Understanding GENETICS

A MOLECULAR APPROACH

Understanding GENETICS

A MOLECULAR APPROACH

NORMAN V. ROTHWELL
Professor Emeritus
Long Island University
The Brooklyn Center

WILEY-LISS
A JOHN WILEY & SONS, INC., PUBLICATION
New York • Chichester • Brisbane • Toronto • Singapore

Address All Inquiries to the Publisher
Wiley-Liss, Inc., 605 Third Avenue, New York, NY 10158-0012

Printed in the United States of America

Library of Congress Cataloging-in-Publication Data
Rothwell, Norman V.
Understanding genetics : a molecular approach / Norman V. Rothwell.
p. cm.
Includes index.
ISBN 0-471-58822-9 (International Ed. ISBN 0-471-59415-6)
1. Human molecular genetics. 2. Human genetics. I. Title.
QH431.R856 1993 92-36402
611'.01816--dc20 CIP

The text of this book is printed on acid-free paper.

To Beverly Logan
An Enduring Inspiration

Contents

Preface

The giant strides that are continuously being made in genetics on the molecular level overwhelm the average well-informed person. They also present a problem to the author of a genetics textbook, who must decide which topics to include at the expense of others. The author is challenged to offer the reader a broad, well-balanced picture of a dynamic field while also keeping the manuscript to a reasonable size. I have tried to accomplish these tasks and to provide a wide audience with a good view of molecular genetics from today's perspective in this extensive revision of *Understanding Genetics*. While the writing assumes a general knowledge of biological principles, several chapters offer a review of classical concepts that are interwoven wherever possible with current information.

Understanding Genetics: A Molecular Approach can well serve undergraduate students in a course that assumes some background in the principles of genetics. It can be of special value to graduate students in a program whose goal is to provide an acquaintance with the direction and methodology of research in molecular genetics. The book focuses on the human genome and can furnish those who have a medical orientation with an insight into approaches that have led to the isolation of genetic regions associated with human afflictions. Attention has been paid to progress at the molecular level in the areas of aging, gene therapy, cancer and mitochondrial inheritance. Considerable treatment has been given to prenatal detection of defective human genes as well as to the recognition of carriers and persons at risk. The treatments of certain other organisms are extensive (*Drosophila*, microorganisms) and are related whenever applicable to the human condition.

To my knowledge, the book brings together for the first time a variety of subjects intimately linked to the advancement of molecular genetics, topics such as the polymerase chain reaction, restriction fragment length polymorphism, chromosome walking and jumping, cosmids, YACs, DNA sequencing, and other procedures used to characterize specific genetic regions in a variety of species. The writing style aims to be relaxed while guiding the reader with the aid of detailed illustrations through topics considered difficult. The questions at the end of each chapter are intended as a study guide for the evaluation of one's grasp of a chapter's content.

Understanding Genetics: A Molecular Approach can be used in a one-semester course or as a guide for two semesters, depending on the specific goals of the instructor and student preparation. It should prove to be a reference work for anyone with a basic science background who wishes to become familiar with the accomplishments of molecular genetics reported in the current literature and the procedures used in the course of the research.

It has been a pleasure to work with the professional staff at Wiley-Liss, Inc., whose efforts made possible completion of this book. Special gratitude is due my editor, Mr. William Curtis, whose constant encouragement and suggestions improved the original manuscript, hastened completion of the project, and made the entire undertaking a gratifying experience.

N.V.R.
Brooklyn, New York

Foundations of Genetics

1

Early genetic studies

1900 was a highly significant year for the science of heredity, for it was then that the work of Gregor Mendel was discovered. Mendel had actually read his paper at Brunn, Czechoslovakia, 35 years earlier, well before any detailed observations had been made on the cell and its contents. This lack of cytological information accounts in part for Mendel's paper lying unappreciated and poorly disseminated among scientists until its significance was pointed out independently by the biologists De Vries, Correns, and von Tschermak. All three worked with plants, and their results confirmed those of Mendel. By 1900, the scientific community was receptive to Mendel's principles because a concept was needed to link inheritance to the observations being made by pioneer cytologists on the cell and its chromosomes. Moreover, Darwin's theory of evolution required an explanation of the source of genetic variation—the raw material of natural selection.

Before 1900, some breeding work had been performed, but it contributed little to an understanding of the mechanism of inheritance. The predominant concept at that time was a blending theory of inheritance that the hereditary material was some sort of fluid, perhaps even blood. The fluids from two parents would come together and blend, forming an inseparable mixture. A contrasting viewpoint conceived of the hereditary material as particulate in nature, composed of particles that were handed down from parents to offspring without losing their original identities. The great significance of the work of Gregor Mendel is that it established the particulate concept of heredity and replaced the blending theory. Unlike most of the other investigators of his time, Mendel was able to obtain clear-cut results after recognizing several requirements essential to genetic analysis.

He chose as his tool the garden pea plant to meet these needs. The plant provided variation, which could be followed from one generation to the next. Mendel was able to study seven characteristics, each one having a pair of clearly defined contrasting traits that could be easily recognized and described, such as tall versus dwarf, green versus yellow (Fig. 1-1). The plant also provided for study a large number of offspring as well as several generations in a short period of time. Mendel recognized the need to work with large numbers so that the results would be scientifically significant. Conclusions that are based on small samples are unreliable and increase the probability that data may be distorted by chance factors. An additional advantage to the pea plant is that it can be quite easily manipulated by the investigator to provide a specific type of mating. Peas are normally self-pollinated, and this feature makes it possible to obtain the equivalent of a cross between two genetically identical individuals, as if an individual were mating with another exactly like itself. In addition, the pea plant can be handled to permit cross-pollination. Consequently a plant from a particular variety can be used as either a male or a female, enabling the investigator to perform crosses in any way desired.

The results obtained before 1900 by many workers had been ambiguous because the species they used as tools in their studies lacked one or more of

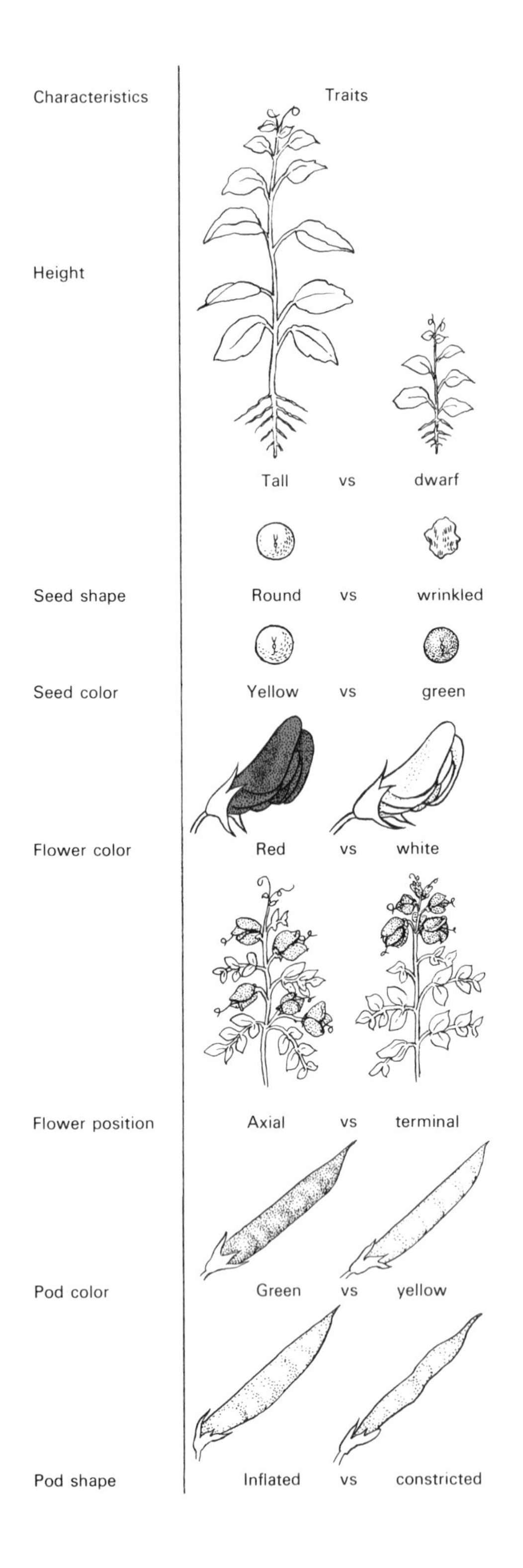

FIGURE 1-1. Seven characteristics in the pea plant followed by Mendel. Each characteristic has two well-defined traits that are easily recognized. This feature permits ready identification of the traits as they are followed from one generation to the next.

the features found in the garden pea. Other organisms (the fruit fly, corn, bacteria, and so forth) that have also contributed to the establishment of fundamental genetic principles possess the same combination of attributes as the pea plant: characteristics that show variations but whose traits are well defined and easily scored, the production of large numbers of offspring in a relatively short generation time, and features that permit crosses to be controlled and manipulated with ease.

Monohybrid crosses and the law of segregation

Studying seven characteristics, Mendel performed monohybrid crosses, matings in which only one pair of alternative traits is being followed. In his study of the inheritance of height, Mendel knew all his tall plants and all the dwarf ones were pure breeding, because he had allowed the two contrasting varieties to self-pollinate for two generations before crossing them. The cross of pure breeding talls with pure breeding dwarfs (the P_1 or first parental generation) produced offspring (the F_1 or first filial generation) that were all tall. The outcome was the same whether the tall plants were used as male or female parents. When he followed the tall F_1 hybrids to the next generation (the F_2 or second filial generation) by allowing them to self-pollinate, he obtained both tall and dwarf offspring in a ratio of approximately 3:1.

Comparable results were obtained with monohybrid crosses for each of the six other characteristics. From his data, Mendel was able to formulate what has been called *Mendel's first law* or the *law of segregation.* Essentially, it states that the hereditary characteristics are determined by particulate units or factors. These units occur in pairs in an individual, but in the formation of germ cells they become segregated so that only one member of a pair is transmitted thorugh any one gamete. Union of gametes at the time of fertilization restores the double number of factors.

Mendel coined the term *dominant* for any trait such as tallness in the pea plant that expresses itself when present with the factor for the contrasting trait and thus seems to dominate it. He used the word *recessive* for any trait, such as dwarfness, that is not expressed when present along with the contrasting dominant trait in the hybrid. Throughout the years, these terms have been widely used to refer to the genetic factors (the genes) associated with the traits as well as to the traits themselves.

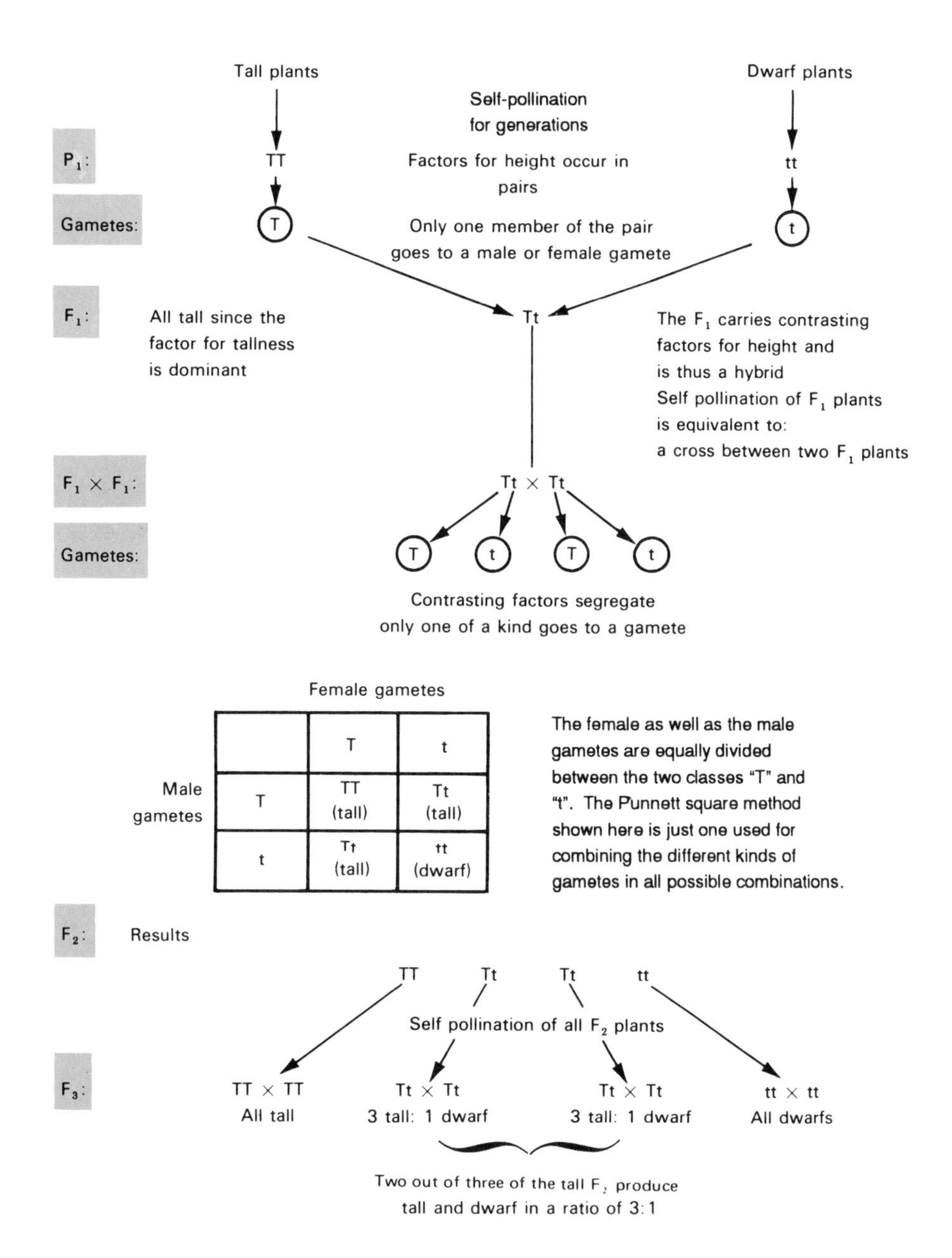

FIGURE 1-2. Monohybrid cross of tall and dwarf pea plants. By convention, the female parent is written on the left. The results of the reciprocal cross are the same.

Let us examine the cross of the tall and dwarf plants, keeping these concepts in mind (Fig. 1-2). Note that when letters are assigned to a pair of contrasting factors (such as those for tallness and dwarfness), the same letter of the alphabet is used (the capital letter for the dominant, the small one for the recessive). Still another accepted way of designating genetic factors will be presented in a later chapter.

In Figure 1-2, the Punnett square or checkerboard method is used to bring gametes together in all possible combinations. Because large numbers of gametes are actually involved in such a cross, the resulting combinations in the F_2 generation occur in a ratio of 1 TT:2 Tt:1 tt. Due to the dominance of tallness, the ratio of different physical types is 3 tall:1 dwarf. However, when the next generation, the F_3, is examined, it can be seen that only one-third of the F_2 tall plants will breed true when self-pollinated. Following selfing, the other two-thirds are shown to be hybrids (Tt), which produce both tall and dwarf offspring in a ratio of 3:1.

Those plants that are pure breeding for tallness (TT) have identical genetic factors for that trait and are said to be "homozygous." The dwarf plants are also homozygotes (tt)—individuals with identical genetic factors for a specific trait. Individuals showing the dominant trait, such as tallness, can be either one of two types, TT or Tt. The latter are not pure breeding and are called *heterozygotes,* meaning that they have a pair of contrasting genetic factors for a particular characteristic. As we will see more clearly later in our discussions, contrasting factors such as these are really different forms of the same gene that produce contrasting detectable effects. Contrasting forms of a gene (such as T vs. t) are known as *alleles.* Any homozygote, TT or tt, has one allelic form of a gene represented twice. On the other hand, the heterozygote (Tt) possesses a pair of contrasting forms (alleles) of the same gene for a given characteristic. The term *allele,* therefore, is used to indicate a specific form of a given gene.

In our present example, we are concerned with the gene for the characteristic height in the pea plant, and we are following the transmission of two specific forms of that gene, the allele T for tallness and the allele t for dwarfness. We cannot tell simply by looking at a tall plant whether it is homozygous or heterozygous. However, we refer to any detectable attribute of an individual such as height as its *phenotype.* The term *genotype* describes the genetic constitution or the kinds of genetic factors carried by an individual. A plant of tall phenotype can have either genotype TT or Tt. However, knowing that an allele is recessive, such as that for dwarfness in the pea, we can deduce the genotype (tt) by noting that the phenotype is dwarf.

Blending inheritance disproved

In the case of flower color in the garden pea, Mendel found that the red flower trait was dominant to that for white. Therefore a cross between homozygous red (RR) and white (rr) plants produces F_1 hybrids (Rr) that are red flowered. Although the white-flowered trait seems to disappear in the F_1, a cross between two F_1 red plants (Rr × Rr) yields an F_2 consisting of red- and white-flowered plants in a ratio of 3:1. The white-flowered plants are identical in phenotype to the original ones in the P_1 and are in no way altered by the presence of the contrasting allele for redness in the F_1. Similarly, the allele for redness has in no way been diluted by the factor for white. This is certainly not to be expected if fluids from two parental sources are mixing; otherwise, the original qualities should become altered in some way, as if paints of different colors had been blended.

The common horticultural plant known as the four-o'clock can also have red or white flowers. However, if red and white plants are crossed, the picture differs from the one presented by Mendel's peas (Fig. 1-3). Both red plants and white ones are homozygous (RR and rr), and the F_1 hybrids formed when they are crossed (Rr) are always pink. When two pink plants are used as parents, the results are always 1 red:2 pink:1 white. The existence of plants with the intermediate pink shade in no way supports the blending concept of inheritance; on the contrary, it strongly supports the particulate concept. Neither the factor for redness (R) nor the one for whiteness (r) becomes altered in any way by being present together in cells of an individual. A cross such as this illustrates Mendel's first law beautifully and differs from that of the pea only in the question of dominance. In the case of the four-o'clocks, neither form of the gene for flower color is dominant to the other. In the hybrid (Rr), each allele expresses itself so that an intermediate effect results. Alleles such as these show a lack of complete dominance. When dominance is not totally complete, the 3:1 phenotypic ratio that typically results from a cross of two monohybrids becomes modified to a ratio of 1:2:1, which is identical to the genotypic ratio that arises from such a cross. This is so, since lack of dominance permits the pink hybrids (Rr) to be distinguished phenotypically, unlike the hybrid pea plants, which appear as red as the homozygotes.

Dominance is not always absolute

Dominant and *recessive* are often used as terms of convenience, but strictly speaking, they are not very precise. One example from the human illustrates this point very well. A certain allele, D, when present in a single dose, can produce a serious skeletal disorder, achondroplasia. Persons heterozygous for this allele (Dd) have a dwarfed stature, short limbs, and a large head with bulging forehead. The allele is generally referred to as a *dominant.* However, persons homozygous for the allele (DD) have a much more abnormal phenotype, and they usually do not survive infancy. Obvi-

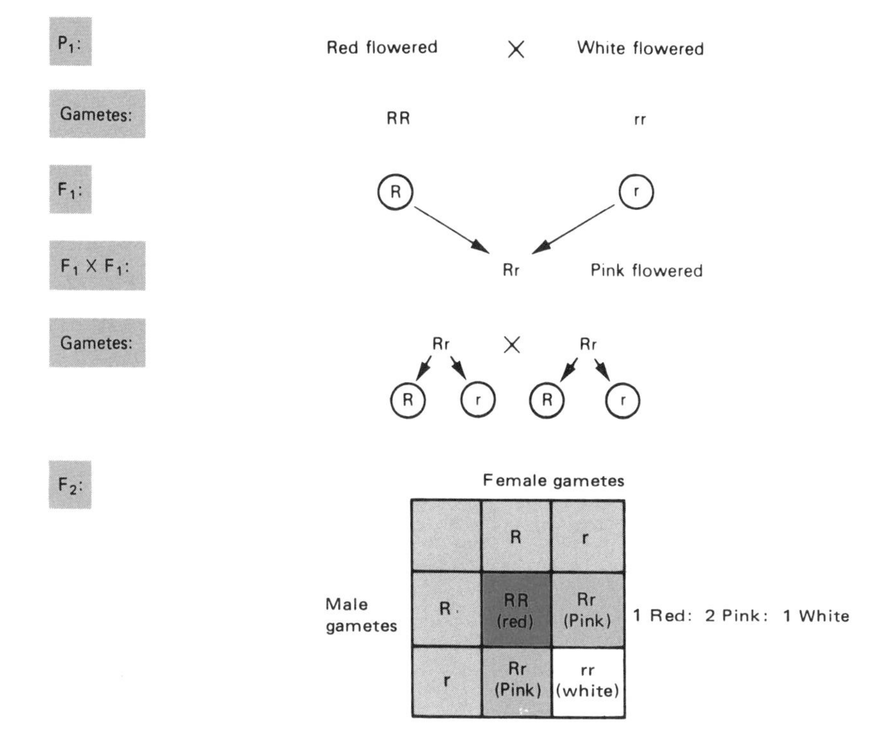

FIGURE 1-3. Incomplete dominance in the four-o'clock. Neither of the two alleles, R or r, is dominant. Consequently the F_2 phenotypic ratio is modified from 3:1 to 1:2:1, which is identical to the ratio of genotypes.

ously, the terms *dominant* and *recessive* are not 100% distinct in this case, because a double dose of the so-called dominant allele produces a far greater effect than a single dose in the heterozygote.

Many examples could be given in which heterozygotes appear on the surface to be identical to homozygotes. Nevertheless, the so-called recessive allele may actually be producing some effect. The expression *incomplete dominance* refers to such situations in which both alleles express themselves in the hybrid but the effect of one allele appears greater than that of the other. In the absence of a refined examination, one allele may even appear to be completely dominant. Another commonly used term is *codominance,* which is also applied to cases in which both alleles produce an effect in the heterozygote. However, in its strict sense, codominance implies that a definite product or substance controlled by each allele can be identified. The genetic factors that govern the A and B blood types in the human are alleles. Each controls the formation of a different red blood cell substance or antigen, antigen A in one case, antigen B in the other. Neither allele is dominant over the other. The individual who is blood type AB is a heterozygote and has both the allele for antigen A and that for antigen B. Both of these substances, A and B, are easily detected in equal amounts in the red cells. This is a clear-cut case of codominance between a pair of alleles.

In other situations it may not be so easy to distinguish codominance from incomplete dominance because use of the proper term depends on our ability to detect in the heterozygote distinct substances controlled by each member of the allelic pair. An example from the human illustrates the difficulty often involved in a precise distinction between codominance and incomplete dominance. Many of us are familiar with the blood disorder known as sickle-cell anemia, a fatal condition in which abnormal hemoglobin is present. It is so named because the red blood cells tend to assume a sickle shape in the blood vessels (Fig. 1-4), resulting in clogging of the capillaries and an assortment of dire symptoms. The disorder stems from the presence in the homozygote of a genetic factor that governs the formation of the abnormal hemoglobin. Those who are homozygous for the normal genetic factor possess only the typical hemoglobin in their red blood cells. Heterozygous persons appear identical to the homozygotes under usual conditions.

For convenience, let us represent the alleles responsible for the normal and the sickle-cell hemoglobins as S and s, respectively. Superficially it would appear that SS and Ss individuals are the same, in contrast to the homozygotes, ss, who present an abnormal phenotypic picture. We might conclude that absolute dominance pertains. However, the heterozygotes, Ss, *are* actually somewhat different from the SS homozygotes. They may be identified when their blood is examined microscopically, because a certain percentage of their red blood cells can be made to sickle by a test that subjects the cells to certain reducing agents capable of reducing the oxygen tension in the immediate environment. Such an effect in the body of the heterozygote would be very unusual. We see, however, that the allele for sickle-cell hemoglobin is not a complete recessive, because the heterozygote can be detected.

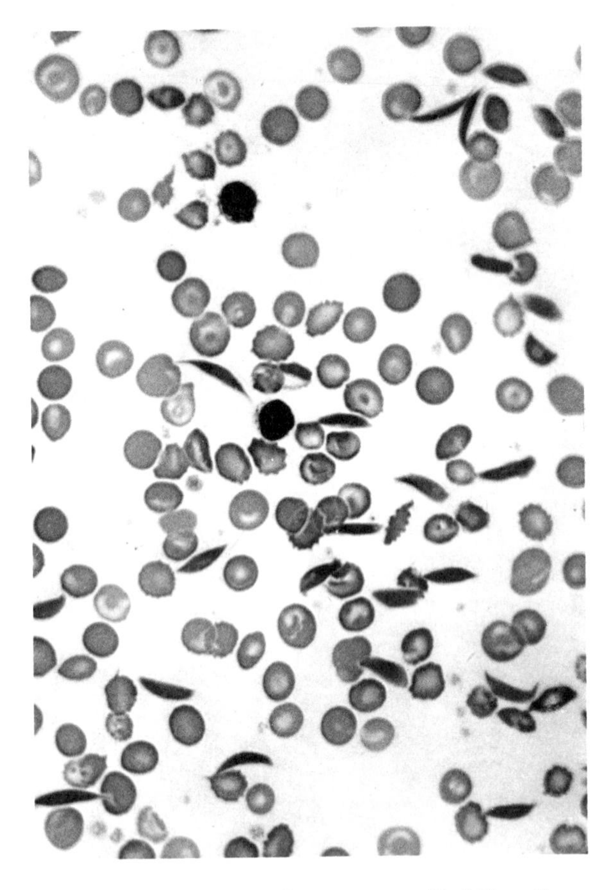

FIGURE 1-4. Blood smear from a person with sickle-cell anemia showing typical sickle-shaped cells. A person homozygous for the allele for normal hemoglobin (SS) will have disk-shaped cells under all levels of oxygen concentration. The homozygote for sickle-cell anemia (ss) carries red blood cells that assume an abnormal sickling shape under the lower oxygen concentrations in the blood vessels. The heterozygotes (Ss) are revealed by a test that subjects a sample of red blood cells to a reduced oxygen level, causing a large number to sickle. Such sickling does not occur in the body of a heterozygote who typically shows no clinical symptoms and may remain undetected. (Courtesy of Carolina Biological Supply Company.)

Is the normal allele S incompletely dominant over the abnormal one, or are the alleles codominant? In the sense that the normal S results in a heterozygote who has normal red blood cells except for those that sickle under test conditions, the allele would appear to be an incomplete dominant. However, refined techniques demonstrate that two different proteins—normal hemoglobin controlled by gene form S and sickle-cell hemoglobin controlled by its allele s—are present in the blood. Based on this criterion, therefore, the alleles are codominants. It is evident that the choice of words may depend on the level at which we describe the phenotype of the heterozygote; it also depends on the techniques available for detecting chemical differences. However, it is important that we avoid getting lost in a maze of terminology. The expression we use may depend on our frame of reference. The alleles for the A and B blood groups are clearly codominants. We may say that the alleles R and r in the four-o'clock are incompletely dominant because the heterozygote is intermediate and no distinct substances specific to each has been recognized. The normal and the sickle-cell alleles may be described as incompletely dominant at one level of reference (a small percentage of sickling) or as codominants at another (presence of two specific protein products). For convenience, at times the allele for normal hemoglobin is referred to as a *dominant.*

Misconceptions to be avoided

To avoid erroneous interpretations of Mendelian principles that have been discussed so far, it is essential to pause and reexamine the implications of certain basic ideas. Several important points concern the concept of dominance, which unfortunately is often misunderstood. Although Mendel showed that the factor for red flower color in the pea plant is dominant over its allele for white, the genetic factors for red and white flowers in the four-o'clock are incompletely dominant. This example was given to illustrate that a genetic factor that produces a certain phenotypic effect (red, in this example) in one species is not the same as one in a different species, even though the two genetic factors appear to produce identical phenotypes (red flower color in both the pea and the four-o'clock). Simply because a particular factor in one group of animals or plants expresses itself as a dominant is no reason to assume that a different factor with a similar or even identical expression must also be dominant in another group. In animals, a factor for black fur may be dominant in one species (as in guinea pigs) but recessive in another (as in sheep).

Another common misunderstanding of dominance concerns the frequency of dominant alleles in a population. One often assumes that simply because a particular factor is dominant it must therefore be more abundant in the population than its recessive counterpart. Implicit in this erroneous idea is the belief that a dominant factor *must* increase in frequency over its recessive allele. A moment's reflection will tell us in the simple case of eye color in humans that this is not so. Although the factor for brown eye color (B) behaves as a dominant to its allele for blue (b), there may be more blue-eyed than brown-eyed people in a population. In Scandinavia this would be so, whereas in Mediterranean countries we expect more people to be brown eyed. The frequency or abundance of an allele is not the same as its degree of dominance. The allele for brown eye color is dominant to the one for blue, which means simply that the factor for brown eyes expresses itself in the presence of its allele, whereas the factor for blue does not. Which allele—the one for brown or the one for blue—will be the more common is a very different matter. The frequency of any allele, dominant or recessive, is related to the survival value it imparts to a population and not just to its dominance or recessiveness.

Nor does the fact that an allele is dominant mean that it is necessarily better or will somehow produce a stronger or more desirable effect. A long list of undesirable dominant alleles in various species could be presented; in humans the following seem to depend on dominants: extra fingers or toes, a form of muscular dystrophy, a type of dwarfism, and a form of progressive nervous deterioration resulting in death. Such conditions are far from desirable and certainly do not make the afflicted person stronger. It is also apparent that the conditions, and hence their associated dominant alleles, are fortunately not as common in the population as the normal conditions and the normal but recessive gene forms. Moreover, these undesirable dominants are not increasing to the disadvantage of the whole population. If their dominance made them increase, everyone would eventually become afflicted.

It is also essential to appreciate the correct use of the Punnett square (Fig. 1-2) and the interpretation of ratios such as 3:1, 1:2:1 or any others. One reason for Mendel's achievement was that he conducted his studies with large numbers of plants. Few of us would expect a cross between two heterozygous tall plants (Tt × Tt) to yield exactly three tall and one dwarf if only four offspring were obtained. The 3:1 ratio (or any other) tells us the probability of getting a certain phenotype or genotype when a particular cross is made. Only after obtaining hundreds of F_2 plants would we expect to approximate a ratio close to the ideal of 3:1. However, the chance of getting a tall plant from any one fertilization when two heterozygotes are crossed is 3 out of 4, and the same value persists regardless of the number of offspring. No matter how many offspring, even after hundreds of them, the 3:1 chance remains the same for obtaining a tall at any one time. Likewise the chance of obtaining a dwarf remains 1 out of 4. It is very important to remember when any ratio is discussed that it does not predict the exact results from a cross but instead gives us the chance at any one event that a certain outcome (tall or dwarf, red or white, boy or girl) will occur.

The testcross and its value

Once he formulated the law of segregation, Mendel performed other types of crosses, one of the first being the testcross. He took his F_1 hybrids and mated them with recessives, such as a cross between hybrid tall plants and dwarf ones (Fig. 1-5). Mendel realized that if the law of segregation was correct, then the monohybrid tall plants should produce two classes of gametes in equal amounts, half with the factor for tall (T) and half with the factor for dwarf (t). A dwarf plant, carrying only the recessive allele for height, can produce only one class of gametes for this characteristic, those carrying the allele t. Consequently, genetic factors for height coming from the dwarf parent will not mask any gene forms for height contributed by the hybrid parent. As a result, it becomes possible to determine the kinds of gametes formed by the hybrid as well as the proportion in which they are formed. To do this it is only necessary to observe the kinds of offspring in the testcross and the frequency with which they occur. As Figure 1-5 shows, half of the offspring from the testcross should be tall and the other half dwarf—a ratio of 1:1. This is so because the two alleles (T and t) are segregating in the hybrid. They are produced in equal amounts, and it is a matter of chance whether a gamete carrying T or one carrying t combines with the gamete from the recessive parent. Mendel performed many testcrosses of monohybrids and obtained results very close to a 1:1 ratio in each case, thus supporting the ideas expressed in his first law. The testcross is such a valuable procedure in genetic analysis that we refer to it frequently throughout the text.

The term *testcross* is often used interchangeably with the expression *backcross*. However, the two do not always mean the same. We have just seen that a testcross is a cross of any individual to one that is recessive for a trait being followed in a cross. A backcross, on the other hand, implies a mating of an individual to a type like one of the parents. In our example (Fig. 1-5) we have both a testcross and a backcross. The monohybrid is crossed to the recessive, but the recessive phenotype is also one of the parental types. If the hybrid had been crossed instead to a homozygous tall (TT), we would have an example of a backcross but not a testcross. Because a backcross may have special value in some genetic analyses, its distinction from the testcross should be recognized.

Dihybrid testcrosses and independent assortment

A question that occurred to Mendel was how two or more pairs of factors might behave in relationship to one another when followed at the same time in a cross. From his monohybrid crosses, Mendel knew that the color of the pea seed can be either yellow or green and that the factor for yellow (G) is dominant over green (g). When plants homozygous for yellow seeds are crossed with those having green seeds, the F_2 generation contains the yellow and green phenotypes in a ratio of 3:1. Likewise, the shape of the seed, round (W) versus wrinkled (w), depends on a pair of alleles with the factor for the former being dominant. The F_2 ratio is thus 3 round to 1 wrinkled. But if both characteristics, seed color and seed shape, are followed at the same time, how will the factors for the two of them behave?

Suppose a plant that is pure breeding for yellow, round seeds is crossed to one from a green, wrinkled variety (Fig. 1-6A). From Mendel's first law, we would expect two factors for each character to be present in each parent. Knowing which genes are dominant, we may represent the genotypes of the parents as GGWW (yellow, round) and ggww (green, wrinkled). The F_1 dihybrid between them will receive one factor for each character from each of the parents and will have the genotype GgWw. The F_1s are all found to have yellow, round seeds. This is not surprising, because yellow and round have been shown to be dominant. The question that now arises is, "What will happen when the F_1 dihybrids are followed further?" Will the factors for yellow and round stay together through later generations because they were together in one of the original parents? The same question arises for green and wrinkled. It is also possible that the original gene combinations are not required to remain together but are free to enter into new combinations.

The most direct way to settle the matter is to take the F_1 dihybrid and perform a testcross, because a testcross can tell us the kinds of gametes produced by an individual and the frequency of the different kinds. As Figure 1-6B shows, the offspring resulting from this testcross are of four types, and they occur with equal frequencies in a ratio of 1:1:1:1. This tells us directly that the dihybrid is forming four different classes of gametes. This in turn means that the factors for seed color and those for seed shape are behaving independently. Just because the factors for yellow and round (G and W) and those for green and wrinkled (g and w) were together in the original parents did not mean they had to stay together. In addition to the old combinations, GW and gw, the F_1 dihybrid also produces new ones, Gw and gW. Nothing seems to tie the factors together; they are free to form new combinations. We say that they "assort independently." As a result, the expected dihybrid testcross ratio is 1:1:1:1, a reflection of the types of gametes and their proportions. We see here, as in the case of the monohybrid, that the testcross parent (double recessive ggww) produces only one kind of gamete in relation to the characteristics we are following. Because all the gametes carry both recessives, g and w, they cover up nothing contributed by the other parent. Thus we can estimate the kinds of gametes and their proportions produced by the other parent simply by counting the kinds of offspring. Figure 1-7 shows one other way to envision the independent behavior of two pairs of alleles in a dihybrid and to determine the kinds of gametes formed.

Independent assortment and the F_2 dihybrid ratio

Knowing that any dihybrid individual forms four different kinds of gametes in equal proportions, we can predict what to expect in a dihybrid cross (one in which two pairs of contrasting factors are being followed) carried through the F_2 generation. We see in Figure 1-8 that the F_2 will fall into four different categories in a ratio of 9:3:3:1. This is to be expected from the independent behavior of the two pairs of alleles. Note that although only four different phenotypes result when dominance is operating in both pairs of factors, there are actually nine different genotypes that

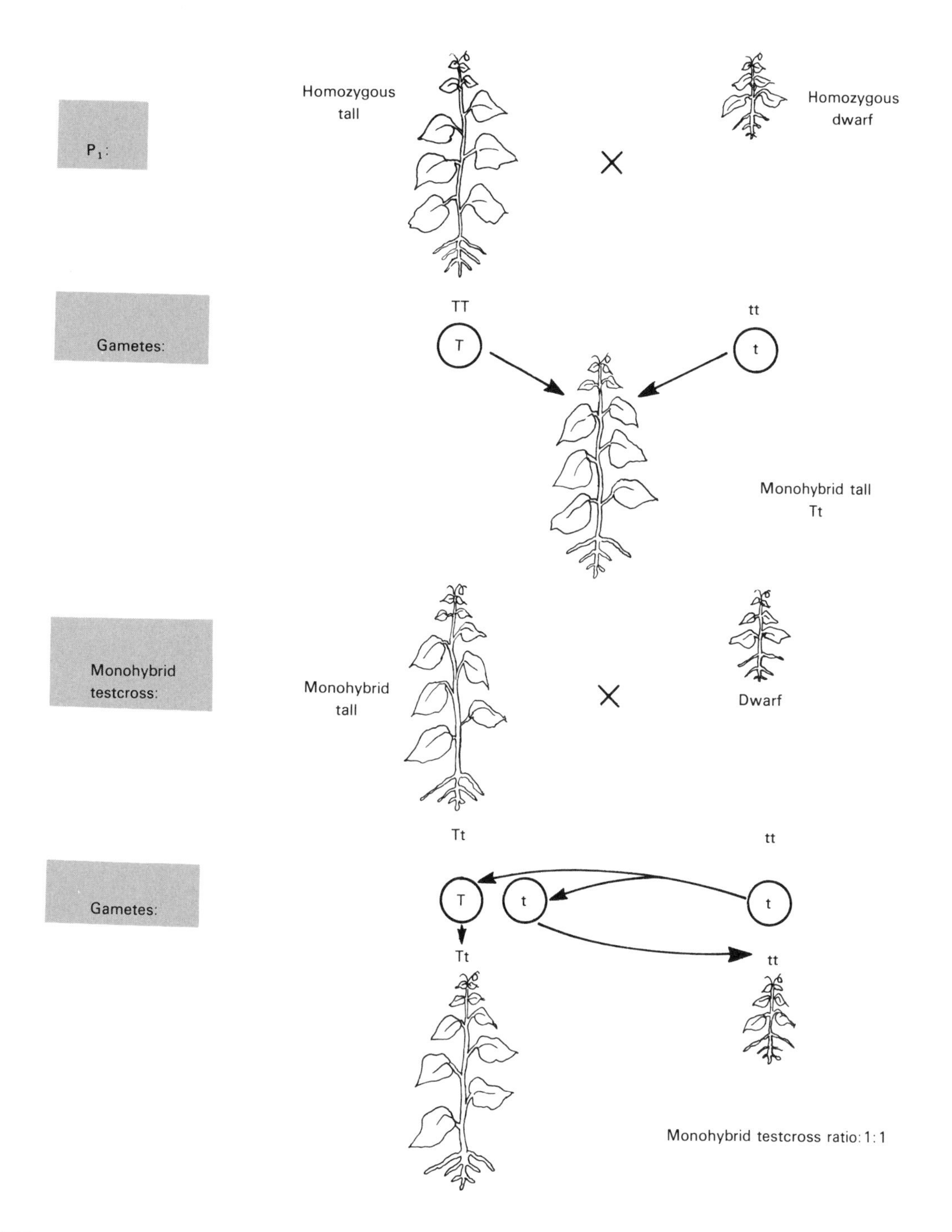

FIGURE 1-5. Monohybrid testcross. The testcross ratio, 1 tall:1 dwarf, is exactly the same as the ratio of the two types of gametes produced by the monohybrid, one kind with allele T, the other with allele t. This is so because the testcross parent (tt) carries only the recessive allele for the character height and cannot mask any factors segregating in the hybrid parent. The testcross shown here is also a backcross, because "dwarf" is one of the P_1 types.

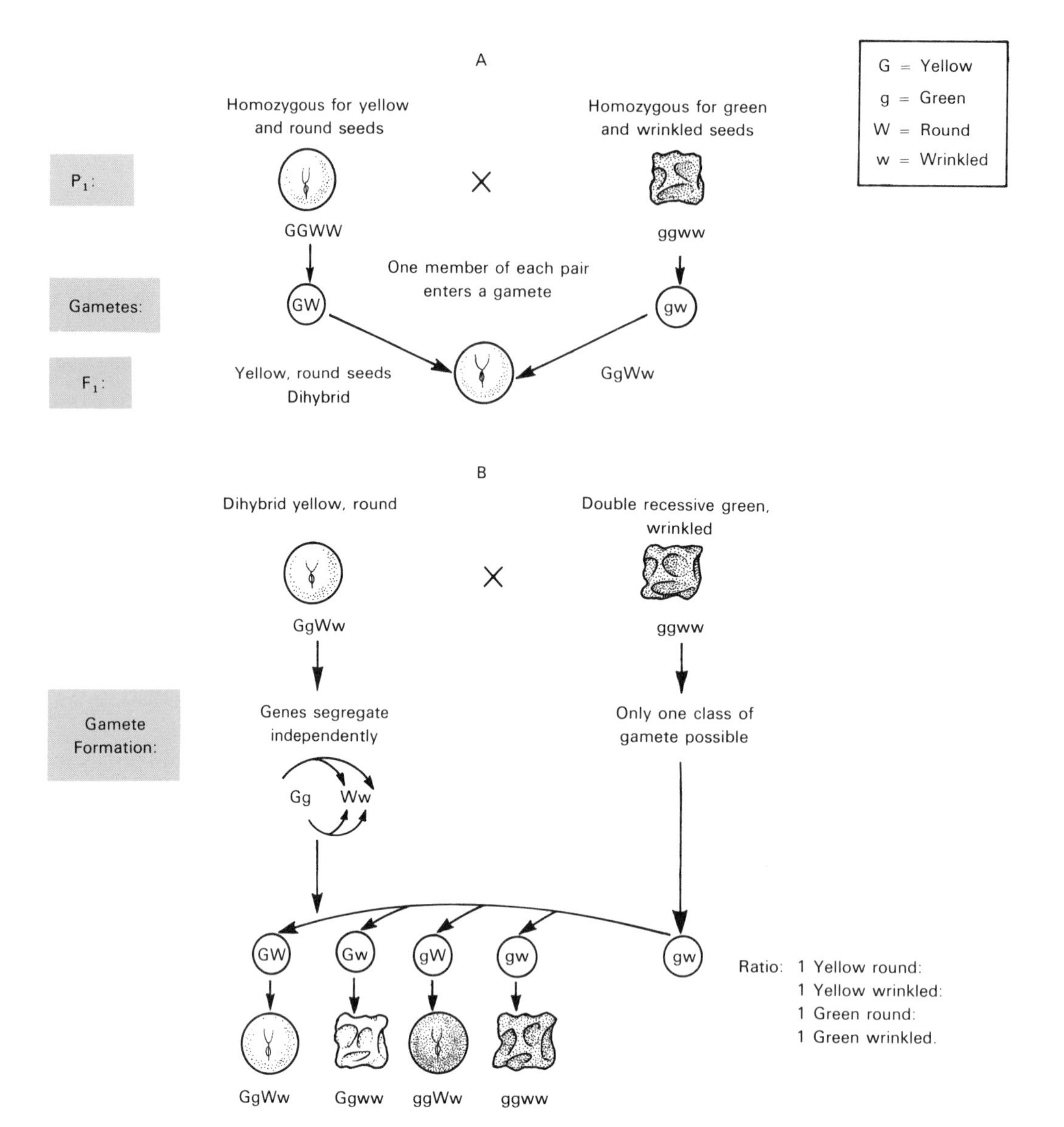

FIGURE 1-6. Origin of a dihybrid and its testcross. **A:** Cross producing a dihybrid. The yellow, round phenotype is to be expected because the factor for yellow (G) is dominant over its allele (g), and the one for round (W) is dominant over wrinkled (w). **B:** The dihybrid is testcrossed. The phenotypic ratio 1:1:1:1 is exactly the same as that in which the four classes of gametes are produced in the dihybrid: 1GW:1Gw: 1gW:1gw.

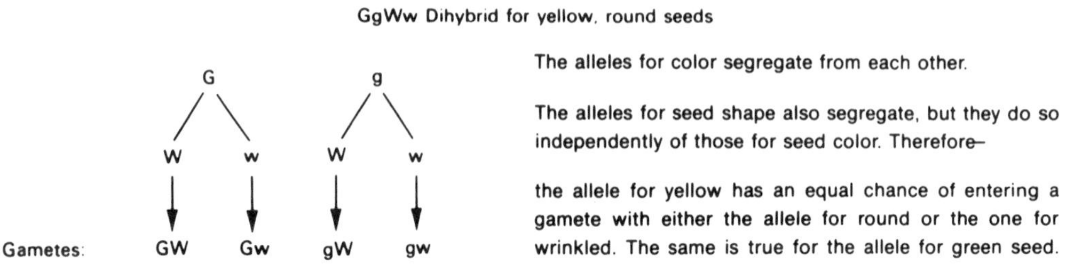

FIGURE 1-7. Branching method. One way to determine the different types of gametes that can be formed by a dihybrid.

arise from independent assortment of the alleles. We can think of a cross between any two dihybrids in the following general way: AaBb × AaBb. When dominance is involved, the expected results are 9 A_B_ (those showing both dominant traits; the dashes indicate that either the dominant or the recessive allele may be present); 3 A_bb (those expressing the dominant A and the recessive b); 3 aaB_ (individuals expressing the recessive a and the dominant B); and 1 aabb (double recessives).

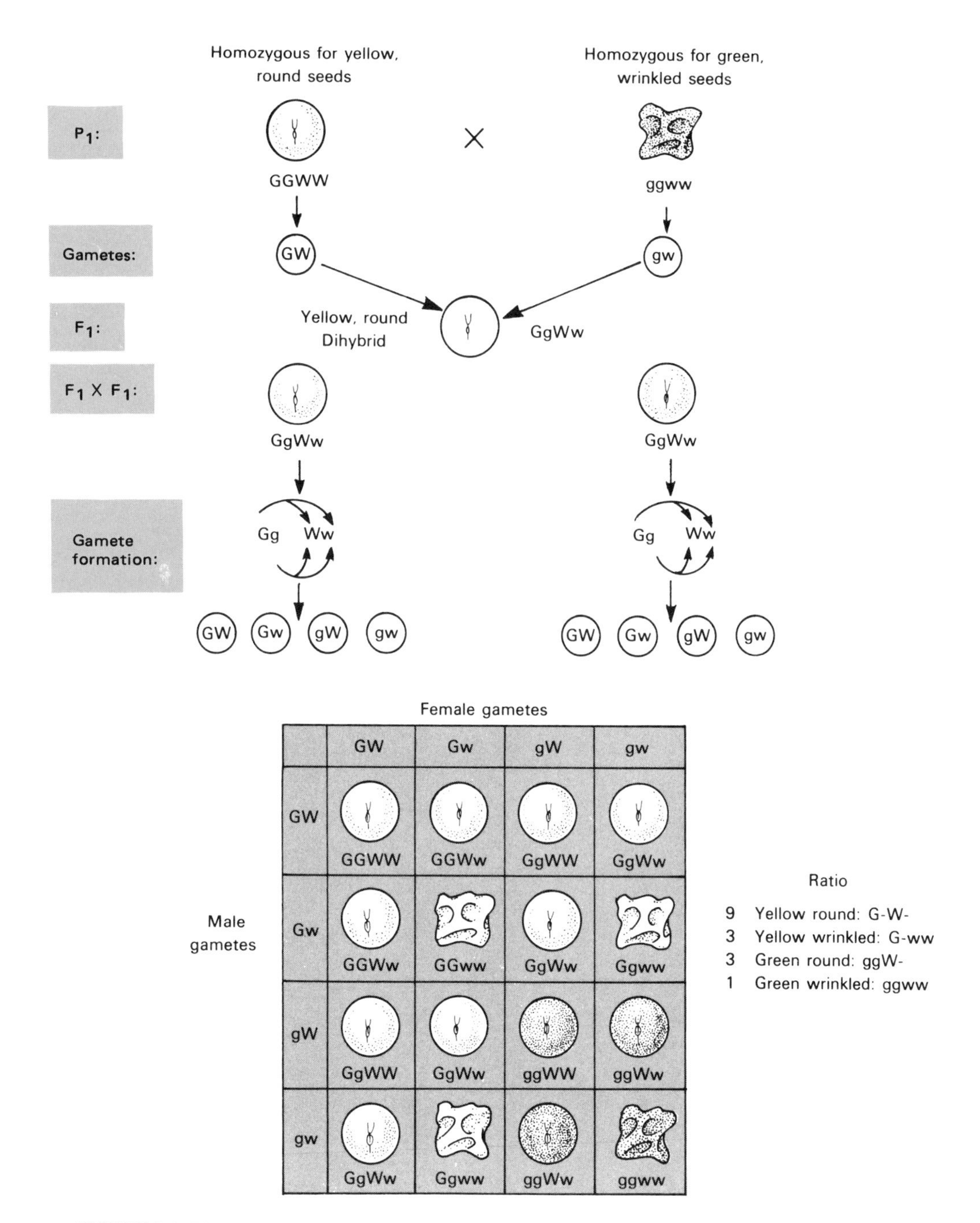

FIGURE 1-8. Dihybrid cross followed through the second generation. Inspection of the Punnett square shows that nine different genotypes are found among the F_2 offspring. Because of dominance in one member of each pair of alleles, only four phenotypes are expressed.

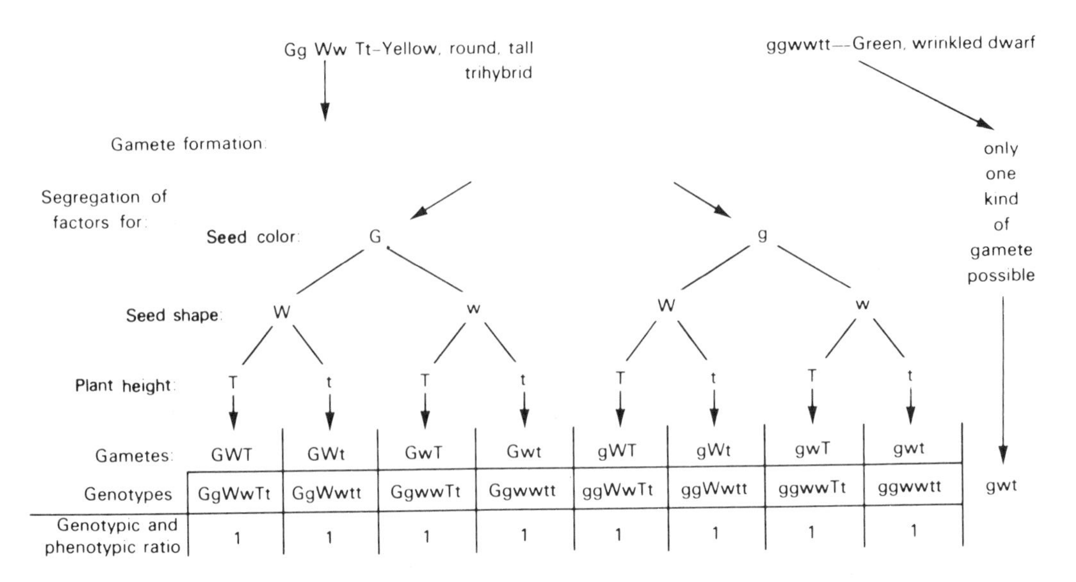

FIGURE 1-9. Trihybrid testcross. The branching method is used here to derive the different kinds of gametes formed by the trihybrid.

Other crosses of pea plants in which the remaining five pairs of alleles were followed produced comparable results and established the 9:3:3:1 ratio as the one to be expected when two dihybrids are crossed and 1:1:1:1 as the dihybrid testcross ratio. From such data, Mendel formulated his second law, *the law of independent assortment,* which states what we have just seen: The members of one pair of factors (alleles) segregate independently of members of other pairs at the time of gamete formation (pair Gg segregates independently of pair Ww). This rule holds for two, three, or more pairs, but we will see that the law pertains only to those genes located on different pairs of chromosomes. The complication of linkage (genes found together on the same chromosome) was not involved in Mendel's crosses to cloud the picture of independent assortment.

Let us ignore linkage for the moment and focus on the enormous amount of variation that can be generated by independent assortment alone. We have seen that a monohybrid such as Tt forms two types of gametes and that a dihybrid such as GgWw forms four. A testcross of a trihybrid (GgWwTt) reveals that it forms eight kinds of gametes (Fig. 1-9). A tetrahybrid would be found to produce 16 different types. The numbers of different possible genetic combinations that can arise from individuals of different degrees of hybridity form a geometric progression that can be summarized by the expression 2^n. Allowing n to represent the number of hybrid gene pairs in an individual, we can easily determine the number of different genetic classes of gametes that can be formed by any hybrid.

Reference back to Figures 1-2 and 1-8 will show that the number of genotypic classes resulting from monohybrid and dihybrid crosses are three and nine, respectively. A cross of two trihybrids produces 27 genotypic classes and that between two tetrahybrids yields 81 types among the offspring. The expression 3^n summarizes this progression, where n again represents the number of hybrid allelic pairs under consideration. Keep in mind that if dominance is not operating for any of the factors in a cross, the number of phenotypic classes is the same as the number of genotypic ones.

We can now consider the possibilities that can arise in the human from independent assortment alone. Humans possess 23 pairs of chromosomes and are highly heterozygous. For any one chromosome pair, the human is hybrid for many, many genes. Keeping the example as simple as possible, let us assume that the human is hybrid for only one pair of alleles on each pair of chromosomes and that there is a lack of dominance. This would mean that any one parent would form 2^n or 2^{23} different kinds of gametes. Among the offspring of any two parents many more genotypic classes are possible, 3^{23}. Actually, this illustration is simplified to the point of absurdity, because countless allelic pairs on any one pair of chromosomes would be undergoing reassortment to increase the number of classes many times more. At any rate, 3^{23}, a figure that defies the imagination, would be the minimal one for the kinds of gametes and phenotypic classes possible from two parents. Keep in mind that we are considering on each chromosome pair just one pair of contrasting alleles showing a lack of dominance. We can well appreciate why no two people on earth, except for identical twins, are genetically identical. Each of us is unique. The chance for another person to be exactly like us, even from our own parents, is so improbable that the idea can be dismissed. Each of us owes our distinct nature in a large degree to independent assortment, which we will see in Chapter 3 is a consequence of chromosome behavior and the nature of the sexual process.

REFERENCES

Dunn, L.C. *A Short History of Genetics.* McGraw-Hill, New York, 1965.

Jordanova, L.J. *Lamarck.* Past Masters Series. Oxford University Press, New York, 1984.

Mendel, G. *Experiments in Plant Hybridization.* Translated in C.I. Davern, ed. *Genetics, A Scientific American Reader.* W.H. Freeman, New York, 1981.

Orel, V. *Mendel.* Past Masters Series. Oxford University Press, New York, 1984.

REVIEW QUESTIONS

1. In tomatoes, round fruit (O) is dominant over oblong fruit (o). Write as much as possible of the genotypes of:

A. A plant from a homozygous round-fruited stock.
B. A plant from an oblong strain.
C. A round-fruited plant that resulted from a testcross.
D. An oblong-fruited plant that resulted from the cross of two round-fruited ones.

2. Using information from Question 1, give the genotypic and phenotypic ratios to be expected regarding fruit shape from each of the following crosses:

A. Homozygous round × heterozygous round.
B. Heterozygous round × heterozygous round.
C. Oblong × oblong.
D. Heterozygous round × oblong.

3. Two short-haired female cats are mated to the same long-haired male. Several litters are produced. Female 1 produced eight short-haired and six long-haired kittens. Female 2 produced 24 short-haired and no long-haired kittens. From these observations, what deductions can be made concerning hair length inheritance in these animals? Assuming the allelic pair S and s, give the likely genotypes of the two female cats and the male.

4. In cattle, the hornless condition (H) is dominant over the horned condition (h).

A. A horned bull is mated to a hornless cow that is heterozygous for the condition. What kinds of offspring are to be expected and in what ratio?
B. If the cow is next mated to a hornless bull that is a heterozygote, what is the chance that the first calf will be horned?
C. Assuming that the first calf born has horns, what is the chance that the second calf will be hornless?

5. In cattle, the alleles for red coat (R) and white coat (r) behave as codominants. Both red and white hair are produced, so that the heterozygote is intermediate, or roan colored.

A. Give the phenotypic and genotypic ratios to be expected among the offspring from a cross of two roan animals.
B. What are the expected genotypic and phenotypic ratios from a cross of a roan animal and a white one?

6. Write as much as possible of the following genotypes:

A. A horned white bull.
B. A hornless roan cow.
C. A hornless red cow whose male parent was horned and roan.

7. In guinea pigs, short hair (L) is dominant over long (l). Black fur (B) is dominant over albino (b). A female from a strain that is pure breeding for black fur and short hair is mated to a male from a strain pure breeding for the albino condition and long hair.

A. What will be the phenotypes of the F_1?
B. If members of the F_1 are mated among themselves, what percentage of offspring can be expected to be homozygous for both traits? Give the genotypes and phenotypes of these homozygotes.

8. In certain breeds of chickens the factor (B) is responsible for the production of black feathers, and its allele (b) produces feathers that are basically white (except for flecks of black). The heterozygote is blue-feathered. Another factor (F) produces straight feathers, whereas its allele (f) results in the frizzled condition, giving brittle feathers. The heterozygote is mildly frizzled. What is to be expected from the following crosses? Give the phenotypic ratios.

A. A black, frizzled hen × a white, straight rooster.
B. A white, frizzled hen × a blue, straight rooster.
C. A blue, mildly frizzled hen × a white frizzled rooster.
D. A blue, mildly frizzled hen × a blue, mildly frizzled rooster.

9. In the human, the allele for normal skin pigmentation (A) is dominant to (a) for the absence of pigment, the albino trait. A certain type of migraine headache (M) behaves as a dominant to the normal condition (m), absence of headache. Write as much as possible of the genotypes of the following three persons:

A. A phenotypically normal individual whose mother was an albino with migraine.
B. A person suffering from migraine who has normal skin pigmentation. Both parents have normal skin pigmentation, and one of them suffers from migraine.
C. An albino person who does not have headaches, whose normally pigmented parents both suffer from migraine.

10. A person who is dihybrid with respect to both the skin pigmentation and headache characteristics marries a person who is heterozygous at the skin pigmentation locus and who does not have headaches.

A. Give the genotype of each and the kinds of gametes each will produce.
B. What kinds of offspring are to be expected and in what phenotypic ratio?

11. The factors for normal hemoglobin (S) and for sickle-cell hemoglobin (s) may be considered codominants. A blood examination can detect the heterozygote, since a small percentage of the cells will show sickling. A man whose cells show no sickling upon examination marries a healthy woman who is found to have a certain percentage of sickle cells.

A. What is the chance of their having a baby with severe anemia?
B. The woman later marries a healthy man whose blood examination reveals sickle cells. They have three children free of sickle-cell anemia. What is the chance that the next child will have the severe disorder?
C. What is the chance that any one of the healthy children is a carrier of the sickle-cell allele?

12. Assume that medical science finds a treatment for sickle-cell anemia. When instituted in infancy, it permits an otherwise doomed person to live a normal life span without any serious effects of the anemia. If two such persons, saved from the fatal effects of the disease, become parents, what will be the genotypes of their offspring? Would they require treatment? Explain. Also give an explanation that would be expected from a supporter of the Lamarckian concept.

13. In the human, free ear lobes (F) are dominant over attached lobes (f). Using this information and that on sickle-cell hemoglobin in Question 11, answer the following:

A. Give the genotypes of the following persons:
(1) A healthy woman with attached ear lobes whose red blood cells show some sickling.
(2) A man with free ear lobes whose mother also had free lobes but whose father had attached ones. This man's red blood cells show some sickling.

B. What kinds of offspring are to be expected if the man and woman marry and in what ratio?

14. The fatal disorder Huntington's disease results in progressive nervous deterioration and death. It usually strikes well after the age of puberty so that afflicted persons may have produced offspring. The genetic factor responsible for the disorder is a dominant one (H), which is very rare in the population and does not seem to be on the increase.

A. Explain why more and more persons do not seem to be suffering from the disorder resulting from this dominant.
B. If a young person had one parent who died of the disorder and another who is apparently free of it, what does he or she know about the chances of developing the disorder?

15. Give the number of phenotypic classes expected from each of the following:

A. Testcross of AaBbCC.
B. Testcross of AaBbCcDd.
C. Testcross of AaBbCcDdEE.

16. Give the number of phenotypic classes to be expected when the following crosses are made and when there is dominance.

A. Cross of two individuals who are AaBbCcDd.
B. Cross of two individuals who are AaBbCcDDEE.
C. Cross of two individuals who are AaBbCcDdEe.

17. Give the number of genotypic classes to be expected in A, B, and C of Question 16.

In the pea plant, tallness (T) is dominant over shortness (t). Yellow seed color is dominant (G) over green (g), and round seed shape (W) is dominant over wrinkled (w).

18. Show the different kinds of gametes that can be formed by individuals of the following genotypes: (1) TtGG, (2) TtGGWw, (3) TtGgWw.

19. What would be the expected phenotypic ratio if each of the three plants described in Question 18 were test-crossed?

20. Suppose plants 2 and 3 in Question 18 were crossed. What would be the expected phenotypic ratio among the offspring?

21. Two pairs of alleles are known to interact in the production of feather color in the parakeet. Allele B is needed for blue pigment, whereas the recessive, b, results in lack of color. The dominant, C, is associated with yellow pigment, whereas its recessive allele, c, also results in lack of color. Blue and yellow pigments together produce green feather color. Give the genotypes of each of the following 4 birds:

A. A green bird that has a white offspring.
B. A yellow bird that has a green parent.
C. A blue bird that has two green parents.
D. A white bird that had a green parent and a blue parent.

Chromosomes and Distribution of the Genetic Material

2

The major categories of cells

All living cells may be classified into two general categories. Those composing the most familiar organisms around us are termed *eukaryotic*. Such cells contain an assortment of intracellular membrane-bound bodies known as *organelles* in which certain specific metabolic processes are carried out. For any kind of living cell to persist from one generation to the next, it must have some mechanism to store and transfer information. All cells have solved this problem by coding information in certain large molecules—the nucleic acids DNA (deoxyribonucleic acid) and RNA (ribonucleic acid). The former is the chemical substance of the gene, and in eukaryotic cells it is confined primarily to the nucleus. RNA, however, occurs as several molecular types widely distributed throughout the nucleus and the cytoplasm. All cells contain ribosomes, each of which is composed of two subunits made up of proteins and RNA. These organelles are quite complex in their organization, and it is through their activities that the messages encoded in DNA and RNA are translated into the diversity of proteins associated with a given type of cell (see Chap. 13).

All cells must have some means to release and transfer energy. In eukaryotes, these vital processes are mediated largely through the activities of certain organelles known as mitochondria, which can be seen with an ordinary microscope but whose details are resolved only by the electron microscope (Fig. 2-1). These bodies contain all the enzymes required for aerobic respiration and the manufacture of ATP (adenosine triphosphate). Energy release through anaerobic processes takes place outside the mitochondria in the cytosol, the fluid portion of the cytoplasm. In addition to mitochondria, many types of plant cells contain chloroplasts, chlorophyll-containing organelles in which energy of the sun may be trapped and converted to ATP.

In contrast to eukaryotic cells, the cells of bacteria and blue green algae do not possess mitochondria, nor do any of those that carry on photosynthesis possess chloroplasts. Their energy capture and transfer depend on modifications of the cell membrane. While they do contain ribosomes and a cell membrane, no membranous organelles such as mitochondria, Golgi, or endoplasmic reticulum are present. Cells such as these are not compartmentalized and are known as *prokaryotes*. Most prokaryotic cells are smaller than eukaryotic ones and fall within the size range of 1 to 10 μm.

Whereas the nuclear region of eukaryotes is bounded by an envelope perforated with pores, the genetic material of prokaryotes is not surrounded by a membranous envelope and is confined to a region called the *nucleoid* (Fig. 2-2). The genetic material of the prokaryotic cell exists as one naked DNA molecule, unlike that of a eukaryotic cell in which the DNA is complexed with proteins to form chromosomes. Moreover, the eukaryotic nucleus contains one or more conspicuous bodies, nucleoli, that are essential for the assembly of the ribosomes (Fig. 2-3).

The organisms used in classic genetic studies were mainly eukaryotic. Therefore we confine our attention to eukaryotes in the next several chapters and will concentrate on prokaryotes in later ones. However, it is essential to keep in mind that any cell, prokaryotic

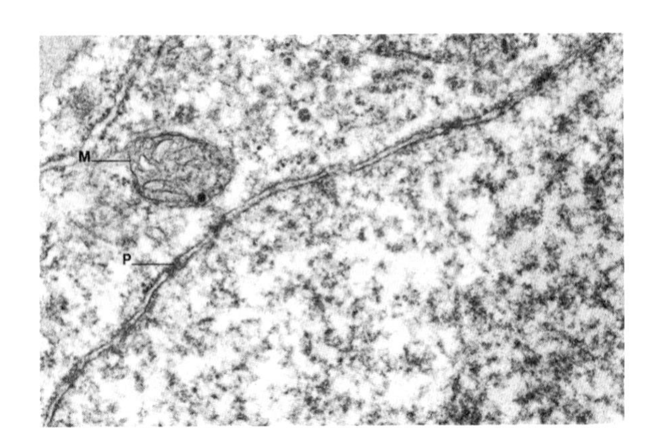

FIGURE 2-1. Electron micrograph of a portion of a cell of the grass *Panicum,* showing mitochondrion (M) in the cytoplasm. Note the pores (P) in nuclear membrane. (Courtesy of R.F. Lewis.)

or eukaryotic, is a dynamic complex system of coordinated interacting components, no single one of which is totally independent. For example, ribosomes depend on DNA, which contains the coded information for both ribosome construction and formation of cellular proteins. Nevertheless, DNA can accomplish nothing without the ribosomes and other components of the cytoplasm. Although DNA contains the information required to perpetuate the cell, genetic information by itself would be meaningless if there were no way for coded messages to be interpreted and put to constructive use.

Chromosome morphology

The name *chromatin* was given to the material in the cell nucleus that has an affinity for certain basic dyes. One of the first changes seen in a dividing nucleus is the gradual condensation of the chromatin into the bodies we recognize as chromosomes (Fig. 2-4A). We now know that chromatin consists of several diverse substances: DNA, the basic histone proteins, and acidic or neutral nonhistone proteins. Therefore, *chromatin* refers to an assortment of macromolecules, whereas chromosomes are the bodies composed of chromatin that are evident at nuclear divisions.

Mitosis is the name given to the nuclear division that ensures that two newly formed cells will contain genetic information identical to that in the parent cell. Mitosis is a fundamental feature of eukaryotes, whereas in prokaryotes the genetic material is distributed in a much simpler fashion. Very characteristic changes in the chromosomes and other parts of the cell accompany the process of mitosis, and names have been assigned to representative mitotic stages: prophase, metaphase, anaphase, and telophase (Fig. 2-4B–E).

At the onset of mitosis, the chromosome is composed of two halves or two sister chromatids. The expression *sister* implies that both structures are identical and have resulted from the replication of an original chromosome strand (Fig. 2-5). This double nature of the chromosome is apparent even at early prophase, the first stage of mitosis. It is a direct consequence of the duplication of the chromosome material that took place during interphase, that portion of the cell cycle between mitotic divisions.

As seen with the ordinary microscope, this double nature appears to extend along the length of the chromosome except at one region that appears constricted. This is the *centromere* region, also known as the *primary constriction,* where the two sister chromatids are held together (Figs. 2-5, 2-6A). The centromere is the dynamic center of the chromosome and is responsible for most of the chromosome movements at mitosis.

The electron microscope shows that a continuous chromatin fiber runs through the centromere and the rest of the chromosome and also reveals that the centromere region contains some kind of structural component whose appearance can vary from one species to the next. This is the kinetochore, the part of the centromere into which spindle fibers become inserted. Each of the two sister chromatids composing a chromosome at prophase and metaphase possesses its own centromere. It is the separation of sister centromeres at mitotic anaphase that triggers the movement of sister chromatids to opposite poles of the spindle. Once separated from its sister, a chromatid then becomes an independent chromosome (Fig. 2-6B).

As is true for the human, the chromosomes typical of a species can differ greatly in appearance (Fig. 2-5). This variation results primarily from differences in both size and centromere position. The centromere may be located more or less in the center of the chromosome, in which case the centromere is in a "median position." Consequently, the arms of such a metacentric chromosome appear approximately equal, and during its movements the chromosome may assume a V shape (Fig. 2-7). If the centromere is just off to one side of the center, it is considered "submedian" in position. This kind of chromosome is submetacentric and will have one arm distinctly shorter than the other; it may resemble an L at anaphase. When the cen-

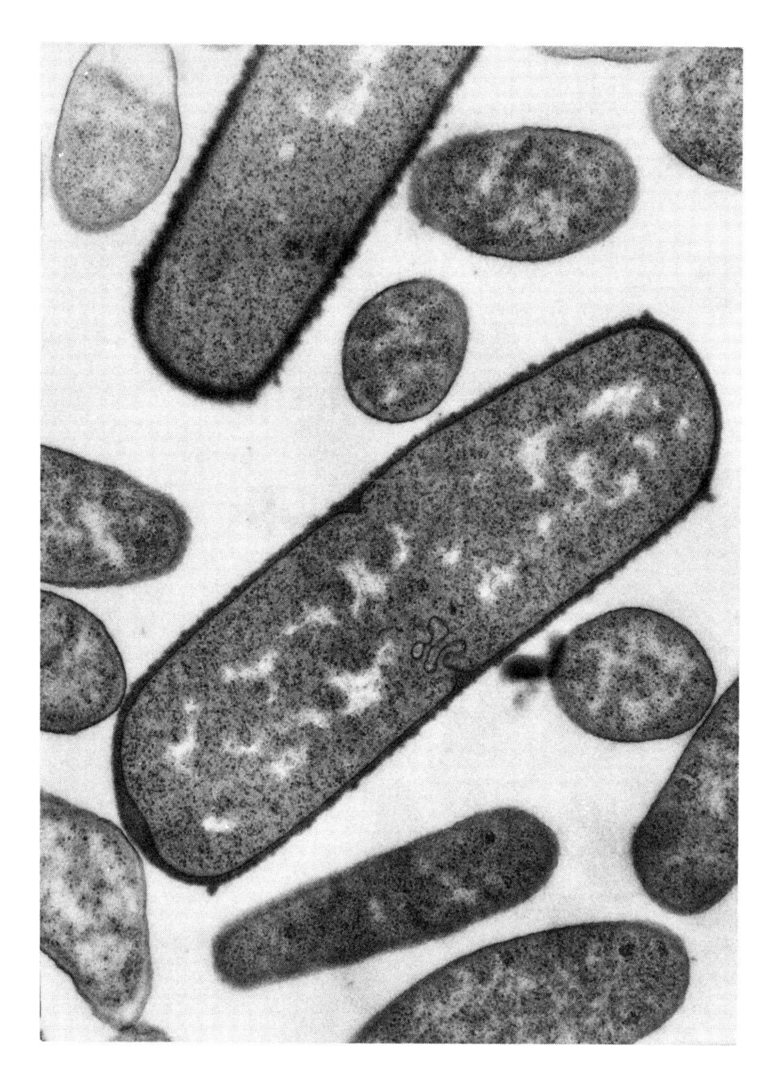

FIGURE 2-2. Electron micrograph of a bacterial cell (*Erwinia* sp.) undergoing division. The cell is packed with ribosomes that appear as deep-staining bodies. An invagination of the cell membrane is obvious. This is the mesosome to which the DNA is attached and which plays a role in DNA distribution as the cell divides. Irregularly shaped nucleoid regions appear as light areas in which the DNA can be seen as thin threads. (Courtesy of R.F. Lewis.)

tromere is way off to one side, giving the chromosome a large arm and one that is very small or inconspicuous, the chromosome is termed *acrocentric* and may have a J or an I shape (see Fig. 2-7). When the centromere appears to be at an extremity and the chromosome has only one evident arm, the expression *telocentric* is commonly used.

In addition to the narrowing attributed to the centromere, some chromosomes possess still other constrictions, known as *secondary constrictions.* These may be so pronounced that nothing more than a filament appears to connect the two portions of the chromosome. In such a case the smaller portion is called the *satellite* (Fig. 2-8). Satellites typically are associated with the formation of the nucleolus. The nucleolus does not simply form anywhere in the nucleus but arises at a special site on a specific chromosome. In a diploid organism, at least two such chromosomes are present. The special sites are the *nucleolar organizing regions,* and these are generally found in the vicinity of the secondary constrictions or satellites. When nucleoli reform at telophase, it is at these sites. They remain attached to the nucleolar organizing chromosomes until the end of the next prophase. Even in the

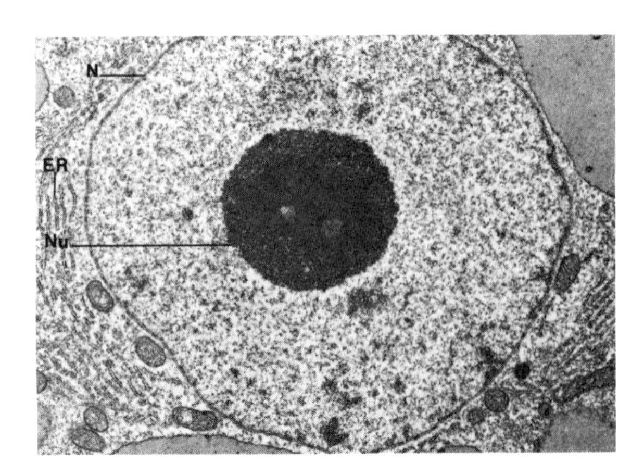

FIGURE 2-3. Electron micrograph of a portion of a *Panicum* cell showing endoplasmic reticulum (ER) covered with ribosomes and the nucleus (N). Note the conspicuous nucleolus (Nu). (Courtesy of R.F. Lewis.)

nondividing cell, the nucleolus remains associated with the chromosomal region on which it was formed. The number of nucleoli present varies from one species to another, depending on the number of nucleolar organizing chromosomes. Because two or more nucleoli may fuse, the number per cell can vary, even within an organism. Although the nucleolus does not influence chromosome movement and is not involved in the segregation of genetic factors, it is essential for the assembly of the ribosomes, as already noted. Therefore, protein synthesis depends on the presence of nucleoli. On the level of molecular events, the nucleolus is a critical part of the cellular machinery essential for the translation of information stored in the genes.

Inspection of chromosomes at metaphase shows that they exist in homologous (corresponding) pairs. This is to be expected, since each chromosome contributed by a female parent should have one that matches it from the male parent. An exception occurs in the case of sex chromosomes, which may be very different in appearance and may contain extensive regions that are not homologous (see Chap. 6). Nevertheless, any sexually reproducing organism will have body cells with two complete chromosome sets, the diploid, or 2n number, which results from the fusion of two sex cells, each containing one complete chromosome set, the haploid, or n number.

Genes are distributed in a linear order along a chromosome. Each gene site is known as a locus, which is typically constant in location. In a diploid cell, the gene residing at a given locus is represented twice such as the gene for height in the pea plant. If the gene occurs in the same allelic form on each homologous chromosome (TT or tt), the individual is homozygous at that locus. On the other hand, if the individual is heterozygous (Tt), then both allelic forms of the gene are present, T on one chromosome and t at the same locus on the homologous chromosome. Any locus, therefore, has contrasting possibilities because it can be occupied by different forms of a gene that can produce very different phenotypic effects (T for tallness and t for dwarfness).

The chromosome and the cell cycle

A single DNA molecule is found in each chromosome or in each chromatid of a replicated chromosome. The genetic content composing all the DNA in a single set of chromosomes, such as a haploid set of 23 human chromosomes, is known as a *genome.* We can think of each chromosome as a vehicle for the distribution of hereditary material at nuclear divisions. A normal cell must contain at least one balanced set of chromosomes. Therefore, when a cell divides, it is critical for each new cell to receive information identical to that present in the parent cell that gave rise to it. Elimination of information from a genome or the addition of extra copies leads to imbalance of the genome and can result in cell abnormality or death. It is the mitotic process that guarantees the transmission of balanced sets of genetic information from one cell generation to the next.

Chromosomes are typically invisible in the nondividing nucleus largely because the chromosome fiber of each is in a highly uncoiled state. However, it is in this greatly elongated condition that chromosomes are most active. When chromosomes are not condensed, major activities involving DNA are taking place on the molecular level. On the other hand, a chromosome in the highly condensed state seen at nuclear divisions is quite inactive in the dynamics of cellular metabolism. Those activities that are evident at division stages are concerned mainly with packaging and distribution of the chromosome to ensure that new cells will carry balanced genomes.

Interphase may be used to denote the state of the metabolic nuclei of cells between divisions. The entire cycle through which dividing cells pass can be separated into two main portions: interphase (the growth period between nuclear divisions) and the actual mitosis or division period, which is referred to as M or D.

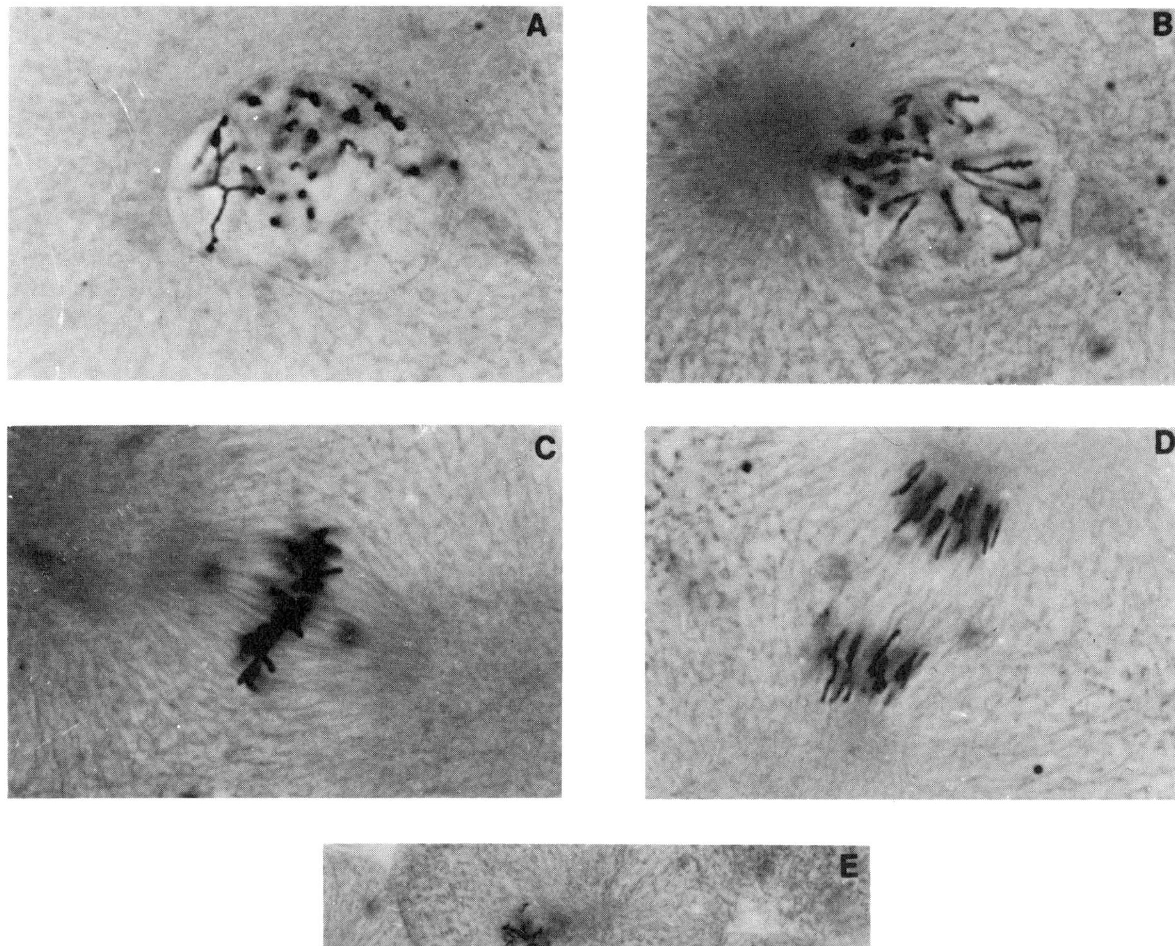

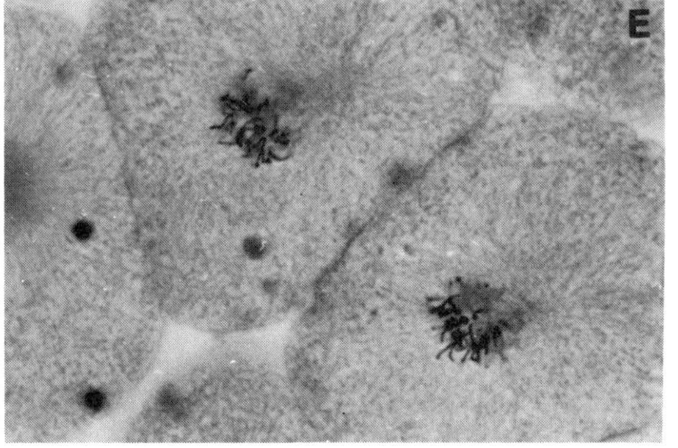

FIGURE 2-4. Cells of whitefish embryo showing changes at mitotic division. (**A**): Nucleus of cell in which the chromatin has started to condense as the cell enters prophase. (**B**): Throughout prophase, chromosomes become more and more evident as a spindle composed of microtubules is assembled. By the end of prophase, the nuclear membrane has broken down and nucleoli have disappeared. The dense region seen here at the left of the nucleus is the centrosome, a clear zone containing rays emanating from the center. The rays plus the centrosome form part of the aster. In the typical animal cell, one aster marks each of the two poles of the spindle at metaphase (**C**). The metaphase chromosomes are found at the midplane of the spindle, the equatorial plate. Anaphase (**D**) is recognized by the movement of chromosomes to opposite poles of the spindle. At telophase (**E**) chromosomes become less and less evident as nuclear membranes and nucleoli reform. Meanwhile, the spindle gradually disassembles. Constriction of the cytoplasm that may begin in anaphase achieves the division into two cells, each with one nucleus.

Early in the 1950s, the technique of autoradiography permitted the recognition of three time periods within interphase itself. It became evident that cells double their DNA content during a certain portion of interphase. This was called the S phase, and all other phases of the cell cycle are defined in terms of it. Immediately preceding the S phase is a period of interphase called G_1; this phase is an interval between the end of the previous nuclear division (M) and the S phase. Following the S phase is another gap period, G_2, before mitosis begins. In various eukaryotic cell types, the following times (in hours), are spent in the various portions of the cell cycle: G_1 (8), S (6), G_2 ($4^1/_2$), M (1) (see Fig. 2-9).

Depending on the cell type and the developmental stage, much variation occurs in the time spent in the phases of the cell cycle, especially G_1. In many animal embryos, cell divisions occur at a very rapid pace. The rate of DNA synthesis accompanying these embryonic divisions is 100 times faster than in adult cells of the same species. Moreover, G_1 may be very brief or even absent. For example, in embryonic cells of the toad *Xenopus* the entire cell cycle is only about 25 minutes long, and G_1 is absent. The S phase is followed by a brief G_2 and then an M phase. Before the M phase is completed, the following S phase is initiated. In contrast, the adult cells of *Xenopus* typically pass through all the phases of the cell cycle, and the S phase itself lasts about 20 hours! Despite such variations, critical events typically taking place in G_1, S, and G_2 appear to be required for the actual division, M, to take place.

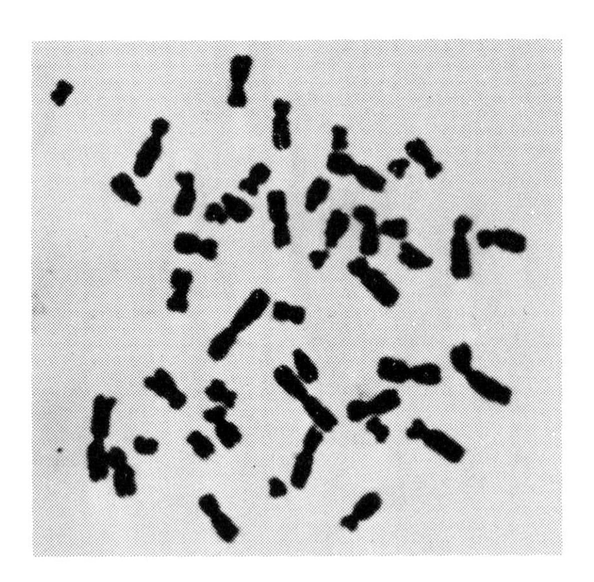

FIGURE 2-5. Human chromosomes in white blood cell from a human male. The cell was actually in metaphase but was swollen by a treatment that separates the chromosomes so that they can be counted and their morphology observed. Note that each chromosome is composed of two chromatids and is double throughout its length. The chromatids are held together at the centromere region, which is evident here as a constricted area. Human chromosomes range in size from large to small and differ in the position of the centromeric region.

While the length of G_1 varies greatly, certain very important key events must take place during that phase in the typical cycle. These entail preparations for the S phase. When G_1 is absent, such processes must occur at some earlier point, such as immediately after M or even before M is completed. DNA replication at S involves very complex interactions at the molecular level (see Chap. 11). It is staggering to realize that, in the human, 46 extended chromosome threads with a total length of about 6 feet must be duplicated while they are packed into a nucleus about 1/2,500 of an inch in diameter! In the G_2 portion of interphase, final preparations are made for the mitotic process. No continued cycling of cells occurs without the S, M, and G_2 phases.

We are well aware that cells can live without dividing and that some cells composing an organism rarely, if ever, undergo division. There is very good evidence that the late G_1 period contains a point, called the *restriction point.* We can think of this as a point of no return. If a cell passes this point, it will enter the S phase and then complete the rest of the cell cycle. Cells that remain arrested at this restriction point, and no longer divide, are often considered to have left the cell cycle and to have entered the G_0 phase.

Studies of the cell cycle are extremely important, not only for the insight they provide on the key mechanisms involved in cell division, but also for the information they provide on the cancer cell, for cancer is a problem of the cell cycle. Although a cancer cell's cycle is not unusual, and is similar to that of any other dividing cell, it keeps repeating the cycle over and over again, as if it had been removed from the constraints imposed on a normal cell, which divides only in harmony with its neighbors. The malignant cell is discussed at greater length in Chapter 18.

Changes at mitosis

The interphase chromosomes begin to contract with the onset of mitotic prophase and become progressively more condensed and conspicuous as this stage progresses (see Fig. 2-4B). By the end of prophase, we can

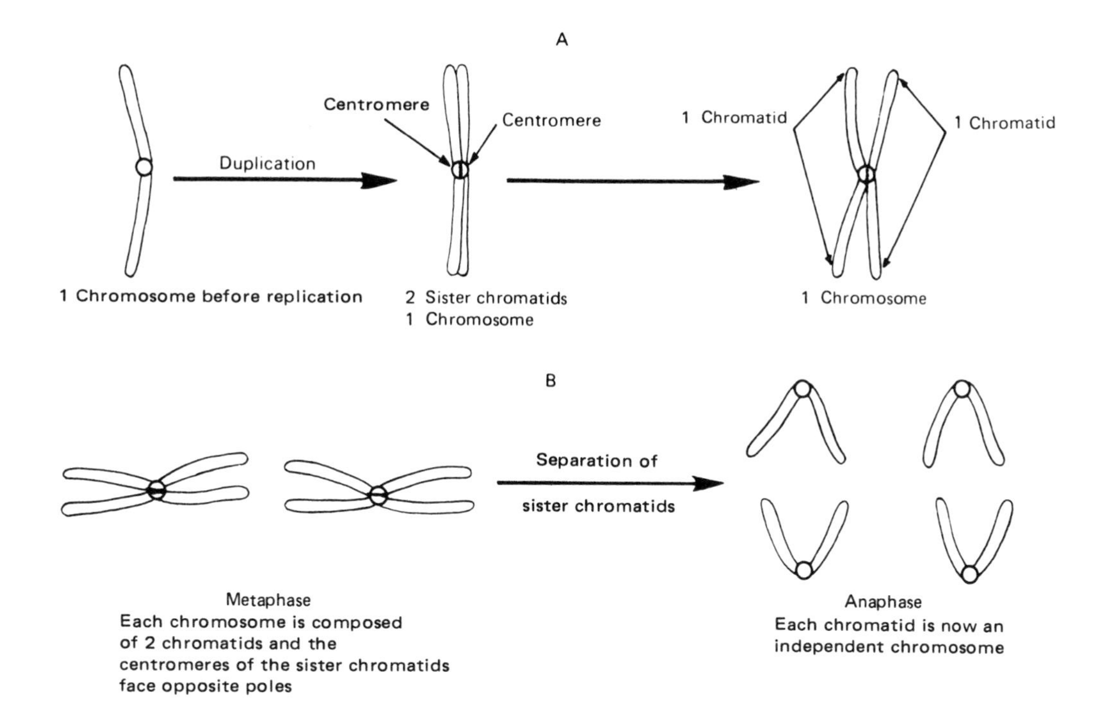

FIGURE 2-6. Chromosome and chromatid. **A:** The chromosome exists as a single structure before it is duplicated. Before the onset of prophase, the chromosome replicates, so that at prophase and metaphase it is composed of two identical sister chromatids, each with its own centromere (represented as circle for emphasis). Both chromatids are held together at the centromere region. **B:** Separation of the centromeres of the sister chromatids initiates anaphase movement. The sister chromatids travel to opposite poles, and each is now an independent chromosome.

distinguish the two chromatids and the centromere position of each chromosome. Meanwhile, throughout prophase the spindle apparatus is forming. Although the spindle includes a number of different elements, it is made up primarily of fibers. These are actually microtubules, hollow cylinders assembled from linear arrays of two very closely related protein molecules, α- and β-tubulin. The spindle begins to form early in prophase in association with the growth of the fibers. By the end of prophase, the nuclear envelope has broken down. The spindle, which is actually made up of two half-spindles, is now seen to consist of fibers extending from two opposite poles (Fig. 2-10A).

In the cells of animals, many fungi, algae, and lower plants, the spindle forms in association with centrioles, microtubular structures that mark the poles. As shown in Figure 2-10A, centrioles always occur in pairs. A pair becomes duplicated in the S phase of the cell cycle, so that two pairs of centrioles are already present at the onset of prophase. Surrounding the centrioles is the centrosome, a clear area that is particularly conspicuous in some cells (Fig. 2-4B,C). The two pairs of centrioles gradually move apart throughout prophase as rays of microtubules emanate from their vicinity, passing through the centrosome. A region composed of a centriole pair, the associated rays, and the centrosome is known as an *aster.* Because two pairs of centrioles are present at the beginning of mitosis, two asters arise, and these come to mark the poles of an astral spindle at the end of prophase.

An animal cell in which the centrioles have been experimentally destroyed still manages to form spindles. Moreover, the cells of higher plants are anastral, lacking asters entirely, yet they nevertheless form mitotic spindles. Such observations leave unclear the exact role of asters and their components in the construction of spindle fibers. When asters are present, they seem to act as focal points for the assembly of microtubules, although they are not essential to their organization. Astral spindles look quite different from the less clearly focused anastral spindles of higher plants (Fig. 2-10A,B). The electron microscope has disclosed at the poles of both astral and

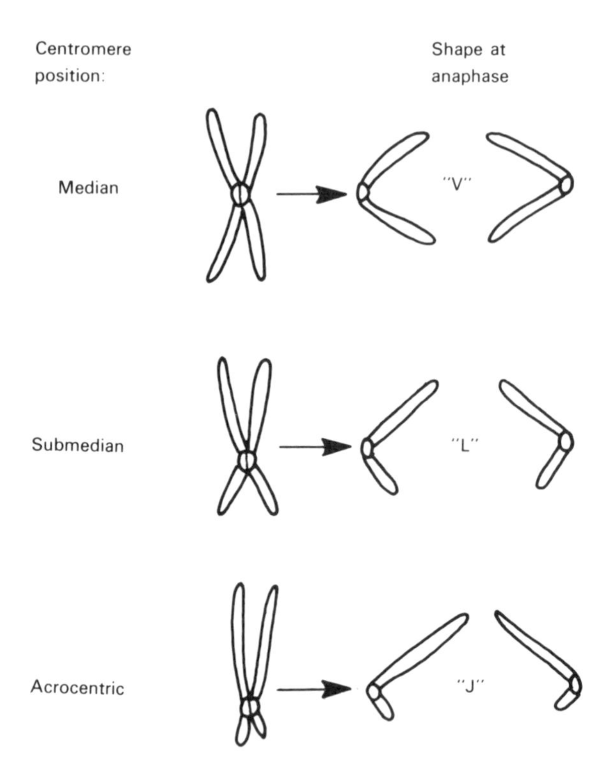

FIGURE 2-7. Chromosome morphology. The shape of the chromosome at metaphase depends largely on the position of the centromere. At anaphase, chromosome shape is the result of the arm length and the fact that the centromere leads in the movement to the poles.

anastral spindles a region that contains lightly staining material from which spindle microtubules appear to grow. This rather hazy region may represent the actual center in the cell that is responsible for the organization of spindle fibers.

New approaches with the electron microscope along with staining refinements are leading to a clearer picture of spindle organization and dynamics during mitosis. The spindle fibers or spindle microtubules can be divided into two major categories based on where their ends terminate. The most numerous of the fibers are the polar fibers, shown in Figure 2-10A,B. These fibers have one end that terminates at a pole of the spindle (at the centrosome of an astral spindle). The other end appears free. Some of these polar fibers are short, but others are long and interdigitate with polar fibers emanating from the other pole of the spindle. The midplane of the spindle (the region lying midway between the poles) is referred to as the *equatorial plate* or *equator of the spindle.*

As prophase progresses, another class of spindle fibers can be recognized. These are the kinetochore fibers and are distinguished from the polar fibers on the basis of their ends (Fig. 2-10C,D). One end of a kinetochore fiber is bound to a kinetochore in the centromere region of a chromosome. The other end of many of these fibers terminates at a pole (or at the centrosome of an astral spindle). Some kinetochore fibers, however, may have the other end free and unbound. Typically, several spindle fibers bind to the

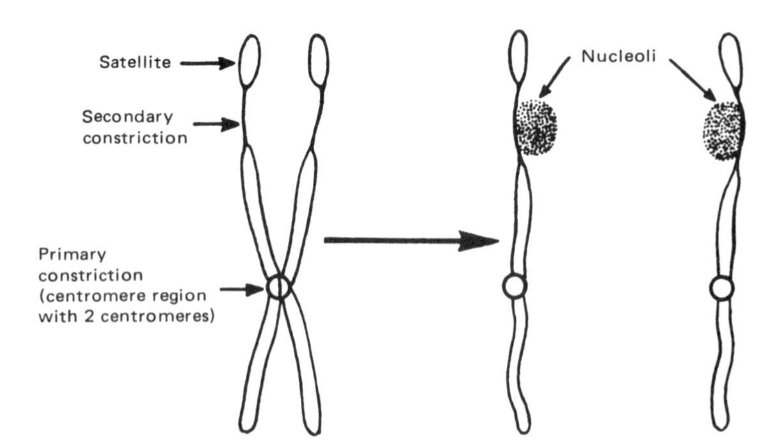

FIGURE 2-8. Secondary constrictions. These occur on some chromosomes and may be so pronounced that a portion of the chromosome appears to be set off as a satellite. These secondary constrictions frequently contain nucleolar organizing regions at which nucleoli reform at telophase. Every typical normal diploid possesses at least two chromosomes with nucleolar organizing regions. All constrictions of the chromosome other than the primary one are known as *secondary constrictions,* even those not associated with nucleolar organizers.

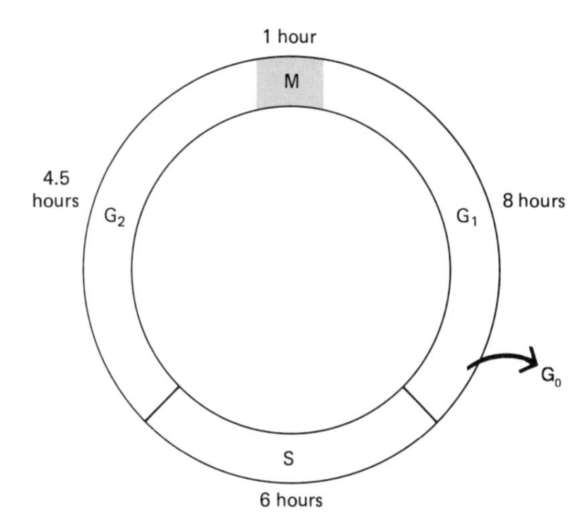

FIGURE 2-9. The cell cycle. The cycle shown is typical for a mammalian cell in culture. Although cell cycles vary from one cell type to the next, that portion of a cycle devoted to mitosis is brief, the major portion being spent in interphase. A cell may enter a G_0 phase in which it no longer undergoes cycling.

very same kinetochore; the number varies from one species to the next. In the human, about 20–40 microtubular fibers are inserted into each kinetochore. In the case of each mitotic chromosome, the kinetochore fibers of one of the sister chromatids at prophase associate with one of the poles of the spindle, and the kinetochore fibers of the other sister associate with the opposite pole (Fig. 2-10D). It is the kinetochore in the centromere region of each chromatid that provides the only connection between the chromosome and the mitotic apparatus.

With the disappearance of the nucleoli from the nucleolar organizing region of the satellited chromosomes and the breakdown of the nuclear envelope, interactions between kinetochores and the attached fibers are responsible for movement of the chromosomes to the middle of the spindle. The arrangement of the chromosomes at the equator marks the beginning of metaphase, the stage at which the chromosomes are in their most compact state (see Figs. 2-4C, 2-10C). At metaphase, it is actually the centromere region of each chromosome that is at the equatorial plate. The arms of the chromosomes project from their centromere regions in various ways because they are being carried along passively by the dynamic centromere regions and therefore do not all lie parallel to the equator as diagrams of metaphase often suggest. At metaphase, the kinetochores of the sister chromatids are facing opposite poles.

One of the most critical events of the mitotic process is the separation of sister chromatids, each of which then becomes an independent chromosome. Their complete separation ends metaphase, and anaphase movement is initiated (Fig. 2-4D). The interactions between the kinetochore and its fibers, which make chromosome movement possible, are still far from clear and are the focus of ongoing research in several laboratories.

Because the two sisters composing a chromosome become independent and move to opposite poles at anaphase, identical genetic material will be found at opposite ends of the cell when this stage is completed (Fig. 2-11). By the end of telophase, the spindle disappears, nuclear membranes and nucleoli are reconstituted, and each new (daughter) nucleus assumes the appearance characteristic of interphase (see Fig. 2-4E). Each chromosome during this part of the cell cycle is single, carrying only one chromatin fiber. If the cell passes through another G_1, S, and G_2, each chromosome will again become composed of two chromatids, each with its own fiber. Note that mitosis is basically a division of the nucleus. Cytokinesis, division of the cytoplasm, may or may not take place. The genetically significant point is that the two daughter nuclei resulting from the mitotic division of a nucleus are identical insofar as their chromosome complements are concerned. Barring rare accidents, mitosis is an exceptionally precise process that ensures genetic continuity from one cell generation to the next and the distribution of balanced sets of chromosomes to all the cells of a multicellular organism.

The human chromosome complement

Treatment of a small sample of blood cells growing in culture with a mitotic stimulator triggers certain leukocytes to undergo mitosis. The dividing cells are then trapped by the application of a drug such as colchicine, a chemical that disrupts the organization of the spindle and prevents anaphase movement. The cells are then killed with a chemical fixative and exposed to a hypotonic solution, which swells them and disperses the chromosomes so that they become separated from one another. Finally, the cells are attached to slides and stained for observation with the microscope. The complement of the human cell can then be photographed (see Fig. 2-5). The term *karyotype* is used to refer to the chromosomes of a single cell or of an individual. Each chromosome may be cut out of a photograph and placed in a sys-

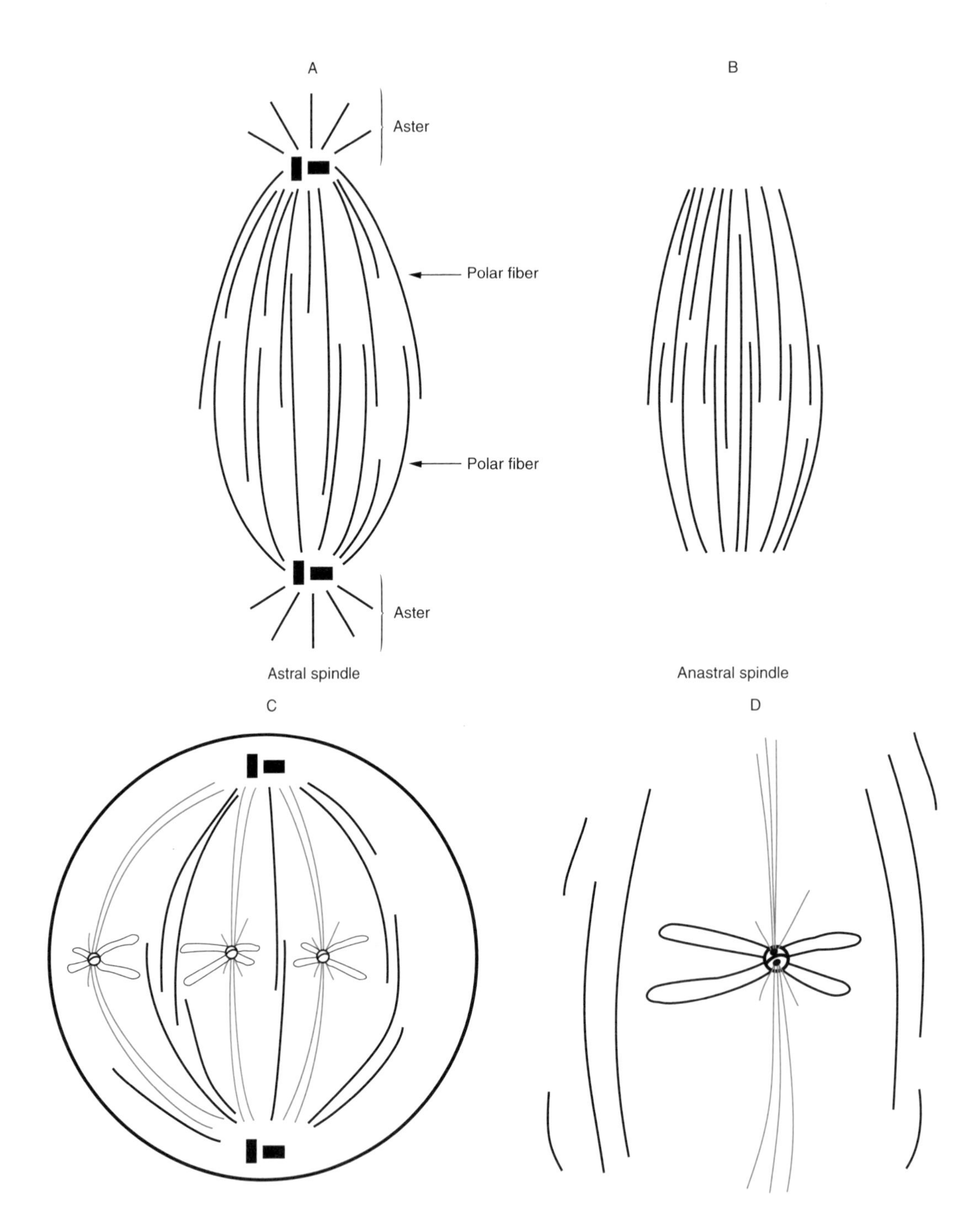

FIGURE 2-10. Organization of the spindle. **A:** A spindle is composed of two half-spindles, each defined by poles from which microtubular fibers emanate. An astral spindle possesses fibers that are focused toward the asters at the poles. An aster includes microtubular rays and a centriole pair surrounded by a clear centromere region. Represented here are polar fibers, each with one end bound at the centrosome region and the other end free. Polar fibers differ in length. **B:** The polar fibers of an anastral spindle are much less focused and lack asters to mark the poles. **C:** By metaphase, kinetochore fibers (red) have arisen. Each fiber in this category has one end bound to a kinetochore. The other end of most such fibers is bound to one of the poles. Some of them do not extend to a pole and possess a free end. **D:** Several spindle fibers bind to a kinetochore. The kinetochore fibers of sister chromatids are associated with opposite poles. (The number of fibers represented was chosen entirely for the sake of clarity.)

tematic arrangement (Fig. 2-12). When a karyotype analysis is done, the chromosomes are assembled into homologous pairs and then arranged in order of decreasing size. In any karyotype preparation, the centromeres are placed on the same line or level so that centromere position and length of chromosome arms can be easily compared from one chromosome pair to the next.

When the chromosome complements of most humans are examined, it is seen that 23 pairs of chromosomes are present and that 22 of these pairs are identical when the chromosome complements of males and females are set side by side (Fig. 2-12). There is, however, a difference in one pair, the pair designated the *sex chromosomes.* It is seen that a female karyotype contains two sex chromosomes of appreciable size and identical appearance. These are the X chromosomes. On the other hand, the male karyotype shows one X chromosome and a very small chromosome, the Y. All the chromosomes in the human complement other than the sex chromosomes are called *autosomes.* The human therefore has 22 pairs of autosomes plus one pair of sex chromosomes (2 X chromosomes in the female and 1 X and 1 Y in the male).

Table 2-1 describes the major groups of human chromosomes as recognized in 1960 at a meeting of cytogeneticists in Denver, Colorado. Following the Denver classification, the autosomes are serially numbered 1 to 22 in descending order of length. The sex chromosomes continue to be called X and Y. Following routine staining procedures, the chromosomes can be easily classified into seven groups. Within a group however, distinctions are frequently difficult or uncertain, as in group C, which contains the X chromosome, and in group G, which includes the smallest chromosomes as well as the Y. Since 1969, techniques have become perfected that clearly reveal distinctions among all the human chromosomes. In the first of these methods, chromosome preparations are exposed to the fluorescent dye quinacrine mustard, which reveals so-called Q bands that form a pattern that is distinct for each chromosome (Fig. 2-13). Moreover, the procedure detects the Y chromosome as a fluorescent body in nondividing male cells. However, fluorescent banding techniques require special optical lenses and have certain limitations. In the 1970s, simpler methods were devised that utilize the common Giemsa stain. These not only permit permanent staining but also reveal distinctive banding patterns.

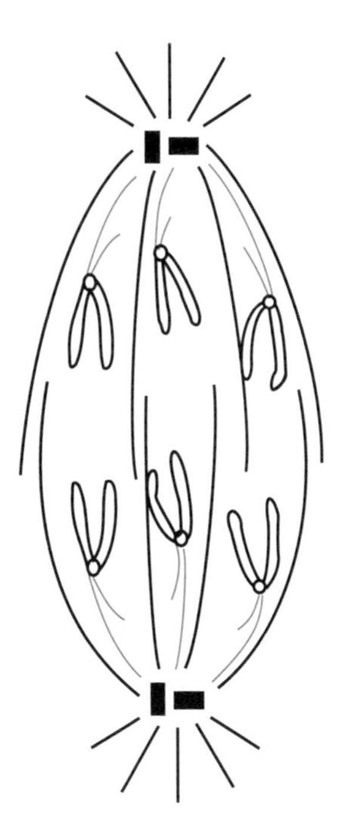

FIGURE 2-11. Anaphase movement. At anaphase, the sister chromatids that compose a chromosome separate, become independent chromosomes, and move to opposite poles as the kinetochore fibers shorten.

One type of procedure, which entails a trypsin-digesting pretreatment followed by Giemsa staining, brings out bands that have been designated *G bands.* These are very similar (although not necessarily identical) to the Q bands; in general, heavily stained G bands correspond to the highly fluorescent Q bands (Fig. 2-14). Another method involving heat denaturation produces a reverse banding pattern that is just the opposite of the Q and G pattern. These so-called R bands stain heavily where the staining is poor in the G band pattern. Those regions that stain poorly in the R band pattern are heavily stained in the G band pattern. R bands can also be brought out when preparations are exposed to certain fluorescent dyes.

Still another banding pattern is seen following a procedure that includes a pretreatment with hydrochloric acid and sodium hydroxide. The C bands that result are very distinct from the other bands we have described and are found only in certain restricted regions of the chromosomes, particularly the centromere region, secondary constrictions, the long arm of the Y chromosome, and the satellites of the G group (Fig. 2-15). These characteristic C bands stain regions that represent a special type of

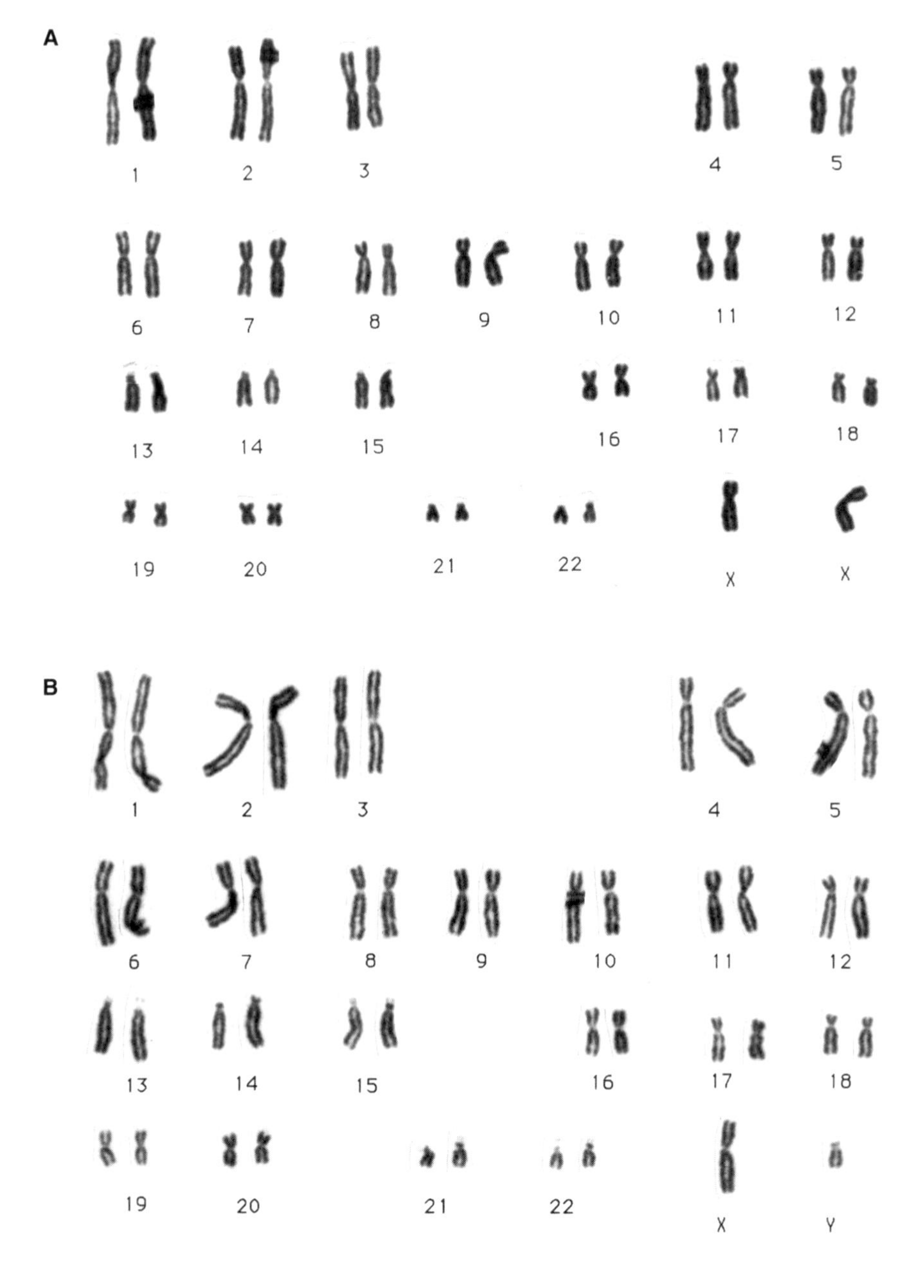

FIGURE 2-12. Mitotic metaphase chromosomes of a human female **(A)** and male **(B)** arranged in homologous pairs. (Courtesy of Steven A. Schonberg, Ph.D., University of California, San Francisco.)

chromatin found along the length of the chromosome. This kind of chromatin tends to remain condensed in the metabolic or interphase nucleus and in prophase, unlike the rest of the chromosome material. Such chromatin is known as *heterochromatin* as opposed to *euchromatin,* which composes most of the chromosomes and tends to be stretched out at interphase and during the early stages of nuclear divisions. In contrast to euchromatin, heterochromatin is late replicating. At the time the genetic material is being replicated at S of interphase, the heterochromatin lags behind the euchromatin. Heterochromatin such as that found at the centromere region and revealed by the C banding procedure is known as *constitutive heterochromatin,* so-called because it is constant in its behavior and always tends to behave differently from

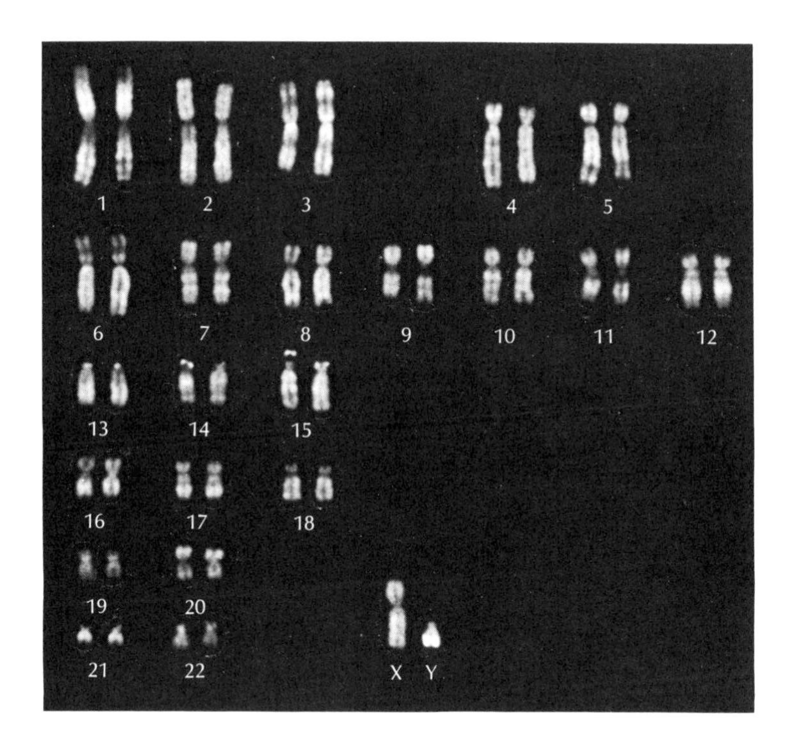

FIGURE 2-13. Karyotype of human male showing Q bands. The sex chromosomes are at the bottom of the karyotype. Note the small, highly fluorescent Y chromosome. (Courtesy of T. Caspersson, Karolinska Institute.)

TABLE 2-1. Classification of human mitotic chromosomes

Group		Description
1–3	(A)	Large chromosomes with approximately median centromeres. The three chromosomes are readily distinguished from each other by size and centromere position
4 and 5	(B)	Large chromosomes with submedian centromeres. The two chromosomes are difficult to distinguish, but chromosome 4 is slightly longer
6–12	(C)	Medium-sized chromosomes with submedian centromeres. The X chromosome resembles the longer chromosomes in this group, especially chromosome 6, from which it is difficult to be distinguished. This large group is the one that presents major difficulty in identification of individual chromosomes
13–15	(D)	Medium-sized chromosomes with nearly terminal centromeres (acrocentric chromosomes). Chromosome 13 has a prominent satellite on the short arm. Chromosome 14 has a small satellite on the short arm. Chromosome 15 was later found to possess a satellite as well. Satellites may be difficult to detect in preparations
16–18	(E)	Rather short chromosomes with approximately median (in chromosome 16) or submedian centromeres
19 and 20	(F)	Short chromosomes with approximately median centromeres
21 and 22	(G)	Very short, acrocentric chromosomes with satellites on the short arms. The Y chromosome resembles these chromosomes and is placed in this group. It cannot be distinguished from the other members in many cases

Source: Reprinted with permission from *JAMA,* 174:159–162, 1960.

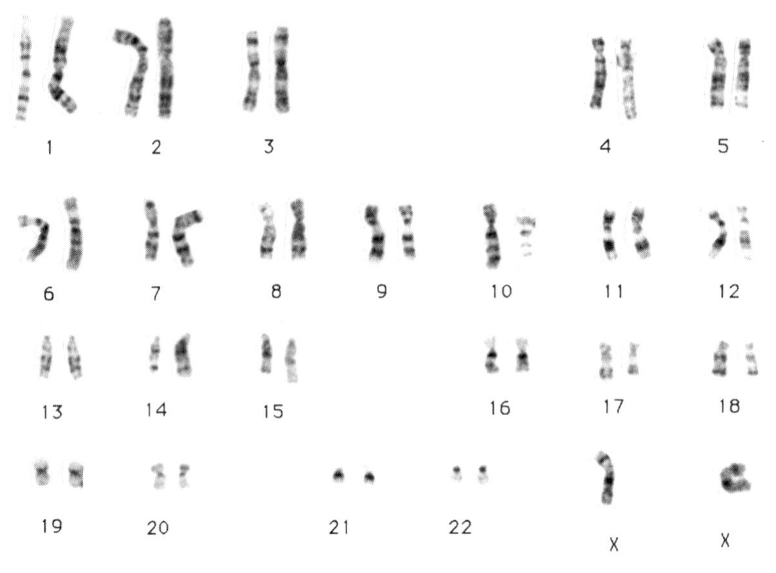

FIGURE 2-14. Karyotype of human female showing G bands. (Courtesy of Steven A. Schonberg, Ph.D., University of California, San Francisco.)

the major portion of the chromatin. We will return to this subject in Chapter 6, where we discuss another type of heterochromatin.

The banding techniques have led to a new era in the investigation of human chromosomes. Various conventions were adopted throughout the years to describe the human haploid chromosome complement at mitotic metaphase. The shorter arm of any chromosome was designated the *p* arm and the longer of the two the *q* arm. When a karyotype is prepared, the chromosomes are arranged so that all p arms are directed upward and the q arms downward. Based on the G banding pattern, 350 bands were recognized at first. Certain bands that are typical features of specific chromosomes were arbitrarily designated *major bands.* These major bands were used along with the centromere and the chromosome ends (telomeres) to identify specific regions on each arm of a chromosome. As Figure 2-16A shows, a region is given a number that reflects its distance away from the centromere toward the end of an arm. Region 7q2 designates that portion of the longer arm of chromosome 7 away from the centromere found between two other regions that are also recognized as "major." Note that within each region both dark and light bands can occur, and these in turn are given specific identifying numbers that also reflect distance away from the centromere in the arm. Thus 7q21 indicates the long arm of chromosome 7, region 2, band 1, a band closer to the centromere than the other band in this region.

Improved techniques continue to increase the resolution of the bands and now reveal 2,000 or more of them in a haploid human chromosome set at late mitotic prophase. As the chromosomes continue to condense from prophase to metaphase, the thinner prophase bands coalesce to produce the fewer but thicker bands of metaphase. Consequently, many bands that have been numbered have now been seen to include smaller subbands. For example, band 1 of region 2 of the long arm of chromosome 7 can now be resolved into three subbands (Fig. 2-16B). Any subband retains the same number as the large band to which it belongs, but it is designated specifically by a number that follows a decimal. This number also reflects the distance of the subband away from the centromere and toward the end of the chromosome arm. The designation 7q21.3 identifies the larger arm of chromosome 7, region 2, band 1, subband 3, the one farthest from the centromere and thus closest to the end of the arm.

Still higher resolution has revealed finer subdivisions of certain subbands. These are added to the subband designation to give two decimal places and are numbered in the same way as the subbands to reflect distance in the direction away from the centromere. As shown in Figure 2-16B, 7q11.22 indicates the long arm of chromosome 7, region 1, band

FIGURE 2-15. Metaphase chromosomes of human female showing C banding. Note the staining of the centromere regions. (Courtesy of Steven A. Schonberg, Ph.D., University of California, San Francisco.)

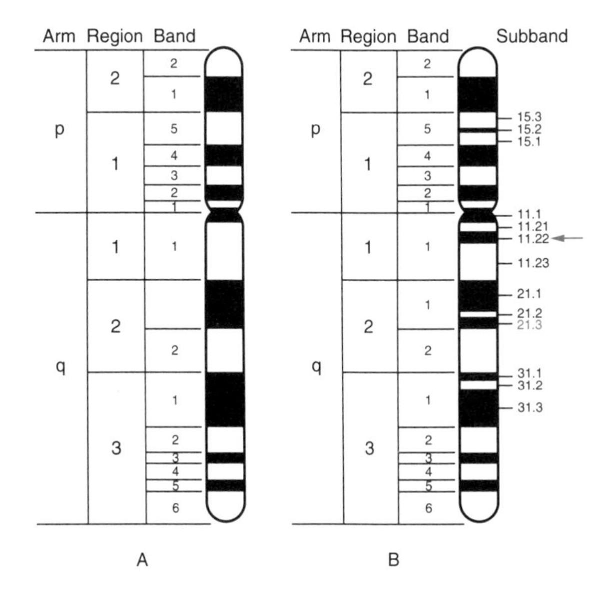

FIGURE 2-16. Designation of G bands on human chromosome 7. The smaller chromosome arm is designated *p* and the larger one *q*. The chromosome regions, bands, and subbands of each arm are numbered to reflect their distance from the centromere. **A:** The designation 7q21 indicates the larger arm of chromosome 7, region 2, band 1 (red number). **B:** Many bands have been resolved into subbands. Band 7q21 can be shown to consist of three subbands, each denoted by a number after a decimal to reflect position in the direction away from the centromere. Thus, 7q21.3 (red number) indicates the subband of band 7q21 which is farther from the centromere than the other two subbands. Certain subbands may in turn be resolved into subdivisions as seen here in the case of subband 11.2. Divisions of the subbands also reflect position in relation to the centromere. The designation 7q11.22 (red arrow) identifies a chromosome 7 segment in the long arm found in region 1, band 1, subband 2, subband division 2.

1, subband 2, second division of subband 2 away from the centromere.

The improvements in staining and band resolution are permitting smaller and smaller chromosome alterations to be recognized and to be related precisely to given sites in an arm. Any alteration resulting in a change in band size, shape, or location may in turn be related to certain genetic effects. In later chapters we will return to this topic of chromosome banding and its application to the localization of genes involved in human afflictions.

Giant chromosomes

One of the many features of the fruit fly *Drosophila* that has contributed to its value as a genetic tool is its giant chromosomes, found in the immense nuclei of the larval salivary gland cells. The larval stage of the fly is concerned primarily with food getting; efficient salivary glands are important to an insect, which literally eats its way through food. These active glands, as well as certain other organs found in some of the dipteran insects (flies), achieve their growth largely through an increase in cell mass and volume rather than an increase in the number of cells. After a certain number of cells is established through mitosis, further cell divisions cease. Growth continues, however, as cells enlarge. As the nuclear size increases, the chromosome threads that represent the chromosomes at metaphase duplicate again and again without an accompanying mitotic division. These extra chromosome sets are probably necessary for the increased activities of cells of greater size and mass. Another unusual feature of these cells is the phenomenon of somatic chromosome pairing. Typically, homologous chromosomes pair only at meiosis (see the next chapter). But, in the cells of the salivary glands, not only do homologous chromosomes pair, but the paired chromosome threads duplicate over and over throughout the larval stage. Approximately 11 rounds of duplication take place, producing thousands of individual threads. The very strong forces of pairing that seem to operate in these cells prevent the threads from separating. The end result is that each nucleus appears to contain a haploid number (four in *Drosophila melanogaster*) of giant chromosomes. Each, of course, is

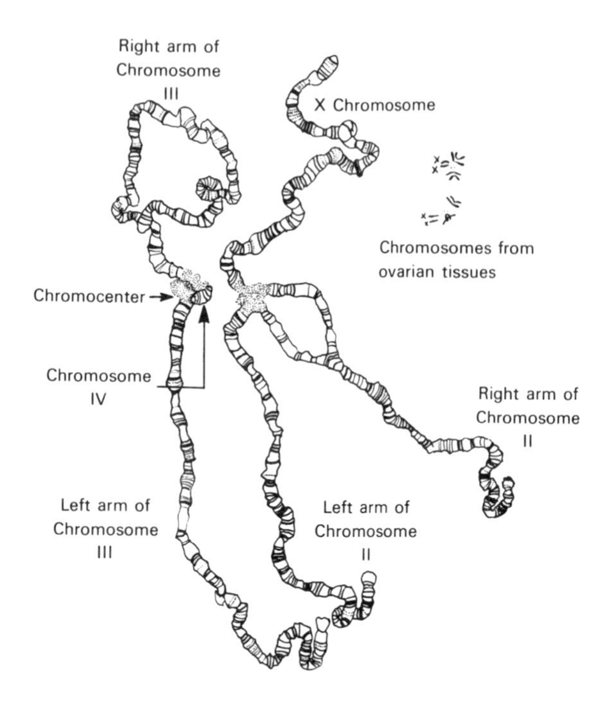

FIGURE 2-17. The salivary gland chromosomes of *Drosophila melanogaster.* (Reprinted with permission from *Hered. J.* 25:464–476, 1934.)

actually composed of the many separate chromosome threads of two homologs.

All the threads held together side by side as a large compound unit form a giant structure called a *polytene chromosome,* that is, a chromosome composed of many threads. This growth of the salivary glands through increasing cell size rather than cell number may be very efficient for the best operation of the enzymatically active cells during the period of active food getting. The glands will finally be resorbed when the larva is transformed into the pupa.

The large salivary cells are not destined to divide again. The large polytene chromosomes are therefore not in any stage of mitosis, but are actually in an extended interphase. They are stretched out almost to their maximal length, a condition that makes them superb subjects for cytogenetic studies. They provide an unusual advantage for assigning genes to definite chromosome regions. When the polytene chromosomes are stained, they display deeper staining regions, bands of varying widths alternating with lighter stained regions, the interbands (Fig. 2-17). The width and arrangement of the bands is so characteristic that a small chromosome segment by itself can be identified by its banding pattern. The deeply staining bands are also called *chromomeres.* There is reason to believe that the chromosome threads are more tightly folded or packed in the region of the bands than in the interbands. As a result, a greater concentration of hereditary material should occur in the band region, and this seems to be the case, as indicated by stains that are specific for DNA. Since thousands of separate chromosome threads are associated side by side to form a polytene chromosome, the bands, or chromomeres, necessarily stain more deeply than the interbands. Specific genes have been associated with specific bands, although it is still unsettled whether one gene or more are located in a distinct band.

The structure of a chromosome may be altered, and a gene may actually be lost or deleted. Such a loss is often detected by the genetic effects that ensue. Observation of the giant chromosomes may then show that a loss of a certain band accompanies the loss of the gene. Other kinds of changes besides gene deletion also alter chromosome structure, such as inversion, in which the order of the genes becomes reversed, or translocations, in which chromosome arms are shifted to new locations (Chap. 10). These changes can be correlated with changes in the bands. This enables the cytogeneticist to assign genes in *Drosophila* to definite chromosome locations. The polytene chromosomes are also of immense value in studies of cell differentiation and gene action (see Chap. 17). We will continue to refer to these giant structures throughout the discussions.

Naming of genes and loci

A few words are in order regarding the conventions used to name genes and their loci on the chromosomes. Although we will use the familiar fruit fly, *Drosophila,* in our examples here, the reasoning applies to other organisms as well.

As commonly used in genetic analysis, the designation *wild* means the standard form typically found in nature. Any inherited departure from the wild may result from a mutation at a certain gene site, a locus on a chromosome. The name given to any mutant stock usually suggests the kind of phenotypic variation from the wild type attributed to the genetic alteration. For example, red eye color in *Drosophila* represents wild type. Gene mutation at one of the loci on the X chromosome can result in the absence of eye pigment. Stocks that are homozygous for this change are designated *white,* and the position or point on the chromosome where the alteration has taken place is named the *white locus.*

The wild gene form for red eye color and its mutant form for white represent a pair of alleles, and either one of the two members may be present at the "white locus" on a particular X chromosome. In elementary discussions, it is common practice to symbolize the dominant member of a pair of alleles by a capital letter and to use a small one for the recessive, such as W and w for red and white eye color, respectively. However, another system of reference is also commonly followed. The plus sign (+) is used to represent any wild trait. The factor for red eye is thus simply designated + (or w^+ for extra clarification), and its allele for white eye remains w. The name given to any genetic locus (in this example, *white*) describes the first mutation or variation detected at that site.

Many recessive mutations in *Drosophila* are known to affect an assortment of other traits besides eye color, and all are named according to this scheme. The recessive *ebony* can result in a fly with a shiny black body instead of the normal gray. The pair of alleles involved is represented by + (or e^+) for wild and e for ebony. The recessive forked, f, can cause shortened, split bristles in contrast to the longer, unsplit ones determined by the wild gene form, + or f^+. Further examples could be given for a long list of recessive alleles. A wild allele corresponds to each mutant one, and every one of the wild alleles is represented as +.

Since many names begin with the same letter, we must often use more than one letter to represent a locus. As e has been used for ebony, it cannot be selected to describe the recessive allele *eyeless,* which results in a reduction of eye size. More than just the first letter is required to stand for eyeless, and the locus is symbolized by ey. The normal eye condition is still + (or ey^+). The method is similar for a long list of others, such as dp (for *dumpy,* a gene form giving reduced wings) and ss (for a recessive, causing reduction in bristle size).

The mutations discussed thus far are all recessive to their wild type alleles (+). Gene alterations that produce a dominant effect, although not as numerous as recessive changes, are responsible for several phenotypic departures from the wild. The naming of dominant gene mutations is identical to that for the recessives. The wild allele, although now the recessive form in such cases, is still represented by +. The mutant allele, however, is symbolized by a capital letter (or an abbreviation beginning with a capital). A frequently encountered dominant genetic effect in the fruit fly is the "Bar eyed" condition in which eye size is narrowed. It is represented by the capital letter B, the wild condition as + or B^+.

The symbol + always represents any wild type allele or standard condition. Any mutant, on the other hand, is designated by a letter or abbreviation. A symbol beginning with a capital tells us that the gene form is dominant to wild as in B (Bar), Cy (curly wings), and Pm (plum, a brownish eye color). The wild alleles are, respectively, B^+, Cy^+, and Pm^+ (or simply + in each case). The system that designates wild by a + has many benefits. These will become very evident in the discussions related to multiple alleles, linkage, and crossing over. At certain times, we may still prefer to use the older scheme of capital and small letters for dominant and recessive when discussing simple monohybrid or dihybrid crosses.

REFERENCES

Alberts, B., et al. *Molecular Biology of the Cell,* 2nd Ed. Garland, New York, 1989.

Brachet, J. *Molecular Cytology, Vol. 1: The Cell Cycle.* Academic Press, New York, 1985.

Broach, J.R. New approaches to a genetic analysis of mitosis. *Cell* 44:3, 1986.

Gorbsky, G.J., P.J. Sammak, and G.G. Borisy. Chromosomes move poleward in anaphase along stationary microtubules that coordinately disassemble from their kinetochore ends. *J. Cell Biol.* 104:9, 1987.

Kavenoff, R., L.C. Klotz, and B.H. Zimm. On the nature of chromosome-sized DNA molecules. *Cold Spring Harbor Symp. Quant. Biol.* 38:1, 1974.

Marx, J. The cell cycle: Spinning farther afield. *Science* 252:1190, 1991.

Mazia, D. The cell cycle. *Sci. Am.* (Jan.) :54, 1974.

Mazia, D. The chromosome cycle and the centrosome cycle in the mitotic cycle. *Int. Rev. Cytol.* 100:49, 1987.

McIntosh, J.R., and K.L. McDonald. The mitotic spindle. *Sci. Am.* 48, 1989.

McIntosh, J.R., and M.P. Koonce. Mitosis. *Science* 246:622, 1989.

Murray, A.W., and M.W. Kirschner. Dominoes and clocks: The union of two views of the cell cycle. *Science* 246:614, 1989.

Murray, A.W., and M.W. Kirschner. What controls the cell cycle? *Sci. Am.* (Mar.) :56, 1991.

Nicklas, R.B. Chance encounters and precision in mitosis. *J. Cell Sci.* 80:283, 1988.

Olson, S.B., R.E. Magenis, and E.E. Lovrien. Human chromosome variation: The discriminatory power of Q-band heteromorphism (variant) analysis in distinguishing between individuals, with specific application to cases of questionable paternity. *Am. J. Hum. Genet.* 38:235, 1986.

Pardee, A.B. G1 events and regulation of cell proliferation. *Science* 246:603, 1989.

Rieder, C.L. The formation, structure, and composition of the mammalian kinetochore and kinetochore fiber. *Int. Rev. Cytol.* 79:1, 1982.

Risley, M.S. (ed.). *Chromosome Structure and Function.* Van Nostrand Reinhold, New York, 1986.

Yunis, J.J. High resolution banding of human chromosomes. *Science* 191:1268, 1976.

Yunis, J.J., R.D. Brunning, R.B. Howe, and M. Labell. High-resolution chromosomes as an independent prognostic indicator in adult acute nonlymphocytic leukemia. *N. Engl. J. Med.* 311:812, 1984.

REVIEW QUESTIONS

1. Name the stage of mitosis at which:

   A. The nucleolus disappears.
   B. Nuclear membranes reform.
   C. Centromeres are aligned on the equatorial plate.
   D. Microtubules are associating to form the spindle fibers.
   E. The nucleolus reforms.
   F. Chromatids composing a chromosome separate and move to opposite poles.

2. The normal chromosome number in a human body cell is 46.

   A. How many chromatids are present at (1) prophase and (2) interphase following DNA replication?
   B. How many chromosomes are present in a nucleus at (1) prophase, (2) telophase, (3) the interphase following DNA replication?

3. In the human:

   A. How many autosomes are present in body cells of a male?
   B. How many autosomes are present in body cells of a female?
   C. How many sex chromosomes are present in a male?
   D. How many sex chromosomes are present in a female?
   E. How many major groups of chromosomes can be recognized on the basis of size and shape?

4. Allow "A" to designate one entire haploid set of human autosomes, and allow "X" and "Y" to represent the sex chromosomes.

   A. How can one represent the chromosome constitution of the body cells of a female in regard to autosomes and sex chromosomes?
   B. Answer Part A for a male.

5. From the column on the right, select the letter of the term that applies best to each of the following statements. A term may be used more than once or not at all.

1. Does not occur in a cell hav-. ing two nuclei.	**A.** Chromomere
2. Its position determines the length of the chromosome arms.	**B.** Cytokinesis
3. Marks the poles of the spindle in some types of cells.	**C.** Chromatin
4. Deeply staining band or region of polytene chromosome.	**D.** Chromosome
5. Holds the chromosome halves together at prophase and metaphase.	**E.** Centromere
6. Nuclear material composed of DNA and associated proteins.	**F.** Centriole
7. Chromosome material that remains condensed in the nondividing nucleus.	**G.** Chromatid
8. Associated with the formation of a nucleolus.	**H.** Genome
9. Separates from its sister at anaphase.	**I.** Heterochromatin
10. The genetic content of a single set of chromosomes.	**J.** Satellite

6. Give at least three features of eukaryotic cells that are absent in prokaryotes.

7. What terms apply to the following definitions:

   A. The specific position on a chromosome that can be occupied by a particular gene.
   B. Chromosomes that correspond in size, shape, and genetic regions.
   C. An individual possessing two identical forms of a gene at a specific genetic region under consideration.
   D. Alternative forms of a gene.
   E. The actual chemical substance composing the gene.
   F. A V-shaped chromosome with its centromere in the middle.
   G. A chromosome composed of many separate chromosome threads that are intimately associated.
   H. The chromosome constitution of a single cell or individual.
   I. That part of some chromosomes that appears as an appendage because of a secondary constriction.
   J. An I-shaped chromosome with one very short arm.
   K. A drug that prevents spindle formation.
   L. The precise period of DNA replication.
   M. Chromosome regions that fluoresce after staining with a fluorescent dye.
   N. Deep-staining chromosome regions occupied by heterochromatin.
   O. The longer arm of a chromosome.

8. Assume that in nine different cells one of the following is experimentally destroyed or eliminated. Indicate whether you would expect mitosis to proceed (+) or to be arrested (−):

   A. Aster
   B. Centriole

C. Kinetochore
D. G_2 phase
E. Polar spindle fibers
F. Centrosome
G. Centromere
H. Cytokinesis
I. S phase

9. A "new" mutation occurs that changes the flower color in a particular plant species from red to lavender. Breeding experiments show that the lavender flower color depends on a dominant genetic factor. The genetic locus involved is named *lavender,* and the first two letters of the name are used to designate the pair of alleles involved in flower color. Using the "+" system of naming alleles, designate (a) a lavender-flowered plant that is homozygous for flower color; (b) a red-flowered plant; (c) a plant with lavender flowers that is heterozygous for flower color.

Chromosomes and Gamete Formation

3

Meiosis and the early meiotic chromosome

Meiosis is the process that reduces the chromosome number from diploid to haploid. Like mitosis, meiosis is also primarily a nuclear event and does not occur among prokaryotes. Although both share certain identical features, several distinctions between them produce very different genetic outcomes. While mitosis preserves the genetic identity between a parent cell and its two daughters, the effect of meiosis is quite the opposite. Not only does it reduce the number of chromosomes by one-half, it also generates new genetic combinations. Meiosis entails two nuclear divisions and therefore typically results in the formation of four nuclei, which are referred to as the *tetrad,* the immediate products of a meiotic process. Since cytokinesis usually accompanies meiotic divisions, a tetrad of haploid cells is the characteristic end product. What these four cells will become (sperm, eggs, spores) depends on the species and sex of the organism.

The first division of meiosis is preceded by an interphase in which G_1, S, and G_2 phases can be recognized. However, the S phase of meiotic interphase requires a much longer time than does the S phase of mitosis. In the newt *Triturus,* for example, 12 hours is typical for the S phase of nonembryonic mitotic cells, but the S phase preceding gamete formation in the male lasts approximately 10 days! Moreover, a very small portion of the DNA may remain unreplicated until the early part of first prophase. It is during first prophase that most of the significant differences between mitosis and meiosis occur. Its duration is quite prolonged compared with that of mitosis and entails pronounced changes in chromosome appearance and behavior. At least five stages are generally recognized. At the earliest of these, *leptonema* (or the *leptotene stage*), the chromosomes are very stretched out, more so than at any other of the meiotic stages. Individual chromosomes cannot be identified, and the nucleus appears to contain a network of thin, entangled threads. The thin chromosome thread is designated the *chromonema* (Fig. 3-1A,B).

Both ends of each chromosome thread are attached to the nuclear membrane. All the changes in chromosome appearance that we see during nuclear divisions are primarily a result of changes in the coiling of the chromosome thread, which often exhibits deep staining regions along its length. These regions give each chromosome the appearance of a string of beads and appear to be coils in the chromosome thread that are constant in their positions along the length of a given chromosome.

The filament at the leptotene stage is so thin that it appears to be single when viewed with the light microscope. However, autoradiographic procedures that follow the incorporation of radioactive building blocks of DNA clearly tell us that duplication took place in the S phase of premeiotic interphase before the onset of leptonema.

Figure 3-1C is a diagram of an imaginary cell from an organism with a diploid chromosome number of four, in which one pair of chromosomes is larger than the other. At leptonema, these two homologous pairs, although long and extended, are separated and are not associated in any prescribed way. The cell in the figure depicts a heterozygous individual, a dihybrid carrying

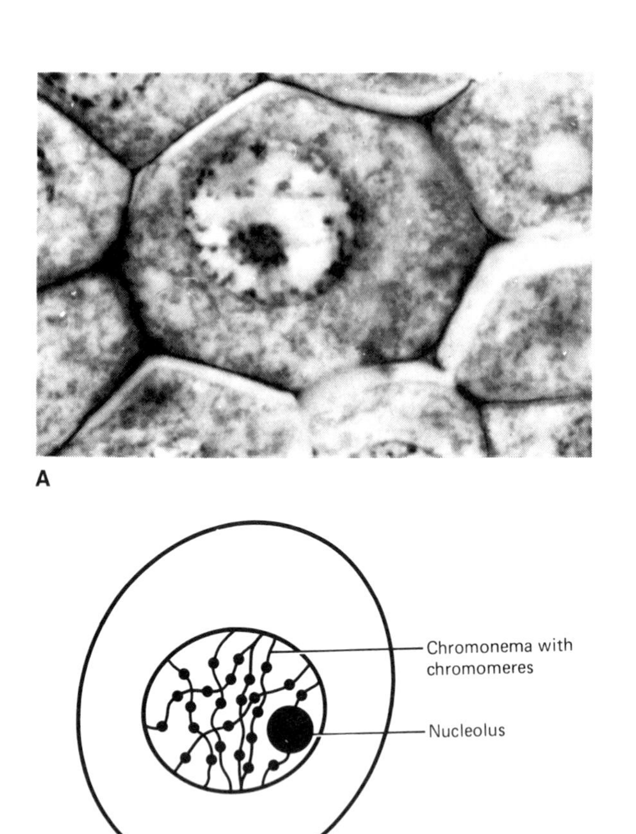

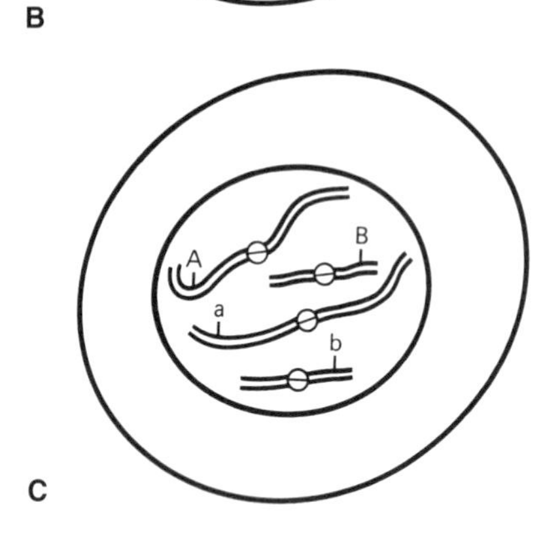

FIGURE 3-1. Leptonema. **A:** Microspore mother cell of the lily. **B:** The chromosomes are in a greatly extended condition and appear as slender threads. **C:** The chromosome number is four. Each chromosome thread is actually double and composed of two chromatids, but this cannot be detected with the light microscope. Two pairs of homologous chromosomes are depicted, one pair with alleles A and a, the other with the alleles B and b. The chromosomes are unassociated and distributed at random in the nucleus.

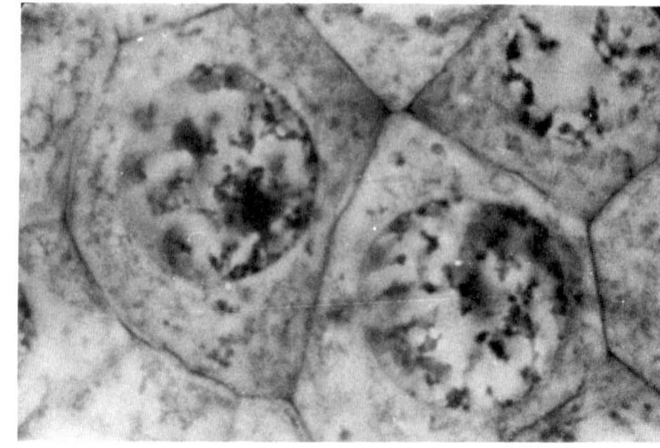

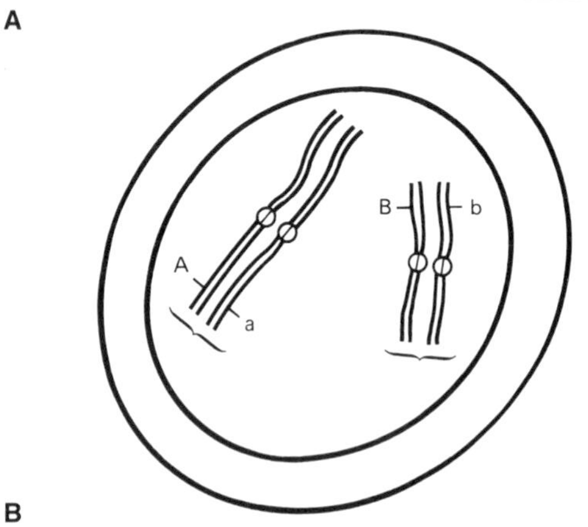

FIGURE 3-2. Zygonema. **A:** Microspore mother cell of the lily. **B:** Each of the four chromosomes in this imaginary cell is double, giving eight chromatids, although this is not evident with the light microscope. Zygonema is the stage of active pairing of the homologous chromosomes, so that two bivalents (brackets) arise in this example. Each of the two bivalents is composed of four threads, and consequently each locus is represented four times: A,A, a,a and B,B, b,b. Each of the two chromatids of a chromosome possesses its own centromere.

two pairs of contrasting genetic factors. These are the alleles A and a, which occupy a locus on the larger of the chromosomes, and the alleles B and b, located at a site on the smaller ones.

The act of synapsis, the precise pairing of homologous chromosomes, marks the end of leptonema and the onset of *zygonema* (the *zygotene stage*), the period of active chromosome pairing (Fig. 3-2A). We can envision the homologous chromosomes associating locus-for-locus throughout zygonema until all the corresponding loci on all the chromosomes are intimately paired. At this stage, the alleles in a heterozygote would be close together, no longer apart in the nucleus (Fig. 3-2B). Although the nature of the synaptic force remains unknown, it serves to bring the homologous chromosomes together in close union, even though they may have been widely separated before the onset of meiosis. A very small amount of DNA synthesis (0.3%) is also known to take place at the zygotene stage.

Unlike the picture at mitotic prophase, separate chromosomes are not distributed throughout the nucleus at zygonema. What appears to be an individual chromosome is actually two, each of which is double. This association of two paired homologs is termed a *bivalent.* (The expression *tetrad,* referring to the four threads present, is also used, but many prefer to restrict usage of this term to the immediate products produced upon the completion of the second meiotic division.) At zygonema, we see that the nucleus contains a number of bivalents that corresponds to the haploid chromosome number for a species (see Fig. 3-2B).

It is important to note that the total number of chromatids in a first meiotic prophase nucleus is the same as in a nucleus undergoing mitotic prophase. At the latter, each chromosome is also composed of two chromatids. The dihybrid represented in Figure 3-2B has eight, the same number that would be present if the cell were undergoing mitosis. However, while mitotic and first meiotic prophase nuclei have the same number of chromosome threads, the distribution of the chromosomes is very different. At mitotic prophase the chromosomes, each composed of two chromatids, are separate and unassociated, whereas at prophase I of meiosis the doubled homologs are paired. It is the latter that are present in the haploid amount.

After zygonema, the chromosome threads become more conspicuous, and we can recognize the next portion of prophase I, known as *pachynema* or the *pachytene stage* (Fig. 3-3A,B). What appeared to be single threads are now revealed to have a double nature, and the two chromatids of each chromosome can be recognized. Pachynema is the stage during which the highly significant process of crossing over takes place, but the visible evidence of this is not apparent until the next stage of prophase I, which will be described shortly.

Some digestion of DNA has been shown to occur during pachynema. The digested portion is replaced, however, by a small amount of DNA synthesis, so the total DNA content of the cell does not change. The significance of this small burst of DNA synthesis is referred to later on in this chapter.

We recognize the next portion of prophase I, *diplonema* (the *diplotene stage*) when the chromosome threads composing each bivalent appear to repel (Fig. 3-4A). Diplonema is often referred to as the stage in which "opening out" occurs, a recognition of the separation of the chromatids along the length of a bivalent. As diplonema proceeds, this becomes more pronounced, particularly at the region of the centromere, where the forces of repulsion appear strongest (Fig. 3-4B). The bivalent association would probably fall apart were it not for certain regions where the homologous chromosomes still remain in contact. These regions are called *chiasmata,* because at each chiasma the threads of separate chromatids appear to cross. Chiasmata are associated with crossing over, the separation of genes linked together on the same chromosome (see Chaps. 8 and 9). Homologous chromosomes exchange segments during meiotic prophase, and the consensus is that each

A

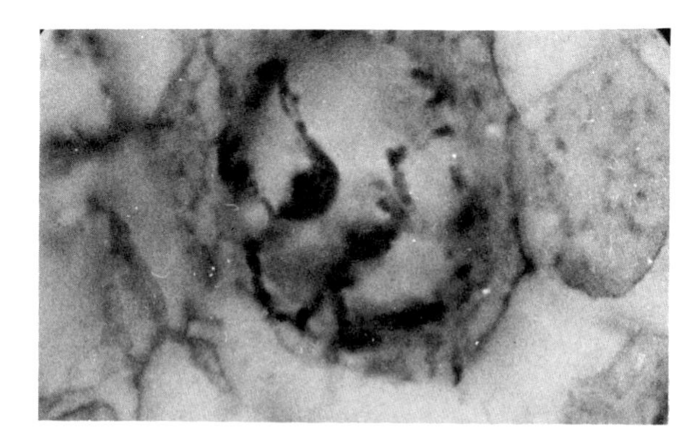

B

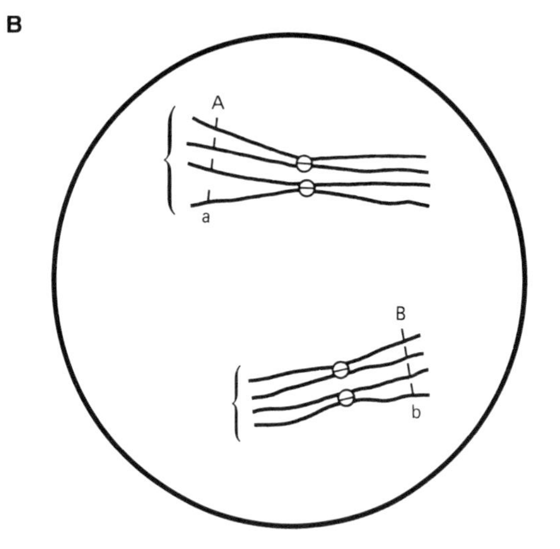

FIGURE 3-3. Pachynema. **A:** Microspore mother cell of the lily. Because thickening of chromosome threads occurs at this stage, the double nature of each thread is evident. **B:** Chromosome number is four and chromatid number is eight in the nucleus of this imaginary cell. The four chromosomes are arranged as two bivalents (brackets).

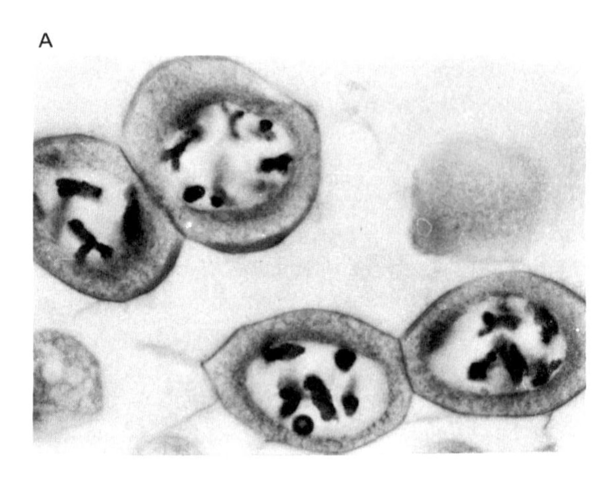

FIGURE 3-4. Diplonema. **A:** Microspore mother cell of the lily. **B:** Imaginary cell. Chromosome number is four. Chromatid number is eight. The number of bivalents is two. The chromosomes composing a bivalent start to repel each other at the beginning of this stage, and the forces of repulsion progress throughout. The chiasmata (arrows) hold the four chromatids together in a bivalent, which would otherwise fall apart. Only the nucleus is represented here.

chiasma represents a point where an exchange between homologs has already taken place. Most evidence supports the concept of an actual breakage of chromosome threads. Following breakage, broken chromatid ends join up reciprocally with chromatid segments of the homologous chromosome (Fig. 3-5). Since the rejoining is reciprocal, the outcome is an exchange of corresponding blocks of genes between homologous chromosomes. Cytologists have determined that this exchange takes place during pachynema before the onset of diplonema when the consequences of the crossover events are seen as chiasmata.

The diplotene stage may be of very long duration in oocytes, meiotic cells leading to formation of eggs. In some species, the chromosomes become very indistinct as the chromatin fibers become less folded and appear to decondense. Synthesis of substances important to the egg as storage products takes place during this time. In some species, diplonema lasts for weeks or even many years, as in the human female.

The last clearly defined stage of prophase I, *diakinesis,* is marked by pronounced changes in the appearance and distribution of the chromosomes whose ends now become detached from the nuclear envelope. The chromosomes contract markedly, and the chiasmata become increasingly evident as diakinesis progresses. The bivalents themselves become widely spaced, making this stage an excellent one for counting chromosomes because the haploid number is clearly indicated by the number of contracted and widely separated bivalents. However, nothing genetically significant is going on to rearrange the genetic material. We can consider this stage to be an exaggeration of diplonema.

Crossing over and the synaptonemal complex

Crossing over is an integral part of the sexual process, and its role in multiplying the number of new combinations of the genetic complement cannot be overemphasized. Intimately related to crossing over is a structure observable with the electron microscope and peculiar to early meiotic chromosomes. This is the synaptonemal complex (SC), a three-layered proteinaceous structure that develops between two homologous chromosomes. The formation of the SC traces back to leptonema, at which time each chromosome thread forms along its length a longitudinal protein core. At zygonema, when the two homologs pair to form a bivalent, the two protein axes are brought together to form the sides or lateral elements of the SC (Fig. 3-6). The central element then proceeds to develop, and by pachynema the SC is complete. Proteinaceous bodies, which are often spherical and are known as recombination nodules, develop along the length of the central element of the SC, whose formation is summarized in Figure 3-7.

The observations all indicate that, although the SC does not initiate synapsis, it does maintain the bivalents

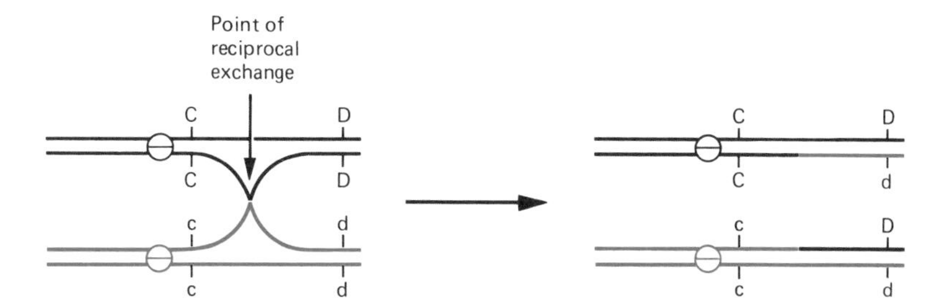

FIGURE 3-5. A single crossover event following genes on the same chromosomes: Cc and Dd. In the bivalent, two of the four threads participate in any *one* crossover event (left). However, more than one crossover event may take place in a bivalent, and any two homologous threads may be involved. In a *single* crossover event, a reciprocal exchange of chromosome segments occurs between two homologous chromatids. After the event, a new combination of alleles results. On the right, we see that in each chromosome one chromatid shows a new arrangement (C-d in the upper; c-D in the lower). This is the direct result of the reciprocal exchange of homologous chromosome segments past the point of the crossing over. (Arrow indicates the point where the reciprocal exchange occurred between the two chromatids.)

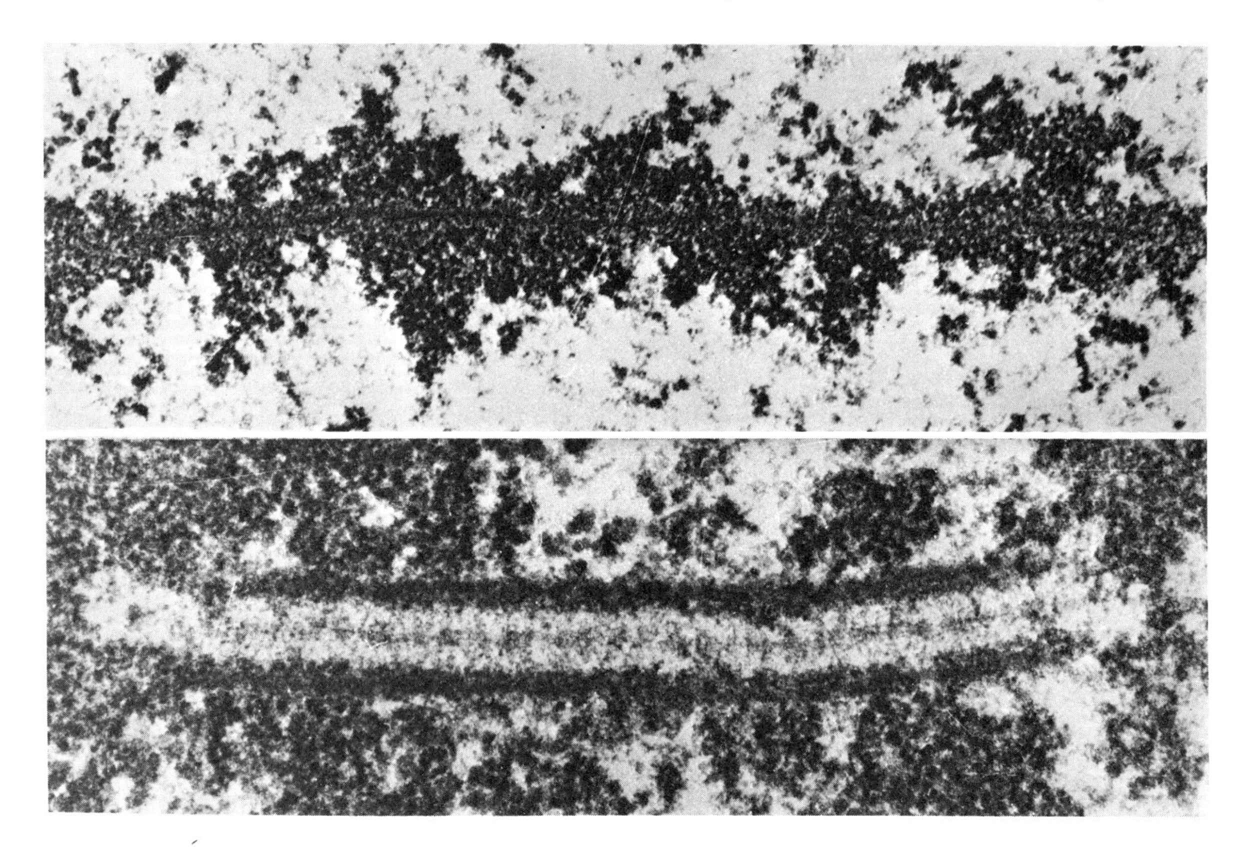

FIGURE 3-6. Synaptonemal complex in the tiger lily. An unpaired chromosome at leptonema produces a proteinaceous axial core (top; ×46,000). When two homologous chromosomes are paired (bottom) the two juxtaposed cores become the lateral elements of the synaptonemal complex. By pachynema the SC is completed, and a central element can be seen between the lateral elements (×80,000). (Courtesy of Dr. P.B. Moens.)

in pachynema by keeping the homologs together in an intimate and precise state of synapsis. In this way, it increases the probability that crossing over will take place. As the homologous regions separate at diplonema, the SC disassembles in a manner that varies with the species. Evidence suggests that the recombina-

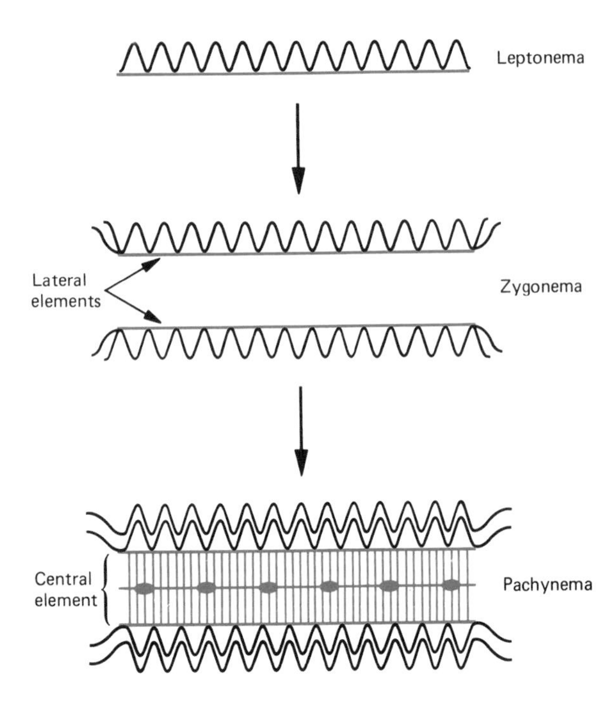

FIGURE 3-7. Development of the synaptonemal complex. At leptonema, each chromosome forms a proteinaceous axis (red) along its length. The chromosome thread is actually double, but the two chromatids are so intimately associated that two chromatids are not evident. At zygonema, the homologous pair and the axial cores of both chromosomes form the lateral elements of the complex. By pachynema, the synaptonemal complex is complete with a central element, along which recombination nodules (red spheres) develop.

tion nodules scattered along the central element of the SC contain enzyme assemblies that are instrumental in bringing about crossing over. Recall that a very small amount of DNA synthesis occurs during zygonema and pachynema. Experiments using radioactive DNA building units show that during pachynema the DNA precursors are taken up by meiotic cells at the site of the nodules. DNA synthesis during pachynema may very well be involved in the process of crossing over, a point that is explored more fully in Chapter 11.

Chromosome behavior and Mendel's laws

As in mitosis, prophase I ends with dissolution of the nuclear membrane, the disappearance of nucleoli, and the movement of chromosomes toward the midplane of the spindle. However, a major distinction exists between the metaphase of mitosis and that of the first meiotic division: In the former, the centromeres of *individual* chromosomes are arranged at the equatorial plate (Chap. 2, Fig. 2-10C), whereas at metaphase I it is the centromere regions of *bivalents* that are found at the equator of the spindle (Fig. 3-8A and B). Thus at mitotic metaphase, *single* chromosomes are oriented on the spindle, whereas *pairs* of chromosomes occupy the equator in the meiotic cycle. The centromere regions of the two individual chromosomes composing each bivalent have been repelling each other and are now clearly directed toward opposite poles of the spindle. The fibers coming from the centromeres of the two sister chromatids of each chromosome are oriented toward the *same* pole (Fig. 3-8C). Contrast this with mitosis, in which the kinetochores of the sisters at metaphase are oriented toward *opposite* poles (see Fig. 2-10C,D).

It is particularly important at this point to note the arrangement of the two pairs of chromosomes in Figure 3-8B. The diagram shows the A allele of one chromosome and the B allele of the other directed to the lower pole and the a and b toward the upper. However, nothing dictates such an orientation requiring factors A and B to go to the same pole. It is equally possible for the chromosome with allele A and the chromosome with allele b to face the same pole and those with a and B to face the opposite one. Assuming that the chromosomes with the A and the B genetic factors came from the maternal parent and those with the a and b factors from the paternal one, we can see that there is no reason to suppose that A and B *must* travel together as well as a and b. The two loci are on completely different chromosomes. How the four chromosomes become arranged in relationship to the poles of the spindle is a matter of chance and forms the foundation of Mendel's law of independent assortment. It is here at metaphase I and the ensuing anaphase I that we find the basis of both of Mendel's primary laws, *segregation* as well as *independent assortment.*

Anaphase I, like anaphase of mitosis, is recognized by the movement of chromosomes to opposite poles of the spindle (Fig. 3-9A,B). Remember, however, that at mitotic anaphase, the kinetochore fibers of the sister chromatids are associated with opposite poles. At the start of anaphase the sisters separate, become independent chromosomes, and travel to the opposite poles of the spindle (see Fig. 2-11). This is not the case at first anaphase of meiosis. Since the kinetochore fibers of the sister chromatids are associated with the same pole, the two sisters composing a chromosome remain associated at the centromere region. The centromeres of the homologs in each bivalent repel each other and travel to opposite poles. Consequently, each chromosome is still composed of two chromatids, which remain associated at the region of the centromere (Fig. 3-9A,B).

Telophase I involves the reformation of nuclear membranes and nucleoli, as in telophase of mitosis (Fig. 3-10A). At meiotic telophase, however, the haploid number of chromosomes is present in the newly formed nuclei. In our example (Fig. 3-10B), each nucleus contains two chromosomes instead of the diploid number of four. However, each of these, unlike those of mitotic telophase, is composed of two chromatids. If we count the chromatid number, in this case four, we see that complete reduction has not yet been achieved. In effect, two genomes are present—two complete sets of genes or genetic information. To bring about the reduction in genome content as well as in chromosome number, the second division of meiosis is required.

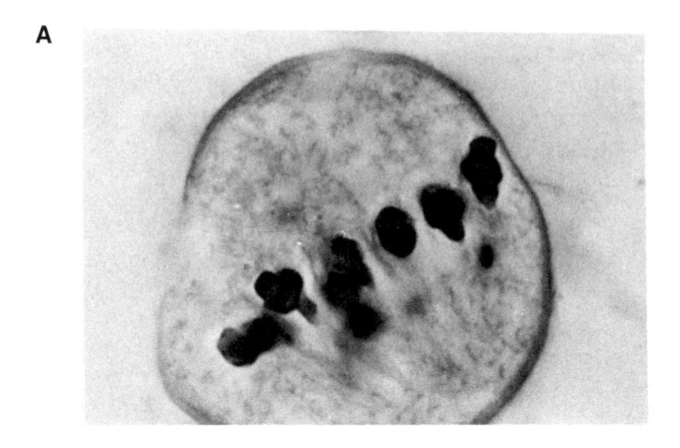

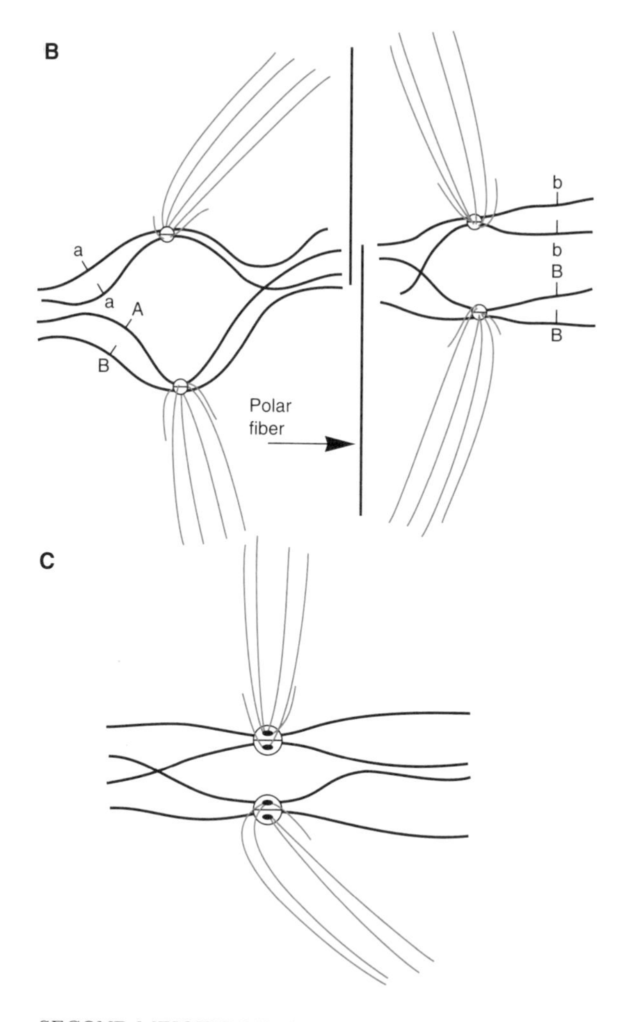

FIGURE 3-8. Metaphase I. A: Meiotic cell in the lily. B: Imaginary cell. Chromosome number is four; chromatid number is eight. The two bivalents are arranged at the equator of the spindle. The chromosomes with alleles A and B are directed toward the lower pole, but it is just as possible to find other arrangements, because the separate nonhomologous chromosomes are not tied together in any way. (Kinetochore fibers are in red.) C: The kinetochore fibers (red) of the sister chromatids are oriented toward the same pole. (The black dots represent the kinetochores in the centromere regions.)

Second meiotic division affects true reduction

Prophase II can begin after a distinct interphase period. In some species, however, there may be an abrupt transition from telophase I to prophase II and in certain cases even to metaphase II. On the surface, second division stages of meiosis resemble those of mitosis (Figs. 3-11A, 3-12A). However, this meiotic division is not a mitotic one. A big distinction relates to the double nature of the chromosomes, which at mitotic prophase results from the duplication of the genetic material at S phase in the preceding interphase. However, no S phase occurs immediately before prophase II. Each chromosome seen at this stage is composed of two chromatids as a result of anaphase I separation (see Fig. 3-9).

It is extremely important to note that the chromatids composing the *mitotic* chromosomes are sisters, meaning they are identical throughout their lengths. Those at prophase II, however, are not truly sisters, because crossing over typically occurs at pachynema to switch chromosome segments between homologs. Only portions of the chromatids of a single chromosome would be identical. We have noted (in Fig. 3-5) that after the point of a crossover a segment of a chromosome is associated with a block of genes from the homologous

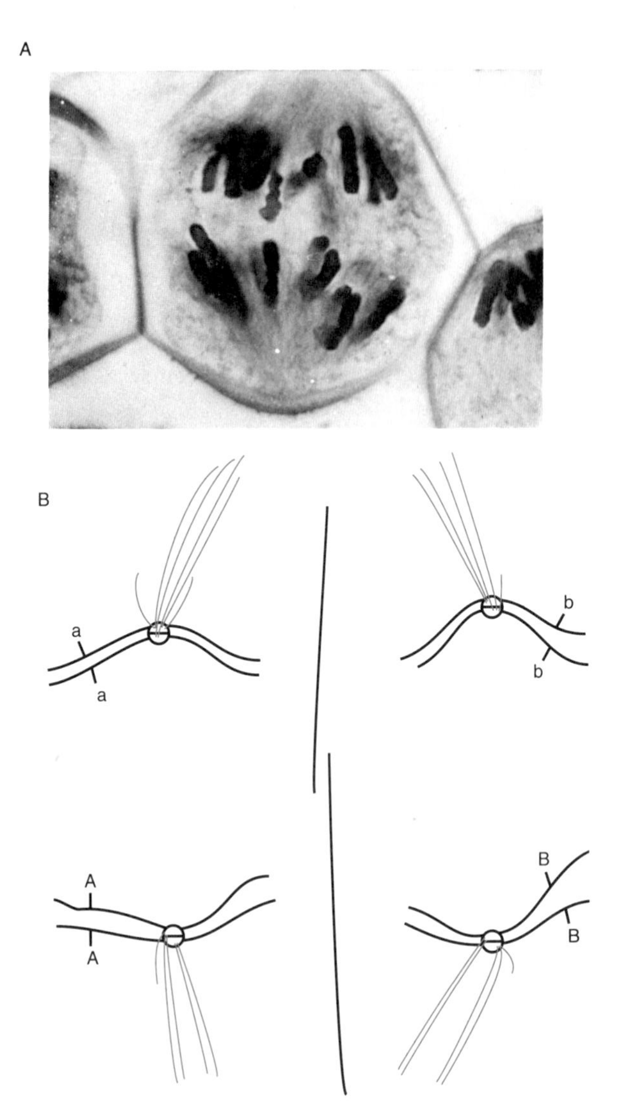

FIGURE 3-9. Anaphase I. **A:** Meiotic cell in the lily. **B:** Imaginary cell. The centromeres of each chromatid in a chromosome remain associated, so that each chromosome moving to a pole is composed of two chromatids.

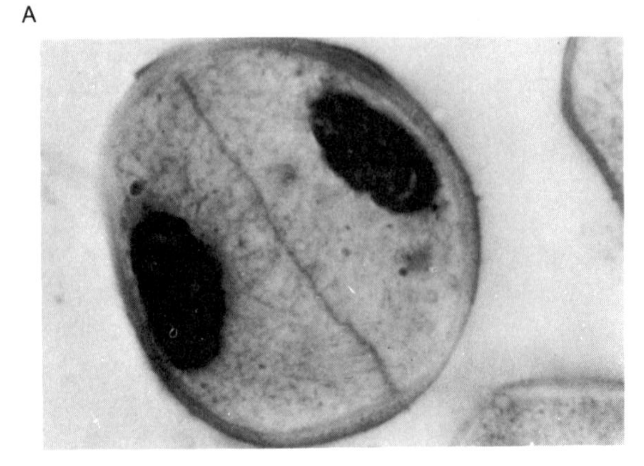

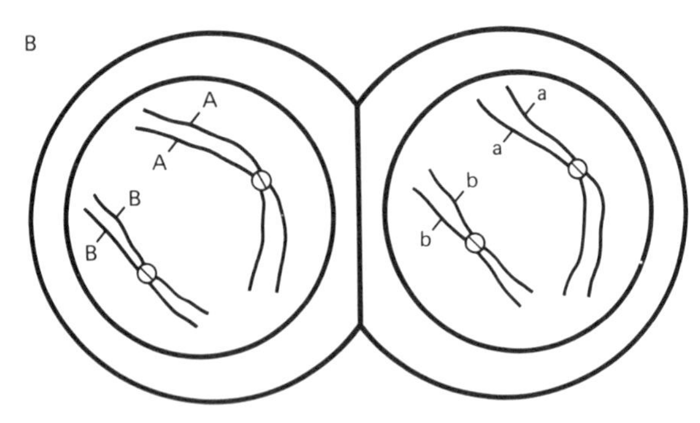

FIGURE 3-10. Telophase I. **A:** Meiotic cells in the lily. At this stage, the nuclear membranes reform. **B:** Chromosome number per nucleus in this imaginary example is two. Chromatid number per nucleus is four. Each sister nucleus is genetically different; the left one is AB, the right ab. Reduction in chromosome number has occurred, but not in genome.

chromosome. Second division of meiosis is necessary to separate the dissimilar chromatids of each chromosome. This critical division will achieve complete reduction in the amount of genetic material and bring about new combinations of genes from the male and female parents of the precedıng generation.

At metaphase II, single chromosomes, each composed of two chromatids held together at the centromere region, are at the equator, and their appearance is similar to that of chromosomes of mitosis (see Fig. 3-11A,B). The kinetochore fibers of the two chromatids are associated with microtubules emanating from opposite poles, just as in the case of mitotic chromosomes (Fig. 2-10C). As in mitosis, separation of the two chromatids of a chromosome occurs at the centromere region, and anaphase II is initiated (Fig. 3-12A,B). *Only* with this separation of the two chromatids is true reduction in the genome achieved. The nuclei of telophase II (Fig. 3-13A,B) contain chromosomes that are now single. Figure 3-13B shows the reduction in both chromosome number and genomic content. The figure also shows that the nuclei of the tetrad are genetically different; two of the resulting cells are AB, having received the chromosome with allele A and the other with allele B, and the other two are ab.

It is essential here to reexamine metaphase I (Fig. 3-8B). It was mentioned that the orientation of the bivalents at this stage is not mandatory. The factors on

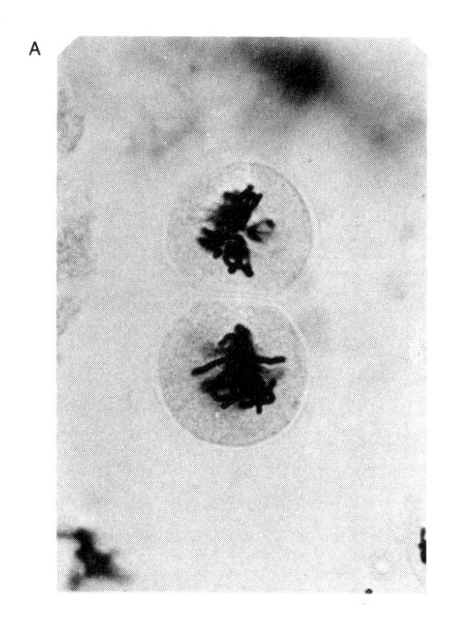

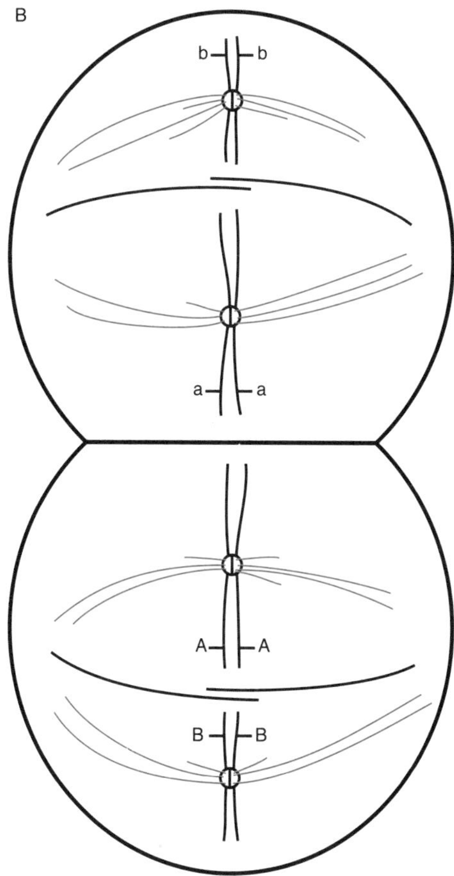

FIGURE 3-11. Metaphase II. **A:** Meiotic cells in the lily. **B:** In this imaginary example the single chromosomes in each sister cell are at the equatorial plate. Contrast this with metaphase I (Fig. 3-8), where bivalents are at the equator.

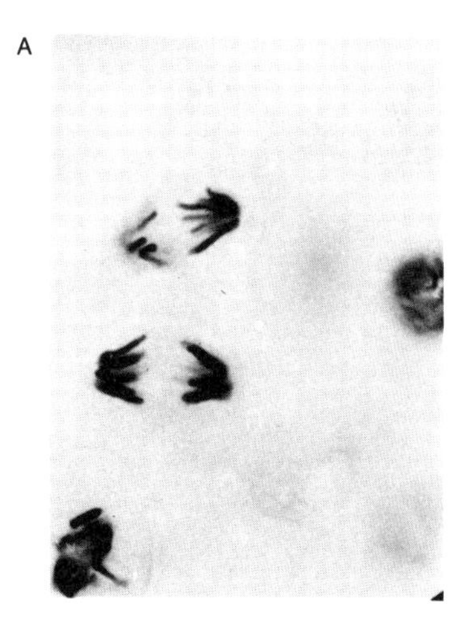

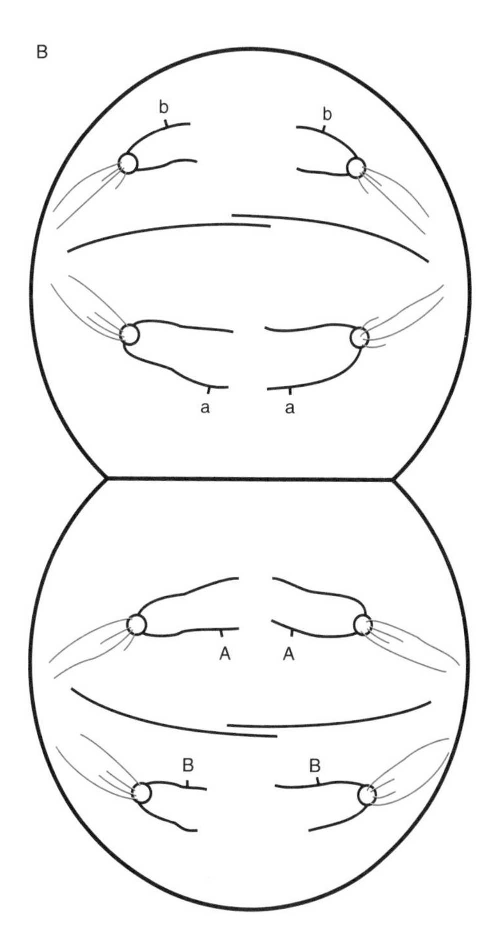

FIGURE 3-12. Anaphase II. **A:** Cells in the lily. **B:** The centromeres of the chromatids now separate. Each chromatid becomes an independent chromosome. Contrast this with anaphase I (Fig. 3-9), where the two chromatids of a chromosome do not separate, and each chromosome therefore remains double.

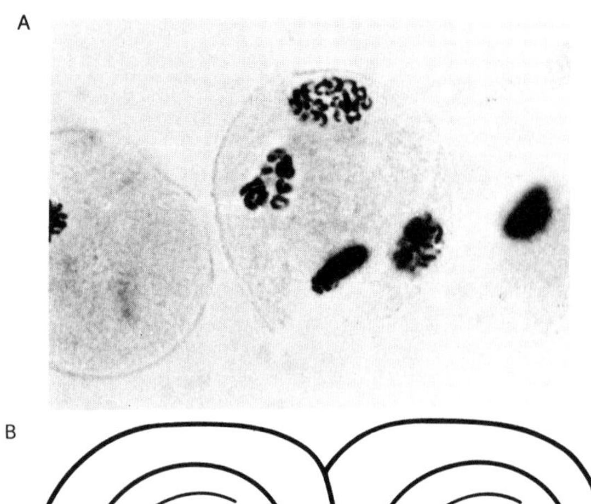

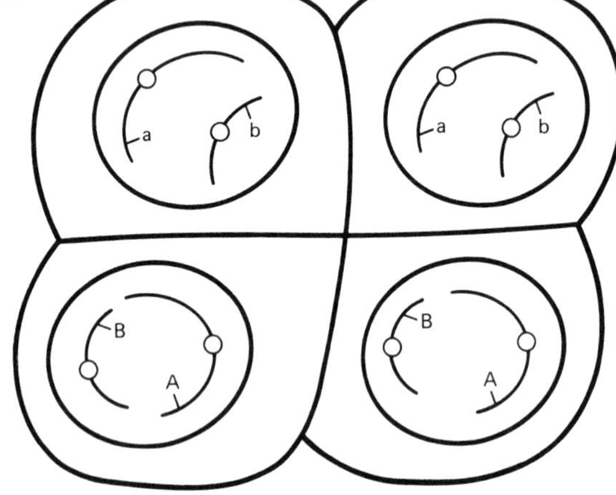

FIGURE 3-13. Telophase II. **A:** Tetrad of microspores in the lily. **B:** The chromosome number per nucleus in our example is two. The chromosome is now single, no longer composed of two chromatids. Each cell of the tetrad has true reduction in both chromosome number and genome. The cells are unalike genetically as a result of segregation and independent assortment. In this example half of the nuclei are genotype AB, and the rest are ab.

separate chromosomes segregate independently. If the A and the b chromosomes face one pole and the a and B chromosomes face the other, the final products at the completion of meiosis would be nuclei with different genetic arrangements from those shown in Figure 3-13B: Ab and aB instead of AB and ab. Since both possibilities can arise with equal frequency as the result of chance chromosome orientation on the spindle, meiosis in the dihybrid AaBb yields four different kinds of gametes in equal proportions: AB, Ab, aB, and ab. Thus the origin and frequencies of different kinds of gametes are determined by the physical events of meiosis, which shuffle the maternal and paternal genetic factors into new combinations. Although the details of linkage and crossing over will be left until later chapters, it should be appreciated at this point that even those genes that are linked on the same chromosome are shuffled as a result of the physical exchange of chromosome segments between homologs (see Fig. 3-5). The overall significance of crossing over, like independent assortment, is the formation of new combinations of genetic material in the gametes.

We see that, unlike mitosis, the nuclei formed from the meiotic divisions of a parent nucleus are very different. Just from independent assortment, almost limitless new combinations of allelic pairs are possible. The meiotic event, along with independent assortment and recombination through crossing over, is the focal point of the sexual process in eukaryotes and has supplied most of the variation for natural selection to work on in the evolution of the diverse forms of life.

Meiosis in the male

In the testis of the mature male animal, a variety of cell types occurs in the wall of the seminiferous tubule. Among these is the spermatogonium, a cell capable of mitotic division or of entering meiotic prophase I. Once it does the latter, it becomes recognized as a primary spermatocyte (Fig. 3-14A,B), a cell destined to complete meiosis with the production of a tetrad of spermatids. Each spermatid then undergoes dramatic cytoplasmic transformations into a sperm (spermiogenesis) without further significant changes in the genetic complement (Fig. 3-15A,B). The entire series of events, beginning with meiosis and resulting in the formation of male gametes, is known as *spermatogenesis,* a process requiring about 64 days in the human (Fig. 3-16).

The stages of first meiotic division during spermatogenesis all take place in the primary spermatocyte. By the time of pachynema, the nuclei of primary spermatocytes of the grasshopper show an identifiable X chromosome (Fig. 3-17). Note that the grasshopper has an unpaired X chromosome and lacks a Y entirely. In this X0 condition, the identification of the X at meiosis is much simpler than in the more common XY situation found in the human and other mammals.

At early prophase stages, the X is more deeply stained and contracted than any other member of the chromosome complement; by diplonema, it is very well defined (Fig. 3-18A,B). The X chromosome in the primary spermatocyte is behaving like heterochromatin, as noted in Chapter 2. When first metaphase is reached, the unpaired X, which has been the most dis-

A

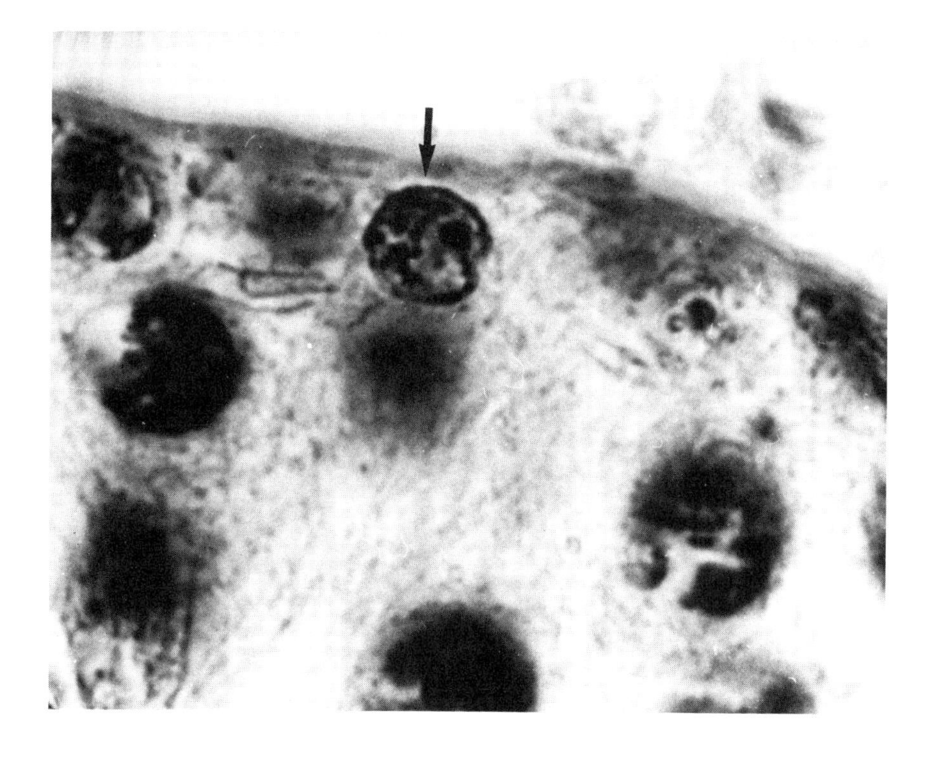

B

FIGURE 3-14. Section of human seminiferous tubule. **A:** The spermatogonium (arrow) is the cell type from which male gametes are derived. Spermatogonia are found just inside the outer membrane of the tubule. Some continue to undergo mitosis, replenishing those spermatogonia that become committed to meiosis. With the onset of first meiotic prophase, a cell becomes a primary spermatocyte (cells lower in figure). **B:** Primary spermatocytes with paired chromosomes.

A

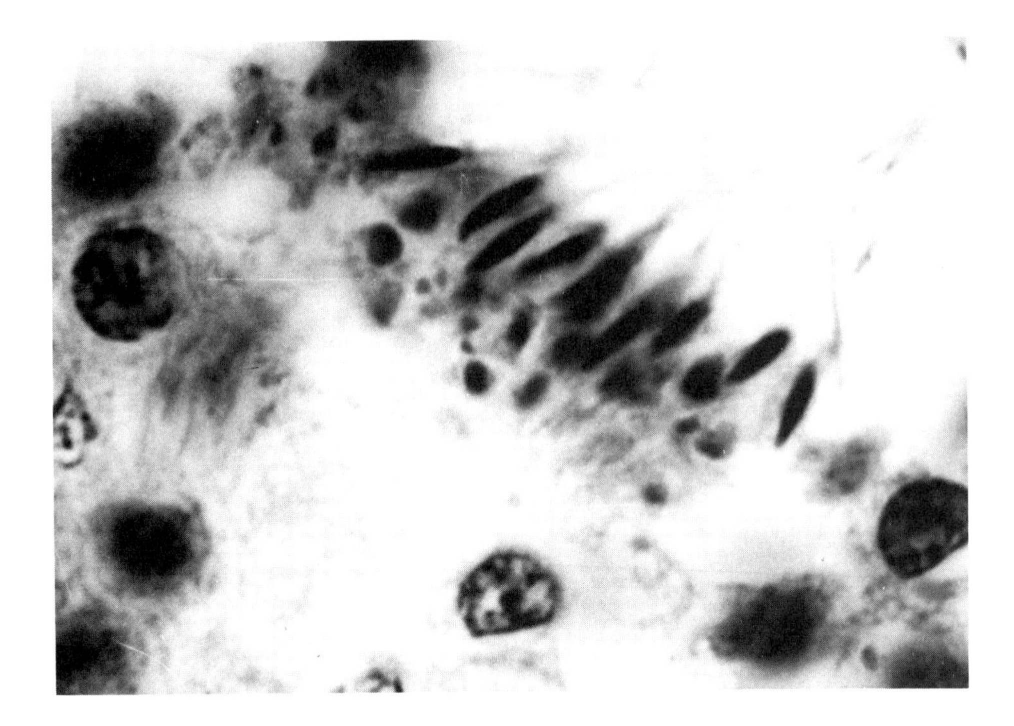

B

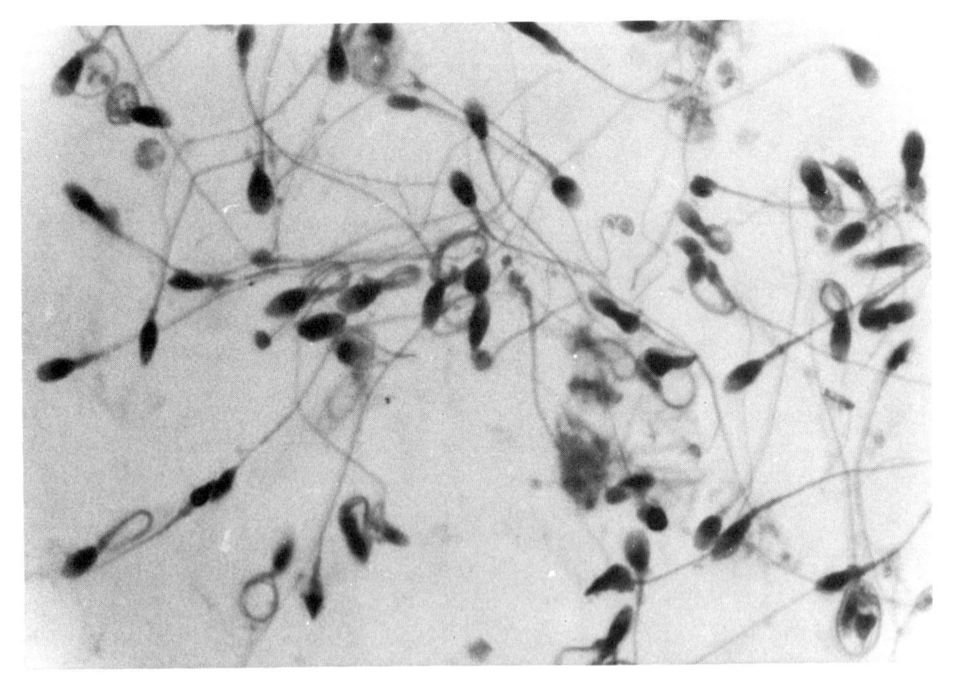

FIGURE 3-15. Sperm maturation in the human. **A:** Each spermatid undergoes marked transformations during which most of the cytoplasm is cast off as the chromatin becomes packed into the head piece. Cast-off cytoplasm can be seen as darkly staining masses in the vicinity of the very deeply staining sperm heads. **B:** From the transformation of the spermatids, mature sperm arise, each with a conspicuous head, neck, and tail region.

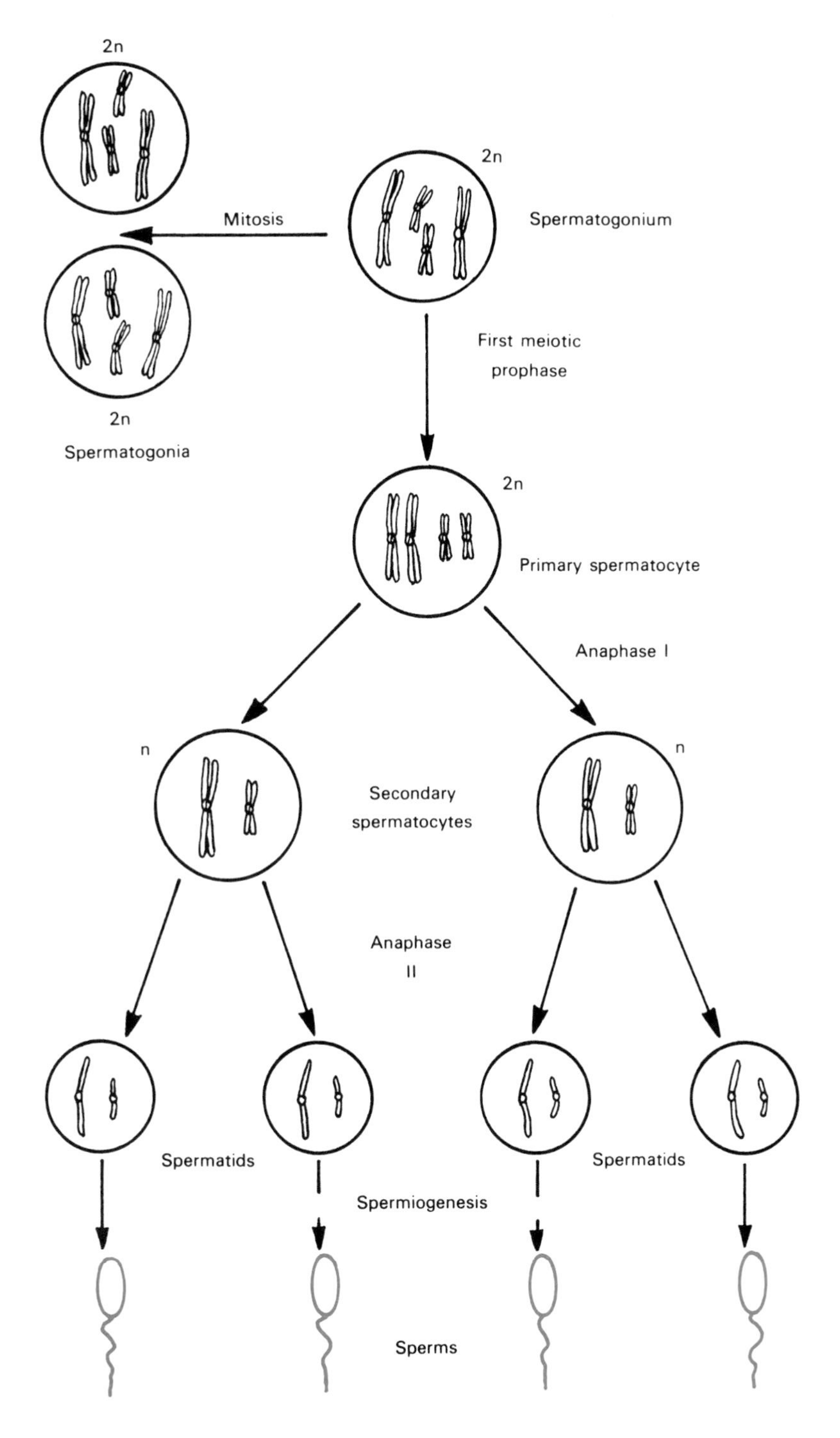

FIGURE 3-16. Diagram of spermatogenesis. Among the various types of cells in the seminiferous tubules is the spermatogonium, which may divide mitotically or undergo the meiotic divisions. The tubules are always filled with cells in various stages of gamete formation. The spermatids are the most numerous. Note that the eventual products of one primary spermatocyte are four functional sperm.

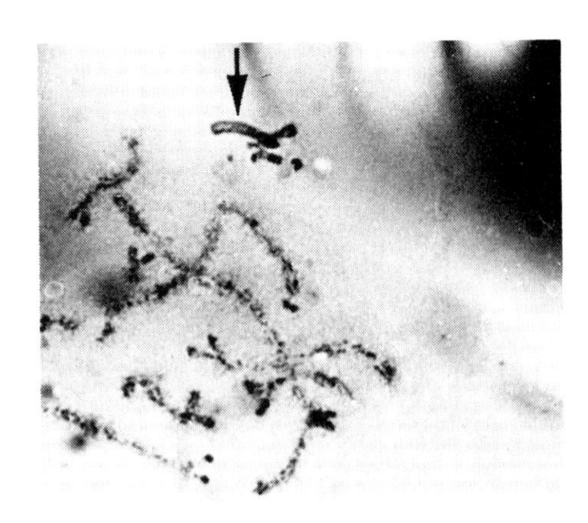

FIGURE 3-17. Pachynema in primary spermatocyte of grasshopper. The X chromosome is much more condensed and stains more deeply than the other chromosomes in the complement (arrow).

A

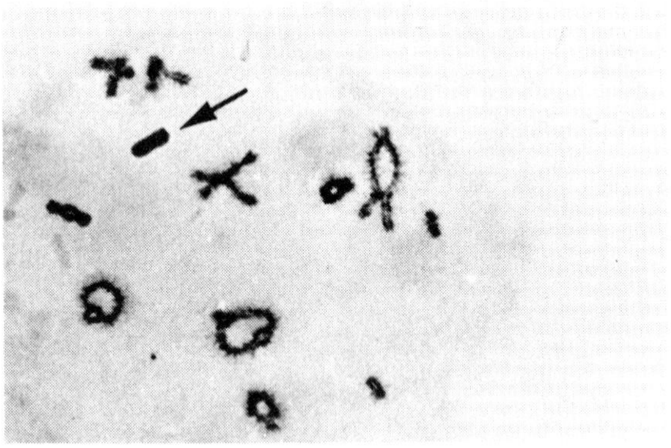

B

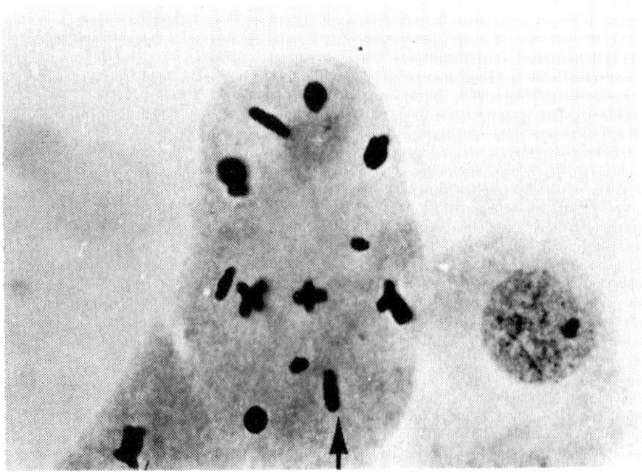

FIGURE 3-18. Primary spermatocytes of grasshopper. These cells show 11 bivalents and an unpaired X (arrow). The X stains more deeply than the other chromosomes at diplonema (**A**). By diakinesis (**B**) all of the chromosomes are highly contracted and react similarly to stains.

tinct chromosome of the complement, begins to appear fuzzy and less intensely stained, resembling the early prophase stages of the rest of the chromosomes (Fig. 3-19). In all mammals, the X chromosome pairs with the Y at early meiotic prophase, but it too shows this difference in stainability as contrasted with the rest of the chromosomes. Such a differential reaction to staining often displayed by sex chromosomes is known as *heteropyknosis*. A difference in staining behavior between the sex chromosomes and the autosomes (all the other chromosomes in the complement) is typical of most animals having the X0 and XY mechanisms of sex determination.

At anaphase I, the more lightly stained X of the grasshopper passes undivided to one of the poles (Fig. 3-20). Following reconstitution of nuclear membranes at telophase I, two cells are formed, secondary spermatocytes. These are quite different genetically (see Fig. 3-13B). A major difference regards the X chromosome. Since it passed undivided to one pole at anaphase I, one of the two secondary spermatocytes in the grasshopper contains one less chromosome than the other. In animals possessing both an X and a Y, one of the two cells contains the X and the other the Y, because the X and Y separate from each other at first meiotic anaphase.

Each secondary spermatocyte formed from first meiotic division undergoes the second stages of meiosis. That secondary spermatocyte with the X produces two spermatids, each with an X. The one in the grasshopper lacking the X gives rise to two spermatids with no sex chromosome, whereas in mammals the Y-bearing spermatocyte produces two spermatids, each with a Y. The meiotic end product of each primary spermatocyte is a tetrad of cells, two of which carry an X and two of which do not. The latter two carry a Y in those males having X and Y chromosomes. The significance of the XY condition in sex determination is discussed in some detail in later chapters.

Meiosis in the female

In females of higher animals, meiosis does not take place continually throughout the age of reproduction as is the case in the male. In female mammals, for example, mitotic divisions of oogonia, the counterparts of the spermatogonia, cease during embryological development (Fig. 3-21). The entire complement of primary *oocytes* is established by the time of birth (Fig. 3-22). A primary oocyte thus remains in an

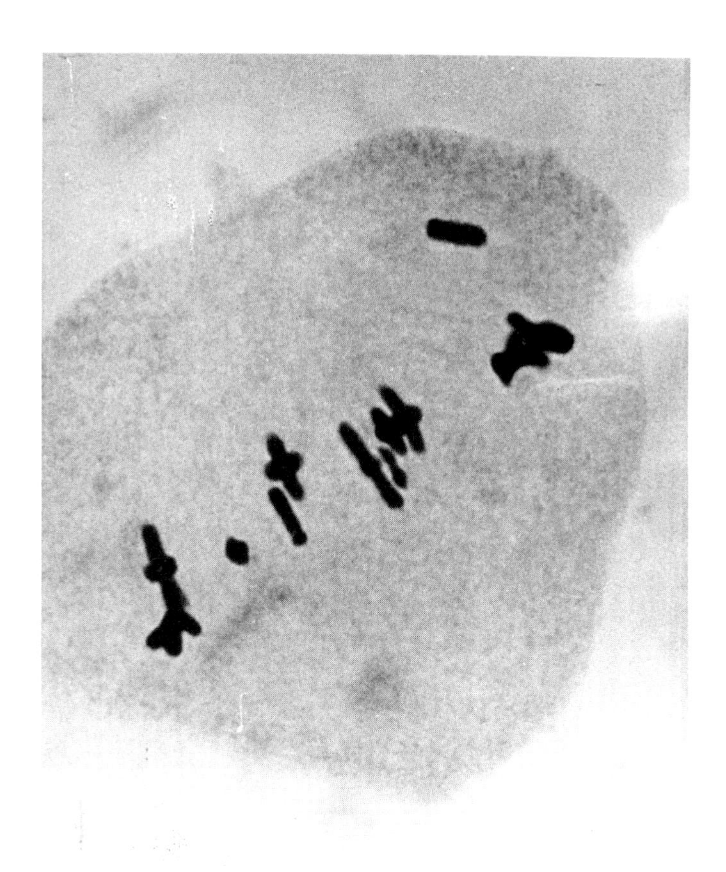

FIGURE 3-19. First metaphase in primary spermatocyte of grasshopper. Note that the X chromosome is off by itself and now stains more faintly than the rest of the chromosomes.

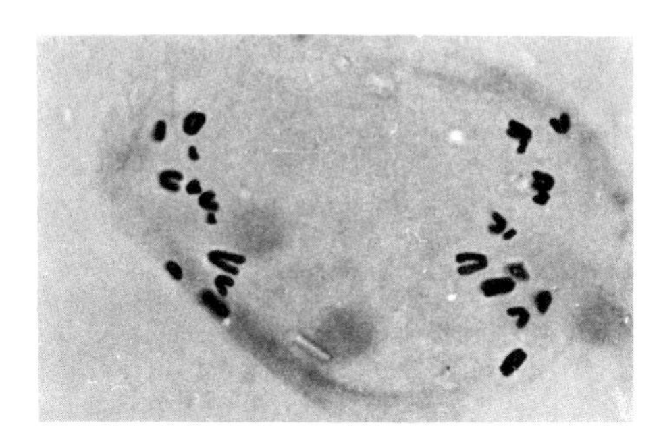

FIGURE 3-20. First meiotic anaphase in the male grasshopper. The chromosomes at each pole will become incorporated in nuclei of cells that develop into secondary spermatocytes. Note that 11 chromosomes are at one pole and 12 at the other, because the latter contains the X chromosome.

extended diplonema of first meiotic prophase for many years. The first meiotic division is not resumed until the egg is about to mature in the follicle at the age of sexual maturity. When the primary oocyte does finally divide, it gives rise to two cells, but these are unequal in size. This occurs because the larger one, the *secondary oocyte,* receives most of the cytoplasm. The smaller cell, the *polar body,* may or may not divide again. The secondary oocyte completes the second meiotic division only if a sperm enters the cytoplasm. The result of the division is again two cells of unequal size, a large one, which will become the egg, and another small polar body. All the polar bodies eventually disintegrate. Throughout oogenesis, the process in the female that produces a gamete from the maturation of an immature germ cell, both X chromosomes stain to the same degree, in contrast to the single X of the male, which shows heteropyknosis in the spermatocytes.

Meiosis in the higher plant

The sexual process undoubtedly arose early in the evolution of eukaryotic cells, long before the divergence of the plant and animal kingdoms. We therefore find comparable meiotic stages in plants and animals. Plant life cycles, however, are very diverse and often quite

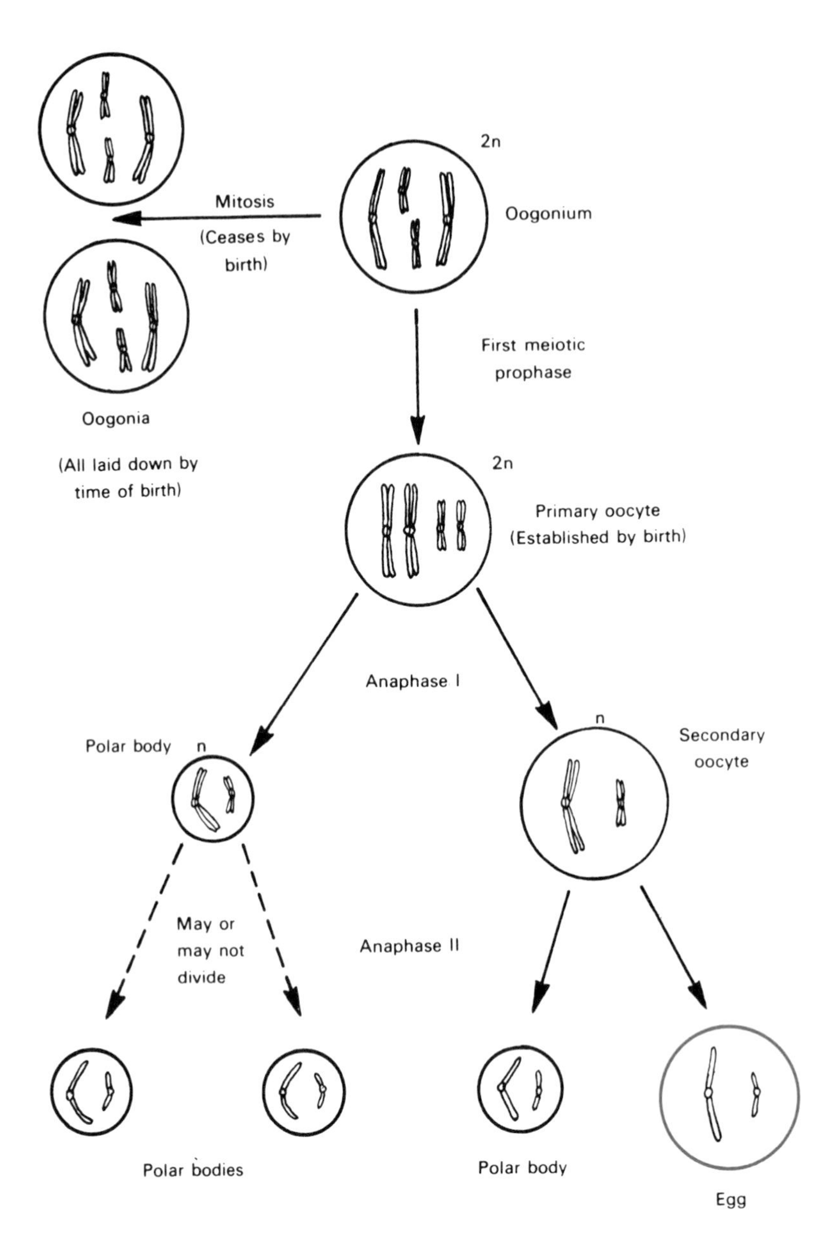

FIGURE 3-21. Diagram of oogenesis. The meiotic stages of oogenesis are comparable to those of spermatogenesis (see Fig. 3-16). The same chromosome behavior is involved, although cytoplasmic events are very different in the two processes. Note that the eventual product of one primary oocyte is one functional egg.

complicated. Meiosis generally takes place in specific plant organs, but the direct result of the process usually is *not* gamete formation as it is in animals. Instead, haploid cells called *spores* are formed. These then undergo a series of mitotic divisions, which eventually leads to the origin of sex cells. In the flowering plant (Fig. 3-23), meiosis occurs in two parts of the flower. On the male side, the anther is the organ involved. Special cells, *microspore mother cells,* arise from the anther wall, and these undergo the meiotic divisions called *microsporogenesis.* (Many of the earlier figures in this chapter show the meiotic divisions and products of microspore mother cells in the lily.) The end product of meiosis of each microspore mother cell is a tetrad of four haploid cells, the *microspores.* Each of the microspores matures into a pollen grain. During this maturation, each haploid nucleus of a microspore divides *mitotically* to produce

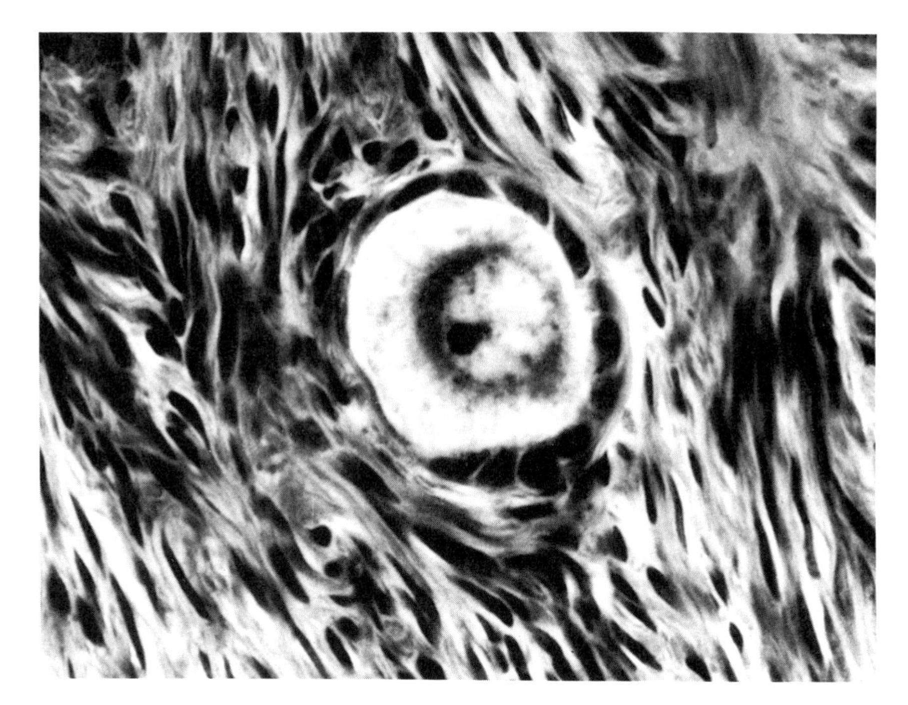

FIGURE 3-22. Section of human ovary with a follicle showing a primary oocyte.

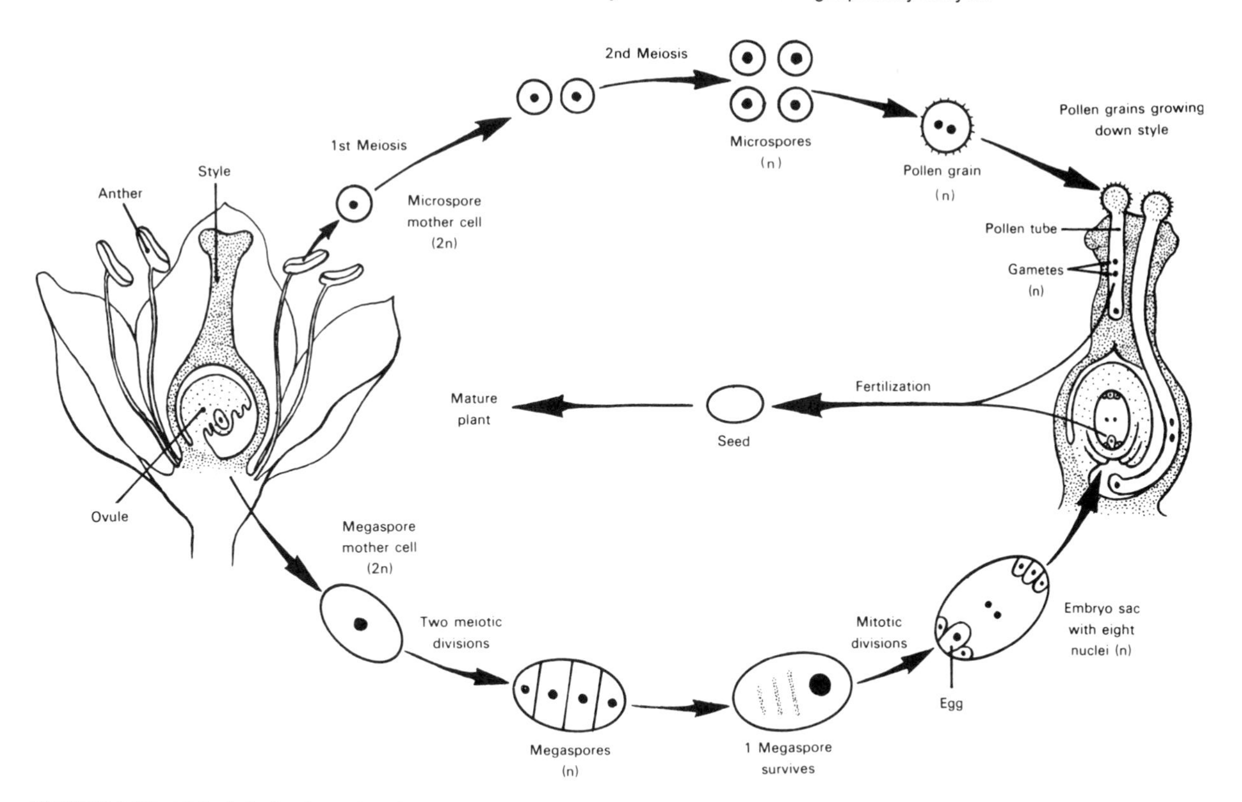

FIGURE 3-23. Meiosis in the flowering plant. Meiotic divisions leading to the formation of male gametes occur in the microspore mother cells of the anther. Four microspores, each of which will become a pollen grain, arise from each mother cell. The haploid nucleus of the pollen divides mitotically and yields two haploid nuclei. One of these divides again to produce two male gametes, one of which will fertilize an egg. The ovary contains ovules, and in each of these, one cell—the megaspore mother cell—undergoes meiosis. Of the four haploid megaspores formed, only one survives. It enlarges to give rise to the embryo sac. Mitotic divisions follow in the embryo sac, and one of the nuclei becomes the egg. Growth of the pollen tube down the style delivers one male nucleus in each pollen tube to an egg in the embryo sac of an ovule.

two haploid nuclei per microspore or pollen grain. Of the two haploid nuclei, one divides again to form two male gametes, one of which fertilizes the egg. Therefore all the stages of meiosis that bring about segregation and independent assortment are completed by the time of microspore formation, well before the production of the male gametes.

Inside the ovary of the flower are found the *ovules,* or immature seeds. In each ovule, *megasporogenesis*—the meiotic divisions that precede the formation of the female gamete—occurs. In each ovule, one cell, the *megaspore mother cell,* enlarges. It undergoes the second meiotic division, with the formation of a tetrad of four haploid *megaspores.* Of these four cells, three disintegrate, leaving a single cell that then enlarges. The nucleus of this remaining megaspore divides mitotically. Typically, two more mitotic divisions follow, forming a group of eight haploid cells. Only one of these is destined to act as the egg and be fertilized if pollination is successful. It is easy to recognize a parallel between megasporogenesis in plants and oogenesis in animals. In both processes, a mother cell (megaspore mother cell or primary oocyte) undergoes meiosis, and only one functional cell results. On the male side, in both plants and animals, each tetrad of cells forms four products (microspores or spermatids) from each of which a functional male gamete is derived.

REFERENCES

Hotta, Y., A.C. Chandley, and H. Stern. Meiotic crossing over in lily and mouse. *Nature* 269:240, 1977.

Moens, P.B. (ed.) *Meiosis.* Academic Press, New York, 1987.

Risley, M.S. The organization of meiotic chromosomes and synaptonemal complexes. In *Chromosome Structure and Function.* M.S. Risley (ed.), p. 126. Van Nostrand Reinhold, New York, 1986.

Stahl, F.W. Genetic recombination. *Sci. Am.* (Feb.):90, 1987.

Stern, H., and Y. Hotta. Biochemical controls of meiosis. *Annu. Rev. Genet.* 7:37, 1973.

REVIEW QUESTIONS

1. Name the precise times in the meiotic cell cycle at which the following pertain:

- **A.** The chromosome threads in the bivalents appear to repel, and chiasmata become evident.
- **B.** Pairing of homologous chromosomes takes place.
- **C.** The chromosomes are greatly contracted, and the bivalents are widely spaced in the nucleus.
- **D.** DNA replication takes place.
- **E.** The chromonema shows obvious chromomeres and is greatly elongated, appearing as a single filament.
- **F.** Crossing over takes place.
- **G.** Chromosomes composed of two chromatids are aligned separately on the equatorial plate.

2. One DNA molecule is present in an unreplicated chromosome. For the human, with a diploid chromosome number of 46, give the number of DNA molecules, the chromosome number, and the bivalent number for a cell or nucleus at each of the following stages:

- **A.** Pachynema
- **B.** Diplonema
- **C.** Diakinesis
- **D.** Telophase I
- **E.** Prophase II
- **F.** Telophase II

3. How can one distinguish between the following cytologically:

- **A.** A cell in mitotic metaphase from one at first meiotic metaphase.
- **B.** A cell at mitotic anaphase from one at first meiotic anaphase.
- **C.** A mitotic prophase from a first meiotic prophase.

4. A cell is dihybrid for the allelic pairs Aa and Bb. The two loci are found on nonhomologous chromosomes. What possible genetic combinations can arise when:

- **A.** This cell undergoes mitosis.
- **B.** A cell of the same genotype undergoes meiosis.

5. A human body cell normally contains 46 chromosomes. Give the number of chromosomes in each of the following:

- **A.** A cell resulting from the mitotic division of a spermatogonium.
- **B.** A primary oocyte.
- **C.** A secondary spermatocyte.
- **D.** The polar body formed along with a secondary oocyte.
- **E.** A spermatid.
- **F.** A cell formed after spermiogenesis.

6. Allow "A" to stand for one set of autosomes in the human, and let "X" and "Y" represent the sex chromosomes.

- **A.** What will be the chromosome constitution of eggs with regard to autosomes and sex chromosomes?
- **B.** Answer Question A in regard to sperms.
- **C.** What will the chromosome constitution in a spermatogonium be?
- **D.** What will the chromosome constitution in an oogonium be?

7. In which of the cell types in Question 5 are bivalents present?

8. **A.** How many sperm or male nuclei will arise from each of the following?

(1) 1,000 primary spermatocytes.
(2) 1,000 secondary spermatocytes.

(3) 1,000 spermatids.
(4) 1,000 microspore mother cells.

B. How many egg cells will arise from each of the following?

(1) 1,000 primary oocytes.
(2) 1,000 secondary oocytes.
(3) 1,000 megaspore mother cells.
(4) 1,000 megaspores that are the immediate products of meiosis?

9. In certain insects such as the grasshopper, which has the X0 condition, the male has only one sex chromosome, an X, whereas the female has two X chromosomes. Assuming the females of such a species have a diploid chromosome number of 12, answer the following questions:

A. How many autosomes will there be in the wing cells of a female?
B. Answer Question A for a male.
C. What will be the total chromosome number in the wing cells of a male?
D. How many chromosomes will there be in the sperm cells that bear an X chromosome?
E. How many chromosomes will there be in the non-X-bearing sperm?

10. In corn, there are 10 pairs of chromosomes in the somatic cells. What would be the expected chromosome number in the following:

A. A pollen tube nucleus.
B. A cell in a petal.
C. A cell in the embryo of the seed.
D. A pollen mother cell.
E. A megaspore.
F. A cell of the embryo sac.

11. Next to each of the following, place the symbol Bo if the item is associated with both mitosis and meiosis; place the symbol Me if it is much more closely associated with meiosis: (1) chromatids, (2) centromere, (3) bivalent, (4) S phase, (5) synaptonemal complex, (6) independent assortment, (7) kinetochore fibers, (8) cytokinesis, (9) chiasmata, (10) Mendel's first law, (11) DNA digestion, (12) sex chromosomes.

Genic Interactions

4

Interaction of several allelic pairs

In the early part of the twentieth century, work with poultry demonstrated that more than one pair of genetic factors can interact to influence feather color. It was found that when certain races of white birds are crossed, the F_1 offspring are not white but colored! The explanation is that two separate pairs of alleles are involved. The dominant genetic factors C and O are both required for pigment formation; their recessive alleles cannot produce pigment. In the absence of either dominant, the feathers are completely white (Fig. 4-1A). Consequently, two different pure breeding races may show the same white phenotype but actually possess different genotypic constitutions. Race A may be white because it is ccOO, whereas race B is white because it is CCoo. A cross between birds from the two races brings together the two different dominant factors, and the offspring can have colored feathers.

The exact color depends on still other genes. Another locus is known that can affect the type of pigmentation. Let B represent the determinant that governs black pigment formation. Its allele b results in white feathers flaked with pigment (white splashed). Because neither factor is dominant, the heterozygote, Bb, is blue feathered. Any bird may be black (BB), blue (Bb), or white splashed (bb), but only if both C and O are present at the other two loci (Fig. 4-1B). For example, the genotype BbCcOo would produce blue feathers, but Bbccoo can result only in white. This example shows clearly that there is not just one single gene for plumage color. At least three separate loci are involved, the interactions of which can produce different effects. This example illustrates the important principle that any characteristic is actually the result of interaction among several genetic elements. In any one specific cross we may be paying attention to just one or two pairs of alleles, but we must not forget that an assortment of other genetic factors is also operating in addition to those under immediate consideration.

Modified ratios

In mice, more than one pair of alleles are known to influence fur color. In addition to the standard grayish color, there are the possibilities of black or white. Two pairs of alleles are known to be involved with respect to these three colors. The expression of gray (wild) requires the presence of a dominant factor, B; black pigmentation depends on its recessive allele, b. However, for any pigmentation at all to develop, gray or black, the dominant factor, C, must be present at another locus. The recessive condition, cc, results in white fur regardless of the allelic forms present at the first locus. Assume that two dihybrid mice are crossed: BbCc × BbCc (Fig. 4-2). Both possess wild coat color, because each carries the dominant factors for gray and for pigment production. Among their offspring, the following can be predicted: 9 B–C– (gray):3 bbC– (black):3 B–cc (white):1 bbcc (white). Notice that the expression of the genotypes is altered by genic interaction and that the expected phenotypic ratio 9:3:3:1 becomes modified to 9:3:4.

This is so because the genotypes B–cc and bbcc cannot be distinguished from each other. Any pigment

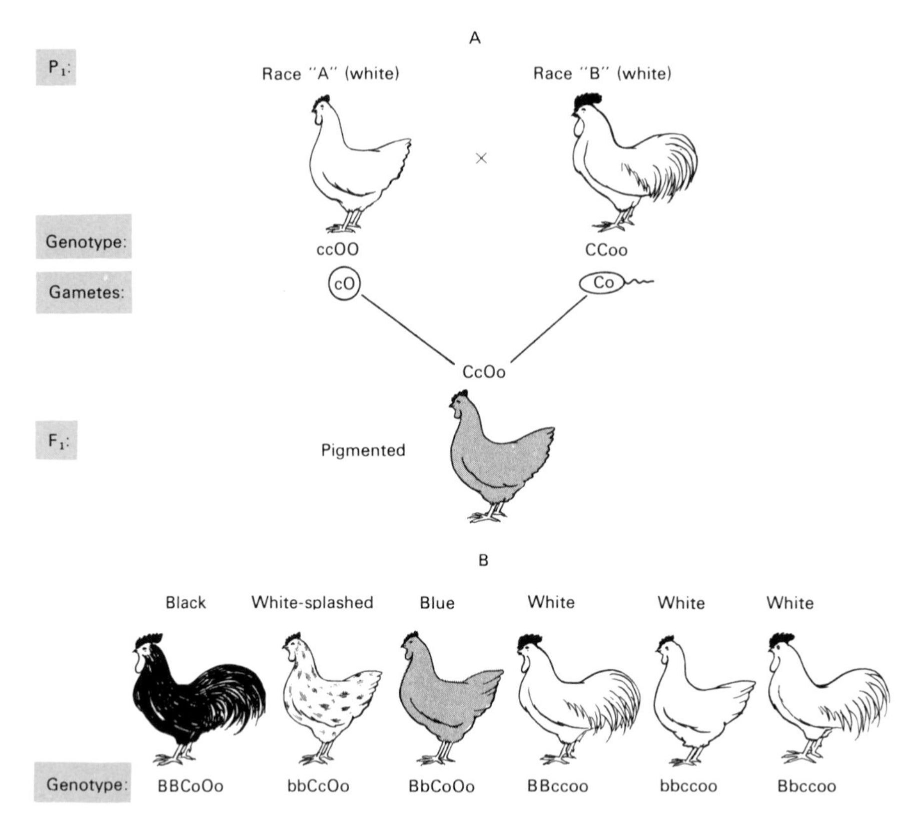

FIGURE 4-1. Gene interaction in poultry. **A:** Pigment formation in the feathers depends on the presence of both dominants, C and O. Races A and B are white because each lacks one of the dominants. The F_1 hybrid between them has colored feathers because it receives a different dominant from each parent. **B:** The specific color of the feathers depends on several pairs of alleles, among them B (black) and b (white splashed). Since there is lack of dominance, the hybrid (Bb) is blue. However, for the pigment alleles to express themselves, the dominant alleles C and O must be present; otherwise a bird will be white, regardless of the genetic factors present at the pigment locus.

in the hair requires the presence of the dominant allele, C. It is as if the c locus can suppress the expression of the pigment factors at the b locus. In one way, this may remind us of *dominance,* but use of that term is restricted to the interaction between a pair of alleles (B is dominant to b). Another word is needed to describe the suppressive influence of any genetic factor on another that is *not* its allele. Such an effect is called *epistasis.* The example of fur color in mice nicely illustrates *recessive epistasis,* because the double recessive condition, cc, is required to mask the expression of genetic factors for pigmentation at the b locus. Epistasis, however, may result from the presence of a dominant allele. Coat color in the dog involves at least two loci. In some varieties, black pigmentation results from the presence of the dominant B; brown pigment depends on the recessive allele, b. But neither color will be expressed if the dominant factor, I (for inhibition of color), is found at another locus on a different chromosome. Figure 4-3 follows the cross of two dihybrid white dogs: BbIi × BbIi. No pigment is found in their hair, because the presence of just one dose of the allele I is sufficient to suppress all pigment formation. Among their offspring, the chances become 12:3:1 for white, black, and brown, respectively.

We have already noted that feather color in poultry requires the dominants C and O; otherwise, no pigmentation is formed in the feather. We see here an example of two genes (one at the c locus and one at the o locus), either of which can suppress the expression of genetic determinants at a third locus, depending on which alleles are present. Birds of the genotypes BbCCoo or BbccOO are white, not blue. Only those birds that have at least one of each dominant allele (C–O–) can form feather pigment (see Fig. 4-1B). Consider a cross of two dihybrids: CcOo × CcOo (Fig. 4-4). Color of some type is expressed phenotypically by each parent, because both dominants are present. Among their offspring, however, pigmented and white occur in a ratio of 9:7, a modification of the 9:3:3:1 ratio. There are only 9 chances out of 16 for any chick to exhibit some type of color, because only 9 out of 16, on the average, will possess both of the required dominant factors. In cases such as this, we say that "duplicate recessive epistasis" is operating, for we have two genes (one at locus c and one at locus o), *either of*

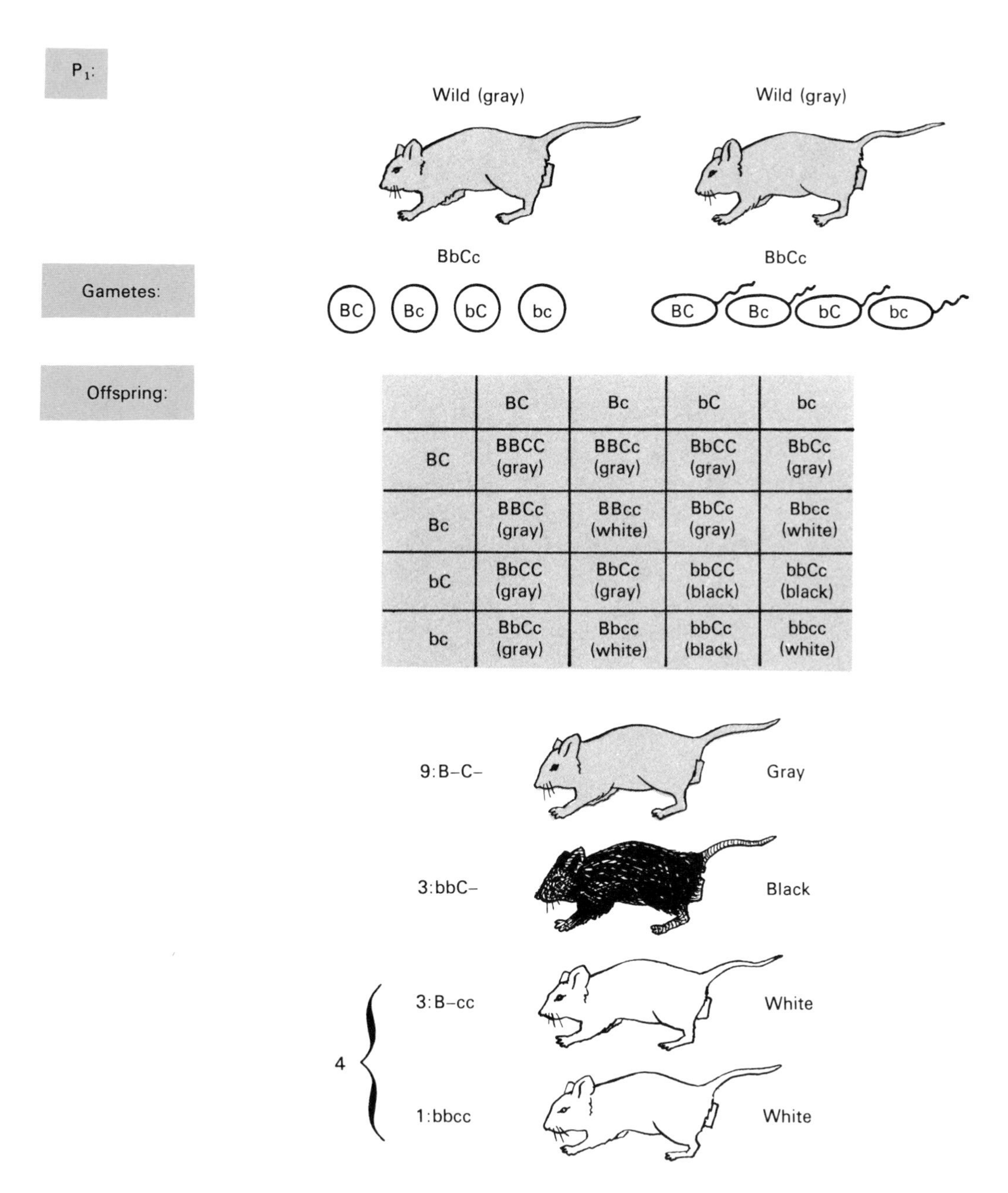

	BC	Bc	bC	bc
BC	BBCC (gray)	BBCc (gray)	BbCC (gray)	BbCc (gray)
Bc	BBCc (gray)	BBcc (white)	BbCc (gray)	Bbcc (white)
bC	BbCC (gray)	BbCc (gray)	bbCC (black)	bbCc (black)
bc	BbCc (gray)	Bbcc (white)	bbCc (black)	bbcc (white)

FIGURE 4-2. Cross of two dihybrid gray mice. The factor B (gray) is dominant to its allele b for black fur. However, no pigment can be produced in the absence of allele C. As a result, two of the four major classes of genotypes cannot be distinguished, and the 9:3:3:1 phenotypic ratio is modified to 9:3:4.

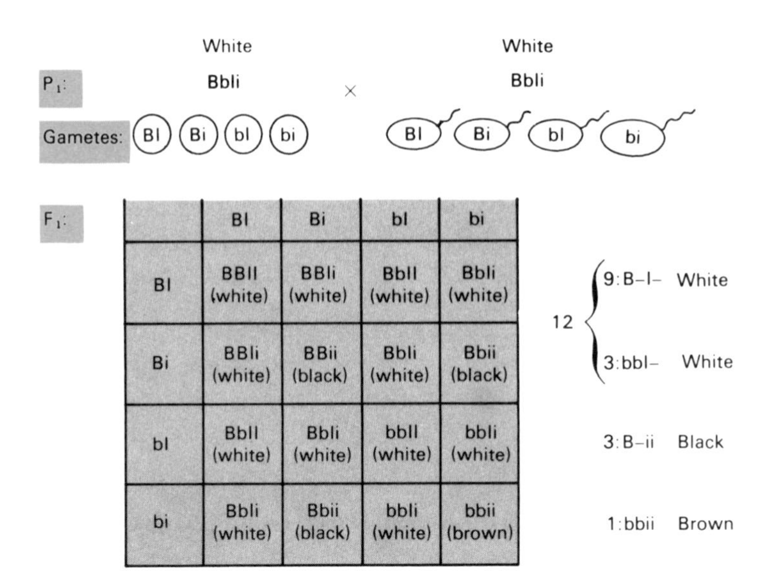

FIGURE 4-3. Dominant epistasis. In certain breeds of dogs, the factor for black fur (B) is dominant to its allele for brown (b). However, neither allele can be expressed if the dominant I is present. Because two of the four major classes of genotypes cannot be distinguished and are classified as white phenotypes, the 9:3:3:1 ratio is modified to 12:3:1. Compare this dihybrid cross with that shown in Figure 4-2, an example of recessive epistasis.

which independently can alter the effects of a second or third one (cc can suppress O– or B–; oo can also suppress C– or B–).

Still other genes are involved in feather color. The allele I at another locus exhibits dominant epistasis and inhibits feather color, just as the allele I in dogs prevented pigmentation of the hair. A pure breeding variety of black birds must therefore have at least the following genotype: BBCCOOii.

The basis of genic interaction

A wealth of molecular biology research has provided insight into the mechanism whereby a gene at one locus can influence the expression of another located elsewhere in the genome. We know that genetic control of metabolism is largely a consequence of the control of proteins. Proteins form the main components of enzymes, all of which are governed by specific genes. Many essential cellular products are the end result of a series of chemical steps, each of which can take place only if the appropriate enzyme is present in an adequate amount. Although Figure 4-5A is a definite oversimplification because it eliminates branches in the sequence, it does enable us to picture the average cellular product as the result of a sequence of steps. No step in the sequence can proceed without the product of a previous step. A breakdown anywhere in the chain prevents the production of the proper end product. Such a breakdown can occur if a step fails to progress because of an enzyme that is defective, absent, or present in insufficient quantities.

We may consider pigment production in the feather of the fowl to depend on a developmental sequence that entails several steps, each controlled by a specific enzyme (Fig. 4-5B). We can see that, if a bird lacks the dominant allele C (Fig. 4-5C) or the dominant allele O, no pigment can form. It does not matter what other genetic factors are present for pigment production if a breakdown in a vital step in the chain occurs. If the chain is not interrupted, pigment will form, but the final color depends on still other genes. Inspection of the sequence of steps shows why a ratio of 9:7 is to be expected after the cross of two dihybrids: CcOo × CcOo. Our knowledge of the classic 9:3:3:1 ratio tells us that, on average, in 9 cases out of 16 both dominants will be present: C–O–. Only these genotypes contain the alleles essential to the completion of the developmental sequence through their control of the needed enzymes. The other seven combinations all lack at least one genetic factor that controls one of the enzymes and hence a step that is essential to a pigment product. If we look at the same chain of reactions (Fig. 4-5C), we see that the presence of allele I blocks completion of the sequence because enzyme i would be lacking. It does not matter if the factors C and O are both present. The presence of allele I, which causes a lack of enzyme i, interrupts the sequence of steps required for the pigment.

Applying the same reasoning to the case of the dihybrid cross in dogs (see Fig. 4-3), we can see that, on the average, 12 offspring out of 16 will contain I and hence will lack enzyme i:

9 I–B–	White because enzyme i is not present
3 I–bb	White because enzyme i is not present
3 ii B–	Black because enzyme i allows pigment to form and the allele B governs the formation of black pigment
1 ii bb	Brown because enzyme i allows pigment to form and the recessive condition bb governs the formation of brown pigment

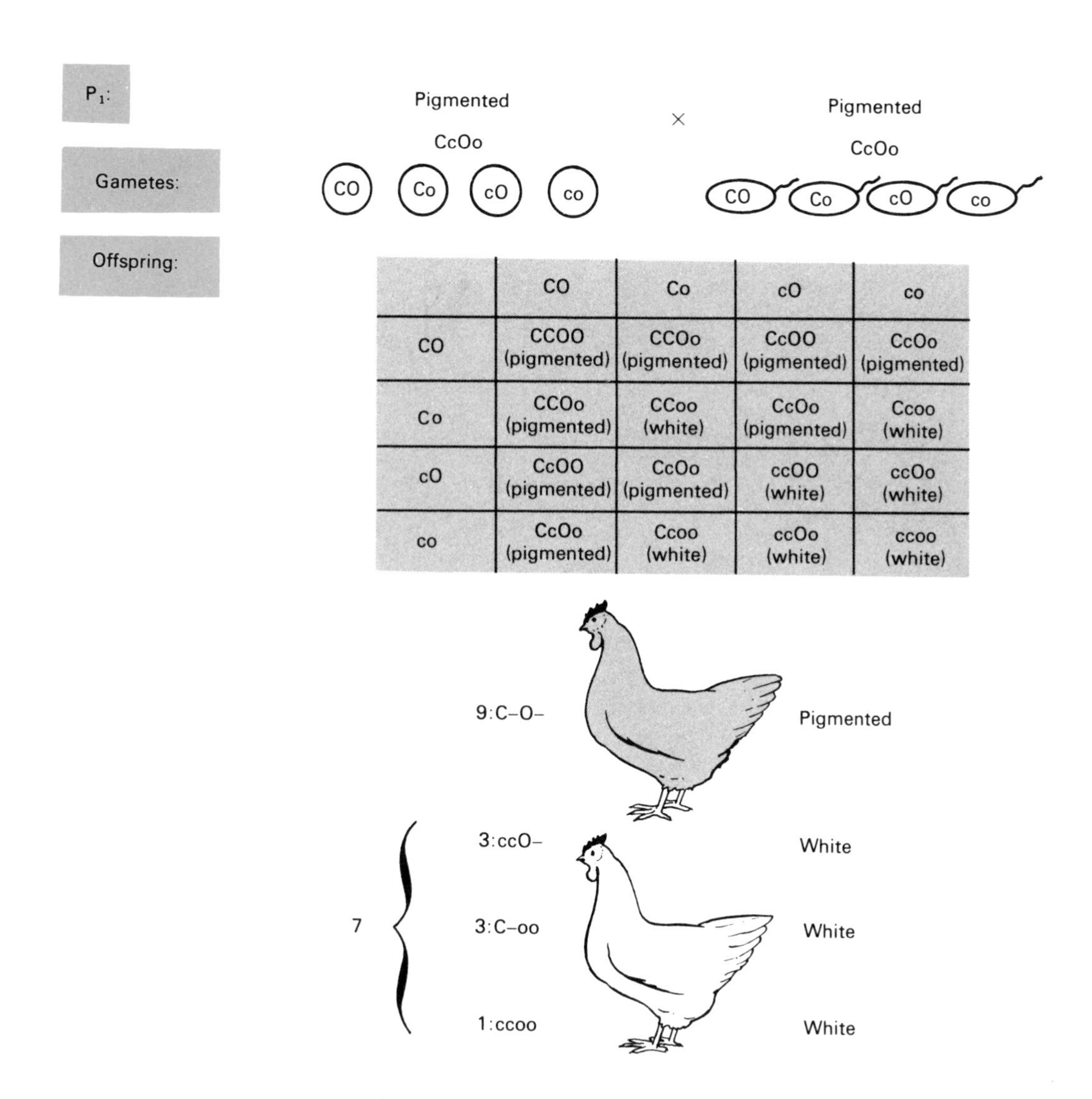

	CO	Co	cO	co
CO	CCOO (pigmented)	CCOo (pigmented)	CcOO (pigmented)	CcOo (pigmented)
Co	CCOo (pigmented)	CCoo (white)	CcOo (pigmented)	Ccoo (white)
cO	CcOO (pigmented)	CcOo (pigmented)	ccOO (white)	ccOo (white)
co	CcOo (pigmented)	Ccoo (white)	ccOo (white)	ccoo (white)

FIGURE 4-4. Duplicate recessive epistasis. In poultry, both dominants (C and O) must be present for any pigmentation of the feather. Either double recessive condition (cc or oo) can prevent the expression of genetic factors at other loci that also influence pigment formation. Because three of the four genotypic classes cannot be distinguished phenotypically, the 9:3:3:1 ratio is modified to 9:7.

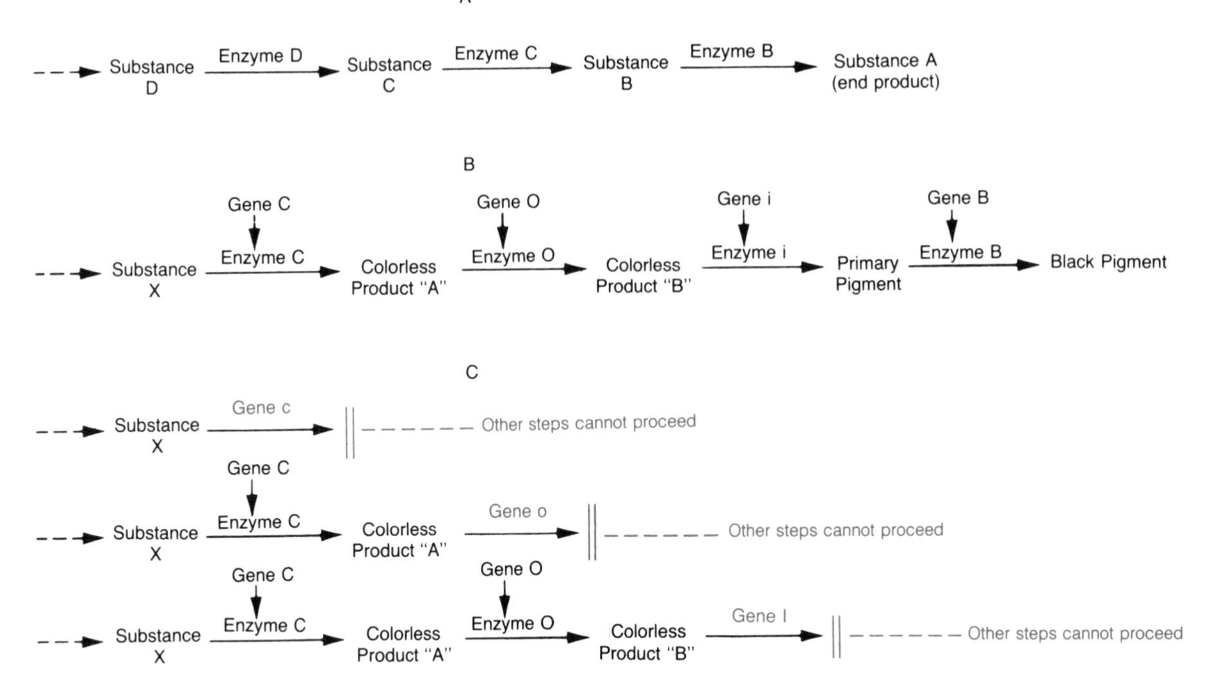

FIGURE 4-5. Gene interaction. **A:** Any cellular product is the result of a series of chemical steps, each catalyzed by a specific enzyme. **B:** A simplified version of pigment formation in the feathers of the common fowl. A specific gene governs the formation of an enzyme at each step along the pathway, leading to the formation of the end product, the pigment. The type of pigment will depend on the enzymes converting the primary pigment. These in turn are controlled by specific genes, such as gene B in this scheme. **C:** Blockage of the pathway leading to pigment formation. The interruption can occur anywhere in the sequence, early (above) or later (below). Absence of a specific gene form (C, O, or i) and the substitution of its allele (c, o, or I) results in deficiency of the enzyme controlled by the gene. The sequence cannot be completed even though all the other required enzymes are available.

Interrelated pathways in metabolism

Few synthetic pathways are as simple as suggested by Figure 4-5A. Most of them are not independent but are linked to others so that any one product in a sequence may be essential to one or more additional pathways (Fig. 4-6). In Figure 4-6, a lack of C substance leads to defects in two end products, A and Z, because both pathways leading to them require C for their completion. Since metabolic pathways are commonly interrelated in this way, it is not surprising to learn that a gene, when studied carefully, is often found to have more than one phenotypic effect. One of these may be more pronounced than the others, but the latter are real and detectable.

The multiple effects of a single gene or allele are termed *pleiotropy.* Although they may seem to be unrelated, study of the chemical or molecular basis of the several effects often reveals a common basis. One excellent example from humans is that of the disorder phenylketonuria (PKU), which is inherited as a simple Mendelian recessive trait. In this unfortunate condition, severe mental retardation is a typical symptom. Affected children, however, also tend to have light hair and light skin pigmentation. Blood and urine analyses reveal abnormally high levels of phenylalanine, an amino acid concentrated in the protein of milk, cheese, eggs, and several other common foods. The high levels of phenylalanine are associated with high levels of another chemical, phenylpyruvic acid, a substance not found in the body fluids of normal individuals.

Any collection of phenotypic effects such as these that defines a clinical condition is commonly called a *syndrome.* The syndrome that describes PKU is now known to result from the pleiotropic effects of a recessive allele that blocks a single step in a metabolic pathway. As Figure 4-7A indicates, protein from the diet is broken down by the cells into its constituent amino acids. The amino acid phenylalanine is normally converted to another amino acid, tyrosine. This single chemical step involves the substitution of an –OH group for –H and requires a specific enzyme manufactured in the liver. The tyrosine that is normally formed is an amino acid linked to other major pathways—one leading to the formation of melanin pigment, one to the formation of thyroxin, and another to complete oxidation to carbon dioxide and water. Tyrosine may also enter cells as a result of its presence in various dietary proteins; all of it is not formed from the conversion of phenylalanine.

The normal picture of the pathways leading from tyrosine may be altered in the absence of the enzyme that forms tyrosine from phenylalanine (Fig. 4-7B). In this case, the latter substance accumulates in the blood. Appreciable amounts of it become diverted into an alternative pathway, which leads to the formation and excessive accumulation of phenylpyruvic acid. The tyrosine required for the pathways that are linked to it must come entirely from preformed tyrosine in the diet. Moreover, excess phenylalanine has been shown to inhibit the activity of the enzyme that converts tyrosine to another substance needed in the pathway leading to melanin production. We can appreciate why victims of the disorder tend to have less pigment in their skin and hair. The mental retardation is a consequence of the high levels of the accumulated phenylalanine and phenylpyruvic acid that forms from it. These substances are toxic to the central nervous system and interfere with the metabolism of the brain cells. The unfortunate, irreversible damage to the nervous system can be prevented if afflicted babies are placed on diets low in phenylalanine. This avoids the accumulation of toxic concentrations of the amino acid and its conversion to phenylpyruvic acid. Fortunately, tests are available for detection of affected infants shortly after birth.

The example of PKU demonstrates several ways in which a gene can bring about pleiotropic effects: by the interruption of a chemical step needed for the formation of a substance that is common to more than one metabolic pathway; through the accumulation of a

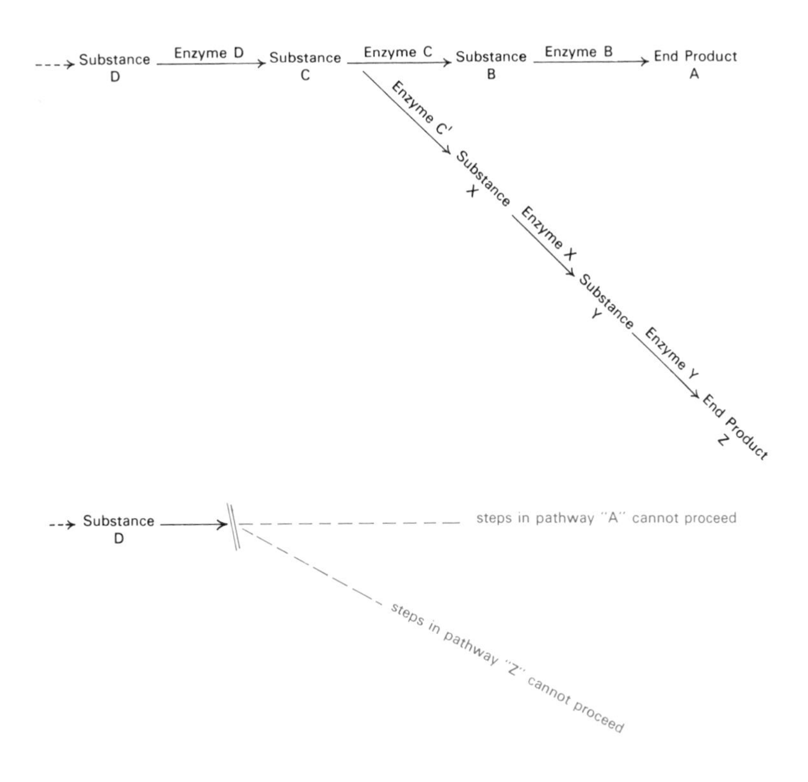

FIGURE 4-6. A branched metabolic pathway. Since most pathways are interrelated (top), a genetic block at one point can have multiple effects. In this example (bottom), a genetic block prevents the formation of substance C, required for the production of two products, A and Z.

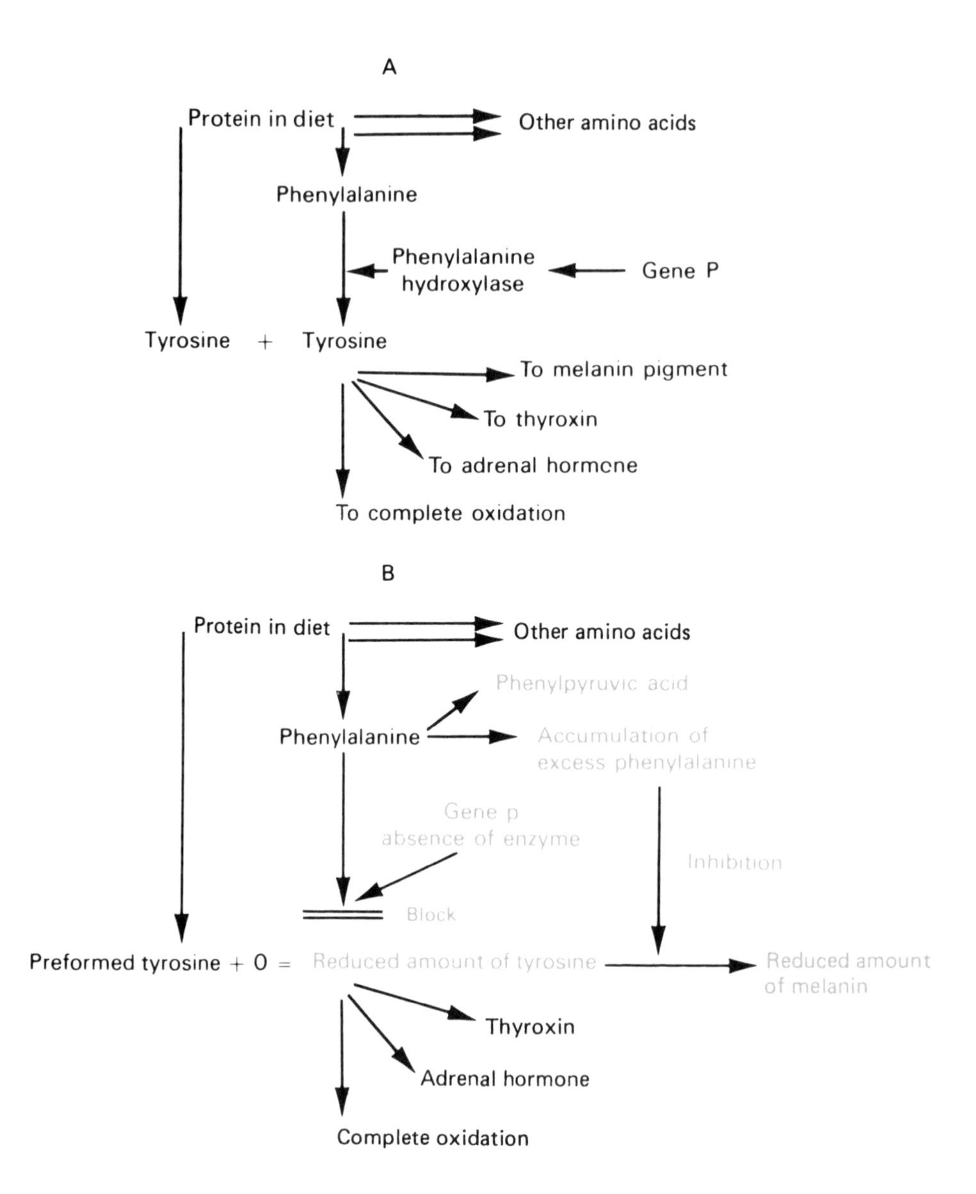

FIGURE 4-7. Interrelated pathways in metabolism of phenylalanine and tyrosine. **A:** Tyrosine normally enters cells through breakdown of dietary protein. In addition, it is formed from phenylalanine by the action of a specific enzyme. The total tyrosine amount then enters other pathways. **B:** In the absence of the enzyme, because of a recessive allele, phenylalanine accumulates, and some is converted to phenylpyruvic acid. The two substances are toxic. The only available tyrosine is preformed and comes from the dietary protein. The excess phenylalanine inhibits an enzyme activity in the pathway leading to the formation of melanin pigment.

metabolite (phenylalanine here) that may be toxic at high levels and may be converted to another toxic product; and through the inhibitory effect of an accumulated metabolite on enzymes in another metabolic pathway. As in the case of PKU, the assorted phenotypic effects of many genes can be traced to a single chemical block in a sequence of steps. Many examples in humans will be encountered in our discussions concerning various other genetic phenomena in addition to pleiotropy.

Variation in gene expression

Certain alleles are very constant in their expression. In the human, for example, the genetic factor S determines the presence of normal hemoglobin (hemoglobin A) in the red blood cells, whereas its recessive allele, s, determines the presence of sickle-cell hemoglobin (hemoglobin S). A person of genotype SS has only hemoglobin A in the red blood cells, and the ss individual has only hemoglobin S. The heterozygote has red blood cells with both types of hemoglobin, A and S. The alleles for the two types can be expected to express themselves whenever they are present in the genotype and to express themselves in the same way.

In contrast with this, we find a large number of other alleles that do not express themselves as predictably. A dominant (P) in humans is responsible for the production of extra digits on the hands and feet, an abnormality known as *polydactyly.* The condition can be present in several members of a family—a parent and one or more children. Occasionally, a phenotypically normal person from such a family marries another normal, unrelated individual but nevertheless produces children with polydactyly. From studying many such cases, it becomes apparent that a person who is heterozygous for the condition (Pp) *may* or *may not* show the trait. The dominant factor for polydactyly is thus not constant in exerting its phenotypic effect. Because it does not express itself in all the individuals who carry it, we say that the allele has *reduced penetrance.* If an allele always expresses itself, as in the case of hemoglobins A and S, we say that it is 100% penetrant. If 10 people have the same genotype (such as Pp for polydactyly), but only 9 out of the 10 show the dominant effect of the mutant factor whereas the tenth appears normal, the allele is said to have a penetrance of 90%. The degree of penetrance varies greatly for different genetic factors. For example, another dominant in humans causes the production of bony projections and has a penetrance of only 60%.

In the fruit fly, a certain recessive, i, causes an interruption in one of the wing veins. When flies with interrupted wing veins are mated, 9 out of 10 of their offspring also show the trait, but the remainder have normal wings. These normal-appearing individuals can be selectively bred together (Fig. 4-8). Among their offspring, 9 out of 10 will show the mutant trait. The remaining normal 10% can be bred further and will again produce 90% mutant progeny and 10% normal. We clearly see from such a study that the 10% that are phenotypically normal individuals are actually homozygous recessives and that the allele for interrupted wing vein has a penetrance reduced to 90%. It is evident that the true genetic constitution of a certain phenotype (e.g., the homozygous recessive genotype of the normal-appearing flies) can remain undetected if selective breeding experiments are not performed. Reduced penetrance is therefore a factor that can complicate genetic investigations and also the interpretation of pedigrees.

In addition to the modification of gene expression that can result from reduced penetrance, another type of variation is frequently encountered. In the case of polydactyly, those persons who *do* express the allele when it is present have an assortment of phenotypes. Some of them may have an extra digit on each hand and foot, others on only one. Moreover, the extra digit can range in development from complete to just a vestige. Obviously, when the allele for polydactyly *is* penetrant, it varies in the kind of phenotypic effects it produces. We say that an allele whose expression varies in degree is of *variable expressivity.* Note the distinction between *penetrance* and *expressivity.* The former measures the ability of an allele to express itself in any way when it is present in the genotype. The latter refers to the kind of phenotype an allele produces when it does express itself (i.e., when it is penetrant). One genetic factor may be 100% penetrant but its expressivity may vary greatly. Another may show reduced penetrance but be very constant in its expression. Besides being 100% penetrant, the factors for hemoglobins A and S do not vary at all in their expressivity.

A good example from the human illustrates an allele with reduced penetrance that, in addition, has a pleiotropic effect with variable expressivity. The dominant factor (O) is associated with *osteogenesis imperfecta,* a condition in which the bones of the body are quite brittle. The allele is also associated with two other effects. When it is present in the genotype, the sclera (the outermost coat of the eye) may be blue, and the bones of the inner ear may be hardened, producing deafness. However, only 9 persons out of 10 who carry the allele express it in any way at all. Such persons may exhibit only one of the three associated effects; others can exhibit all three or any combination of two (Fig. 4-9A). Moreover, when any one of the traits is expressed, its severity varies markedly from one individual to the next. The effect on the bones may be very pronounced, subjecting such persons to multiple fractures, whereas the condition is so mild in others that it almost

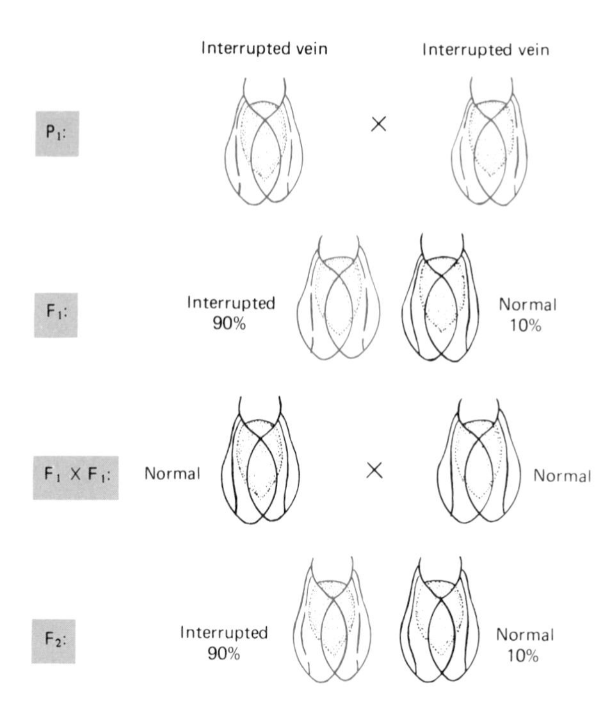

FIGURE 4-8. Reduced penetrance. When flies with the mutant trait interrupted wing vein are mated, 10% of their offspring appear normal. When these are mated to each other, 90% of their offspring are mutant and 10% normal. Further breeding of these latter types produces the same results. All of the flies, those normal in phenotype as well as those expressing the mutant trait, are homozygous recessives for interrupted wing vein (ii). Therefore, the recessive genotype is only 90% penetrant.

escapes detection. The sclera color of those who show any discoloration may range from a tinge of blue to almost black (Fig. 4-9B).

Genetic basis for variation in gene expression

How can we explain the basis of the phenomena of reduced penetrance and variable expressivity, which can cause marked variations in gene expression? In our studies we have encountered several examples of genic interaction in which one gene alters the expression of another. In the simple example of complete dominance of the allele for tallness in peas, the recessive allele does not express itself at all in the heterozygote (Tt). In a sense, the recessive (t) is not penetrant when its allele (T) is present. When epistasis is operating, an allele may not be penetrant because of the presence of another genetic factor at another locus (feather color is not expressed in the genotype (CCII). In these examples, we clearly see that the penetrance of an allele may be the direct result of the influence of some other genetic factor, either its own allele (in the case of dominance) or a nonallelic factor (as in epistasis). Therefore reduced penetrance should not be regarded as any sort of contradiction to the principles of heredity. It is an expected consequence of the role of the rest of the genotype in the expression of any form of one specific gene.

Similarly, variation in the *kind* of expression an allele has when it *is* penetrant often depends on the presence of certain other genes. Many examples relating to spotting have been studied in cattle and rodents. In mice, white spotting may depend on the presence of a recessive (s). However, although spot-

FIGURE 4-9. Penetrance, pleiotropy, and variable expressivity. **A:** The dominant allele, O, for osteogenesis imperfecta can affect the bones, inner ear, and eye (top). However, a person who carries the allele and expresses it may show only one abnormality, such as affected eyes. Offspring of such a person who receive the allele may express all three defects, none, or any combination of defects (middle). Since the allele is only 90% penetrant (bottom), a person carrying it may not express it at all in any detectable way. However, offspring of such a person may show none, all, or any combination of abnormalities. **B:** A person carrying the allele O may not exhibit any discoloration of the sclera of the eye. Among those who express this trait, there is variation in the shade of the sclera from one person to the next.

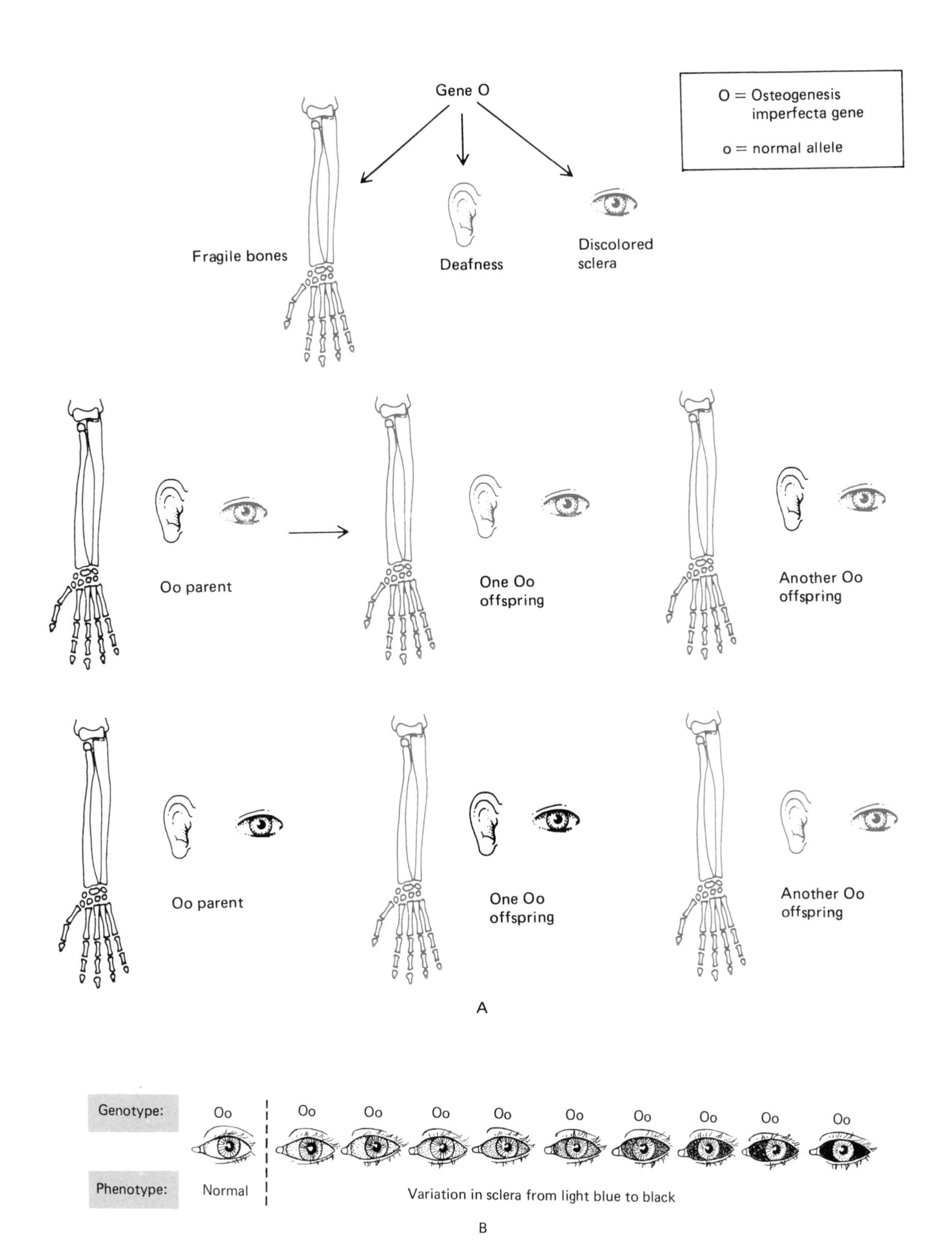
Gene O
O = Osteogenesis imperfecta gene
o = normal allele
Fragile bones
Deafness
Discolored sclera
Oo parent
One Oo offspring
Another Oo offspring
Oo parent
One Oo offspring
Another Oo offspring
A
Genotype:
Oo
Oo
Oo
Oo
Oo
Oo
Oo
Oo
Oo
Oo
Phenotype:
Normal
Variation in sclera from light blue to black
B

ting may be expressed in the homozygous animals (ss), the size of the white areas can range from tiny points to large spots to a coat that is entirely white. The amount of whiteness depends on a host of genes, each with a small but definite effect, interacting with the allele for spotting (s). The term *modifier* (or *modifying gene*) is usually reserved for any such genetic factor whose effect is to alter, in a small, quantitative way, the expression of another genetic factor. The cumulative or added influence of modifiers can be very significant. If different modifiers are present in different individuals who possess the same "main allele," the expression of the latter can vary greatly from one individual to the next.

Certain modifiers have such an extreme effect that they completely prevent the expression of some other allele. One such case in the fruit fly involves the size and shape of the wing. Flies homozygous for the recessive allele vetsigal (vgvg) have reduced and distorted wings. However, the mutant condition may not appear if a certain other allele (called a *dimorphos*) is present at another locus. We designate the latter a *suppressor,* meaning that it is a genetic factor that can prevent the expression of a mutant allele. Some suppressors behave as dominants, others as recessives. Suppressors as well as modifiers with small effects can confuse the genetic picture if we are unaware of them. Their recognition enables us to appreciate part of the basis for penetrance and expressivity, because modifiers and suppressors, along with other genes composing the total genotype, can interact to produce variation in the expression of any gene.

The environment and gene expression

The activities of all the genes take place in the cellular environment, which in turn can be influenced by the external one. Any characteristic depends on a certain environment as well as on a certain genotype for its typical expression. The environment has been shown to be a critical factor in the degree of penetrance and expressivity of many alleles. In the fruit fly alone, temperature changes are known to be able to change the penetrance of many alleles from 0% to 100%. Similarly, the expressivity of an allele can vary greatly with temperature. The genotype that produces blisters on the wings is much more extreme in its expression at 19° than at 25°C. Not only temperature, but all aspects of the environment must be considered in relation to their influence on gene expression. Some of the variation in the expression of polydactyly is undoubtedly environmental. The fact that any one person showing the trait may have a complete extra digit on one hand and none at all or just a protuberance on the other argues that environmental influences are operating. We are seeing here an example of variable expressivity in *one* individual. Genetic factors for extra digits are also known in animals in which there is also evidence for an effect of maternal age on the penetrance of the specific factors.

Nutrition or diet is another critical environmental factor to consider. Recall that the dire effects of phenylketonuria are expressed when there is an accumulation of phenylalanine. A diet that eliminates much of this amino acid is a factor in preventing the mental retardation that full expression of the allele for PKU can cause.

An excellent example demonstrating the interaction of the genetic component with environmental factors centers around variations in a gene coded for a particular enzyme, glucose-6-phosphate dehydrogenase (G6PD). The gene is sex linked, located on the X chromosome. The enzyme that it specifies occurs in all kinds of tissues, including red blood cells, where its role is of special importance. Several different mutant forms of the gene are known, and these typically cause the activity of the enzyme to be lowered. Ordinarily a defect in G6PD is not of any consequence, since the enzyme is called upon to carry out a biochemical step that is of relatively minor importance. The step that it performs is linked to a reaction that brings about the reduction of glutathione. Reduced glutathione is needed to keep the cell membrane intact, but only a low activity of G6PD is needed for this purpose, and so a person with a mutant allele is not usually at a disadvantage. However, if a situation arises that favors oxidation in the cell, as in the presence of peroxides, reduced glutathione levels may fall. If normal enzyme with full activity is present, the situation is readily handled, but mutant enzymes with lowered activity may not meet the challenge, and the level of reduced glutathione may drop. Since mature red blood cells are anucleate, they cannot synthesize more enzyme and depend solely on that made when they were immature. Consequently, an inefficient enzyme may fail to maintain the membranes of the red blood cells, which then break down, bringing about an attack of acute hemolytic anemia that can prove fatal.

Several environmental factors are well known to cause oxidation of reduced glutathione and thus spark an episode of red blood cell disruption in persons who carry a mutant form of the gene for G6PD. The condition, known as *favism,* is a hemolytic response that can be triggered by eating fava beans. Susceptible people are also vulnerable to red blood cell instability following exposure to oxidizing substances such as certain sulfa drugs, chloramphenicol, and the naphthalene found in mothballs. In those individuals with a different mutant form of the sex-linked allele, exposure to the antimalarial drug primaquine can bring about red blood cell breakdown. They are also susceptible to naphthalene and to certain sulfa and other drugs.

It is significant that those persons showing sensitivity are otherwise completely normal. Only an encounter with the specific substance reveals any difference between them and the majority of the population. The reaction of those who are sensitive emphasizes the importance of caution when experimenting with drug intake for any reason. Wide differences in response resulting from genetic factors could have disastrous effects. This is seen most vividly in the violent reaction to barbiturates, which can cause the death of susceptible persons. Such people have a disorder known as *porphyria,* which results from a defect in an enzyme that renders them incapable of properly metabolizing a component of hemoglobin, porphyrin. The disorder is transmitted as a dominant and is caused by a genetic alteration in a specific gene found on an autosome (a chromosome other than a sex chromosome). Persons with porphyria accumulate porphyrins and excrete some of the excess into the urine that turns a wine color following its exposure to light. The high levels of porphyrin may cause a person in his or her middle years to suffer episodes of intense abdominal pain, delirium, mental seizures that mimic schizophrenia, and various other symptoms. In addition to barbiturates, reactions can be set off by certain other components found in sedatives and anesthetics as well as by certain types of infection and insufficient food intake.

A second form of porphyria is due to a defect in still another enzyme. It also behaves as a dominant trait, but stems from an alteration in another autosomal gene. In addition to the symptoms already described that can be induced by drugs, persons with this form of the disorder also suffer from a variety of skin problems and are unusually sensitive to sunlight.

The realization that persons in the population vary in their response to certain substances as the result of enzyme variations due to genetic factors has been largely responsible for the development of the discipline of pharmacogenetics. This area concerns genetic variations in response to drugs and additives, many of which can be life threatening.

Importance of both environmental and genetic factors

A frequently asked question is, "Which is more important, heredity or environment?" Concern with environmental influence on heredity has persisted from the earliest days of genetics. Many have thought that the environment, acting over long periods of time, could change the genes and direct the course of evolution. A classic experiment, performed in 1909, shows the fallacy of this idea. It was known that black coat color in guinea pigs depends on a dominant (A) and an albino coat on its recessive allele (a). Geneticists removed the ovaries from a white animal and transplanted into it ovaries from a young black guinea pig. The albino female, bearing ovaries from the black one, was mated to an albino male (Fig. 4-10). If whiteness of the female can affect the genes in the transplanted ovaries, we might then expect the offspring to be white or some intermediate shade. However, in three separate litters, the offspring produced were completely black. Even though the body of the host mother was white, the cross is genetically AA (black female) × aa (white male). This is an elegant demonstration that the body cells do not affect the transmission of genetic factors to the next generation. The genetic factors for black fur in the gametes of the transplanted ovary were in no way altered just because they were housed in a white body.

This example from the guinea pig illustrates the distinction between phenotype and genotype. It shows that the phenotype is not transmitted. The white phenotype of the host mother was not passed to her offspring. What was actually passed down was the genetic constitution found in the two gametes that combined at fertilization to establish a genotype, a combination of genetic factors from both parents. This genotype sets a potential, but the phenotype that is actually realized depends on the interaction of all the genes and the environment. Whether individuals burdened with the recessive genotype for phenylketonuria escape the most dire effects of the disorder depends on the environment, their diet. Nevertheless, those who are placed on the proper diet as infants and who come to lead an existence indistinguishable from that of other persons still carry the recessive genotype for the disorder. The effect of the proper environment on their bodies has in no way changed the defective alleles they carry. Two such parents can produce only phenylketonuric offspring who also have the metabolic defect and whose chances for a normal life will again be determined by the environment, a special diet in infancy.

When dealing with very complex characteristics such as body size or intelligence, the interactions between heredity and environment become so complex that they cannot be untangled for precise measurement. But we do know that the expression of the best genotype, let us say one that makes possible the expression of great intelligence, can be suppressed by factors in an unfavorable environment. On the other hand, the most favorable environment cannot produce a genius if the hereditary endowment is lacking.

Genes that influence survival

We have encountered thus far some of the ways in which genic interaction results in modifications of classic Mendelian ratios, such as 9:3:3:1 to 9:3:4 or 9:7. Expected ratios can be altered by still another factor, which can be illustrated by a familiar example in mice. A cross between any yellow mouse and a gray one always produces yellow and gray offspring in a ratio of 1:1. This makes us suspect that all yellow mice are heterozygotes because the ratio of 1:1 is expected when a monohybrid is testcrossed. The truth of this hypothesis is borne out by following crosses between any two yellow mice (Fig. 4-11). The outcome of such a cross is always two yellow to one gray. And these yellow animals always give the typical 1:1 testcross ratio when crossed to grays.

Evidently some factor is preventing the birth of homozygous yellow animals. Examination of gravid females reveals that these homozygotes are produced but they die as embryos while still in the uterus. The factor for yellow fur (Y) is responsible for killing the individual when it is homozygous, in contrast to its allele for gray. We say that it is a *lethal* allele, meaning that it is responsible for the death of the carrier. Yellow is a *complete lethal,* because it kills the individual before reproductive age. Such lethal alleles are by no means exceptional and must always be considered in populations of plants and animals. Indeed, when a gene mutation arises, it is much more apt to produce an allele that has a harmful or a lethal effect than one

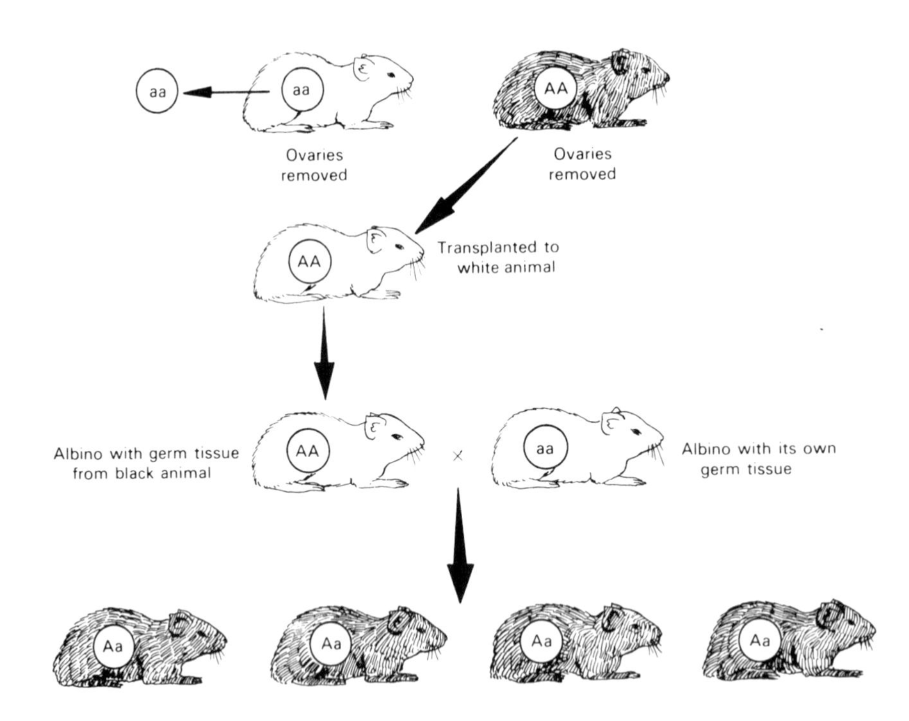

FIGURE 4-10. Ovarian transplant in the guinea pig. An albino animal lacks the dominant required for black pigment formation. When germ tissue from a black animal is transplanted into an albino, the body of the latter in no way alters the germ cells that are being sheltered. A cross of this animal with an albino will result only in black offspring, because the cross is AA × aa. Such experiments demonstrate that the gametes link one generation to the next and that the genetic information they contain is not changed by the body cells.

that is neutral or conveys a benefit. In the fruit fly, invisible lethal mutations outnumber those that produce a detectable phenotypic effect by 10 to 1! The allele for yellow coat in mice is pleiotropic; it has a dominant effect on fur color but a recessive one on viability. Therefore we can look at the factor for yellow in another way and call it a *recessive lethal.* Many lethals produce no pronounced effect at all on the phenotype, but they may make their presence known by a decrease in the lifespan or by the very early elimination of the carrier. Some geneticists calculate that each human carries, on the average, the equivalent of three lethal alleles.

How can an allele have a killing action? A glance at Figure 4-6 reminds us that metabolism is the result of many interlocking pathways. A mutation can produce an allele that proves to be lethal because it interferes with a pathway leading to normal development of some essential organ such as the heart, resulting in the death of an embryo or fetus. Tay-Sachs disease (ganglioside lipidosis) is one example of a human lethal whose genetic basis has been clarified. This fatal condition is expressed in children who are homozygous for a certain recessive. The mutant allele is found with a higher incidence among Jews of northern European origin than in other groups. The severe symptoms (blindness, growth retardation, loss of motor coordination) typically end in death by the age of 4 years. The severity of the disorder follows from the abnormal accumulation in brain cells of a certain type of lipid, a ganglioside. Its buildup produces actual swelling of separate nerve cells that in turn causes

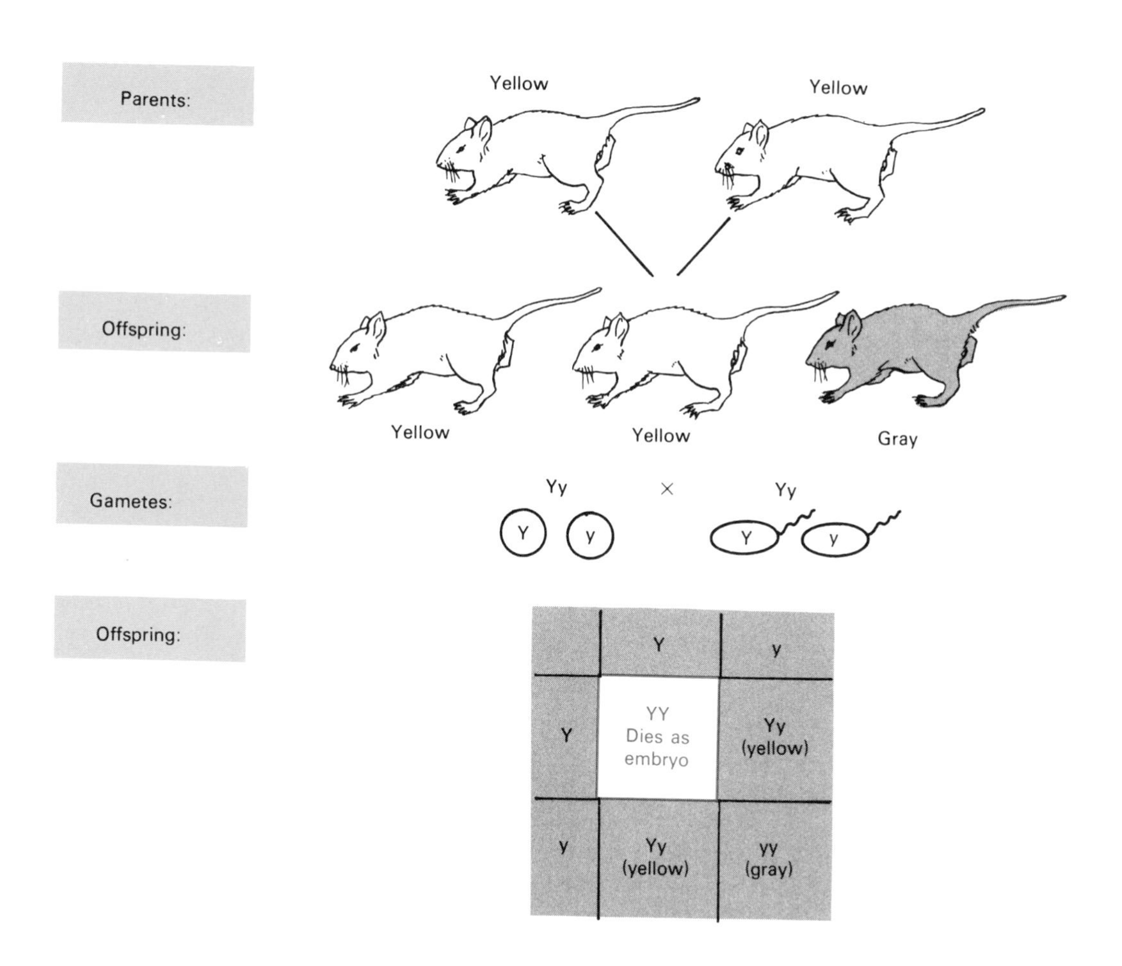

FIGURE 4-11. Recessive lethal in mice. When two yellow mice are crossed, the offspring occur in a ratio of 2 yellow:1 gray (above). The explanation is that all yellow animals are heterozygotes. Although the yellow phenotype is inherited as a dominant, the allele that is responsible behaves as a recessive lethal. Therefore no homozygotes for the allele for yellow fur survive.

swelling and malfunction of the brain. We now know that this lethal condition is associated with a blockage of a critical step in the normal breakdown of the particular ganglioside. Victims of Tay-Sachs disease show an abnormality in the enzyme hexoseaminodase A (Hex A). This enzyme is actually composed of two different polypeptide chains, α and β, encoded by genes on chromosomes 15 and 5, respectively. Only β-chains are found in normal amounts in persons with the disorder. It is this deficiency of the enzyme Hex A that blocks the normal breakdown of the ganglioside in brain cells and leads to its accumulation. Parents of afflicted children are able to handle the lipid normally, but a blood test shows them to possess less than the typical level of α-chains of Hex A. Prospective parents who show such a deficiency run a risk of 1 in 4 of producing a child with Tay-Sachs disease. It is possible to detect such a child in utero. The cells that are shed into the amniotic fluid that surrounds the fetus are fetal in origin and a sample can be obtained by amniocentesis. The fluid is withdrawn by a hypodermic needle inserted through the pregnant woman's abdomen and uterine wall. Cultures are made of cells in the fluid, and, if they reveal that α-chains of Hex A are missing, steps can be considered to prevent the birth of a seriously afflicted infant.

The Tay-Sachs allele and the yellow allele in mice are examples of complete lethals, those that remove the individual before reproductive age so that they leave no offspring. Some lethals exert their killing action later in the life cycle, after the carrier of the allele may have left offspring. In the human, the dominant factor for Huntington's disease (see Chap. 20), a fatal deterioration of the nervous system, does not usually express itself before the age of 30. Such genetic determinants, which can result in death but which permit the carrier to live to reproductive age, are often grouped as *sublethals*. There are actually no sharp boundaries between the stages of the life cycle at which different lethals can act to decrease the chances of survival.

REFERENCES

Beadle, G.W., and E.L. Tatum. Genetic control of biochemical reactions in *Neurospora*. *Proc. Natl. Acad. Sci. U.S.A.* 27:499, 1941.

Eaton, G.J., and M.M. Green. Implantation and lethality of the yellow mouse. *Genetica* 33:106, 1962.

Kidd, K.K. Phenylketonuria: Population genetics of a disease. *Nature* 327:282, 1987.

Ledley, F.D., H.L. Levy, and S.L.C. Woo. Molecular analysis of the inheritance of phenylketonuria and mild hyperphenylanemia in families with both disorders. *N. Engl. J. Med.* 314:1276, 1986.

McKusick, V.A. *Mendelian Inheritance in Man: Catalogs of Autosomal Dominant, Autosomal Recessive, and X-Linked Phenotypes,* 8th Ed. The Johns Hopkins University Press, Baltimore, 1988.

Scriver, C.R., A.L. Beaudet, W.S. Sly, and D. Valle (eds.) *The Metabolic Basis of Inherited Disease,* 6th Ed. McGraw Hill, New York, 1989.

Scriver, C.R., S. Kaufman, and S.L.C. Woo. Mendelian hyperphenylanemia. *Annu. Rev. Genet.* 22:301, 1988.

Sutton, H.E. *An Introduction to Human Genetics,* 4th Ed. Harcourt Brace Jovanovich, New York, 1988.

REVIEW QUESTIONS

1. A rare allele for extra digits in the human behaves as a dominant (E) to the normal (e). A man with an extra toe on each foot marries an unrelated woman with the normal condition. The first child has an extra finger; the second child is phenotypically normal. In time, the normal offspring marries a normal, unrelated woman and produces a child with an almost complete extra toe on one foot and the normal number of toes on the other. Offer an explanation, and give the probable genotypes of the persons mentioned.

2. In the human, study of family histories indicates that 4 out of 10 persons who carry a certain rare dominant allele for a skeletal defect do not express it. What would be the penetrance of this allele?

3. In the human, a certain allele results in severe anemia plus damage to the kidneys and the central nervous system. What genetic phenomenon is illustrated by the expression of this allele?

4. In sweet peas, the two allelic pairs C, c and P, p are known to affect pigment formation in the flowers. The dominants, C and P, are both necessary for colored flowers. Absence of either results in white. A dihybrid plant with colored flowers is crossed to a white one heterozygous at the c locus.

A. What are the genotypes of these two plants?

B. What kinds of flowers, colored or white, should be expected from the cross, and in what ratio?

5. Assume that another allelic pair in sweet peas also affects pigment formation in addition to the genes mentioned in Question 4. The presence of the dominant, R, is required for red flowers. Its recessive allele, r, produces yellow flowers. What would be the phenotypes of the following plants in relation to flower color?

A. CcPpRr.
B. CcppRR.
C. CcPPrr.
D. ccPPRR.

6. In a certain breed of dog, the dominant, B, is required for black fur; its recessive allele, b, produces brown fur. However, the dominant, I, is epistatic to the color locus and can inhibit pigment formation. The recessive allele, i, on the other hand, permits pigment deposition in the fur. Describe the phenotypes of the following sets of parents, and the results of their mating.

A. bbii × BbIi.
B. bbIi × Bbii.
C. bbIi × BBIi.

7. In the human, the dominants, D and E, are both required for normal development of the cochlea and the auditory nerve, respectively. The recessives, d and e, can result in deafness because of impairment of these essential parts of the ear. Give the phenotypes of each of the following sets of parents and the chance of a deaf child being born as the first offspring.

A. DDee × ddEE.
B. DdEE × DDEe.
C. DdEE × DdEe.
D. DdEe × DDEe.

8. In poultry, the shape of the comb varies greatly and involves at least two pairs of alleles. The allele R can result in rose comb, and the allele P can result in pea-shaped comb. If both of these dominants are present, genic interaction produces a walnut comb. When a bird is carrying both recessives, r and p, in the homozygous condition, single comb type results. Give the type of comb of each of the following pairs of birds and the phenotypic ratio to be expected among their offspring as to comb type.

A. rrPP × RRpp.
B. RrPp × RrPp.
C. RrPp × rrpp.

9. Using the information on comb shape given in Question 8, deduce the genotypes of the parents in each of the following crosses.

A. A walnut bird crossed with a single-comb bird. These parents produce three walnut, four pea, two rose, and three single-comb birds.
B. A rose-combed bird crossed with a pea. This parental combination gives rise to rose and walnut in a ratio of 1:1 after several matings.
C. A rose-combed bird crossed with a walnut. These parents produce three walnut, five rose, two pea, and one single-comb birds.

10. In poultry, the allele B can result in black feathers. Its allele b can produce white-splashed feathers. The heterozygote is blue. A bird with black feathers and pure breeding for rose comb is crossed with a bird that is white splashed and pure breeding for pea comb. (Use the information in Question 8 for comb shapes.)

A. Diagram the cross. What would the F_1 be like?
B. What kinds of birds are to be expected from crossing F_1s with birds that are black and have single combs?

11. In poultry, the factors C and O are both required for any color at all in the feathers. Homozygous recessiveness at either locus will result in white. In each of the following three cases, give the genotypes of the parental birds and their offspring. (Use the information on color in Question 10.)

A. Two completely white birds that produce an F_1 that is black.
B. Two black birds that produce 54 black birds and 42 white ones.
C. Two blue birds that produce 16 black, 37 blue, and 14 white-splashed birds.

12. Two pairs of alleles are involved in cyanide production in white clover. Some strains have a high cyanide content, others a low one. When low strain A was crossed to low strain B, the F_1 all had a high cyanide content. When the F_1 plants were crossed among themselves, the following F_2 resulted: high cyanide content, 450; low cyanide content, 350. Offer an explanation for these observations.

13. Two different highly inbred strains of chickens both have feathers on the upper portion of their legs. When birds from the two different strains are crossed, the F_1s all have feathered legs. In a typical case, crossing the F_1s to each other gave 323. Of these 301 had feathered legs, and the remainder were unfeathered. When the unfeathered ones were crossed among themselves, only birds with unfeathered legs resulted. Give a likely explanation.

14. Very little hair is found on a Mexican hairless dog. A cross between a Mexican hairless and a dog with a typical coat of hair usually produces litters of pups in which half of the animals are hairless and the other half have hair. On the other hand, a cross between two Mexican hairless dogs tends to produce litters in which two thirds of the pups are hairless and one third have hair. However, in addition to these surviving puppies, usually some are born dead. These appear hairless and occur in about the same frequency as the pups with hair. Offer an explanation and represent the genotypes of the different kinds of animals using any symbols you choose.

15. Several loci affect hair development in dogs. The allele for wire hair (W) is dominant to its allele for straight hair (w). Allow H to represent hairlessness in the Mexican hairless dog and its recessive allele h to represent the gene for typical hair growth in other breeds. The Mexican hairless dog is homozygous for straight hair, but, because of the epistatic effect of H, alleles for hair formation cannot be expressed. Write genotypes of the following animals with respect to the two loci mentioned.

A. A Mexican hairless dog.
B. A dog with straight hair.
C. A dog from a strain that is pure breeding for wire hair.

16. Using information in Question 15, give the results of a cross between the following:

A. A Mexican hairless dog and a dog from a strain that is pure breeding for wire hair.
B. Two animals of the first generation resulting from the cross given in A. (Cross only the animals that have different genotypes.)

17. Assume that in a disease-resistant variety of plant the recessive lethal w occurs with a high frequency, and a considerable proportion of the plants die because they lack the dominant allele W and cannot produce chlorophyll. A heterozygote with one dose of the dominant allele cannot be distinguished from a plant homozygous for the dominant. Suggest a program by which the undesirable recessive might be eliminated from the variety, considering that the plants are exclusively cross-pollinated.

18. In guinea pigs, black fur, A, is dominant over the albino condition, a, which gives white fur. A black female from a pure-breeding strain carries transplanted ovaries from an albino female. The albino female in turn received the ovaries from the homozygous black animal.

A. What are the expected results among the offspring when the black female is crossed to a black male that had a white parent?
B. What would be expected from this cross if the black female had also had a white parent?
C. What results are to be expected if the albino female is crossed to the same male?

19. The dominant genetic factor P is required for the formation of an enzyme needed to convert phenylalanine to tyrosine. The recessive allele, p, results in a lack of the enzyme, and as a consequence phenylketonuria (PKU) can ensue.

A woman who was spared the dire effects of PKU as a result of a special diet during infancy and childhood is married to a man who does not have PKU but does carry the recessive allele. During her pregnancy, the woman is placed on a special diet once more, since excess phenylalanine and other toxic products can cross the placenta in a PKU woman and result in mental retardation in the newborn. What genotypes are to be expected among the offspring in this case, and what would be their phenotypes?

20. Suppose the woman in Question 19 is not placed on the diet during pregnancy. What would be expected among the offspring?

Alleles, Antigens, and Antibodies

5

Allelic forms of a gene

In *Drosophila,* a gene located at a certain locus on chromosome II influences the size and shape of the wing, which is normally long and smooth in outline. A mutation occurred at this locus that produced an alternative form of the gene that causes the wing to become nonfunctional and reduced in size. Following the system discussed in Chapter 2, the locus and gene were named *vestigial,* or *vg,* because this was the first mutation found at this site, and it proved to be recessive to its wild allele, vg^+.

Later, a second mutation arose in the same gene, and this produced still another gene form. This mutation was referred to as *antlered,* because it produces a wing form suggestive of antlers. A cross between pure breeding wild flies and antlered ones shows that antlered is recessive to the wild-type allele vg^+. Since antlered is a mutation at the vestigial locus, we designate it vg^a. A cross between antlered and vestigial flies produces an F_1 with wings intermediate between the two mutant types. This shows us clearly that antlered is allelic to vestigial. If antlered had been a recessive mutation at a locus other than vg, a cross between vestigial and antlered would be wild, since each mutant stock would contribute the normal allele that the other lacks. But this did not happen. The appearance of flies with distorted wings in the F_1 tells us that antlered and vestigial did not complement each other. The F_1 flies do not have a wild gene form for vestigial and a wild gene form for antlered.

Only one gene is involved, and in the cross between the two mutants the F_1 carries two mutant allelic forms of this one gene (Fig. 5-1). Thus when we symbolize the allele for antlered, which is a second mutation at the vestigial locus, we keep the same base, vg, and add a superscript, giving us vg^a. We are now recognizing the existence of three alternative forms of the gene at the vestigial locus, vg^+, vg^a, and vg. Many genes exist in various forms, and we refer to each set as a series of multiple alleles, three or more forms of one gene associated with a specific locus.

Multiple allelic series are known in all groups of organisms. The number of alleles composing a multiple allelic series can be quite large. For example, on the X chromosome of the fruit fly at the white (w) locus, 12 alleles are known that influence eye color, producing a spectrum of pigmentation from white through shades that deepen to the normal dark red of wild type. The mutations have been named to suggest their phenotypic effects, such as white, ivory, pearl, apricot, and cherry (w, w^i, w^p, w^a, and w^{ch}). A heterozygous female has an eye color intermediate between that of the homozygotes.

The occurrence of numerous mutations at a locus also shows us that, although many changes can occur within a gene and produce other phenotypic effects, the *primary* effect is confined to the one characteristic. Thus we can find many alterations within the white gene that affect eye pigment in different ways, but none of these has a pronounced influence on another characteristic, such as wing shape. This makes perfect sense, because a gene at any locus has a primary effect in a certain chemical pathway as a result of its control of a specific metabolic step (see Fig. 4-5A, B).

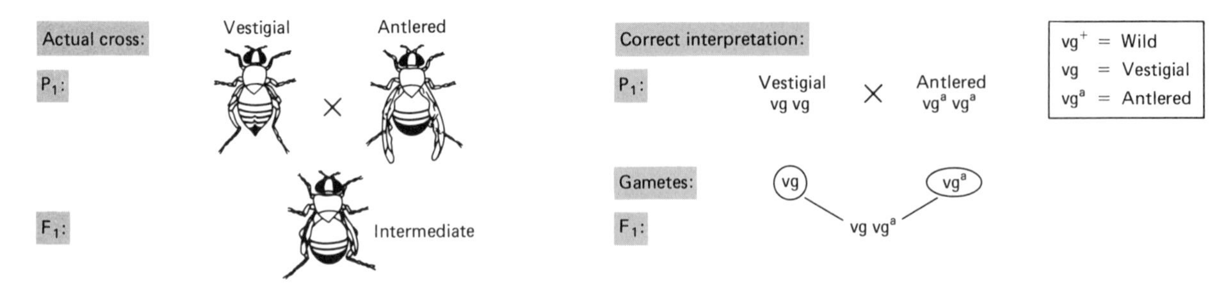

FIGURE 5-1. Allelism. When a cross is made between the mutants vestigial and antlered, the F_1 flies are mutant. This indicates that the F_1 did not receive a wild-type allele from each parent. Complementation did not take place. Vestigial and antlered are variant forms of the same gene that affects the wing. Neither vestigial nor antlered winged flies possess a wild form of the gene vg^+, and thus the F_1 is mutant. Because the antlered effect is the result of a mutation at the vestigial (vg) locus, a symbol (vg^a) must be used to acknowledge this.

More than two allelic forms are known for each of the two human genes discussed a bit earlier: the one for the enzyme G6PD and the gene for the alpha-chains of Hex A, a deficiency of which results in Tay-Sachs disease. In the case of the latter, mutations are known to occur at different sites in the gene on chromosome 15. Most Tay-Sachs patients are homozygous for one of the alleles, but some have been found to carry two different mutant alleles of the gene. Such an individual is unable to produce any normal alpha-chains, since the two defects are at the same gene locus and complementation cannot occur, as is the case in the cross between vestigial and antlered in the fruit fly (Fig. 5-1).

The gene for G6PD actually occurs in 300 different variations! However, the two that occur most commonly are the ones associated with favism and primaquine sensitivity. The allele responsible for the former is the more severe, producing effects in red blood cells and other tissues as well. The one for primaquine sensitivity is expressed mainly in red blood cells, since the G6PD variant associated with this allele has a higher activity than does the variant causing favism. Most affected persons are males, since the gene is sex linked (for details, see Chap. 6), although heterozygous females carrying one normal and one of the mutant alleles show intermediate responses. The favism allele is very widespread and occurs in a high frequency in some Mediterranean populations (e.g., Greece, southern Italy) and even in regions of India. The allele for primaquine sensitivity is found primarily among blacks of African descent and affects about 15% of black American males.

Antigens, antibodies, and the ABO blood grouping

In humans, about 20 systems of blood groups have been recognized, and some of these are inherited on the basis of multiple allelic series. The ability to classify the blood of any individual into the various groupings depends on the presence or absence of specific antigens on the surface of red blood cells. Antigens are large molecules that are often proteinaceous, at least in part. They can elicit the formation of specific antibodies, that is, large protein molecules (immunoglobulins), with which they react to produce clumping or lysis of the cells. The presence of a specific antigen on the red blood cells A, B, or both A and B defines the corresponding blood type within this major grouping (Table 5-1). Cells lacking both antigens are classified as type O. Moreover, if an antigen is absent from the cells, the serum has the corresponding antibody. Since antigen A reacts with antibody a (and likewise B with b), knowledge of the ABO type is crucial for successful blood transfusions. Any transfusion is incompatible if the cells of the donor are clumped by serum of the recipient. Persons of type A have antibody b in their sera. The millions of cells from a B donor would be clumped on contact with the b antibody and would clog the smaller blood vessels, resulting in severe illness or death. Only type O can donate blood to the other types with any degree of safety. Although both types of antibodies, a and b, are present in the serum of type O and will react with type A and B cells, the serum of the O donor would be diluted by the A, B, or AB recipient so that a severe effect is avoided.

The A and B antigens are short carbohydrate molecules, each one a chain composed of only six sugars. The short antigen molecule protrudes from the surface of the red blood cell and is anchored to a lipid or protein molecule embedded within the membrane. The A and B molecules are nearly identical, differing only in the last sugar residue at the end of the chain. The terminal galactose in B antigen differs from the terminal *n*-acetylgalactosamine of the A antigen with respect to just one chemical grouping. This slight difference in the terminal sugar is what accounts for the antigenic specificity of each molecule. It is the protruding sugar, not the base to which it is embedded, that reacts specifically with an antibody. This antigenic specificity is under the genetic control of the gene at the ABO locus, located on the long arm of chromosome 9. The five-sugar substrate to which the terminal sugar is added is known as the H substance, or the H antigen, since it too can react with the appropriate antibody against it. The H antigen is under the control of a gene other than the one at the ABO locus. The H substance is thus essential to the formation of both A and B antigens, since it provides the five-sugar chain to which the sixth sugar can be attached. If no sixth sugar is added, the H substance remains unaltered, and the blood is considered type O, since neither A nor B antigen is present on the cell surface.

In the simplest state, three different alleles occur at the ABO locus (Fig. 5-2). The I^A and I^B forms of the gene determine which specific sugar is placed in the terminal position on the H substance sugar chain. Each of these two alleles is dominant to the allele, i, for blood type O, the absence of A and B antigens (Table 5-2). However, a person of blood type O does produce H antigen in its unaltered form on the cell surface, and this can be detected in such an individual by the appropriate antigen–antibody test. The A, B, and H antigens also occur on cells other than the red blood cells.

The I^A and I^B alleles provide an excellent example of codominance, because the heterozygote, I^AI^B, produces both the A and B antigens. The number of alleles at the ABO locus is now known to include several subgroupings, including A_1 and A_2. The I^{A1} allele acts as a dominant to I^{A2}. If we just consider the alleles I^{A1}, I^{A2}, I^B,

TABLE 5-1. General scheme of antigens and antibodies in the ABO system

Blood type	A and B antigens in red blood cells	a and b antibodies in serum
A	A	b
B	B	a
AB	A and B	None
O	None	a and b

and i, we see that different subgroupings of types A and AB are possible (Table 5-3). Six different types can be recognized, distinguished from one another on the basis of antigen–antibody reactions. As additional subgroupings of A and B are recognized, the number of ABO types increases accordingly.

Ignoring the subgroupings for the moment, let us see how knowledge of the inheritance patterns of the blood groups can aid in a case of disputed paternity. Suppose a man is suing his wife for divorce and accuses another man of being the father of the wife's child. Blood typing reveals the following: wife, type B; baby, type AB; husband, type O; the other man, type A. Our genetic knowledge tells us that the husband's suspicion is well-founded. He cannot be the baby's father because he is blood type O and therefore cannot contribute an allele for the production of either A or B antigen. The baby must have received an allele for antigen A from the male parent. In reaching a decision regarding the correct male parent, however, caution must be exercised. Although the accused man possesses blood type A, an appropriate type for the baby's father, this by itself cannot incriminate him, for many men have blood type A. Moreover, the father could also be type AB, because such a person can contribute either the allele A or the allele B to the offspring. So we see that the genetics of blood types may eliminate as a possible parent one who lacks the allele required for the production of a specific antigen; however, it cannot prove a person *is* the parent in question just because he has the suitable blood grouping.

In our example, suppose that subgroupings of A were identified and that the baby was found to be A_1B and the man A_2. Since an A_2 person (see Table 5-3) can be only genotype $I^{A2}I^{A2}$ or $I^{A2}i$, he cannot give the offspring the I^{A1} allele. The man would thus be exon-

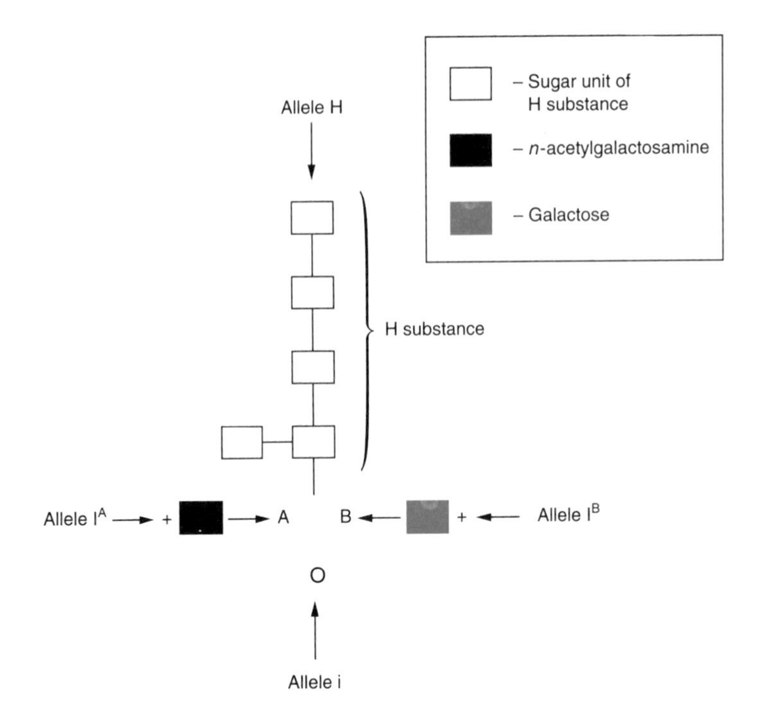

FIGURE 5-2. Interactions of alleles determining ABO blood type. Allele H controls the formation of H substance, a five-sugar unit. Alleles I^A and I^B at the ABO locus are coded for enzymes that determine what sixth sugar is to be added to the sugar chain. These two alleles for the A and B antigens specify different terminal sugars, resulting in the formation of A and B antigens from H substance. Allele i does not bring about the addition of a sixth sugar to the chain, and thus H substance remains unchanged and type O blood is recognized, since neither A nor B antigen is present.

erated on the basis of this more detailed blood typing. In this way, a person can be eliminated as a possible parent as more detailed blood groupings are examined. Although the probability of parenthood becomes greater as more and more subgroupings between child and suspected parent agree, the evidence by itself is still inconclusive.

We have just noted that a person of blood type AB cannot have a parent who is type O. However, extremely rare cases have been described of individuals who carry an allele for antigen A or B but who do not express them. A person of genotype I^AI^B could actually appear to be type O! As we noted in Figure 5-2, the ability to produce A or B antigen at all depends not only on the ABO locus but also on the locus where the gene for H substance resides. The presence of the dominant allele H at this site determines the synthesis of substance H, the substrate from which both A and B antigens are made under the guidance of alleles I^A and I^B. Since most persons possess the dominant allele H, they produce the appropriate A and B antigens when the corresponding alleles are present in the genotype.

Unusual individuals have been found who possess the genotype hh and who therefore cannot produce H substance. Such a person not only tests negative for H antigen but also appears to be blood type O regardless of the actual ABO genotype (Fig. 5-3). This illustrates an example of the Bombay phenomenon, named for the city where the first case was reported. Although very rare, the Bombay phenomenon is an excellent example in the human of epistasis and underscores the caution that must be exercised before reaching conclusions on the basis of a single pedigree.

Another locus, also distinct from the ABO locus, is the secretor locus. About 78% of the population possesses the dominant secretor allele, Se, which is responsible for the secretion in a water-soluble form of antigens A, B, and H in a number of body fluids, such as saliva and tears. Secretion of the antigens does not take place in those homozygous for the recessive, se. Some H substance is present on the cell surfaces of

TABLE 5-2. General scheme of genotypes in the ABO system

Blood type	Possible genotypes	A and B Antigens in red blood cells
A	I^AI^A, I^Ai	A
B	I^BI^B, I^Bi	B
AB	I^AI^B	A and B
O	ii	None

TABLE 5-3. Two subgroupings of type A and possible blood groups

	Blood group
$I^{A1}I^{A1}$	A_1
$I^{A1}i$	A_1
$I^{A1}I^{A2}$	A_1
$I^{A2}I^{A2}$	A_2
$I^{A2}i$	A_2
I^BI^B	B
I^Bi	B
$I^{A1}I^B$	A_1B
$I^{A2}I^B$	A_2B
ii	O

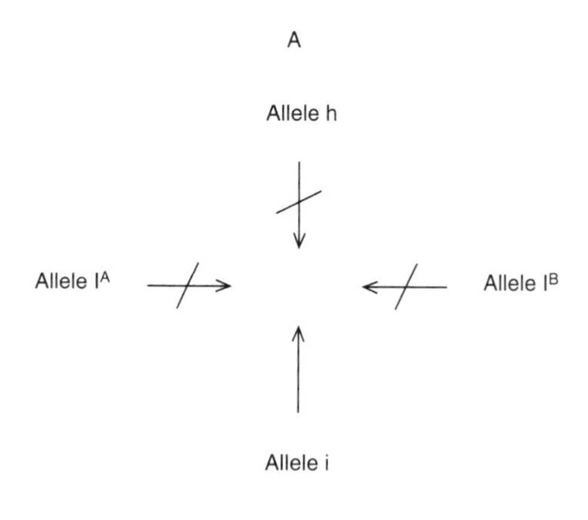

B

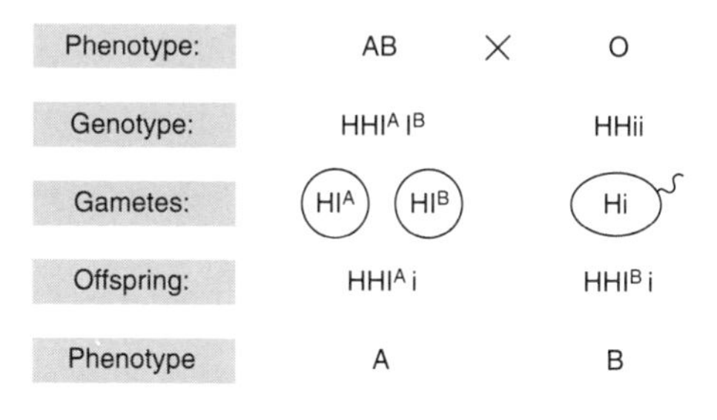

FIGURE 5-3. The Bombay phenomenon. **A:** In the absence of allele H, H substance cannot be formed from precursor material. Consequently, the alleles for A and B antigens cannot be expressed, since there is no substrate for the enzymes that they determine to act upon. A person lacking H antigen will test as type O regardless of which alleles are present at the ABO locus. (cf. Fig. 5-2). **B:** Most persons are homozygous for allele H. In a typical mating between persons of AB and O blood types (left), the offspring will be either type A or B. A very rare person (right) may be homozygous for the recessive allele h and will manifest the Bombay phenomenon. Although phenotypically type O, the person has the genetic information to make both antigens A and B if H substance is present. Since the other parent carries the H allele, the offspring will be either type A or B, and the most unusual situation will arise in which two parents of type O blood have offspring who are type A or B.

persons who are types A, B, and AB and can be secreted by them if they carry allele Se, since not all of the H substance is converted into A and B antigens. Table 5-4 summarizes the situation found in secretors and nonsecretors.

Antigens and the Rh blood grouping

The discovery of the Rh blood grouping occurred in 1940, when it was demonstrated that both human and rhesus monkey blood contain an antigen that had been

TABLE 5-4. Antigens A, B, and H in secretors and nonsecretors

	Antigens of secretors		Antigens of nonsecretors	
Blood group	Red blood cells	Body fluids	Red blood cells	Body fluids
O	H	H	H	None
A	A	A and H	A	None
B	B	B and H	B	None
AB	A and B	A, B, and H	A and B	None

undetected up to that time. The antigen was designated Rh to denote its presence in the rhesus monkey. Those persons with the Rh factor were designated Rh^+. The 15% of the white population whose blood did not react with Rh antibodies were termed Rh^- because the antigen or factor was absent from their blood. The inheritance of the Rh type seemed to depend on a single pair of alleles: the dominant R for the presence of the antigen and the recessive r for its absence. Unlike the ABO system, no naturally occurring Rh antibodies of any kind are formed in the blood. The Rh^- person (rr) is not born with antibodies against the Rh antigen, nor is the Rh^+ person born with antibodies against Rh^- blood. However, Rh^- individuals, it was found, could form antibodies against Rh antigen if they received an injection of Rh^+ blood. These antibodies can clump cells containing Rh antigen once the antibodies are formed by the Rh^- person in response to the presence of Rh^+ blood in the bloodstream. It soon became evident that the Rh factor is just as important as the ABO type in transfusions. Rh^- persons may be able to tolerate one transfusion of positive blood, because they contain no preformed antibodies. Once sensitized, however, they will produce a high level of Rh antibodies on a second transfusion, and the reaction with Rh^+ antigens in red blood cells can lead to fatal results.

Not only did the Rh type demand consideration in transfusions, it also became implicated in certain cases of blood incompatibility between mother and child. Some babies are born severely jaundiced and seriously anemic. This condition, erythroblastosis fetalis, is the consequence of fetal red blood cell destruction, which not only produces an oxygen defect but also causes clogging of the blood vessels of the liver with damaged red blood cells and the entry of bile pigments into the blood. In very severe cases, the child may suffer permanent damage. Death may even occur shortly before or after delivery. The destruction of the baby's red blood cells results from an antigen–antibody reaction in the child's bloodstream. We have noted that a person can become sensitized to Rh antigen after a transfusion. A high antibody level can be reached in an Rh^- woman who has been thus sensitized. If her husband is Rh^+, then his genotype may be either RR or Rr. If he is heterozygous, then there is a probability of 1/2 that the dominant allele will be transmitted. If the offspring does inherit the dominant allele and is thus Rh^+, any antibodies against the Rh antigen will react with the blood of the fetus *if* they pass the placental barrier from mother to child because of some defect in the placenta (see later). Thus an Rh^- mother, already sensitized, can possibly pass antibodies to her offspring in utero. Serious consequences can then follow (Fig. 5-4A).

Unfortunately, this is not the only way in which an incompatible Rh reaction can come about (Fig. 5-4B). In an Rh^- woman carrying an Rh^+ child, it is possible for some blood from the fetus to leak into the circulatory system of the mother. Once this happens, the Rh^- mother may become sensitized, just as if she had received an injection of Rh^+ blood. The antibodies formed by the Rh^- mother may in turn cross the placental barrier and cause their damaging effects. Although Rh incompatibilities between mother and child are certainly a matter for concern, knowledge of the subject should be used properly. Trouble *may* ensue if the mother is Rh^- and the baby Rh^+. This follows only from the mating between an Rh^+ man and an Rh^- woman. If the woman is positive and the man negative, there is no

FIGURE 5-4. Rh incompatibilities. **A:** An Rh^- woman and an Rh^+ man who is heterozygous have a 50:50 chance of having an Rh^+ child (above). If the child is Rh^+ and the mother is sensitized by a previous transfusion, Rh antibody formation may be evoked, and antibodies may at times pass the placenta (below). Consequently, an Rh^+ child carried by the previously sensitized mother can suffer from erythroblastosis. **B:** A woman not carrying antibodies against Rh^+ blood may become sensitized by an Rh^+ offspring, but usually only at the time of birth. The mother will usually not form a sufficient number of antibodies to affect the first Rh^+ child, even if she becomes sensitized by the pregnancy. An Rh^- woman already sensitized by a previous pregnancy with an Rh^+ child can respond rapidly to antigen from a second Rh^+ fetus (right). Antibody production can be high enough to affect the second Rh^+ child.

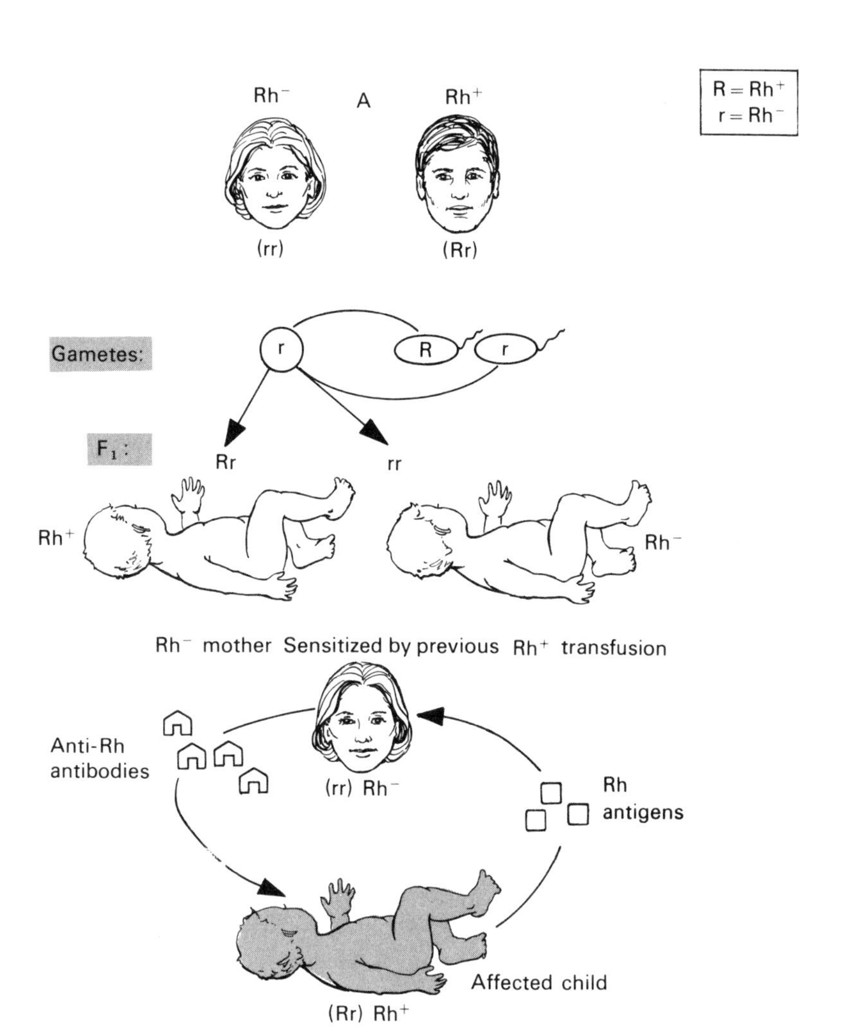
R = Rh⁺
r = Rh⁻
Rh⁻
A
Rh⁺
(rr)
(Rr)
Gametes:
r
R
r
F₁:
Rr
rr
Rh⁺
Rh⁻
Rh⁻ mother Sensitized by previous Rh⁺ transfusion
Anti-Rh
antibodies
(rr) Rh⁻
Rh
antigens
Affected child
(Rr) Rh⁺

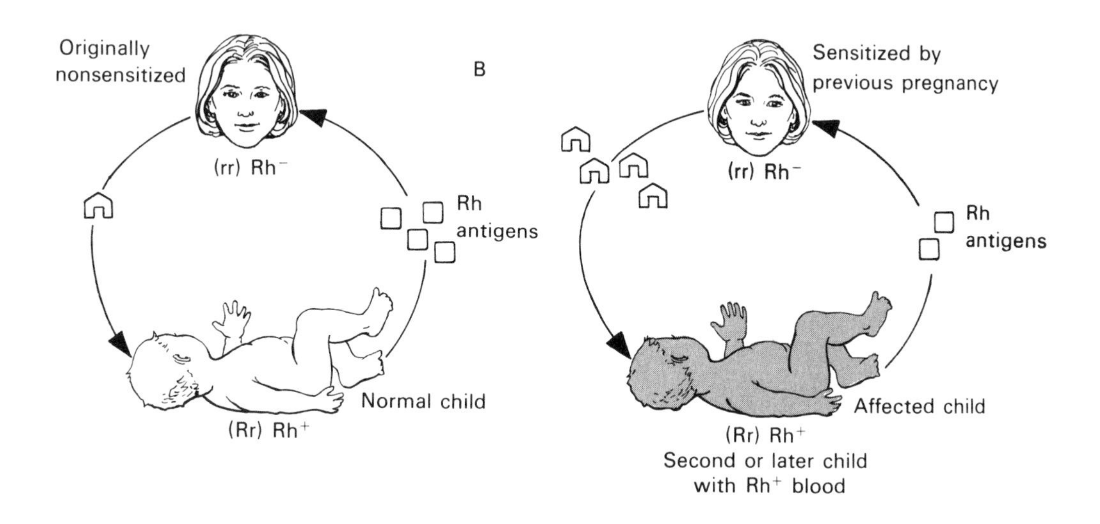
Originally
nonsensitized
B
(rr) Rh⁻
Rh
antigens
Normal child
(Rr) Rh⁺
Sensitized by
previous pregnancy
(rr) Rh⁻
Rh
antigens
Affected child
(Rr) Rh⁺
Second or later child
with Rh⁺ blood

problem, because an Rh^- baby carried by an Rh^+ woman cannot cause the mother to form antibodies that will react with Rh^- blood. But even if the woman is Rh^- and her mate is Rh^+, there still may be no problem. If the male is heterozygous (Rr), there is a 50% chance that the child will be Rh^+. If the child is Rh^+, he or she will most probably not be affected if this is the first Rh^+ born to an unsensitized Rh^- mother. This is so because most of the transfer of fetal cells to the maternal circulation occurs late in the pregnancy, at or near the time of birth. And we must keep in mind that an Rh^+ offspring does not necessarily sensitize the mother. The placenta is normally an effective barrier between mother and child. In most cases, no exchange of blood occurs between mother and child. Erythroblastosis fetalis is actually uncommon. This means that an Rh^- mother may have one or more Rh^+ children and encounter no difficulties. So we must apply the genetic knowledge intelligently. Today, with modern treatment, an Rh^- mother who has borne an Rh^+ child may never become immunized, since immunization to the Rh antigen on the fetal cells can almost always be prevented. This is accomplished by injecting preparations of antibody to the Rh^+ antigen into the Rh^- mother shortly after the birth of an Rh^+ child. These antibodies remove any Rh^+ fetal cells that may have entered the mother's bloodstream. Thus, since these fetal cells cannot act to stimulate antibody formation in the mother, the mother is not sensitized to Rh^+ cells. Proper medical counseling can avoid unnecessary alarm and prevent any unfortunate consequences of Rh incompatibility.

Not long after its discovery, the Rh grouping was shown to have a more complex genetic basis than we have just described and to involve more than the one pair of alleles, R and r. With the improvement of techniques for the detection of antigenic differences, it became clear that not all Rh^+ individuals are the same. As shown by antibodies produced against Rh^+ blood in experimental animals, there are actually several different Rh^+ types. These various Rh^+ individuals possess different combinations of antigens, which may be detected through antigen–antibody reactions. Indeed, Rh^- people were also found to vary. The many Rh variations stem from the fact that several distinct Rh antigens occur in humans, not just one or two. Even Rh^- persons possess Rh antigens that can bring about antibody production. But the most commonly occurring antigens in Rh^- people are not responsible for the Rh incompatibilities we have just discussed.

To account for these many Rh types, two ideas have been proposed, which has caused some confusion in terminology. Essentially, one concept, that of Wiener in the United States, pictures the Rh genetic region as a complex locus on the chromosome (Fig. 5-5). We may think of it as a stretch of DNA. Changes at different points within this DNA segment result in the production of different gene forms or alleles, which form part of a multiple allelic series. According to the Wiener theory, at least eight different Rh alleles exist. Each allele controls the formation of a certain *combination* of antigens (Table 5-5). An individual can have any two of the eight alleles and is Rh^+ or Rh^-, depending on the genes he or she carries (Table 5-6).

A group of English hematologists, Fisher, Race, and Sanger, interpret the situation somewhat differently. According to their hypothesis, the Rh region on the chromosome includes not just one but three loci (Fig. 5-6). Each locus is the site of a pair of alleles: Cc, Dd, and Ee. Eight different combinations of the six alleles are possible on any one chromosome: CDE, CdE, and so on. Five of the six alleles (all but the allele d) control distinct antigens, and each of these has been detected serologically by antigen–antibody reactions. Antigen d has never been demonstrated, and therefore no antibody d has been produced. Because eight combinations of alleles are possible, there are different combinations of antigens, one for each arrangement of alleles (Table 5-7). The three loci are so closely linked that crossing over in the cde region of the chromosome is very rare. Therefore any one of the allelic combinations (CDE, cDe, etc.) would usually stay linked and thus be inherited together as a unit. In essence, each combination appears as if it were one allele, even though three are actually present at each Rh region on the chromosome. Based on this idea, any person possesses six Rh alleles, any two combinations of three each, such as CDE on one chromosome and cdE on the other. The genotype of such a person could be represented as CDE/cdE (see Table 5-6).

No matter how we view the Rh mechanism, either as one locus with a series of multiple alleles or as three tightly linked loci, most of the Rh alleles exhibit codominance. As Tables 5-5, 5-6, and 5-7 indicate, each allele, when present (except for allele d), is expressed and causes the production of its specific antigen. The antigens are in turn detected by reactions with their corresponding antibodies. The maternal–fetal Rh incompatibility is mainly the result of the presence of D antigen, which is a very strong stimulator of antibodies. The Rh^- condition is defined as the absence of D anti-

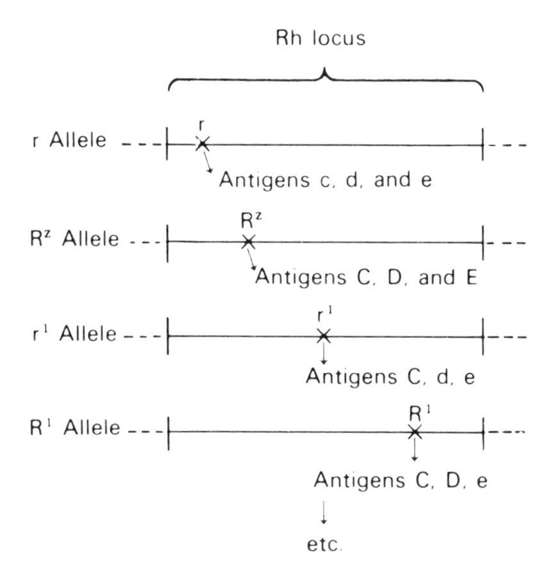

FIGURE 5-5. The Wiener concept of the Rh region. According to this interpretation, a series of multiple alleles exists at one locus. Each allele controls not one but a combination of antigens. For example, the r allele would govern the formation of antigens c, d, and e, which would react with the corresponding antibodies. A total of eight different alleles exists, each controlling a different combination. Only four are depicted here. Actually, the CDE designation shown here is employed in the other system (Fig. 5-6) rather than in the Wiener. However, the antigens and antibodies correspond in both, and the one designation is used here for the sake of clarity.

TABLE 5-5. The eight different Rh alleles according to Wiener[a]

	Different alleles	Antibodies formed against antigens governed by the allele				
		Anti C	Anti D	Anti E	Anti c	Anti e
Rh^-	r				+	+
	r′	+				+
	r″			+	+	
	r^y	+		+		
Rh^+	R^o		+		+	+
	R^1	+	+			+
	R^2		+	+	+	
	R^Z	+	+	+		

[a]Each allele is responsible for the production of a combination of two or more antigens, and these can cause the formation of corresponding specific antibodies (anti-d serum not available).

TABLE 5-6. Some common Rh^- and Rh^+ genotypes[a]

Phenotype	Genotype: Wiener	Genotype: Fisher	Antibodies
Rh^-	r/r	cde/cde	c, e
Rh^-	r″/r	cdE/cde	c, E, e
Rh^+	R^o/r	cDe/cde	c, D, e
Rh^+	R^1/R^1	CDe/CDe	C, D, e
Rh^+	R^2/r	cDE/cde	c, D, E, e
Rh^+	R^1/R^2	CDe/cDE	C, c, D, E, e
Rh^+	R^1/r	CDe/cde	C, c, D, e

[a]Any person, according to Wiener, can have any combination of the eight alleles, one allele on each chromosome. According to Fisher, any person can have a combination of any three alleles on one chromosome and a combination of any three on the other. No matter how the situation is viewed, the individual forms antibodies against specific antigens. Antigen D is the one most commonly involved in Rh incompatibilities. The Rh^+ condition is defined as *presence of D antigen.* (Anti-d serum is not available.)

gen, because that antigen is the most important one in the Rh incompatibilities.

Antigens C and E are not as antigenic as is the D factor, but at times they may cause a production of antibodies in the maternal bloodstream. Moreover, antigens C and E usually occur along with antigen D. In most cases, c factor does not cause a problem, since it is not a potent antigen, but at times it may be responsible for an Rh incompatibility, for example, if the mother is genotype CDe/CDe and the father is CDe/cde.

Whether the Wiener or the Fisher concept is correct is still unsettled, and there are good arguments in favor of each. Nevertheless, we know today that the genetic region associated with the Rh blood grouping is located at a specific region on chromosome 1. To explain a very complex story in the simplest way, we can think of the Rh grouping based primarily on a pair of alleles, D and d, because it is the D antigen that must be considered primarily when Rh types differ.

Our discussion of blood groupings up to this point has by no means exhausted the number identified. Very well known is the MN blood grouping. Each of us falls into one of three classifications: M, N, or MN, due to the possession of M or N antigen in the red blood cells. The average person is unaware of his or

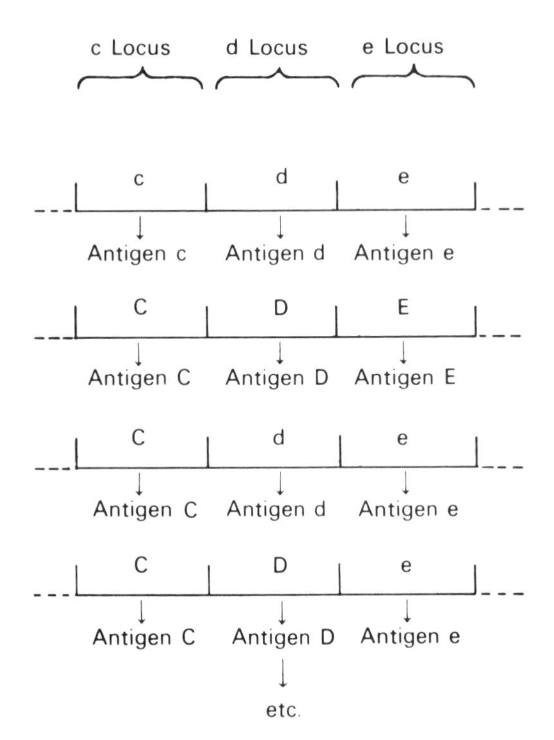

FIGURE 5-6. The Fisher concept of the Rh region. According to this theory, the Rh region includes three very closely linked loci that rarely are separated by crossing over. At each locus a pair of alleles exists, and each allele controls the formation of a specific antigen. On any one chromosome, any combination of three of the six alleles can occur. A total of eight different combinations of three is possible. Contrast this figure with Figure 5-5.

TABLE 5-7. The eight combinations of the Rh alleles according to Fisher[a]

	Rh alleles combinations	Antibodies formed against antigens governed by genes at the loci				
		Anti C	Anti D	Anti E	Anti c	Anti e
Rh^-	c d e				+	+
	C d e	+				+
	c d E			+	+	
	C d E	+		+		
Rh^+	c D e		+		+	+
	C D e	+	+			+
	c D E		+	+	+	
	C D E	+	+	+		

[a]Each combination of three alleles is strongly linked and behaves as if it were one allele. Each combination of three, according to this hypothesis, corresponds to one of the alleles, as pictured by the Wiener concept (see Table 5-5).

her MN type, because the MN antigens play no role in blood transfusions. Antibodies to M or N antigens do not occur naturally in the blood serum. Such antibodies, however, can be elicited by injecting human blood into an experimental animal such as a rabbit. Antibodies to M or N form in the animal's serum and, when mixed with human blood, can reveal the presence of the antigens in the human red blood cells, as they cause the latter to clump. At its simplest, the formation of M or N antigens depends on a pair of alleles, L^M and L^N. The two forms exhibit codominance, so that three possible genotypes and phenotypes are recognizable: L^ML^M (type M), L^ML^N (type MN), and L^NL^N (type N). These react specifically with antibodies M, M and N, and N.

Several other blood groupings are much less familiar to most of us but may be encountered in specific situations. To mention just a few, there are the Kell, the Kidd, the Duffy, Lutheran, and Lewis groupings. The names usually refer to the family in which the grouping was first detected. For example, the Kell factor (K) denotes an antigen found in most persons. Women who are double recessive (kk) may give birth to erythroblastotic Kell positive children. As in the case of the Rh^- mothers, the Kell negative women may produce antibodies (anti-Kell) that react with the Kell antigen in the fetal circulation. These many additional blood groups can also be of aid in legal matters. They also demonstrate how unique each of us is. The number of possible combinations of the many blood group loci, some of which are multiple allelic, defies comprehension.

Antigens, transplants, and the immune system

Intimately related to the topic of blood groups is the subject of tissue grafting and organ transplants. We might expect other tissues as well as blood to contain distinct antigens that react with specific antibodies. And indeed, the rejection of an organ such as the kidney or the failure of a skin graft to take are examples of immunological responses. All cells possess antigens, which vary from one individual to the next and may cause the rejection of transplants. Such factors are called *histocompatibility antigens,* and the genes that control them are *histocompatibility genes* (H genes). The antigens, as well as those associated with the blood types, are present in the cell membrane, where they may play some role in its structure and biological properties.

The most thorough studies on the genetics of grafting have been performed with mice. However, the same principles are believed to apply to other species, including the human. Mice within one family line may be mated for generations to produce strains that are so inbred that they are homozygous for an appreciable number of alleles. Valuable genetic information on tissue compatibility factors and their role in tissue transplants may be gained from studying graft tolerance within and between the inbred strains. Such work gives strong evidence for the following concept: A recipient will accept tissue from a donor if the recipient possesses the same kinds of histocompatibility alleles as the donor. Conversely, if the donor, and hence any tissue from the donor, possesses H alleles that are *not* present in the receiver, the graft will be rejected.

Table 5-8 represents a general summary depicting results from two inbred strains, A and B, in which two pairs of histocompatibility alleles are being followed. As was true for the ABO and Rh alleles, the H alleles are codominant in their expression. It is obvious from Table 5-8 that within an inbred strain tissue can be successfully grafted and accepted because there is no difference between donor and recipient in the kinds of H alleles they possess. When the two inbred strains are crossed, their F_1 offspring contain H alleles from each parent. These hybrids have all the alleles that are typical of each strain. Thus they can accept grafts from either parent type. However, the F_1 cannot act as donor to either inbred parental strain, because the F_1 has some alleles not found in strain A or B. If the F_1s are interbred to produce F_2 progeny, the latter may have various new combinations (3^2) of the two pairs of alleles, such as $A_1A_1B_1B_2$. But no matter what the new combination in the F_2 animals, the F_1s will be able to accept grafts from them. This is so, because the F_1s will contain *all* the types of alleles present in any F_2. The reverse, however, is not necessarily true, because independent assortment will scramble the H alleles into a variety of new combinations (nine in this case). The only F_2 animals capable of accepting grafts from F_1 animals must be dihybrids, just as the F_1s, and they are in the minority.

This simple example illustrates the principle behind graft tolerance, but it is an oversimplification. In mice, about 30 separate loci are associated with histocompatibility genes. The number of different possible combinations becomes staggering. Moreover, some of these loci exert a greater effect than others. In mice, there is a genetic region known as the *major histocompatibility complex* that is the main one in the sense that the antigens it controls can bring about a very strong immunological response, much more so than other loci. This region, also called the *H-2 complex,* has been studied in detail and has been shown to involve several very closely linked loci adjacent to each other.

In addition to circulating antibodies, which form in the body of a mammal in response to foreign antigens, there is another immune response—one that plays a major part in actual rejection of foreign tissue received in a transplant. This is the cellular response, which entails those particular lymphocytes, the so-called T lymphocytes (or T cells), that have become differentiated in the thymus gland. The T cells can recognize foreign antigens on the surfaces of cells in tissue grafted to a host. A T cell has antibody molecules embedded in the surface of its cell membrane, and these apparently enable T cells to respond to the presence of antigens on the surfaces of cells in foreign tissue.

In addition to the mouse, other mammals have also been shown to possess a major histocompatibility complex (MHC), a genetic region responsible for the strongest immune response. The MHC in mammals is the most variable system or genetic complex known, capable of yielding a vast number of different allelic combinations and thus generating an immense amount of individual variation. In the human, this MHC has been designated the *HLA complex.* All of the exceedingly varied cell surface antigens determined by the HLA complex are glycoproteins (proteins containing small amounts of carbohydrate). These antigens fall into two categories. Those in class I are found on all nucleated cells, including white blood cells, while the antigens in class II occur mainly on certain cells of the immune system (some T cells, B lymphocytes, macrophages). Each antigen is composed of two closely associated polypeptide chains, chains made up of linked amino acids. These two units, the alpha (α)-chain and the beta (β)-chain, differ in their amino acid composition, and each antigen molecule is known as a *heterodimer.* The α-chain of class I antigens varies greatly in its amino acid composition, but the β-chain is constant in structure. Among class II antigens, the great variation that is present is the result of differences in the β-chain.

Those genes coded for the class II antigens of the HLA complex are found in a position proximal to

TABLE 5-8. Grafting compatibilities between animals of various genotypes

Donor	Donor's genotype	Recipient	Recipient's genotype	Ability to accept grafts
Strain A	$A_1A_1B_1B_1$	Strain A	$A_1A_1B_1B_1$	+
Strain B	$A_2A_2B_2B_2$	Strain B	$A_2A_2B_2B_2$	+
F_1 hybrid	$A_1A_2B_1B_2$	Strain A	$A_1A_1B_1B_1$	–
F_1 hybrid	$A_1A_2B_1B_2$	Strain B	$A_2A_2B_2B_2$	–
Strain A	$A_1A_1B_1B_1$	F_1 hybrid	$A_1A_2B_1B_2$	+
Strain B	$A_2A_2B_2B_2$	F_1 hybrid	$A_1A_2B_1B_2$	+
F_2 hybrid	9 Types	F_1 hybrid	$A_1A_2B_1B_2$	+
F_1 hybrid	$A_1A_2B_1B_2$	F_2 hybrid	9 Types	– (In most cases)

(toward) the centromere on the short arm of chromosome 6 (Fig. 5-7A). At least three closely associated class II genes have been recognized: DR, DQ, and DP. Some of these genes that code for the class II antigens can be found in more than one copy on a chromosome. Three class I genes of the HLA complex are located in a position distal to (away from) the centromere. These genes, B, C, and A, are coded for the α-chains of class I antigens and by convention are written in the order in which they occur in their region of chromosome 6. The gene for the constant β-chain is found on chromosome 15 and will not be discussed further. Between the class I and class II regions of the HLA complex are at least four other genes that are important in the immune response to foreign antigens. These code for proteins that cause lysis when they interact with cells that have been marked by antibodies. Two other loci that are coded for an enzyme also occur in this part of the complex, which is often called the *HIII region.* It is only the class I and II genes, however, that are pertinent to our immediate discussion.

To appreciate the antigen variability found among humans that is made possible by the class I and class II genes, it is necessary to consider the following. Many alleles of the three class I genes are known: about 40, 10, and 20 for the B, C, and A genes, respectively. For the DR, DP, and DQ class II genes the allelic varieties are about 15, 7, and 4, respectively. Any given allele is detected by the appropriate antibody that reacts with the specific antigen for which it is coded. It is expected that even more alleles at the different loci in the HLA complex will be recognized as more antibodies against the as yet undetected antigens become available.

Because the antigen-determining loci in the HLA complex are so close together on chromosome 6, their separation as a result of crossing over is rare. Therefore, the combination of alleles that occur together on a chromosome tends to persist and to be transmitted intact from one generation to the next. For example, the class I allelic combination A7B15C5 behaves as a single unit and is termed a *haplotype* (Fig. 5-7B). Each person carries two haplotypes for the class I and the class II genes, one on each chromosome 6. For the sake of simplicity, we can consider each haplotype (a given combination of alleles) as a single genetic entity (say, H1, H2, H3, etc.) and think of any individual as being heterozygous at the HLA region for a pair of units such as H1/H2. Considering just the class I genes, we see that the number of different possible haplotypes becomes immense, about 8,000 (the product of the number of different alleles of genes B, C, and A: $40 \times 10 \times 20$). The large number of haplotypes makes it extremely likely that most people are heterozygous at the HLA region. Since any person possesses two haplotypes, the number of possible genotypes becomes staggering. The chance that any two unrelated persons taken at random will have the same genotype in respect to these haplotypes and hence the antigens they determine is well under 1 in 10,000. We can thus appreciate why the HLA antigens are now much more important in settling legal matters than are the antigens for the blood groupings. Since any given haplotype is not likely to be widespread in a population sample, this greatly narrows the chance that a given person is the parent in a particular case.

Considering the HLA complex in the simplest way, we can still understand why most grafts and transplants between individuals are not tolerated (Fig. 5-8). A mating between persons of genotypes H1/H2 and H3/H4 results in the combinations H1/H3, H1/H4, H2/H3, and H2/H4. None of the children will be able to accept a graft from either parent and vice versa. There is a 25% chance, however, that two sibs will possess the same haplotypes, such as two who are H1/H4 in our example. This means that two brothers or sisters may be compatible for haplotypes of the HLA region and may possibly exchange tissue.

Both class I and class II antigens play a major role in kidney transplants. Except for identical twins, the greatest chance of a successful transplant is one made between sibs due to the 25% chance that two of them will be identical for haplotypes at the HLA

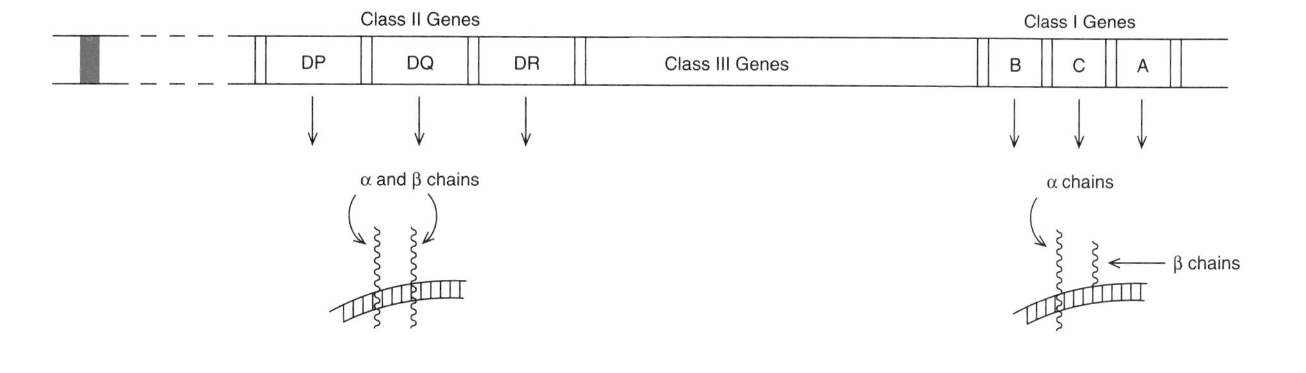

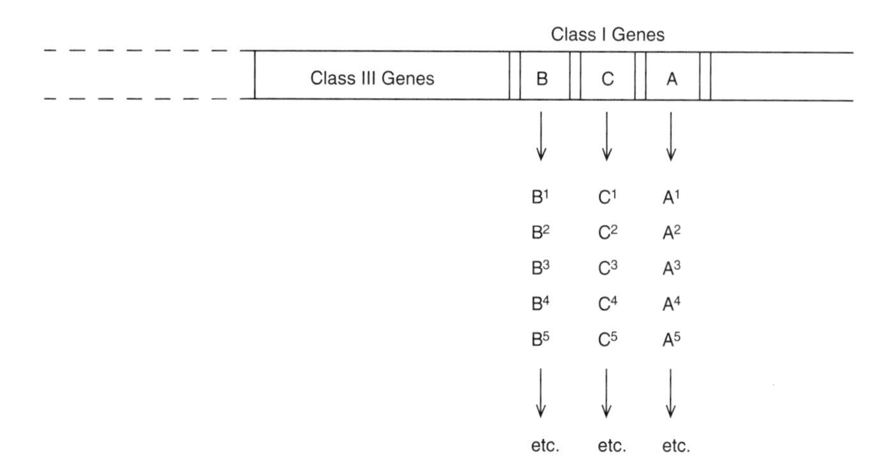

FIGURE 5-7. Genes of the HLA complex. **A:** Class II genes code for both the α- and β-chains of antigens found on the surfaces of membranes of certain types of white blood cells. The genes occur on chromosome 6 in a cluster located toward the centromere (red). Several such genes are found in the region, and three distinct ones have been recognized, one or more of which may occur in more than one copy. The exact gene order has not been established. Class I genes occur in another cluster farther away from the centromere. This group includes genes B, C, and A, in that order. These genes code for the variable α-chain of class II antigens, which are found on the surfaces of all kinds of nucleated cells. The invariable B chain is coded by a gene on chromosome 15. Class III genes occur between the other two and code for proteins that are not cell surface antigens. **B:** Class I genes exist in several allelic forms. (Simplified designations are used here.) The many allelic possibilities at each site make possible many different kinds of α-chains. This generates the variability of the class I antigens. Since the genes in the complex are so close together, any one combination of genes, a haplotype, tends to stay together and be inherited as a genetic unit. A similar situation pertains to class II genes, which also exist in more than one allelic form and which yield combinations that behave as haplotypes.

region. To find such compatibility in any unrelated person is most unlikely. Close relatives may share one haplotype in common and are second in choice to sibs when searching for a donor. Bone marrow transplants have been quite successful and also depend on compatibility at the HLA complex. In any transplant, the ABO blood type must also match. You will recall that these antigens appear on cell types other than the red blood cells. Moreover, still other antigens controlled by genes at additional chromosome locations can also play roles in the acceptance or rejection of a transplant. This becomes apparent in unsuccessful grafts that survive for just a few weeks even though the tissue exchange was between sibs who matched for both the HLA complex and the ABO blood grouping.

An interesting point comes to light when we consider the frequencies of the different class I and class II alleles in the population. Although crossing over is infrequent among the tightly linked genes in the HLA complex, it is still to be expected that the frequencies of the various haplotypes will be the product of the separate allele frequencies in the population. For example, if the frequencies of certain C and A alleles in the population are 5% and 20%, respectively, then they would be expected to occur together with a frequency of approximately 1% in the population (0.05 × 0.20). However, this does not prove to be the case. Certain combinations of alleles are more frequent in the population than would be expected due to chance (say that it is 5% in our example), whereas certain other alleles do not appear to associate, and thus certain haplotypes are much less frequent than one would expect on the basis of chance. The alleles of the complex are thus in *linkage disequilibrium*; the frequencies of the combinations in which they occur in the population to form haplotypes are not random. It is unknown why certain combinations are more frequent than others. Perhaps such haplotypes give an advantage to individuals carrying them, and they are thus selected, or else the population may not have had time as yet to come to equilibrium for the haplotypes and what we observe now will continue to change until the expected values are reached.

Very interesting is the observation that certain antigens of the HLA system are associated with certain diseases. For example, a person with one of the class I antigens (B27) is 87 times more likely than a person without it to develop ankylosing spondylitis, an inflammatory condition of the sites of attachment of ligaments to bones. Other specific HLA alleles are associated with such diseases as rheumatoid arthritis, juvenile diabetes mellitus, multiple sclerosis, myasthenia gravis, and systemic lupus erythematosus. While the reason for such associations is unknown, there is a hint that a common basis may exist since most of the diseases involve the human immune mechanism in one way or another.

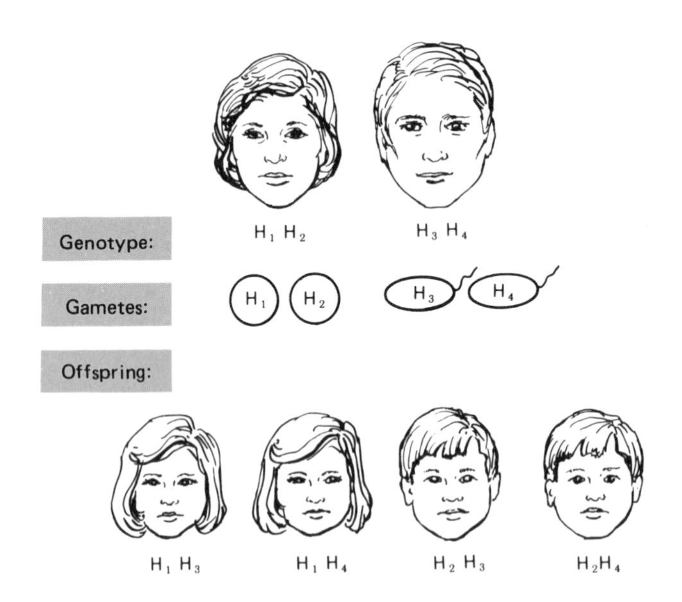

FIGURE 5-8. Tissue incompatibility at the HLA complex in humans. Any one person is likely to be heterozygous at the HLA region, carrying two different haplotypes. The parents in this case, as would probably be true of any two unrelated persons, do not possess both haplotypes in common and would be unable to exchange grafts. Among their children, diverse types occur. These cannot give grafts to or accept grafts from either parent, because there will be a difference in one haplotype. For the same reason, children of different genotypes cannot exchange grafts. However, two children can occur in the family who carry the the same haplotypes, and these may be able to exchange tissue. However, still other genetic regions exert an influence; differences at these will reduce the chances of free exchange, even if compatibility exists at the HLA complex.

A tantalizing question relates to the actual function of the HLA locus and the many antigens that it governs. The function certainly has nothing to do with grafting and transplants, since these are recent devices conceived by the human and are not natural biological processes. While most suggestions for the role of the complex are not well supported by evidence, a good case can be made for its role in the regulation and refining of the immune response. There is also reason to believe that the evolution of the great complexity of the immune system has resulted largely from the value that it imparts in the recognition of those cells in the body that suddenly undergo malignant transformation. Any new cancer cell has alterations on its surface and is recognized as foreign. The origin of a cancer cell probably entails a vast assortment of alterations, each quite distinct and requiring a specific cellular response if it is to be eliminated.

Good illustrations of the importance of the immune response are seen in those genetic diseases in which there is a decreased efficiency in the ability to respond to foreign antigens. Several types of inherited immune-deficient disorders exist, and these vary in response to treatment and in genetic basis. In those disorders known as the *agammaglobulinemias,* no circulating antibodies are found in the blood. In one rare condition of this type, a sex-linked recessive interferes with the maturation of B lymphocytes, those white blood cells that manufacture immunoglobulins. All individuals afflicted are males, and they usually show symptoms before they are 1 year old. Since T cells are present, these individuals can ward off some viral infections, but they are completely vulnerable to bacteria and require antibiotics as well as antibody injections to prevent constant attacks.

Especially severe are those rare immune disorders classified as SCID (severe combined immunodeficiency disease). In these conditions, both B and T cells are lacking. Afflicted babies usually die in infancy, since they cannot develop resistance to either infectious viruses or bacteria. Bone marrow transplants have proved effective in some cases. SCID is transmitted as a sex-linked recessive about half the time and as an autosomal recessive in the rest of the cases. Autosomal forms of the disorder are now known to result from certain specific enzyme deficiencies to which lymphocytes appear to be especially vulnerable. In one autosomal form of SCID, the enzyme defect is related to a gene on chromosome 14. In a second form, the enzyme defect is associated with a gene on chromosome 20. Now that these associations are known, there is hope that recombinant DNA technology (see Chap. 19) will be able to alleviate the dire effects of some SCID cases. We discuss the intricacies of the immune system and the virus (HIV) responsible for the grave immune disorder AIDS (acquired immune deficiency syndrome) in later chapters.

REFERENCES

Aufray, C., and J.L. Strominger. Molecular genetics of the major histocompatibility complex. *Adv. Hum. Genet.* 15:197, 1986.

Dausset, J. The major histocompatibility complex in man: Past, present, and future concepts. *Science* 213:11469, 1981.

Hames, B.D., and D.M. Glover (eds.). *Molecular Immunology.* IRL Press, Washington DC, 1989.

Hood, L.E., M. Steinmetz, and R. Goodenow. Genes of the major histocompatibility complex. *Cell* 28:685, 1982.

Mange, A., and E.J. Mange. *Genetics: Human Aspects,* 2nd ed. Sinauer, Sunderland, MA, 1990.

REVIEW QUESTIONS

1. S_1, S_2, S_3, and so on are sterility alleles that form a multiple allelic series in tobacco. A plant cannot be homozygous for any one of them. Pollen will abort if it carries a sterility allele possessed by a plant used as an egg parent. What will be the constitutions of the F_1 plants regarding the sterility alleles in each of the following crosses?

A. S_1S_2 egg parent × S_4S_5 pollen parent.
B. S_3S_4 egg parent × S_4S_5 pollen parent.
C. S_1S_2 egg parent × S_1S_2 pollen parent.
D. S_4S_5 egg parent × S_3S_4 pollen parent.

2. In the rabbit, a multiple allelic series affects the color of the fur. The wild-type allele (c^+) for dark fur is dominant to all the other alleles in the series. The allele c^h for Himalayan is dominant to albino (c). However, the allele for chinchilla coat color (c^{ch}) gives light gray color when present with either the allele for Himalayan or albino. Give the genotypes and the phenotypes of the offspring from the following crosses.

A. A Himalayan animal whose male parent is albino × a chinchilla.
B. An animal with wild-type fur that has a chinchilla parent × an albino.
C. A light gray animal that has an albino parent × a light gray that has a Himalayan parent.
D. A rabbit with wild-type fur that has an albino parent × a Himalayan that also has an albino parent.

3. Give the probable genotypes of the following pairs of parents with respect to the A, B, O blood type alleles.

A. An A type parent and an O who have one child who is type O.
B. An AB type parent and an A who have one child who is type B and one who is type A.
C. A type A parent and a type B who have eight children who are type AB.

4. Two subgroups of type A blood can be recognized. The allele A_1 is dominant to A_2 and to O. A_2 is also dominant to O. Give the results of the following crosses:

A. An A_1 type person who has an O parent × an A_2 who has an O parent.
B. An A_2 type person who has an O parent × an A_1B.
C. $A_1B \times A_2B$.
D. $A_1B \times O$.

5. A woman is blood type B. Man 1 is type O. Man 2 is A_2. The woman's baby is type A_1B. Which man could be the father? Explain.

6. The secretor trait (Se), which enables A and B antigens to be secreted into the body fluids, is dominant to nonsecretor (se). Give the blood type and indicate whether the saliva will contain A or B antigens in the offspring from each of the following parents:

A. I^AI^BSeSe × iisese.
B. I^AI^BSese × iiSese.
C. I^AiSese × I^BiSese.

7. The allele H is required for the completion of A and B antigens. In its absence, the allele h cannot direct the formation of A and B antigens. The h allele is thus epistatic to the ABO locus and can cause a person to be blood type O regardless of the presence of the A and B alleles. Give the blood types expected among the offspring in each of the following matings, as well as the ratio in which they would occur:

A. I^AI^BHh × I^AI^BHh.
B. iiHh × I^AiHh.
C. I^AiHh × I^Bihh.

8. For each of the following four sets of parents, give the ABO blood types that can occur among the offspring and the proportion in which they can be expected. Indicate which, if any, of the antigens involved will be secreted into the body fluids.

A. iiSeSeHh × iiSeSeHh.
B. I^AiSeSeHh × I^BiSeSehh.
C. iiSeseHh × iiSeseHh.
D. I^AI^AseseHh × I^BI^BSeseHh.

9. A woman is blood type A MN. Her baby is type O N. Which of the following two men is the father? Explain.

A. Man 1: type A M.
B. Man 2: type O N.

10. Considering Rh^+ blood to be associated with the dominant, D, and Rh^- with its recessive allele, d, give the possible offspring from the following matings:

A. An O Rh^- woman (who has an Rh^+ parent) and an O Rh^+ man (who has an Rh^- parent).
B. An A Rh^+ woman and a B Rh^+ man, both of whom have one Rh^- parent.
C. A woman of type O who suffered erythroblastosis fetalis and a man who is AB negative.

11. A woman is blood type O MN Rh^+. Her baby is O MN Rh^-. Man 1 is A N Rh^+. Man 2 is O MN Rh^-. Which of the two could be the father?

12. Give the ABO blood grouping of each of the following persons and tell which of the pertinent antigens will be secreted into the body fluids.

A. iiSeseHhL^ML^N.
B. I^AI^AseseHhL^NL^N.
C. I^AI^BSeSehhL^ML^N.
D. I^AI^BseseHHL^ML^N.
E. I^AI^BSeSehhDDL^NL^N.

13. Of the following matings, which run a risk of producing a baby with erythroblastosis fetalis?

A. Female $\frac{C\,D\,E}{C\,D\,E}$ × Male $\frac{c\,d\,e}{c\,d\,e}$.

B. Female $\frac{c\,d\,E}{c\,d\,e}$ × Male $\frac{C\,D\,e}{c\,D\,e}$.

C. Female $\frac{C\,d\,e}{c\,d\,e}$ × Male $\frac{c\,D\,E}{C\,d\,e}$.

D. Female $\frac{C\,d\,E}{C\,d\,e}$ × Male $\frac{C\,d\,E}{C\,d\,e}$.

14. Three genes in the HLA complex, designated B, C, and A, code for histocompatibility antigens. Each gene occurs in many allelic forms. Assume that a woman and her husband have the following genotypes, respectively, for the antigen-determining genes:

$$\frac{B_2C_1A_2}{B_1C_2A_1} \text{ and } \frac{B_4C_3A_4}{B_3C_4A_3}.$$

A. Give the genotypes possible among the offspring, ignoring crossing over.
B. Can the children donate to or accept tissue from either parent?
C. How would you answer the preceding question if one of the children were a boy with agammaglobulinemia who could not produce T lymphocytes?

15. Following are the phenotypes of members of a family with respect to histocompatibility antigens. Answer the questions regarding this family:

Mother: $B_8B_9A_1A_2$
Father: $B_6B_8A_3$

Child 1: $B_8B_9A_2A_3$
Child 2: $B_6B_8A_1A_3$
Child 3: $B_6B_9A_2A_3$

A. Give the genotypes of all these persons.

B. Suppose another child is born of the following type: $B_8A_2A_3$. Give this child's genotype and explain its occurrence.

16. In the human, there are two different genetic blocks in the pathway leading to melanin pigment formation. One of them, a, interrupts a step that prevents the formation of the immediate precursor of melanin. The second block, b, prevents the conversion of the precursor to melanin pigment. Each of the two types of albinism that result from these blocks is a recessive trait.

A. What kinds of offspring are to be expected if an albino person of type a has children with an albino person of type b?

B. What is the chance that two persons of the children's genotype will produce an albino at any birth if two such persons marry?

17. A family of four is tested for the presence of class I antigens (B, C, A) of the HLA complex. The results are given below. Assuming no recombination, deduce the haplotypes of the father, mother, and two children.

	Antigen types present		
	B	C	A
Father	5, 40	10	8, 20
Mother	17, 27	2,10	11, 20
Child 1	27, 40	2,10	11, 20
Child 2	17, 40	10	20

Sex and Inheritance

6

Sex chromosomes

Before 1910, pioneer cytologists had worked out the chromosome arrangement in many insects. They recognized that chromosomes occur in pairs but that the two sexes often differ with respect to one of them. The common fruit fly, *Drosophila melanogaster,* was found to possess four pairs of chromosomes (Fig. 6-1A). Cells from females show one pair of rod-shaped chromosomes (pair I, which is acrocentric), two pairs of V-shaped chromosomes (pairs II and III with median centromeres), and a very small pair designated the *dot chromosomes* (pair IV with centromeres near one end). In the female, the two members of each pair appear identical, but this is not so in the male. The difference is found in pair I. Males have only one rod-shaped chromosome instead of two but carry a hook-shaped or J-shaped chromosome, which is absent in the normal female. Similar distinctions are now well known in other species as well.

The two members of the chromosome complement that differ between the sexes are called the *sex chromosomes.* The chromosome found singly in the male and paired in the female is the X, and the one confined to the male is the Y. As noted in earlier chapters, all the chromosomes of the complement other than the sex chromosomes are known as the *autosomes. Drosophila* possesses three pairs of autosomes and one pair of sex chromosomes, two Xs in the female and an X and a Y in the male. Such a chromosome picture is known as an *X–Y* condition and implies that the male has dissimilar sex chromosomes and will produce gametes that differ in their sex chromosome content. For example, male fruit flies form sperms with three autosomes, but half of the gametes carry an X and the other half a Y. Since his gametes fall into two distinct classes, we say that the male is the *heterogametic sex.* The female is termed *homogametic;* all of her eggs are alike in the type of sex chromosome they contain (Fig. 6-1A).

However, the X–Y condition is not universal. Cytological examination shows that cells of one sex can contain one chromosome more than the other. In such species, the sex chromosome is completely unpaired. Since a Y is absent, the X has no pairing partner. We followed the behavior of an X in this so-called *X–O condition* during our discussion of spermatogenesis (see Chap. 3). The X–O arrangement is found in many insects and various other groups as well. In species such as the grasshopper, the male is again the heterogametic sex, because the sperms are of two kinds in relation to their sex chromosome content: Half contain one more chromosome than the other. In some organisms with the X–O mechanism, it is the female sex that is heterogametic (XO) and the male that is homogametic (XX). Indeed, the females of certain species may contain an X and a Y, whereas the males are XX. Such an arrangement is found in moths and butterflies and is typical of birds. This is designated *W–Z,* to imply that the female (ZW) forms two classes of sex cells, one with the large Z and one with the small W. The female is thus the heterogametic sex. The male is homogametic (ZZ), possessing two large Z chromosomes. This is the reverse of the more common and familiar X–Y arrangement found in *Drosophila* as well as in the human. In our species, we find 22 pairs of autosomes and one pair of sex chromosomes (two Xs in the female; one X and one Y in the male).

Geneticists recognized that the presence of sex chromosomes could account for the determination of equal numbers of male and female offspring. It could thus explain what appears to be a monohybrid 1:1 testcross ratio. In an animal with the X–Y mechanism, the eggs are all X bearing and are fertilized by either X-bearing or Y-bearing sperms. Since the male produces these two different classes of gametes in equal amounts, it is a matter of chance whether any one egg is fertilized by an X-containing sperm or by one with a Y. The outcome is male and female offspring in equal proportion (Fig. 6-1B).

Genes on the X chromosome: sex linkage

Work with the fruit fly by T.H. Morgan and his associates at Columbia University began in 1909 and contributed a wealth of information that is basic to an understanding of sex determination. The investigations of this outstanding group also uncovered other fundamental principles that established genetics firmly as a scientific discipline. One of the early discoveries was that the X chromosome carries genes that influence characteristics other than sex. A mutation arose affecting the eye color of *Drosophila,* changing it from the red found in nature to unpigmented or white. The white-eyed fly that arose was a male, and it was crossed to a normal red-eyed female. All of the offspring were red eyed, suggesting that the genetic determinant for white eyes was recessive to the one for red. A cross between males and females of the first generation produced red-eyed and white-eyed flies in the familiar Mendelian ratio of 3:1. The results indicated that a pair of alleles was involved, W for the dominant red condition and w for the recessive white. However, there was something unusual about this 3:1 ratio. All of the flies with white eyes were males (Fig. 6-2A)! Half of the males had red eyes, but *all* of the females were red eyed. Morgan realized that such results could be explained by assuming that the locus determining red versus white eye color is on the X chromosome (Fig. 6-2B). The females used in these crosses would have two X chromosomes, each bearing the dominant W. They would thus be homozygous for red eye color and can be represented simply as WW. The white-eyed male, however, has but one X chromosome and therefore has the eye color locus represented only once. The Y chromosome would not carry a locus for any eye color trait. The white-eyed male would thus have an X chromosome with the recessive allele w

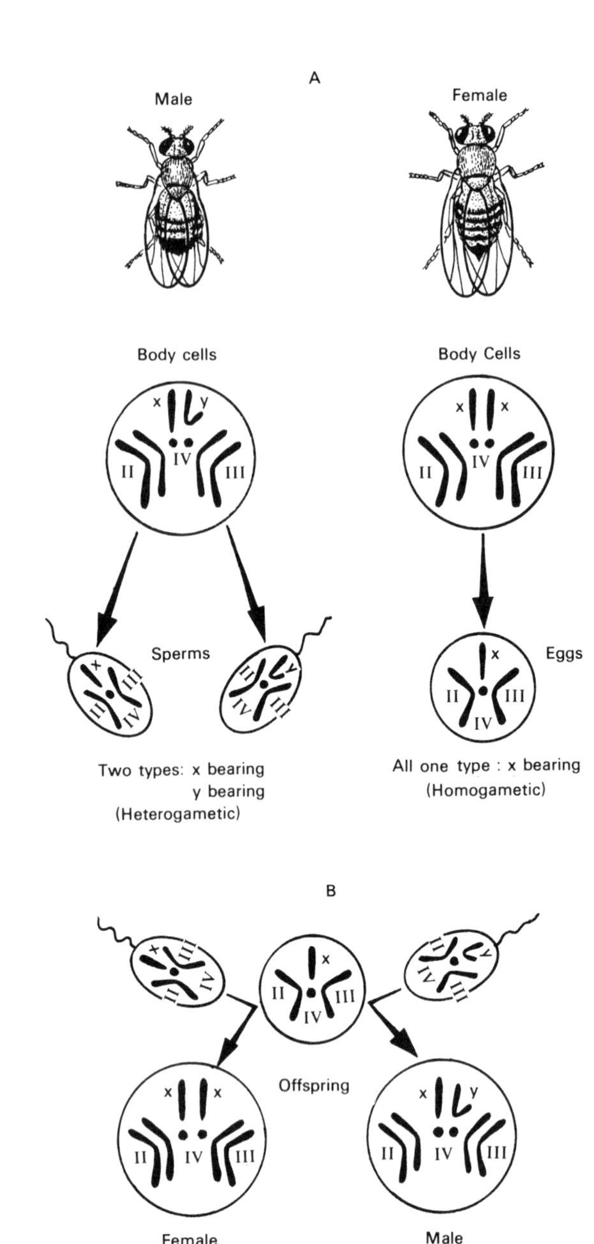

FIGURE 6-1. Sex determination in the fruit fly. **A:** The male and the female both possess two sets of autosomes (chromosomes II, III, and IV) in their body cells. They differ regarding the sex chromosomes: The male carries 1X + 1Y, whereas the female carries two Xs, also called chromosome I. All of the eggs produced by a female are alike in that they are X bearing. The male forms two classes of sperms and is designated the *heterogametic* sex. **B:** There is an equal chance for any egg to be fertilized by an X-bearing or a Y-bearing sperm. Consequently, male and female offspring occur in equal numbers.

and a Y chromosome with no corresponding eye color locus. The genotype of such a fly can be represented as wY. The Y chromosome must be shown when the genotype is written, to indicate that the male is heterogametic and carries only one factor for white eye color. We cannot say that the male is either *homozygous* or *heterozygous* for eye color, because, by definition, both of these terms imply that a particular gene is represented twice for a particular character. Since only one dose of a gene can be present in the case of an X-linked gene in the heterogametic sex, we say that the individual is *hemizygous.* Figure 6-2B shows why all of the F_2 white-eyed flies must be males after a cross between a red-eyed female and a hemizygous white-eyed male. Moreover, all of the red-eyed females in the first generation must be heterozygous, Ww. Mating them with white-eyed males produces red-eyed and white-eyed flies of both sexes (Fig. 6-3). Matings of this type yield white-eyed females that can then be crossed to red-eyed males. This is the reciprocal of the original cross (red-eyed female × white-eyed male). Figure 6-4A shows that the results of this reciprocal cross are very different from the original mating. In the first generation, all of the females are red-eyed like the male parents and all of the males white-eyed like their mothers. This is to be expected if the gene for eye color is on the X chromosome (Fig. 6-4B). A *crisscross* pattern of inheritance is seen, meaning that a phenotype present in the female parent appears among all the sons, whereas the contrasting phenotype of the male parent shows in all the daughters. Note from Figure 6-4B that the cross of the F_1 red-eyed females with their white-eyed brothers is really the same as the cross diagrammed in Figure 6-3, which first yielded white-eyed females.

Further work with *Drosophila* revealed genes for still other characteristics that do not influence sex (body color, wing and eye shape) but that nevertheless are associated with the X chromosome. Such genes—whose loci are found only on the X and not on the Y—are said to be *sex linked.* Actually, the first case of sex linkage had been found in 1906 by Doncaster, who was following inheritance of body color in the moth *Abraxis.* The genetic results suggested that the female is the heterogametic sex and the male is homogametic. This was later verified cytologically. Knowing now that the female is indeed heterogametic in this species, we can diagram sex-linked crosses in the moth, as shown in Figure 6-5. Notice that the reciprocal crosses are different, and a crisscross pattern occurs when dark-bodied females are mated with light-bodied males. Such results strongly indicate sex linkage. The fact that the sex chromosome arrangement is opposite to the one found in *Drosophila* is no contradiction to the concept of sex linkage. However, this feature of the moth made it difficult at the time to compare it with well-known organisms having the X–Y arrangement and in which the complications of sex linkage were being worked out. As noted, we now know that this type of pattern seen in *Abraxis,* in which the female is heterogametic, is typical of birds and certain other groups.

Sex linkage in humans

After the discovery of sex linkage in *Drosophila,* attention was called to similar inheritance patterns for certain human traits. Among the most familiar are red–green color blindness and hemophilia, both of which behave as sex-linked recessives. Approximately 8% of American males have a defect in color recognition, whereas the trait in females is well under 1%. We can easily understand this once we realize that the loci for the most frequent types of color blindness are on the X chromosome. The normal ability to perceive color depends on several genes, X linked and autosomal. These genes are responsible for the production of specific proteins that are sensitive to certain wavelengths of light. The three different types of visual pigments, red, green, and blue, are found separately in the cones, light-sensitive cells of the retina. About three-fourths of all color-blind persons carry a sex-linked recessive that causes a defect in the green-sensitive pigment. Consequently, difficulty is encountered in recognizing certain shades of red and green. This is known as the deutan type of color blindness. Less common is the protan type, which affects about 2% of the males in the population. This sex-linked recessive causes a defect in the pigment found in those cones sensitive to red, again resulting in poor recognition of certain red and green shades. Since the effects of the two genes are similar, it was at first thought that only one sex-linked recessive was responsible, but more refined observations led to the recognition of two genes on the X that affect red and green color perception. The protan and deutan genes are now known to reside in a band at the tip of the long arm of the X (Xq28).

Each of the two genes, the deutan and protan, has been found to occur in three allelic forms. In addition to the alleles for normal and defective pigments, a third less common one exists that results in complete absence of pigment, the green-sensitive one in the case of the deutan gene and the red-sensitive one in the case of the protan gene. Complete absence of either pigment causes a more pronounced color vision defect than does the presence of pigment in defective form. We see here examples of X-linked multiple

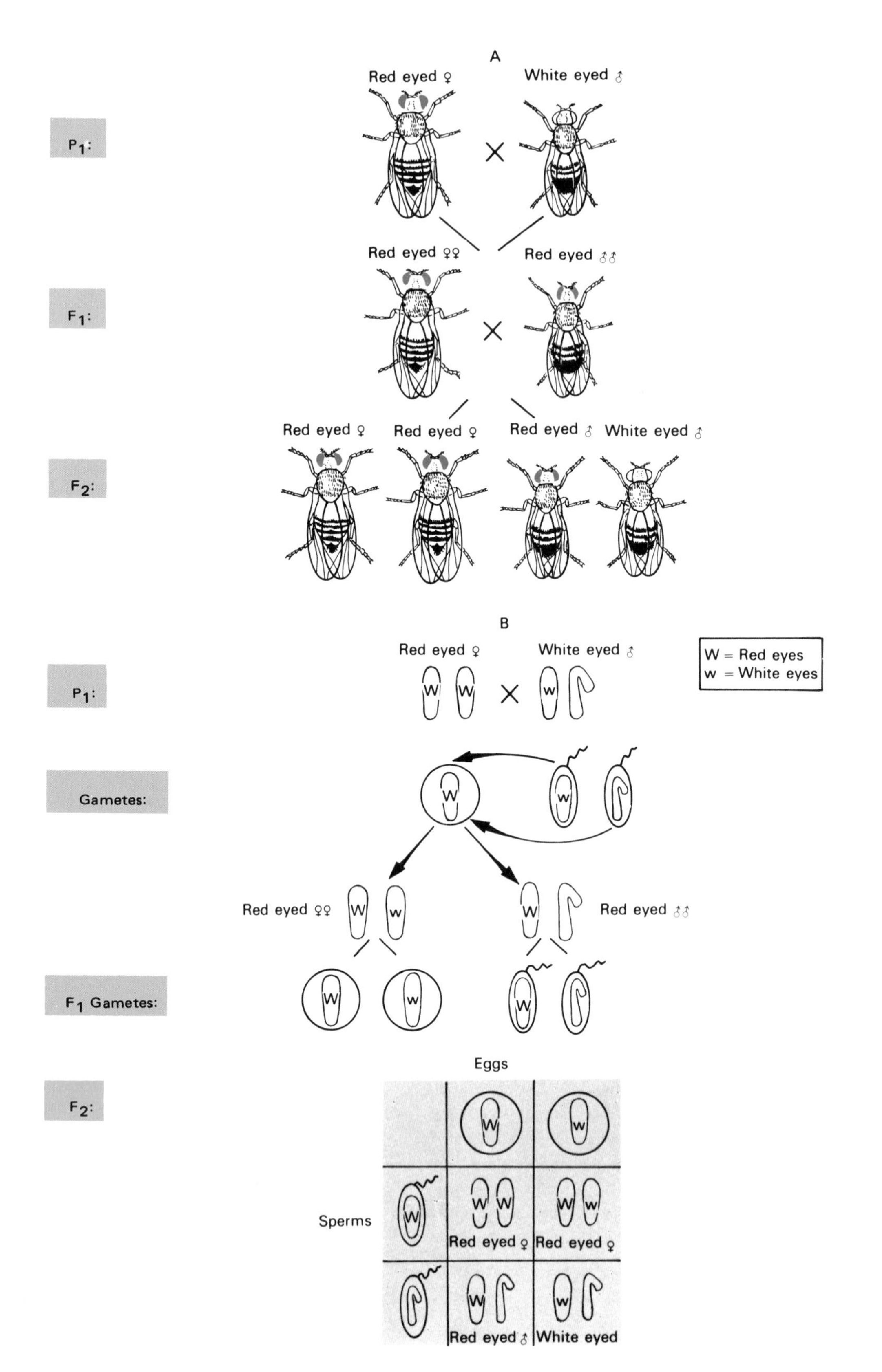
A
P1:
Red eyed ♀
White eyed ♂
F1:
Red eyed ♀♀
Red eyed ♂♂
F2:
Red eyed ♀
Red eyed ♀
Red eyed ♂
White eyed ♂
B
P1:
Red eyed ♀
White eyed ♂
W = Red eyes
w = White eyes
Gametes:
Red eyed ♀♀
Red eyed ♂♂
F1 Gametes:
F2:
Eggs
Sperms
Red eyed ♀
Red eyed ♀
Red eyed ♂
White eyed

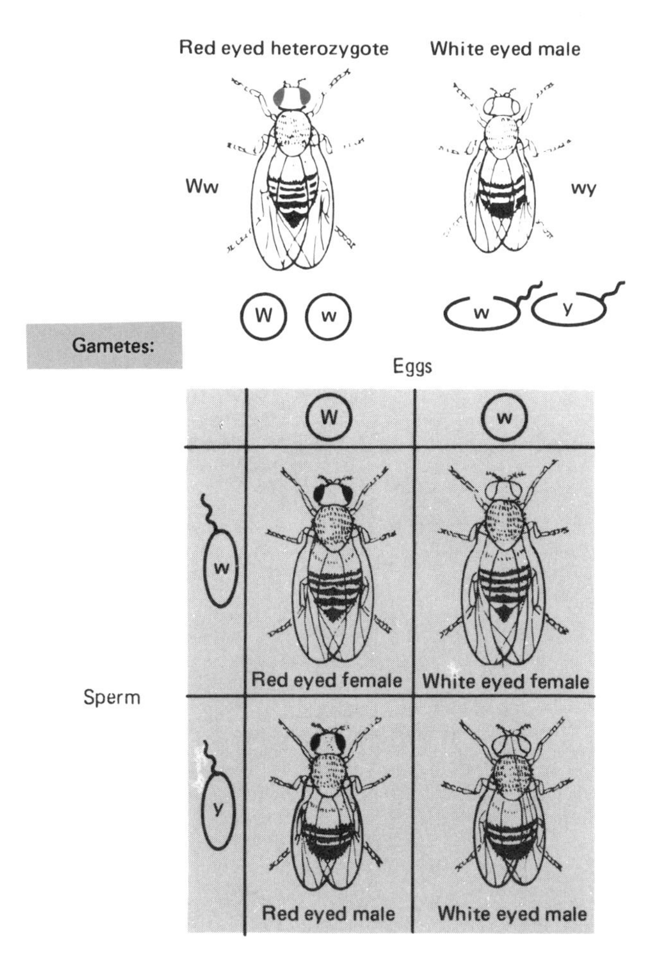

FIGURE 6-3. Derivation of white-eyed females. Red-eyed heterozygotes are obtained from a cross such as that depicted in Figure 6-2. When these F_1 females are crossed with white-eyed males, half of the female as well as the male offspring will be white eyed.

allelic series in the human. In the case of each gene, the allele for normal pigment is dominant to both mutant forms, while the allele for defective pigment is dominant to the one for its complete absence.

FIGURE 6-2. Inheritance of white eye color in *Drosophila.* **A:** A cross of a female from a red-eyed stock with a white-eyed male yields an F_1 in which all are red eyed. A mating of the F_1 gives an F_2 in which red-eyed and white-eyed flies occur in a ratio of 3:1. However, the white-eyed flies are always males. **B:** The results of the preceding cross are explained by the fact that the locus for white eye color is on the X chromosome. No corresponding locus is found on the Y. All of the F_2 females are red eyed because they receive an X from the F_1 males, and these all carry the dominant allele. The males of the F_2, like any males with an X–Y mechanism, receive the X from their mothers, not their fathers. Half of these carry the recessive, w. The Y each male receives from the male parent carries no locus for eye color. Therefore the males express any factors on the X chromosome.

Advances in molecular biology (see Chap. 19) have permitted the isolation of several human genes, among them three responsible for the color-sensitive pigments green, red, and blue. Analysis of the DNA of each gene has shown that the X-linked deutan and protan genes (for green-sensitive and red-sensitive pigments, respectively) are almost identical and code for light-absorbing proteins that are about 99% the same! The two genes occur very close to each other on the X chromosome and are believed to represent the results of duplication of an ancestral gene during mammalian evolution in the not so distant past. This concept is supported by the further discovery that some males have the deutan gene represented two or even three times on the X chromosome! (We will discuss the importance of duplication in the evolution of genes in more depth in Chapter 10.)

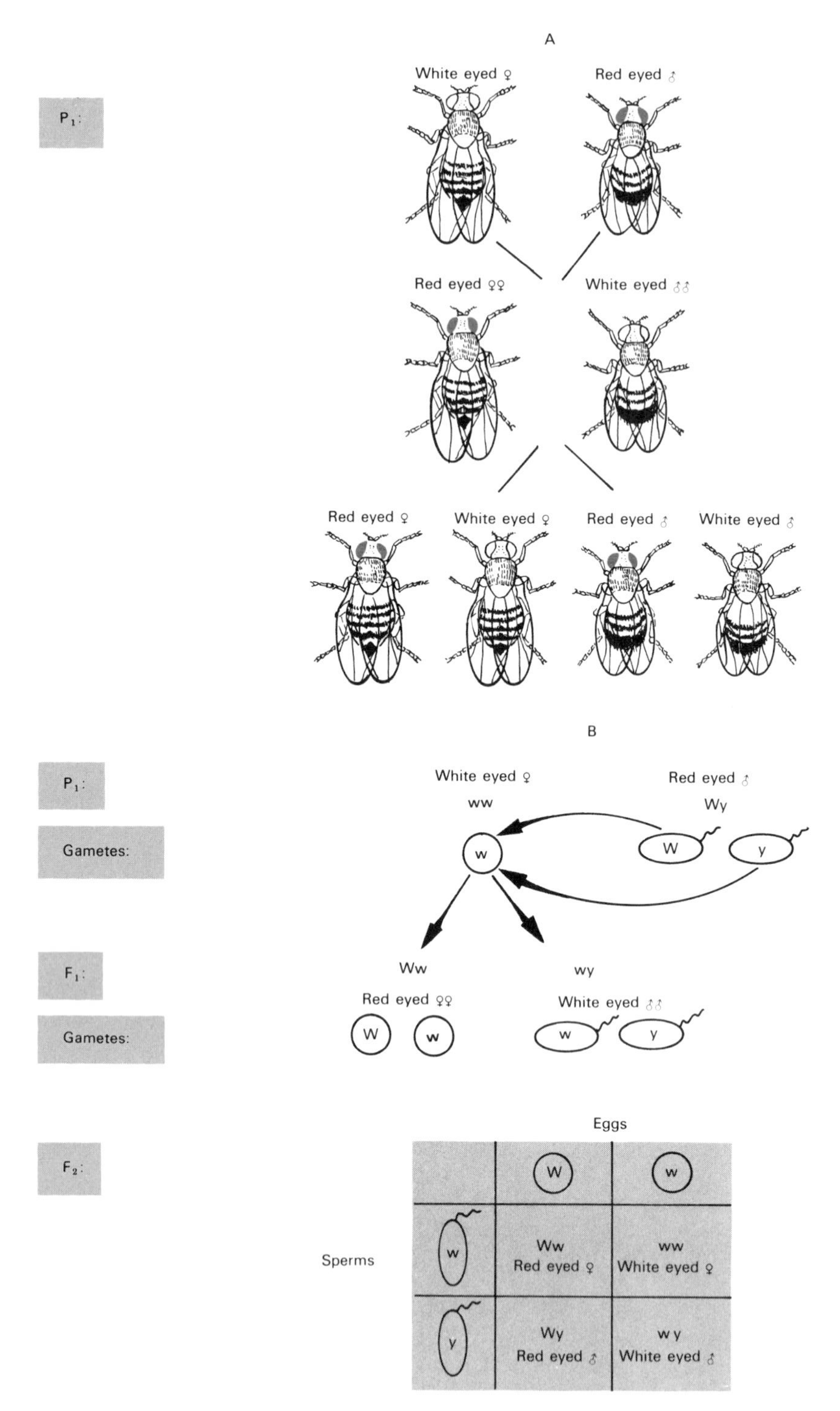

FIGURE 6-4. White-eyed female *Drosophila* × red-eyed male. **A:** The results are very different from the reciprocal cross (Fig. 6-2). In the F_1 a crisscross pattern occurs. The trait seen in the female parent occurs in the male offspring; the trait expressed by the male parent is found in the female offspring. **B:** The diagram of the cross shows that in the case of a sex-linked allele, the males will express the trait that the gene governs, regardless of dominance. This is so because males receive the Y chromosome from their fathers, and thus have no homologous region to mask any factors on the X contributed by their mothers.

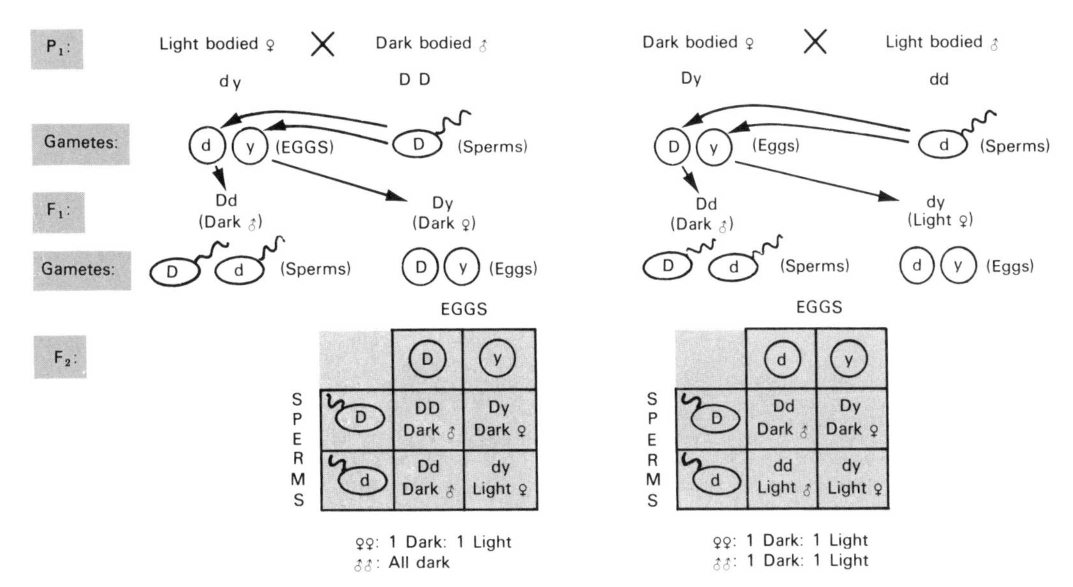

FIGURE 6-5. Sex-linked inheritance in the moth *Abraxis.* A pair of alleles, D (dark body) and d (light body), is associated with the X chromosome. The reciprocal crosses diagrammed here produce very different results. The crisscross pattern (right) indicates sex linkage. In moths, the female is the heterogametic sex. However, the principles of sex linkage remain exactly the same as in the more familiar situation in which the male is heterogametic. Compare this figure with the crosses in Figures 6-2, 6-3, and 6-4.

The third color gene that was isolated and analyzed codes for blue-sensitive visual pigment, which is not as similar to the red- and green-sensitive ones as they are to each other. This gene has been found to reside on an autosome, specifically the long arm of chromosome 7.

Since the loci for recessives responsible for red–green color blindness are sex linked, we can understand why this condition is much more common in the male. Any female of normal chromosome constitution carries two X chromosomes. For her to express red–green color blindness (let us say of the more common deutan type), her father must show the trait. The mother must show deutan color blindness herself or at least be a heterozygote. This is so because the daughter must receive a deutan color-blind recessive from each parent to express the trait (Fig. 6-6A,B). The same reasoning applies to the protan type of red–green color blindness.

It should be apparent also that a male with the normal XY chromosome constitution will always express any gene that is sex linked because he has only one X chromosome. Being hemizygous, he cannot carry the normal allele and the defective form at the same time. Therefore, any sex-linked allele that expresses itself in a heterogametic male must have come from the female parent. In humans, a son receives the Y from his father; this is a requirement for maleness. Moreover, all the sons of a woman who expresses a sex-linked recessive, such as color blindness, would be affected, even if her mate is normal (Fig. 6-6C). The cross of a color-blind woman and a normal man results in a crisscross pattern and is comparable to a mating in *Drosophila* between a white-eyed female and a red-eyed male.

The sex-linked recessive responsible for the lethal effect of hemophilia, a condition in which the blood fails to clot normally, is of special interest for both its consequences for the sufferer and its historical import. The affliction, which has been recognized for hundreds of years, received attention when it appeared among European royalty during the reign of Victoria of England. There is little doubt that Queen Victoria was heterozygous for the recessive, Hh. The probability is great that a mutation from the normal allele (H) to a recessive form (h) was carried in either the sperm from her father or the egg from her mother. At any rate, of the queen's nine children, several definitely received the recessive allele from her. One of her sons was consequently a hemophiliac, and some of her daughters were carriers who were responsible for passage of the hemophilia factor through several royal

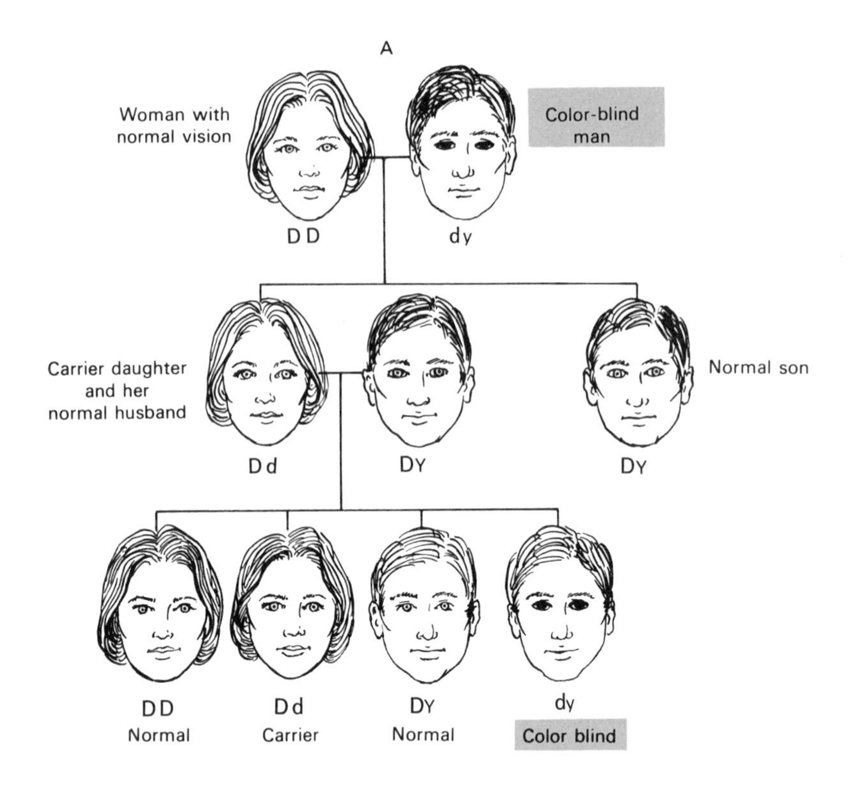

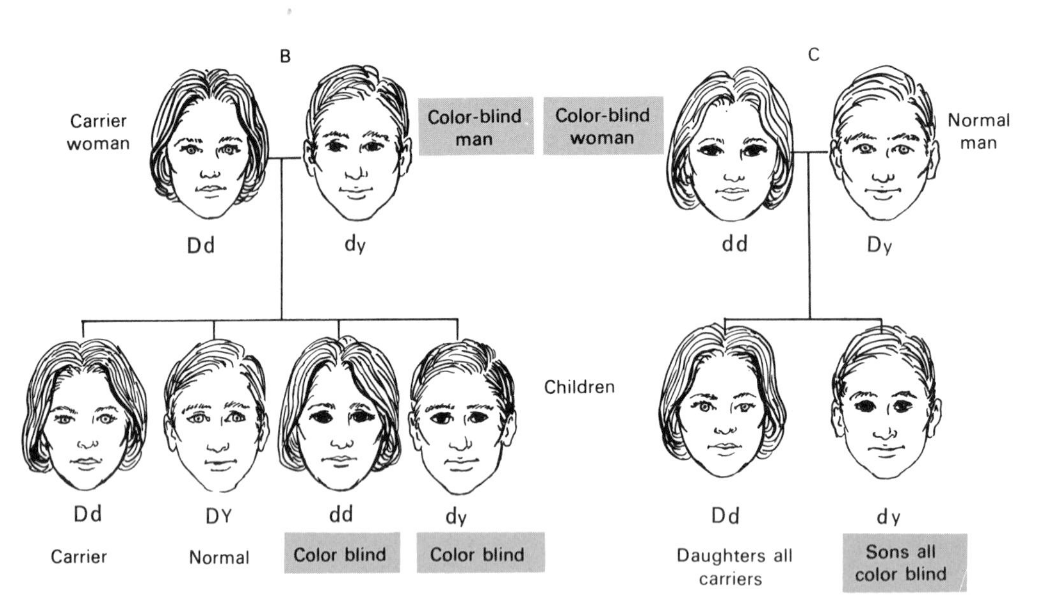

FIGURE 6-6. Color-blind inheritance. **A:** The locus for green color perception is on the X chromosome. D represents the dominant allele for normal color distinction and d its recessive allele for deutan color blindness. The trait passes from a color-blind man to half of his grandsons by way of his carrier daughters. Since a male contributes his Y chromosome to his sons and not the X, the sons of a color-blind man cannot receive the recessive from their father. **B:** For a color-blind female to arise, her mother must at least be a carrier and her father must be color-blind. The color-blind female receives from each parent an X that carries the recessive allele. **C:** All of the sons of a color-blind woman must be color-blind because a male receives his X from his mother. The daughters of a color-blind woman will be carriers with normal vision if the male parent has normal vision.

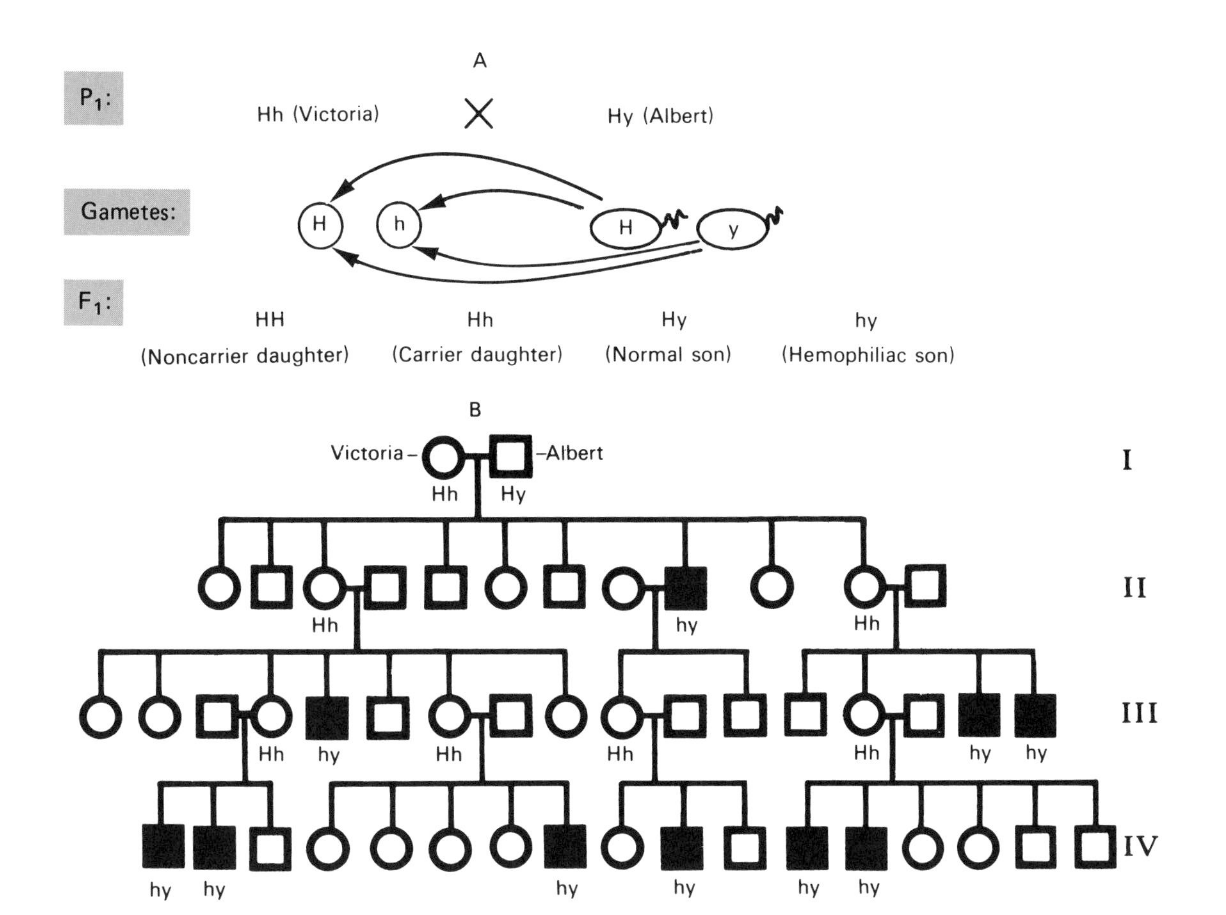

FIGURE 6-7. Inheritance of hemophilia through Queen Victoria. **A:** Hemophilia among European royalty can be traced back to Queen Victoria. The trait was not found among her ancestors, but she was undoubtedly heterozygous for the defective allele. Since hemophilia is sex linked, we would expect her to pass the allele to half of her sons and half of her daughters. **B:** When the actual pedigree of Victoria and her descendants is studied, the transmission of hemophilia is easily understood on the basis of the expectations summarized in A. In the pedigree, circles represent females, and squares are males.

houses, even down to the present. This is easy to understand from what we know about the inheritance of sex-linked recessives (Fig. 6-7A,B). We would expect from such a mating that half of the daughters would be carriers and half of the sons would express the factor for hemophilia.

Today we know a great deal about the molecular basis of hemophilia. We may think of blood clotting as the end result of a series of reactions requiring various enzymes and cofactors. The end result of this series of interactions is the formation of insoluble fibrin from soluble fibrinogen. This last step, which produces the final fibrin clot, requires the action of the enzyme thrombin. The many protein substances (enzymes and cofactors) leading to the activation of thrombin are governed by genes. A defect in any one of these genes leads to defective clotting, since a required enzyme or cofactor would be either defective or absent.

Hereditary disorders have been described in which one or more of these substances is deficient in the blood plasma. It is therefore not surprising that more than one kind of hemophilia is known. About 85% of the cases, designated hemophilia A, result from a deficiency of a protein cofactor known as factor VIII. The gene for factor VIII resides on the X chromosome. Factor VIII is needed for full activity of one of the enzymes, factor IX, in the series of events leading to the activation of thrombin (Fig. 6-8). Absence of functional factor VIII interrupts the steps leading to the activation of thrombin, and consequently fibrin cannot form.

The remaining 15% of persons with hemophilia suffer from hemophilia B, which results from a defect in another X-linked gene that governs the formation of factor IX. In its active form, factor IX behaves as an enzyme to activate the next enzyme (factor X) in the reaction series. Factor VIII (the substance lacking in hemophilia A) is a cofactor required to work along with active factor IX in activating factor X.

The techniques of molecular biology (see Chap. 19) have made it possible to isolate the genes coded for factors VIII and IX and also to characterize factor VIII. This substance has been shown to be a large protein, approximately 2,332 amino acids long. It is also known that on the molecular level more than one kind of defect can occur in the factor VIII gene, and hence in the factor VIII protein. Different defects are associated with different degrees of severity in cases of hemophilia A. The information and techniques on the molecular level provide hope that factor VIII may soon become available on a wide scale for sufferers of hemophilia A and enable them to avoid the dangers entailed in transfusions of factor VIII from pooled blood.

For years it was thought that hemophilia could not occur in females, because the homozygous condition hh was believed to be lethal to the embryo. We are now aware of a few authentic cases of female bleeders. As might well be expected, they are very rare. As in the case of color blindness, the female parent of an afflicted girl must at least be heterozygous for the condition, and the father must actually suffer from the recessive disorder (see Fig. 6-6). Until the 1960s, most males with hemophilia did not survive to reproductive age. The few who did manage to have offspring would almost always have mated with women who are homozygous normal HH, because the recessive allele involved is not common in the population. When a rare mating between a hemophiliac and a carrier female does take place, an afflicted daughter can be produced.

Sex and genic balance

Investigations with *Drosophila* in Morgan's laboratory produced other findings that are fundamental to an understanding of genes and chromosomes. Crosses involving various unusual flies gave unexpected results that led to the discovery of nondisjunction, the failure of chromosomes to separate at anaphase of either meiosis or mitosis. Figure 6-9A shows the outcome when a certain white-eyed female was crossed to a red-eyed male. This odd female produced some unexpected white-eyed daughters and red-eyed sons. Figure 6-9B,C gives the explanation for these peculiar results. Failure of the two X chromosomes to separate occasionally at anaphase I of meiosis in the unusual white-eyed female leads to the formation of abnormal eggs, those with two X chromosomes and those with no sex chromosomes at all. Offspring with atypical chromosome compositions result from fertilization of these abnormal eggs. Since at least one X is required for life, no flies develop from zygotes lacking this chromosome entirely. However, females with three X chromosomes, females with a Y, and sterile males lacking a Y can arise. The discovery of the latter two types led to the realization that the presence of a Y chromosome in a zygote does not mean that a male fly will necessarily develop. In *Drosophila,* the Y is not required for life or even for maleness, because a fly can be male without a Y or can be female with one! Although the Y is needed for fertility and thus for the survival of the species, it is not essential for life or for sex determination of any one individual. In the

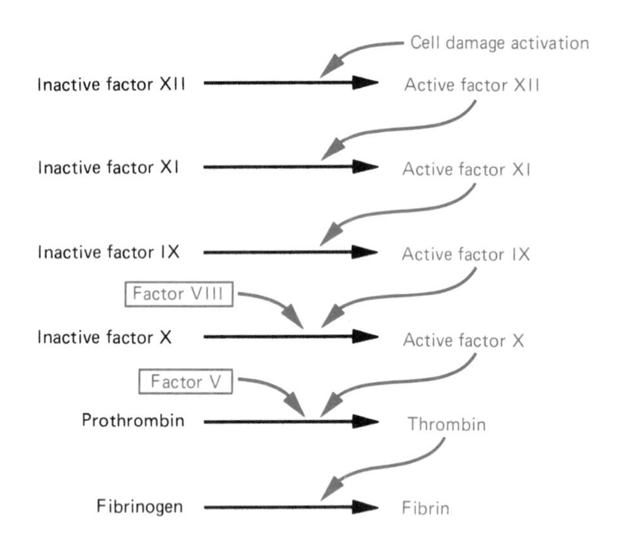

FIGURE 6-8. Some steps in blood clotting. Normal coagulation involves a series of reactions leading to the formation of insoluble fibrin. Many of the steps in the chain entail the conversion of a gene product to its active enzymatic form by the enzymatic action of the active product of a previous step. Cofactors (e.g., factors VIII and V) are required to work along with some of the active enzymes. Factor VIII, absent in hemophilia A, is a cofactor required along with factor IX to activate factor X. If factor VIII is absent as a result of a sex-linked recessive, the sequence is interrupted, and the end result is defective clotting. In hemophilia B, factor IX is deficient.

fruit fly, sex has been shown to depend not on one or two chromosomes but rather on a balance between autosomes and X chromosomes. When the balance is two X chromosomes to two sets of autosomes, a female develops, whereas one X to two sets of autosomes determines a male.

This concept of genic balance in sex determination was confirmed by later work with other exceptional flies, those with entire extra sets of chromosomes. Individuals with three or four chromosome sets instead of the normal two are *triploids* and *tetraploids,* respectively. The presence of more than two homologous chromosomes in the cells of such flies leads to irregular chromosome separation at meiosis. Consequently, when these flies are bred, their offspring can have atypical chromosome numbers, having different combinations of autosomes and X chromosomes. Normal diploid males and females possess two sets of autosomes. As shown in Figure 6-1B, one set of autosomes in *Drosophila* includes two different V-shaped chromosomes (II and III) and a dot chromosome (IV). Two of these sets of autosomes plus one X produce a male, whereas two sets plus two Xs result in a female. Unusual flies that develop from mating triploids and tetraploids show departures from this normal relationship of autosomes to X chromosomes. Studies of their cytology have indicated that genes for maleness and femaleness are distributed throughout all the autosomes and the X. As the summary of results shown in Figure 6-10 shows, the X contains genes with a strong female-determining tendency. One set of autosomes leans in the male direction as a result of the many genes for maleness scattered throughout. However, the strength of one X toward the female side is greater than the strength of one set of autosomes in the male direction. They are not 1:1; rather, it seemed that the femaleness of an X was about 1.5 to the maleness of *a set* of autosomes (1.5:1). This should not be construed to mean that there is any kind of fluid or substance that has male- or female-determining properties. The ratio 1.5:1 only symbolizes the comparative strengths of the two major factors (one X vs. one set of autosomes) in sex determination.

The observations indicate that the sex of a fly is determined by a balance between many genes distributed throughout the autosomes and those found on the X. The absolute number of chromosomes is not the sole factor, because *any* fly will be a female when the ratio of femaleness to maleness is 3:2 (which in turn results from a ratio of one X to one set of autosomes). It does not matter if four sets of autosomes and four Xs are present or if there is only one set of autosomes and one X. The ratio is still the same (1:1 between X and autosomes). Any departure from a certain balance, however, produces a fly that is abnormal in some way. The presence of just one extra X tips the balance heavily in the female direction (4.5:2), giving a fly with exaggerated secondary sex characteristics—a superfemale or metafemale. Such a fly is weak and usually dies.

A comparable story holds for the determination of maleness. A ratio of one X chromosome to two sets of autosomes (a female:male ratio of 1.5:2) will always result in a male regardless of the presence or absence of a Y. A supermale or metamale, also generally weak, results from the extra strength of maleness provided by one extra set of autosomes. When the female:male ratio is equal (3:3), a sterile intersex, completely intermediate in secondary sex characteristics, is produced.

The involvement of genes on the autosomes is very dramatically shown by a rare recessive autosomal allele in *Drosophila* known as transformer (tra). This is so named because a double dose of this recessive causes flies that have two X chromosomes and two sets of autosomes (and therefore should be normal females) to be phenotypically male (Fig. 6-11)! Such recessives cannot be distinguished from normal flies by observation; however, the transformed flies are sterile. Because the recessive (tra) is rare in fly populations, most females are Tra Tra XX and most males Tra Tra XY. This is another good example of the interaction of many genes in the determination of a phenotypic characteristic, in this case sex.

Some sex chromosome anomalies in humans

Genic balance plays an important role in human sex determination, but there are distinct differences from the situation in the fruit fly. The Y chromosome of *Drosophila* appears to be devoid of genes for male development. Thus an XO individual is male, and an XXY is female. However, such is not the case in humans and other mammals where an intact Y chromosome is essential for male attributes. Examples of XO and XXY persons are well known. Those of the former constitution have a chromosome number of 45 and are sterile females who display the various abnormalities of Turner syndrome, a condition that affects the development of both sexual and other bodily char-

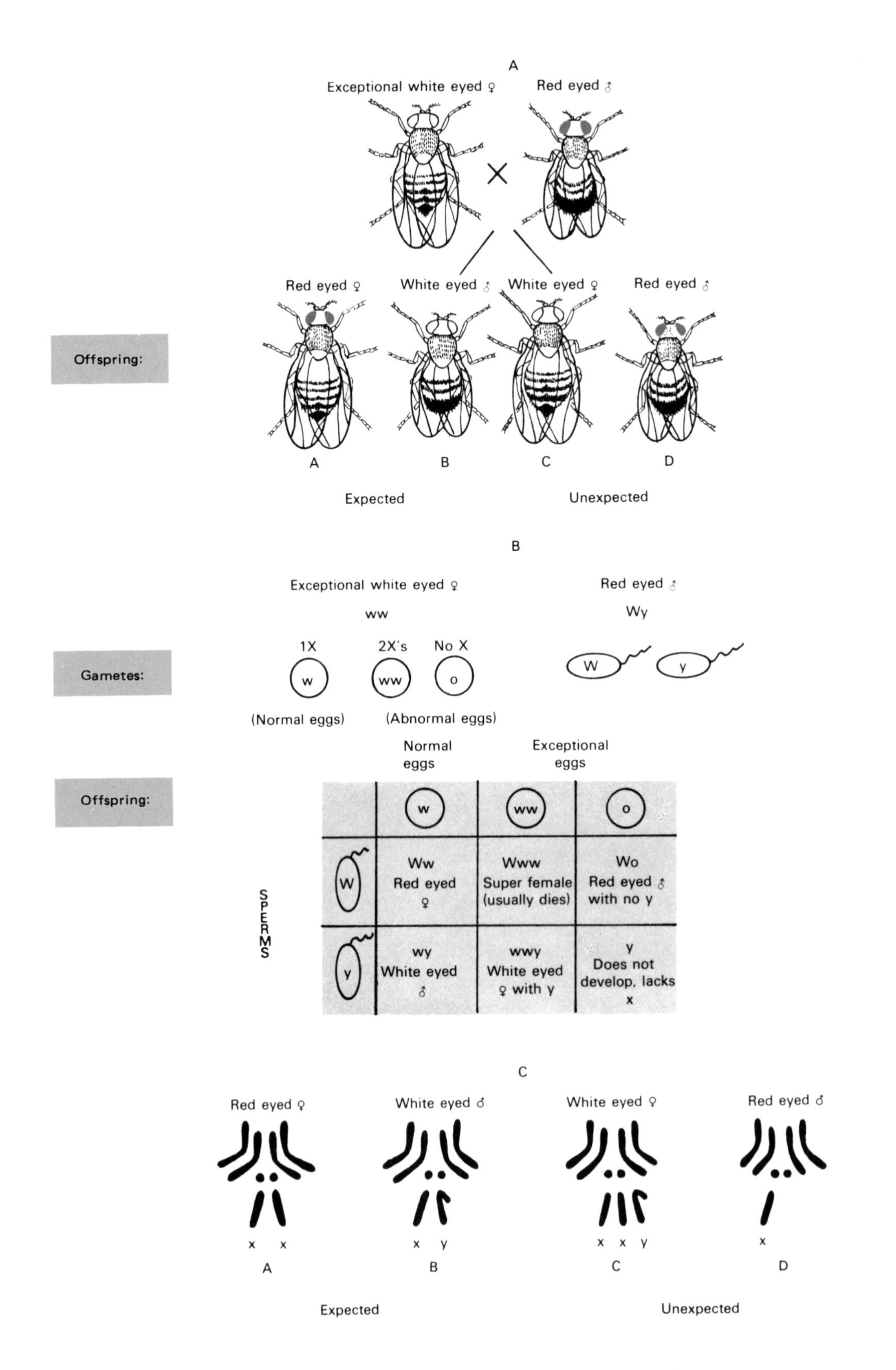
A
Exceptional white eyed ♀
Red eyed ♂
Offspring:
Red eyed ♀
White eyed ♂
White eyed ♀
Red eyed ♂
A
B
C
D
Expected
Unexpected
B
Exceptional white eyed ♀
ww
Red eyed ♂
Wy
Gametes:
1X
2X's
No X
w
ww
o
W
y
(Normal eggs)
(Abnormal eggs)
Normal eggs
Exceptional eggs
Offspring:
SPERMS
Ww Red eyed ♀
Www Super female (usually dies)
Wo Red eyed ♂ with no y
wy White eyed ♂
wwy White eyed ♀ with y
y Does not develop, lacks x
C
Red eyed ♀
White eyed ♂
White eyed ♀
Red eyed ♂
x x
x y
x x y
x
A
B
C
D
Expected
Unexpected

acteristics. A Turner female, found in about 1 in 2,000 female births, is usually of short stature, possesses a webbed neck, and has undeveloped ovaries, an immature uterus, cardiovascular defects, and other somatic aberrations (Fig. 6-12A,B). Persons who are of the chromosome constitution XXY are definitely male, but they exhibit an assortment of deviations from normal that describes Klinefelter syndrome. A Klinefelter male arises in about 1 out of 600 male births, a higher figure than that for Turner syndrome. These males typically show some breast development, small testes, sparse body hair and some mental deficiency (Fig. 6-13). Klinefelter persons are almost always sterile. The Y chromosome in other mammals plays a role in maleness similar to that in humans. In cats, XXY animals occur and have been found to be the sterile equivalents of human Klinefelter males.

Persons exhibiting Klinefelter syndrome have been identified who have chromosome pictures much more complex than the XXY karyotype shown in Figure 6-13B. Individuals of XXXY, XXXXY, XXYY, and XXXYY chromosome constitutions have been found. On average, the XXY male has lower intelligence than the XY male. Mental deficiency and body abnormalities appear to be more pronounced in those Klinefelter males with three and four X chromosomes.

Other kinds of atypical sex chromosome conditions have been recognized in the human, among them the XYY constitution with an incidence of about 1 per 1,000 males. XYY males are taller as a group than XY males, and many of them are below average in intelligence. Cytological studies conducted on inmates of prisons and mental institutions revealed a significantly high proportion of men with an extra Y chromosome, particularly among those over 6 feet tall. The data suggest that an XYY male faces a greater risk of confinement in an institution than does the XY person. However, an extra Y chromosome is present in some males who appear to be normally adjusted and who are indistinguishable from other males in the population. Proper evaluation has yet to be made of the complex of environmental and biological factors that may trigger aggressive behavior. At this time, the reasons for the apparently higher risk of confinement for the XYY male remain unclear. It is clear, however, that XYY males are fertile and that their offspring are usually typical XX females or XY males. Some unknown factor at meiosis seems to prevent an extra Y from becoming part of the chromosome constitution of a functional sperm in the XYY male.

It should be appreciated at this point that the XO, XXY, XYY, and various other unusual sex chromosome constitutions can be explained on the basis of nondisjunction at the first or second meiotic division in one of the parents. At zygonema of meiosis in the male, the X and Y chromosomes form a pairing association referred to as the *sex bivalent* (Fig. 6-14). The pairing does not affect the entire length of the chromosomes but is confined to ends of the short (p) arms of each (about one-fourth of Xp and most of Yp). As we will discuss later on, some homology between the X and Y chromosomes is found in a very small portion of this pairing region. The two chromosomes normally separate at first anaphase to produce two cells: one with the X, the other with the Y (see Chap. 3). Each of these secondary spermatocytes in turn undergoes the second meiotic division. The final outcome is X- and Y-containing sperm in equal numbers (Fig. 6-15A).

Failure of separation of the two chromatids of the Y chromosome at the second division will produce a spermatid with two Y chromosomes and a spermatid with no sex chromosome at all. The former is responsible for the production of the XYY males just discussed. The latter can be responsible for an XO person if it fertilizes a normal, X-bearing egg (Fig. 6-15B). Consider the effects of nondisjunction of the X and Y at first meiotic division (Fig. 6-15C). Sperm of the constitutions XY and O can be produced. If normal, X-bearing eggs are fertilized by such gametes the result is XXY (Klinefelter) and XO (Turner). Nondisjunction can also involve a female parent and give rise to abnormal eggs of the constitutions XX and O (Fig. 6-15D). Fertilization of the exceptional eggs by nor-

FIGURE 6-9. Unexpected results in *Drosophila*. **A:** When a certain exceptional white-eyed female was crossed with an ordinary red-eyed male, the results yielded the expected red-eyed females (A) and white-eyed males (B). However, a small number of unexpected white-eyed females (C) and red-eyed males (D) occurred. **B:** The unusual results can be explained by assuming that in a certain percentage of oocytes in the exceptional female the two X chromosomes do not separate at meiosis but move to the same pole. Thus, in addition to the normal eggs, two exceptional types would arise: one with two Xs and the other lacking an X entirely. When fertilized by normal sperms, these exceptional eggs would give rise to exceptional zygotes. **C:** Examination of the offspring cytologically showed that the expected flies (A and B) possessed the normal chromosome complement. The unexpected females (C) carried a Y; the red-eyed males (D) lacked a Y and were sterile. No flies were found with three Xs. These were discovered later and are unusual females (superfemales) that are weak and tend to die. Flies lacking an X are never found, as at least one X is required for normal development.

IX = ♀ Determining strength of 1.5

1 Set of autosomes = ♂ determining strength of 1.0

Number of x's	Number of sets of autosomes	Ratio femaleness:maleness	Sex
1 (1.5)	2 (1)	1.5: 2	Male
2 (1.5)	2 (1)	3.0: 2	Female
3 (1.5)	3 (1)	3.0: 2 (4.5:3)	Female
4 (1.5)	4 (1)	3.0: 2 (6.0:4)	Female
2 (1.5)	3 (1)	3.0: 3.0	Intersex
3 (1.5)	2 (1)	4.5: 2.0	Superfemale
1 (1.5)	3 (1)	1.5: 3.0	Supermale

All ♀ since ratio of x to sets of autosomes is the same in each case: 1x to 1 set of autosomes

BUT

Adding a single x tips the balance toward femaleness

Adding a set of autosomes equalizes strength of male and femaleness

In both of these the ratio of x to sets of autosomes departs from the balance of 1x to 1 set of autosomes

FIGURE 6-10. Genetic balance and sex determination in *Drosophila*. A female results whenever one X chromosome is present for each set of autosomes. A male develops when one X occurs with two sets of autosomes. Any departure from these ratios results in flies that are in some way abnormal in sexual features.

mal sperm can also produce zygotes that are unbalanced, among them XXY and XO types as well as XXX. The latter constitution occurs in 1 out of 1,600 female births. Many such females have normal intelligence, but a slightly higher frequency of XXX persons is found in mental institutions. Triple-X women are fertile and usually give birth to offspring with normal chromosome constitutions. Evidently those nuclei that are XX in constitution following meiosis in an XXX female are lost, perhaps cast off into a polar body. As in the case of XYY males, some factor appears to be operating in XXX females to prevent the extra chromosome from entering a functional gamete. Very rare XXXX women have been found, and these individuals suffer from pronounced mental deficiency as well as various physical abnormalities, although their sexual development appears normal.

Mosaicism

The fact that individuals with only one Y chromosome but two, three, and even four Xs are still males shows the importance of the Y in determining the secondary sex characteristics in humans. This is quite a contrast to the situation in *Drosophila* and is even more impressively illustrated by comparing certain other chromosome anomalies in the fruit fly and the human. In the fly, certain mishaps that affect the sex chromosomes occasionally take place during embryological development. At rare times during the mitotic divisions in some female embryos, one of the X chromosomes may lag and fail to reach the pole at anaphase (Fig. 6-16A). It may be left out in the cytoplasm where it disintegrates. The result is one cell with two sets of autosomes plus two Xs and another cell with two sets of

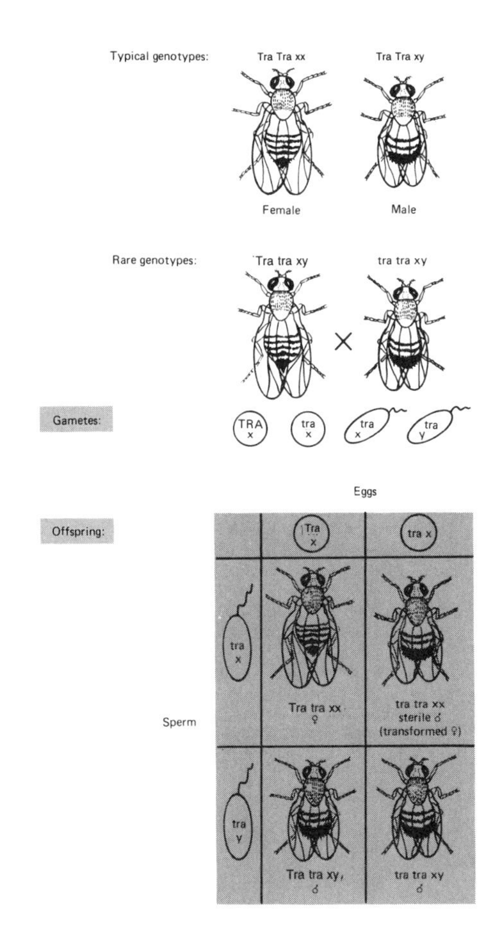

FIGURE 6-11. The transformer allele in *Drosophila.* The recessive tra can transform XX flies, which would normally be females, into sterile males. Most flies (above) are homozygous for the dominant allele (Tra), so that the ratio of X chromosomes to autosomes usually determines the sex. When the recessive tra is present, however, it may upset the sex ratio (below). In this case, a female carrying the recessive is crossed with a male that is homozygous for it (the recessive does not affect the fertility of X–Y males). Among the offspring, half of the expected females are transformed to sterile males. An autosomal gene can thus clearly influence sex.

autosomes plus one X. In *Drosophila,* this means that some cells will be genetically female and others male. Flies such as these actually arise and are called *gynandromorphs,* mosaic individuals with discrete female and male body segments. If the loss of the X takes place at the first division of the zygote, the fly will be a bilateral gynandromorph, male on half of the body, female on the other (Fig. 6-16B). If the mishap occurs later, smaller patches of male tissue will be present among the female background.

Mosaic persons, those who are mixtures of different cell lines, are well known. Actually about 15% of persons with Turner syndrome prove to be mosaics, persons with two distinct cell lines. The karyotype of some of these persons can be represented as X/XX, since one population of cells contains only one X chromosome whereas the other has the normal two. Other persons have the karyotype X/XXX, one of the cell lines with three X chromosomes, the other line with only one. Some Turner persons have been found whose karyotype is X/XY. The various mosaic individuals differ in their phenotypes. This is to be expected, because the mitotic accident resulting in the loss of one X chromosome from one cell line can occur at any time during prenatal development. The earlier the accident, the greater the number of XO cells and the closer the phenotype to that typical of Turner syndrome. Mosaicism accounts for about 15% of the cases of Klinefelter syndrome. Among them are found such karyotypes as XY/XXY and XX/XXY. Again, the mosaics vary in phenotype, depending on the number of unusual cells and their location in body tissues.

It must be pointed out that these mosaic persons are *not* gynandromorphs. No discrete patches of definite male or definite female tissues are seen in any kind of human mosaic. The mosaics may show a range of abnormal phenotypes, but hormonal activity in mammals will not allow the development in any one of them of two distinct classes of cells, those with typical female features intermingled with those that are obviously male.

Departures from the normal number of chromosomes because of nondisjunction and other mishaps can affect autosomes as well as sex chromosomes. Moreover, any chromosome may become altered in its structure. These and other chromosome anomalies are pursued further in Chapter 10.

Incomplete sex linkage

In addition to X-linked genes, others must be considered in relation to the sex chromosomes. We have defined sex-linked genes as those that are found only on the X chromosome. We have also noted that the X and the Y chromosomes pair at first meiotic division. This implies that some homologous segments may be present in each. In the fruit fly, a locus named *bobbed,* which affects bristle length, is located at the end of the X chromosome near the centromere. A homologous locus, however, also occurs on the Y. This means that a female fly has two bobbed loci present (one on each X), and so does the male (one locus on the X, the other on the Y). Thus a male, like a female, can be heterozy-

A

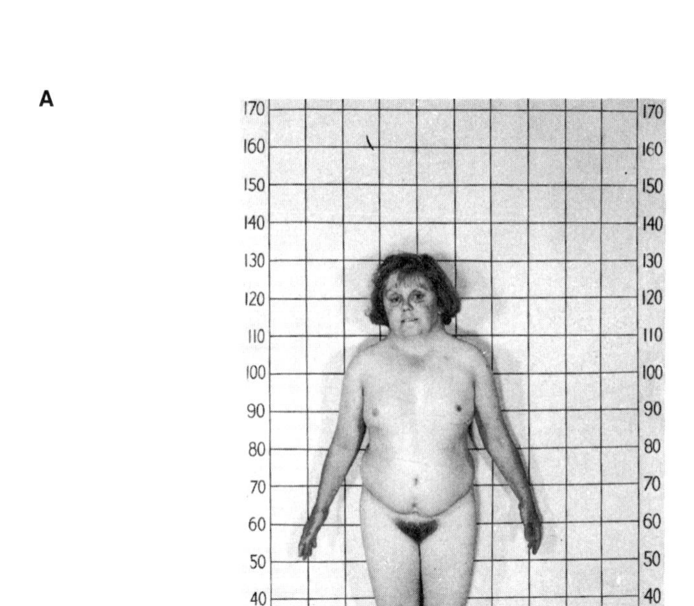

B

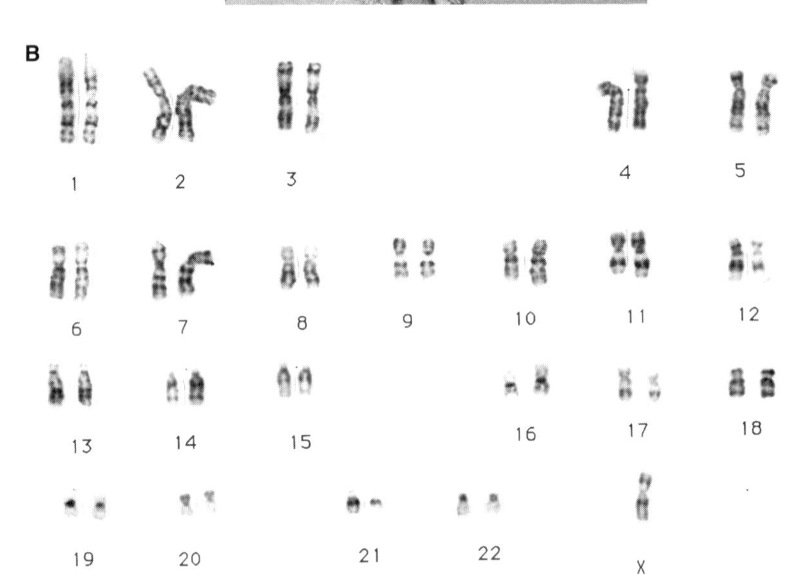

FIGURE 6-12. Turner syndrome. **A:** Patient. (Reprinted with permission from V.A. McKusick, *Medical Genetics* 1958–1960. C.V. Mosby, St. Louis, 1961.) **B:** Karyotype of a patient with Turner syndrome. The chromosomes are G banded. Note that only one sex chromosome, the X, is present. (Courtesy of Steven A. Schonberg, Ph.D., University of California, San Francisco.)

gous or homozygous at this locus. The wild allele (+) is needed for long bristles; its recessive allele, bb, can result in short ones. Alleles such as these are termed *incompletely sex linked*, meaning they occur on both the X and the Y, in a region of homology between the X and Y chromosomes. This is in contrast to sex-linked alleles, which occur only on the X. The cross diagrammed in Figure 6-17 brings out some of the unique features of a trait that is inherited on the basis of an incompletely sex-linked gene.

As noted earlier, pairing regions exist on the short arms of both human sex chromosomes, the X and the

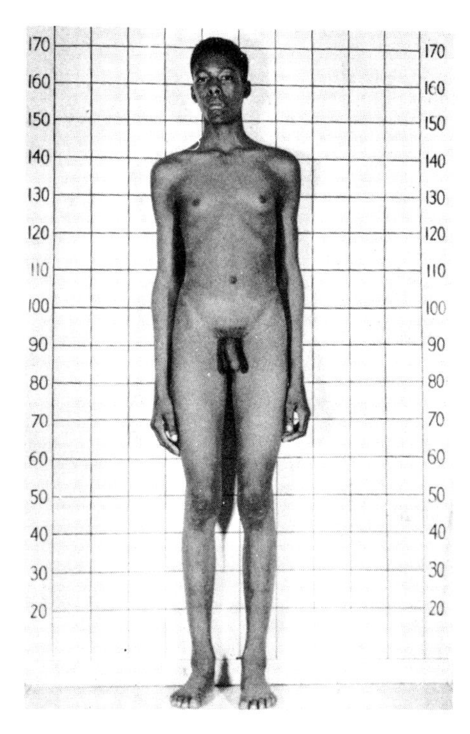

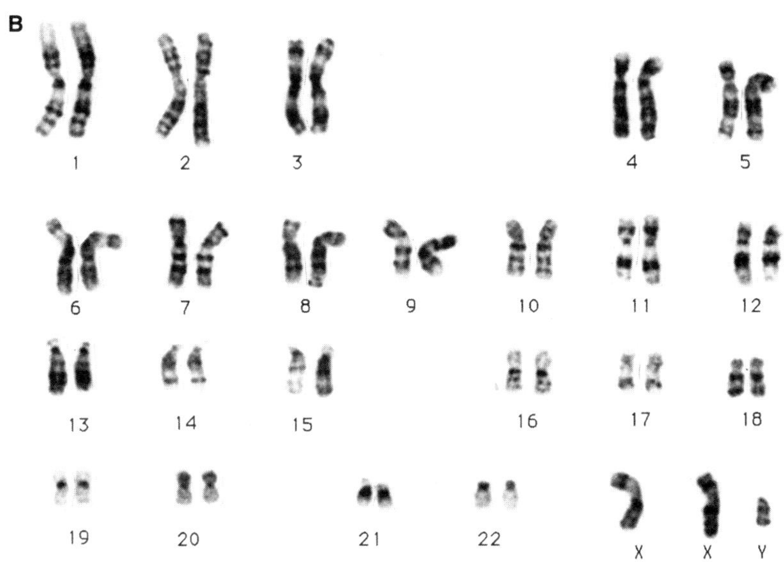

FIGURE 6-13. A: Patient showing features of Klinefelter syndrome. (Reprinted with permission from V.A. McKusick, *Medical Genetics* 1958–1960. C.V. Mosby, St. Louis, 1961.) **B:** Karyotype of a patient with the syndrome. The chromosomes are G banded. Note that two X chromosomes are present in addition to the Y chromosome, which is grouped with them. (Courtesy of Steven A. Schonberg, Ph.D., University of California, San Francisco.)

Y. These regions enable the X and the Y to associate at meiosis in the male (Fig. 6-14) and to segregate in an orderly fashion. The pairing regions contain tiny segments at the tips of the X and the Y that are homologous (Fig. 6-18). Within these segments, one crossover event regularly occurs at first prophase in the male. Since any genes located in this homologous segment are present in two doses in the male as well as

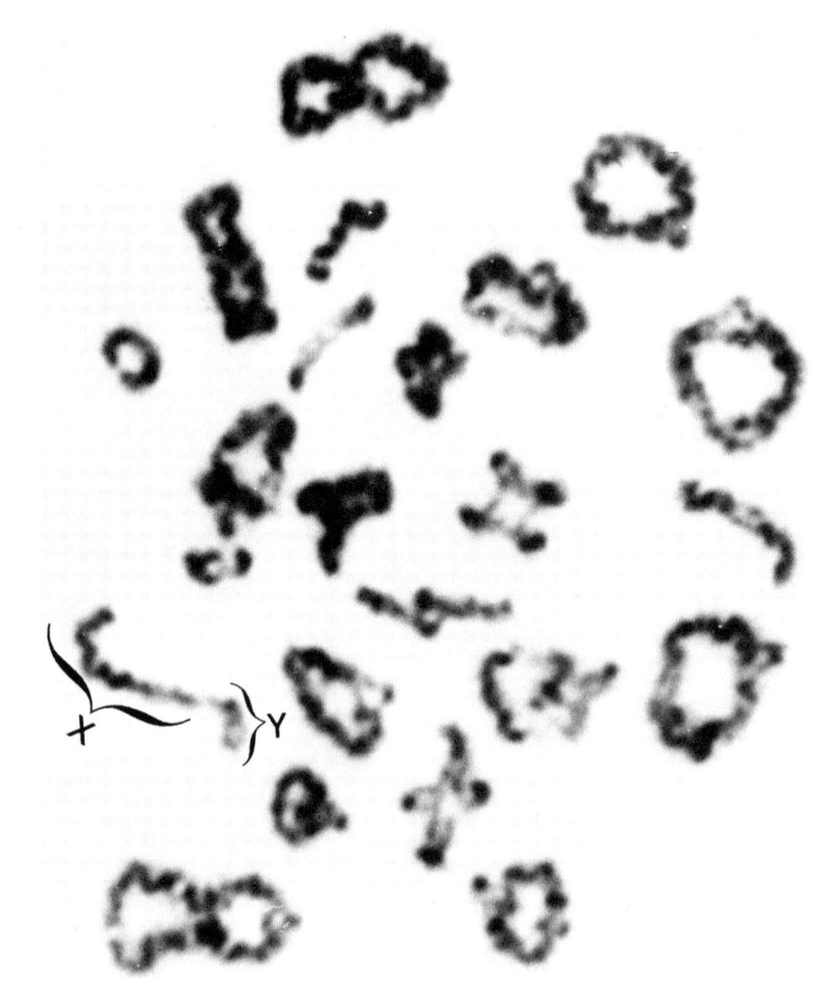

FIGURE 6-14. Meiotic chromosomes of human male at diakinesis. The tiny Y chromosome appears attached terminally to the much larger X. (Reprinted with permission from W.V. Brown, *Textbook of Cytogenetics.* C.V. Mosby, St. Louis, 1972. Courtesy of Dr. J. Melnyk, City of Hope Medical Center, Duarte, CA.)

in the female, their transmission pattern would resemble somewhat that of autosomal genes, as is the case for the gene bobbed in the fruit fly (Fig. 6-17). Therefore the segment is often referred to as *pseudoautosomal.* The only gene identified thus far in this segment in the human is one coded for a cell surface antigen. The remainder of the pairing region is an almost entirely nonhomologous segment in which crossing over does not regularly take place. However, on rare occasions this does occur and can account for the origin of the unusual sterile XX males and XY females discussed a bit later in this chapter.

FIGURE 6-15. Normal and abnormal separation of the sex chromosomes at meiosis. **A:** In the male, normal disjunction of the X and Y chromosomes at first meiotic division, followed by a normal second meiosis, yields normal X-bearing and Y-bearing sperm. **B:** Failure of separation of the two chromatids of the Y chromosome at second meiosis yields sperm of exceptional chromosome constitution, one with two Y chromosomes and the other with no sex chromosome at all. **C:** Nondisjunction of the X and Y chromosomes at first meiotic division yields sperm of unusual chromosome constitution, those that contain both an X and a Y and those that contain no sex chromosome at all. **D:** Nondisjunction of the two X chromosomes in the female may also occur, resulting in eggs of exceptional chromosome constitution.

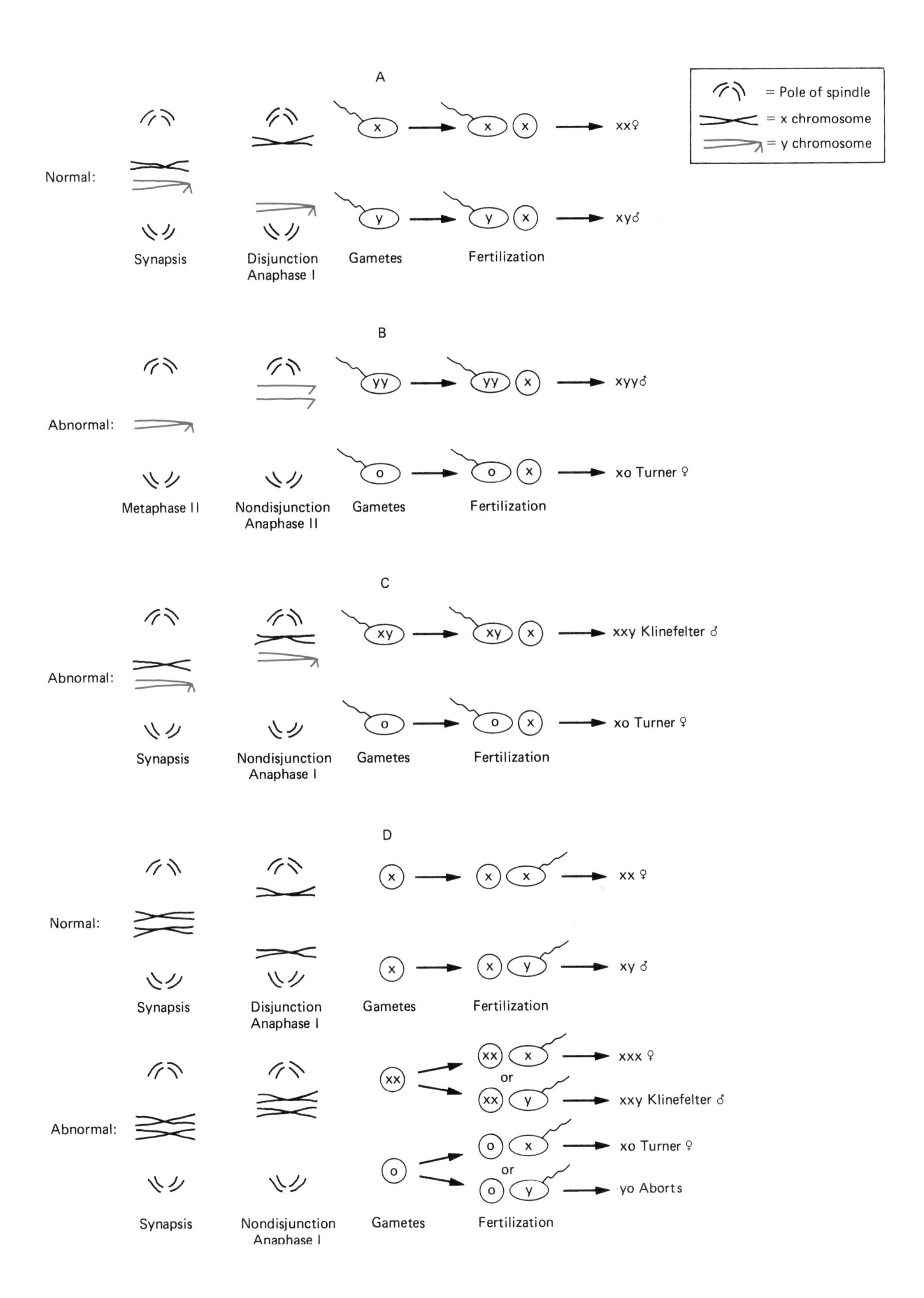
A
= Pole of spindle
= x chromosome
= y chromosome
Normal:
x
x
x
xx♀
y
y
x
xy♂
Synapsis
Disjunction
Anaphase I
Gametes
Fertilization
B
Abnormal:
yy
yy
x
xyy♂
o
o
x
xo Turner ♀
Metaphase II
Nondisjunction
Anaphase II
Gametes
Fertilization
C
Abnormal:
xy
xy
x
xxy Klinefelter ♂
o
o
x
xo Turner ♀
Synapsis
Nondisjunction
Anaphase I
Gametes
Fertilization
D
Normal:
x
x
x
xx ♀
x
x
y
xy ♂
Synapsis
Disjunction
Anaphase I
Gametes
Fertilization
Abnormal:
xx
xx
x
xxx ♀
or
xx
y
xxy Klinefelter ♂
o
o
x
xo Turner ♀
or
o
y
yo Aborts
Synapsis
Nondisjunction
Anaphase I
Gametes
Fertilization

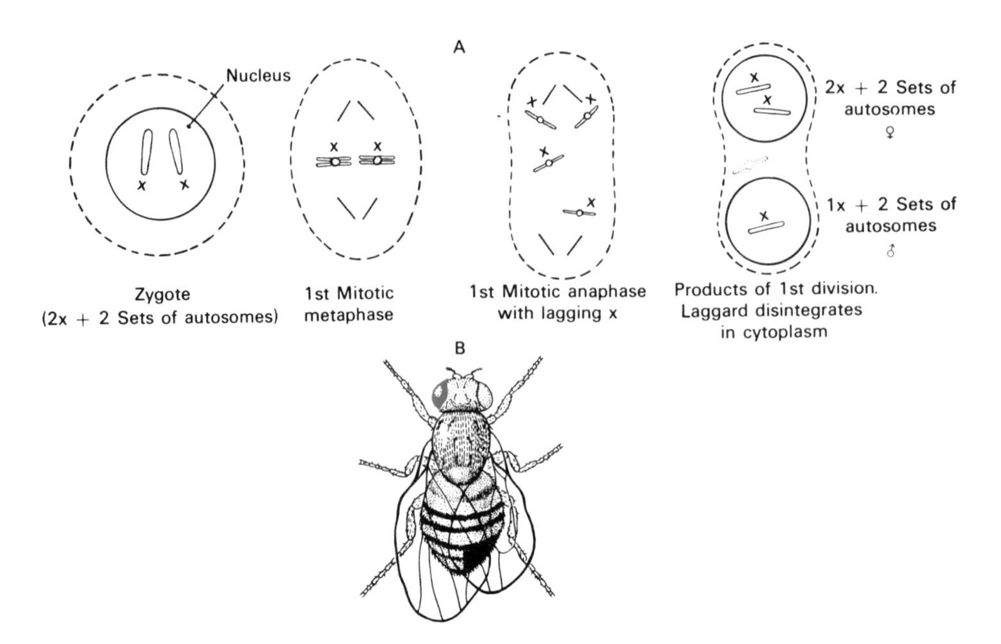

FIGURE 6-16. Gynandromorph origin in *Drosophila.* **A:** A zygote with two X chromosomes plus two sets of autosomes normally develops into a female. However, at mitotic divisions during embryology, one X chromosome at rare times may lag and be eliminated. The outcome is two cells, one with a nucleus balanced in the female direction and the other with a nucleus balanced in the male direction. If the mishap occurs at the first division of the zygote, a fly that is male on one side of its body and female on the other can arise. **B:** A bilateral gynandromorph that is male on the right side of the body and female on the left. The eye color on the male side is white, because only one X is present, and this carries the recessive for white eye color (genotype wO). The X that was lost carried the dominant allele for red color. On the female side, both Xs are present, one with the dominant, the other with the recessive (genotype Ww), and so the eye is red. (Reprinted with permission from E.J. Gardner, *Principles of Genetics,* 4th ed. Wiley, New York, 1972.)

The absence of extensive regions of homology between the X and the Y chromosomes would appear to be advantageous and to preserve the genes on the X as a single unit as well as those on the Y as a different unit. If the X and Y did possess extensive regions of homology, crossing over could occur freely between the two of them at meiosis. As a result of exchange of genetic segments along the length of the two sex chromosomes in the male, genes for maleness and those for femaleness could become distributed on both the X and the Y, and this would interfere with clear-cut segregation of the genes for sex determination. It seems likely that, during the evolution of the mammalian sex-determining mechanism, natural selection favored any changes that made the X and Y chromosomes less homologous. Some of the homology that still remains may facilitate the pairing of the X and the Y and their orderly segregation at first meiotic division of spermatogenesis. Crossing over between the X and the Y could be limited further by the condensation of the X, which takes place at early meiotic stages in the male (Fig. 3-17).

Sex chromosomes and sex determination in mammals

In the human embryo, the gonads are first represented by the gonadal ridges, which appear about 30 days after zygote formation. These are capable of developing into either testes or ovaries. Thickenings appear in the ridges that indicate the positions of the Wolffian and Müllerian duct systems. The former have the potential to develop into the male duct system, the latter into the female duct system. Normal differentiation is triggered by the sex chromosome constitution of the cells of the bipotential primitive gonad.

An intact Y chromosome is essential for the development of a normal testis; in its complete absence no testis will form. A testicular organizing substance is believed to exist, and much research has been

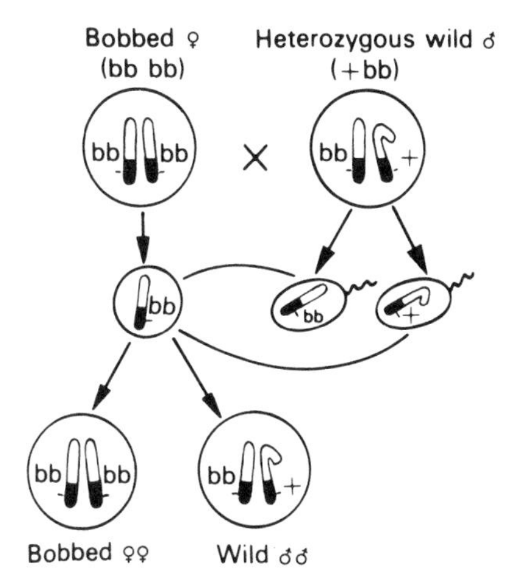

FIGURE 6-17. Incomplete sex linkage in *Drosophila.* The locus bobbed, which affects bristle length, is found on both the X and the Y chromosomes. In a cross between a heterozygous male carrying the dominant wild allele on the Y and a female homozygous for the recessive allele, the normal trait is passed from father to sons. All of the daughters will show the mutant phenotype. This is in contrast with a sex-linked gene. If the locus were on the X chromosome alone and not on the Y, a crisscross pattern would occur. The inheritance of an incompletely sex-linked gene resembles that of one located on an autosome.

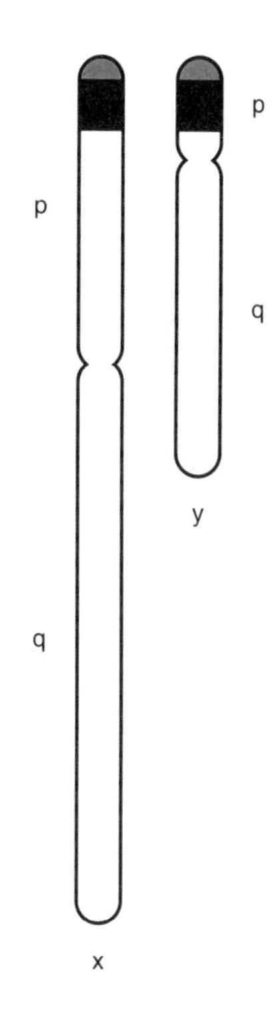

FIGURE 6-18. Pairing regions of the X and Y chromosomes. Each sex chromosome contains a pairing region at the end of its p arm (red and black segments). At the extreme tip of each pairing region is a tiny segment (red) that contains homologous genetic material. In this segment a crossover event regularly takes place at meiosis. The remainder of the pairing region (black) is essentially nonhomologous, although a crossover event may take place in it on rare occasions.

directed toward its recognition. If a normal Y is present, testes develop, even in persons with unusual karyotypes (XXY, XXXY, XXYY, XXXXY). The testis of the fetus secretes substances critical to normal male development. Among these is MIH, Müllerian-inhibiting hormone, an inhibitor of the Müllerian system that otherwise will evolve into fallopian tubes and uterus. Testosterone, the major male hormone, then guides the Wolffian duct system to form the prostate, seminal vesicles, and tubes of the male system. A conversion product of testosterone, DHT (dihydroxytestosterone), organizes development of the penis and scrotum (Fig. 6-19A).

In the absence of a Y chromosome and male hormones, as in the XX condition, no testis differentiation can be triggered, and the Wolffian duct system regresses. Since no Müllerian inhibitor is present, the Müllerian ducts are free to realize their potential in the female pathway (Fig. 6-19B). The primitive gonad in the XX individual differentiates into an ovary. Oogonia arise in the cortex of the gonads, and primary oocytes form before birth.

The X–Y mechanism makes possible equal numbers of male and female zygotes. However, the human sex ratio at birth is slightly in favor of males: 106 boys to every 100 girls. Various interpretations have been offered for this slight excess of males at birth, but the topic remains a matter of debate. Regardless of the correct explanation, the ratio of males to females at the age of sexual maturity is 1:1.

The Y chromosome and the H–Y antigen

A gene that occurs exclusively on the Y chromosome is said to be *holandric.* Such a Y-linked gene occurs only in males and is transmitted from father to son. In *Drosophila,* genes that affect sperm motility are holandric, although the Y chromosome in the fruit fly is not necessary for secondary male characteristics. In the mammal, however, there must be a gene on the Y required for the development of a testis, which forms very early in development and soon produces the hormones for masculinity.

An interesting discovery in mice has supported the concept that a testicular organizing factor occurs on the Y. These animals can be inbred for many generations to produce isogenic strains, populations within which all individuals have essentially the same genotype. Nevertheless, within an isogenic strain a genetic

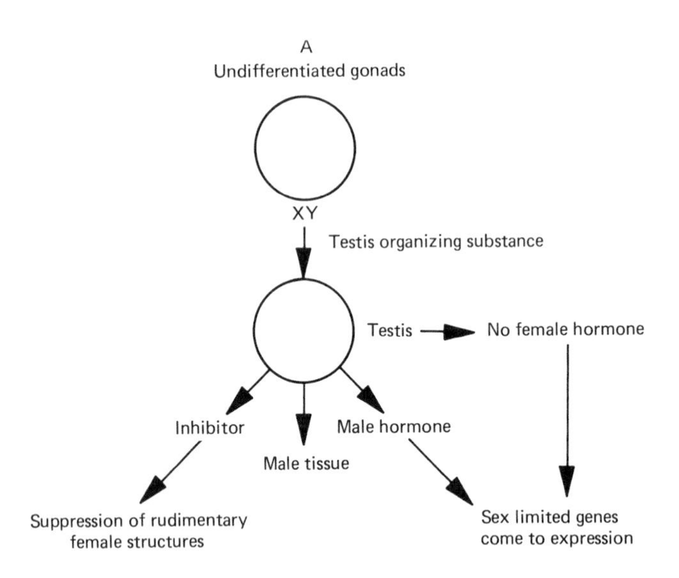

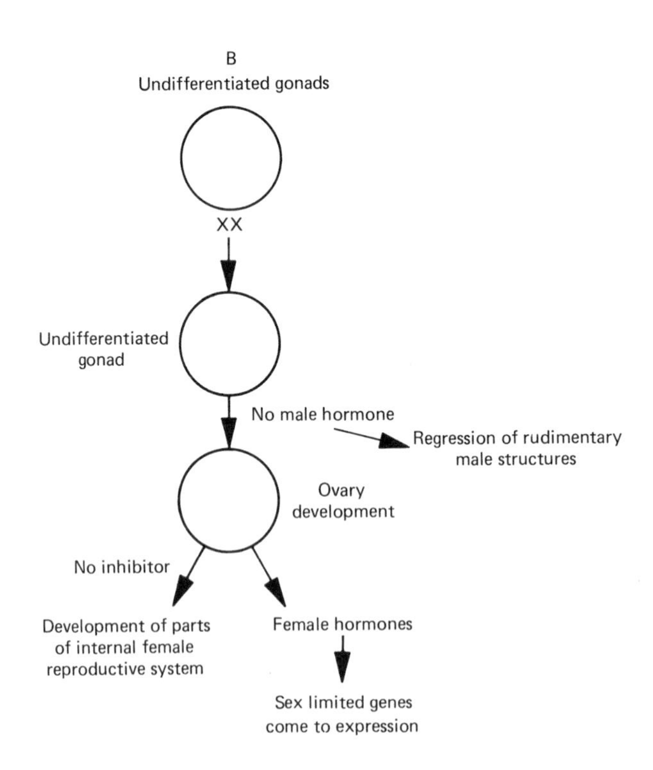

FIGURE 6-19. Summary of the interaction of genetic constitution and the hormonal environment in sex determination. **A:** Development in the male direction. **B:** Development in the female direction.

difference persists between males and females, since the sexes differ in respect to the sex chromosomes. A male possesses all the alleles found in a female of the same strain. However, any genes on the Y are present only in the male. In 1955, it was demonstrated that skin can be successfully transplanted from female to male mice of the same inbred strain; however, the reverse does not hold. Male skin transplanted to females of the same strain is eventually rejected (Fig. 6-20). Serum taken from females that have already rejected male tissue contains antiserum capable of killing male cells. From such observations, it was concluded that the Y chromosome carries a gene or genes responsible for antigenic determiners on male tissue. The Y-linked region or locus was named *H–Y,* and the term *H–Y antigen* was given to the transplantation antigen on male tissue. This antigen has been found on all normal male tissue of the human and other mature male mammals thus far tested. Men with an XYY genetic constitution possess more H–Y antigen than do XY males.

The findings led to the proposal that the H–Y antigen in mammals is the male-determining substance required for testis organization and that the gene on the Y associated with the antigen is the one responsible for the organization of the testis and hence the determination of maleness. However, later studies (discussed in the following section) showed that the H–Y antigen gene is not the one required for testis organization.

The testis determining factor

The search for the location of the gene responsible for the testis determining factor (TDF) focused on sex-reversed individuals, males with an XX and females with an XY karyotype. XX men, who occur with a frequency of 1/20,000 males, present a normal appearance but prove to be sterile. The fact that there has been testis development suggests that a portion of the Y that carries the TDF gene and that is too small to detect with the ordinary microscope is present somewhere in the genome, probably on an X. Conversely, XY females must be missing that part of the Y that determines maleness, the part that contains the TDF gene. Such females are sterile, usually have poor breast development, and fail to menstruate.

To determine whether these ideas are correct, a team of investigators turned to the tools of molecular biology. These procedures are discussed in some detail in Chapters 19 and 20. At the moment, it is sufficient to understand that DNA probes were utilized in the search for the TDF segment of the Y. A DNA probe is essentially a radioactive piece of genetic material that will bind to a DNA segment that resembles it closely. In this way, it can identify and thus locate the presence of a piece of DNA in the entire genome that is identical or very similar to it. Conversely, if a complementary piece of DNA is absent, the specific probe will not bind to any pieces or portions of the DNA to which it is exposed. Probes were obtained that were actual pieces of the Y chromosome DNA. If the DNA from XX men bind to any Y DNA probe, this means such men must carry a piece of the Y in their DNA. It was found that the XX men do indeed carry a part of the Y and that the amount carried varies from one case to the next. The piece present in these men is normally located on the short arm of the Y. Since this must therefore be the location of the TDF gene, it became clear that the H–Y antigen is not the sex determining factor, since the locus of the H–Y gene resides on the long arm of the Y.

A search was then made for the smallest DNA piece that all XX men have in common. The short arm of the Y was divided into more than a dozen segments. Using probes for these segments, the research team was able to narrow the location of the gene down precisely. The DNA of XY females was also screened to determine which piece might be missing. It turned out that some of them had almost the whole Y chromosome, and yet they were still basically female! All XY females, however, proved to have a certain piece missing that was present in all the XX men.

One XY female in particular led a team of investigators to believe they had finally found the testis determining gene. This female was missing only a tiny Y chromosome segment. Successful cloning of this piece (see Chap. 19) revealed that it contained a gene coded for a protein product. The structure of the protein was deduced to be that of a type known as a zinc finger protein. Proteins in this class contain amino acid loops held together by chemical bridges. Such a structure suits a protein well for binding to the genetic material and hence regulating gene activity. The gene coding for this particular protein (the presumed testis determing gene) became known as *ZFY* (*zinc finger Y*).

Further studies conducted largely by two British teams have shown that the ZFY gene is not the primary one determining sex. XX males were found who lacked ZFY but who had inherited another tiny piece located close to it. This small region was found to contain a single gene that has been designated *SRY* (*sex determining region Y*). A comparable region known as *Sry* has been recognized in mice and several other mammals. XY mice were found that lacked the Sry gene and that were biologically female. The Sry gene has also been shown to be active in the proper tissue and at the proper time, only as testes develop in the embryo. A very crucial experiment has been performed that strongly implicates SRY (and the corresponding Sry) as the testis determining gene. This has been the insertion of a very small DNA segment that includes Sry into mouse embryos that carry two normal X chromosomes. Some of these developed into male animals having normal testes and normal male behavior.

While the testis determining gene appears to have been found, many questions regarding normal sex differentiation remain. First, SRY appears to act only as a switch. It is known that testes can develop in females of normal XX sex chromosome constitution as a result of mutations in other parts of the genome. Normally, the SRY gene, present in the Y, triggers the change in the undifferentiated gonadal ridges that leads down the male pathway. Other genes that lie outside the Y and that are also present in females then come into play to bring about differentiation of the ridges into male gonads. Still to be answered are "What turns the SRY gene on at the proper time in the appropriate cells?" "What are the other genes that SRY triggers?"

Recognition of the pairing regions found on the X and Y chromosomes permit an explanation of the origin of XX males and XY females (Fig. 6-21). The chromosome region containing the SRY gene is located in the nonhomologous pairing region of the Y. While crossing over is usually confined to the homologous segments, an unusual crossover event may take place in the nonhomologous segments, resulting in a transfer to the X chromosome of genetic material that is typically confined to the Y and that contains the SRY gene. Support for this concept of such a transfer from Y to X chromosome during first meiotic prophase in the male comes from investigations utilizing tools of molecular biology. These have shown that XX males

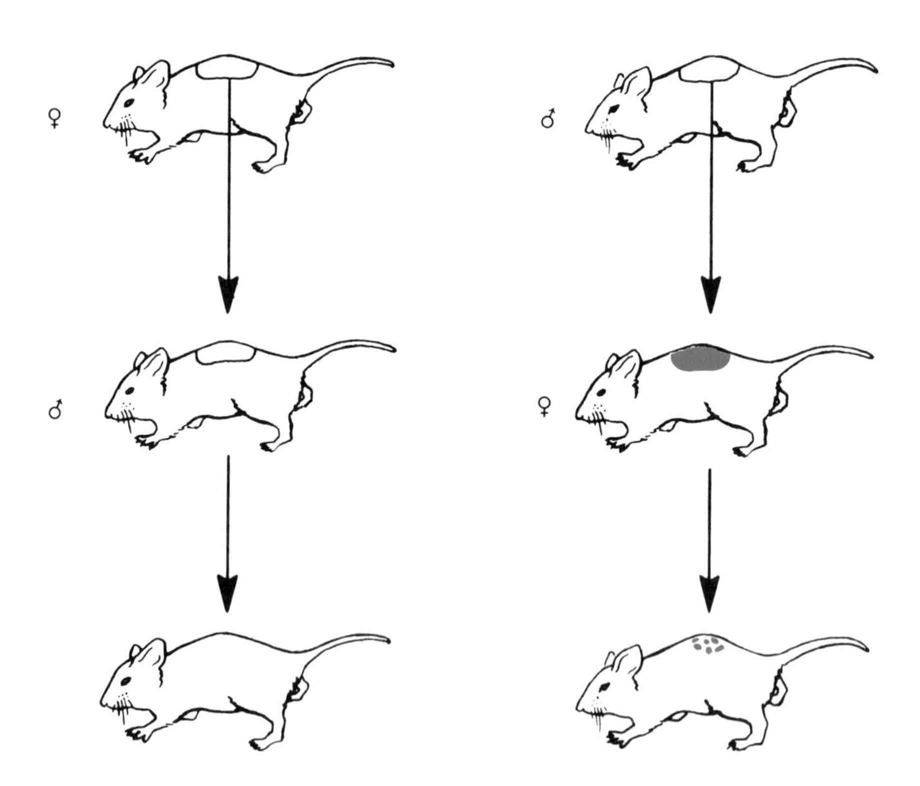

FIGURE 6-20. Tissue transplants in isogenic strains of mice. Skin taken from a female (left) can be transplanted into a male of the same strain with acceptance of the transplant. Skin from a male, however (right), transplanted into a female of the same strain is eventually rejected.

carry a portion of the short arm of the Y bearing the SRY region on that X chromosome that they inherited from their fathers, not from their mothers.

Normal sexual development is indeed a complex process that depends on genes located on the autosomes and both sex chromosomes as we will now see in the examples that follow. During the embryonic development of certain XY individuals, testis formation is induced by a normal Y chromosome, and normal amounts of male hormones and MIH are produced. Consequently, no internal parts of the female reproductive tract are formed, and the stage is set for normal male development. However, a specific X-linked recessive allele can prevent the formation of appropriate receptors on cells of those tissues that normally respond to male hormones. Although the testes are normal and functional in their production of male hormones, they do not descend into a scrotum. While Müllerian ducts have regressed, male parts cannot develop from the Wolffian system in the absence of the required receptors to male hormones. Due to the appearance of the external genitalia, these XY persons are raised as females. At puberty, they actually develop breasts and other secondary female sex characteristics despite the fact that normal amounts of testosterone are present! This comes about as a result of the female hormone estrogen, which arises from metabolism of testosterone in the bloodstream. This condition, known as the *testicular feminization* or the *androgen insensitivity syndrome,* also occurs in the mouse. Since affected individuals are always sterile, the condition cannot be transmitted by them but is passed to male offspring by females who carry the recessive allele on one of the X chromosomes.

A different sex-linked recessive acts in a way similar to that described above but results in partial instead of complete insensitivity to male hormones. XY persons in these cases have external genitalia that are partly male and partly female. Individuals with ambiguous external genitalia but who nevertheless possess either a testis or an ovary are frequently referred to as pseudohermaphrodites. A very dramatic form of this condition is due to an autosomal recessive and is associated with the XY karyotype. Absence of the appropriate dominant allele prevents the conversion of testosterone to DHT. Since a normal testis has been triggered by the Y and normal receptors to male hormones are present, the Wolffian duct system develops properly into parts of the male reproductive tract. MIH prevents development of internal female parts. However, since DHT is absent, the penis and scrotum do not develop normally, and the external genitalia appear more female than male. Such persons are raised as females. However, puberty brings about increased levels of testosterone, which trigger a dramatic virilization. Testes descend into a developing scrotum, and a functional penis arises from what appeared to be a clitoris. Beard development and other normal changes associated with masculinity occur, including the ability to produce sperm. This is a very rare condition known primarily in a few Caribbean villages where a high incidence of marriage between related persons has taken place. Nevertheless, this and other unusual conditions serve well to emphasize that normal sexual development is the result of the balance and interaction of many genes located throughout the genome, only a few of which have been discussed here.

Sex chromatin and the Lyon hypothesis

For many years, biologists wondered about the fact that in mammals the female carries a double dose of all the genes on the X chromosome while males carry a single dose. This would imply that a female homozygous for the dominant allele for the antihemophilia factor has twice as much of factor VIII as does a normal male. The idea is disturbing when considered in relation to the concept of genic balance, particularly in view of the fact that the X chromosome is of appreciable size, containing about 6% of the genetic material in a haploid chromosome set. In contrast to the X-linked loci, all autosomal loci are represented twice in both sexes. It seemed to geneticists that something must compensate for the difference in dosage of the sex-linked genes to preserve genic balance.

An answer to this problem dates back to 1949 with the discovery by Barr of a nuclear body found in the neurons of the female cat. This small intranuclear entity was completely absent from neurons of the male (Fig. 6-22). Examination of other tissues and other species showed that a constant distinction can be made between nuclei of cells taken from females and from males. A simple way to demonstrate this in the human is by scraping epithelium from the buccal mucosa and staining it with some common dye. The nuclei of most of the cells from a female show a small rod-shaped structure, more deeply stained than the surrounding chromatin and usually located at the periphery of the nucleus (Fig. 6-23A). Comparable cells from a normal male show no such body, which has been designated the *Barr body* or the *sex chromatin.*

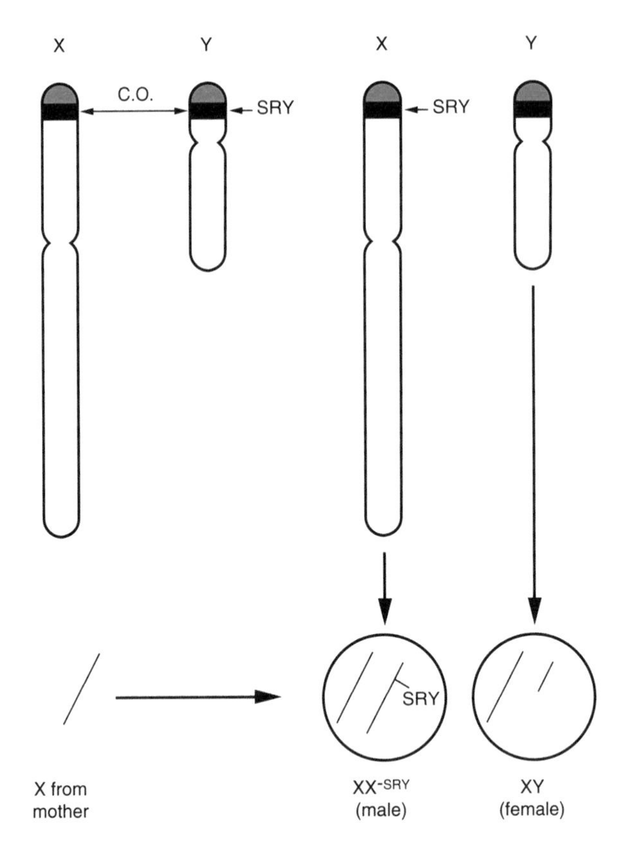

FIGURE 6-21. Origin of sex-reversed individuals. One way in which exceptional XX males and XY females arise results from an unusual crossover event (double-headed arrow) in the nonhomologous pairing segments (black) of the X and Y. The homologous segments (red) are at the extreme tip of each chromosome. The SRY gene resides in the nonhomologous segment. Following a crossover event that occurs on the side of SRY toward the centromere, the gene may be transferred to the X, leaving the Y without SRY. If a sperm carrying the X with the SRY gene fertilizes a normal egg, an XX individual arises who possesses an SRY derived from a Y chromosome and who would be male. The Y lacking the SRY gene, when combined in a zygote with a normal X, gives rise to an XY female.

In certain white blood cells, polymorphonuclear leukocytes, a "drumstick" can be recognized (Fig. 6-23B). This is seen as a small appendage attached by a filament to one of the lobes of the nucleus. Although they can be identified in only about 2% of the leukocytes from a female, drumsticks are entirely absent from cells derived from a male (Fig. 6-23C).

The discovery was then made that drumsticks and Barr bodies are absent from females with Turner syndrome. On the other hand, cells from XXY males contain one Barr body or one drumstick; males who are XXXY and XXXXY show two and three, respectively. Cases were found of women with multiple X conditions: XXX and XXXX. These persons were found to have two and three sex chromatin bodies, respectively. It became apparent that the number of sex chromatin bodies and drumsticks was one less than the number of Xs present in the cells of the individual.

Such observations led to the formulation of a theory that accounts for the relationships between the number of sex chromatin bodies and the number of Xs. More important, it suggests a mechanism that compensates for the presence of different dosages of X-linked genes in males and females. Dr. Mary Lyon has presented some of the best arguments for the concept. According to the Lyon hypothesis, all normal females are sex mosaics, composed of different cell lines insofar as the X chromosomes are concerned. Only one X is functioning entirely in a given cell; the other is present largely in an inactive state. Which specific X is active in a cell—the one from the maternal or the one from the paternal parent—is a matter of chance. Any female has the X she received from her mother functioning in some cells and the one from her father in others. Which X will be the active one in a cell is believed to be decided during embryological development.

Several lines of evidence support the Lyon hypothesis. It will be recalled from the discussions of mitosis and meiosis that chromosomes in a compact state, as at the time of nuclear divisions, actually do not participate in cellular activities other than those concerned with the division process itself. Evidence from the molecular level has shown that DNA in its compact state is relatively inactive in directing the synthesis of other molecules. Cells can be taken from human females and maintained for study in tissue culture. Observations have revealed that at early mitosis one of the two X chromosomes in any cell tends to be more compact and to stain more deeply than the other. It also tends to undergo synthesis of its DNA later. All in all, the behavior of one of the Xs at mitosis has been shown to be distinctly different from the other members of the chromosome set, and some of the features it displays are typical of chromatin, which is relatively inactive. In the earliest cell divisions of the female embryo, both X chromosomes behave in the same fashion. However, at approximately 16 days, in the late blastocyst stage, sex chromatin can be recognized in all kinds of cells, somatic as well as those destined to become germ cells. However, in the latter, reactivation of the condensed X occurs so that two active X chromosomes are present as meiosis begins. During prophase of oogenesis, both X chromosomes appear identical and react the same way to stains. During first meiotic prophase, they pair and engage in crossing over. All the observations indicate that both X chromosomes are equally functional in cells of the very early embryo but that some mechanism becomes operative at the blastocyst stage, keeping one X (or a large part of one) in a permanently inactive state in somatic cells by condensing the chromatin. The Barr body is a visual manifestation of this. However, in the germ line, reactivation occurs in the condensed X, rendering both X chromosomes equally active.

You will recall that in the male the X chromosome is condensed during early meiotic stages. This behavior of the X in the male may be designed to discourage crossing over between X and Y chromosomes and may not represent the same process operating to condense one X in somatic cells of the female.

In Chapter 2, we encountered examples of chromatin that always remain condensed, the constitutive heterochromatin. We see here in the female that, whereas an X chromosome may become condensed and behave as heterochromatin, it may be reactivated in a germ cell. Moreover, an X that does become inactivated is not the same in every somatic cell of the embryo or fetus. It is the paternally derived one in some cells and the one maternally derived in others. This means that the same kind of chromatin can act as heterochromatin in certain cells and as euchromatin in others. The term *facultative heterochromatin* is used to distinguish the kind of heterochromatin exemplified by the X chromosomes from the constitutive heterochromatin, which is constant from cell to cell.

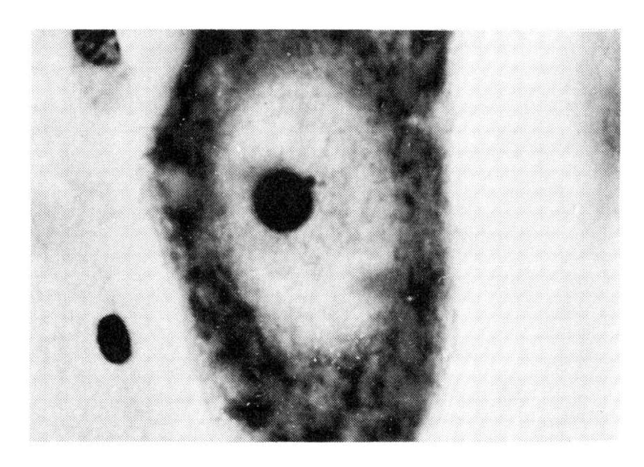

FIGURE 6-22. Barr body in neuron of the female cat (× 1,050). This body, which is very evident in the nucleus of the female, is often found in the vicinity of the nucleolus (the large intranuclear body). It is entirely absent from the nucleus of the male.

Barr bodies and drumsticks show various parallels that indicate they are both visible manifestations of the same phenomenon, a compacted X, which is thus in an inactive state. The fact that sex chromatin is not seen in every cell of a female (drumsticks in only about 2% of the polymorphonuclear leukocytes; Barr bodies in 90% of the buccal cells) is believed to result from its orientation in a given nucleus. The extreme lobing of the polymorphonuclear leukocyte nucleus could affect visualization of the drumstick.

Further evidence for the Lyon hypothesis

Excellent support for the Lyon hypothesis comes from chemical studies of individuals known to be of different genotypes for sex-linked genes, such as the one for hemophilia. For example, normal homozygous women (HH) and normal men (HY) have the *same* amount of antihemophilia protein in the blood, even though the gene dosage is different. The female carrier (Hh) is almost always without symptoms of the disorder, but blood analysis shows that the concentration of antihemophilia protein varies from one carrier to the next. A few of them have about the same amount of the clotting factor as does any normal male or homozygous female, but in a few the essential substance is very low. Most carriers have about half that of the average person. This is precisely what one would expect on the basis of the Lyon hypothesis, which predicts that which X chromosome remains inactive in a given cell is determined by chance. Consequently, in a carrier female (Hh), approximately half of the cells, on average, should have the maternal X with the normal allele operating (Fig. 6-24). In the rest of the cells, the paternal X with the defective allele is being expressed, and thus only half the normal amount of the clotting factor is produced. In a small number of cases, however, by chance the majority of cells have the maternal chromosome operating. An almost normal level of the antihemophilia protein is present as a result. Likewise, in a few carriers the paternal X is active in most of the cells, by chance, and consequently a low level of the factor is present and some symptoms of hemophilia may be manifest.

Another product associated with a gene on the X chromosome is the enzyme glucose-6-phosphate dehydrogenase (G6PD), discussed in Chapter 4. The effects of a deficiency of this enzyme can be severe following exposure to certain drugs. Since males are hemizygous for X-linked genes, they either respond to an environmental trigger such as primaquine or they have no response at all if they carry the normal allele. Heterozygous females, on the other hand, typically have an intermediate response to the stimuli. However, they show different degrees of response, as was true for carriers of the hemophilia allele, since in some of them the X with the defective allele is inactivated in the majority of cells, whereas in others it is the X with the normal allele that is not being expressed in most cells.

Single cells can be isolated from a cell population in tissue culture, and any cell can then be allowed to divide to produce a clone, a population of genetically identical cells. In the case of the females heterozygous at the G6PD locus, two types of clones can be established: those from which only normal enzyme can be isolated and those from which only the variant form can be retrieved (Fig. 6-25). No clones are ever found that produce enzyme levels intermediate between the normal and the variant or that produce both types of the enzyme. Cultures of cells taken from any given male, however, yield only one kind of population, a clone with the normal enzyme or a clone with the variant. This shows that no mixture of cell types is found in the male with respect to the gene or the enzyme.

Moreover, when persons of normal karyotypes, XX and XY, are compared with XXX and XXY persons, it is found that the amount of G6PD is the same in their cells, indicating that only one X chromosome is active in a cell despite the presence of additional ones.

Many other examples can be given in the human for other X-linked loci, which, in each case, are also associated with a particular product. The situation is similar to the one just described, and all therefore

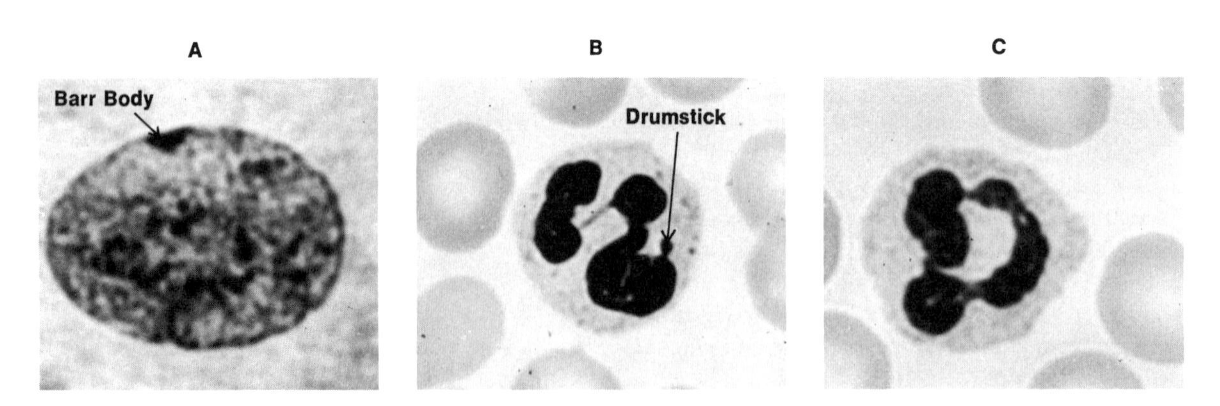

FIGURE 6-23. Sex chromatin in humans. **A:** Barr body in nucleus of a squamous epithelial cell from the buccal mucosa of a female. **B:** The drumstick in a polymorphonuclear leukocyte of a female. **C:** Polymorphonuclear leukocyte of a male. (Courtesy of Carolina Biological Supply Company.)

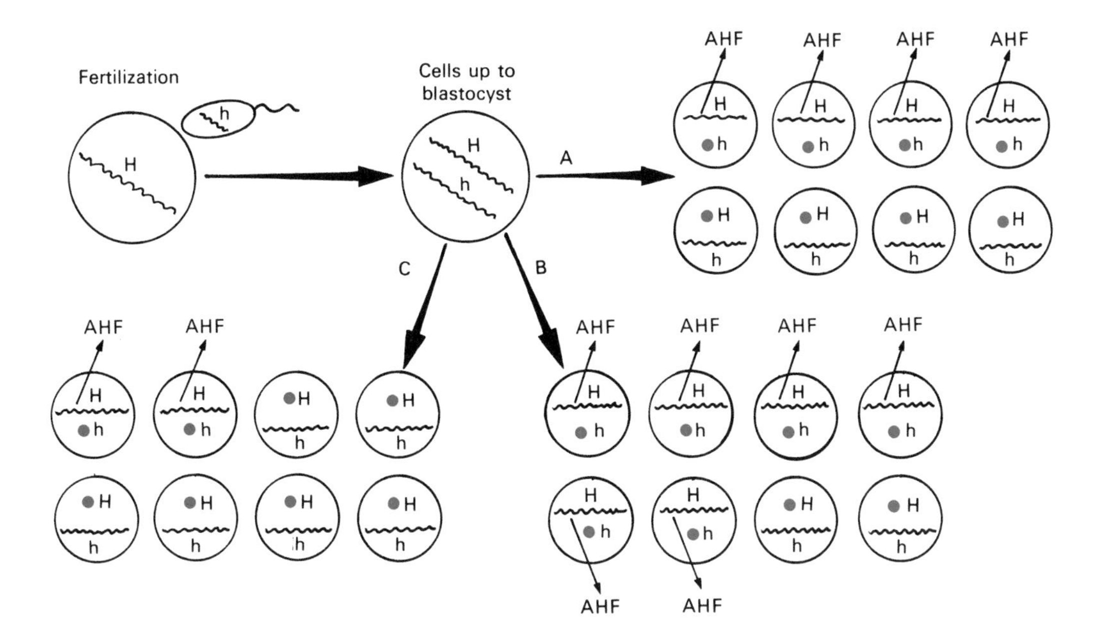

FIGURE 6-24. X inactivation and the AHF. A female carrier of hemophilia can arise if a sperm carrying an X chromosome with the recessive unites with an egg carrying an X with the dominant allele responsible for the production of AHF. Both X chromosomes are active in each cell up to the time of the blastocyst, when one of them becomes inactive (arrow A). The inactive X may be visualized as condensed chromatin, a Barr body, or sex chromatin. It is chance whether the maternally derived X or the paternally derived one remains active. On the average, a carrier will have half of her cells with an active X carrying the dominant, and thus half the normal amount of AHF will be present. In a smaller number of cases (arrow B), the maternal X will be active in most cells, and the amount of AHF will approach that of a normal homozygote or of a normal male. Conversely (arrow C), in a few carriers the paternal X will be active in most cells, and the amount of AHF will be much lower than the normal.

support the premises of the Lyon hypothesis. A final example from a natural hybrid gives even more direct evidence for two aspects of the Lyon hypothesis. A female mule is a hybrid resulting from the cross of a horse and a donkey. The X chromosome of the horse has a median centromere and is quite readily distinguished from the X of the donkey, which has its centromere close to one end. Moreover, the G6PD enzyme from the horse can be distinguished electrophoretically (see Chap. 14) from that of the donkey. Cells can be isolated from the female mule and raised in culture from which two types of clones can be isolated. In one kind it can be demonstrated that the donkey X is late replicating (and thus is the inactivated X). In such a clone, only the horse G6PD is present. In a clone that produces only donkey G6PD, the X chromosome of the horse is late replicating and inactive. We see in this example that not only are two populations of cells present in the female, but the presence or absence of a given product is associated with the inactivation of a specific chromosome.

The precise mechanism that causes inactivation of one X chromosome, and thus compensates for the different dosages of X-linked alleles in males and females, is a matter of ongoing research (Chap. 17) on the molecular level. However, it has undoubtedly been perfected by natural selection, along with the evolution of the X and Y chromosomes, to ensure genic balance in both sexes. According to Ohno, the X-inactivation mechanism has preserved the ancestral X chromosome intact throughout the evolution of mammals. While the Y has become miniaturized and varies appreciably from one species to the next, the evolution of the X has been very conservative. Several good arguments support Ohno's law of the conservation of the X chromosome. For example, the X of all mammals is about the same, a good-sized chromosome making up about 5% of the genome and having the same pattern of bands. Even the X chromosome of the kangaroo is similar to that of placental mammals. The X chromosome of the gibbon appears to be very similar to that of the great apes and the human, although the autosomes of the gibbon have experienced much evolutionary change. No exception has been found in mammals regarding genes that are sex linked. Demonstration of sex linkage for a given gene in any other mammal is a strong indication that the comparable gene will be sex linked in the human. In spite of the vast amount of speciation and chromosome change that has occurred in mammals in more than 2 million years, changes in the X chromosome appear to have been highly conservative.

As a result of deactivation of one X chromosome in the female, both sexes are, in effect, hemizygous in any one cell for genes on the X. It should be noted, however, that the deactivation of one X in the XX female must be partial and not entire. The XX female is certainly quite distinct from the Turner female, who is truly hemizygous for genes on the X. Moreover, it is known that two X chromosomes are essential for female fertility. Infertility results in cases in which just a portion of one X chromosome is missing.

Studies have shown that certain X-linked genes close to the tip of the p arm are not inactivated. These include the gene responsible for the Xg blood type, one associated with the production of an enzyme required to prevent ichthyosis (an abnormal skin condition, and one associated with a cell surface antigen). The various observations indicate that at least some segment of both X chromosomes remains active in XX females, and the devices that keep large parts of one X chromosome permanently inactive nevertheless permit some genetic material to remain active in that same X chromosome.

Sex-influenced alleles

Many species are known in which the manner of expression of certain alleles is affected by the sex of the carrier. Such factors are not necessarily located on the sex chromosomes; most of them are actually autosomal. Among these are various alleles whose expression of dominance depends on the sex of the individual. For example, in certain breeds of spotted cattle, the colored regions on the body may be red or mahogany, a deep reddish brown. When a mahogany female is crossed with a red male (Fig. 6-26A), the results are red-spotted females and mahogany-spotted males. This crisscross pattern suggests a sex-linked recessive in which the female is double recessive and the male hemizygous.

If the red females are mated to their mahogany brothers, however, we see that this is not so. Among the male offspring, mahogany and red occur in a ratio of 3:1. The two phenotypes also show in the females, but the ratio is the reverse: three red to one mahogany. We can recognize the characteristic monohybrid ratio within the females as a group and within the males as a group. The results suggest that the allele for mahogany (M) is dominant in the male and recessive in the female, whereas its allele for red (m) is dominant in females and recessive in males (Fig. 6-26B). We could just as well have written M for red and m for mahogany. It makes no difference as long as we keep in mind the distinction in dominance between the sexes. Therefore the heterozygote Mm will be mahogany if male and red if female.

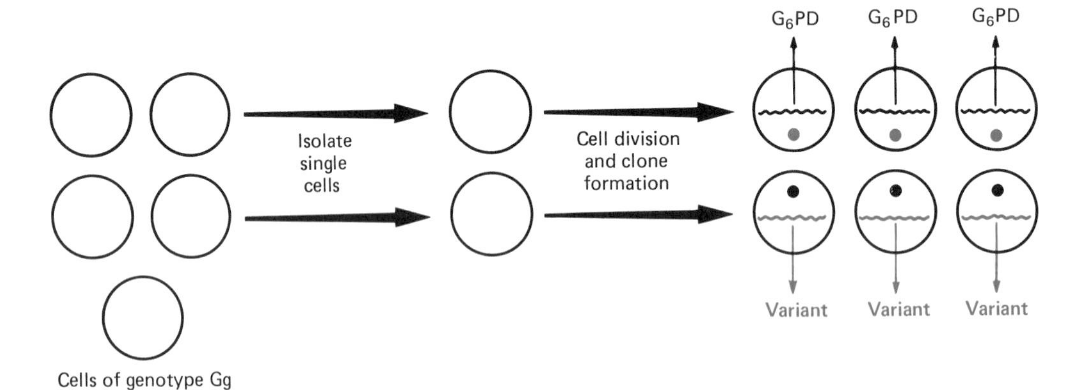

FIGURE 6-25. G6PD and skin cells. Single cells can be isolated from heterozygotes who carry the sex-linked recessive allele (g) for a variant form of G6PD. Single cell isolates can be allowed to form clones. Any one clone of cells is found to produce only the normal form of the enzyme or the variant, but not both. This indicates that the X that has the allele for the normal enzyme form (G) is active in cells of clones producing the normal enzyme and that the X with the allele for the variant form (g) is inactive. Conversely, a clone in which the variant allele is active contains only the variant enzyme form since the X with the standard form is inactive in these cells.

Alleles like these, whose degree of dominance is determined by the sex of the individual carrying them, are called *sex influenced.* Many other such alleles are known in animals. One is the presence or absence of horns in some breeds of sheep, in which the horned condition behaves as a dominant in males but as a recessive in females; the hornless state is dominant in the female sex but recessive in the male.

Although other genetic interpretations are possible, a trait that is of concern to many of us, pattern baldness, appears to be sex influenced. A pair of alleles seems to be involved, B for bald and b for nonbald. According to this interpretation, simple monohybrid inheritance operates, but the allele for baldness (B) behaves as a dominant in males and a recessive in females. Therefore a person of the genotype Bb is bald if male but nonbald if female. The alleles at the pattern baldness locus interact with other genes that affect the presence of hair on the head. The particular pattern in which the hair is lost, as well as the time of onset, entails the interaction of many genes with the environment. The internal or hormonal environment exerts effects as pronounced as those outside the body. A woman may become bald if she receives the bald-determining allele B from both parents. However, the onset of hair loss in a woman of genotype BB does not occur until late in life because of the effects of her hormonal environment.

The dominance difference typical of sex-influenced alleles is mainly the result of hormonal interaction with the genotype. The male hormone is responsible for the expression of the bald-determining allele B when it is present in just a single dose. If a male must take testosterone for some medical reason, the hormone level rises and can result in hair loss in a man who has not yet expressed the trait. Conversely, therapy may prescribe the taking of a female hormone by a man. Accompanying the consequent decrease in male hormone level, a balding man may start to resume hair growth along with an increase in breast size.

Sex-limited genes and the hormonal environment

The effects of the internal environment are most evident in the case of *sex-limited genes,* so-called because their expression is confined to just one sex. Primary examples of sex-limited genes are those genetic factors responsible for obvious secondary sex characteristics. Once the male and female gonads form, they produce their respective hormones. These in turn interact with the genotype, not only to influence the degree of dominance of certain alleles but also to suppress completely the expression of others. Genes for secondary sex characteristics such as breast development are inherited from both parents, male as well as female. The kind of beard a male develops is determined equally by contributions from both his mother and father. The genetic potential for beard development requires the presence of a certain level of male hormone to trigger it to expression. Normally, this potential lies dormant in women. However, a tumor of the adrenal, a gland that normally produces some male hormone in both sexes, can bring about an excess level of male hormone. In a woman with such a condition, the genes she carries for beard growth are called to expression, and a woman with a phenotypic feature typical of a male results. The taking of estrogen by a man can cause breast enlargement, a phenotype usually limited to females. However, it is not only the presence of a specific hormone that is required for the expression of sex-limited genes. Instead, the absence of a given hormone allows some of them to be expressed. Although the female hormone is needed for breast enlargement, it is not required for female voice. The latter develops in the absence of male hormone. A male who is deprived before puberty of his main source of male hormone as a result of castration will develop neither a beard nor a deep male voice since both features require male hormone.

A man who undergoes "sex change" surgery remains genetically male. Having only one X chromosome, such a person can never show a Barr body or a drumstick. Beard growth and the development of the Adam's apple have both been triggered by male hormone and can be altered following castration only by surgical and cosmetic means. Voice will remain essentially masculine, and breast development will require continued intake of estrogen. The female-to-male transformation also requires surgical procedures. Male hormones must be administered to achieve beard growth and the expression of other male secondary sex characteristics. However, the XX chromosome constitution remains unaltered and sex chromatin continues to be present in the body cells.

Some of the most striking examples of sexual dimorphism are found among various groups of birds. The male often appears so different from the

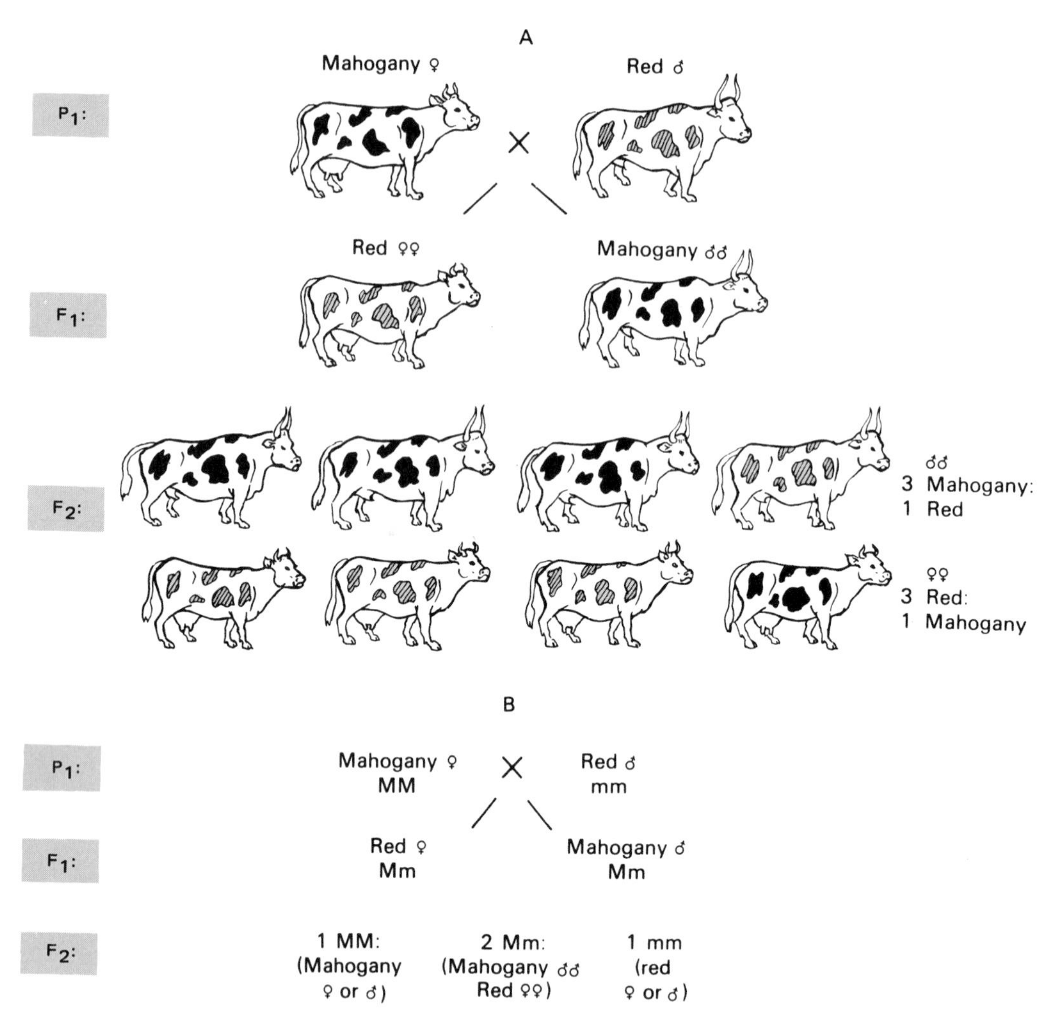

FIGURE 6-26. Sex-influenced inheritance in cattle. **A:** A cross of a mahogany-spotted female and a red-spotted male produces an F_1 of red-spotted females and mahogany-spotted males. The F_2 results indicate that the locus involved is not sex linked but that expression of the genes for coat color is influenced by an individual's sex. **B:** The factor for mahogany (M) behaves as a dominant in males and a recessive in females. Its allele for red color is dominant in females and recessive in males. Therefore heterozygotes (Mm) will differ in color depending on their sex.

female that the novice bird watcher may classify them in separate species. Sex-limited genes are the basis for these striking distinctions. The common fowl is a group that exhibits variation in its plumage pattern. The hen-feathered plumage is quite distinct from the cock-feathered phenotype. Both hen and cock feathering can occur in roosters, but only the hen-feathered phenotype is expressed in females (Fig. 6-27). The allele for hen feathering (H) acts as a dominant in males, whereas its allele for cock feathering (h) behaves as a recessive. But the allele h cannot be expressed at all in females, which are always hen feathered regardless of the genotype.

If the gonads are removed from a young bird, depriving it of either sex hormone, the genotypes HH and Hh are expressed phenotypically as cock feathering in both sexes! This indicates that the factor H prevents cock feathering if either the male or the female sex hormone is present. Its recessive allele, h, when homozygous, allows the cock-feathering pattern to develop, but the female hormone does not permit the expression of that phenotype. Therefore, two birds of the genotype hh may show different kinds of plumage. It is expressed phenotypically as cock feathering in the absence of female hormone but gives hen feathering in the presence of the female hormone. We see here clearly that

FIGURE 6-27. Sex-limited inheritance in the fowl. The dominant H causes the male to have hen feathering. Its recessive allele h is responsible for cock feathering in males. Females, however, will not express the allele h. They will be hen feathered regardless of the presence or absence of allele h.

the potential set by a normal genotype for the formation of a fertile individual showing typical secondary sex characteristics can be realized only in the setting of a conducive environment, internal as well as external. Sex is but one of the many phenotypic characteristics of an organism that depends for its normal expression on both the entire genotype and the background in which this genotype is allowed to express itself.

REFERENCES

Anderson, M., D.C. Page, and A. de la Chapelle. Chromosome Y–specific DNA is transferred to the short arm of X chromosome in XX males. *Science* 233:786, 1986.

Anneren, G.M., et al. An XXX male resulting from paternal X–Y interchange and maternal X–X nondisjunction. *Am. J. Hum. Genet.* 41:594, 1987.

Brownlee, G.G., and C. Rizza. Clotting factor VIII cloned. *Nature* 312:307, 1984.

Cervenka, J., D.E. Jacobson, and R.J. Gorlin. Fluorescing structures of human metaphase chromosomes. Detection of "Y" body. *Am. J. Hum. Genet.* 23:317, 1971.

Charlesworth, B. The evolution of sex chromosomes. *Science* 251:1030, 1991.

Cherfas, J. Sex and the single gene. *Science* 252:782, 1991.

Conley, M.E., et al. Expression of the gene defect in X-linked agammaglobulinemia. *N. Engl. J. Med.* 315:564, 1986.

de la Chapelle, A. Sex reversal: Genetic and molecular studies on 46, XX and 45, X males. *Cold Spring Harbor Symp. Quant. Biol.* 51:249, 1986.

Eicher, E.M., and L.L. Washburn. Genetic control of primary sex determination in mice. *Annu. Rev. Genet.* 20:327, 1986.

Epstein, C.J., S. Smith, B. Travis, and G. Tucker. Both X chromosomes function before visible X chromosome inactivation in female mouse embryos. *Nature* 274:500, 1978.

Federman, D.D. Mapping the X-chromosome: Mining its p's and q's. *N. Engl. J. Med.* 317:161, 1987.

Gartler, S.M., and A.D. Riggs. Mammalian X-chromosome inactivation. *Annu. Rev. Genet.* 17:155, 1983.

Goodfellow, P.J., S.M. Darling, N.S. Thomas, and P.N. Goodfellow. A pseudoautosomal gene in man. *Science* 234:740, 1986.

Goodfellow, P., S. Darling, and J. Wolfe. The human Y chromosome. *J. Med. Genet.* 22:329, 1985.

Gordon, J.W., and F.H. Ruddle. Mammalian gonadal determination and gametogenesis. *Science* 211:1265, 1981.

Grant, S.G., and V.M. Chapman. Mechanisms of X-chromosome regulation. *Annu. Rev. Genet.* 22:199, 1988.

Graves, J.A.M. The evolution of mammalian sex chromosomes and dosage compensation: Clues from marsupials and monotremes. *Trends Genet.* 3:252, 1987.

Haseltine, F.P., and S. Ohno. Mechanisms of gonadal differentiation. *Science* 211:1272, 1981.

Hoshijima, K., K. Inoue, I. Higuchi, H. Sakamoto, and Y. Shimura. Control of doublesex alternative splicing by transformer and transformer-2 in *Drosophila. Science* 252:833, 1991.

Koo, G.C. et al. H–Y antigen expression in human subjects with testicular feminization syndrome. *Science* 196:655, 1977.

Lawn, R.M., et al. Cloned factor VIII and the molecular genetics of hemophilia. *Cold Spring Harbor Symp. Quant. Biol.* 51:365, 1986.
Lawn, R.M., and G.A. Vehar. The molecular genetics of hemophilia. *Sci. Am.* (Mar.) : 48, 1986.
Lyon, M.F. Possible mechanism of X chromosome inactivation. *Nature* [*New Biol.*] 232:229, 1971.
Martin, G.R. X-chromosome inactivation in mammals. *Cell* 29:721, 1982.
McKusick, V.A. The royal hemophilia. *Sci. Am.* (Aug.):88, 1965.
McLaren, A. Sex determination in mammals. *Trends Genet.* 4:153, 1988.
Mittwoch, U. Males, females, and hermaphrodites. *Ann. Hum. Genet.* 50:103, 1986.
Mollon, J.D. Understanding colour vision. *Nature* 321:12, 1986.
Nathans, J. The genes for color vision. *Sci. Am.* (Feb.) : 42, 1989.
Ohno, S. Evolution of sex chromosomes in mammals. *Annu. Rev. Genet.* 4:495, 1969.
Page, D.C. Sex reversal: Deletion mapping of the male-determining function of the human Y chromosome. *Cold Spring Harbor Symp. Quant. Biol.* 51:229, 1986.
Page, D.C., L.G. Brown, and A. de la Chapelle. Exchange of terminal portions of X- and Y-chromosomal short arms in human XX males. *Nature* 328:437, 1987.
Roberts L. Zeroing in on the sex switch. *Science* 239:21, 1988.
Simpson, E., et al. Separation of the genetic loci for the H–Y antigen and for testis determination on human Y chromosome. *Nature* 326:876, 1987.

REVIEW QUESTIONS

1. In the human, the autosomally linked allele A is required for normal skin pigmentation. Its recessive allele is associated with the albino condition a. Several pairs of alleles affect color vision. One of the X-linked recessives responsible for red–green color blindness may be represented as p. Its dominant allele P is required for normal color vision. Write as much as possible of the genotypes of the following three persons:

A. A woman with normal vision and normal skin pigmentation whose father is a color-blind albino.
B. A normally pigmented man whose mother is color-blind and whose father is albino.
C. A man with normal vision and normal skin pigmentation whose father is a color-blind albino.

2. Using the information in Question 1, show the different kinds of gametes that can be formed by the following three persons:

A. An albino woman who has normal vision but carries the allele for color blindness.
B. A normally pigmented man who carries the allele for albinism and is color-blind.
C. A normally pigmented woman with normal color vision whose father was a color-blind albino.

3. The autosomal recessive (a) described in Question 1 is associated with oculocutaneous albinism, a condition in which pigment is absent from the skin and eyes. A sex-linked recessive (o) produces ocular albinism in which pigment is absent *only* from the eyes. The dominant (O) is required for pigment deposition in the eyes. For each of the following two matings, give the genotypes and the phenotypes to be expected among the children:

A. A woman with oculocutaneous albinism who is not carrying the sex-linked recessive and a man with ocular albinism who carries no recessive at the autosomal locus.
B. A woman with no pigment in her eyes but is homozygous for the dominant allele at the autosomal locus and a man who has the oculocutaneous albino condition but has the dominant sex-linked allele.

4. In the human, a certain rare sex-linked recessive, i, can result in a cleft in the iris of the eye. Its allele I is required for a normal iris. What would be the results of the following three crosses?

A. A man with cleft iris and a woman who carries only the normal allele.
B. A woman with cleft iris and a man with normal iris.
C. A woman who is heterozygous for the iris character and a man with cleft iris.

5. The dominant allele M is associated with a certain form of migraine headache. Its recessive allele m is required in homozygous condition for absence of headache. This pair of alleles is autosomal. Using this information and that given in Question 4 on cleft iris, consider the following: A woman who suffers from no apparent afflictions takes her daughter to a physician because the young girl is suffering from migraine. The doctor notices that the girl has a cleft iris. From just this information, write what the doctor knows about the genotypes of the girl, her mother, and the girl's father.

6. Using the information given in the questions on cleft iris and migraine, give the expected phenotypic ratio among the children from the following cross: mmIi × MmiY.

7. The rare autosomal recessive t results in total color blindness. The dominant allele T is required for development of rods and cones. Suppose a red–green color-blind woman (due to the sex-linked recessive p noted in Question 1) is homozygous for the dominant autosomal color-vision allele. Her husband is totally color-blind but carries the dominant sex-linked allele.

A. Diagram the cross and give the expected F_1.
B. Assume a woman and a man with genotypes like the above F_1 have children. What types are to be expected and in what proportion?

8. In the human, the sex-linked dominant allele R produces a type of rickets, a skeletal defect. The recessive allele r is associated with normal skeletal development. Give the results expected from the following crosses:

A. A man with rickets and a woman who does not have the condition.
B. A woman who has rickets but whose father did not and a man who does not have the condition.

9. In chickens, the Bar feather pattern is due to a sex-linked dominant allele, B, whereas its recessive allele, b, results in non-Bar. Since the female is the heterogametic sex in birds, what would be the results of the following two crosses? (Give the genotypes and phenotypes of the offspring.)

A. A Bar hen and a non-Bar rooster.
B. A non-Bar hen and a Bar rooster whose female parent was non-Bar.

10. A. In mice, as in all mammals, the male is the heterogametic sex. Assume that a sex-linked lethal trait is present in a strain of animals and that this causes the death of the late embryo. How would this affect the sex ratio?

B. Answer the same question if a sex-linked lethal were present in a strain of chickens.

11. In the human, a certain sex-linked recessive trait causes a child to exhibit very abnormal behavior in childhood. An assortment of bodily upsets occurs, and the affected child dies. Only males suffer from this genetic disorder; it has never been reported in a girl. Explain.

12. In cats, the factor B (for black fur) is codominant with its allele (b) for yellow fur. The heterozygote is tortoise. The pair of alleles is sex linked. In the clover butterfly, the autosomal factor W (white) is dominant over w (yellow), but the expression of the dominant is limited to females. From a distance, you observe a tortoise kitten and a yellow one playing with a white butterfly. What can you say regarding the sex and the genotypes of the cats and the insect?

13. A noncolor-blind woman is carrying the sex-linked recessive, p, noted in Question 1, which can result in a type of red–green color blindness. Assume that nondisjunction of the X chromosomes occurred at first meiosis in an oocyte of this woman. Give the kinds of offspring that could result if one or the other of the possible exceptional eggs is fertilized by sperm from a noncolor-blind man.

14. Give the possible results if normal eggs resulting from a noncolor-blind carrier female (Pp) are fertilized by unusual sperm resulting from nondisjunction of the X and Y at first meiosis in a noncolor-blind male.

15. An X-linked dominant allele, H, is required for the production of an enzyme, HGPRT, normally present in human cells. Its allele, h, results in lack of the enzyme and the elevation of uric acid in the blood. Suppose 100 fibroblast cells growing in culture are tested for this enzyme from persons of the following genotypes. In each case, how many cells on the average would be expected to show the presence of the enzyme?

A. HH
B. HY
C. Hh
D. hy
E. hh
F. HhY
G. HO
H. HYY

16. The multiple alleles w (for white eye), w^a (for apricot eye color), and w^{ch} (for cherry or light red) are found at the w (white) locus on the X chromosome of *Drosophila.* Heterozygotes for these alleles have intermediate eye colors, since the alleles in the series are incompletely dominant to each other. However, each allele is recessive to the wild-type allele (w^+) for red eye color. Give the results expected from the following crosses:

A. An apricot female × a white male.
B. An apricot female × a cherry male.
C. A cherry female × a red male.
D. F_1 females from the cross in C × an apricot male.

17. A certain bull is considered a prizewinner on the basis of his musculature and other fine anatomical points. However, when he is mated with cows the female offspring sired by him produce much less milk than their mothers. Explain.

18. Consider pattern baldness in the human to result from the expression of the autosomal factor (B) as a dominant in males and a recessive in females. Its allele, b (nonbald), behaves as a dominant only in females. What would be the results of the following matings:

A. A nonbald man and a nonbald woman who is heterozygous for the pair of alleles.
B. A nonbald man and a woman who became bald later in life.
C. A bald man whose father never became bald and a nonbald woman who is a homozygote.

19. Consider the alleles for baldness in Question 18 along with the sex-linked alleles that affect color vision, P and p (Question 1). Give the results of a mating between a nonbald color-blind man and a nonbald woman with normal vision whose father was color-blind and whose mother became bald.

20. The genetic factor (L) for long index finger behaves as a dominant in females, while its allele (l) for short index finger behaves as a dominant in males. This pair of alleles is autosomal. Give the genotypes of 1) a female with short index finger and 2) a male with long index finger. Give the genotypes and phenotypes of 3) their daughter and 4) their son.

21. According to one idea, the autosomal genetic factor, E, for early baldness is dominant to its allele, e, for no early loss of hair. The penetrance of the allele for early baldness is 100% in males but 0 in females. What would be the expected phenotypes among the offspring of a man and a woman, both with genotype Ee?

22. In each of the following cases, assume that the donor and recipient are compatible regarding most antigens that would affect the ability to accept a skin graft. Indicate whether a skin graft could be accepted (+) or rejected (–) on the basis of H–Y antigen or its equivalent.

	Donor	Recipient
A.	XY	XXY
B.	XXY	XX
C.	XXY	X0
D.	X0	XXY
E.	XXX	XYY
F.	XY female (SRY^-, xq intact)	XX female
G.	XX male (SRY^+, xq deleted)	XX female
H.	XY male	XY female (SRY^-, xq intact)
I.	XX female	XY female (SRY^-, xq intact)

23. Suppose a 9-year-old girl needs a kidney transplant and that it is not possible to check for tissue antigens. Using information in Chapters 5 and 6, rank the following individuals in order of preference as potential donors: older sister, sister who is fraternal twin to the girl, younger brother, unrelated 9-year-old girl, aunt, mother.

24. Analysis of a buccal smear from a patient reveals the presence of three Barr bodies. Which of the following is indicated? Select the correct answer(s):

A. The patient is probably suffering from Turner syndrome.
B. The patient is probably a gynandromorph.
C. The patient is probably an XXY Klinefelter male.
D. The patient will probably show three drumsticks in a blood smear.
E. The patient will probably show four X chromosomes in a karyotype analysis.

25. Deutan color blindness is associated with allele d, which results in defective visual pigment and is recessive to allele D for normal pigment. A third form of the gene, d^a, is also recessive to the normal allele and results in complete absence of visual pigment. Assume that alleles d and d^a are codominant with each other, and answer the following:

A. A woman with normal vision is a carrier of the allele for defective pigment. Her husband is color-blind, since he carries the allele for complete absence of visual pigment. Give the genotypes and phenotypes that are possible among their children.
B. A certain color-blind woman is found to carry both defective alleles for deutan color blindness. Her husband has normal vision. What genotypes and phenotypes are to be expected among their offspring?

Continuous Variation

7

Continuous versus discontinuous variation

The basic principles of genetics are based on traits that are very well defined. Mendel's pea plants were definitely either tall or short; flowers were either red or white. Morgan's fruit flies also possessed many clear-cut phenotypic features that could be easily followed from one generation to the next (red v. white eye color; short wing v. long wing). Characteristics such as these (height in peas; eye color in the fly, etc.) are said to show discontinuous variation, because contrasting traits can be recognized and easily assigned to distinct categories. In contrast are those characteristics that exhibit continuous variation, in which the phenotypic differences slowly intergrade. Height, weight, or length and breadth of an organ usually describe a continuous distribution from one extreme to the other. No sharp distinctions are found among the phenotypes to permit the recognition of clear-cut classes. Moreover, extra effort must be taken to control environmental effects before the genetic basis of such characteristics can be analyzed, and special methods are needed to describe the variation that is seen because the expression of the characteristic intergrades from one individual to another. This kind of variation is also referred to as *quantitative variation,* and the characteristics that display it are frequently called *quantitative characteristics.* These expressions recognize the fact that the variation shown can be measured and expressed in mathematical terms.

One of the first clear-cut demonstrations that a quantitative characteristic depends on the interaction of several genes involved color inheritance in wheat. Grain color intensity may differ among races and may vary from white through several intermediate shades to red. When individuals from a pure breeding red race are crossed with those from a white-grained one, the F_1 plants are intermediates. In the F_2, however, various shades occur among the grains, from full red to white. Among the F_2 individuals, 1 out of 64 is white grained and about 1 out of 64 is red, like the red and white of the P_1 parents. These results can be interpreted on the basis of three pairs of alleles that assort independently (Fig. 7-1). Each of the factors for redness, R_1, R_2, and R_3, contributes an equal amount of pigment to the phenotype. The effects are cumulative and can be added together. The alleles of the pigment genes, r_1, r_2, and r_3, contribute no pigment, and thus an individual homozygous for all three is white grained.

Other such examples of quantitative inheritance have formed the basis for the theory of multiple-factor inheritance. Actually, no new concept is introduced, because the multiple-factor idea explains continuous variation in terms of Mendelian genetics. However, characteristics that exhibit continuous variation typically involve the interaction of more than two pairs of alleles, as well as a pronounced effect by environmental influences. And the effects of the different alleles are quantitative; they may be added up, in contrast to pure dominance or lack of dominance. Multiple factor inheritance, also known as *quantitative* or *polygenic inheritance,* implies that several pairs of alleles are interacting and that each has a similar and measurable effect on a characteristic, which can also be influenced by the environment.

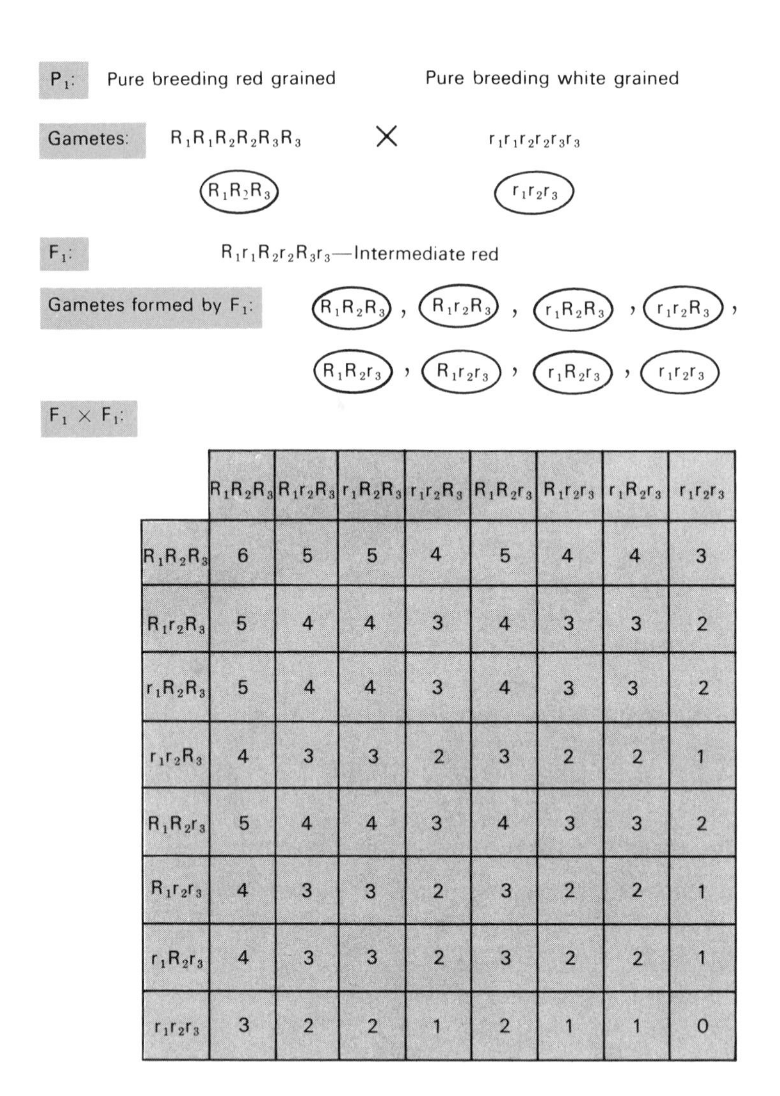

	$R_1R_2R_3$	$R_1r_2R_3$	$r_1R_2R_3$	$r_1r_2R_3$	$R_1R_2r_3$	$R_1r_2r_3$	$r_1R_2r_3$	$r_1r_2r_3$
$R_1R_2R_3$	6	5	5	4	5	4	4	3
$R_1r_2R_3$	5	4	4	3	4	3	3	2
$r_1R_2R_3$	5	4	4	3	4	3	3	2
$r_1r_2R_3$	4	3	3	2	3	2	2	1
$R_1R_2r_3$	5	4	4	3	4	3	3	2
$R_1r_2r_3$	4	3	3	2	3	2	2	1
$r_1R_2r_3$	4	3	3	2	3	2	2	1
$r_1r_2r_3$	3	2	2	1	2	1	1	0

FIGURE 7-1. Quantitative inheritance in wheat. Kernel color depends on three pairs of alleles. Each genetic factor that contributes to pigment formation (R_1, R_2, and R_3) adds an equal dosage. Their alleles (r_1, r_2, and r_3) contribute nothing to pigment formation. The trihybrid carries three alleles for pigment and produces kernels intermediate in color between the parents. The F_1s form eight classes of gametes. When these combine in all possible combinations, a range in shade is found. The Punnett square shows only the number of effective pigment alleles carried in the offspring. Only 1 out of 64 possesses six pigment factors, and only 1 out of 64 carries none at all. All of the other offspring vary in shade between the original (P_1) parents.

The example of grain color in wheat brings out several important points. Many nineteenth-century biologists might have thought that blending inheritance would explain the results. However, although the F_1 has only one phenotypic class (intermediate to the parents), the F_2 results clearly show that blending has not occurred because the extremes (white and red) again appear unchanged. The range of variation seen in the F_2 generation indicates that genes are segregating and that recombination is taking place. It is also evident that there is not one gene for one character. Grain color depends on the interaction of several genes and their cumulative effect. Moreover, in the F_2 environmental factors may operate to eliminate sharp distinctions among the major color groupings. The cross also reveals that two individuals of the same phenotype may have different genotypes. For example (Fig. 7-1), the genotypes $R_1R_1R_2R_2r_3r_3$ and $R_1r_1R_2r_2R_3R_3$ are in the same general color class because each has four pigment alleles, but the two are quite different genetically.

Quantitative inheritance and skin pigmentation

Not long after its formulation, the multiple-factor hypothesis was applied to skin color differences in humans. All normal individuals possess skin pigment, which is found in the melanocytes (pigment cells) of the live layer of the epidermis. The skin pigment is melanin, a dark brown product that combines with protein to form pigment granules in the melanocytes.

Melanin formation depends on the oxidation of the amino acid tyrosine, and the amount of melanin in the melanocytes is responsible for varying shades of skin color. A certain autosomal recessive allele may block melanin formation and result in albinism, a condition characterized by an abnormally low amount of pigment. This allele is not confined to any one group. Among humans, however, the normal range of skin shades varies from one population to the next and even within so-called black and white populations. In the earliest studies of inherited pigment differences in humans, the methods used to measure skin color were very limited. Nevertheless the observations could be explained on the basis of quantitative inheritance, in which approximately two pairs of independent alleles are involved. Recently, more accurate methods using spectrophotometry have been employed to estimate the degree of pigmentation. The equivalents of F_1 and F_2 testcross progeny between blacks and whites have been followed. Although still limited, the results suggest that three or four pairs of alleles are operative in skin color differences. Although the present interpretation may still be an oversimplification, there is no doubt that these color variations have a polygenic basis and that only a relatively small number of genes is directly involved.

For the sake of simplicity, we will assume that two pairs of alleles provide the basis for skin shades in humans (Fig. 7-2). An extreme white person would contain no effective pigment alleles above a certain basic level. The extreme dark parent would have four pigment alleles in our simplified illustration. The principles demonstrated by the red pigment of the wheat grains also apply here to melanin pigmentation of skin. As in the case of grain color, environment also influences gradations of shading in humans, so that no distinct steps exist from shade to shade. Again, individuals of the same phenotype may have very different genotypes. For example (Fig. 7-2), persons of the following three genotypes would have the same intermediate phenotype, ignoring the environment: 1) AaBb, 2) AAbb, and 3) aaBB. Certain consequences ensue from this. A cross of two persons of genotype 1, the dihydrid AaBb, will produce offspring that show a range of variation, indicated in Figure 7-2. On the other hand, a cross of a person of genotype 2 with one of genotype 3 (Fig. 7-3A) can give rise only to offspring of the same shade as the parents. So we would expect color variation to be great among the offspring in certain families in which the parents are intermediate in degree of pigmentation. In other families, little or no variation in shade would be expected from intermediate parents. All observations suggest that this is the case. It should also be evident that a mating between a person of any shade with a white individual cannot produce any children who are darker than the more pigmented parent (Fig. 7-3B).

Quantitative inheritance and deep-seated characteristics

Although skin and grain pigmentation represent rather superficial characteristics, a moment's reflection tells us that the most deep-seated characters in any species are ones that show continuous variation and are quantitative in nature—such features as height, weight, length, and breadth of limbs or organs. Although the details of the inheritance of such characteristics are unknown for most species, there is little doubt that they have a polygenic basis. Indeed, polygenic differences undoubtedly provide the distinctions between closely related species. An accumulation of many genetic differences that influence basic physiology may affect the ease with which members of separated groups can mate and thus lead to genetic isolation between them, even though both arose from one original population.

We still know little about the inheritance of such very complex characteristics as behavior, intelligence, and personality. We actually know much more concerning the genetic basis of the more superficial characters of living things than we do about those that are much more fundamental. The reason for this soon becomes obvious. Figure 7-1 shows a Punnett square for the superficial color characteristic in wheat grains in which only three pairs of alleles apparently play a major role. Yet segregation of just these few pairs produces an array of types. Only 1 individual out of 64 has the same genetic combination for pigmentation as does one of the parents, $R_1R_1R_2R_2R_3R_3$ or $r_1r_1r_2r_2r_3r_3$. More complex characteristics must involve many more genes than this. Table 7-1 shows

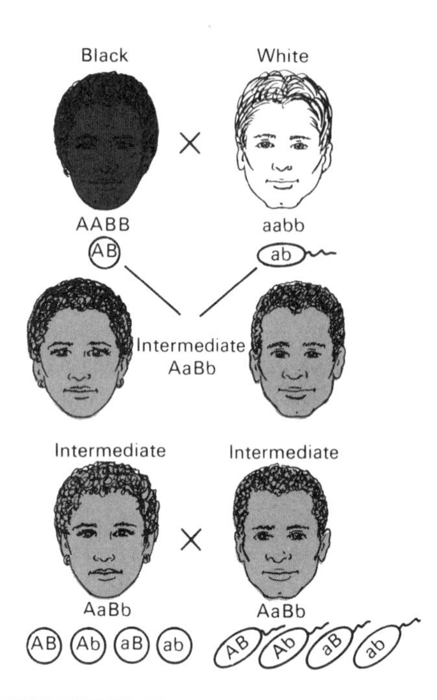

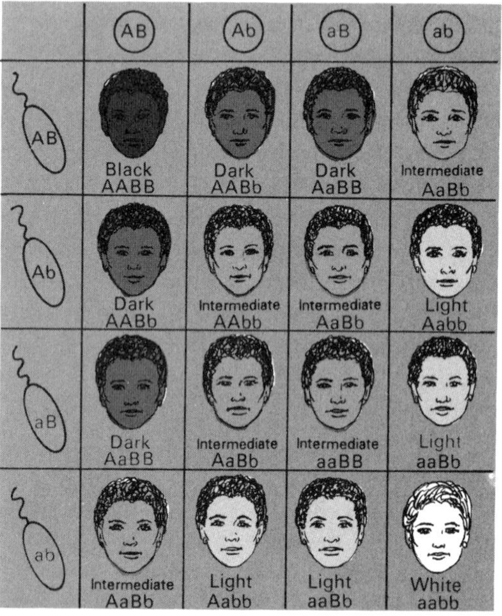

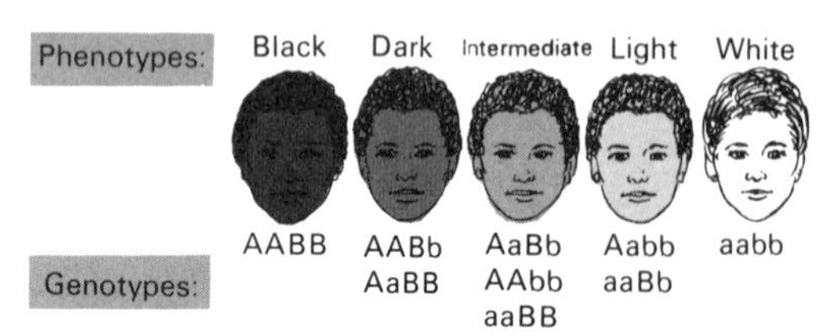

FIGURE 7-2. Inheritance of skin color. A mating between a black and a white person produces offspring intermediate in shade. Two parents of the genotype shown here, who are intermediates, will have offspring who vary in color from black, through various lighter shades, to white. (Probably three or four pairs of alleles are involved, but the principle is the same, as illustrated on the basis of two.)

that with just five genes (pairs of alleles) so many new combinations are possible that only a minute number (1/1,024) would resemble one of the original parents. This is so because many pairs of alleles are segregating and forming an assortment of new combinations. We would certainly expect the genetic basis of a very complex character like intelligence to involve more than 20 different allelic pairs. The number of combinations possible becomes staggering. And the expression of the many genotypes is modified further by the influence of the environment. Moreover, for the sake of simplicity, we have assumed that the effects of each pair of alleles are equal and additive. We now know that this is not so in most cases of polygenic inheritance. Some alleles contribute more of an effect than others to a quantitative character. We have also ignored the possible involvement of dominance and epistasis. It is no wonder that the genetic analysis of continuous variation has progressed more slowly than that for discontinuous variation. However, we do have approaches to the study of polygenic inheritance. Although certain assumptions must be made when applying them, they can still give some idea regarding the number of pairs of alleles involved and their quantitative contribution to the characteristic.

Estimation of allelic differences and their quantitative contribution

To approximate the number of pairs of alleles we turn our attention to the fraction of F_2 offspring that expresses a phenotype as extreme as one of the parents (Table 7-1). For example, suppose two races of tobacco plants are crossed (Fig. 7-4A). One race has long flowers in which measurements range from 88 to 100 mm, with a mean or average of 93 mm. The short flowers of the second race range from 34 to 42 mm, with a mean of 39 mm. The F_1 offspring are approximately intermediate in flower length, ranging from 55 to 70 mm, with a mean of 66 mm. When the F_1 are crossed, the 800 F_2 progeny show a much greater range of variation than did the F_1 (from 42 to 91 mm). Of the 800 F_2 offspring, only 12 were in the size range of the longer parent and 10 in the range of the shorter one. The data from the cross indicate several things and suggest polygenic inheritance. On such a basis, we would expect two pure-breeding races to produce an intermediate F_1 when they are crossed. The much greater variation among the F_2 than the F_1 offspring results from the segregation of several allelic pairs and the formation of many new genetic combinations. The 12 largest ones (12 out of 800, or approximately 1/66) and the 10 shortest (10 out of 800, or 1/80) suggest that three pairs of alleles are segregating, because three allelic pairs will produce a fraction of approxi-

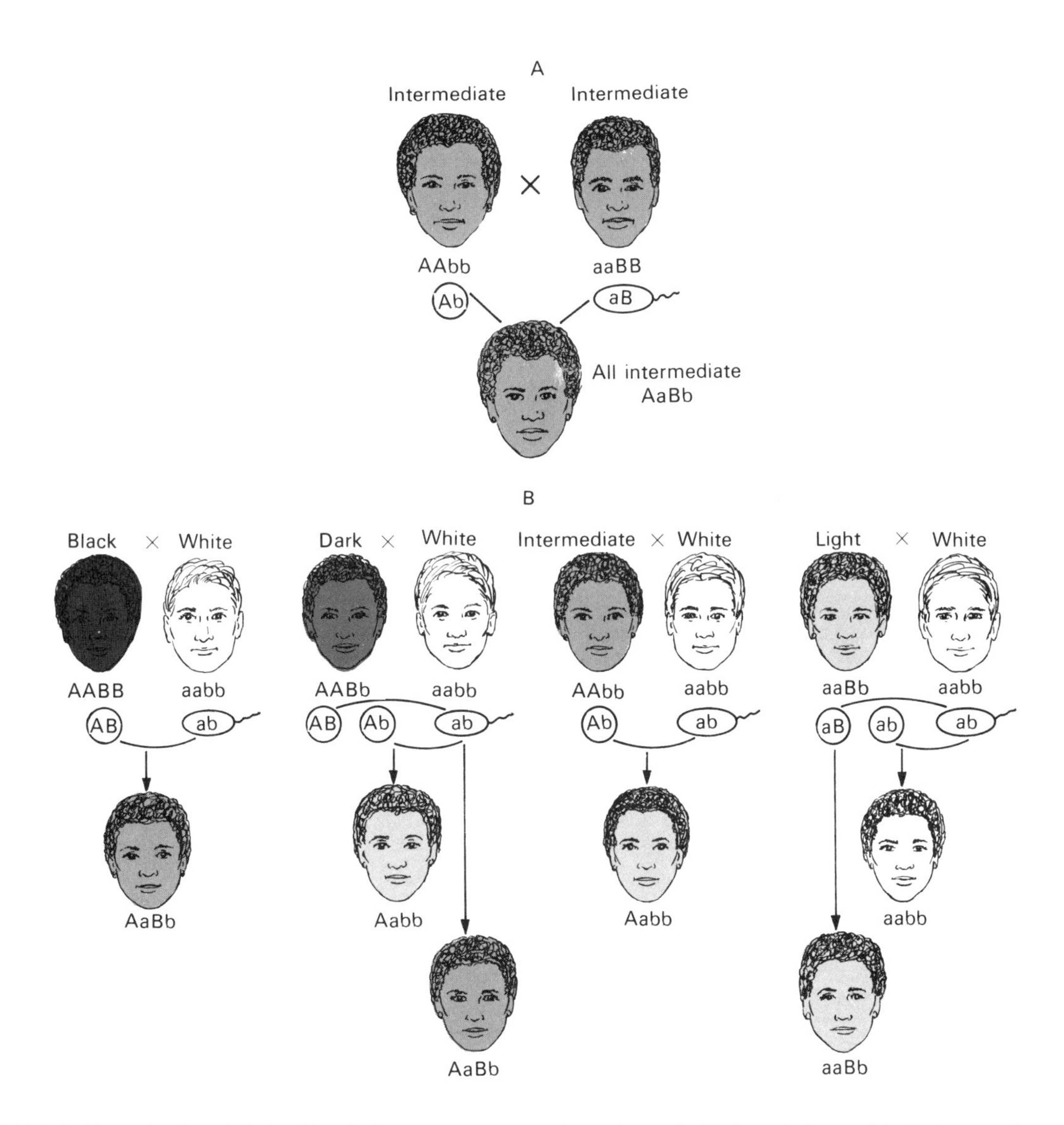

FIGURE 7-3. Skin color in certain families. **A:** Some parents who are intermediate in shade give rise to offspring all of whom are also intermediate. In such cases, each parent is homozygous for pigment alleles at one of the loci. In this example, they happen to be homozygous at different loci, but intermediate offspring would also result if the genotypes of both parents were either AAbb or aaBB. These results should be contrasted with those in Figure 7-2. **B:** A cross of a person of any shade with a white individual can never give an offspring darker than the nonwhite parent. This is so because the white contributes no effective pigment alleles that can add to those that may be donated by the darker parent. (Only some of the possible combinations are represented here, but the principle is the same for all of them.)

TABLE 7-1. Fraction of offspring showing one parental extreme[a]

Allelic pairs	Different kinds of gametes	Fraction showing one parental extreme
1	$2^1 = 2$	1/4
2	$2^2 = 4$	1/16
3	$2^3 = 8$	1/32
4	$2^4 = 16$	1/64
5	$2^5 = 32$	1/256
6	$2^6 = 64$	1/1,024
7	$2^7 = 128$	1/4,096
8	$2^8 = 256$	1/16,384
9	$2^9 = 512$	1/262,144
10	$2^{10} = 1024$	1/1,048,576

[a]This fraction indicates the number of allelic pairs involved in the inheritance of the characteristic under consideration.

mately 1/64 (Table 7-1), which is as extreme in phenotype as one of the parents.

We can now consider the small-flowered P_1 parent to have a genotype in which the minimal effect on flower length, an average of 39 mm, is being expressed by an unknown number of genes (Fig. 7-4B). The same minimal effect on length is also being exerted in the long-flowered parent. The length in millimeters above the minimum (the smaller average of 39 mm) represents the effective genetic contribution of other genetic factors. It is these that differ between the long- and the short-flowered races to cause a difference in flower length. So the lower average length (representing the minimal or basic length) is subtracted from the larger average. The 54 mm that remains is the effective difference between them. Since six genetic factors (two each of alleles A, B, and C) contributed to that amount, each allele donates an average of 9 mm to petal length. The F_1 received three alleles from the larger parent that contribute an effective difference. The small parent gave no alleles that add to the difference. Therefore the F_1 offspring would have a mean value approximately midway between the two parental means.

To study continuous variation in this way, environment must be controlled as closely as possible because the length of any organ can certainly be influenced by environmental factors. Two individuals of the identical genotype could have very different phenotypes simply as a result of these factors. To estimate with any degree of accuracy the number of effective alleles that are operating in a specific case, large numbers of individuals must be raised. When five or more segregating pairs of alleles are involved, it may be impossible in many cases to obtain one of the parental extremes. Moreover, in this example of flower length we assumed equal effects of each contributing allele, and we know this may not be so. Despite these limitations, however, some knowledge may be gained concerning the genetic basis for a character difference between two different stocks or varieties. It must also be emphasized that estimating the number of gene differences in this way does not mean that these are the only genes affecting the character. In our example, the smaller P_1 parent certainly has a flower length, and we noted that an unknown number of genes contribute to it, as it does for the same amount in the longer parent. The same holds in the case of pigment differences in humans. Normal whites, as noted, possess melanin pigment. Considering the average white genotype to be minimal, we try to estimate the number of effective allelic differences between white and black populations, just as we approximated that three genes are involved in producing the differences between the large- and small-flowered races.

Transgressive variation

Let us next examine a cross that produces results that may at first seem bizarre but can easily be explained on the basis of polygenic inheritance. Suppose that each of two inbred varieties of corn has a mean value of 64 inches in height. Suppose that the two varieties are crossed to obtain greater disease resistance and that height is not being considered. As it happens, the F_1 offspring also average approximately 64 inches in height (Fig. 7-5A). When the F_2 progeny arise, however, a spectrum of sizes is found. Of 2,000 F_2 plants, 7 are only about 32 inches in height, whereas 9 reach 96 inches. The unexpected results can be readily attributed to the segregation of several pairs of alleles in the F_1. Apparently, the P_1 plants, although homozygous for effective alleles that influence height, were not extremes. The possible extremes appeared in the F_2 after segregation and independent assortment in the F_1.

We can now estimate the number of effective pairs of alleles segregating for height, and we can approximate their contribution, again making the assumptions we did previously (Fig. 7-5B). Taking 32 inches as the minimal height resulting from factors common to both races, the effective genetic difference becomes

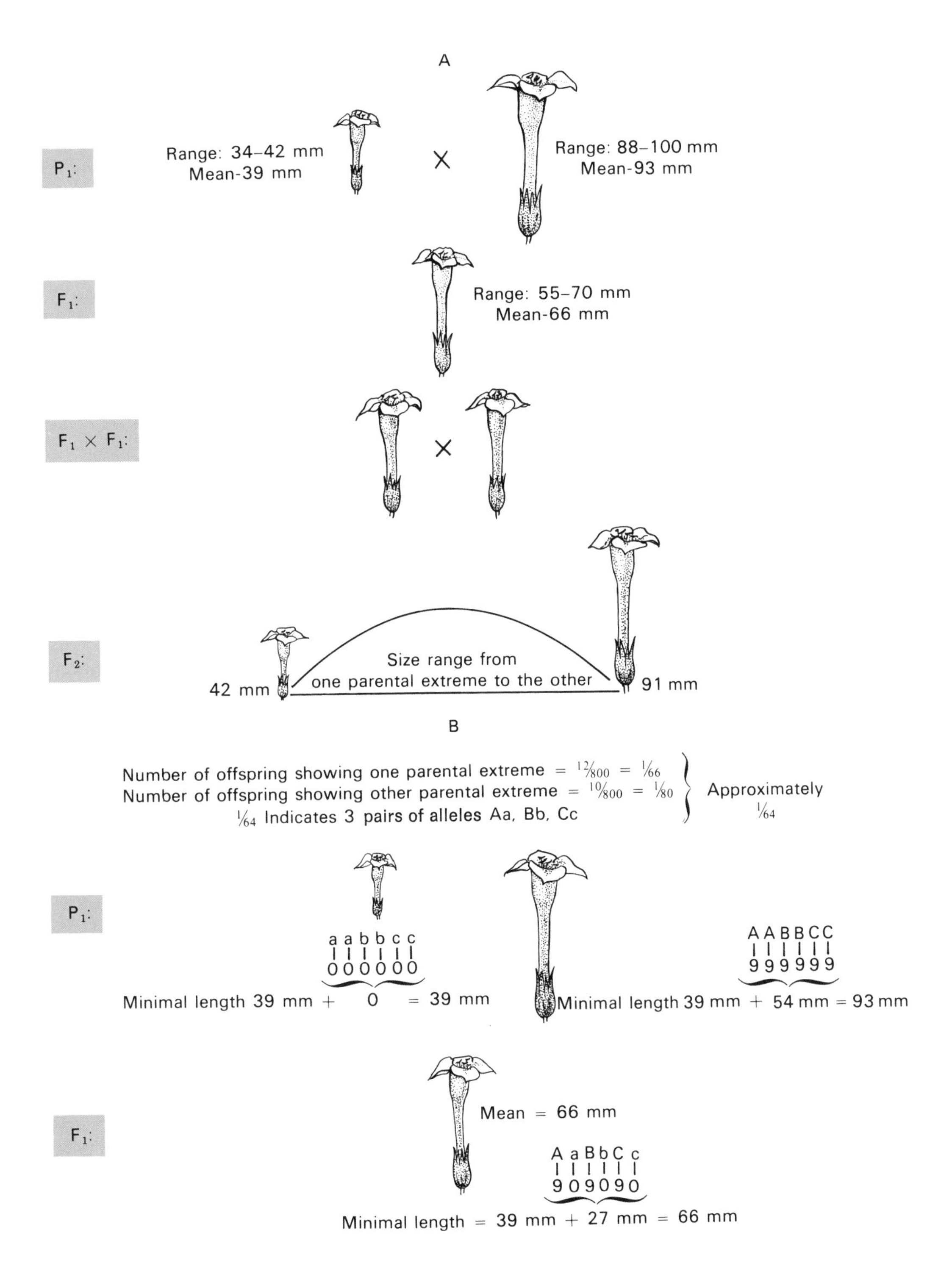

FIGURE 7-4. Quantitative inheritance in tobacco. **A:** When a short-flowered race is crossed with a long-flowered one, the F_1 has a mean about intermediate between the parents. A range in flower size is seen in the P_1 and the F_1. When two F_1 plants are crossed, the F_2 show a range in flower size that greatly exceeds the size range in millimeters shown in the F_1 or P_1. Some of the F_2 plants have flowers as short as those found in the short parental race and some as long as those in the longer parental race. Most of the F_2s have flowers that range in length between these extremes. The observations indicate that several pairs of alleles that contribute to flower length segregated in the F_1 to produce the variation seen in the F_2. **B:** Calculation of effective allele differences involved in the cross. The genotype of the shorter race can be represented as aabbcc and the longer as AABBCC, because three genes are indicated from the crossing data. The six factors in the longer parent contribute to the difference between the two mean lengths, a difference of 54 mm. Therefore each allele must add 9 mm.

96 – 32 inches = 64 inches (the average largest height minus the smallest). The number of F_2 plants showing parental extremes (9/2,000 and 7/2,000) is close to 1/256 and suggests that four pairs of alleles contribute to this difference of 64 inches, meaning that four genes are affecting height. This means that each allele in the largest individuals on the average contributes approximately 8 inches to height above the minimal of 32. There is actually no basic difference between this and the preceding examples. The only variation is that we did not start out with either possible extreme phenotype in the original parents. We see from this example that offspring can exceed the measurements of the P_1 parents. Such a distribution, in which the offspring exceed the parental measurements, is called *transgressive variation.* It can explain cases in which progeny may be much taller or heavier than their parents, both of which may have more modest measurements.

Studying behavior

Behavior is the most complex phenotype that can be studied in both human and other animals. Not only does behavior involve all aspects of the individual, it also varies greatly and is always changing since it involves a dynamic interaction with the environment. Most behaviors are not discontinuous traits but instead exhibit continuous variation. Most are not caused by a single gene but result from the interaction of many genes, each producing a small effect and interacting with the environment.

The human brain is the most complex of all organs, and its intricacies have enabled the human to overcome environmental challenges to a degree unequaled by any other animal. The human is far less restricted than any other species by problems posed by the environment, and this is largely the result of the superior mental abilities of *Homo sapiens.* The human not only copes with existing environmental problems but also changes the environment to suit immediate needs. In so doing, people influence the course of their own evolution as well as that of other species. Since human behavior has the potential to influence all life forms, any information that provides an insight into the nature of this most important characteristic has a decided value. A moment's consideration tells us that many of the most serious problems facing society today are behavioral. These include mental illness, mental retardation, alcoholism, and other substance abuse. Such a list underscores the need for studies into the complexities of human behavior.

In the same way that quantitative genetics applies to studies of such complex characteristics as height, weight, and form, it also appears to apply to animal behavior. Actually, the application of genetics to behavior is not a new approach. It has been carried on for centuries in the breeding of animals for their behavior as well as for their physical qualities. We need think only of the many breeds of dogs that differ in their behavior as well as in their appearance. Such differences seen in just a single species reflect the power of artificial selection over the course of centuries.

Controlled laboratory experiments also demonstrate the effects of selection on animal behavior and reveal the involvement of genetic factors on this complex characteristic. In one type of experiment, the behavior of mice was studied under conditions known to be extremely unpleasant to them, a brightly lighted environment. The animals were selected for their activity, high or low, in such an adverse setting. Over the course of 30 generations, highly active animals were selected from the population and bred together. The same was done for mice displaying low activity in the lighted environment. As a control, animals were randomly selected from the population and bred together without regard to their activities. At the completion of the study two lines of mice could be demonstrated in which the behaviors were distinct and showed no overlapping: a line displaying low activity and one with high. Similar results have been obtained for still other behaviors in mice such as amount of sleep time, aggressiveness, and various kinds of learning. Selective breeding experiments have also been applied to rats and even to the fruit fly and have been shown to influence behavior in these very different species.

Such studies demonstrate various important points regarding animal behavior. While they show that there is indeed a genetic influence on behavior, they also tell us something about the extent to which the genetic component operates. This brings us to the concept of a measure known as *heritability.* When we observe any deep-seated characteristic such as height, we notice that it varies from one individual to the next in the population. An important question that arises is, "How much of the variation shown in the expression of a complex characteristic is due to genetic factors?" Heritability is a measure of that percentage of the variation shown in the expression of a characteristic that is due solely to genetic factors. We can think of it as the degree to which a trait is genetically determined. If all the variation displayed can be shown to be due entirely to genetic factors, then the heritability for that feature

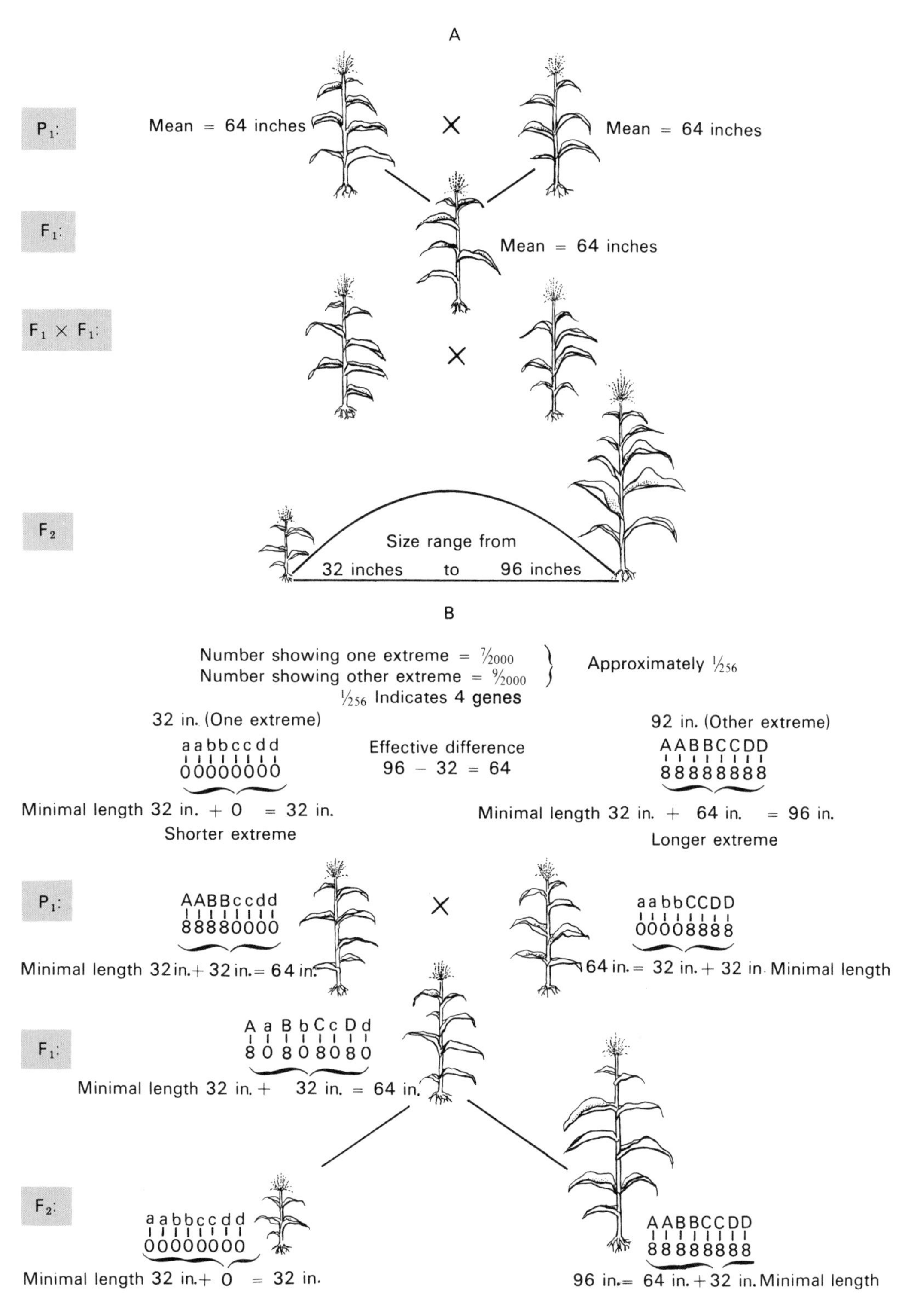

FIGURE 7-5. Transgressive variation in corn. **A:** In certain crosses, two plants of the same height yield an F_1 that is also similar in height to the parental strains. A cross of these F_1s, however, may yield an F_2 that varies greatly and in which some of the individuals exceed the measurements of the original parents. **B:** This transgressive variation results when the P_1s are themselves not the extreme types. The number of pairs of alleles involved and the effective contribution of each can be calculated (see the text for details).

would equal 1. If the heritability were 0.5, then half of the variation would be due to genetic influences and the other half to environmental factors. A heritability of 0 would tell us that all the observed variation in the characteristic is environmental. Estimates of heritability in most selection studies such as the one described above for the mice give values that are usually less than 50%. This implies that while genetic influences can account for major differences between the two lines of mice after several generations of selection, most of the variation in behavior seen in the population is not due to genetic factors.

The results of selection studies also tell us that many genes influence behavior. This is seen in the fact that even though intense selection continues for a particular type of behavior, that behavior reaches no upper limit. For example, selecting mice for higher and higher activity under adverse conditions can keep producing lines of animals with greater and greater activity, even though the low lines have reached the limit of minimal activity. If only a few genes were responsible for the genetic effects on the behavior under study, it would take only a few generations of selection to establish two set lines, high and low. Since selection results in increasing divergence between the lines, we must conclude that many genes are involved. Behavioral differences between inbred strains of animals that are raised under the same laboratory conditions have also been studied. Each inbred strain results from brother–sister matings for a minimum of 20 generations and must consist of individuals that are homozygous at most of their loci. When behavior differences between the inbred strains are compared, a significant genetic influence is revealed, but again those genetic factors do not explain most of the observed variance. Moreover, the studies also show that the behavioral differences are not due to a single gene.

The finding that many genes are involved in the genetic component that influences behavior does not mean that certain single genes are incapable of producing a pronounced effect on behavior. This is clearly seen in the case of an aberrant form of behavior in mice that has been well understood for many years. This is the waltzing trait, characterized by pointless, irregular twirling motions that leave an animal exhausted. When a cross is made between waltzers and mice from a wild stock, the F_1 generation exhibit normal behavior, but the F_2 produces normals and waltzers in a ratio of 3:1. Clearly, the trait is inherited as an autosomal recessive (Fig. 7-6). The bizarre behavior has been shown to result from a nervous disorder associated with deafness and abnormalities of the inner ear. In a waltzer, there is a lack of genetic information required for normal development of the inner ear and associated nervous tissue. The block to development manifests itself in the waltzing behavior. We will discuss later single genes that may play a significant role in the aberrant behavior of certain humans. However, as will become clearer in the discussions that follow, the evidence indicates that such genetic factors do not play a significant role in the ordinary or normal behavioral activity observed in a population.

Behavioral studies in humans

How can we proceed to study human behaviors, since we cannot use selection studies or inbred strains? Research relies heavily on family studies. Comparisons are made between parents and children, between sisters and brothers, and between adoptive individuals and other family members. Especially valuable are twin studies, particularly those dealing with identical twins. Identical twins are monozygotic (MZ) and therefore have identical genotypes. On the other hand, fraternal twins are dizygotic (DZ) like any other brothers and sisters and have on the average 50% of their genes in common. Since the genotypes of identical twins are the same, they provide a superb opportunity to help evaluate the relative roles of heredity and environment in the expression of a trait. A truly unique opportunity is provided by those very rare MZ twins who have been raised apart since early infancy or childhood. Since the environmental effects would be more varied on twins reared apart than on those raised together, the contribution of environmental factors to the expression of a trait can be revealed. A pair of identical twins thus makes it possible to study the expression of the same genotype in a similar environment (MZ twins reared together) and the expression of the same genotype in different environments (MZ twins reared apart).

Studies of DZ twins and ordinary brothers and sisters reared together and reared apart are compared with those of MZ twins. Two identical twins should always show the same trait if that trait is based entirely on hereditary factors. Blood type provides such an example. MZ twins always agree in their blood groups (ABO, Rh, MN, etc.). We say that the two members of a twin pair are concordant if they are alike in regard to a specific trait such as ABO blood type. They are also considered concordant if both do not exhibit a given trait (neither is an albino). On the other hand, members of a twin pair are said to be discordant if they differ with respect to a given trait. If one pair of identical twins in a study is discordant, the operation of environmental factors in the expression of the trait is indicated (one is an alcoholic whereas the other is a teetotaler). If both MZ and DZ twins are very similar in concordance value for a certain trait, then a strong environmental influence is indicated, since the members of each DZ twin pair have only about 50% of their genes in common. If the trait is based largely on genetic factors, the MZ twin pairs should have a much higher concordance value. We can therefore use the extent to which twins, identical and fraternal, differ in their concordance values as a sort of measure of the effects of the environmental and genetic components in the expression of a characteristic or a trait. From studies of this type, heritability values may be estimated. Suppose 50 pairs of MZ twins are studied in relation to a certain trait and that 40 of the pairs (80%) show concordance. A study of 50 pairs of DZ twins for the same trait shows that 10 pairs (20%) are concordant. This would provide evidence for a strong hereditary component. If both groups, MZ and DZ, showed about the same amount of concordance, for example 70% (or conversely the same amount of discordance), then a greater influence of the environment and lower heritability are indicated. (Such studies have indicated that in the case of height, heritability is in the vicinity of 90%.)

One of the most studied characteristics in the human is that of general knowledge or perception as measured by performance on an IQ test. Studies of identical and fraternal twins (10,000 pairs) have been carried out. They have included estimates of IQ correlation between twins reared together and apart and for parents and their adoptive children. The results are consistent for a heritability value of 0.50 for IQ. However, errors could be as high as 20%. The conclusion is that one can only say with confidence that heritability of IQ performance is between 30% and 50%.

In the human, studies provide no evidence for the effects of any major single gene on learning abilities. For example, no clear-cut evidence has been found for any single gene with a highly pronounced effect on reading ability. This does not mean that no genes exist that can significantly impair normal learning development in a person. Indeed, more than 100 are known that can lower IQ scores, but the alleles are rare in the population, where they account for only a minor amount of IQ variance. The fragile X syndrome, for example (see Chap. 20), is the source of some mental retardation in males. However, it cannot account for much of the IQ variance in the population as a whole, since it occurs in less than 1 in 1,000 males, and the intelligence of many persons with the fragile X is within the normal range.

Personality is another deep-seated human characteristic that has been examined in twin and adoptive studies. Aspects that have been investigated include neurotic behaviors and sociability. Again, there is no evidence for the role of any single major gene, and estimates of heritability give a value in the range of 30%–50%.

Behavioral disturbances in humans

One major area of behavior genetics is concerned with mental disorders such as schizophrenia and the depression syndromes. Studies were made of 18,000

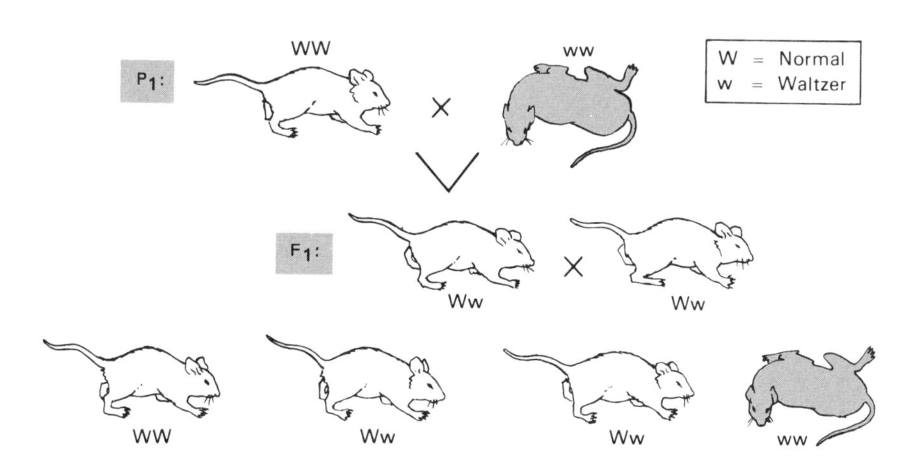

FIGURE 7-6. Waltzing behavior in mice. This aberrant waltzing trait is inherited as a simple Mendelian recessive.

relatives of schizophrenics. These were all persons who had about 50% of their genes in common with the disturbed individuals (sibs, parents). It was found that these relatives with a genetic relatedness of 0.5 to the schizophrenics showed a risk that was eight times greater for the disorder than was the case for others in the general population. Twin and adoption studies suggest strongly that familial schizophrenia is the result of hereditary factors rather than the effects of a common environment. A twin study of veterans of World War II entailed 164 pairs of identical twins and 268 pairs of fraternal ones. Concordance rates were found to be 30.9% and 6.5%, respectively. Adoption studies support the importance of the role of heredity in schizophrenia. However, they also show that nongenetic factors are critical as well. While a risk of 30% for the identical twin of a schizophrenic is much higher than the 1% risk for a person in the general population, it still indicates that there is no one gene or major hereditary component that will guarantee that the other twin will be affected as well. Nonhereditary factors play a decided role.

Manic depression was reported to be linked to a dominant allele on chromosome 11 in certain families. However, newer data have rendered this report inconclusive. There is no convincing evidence at the moment for the role of any one major gene in this syndrome. All the results of research on behavior, including animal studies and those concerned with mental illness in the human, suggest a genetic component that involves multiple genes rather than one or two major ones. They also indicate that nongenetic sources of variance are at least as important as are the genetic ones. Again, this does not mean that certain rare alleles with a pronounced effect on behavior are nonexistent, as the following examples will demonstrate.

We have discussed (see Chap. 4) the autosomal recessive responsible for phenylketonuria, a condition that may cause severe mental retardation. However, dietary restrictions can prevent the accumulation in the blood of substances toxic to the developing nervous system of a baby with the affliction. We see here the interaction of environmental and genetic components in the expression of a phenotype. An unfortunate behavioral outcome may be avoided if the proper environment (low levels of dietary phenylalanine) is present at a crucial time. The mental retardation cannot be prevented if the dietary regimen is initiated too late. Nor will higher levels of phenylalanine in the diet impair the mental functions of individuals homozygous for the allele who were given the proper diet in infancy. As in many cases of phenotypic expression of a genotype, behavioral or otherwise, the environmental influence at a specific period of development may be critical.

One of the most bizarre behavior disturbances in the human is associated with a single genetic defect, a sex-linked recessive. This is a type of cerebral palsy known as the Lesch-Nyhan syndrome in which afflicted children, all boys, develop a compulsion for self-mutilation. While seeming to wish protection against his own self-destructive behavior, a Lesch-Nyhan child will nevertheless bite his tongue, lips, and fingers. Such an unfortunate boy must be fitted with restraints to protect him from inflicting damage on his own body. He also demonstrates very aggressive behavior toward others and when upset will kick and bite those caring for him and additionally may vomit on them. The responsible sex-linked recessive has been shown to produce a molecular defect, the deficiency of a certain enzyme required for normal purine metabolism. Lacking the enzyme, a person accumulates products that produce a high metabolic rate. Lesch-Nyhan victims typically die before puberty, usually as a result of kidney damage and frequently after developing gout. Persons who are gout sufferers also have a biochemical upset similar to that found in the Lesch-Nyhan syndrome; however, they do not experience such nervous derangements. The syndrome is extremely relevant to studies of aggressive behavior and compulsions of various kinds, and a complete understanding of it could yield valuable information on behavior disturbances in general, the biochemical upsets that accompany them, and the genetic and environmental factors that may interact to trigger them.

There are other single genetic alterations that are associated with mental disturbances (fragile X syndrome, for example). There is also reason to consider the presence of an extra Y chromosome as a factor predisposing a male to antisocial behavior. Nevertheless, while it is possible for any one kind of genetic alteration to upset the development of normal behavior, the normal range of variation seen in a population is the result of the interaction of the environment and many genes, each with a small effect. A rare allele or chromosome alteration may disrupt normal behavior development in a few individuals and is more easily noticed than is one producing a small effect, but it is in reality only one factor in the total picture. Most alleles that influence behavior probably work in small ways, but any one of them does not show a pronounced effect on a few individuals. Research in the genetics of behavior makes it clear that nongenetic sources of variation are very important, since genetic variance rarely accounts for as much as half of the variance of behavioral traits. In other words, the heritability seldom exceeds 50%.

Species-specific behavior

Arguments have continued for years over the inheritance of "innate" behavior, that behavior so typical of a species that it may seem inborn and unalterable. Immediately after hatching, the duckling typically follows its mother faithfully; a female rat properly attends to her offspring after their birth. These animals seem "to know" to do such things without having been taught in any way. For a few specific behavior traits in some animals, the genetic basis has been determined. A good example occurs in honeybees.

Bees are particularly interesting in behavior studies because workers and queens, which are genetically similar, display very different types of behavior as a result of environmental influences operating during larval development. In addition, there are great differences in behavior among strains of bees. Some colonies exhibit "hygienic" behavior, which means that the workers remove dead larvae and pupae from the cells of the hive, an act requiring uncapping the cell and physically removing the dead remains. These hygienic colonies are also resistant to a bacterial infection, American foulbrood, and their resistance stems from the removal of the dead bees, which may harbor the responsible agent. Other bee colonies are "nonhygienic"; no attempt is made by the workers to remove the dead from their cells. Such colonies are thus sensitive to the bacterial disease.

A cross between hygienic and nonhygienic strains produces an F_1 in which the workers display nonhygienic behavior. It will be recalled that worker bees are sterile females; only the queen engages in sexual reproduction. The drones, or males, are all haploid and develop parthenogenetically from unfertilized eggs. Male offspring of the F_1 queen (whose workers are nonhygienic) can be mated to queens from other colonies, which are inbred or pure-breeding for either the hygienic or nonhygienic behavior. When these drones are mated to inbred hygienic queens (Fig. 7-7), the results are a 1:1:1:1 ratio of hygienic bees: nonhygienic bees:bees that just uncap the cell but leave the remains:bees that do not uncap cells but remove remains when the uncapping is performed for them. This typical testcross ratio clearly indicates a two-gene difference between hygienic and nonhygienic behavior.

The example of behavior in bees illustrates a very significant point. The behavior of the hygienic strain is of definite value to the colony. It is important to bear in mind that the general behavior that is typical of a species is a phenotypic characteristic that has undergone a long history of natural selection. Behavioral patterns have been selected for their survival value, just as any other part of the total phenotype.

The very fact that one can recognize behavior that is typical of a species indicates that there is some genetic basis for it. However, a characteristic behavior pattern is not simply encoded in a zygote's DNA, guaranteed to emerge intact at a given moment. Environmental influences, many of them subtle, must be considered before definitive statements can be made about innate behavior. A good example from the rat serves to illustrate the interplay of environmental stimuli with the genetic component in shaping a species-specific characteristic.

The female rat apparently does not need to learn to act as a mother. Normally, she cares for her offspring in her very first litter, even without prior experience or opportunity to observe another mother rat. On the surface, it appears that this maternal behavior is inherited as such and emerges at the proper moment. However, it can be demonstrated that a female rat will show hostility toward her offspring, avoid them or even kill them if she is deprived of certain stimuli. Normally, a rat licks and smells its own body as it develops into an adult. If a newborn female is fitted with a collar of sufficient size, she will be unable to examine herself in this way. When she bears her young, even after removal of the collar she will fail to display the species-specific maternal behavior. This example shows that a normal environmental stimulus may be something quite unsuspected. "Seeing" maternal behavior is not a necessary prerequisite here. Quite a different experience is required: familiarity of the female rat with her own body. Various other data gathered from animal studies caution against reaching premature conclusions about innate animal and human behavior.

Human babies, although quite helpless immediately after birth, can start to speak a language at a certain age, something no other animal can do, no matter how precocious it is shortly after its birth. Children have the potential to speak because the DNA transmitted to them has the encoded information required to equip them with a complex brain, a voice box, and other necessary anatomical structures. Animals without the set of genetic instructions needed to guide the orderly construction of the features essential to speech

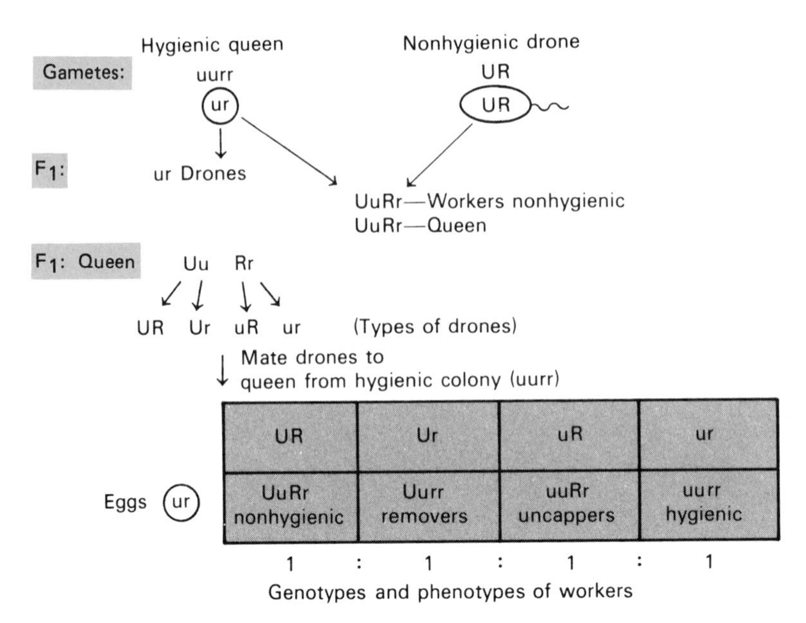

FIGURE 7-7. Inheritance of hygienic behavior in worker bees. Workers are sterile females, but certain of their behavioral characteristics can be studied genetically by crossing queens and drones of various stocks. The male, or drone, develops from an unfertilized egg of the queen. Thus males are haploid and in effect represent the kinds of gametes produced by the queen. The difference between hygienic and nonhygienic behavior in the workers has been shown, as follows, to be based on two pairs of Mendelian alleles. If a queen from a hygienic colony is mated to a drone from a colony that is nonhygienic, all the F_1 workers exhibit nonhygienic behavior. The F_1 drones are all haploid and reflect the genotype of the hygienic queen. The F_1 queen and workers are dihybrids, but only the workers exhibit the behavior characteristic under study. Because the workers are sterile, only the F_1 queen can be crossed. The dihybrid queen will produce four kinds of gametes. Thus four different kinds of drones represent these gametes. If these drones are mated to a queen from a hygienic colony, four different kinds of workers result, in the ratio 1:1:1:1, showing that a two-gene difference is involved. (If the drones from the F_1 queen are mated with a queen from a nonhygienic colony, all the workers will be nonhygienic, because the two dominant alleles, U and R, would be contributed by the nonhygienic queen.)

will never utter an intelligible word. Of course, there is no gene for "speaking" any more than there is a special gene for intelligence. Despite their structural equipment, children will begin to speak at a certain period in their development only if their environment surrounds them with the proper sound incentives. Whether they will speak English, Chinese, or some other tongue depends on the sounds they hear. Their genetic endowment empowers them with the remarkable ability to assemble separate sound symbols and arrange them in meaningful order. Without ever being taught the rules of grammar, they become able to understand the meaning of sentences and to express themselves by manipulating the symbols of speech.

The magnitude of this feat is appreciated when we note that those children surrounded by the unrelated sound stimuli of more than one language can, without difficulty, sort them out in their proper relationships and thus achieve expertise in more than one tongue. However, at about the age of sexual maturity this talent begins to fade, so that mastery of a language from that time on becomes an effort. It is the rare adult who achieves the perfection and facility with a newly studied tongue that a child acquires so effortlessly. Thus there is a critical period during development when environmental factors interact with those genetically determined. The former are the proper sounds, the latter the structures needed to receive, interpret, and associate symbols. Should a child be deprived of the proper stimuli at a critical developmental stage, he or she may never acquire the same language facility as one who was surrounded by the sounds that are meaningful in the cultural environment. Therefore group members who are isolated from infancy from appropriate language stimuli would be unable to communicate verbally in any known language. After a series of generations, we might expect descendants of this still isolated group to have developed some means of verbal communication that follows certain rules that give it a structure and grammar, for speaking a language is an aspect of behavior, a potential that is peculiar to the human zygote. Moreover, it is a phenotypic character with survival value. Speech enables the members of the group to transfer an ever-growing amount of cultural information from one generation to the next. This enables the members, building on information from the past, to continue to devise new ways of coping with different environments, which enables their species to survive under a range of diverse conditions. Natural selection has undoubtedly favored this aspect of human behavior, which helps set humans apart from other creatures.

Of course, a spoken language is by no means the only type of communication. An individual born with a defective vocal apparatus can still communicate by symbols, as can a person deaf from birth. The latter, although possessing perfect vocal cords, will never speak with the clarity of the more fortunate person who has been able to receive the appropriate sound stimuli during development. The complex interaction of hereditary and environmental elements is very evident here. Hereditary deafness can prevent the child from reacting with the appropriate environmental factors. However, if the infant is genetically normal otherwise, such a child will have the kind of nervous system that enables him or her to associate and manipulate nonverbal abstract symbols that can be substituted for the spoken word. On the other hand, deafness or faulty speech may be environmentally caused. The proper genetic information coded in DNA may be present in all the cells, but suppose a virus crosses the placenta at a critical point in embryonic development. Its interference with the formation of a normal anatomy can prevent the individual from ever speaking clearly, even though the person is later given the most conducive environment.

While a genetic component certainly exists for species-specific behavior, we must recognize the advantage the human has in the flexibility that the human genetic endowment permits in behavior. It would seem less wise, on the basis of highly debatable arguments, to assert that certain types of behavior are programmed or inherent than to appreciate the *range* of human responses made possible by the very plasticity that the genetic endowment allows. A fuller understanding of the subtle interplay between genetic factors and environmental influences in the establishment of human behavior can lead to a deeper appreciation of differences among groups and individuals. It may also offer clues to the alleviation of serious problems that can arise from certain forms of human behavior.

REFERENCES

Bouchard, T.J. Jr., D.T. Lykken, M. McGue, N.L. Segal, and A. Tellegen. Sources of human psychological differences: The Minnesota study of twins reared apart. *Science* 250:223, 1990.

Holden, C. The genetics of personality. *Science* 237:598, 1987.

Hunt, E. On the nature of intelligence. *Science* 219:141, 1983.

Kidd, K.K., and L.L. Cavalli-Sforza. Analysis of the genetics of schizophrenia. *Soc. Biol* 20:254, 1973.

Koshland, D.E., Jr. Nature, nurture, and behavior. *Science* 235:1445, 1987.

Kuehn, M.R., A. Bradley, E.J. Robertson, and M.J. Evans. A potential animal model for Lesch-Nyhan syndrome through introduction of HPRT mutations into mice. *Nature* 326:295, 1987.

Marx, J. Dissecting the complex diseases. *Science* 247:1540, 1990.

Plomin, R. The role of inheritance in behavior. *Science* 248:183, 1990.

Rothenbuhler, W.C., J. Kulincevic, and W.E. Kerr. Bee genetics. *Annu. Rev. Genet.* 2:413, 1968.

Sherrington, R., et al. Localization of a susceptibility locus for schizophrenia on chromosome 5. Nature 336:164, 1988.

Smith, J.M. The evolution of behavior. *Sci. Am.* (Sept.):176, 1978.

Stern, C. Model estimates of the number of gene pairs involved in pigmentation variability of the Negro-American. *Hum. Hered.* 20:165, 1970.

Winston, M.L., and C.D. Michener. Dual origin of highly social behavior among bees. *Proc. Natl. Acad. Sci. USA* 74:1135, 1977.

REVIEW QUESTIONS

In Questions 1 through 4, assume that two pairs of alleles, Aa and Bb, form the basis of skin pigmentation differences in humans and are responsible for the recognition of five classes: black, dark, intermediate, light, and white.

1. Give the possible genotypes of each of the following:

A. A dark-skinned person.
B. An intermediate who had a white parent.
C. A white who had one intermediate parent.

2. A white person and one who is intermediate in skin color produce nine children, all of whom are light skinned. What are likely genotypes of the two parents?

3. Give the phenotypes and genotypes possible from a cross of an intermediate person who is heterozygous at both loci and a white person.

4. Two persons of intermediate skin color produce seven children, all of whom are about the same in skin color as the parents. However, two other persons of intermediate skin color produce five children who vary greatly from one another, ranging from white through dark. Give the most likely genotypes of the two sets of parents.

5. Assume that in a particular variety of wheat the color of the grains varies from deep red through intermediate shades to white. Six pairs of alleles are involved: R_1r_1, R_2r_2, and so on. Each pigment-contributing allele, R_1, R_2, and so on adds an equal dosage, whereas the alleles r_1, r_2, and so on contribute nothing to pigmentation. Plants from a pure-breeding strain that has deep red grains are crossed to a strain with white grains. The F_1s are intermediate in color.

   A. When F_1s are crossed to white-grained plants, how many shades of grain are possible among the offspring?
   B. When the F_1s are crossed among themselves, how many of the offspring can be expected to be white grained and how many deep red?

6. In one variety of rabbit the ear length of the animals averages about 4 in. In a second variety, the average is 2 in. The ear length of hybrid animals obtained after crossing the two varieties averages 3 in. The hybrids, when crossed among themselves, produce offspring whose ear lengths vary much more than that of the F_1 hybrid parents. Of 496 F_2 animals, two have ears that measure 4 inches and two have ears about 2 inches long. How many pairs of alleles would you say seem to be involved? How much does each effective allele contribute to length of ear?

7. In one variety of oats, the weight yield of grain per plant is 10 g, whereas in another variety it is only 4 g. Hybrids between the varieties give an average yield of about 7 g per plant. When the hybrids are crossed to each other, the yield varies greatly. About 4 plants in 253 yield 10 g each and about the same number yield about 4 g each. How many pairs of alleles appear to be involved in weight yield per plant? What contribution does each effective allele make to weight yield?

8. A race of inbred plants has a mean petal length of 12 mm. A second race from another location has the same mean petal length. When the two races are crossed, the F_1s also have a mean petal length of 12 mm. However, the F_2 generation derived from crossing the F_1s to one another shows a very wide spread in petal length. About 3 out of 770 have a length as small as 8 mm, and about 3 out of 770 have a length as long as 16 mm. How many allelic pairs seem to be involved in the difference in petal length? How much does each effective allele contribute to length of the petal?

9. Assuming that height difference in the human depends on four pairs of alleles and that environmental factors are constant, explain why two parents of average height produce children much taller than they.

10. **A.** Diagram a cross between a queen bee from a nonhygienic hive and a drone from a hygienic colony. Give the genotypes of the F_1 queen and the workers. What is the workers' behavior?

    B. What are the genotypes of the drones produced by the P_1 queen?
    C. What are the genotypes of the drones produced by the F_1 queen?
    D. The drones from the F_1 queen are mated to a queen from a nonhygienic colony. Give the genotypes and the behavior of the resulting workers.

Select the correct answer or answers, if any, in the following:

11. Species-specific behavior in higher animals:

    A. Is a phenotypic characteristic.
    B. Has evolved under the force of natural selection.
    C. Depends on environmental stimuli for its normal development.
    D. Requires that the individual be taught by observing the typical behavior.
    E. Is inborn in those cases in which the animal displays the typical behavior pattern without ever having seen it.

12. All mammals possess the equivalent of the HLA complex found in humans (see Fig. 5-7). This genetic region includes several very closely linked genes, each of which exists in many allelic forms. The extreme variability at this region accounts for the fact that in mammals graft rejection occurs when tissue is transplanted from one unrelated individual to another. In the case of the cheetah, any animal will reject a skin graft from the domestic cat. However, any cheetah will tolerate a skin graft from any other cheetah, even though the animals may be from different areas. Explain.

13. Suppose that a study of MZ and DZ twins give the concordance values shown below for each of the different traits under study. Evaluate each as to the importance of hereditary or environmental factors in the expression of each trait.

	MZ	DZ
1. Susceptibility to a certain virus	90%	78%
2. Development of a certain bone disorder	88%	20%
3. Development of a certain tumor	27%	3%

Genes on the Same Chromosome

8

Independent assortment versus complete linkage

Shortly after Morgan undertook his classic studies with *Drosophila,* he showed that two or more pairs of alleles, and thus two or more genes, can be associated with the same chromosome. Let us examine certain crosses in the fruit fly to illustrate the nature of linkage and its genetic implications. Suppose a cross is made between two mutant stocks: "eyeless" (reduced eye size) and "black" (black body instead of gray). Both of the mutant alleles are inherited as autosomal recessives. The F_1 flies are found to have the wild phenotype, large eyes, and gray body (Fig. 8-1). A testcross of these F_1s produces the typical dihybrid testcross ratio 1:1:1:1, expected on the basis of independent assortment. The testcross reveals the kinds of gametes produced as well as the proportion in which they were formed. The results of this cross tell us that the eyeless locus and the one for black are not in any way linked or tied to each other but are free to segregate independently at gamete formation.

Let us next follow the allele "black" (b) along with another recessive, "purple" (p, for dark eye color as opposed to red). Treating the cross between black-bodied and purple-eyed flies exactly as in the preceding case, we might expect again all wild flies in the F_1 and a testcross ratio of 1:1:1:1. After such a cross, the F_1 flies are all wild as expected, but a testcross of F_1 males with double recessive females (purple-eyed, black-bodied) gives quite different results (Fig. 8-2). Instead of four kinds of offspring in equal proportions, we find only two types in a ratio of 1:1, purple and black, which are identical to the original P_1 parental types. Independent assortment has not occurred, and the reason is that the loci p and b are not found on different chromosomes but happen to be located on chromosome II and are thus linked. As Figure 8-2 shows, a cross must be diagrammed a bit differently when linkage is involved. It is essential to indicate which specific allelic combination is present on a given chromosome, because those alleles tend to stay together and do not assort independently.

The cross shows that each P_1 parent has, on each chromosome, a recessive in combination with a dominant and that the combination stays together throughout the testcross. The bar or line represents two chromosomes. The combination of alleles above the line tends to remain together, as does the combination below it. We see that pb^+ and p^+b entered the cross together through the P_1 parents. The combination did not come apart but remained together among the testcross progeny. The testcross clearly tells us that the two genes are completely linked, because only the old P_1 combinations are formed (purple eyes with wild body color and red eyes with black body).

Now let us make the cross in a slightly different way. In Figure 8-3, the P_1 parents are purple, black (double recessives), and homozygous red eyed, gray bodied (wild eye, wild body). Again we see that independent assortment does not occur. The combination of alleles in the original P_1 parents stayed together and remained completely linked through the testcross. A very significant difference between this mating and the previous one (Fig. 8-2) is seen in the combination of the alleles. Note that in both cases the F_1 hybrids are

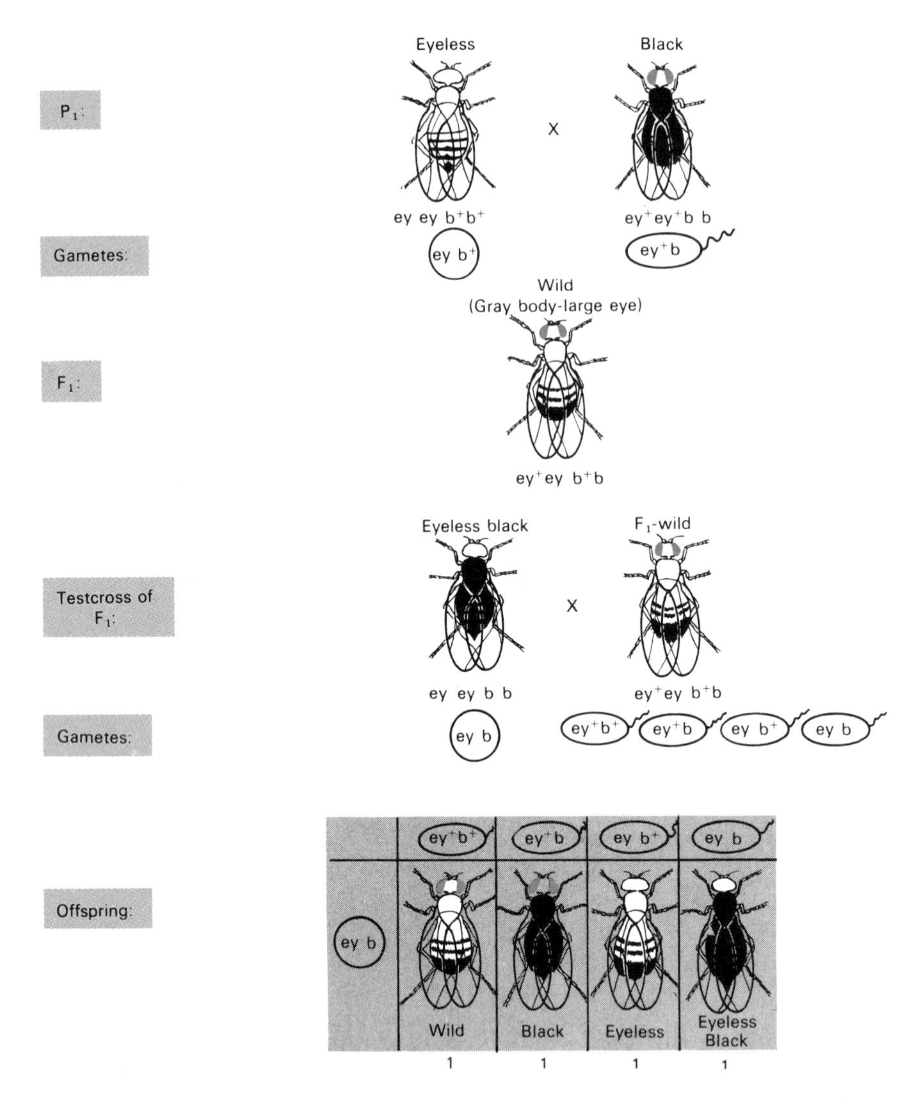

FIGURE 8-1. Dihybrid testcross. The eyeless and the black traits are inherited as autosomal recessives. When the F_1 dihybrid is testcrossed to an eyeless black fly, the offspring occur in the expected ratio of 1:1:1:1.

wild in phenotype, but the allelic arrangement is very different. In the first case (Fig. 8-2), one chromosome has the combination of a recessive with a wild (pb^+). The homologous chromosome has the reciprocal of this, wild with recessive (p^+b). This arrangement in the F_1 dihybrid, p^+b/pb^+, is called the *trans* or *repulsion arrangement* and results from a cross between two individuals carrying different recessives on the same chromosome. This trans arrangement is quite different from the *cis* or *coupling arrangement* shown in Figure 8-3. In this case, the dihybrid contains both recessives on one chromosome and both dominants on the other: p^+b^+/pb. This results from the cross of any individual that is double recessive at two linked loci with an individual homozygous for the corresponding dominant alleles.

Figures 8-2 and 8-3 clearly show that although both dihybrids, cis and trans, are identical phenotypically, they represent very different arrangements of alleles and produce dissimilar results in the testcross. In each case, however, the testcross offspring are like those of the P_1 parents in phenotype. At times it may be very important to know whether two dihybrids showing the same phenotype have the same allelic arrangement, cis or trans, because the two types differ greatly when they are bred. The testcross is the most direct way to

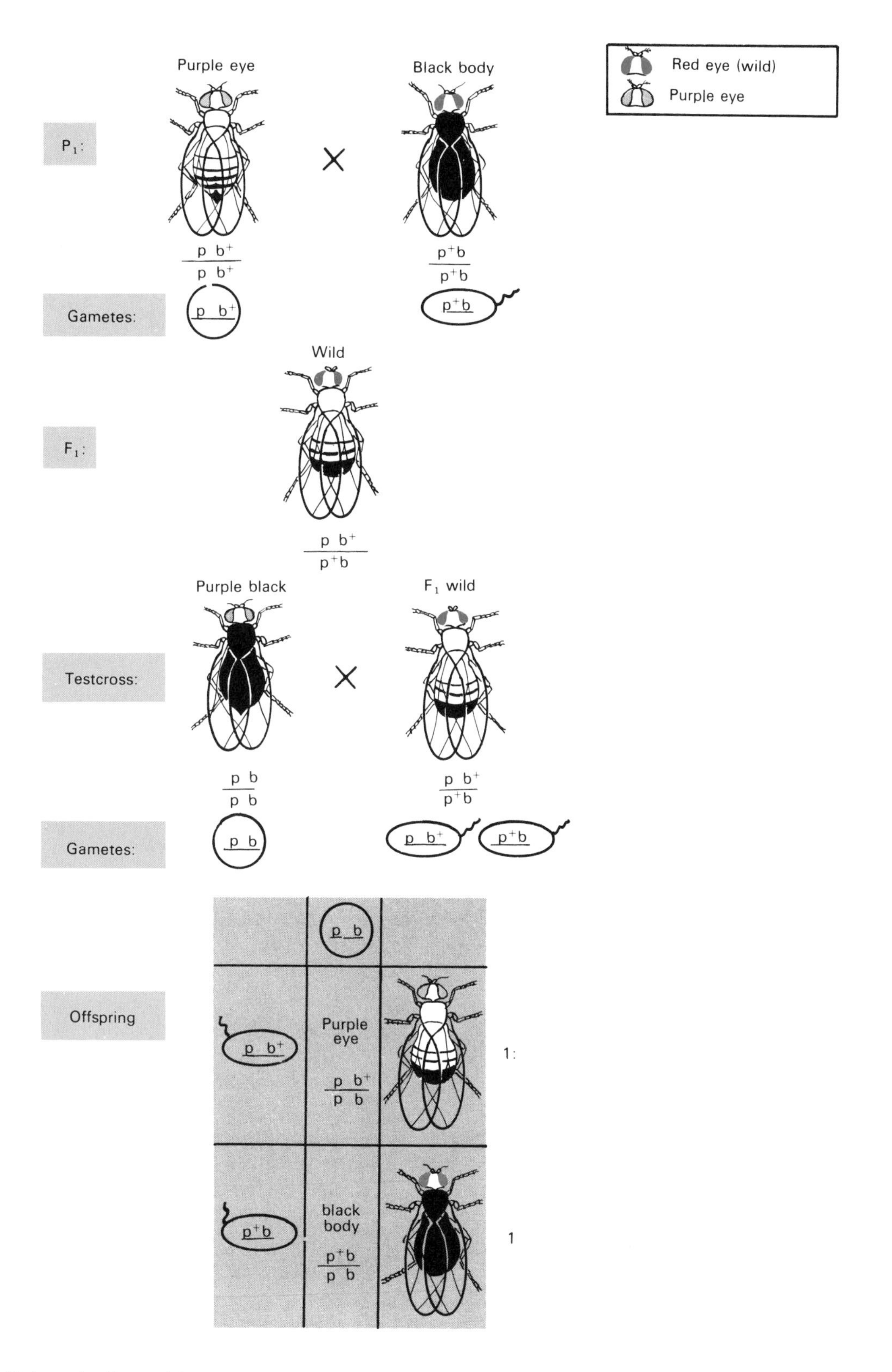

FIGURE 8-2. Linkage in *Drosophila.* The gene for eye color and the gene for body color are on the same autosome. Therefore the combination of alleles found in the P_1 parents tends to remain together. When a wild dihybrid male is testcrossed, only two types of offspring result, and these are identical to the P_1 parents. The F_1 male forms only two classes of gametes because the two genes are completely linked.

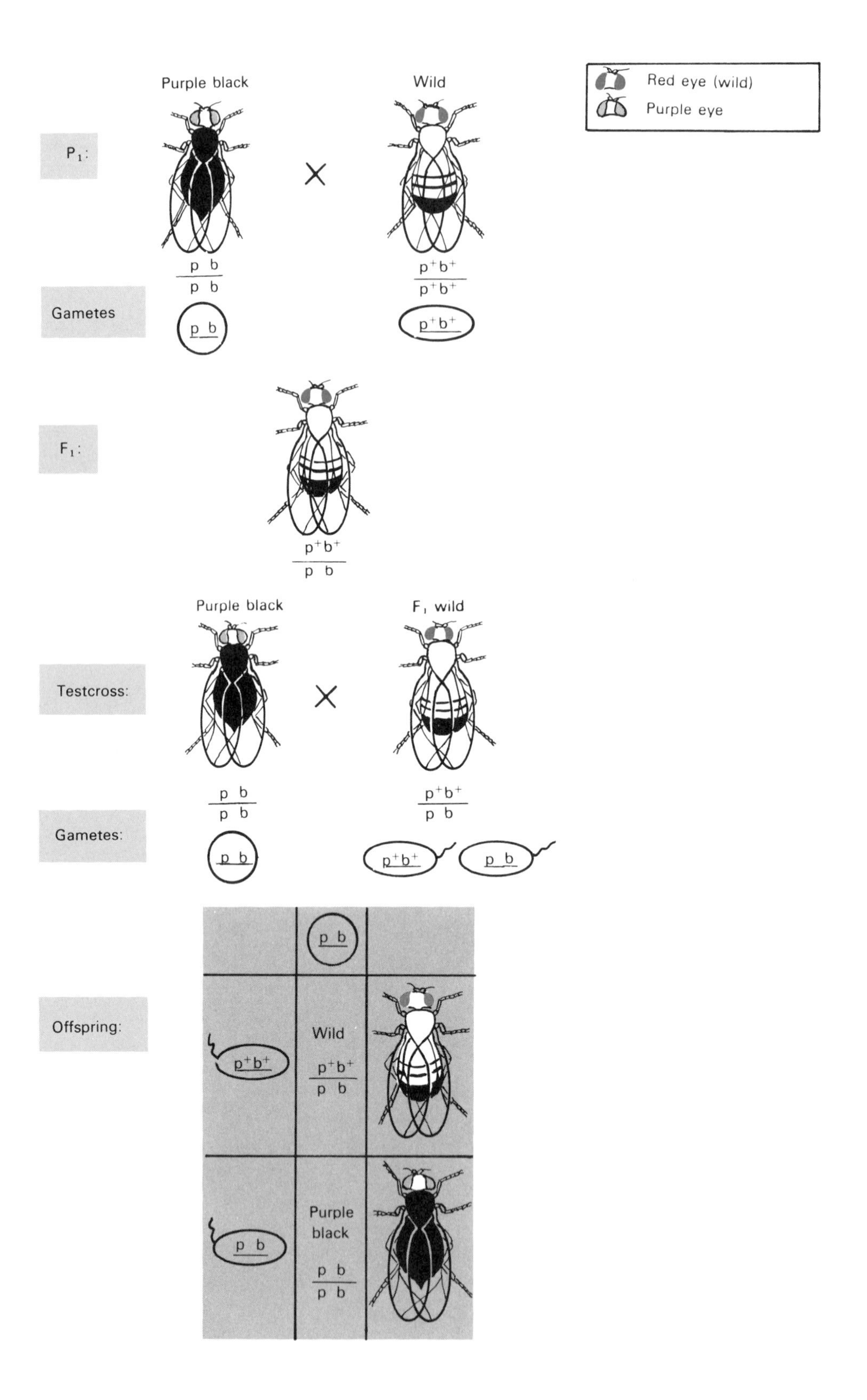
Red eye (wild)
Purple eye
P1:
Purple black
Wild
p b / p b
p+b+ / p+b+
Gametes
p b
p+b+
F1:
p+b+ / p b
Purple black
F1 wild
Testcross:
p b / p b
p+b+ / p b
Gametes:
p b
p+b+
p b
Offspring:
p b
p+b+
Wild
p+b+ / p b
p b
Purple black
p b / p b

determine which of the two arrangements is present in a dihybrid.

Recombination and its calculation

In both of the testcrosses we have been discussing (Figs. 8-2 and 8-3), dihybrid males were selected and crossed with double recessive females. We now return to these crosses, but this time F_1 dihybrid females are mated with double recessive males. As Figures 8-4 and 8-5 show, testcrosses of the F_1 females give not just two types of offspring, as in the case of the F_1 males, but four kinds! This means that four different kinds of gametes were formed by the F_1 females; however, the numbers show that independent assortment did not occur. It is the old (P_1 or parental) combinations that are the most common. The new combinations, the recombinants, are in the minority. Thus, although there was a definite tendency for the parental allelic combinations to remain intact, a reassortment did occur in a smaller percentage of the gametes. These new combinations of linked genes result from crossing over, an event that entails a reciprocal exchange between two homologous chromosome segments at early meiotic prophase (see Fig. 3-5). In this example, the percentage of recombination or the percentage of crossover gametes is 6%, and consequently the strength of linkage is 94%. The recombinant gametes are responsible for the recombinant phenotypes seen in the testcross.

Crossing over does not occur at all in the male fruit fly, in which the strength of linkage is normally 100%. But it must be kept in mind that in most species crossing over is typically characteristic of meiosis in both sexes. The absence of crossing over in the male fruit fly is an exception. This peculiarity, however, is a valuable one when working with *Drosophila,* for it permits the detection of linkage with relative ease. In most organisms, however, crossing over, and the recombination it generates, is a normal and integral feature of gamete formation in both the male and the female.

FIGURE 8-3. Linkage in *Drosophila.* The principle in this cross is identical to that shown in Figure 8-2. However, the F_1 dihybrid males in the two crosses must be compared. They are both wild in phenotype, but their allelic arrangements are different. In this case, the alleles are in the cis arrangement as opposed to the trans in the other cross. The testcrosses give very different results, but in each case the testcross offspring are identical phenotypically to the P_1 parents. No new combinations of alleles are formed.

Figures 8-4 and 8-5 also contrast the results of testcrossing F_1 females with the alleles in the trans arrangement and F_1 females with the alleles in the cis arrangement. Note that the old or parental combinations among the offspring of the trans females [(pb^+/pb) and (p^+b/pb)] (Fig. 8-4) are the new combinations among the offspring of the cis females (Fig. 8-5). Likewise, the old combinations occurring among the cis females [(p^+b^+/pb) and (pb/pb)] are the recombinants or new combinations among the progeny of the trans females. This is exactly what is expected, since the gene affecting body color and that affecting eye color are linked together on the same chromosome. They tend to remain together in the original P_1 arrangement and thus result in a larger number of parental types among the testcross offspring.

The following example illustrates how the amount of recombination between two linked alleles can be easily calculated from an inspection of raw data obtained in a crossing experiment. A recessive allele of another gene is known to be linked to the one responsible for black body in *Drosophila.* Its effect is to produce a bent wing in the homozygous recessive condition. The allelic pair is "a^+" (for normal wing) and "a" for arc (bent wing). A cross between arc females and black males produces wild-type progeny (Fig. 8-6). Note that they have the alleles in the trans arrangement. When the F_1 females are testcrossed, four phenotypic classes can be recognized. To determine the amount of recombination, the data are inspected to distinguish the parental types from the recombinants, those with new allelic arrangements. The latter are quickly recognized by their phenotypes (wild-type flies and those that are both black and arc) and are the less numerous types. The amount of recombination is estimated simply by calculating the percentage of these crossover offspring among the total number of progeny produced (1,020/2,392 = 0.426 = 42.6%). This example may be used as a guide to estimate the crossover value between any two linked autosomal genes.

In the fruit fly, however, we must remember that the F_1 males will give a very different result in a testcross because of the absence of crossing over (Fig. 8-7). Because linkage strength is 100%, the genes a for wing shape and b for body color will not separate. Therefore half of the flies will show the arc phenotype, and the remaining half will be black. These phenotypes will be about equally distributed among the progeny. It should be apparent that the F_1 *Drosophila* males cannot be used in a testcross to estimate the crossover amount,

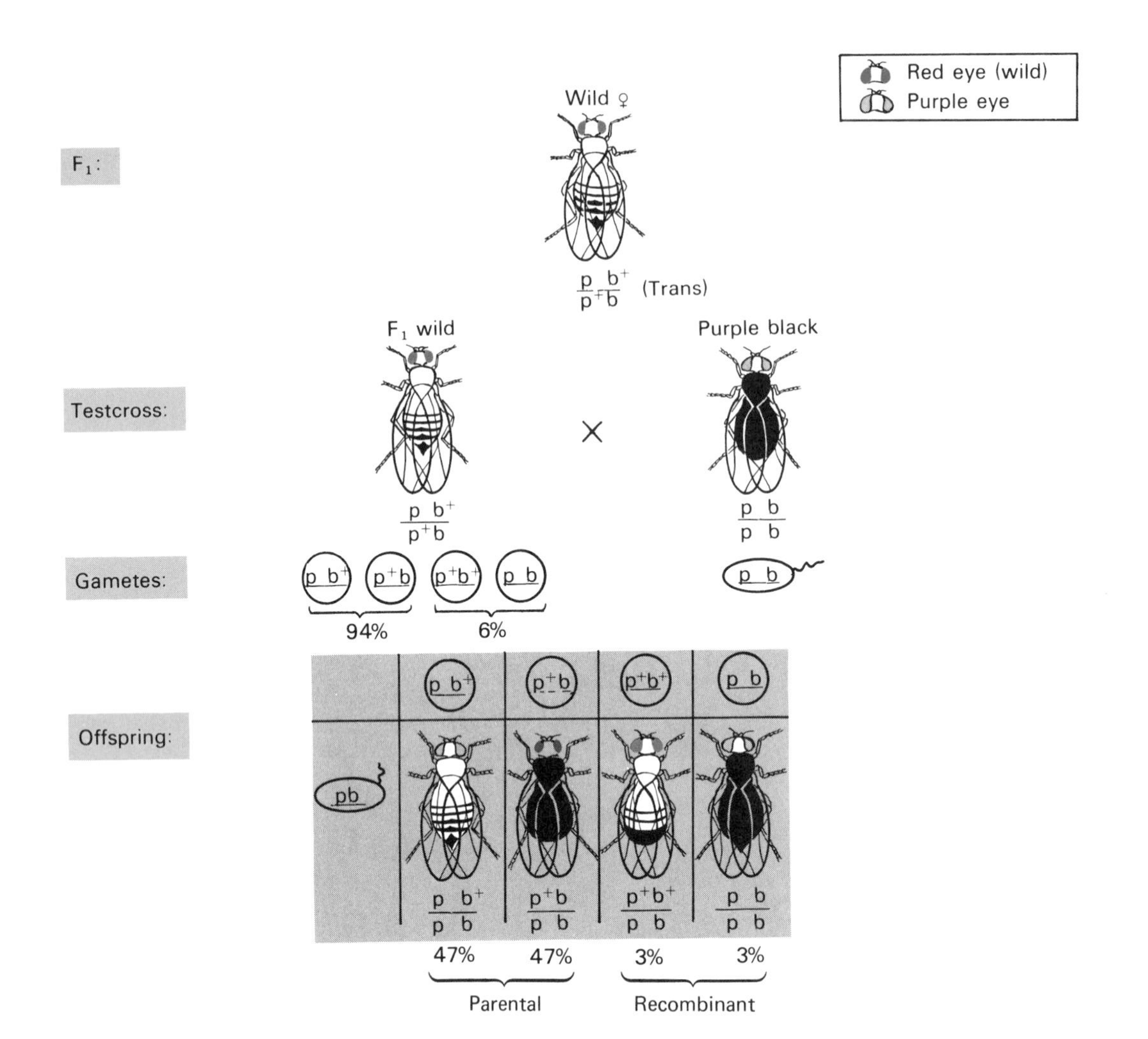

FIGURE 8-4. Testcross of F_1 female, trans arrangement. When a dihybrid female is testcrossed, the linked genes separate or engage in crossing over. The old combinations are more common than the new ones. Four types of testcross progeny result, but they occur in unequal proportions, which reflect the amount of crossover gametes (6% here). This testcross contrasts with the testcross of the F_1 male shown in Figure 8-2.

because a ratio of 1:1 will always result regardless of which genes are followed. As mentioned, this absence of crossing over in the male fruit fly is very valuable in showing that two genes are linked and is one of the many advantages of *Drosophila* as a tool in pioneer genetic studies.

Linkage and the F_2 generation

The discussion thus far has stressed the use of the testcross in the detection of linkage and the amount of recombination. Its advantage is clearly seen when we consider a cross that follows two linked genes through the F_2 generation. Suppose we are working with corn plants and suspect that two particular loci are linked. The alleles of one of the genes affect the leaves (cr^+, the dominant for normal, and cr, its recessive allele for crinkly leaves); the alleles of the other influence height (d^+ for normal and d for dwarf). Assume that the actual amount of recombination that can take place between these genes is 20% but that we are unaware of this at the outset. Instead of making a testcross, we decide to follow the plants through the F_2 generation and to estimate the crossover percentage from the F_2 data. Figure 8-8A represents the cross of a crinkly plant with a dwarf. Since we are not dealing with *Drosophila,* each parent (instead of just the female) produces crossover gametes as well as the parental types. When combining the gametes in all possible combinations, we may use the familiar Punnett square method shown in Figure 8-8A, but we must remember that the gametes are not being formed in equal numbers, as is true in independent assortment. Therefore, we must indicate the frequency of each type of gamete. Each frequency in turn indicates the probability of the occurrence of that kind of gamete. When we combine two gametes, we must therefore multiply the two separate probabilities to indicate the chance of the two coming together at any one fertilization. This is no different from what we have been doing all along in problems dealing with independent assortment, in which the frequencies of each type of gamete have been equal.

We can see from the diagram of the cross (Fig. 8-8A) that on the basis of 100%, the four different phenotypes would occur in a frequency of 24:24:51:1. If we depended on such an F_2 ratio to tell us the amount of recombination, we could often be misled by fluctuations due to chance or to such factors as lethals and genic interactions. But even at its simplest, the F_2 ratio would usually require intricate calculations to estimate

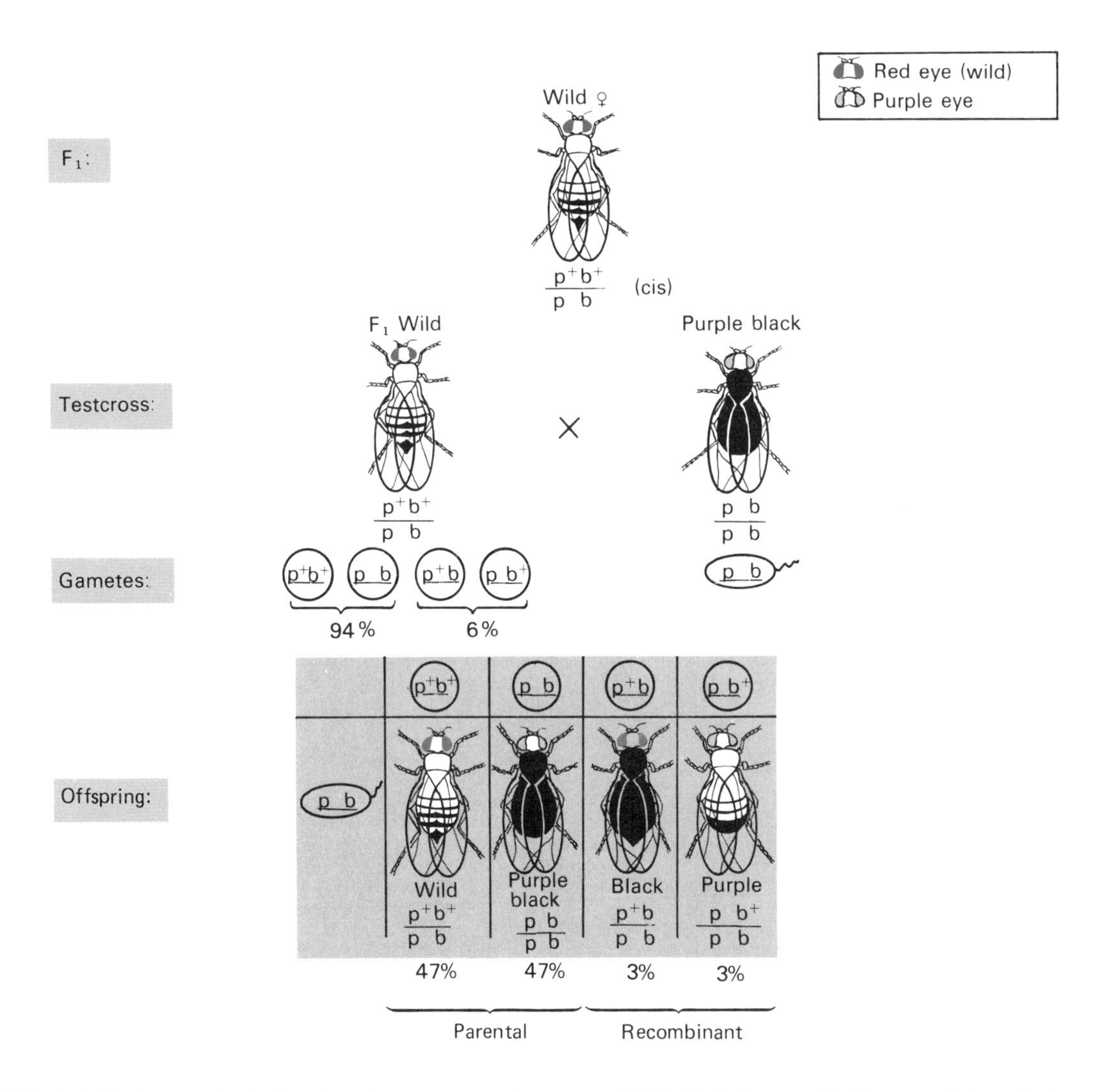

FIGURE 8-5. Testcross of F_1 female, cis arrangement. Because crossing over occurs, four types of gametes are formed, but the old (parental) combinations are the most common. These results are just the reverse of those shown in Figure 8-4, but in both cases it is the parental types that are more common. This cross should be contrasted with the one in Figure 8-3, in which the corresponding F_1 male is testcrossed.

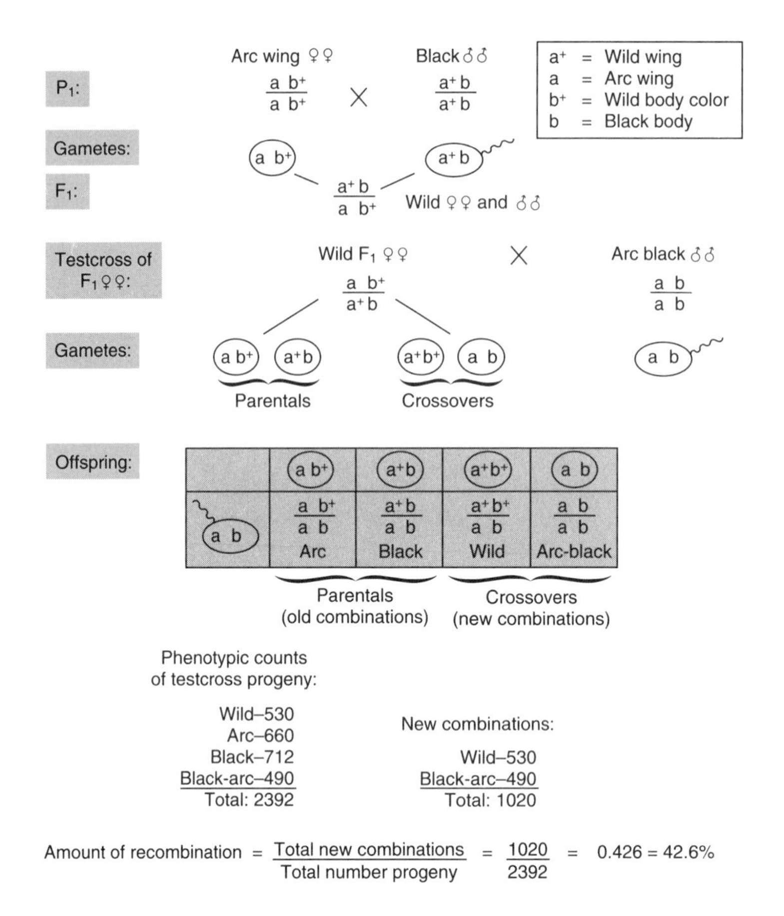

FIGURE 8-6. Calculation of the amount of recombination. To calculate the amount of recombination between two autosomally linked genes, the F_1 dihybrid is testcrossed. The crossover types and the parentals are easily distinguished phenotypically. Crossover percentage is calculated by dividing the total number of testcross offspring into the number of new combinations.

the percentage recombination. Obviously, the testcross is the more direct route (Fig. 8-18B).

Genes linked on the X chromosome

There is a qualification to the statement that using the F_2 is not the best way to estimate the amount of recombination. This pertains to F_2 ratios obtained from genes linked together on the X chromosome. A cross between two stocks, mutant for different sex-linked alleles, produces an F_2 generation that is particularly valuable because it can be used to determine the amount of recombination. The F_2 generation in a sex-linked cross may be used for this purpose, because at least half of the F_2 in a sex-linked cross are always testcross progeny. This statement can be appreciated if we recall that sex-linked genes are carried on that part of the X that has no homologous region on the Y chromosome. Consequently, all XY males must express the X-linked alleles and thus will be testcross offspring for any sex-linked genes. This is so, because the Y does not mask any alleles on the X contributed by the female parent. We may, in this sense, consider the Y chromosome devoid of genes being followed on the X.

Figure 8-9 illustrates linkage between two recessives on the X of *Drosophila:* w (white eyes) and ct (cut wings). Note that the alleles in the F_1 female are in the trans arrangement. When allowed to mate with the F_1 males, these females produce crossover as well as parental gametes. The males, on the other hand, contribute only a Y chromosome and an X with the parental combination. The figure shows that the F_2 female progeny fall into only two phenotypic classes, wild and white. However, four different classes can be recognized among the F_2 males, because their fathers

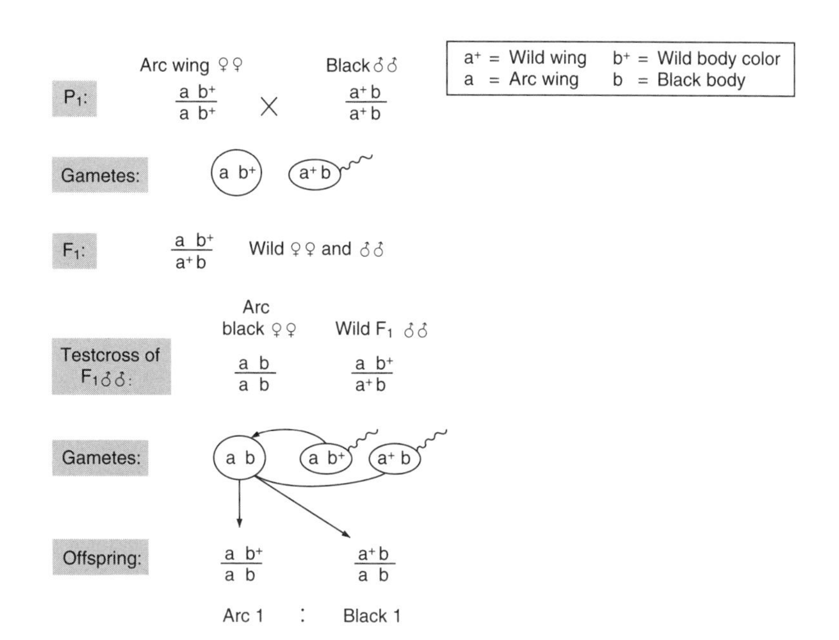

FIGURE 8-7. Testcross of dihybrid *Drosophila* males. The testcross of a male that is dihybrid for two autosomally linked genes yields only the parental combinations in a 1:1 ratio. Linkage strength is 100% in the male fruit fly. Thus crossover percentage cannot be determined using male *Drosophila* in a testcross. Compare this with the testcross of F_1 females in Figure 8-6.

contributed the Y chromosome to them. They are thus testcross offspring, having received from their male parents nothing to mask any recessives on the X from their mothers. The F_2 females clearly are not testcross progeny; they received an X chromosome from their fathers that carried a dominant (ct^+), which did mask the presence of the recessive.

We can ignore these female offspring and concentrate only on the different types of males to estimate the crossover frequency. Suppose the mating between "white" females and "cut" males, which is diagrammed in Figure 8-9, produced 1,000 flies in the F_2 generation, distributed as follows:

Females		Males	
Wild	258	White	191
White	250	Cut	198
	508	White, cut	45
		Wild	38
			472

The F_1 and the F_2 data tell us that the genes are sex linked, because the expression of the genes is seen to depend on the sex of the offspring; only the males are mutant in the first generation; only the males show the cut trait in the F_2. To calculate the crossover frequency, we pay attention only to the F_2 males, the testcross progeny. The parental and crossover combinations among them are recognized, and their numbers are recorded, as follows:

Parentals		Crossovers	
White (wct^+)	191	Wild (w^+ct^+)	45
Cut (w^+ct)	198	White, cut (wct)	38
	389		83

The percentage of recombination is then obtained by dividing the total number of male crossover types by the total number of male offspring: crossover frequency = 83/472 = 17.6%.

We must remember that when experimental crosses are performed they are done routinely in reciprocal, to reveal any differences that might be related to sex. The student should diagram the reciprocal of the cross followed here: Cut females × white males. Although the F_1 male offspring differ phenotypically from those F_1 males diagrammed in Figure 8-9, the F_2 results among the males are the same. This is so because the F_1 females are identical genotypically in both crosses. In actuality, comparable figures will be obtained if the reciprocal matings are performed under similar conditions.

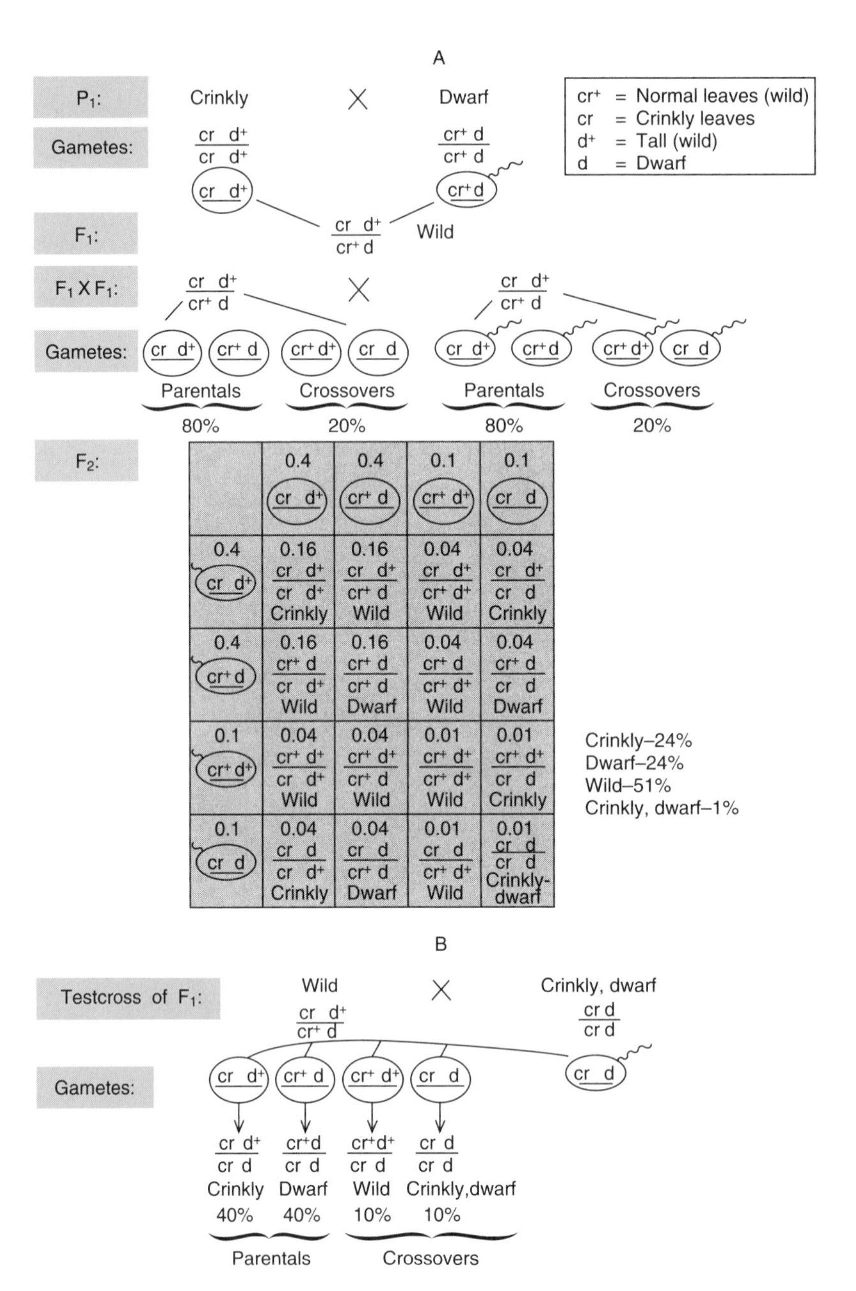

FIGURE 8-8. Linkage and the F_2 generation in corn. **A:** The amount of recombination between cr and d is 20%. Each parent forms crossover as well as parental types of gametes. If the amount of recombination were unknown to begin with, its calculation on the basis of the F_2 data would be difficult. **B:** A cross of any F_1 hybrid to a double recessive reveals directly the number of new combinations (wild and crinkly, dwarf), and the percentage of recombination is readily determined.

Although males are always testcross progeny in the case of sex-linked genes, females may also be testcross offspring, depending on the male parent. Consider the cross shown in Figure 8-10. In the F_2 generation, all of the offspring, female as well as male, are testcross progeny. This is so because the X chromosome of the F_1 males contains both recessives. Consequently, these F_1 males contribute no sex-linked dominants to any of the offspring. Therefore, no factors mask the effects of any sex-linked alleles coming from the F_1 females. The diagram in Figure 8-10 shows that all of the possible phenotypes are represented equally among the females and

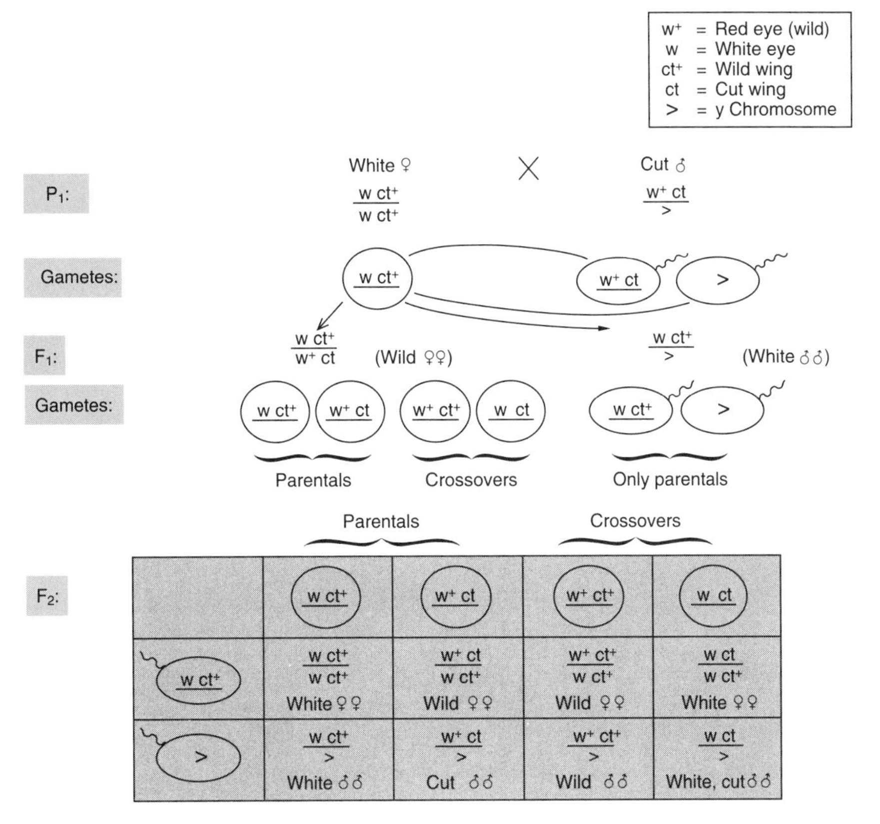

FIGURE 8-9. Genes linked on the X chromosome. When genes on the X are being followed, the F_2 generation may be used to estimate the amount of recombination. This is so because the Y chromosome in the male masks no factors on the maternal X. Therefore all the F_2 males are testcross offspring, and they can always be used in such crosses to estimate the percentage of recombination. In the cross shown here, *only* the F_2 male offspring can be used because the F_2 females received a dominant allele from the F_1 male parent. This masks recessives on the maternal X chromosome.

the males. Since they are all testcross offspring, every one of the progeny, male and female, can be counted to determine the crossover percentage. In contrast, the reciprocal mating (homozygous wild P_1 females × white, cut males) gives very different results. The student should diagram this cross to be convinced that in this case only the F_2 males can be used in the counts to determine the crossover frequency. This is so because the F_1 males contribute an X to their daughters that contains both of the dominant wild-type alleles.

It soon became evident in *Drosophila* that the sex-linked genes fall into one large group and the autosomal ones into three, two large and one very small. This is to be expected since a haploid set of fruit fly chromosomes consists of a large X, two large autosomes, and one very small "dot" chromosome. If genes are indeed on the chromosomes, the number of linkage groups and their relative sizes should correspond to chromosome number and size. The same correspondence has now been established for all other species whose genetics and cytology have been studied.

Genes that are linked tend to stay together, but crossing over makes it possible for them to separate and to enter into new combinations. If it were not for crossing over, all allelic combinations on one chromosome would be forced to remain together. The only way a new combination could be formed would be through the slow chance process of gene mutation. Meiotic crossing over, however, brings about the new combination quickly. Although alleles of linked genes may not assort independently, crossing over extends recombination down to the level of the chromosome. Without crossing over, much of the value of the sexual process—the formation of new gene combinations—would be lost. Evolution would be greatly slowed down in those species lacking it, because fewer variants would be available for the operation of natural selection. Crossing over probably arose long before independent assortment—when only one pair of chromosomes was found in living things. It would have provided the first mechanism to ensure new gene combinations among the earliest cells.

Chromosome mapping

All the data on linkage and crossing over assembled in Morgan's lab indicated that the units of heredity are located in a linear order on the chromosome rather than in some complex arrangement such as one on top of the other.

It was reasoned that, if a linear order is the case, one should be able to determine the relative positions of genes on the chromosomes by studying strengths of linkage and the amount of recombination. When a particular kind of cross is repeated under the same experimental conditions, the amount of recombination between two genes is found to be constant. The frequency of crossing over between a particular pair of genetic loci is apparently not haphazard. Crossover values for different pairs of linked genes, however, may be quite dissimilar. Consider the two hypothetical crosses in Figure 8-11. There would be a crossover value of 10% for p and l and a different constant figure, 20%, for l and c. We can interpret such results to reflect distances among these three genes, assuming a linear order. For if crossing over does entail some kind of breakage and reunion of chromosome threads, it is obvious that the closer together two genes are, the less the chance is that a random break will occur between them. So if l is closer to p than it is to c, there will be fewer chance breaks between p and l. Thus they will tend to stay together more than will l and c when a single break takes place.

Tentatively, let us arrange the three genetic loci in the following order: p, l, c. The reasoning here is that

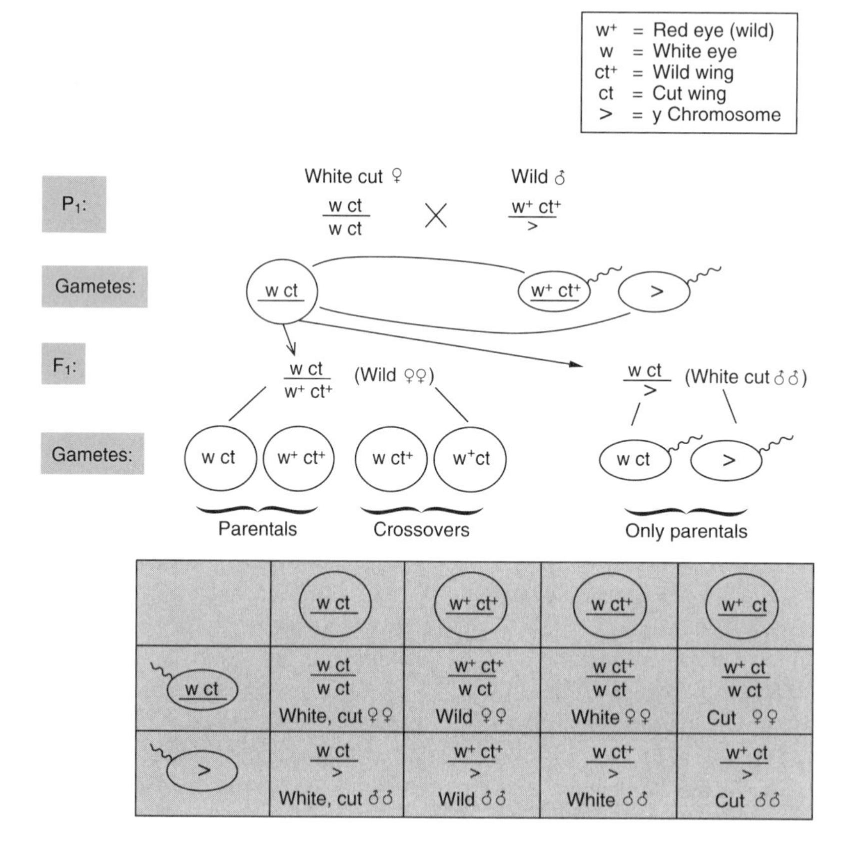

FIGURE 8-10. F_2 females and males as testcross progeny. In this example, the F_2 females receive only recessives from the F_1 male parent. They are thus testcross offspring in the case of the X-linked genes. All of the offspring in this cross may be used to estimate the amount of recombination. Contrast this with Figure 8-9.

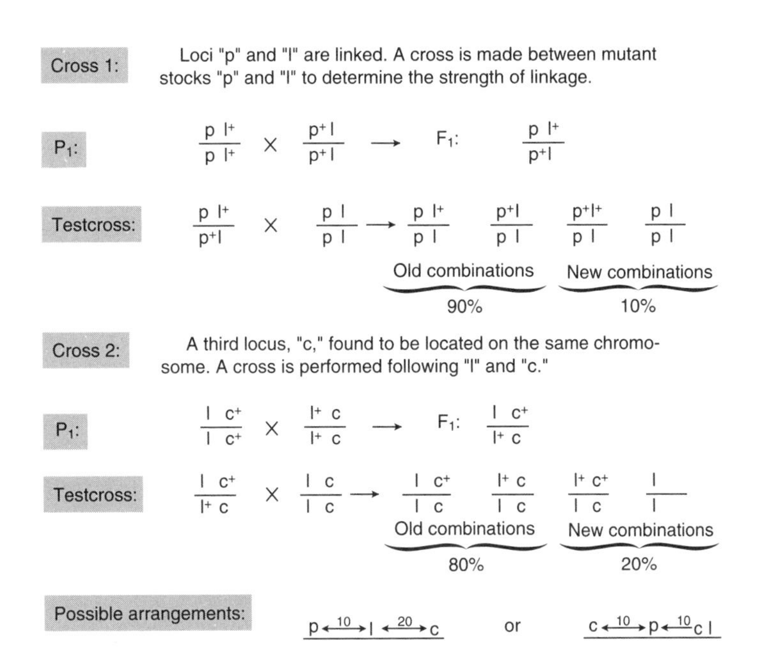

FIGURE 8-11. Determining gene arrangement. In cross 1, the value of 10% indicates the amount of recombination, the frequency with which the loci p and l enter into new combinations. Similarly, in cross 2 the value of 20% indicates the amount of crossovers between l and c. The crossover frequencies reflect the distance between the loci. Thus l and p with a 10% crossover value (90% old combinations) must be closer to each other than l and c with 20% recombination (80% parentals). From the data, two map arrangements are possible. The correct order can be established only by making a cross between stocks p and c to determine the amount of crossovers between them and hence their map distance.

the number of new combinations observed in a testcross stems from the amount of crossing over and that this, in turn, reflects the distance between the genes. If these new combinations can be used to indicate distance, what sort of units should be used as a measure? H.J. Muller suggested taking the percentage of new combinations and using them directly as a kind of measure of distance simply by converting the percentage into crossover units. So we would say that genes p and l are 10 units apart, whereas l and c are separated by 20 units. A moment's consideration of our tentative arrangement of the three genes (p, l, c) tells us that the gene order is not justified from the information at hand thus far. It is equally possible that the correct order is c, p, l. Provisionally, we can only say that l and c are twice as far apart as p and l. However, c could be on either side of p. Either of the two map arrangements shown in Figure 8-11 is feasible.

Obviously, we must find the number of new combinations between p and c. We cannot place any three genes (1, 2, 3) in the proper relative positions if we find only the crossover values between genes 1 and 2 and between genes 1 and 3. We must also determine the number of new combinations between genes 2 and 3 in order to arrange the loci on the chromosome with any degree of accuracy. Suppose a cross between p and c gives a value of approximately 30%. This would indicate that the first supposition was correct (the left arrangement in Fig. 8-11). A value close to 10%, on the other hand, would indicate that c is to the left of l (the right arrangement, Fig. 8-11).

Assuming that the correct order is found to be p, l, c, we can appreciate the physical basis for the different crossover values from the illustrations in Figure 8-12. Notice that in all the illustrations of crossing over each chromosome is represented as a double structure. Chromosomes are depicted in this way because genetic analyses of such organisms as *Neurospora* and *Drosophila,* as well as studies of the cell cycle, indicate that at the time of crossing over each chromosome thread is already replicated. Each chromosome would thus be composed of two chromatids, and crossing over would take place in the four-strand stage, which simply means that four chromatids are present when it occurs. Any one crossover event, however, involves only two of the four strands, and nothing dictates that it must always be the same two. In the illustrations, strands 2 and 3 were selected as the crossover chromatids only for the sake of diagrammatic clarity. This should not be construed to mean that only these two can participate. Strands 1 and 3, 1 and 4, or 2 and 4 could have engaged equally well.

Although the reasoning discussed in Figure 8-12 seemed to hold, it was soon realized that certain complications could enter the picture. For example, if the gene order is actually p, l, c, and p and c are truly 30 units apart, it might be possible for crossing over to occur simultaneously at two points between p and c, that is, to get double crossing over between them.

Let us examine the effects of such an event on the relationship between these two genes and compare them with the results of a single crossover event. If just genes p and c are being followed, any double crossover between them would produce the results in Figure 8-13A. Compare these with the results of the single crossover event in Figure 8-13B. We see that the single crossovers produce detectable new combinations between p and c. In the case of the doubles, however, p and c are not placed in a new relationship to each other but still occur together on the same chromatid. Therefore there would be no way to recognize the fact that two crossovers, not just one, had occurred between them. The doubles would be classified along with the parentals, the types with the old combinations. Because we would be losing sight of two single crossovers for every double that was not detected, p and c would appear to us to remain together in the parental combination more often than they actually do. Since the chromosome map is based on the number of new combinations that we can observe, the genes p and c would appear to be more strongly linked and hence closer together than they really are.

Trihybrid crosses and mapping

How can we overcome this tendency to underestimate the true number of crossovers? The problem is quite easily solved by making a trihybrid or three-point testcross. Instead of following just two genes at a time in the testcross, we consider three. Trihybrid crosses are especially valuable in chromosome mapping and are routinely performed to ascertain gene locations. The same basic procedures are followed when working with microorganisms or with higher species. Because chromosome mapping is so critical to genetic analysis, the steps and the reasoning behind them are presented here in considerable detail.

If we have established the gene order as p, l, c and know that l is between the other two, we can use this knowledge to detect any double crossovers between the more distant genes, p and c. Now consider a double crossover when a trihybrid is being followed (Fig. 8-14). We see that on either side of gene l there are two

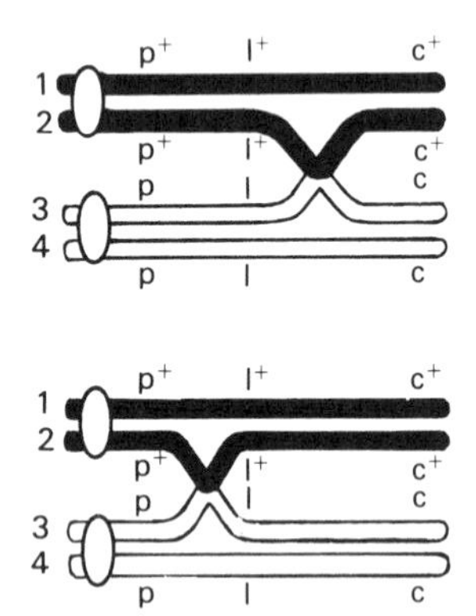

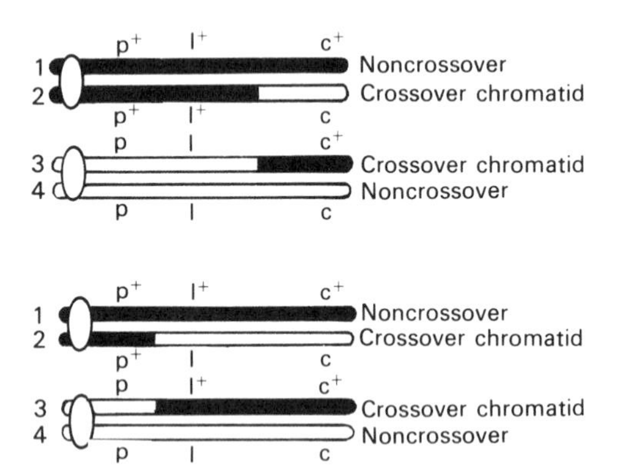

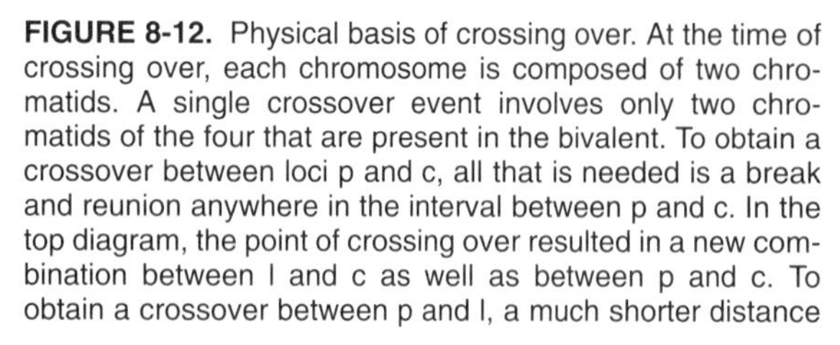

FIGURE 8-12. Physical basis of crossing over. At the time of crossing over, each chromosome is composed of two chromatids. A single crossover event involves only two chromatids of the four that are present in the bivalent. To obtain a crossover between loci p and c, all that is needed is a break and reunion anywhere in the interval between p and c. In the top diagram, the point of crossing over resulted in a new combination between l and c as well as between p and c. To obtain a crossover between p and l, a much shorter distance is involved. Imagine that, for a crossover event to occur within this shorter interval, the threads must bend more sharply. The lower diagram shows a crossover between p and l. Note that in this case and also in the one shown above, there was crossing over between p and c. This is true simply because any one break between p and l must also be a break between p and c. Thus, as a result of their greater distance apart, we would expect more crossing over between p and c than between p and l.

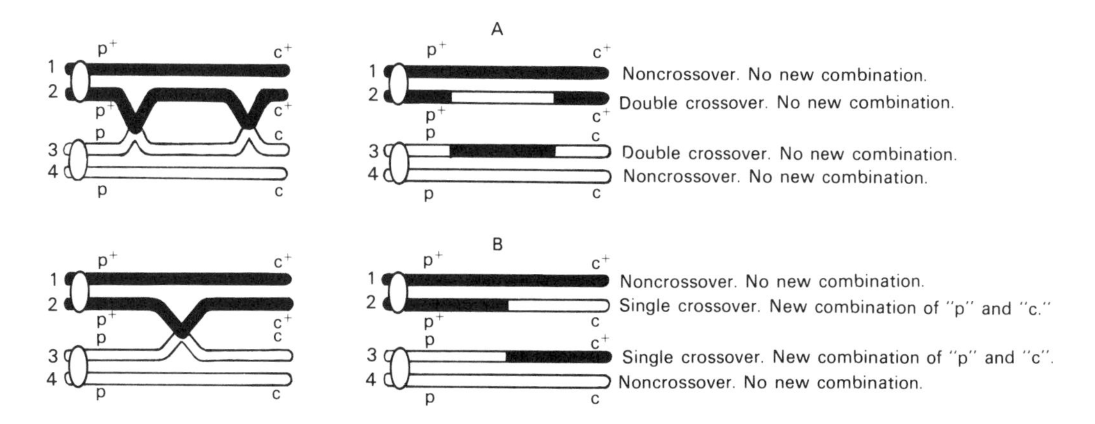

FIGURE 8-13. Results of double and single crossing over between two genes. **A:** A double crossover event between p and c yields no new combinations of alleles. Therefore, if only these two loci are being followed in a cross, we would lose sight of two single crossovers each time double crossing over took place. **B:** A single crossover event between p and c can be readily detected by the new combination among the offspring.

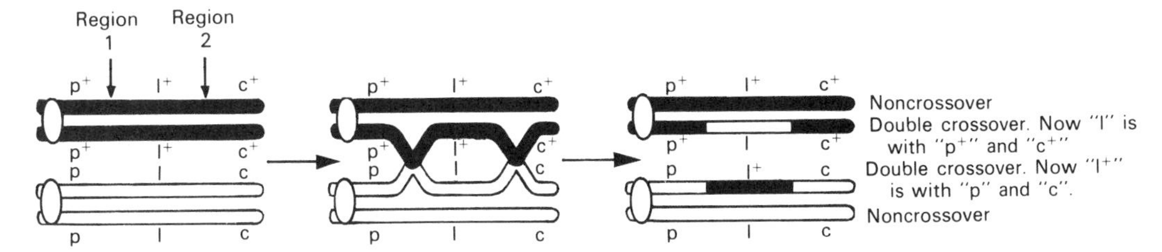

FIGURE 8-14. Double crossing over in a trihybrid. A double crossover event involves two single events. There are two regions therefore where single crossing over can take place. A double crossover in a trihybrid can be detected, because the "middle" gene will assume a new relationship with those on either side of it. It will appear to have switched position. The alleles on either side of it will remain in the old combination.

chromosome regions, 1 and 2, where single crossover events can occur. If two singles occur together to give a double, then no new combinations between p and c are produced. However, we now have a way to detect these doubles. For when a double crossover event occurs, the gene that is more central, gene l in this case, enters into a new combination with the genes on either side of it. Note the genic parental arrangements in the trihybrid parent. On the one chromosome, the three wild alleles of the three genes are together: $p^+l^+c^+$. On the other, the three mutant alleles are linked: plc. As a result of the double crossover event, we now have p^+lc^+ and pl^+c. It appears to us that l, the gene in the middle, has shifted its position. The apparent shift has resulted from the occurrence of two crossover events, one on either side of l. With the trihybrid testcross we can now recognize double crossovers that would otherwise be lost, and we add these to the number of singles to give the true amount of crossing over between p and c. These two genes on either side of l will consequently appear farther apart than they otherwise would. We see from this example that the number of new combinations is not the sole criterion for an estimation of the amount of crossovers between two distant genes. Whenever possible, three closely linked loci should be followed to arrive at a more accurate estimation of map distance.

Let us now use a trihybrid testcross to map three hypothetical genes that are autosomally linked in *Drosophila.* We will call the three genes b, n, and t. Nothing is known about these three sites except that they are linked and the wild-type alleles, b^+, n^+, and t^+, are dominant. We do not even know the precise order of the three loci. Since we do not know which gene is the one in between, we cannot say which is the correct way to illustrate the gene arrangement when we diagram the cross. However, if we are consistent, it does not really matter. To start with, we may select the order b, n, t. This may or may not be accurate, and we could have chosen another arrangement to work with.

Data from the actual cross will tell us later which arrangement is actually correct.

Suppose a cross is made between homozygous wild females and triple mutant males. Having decided to use the order b, n, t, we can depict the trihybrid cross and the testcross as shown in Table 8-1. Eight different categories of offspring can be recognized phenotypically on the basis of the combinations of the wild and mutant traits. Eight phenotypic classes are expected, because a trihybrid forms eight classes of gametes ($2^3 = 8$). Remembering the meaning of a testcross, we know that these eight classes actually represent the eight different kinds of gametes produced by the trihybrid and that the data reflect the relative frequency in which they were formed. Therefore, we can use the testcross data to map the three genes, for now we have a way to estimate how often the linked genes separated and entered into new combinations during gamete formation. The steps and reasoning go as follows (follow Table 8-1):

1. Since linked genes tend to stay together, we would expect the noncrossover types to be the most common. The phenotypic classes that are the most numerous among the progeny should represent the parental-type gametes, those in which no crossing over has been involved. The data show us these types, and we record them as the parentals: $b^+n^+t^+$ (298) and bnt (289).

2. Single crossovers would be less common than parentals, but doubles would be even fewer and can be easily identified. So we record the least numerous classes: bn^+t^+ (15) and b^+nt (21).

3. Once both the parental and double crossover classes have been recognized the correct order of the three genes is readily determined. We have seen that when a double crossover event takes place the gene in the middle appears to have switched its position in relation to the two on either side of it. A comparison of the two classes, the parentals and the doubles, indicates that b is the gene that is in between. We see that the three wild alleles went into the cross together on one chromosome and the three recessives on the other. In one group of doubles (bn^+t^+) two of the wild types are still together, and in the other group (b^+nt) two of the recessives are still in the parental arrangement. It is alleles of gene b that do not appear in the original combination, and we conclude that gene b must be between the other two.

4. We next rewrite the trihybrid parent to show the proper gene arrangement. It makes no difference whether we write $n^+b^+t^+$/nbt or $t^+b^+n^+$/tbn, as long as the proper gene is in the inside position. If this seems confusing, remember that we are primarily concerned with distances. As long as the relationships between the genes are preserved, the sequence selected is of no consequence, just as it would not matter if a road map were turned upside down when it was read. We will choose arrangement $n^+b^+t^+$/nbt to stand for the trihybrid. To convince yourself that the other possibility is equally valid, you should later redo this cross using the alternative order; you will obtain the same results.

5. Looking at the correctly written trihybrid parent, we can see that there are two places where crossing over can occur to rearrange the three linked genes: one between n and b and the other between b and t. For reference, these regions are designated region 1 and region 2, respectively.

6. A break in region 1 will produce the single crossovers n^+bt and nb^+t^+. The data show that they amount to 120 and 109, respectively, giving a total of 229 crossovers in that region. The singles in region 2 are n^+b^+t (69) and nbt^+ (71), totaling 140. It is critical to understand that these single crossovers, which can be recognized from the data, do not represent all the crossovers in regions 1 and 2. Double crossing over has occurred; this means that there were additional breaks in both regions. These cannot be detected by considering the singles alone. Since one double masks the effects of two single crossover events, the genes n and t will appear closer together than they should on the final map if we fail to take the doubles into consideration. However, we expressly made a triple testcross to detect these doubles, and we identified (step 2) 36 of them among the progeny. A glance at the trihybrid (Table 8-1) shows that, to get the classes n^+bt^+ and nb^+t, a crossover event must have occurred in *both* regions 1 and 2. Since this happened 36 times in region 1, we must add this figure to the total of singles we have already identified in this region. Instead of a total of 229 crossovers, there were actually 229 + 36, or 265. Exactly the same reasoning applies to region 2. Again, 36 must be added, this time to 140, giving the correct value of 176 for the number of crossovers in this region. The value of the triple testcross should now be quite apparent. It enables us to detect in two regions of the chromosome those single crossovers that otherwise would go unobserved. We must not forget to add the doubles identified in any three-point testcross to each set of singles, those in region 1 and those in region 2 as well.

7. All that remains to be done to obtain crossover values among the three genes is to take the total

TABLE 8-1. Trihybrid testcross data and construction of a map segment

$$P_1 \frac{b^+\ n^+\ t^+}{b^+\ n^+\ t^+} \times \frac{b\ \ n\ \ t}{b\ \ n\ \ t} \longrightarrow F_1 \frac{b^+\ n^+\ t^+}{b\ \ n\ \ t}$$

Testcross $\frac{b^+\ n^+\ t^+}{b\ \ n\ \ t} \times \frac{b\ n\ t}{b\ n\ t}$

Testcross progeny	
b^+ n t	21
b n^+ t	120
b n t	289
b^+ n^+ t	69
b n t^+	71
b^+ n^+ t^+	298
b^+ n t^+	109
b n^+ t^+	15
Total	992

Construction of Map Segment From Testcross Data

Parentals	
b^+ n^+ t^+	298
b n t	289
Total	587

Doubles	
b n^+ t^+	15 Doubles
b^+ n t	21
Total	36

Alleles at b appear to have switched position. The b locus must be between n and t. One correct way to represent the trihybrid parent is

$$\frac{n^+\ b^+\ t^+}{n\ \ b\ \ t}$$

Crossing over may occur in the segment between n and b; it may also occur in the setment between b and t. These regions can be called, respectively, region 1 and region 2, and the trihybrid written as

$$\overset{\quad 1 \qquad 2}{\frac{n^+ \quad b^+ \quad t^+}{n \quad\ b \quad\ t}}$$

Single crossover region 1

$n^+ \times b^+\ t^+$ / n b t →

n^+ b t	120
n b^+ t^+	109
Region 1 crossovers	229
Total region 1	36
	265

Crossover regions 1 and 2

$n^+ \times b^+ \times t^+$ / n b t →

n^+ b t^+	15
n b^+ t	21
Doubles total	36

Single crossover region 2

$n^+\ b^+ \times t^+$ / n b t →

n^+ b^+ t	69
n b t^+	71
Region 2 crossovers	140
Total region 2	36
	176

Amount of region 1 crossovers $= \frac{265}{992} = 26.7\%$ Amount of region 2 crossovers $= \frac{176}{992} = 17.7\%$

Map segment n 26.7 b 17.7 t

crossover figure for each single region and divide it by the entire number of offspring. The respective percentages, 26.7% and 17.7%, are simply translated into map units.

Factors influencing crossover frequency

It must be kept in mind that while a road map reflects precise distances between points, this is not true of a genetic map. Although the number of map units between two genes does indeed depend in part upon distance, we must not think of the units in terms of accurate spatial measurements. Various factors that influence crossover frequency have come to light throughout years of genetic studies. Moreover, the ease with which crossing over occurs along the length of a chromosome is not uniform; there is more crossing over in some regions than in others. For example, the area around the centromere engages in less crossing over than most of the rest of the chromosome. As a result of its influence, genes in the vicinity of the centromere tend to appear more clustered than do those found elsewhere.

In addition to the centromere, certain other chromosome regions are found in which crossing over may be cut down and others in which it seems to take place with ease. This was established many years ago in *Drosophila* (see Chap. 10) and is now known to be true in the human and other species. Many human genes have been assigned to specific bands on the chromosomes largely through the use of techniques that will be discussed in later chapters (see Chaps. 9, 20). Recombination values have been derived from family studies and have permitted the construction of genetic maps for some human chromosome regions. It is now possible to compare two kinds of human maps: the genetic map expressed in map units based on recombination values and the cytological map based on assignments of genes to specific bands. The two maps prove to be colinear: The order in which the genes occur is the same. It soon becomes apparent, however, that the distances among the genes on the cytological map are not accurately reflected by the distances expressed in map units. For example, two genes, let us say A and B, may be separated by one band, whereas genes C and D are separated by four such bands and are thus much farther apart than are A and B on the cytological map. However, the genetic map may show that A and B are 20 map units aparts, whereas C and D are separated only by 10 units. This discrepancy is a direct reflection of the tendency for crossing over to occur with a higher frequency in some chromosome regions than in others. Reasons for these differences in various species are unknown.

Crossing over may be influenced by environment, both external and internal (hormonal). The latter effect is clearly seen when crossover frequencies are compared between the two sexes. Crossing over does not normally occur in the male fruit fly or in the female silkworm. Female mice show about 25% more crossing over than do males. Recombination frequencies in humans suggest a 40%–50% excess in females over males. It appears that when a difference in crossover frequency between the sexes occurs in a species, it is the heterogametic sex that shows the lower amount. (In the silkworm, the female is heterogametic.)

Many years ago Bridges found a definite effect of age on crossover frequency in *Drosophila.* More crossing over takes place in young females and tends to decrease with age of the fly. In the human, the effect of age on crossing over has not been studied sufficiently to give an accurate appraisal. However, counts of the number of chiasmata in meiotic cells of males show an increase with age, whereas in females there is a reduction. The few chiasmata observed in older oocytes seem to be associated with an increase in the number of unpaired chromosomes. This in turn could reflect the effect of maternal age on the frequency of nondisjunction.

Temperature must be controlled when precise measures of crossing over are desired in *Drosophila.* Departures from the normal temperature range (extremes of heat or cold) tend to raise the frequency of crossing over. Radiation has a pronounced effect on the amount of crossing over and has been shown to stimulate it significantly in the fruit fly. Both temperature and radiation seem to encourage crossing over in regions where it is normally reduced, such as in the vicinity of the centromere. They may even induce it in the male fruit fly and stimulate it to occur in body cells where it is normally inhibited.

Certain divalent cations such as Ca^{2+} and Mg^{2+} have been shown to alter the amount of crossing over. In one experiment, flies raised for 4 days on a diet with an excess of calcium exhibited less crossing over than those raised on a normal food medium. The decrease persisted for more than a week. When metallic ions were reduced below that of normal, the crossover frequency increased significantly.

Alterations in chromosome structure (see Chap. 10) may decrease the amount of crossing over. The presence of a deletion, for example, in one chromosome may interfere with crossing over in the chromosome parts that do pair properly. Different kinds of alterations in chromosome structure all seem to cause a decrease in crossover frequency, as if they were somehow interfering with the normal forces of attraction between the homologous chromosomes.

Use of a chromosome map

Since crossing over can be influenced in so many ways, crosses that are used to generate genetic maps should be carefully controlled. The amount of crossing over is constant for a given species under a prescribed set of conditions. Therefore a map can be used to make predictions when experimental matings are performed properly. If we can use a genetic map to tell us the amount of recombination to be expected in different chromosome regions, we should be able to predict from it the number of double crossovers.

The map gives us the probabilities of the single events expressed in map units. Since a double crossover is the result of two singles happening simultaneously, we should be able to estimate the probability of a double in a certain chromosome region simply by multiplying together the separate chances for the singles. In our example, the probability is 26.7% that a crossover will occur alone in region 1 and 17.7% that one will take place in region 2. Since the chance for two independent events happening together is the product of their separate probabilities, we simply multiply the two to find the chance of the double event ($0.267 \times 0.177 = 0.047 = 4.7\%$). The map as a table of probabilities gives the percentage of doubles to be expected in that particular chromosome region. But let us see if the results of the cross bear this out.

We obtained 992 offspring (refer to Table 8-1). By our calculations, of these 4.7% should be double crossover progeny. Therefore approximately 47 individuals should be distributed between the types n^+bt^+ and nb^+t (992×0.047). The data show that actually only 36 of these types arose from this cross. We might conclude that this lower figure is just the result of chance in this particular instance. However, if we repeated the mating many times, we would continue to find fewer double crossover types than the number predicted from the map. Finding a number of doubles that is smaller than the amount indicated by the map is typical rather than exceptional. This suggests something about the single crossover events themselves. They must not be independent, for if they were, their product should be a reliable prediction of the actual number of doubles that will be obtained in a cross. We are forced to conclude that the occurrence of one crossover event can influence the occurrence of another in the same chromosome. This is particularly true within short regions of the chromosome, where the influence of one crossover on another may be quite pronounced. The term *interference* describes the effect crossing over in one chromosome region exerts on the probability of a crossover event in an adjacent region.

This interference varies from one part of the chromosome to another. For a more accurate estimate of the amount of double crossovers that will arise in a cross, interference must be taken into consideration. This factor is measured by the coefficient of coincidence. This is nothing more than an expression of the ratio of double crossovers actually obtained to those predicted from the map (Table 8-2). We expected approximately 47 doubles but found only 36. The coefficient of coincidence equals the actual number of doubles over the expected number, which in this example is $36/47 = 0.76$. This coefficient tells us that we must modify the number of double crossovers predicted from the map. We expected 47, but, knowing the coefficient of coincidence, we multiply this figure by 0.76. This in turn indicates that we should expect only 36 doubles, which is what we actually obtained from the cross. The coefficient of coincidence varies between zero for very short distances (meaning that no doubles at all are to be anticipated) and 1 (meaning that the expected amount is simply the product of the two singles, and no interference is operating). It is typical for the value of the coefficient to be between the two extremes.

Now let us see how we can use the map to make further predictions. Referring back to the map segment in Table 8-1, we see that we can expect 26.7% crossovers between genes n and b. Similarly, for region 2 we can predict 17.7% crossovers between b and t. Therefore, if 1,000 offspring result from a mating, 177 of them ($17.7\% \times 1{,}000$) should represent crossovers between b and t. However, remember that the crossover amount is not equal to the number of new combinations, since any one double crossover event between two genes will regenerate the old combinations. But knowledge from our map segment can aid us in this regard. We have just calculated that 36 double crossovers are to be expected, taking into consideration the coefficient of coincidence. This figure, 36, was added to each set of singles when the genes were being mapped, to give a more accurate estimate of the true amount of crossovers. Now, however, we must subtract the 36 from our expected new combinations, since we will get 36 fewer new combinations in each region (1 and 2) than the distance indicates in crossover units. On the basis of 1,000 offspring, therefore, we would expect, in region 1, 231 new combinations ($26.7\% \times 1{,}000 - 36$) and, in region 2, 141 new

TABLE 8-2. Calculation of coefficient of coincidence based on data in Table 8-1

Map segment

n ___26.7___ b ___17.7___ t

Number of doubles obtained in actual cross = 36

Amount of double crossovers expected from map segment = crossover probability in region 1 × crossover probability in region 2:

$$0.267 \times 0.177 = 0.047 = 4.7\%$$

Number of doubles to be expected on basis of 992 offspring:

$$992 \times 0.047 = 47$$

$$\text{Coefficient of coincidence} = \frac{\text{no. of doubles actually obtained}}{\text{no. of doubles predicted from map}} = \frac{36}{47} = 0.76$$

combinations (17.7% × 1000 − 36). What we are doing is using the map to make predictions in terms of probabilities, and that is basically how a chromosome map serves us.

When mapping genes on the X chromosome, the procedure followed is essentially the same as that for mapping autosomal genes. One need only remember that males are always testcross progeny in the case of genes on the X chromosome. The F_2 females, on the other hand, may or may not be. Females are testcross offspring in regard to the X chromosome if their fathers express all the X-linked recessives being followed in the cross (refer back to Figs. 8-9 and 8-10). Therefore, depending on the specific cross, one can at times use both male and female F_2 progeny to calculate map distances among genes on the X. At other times only the males are considered.

Map length, multiple crossovers, and recombination

Chromosome maps have been constructed in considerable detail for certain organisms, particularly *Drosophila, Neurospora,* the mouse, and corn. Figure 8-15 shows the location of some of the better-known loci in the fruit fly. When detailed chromosome maps are made, they are usually built up by determining the crossover values between genes that are closely linked. Trihybrid testcrosses are used to establish gene order, but, if genes are a considerable distance apart, multiple crossovers (doubles, triples, and even higher degrees) may occur between them to obscure the true crossover value. Therefore more accurate map units are determined by studying recombination values between genes close enough to minimize the effects of these factors.

Suppose a gene order such as the following has been established by trihybrid testcrosses: f, g, h, i, j, k, l. The map would then be more precisely built up in map units by finding the crossover value between h and i, then between h and g, g and f, and so on. Obviously, this would be more accurate than just relying on recombination values between f and l to reflect the number of map units between these two loci, because multiple crossovers could occur in the f–l interval. To have a more reliable estimate of the distance between f and l in map units, we would add up the small crossover values between the genes in the f–l interval.

It soon becomes apparent from glancing at chromosome maps that they can include more than 50 map units, even more than 100 (note the two large chromosomes of *Drosophila;* Fig. 8-15). When estimating map lengths, the relationship between chiasmata and crossover events is taken to be 1:1. Cytogenetic analyses have continued to show a good correspondence between the amount of crossing over detected genetically by observation of the amount of recombination and the number of chiasmata seen with the microscope. A higher chiasma frequency at meiotic prophase is associated with a higher frequency of genetic recombination. Moreover, in those species with large chromosomes the number of chiasmata present in the bivalents can be determined more accurately than in species with tiny meiotic chromosomes, such as *Drosophila.* In corn, for example, which has large chromosomes, a very good correlation exists between the amount of genetic recombination and the chiasma frequency. An average of one chiasma in a bivalent would therefore reflect a map length of 50

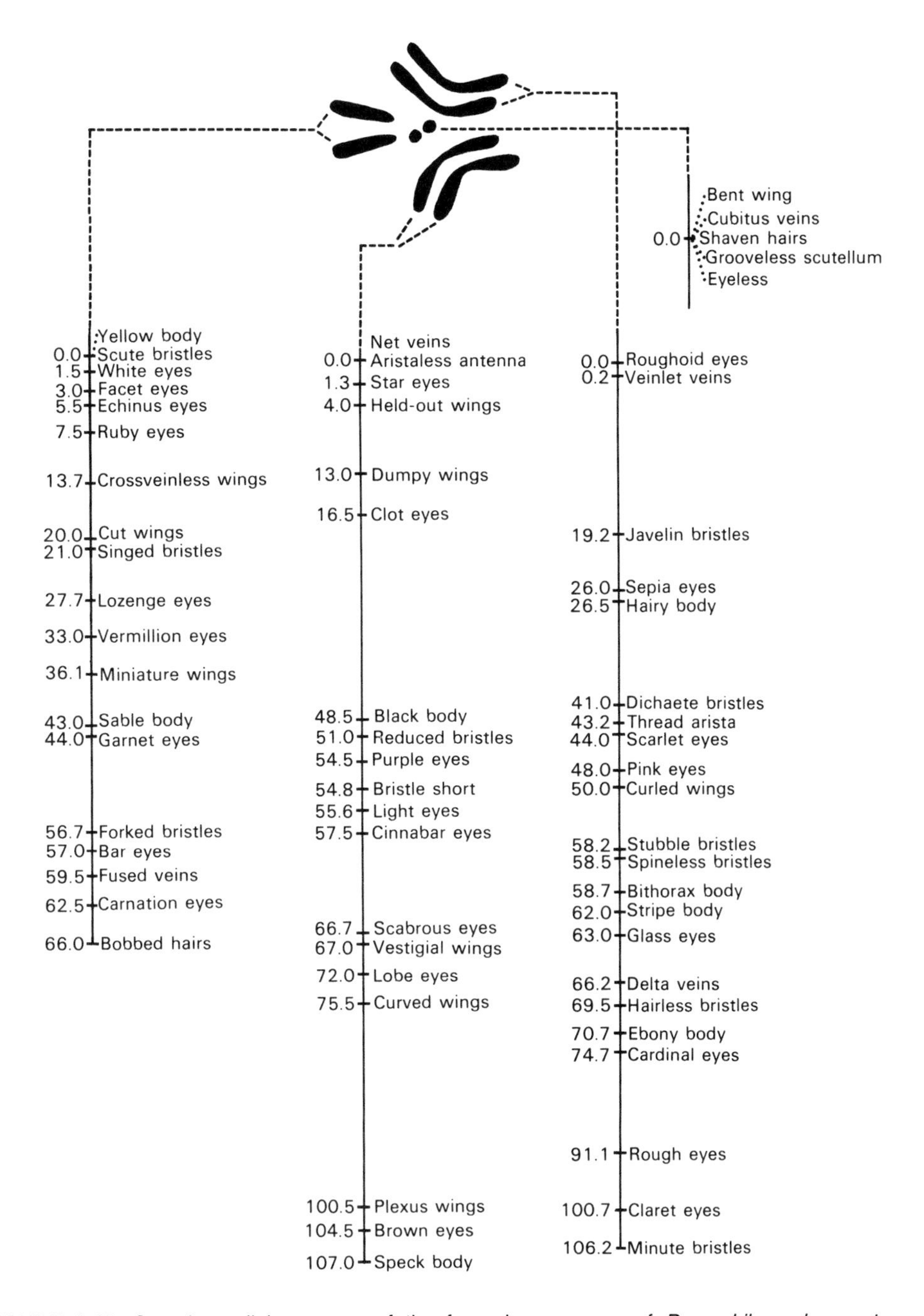

FIGURE 8-15. Genetic or linkage map of the four chromosomes of *Drosophila melanogaster.* (Reprinted with permission from E.W. Sinnott, L.C. Dunn, and T. Dobzhansky, *Principles of Genetics.* McGraw-Hill, New York, 1958.)

units. This would be so because genetic recombination between the pairs of genes located at the opposite ends of the chromosome would take place 50% of the time. This follows from the fact that crossing over is the event that takes place in the bivalent to produce recombinant, or crossover, gametes. If a crossover event occurred between two linked genes in 100% of the meiotic cells, half of the resulting gametes would be recombinants and half would be noncrossovers. This is so because only two of the four chromatids in a bivalent participate in any one crossover event (Fig. 8-16A). An average of two, three, or four chiasmata per bivalent would give, respectively, chromosome map lengths of 100, 150, and 200 units.

What does this mean in terms of the amount of recombination to be expected between very distant genes? Does this mean that 60% or even higher amounts of recombination will occur? The answer is that the number of recombinants cannot exceed 50%. A moment's reflection tells us that 50% recombination is equivalent to independent assortment. If one crossover between two linked genes, a and b (Fig. 8-16A), occurs in every meiotic cell, that means equal numbers of parentals and crossover gametes will form. Indeed, as noted before, if genes are far enough apart on a chromosome then this tends to occur, and we may not be able to tell at first that the two genes are linked; they would be assorting independently, as if they were on separate chromosomes (Fig. 8-16B).

The same principles of map construction used in lower forms can be applied to higher groups. Well over 200 genetic loci have been assigned to specific chromosomes in the human, and the information is being used in a very practical way (see Chap. 9). An estimate has been made of the map length of the human X chromosome and the total map length of all the autosomes. This has been reasoned from studying the frequency of chiasmata in meiotic cells of testicular tissue. Recall that a single chiasma is equivalent to a crossover value of 50%. Counts of chiasmata average about 552 for all the autosomal bivalents in a meiotic cell. This gives a total map length for all the autosomes of 2,600 units. Since the human X chromosome is about 6% the length of one total haploid set of autosomes, its genetic length has been calculated to be at least 160 map units. The figure cannot be considered precise, however, because chiasmata counts were made in male meiotic cells and on autosomes. It is very difficult to obtain meiotic cells from a female, in contrast to a male. Apparently, less crossing over occurs in the human male, as reflected in less recombination than in the female. Moreover, males do not have two X chromosomes so that the number of chiasmata in a bivalent composed of two Xs has not been directly observed but only estimated from the number occurring in pairs of autosomes. Therefore, the true

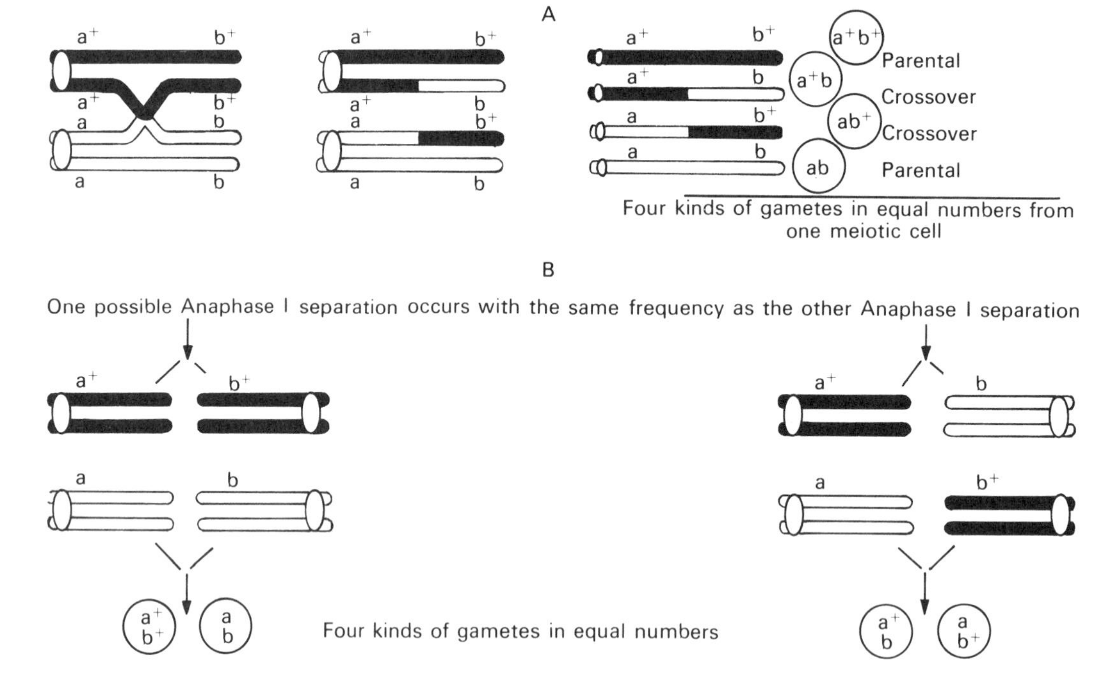

FIGURE 8-16. Effects of a high frequency of single crossovers. **A:** If crossing over occurs between two genes in a meiotic cell, four different kinds of gametes are formed. Only two of the four chromatids participate in any one crossover event. If one crossover event occurs in every meiotic cell between a and b, it will produce recombinants in a frequency of 50%, and the loci will appear to be assorting independently. **B:** The diagram shows independent assortment, assuming a and b are on separate chromosomes. The two possible anaphase separations occur with equal frequency. The result is four kinds of gametes in equal numbers. It can be seen by comparing diagrams A and B that a single crossover event in every cell at meiosis will have the same effect as independent assortment.

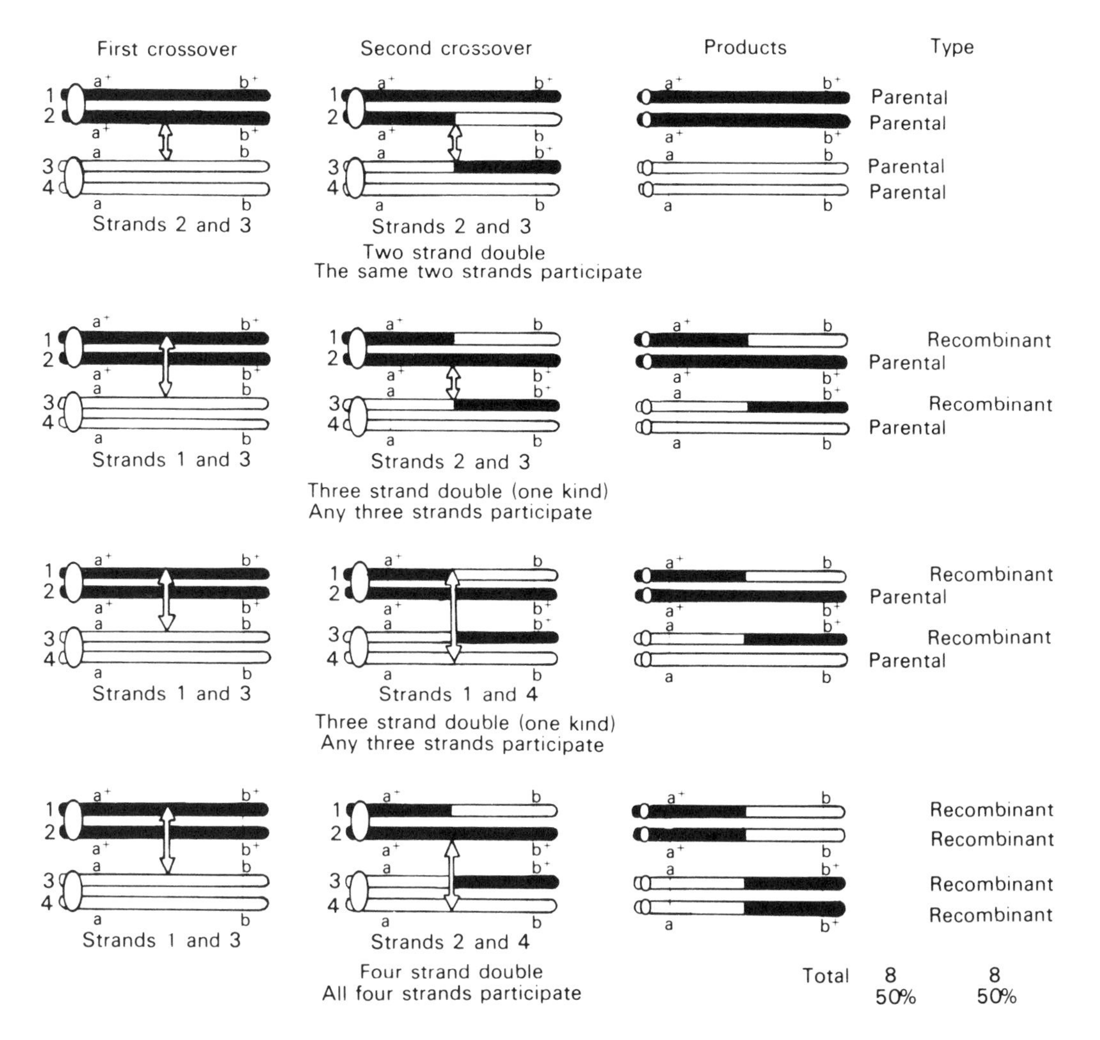

FIGURE 8-17. Types of double crossovers. Double crossing over can involve any of the four chromatids and may take place in a variety of ways. This figure illustrates just one of these. A three-strand double can always occur in two different ways. Since double crossing over is random, the two-, three-, and four-strand doubles occur with a frequency of 1:2:1, respectively, as indicated here. The number of recombinant types, however, cannot exceed 50%. (The second crossover event usually does not occur at exactly the same position as the first. The diagram here clarifies events between loci a and b.)

value of the X chromosome in map units may be greater, perhaps over 200 units. Nevertheless, the length of the X in map units is seen to be sufficient to account for the fact that certain genes known to be X-linked still appear to be undergoing independent assortment. Students of human genetics employ the term *synteny* to describe the situation in which two loci have been assigned to the same chromosome but still may be separated by such a large distance in map units that genetic linkage has not been demonstrated. The failure to show linkage results from the fact that crossing over usually separates the two distant loci.

The complete absence of crossing over in the male fruit fly gives a great advantage in showing that even widely separated genes are located on the same chromosome. Remember that only two of the four chromatids participate at any one crossover. Therefore, even if a crossover event were to occur in every meiotic cell, there would still be only 50% recombination. One might believe that multiple crossovers could increase the recombination value above 50%. A consideration of the different kinds of double crossovers (Fig. 8-17) may clarify this. It was stated in our discussion that a double crossover can involve any of the four strands.

We have considered, for the sake of clarity, only two-strand doubles, those double crossovers in which the same two strands participate in each of the two crossover events. Figure 8-17 shows that three- and four-strand doubles may also take place. Because they are independent events, the doubles occur in a random frequency of 1:2:1 (one two-strand double:two three-strand doubles:one four-strand double). If the numbers of parental and recombinants are added up, it is seen that they total 50:50. Again the recombination value does not exceed 50%, even on this basis.

What about even higher numbers of crossover events between two genes? The greater the distance between two genetic loci, the greater is the chance for multiple crossing over to occur between them. Those multiple crossover events with an even number of exchanges between two loci produce only the parental combinations, whereas those with an odd number of exchanges yield recombinants. When two genes are very far apart on the same chromosome, the frequency of even-numbered crossovers will be about the same as the frequency of the odd-numbered ones. Consequently, the two linked genes will appear to be assorting independently and hence to be unlinked. Nevertheless, recombination frequencies cannot exceed 50%. In certain cases, such as that of "black" and "arc" in *Drosophila,* which are over 50 map units apart, the amount of recombination is actually less than 50%. Therefore one cannot always look at a map and translate the distance indicated by map units into the amount of recombination to be expected. The number of map units corresponds closely to the amount of recombination only for those genes that are relatively closely linked, usually in the vicinity of 20 map units or fewer. The following chapter pursues further aspects of crossing over, with attention to its practical applications.

REFERENCES

Allan, G.E. *Thomas Hunt Morgan: The Man and His Science.* Princeton University Press, Princeton, NJ, 1979.

Hotchkiss, R.D. Models of genetic recombination. *Annu. Rev. Microbiol.* 28:445, 1974.

Morton, N.E., and G.A. Burns. Report of the committee on the genetic constitution of chromosomes 1 and 2. *Cytogenet. Cell Genet.* 46:102, 1987.

Muller, H.J. The mechanism of crossing over. II. *Am. Nat.* 50:284, 1916.

Nicklas, R.B. Chromosome segregation mechanisms. *Genetics* 78:205, 1974.

Stern, H., and Y. Hotta. Biochemical controls of meiosis. *Annu. Rev. Genet* 7:37, 1973.

Stern, H., and Y. Hotta. DNA metabolism during pachytene in relation to crossing over. *Genetics* 78:227, 1974.

Sturtevant, A.H. The linear arrangement of six sex-linked factors in *Drosophila,* as shown by their mode of association. *J. Exp. Zool.* 14:43, 1913.

von Wettstein, D. The synaptonemal complex and four-strand crossing over. *Proc Natl. Acad. Sci. USA* 68:851, 1971.

REVIEW QUESTIONS

1. Assume that the two allelic pairs r^+, r and s^+, s are so closely linked that crossing over can be ignored. What kinds of gametes can be formed by the following?

A. A double heterozygote with the alleles in the trans arrangement.
B. A double heterozygote with alleles in the cis arrangement.

2. Assume that the two allelic pairs t^+, t and u^+, u are linked and that the frequency of crossovers between the two loci is 30%.

A. What kinds of gametes will be formed by a dihybrid with the genes in the trans arrangement, and what would be the expected frequency of each type considering crossing over?
B. Answer the same question for a dihybrid with the alleles in the cis arrangement.

3. Assume that a double crossover takes place in a meiotic cell of the dihybrid in Question 2B. What would be the combination of the alleles resulting from the double crossover between t and u?

4. In the fruit fly, the allele for purple eye color (p) is recessive to its allele for red (p^+). The allele for vestigial wings (vg) is recessive to the wild allele for normal wings (vg^+). The two genes are autosomally linked. Females from a purple stock are crossed to males from a vestigial stock. The F_1 flies are all wild (red eyes and normal wings). F_1 females are testcrossed with the following results:

Purple	210
Wild	40
Vestigial	215
Purple, vestigial	35

A. Diagram the testcross.
B. Calculate the amount of recombination.

5. Give the results to be predicted if an F_1 dihybrid male from the cross described in Question 4 is testcrossed, assuming 500 offspring.

6. In tomatoes, round fruit (o^+) is dominant to long (o). Simple flowering shoot (s^+) is dominant to branching flowering shoot (s). Plants from two different pure-breeding varieties are crossed. One variety bears round fruit and has branched flowering shoots. The other variety has long fruits and simple flowering shoots. F_1 plants were testcrossed, and the following progeny were obtained:

Round, simple	23
Long, simple	83
Round, branched	85
Long, branched	19

A. Diagram the testcross.
B. Calculate the amount of recombination.

7. Using information from Question 6, assume that two dihybrids with the alleles in the cis arrangement are crossed to each other.

A. Give the kinds of gametes and the frequencies of each expected from each parent.
B. Give the phenotypes and the expected frequencies from a cross of two such hybrids with the cis arrangement.
C. Assuming 1,000 plants are obtained in this cross, how many of them would you expect to have the combination long fruits and simple flowering shoots?

8. In chickens, the dominant allele I prevents pigment formation so that the feathers are white. The recessive allele (I^+) permits the feathers to be colored. This pair of alleles is autosomally linked to a pair that influences the development of the feather. The allele F^+ results in normal feathers, but the allele F causes brittle feathers when in the homozygous condition. However, there is no dominance, so that the heterozygote has mildly brittle feathers.

A hen with colored, brittle feathers is mated several times to a rooster with white, normal feathers. Over three dozen eggs are produced that yield chicks with white, mildly brittle feathers. Diagram the cross, giving the genotypes of the parents and the F_1.

9. F_1 hens from the cross in Question 8 are crossed to roosters having colored, normal feathers. The following chicks are obtained:

White, mildly brittle	17
Colored, mildly brittle	64
White, normal	61
Colored, normal	15

A. Diagram this cross.
B. Give the genotypes of all the kinds of birds.
C. Calculate the amount of recombination.

10. In the fruit fly, the allele w^+ for red eye color is dominant to its allele for white eye, w. This pair of alleles is sex linked, as is another pair that influences body color: the dominant for gray body (y^+) and the recessive for yellow (y). Females from a pure-breeding, white-eyed, gray-bodied stock are crossed to males from a stock pure breeding for red eye and yellow body. Give the genotypes of the P_1 and the genotypes and phenotypes of the F_1.

11. F_1 females from the cross described in Question 10 are crossed to white-eyed, gray-bodied males. From the results presented below, calculate the amount of recombination between alleles at the white (w) locus and alleles at the yellow (y) locus.

Females:	Red-eyed, gray-bodied	340
	White-eyed, gray-bodied	350
Males:	White-eyed, gray-bodied	346
	Red-eyed, yellow-bodied	324
	Red-eyed, gray-bodied	9
	White-eyed, yellow-bodied	6

12. In the human, two pairs of alleles d^+, d and p^+, p are associated with color perception. The recessives, d and p, can result in color blindness. Both loci d and p are found on the X chromosome. Linkage is so close that the amount of crossing over is very low.

A. Ignore crossing over, and diagram a cross between a color-blind woman who is homozygous for the dominant allele at the p locus and a man who is color blind because of the recessive he carries at the p locus. Show the expected offspring.
B. Assume that a dihybrid female with the color perception alleles in the trans arrangement marries a man who carries neither recessive. Diagram the cross and show the offspring, ignoring crossing over.
C. In the last case, show the expected results if crossing over occurs.

13. The dominant allele o^+ is required for pigment deposition in the iris of the human eye. Its recessive allele, o, is responsible for ocular albinism. The locus is found on the X chromosome, as is the d locus associated with color blindness that was mentioned in Question 12.

A. Assuming no crossing over, what are the possible results of a cross between a woman with ocular albinism who is homozygous for the allele for normal color vision and a man who has a pigmented iris but is color blind because of the recessive d.
B. Assuming no crossing over, give the results of a cross between a woman who is dihybrid at the albinism and color-vision loci (cis arrangement) and a man with both normal traits.

14. The amount of recombination between the linked genes e and s is found to be 11%. The gene r is found to be linked to e and s. Alleles of the r gene undergo recombination with those of the s gene with a frequency of 7% and with those of the e gene with a frequency of 4%. What is the order of these genes on the chromosome and the distance between them in map units?

15. Answer the questions given the following portion of a chromosome map:

d 14 m 18 p

A. The double crossover value in this region is found to be 1.5%. What is the coefficient of coincidence?

B. Alleles of the s gene are found to be linked to alleles of the genes shown on the map. A crossover value of 13% is found between the s and the p genes. Where should the s gene be placed on the map?

16. In corn, the genes v, b, and l are linked. Using the following data summarizing the results of a trihybrid testcross, do the following:

A. Give the correct genotype of the trihybrid parent showing the correct sequence of genes.
B. Construct a map segment showing the distances among the genes.
C. Calculate the coefficient of coincidence.

v^+ b^+ l	304
v^+ b l	119
v b l	18
v b^+ l	70
v^+ b l^+	64
v^+ b^+ l^+	22
v b^+ l^+	108
v b l^+	295

17. In the Oriental primrose, short style (l^+) is dominant over long style (l). Magenta flower color (r^+) is dominant over red (r), and green stigma (g^+) is dominant over red (g). A trihybrid testcross is performed. From the following data, construct a map giving the positions of the three genes and the distances involved:

l^+ r^+ g	292
l^+ r^+ g^+	153
l r g^+	286
l r g	139
l^+ r g^+	36
l^+ r g	22
l r^+ g	40
l r^+ g^+	18

18. In *Drosophila,* assume that the three genes x, y, and z are linked and that the alleles x^+, y^+, and z^+ are dominant to x, y, and z. Crosses of trihybrid females and males showing the three dominant traits x^+, y^+, and z^+ give the following results:

Females:	x^+ y^+ z^+	1012
Males:	x^+ y^+ z^+	30
	x^+ y^+ z	32
	x^+ y z^+	441
	x^+ y z	2
	x y^+ z^+	3
	x y^+ z	430
	x y z^+	27
	x y z	38

Based on these data, do the following:

A. Give the genotypes of the P_1 males and females, showing the correct sequence of the genes.
B. Construct a map giving the locations and distances.
C. What can you say about the interference in this map region on the basis of the data?

19. The following is a map segment showing the location of loci r, s, and t:

r 12 s 10 t

The coefficient of coincidence in the region is equal to 0.5. The following cross is made in the fruit fly:

$$\frac{r^+s^+t}{rst^+} \times \frac{r^+s^+t^+}{\text{Y chromosome}}$$

Answer the following based on the expectation of 4,000 offspring:

A. How many will be wild-type females ($r^+s^+t^+$)?
B. How many males will be of the constitution r^+st?
C. How many males will be r^+s^+t and rst^+?

20. In the fruit fly, the allele for sepia eye color (se) is recessive to the allele for red (se^+). Straight wings (cu^+) are dominant over curled (cu). The two genes, se and cu, are linked on an autosome, chromosome III. The amount of recombination between them is 24%. A cross is made between a female from a sepia-eyed stock and a male from a curled stock. The F_1s are wild (red-eyed with straight wings). They are crossed to one another and produce an F_2.

A. What kinds of offspring will appear in the F_2 and in what ratio?
B. The se gene is also linked to the e gene (for ebony body). The allele e is recessive to e^+ for gray body. Diagram a cross between a female from a sepia stock and a male from an ebony stock. Can you tell the amount of recombination between the two genes on the basis of the F_2 results? Explain.

21. Crossing over is the event that produces crossover or recombinant gametes. Suppose a and b are linked and a dihybrid is carrying the combination of alleles in the cis arrangement:

$$\frac{a^+\ b^+}{a\ \ b}$$

Assume that in every 10 meiotic cells crossing over takes place in 5 cells.

A. What would be the percentage of crossover gametes and the percentage of noncrossovers?
B. Suppose a crossover event occurs between a and b in every meiotic cell. What would be the amount of recombination?

Further Aspects of Linkage and Crossing Over

9

Family pedigrees and linkage

Until 1967, the study of family pedigrees was just about the only method available for recognizing human linkage groups. Linkage of genes to the X chromosome can be established much more readily than the linkage of two or more genes on an autosome. Consequently, the first advances in assignment of loci were made with the X. Linkage of genes to the X chromosome is more easily detected because pedigree analysis may show more affected males or a crisscross pattern of inheritance. More than 100 human genes have been found to be sex linked. The actual map distances have been calculated for some of these through analyses of many pedigrees and the application of the fundamental principles of sex linkage and genetic recombination. For example, knowing that the locus for deutan color blindness and that for the production of the enzyme glucose-6-phosphate dehydrogenase (G6PD) are sex linked, one can proceed to study family pedigrees that include doubly heterozygous mothers and their sons. The pedigree must be complete enough to establish whether the alleles in any heterozygous woman are in the cis or the trans arrangement. Such information can be obtained if the pedigree also includes information on the woman's father. If the latter, for example, is color-blind and cannot produce normal G6PD, whereas the woman expresses both dominant traits, then we know that her genotype is

$$\frac{gd^{+}d^{+}}{gd^{-}d}$$

We can deduce this because her father can only give her an X chromosome carrying both recessives.

Similarly, we might reason that another woman carries these alleles in the trans arrangement. The pedigrees are then examined with respect to the sons of such women for whose genotypes the cis or the trans arrangement is known. When this was actually done, it was estimated that about 1 son in 20 of doubly heterozygous mothers showed new combinations of the alleles, meaning that recombination occurred in about 5% of the cases (Fig. 9-1). The two loci can be considered roughly 5 map units apart. Studies of this type have established the map distances for several X-linked genes. The loci for hemophilia A and protan color blindness are so closely linked to those for deutan color blindness and G6PD production that the precise map distances in the linkage group are still unknown.

Some genes known to be located on the X chromosome, however, seem to be assorting independently from other genes whose loci are also known to be X linked. The reason for this is undoubtedly the length of the X (see Chap. 8); thus two loci that are appreciably far apart are frequently separated by crossover events and so appear to be assorting independently. Many students of human genetics would say that two such genes are syntenic (on the same chromosome) but that linkage has not been established between them. In our discussions, we will continue to use the term *linkage* in the classic sense of two genes being linked if they are on the same chromosome.

In contrast to the X chromosome, when working with family histories it is much more difficult to establish that two or more loci are linked on an autosome. Even when this is accomplished, pedigree analysis

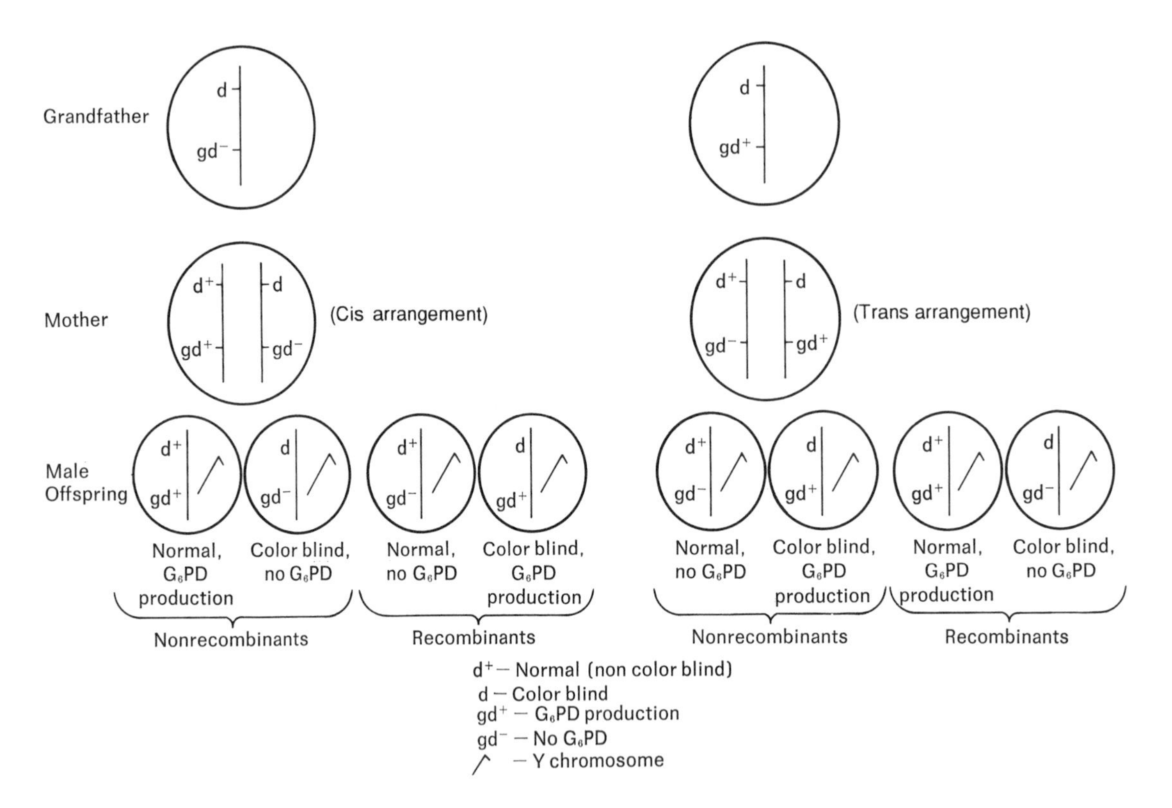

FIGURE 9-1. Uses of family pedigrees in linkage analysis. To ascertain the amount of crossovers between the locus for deutan color blindness and the locus for normal G6PD production, pedigrees must be complete enough to establish whether the doubly heterozygous mothers of sons being studied carry the genes at these loci in the cis or in the trans arrangement. This requires information on the maternal grandfathers. Once the arrangement is determined, a count of the sons is made to determine how many are recombinants and how many nonrecombinants. In this case, 5% of the sons of doubly heterozygous mothers were found to be recombinants, giving a map distance of 5 units between the two loci.

does not allow assignment of the linkage group to a specific autosome. The most useful pedigrees are those in which one parent is doubly heterozygous and the other doubly recessive (a testcross). Again, it is necessary to know whether the double heterozygote has the alleles under consideration in the cis or trans arrangement for precision. Today, family pedigrees continue to be used in the assignment of loci to linkage groups and in the mapping of chromosomes, but complementing this is the important method of somatic cell genetics that has greatly accelerated the assignment of genetic loci in the human (discussed later in this chapter).

Application of linkage knowledge to the human

Since the hemophilia allele is sex linked, we know that there is a 50% chance that a woman carrying it on one of her X chromosomes will pass it to a son who will be afflicted with the disorder. Knowledge of the close linkage now known to exist between the hemophilia A and G6PD loci can be put to excellent use to enable us to narrow down the chances for such a woman to have an affected offspring. Suppose it has been established that a woman known to be a carrier for hemophilia received the defective allele from her hemophilic father. She also received from him the allele enabling her to produce normal G6PD (Fig. 9-2A). Her mother donated an X chromosome carrying the normal allele for blood clotting as well as the recessive for lack of normal G6PD. The woman in question therefore carries the two pairs of alleles in the trans arrangement. Since the two genes are less than 1 map unit apart, we know that the parental combinations will tend to remain together in the preponderance of the woman's gametes. Assuming for the sake of simplicity that they are 1 map unit apart, we can predict that 99% of the gametes will be parentals and only 1% recombinants (Fig. 9-2B). Since the hemophilia allele is carried along on the same X as the allele for normal enzyme production in half of the parentals, the chances are great that any son who produces normal G6PD will suffer from hemophilia. A nonafflicted son will most likely produce variant or no G6PD. Fetal cells may be obtained during pregnancy by amniocentesis. If the karyotype proves to be that of a male and if there is variant or no G6PD production, the chance is very high, 49.5 of 50, or 99%, that the woman is carrying a baby free of the disorder. This is so, since 50% of the gametes carry the gd^- allele, and they are of two types. One type carries the normal h^+ in combination with gd^-, and it occurs with a frequency of 49.5% (a parental). The other combination, h with gd^-, is a rare crossover with a frequency of only 0.5%. Therefore, only 0.50/50 (1%) of the gd^- sons in such a case would also carry the hemophilia allele. We see here clearly the value of linkage in enabling prenatal diagnosis to be more exact in estimating risk of a disorder.

Myotonic dystrophy was the first serious human affliction to be related to a gene that is a member of an autosomal linkage group. The linkage relationship was established years before the gene was assigned to chromosome 19. The dystrophy locus was found to be linked to two others, one that is responsible for the blood antigens of the Lutheran grouping and the other that determines the secretor trait, the possession of an enzyme resulting in the secretion of red blood cell antigens into various body fluids (see Chap. 5). When the secretor allele (Se) is carried by a fetus, the antigens are secreted into the amniotic fluid (see Table 5-4). This fact, plus that of close linkage between the secretor and dystrophy loci, has a valuable application, because the dominant dystrophy allele has reduced penetrance and delayed onset of expression. Moreover, it produces no detectable product in body fluids and therefore lies undetected until it strikes a victim (see Chap. 20 for molecular studies of myotonic dystrophy).

Fortunately, the knowledge of linkage can guide us in some cases. The secretor gene can be used as an excellent marker, since the expression of either of its alleles can be easily detected (secretion of antigens vs. no secretion). Approximately 10% recombination takes place between the secretor and dystrophy loci.

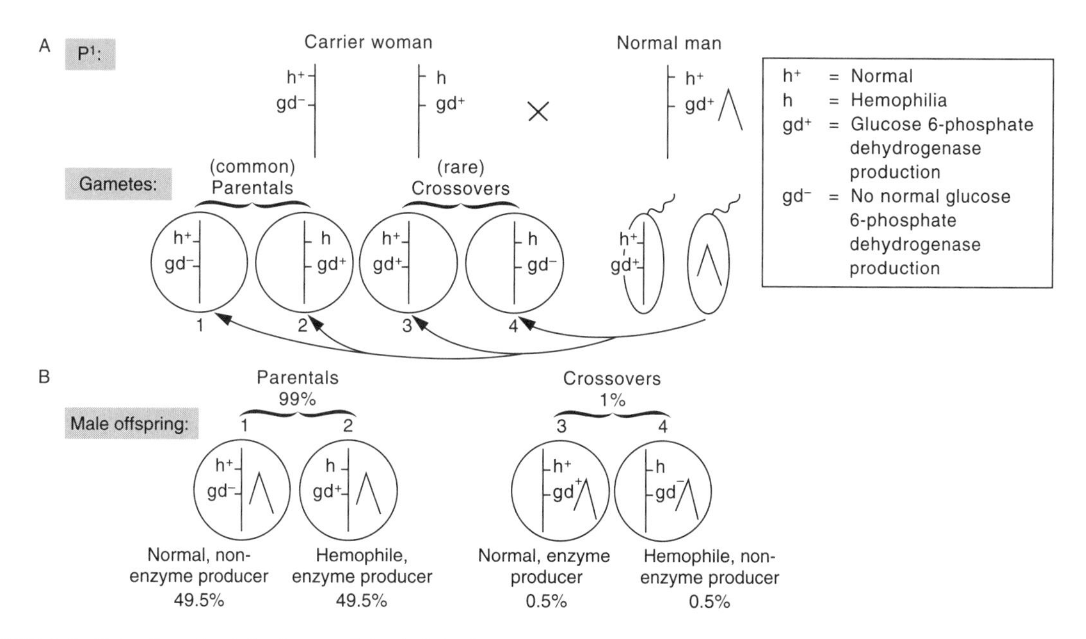

FIGURE 9-2. Detection of a defective sex-linked allele. In this pedigree, the P_1 female parent is known to have received an X chromosome from her father who carries the recessive hemophilia allele and the dominant for the production of G6PD. The X she received from her mother carries the contrasting gene forms. This information, plus the fact that the two loci are tightly linked on the X, can be put to practical use in prenatal diagnosis. Any male offspring receives one X from his mother, making the chance 50% that any son of the P_1 woman will have hemophilia. The enzyme locus can be used as a marker to check for the presence in utero of the defective clotting allele. Because the two loci are closely linked, the crossover eggs will be much rarer than the parentals. Therefore, if the amniotic cells indicate a male offspring and if the normal enzyme is not present, the chances are 99% that the unborn son has received the parental X carrying the normal allele for blood clotting.

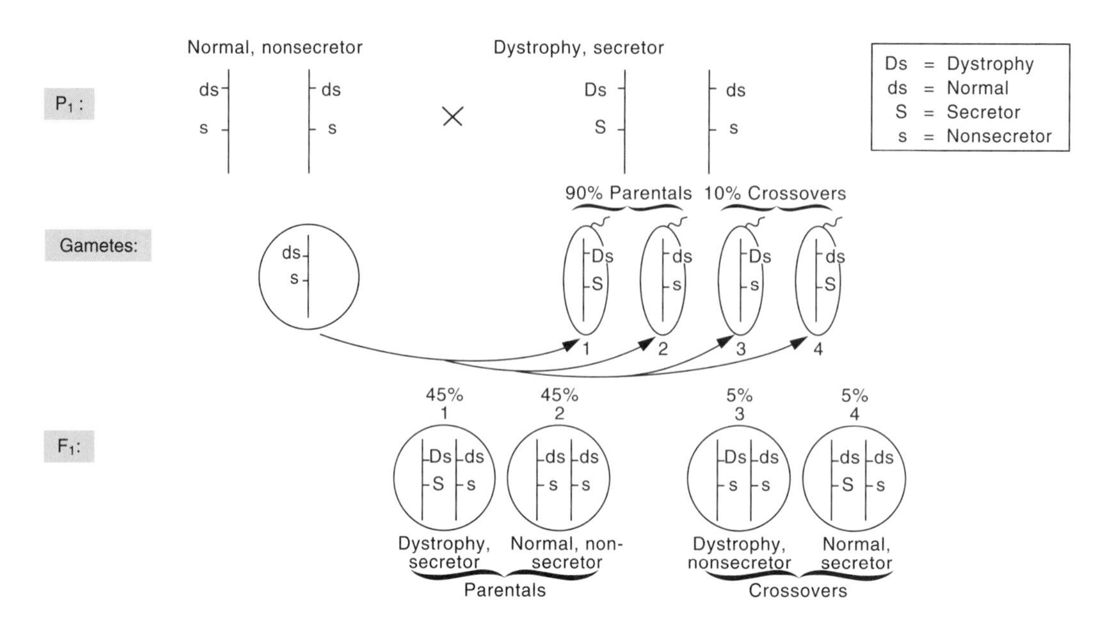

FIGURE 9-3. Detection of a defective autosomal allele. The locus associated with myotonic dystrophy and the secretor locus, which determines the ability to secrete certain antigens into body fluids, are autosomally linked. In this case, the male parent is known to carry the dominant factors for dystrophy and for the secretor ability on one chromosome; the recessive alleles are on the other. His wife is homozygous recessive. Since only 10% recombination takes place between the two genes, the crossover gametes produced by the male would be much less common than the parentals. If the amniotic fluid reveals blood antigens, this means the secretor allele is present and the embryo probably carries the dominant for dystrophy.

Suppose a man is known to carry the dominant secretor allele on the same chromosome as the dominant for dystrophy (Fig. 9-3). The man's wife is found to be a nonsecretor. Since she is phenotypically normal, she is not carrying the dominant dystrophy allele. If a couple such as this is concerned about the possibility that their unborn child will be afflicted with the disorder, amniotic fluid can be withdrawn. If it gives a positive reaction for the presence of red blood cell antigens, this means that the secretor allele from the male parent is present. Since this allele is closely linked to the locus for dystrophy, only a small probability exists for crossing over to take place. Therefore, the probability is very high (90%) that the fetus carries the harmful allele. If, on the other hand, the fluid tests negative for the antigens, this means that the offspring received from the father the chromosome that carries the nonsecretor allele. Since this is closely linked to the normal allele, the chances are 90% that the dystrophy allele is absent.

Without the knowledge of linkage between the dystrophy and secretor loci, we would only know in such a case that the chance is one-half for the dominant disease allele to be transmitted to an offspring. This and the previous example illustrate well the value of the role of suitable markers along with linkage information in aiding the genetic counselor to make more precise estimations of the chances that an undesirable allele is present.

The value of marker genes

Techniques employed in molecular biology are rapidly providing a host of chromosome markers and are thus facilitating the establishment of linkage relationships between disease genes and specific chromosome sites (see Chap. 20). This in turn is making it possible to isolate genes associated with serious human afflictions and is also permitting more and more precise estimates to be made of the degree of risk for persons in families afflicted with certain disorders. Since the ability to recognize genetic markers is in large part responsible for the burst of activity in locating and isolating genes as well as in aiding genetic counseling, it is important to understand and appreciate the value of a suitable genetic marker.

Keep in mind that if a gene is to be useful at all as a marker, it must exist in more than one allelic form and must be associated with some detectable phenotypic effect. Typically, this is the production of a protein or polypeptide. In the case of certain genes, one allelic form may be associated with a variant form of a protein in contrast to the typical protein associated with the standard or wild form. This was the situation with respect to the G6PD gene discussed in the previous section. In other cases, the presence or absence of some product or products may be associated with different allelic forms of a gene. The secretor gene has value as a marker, since its two allelic forms are associated with the presence or absence of blood antigens in body fluids. These two contrasting phenotypic effects, each associated with a different version of the secretor gene, permitted an estimate to be made of the chance that the defective dystrophy allele was present.

Obviously, if no variation at all exists in a gene, it is of no use whatsoever as a marker. It would be impossible to detect linkage between an invariant gene and a gene at any other locus in the genome (Fig. 9-4A).

If a gene exists in two allelic forms (such as A vs. a), and is associated with two versions of a protein, it may be useful as a marker. In a family pedigree, it may be possible to ascertain that an inherited disease is being transmitted along with one of the alleles (A or a) of a gene. As shown in Figure 9-4B, the presence of gene form A is detected by the presence of protein A, and this in turn indicates the presence of the disease-causing dominant, allele D.

However, if a gene exists only in two versions, its value as a marker may become somewhat limited, especially if one of the allelic forms occurs at a much higher frequency than the other in the population. If form A, for example, is very common, the chances of heterozygosity are decreased. No information may be gained in certain families where determination of risk is to be made. The situation would often be the same as that depicted in Figure 9-4A.

Those genes that are particularly valuable as markers are those that exist in more than two allelic forms (form a series of multiple alleles). The several different versions of such a gene may be associated with slightly different versions of a protein. Detection of the latter increases the chances of establishing linkage between it and some other gene. A marker gene with multiple allelic forms makes it much easier to estimate risk in pedigrees, since the chance that any one person is heterozygous for different versions of the marker are greatly increased (Fig. 9-4C).

Until recently, only about 30 markers of any great value were available to detect linkage to a locus associated with a disease. These valuable markers exist as multiple alleles, each series associated with a protein polymorphism, different versions of a given protein product. Due to the scarcity of suitable markers, genes related to many genetic disorders have not been associated with any other genes. As a result of the failure to establish linkage relationships, the chromosome locations of many serious disease genes have remained unknown. This lack of information of linkage has also prevented the determination of more precise estimates of risk. Fortunately this rather bleak picture is now rapidly changing due to the methodology of molecular biology. In Chapter 20, we will become acquainted with valuable chromosome markers that are enabling us to pinpoint the location of many defective genes. As a result, isolation of human disease genes is now progressing in addition to improved estimations of risk in the cases of many disorders in which the product of the gene in question remains unknown.

Somatic cell hybridization

Mapping by somatic cell hybridization entails the fusion of somatic cells of two different species. When grown together in cultures, cells from different sources fuse on rare occasions. It was found that the frequency of cell fusions can be greatly increased by adding to the culture certain chemicals such as polyethylene glycol or by adding Sendai viruses that have been inactivated by ultraviolet light. Such agents can alter the properties of cell membranes in such a way that fusion between cells is enhanced, even fusions between cells of very different species, such as the mouse and the human. The human cell types commonly employed in such a procedure are the fibroblast, obtained from bits of skin growing in tissue culture, and the white blood cell.

After fusion occurs between cells from different genetic sources, a cell is formed containing two nuclei. Such a hybrid cell containing two genetically different nuclei is called a *heterokaryon.* When the heterokaryon undergoes mitosis, the chromosomes from the two nuclei align on the single spindle. Two hybrid cells are produced, each with one nucleus containing chromosomes derived from the two different species. However, a problem remains after cell fusion has occurred; this is the need to select out only the hybrid cells that are growing amid all the other kinds of cells in the culture—those that are unfused and those that have resulted from the fusion of two cells of the same origin, such as two human cells and two mouse cells. The problem has been essentially solved

A, a, a^1, a^2, a^3, a^4 – alleles associated with different proteins
D – dominant disease allele
d – normal recessive allele

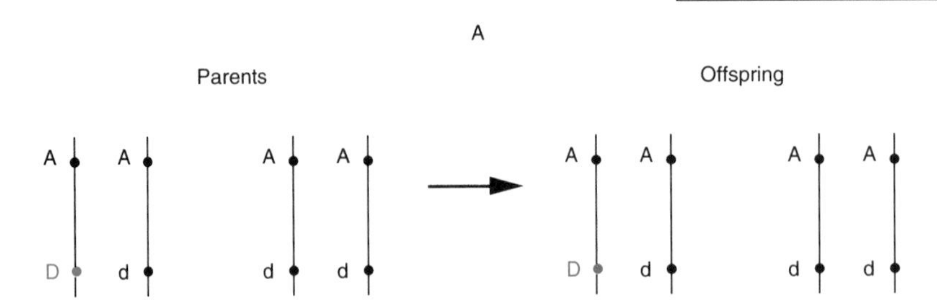

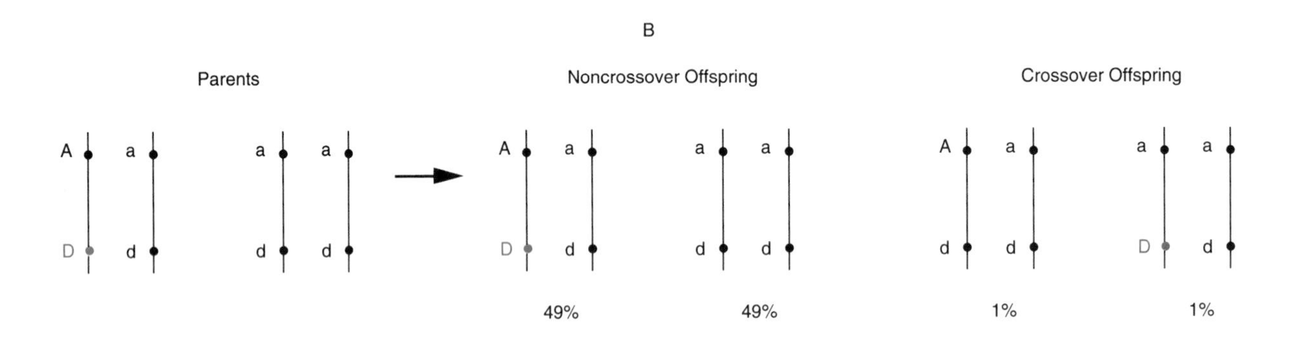

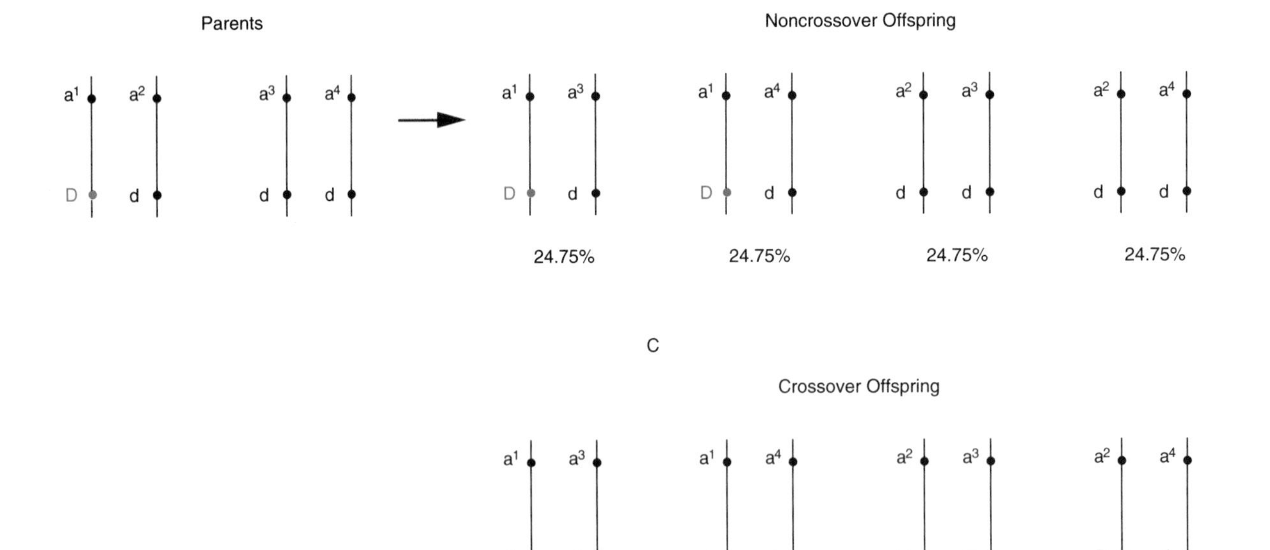

by the development of a technique that facilitates the screening out of the hybrid cells.

To appreciate the basis of the procedure, a few essential points must be grasped concerning the growth requirements of cells. Among other things, a cell must be able somehow to obtain the necessary materials to construct its nucleic acids if it is to multiply in a cell culture. The hereditary material, DNA, contains units called *purine* and *pyrimidine nucleotides* (see Chap. 11). These nucleotide requirements may be met by two general pathways in a cell (Fig. 9-5). Through the *de novo pathway,* a cell can build up its purine and pyrimidine nucleotides starting only with very simple materials. The other general pathway is referred to as the *salvage pathway,* so called because a cell following this route can use more complex substances, such as breakdown products of nucleic acids, from which it can salvage the required purine and pyrimidine components of nucleic acids.

Certain mutant cells, however, are unable to use this salvage pathway. One such cell type lacks the enzyme hypoxanthine phosphoribosyltransferase (HPRT) and is designed $HPRT^-$. (It is this enzyme that is deficient in the Lesch-Nyhan syndrome discussed in Chapter 7.) The $HPRT^-$ cell cannot form purines from substances in the salvage pathway and must therefore make them de novo. Other mutant cell types are unable to carry out another aspect of the salvage pathway, because they lack the enzyme thymidine kinase (TK). These TK^- cells can make the necessary pyrimidine-containing nucleotide (thymidylic acid) along the de novo pathway, but they cannot bring about the essential step (adding a phosphate) required to form thymidylic acid from thymidine (a precursor that lacks the phosphate).

This knowledge of the cell pathways has been put to good use to screen out only hybrid cells on a selection medium known as the HAT medium. The medium is so called because it contains hypoxanthine (which can be used by $HPRT^+$ cells as a purine source using the salvage pathway), aminopterin (an antimetabolite that inhibits the de novo pathway), and thymidine (which can be converted to thymidylic acid by TK^+ cells in the salvage pathway).

Now let us suppose that the human cell line used in a cell hybridization procedure is $HPRT^-$. This population cannot grow on the HAT medium, because its cells depend on the de novo pathway to form purines, and the aminopterin in the medium prevents this. Only a cell that is $HPRT^+$ can use the hypoxanthine. If the other parental cell line, the mouse line, is TK^-, it is prevented by the aminopterin in the HAT medium from forming the thymine-containing nucleotide thymidylic acid along the de novo pathway. Lacking the required enzyme, it cannot use the supplied thymidine using the salvage pathway. The only cells that can survive on the HAT medium are those that are $HPRT^+/TK^+$. Only the hybrid cells between the two species will have both of the metabolic requirements to grow on the HAT medium.

As the human–mouse hybrid cells divide, there is a pronounced tendency for the human chromosomes to be lost at cell divisions while the full complement of the rodent chromosomes is maintained (Fig. 9-6). Eventually cells are produced with just a small number of human chromosomes. Such single cells can be isolated to establish stable hybrid cell lines from which no further chromosome loss occurs. Each cell line constitutes a clone, since the members of the particular line arose from one cell and must therefore be genetically identical. The human chromosome loss that takes place is a random one so that the established stable clones differ with regard to the particular combinations of human chromosomes that they possess. This very fact has been put to good use in associating a particular trait with a specific human chromosome.

FIG. 9-4. Value of a gene as a marker. **A:** Gene A is of no value as a marker, since it exists in only one form. Protein A associated with the gene is present in all persons. No detection of linkage and the disease gene is possible. Crossover offspring would be the same as the noncrossovers. **B:** If a gene exists in two allelic forms, it may be possible to detect linkage between it and another gene in certain pedigrees. In this pedigree, protein A is associated with persons expressing a genetic disease, which is transmitted as a dominant. Once the amount of recombination has been estimated from studies of various family histories, an estimate of risk may be possible. In this case, assuming that the amount of recombination has been found to be 2% between A and D, then an offspring producing A protein (and hence carrying version A of the marker) would be at 98% risk of having the disorder. **C:** When a marker exists in several allelic forms, the chances are increased that any two persons will be heterozygous for different alleles of that gene. This makes it much easier to deduce that a defective allele is being transmitted in a family along with a specific variant protein coded by an allelic form of the marker. In this case, the defective allele is being transmitted along with version a_1 of the gene, and its presence is in turn detected by presence of a_1 protein. If the amount of recombination between the two genes is found to be very low, say 1%, we know that the marker gene lies close to the gene associated with the disease. In this case, it would be possible to say that an offspring lacking a_1 protein would have only a 1% chance of carrying the defective allele. In the absence of linkage information, all that could be said is that any offspring has a 50% chance of receiving the defective allele from a parent.

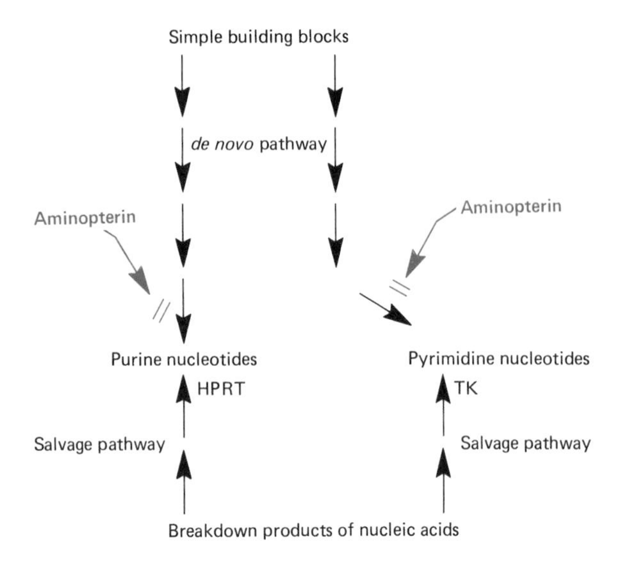

FIGURE 9-5. Summary of de novo and salvage pathways in synthesis of nucleotides. A cell requires purine and pyrimidine nucleotides to construct its DNA. A cell can form these in two general ways, by starting with a very simple compound (de novo pathway) or by using more complex substances resulting from nucleic acid breakdown (salvage pathway). The enzymes HPRT and TK are needed in the salvage pathway. A cell lacking either of these enzymes ($HPRT^-$ or TK^-) would be unable to perform a necessary step in the salvage pathway and would thus depend on the de novo pathway. The antimetabolite aminopterin can block steps in the de novo pathway, thereby preventing the construction of DNA requirements along this route (see text for details).

Actually the first gene to be assigned to a chromosome by the somatic cell procedure was the TK gene, the one responsible for the production of thymidine kinase, which we have noted is needed for growth on the HAT medium. The human line used in the hybridization was TK^+, and the mouse line was TK^-. No clones were established that lacked human chromosome 17. In all stable, established cell lines, human chromosome 17 was present along with the human enzyme thymidine kinase. Therefore chromosome 17 must carry the genetic factor responsible for the enzyme.

The cell hybridization procedure, to be useful in assigning a gene to a chromosome, depends on the ability of the parental genes to be expressed in the hybrid cell. This appears to be the case for many genes in the human–rodent hybrid cells. An added advantage is that a given enzyme found in both the mouse and the human (e.g., G6PD), although it is very similar and performs the same function in both species, exists in different forms—a human form and a rodent form. This allows us to identify a specific enzyme in a cell population. The cell hybridization method also permits assignment of genetic loci in the human that are not associated with the production of a biochemical product. For example, loci affecting the susceptibility of cells to certain viruses and drugs can be assigned by the procedure. Extremely important is the fact that the association of two or more genes with the same chromosome by cell hybridization permits the identification of linkage groups. The genes for HPRT, G6PD, and the enzyme phosphoglycerate kinase were all shown to be linked on the X chromosome. The three enzymes are always present in those clones with an X and are absent when the X is not present in a cell line. The locus for the gene responsible for G6PD was already known to be X linked from pedigree analysis.

Stable hybrid clones covering all of the human chromosomes are available for use in assigning an unmapped locus to a chromosome. Each clone has its unique combination of human chromosomes. Let us suppose we wish to assign the locus of a gene associated with a certain enzyme (enzyme A) to a given chromosome. We have available the three clones A, B, and C (Table 9-1). We find that human enzyme A is detectable in clones A and B but absent from C. We can now deduce that the genetic locus for the enzyme is on chromosome 2. We can easily see this from Table 9-1 by comparing the enzyme pattern (whether it is + or – in a clone) with the chromosome pattern (whether a specific chromosome is present or absent in a clone). It is the vertical columns in each case that we compare. We can quickly see that the vertical pattern + + – for the enzyme is the same as the vertical pattern in column 2 for the chromosomes. None of the other vertical columns are + + –. This is so because clone C lacks chromosome 2. Chromosome 1 cannot be associated with enzyme A, since that chromosome is present in clone C, which lacks the enzyme. Chromosome 2 is the only other chromosome that clones A and B have in common.

Microcell transfer and somatic cell hybridization

A variation of the somatic cell hybridization procedure makes it possible to introduce just a few human chromosomes or even just a single intact one into a recipient cell. This method, known as *microcell transfer,* entails the treatment of cells with high doses of the mitotic inhibitor colcemid, a derivative of colchicine.

Exposure to the drug inhibits the assembly of microtubules and blocks the metaphase stage of the mitotic cycle. Since no functional spindle can arise, the chromosomes become scattered throughout the cell. When the cells eventually leave the M phase of the cell cycle, nuclear membranes reform around scattered chromosomes. As a result, the cell comes to possess several micronuclei containing single or small clusters of chromosomes. Another chemical, cytochalasin B, is then applied to the multinucleate cells. This substance is a fungal metabolite that can prevent cytokinesis and that is able to cause nuclei to bulge or extrude from cells. When centrifugation is carried out in the presence of this drug, the protruding nuclei break off from the multinucleate cells on a thin stalk of cytoplasm. The small micronuclei, each surrounded by a thin layer of cytoplasm, are known as *microcells* and can be collected from the bottom of the centrifuge tube. A microcell can then be fused to recipient cells in a manner similar to that described for whole cells. The transfer of just a single human chromosome at a time into a recipient cell by the microcell transfer method has yielded valuable information on the chromosome locations of certain very important genes, such as those involved in the suppression of tumors (see Chap. 20).

It is important to note that somatic cell hybridization does not give the same kind of linkage information as pedigree analysis. The somatic cell procedure tells us that two or more genes are syntenic, associated with the same chromosome. It tells nothing, however, about the distance between two loci. We rely on classic genetic linkage analysis to give us this information. The two methods actually complement each other. For example, if the family method shows that two or more loci are autosomally linked and a certain map distance apart, the cell hybridization method may establish the exact chromosome to which one of the members of the linkage group belongs because the one gene is associated with a product (say an enzyme) that can be followed in somatic cell studies. The locus for the ABO blood grouping does not produce a detectable product in cell culture and thus cannot be assigned to a chromosome using the cell fusion procedure. However, family studies showed the ABO locus to be linked to a locus that controls the production of an enzyme that is detectable in cell cul-

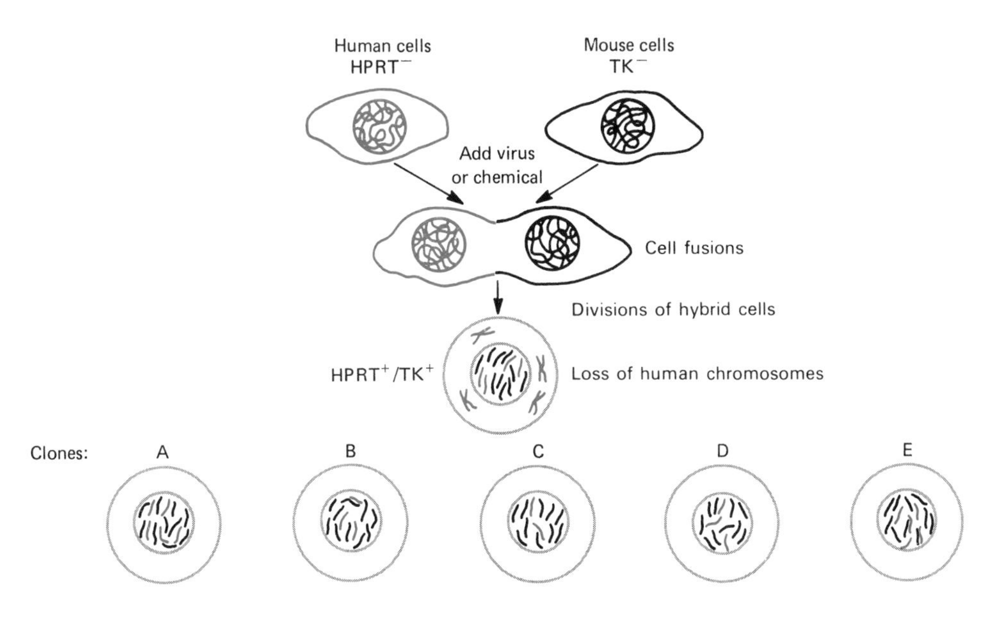

FIGURE 9-6. Somatic cell hybridization. Hybrid cells may be formed from the fusion of human and mouse cells. Only hybrid cells in this example can grow on the HAT medium, since they are $HPRT^+/TK^+$. In the subsequent divisions of the hybrid cells, the human chromosomes tend to become lost. Cells can eventually be isolated that contain just a few human chromosomes. Such cells may give rise to various stable cell lines, each line carrying a different combination of human chromosomes.

TABLE 9-1. Human chromosome content of three mouse–human hybrid clones

	Human chromosome content							
Hybrid clone	1	2	3	4	5	6	7	8
A	+	+	+	+	–	–	–	–
B	+	+	–	–	+	+	–	–
C	+	–	+	–	+	–	+	–

+ , present; – absent.

tures. Since we can show that this locus is associated with chromosome 9, the ABO locus to which it had been shown to be linked by family studies could be assigned to chromosome 9.

Somatic cell hybridization has been very useful in assigning genes to specific autosomes, a most difficult task if attempted on the basis of family histories alone. Well over 1,000 genes have now been assigned to autosomes, a feat made possible largely by somatic cell hybridization. It has also shown that the pictures of synteny in the human and in the chimpanzee are very similar, the same groups of genes being associated together with a specific chromosome and falling into a group. The human and the mouse, however, differ in this way. A very interesting point, however, is that in all mammals the same genes are X linked (recall Ohno's law, in Chap. 6).

Precise assignment of genetic loci to specific arms or parts of chromosome arms has also been achieved for some genes. This has been made possible by studying human cells in which chromosome breakage and rearrangement of some kind have occurred (see Chap. 10). For example, a piece of one chromosome may be shifted or translocated to another chromosome. The piece may be identified by studying the pattern of chromosome bands (see Chap. 2). If it is found that a band or bands are missing from one chromosome and have been inserted into another, cell hybridization may show that a certain gene product, usually associated with a given chromosome, is now associated with another chromosome. The knowledge that the band is missing from a specific location in the first chromosome and is now associated with a new position in another chromosome tells us that the locus associated with the product is normally found in the region of that band on the normal chromosome.

Deletions, chromosome aberrations in which segments of chromosomes are actually missing, have similarly been used to assign genetic loci to specific locations on given chromosomes. A given chromosome band or a portion of a chromosome arm may be missing in some cell lines. If absence of the band can be associated with absence of a product known to be controlled by a locus assigned to that chromosome, and if presence of the band is associated with presence of the product, then the particular genetic locus can be related precisely to that portion of the chromosome. For example, a person who was heterozygous for Rh blood antigen (Rh^+ is dominant over rh^-) was found to be producing both Rh^+ and rh^- red blood cells. The person was then found to have a cell line with a deletion in chromosome 1 in the short arm, away from the centromere. Thus the Rh locus, known to be on chromosome 1, was assigned to that specific region.

Somatic crossing over

We have noted that radiation can stimulate crossing over in somatic cells; this exceptional event has been found to occur in *Drosophila* and in many fungi. Let us consider the consequences of somatic crossing over using a hypothetical example from humans (refer to Fig. 9-7). Suppose a female zygote is formed that is dihybrid for the hemophilia trait and for the ability to produce normal G6PD dehydrogenase. The alleles are carried in the trans arrangement. An individual of this genotype typically expresses the normal blood condition and is able to manufacture the enzyme. At birth, every cell of the body should be identical, because normal mitotic divisions guarantee equal distribution of genetic material to daughter cells.

Figure 9-7 shows how somatic crossing over can upset this. In rare cells, the phenomenon of somatic pairing may take place spontaneously or as the result of exposure to radiation. A crossover may then occur between the centromere and the two loci we are following. This crossover event produces two metaphase chromosomes whose chromatids are not identical. Therefore, unlike a normal mitosis, the two chromatids that separate from each other and move to opposite poles at anaphase will be different.

One outcome is that two new cell lines can rise from the one original genotype. The individual would be a mosaic, a mixture of wild-type cells, of cells homozygous recessive for the hemophilia trait, and cells that lack the ability to synthesize normal G6PD. The extent of the mosaicism depends on how early in development the somatic crossover occurs. The earlier it takes place, the larger will be the number of cells in the two new cell lines, because the recombinant cells will have undergone more mitotic divisions before completion of embryology.

We see from this example that somatic crossing over can uncover recessives in the body, leading to patches of mutant cells scattered among normal cells. Obviously, this can produce undesirable effects. It would therefore be advantageous for crossing over to be suppressed in body cells. It is logical to assume that in the evolution of living things crossing over provided the first means of recombination, since the simplest diploid cells must have had only one pair of homologous chromosomes. And so the most primitive kinds of diploids must have relied on crossing over for recombination. As the diploid stage evolved and the body became differentiated between germ cells and somatic cells, strategies must have evolved to allow crossing over in the germ line, where it was needed to provide new combinations of linked alleles, but to prevent it in body cells where the effects could be harmful because of the production of mosaicism. Certainly there are devices that prevent crossing over at meiosis in one sex of certain species, such as the fruit fly and the silkworm. It seems likely that comparable genetic controls must have evolved to prevent its occurrence at mitotic divisions, to ensure a soma composed of cells with the identical genetic information.

Importance of crossing over to evolution

We can aptly conclude this chapter with a consideration of the important role of crossing over in evolutionary progress. An organism that is completely asexual would not have the mechanism of meiosis to bring about new allelic combinations through independent assortment. The only way a new combination can result is through gene mutation. Consider the two diploid members of the same species, with a chromosome number of 4, which are represented in Figure 9-

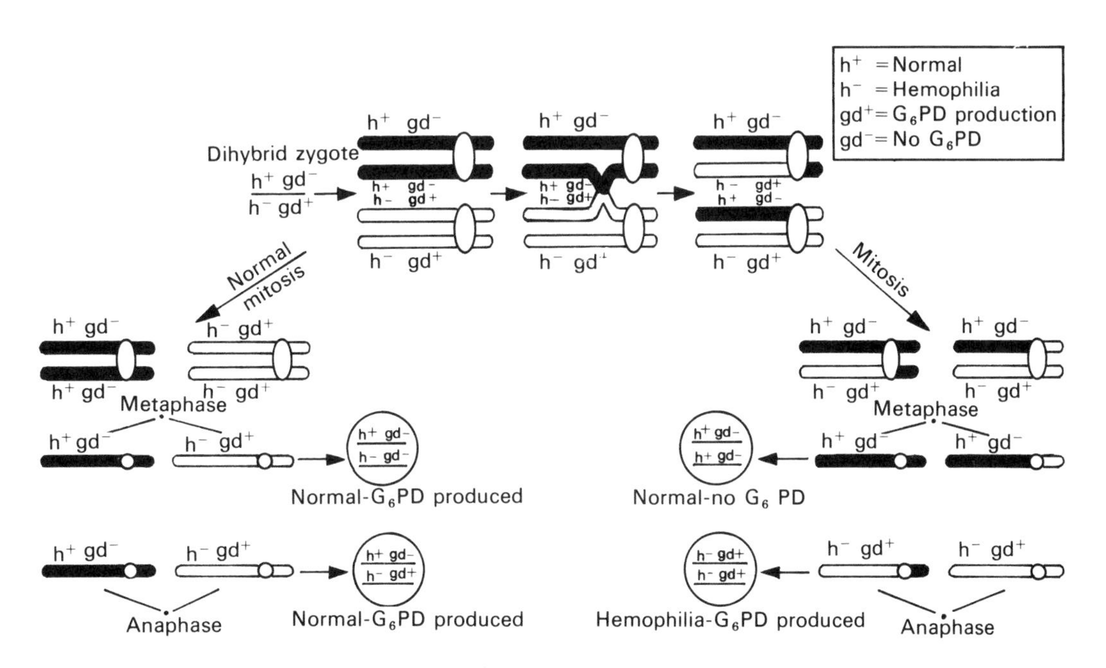

FIGURE 9-7. Effects of somatic crossing over. In this hypothetical example, a zygote is considered to be dihybrid at the hemophilia locus and the locus that governs the production of G6PD. Normally (left) mitotic divisions follow that ensure that every cell of the body will be identical and have the potential to make antihemophilia factor (AHF) and the enzyme G6PD. On very rare occasions, pairing of homologous chromosomes occurs in the body cells. If this is followed by a crossover event, the two resulting chromosomes carry chromatids that are not identical. These chromosomes may arrange themselves on the spindle in such a way that the products of the ensuing mitotic divisions are different. In this case, a cell results that can make AHF but cannot make normal G6PD. The other cell cannot make AHF but can produce the enzyme. Since most of the mitotic divisions will be normal, the majority of cells in the body in this case will have the potential to manufacture both AHF and G6PD. The individual will be mosaic, composed of three lines of cells. The earlier in development the somatic crossing over takes place, the larger the number of exceptional cells and the extent of the mosaicism.

8. One of them expresses the genotype ab^+, the other the genotype a^+b. The offspring will be identical to each parent: all ab^+ in the one case, all a^+b in the other. If two diploid organisms of these same genotypes were sexual, they would be able to produce haploid gametes that could unite to form dihybrids. These, in turn, by independent assortment can give rise to four kinds of haploid gametes. A cross between the two dihybrids will produce the familiar dihybrid ratio 9:3:3:1. We now have *four* different phenotypes (2^2) and nine different genotypes (3^2). Without the sexual process, there would be just two phenotypes and two genotypes, one for each of the cell lines. Much more variation is possible through sexual reproduction and the independent assortment that meiosis makes possible.

Now let us consider the importance of crossing over as a source of variation that greatly supplements that contributed by independent assortment. Referring back to Table 8-1, which diagrams a trihybrid testcross, it can be seen that if no crossing over takes place, then only two types of gametes will be formed—the old combinations. This means that the main advantage of sexual reproduction, the production of new combinations of alleles, would be lost for genes linked on the same chromosome. However,

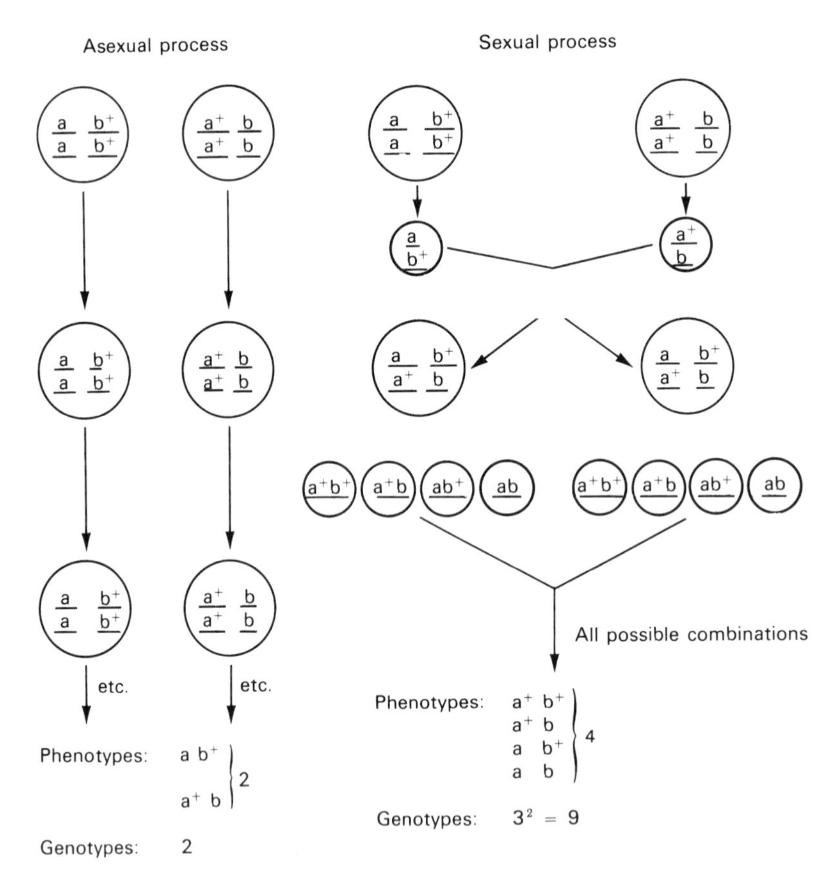

FIGURE 9-8. Asexual and sexual reproduction compared. If members of a species are completely asexual, they continue to reproduce their own type indefinitely (left). With two pairs of alleles, and assuming dominance, only two phenotypes and two genotypes are possible in the population. This will remain so until rare spontaneous mutation causes one of the genes to mutate. In contrast (right), the sexual process generates new combinations swiftly. The genetic material of the two types is brought together. With just two pairs of alleles, four phenotypes and nine genotypes are possible. These different combinations can vary in the advantages they give. Natural selection favors those that are more beneficial. Because the sexual process brings about more variation than the asexual in a given length of time, those lines that are sexual have evolved more quickly than those that are exclusively asexual. Crossing over speeds the process up by contributing even more to the variation effected by independent assortment.+, present; –, absent.

if all possible crossovers occur among the three genes, then eight different types of gametes are formed, the two parentals plus the six crossover types. It is evident that crossing over, along with independent assortment, contributes greatly to the variation generated by the sexual process. With crossing over, recombination is extended down to the level of the chromosome.

Inasmuch as crossing over typically occurs at every meiosis, new gene combinations are continually being formed. Crossing over, therefore, speeds up the production of new combinations of genetic material. Its contribution to the pool of variations in humans can also be appreciated by referring back to Chapter 1. When discussing new combinations provided through independent assortment, we noted that 2^{23} types of gametes would be a minimal number for a human. It should now be evident to us that each of the 23 pairs contains thousands of genes and that much heterozygosity is present, because the human is highly hybrid as a result of outbreeding. Since crossing over can occur at any point along a chromosome, the number of new combinations that can be generated is staggering. Each crossover point doubles the number of types of gametes. The new allelic combinations make possible a vast assortment of phenotypes. These are the variations on which the evolutionary force of natural selection can operate. The greater the number of variations, the greater are the chances for types better adapted to different sets of environmental conditions. The larger the number of different types, the greater is the chance for superior forms to occur in a given period of time and thus the faster the rate of evolutionary change. Therefore it is not surprising that sexual organisms have provided the main branches leading to the evolution of the diverse kinds of living species. The contribution of crossing over to the pool of variation, by extending recombination to linked alleles, has been a major factor in the evolution of higher forms of life and cannot be overestimated.

REFERENCES

Caskey, C.T., and G.D. Kruh. The HPRT locus. *Cell* 16:1, 1979.

Creagan, R.P., and F.H. Ruddle. New approaches to human gene mapping by somatic cell hybridization. In *Molecular Structure of Human Chromosomes.* J.J. Yunis (ed.), Academic Press, New York, 1977.

Creighton, H.S., and B. McClintock. A correlation of cytological and genetical crossing over in *Zea mays. Proc. Natl. Acad. Sci. USA* 17:492, 1931.

D'Eustachio, P., and F.H. Ruddle. Somatic cell genetics and gene families. *Science* 220:919, 1983.

Ephrussi, B., and M.C. Weiss. Hybrid somatic cells. *Sci. Am.* (April): 26, 1969.

Harper, P.S., M.L. Rivas, and W.B. Bias, et al. Genetic linkage confirmed between the locus for myotonic dystrophy and the ABH-secretion and Lutheran blood group loci. *Am. J. Hum. Genet.* 24:310, 1972.

Harris, R. Genetic counselling and the new genetics. *Trends Genet.* 4:52, 1988.

Mayo, O. The use of linkage in genetic counseling. *Hum. Hered.* 20:473, 1970.

McKusick, V.A. Mapping and sequencing the human genome. *N. Engl. J. Med.* 320:910, 1989.

McKusick, V.A. Genetic nosology: Three approaches. *Am. J. Hum. Genet.* 30:105, 1978.

McNeill, C.A., and R.L. Brown. Genetic manipulation by means of microcell-mediated transfer of normal human chromosomes into recipient mouse cells. *Proc. Natl. Acad. Sci. USA* 77:5394, 1980.

O'Brien S.J., and W.G. Nash. Genetic mapping in mammals: Chromosome map of the domestic cat. *Science* 216:257, 1982.

Ruddle, F.H. Linkage analysis in man by somatic cell genetics. *Nature* 242:165, 1973.

Ruddle, F.H., and R.S. Kucherlapati. Hybrid cells and human genes. *Sci. Am.* (July) :36, 1974.

Stern, C. Somatic crossing over and segregation in *Drosophila melanogaster. Genetics* 21:625, 1936.

REVIEW QUESTIONS

1. In the human, the allele for normal blood clotting (h^+) is dominant over the recessive allele (h), which results in hemophilia. The allele for a certain protein that can be detected in the blood (m^+) is dominant over its allele (m) for the absence of the protein. Both allelic pairs are X linked. A man with hemophilia has a daughter who is concerned about her chance of bearing a child with the disorder. It is found that the man does not have the blood protein but that his wife, the woman's mother, does. The concerned daughter also has the protein. She is married to a man without hemophilia who also has the protein. Prenatal analyses show that the daughter is carrying a male fetus that also carries the blood protein. What is the chance that the baby will have hemophilia, considering that a chromosome map shows the m and the h loci to be 13 map units apart on the X chromosome?

2. Assume that in the human a dominant allele (A) is responsible for a fatal nervous deterioration, which usually has its onset after age 20. The normal condition depends on the recessive allele (a). The presence of the dominant allele is associated with no detectable product before the onset of the disease. Assume that the locus associated with the disorder is

on an autosome and is found to be linked to a locus associated with a specific enzyme that can be detected in utero. The allele for enzyme production (E) is dominant over the recessive allele for absence of enzyme (e). Pedigree analyses indicate that the two loci, A and e, are 10 map units apart.

A man with the nervous disorder is also found to be an enzyme producer. Family history reveals he is a dihybrid with the alleles in the trans arrangement. The man's normal wife is not an enzyme producer. During her pregnancy, amniocentesis shows that the offspring is not an enzyme producer. What is the chance that the fetus carries the gene for the nervous disorder?

3. The ability to produce normal G6PD depends on a sex-linked dominant allele gd^+. The recessive gd^- produces defective enzyme. Also sex linked is the dominant allele P, which is necessary for normal color vision. The recessive allele, p, produces protan color blindness. The loci for the enzyme and for color vision are 6 map units apart. A dihybrid woman with alleles in the cis arrangement marries a man with normal vision who is a producer of normal G6PD. Diagram the cross and show the possible offspring and their expected frequencies.

4. In *Drosophila,* the loci for yellow body (y) and distorted or singed bristles (sn) are sex linked. The alleles for gray body (y^+) and normal bristles (sn^+) are dominant to their recessive alleles, y and sn. The arrangement on the chromosome, where the circle indicates the centromere, is as follows:

———0——sn————y——

A dihybrid zygote is formed with the alleles in the trans arrangement. Assume that somatic synapsis takes place and that it is followed by crossing over between sn and the centromere in one of the cells of the developing fly. What will be the consequences?

5. In the fruit fly, the locus for dumpy wings (dp) is on Chromosome II. The locus for sepia eyes (se) is on Chromosome III, and the locus for the eyeless trait (ey) is on Chromosome IV. The traits dumpy, sepia, and eyeless are recessive to the wild-type or normal traits. Suppose a mutation arises in a wild-type laboratory stock of flies and causes the development of extra hairs on the body. The genetic factor causing hairiness (h) is found to be recessive and not sex linked. The following three crosses are made and followed through the second generation: 1) hairy × dumpy; 2) hairy × sepia; 3) hairy × eyeless. The results of the F_2 are as follows: cross 1, 9 wild; 3 hairy; 3 dumpy; 1 hairy, dumpy; cross 2, 2 wild; 1 hairy, 1 sepia; cross 3, 9 wild; 3 hairy; 3 eyeless; 1 hairy, eyeless. What conclusions can be reached on the assignment of the h locus to a chromosome? Explain.

6. Human cells are fused with those of a mouse, and three clones of hybrid cells are derived with the human chromosome content given in the following panel (+, presence of a chromosome; –, absence). The clones are checked for the presence or absence of three enzymes, X, Y, and Z. Enzyme X is found to be present only in clone A. Enzyme Y is in clones A and B, whereas enzyme Z occurs in all three. Associate each enzyme with a specific chromosome.

	Human chromosomes						
Clone	1	2	3	4	5	6	7
A	+	–	+	–	+	–	+
B	+	+	–	–	+	+	–
C	+	+	+	+	–	–	–

7. Assume that production of a certain enzyme in the human has been associated with chromosome 1 through somatic cell hybridization. However, studies of cells from one person indicate that the enzyme is associated with chromosome 5. Explain this and how one might get evidence for it.

8. The recessives a, b, and c are sex linked in *Drosophila.* A cross of two parental strains results in the following F_1:

$$\frac{a^+b^+c^+}{abc} \quad \text{and} \quad \frac{abc}{Y}$$

If these two F_1s are crossed, which of the following would be true:

A. The amount of recombination cannot be estimated from the resulting F_2.
B. The amount of recombination can be determined from the F_2 from the male offspring alone.
C. The amount of recombination can be determined by considering all of the resulting offspring.
D. The alleles a and b are in the trans arrangement in the female F_1s.
E. Crossing over can be determined only from the F_2 females.

Changes in Chromosome Structure and Number

10

The importance of chromosome anomalies

Sudden inheritable changes are often called *point mutations,* since they represent modifications at specific points along the DNA. These are the alterations that lead to the origin of alleles, alternative forms of a gene. Point mutations are not, however, the only kind of modification that can bring about inheritable variation. Some changes alter the very morphology of the chromosome by modifying its structure in some way. Others entail the actual loss or gain of chromosomes. Although variations of this kind can be classified as *mutations,* the term is usually reserved for point mutations. The more gross modifications are generally referred to as *chromosome aberrations* or *anomalies.*

Chromosome anomalies in the human are of special concern to the physician, genetic counselor, and certain family groups. The frequency of newborns with defects attributable to some kind of chromosome anomaly is about 1 in 440. It has been estimated that among normal, fertile men, 10% of the sperm is abnormal due to some kind of chromosome anomaly. Of the successful fertilizations that do take place, at least 20% will result in spontaneous abortuses before the completion of pregnancy, and about half of these entail some kind of chromosome aberration! The figures given are undoubtedly conservative, since those spontaneous abortions occurring very early can go unrecognized. According to some estimates, at least one-half of successful fertilizations are eventually lost. Moreover, certain radiations are known to be potent producers of chromosome aberrations, and the list of chemicals found to alter chromosomes continues to grow. Such factors must be considered as routes through which additional chromosome rearrangements enter the population. We can well appreciate from their wide occurrence that a knowledge of chromosome aberrations is essential to an understanding of a host of genetic phenomena in humans and other species.

Inversions and their consequences

One main type of structural chromosome change that can occur spontaneously or that can be induced by radiations and chemicals is the *inversion.* As the name implies, a stretch of the chromosome becomes turned around (Fig. 10-1A). This requires two breaks followed by a healing of the breakage points. Note from Figure 10-1A that when an inversion arises the interstitial segment (cdef) does not attach to either of the original chromosome ends (a or g). The reason is that the ends of a chromosome are polarized and do not attach to other segments. These two extremities of each chromosome are called *telomeres* and are often described as being "nonsticky." In contrast, the chromosome segments between the telomeres are "sticky," since they can attach to one another, but not to the telomeres.

An inversion can arise in a somatic cell or in a cell in the germ line. Assume that a sperm contains a chromosome with an inverted segment and that it fertilizes an egg lacking the inversion (Fig. 10-1B). Any individual arising with a normal chromosome and with one that has been altered in some way is called a *structural*

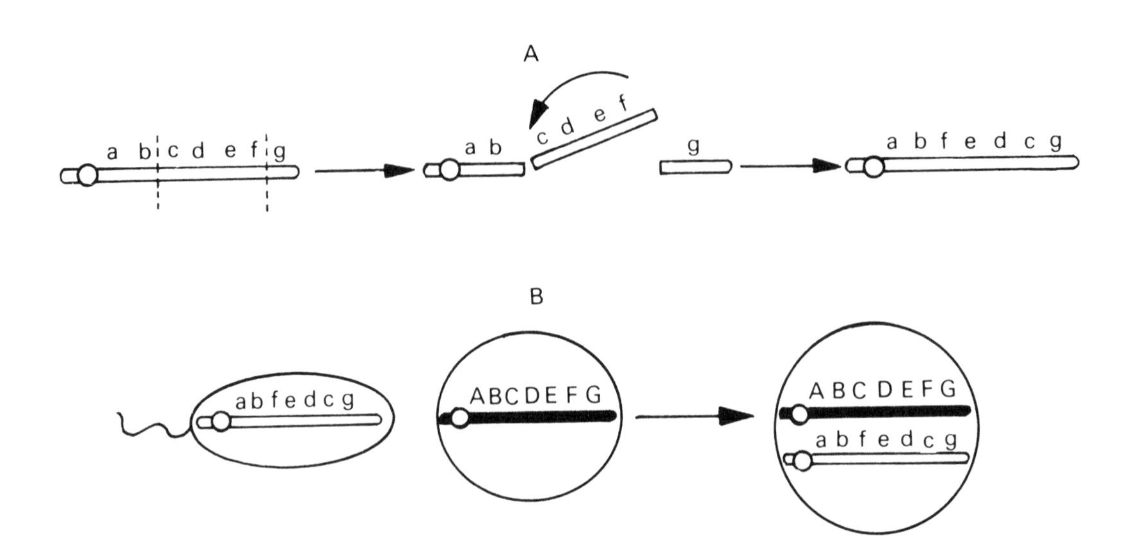

FIGURE 10-1. The inversion. **A:** Origin of an inversion after two breaks in a chromosome. **B:** A sperm carrying an inversion fertilizes an egg in which the homologous chromosome has the normal gene sequence. The result is a zygote that is an inversion heterozygote. (The centromere is represented by an open circle.)

heterozygote. The inversion heterozygote in this example is genically balanced, because there is no change in the kind or number of genes despite the change in gene order. Generally such an individual is normal and can live to maturity without any ill effects from the anomaly. However, most inversions are weeded out of populations, and the reason for this becomes apparent when the meiotic picture is considered. Two homologous chromosomes have pairing difficulties if one member contains a segment that is reversed. The forces of synapsis are powerful enough to ensure pairing of homologous regions if at all possible, even if the chromosomes must undergo contortions. In the inversion heterozygote, the pairing is accomplished through the formation of a loop in one of the chromosomes, either the normal or the inverted one (Fig. 10-2A). The two chromosomes may separate normally at anaphase I, and there would be no complications. However, if the inversion is large enough, a crossover event is certain to occur in some cells. In Fig. 10-2A, a crossover event is indicated in the loop between loci d and e, and its consequences are seen. As a result of crossing over anywhere within the inverted segment, two chromosomes are formed that are very abnormal in structure. One of them possesses two centromeres (a dicentric chromosome); the other lacks a centromere (an acentric). Both are also genically unbalanced; certain loci are represented more than once in each, whereas other loci are missing (in the dicentric, the a and the b loci are present twice but g is absent; the acentric has two doses of the g locus but lacks the a and the b).

If the acentric and the dicentric separate at first anaphase and move to opposite poles, the two nuclei that form will be genically unbalanced. But the affected chromosomes are apt to be involved in still further cytological problems. The dicentric will be attracted to both poles, the acentric to neither. Since the centromere is the dynamic center of the chromosome, the dicentric may be stretched as the two centromeres travel to opposite poles. At the level of the light microscope, this may be visualized as a bridge between the poles (Fig. 10-2B). Next to it may be seen a fragment, actually the acentric chromosome, that is unable to move at all and that will disintegrate in the cytoplasm after being left out of a nucleus. The stretched chromosome may break anywhere along its length.

In any case, genetically unbalanced cells will arise, and the consequence is a certain proportion of unbalanced eggs or sperm. If the gametes survive, any offspring resulting from them would be abnormal in some way or would fail to survive due to the imbalance. In general, inversion heterozygotes suffer from some decrease in fertility as a result of the acentrics and dicentrics that arise following crossing over in the inverted segment and are at a reproductive disadvantage, leaving fewer offspring in the long run than do individuals with normal chromosome structure. Consequently, natural selection tends to eliminate inver-

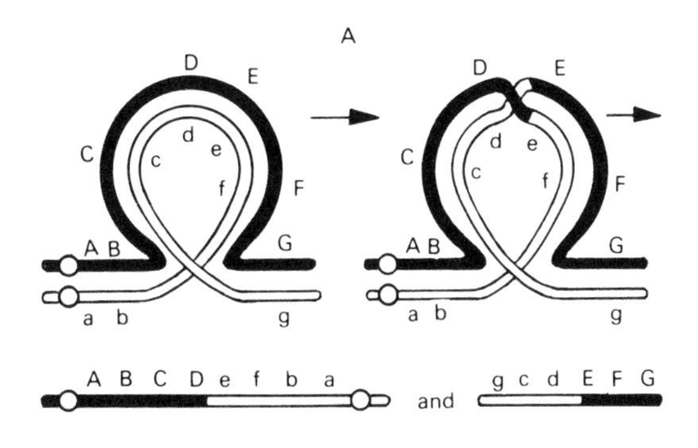

B

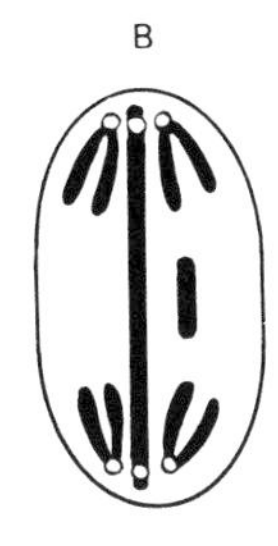

FIGURE 10-2. Pairing and crossing over in an inversion heterozygote. **A:** Loop formation is shown followed by crossing over. A crossover event between loci d and e (or any other two loci in the inverted segment) results in the origin of a dicentric chromosome and one that is acentric. Both are genically unbalanced. For the sake of clarity, each chromosome is represented as being single. Actually, at the time of crossing over, each is double, but only two chromatids engage in any one crossover event. **B:** An inversion may be detected with the microscope, because anaphase I cells will show a bridge and a fragment. These represent the dicentric, being pulled to both poles, and the acentric, which cannot move at all.

sions from a population. A tiny inversion has a greater chance of persisting than does a larger one, because the probability of crossing over within it is small.

Inversions in *Drosophila*

The outstanding exception to a low level of inversions in the wild is found in the genus *Drosophila.* In many fruit fly populations, inversions of appreciable size are very common and may actually be a normal part of the hereditary composition of the population. How can these persist without causing a reduction in fertility? We have noted that the fruit fly is unusual in that normally no crossing over occurs in the male. Therefore, the complications discussed earlier are not encountered, and no reduction in male fertility ensues from the inversion. In the female, however, crossing over is the rule at meiotic prophase. Still, female fertility is not reduced by the presence of the inversion. Remember that crossing over takes place in the four-strand stage. Only two of the chromatids of the bivalent participate in any one crossover event. In the case of an inversion, two chromatids usually are not involved in a crossover event in the inverted segment and are normal in their genic balance. The abnormal, dicentric chromosome is held back as a result of the chromatid bridge, and the acentric does not move at all and disintegrates in the cytoplasm. Consequently, one of the noncrossover chromatids (either the one with the inversion or the one with the normal order) completes meiosis and enters the egg (Fig. 10-3A). The overall effect of the chromatid bridge in the females of certain animal species, such as those of *Drosophila,* is to prevent the dicentric chromatid from entering the egg. The balanced chromatid carrying the inversion has just as much chance as the standard one to enter the egg. Since neither the female nor the male fruit fly experience a reduction in fertility, inversions can therefore accumulate in populations.

Inversions can actually contribute an advantage to a *Drosophila* population. The reason for this can be appreciated by considering the consequences of crossing over within the inverted segment (Fig. 10-3A). A single crossover event brings about the recombination of linked alleles, but in the case of the inverted region the recombination is associated with abnormal chromatids that do not enter the egg of the fruit fly. Only the old combinations (the chromatids with the standard arrangement and the one with the original inversion) are preserved to enter the gamete. Therefore the effect of the inversion is a decrease in the amount of recombination among genes in the inverted region. Crossing over and the recombination it generates are highly desirable aspects of the sexual process and have enhanced the variation available for natural selection. However, the coin has a reverse side (Fig. 10-3B). It is quite possible for a group of certain closely linked alleles to constitute a desirable combination that confers a distinct advantage. These can be destroyed by recombination, but the presence of an inversion ensures that the combination will persist from one generation to the next.

In some natural populations of fruit flies, nearly every individual is heterozygous for one or more inversions on each of its chromosomes. Studies of populations through several years have shown that the frequency of certain specific inversions varies characteristically with the season. Controlled laboratory conditions with fly populations have shown that similar variations in the frequency of a particular inversion type occur with temperature changes that duplicate those in nature. The evidence strongly indicates that a given inversion by preserving a certain combination of alleles confers a specific advantage. It is suspected that small inversions in many plant and animal groups may also provide a mechanism to preserve combinations of alleles favorable under a variety of environmental conditions.

Different types of inversion

Two major categories of inversions are recognized. In our discussion thus far, both breaks have been in one chromosome arm, to one side of the centromere. Such inversions are called *paracentric.* These are in contrast to *pericentric inversions,* those in which the breaks occur on either side of the centromere (Fig. 10-4A). The details of synapsis and loop formation are the same for the pericentric heterozygote as for

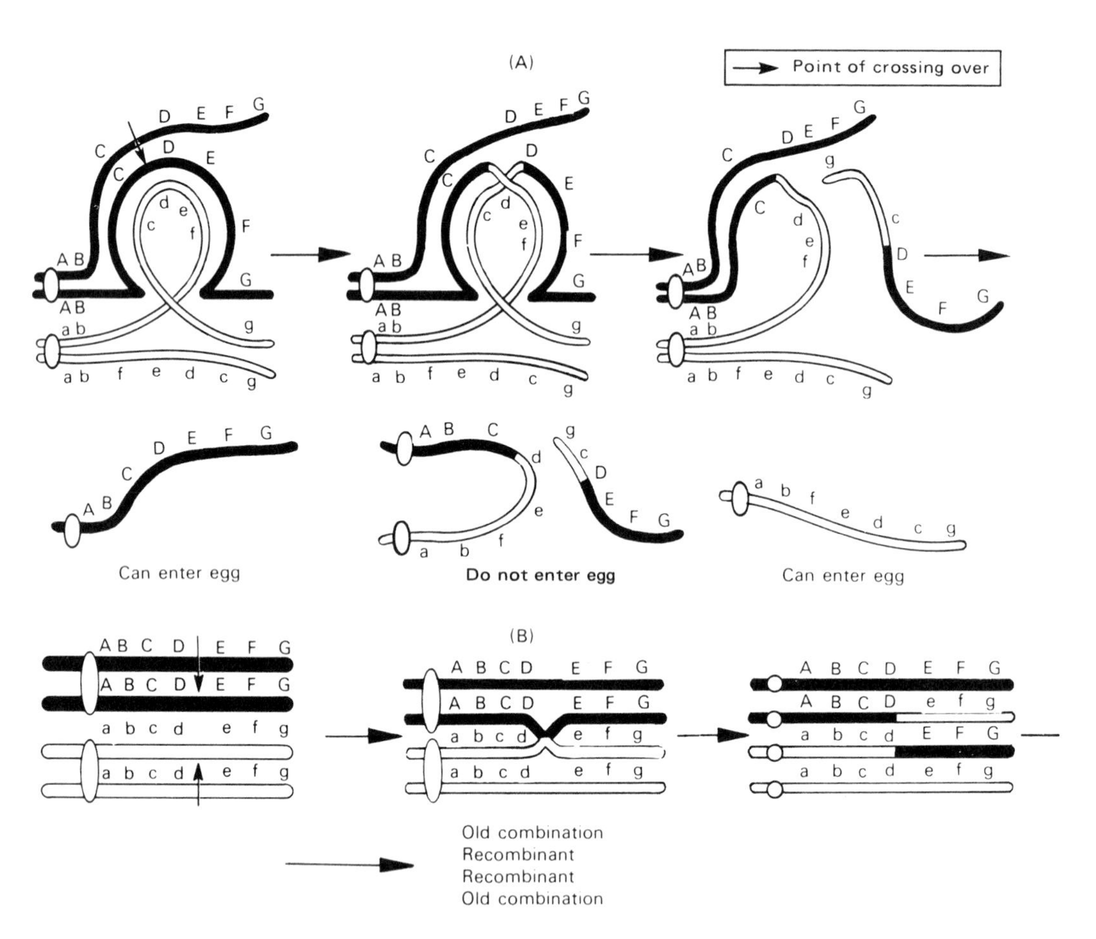

FIGURE 10-3. Bridge formation at oogenesis. **A:** Only two of the four chromosome strands participate in any single crossover event. As a result of the formation of the bridge, the defective dicentric strand is held back at the end of the first meiosis. The strands of normal morphology (those with just one centromere) become directed toward the "outside" of the cell. Upon completion of second meiosis, one of the outside nuclei becomes the nucleus of the egg. The defective dicentric and acentric chromosomes are never oriented toward the outside and do not enter the egg. Note that the egg will therefore contain one of the parental gene combinations, CDEF (the noninverted arrangement) or fedc (the inversion). Since only the parental combinations are present, the chance remains good that many individuals will be heterozygous for the inverted region and also heterozygous for the genes in the region. **B:** Suppose the gene combinations CDEF and cdef (in any order) give an advantage when present as a unit on a chromosome. If no inversion is present, any single crossover in the cdef region can break up the combination. The inversion, therefore, as shown in A, tends to preserve it. Without the inversion, there is also a greater chance for genes in the region to become homozygous.

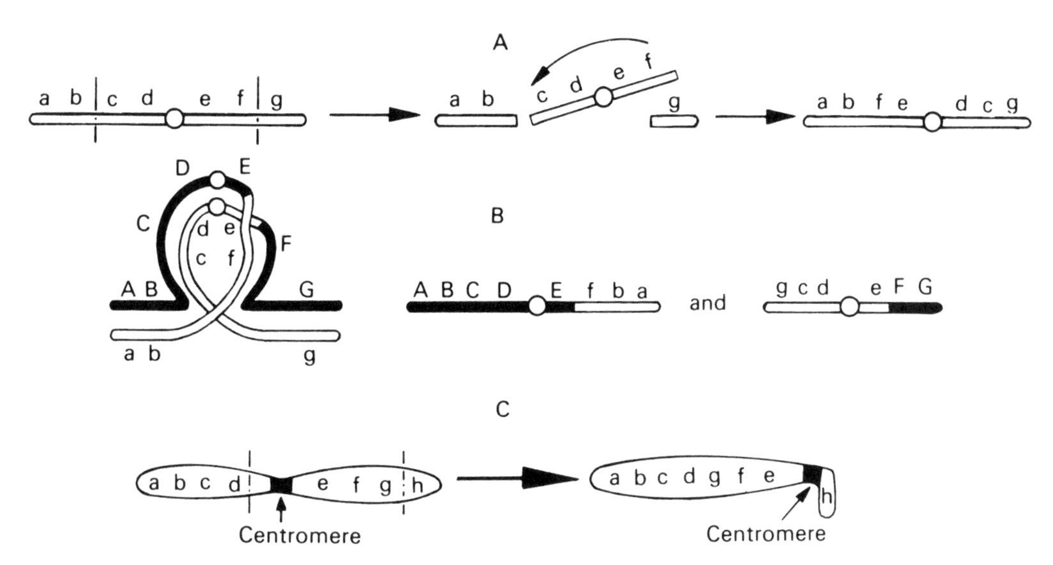

FIGURE 10-4. Pericentric inversions. **A:** Note that each break is on either side of the centromere. Contrast this with the paracentric inversion (Fig. 10-1A) in which both breaks are on the same side of the centromere. **B:** Crossing over anywhere in the inverted region in the case of a pericentric inversion will produce two chromatids, each normal morphologically (with one centromere each) but each genetically unbalanced. Note that in one chromatid loci a and b are represented twice and g is absent. The other has a double dose of g but lacks a and b. Only two chromatids are shown, for the sake of clarity. The other two would not participate in this crossover event and would be balanced parental types. However, because there is no bridge to hold back the unbalanced ones, the egg nucleus may receive a defective chromosome, and thus a decrease in fertility ensues. **C:** If the two breaks are not equidistant from the centromere, the centromere position shifts. In this case, it is shifted closer to one end, yielding an acrocentric chromosome from a metacentric one.

the paracentric, but a large difference is seen when the products of crossing over are examined (Fig. 10-4B). No dicentrics or acentrics are formed. While the two resulting crossover chromatids are normal morphologically, each is genically unbalanced. Since there is no dicentric chromatid, no chromatid bridge forms at anaphase. The two unbalanced chromatids are not held back at meiosis and have as much chance to enter the egg as do the balanced ones. The consequence is some reduction in female fertility, even in *Drosophila.* Therefore, we find pericentric inversions to be uncommon in fruit fly populations. However, small pericentrics have a chance to accumulate. From Figure 10-4C, it can be seen that if the two breaks on either side of the centromere are not equidistant, the centromere position will shift. Small centromere shifts resulting from pericentric inversions have almost certainly operated in *Drosophila* during evolution of the various species to alter chromosome morphology.

Once an inversion has occurred in a chromosome, additional ones may arise later, and each of these would usually involve different points of breakage. A second inversion in a chromosome may even have one point of breakage in a segment that has already been reversed by a previous inversion (Fig. 10-5A). Such overlapping inversions have occurred in the evolution of *Drosophila* populations. Within a fruit fly species, certain races have been shown to differ by two or more inversions (Fig. 10-5B).

In the human as well as other species, a pericentric inversion in a chromosome can be recognized by a shift in centromere position along with a change in the length of the two chromosome arms. Paracentric inversions in the human have been especially difficult to identify as well as those pericentric inversions in which the two breaks are equidistant on either side of the centromere. However, refinements in chromosome banding techniques are making it possible to detect different kinds of inversions much more readily in the human by revealing alterations in band pattern. Pericentric inversions are now known for almost all the human chromosomes, but so far paracentrics have been found on fewer than 10 of them. The data indicate that certain chromosome regions may be more susceptible to chromosome breakage and the production of inversions. With the improvements in detection of inversions, it may be possible to examine more thoroughly karyotypes of persons in families with a history of congenital disorders to determine whether any association exists between inversions and certain complex defects.

Deletions and their effects

The deletion, the very first type of structural chromosome change to be identified, was recognized in *Drosophila* in 1915. A deletion can follow a double break in a chromosome and entails the actual loss of a chromosome segment so that a stretch of genetic material is missing (Fig. 10-6). The affected chromosome actually becomes smaller as a result of the loss, but a size change may not be obvious when the eliminated piece is minute. A deletion is apt to cause some genic unbalance, and unexpected genetic effects usually ensue. For example, in the fruit fly the deletion of a sex-linked locus that affects eye development causes a nick in the wing tip. We see here a good example of the fact that a gene may be pleiotropic and thus have more than one effect, even on different body parts. As a consequence of a deletion, a recessive allele on one chromosome may unexpectedly

A

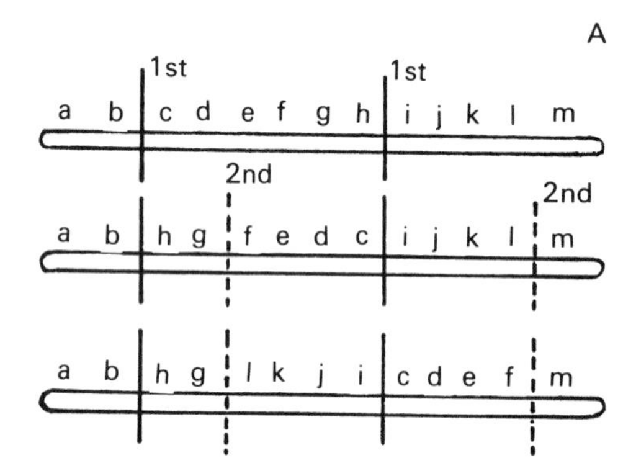

B

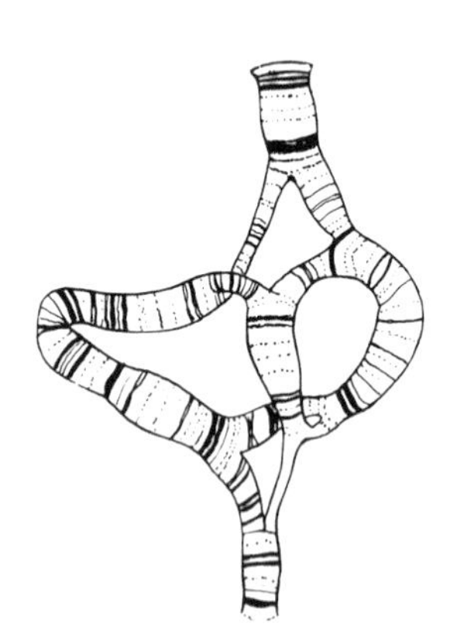

FIGURE 10-5. Simple and complex inversions. **A:** Origin of an overlapping inversion. Assume 1 is the standard order and that the first breakage points result in an inversion. When another inversion occurs, the second points of breakage can involve loci that were inverted in the first rearrangement. **B:** Simple loop (left) and compound one seen in the salivary gland chromosomes of *Drosophila azteca*. If two races are crossed and a simple loop forms in a chromosome pair, we know that the races differ by a single inversion. If a compound loop forms, we know there are overlapping inversions. By studying the loops and their complexities, the cytogeneticist can tell by how many inversions two races may differ. Relationships among races may be worked out by making use of this reasoning. (Reprinted with permission from *J. Hered.* 30:3, 1939.)

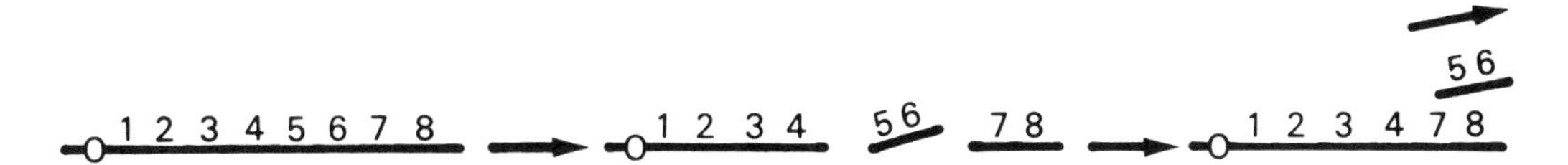

FIGURE 10-6. Formation of a deletion. Two breaks may arise in a chromosome, followed by loss of the interstitial piece. The portions with the centromere and telomeres rejoin.

express itself in a heterozygote after the loss of the dominant factor on the homolog. This phenomenon is known as *pseudodominance,* the expression of a recessive allele as a consequence of the elimination of the dominant by a deletion.

When a deletion becomes homozygous, it usually prevents survival and thus behaves as a recessive lethal. A deletion in an X chromosome can prevent development of males in species with the X–Y method of sex determination, since the hemizygous sex has only one X chromosome and would lack vital genetic material if a piece of the X were missing. Females homozygous for a deletion on the X typically fail to survive. If their effect on genic balance is not too severe, small deletions, autosomal as well as sex linked, may accumulate in a population to some extent. But when they become homozygous, they usually prevent development and act as recessive lethals. If a deletion is of sufficient size or includes certain critical genes, it may also act as a dominant lethal and kill the heterozygote.

Detailed studies of different deletions in the mouse and *Drosophila* have illustrated the crucial role of various genes at specific times in differentiation. A certain gene product may be required at a specific time for a critical step in development. If it is absent, a crisis ensues that can terminate development and prove lethal. Moreover, a deletion of any size can come to exist in the heterozygous state only if it can be transmitted through haploid cells. Deletions frequently prove lethal to the gametes of an animal, the mouse, for example. The haploid (gametophyte) generation of plants is especially vulnerable to deletions. Those genes that are essential to the formation of viable gametes appear to be located at various sites throughout the genomes of these species, which, unlike the fruit fly, cannot tolerate deletions of any substantial size in their sex cells or cells of the haploid generation.

The most common deletion in the human occurs in about 1 in 50,000 live births and entails loss of genetic material from the short arm of chromosome 5. The size of the deletion can vary, but it usually includes a small segment of band 5p15 and is associated with cri du chat (cat cry) syndrome, so named for the cry produced by affected infants. In addition to various characteristic facial features (Fig. 10-7), those with the deletion suffer from growth defects and mental retardation. Deletion of material from the short arm of chromosome 4 is the next most frequent human deletion and produces effects so severe that the babies die in infancy.

Most cases of Duchenne muscular dystrophy are now known to be associated with extensive deletions of segments of the X chromosome (see Chap. 20 for details). Of prime importance to cancer research is the accumulation of data implicating various chromosome deletions in the malignant transformation of cells. Retinoblastoma, an eye tumor in children, and Wilm's tumor, a kidney tumor in children, are two examples and are associated with deletions in chromosomes, 13 and 11, respectively. We will return to these as major topics in Chapter 18.

The importance of deletions in the assignment of human genes to specific chromosome regions was noted in Chapter 9. Deletion mapping in the human, which involves techniques of chromosome banding and somatic cell hybridization, along with pedigree analysis, has been very valuable in the localization and ordering of genes on a given chromosome.

Duplications and their effects

Another kind of structural chromosome change that can produce unexpected results is the duplication, a condition in which a portion of a chromosome is present in excess. If a region is present three or more times instead of the normal two in a somatic cell, it is said to be *duplicated.* One way in which duplication can arise is shown in Figure 10-8A,B. A simple translocation, or shift (details are given later), may place a chromosome segment in a new location in the chromosome. The chromosome containing the shift would still be balanced, and so there probably would be no unusual effects in an individual carrying the

A

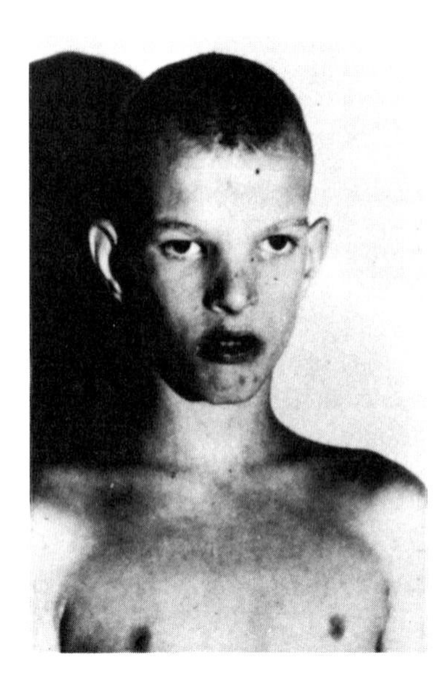

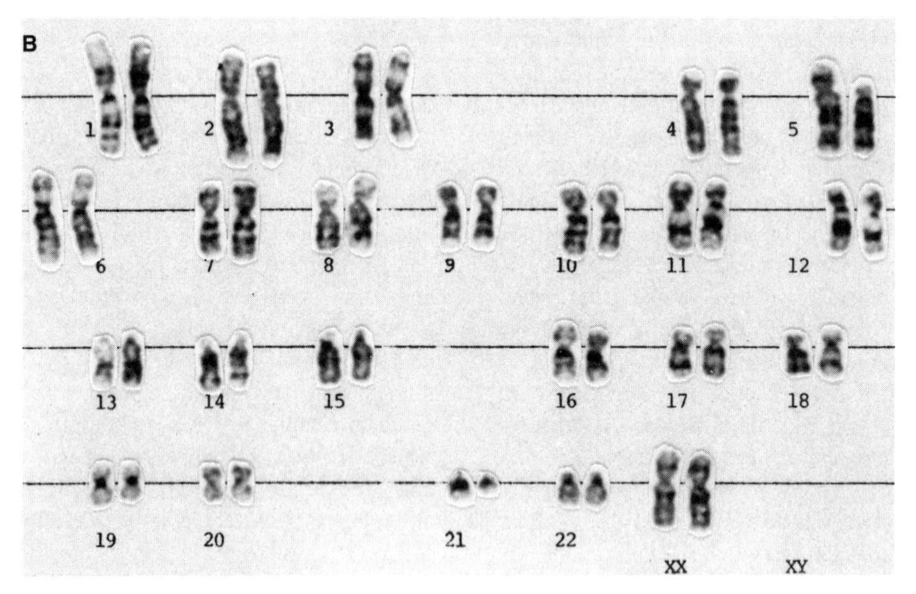

FIGURE 10-7. Cri du chat syndrome. **A:** Child showing features of the syndrome. (Reprinted from D. Bergsma (ed.). *Birth Defects: Atlas and Compendium,* The National Foundation—March of Dimes, White Plains, New York, with permission of the editor and contributor.) **B:** Karyotype of female with cri du chat syndrome. Note deletion in the short arm of one member of the chromosome 5 pair. The chromosomes are G banded. (Courtesy of George I. Solish, M.D., Ph.D., Jyoti Roy, M.S., Thomas Mathews, M.S., Long Island College Hospital, Cytogenetics Lab.)

altered chromosome along with one with the normal gene sequence.

Figure 10-8 shows that pairing problems will arise at meiosis. Two loops can form to permit synapsis of most of the homologous regions. Crossing over anywhere between the loops, however, leads to two altered chromosomes, one with a deletion and one with a duplication. The chromosome with the latter, as in the case of a deletion, can cause phenotypic effects because of an upset in genic balance. If it enters a gamete that then combines with a normal gamete, the zygote will contain some genes in three doses instead of two (the loci FG in Fig. 10-8C). The outcome depends on the size of the duplication and on the specific genes involved. In some instances, a dose of two recessive alleles along with a dominant results in the expression of the recessives. For example, a female *Drosophila* with two doses of the sex-lined recessive v (vermilion eye color) and one of the wild allele for red eye, v^+, has vermilion eyes. The genotypes v^+vv and vv are expressed the same phenotypically. In contrast, many examples could be given in which one dose of the dominant expresses itself over two doses of the recessive. In still other cases, intermediate mutant effects of some types are produced. An unsuspected duplication can very well lead to puzzling effects in certain matings.

Position effects

The unexpected genetic effects of a duplication are well illustrated by the bar mutation in *Drosophila*. The effect of bar is to decrease the number of facets in the eye so that it resembles a band or bar. The condition is inherited as a sex-linked incomplete dominant and was thoroughly studied in the laboratory of T.H. Morgan. Various matings involving bar-eyed effects were correlated with cytological examinations of banding patterns of the giant salivary gland chromosomes (see Chap. 2). The results revealed that the dominant bar effect is actually the result of a duplication of the bar

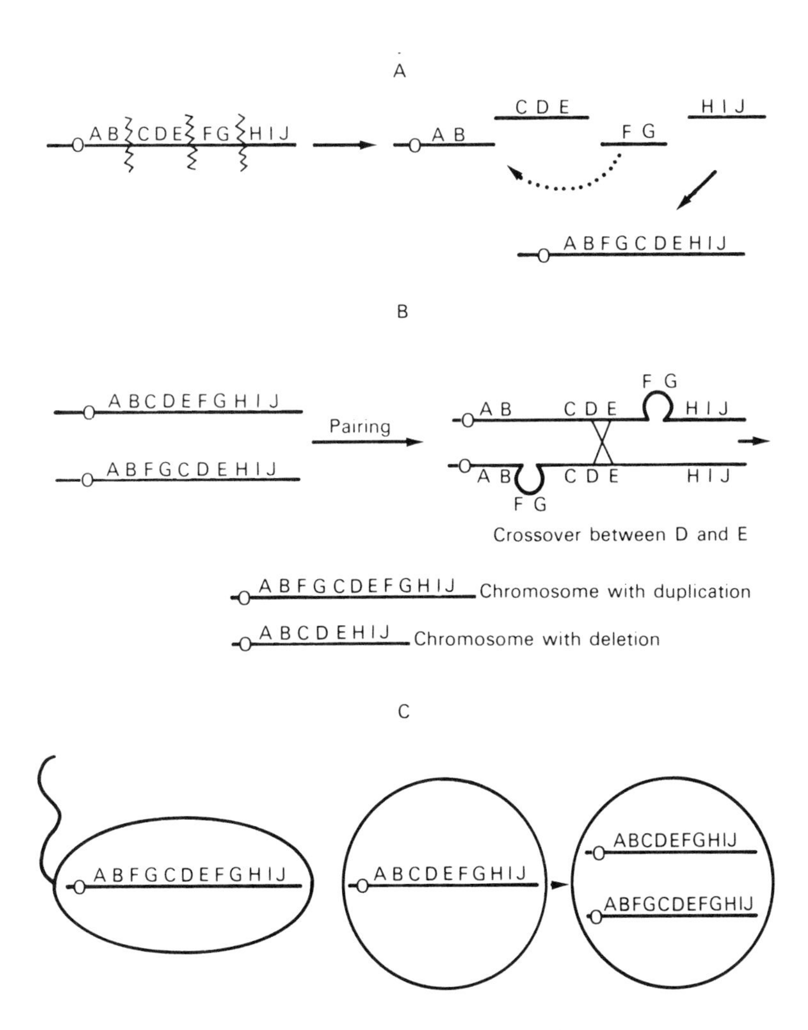

FIGURE 10-8. Origin of a duplication. **A:** A shift of a genetic region takes place after three breaks in one chromosome. **B:** At meiosis, the chromosome with the shift cannot pair exactly throughout its length with its homolog. A buckle forms in each chromosome to bring together most of the homologous regions. A crossover event anywhere between the loops results in a chromatid with a deletion (FG) and another with a duplication. **C:** A gamete with the duplication fertilizes a normal gamete. The zygote has one genetic region present three times. The region in excess (FG) is said to be duplicated.

locus on the X chromosome and is not a point mutation. Every X chromosome in a pure breeding stock of bar-eyed flies would carry two doses of bar adjacent to each other. Bar-eyed female and male flies would be, respectively, BB/BB and BB/Y in contrast to flies with normal eye shape, B/B and B/Y. The flies with normal eyes have just one dose of bar on each X. It is the duplication, two adjacent bar genes on an X, that produces the mutant phenotype.

Examination of flies showing different degrees of the bar effect (wider vs. more narrow eyes) revealed different genotypes in respect to the bar locus and led to a very important genetic discovery. Not only was the number or doses of the bar locus found to be important in affecting eye size, but the way in which the genes are arranged was also shown to play a role. For example, females known to possess three bar genes next to one another on one X chromosome and just one bar on the other (BBB/B) have much narrower eyes than do the typical bar females with two doses of bar on each X (BB/BB). This is so despite the fact that both types of flies possess four doses of the bar gene. This type of phenotypic effect resulting from the position of one gene with respect to another was named *position effect.* In this case, the arrangement of three genes together on one chromosome along with a single dose on the homolog differs from that of two genes together on each homolog. Position effects have now been demonstrated in organisms other than the fruit fly and must be taken into consideration when any structural heterozygote is studied.

Not all genes show position effects when placed in new locations; the particular gene involved and the new location both play a role. Very interesting effects are produced when certain genes are shifted to the vicinity of heterochromatin (constitutive and facultative heterochromatins are discussed in Chaps. 2 and 6). The majority of those genes familiar to us are located in euchromatic regions. It has been clearly demonstrated in the fruit fly that a dominant allele from the euchromatin placed next to heterochromatin tends to lose its dominant effect so that mosaic or variegated phenotypes arise (Fig. 10-9A). It is as if the heterochromatin were rendering the wild allele inactive in some cells, permitting the recessive to be expressed. If exposed to enough heterochromatin, a dominant allele may be unable to express itself in any of the appropriate cells. The dominant itself has not been structurally altered in any way when a loss of dominance occurs. In several cases, a rearranged dominant associated with a variegated effect has been restored to its normal location in the euchromatin. It then acts again as a full dominant, illustrating that it was the neighboring heterochromatin and not a gene mutation that was responsible for the variegation.

Not only can one allele at a time show the variegation effect. If a transposed euchromatic segment contains two or more alleles, then more than one of them can be influenced (Fig. 10-9B). The allele closest to the heterochromatin would show the greatest effect. The influence of the heterochromatin decreases with its distance from a particular euchromatic gene. If a more distant allele in a group is affected, those between it and the heterochromatin are certain to be affected also. This illustrates the spreading effect, the tendency of more than one allele to be repressed by the presence of heterochromatin. The closer the allele is to the heterochromatin, the greater is its chance of being influenced.

Position effect has been demonstrated in the mouse. A variegated effect and a spreading effect can result from translocations that place X-derived heterochromatin next to autosomal genes found in the euchromatin. Both facultative and constitutive heterochromatin can apparently bring about a position effect. The assortment of effects that can arise from duplications and from simple changes in gene position call attention to the complications that must be considered in cytogenetic analyses.

Translocation

A well-known chromosome aberration is the translocation, an alteration in which chromosome segments, or even whole arms, are transported to new locations. A simple type of translocation mentioned earlier, commonly called a *shift,* involves three breakage points. All three may be in one chromosome (Fig. 10-10A). Other simple shifts may involve two breaks in one chromosome and a single break in a nonhomolog (Fig. 10-10B). The segment from chromosome 1 is then inserted into chromosome 2. Some consequences of shifts were discussed in the preceding sections.

A shifting of chromosome segments is termed a *reciprocal translocation* when a mutual exchange occurs between two nonhomologous chromosomes. Reciprocal translocations have been studied extensively in many plant and animal species. Figure 10-11A illustrates a reciprocal translocation in which whole arms are exchanged. Assume that the alteration occurs during spermatogenesis, so that a sperm with the chromosome alteration fertilizes a normal

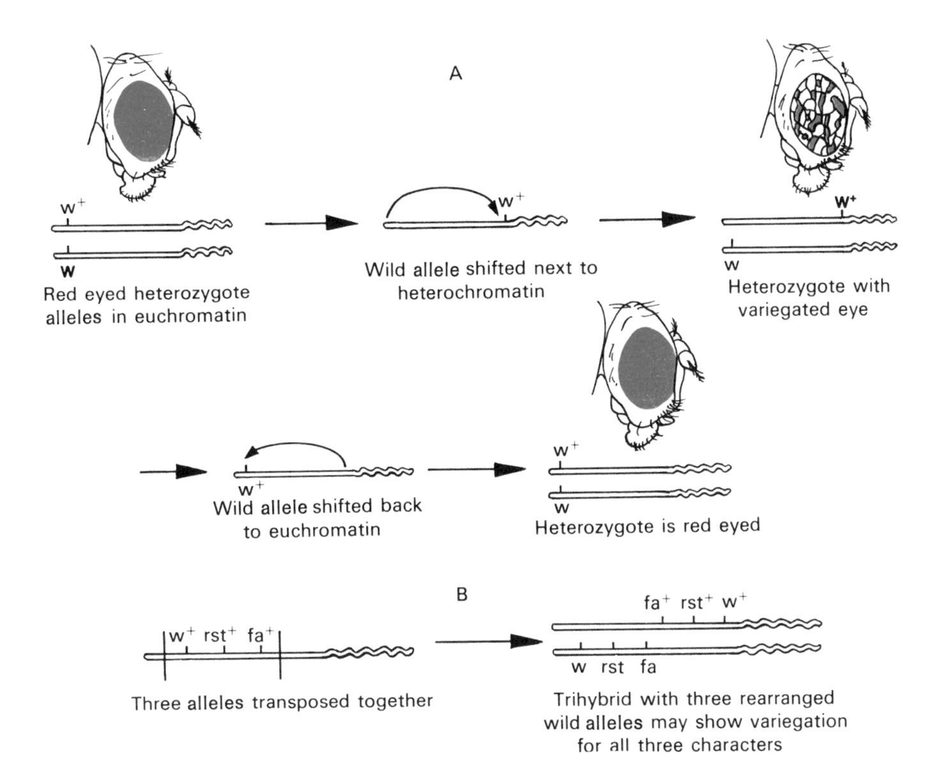

FIGURE 10-9. Heterochromatic position effect (wavy line indicates heterochromatin). **A:** Wild-type allele placed next to heterochromatin in the same or in a different chromosome after a chromosome alteration can result in failure of the wild allele to express itself in some cells. When returned to a normal euchromatic region after a second chromosome alteration, the wild allele again expresses itself in every cell. **B:** Spreading effect is seen when more than one gene is shifted to a region of heterochromatin. If the gene farthest away is influenced (facet, in this case), all those intervening up to the heterochromatin will also show the position effect.

egg. The result is a translocation heterozygote that, as in the case of the inversion, would be genically balanced and probably normal phenotypically. Like the inversion heterozygote, the individual with the translocated chromosomes would encounter problems during synapsis at first meiotic prophase. The forces of attraction may achieve pairing by the formation of atypical chromosome associations. When whole arm translocations are present, a characteristic cross figure forms that can be easily seen at pachynema in certain favorable material (Fig. 10-11B). All the homologous segments are paired, and crossing over may occur. The chief problem in the translocation heterozygote does not concern crossing over but rather disjunction, the manner in which the chromosomes will separate at anaphase. The association of four chromosomes at first meiotic division is termed a *tetravalent,* to distinguish it from the normal meiotic bivalent. At diplonema and diakinesis, the cross figure opens out into a circle as the forces of repulsion become operative. By first metaphase, the circle of four will be at the equatorial plate. The manner in which the chromosomes of the tetravalent disjoin determines the genic balance of the resulting gametes. Inspection of Figure 10-11C shows that if any adjacent chromosomes in the circle move to the same pole, unbalanced gametes arise. These contain extra doses of some genes and lack others entirely. However, if alternate chromosomes in the circle move to the same pole, balanced gametes are formed. Of those arising from a single meiotic cell, half carry the normal chromosome arrangement, and half have the translocated chromosomes. The end effect, then, of the reciprocal translocation is to produce a reduction in fertility in the structural heterozygote. Usually, the unbalanced gametes will not survive, or they will give rise to abnormal offspring.

In some cases, more than two homologous chromosomes may engage in reciprocal translocations. This increases the size of the circle or even the number of circles seen at meiosis, depending on how the arms have been distributed. For example, in the cultivated plant *Rhoeo discolor* all the chromosome arms in the complement have been shifted. The result is a multivalent association that includes all 12 chromosomes (Fig. 10-12). At times a chain rather than a circle forms in some meiotic cells. The overall effect of the large number of reciprocal translocations in this plant is to produce a certain number of unbalanced gametes because of irregular chromosome distribution at anaphase.

As is true of inversions, reciprocal translocations do not tend to accumulate to any extent in natural populations, but again there is one striking exception to the rule. This is found in the plant *Oenothera,* the evening primrose, in which the size and shape of the chromosomes favor alternate disjunction at meiosis, thus permitting the formation of balanced gametes.

Translocations and Down syndrome

Translocations in the human occur in about 1 in 500 live births, and their consequences are of great concern. As will be discussed in more detail later in this chapter, Down syndrome (Fig. 10-13) typically results from the presence of one extra chromosome, chromosome 21, because of nondisjunction at meiosis. Consequently, the Down syndrome individual usually has 47 chromosomes. However, this is not the only arrangement in those with this disorder; some have 46 chromosomes, the normal number. Karyotype analyses of such persons reveal that a structural chromosome change has altered the appearance of one of their chromosomes, which now appears longer than usual as a result of a translocation. Most translocations associated with Down syndrome involve the translocation of 21q (the long arm of chromosome 21) to one of the three members of the D group. Of the D translocations 60% involve chromosome 14; the rest are associated with chromosome 13 or 15. Let us examine a case in which chromosome 15 happens to be the chromosome involved.

Two normal chromosomes 21 are present, but a change is evident in pair 15 in this case. One of the members is normal, but the other is conspicuously larger and does not match its homolog. Therefore, the normal 15 remains unmatched. The change here is explained as the result of a translocation between chromosomes 21 and 15 (Fig. 10-14A). It can be seen in the figure that the new chromosome at the left contains most of the material from the two original chromosomes, 21 and 15. The tiny chromosome that forms may get lost. This kind of structural chromosome alteration that occurs in Down syndrome is an example of a *Robertsonian translocation.* In this type of translocation, the two nonhomologous chromosomes involved are acrocentrics (as is the case for members of the D and G groups), and the two chromosome

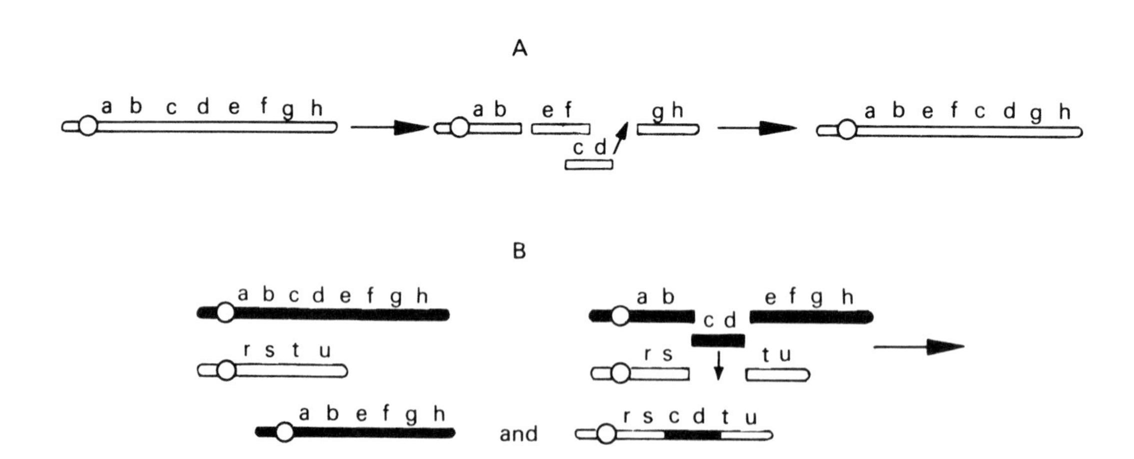

FIGURE 10-10. Shifts. **A:** A shift involving three breaks in one chromosome. The genes c and d are switched to a new location in the same chromosome. **B:** A shift involving breaks in two nonhomologous chromosomes. The two genes c and d are now switched to a new location in a different chromosome.

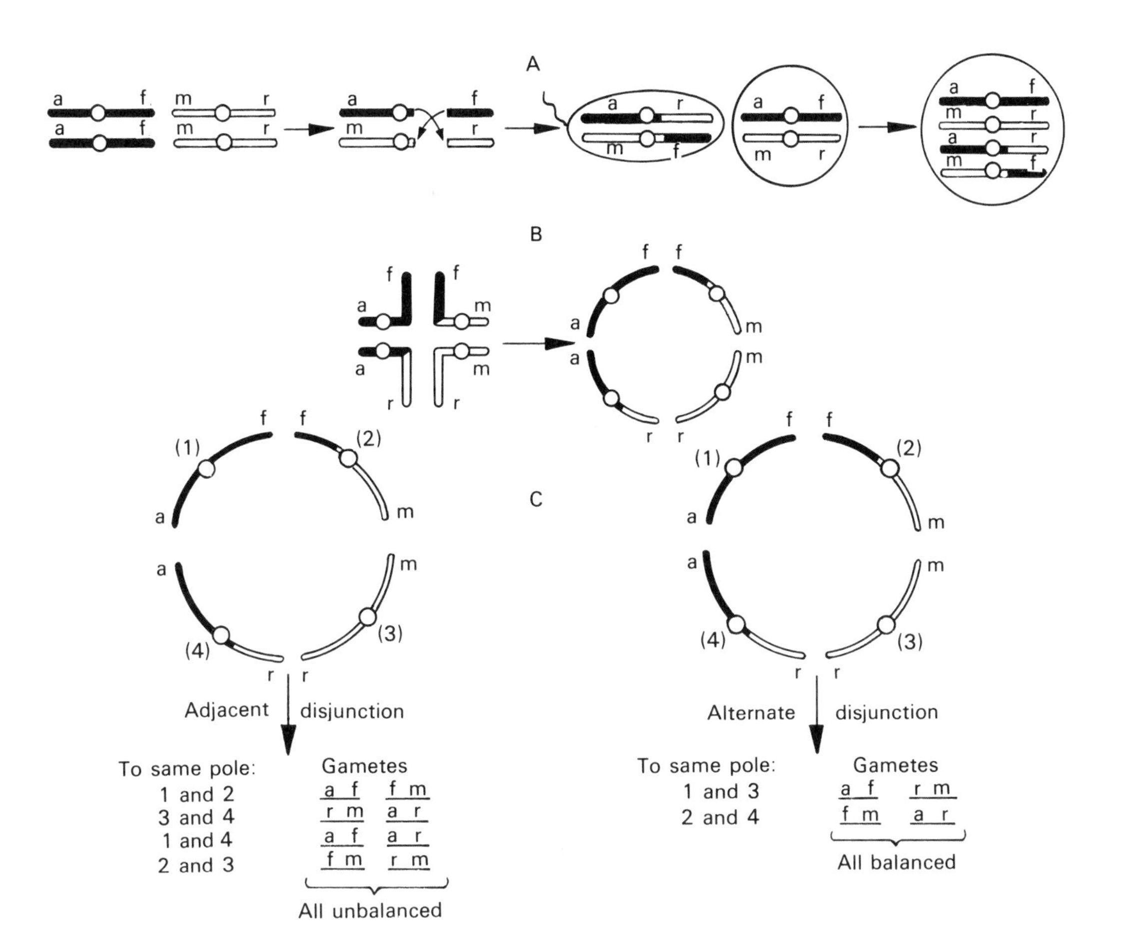

FIGURE 10-11. Reciprocal translocation and its consequences. **A:** Two pairs of nonhomologous chromosomes engage in a reciprocal translocation. This follows two breaks, one in each of two nonhomologous chromosomes, shown here as occurring next to the centromere. If a sperm carrying the translocation fertilizes a normal egg, the result is a translocation heterozygote. (Only the end genes are labeled in each arm, for clarity.) **B:** Since arms have been shifted between nonhomologous chromosomes, the pairing of homologous arms is achieved by formation of an association of four chromosomes, a tetravalent, seen as a cross in early meiosis and then as a circle by late diplonema. **C:** Adjacent disjunction (left) may occur at anaphase. If chromosomes 1 and 2, for example, move to the same pole, the gametes will be unbalanced. Some genes will be represented twice (those in arm f) and others will be completely absent. The other adjacent combinations are also seen to be unbalanced. If alternate disjunction occurs (right), the gametes will be balanced. Half of them will carry the original arrangement and the other half the translocation.

breaks occur very close to the centromere regions. The two short arms would be largely heterochromatin, because this type of chromatin surrounds the centromere. The small chromosome that arises would therefore not carry any genes essential to life, so its loss would not prove lethal. The two heterochromatic arms could even be lost without actually joining to form the tiny chromosome at all.

Let us say that an individual inherits a 15/21 translocation through a sperm (Fig. 10-14B). He or she would receive from the female parent a normal 15 and a normal 21. However, the person would have a chromosome total of 45 instead of the normal 46 and can be easily identified by karyotype analysis. The reduced number is present because this individual does not possess a pair of 21 and a pair of 15. Instead, there is a normal 15, a normal 21, and the altered chromosome, 15/21. Since the essential genetic materials of chromosomes 15 and 21 are present, the person would exhibit no ill effects and might be considered completely normal until a karyotype analysis revealed the aberrant chromosome number.

The meiosis in such a person would be quite irregular, because the translocated chromosome, 15/21,

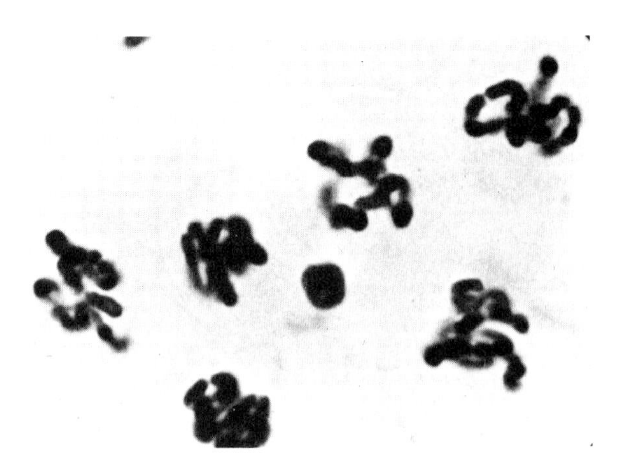

FIGURE 10-12. Multivalent associations in *Rhoeo discolor* as a result of translocations.

can synapse with both the normal 21 and the normal 15 (Fig. 10-14C). A trivalent association of three chromosomes may be seen at meiosis. The segregation of these three will not be regular and can result in the formation of gametes of different constitutions. Of these, some will be normal, containing one chromosome 21 and one chromosome 15. But an appreciable amount will contain number 21 and the translocated chromosome, 15/21. When a gamete of this sort combines with a normal one, chromosome 21 is in effect present three times in the zygote (Fig. 10-14D). The chromosome number of such an individual would be normal (46), but he or she would be indistinguishable from those sufferers of Down syndrome, who have 47 chromosomes. This is so because both classes of victims have most of chromosome 21 in three doses. Down syndrome arises with an unfortunate frequency (approximately 1 in 700 newborn); it therefore accounts for a large number of mentally defective children. Most of them possess 47 chromosomes and have arisen as a result of nondisjunction in the female parent, but a percentage arises, as discussed here, through a parent who is a carrier of a translocation. Identification of translocation carriers by karyotype analysis is important when the occurrence of Down syndrome appears to run in a family, since the heterozygous carrier of the translocation can continue to transmit both gametes that may give rise to Down children and the carrier condition to other children who may do the same (Fig. 10-14C). Identification and proper genetic counseling of carriers can help to reduce the incidence of afflicted children. Those cases of Down syndrome resulting from the presence of one extra chromosome 21 due to nondisjunction rather than a translocation involving a D and a G group chromosome (Fig. 10-13) are not familial, because the aberration arises in meiotic cells of the immediate parent of the Down individual, who almost always fails to reproduce.

Translocations, cancer, and other human disorders

As noted in Chapter 9, studies of cells with translocations are especially valuable in the assignment of a given gene to a localized region of a chromosome. Any of the arms of the 46 human chromosomes may be involved in translocations that occur in about 1 out of 500 births. Those heterozygotes with balanced translocations (for example, in Fig. 10-14B) may have some reduction in fertility but otherwise suffer no ill effects. However, unbalanced translocations producing trisomic conditions are associated with mental retardation and serious phenotypic effects and account for a significant number of spontaneous abortions.

Very interesting are those translocations in the female in which a piece of an X chromosome becomes inserted into one of the autosomes. Although no genes are missing since the translocation is balanced, the effect mimics that of a deletion. This stems from the surprising fact that the intact X, the one that did not lose a piece to an autosome, becomes preferentially inactivated in all the somatic cells! The X with the missing portion is the one that remains active. This means that such a female can express a sex-linked recessive trait. When a particular band of the short arm of the X is translocated, the female suffers from Duchenne muscular dystrophy, which occurs much more frequently in males and which is associated with a deletion in their one X chromosome. These unusual females played an important role in the identification of the Duchenne muscular dystrophy gene, which is discussed in some detail in Chapter 20.

A most significant fact is the association of translocations with certain cancers. For many years, only one cancer, chronic myelogenous leukemia (CML), had been found to be associated with a consistent genetic defect. Leukocytes of affected individuals typically contain a chromosome 22 in which more than one half of the long arm is missing. Such a chromosome was designated the *Philadelphia chromosome* and was originally thought to be due to a simple deletion. However, chromosome banding revealed that the Philadelphia chromosome is actually the result of a reciprocal translocation between chromosomes 22 and 9. About 10% of CML patients lack the translocation, but some believe that these persons may have a different form of the disease. In CML, chromosomes 22 and 9 appear to break consistently at specific subbands. It has been established that the translocation entails the mutual interchange of the ends of the long (q) arms of the two chromosomes. The sizes of the pieces that are

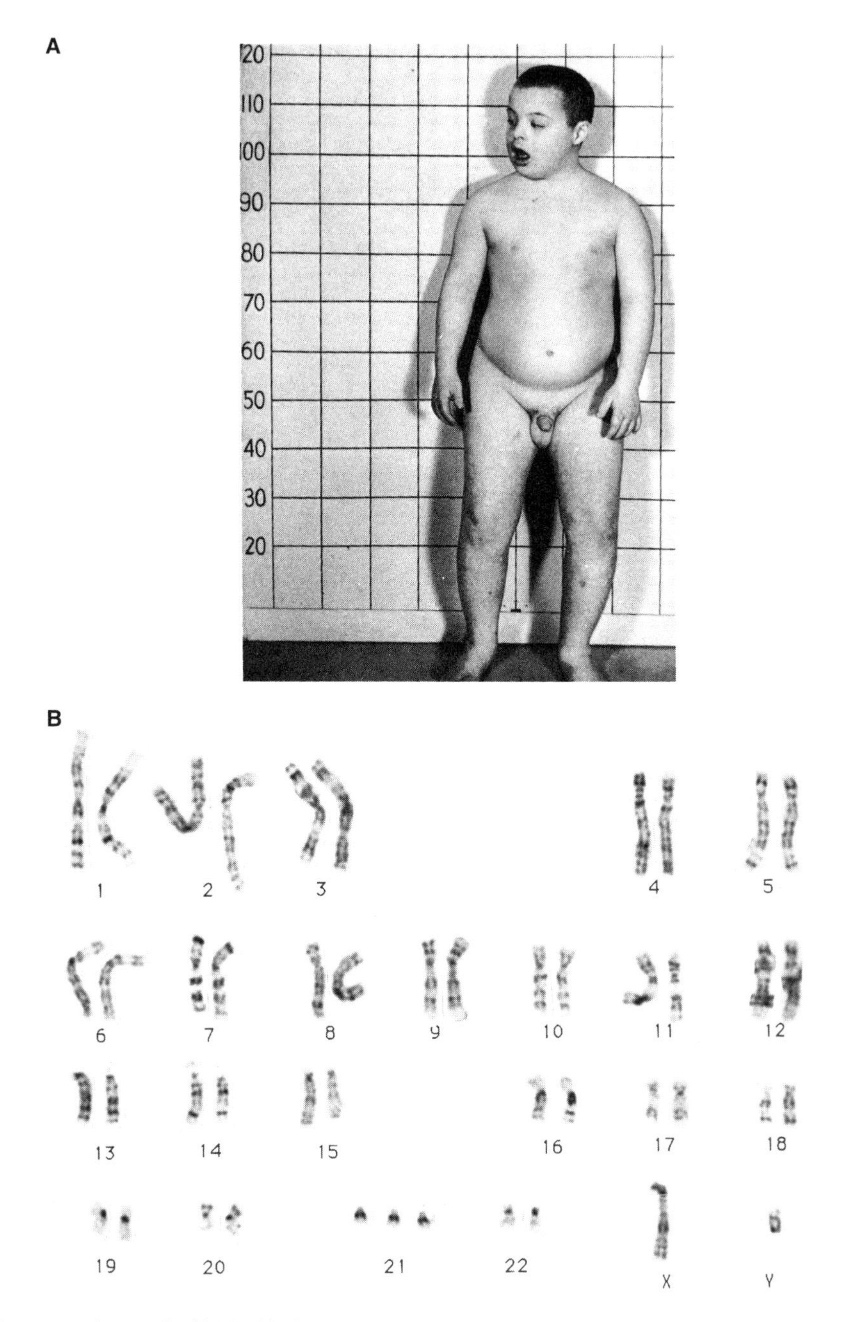

FIGURE 10-13. Down syndrome. **A:** Child with characteristic features. (Reprinted with permission from V.A. McKusick, *Medical Genetics* 1958–1960. C.V. Mosby, St. Louis, 1961.) **B:** Typical karyotype of a male with Down syndrome showing 47 chromosomes. The extra chromosome is a number 21 of the G group. The chromosomes are G banded. (Courtesy of Steven A. Schonberg, Ph.D., University of California, San Francisco.)

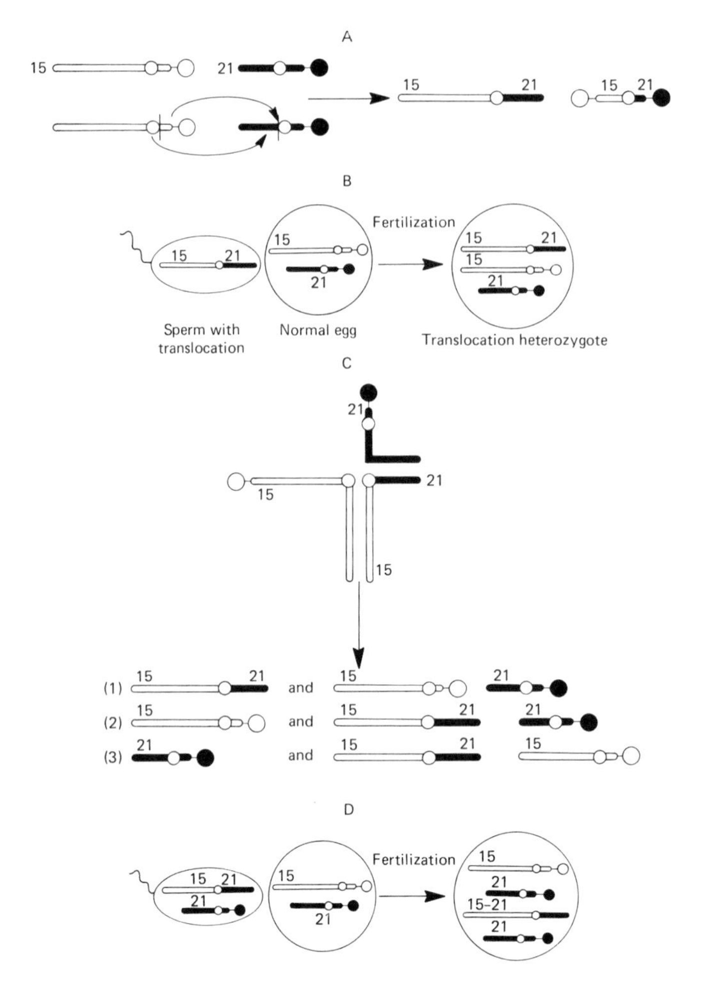

FIGURE 10-14. Translocation in the human between chromosome 21 and a member of the D or G group. In this example, chromosome 15 of the D group is depicted. **A:** Arms are exchanged between chromosomes 15 and 21. The small chromosome that may arise would contain little genetic material. The larger chromosome that arises contains most of the original chromosomes 15 and 21. **B:** A sperm carrying the translocated chromosome fertilizes a normal egg. The zygote that results will be essentially balanced, because two chromosomes 15 and two chromosomes 21 are, in effect, present. The chromosome number is one fewer than normal because 15 and 21 are joined as a result of the translocation. **C:** A trivalent forms at meiosis in the translocation heterozygote. This brings about pairing of homologous regions, but it does not guarantee balanced segregation. Because any one of the three chromosomes can move by itself to one pole, several possibilities follow. Segregation 1 will produce gametes that will pass down the translocation; it will also produce gametes that are normal. Segregation 2 will produce a gamete lacking chromosome 21 (a lethal condition) and a gamete that can result in Down syndrome, as shown in D. The third segregation produces gametes that are very unbalanced. One type lacks chromosome 15. The other has it present twice. The unbalance proves lethal. **D:** A sperm formed in a translocation heterozygote as a result of segregation of type 2 (C) fertilizes a normal egg. The zygote has three doses of chromosome 21 and will develop Down syndrome. The chromosome number, however, will be normal.

switched are unequal. Consequently, a longer chromosome 9 and a shorter 22, the Philadelphia chromosome, are produced. The improvements in staining that now permit high resolution of chromosome bands have revealed that most malignant tumors have characteristic chromosome alterations. In most leukemias and lymphomas, as well as in some carcinomas, a reciprocal translocation is present in the malignant cells. Malignant cells in Burkitt's lymphoma, a very aggressive cancer, contain a reciprocal translocation involving chromosome 8 and usually chromosome 14.

Although a particular chromosome aberration may be characteristic of a given malignancy, as in the case of CML, this does not mean that the alteration is necessarily specific for that one disorder. For example, the translocation characteristic of CML is also found in acute lymphocytic leukemia and acute myelogenous leukemia. The same situation pertains to various other malignancies in which several different types of cancers exhibit the same chromosome aberration. On the other hand, in certain cancers the specific chromosome defect appears to be unique to that one disorder.

Characteristic chromosome aberrations are present early in the development of a malignancy; however, further chromosome aberrations appear as the disease progresses. These secondary changes are nonrandom. For example, among changes in the white blood cells of persons with CML are the gain of extra Philadelphia chromosomes and extra chromosomes 8 or 17. This changing cytological picture of the white blood cells is correlated with changes in the leukemia from chronic to acute. Secondary characteristic chromosome abnormalities often involve duplications of certain chromosomes or chromosome segments and are found in many malignancies. Perhaps these changes confer an advantage on the tumor cells, enabling them to proliferate and spread to other tissues more rapidly. The involvement of structural chromosome changes in the malignant transformation of cells and CML translocation at the molecular level are discussed more fully in Chapter 18.

Chromosome alteration and evolution

There is no question that structural changes have altered the size, shape, and number of chromosomes during the evolution of many plant and animal groups. In *Drosophila,* it is evident from genetic and cytological observations that paracentric inversions and whole-arm translocations have been involved in the differentiation of species within the genus (Fig. 10-15A,B). Pericentric inversions are undoubtedly responsible for the shifting of the centromere, bringing about a change in chromosome morphology.

The evolution of human chromosomes can now be studied in a more precise way. Chromosome banding is especially valuable in revealing details of chromosome morphology, permitting comparisons of chromosomes between the human and other species. Detailed karyotype analyses can give some insight into possible relationships among different groups and any chromosome changes that may have arisen as they diverged. For example, the chimpanzee's closest ancestor was probably shared with the human, as is strongly suggested by similarities in their DNA and proteins. The chromosome number of the chimpanzee is 48; one more pair of autosomes is present than in the human. However, this difference in number can be attributed to a translocation involving two acrocentric chromosomes. In the translocation, the two larger arms of two nonhomologous chromosomes would have become associated, much as shown in Figure 10-14A. The banding pattern suggests that human chromosome 2 has arisen in this way from two nonhomologous ancestral chromosomes, each with one very short arm. Many other differences between the chromosome complements of the human and the chimpanzee can be attributed to pericentric and paracentric inversions.

Duplications are considered to have played a very important role in the evolution of living things, since they increase the number of genes in the chromosome complement. Once extra amounts of a genetic region are present, the stage is set for the originally identical genes to diverge. This is true because each of the duplicated regions can undergo its own subsequent course of independent mutation. As different mutations, and hence different gene changes, continue to arise and as the force of natural selection operates, what were once identical gene forms may eventually be recognized as two very different genes, each controlling the formation of a somewhat different product. There is good evidence for this in the fruit fly, mouse, and human. In these species, there are various cases of very close linkage among genes that determine the amino acid composition of similar protein chains. If the two protein chains controlled by two separate genes are much alike in amino acid sequence, this means that the two genes involved must also be quite similar, since their information contents are similar.

You will recall from Chapter 6 that the X-linked genes for green (deutan) and red (protan) visual pigments lie very close to each other, are almost identical, and encode almost identical pigments. Some males have been found to have more than one copy of the green pigment gene. There is little doubt that the protan and deutan genes represent recent duplications of an ancestral gene that have not yet diverged to any extent as a result of gene mutation.

Evidence is suggestive that the genes encoded for the various globin chains of hemoglobin are all descended from a common ancestral gene and that a series of duplications, followed by mutation, has resulted in the present families of genes associated with the globin polypeptides (Chap. 20).

Aneuploidy in the human

In addition to those aberrations that change the structure of individual chromosomes, other anomalies alter the actual chromosome number of an organism. Any variation from the normal number is termed *heteroploidy.* A heteroploid organism may differ from a typical member of its species by containing one or a few chromosomes more or even one or a few less than the number typical for the species. This situation is termed *aneuploidy,* a condition in which the chromosome number differs from the normal diploid by less than an entire set (2n – 1; 2n + 1; 2n – 2, etc.).

We would expect aneuploidy to lead to genic unbalance, and we have already encountered several examples of this in humans in respect to the sex chromosomes (e.g., Turner and Klinefelter syndromes). Down syndrome (Fig. 10-13), resulting from an extra chromosome 21, is a good illustrative example of the serious consequences of the unbalance due to the presence of an extra autosome. Individuals with one extra chromosome are called *trisomics,* meaning that their body cells contain three doses of a particular chromosome. Trisomics such as a child with Down syndrome may result from nondisjunction at meiosis in one of the parents (Fig. 10-16). If two chromosomes fail to separate at anaphase I of meiosis, one-half of the gametes will contain an extra chromosome and half will have one chromosome less than normal. A gamete of the former type, when it combines with a normal gamete, produces a zygote that is trisomic. The zygote receiving only one member of a chromosome pair is designated a *monosomic* and will not survive.

Over 93% of the cases of Down syndrome result from trisomy for chromosome 21 as a consequence of nondisjunction at meiosis. There is an interesting sidelight to the story of this chromosome, which came to be shown by improved staining techniques to be the

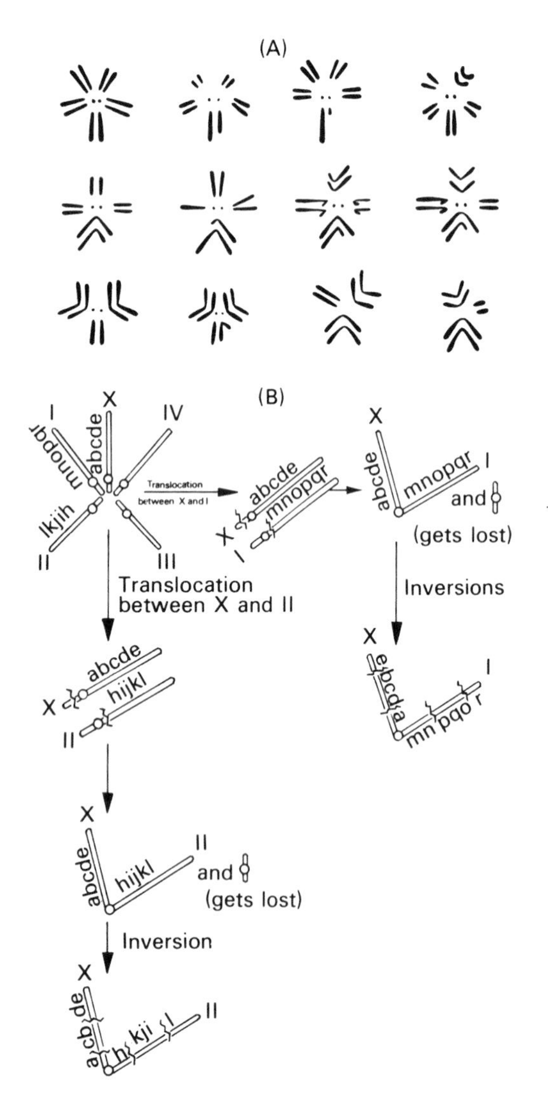

FIGURE 10-15. A: Chromosome complements of some species of *Drosophila*. The karyotypes are of males, and the X and Y chromosomes are the two lower ones in each case. The different chromosomes (metacentric, acrocentric) appear as arrangements of rods, Vs, and Js. This can be explained on the basis of shifting of chromosome arms and segments during evolution of the species by translocations and inversions. The species shown are (1) *D. virilis,* (2) *D. funebris,* (3) *D. repleta,* (4) *D. monata,* (5) *D. pseudoobscura,* (6) *D. miranda,* (7) *D. azteca,* (8) *D. affinis,* (9) *D. putrida,* (10) *D. melanogaster,* (11) *D. willistona,* and (12) *D. prosaltans.* **B:** Alterations in karyotype. In *Drosophila,* chromosome arms may have been shifted about in a fashion similar to this. From an ancestral type having five acrocentric chromosomes, metacentrics can arise. The X is shown here undergoing a reciprocal translocation with chromosome I in one case and with chromosome II in another. The tiny chromosome may be dispensable, because it contains large amounts of heterochromomatin, which is found around the centromere. Eventually two species may be derived in which certain genes are sex linked in one but not in the other. However, a group of genes would be similar and sex linked in both species. These genes would be found in the same arm. This arm would correspond to the single arm in a species having an X with just one prominent arm, because this arm traces back to the same ancestral X. The order of genes in the arms may differ because of different inversions as the species evolved separately.

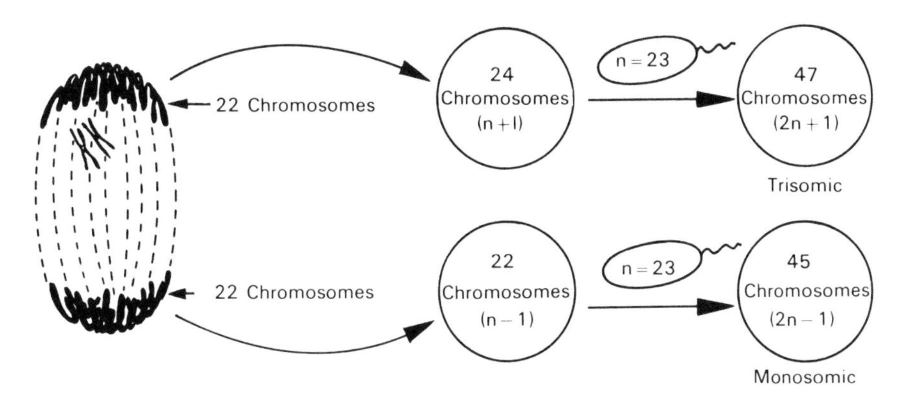

FIGURE 10-16. Nondisjunction of an autosome at meiosis. Two homologous chromosomes (assume chromosome 21 in the human) fail to separate at first anaphase. They move to the same pole and enter the same nucleus. A gamete derived from this will contain two chromosomes 21 instead of just the normal one. A normal gamete from the other parent will contribute one chromosome 21. The resulting zygote is thus trisomic, carrying three instead of two. A gamete derived from the other nucleus at meiosis (bottom in figure) will lack chromosome 21 entirely. After fertilization, the zygote will be monosomic, having just a single chromosome 21, which was contributed by the balanced gamete.

smaller of the two autosomes in group G. You will recall that the chromosomes in any group are numbered according to decreasing size. This means therefore that the Down chromosome should technically be number 22, not 21, as designated for many years. However, inasmuch as Down syndrome had been called *trisomy 21* for so long and the terminology is so well established in the medical literature, geneticists agreed to keep the designation 21 for the Down chromosome in order to avoid confusion. Although the other G group chromosome is actually the larger of the two, it has arbitrarily been designated 22. The fact that chromosome 21 is actually the smallest of the human chromosomes may be an important part of the reason that Down syndrome is the most common of the trisomies among newborns. It has been estimated that chromosome 21 contains only 1.5% of the total amount of genetic material in a human chromosome complement. Recent advances in molecular biology are permitting recognition of the precise region on chromosome 21 that is responsible for Down syndrome and raise hopes that some day the condition may be prevented or treated more effectively (Chap. 20).

Nondisjunction of chromosome 21 can take place at a mitotic division. Let us examine the consequences of such an event. If it should occur at the first mitotic division of the zygote, the result would be two cells, one trisomic for chromosome 21 and the other monosomic, having only one copy of 21. Since cells monosomic for an autosome usually die, the remaining cell is left to give rise to a Down individual, all of whose cells will be trisomic for chromosome 21. However, if the nondisjunction takes place at a later mitotic division, normal diploid cells will be present along with the surviving trisomic cell. The result is now an offspring who is mosaic for trisomy 21. Approximately 1%–2% of Down persons are mosaics possessing two types of somatic cells, those with the normal diploid number and those that are trisomic for chromosome 21. The extent of the symptoms depends on when in development the nondisjunction took place as well as on the tissues affected. (You will recall from Chapter 6 that mosaicism involving the sex chromosomes is well known.)

The incidence of Down syndrome has been correlated with the age of the mother. Women over 40 have a much higher chance of bearing a Down syndrome child than do younger mothers. The age of the father does not seem to be a factor. It should be noted that the incidence of Down syndrome resulting from translocations is not correlated with maternal age. There is reason to believe that chromosomes of oocytes are much more subject to nondisjunction the longer they remain in first prophase of meiosis. Since the first meiotic division in the human oocyte is not completed until the follicle matures, this means that an oocyte has been lying in a sustained meiotic prophase for perhaps decades by the time meiosis is resumed. Spermatogonia, on the other hand, provide a supply of spermatocytes that are continually undergoing spermatogenesis throughout adult life. However, nondisjunction can occur in a male, and it has been estimated from studies utilizing molecular methods that as many as 25% of Down cases may have resulted from an extra chromosome 21 derived from the father.

Moreover, the mosaicism mentioned above can cause problems both in diagnosis of Down syndrome as well as in detection of its mode of transmission. For example, two phenotypically normal fathers produced children with Down syndrome. After cells were taken from various tissues and then grown in culture, both men were found to be mosaics for chromosome 21. In one man, the extra chromosome was found in skin and testicular fibroblasts but not in leukocytes. In the other man, mosaicism for chromosome 21 was found among the leukocytes. In both cases, the Down aberration was coming through the male parent. All of the persons involved were under age 20. Such examples indicate the need to consider both parents in cases of suspected inherited anomalies and to look for mosaicism in different tissues.

It is interesting to note that trisomy 21 occurs in the chimpanzee. This human condition results in a syndrome closely resembling Down syndrome.

Trisomy 22 has been recognized in persons through chromosome banding techniques that permit accurate distinctions to be made between chromosomes 21 and 22. Trisomy 22 is associated with mental and growth retardation, as well as with cleft palate and an assortment of other disorders. The survival rate of the afflicted individuals appears to be high.

A few other human trisomies are known that also permit survival beyond birth. However, infants trisomic for chromosome 8, 9, 13, or 18 suffer severe abnormalities of the skeleton and brain in addition to a variety of other disturbances. The unbalance resulting from the excess genetic material is so severe that death usually ensues in early childhood.

The autosomal trisomies we have mentioned are those that permit survival beyond birth. However, tissues from abortuses have shown that nondisjunction can involve any chromosome. Most trisomies are apparently lethal to the embryo or fetus. Actually, trisomy as a class of aberrations is the most frequent of the chromosome anomalies found among abortuses, occurring in over 50% of them. The lethal risk for trisomy 22 is evidently greater than that for trisomy 21, which is more likely to survive the period of fetal development and is thus more common among trisomic individuals than trisomy 22. A maternal age effect is also associated with the other autosomal trisomies in addition to trisomy 21.

Those aneuploid persons with an XXXY or XXXXY condition (see Chap. 6) are trisomic and tetrasomic, respectively, for the X chromosome. Although they have the abnormalities of Klinefelter syndrome, they nevertheless survive and are in no way as defective as those individuals trisomic for one of the autosomes. This ability to tolerate extra doses of the X may be possible as a consequence of some deactivation of those Xs in excess of one. Remember that one, two, and three Barr bodies are present, respectively, in persons who are XXY, XXXY, and XXXXY. A maternal age effect is also associated with Klinefelter syndrome.

As noted, monosomic cells lacking an autosome are produced as a result of nondisjunction. However, such cells do not survive in the human. There is no clear-cut example in the human of an individual who is monosomic for an autosome. Autosomal monosomy is even rare among abortuses. Apparently the effect is so lethal that almost all are lost, even before implantation of the embryo can take place.

However, individuals monosomic with respect to the sex chromosomes are well known. Persons with one X chromosome and no other sex chromosome (XO chromosome constitution) show the assortment of characteristics that describe Turner syndrome (Fig. 6-12A), a condition that can arise from nondisjunction during parental gamete formation (Fig. 6-15). No maternal age effect has been associated with Turner syndrome.

Monosomy in humans causes a greater degree of unbalance than does the presence of an extra chromosome. Only those monosomics who have a missing sex chromosome (Turner females) can apparently survive to birth, but a very large proportion of such XO embryos evidently do not reach full term. Turner syndrome births are much less common than are Klinefelter syndrome ones. The XO chromosome abnormality is actually the most common chromosome aberration among abortuses. The number of XO zygotes that manage to develop and survive to birth has been estimated to be only 1%. Apparently 99% are aborted as a result of prenatal disturbances, again indicating the sensitivity of human cells to monosomy.

Polyploidy in plants

Plants tolerate departures from the typical chromosome number much more readily than do animals. Various plant species have provided valuable information on the cytological and phenotypic effects of aneuploidy. Moreover, plants have the ability to tolerate the addition of entire chromosome sets to the typical diploid amount. Such heteroploid individuals with entire extra sets are called *polyploids*. For example, if a diploid number of 10 chromosomes composed of two haploid sets of 5 is typical, individuals with 1 or 2 extra sets having numbers of 15 and 20 would be polyploids known as *triploids* and *tetraploids*, respectively.

Polyploidy is very characteristic of flowering plants and has played an important role in the evolution of many plant groups, including a number of valuable crop plants such as wheat and cotton. When we examine a list of chromosome numbers of species with a plant genus, we often find that the numbers are multiples of a basic one. For example, in different wheats, 2n = 14, 28, 42 (multiples of 7); in *Chrysanthemum*, 2n = 18, 36, 54, 72, 90 (multiples of 9). We can explain such observations on the basis of polyploidy; the groups with the higher numbers have been derived from those with lower ones through the multiplication of whole chromosome sets.

One of the main types of polyploidy is *autopolyploidy*, a condition in which the extra chromosome sets have originated from within the species. A plant species with a diploid number of 10 would have two haploid sets, AA. If one or two extra sets are present, we would have an autotriploid (AAA) and an autotetrapaloid (AAAA). The origin of the extra sets could be some failure during meiosis. Accidental disruption of the spindle may prevent anaphase movement and give rise to an unreduced gamete (AA). Union of such a gamete with a normal one (A) would produce the autotriploid (AAA; Fig. 10-17). Fusion of two diploid gametes produces the autotetraploid. The factor responsible for autopolyploidy could occur in somatic tissue as well. Anaphase of mitosis could fail; the nuclei of two diploid body cells could fuse. The result would be a tetraploid cell (AAAA). If such a somatic cell goes on to divide, it could give rise to an entire tetraploid plant. Any flowers on a tetraploid branch would produce unreduced gametes (AA), and, through self-fertilization, tetraploid offspring would arise.

Tetraploids may also be induced artificially by a variety of treatments such as application of the drug colchicine. This alkaloid, which is used medicinally to relieve the pain of gout, upsets the organization of the microtubules composing the spindle and thus prevents anaphase movement. All the chromosomes become incorporated into one nucleus, and a tetraploid cell is formed. Since colchicine does not interfere with the duplication of the chromosomes, its presence during a series of chromosome cycles can lead to high levels of ploidy. Solutions of the drug may be sprayed on buds; seeds or roots may be dipped or soaked for several hours, but, since colchicine is very toxic, concentration and exposure time must be carefully controlled. Some polyploids may encompass certain features that make them more desirable than the diploids from which they were derived, such as being larger or more vigorous. The polyploid tends to grow more slowly than the diploid and may flower later over a longer period of time.

Cytologically, most autopolyploids have meiotic irregularities because of the presence of extra homologous chromosomes. This is well demonstrated by the triploid, which is usually highly sterile (Fig. 10-18). Any three homologous chromosomes can enter into a trivalent association, or they can form a bivalent and a univalent. It is even possible that no pairing will be established and that three univalents will

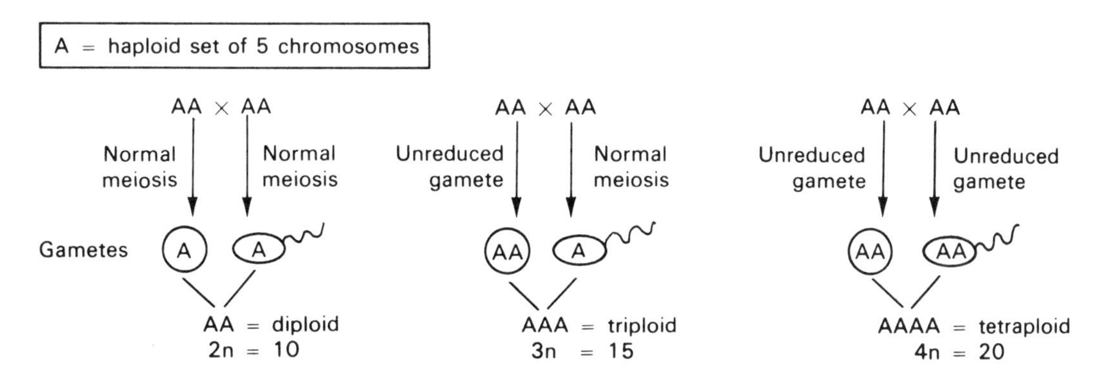

FIGURE 10-17. Diploidy and autopolyploidy. The extra set or sets of chromosomes in an autopolyploid are derived from within the one race or group, so that more than two completely homologous sets are present.

result at first meiotic prophase. There is no mechanism to ensure an anaphase separation in which two chromosomes of each kind will move to one pole and one of each kind to the other. Such a separation would produce balanced diploid and haploid gametes. However, random segregation of the members of each group of three homologs usually occurs, so that only a few gametes have a balanced combination of chromosomes. This in turn means that the gene dosage is unbalanced in most of the cells after meiosis. Consequently, the pollen of triploids is often very sterile, and seed set is very low. The same sort of meiotic irregularity accounts for the reduced fertility characteristic of polyploids with high but odd chromosome numbers (5n, 7n, etc.). This sterility has commercial value in plants such as the watermelon, where the almost seedless fruit is more desirable than that of the fertile diploid. To maintain sterile triploids, the plant breeder must propagate them asexually by means of leaf or root cuttings; seeds that will grow into triploids can also be obtained by crossing a tetraploid with a diploid.

In the autotetraploid, there is a greater chance than in the triploid for the formation of balanced gametes. Since four chromosomes of each kind are present, multivalents composed of an association of four are seen (tetravalents). Trivalents and univalents are also possible. But there is a higher probability of bivalent formation in the tetraploid (or any polyploid with an even chromosome number). Therefore anaphase segregation results in a much larger proportion of cells that have balanced gene combinations. In certain species, such as the Jimson weed Datura, the autotetraploids suffer little decrease in fertility.

The second main type of polyploidy is *allopolyploidy*, a polyploid condition in which the chromosome sets are derived from different races or species. For an allopolyploid to arise, hybridization must take place between two genetically different groups. One criterion that distinguishes two groups as separate species is their inability to cross and produce fertile offspring. If two groups are able to cross at all, they must share sufficient genetic similarities to permit any degree of gene exchange to occur. Some may be true species in every sense of the word, so that production of offspring between them on the diploid level rarely, if ever, occurs. Others may not be as genetically isolated but may actually be subspecies. In other words, no sharp distinction exists between autopolyploidy and allopolyploidy, because the distinction between species is often blurred.

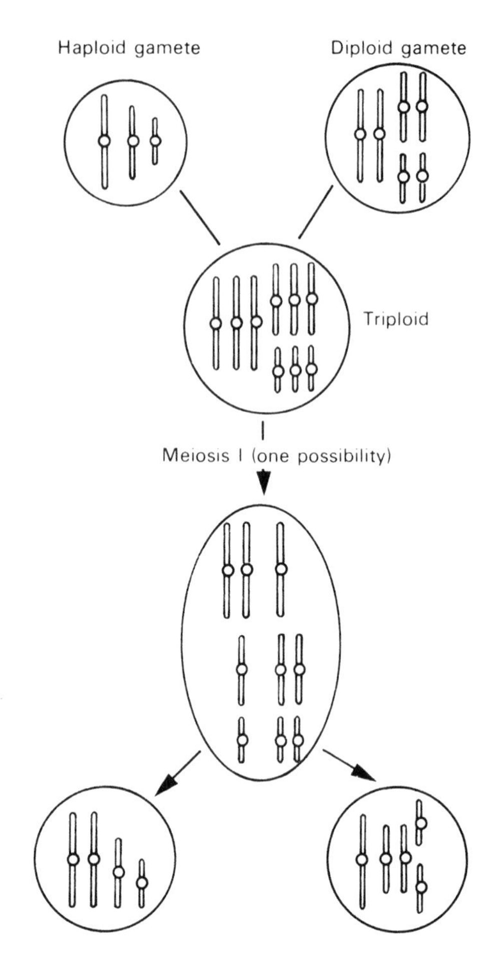

FIGURE 10-18. Irregular segregation in a triploid. Since three of each kind of chromosome are present, trivalents and univalents may occur at meiosis, and segregation will tend to be irregular. Balanced gametes would contain two of each kind of chromosome or one of each kind. Most cells, however, will contain unbalanced combinations. For example, the gamete at the left has one extra long chromosome. The gamete at the right has two extra chromosomes, one of medium size and a small one; if it possessed another long one, it would be a balanced diploid gamete.

Various lines of cytogenetic evidence leave no doubt that allopolyploidy has played a major role in the evolution of plants. Following hybridization between two different but genetically related groups, a highly infertile hybrid may arise that in rare instances produces unreduced gametes containing the entire haploid chromosome complements of both parental species. Fusion of these unreduced gametes can lead to the origin of an allopolyploid with two sets of chromosomes from each parental species. Figure 10-19 depicts an experimental cross that illustrates some of the events leading to the production of an allopolyploid.

Allopolyploids can be produced experimentally with the aid of colchicine, which is applied to the diploids to obtain diploid gametes. At times the polyploid possesses an advantageous combination of characteristics that enables it to thrive under conditions that would be unfavorable for either of the diploid parents. The polyploid may prove to be more robust than either diploid and may actually replace them. In nature, allopolyploidy has played a more significant role than autopolyploidy in the history of major plant groups. As Figure 10-20 shows, a cross between the allotetraploid and either one of the parents yields triploid combinations with the ensuing formation of unbalanced gametes. Thus does allopolyploidy provide one mechanism for the evolution of new species.

Somatic cell hybridization and plant hybridization

Besides their application to linkage studies (see Chap. 9), techniques that apply cell fusion to plants offer exciting possibilities for the derivation of superior hybrids. Mature, fertile, interspecific hybrids have actually been obtained through cell fusion methods that completely bypass the normal, sexual process. One such feat involved two wild tobacco species, *Nicotiana glauca* (2n = 24) and *Nicotiana langsdorfii* (2n = 18). Leaf cells were subjected to enzyme digestion to free them from their cellulose walls. More than 10^7 protoplasts of each species were plated in a solution that stimulated them to fuse (Fig. 10-21). Approximately 25% of the cells underwent fusion in culture. Some fusions were between two cells of the same species, but others were hybrid fusions. These latter cells were then selected by plating all the cells on a medium that permits growth only of those containing the genetic information of both parent species. After further culture, the isolates developed rudimentary shoots and leaves but failed to form roots. However, when grafted to other tobacco roots, they developed into mature plants that were identical in chromosome number (2n = 42) and in all other characteristics to hybrid plants produced by conventional cross-pollination. Moreover, their seeds gave rise to offspring identical to them in all ways, illustrating that these "fusion hybrids" can breed true.

As we have just seen, in the typical production of an interspecific polyploid such as this, chromosome doubling is required to produce fertile hybrid offspring from highly infertile F_1 plants. The cell fusion method, called *parasexual hybridization* because the normal sexual process is circumvented, avoids this difficulty and others involved in crossing species. The parasexual process holds great promise as a means of combining the genetic information from plant species that would otherwise be unable to cross because of reproductive barriers.

Hybrids derived parasexually may very well enable the plant breeder to produce vigorous new species

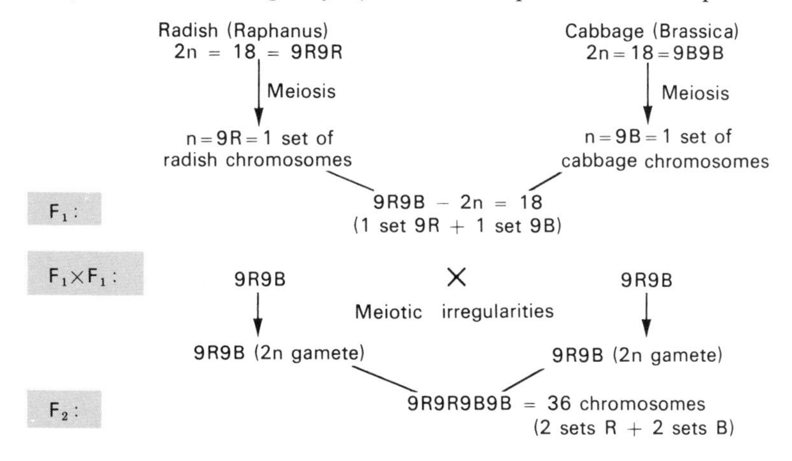

FIGURE 10-19. Origin of an allotetraploid from a cross of radish with cabbage. One haploid set of radish chromosomes is represented by 9R and one haploid set of cabbage by 9B. The irregular meiotic behavior in the highly infertile F_1 on rare occasions produces a gamete that is diploid. If two of these unite, an allotetraploid arises that is fertile, because each parental set (9R and 9B) is represented in the diploid condition and can form bivalents. The allotetraploid was named *Raphanobrassica.*

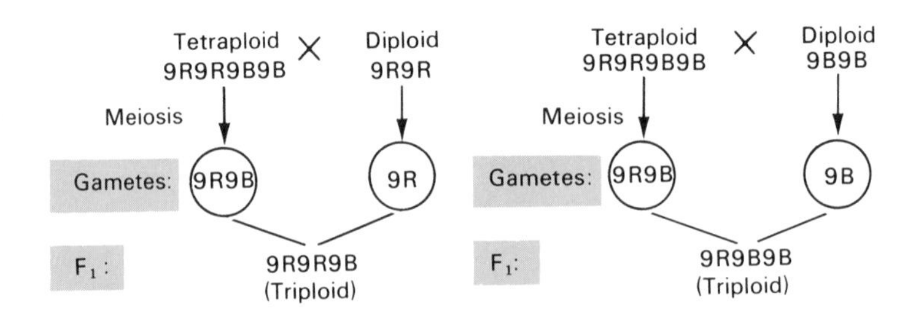

FIGURE 10-20. Reproductive isolation of the allotetraploid. If an allopolyploid such as *Raphanobrassica* (see Fig. 10-19) is backcrossed to either parent species (radish or cabbage), the F_1 is a sterile triploid. The backcross to the radish parent (left) leaves the nine cabbage chromosomes unpaired. Similarly, the cross to the cabbage parent leaves nine unpaired radish chromosomes. Irregular segregation of the univalents causes

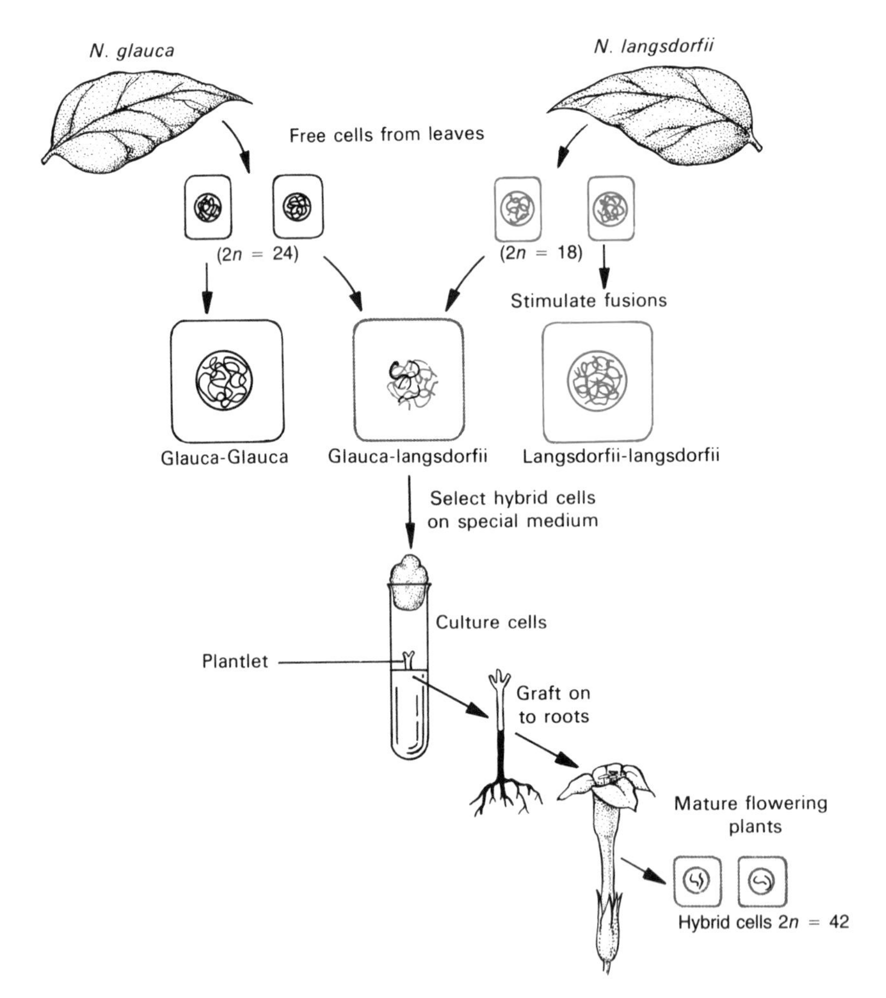

FIGURE 10-21. Parasexual hybridization in tobacco. Cells of tobacco leaves are enzyme treated to release free cells. Cells from two different species that differ in chromosome number are cultured together. Cellular fusions occur; some of these are fusions between the same kinds of cells; others are hybrid cells between the species. These are selected by the growth medium. The hybrid cells develop into plantlets lacking roots. The small plants are grafted onto other roots and eventually develop into mature plants bearing flowers and producing seeds. The chromosome numbers as well as all the other characteristics of the plants derived in this way are identical to the interspecific hybrid produced by crossing the two species and doubling the chromosome number to obtain a fertile polyploid.

combining the most desirable features of both parents, such as increased disease resistance or increased yield of an edible product. Attempts to produce plant hybrids through cell fusion depend largely on selective techniques that make possible the isolation of only the hybrid cells that would be scattered among the total percentage of cells fused. It seems quite likely that this problem can be solved once differences in growth requirements between the parental species are fully known. The parasexual process may prove to be a very valuable method to increase food production in needy areas by the derivation of hardier and more nutritious plant species.

Polyploidy in animals

Since polyploidy is so prevalent in the plant kingdom, we might expect it to have been important in animal evolution as well. Polyploidy does occur normally in certain tissues of the body, such as in the liver, but actually in humans and other mammals polyploidy exerts a lethal effect when present in all cells. The consequences of tetraploidy are apparently more severe than those of triploidy. The frequency of spontaneously aborted embryos that are tetraploid is quite low, indicating that the zygotes do not usually develop very far; no such fully developed fetuses have been recovered. However, nearly 4% of all spontaneous human abortuses show triploid cells, making triploidy the most common class of chromosome anomalies in spontaneous abortuses after trisomies. Triploid embryos usually develop further than those with tetraploid cells and may approach the development of a full-term fetus. A very small number of triploids have actually lived for a few hours, and a few have even survived for several years, but the latter all proved to be mosaic for the triploidy. Examples of tetraploid cells in surviving humans have been extremely rare, and these cases all involved mosaicism.

Polyploid animals have been produced experimentally in some species, such as *Drosophila* and the silkworm. However, when the major animal groups are surveyed, it becomes evident that naturally occurring polyploids within them are uncommon. One main reason is that animals, unlike plants, usually have separate sexes. Normal sex determination requires a balance among many genes in the genetic complement. Polyploidy would upset this balance and result in sterile individuals, such as intersexes in the fruit fly. We need only recall that one X chromosome in mammals is sufficient to trigger testis formation and that aneuploid persons with different doses of X and Y chromosomes such as XXY, XXYY, and XXXY are sterile.

REFERENCES

Arrighi, F.E., and T.C. Hsu. Localization of heterochromatin in human chromosomes. *Cytogenetics* 10:81, 1971.

Astaurov, B.L. Experimental polyploidy in animals. *Annu. Rev. Genet.* 3:99, 1969.

Beatty, R.A. The origin of human triploidy: An integration of qualitative and quantitative evidence. *Ann. Hum. Genet.* 41:299, 1978.

Boue, A., J. Boue, and A. Grapp. Cytogenetics of pregnancy wastage. *Adv. Hum. Genet.* 14:1, 1985.

Caspersson. T., et al. Identification of the Philadelphia chromosome as a number 22 by quinacrine mustard fluorescence. *Exp. Cell Res.* 63:238, 1970.

Croce, C.M., and G. Klein. Chromosome translocations and human cancer. *Sci. Am.* (Mar) :54, 1985.

Epstein, C.J. Mechanisms of the effects of aneuploidy in mammals. *Annu. Rev. Genet.* 22:51, 1988.

Hamerton, J.L. Cytogenetic disorders. *N. Engl. J. Med.* 310:314, 1984.

Hassold, T.J. Chromosome abnormalities in human reproductive wastage. *Trends Genet.* 2:105, 1986.

Hassold, T.J., and P.A. Jacobs. Trisomy in man. *Annu. Rev. Genet.* 18:69, 1984.

Hook, E.B. Behavioral implications of the human XYY genotype. *Science* 179:139, 1973.

Hook, E.B., and P.K. Cross. Rates of mutant and inherited cytogenetic abnormalities detected at amniocenteses: Results on about 63,000 fetuses. *Ann. Hum. Genet.* 51:27, 1987.

Kaiser, P. Pericentric inversions: Problems and significance for clinical genetics. *Hum. Genet.* 68:1, 1984.

Le Beau, M.M., and J.D. Rowley. Chromosomal abnormalities in leukemia and lymphoma: Clinical and biological significance. *Adv. Hum. Genet.* 15:1, 1986.

Mules, E.H., and J. Stamberg. Reproductive outcomes of paracentric inversion carriers. Report of a liveborn dicentric recombinant and literature review. *Hum. Genet.* 67:126, 1984.

Niebuhr, E. The cri du chat syndrome. *Hum. Genet.* 44:227, 1976.

Niikawa, N., and T. Kajii. The origin of mosaic Down syndrome: Four cases with chromosome markers. *Am. J. Hum. Genet.* 36:123, 1984.

O'Riordan, M.L., et al. Distinguishing between the chromosomes involved in Down's syndrome (trisomy 21) and chronic myeloid leukemia Ph' by fluorescence. *Nature* 230:167, 1971.

Patterson, D. The causes of Down syndrome. *Sci. Am.* (Aug) :52, 1987.

Solomon, E., J. Borrow, and A.D. Goddard. Chromosome aberrations and cancer. *Science* 254:1153, 1991.

Stebbins, G.L. *Chromosomal Evolution in Higher Plants.* Addison-Wesley, Reading, MA, 1971.

Yunis, J.J. (ed.). *New Chromosomal Syndromes.* Academic Press, New York, 1977.

REVIEW QUESTIONS

1. Let the following represent a pair of homologous chromosomes in an inversion heterozygote. (The 0 represents the centromere.)

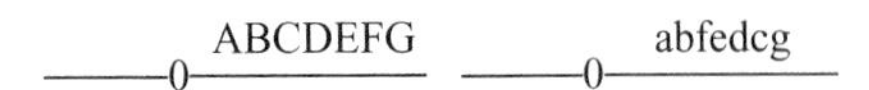

Assume that a crossover takes place between the c and the d loci at meiosis. Give the two products that will result from the crossover event.

2. In the inversion heterozygote in Question 1, what will the two products be, assuming a double crossover occurs between the same two strands—the first one between c and d and the second between e and f?

3. Let the following represent a pair of homologous chromosomes in another inversion heterozygote. (The 0 represents the centromere.)

HIJ —0— KLMN   hlk —0— jimn

Assume that crossing over takes place between loci k and l. Give the two products resulting from the crossover event.

4. In the inversion heterozygote in Question 3, what will the two products be, assuming double crossing over between the same two strands, one crossover event between i and j, the other between k and l?

5. Assume that an individual has one chromosome with the gene order as follows:

—0— RSTUVWXYZ

In the homolog, a shift resulted in the insertion of the segment TUV between X and Y. Show how this situation can lead to the origin of a chromosome with a duplication and one with a deficiency.

6. Following are two pairs of homologous chromosomes with the centromeres in a median position. Assume a reciprocal translocation occurs so that whole arms are shifted between two of the nonhomologous chromosomes following the breaks whose positions are indicated:

g —0— h   m —0— n

g —0— h   m —0— n

   A. What is the composition of a translocation heterozygote carrying the two pairs of chromosomes?
   B. If adjacent disjunction occurs, what are the two kinds of gametes possible for each of the two possibilities?
   C. What kinds of gametes are possible from alternate disjunction?

7. Assume that a deletion arises in the X chromosome of the mouse and removes a gene required in at least single dose for the survival of the embryo.

   A. What would be the result of crossing a female carrying the deletion and an ordinary male?
   B. Assume that the deletion of an autosomal gene required in single dose accumulates in a population. What will be the outcome when carriers of the deletion mate with each other?

8. Several cases of reduction of eye size to a narrow bar in *Drosophila* have been related to removal of the Bar locus from its normal location on the X to tiny chromosome IV, which is rich in heterochromatin. Different doses and locations of the Bar locus are depicted in the following. (B represents one dose of the Bar locus. X and IV represent the X and fourth chromosomes, respectively.) Types 1 and 2 flies have normal eye size. Types 3, 4, and 5 have the same narrow eye. Type 6 has an eye size much narrower than even these. Based on the following picture, explain the eye size differences.

1	2	3	4
X B	X B	X ___	X B
IV___	X B	IV B	IV B
IV___	IV___	IV___	IV___
	IV___		
(Male)	(Female)	(Male)	(Male)

5	6
X B	X ___
X B	IV B
IV B	IV B
IV___	
(Female)	(Male)

9. Assume that a certain animal species has a haploid chromosome complement that consists of the five chromosomes designated I, II, III, IV, and V. Give the chromosome constitution of a body cell of an individual which happens to be:

   A. Trisomic for II.
   B. Triploid.
   C. Monosomic for IV.
   D. Tetrapolid.

10. Species A has a diploid chromosome number of 24; so does another plant, species B. After many attempts a highly infertile F_1 is produced. From this F_1, a fertile allotetraploid is eventually derived.

   A. Give the chromosome number of the F_1 plant.
   B. Give the chromosome number of the fertile derivative.

11. Species C has a diploid chromosome number of 30, whereas that for species D is 20. A highly infertile F_1 arises from a cross between the two species, and eventually a fertile allopolyploid is derived.

   A. Give the chromosome number of the F_1 plant.
   B. Give the chromosome number of the allopolyploid.

12. The same two species discussed in Question 11 are manipulated in a parasexual hybridization procedure in which cells freed from leaves are used to obtain a hybrid. Briefly outline the procedure used, and give the chromosome number of the hybrid obtained from the two species.

13. In the course of several studies of plant populations, the following cytological pictures were encountered in individual plants. In each case, the somatic chromosome number was 24. Present a likely explanation for each picture:

    A. Ten bivalents and a circle of four chromosomes at metaphase I of meiosis.
    B. Bridges and fragments in pollen mother cells at anaphase I.
    C. Two tetravalents and eight bivalents at metaphase I in pollen mother cells.
    D. Five trivalents, three bivalents, and three univalents in pollen mother cells at metaphase I.
    E. Twelve bivalents, one of which shows a distinct buckle at late prophase I of meiosis.
    F. Two tetravalents, six bivalents, one trivalent, and one univalent at diakinesis.
    G. Eleven chromosomes composed of two chromatids in a cell undergoing metaphase II.
    H. Twenty-four chromosomes composed of two chromatids in a cell undergoing second meiotic division.

14. Assume plant species 1 has a diploid chromosome number of 16. Give the following chromosome numbers:

    A. An autotriploid derivative of species 1.
    B. A trisomic member of species 1.
    C. An allotetraploid derived from a cross with species 2, which has a haploid chromosome number of 8.
    D. A monosomic member of species 1.
    E. A plant derived from species 1 and species 2 using somatic cells and the technique of parasexual hybridization.

15. Give at least one way to account for the origin of humans with the following chromosome constitutions:

    A. XXY.
    B. XO.
    C. XYY.
    D. Only one normal metacentric chromosome 1 but with an unusual chromosome the size of chromosome 1 but acrocentric.
    E. Three complete sets of chromosomes.
    F. Only one chromosome 21 and only one chromosome 13 plus one unusually large chromosome.
    G. XXX.
    H. XX/XO mosaic.
    I. XXY/XX mosaic.
    J. XY/XYY/XO mosaic.

16. Match the disorder in column A with the letter of an appropriate chromosome anomaly in column B (there may be more than one possible choice in a given case, and an item in column B may be used more than once):

A	B
____ Turner syndrome	1. Polyploidy
____ Klinefelter syndrome	2. Aneuploidy
____ Down syndrome	3. Monosomy
____ Cri du chat syndrome	4. Triploidy
____ Philadelphia chromosome	5. Inversion
____ XYY male	6. Translocation
	7. Deletion
	8. Nondisjunction

17. Demonstrate or explain how a metacentric chromosome can be derived from two acrocentric chromosomes with centromeres very near the end of a chromosome arm.

18. Occasionally males with Klinefelter syndrome are found to have an XX sex chromosome constitution, with no Y chromosome present! Explain, using information on the sex chromosomes in Chapter 6.

Chemistry of the Gene

11

The *Pneumococcus* transformation

Of all the materials found in cells, only proteins seemed complex enough to compose the countless different kinds of genes in the biological world. In 1944, a landmark was reached in biology with the announcement that DNA and not protein might be the genetic substance. The story actually began in 1928 with the English bacteriologist Griffith, who was studying the pneumonia organism *Streptococcus pneumoniae.* This bacterium, which was then known as *Pneumococcus,* is a spherical cell that typically occurs in pairs surrounded by a sizeable common capsule. Such encapsulated cells are virulent and can kill laboratory animals. On agar plates, they form glistening, mucoid colonies described as *smooth.* On occasion smooth (S) cells can change genetically and lose the capsule. These cells are now harmless, because they are vulnerable to attack by the body's immune defenses. On agar, their colonies appear granular and are termed *rough* (R) to contrast them with the smooth (S), encapsulated type.

When injected into an animal, S cells elicit the formation of specific antibodies, because the capsule, a polysaccharide, acts as an antigen. S cells fall into different strains, each of which brings about the formation of a distinct type of antibody. The different kinds of virulent cells can be designated types I, II, III, and so on in recognition of their specific antibodies. These types are genetically distinct, each being inherited from one cell generation to the next. When a virulent cell mutates to rough, the resulting R cells on rare occasions mutate back (revert) to the virulent S form. When this takes place, however, the reversion is always to the S form from which the R types were originally derived. Descendants of a rough cell that arose from type I would revert only to S type I, never to III or to some other type.

In the course of his experiments, Griffith injected mice with virulent cells from smooth colonies. As expected, the animals died, and autopsy revealed the presence of living, smooth cells. When he injected them with rough cells, they survived as again expected, because R cells are not virulent. Moreover, injection of animals with virulent S cells destroyed by heat caused no deaths, because killed bacteria cannot multiply and produce toxin.

Griffith also injected mice with two kinds of cells: a small number of living R cells derived from type II and heat-destroyed S cells, type III. The expectation was that the animals would live, because neither living R cells nor heat-killed S ones alone can cause disease. However, many animals died and were found to contain living cells that produced smooth colonies, type III (Fig. 11-1)! Such cells can indeed kill, but how did they arise? Somehow the living R cells had been "transformed" and took on the specific capsule type of the killed virulent cells. In this phenomenon, known as the *Pneumococcus transformation,* R cells derived from any S type can acquire the capsule of any other S type cell. Somehow R cells can be changed genetically through incubation with nonliving S cells. Something was causing the genetic change, and this was termed *TP* (*transforming principle*). Later work by others showed that transformation could be accomplished in the test tube without the presence of an animal.

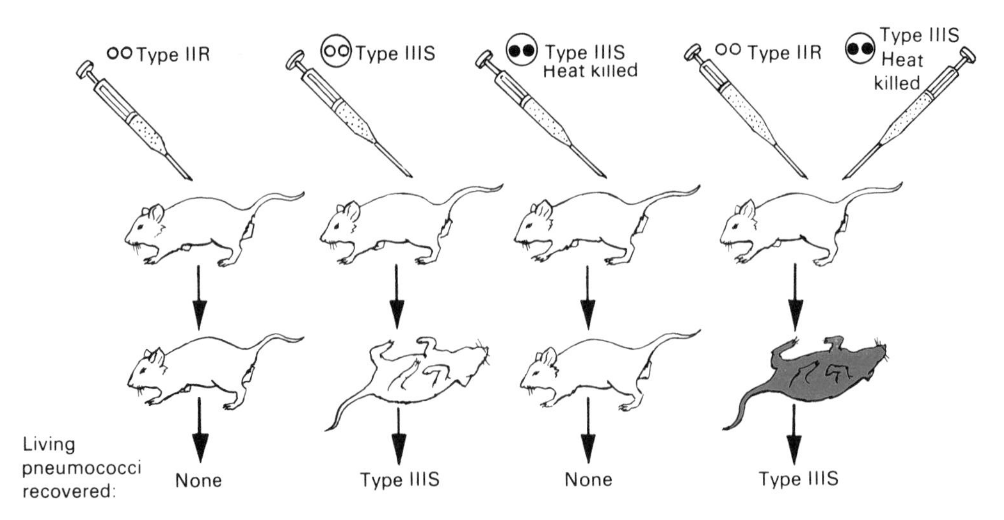

FIGURE 11-1. *Pneumococcus* transformation. Mice injected with rough cells alone or heat-killed cells alone survive as expected, because the rough are not virulent and dead cells should not kill. However, a combination of living rough and heat-killed smooth cells can cause the death of an animal. Living smooth cells can be recovered after autopsy, and they are always found to be of the same bacterial type as the heat-killed smooth.

Employing refined chemical procedures, Avery, McCarty, and MacLeod of Rockefeller Institute obtained small amounts of highly purified TP from SIII cells. This TP extract was very active in transforming RII cells to SIII. To determine its specific composition, Avery and his team proceeded to analyze the active TP using an assortment of exacting chemical tests. The final preparation proved to be rich in DNA. The transforming ability of the TP was not affected by exposing it to protein-digesting enzymes or to RNAase treatments that would destroy any protein or RNA still present. However, exposure to the enzyme DNAase caused loss of the preparation's transforming nature, leaving little doubt that the power resided in DNA. Therefore, since pure DNA by itself could effect a genetic change, the gene for capsule type in the pneumonia organism must be composed of this substance.

The Hershey-Chase experiment

By the early 1950s, additional lines of evidence had accumulated that also implicated DNA as the genetic material. In their classic experiment performed in 1952, Hershey and Chase selected for study the bacterium *Escherichia coli* and the T/2 phage (bacterial virus) that attacks it. From electron microscope studies, they knew that the virus is composed of a protein envelope surrounding a DNA core (Fig. 11-2). Such a structure is characteristic of these T-even phages that reproduce only in *E. coli* cells. Infected cells eventually burst, releasing phage particles identical to the original infecting viruses. The T/2 virus has genetic continuity and must provide a cell with the information needed to make more virus.

Hershey and Chase, knowing that the phage attached to the bacterial cell wall as if it were injecting something into the cell, designed an experiment to determine exactly what this might be (Fig. 11-3). They raised one group of bacteria (group A) on a medium containing radioactive sulfur, ^{35}S. Since sulfur is a protein component, all the bacterial protein became labeled. Viruses were then allowed to infect these bacteria. The new particles released upon burst of the cells would all contain radioactive protein envelopes that had been assembled in the labeled bacterial cells.

A second group of bacteria (group B) was grown on a food source supplied with radioactive phosphorus, ^{32}P. Phosphorus, a principal ingredient of DNA and RNA, does not become incorporated into any protein but only into the nucleic acids. The group B phages were then allowed to infect the labeled cells. The released phage particles would have radioactive phosphorus incorporated into their DNA, which would consequently be tagged or labeled. Hershey and Chase now had two groups of viruses—group A, with protein coats tagged with radioactive sulfur, and group B, with DNA labeled with radioactive phosphorus. The group

A

Head with DNA

Sheath

Core

Plate

Tail fiber

B

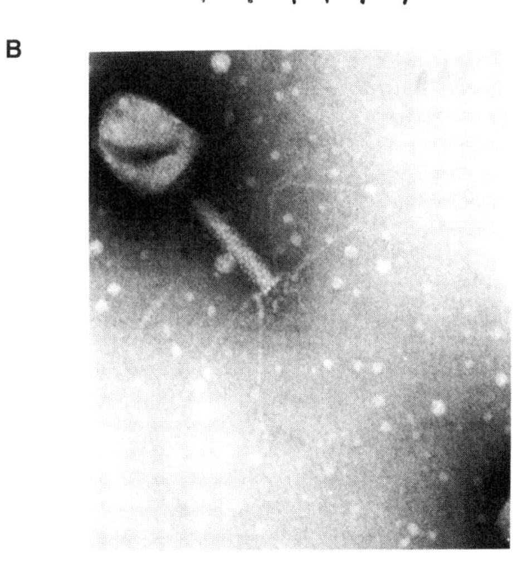

FIGURE 11-2. T-even viruses. **A:** Diagram of some features of the virus. A contractile sheath surrounds the core. Except for DNA in the head, all of the other parts are protein. **B:** Electron micrograph of phage T4. (Courtesy of Dr. James G. Wetmur.)

A phages were then allowed to infect ordinary *E. coli.* A few minutes after infection, the bacterial cells were placed in a blender. The tiny cells cannot be injured by the blades, but any particles adhering to their walls can be torn away. After exposure to the blender, the cells were separated from the fluid medium by centrifugation, to give a cellular fraction and a fluid fraction. Both fractions were then tested for radioactivity of ^{35}S. It was found that almost all of the activity was in the fluid fraction. This meant that only a little of the sulfur was associated with the cells.

Group B phages with tagged DNA were also allowed to infect ordinary *E. coli.* The identical procedure was followed to give a cellular fraction and a fluid fraction. Both were then tested for the labeled DNA. It was found that almost all of the radioactivity was with the cells, in contrast to the results with group A. In other words, most of the DNA of the phages entered the bacteria, whereas most of the protein remained outside in the fluid. It was also shown that the small amount of protein that was found with the cells did not appear in the new viruses, which arose if the cells were allowed to burst. The protein tended to disappear and not enter into the construction of new viruses from the infected cells. These results are strongly in favor of DNA as the genetic material of the phage that bears instructions for the manufacture of new virus particles.

Chemistry of nucleic acids

The first studies of nucleic acids go back to a Swiss biochemist, F. Miescher. Working in 1869 with salmon sperm and with the nuclei of white blood cells, Meischer isolated a substance that was unusually acidic and contained a high amount of phosphorus. Meischer named it *nuclein,* but this was later changed to *nucleic acid* by the biochemist R. Altman, who purified the substance and showed that it was composed of sugars and compounds called *nitrogen bases.* Later investigations revealed that two kinds of nucleic acids occur in cells, deoxyribonucleic acid (DNA) and ribonucleic acid (RNA), and that the nuclein of Meischer was DNA.

An excellent source of animal DNA proved to be the calf thymus gland. Hydrolysis showed that DNA is constructed from (1) phosphoric acid; (2) the five-carbon sugar D(-)2-deoxyribose; (3) the purine nitrogen bases adenine and guanine; and (4) the pyrimidine nitrogen bases thymine and cytosine (Fig. 11-4A).

A rich source of RNA proved to be yeast cells. Upon its hydrolysis, RNA yielded the same kinds of products as DNA. One major distinction between the two lies in the sugar component, and a second lies in one of the pyrimidine bases (Fig. 11-4B). The sugar of RNA, D-ribose, is slightly different from deoxyribose. At its carbon 2 position, deoxyribose has one less oxygen than ribose and therefore does not possess a hydroxyl group at that site. The pyrimidine base thymine does not typically occur in RNA.

Instead, it is replaced by the pyrimidine uracil, which is not found in DNA. Another distinction between the two nucleic acid types is their cellular distribution. DNA is confined almost exclusively to the nucleus in eukaryotes, whereas RNA has a wide distribution in the nucleus and cytoplasm.

Chemical analysis showed DNA and RNA to be constructed of repeating units, nucleotides. Each DNA nucleotide is a combination of a phosphoric acid, a deoxyribose sugar, and one of the nitrogen bases, either a purine or a pyrimidine. Therefore, four different kinds of nucleotides are possible in DNA, depending on the nitrogen base (Fig. 11-5A): adenylic acid, guanylic acid, thymidylic acid, and cytidylic acid. DNA is thus a polynucleotide in which the individual nucleotides are connected through bonds between the sugar and phosphoric acid components (Fig. 11-5B). The nitrogen bases in a strand of DNA are not joined to one another but are linked to the sugar. As the figure indicates, the bases provide the only distinction among the nucleotides. In any stretch of DNA, one kind of nucleotide need not be

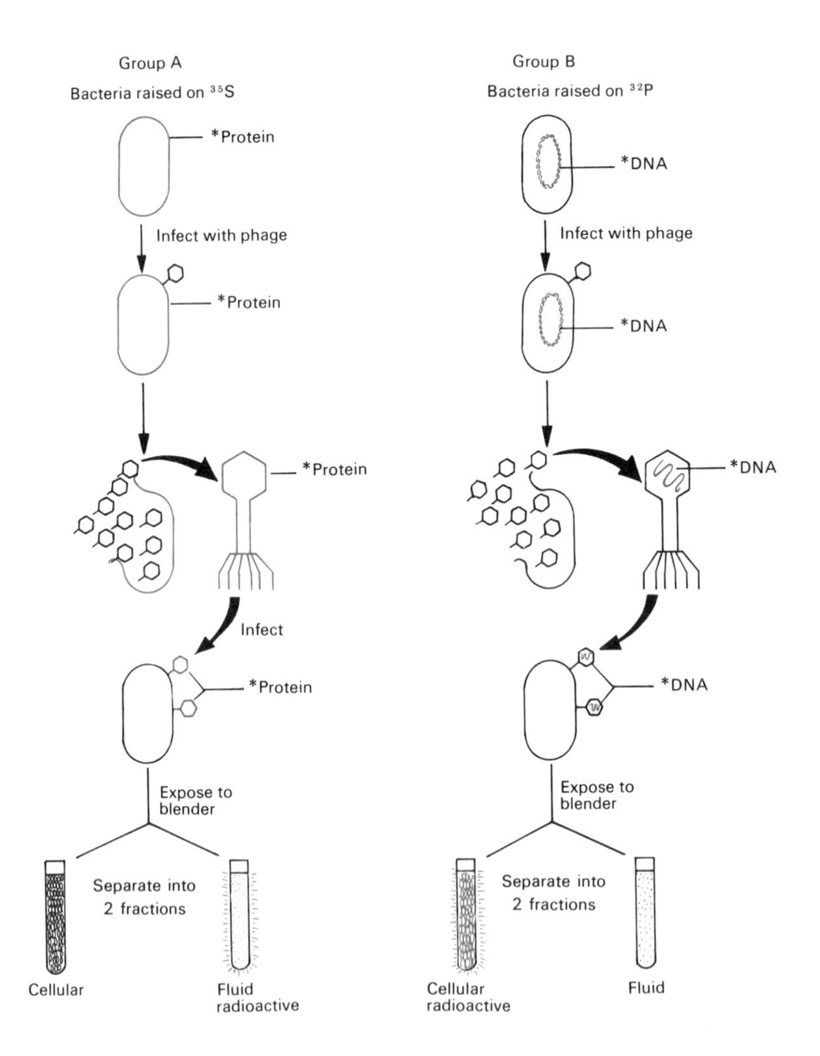

FIGURE 11-3. The Hershey-Chase experiment. Bacteriophage with radioactive (*) protein coats were obtained by infecting bacteria whose protein was all radioactive. Phages with radioactive (*) DNA were obtained by infecting another group of bacteria whose DNA was all radioactive. The phages whose protein was tagged (group A) and those whose DNA was tagged (group B) were allowed to infect ordinary bacteria. Shortly after infection, the cultures were separated into a cellular fraction and a fluid one. The radioactivity in group A was associated with the fluid, meaning that the labeled protein of the infecting phage did not enter the cells. In group B, the cellular fraction was labeled, indicating that the labeled DNA did enter.

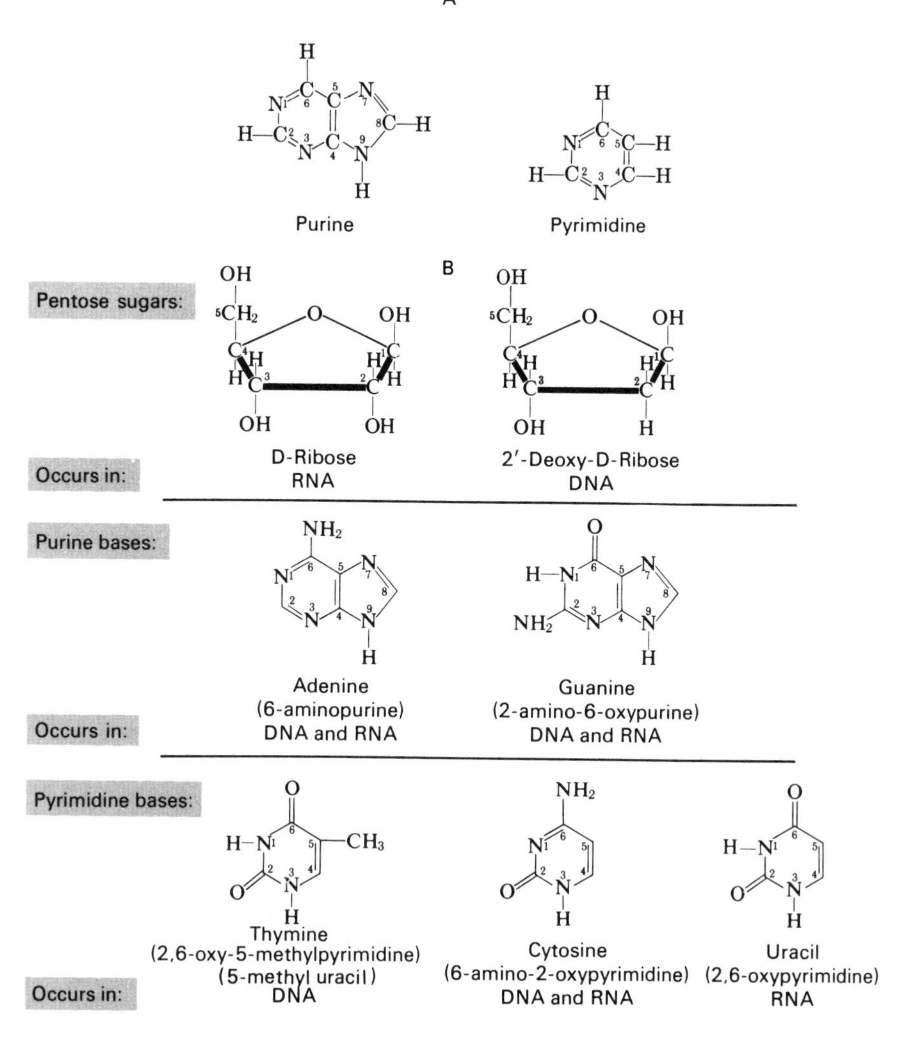

FIGURE 11-4. Sugars and bases in the nucleic acids. **A:** Structure of purine and pyrimidine, the parent compounds of the nitrogen bases in DNA and RNA. **B:** Structure of sugars, purines, and pyrimidines in DNA and RNA. Note that although both types of nucleic acids contain similar building units, an important difference exists in the sugar component and with respect to two pyrimidine bases.

followed by any particular nucleotide. As DNA from various sources was analyzed, no restriction seemed to be imposed on the sequence of nucleotides within the DNA molecule.

Note (Figs. 11-4 and 11-5) that the sugar component of the nucleic acids exists in the furanose form, in a ring of five instead of six as in the pyranose form. The sugar may occur in the cell in association with one of the bases without being linked to a phosphoric acid. Such a combination of a pentose sugar and a purine or pyrimidine base is called a *nucleoside.* Since a nucleoside is equivalent to a nucleotide devoid of the phosphate component, four different kinds correspond to the four nucleotides of DNA: adenosine, guanosine, thymidine, and cytidine (Fig. 11-6A). Uridine is the nucleoside in RNA that corresponds to the RNA nucleotide uridylic acid (Fig. 11-6B). Chemically, the nucleotides are phosphoric

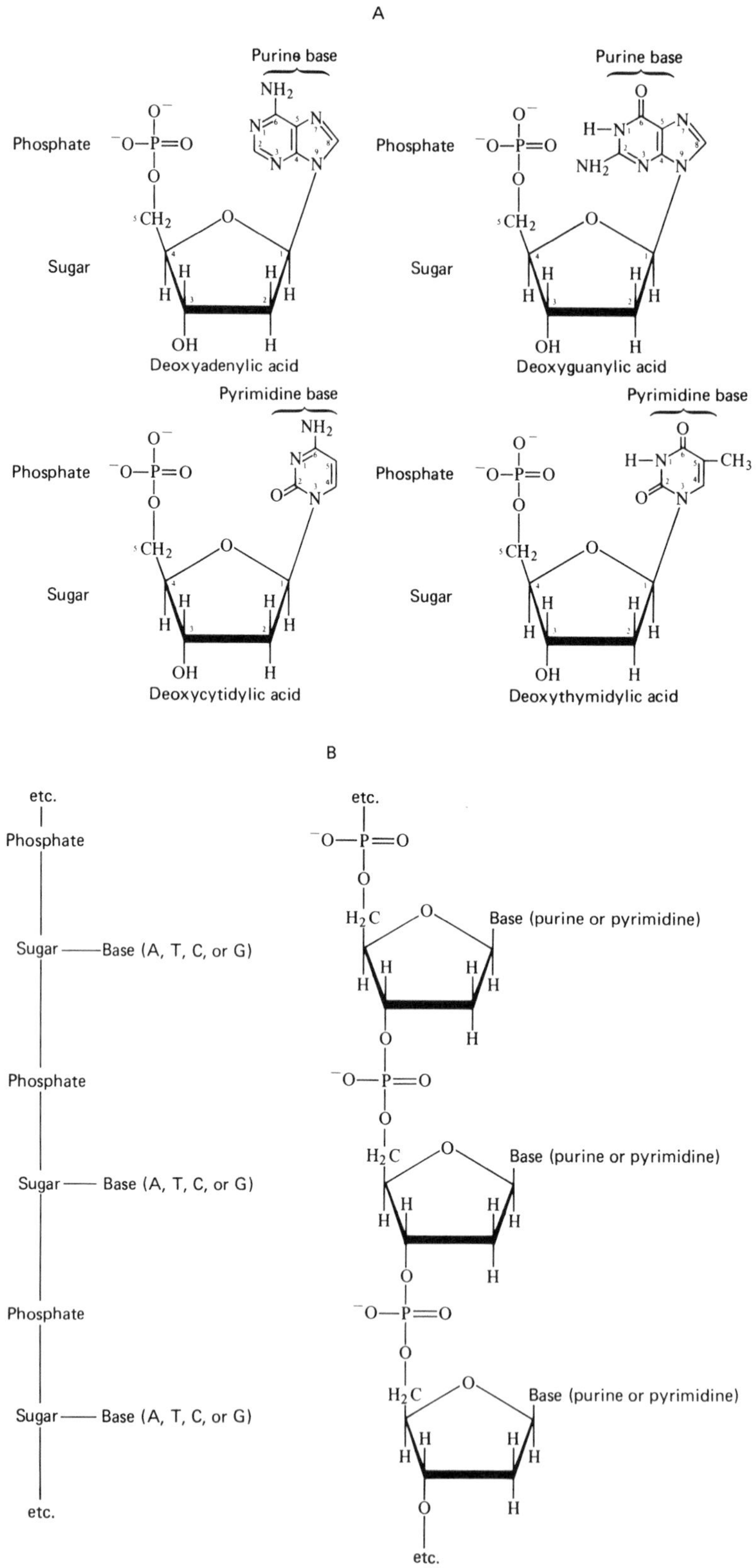

A

Purine base

Phosphate

Sugar

Deoxyadenylic acid

Purine base

Phosphate

Sugar

Deoxyguanylic acid

Pyrimidine base

Phosphate

Sugar

Deoxycytidylic acid

Pyrimidine base

Phosphate

Sugar

Deoxythymidylic acid

B

etc.

Phosphate

Sugar — Base (A, T, C, or G)

Phosphate

Sugar — Base (A, T, C, or G)

Phosphate

Sugar — Base (A, T, C, or G)

etc.

etc.

Base (purine or pyrimidine)

Base (purine or pyrimidine)

Base (purine or pyrimidine)

etc.

esters of the nucleosides. It should be noted that nucleotide structure is found not only in nucleic acids but also in several biologically important compounds, such as certain coenzymes and the universal energy source adenosine triphosphate (ATP).

One theory in the 1940s proposed that DNA consists of repeating tetranucleotide units, each being a sequence of four nucleotides in which the bases adenine (A), guanine (G), cytosine (C), and thymine (T) are always represented. Consequently, the four kinds of bases would occur in equal amounts. Chargaff and his associates showed this is not the case. DNA does not contain equal amounts of the four different bases. Nothing restricts the sequence of the nucleotides. Moreover, the proportion of bases varies greatly with the source of the DNA. But one very important fact came to light from Chargaff's analyses. It was found that the amount of A was always equal to the amount of T. Similarly, the amount of G equaled that of C. Therefore, the A:T ratio is 1 and so is the G:C ratio.

However, the amount of A + T is not necessarily equal to the amount of G + C. Indeed, wide differences were found in these proportions. Some types of DNA have much more A + T than G + C, whereas an abundance of G + C over A + T is found in the DNA from other species (Table 11-1). Thus the makeup of the DNA was shown to vary from one species to another in its base composition, but the different tissues of any one species showed the same kind of DNA. This, of course, made sense. We would expect the hereditary material to differ among the diverse forms of life but not from one cell type to another within an organism or a species.

The base composition of the DNA is therefore a characteristic of any particular species. Each species has its typical A + T:G + C ratio. However, when the base composition of a species is discussed, rather than talking in terms of the A + T:G + C ratio, reference is usually made to the "G + C content" or the "percent GC content" for short. If it is said that an organism has a G + C content or percent GC content of 40%, this is taken to mean G = 20%, C = 20%, A = 30%, and T = 30%.

A striking finding is that the percent GC content of lower forms (prokaryotes, for example) shows great variability between species, perhaps 25% in one and 75% in another. With evolutionary progression, the percent GC appears to vary less and less within the divisions of plants and animals and to be a

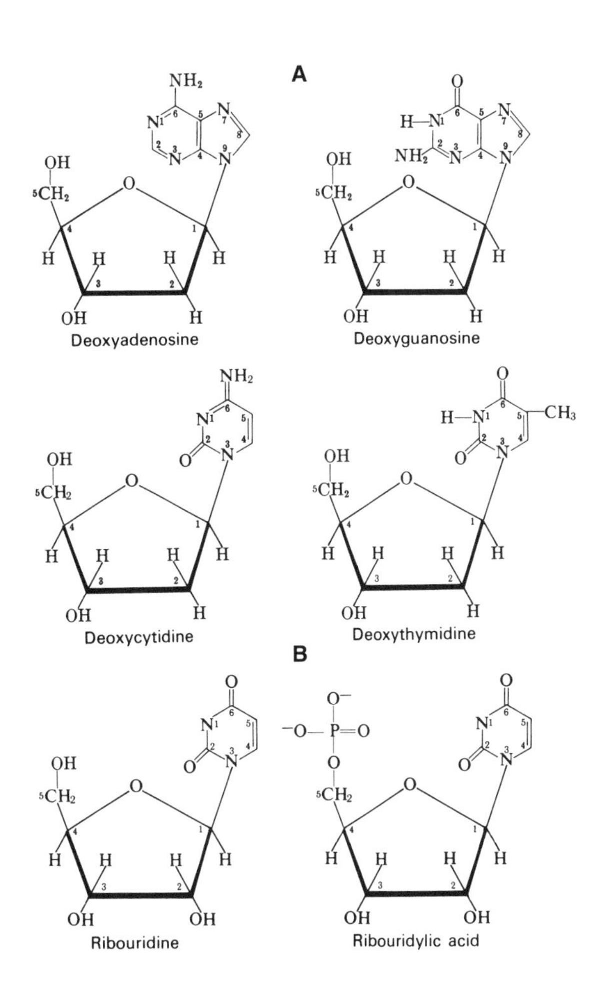

FIGURE 11-5. Nucleotides in DNA. **A:** Four different nucleotides occur in DNA. Each is a combination of deoxyribose sugar joined to a phosphate and to one of the four bases: adenine (A), guanine (G), cytosine (C), or thymine (T). **B:** DNA is a polynucleotide. The nucleotides are attached to each other through the sugar and phosphate components. Note that the bases are not connected to the phosphate but are linked only to the sugar.

FIGURE 11-6. Nucleosides in nucleic acid **A:** A nucleoside is a combination of a purine or pyrimidine base with a pentose sugar. The four kinds in DNA are deoxyribonucleosides and correspond to the nucleotides (Fig. 11-5), which contain a phosphate group linked to the sugar. **B:** Nucleosides and nucleotides of RNA correspond closely to those of DNA. One major difference is in the sugar. The ribose of RNA contains an oxygen at position 2 in the sugar ring that is absent from that position in DNA. Another distinction is that in RNA the nucleoside ribouridine and the nucleotide ribouridylic acid occur in place of thymidine and thymidylic acid.

TABLE 11-1. Base composition of DNA from various sources[a]

DNA source	Adenine	Thymine	Guanine	Cytosine
Human (liver)	30.3	30.3	19.5	19.9
Human (sperm)	30.7	31.2	19.3	18.8
Human (thymus)	30.9	29.4	19.9	19.8
Bovine sperm	28.7	27.2	22.2	21.9
Rat bone marrow	28.6	28.5	21.4	21.5
Salmon	29.7	29.1	20.8	20.4
Sea urchin	32.8	32.1	17.7	17.7
Wheat germ	27.3	27.2	22.7	22.8
Yeast	31.7	32.6	18.3	17.4
Escherichia coli	26.0	23.9	24.9	25.2
Streptococcus pneumoniae	30.3	29.5	21.6	18.7
Mycobacterium tuberculosis	15.1	14.6	34.9	35.4
Vaccinia virus	29.5	29.9	20.6	20.3
Bacteriophage T2	32.5	32.5	18.3	16.7[b]
Bacteriophage T7	26.0	26.0	24.0	24.0

[a]Note that although both the A:T and the G:C ratios tend to equal 1, the proportion of A + T/G + C varies from one species to the next. Some have little G + C (sea urchin), whereas others have a great deal (*M. tuberculosis*).
[b]Has hydroxymethylcytosine in place of cytosine.

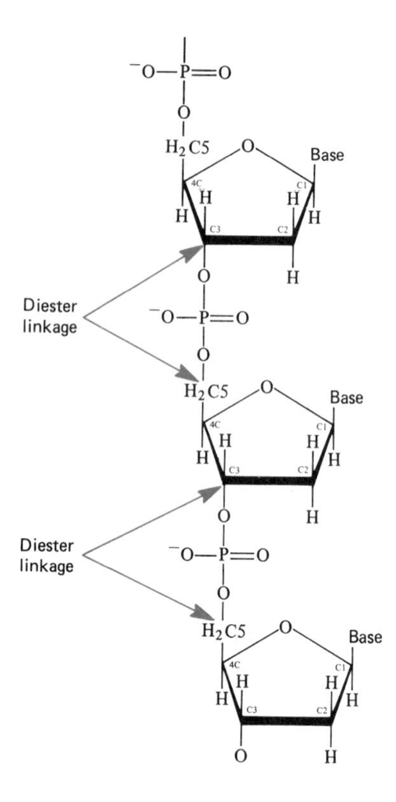

FIGURE 11-7. The linkage between nucleotides. Note that carbons 3 and 5 of the sugar are involved in the formation of a linkage with the phosphate component. A phosphoric acid has formed one ester bond with the -OH associated with the carbon in position 3 of the one sugar and another ester bond with the -OH associated with the carbon in position 5 of the adjacent sugar. (Refer also to Fig. 11-5B, which does not show the C atoms in the sugar.)

more fixed value. In mammals, the overall average value is about 40%, one species of mammal showing a similar value to another. The reason for this stability of G + C value in higher life forms and its evolutionary significance are unknown.

More detailed chemistry studies showed exactly how the nucleotides are linked to one another through the sugar and phosphate (Fig. 11-7). The number 1 carbon in the sugar is attached to the nitrogen base. The second carbon is the one lacking the oxygen. Carbon 4 is completely in the ring. Carbons 3 and 5 are involved in the internucleotide linkages. The -OH group of carbon 3 in the sugar component of one nucleotide forms an ester linkage with the phosphoric acid. The phosphate in turn forms another ester linkage with the -OH group of carbon 5 in the adjacent nucleotide. Thus the phosphoric acid engages in the formation of a double ester. It is these diester linkages that hold together the sugar–phosphate backbone. Because positions 3 and 5 of the sugar are involved, we say the nucleotides are joined by 3′–5′ phosphodiester linkages.

Physical structure of DNA and the Watson-Crick model

By the early 1950s a great deal was known about the chemistry of DNA, but this told little about the architecture of the molecule. Important data on the physical structure of DNA were obtained in the laboratory of Maurice Wilkins and his group at King's College. The information was primarily in the form of X-ray diffraction studies of fibers of DNA. In diffraction studies, the material is subjected to a beam of X-rays. Assuming the material is composed of molecules and atoms arranged in a random or haphazard manner, the rays, in their passage through the material, will not be diverted in any characteristic fashion. If they pass through the material and fall on a photographic plate, they will not describe any definite pattern, and when the plate is processed, little more than a diffuse blackening may be seen. If, however, there is a definite arrangement of the parts within the molecule, the rays will be deflected more in some directions than in others. As the arrangement repeats itself, the photosensitive emulsion may be bombarded over and over again in certain areas. This will heighten the darkening in these areas, with the result that certain dark spots may repeat and form a pattern (Fig. 11-8). Studies of such patterns can therefore tell a great deal about the spatial organization within a molecule and the repetition of the units that compose it. Diffraction studies by Wilkins's group showed that the DNA molecule is not haphazard in arrangement but possesses a definite organization in which the parts repeat themselves. The helical form of the DNA molecule is indicated in Figure 11-8 by the pattern in the form of a cross in the center of the photograph. The figure represents a photograph taken by Rosalind Franklin, Wilkin's associate, and was pivotal in establishing the helical form of DNA. The pronounced darkenings at the top and bottom indicate that the base pairs are stacked in a regular arrangement perpendicular to the axis of the molecule. Using the information from chemical analyses along with the diffraction photographs, Watson and Crick were able to propose a model of the structure of the DNA molecule. The Watson-Crick model is perhaps the greatest landmark in modern biology. It has given rise to the discipline of molecular biology and has provided a deep insight into the very nature of gene action.

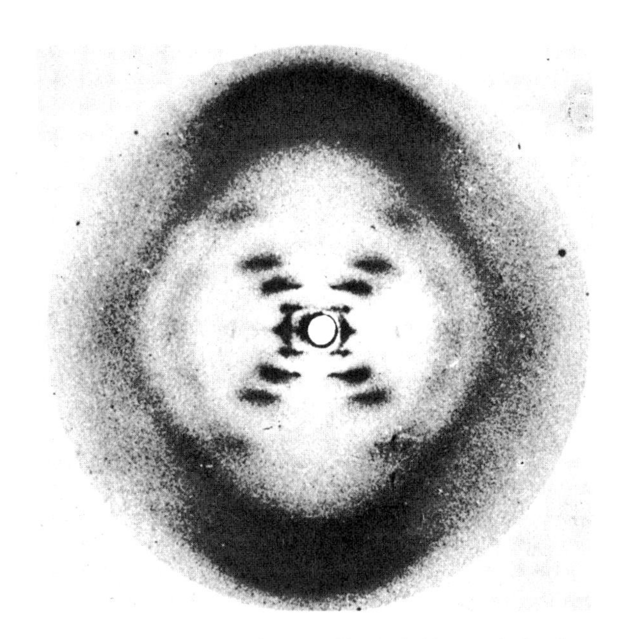

FIGURE 11-8. X-ray diffraction. (Copied with permission from J.D. Watson and F.H.C. Crick, *Cold Spring Harbor Symp. Quant. Biol.* 18:123–131, 1953.)

The diffraction photographs of Wilkins and Franklin revealed patterns that showed that the DNA molecule is composed not of just a single long strand of nucleotides but rather two strands bonded together in some way. Moreover, the two polynucleotide chains are not simply stretched out but are twisted in a spiral, much as the railings of a spiral staircase. The DNA molecule is in the form of a double helix (Fig. 11-9), having a diameter of about 20Å. One complete turn of each chain is 34 Å, and this includes 10 nucleotides. Therefore the distance between one nitrogen base and the next one on a chain is 3.4 Å. The chemical information told Watson and Crick that the nitrogen bases in DNA are in the ratios: A:T = 1 and G:C = 1. Added to what was known about the physical structure of the DNA, this information suggested to them that preferential base pairing occurs in the molecule and that this pairing is responsible for holding the two chains together. The purine adenine would thus pair preferentially with the pyrimidine thymine, whereas the purine guanine would pair with the pyrimidine cytosine (Figs. 11-10 and 11-11). The two bases composing each base pair would be held together by hydrogen bonds, much weaker attractive forces than covalent bonds, which entail the sharing of electrons between atoms.

Knowledge of the molecular structures of the four bases indicated that two hydrogen bonds could hold the AT pair together, whereas three would be formed in the GC pair. As a consequence of this hydrogen bond formation between the base pairs all along the molecule, the two polynucleotide chains would in turn be held together. The base pairs extend from the sugar–phosphate backbone and are stacked up inside the double helix, resembling the stairs in the analogy of the spiral staircase. The two chains of the molecule run in opposite directions with respect to the attachment between sugar and phosphate groups. Recall that 3′–5′ phosphodiester linkages hold the nucleotide units together in a chain. If we start at the "top" of a DNA molecule, the orientation of one chain would be 3′–5′–3′–5′– and so on. The other chain would be in reverse order and would run 5′–3′–5′–3′– and so on. As a result of this orientation, the molecule looks the same from the top as from the bottom: A 3′ chain and a 5′ chain terminate at each end (Fig. 11-11). Thus

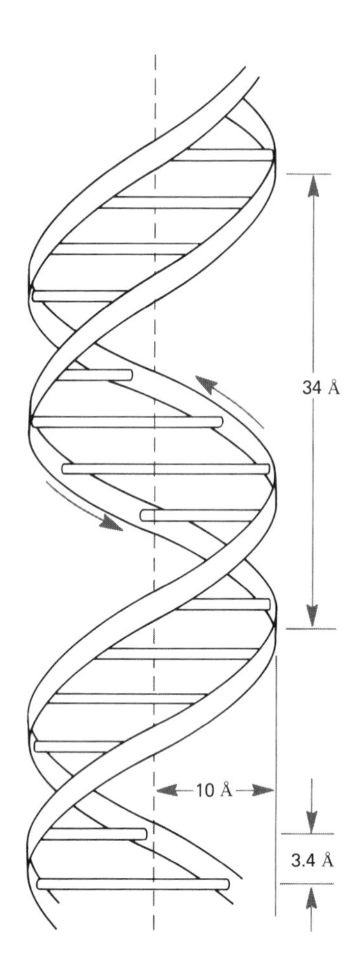

FIGURE 11-9. The Watson-Crick double helix. According to the model, the DNA is composed of two polynucleotide chains twisted in a spiral. The sugar and phosphate components provide the backbone of the chains. These in turn are held together by pairing between the nitrogen bases (see Figs. 11-10 and 11-11). The bases would be stacked inside the double helix, much like stairs in a spiral staircase. Note that a large groove and a small one are present.

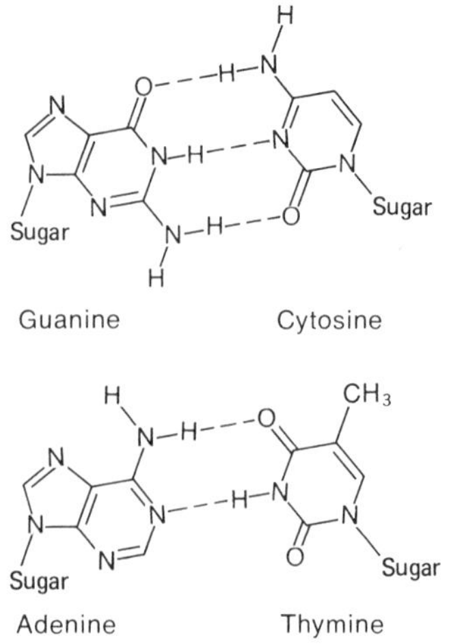

Guanine Cytosine

Adenine Thymine

FIGURE 11-10. Preferential base pairs in DNA. Adenine on either one of the two chains pairs with thymine; cytosine pairs with guanine. Note that two hydrogen bonds (dashed lines) hold the AT pair together, whereas three are involved in the GC pair.

each end of the molecule has one 5′ or phosphate terminating chain and one 3′ or –OH terminating chain. The double-helical DNA is a molecule with a right-handed twist. Note (Fig. 11-9) that it also possesses two grooves—a narrow minor groove and a wider major groove.

The structure of the DNA molecule, as proposed by Watson and Crick, was constructed through careful, painstaking correlation between chemical and crystallographic data. All of the suggestions in the model were in complete agreement with established chemical and physical principles. One critical feature of the model is that the two polynucleotide chains are not identical because of the preferential base pairing. Not only do the chains run in the opposite direction, but, most important, they are complementary, the A and C in one strand corresponding to a T and G in the other. Since the model imposes no restrictions on the sequence of the bases in any one chain, A could be followed by T, G, C, or another A. However, if one knows the sequence of bases along a length of one of the chains (say, ATCCAG), one can be certain of the sequence of bases occurring along the complementary chain (TAGGTC).

As research on the DNA molecule continued, it became apparent that DNA can exist in more than one form. The one presented in Figure 11-9 and described by Watson and Crick is known as the *B form,* the form in which most cellular DNA is believed to exist. Other forms of DNA and their significance are discussed in Chapter 17 in relation to the topic of gene regulation.

At this point, a few remarks are in order concerning the need to designate the polarity of a written sequence of bases in a DNA strand. When just a portion of a single DNA chain is being written, the polarity must be designated in some way, since the two DNA chains of a double helix are antiparallel (Fig. 11-11). A convention is needed to clarify the polarity of a written sequence such as TACG. One can write TACG as pTpApCpG with the understanding that the 5′ end of the chain is to the left and that each p between the

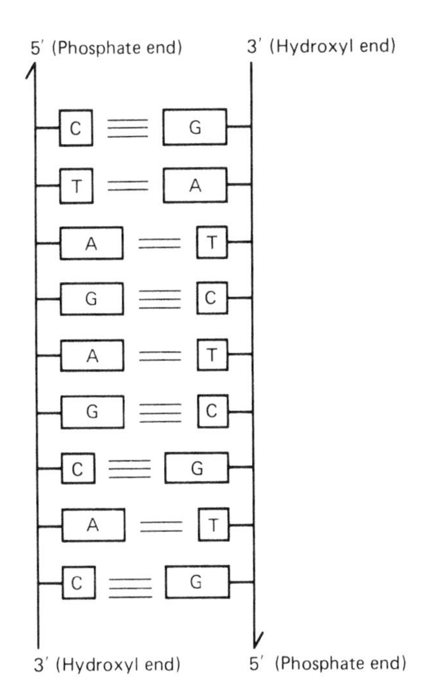

FIGURE 11-11. Diagram of the DNA double helix. Note the difference in orientation of the two chains: One runs in the order 3′–5′ and the other in the order 5′–3′. It can also be seen that the two chains are complementary, not identical.

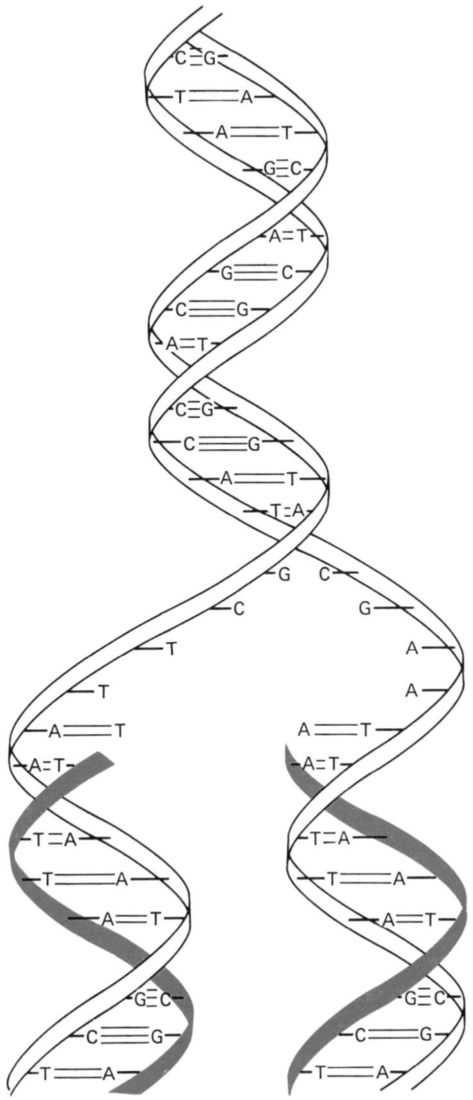

bases represents a phosphodiester linkage. Another way to represent the same sequence is p5′–TACG–3′–OH, again with the 5′ end to the left. The writing can now be simplified, however, to TACG, since by convention the 5′ end of a written sequence is taken to be toward the left unless otherwise indicated.

FIGURE 11-12. Self-duplication of DNA. During replication, the hydrogen bonds between the strands are dissolved. Each original strand remains intact and acts as a pattern for the assembly of a complementary chain (new strands are shown in red).

Implications of the Watson-Crick model

Although it was well-known that the genetic material undergoes self-duplication, there was no well-founded suggestion as to how this might take place. The Watson-Crick model offered a valid basis for this. According to the model, at the time of replication the hydrogen bonds between the bases of the two complementary chains dissolve (Fig. 11-12). The two chains then unwind and separate. Each strand maintains its integrity in the process and does not break down in any way. Instead, each one acts as a template or pattern for the assembly of another strand, one that is complementary to it. These complementary strands are constructed from their building blocks, the nucleotides present in the cell. As the sugar–phosphate backbone is assembled, each nitrogen base in the original strands attracts the complementary one. The purine adenine (A) in one chain attracts the pyrimidine thymine (T) in the cellular environment. Likewise, T in the other original chain attracts nucleotides of A in the cell. This complementary attraction by all the bases in each of the two original strands, along with the assembly of the sugar–phosphate backbone, results in the construction of two new chains, each complementary to the original "old" ones. The overall effect is two double-helical DNA molecules, each identical to the original.

This fits perfectly with what we would expect if the genetic material remains identical from one cell generation to the next, for the genetic material must contain a store of information. The Watson-Crick model has important implications on this extremely critical point. Since the model assumes no restriction on the sequence of base pairs, the genetic information could somehow be stored in the form of a code. The code cannot reside in the sugar or phosphate, because these are the same from one nucleotide to the next; only the bases can differ. The sequence of bases could then provide the basis of a code. If the sequence is ATT along one stretch of the DNA, this could mean something different from the sequence GTC. A gene, a segment of double-stranded DNA, would then contain information coded in the sequence of its base pairs. And one gene or one allele would differ from the next because of the difference in the sequence. Since a gene is composed of many nucleotides, it would include many base pairs. The unrestricted sequence allows a limitless number of differences and could account for the endless number of different genes found in living species.

But it is known that the genetic material can suddenly change as the result of a spontaneous gene mutation. This implies a change in the coded information. Chemical analysis of DNA showed that the molecule is tautomeric. This means that the molecule undergoes occasional shifts in the positions of certain protons within it. One form of the molecule would be the most common, but at any one moment a hydrogen can change position so that a molecule could be in its rarer form or state (Fig. 11-13A). This rare state might occur at the time the gene is replicating or building more of itself. Since the tautomeric changes can affect hydrogen atoms in the purine and pyrimidine bases, the change could alter the pairing preferences of the bases (Fig. 11-13B).

Let us assume that in a stretch of DNA a shift occurs in the position of a hydrogen in one of the pyrimidine bases, thymine. As a consequence of the shift, it may not be able to pair with its normal partner, adenine, but instead, pairs with guanine (Fig. 11-14). The next time the molecule undergoes replication, there would be no reason to expect a tautomeric shift, because this would be a rare event in any case and would not have to affect the same bases. Normal bonding preferences would therefore be expected at the next replication. However, because G has been inserted in the one strand (the right one in the figure) as a result of the first error, C will now be placed in the corresponding position on the new complementary strand. Instead of the original base pair sequence AT, TA, CG, we now have AT, CG, CG. This TA pair has been replaced by a CG. If the base pairs do provide the basis for a code of stored information, the code would now be changed. The new sequence of base pairs could mean something very different from the original message. This model could also explain the dire effects of a deletion in which a part of the chromosome is lost. If a sequence of base pairs is omitted, then part of the message would be missing. Any change in the structure of the chromosome could alter the base sequence and thus upset the genetic message.

If the Watson-Crick model is correct, then the coded information in the DNA must somehow be transferred to other parts of the cell. Obviously, a code is of no use if its information cannot be communicated. The model triggered a wealth of investigations on the mechanism of information transfer from nucleus to cytoplasm, a most important topic, which is the subject of Chapter 12.

Experimental testing of the Watson-Crick model

One very admirable quality of the Watson-Crick model is that its formulation permits tests of its validity. Throughout the years, the model has passed the scrutiny of many investigations, the results of which support all the features of the model. In 1958, Meselson and Stahl presented strong support for the concept of replication as pictured by Watson and Crick. According to the model, DNA replication is semiconservative. This means that the entire molecule does not remain intact when new DNA is being formed; instead, the two strands come apart. However, each strand is preserved intact, as shown in Figure 11-12; it does not dissociate in any way. Besides this semiconservative concept of replication, two other possibilities

FIGURE 11-13. Tautomerism and formation of exceptional base pairs. **A:** Each of the four bases in DNA typically exists in a common state. However, hydrogen atoms may at times shift position, so that a rare state may exist at a particular time. For example, adenine bears a $-NH_2$ group. At rare times, one of the H atoms shifts so that the $-NH_2$ becomes $-NH$, and the H is in a new position. **B:** Such tautomerism can change pairing preference. For example, the shift that can occur in adenine now allows it to bond with cytosine. Shifts in the other bases cause similar changes in pairing preference.

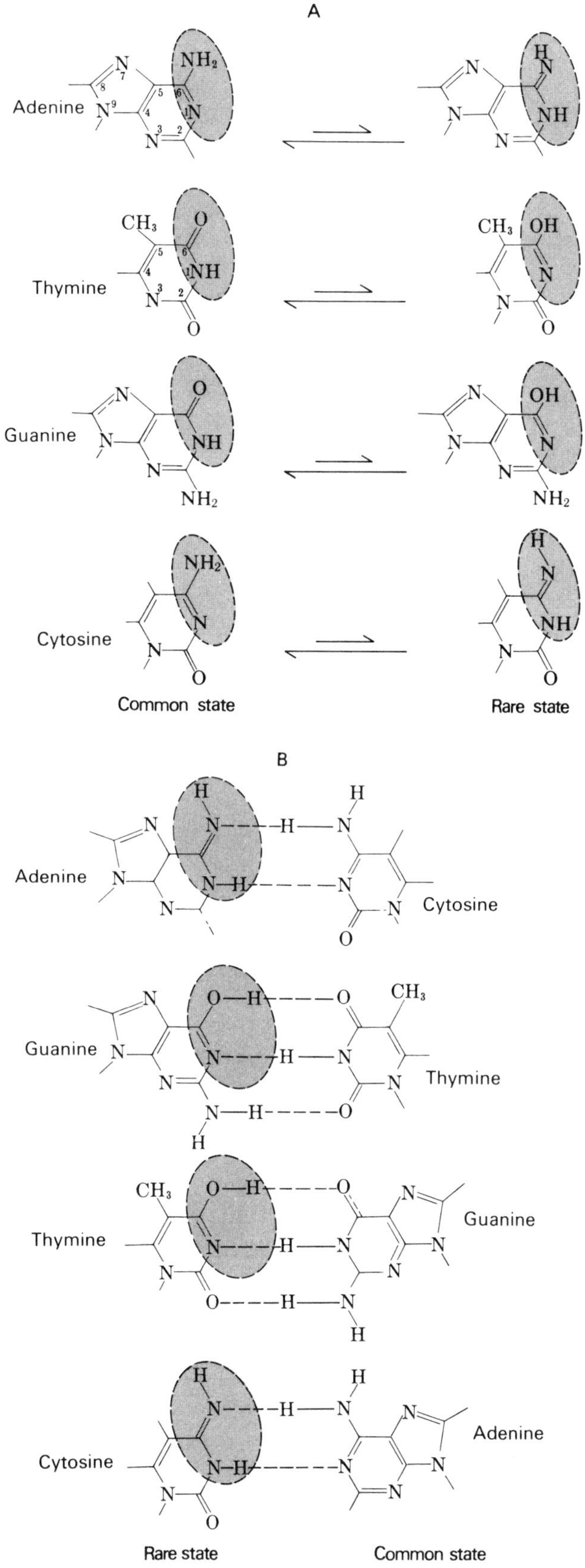

A

Adenine

Thymine

Guanine

Cytosine

Common state

Rare state

B

Adenine

Cytosine

Guanine

Thymine

Thymine

Guanine

Cytosine

Adenine

Rare state

Common state

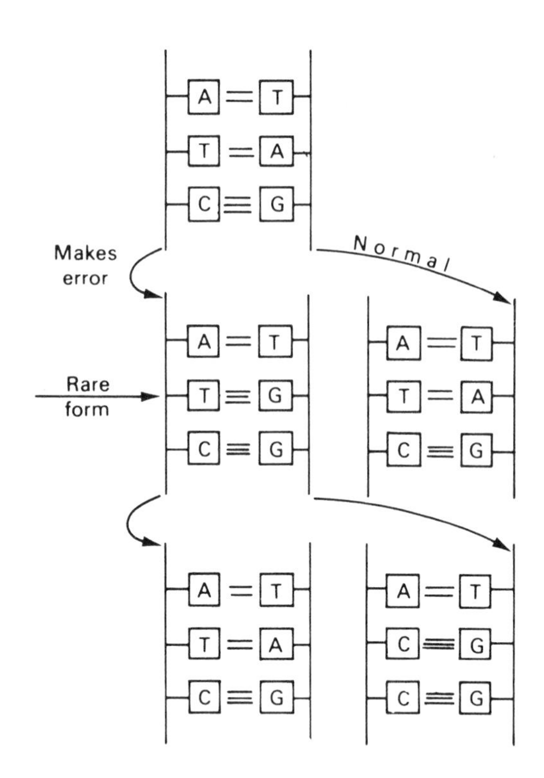

FIGURE 11-14. Tautomerism and mutation. A sequence of normal nucleotide pairs (above) is typically found in a stretch of DNA. If a base should exist in its rare form (middle) in one of the strands at the time of replication, an exceptional base pair may form. At the next replication (below), the bases in each strand would usually behave normally and attract their complementary bases. However, because a substitution was made at the previous replication, a new base pair, CG, now substitutes for the original, TA, producing a mutation.

can be imagined: conservative (the two strands stay together, and a new double-stranded molecule is built up next to them) and dispersive (the two original strands break down and entirely new strands are constructed from these and other precursors in the cell).

To gain information on this critical point, Meselson and Stahl employed a procedure now routine in laboratories—density gradient ultracentrifugation. The technique is based on the knowledge that a solute (e.g., sugar, salt), when spun at a high speed in a centrifuge, becomes distributed from one end of the tube to the other. It eventually reaches an equilibrium, at which time the molecules are most dense at the centrifugal end and gradually decrease in density toward the centripetal end. Such an equilibrium density gradient has practical laboratory application. A mixture of substances that are very similar in density may be separated by this procedure, for when a mixture is spun in the appropriate solution a density gradient is established in the centrifuge tube. The substances composing the mixture tend to separate out at characteristic levels in the tube. This dissociation occurs because a substance settles at a location where the density of the solute molecules matches its own density. The technique is so sensitive that materials of very similar densities can be separated by using the proper solute. For example, a mixture of DNA molecules of different densities can be separated in a gradient of cesium chloride (Fig. 11-15). Under laboratory conditions, a given kind of DNA can be made heavier than normal by supplying organisms with a medium containing only heavy nitrogen, ^{15}N, an isotope of the common ^{14}N. If "heavy" DNA (heavy because its bases contain ^{15}N) and light DNA are mixed, they can be dissociated in a cesium density gradient. This is possible, because they come to equilibrium and form distinct bands at slightly different locations in the centrifuge tube.

In their experiment (follow Fig. 11-16), Meselson and Stahl employed the bacterium *E. coli.* The organism was grown on a source of labeled, heavy ^{15}N for several generations until all of its DNA was heavy. When extracted from the cells, this heavy DNA settled out in the density gradient at a characteristic position. Meselson and Stahl removed samples of bacteria from the heavy cultures and placed them on a medium containing ordinary ^{14}N. They tested the DNA of some of these immediately, without allowing any cell division (0 generations). Some samples were allowed to divide once (one generation), others two and three times (two and three generations), before their DNA was extracted. As Figure 11-16 shows, all of the DNA from the 0 generation bacteria was heavy, as expected.

It was reasoned that if the Watson-Crick model was correct, each of these heavy strands would separate during the first round of cell division. Each could only direct next to it the construction of a light strand, because only light nitrogen would now be present. The DNA of the first-generation cells would thus contain DNA that is hybrid, intermediate in density between double-stranded heavy and double-stranded light molecules. It should therefore settle out at a location in the centrifuge tube distinct from that of the other two.

This is what was found, a single band at a position characteristic of neither ^{14}N nor ^{15}N. The DNA was then tested from the bacteria that had undergone two divisions. According to the Watson-Crick model, each strand will maintain its integrity. But under the conditions of the experiment, each strand can order the construction of only light strands. The original heavy ones build up light strands, as do the light ones made during the first division. As a result, half of the cells will have DNA with two light strands and half will have DNA that is hybrid ($^{14}N/^{15}N$). A perfect correlation was found between the prediction and the experimental results, for now *two* bands of *equal* size were present in the density gradient: one at the level of the hybrid DNA and the other at a less dense position, that typi-

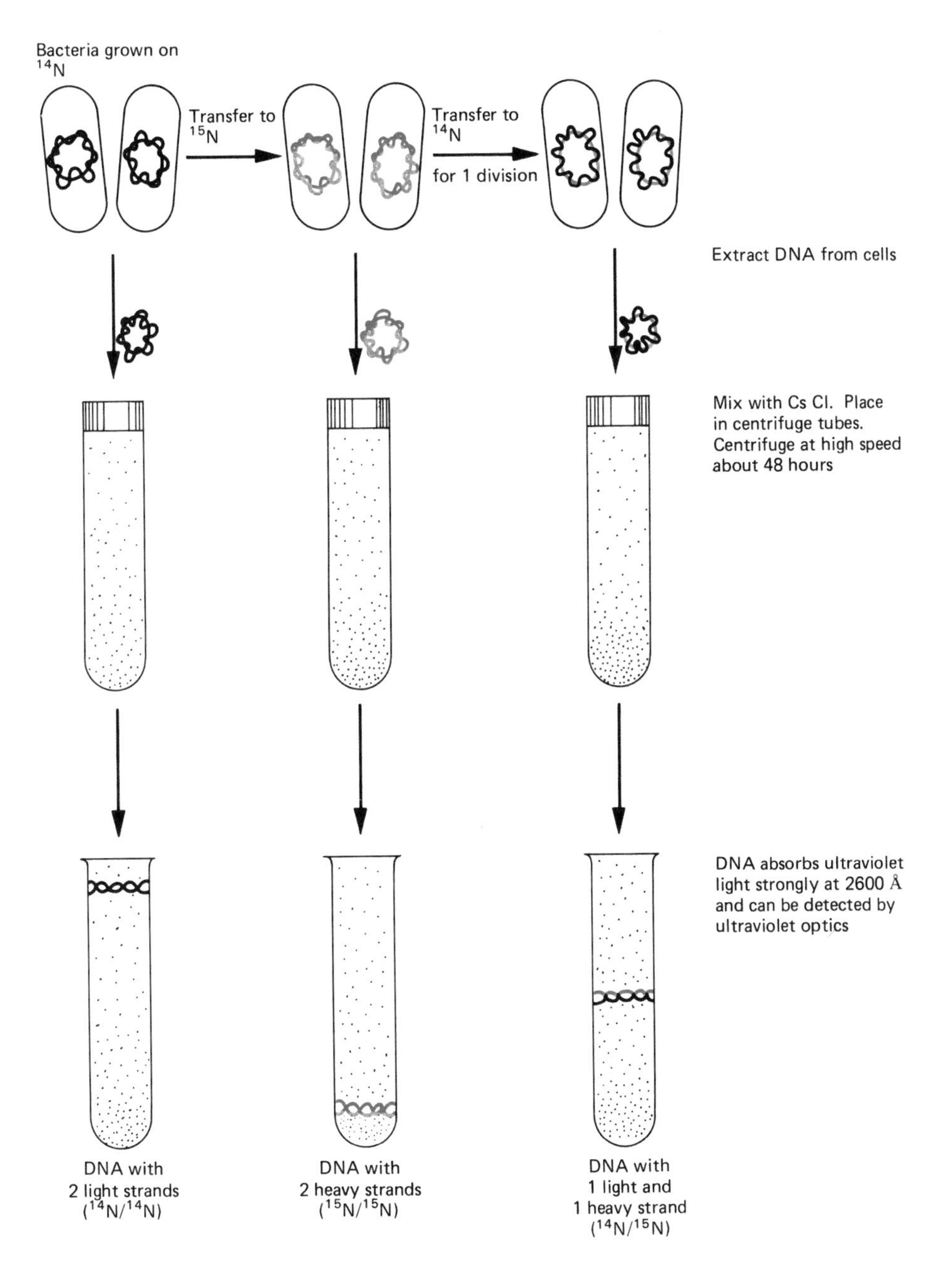

FIGURE 11-15. Density gradient ultracentrifugation. Substances that are very similar in density may be separated by this sensitive method because each will tend to settle where its density matches that of a solute. Different types of DNA can be separated, such as the three kinds indicated here: one type composed of two light chains containing ^{14}N, another composed of two heavy chains with ^{15}N, and an intermediate DNA made of one light and one heavy. (See text for details, and compare with Fig. 11-17.)

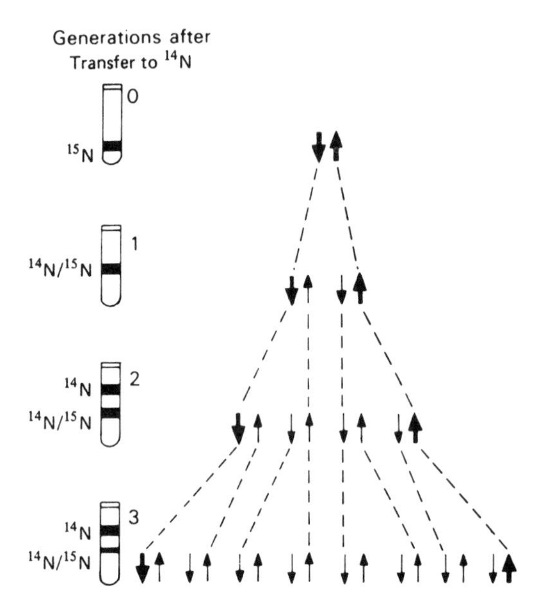

FIGURE 11-16. Meselson and Stahl experiment. In this highly schematic diagram, a pair of reversed arrows represents a double helix. It can be seen that if the cells are grown on medium with heavy (^{15}N) nitrogen, all of the cellular DNA is heavy if no division is permitted (0 generation). As the number of cell divisions increases, following transfer to ^{14}N medium, the number of cells with light strands increases. Note that the strands maintain their integrity. Any strand formed in one generation acts as the template for a new light strand in the next generation. In the third generation, the hybrid ($^{14}N/^{15}N$) band is narrower than the ^{14}N, because only light strands can be made. Still, the heavy strands continue to act as templates.

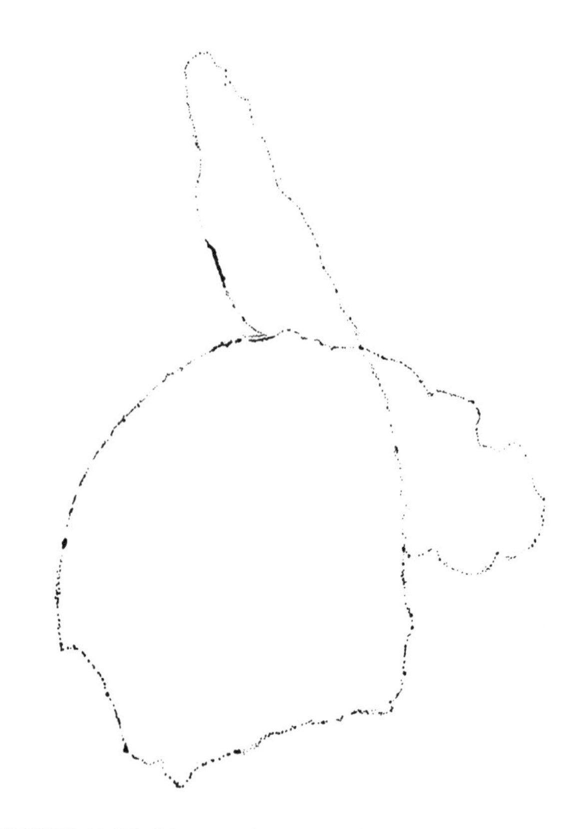

FIGURE 11-17. Diagram of an autoradiograph of the *E. coli* chromosome. The chromosome is actually in the act of duplicating. The DNA is from cells that have been grown in the presence of radioactive thymidine for two generations (see text for details). (Drawn after J. Caims, *Cold Spring Harbor Symp. Quant. Biol.* 28:43, 1963.)

cal of light DNA. The third-generation DNA again gave exactly what would be expected from the Watson-Crick model. Hybrid and light strands are again both present, because the original heavy ones continue to act as templates. However, cells with hybrid DNA are fewer in proportion to cells with completely light DNA, since only light strands can be constructed. Observations clearly showed that the light band in the density gradient was conspicuously thicker than the slightly heavier hybrid one. If replication were conservative or dispersive, such results could not be explained. Replication as proposed by Watson and Crick offers the only answer.

According to the model, as replication takes place the two chains separate from each other while each base attracts the complementary one. It is conceivable that chain separation is unnecessary for the construction of two new chains. It is also possible that both strands must separate completely along their lengths and exist entirely as single chains at the time of replication. The work of Cairns with *E. coli* DNA provided details on these aspects of the duplication process. Cairns was a pioneer in the study of the "chromosomes" of bacteria. His superb isolations of bacterial DNA showed that the DNA of *E. coli* is in the form of a circle (Fig. 11-17). Bacteria and other prokaryotes do not possess chromosomes in the true structural sense of the word as used for eukaryotes. The genetic material of these simple organisms contains naked strands of DNA that are not complexed with histone protein. Much less DNA is found here than in the true chromosomes of eukaryotes. When extracted from a cell, the prokaryotic DNA can often be followed through its entirety. The overall simplicity of this DNA provides a system for detailed study that is extremely difficult in eukaryotes.

To follow the replication of DNA in *E. coli,* Cairns used the technique of autoradiography. When cells of any kind are grown in a medium containing thymidine, this nucleoside is incorporated only into the DNA. The thymidine can be rendered radioactive if it contains tritium (a heavy isotope of hydrogen). As it disintegrates, tritium emits β particles (electrons). If these fall on a photographic emulsion, the film will contain darkened spots when it is processed. Cairns grew *E. coli* cultures in a medium of tritiated thymidine for different periods of time. He extracted the DNA and covered it with an emulsion that was eventually developed in a fashion similar to that of any photographic film. The dark spots on the resulting autoradiograph indicate the breakdown of radioactive atoms in the DNA. The more such atoms exist in one location, the more spots there are because of the greater number of β rays affecting the emulsion. Thus, the number of spots or their density would be proportional to the number of radioactive nucleotide chains in the DNA molecule. If two radioactive chains are present at a certain location and only one at a second location, the former gives off twice as many emanations as the latter. The darkening is thus twice as dense.

Cairns compared autoradiographs from bacteria that had grown in labeled thymidine for different lengths of time. Particularly significant were autoradiographs from those cultures of cells that had completed one round of DNA replication in the labeled medium and were now undergoing a second replication in the presence of the label. On each of these, Cairns was able to identify a region that appeared as a fork (Fig. 11-18). This single fork appeared at different positions, depending on the length of time the cells had grown in the presence of the radioactive thymidine.

The observations can be readily interpreted according to the Watson-Crick model. The fork marks the point in the DNA where replication is taking place. If one round of replication is completed, the DNA is composed of one labeled and one nonlabeled strand. As a second round starts, the two strands unwind, and each proceeds to direct the formation of a radioactive partner. However, one strand has two labeled chains, the other only one. Consequently, a "Y" figure is produced on the emulsion at this point, but one arm of the Y has a density of dots twice that of the base of the Y and the other arm.

Cairns was able to time the movement of a single replicating site from the initiation point around the circular chromosome by comparing autoradiographs from cultures grown for different lengths of time in the labeled thymidine. A fork was estimated to travel approximately 20–30 μ/min. It was later established that replication of the DNA in the *E. coli* chromosome is initiated at a single site, as indicated by Cairn's original observations, but that two growing points are established at that site. Replication is then bidirectional; these two replicating sites move in opposite directions around the circular bacterial DNA until they eventually converge and fuse. The observations clearly

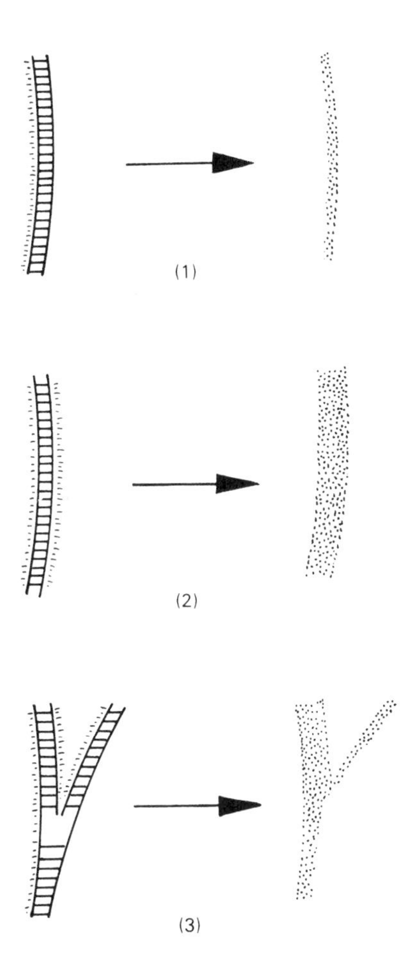

FIGURE 11-18. Point of replication in DNA. **1:** If one chain of the double helix is labeled with tritiated thymidine, dark spots will appear on the autoradiograph because of the emission of electrons as the label decays. **2:** If both chains are fully labeled, the density of dots on the autoradiograph will be twice as great as in the case of the half-labeled DNA. **3:** Half-labeled DNA that is undergoing a second round of replication in labeled medium will show a region that appears like a fork with one arm having a density of dots twice that of the other. In the denser arm, two radioactive chains are present (the original and the new one being constituted). In the other arm, only the new chain will be radioactive, because it is being formed in a labeled medium on a nonlabeled template.

indicate that chain separation goes hand in hand with replication of the DNA strands much as predicted by the Watson-Crick model.

In addition to bacteria, the genomes of many viruses, as well as those of organelles (see Chap. 21), are known to occur in the form of circular DNA molecules. When the template is circular, as in the case of *E. coli,* the replicating molecule, when viewed with the electron microscope, resembles the Greek theta, θ. This is the result of the formation of an eye, which becomes progressively larger as replication proceeds (Fig. 11-19). Another way that circular DNA replicates is discussed in Chapter 15.

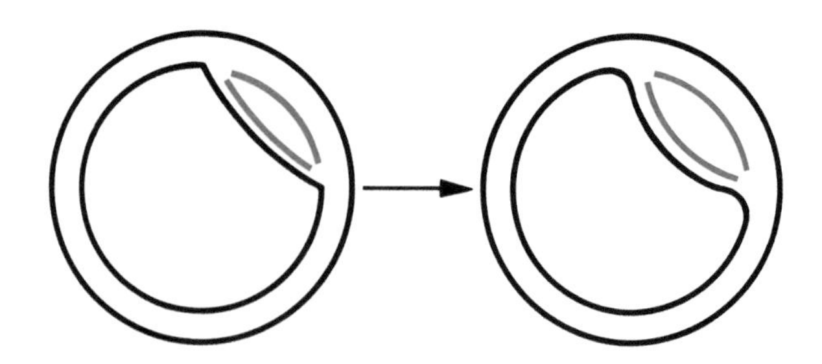

FIGURE 11-19. Replicating circular DNA. A replicating circular DNA molecule resembles the Greek theta (θ), since the region in which replication has taken place appears as an eye. Such a molecule is often referred to as a *theta structure* or *theta molecule.* The eye becomes progressively larger as the newly synthesized DNA (red) continues to lengthen.

The DNA of higher organisms, including the human, has been studied in a similar manner by refined autoradiography and electron microscopy. In eukaryotes, however, there are many initiation points and thus many replication forks—hundreds per molecule—along the length of the DNA. As in *E. coli,* replication proceeds bidirectionally from any one initiation point (Fig. 11-20). The replication forks that originated from different initiation points eventually meet. In electron micrographs, replicating regions of eukaryotic DNA molecules show several loops or eyes along the molecule, in contrast with viral and bacterial DNA, which shows only one, since typically only one initiation point exists in the DNA of bacteria and viruses. In eukaryotes, the loops become larger as the smaller loops fuse. Eventually, at the end of replication, two linear, double-helical DNA molecules result.

We see that the entire eukaryotic chromosome includes several units of replication, each with an initiation site or point of origin of DNA replication and two termination sites on opposite sides of the origin. Such a unit of replication has been termed a *replicon.* On the other hand, a chromosome of a virus or of a prokaryote usually consists of but a single replicon. In later chapters, we will learn that replicons that reside outside the chromosome can exist in a cell.

The work of Cairns and that of Meselson and Stahl are only two of the many experimental approaches that have tested and supported implications of the Watson-Crick model. Some of the other approaches will be encountered as we pursue discussions of gene action at the molecular level.

Other aspects of semiconservative replication

In the 1950s, Kornberg isolated from *E. coli* an enzyme, DNA polymerase I, which was thought to be the enzyme of replication. DNA polymerase I is able to catalyze the joining together of nucleotides and in this way to build up a DNA chain complementary to the template chain. However, the enzyme is only able to recognize the sugar–phosphate portion of the nucleotide. Preformed DNA must be present to act as a template to establish the sequence of the four kinds of nucleotides, as depicted in the Watson-Crick model. Another feature of the DNA polymerase is that it increases the length of a growing chain only in the 5′ to 3′ direction. This means that, as a chain grows, the 5′ end of the nucleoside triphosphate being added is joined to the 3′ end (the –OH end) of the nucleotide already in the chain (Fig. 11-21). In other words, a nucleotide just inserted would be found at the 3′ end of the growing chain. The first nucleotide in the growing chain would have a free 5′ end, or phosphate end. Recall that the two complementary chains in DNA show opposite polarity, one running 5′ to 3′ and the other 3′ to 5′ along the length of the molecule (Figs. 11-11, 11-21). Nevertheless, we will see in a moment that, as the double helix unwinds, replication in both newly growing strands is 5′ to 3′.

Since DNA polymerase requires a 3′–OH end, the enzyme by itself cannot initiate the joining of two nucleotides. A primer must be present—a molecular segment that provides the free 3′–OH for the internucleotide linkage. So both a template and a primer are required for DNA polymerase activity (Fig. 11-22).

It was found that DNA polymerase I had still another catalytic property in addition to its synthetic ability. Besides adding nucleotides in a 5′ to 3′ direction, it can actually remove nucleotides, one at a time, from the end

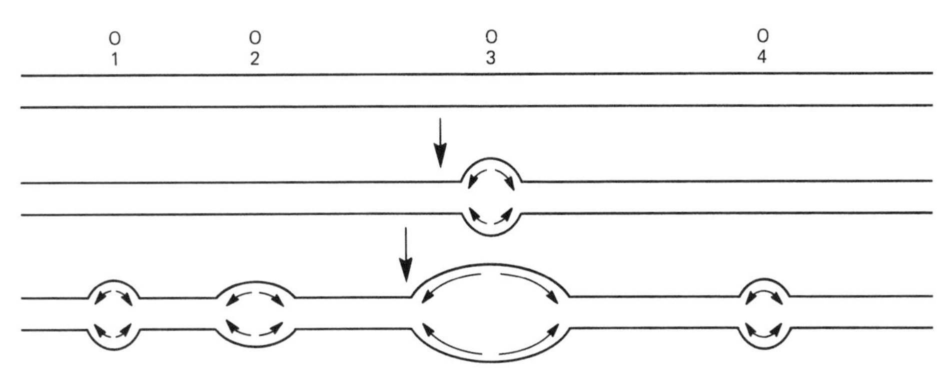

FIGURE 11-20. Replication forks in eukaryotes, In eukaryotes, replication begins at numerous origin points (e.g., O1, O2, O3) and then proceeds bidirectionally by two replication forks from each point of origin. Initiation of replication may be staggered; replication may start from one origin point well in advance of another. Eventually, replication forks that arose from different origins meet, and fusion of loops occurs.

of a DNA chain. It thus possesses exonuclease activity. It can catalyze the removal of nucleotides, however, from either the 3′ or the 5′ end of the chain. Thus its exonuclease activity is both 5′ to 3′ and 3′ to 5′.

In addition to exonuclease activity, many enzymes possess endonuclease activity and have the ability to attack internal bonds in the DNA, thus producing nicks in DNA strands. We will encounter endonucleases in many of our following discussions.

Many years after the discovery of DNA polymerase I (Pol I), a mutant strain of *E. coli* was isolated in which very little of the enzyme was present. Nevertheless, the mutant was still able to synthesize DNA at a normal rate. From this mutant strain, two other DNA polymerases were isolated in 1972, DNA polymerase II (Pol II) and DNA polymerase III (Pol III). In nonmutant *E. coli* cells, the high activity of Pol I obscures the activities of the other two polymerases, making them difficult to detect. Like Pol I, they require a template and can assemble DNA nucleotides only in a 5′ to 3′ direction on the end of a primer. They also possess exonuclease activity, but Pol II differs from DNA polymerases I and III by lacking such activity in the direction 5′ to 3′. Studies of *E. coli* mutants reveal that Pol I is not the primary enzyme of DNA replication but plays an accessory role that will be described shortly. The role of Pol II has not as yet been defined and is apparently of minor importance. It is Pol III that is the primary DNA polymerase of replication. Mutants deficient in this polymerase are unable to link nucleotides together to form long DNA chains. Pol III is composed of seven different polypeptide chains, each of which is required for the enzyme to function.

Since all known DNA polymerases can assemble DNA chains only in a 5′ to 3′ direction, certain problems become apparent. How do the necessary primers arise to provide 3′ –OH ends for the construction of new chains on the template strands? Moreover, one new chain that is assembled on its template must run in the direction 5′ to 3′, whereas the other must run 3′ to 5′ (Fig. 11-23A). How can assembly of both strands take place in the same required direction, 5′ to 3, if they have opposite orientations?

Answers to these questions were provided by several discoveries that have filled in many details of DNA replication. Figure 11-23B shows that one of the newly growing chains, called the *leading chain,* is synthesized in the direction toward the fork. This is so because any free 3′ –OH ends are oriented in that direction. The overall synthesis of this chain is in the direction 5′ to 3′, and it takes place in a continuous fashion. The situation is quite different for the other growing chain, the *lagging* or *retrograde chain.* Its overall direction of growth is 3′ to 5′, opposite that of the leading chain. Its synthesis is not continuous but takes place by the assembly of short DNA chains, each about 1,000 nucleotides long. These are called *Okazaki fragments* after their discoverer. Note (Fig. 11-23B) that synthesis of the fragments takes place in the direction away from the fork and is a consequence of the orientation of the free 3′ –OH ends. The Okazaki fragments will become joined to form a complete chain as a result of enzyme activity that is described later. It is important to realize that while the overall direction of growth of the leading and lagging chains is opposite each other, DNA synthesis is occurring entirely in the direction 5′ to 3′, in construction of the Okazaki fragments as well as of the continuous chain. Moreover, synthesis of the Okazaki fragments does not begin just anywhere but is initiated at definite sites on the template strand. Consequently, fragments are formed that have a typical or average length for a specific DNA. Nor do replication forks just arise at any point. Their origins are distinct from those of the Okazaki fragments and are the origin points discussed in the previous section (Fig. 11-20). At an origin site,

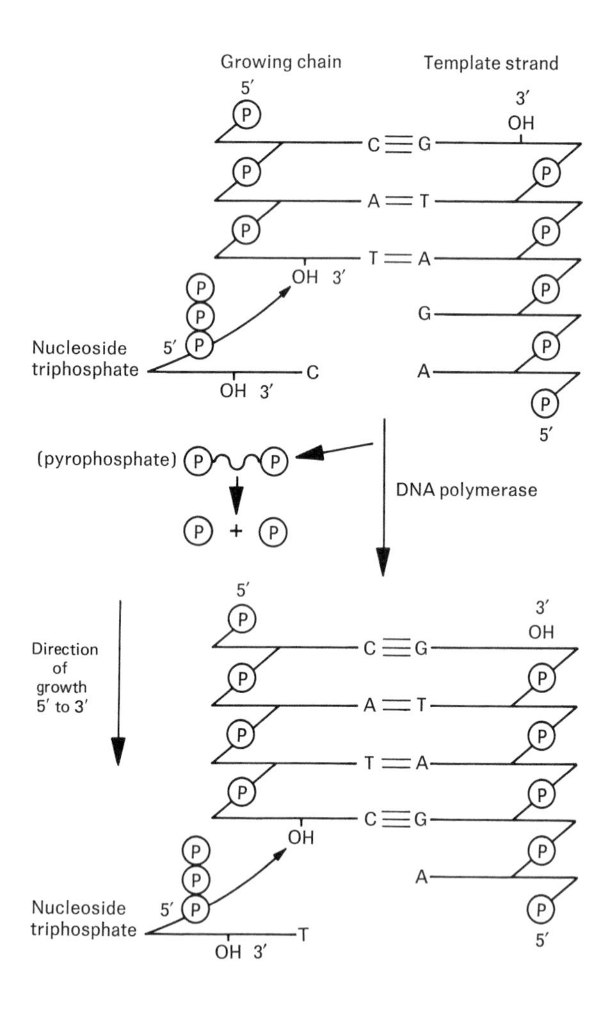

FIGURE 11-21. Growth of DNA chain in the 5′ to 3′ direction. Each nucleotide added to a growing DNA chain is in its triphosphate form and is inserted by the union of its 5′ (PO_3) end, with the 3′ (OH) end of the nucleotide already in the chain. A pyrophosphate is produced and is hydrolyzed to two phosphates in the elongation process, which is catalyzed by DNA polymerase, working only in the 5′ to 3′ direction.

leading strand synthesis commences before that of the lagging strand, making the naming of the two growing strands quite appropriate.

A most important discovery provided some answers regarding the primers needed to provide the free 3′ –OH ends at the replication fork. It has been shown that RNA synthesis is required for the replication of DNA. Enzymes recognize the specific points where DNA synthesis is to be initiated, such as those where Okazaki fragments are started. These enzymes, known as *primases,* then lay down short RNA chains, each about 10 nucleotides in length, to serve as primers for DNA synthesis. In *E. coli,* the primase is a polypeptide chain that by itself is not very efficient but that functions at maximal level when complexed with about six other polypeptides. The whole assemblange of polypeptides has been named a *primosome.* The entire complex moves along a lagging chain template in the direction toward the fork (Fig. 11-24). This direction is actually opposite to that in which synthesis of the RNA primer is occurring. Perhaps this accounts for the fact that the primers are so short. As the replication fork continues to move, the primosome moves along the template strand to recognize the next initiation site and to begin synthesis of the next short RNA primer for an Okazaki fragment. While the functions of those chains associated with the primase in the primosome are not known, they are thought to be involved in movement and recognition of the various initiation sites. Additional proteins are also typically found at a replication fork. Some of these ensure that an RNA primer will not come off a DNA template at an initiation site. Others prevent the strong tendency of single DNA strands to reassociate, a feature that would impede replication.

Once RNA primers are assembled, DNA nucleotides are added to them in the usual 5′ to 3′ direction. As replication proceeds, it becomes necessary to excise the RNA primers. This removal can be accomplished by the exonuclease activities of the DNA polymerases. Pol I is now known to be the major enzyme in removing primers and in filling in the gaps between Okazaki fragments. It does this by adding nucleotides to the 3′ end of a fragment using the parental strand as a template while at the same time removing the RNA primer ahead of it through its 5′ to 3′ exonuclease activity. Note (Fig. 11-25) that, although Pol I can fill in gaps between fragments, it cannot join the 3′ end of the very last nucleotide of the "new" segment to the 5′ end of the first nucleotide of the following segment. Conse-

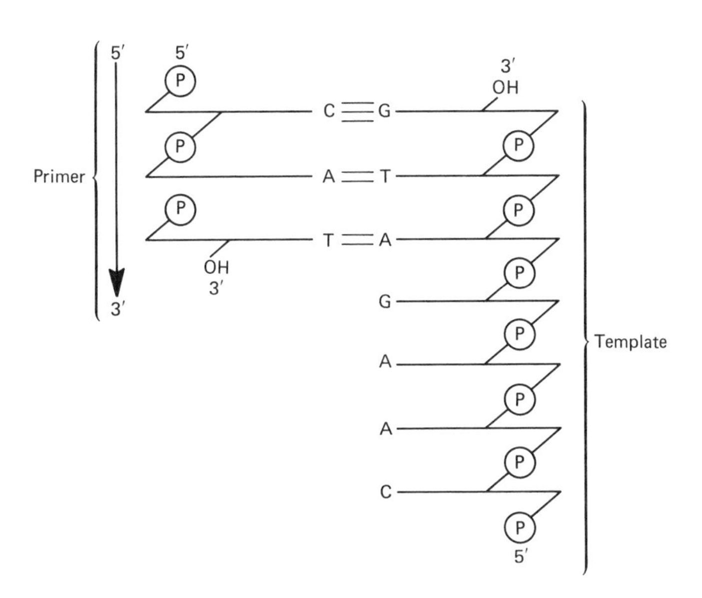

FIGURE 11-22. Template and primer in replication. DNA polymerases require a template strand to determine the specific sequence of nucleotides in the growing strand. In addition, they require a primer strand, a short segment of nucleotides to provide a free 3′ -OH end. Without a primer, a DNA polymerase cannot initiate the linking of two nucleotides. The primer strand is increased in length by the addition of nucleotides in the 5′ to 3′ direction. (Compare with Fig. 11-21, which shows growth of the primer.)

quently, a nick will be present at each site where a gap has been filled. These nicks become sealed through the activity of still another enzyme, DNA ligase, which can join the two ends together by forming a phosphodiester linkage. If a mutant cell lacks the ligase, Okazaki fragments accumulate, because they cannot be joined together.

Another important function has been attributed to the DNA polymerases. This is their ability to act as proofreaders through their possession of 3′ to 5′ exonuclease activity. This guarantees a high degree of accuracy for DNA replication. Any nucleotide that has been incorporated erroneously into a growing chain does not pair properly with the nucleotide in that position on the template strand. A polymerase will add a nucleotide to a growing chain only if the "last" base is correctly paired. Once the wrong nucleotide is removed, the enzyme again exerts its polymerase activity and links the correct nucleotide into position (Fig. 11-26).

Although most of the details of replication have been assembled from experiments with prokaryotes and viruses, replication in eukaryotes has been found to entail many similarities. Okazaki fragments have been demonstrated in eukaryotes, but these are shorter than the prokaryotic ones, consisting of approximately 100–200 nucleotides. RNA primers and ligase activity have been found. Moreover, three DNA polymerases are known, and they have been designated α (alpha), β (beta), and γ (gamma). All three have the ability to increase the length of a DNA chain, but again only in the direction 5′ to 3′. The α polymerase is believed to be the primary one in DNA replication in the eukaryotic cells just as Pol III is the major one in prokaryotes. The β enzyme may play roles in gap filling or repair. It is known, however, that the γ polymerase is found only in mitochondria, where it is required for the replication of the DNA of that organelle. (It is possible that it may also occur in chloroplasts.)

Before DNA replication can begin, the two strands of the double helix must unwind to permit the formation of a fork. Separation of the strands requires a class of catalysts known as the *unwinding enzymes* or helicases. A helicase is able to bind to a single-stranded portion of the DNA as well as to ATP from which it acquires energy. An energized helicase is then able to move along the DNA to ensure that strand separation keeps taking place at the proper rate. There

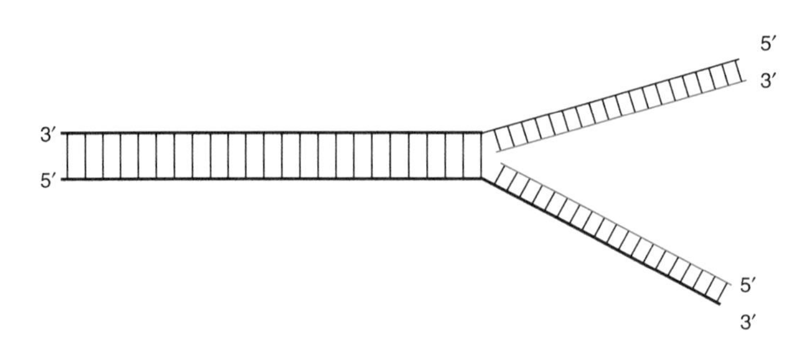

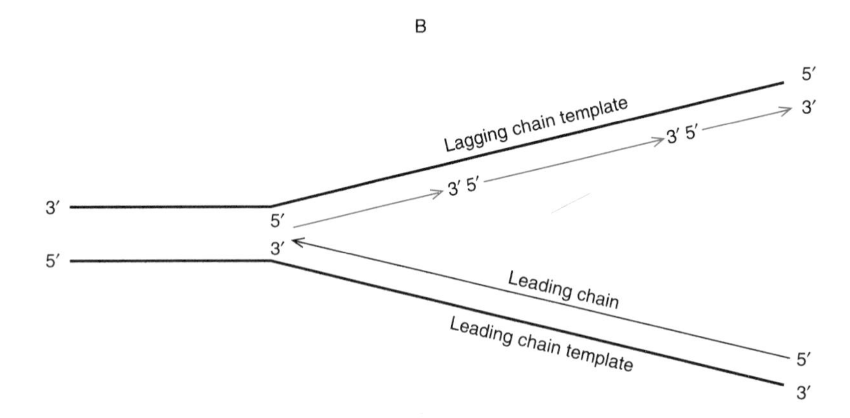

FIGURE 11-23. Assembly of DNA at replication. **A:** The two new DNA chains (red) that are assembled during replication are oriented in opposite directions. One of them (top) runs in the direction 3′ to 5′ toward the fork, whereas the other (bottom) runs in the direction 5′ to 3′. **B:** DNA synthesis is continuous only in relation to one of the new chains, the leading chain (below), which is assembled in the direction toward the growing fork due to the orientation of the nucleotides already in the chain. (Compare with Fig. 11-22.) The DNA is synthesized in a discontinuous fashion by the formation of Okazaki fragments (red) on the template of the lagging strand. These fragments will later become joined to form the completed lagging chain. This strand is synthesized in a direction away from the fork due to the orientation of the 3′ ends of the nucleotides. The overall growth of the leading chain is in the direction 5′ to 3′, whereas it is 3′ to 5′ for the lagging chain. Nevertheless, all DNA is assembled in the direction 5′ to 3′, even in the synthesis of the Okazaki fragments.

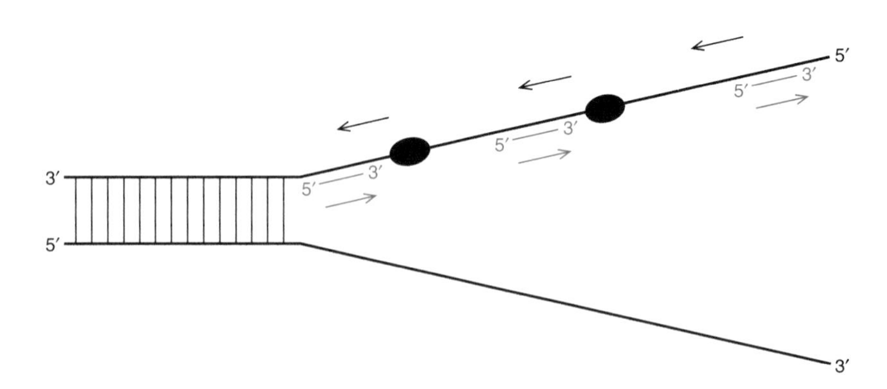

FIGURE 11-24. Formation of RNA primers. As the primosome moves down the template of the lagging strand, it recognizes sites for the initiation of DNA synthesis and lays down short RNA primers (red). The direction of synthesis of the primers (red arrows) is opposite that of the movement of the primosome (black arrows).

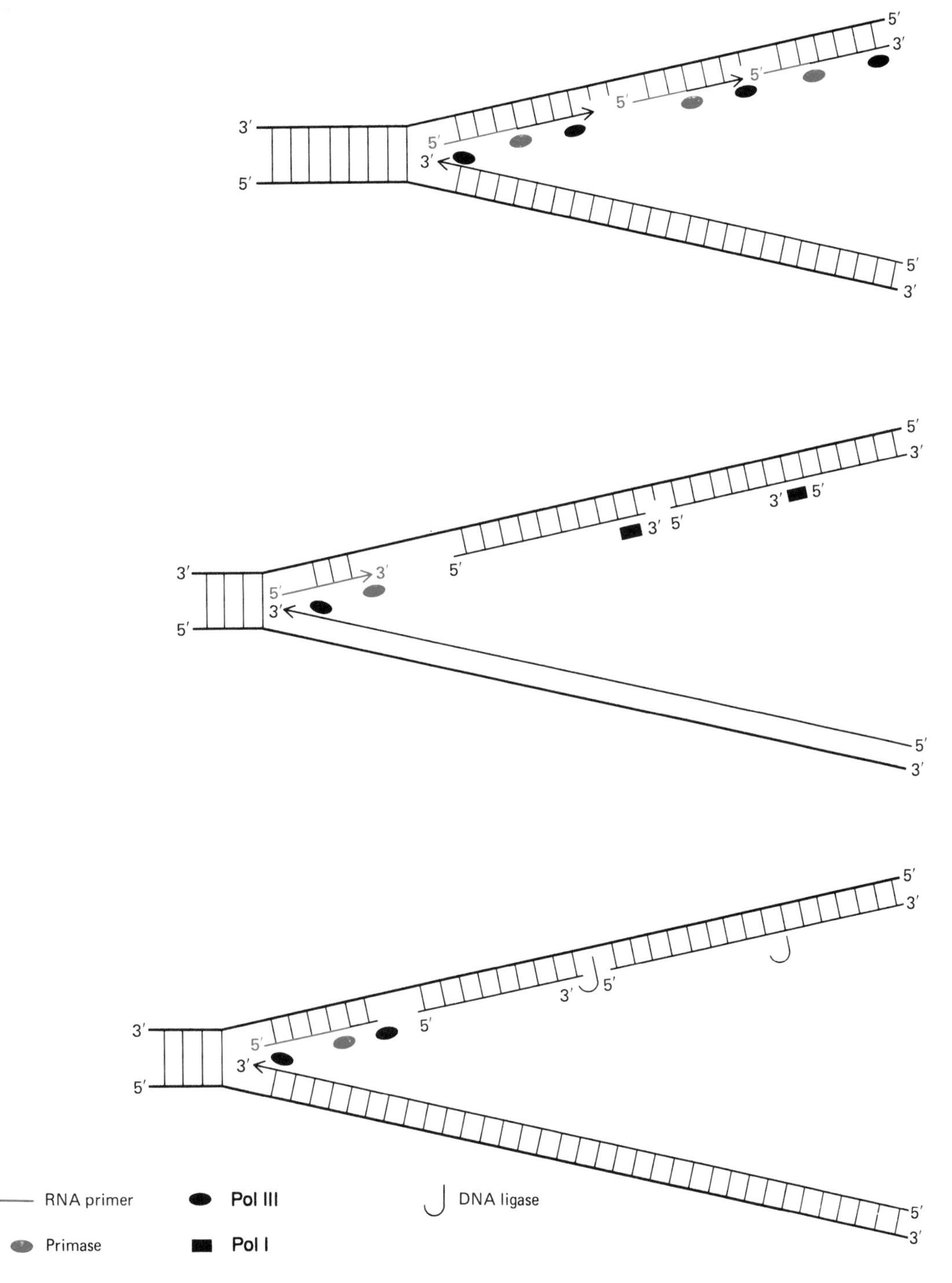

FIGURE 11-25. RNA primer and replication. **Top:** Each short primer provides a free 3′ end for Pol III, which adds DNA nucleotides to the primer in the direction 5′ to 3′. **Middle:** RNA primers are later excised by the exonuclease activity of Pol I, which fills in the gaps resulting from primer removal. **Bottom:** Since no DNA polymerase can seal nicks left between DNA segments after the gaps are filled, DNA ligase is called upon to perform the function.

are at least two different kinds of helicases. One of these binds to the template for the leading strand and moves in the direction 3′ to 5′, whereas the other binds to the template of the lagging strand and moves 5′ to 3′ (Fig. 11-27A). When the two DNA strands separate, they will tend to reassociate as the helicases move along the DNA molecule. Such a reassociation would interfere with replication, but this is prevented by the single-stranded DNA-binding protein (SSB). Each SSB is a tetramer composed of four identical polypeptide chains. The SSBs are not enzymes, however, and they do not bind to ATP. As one SSB binds to a section of single-stranded DNA, it facilitates the binding of another one to an adjacent spot. When a portion of a single strand of DNA is covered with SSBs, it is rigid, free of loops, and does not curve back upon itself to form base pairs. Such an extended condition is necessary for replication to proceed (Fig. 11-27B). The SSBs disassociate from the DNA chains upon completion of replication in that region of the molecule, thus permitting the reestablishment of double helixes.

Contrary to most of its depictions, naturally occurring DNA does not occur as a long, extended, linear double helix. Instead, the DNA is organized into loops and is twisted around upon itself to form a supercoil (Fig. 11-28). In our discussions of DNA, we will encounter enzymes that fall into a class known as the topoisomerases. Such enzymes are able to convert DNA from one topological form to another. One topoisomerase, DNA gyrase, is needed at a replication fork. This enzyme possesses endonuclease activity and is required to relax the double helix in front of the fork. It does this by nicking a DNA strand. This enables the free end to rotate around the other chain, which performs the role of a swivel. As the DNA becomes relaxed following the removal of certain supercoils, the DNA gyrase reseals the nick that it made in the DNA strand (Fig. 11-27B). The assemblage of the various different proteins involved in replication at the site of the replication fork is sometimes referred to as a *replisome.* This complex includes the DNA polymerases, helicases, primases, topoisomerases, DNA ligase, and others.

A burst of experimentation that continues to the present day was set off by the announcement of the Watson-Crick model. The ongoing research has supplied us with a wealth of detail on the interplay of molecules involved in duplication of DNA. We can expect such effort to continue to fill in those gaps that still remain in our knowledge of the marvels of replication.

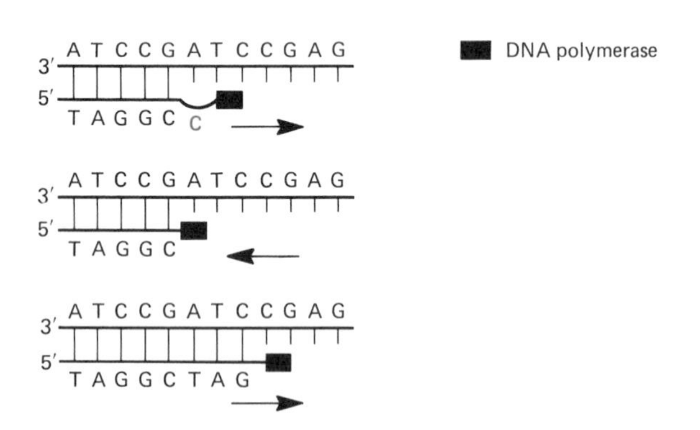

FIGURE 11-26. Proofreading activity of DNA polymerase. An incorrect nucleotide is added to a growing chain (above). The 3′ end of the chain cannot serve as a primer, because the incorrect nucleotide cannot pair properly with the nucleotide on the template. The DNA polymerase moves in reverse direction (middle) and removes the error through its 3′ to 5′ exonuclease activity. Following excision, the polymerase moves forward again (bottom) and continues to add nucleotides to the growing chain in the direction 5′ to 3′.

The DNA molecule and crossing over at the molecular level

Biologists have long recognized the importance of crossing over in the evolution of sexual species, and speculations on the mechanism whereby it occurs have fascinated cytogeneticists for years (Chaps. 8 and 9). With the advent of molecular biology and the wealth of information that has accumulated on the structure of the DNA molecule and its replication, new approaches have been provided to help solve the riddle of the mechanism of crossing over. In the long run, it is events at the molecular level that must be sorted out before any real understanding of crossing over at the level of the chromosome can be gained.

Several models have been proposed to explain crossing over on the molecular level. One that is compatible with various genetic analyses as well as with data from molecular biology is the Holliday model, which has undergone several modifications since its proposal in 1964. Figure 11-29 summarizes some of the major points in the Holliday model. Figure 11-29 (step 1) shows two single, homologous DNA duplexes. Each duplex represents one duplex DNA of a single chromatid present in a bivalent. These two

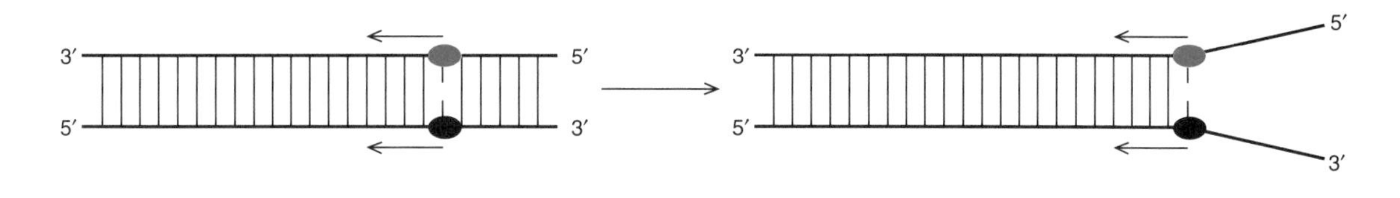

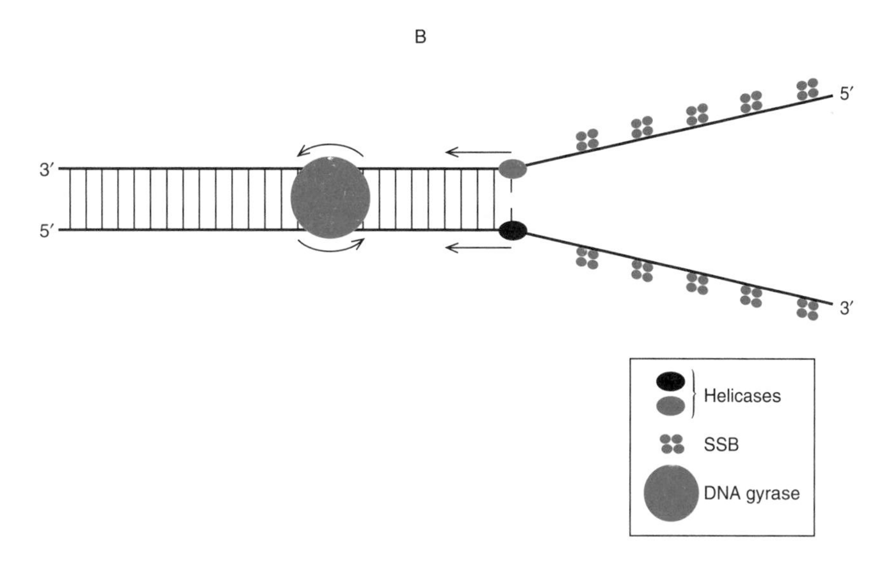

FIGURE 11-27. Opening the double helix. **A:** One type of energized helicase (black) binds to the template for the leading strand and moves in the direction 3′ to 5′. A second type (red) binds to the template for the lagging strand and moves in the direction 5′ to 3′. Action of the helicases triggers strand separation. **B:** The tendency for DNA strands to reassociate is prevented by SSBs, which bind to the single strands and maintain them in a rigid, linear state conducive to replication. DNA gyrase relaxes the double helix in front of the fork. It produces a nick in one of the strands, permitting rotation of the nicked strand. The gyrase later reseals the nick.

nonsister duplexes are the ones that will participate in crossing over. Therefore the duplexes that are sisters to each of these are not shown, because only two of the four chromatids in a bivalent are involved in any one crossover event (Chap. 8). Two pairs of genetic factors are being followed in the given example. One parental duplex carries the linked marker genes A and B; the other duplex carries the corresponding alleles a and b. Before recombination can take place, the two homologous duplexes involved in crossing over must recognize each other. They must then arrange themselves in precise alignment. Otherwise a crossover event would result in duplication or deletion of genetic material.

Step 2 in Figure 11-29 shows two endonuclease-generated nicks that affect strands of the same polarity. Each nicked strand separates from its complementary strand as unwinding takes place at the site of the nick. Each released strand now leaves its complementary partner, invades the homologous duplex, and associates with its complementary strand in the homologous duplex (Fig. 11-29, step 3). The strands are held together at first by hydrogen bonds, but the nicks in the strands soon become sealed by DNA ligase activity. Each strand that has crossed over continues to migrate, as more regions of pairing occur between the nonsister strands (Fig. 11-29, step 4). This is called *branch migration,* and its effect is to increase the regions of pairing between the two homologous strands that were originally parts of two separate double helixes. Note that each double helix now has a region at the point of exchange that is *heteroduplex* and that branch migration increases the extent of the heteroduplex regions. The heteroduplex DNA is represented here by the segments in which pairing occurs between complementary single strands with different origins. In the heteroduplex regions, the two strands came from different original (or parental) DNA molecules.

Note also that the two original duplexes are now joined together into one molecule as a result of the strand exchange. They are held together by the two single strands that crossed over to pair with their complements in the homologous duplex. This joint molecule must be resolved into two separate duplexes by cutting in some way. One way in which this can take place is shown in step 5 of Figure 11-29, where the two strands *opposite* the crossover strands are now nicked. Some digestion of DNA may occur. A rotation of the nicked strands then takes place. Gaps are filled in by DNA polymerase action, and nicks are sealed by DNA ligase. The result is two recombinant chromatids, Ab and aB (Fig. 11-29, step 6).

According to the model, it is also possible for the nicks that resolved the joined duplexes to break the *same* two strands that were nicked at the outset, leaving the other two strands completely unaffected. It can be seen from step 7 of Figure 11-29 that if this is the case, then the two resulting chromatids will *not* be recombinant with respect to the two marker genes on either side of the exchange. Each retains the parental arrangement, AB and ab. However, it is important to note that in both cases the duplexes contain DNA regions that are heteroduplex. The heteroduplex DNA is represented here by those segments in which pairing occurs between complementary strands that had *different origins*. The two paired DNA strands in these regions originated in different parental DNA molecules. As indicated in the depiction of the Holliday model, heteroduplex DNA arises following crossing over, whether all four strands of the two duplexes experience endonuclease-generated nicks or the same two strands are nicked again (Fig. 11-29, steps 6 and 7).

The Holliday model is compatible with a vast amount of genetic data. Moreover, many of its features are supported by studies of *E. coli* mutants in which genetic recombination is absent or greatly reduced. As a result of studying such mutants, investigators have identified various proteins that appear to be required for recombination. Among these are SSB proteins, which, as noted in the previous section, maintain single strands at the replication fork, and the RecA protein. This RecA protein possesses several remarkable properties. It can promote pairing between two homologous DNA molecules and bring them into alignment, a prerequisite for an exchange of strands between two duplexes. Like the SSB proteins, it too can bind tightly to single DNA strands. It also has the ability to hydrolyze ATP with the release of energy and can apparently use this ability to enable a single DNA strand of one double helix to invade another double helix and to base pair with the complementary strand in the other helix, much as depicted in the Holliday model. The RecA protein can also bring about branch migration, as pictured in the model.

The RecA protein and the SSB proteins must act together to bring about the pairing reaction, since a mutant bacterial cell that lacks just one of the two types of proteins has a greatly reduced amount of recombination. Many details on recombination are yet to be resolved, and further modifications of the Holliday model may yet be made. However, the riddle of crossing over is certain to be solved as refined techniques in molecular biology continue to provide an insight into the details of this process, which is essential to all sexual species.

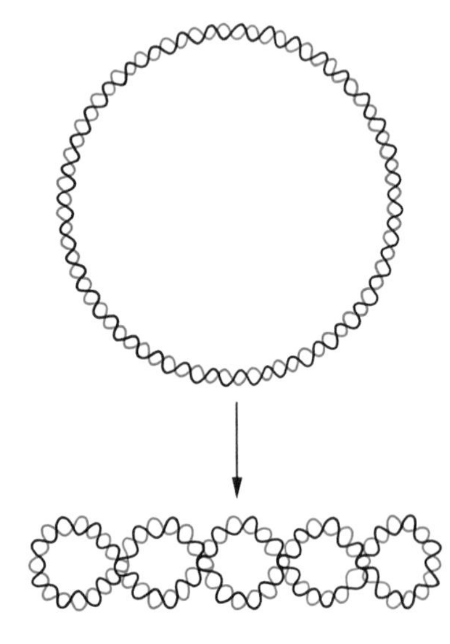

FIGURE 11-28. Supercoiling of DNA. Naturally occurring DNA exists as a circular molecule or as a series of large loops that are held intact and thus have the form of circles. A circular molecule (above) can twist around about itself and assume a supercoiled configuration (below).

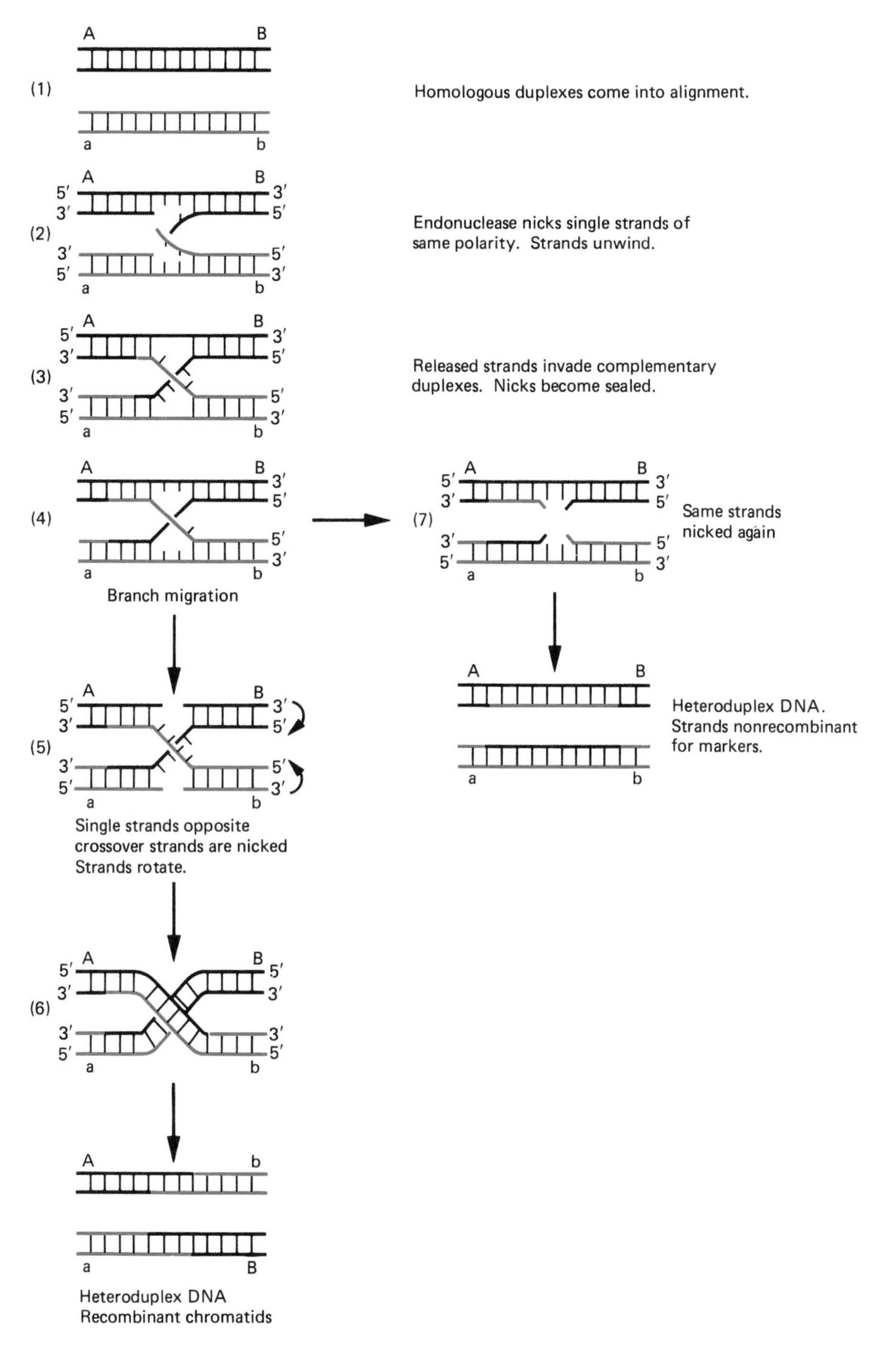

FIGURE 11-29. Summary of steps in the Holliday model of crossing over. (From *Cold Spring Harbor Symp. Quant. Biol.* 43. 1979.)

REFERENCES

Avery, O.T., C.M. Macleod, and M. McCarty. Studies on the chemical nature of the substance inducing transformation of pneumococcal types. *J. Exp. Med.* 79:137, 1944. Reprinted in *The Biological Perspective, Introductory Readings.* W.M. Laetsch (ed.), p. 105. Little, Brown, Boston, 1969.

Brutlag, D., and A. Kornberg. Enzymatic synthesis of deoxyribonucleic acid: A proofreading function for the 3′ to 5′ exonuclease activity in DNA polymerases. *J. Biol. Chem.* 247:241, 1972.

Cairns, J. The bacterial chromosome. *Sci. Am.* (Jan.):36, 1966.

DePamphilis, M., and P.M. Wassarman. Replication of eukaryotic chromosomes: A closeup of the replication fork. *Annu. Rev. Biochem.* 49:627, 1980.

Hershey, A.D., and M. Chase. Independent functions of viral protein and nucleic acid in growth of bacteriophage. *J. Gen. Physiol.* 36:39, 1952. Reprinted in *Papers on Bacterial Viruses,* 2nd ed., G.S. Stent (ed.), p. 87. Little, Brown, Boston, 1965.

Holliday, R. A mechanism for gene conversion in fungi. *Genet. Res.* 5:282, 1964.

Huberman, J.A., and A.D. Riggs. On the mechanism of DNA replication in mammalian chromosomes. *J. Mol. Biol.* 32:327, 1968.

Kornberg, A. *DNA Replication.* W.H. Freeman, San Francisco, 1980.

Kornberg, A. *1982 Supplement to DNA Replication.* W.H. Freeman, San Francisco, 1982.

Kriegstein, H.J., and D.S. Hogness. Mechanism of DNA replication in *Drosophila* chromosomes: Structure of replication forks and evidence for bidirectionality. *Proc. Natl. Acad. Sci. USA* 71:135, 1974.

Laskey, R.A., M.P. Fairman, and J.J. Blow. S phase of the cell cycle. *Science* 246:609, 1989.

Masters, M., and P. Broda. Evidence for the bidirectional replication of the *Escherichia coli* chromosome. *Nature [New Biol.]* 232:137, 1971.

Meselson, M., and C.M. Radding. A general mode for genetic recombination. *Proc. Natl. Acad. Sci. USA* 76:2615, 1978.

Meselson, M., and F.W. Stahl. The replication of DNA in *Escherichia coli. Proc. Natl. Acad. Sci. USA* 44:671, 1958. Reprinted in *Biochemical Genetics,* G.L. Zubay (ed.), p. 397. Little, Brown, Boston, 1966.

Mirsky, A.E. The discovery of DNA. *Sci. Am.* (June):78, 1968.

Ogawa, T., and T. Okazaki. Discontinuous DNA replication. *Annu. Rev. Biochem.* 49:421, 1980.

Radding, C.M. Homologous pairing and strand exchange in genetic recombination. *Annu. Rev. Genet.* 16:405, 1982.

Wang, J.C. DNA topoisomerases. *Sci. Am.* (July):94, 1982.

Watson, J.D., and F.H.C. Crick. Molecular structure of nucleic acids. A structure for deoxyribose nucleic acid *Nature* 171:737, 1953. Reprinted in *Classic Papers in Genetics,* J.A. Peters (ed.), p. 241. Prentice-Hall, Englewood Cliffs, NJ, 1959.

Watson, J.D, and F.H.C. Crick. Genetical implications of the structure of deoxyribonucleic acid. *Nature* 171:964, 1953.

Watson, J.D., N.H. Hopkins, J.W. Roberts, J.A. Steitz, and A.M. Weiner. *Molecular Biology of the Gene,* 4th ed. Benjamin Cummings, Menlo Park, CA, 1988.

REVIEW QUESTIONS

1. In each of the following, DNA preparations from a donor bacterial strain are incubated with living cells of a recipient. What would be the nature of any transformed cells with respect to each of the specific traits being studied?

   A. Donor: capsulated, type III. Recipient: rough, type V.
   B. Donor: streptomycin sensitive. Recipient: streptomycin resistant.
   C. Donor: penicillin resistant. Recipient: penicillin sensitive.

2. Bacteria were infected with phages labeled with ^{35}S and then whirled in a blender. Cells were then separated from the fluid medium. Cells and fluid were next tested for radioactivity. Where will most of it be found and why?

3. In the Hershey-Chase experiment, suppose labeled carbon, ^{14}C, had been used in place of ^{35}S and ^{32}P. What would have been the results? What conclusions could one have reached? Explain.

4. Of what building units is each of the following composed?

   A. A nucleotide.
   B. A nucleoside.

5. How does DNA differ from RNA with respect to the following:

   A. Its base content.
   B. The chemistry of its sugar component.
   C. Its cellular distribution.

6. Four samples of double-stranded DNA are analyzed, and the following information is obtained:

Sample 1	15% thymine
Sample 2	20% guanine
Sample 3	30% adenine
Sample 4	40% cytosine

   For each of the samples, what can you predict as likely percentages for all of the bases? Is it possible that any of these samples represents DNA from the same source?

7. Consider the nucleotide sequence, TAG. Assume at the time of DNA replication that A in this strand cannot pair with its normal partner but pairs with C instead. At the replication following this rare event, what will be the resulting sequence in the two new DNA molecules?

8. Using the following segment of a DNA chain, answer the questions below:

5′ ACACCCTTTACAAAT 3′

A. What is the orientation and base sequence in the complementary strand?
B. What is the A + T/G + C ratio in the intact double helix?
C. In the intact double helix, what is the A/T:G/C ratio?
D. In the intact double helix, what is the ratio A + G/C + T?

9. Following is the nitrogen base composition of DNA from three sources. For each, calculate 1) A + T/G + C; 2) A + G/T + C; 3) A/T; 4) G/C; 5) %G:C.

	Base (and percentage of total nitrogen bases)			
DNA source	Adenine	Thymine	Guanine	Cytosine
Human liver	30.3	30.3	19.5	19.9
D. melanogaster	30.7	29.4	19.7	20.2
Yeast	31.3	32.9	18.7	17.1

10. The two strands of a DNA molecule are separated, and one of them is analyzed for its A + T:G + C ratio. This is found to be 0.2.

A. What is the A + T:G + C ratio in the other strand?
B. Suppose analysis of a single DNA strand shows the A + G:T + C ratio to be 0.2. What is the A + G:T + C ratio in the complementary strand?
C. What is the ratio of A + G:T + C in the intact double-stranded DNA in the case of the two DNAs described in A and B?

11. Suppose analysis of a single DNA strand gives an A + T:G + C ratio of 0.5.

A. What ratio is expected in the complementary strand?
B. What is this ratio for the whole molecule?

12. If the A + G:T + C ratio in one strand of DNA equals 0.5, what is expected for this ratio in the complementary strand?

13. Consider four separate nucleoside triphosphates—one with the base adenine, one with cytosine, one with guanine, and one with thymine. Each is inserted into a growing DNA chain next to a template through the action of DNA polymerase. The nucleoside triphosphate with thymine becomes the first residue at the beginning of the chain, the one with adenine the second, and that with cytosine the third; the one with guanine is the last residue in the chain. Answer the following:

A. In the new chain formed, which base will be in the nucleotide that has a free–OH end?
B. In the new chain, which base will be in a nucleotide with a free phosphate?
C. Which end of the second nucleotide will be joined to that of the third one that is added to the growing chain?
D. In the template strand, what will be the base in the nucleotide with the free–OH end.

14. Suppose that DNA did not replicate semiconservatively, as proposed by Watson and Crick. Assume that the double helix stays intact and a double helix composed of two new strands is formed at replication. This would be a conservative rather than a semiconservative method of replication. If the DNA replication were conservative, what would have been the observations in the density gradient experiment of Meselson and Stahl?

15. The two chains composing a DNA double helix unwind, and a replication fork is formed. Chain A is presented to the DNA polymerase in the direction 3′ → 5′. Its complementary chain, chain B, is presented in the direction 5′ → 3′. Answer the following:

A. On which of the two chains will DNA synthesis occur in a direction toward the fork?
B. On which chain will the free–OH ends of replicating DNA be oriented toward the fork?
C. On which chain will the PO_3 end of the RNA primer be directed toward the fork?
D. On which chain will the Okazaki fragments have –OH ends directed away from the fork?

16. Suppose that both strands of a DNA molecule are labeled with tritiated thymidine. This DNA is allowed to undergo another round of replication in labeled medium.

A. What will be the appearance of the Y-like fork on an autoradiograph of this DNA undergoing another round of replication in the labeled medium?
B. Suppose the DNA that is fully labeled in both strands is allowed to undergo another round of replication in unlabeled medium. What will be the appearance of the Y on an autoradiograph of the DNA undergoing another round of replication in the unlabeled medium?

17. Match the enzymes given on the left with the aspects of DNA listed on the right. More than one item may match some of the enzymes, so an item may be used more than once.

A. DNA gyrase
B. DNA ligase
C. DNA polymerase I
D. DNA polymerase II
E. DNA polymerase III
F. Primase
G. Helicase
H. Rec A

a. Extends DNA chains in a 5′to 3′ direction.
b. 5′ to 3′ exonuclease activity.
c. The primary enzyme of DNA replication.
d. Assembles RNA in a 5′ to 3′ direction.
e. Needs a primer and a template to act.
f. Hydrolyzes ATP to unwind DNA.
g. Is main enzyme in removing RNA primer.
h. Is needed to join Okazaki fragments.
i. Brings about pairing between DNA molecules during a crossover event.
j. A proofreading enzyme.
k. Nicks a DNA strand in front of a fork.

Information Transfer

12

Gene control of protein synthesis

By 1950, a wealth of research with microorganisms led geneticists to conclude that genes control the formation of proteins. Many examples accumulated showing that gene mutations can block steps in chemical pathways through their effects on specific enzymes, the proteinaceous catalysts of cells. The first gene in the human to demonstrate a relationship with a protein concerned the effect of the recessive allele for sickle cell anemia on hemoglobin. This protein contains two kinds of chains, each composed of more than 140 amino acids (details given in Chap. 20). The defective sickle cell hemoglobin departs from the normal in the replacement of just a single amino acid (glutamic acid in normal hemoglobin) by another (valine in sickle cell hemoglobin) at one position in one of the chains. Comparable alterations in the hemoglobin molecule were then shown to be related to other kinds of hereditary anemias.

An argument for a relationship between DNA and protein can be made on the basis of the physical structure of the two kinds of molecules. Although DNA is a double helix, each of its two strands is composed of nucleotide units assembled in a linear order along the length of the molecule. Classic data on linkage and crossing over established that genetic maps are also linear. This concept of one gene following another along the length of the chromosome fits in well with the molecular structure of the gene, a segment of a linear molecule. So the linear genetic map corresponds to the linear sequence of nucleotides composing DNA. Furthermore, protein molecules are also essentially linear, consisting of 20 different kinds of amino acids joined together in a linear order.

Although the various amino acids differ in their complexity, they possess certain features in common, with the exception of proline (Fig. 12-1). All contain a first carbon, known as the α-carbon, to which characteristic parts are bonded. One of these is a hydrogen atom, a second is a free α-carboxyl group (–COOH), and a third is a free α-amino group (–NH2). (In proline, which strictly speaking is an imino acid, the amino group is found in a ring.) It is the fourth position attached to the α-carbon of each amino acid that distinguishes one from the other. This is the R group (R designates radical). The R group may represent nothing more than a hydrogen atom, as in glycine, the simplest of the amino acids. It may, however, be a more complex side grouping, as in tyrosine.

When the amino acids unite in the formation of a protein, it is their carboxyl and amino groups that play the most important role, for the separate amino acids are joined by the reaction of the carboxyl group of one with the amino group of another (Fig. 12-2). A molecule of water is evolved, and a bond is formed, known as the *peptide linkage*. As a result of these linkages, short, linear stretches may form, composed of two to several linked amino acids. These are called *peptides*, a dipeptide consisting of two linked amino acids, a tripeptide of three, and so on. A polypeptide is a chain composed of many linked amino acids. Proteins, which can be composed of one or more polypeptide chains, can become very complicated and diverse and are the most complex of all the chemical substances in the cell. Twenty kinds of amino acids arranged in dif-

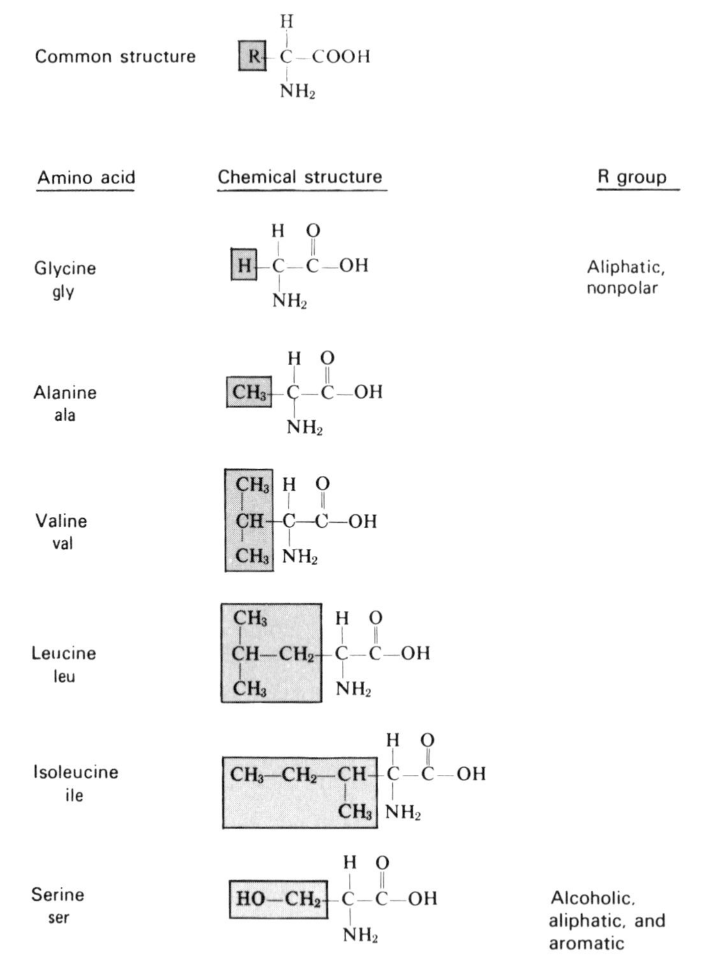

FIGURE 12-1. The 20 common amino acids. Amino acids possess a carboxyl group, an amino group, and an R group (outlined in the figure). It is the R group that is the distinguishing feature among the amino acids. Its complexity varies from a single H, as in glycine, to the complex groups seen in the aromatic amino acids. (Figure 12-1 continues on following pages.)

ferent ways in chains of varying lengths permit numerous varieties. In addition, a complex R grouping on one amino acid may react with the R grouping on another amino acid in the same chain or in an adjacent one. These interactions can produce complex foldings of the separate polypeptide chains. All of this makes possible a diversity of protein molecules that differ not only in their amino acid content but in their three-dimensional architecture as well. But no matter how complex the protein, it is still a linear molecule in the topographical sense; for if we unfold a protein, we see that it is fundamentally a series of amino acids joined together by peptide linkages. We see, therefore, a correspondence from the linear genetic map to the linear DNA molecule to the linear protein molecule.

Based on such considerations, it seemed quite likely that the specific linear order of the amino acids in a protein might be determined by the specific linear order of the nucleotides in a gene. If information on protein structure is coded in the DNA it becomes necessary to explain how this information is transferred to the cytoplasm to the sites of protein assembly, the ribosomes (refer to Fig. 2-3). It is conceivable that the ribosomes themselves contain information determining amino acid sequence. Moreover, ribosome composition is a complex of protein and RNA, and most of

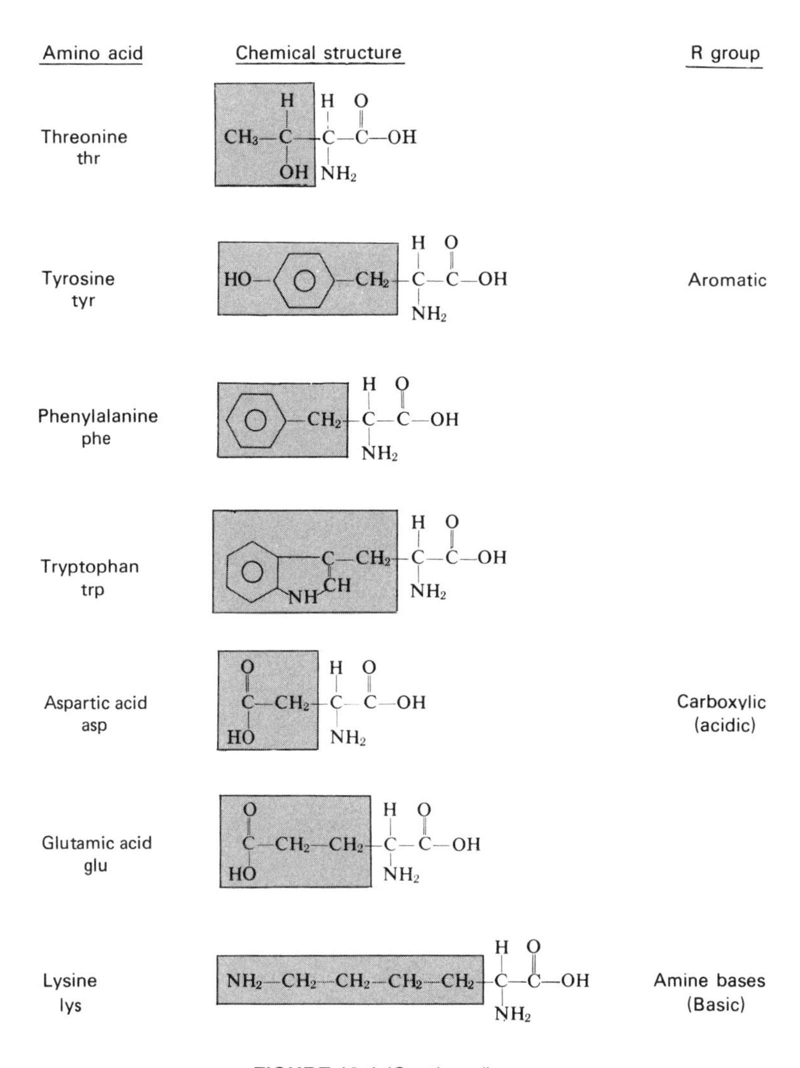

FIGURE 12-1 (Continued)

the cellular RNA is actually located in the ribosome fraction. It was therefore suspected that RNA molecules were intimately involved in information transfer.

The close similarity between the DNA and RNA molecules also suggested that RNA could somehow carry information coded in DNA. A key to the solution of the problem of information transfer was provided through research conducted at Harvard University by Hoagland and his associates. The important discovery was made that each amino acid, before engaging in protein synthesis, attaches to a special kind of RNA that is not the RNA of the ribosome. This special RNA makes up a small portion of the soluble part of the cytoplasm and was originally called sRNA but is now popularly known as *transfer RNA* (tRNA). Once each amino acid is attached to a tRNA, it is transported to the ribosome, where it may participate in protein formation.

The tRNA was shown to be more than a mere vehicle that carries amino acids. Since there are 20 types of amino acids that take part in protein synthesis, there must be at least 20 different kinds of tRNA, one for each type of amino acid. The structure of certain tRNA molecules was worked out in the late 1960s. It was shown to be a single-stranded RNA and to have a cloverleaf shape (details are given in Chap. 13). An amino acid in the cell, once it is joined to its specific tRNA, receives an identity. After attachment to its tRNA, the amino acid is placed in a specific position in a growing polypeptide chain. Without being identified in such a way, the amino acids could not be

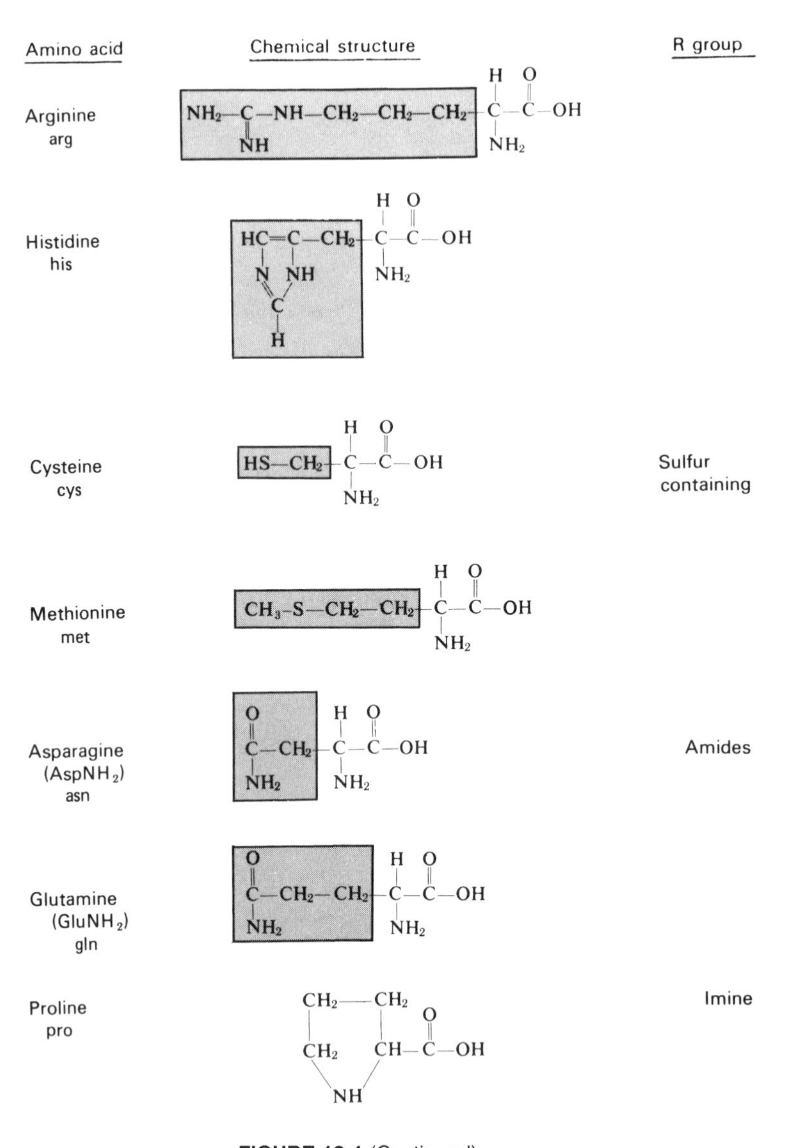

FIGURE 12-1 (Continued)

assembled into any specific sequence. Therefore, the binding of each amino acid to its identifying tRNA is a critical event in protein synthesis, which was shown to entail a series of steps.

Before any free amino acid in the cell can be bound, it must be activated so that it can react with tRNA and also have the energy needed to form peptide linkages with other amino acids (Fig. 12-3A). The energy required to activate an amino acid is supplied by ATP, the universal energy donor of the cell. The transfer of energy from ATP to an amino acid requires a specific enzyme that can recognize both the amino acid and the ATP. Since there are 20 different kinds of amino acids, there are 20 varieties of the enzyme, one for each specific amino acid. These essential enzymes are called *aminoacyl-tRNA synthetases*. Not only are they necessary for the activation of the amino acids, they are also responsible for joining the energized amino acids to their appropriate tRNAs (Fig. 12-3B). Each synthetase thus has two critical roles to perform: 1) to make the specific amino acid active by effecting the transfer of energy from an ATP molecule; and 2) to join the activated amino acid to its proper tRNA by way of a high-energy bond (an aminoacyl bond). Once it is coupled to its tRNA, the amino acid is carried to the ribosome. Using energy from the aminoacyl bond, it can now form peptide linkages with other amino acids that have undergone the same sequence of events. The tRNA drops away and can recycle to pick up another specific amino acid.

Transcription and messenger RNA

A series of brilliant experiments performed by different teams of investigators filled in more parts of the picture of information transfer. One line of research pursued the synthesis of RNA from its building units, the four different kinds of ribonucleotides. This kind of study was advanced by the isolation of the enzyme RNA polymerase from the bacterium *Escherichia coli.* It was found that this catalyst can stimulate the synthesis of RNA in vitro. Whole cells were not necessary for synthetic RNA to be made, but certain requirements had to be met (Fig. 12-4). Among these was the need for the ribonucleosides to be present as triphosphates, their activated or energized form. If any RNA is to be made in the test tube, it is also essential for some DNA to be present. The reason for this is that the enzyme RNA polymerase cannot link the nucleotides together on its own. DNA is required as a blueprint or template; RNA formation by the polymerase depends on DNA. The RNA that is synthesized is complementary to the DNA that is present and that serves as the template in the RNA formation. It has now been established that in a cell the double helix unwinds, not always to engage in replication, but to permit the formation of RNA, which is complementary to it (Fig. 12-5). For example, along one strand the enzyme RNA polymerase, using the DNA as a guide, assembles the RNA precursors (the ribonucleoside triphosphates) into an RNA strand complementary to the DNA strand. The bases along a stretch of DNA (CAATG, for example) pair off with the complementary ones of the ribonucleosides (GUUAC). This process is called *transcription,* the formation of RNA complementary to, and under the direction of, a segment of DNA. The RNA formed after transcription closely resembles the DNA stretch that acted as a template. The sequence of bases in the DNA is reflected in the sequence of the complementary bases in the RNA.

It must be noted that RNA as well as DNA has a definite orientation, a 5′ and a 3′ end, because of the ester linkages between the phosphate and the sugar. This was discussed with respect to DNA, as was the fact that the assembly of a DNA strand occurs in the 5′ to 3′ direction (see Chap. 11). Similarly, the

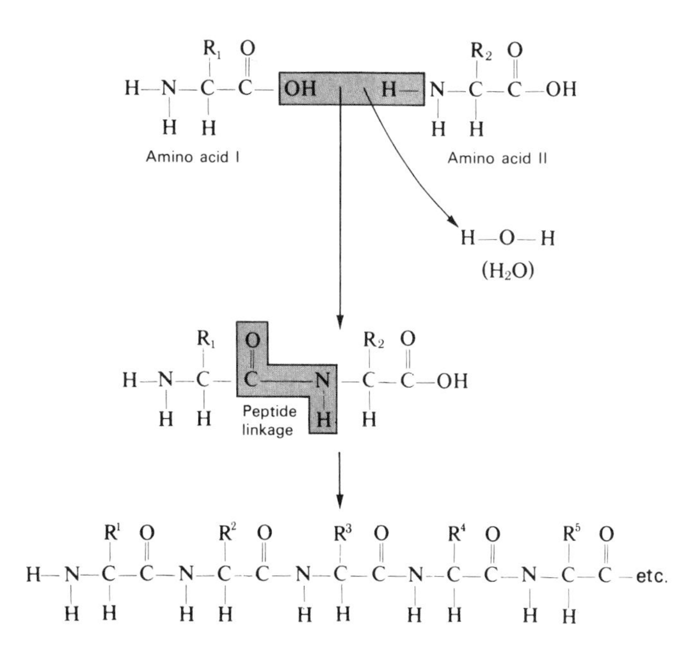

FIGURE 12-2. The peptide linkage. The backbone of the protein molecule consists of amino acids joined together through peptide linkages. Reactions may take place between various R groups of the many amino acids, so that the molecule can come to assume a complex architecture. Still, when unfolded, the protein is seen to consist of amino acid units in a linear order joined by peptide linkages (below).

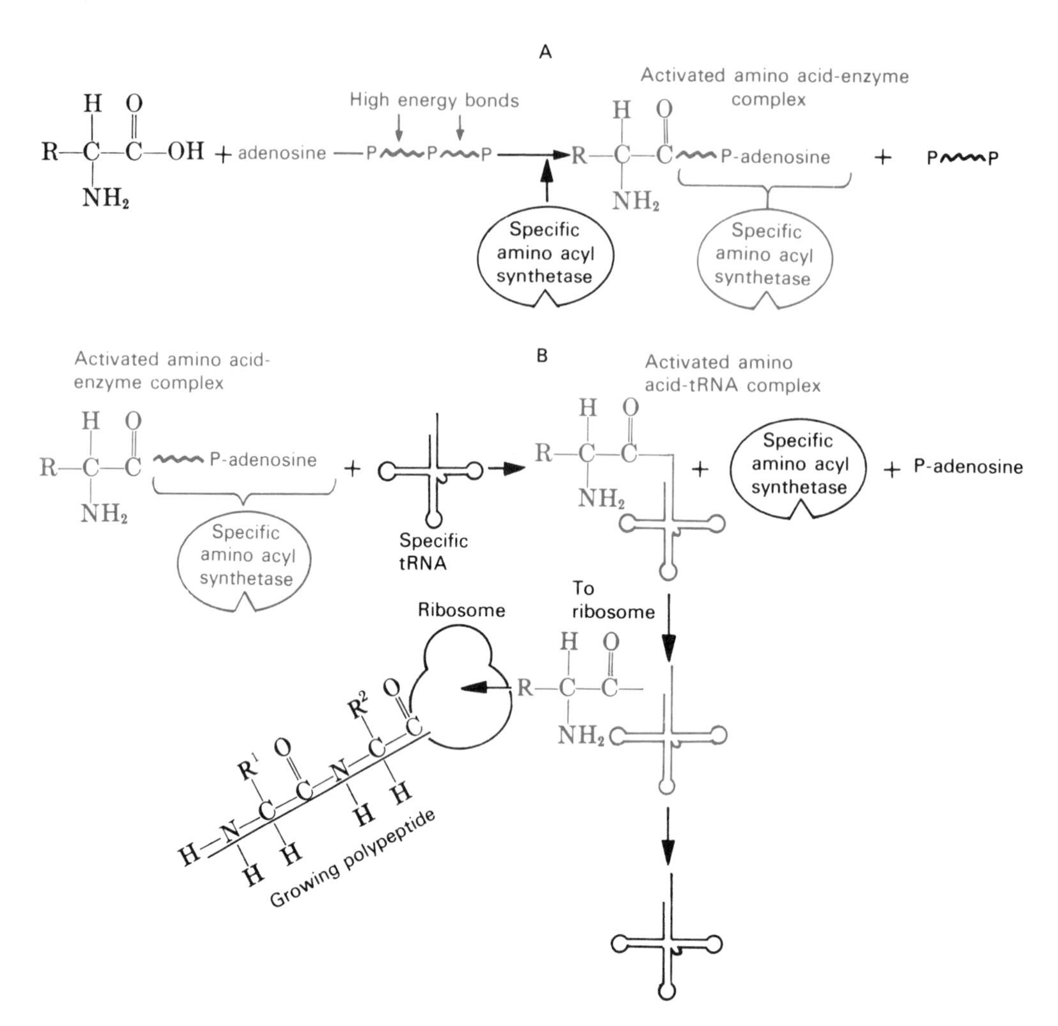

FIGURE 12-3. Activation of an amino acid and its attachment to tRNA. **A:** An amino acid reacts with the high-energy compound ATP. This is brought about by an enzyme specific for the particular amino acid. The enzyme forms a complex with the amino acid in which a high-energy bond is transferred to the amino acid from the ATP. Two phosphates are released. **B:** The activated amino acid–enzyme complex possesses sufficient energy to react with a tRNA and to transfer the amino acid component to the tRNA. Each type of amino acid has a specific type of tRNA with which it reacts. After the reaction, the enzyme and adenosine are released. The tRNA, carrying its specific amino acid, can now move to the ribosome where the amino acid will be placed in a specific site in a growing polypeptide chain. The amino acid, using energy from the amino acyl bond, can form peptide linkages. The tRNA is then free to recycle and to react again with its specific kind of amino acid. Until an amino acid reacts with its kind of tRNA, it has no identity and cannot react with the ribosome.

nucleotides in RNA are joined in the 5′ to 3′ direction; the 5′ end of the nucleotide being added is linked to the 3′ end of the nucleotide in the growing chain. Therefore the polarity of RNA formed in transcription is just the opposite of that of the DNA template on which it was assembled. For example, the DNA sequence reading 3′–CAATGT–5′ would form a transcript with the sequence 5′–GUUACA–3′ (Fig. 12-5).

If the base order in DNA corresponds to coded information, then the RNA contains information that is stored in the DNA. The RNA formed in this way could therefore be involved in the transfer of information. But exactly what kind of role does this RNA play in the cell? Is this the RNA of the ribosomes (rRNA), or is it still another kind of RNA that is distinct from the rRNA and tRNA?

Answers came from research using *E. coli* and certain phages that attack it (T/2 and T/4). It had been established that, once the DNA of the virus enters the bacterium, the host is directed to make protein that is

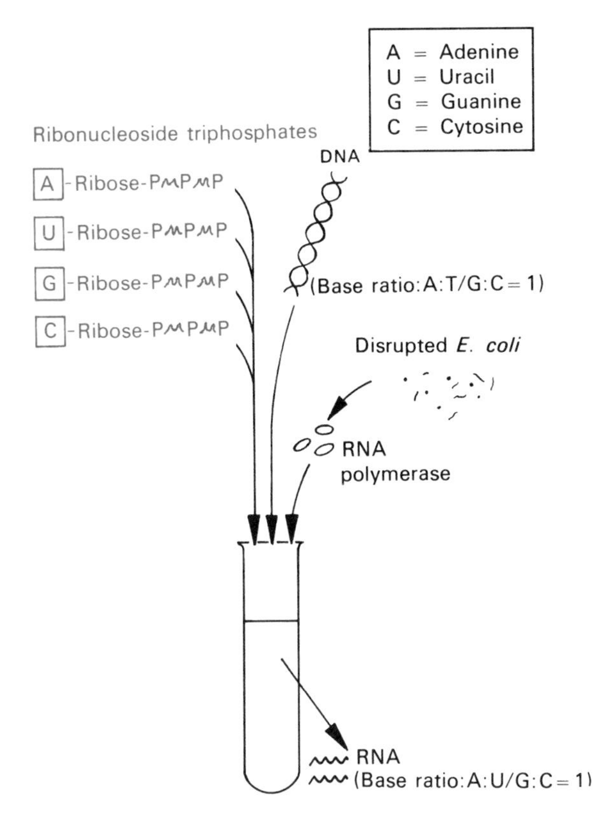

FIGURE 12-4. In vitro synthesis of RNA. Cell-free synthesis of RNA can be achieved in the presence of several basic ingredients. Among these are the building units of RNA in their energized triphosphate form, DNA to act as a template in their assembly, and RNA polymerase needed as the catalyst. The RNA formed will possess the same base ratio as the DNA that served as the template.

not normally found in the cell. Clearly, the virus can instruct the host cell to make products specific to the virus. Therefore it must have some mechanism that enables it to convey the necessary information to the host for the synthesis of the viral necessities. Studies of virus-infected bacteria revealed that immediately after entry of the virus DNA, an RNA appears (Fig. 12-6). This RNA was found to resemble the DNA of the virus, *not* the DNA of the host.

Another very important fact emerged from studies of infected bacterial cells. It was demonstrated that after viral infection *no* new ribosomes are formed in the cell; those present in the cell *before* infection are retained. The new RNA formed under the direction of the virus DNA was shown to associate with the ribosomes of the cell. Newly formed protein, protein specific to the virus, can then be detected in association with the ribosomes. No doubt remains that the bacterial ribosomes can somehow synthesize viral protein by assembling amino acids in the proper order. To do this, the ribosomes must require direction in the form of a set of instructions. This is apparently provided by RNA—an RNA different from the ribosomal and the transfer RNAs. This special variety was named *messenger RNA (mRNA)* because it contains a message that directs the ribosomes to assemble amino acids in an order specific for a given kind of protein. Without the mRNA, the ribosomes can do nothing.

Evidence for a triplet code

The process of transcription, the formation of coded information in RNA from a DNA template, is essential for the orderly transfer of genetic instructions. In transcription, passage of information occurs from one kind of nucleic acid to another—from DNA to RNA. To be meaningful, this information must now undergo *translation:* The information coded in the RNA must be read and finally transferred to protein. Before the mechanism involved in translation can be understood, we must decipher the code that resides in the DNA. Any such code must depend on the sequence of the bases in the DNA molecule. If this is so, then the bases must somehow correspond to the amino acid units that compose protein.

It is obvious that one base cannot correspond to one amino acid, because there are only 4 bases to designate 20 amino acids. Perhaps a sequence of two bases, a doublet such as CA, indicates an amino acid. However, if 4 different entities of any kind are involved, such as 4 different bases, and if 2 of these are taken at a time, only 16 different permutations or arrangements of the 4 (4^2) are possible. This is 4 fewer than the needed 20. If the code is based on triplets, three bases taken at a time, then 64 (4^3) distinct arrangements can be realized. This is many more than necessary.

Many ideas were offered on this point; among these was the suggestion that only 20 triplets designate amino acids, one triplet for one amino acid. The remaining triplets would be meaningless or nonsense. Another idea was that the code is *degenerate;* most or all of the triplets would represent amino acids. In a degenerate code, two or more triplets could code for the same amino acid.

One of the first clear lines of evidence that the code *is* actually based on triplets, adjacent groups of three nucleotides, came from the work of Crick and his colleagues. Crick favored the idea that the simplest explanation might be the correct one. He suggested that the message coded in the gene is based on triplets and that the message is read three nucleotides at a time from a definite starting point. Crick's idea proved to be correct and was supported by a host of later investigations that established the foundations of our knowledge of transcription and translation at the molecular level. We know that only one of the two DNA strands of a gene undergoes transcription to form a messenger RNA transcript coded for the amino acid sequence of a polypeptide chain. This message is read in triplets, and its beginning is in the form of a specific triplet that indicates the start of the message, and hence the very first amino acid in the polypeptide (Fig. 12-7). The reading of triplets continues up to the endpoint of the message, indicated by another specific triplet. The sequence of triplets in the message constitutes a *reading frame,* which is a reflection of the complementary reading frame in the DNA.

To obtain information on the nature of the genetic code, Crick employed *E. coli* and phage T/4. The wild type bacterial virus can infect E. coli and cause cell lysis. On an agar plate, the lysis is evident as plaques, clear areas where cells have been destroyed. The size of the plaque is characteristic of wild T/4. Several types of mutations may affect the ability of the virus to produce lysis so that mutant plaques, or even no plaques at all, form on an agar plate.

Crick studied a number of different mutations in a gene (the B gene of the virus) that affects plaque formation. The mutations were induced by mutagens, and the separate mutation sites within the gene were mapped from recombination data (crossover frequencies).

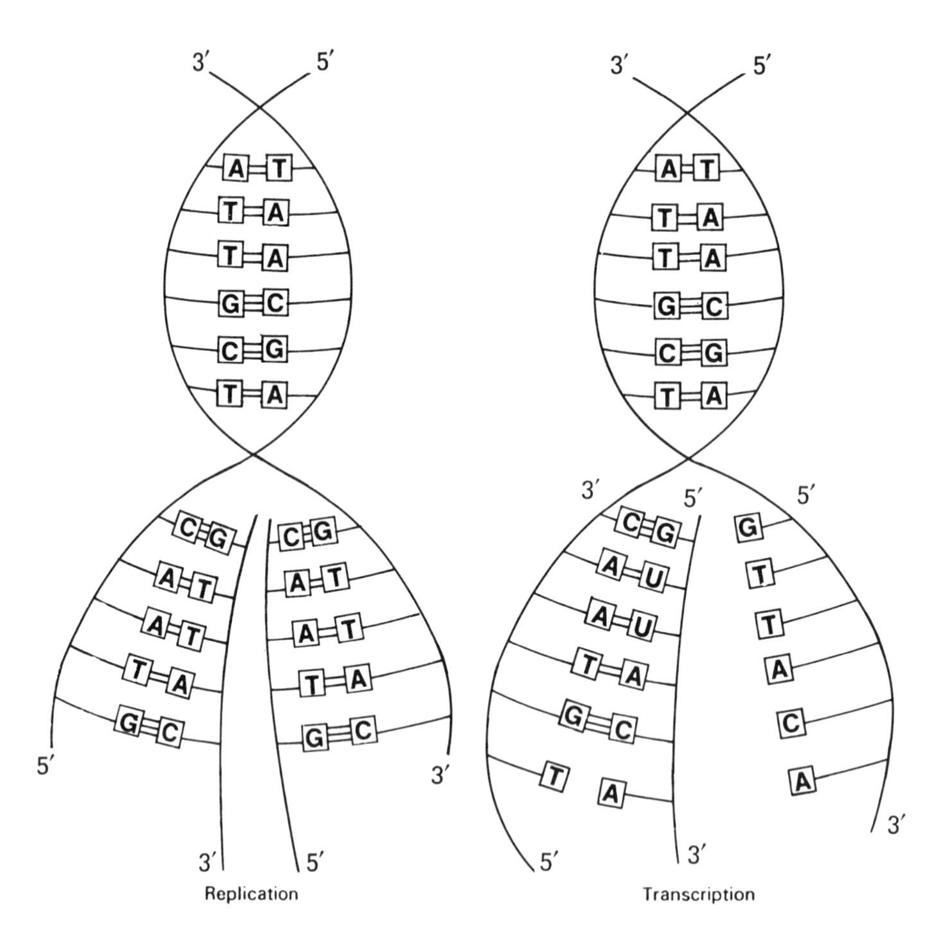

FIGURE 12-5. Self-replication and transcription. Self-replication (left) is necessary to ensure continuity of the genetic information from one cell generation to the next. However, this information must be transmitted to the cell if it is to be of any consequence. A critical step in this transfer is transcription (right). A portion of the double helix unwinds and acts as a template for the formation of a stretch of RNA. The RNA resembles the DNA closely, but one major difference is the replacement of T by U in the RNA. Also, the polarity of the RNA is opposite that of the DNA template.

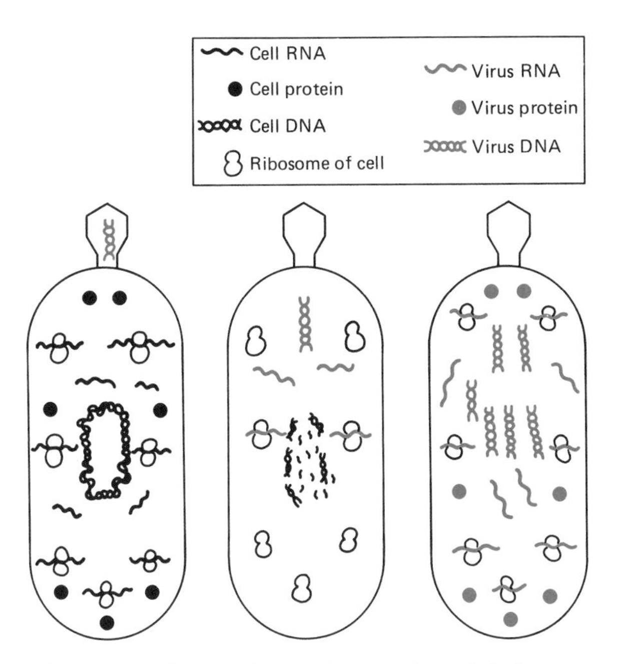

FIGURE 12-6. Phage infection of a bacterial cell. Before the phage injects its DNA into the bacterial cell (left), the DNA of the cell is engaged in transcription, forming RNA that resembles the DNA of the cell. This RNA associates with the ribosomes of the cell to form necessary cell proteins. Shortly after the viral DNA enters the cell (middle), a new type of RNA appears. This is a transcript of the viral DNA, and it associates with the ribosomes of the cell. The cellular DNA decomposes; RNA and protein synthesis typical of the cell cease to take place. Shortly thereafter (right), more viral DNA and viral RNA appear in the cell. Viral-specific protein also is synthesized. No new ribosomes are made; the viral mRNA associates with the ribosomes that were originally present.

In the studies, acridine dyes were employed as mutagens. These chemicals, by actually inserting into the DNA, can cause deletions and duplications of one or a few base pairs (see Chap. 14). Most other mutagens typically produce their effect by causing base pair substitutions in the DNA. After exposing the T/4 virus to acridines, it was noted that the different mutations within the gene often resulted in complete loss of gene function. The sites of various acridine-induced mutations were mapped. This made it possible to infect bacteria with viruses mutant at two sites. From such a mixed infection, offspring viruses could be obtained. Among these progeny viruses, recombinants could be formed; some would be doubly mutant, others wild (Fig. 12-8). The doubly mutant ones could then be studied for their ability to produce plaques on *E. coli.* It was found that certain double mutants were nonfunctional, whereas others retained some function or behaved as if they were wild. It was possible to classify the single mutants into two categories: + and –. A combination of a + and a – could suppress the mutant effect, whereas two + or two – mutations in one virus DNA would be completely mutant.

Such results can be easily explained according to Crick's hypothesis and fit in perfectly with our present knowledge. As Figure 12-7 shows, the reading frame starts from a beginning triplet, and the message in the mRNA then proceeds to be read three bases at a time, each triplet designating an amino acid. In the absence of mutant sites within the gene, and hence within the mRNA, a wild-type polypeptide or protein is formed. Allowing + mutations to stand for added bases, we can see that the addition of just one nucleotide will upset the reading of the gene (Fig. 12-9A). After the point of insertion, the reading is out of phase, because there is nothing within the gene to set the triplets apart. The reading of triplets from the point of insertion on is therefore changed. Many of the triplets now designate amino acids that should not be placed in particular positions in the protein chain. The result is an altered protein and a mutant effect. The same sort of reasoning applies to a – or deletion mutation, in which the reading is again out of phase from the point of alteration where the base is deleted (Fig. 12-9B). However, a combination of a + and a – mutation can bring the reading back into phase again (Fig. 12-9C). Therefore, if the stretch between the + and the – is not too long, wild-type function might be restored. A combination of three + or three – mutations would also result in a return of function if enough of the correct message is in phase. However, two + or two – mutations together would not restore correct reading to the frame, and gene function would be lost (Fig. 12-9D).

Various combinations of mutants were made, and results were all compatible with Crick's hypothesis. They also demonstrated that at least two different classes of gene mutation can be recognized. There are the familiar kinds of mutations in which a single base pair is changed, rather than removed or duplicated (Fig. 12-9E). These are designated *missense mutations.* Although they may cause a mutant effect, the gene often retains some of its function. Except for one triplet, the rest of the gene reads correctly. Therefore only one amino acid is changed in the protein dictated by this mutant gene. Missense mutations can also mutate back to wild on rare occasions.

In contrast with these are the *frame shift mutations,* illustrated by the + and – alterations. In this kind of mutation, a base is added or deleted, with the conse-

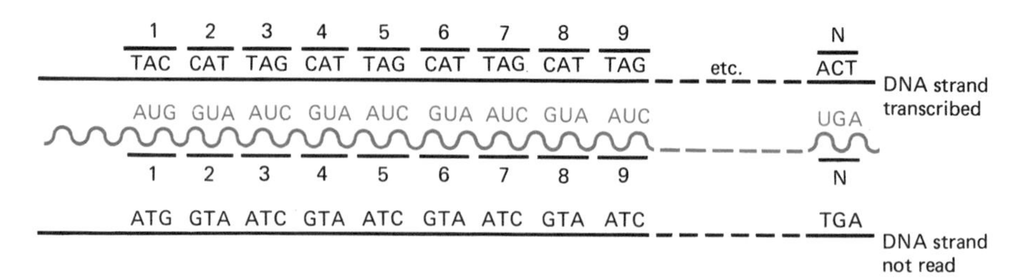

FIGURE 12-7. Reading of a gene. The messenger RNA (red) contains a sequence of nucleotides coded for an amino acid sequence that constitutes a reading frame. There is a definite starting point, a first triplet (AUG), which indicates where the message begins, and the reading proceeds triplet after triplet from that point. The reading will come to a halt when another specific triplet (UGA) indicates the endpoint of the message. The reading frame in the messenger RNA corresponds to the complementary one in the template strand that was transcribed.

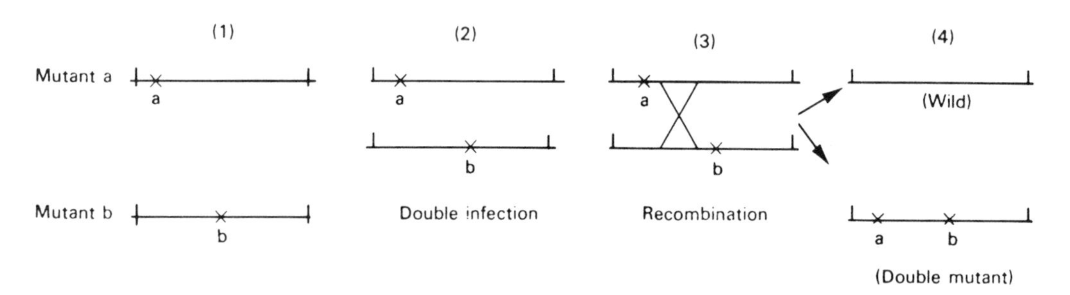

FIGURE 12-8. Intragenic recombination. After exposure to a mutagen such as an acridine, mutations may arise in the particular gene under study. The location of each separate mutation can be mapped (1). A bacterium may be infected with two types of virus (2). A process akin to crossing over between the virus particles can take place (3). Among the progeny viruses, wild types and double mutants can be recognized (4).

quence that the reading of the gene may be shifted over a long stretch of nucleotides. Any protein associated with this type of genetic alteration would thus possess a long stretch of amino acids different from the normal and would very likely be nonfunctional. Spontaneous revertants of the + and − frame shift mutations would not be expected, and this seems to be the case. All the experimental evidence supported Crick's concept of a gene that is read in groups of three nucleotides from a fixed starting point.

According to this model, the portion of the gene that is coded for a polypeptide is "commaless"—it contains within it no punctuation or signals to indicate a stop to translation and then another start. However, some signals must exist to bring translation to a halt. This is necessary because the messenger RNA contains a trailer sequence at its 3′ end that is not coded for the polypeptide and must not be read. Later in this chapter, we will discuss these signals as well as those found at the end of genes that bring transcription to a halt to prevent the reading of one gene into the next.

Codons and their assignments

As supporting evidence for the triplet nature of the code was assembled, important investigations were conducted to elucidate the very meaning of the 64 triplets. Which triplets stand for which amino acids, and which of them are nonsense? The pioneer studies in deciphering the genetic code were carried out by Nirenberg and Matthei, who reported a breakthrough in 1961. The logic behind the work was to provide the protein-making machinery of the cell with a *known* genetic message. Any protein made would then reflect the known information that had been supplied by the investigator. Nirenberg and Matthei employed cell-free test-tube systems containing components from ruptured bacteria plus mixtures of amino acids. For any polypeptide to be made in vitro, certain essentials are required. Besides the amino acids, all of the parts of the system needed to assemble them into protein must be present (Fig. 12-10). This includes an energy source, ribosomes, the various tRNAs along

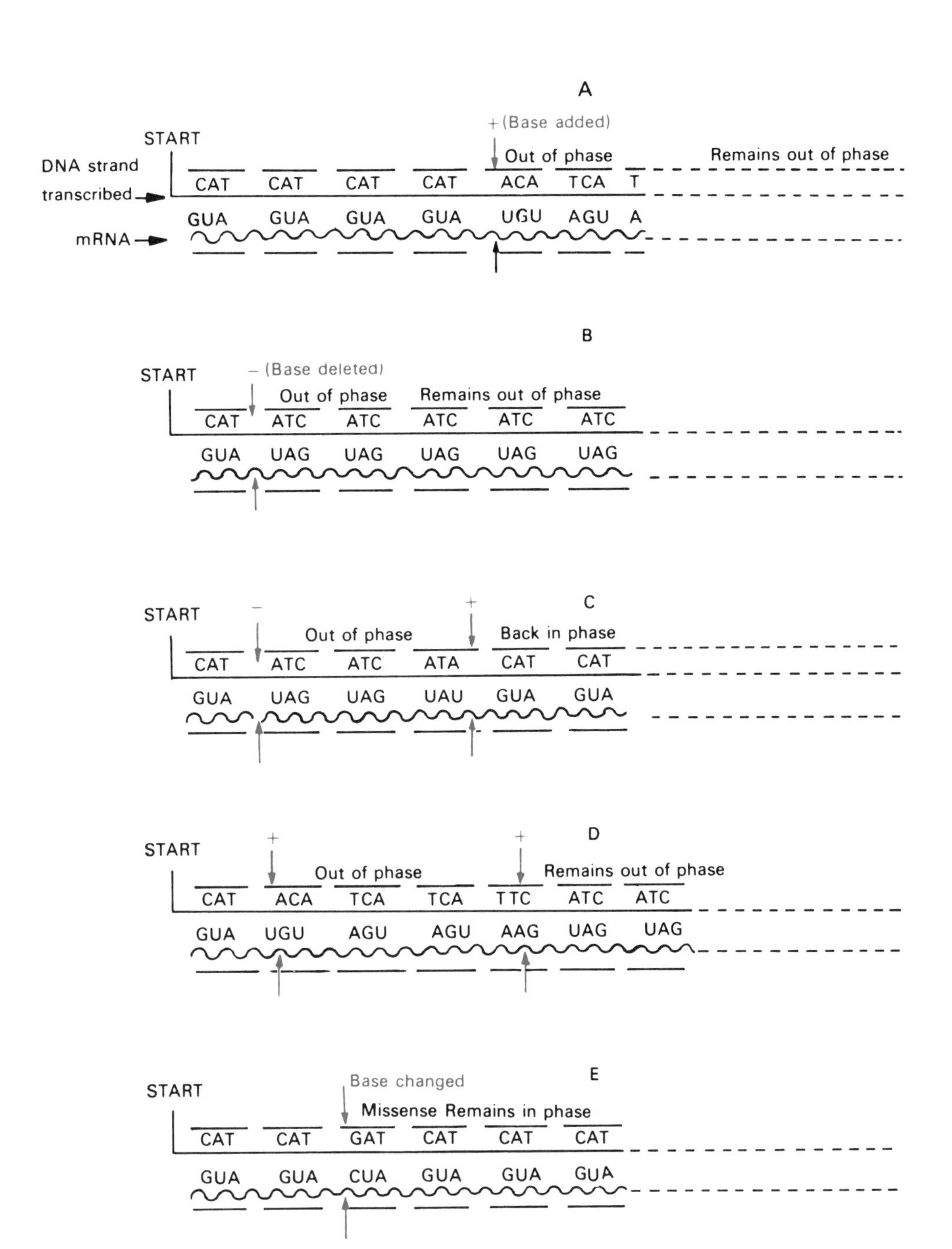

FIGURE 12-9. Types of mutations and their effects on the reading of the gene. Assume that a portion of a reading frame in the template strand is CAT, CAT, CAT, and so on. The addition or deletion of a base (A,B) will cause the reading frame to go out of phase, because the bases are read in triplets starting from a fixed point within the gene. Note that the mRNA that is transcribed from the DNA will reflect the change and will also have stretches that are out of phase. A combination of a + and a – mutation (C) can bring the reading frame back into phase. If the stretch that is out of phase is not too extensive, the gene may show some function. A combination of two + mutations (D) cannot bring the frame back into phase. This is also true of a combination of two – mutations. Most spontaneous mutations are probably the result of a base substitution (E) rather than the addition or deletion of a nucleotide. Except for one triplet, which codes for one "incorrect" amino acid, all the bases are in phase and would be read correctly. The gene would retain some function. These missense mutations can revert spontaneously to normal if a later mutation causes the original base to be inserted for the substituted one. A + or a – mutation (A–D), on the other hand, would not be expected to revert spontaneously with any ease, because a whole nucleotide must be either eliminated or added rather than just a base substituted.

with the corresponding synthetases, and a coded mRNA. The ability to manufacture a synthetic messenger RNA with a known sequence of nucleotides provided the critical key to deciphering the genetic code. We noted that the transcription of DNA into mRNA by the enzyme RNA polymerase requires a DNA template. Fortunately, in the mid-1950s Ochoa and Grunberg-Mangano had isolated from *E. coli* still another enzyme, polynucleotide phosphorylase, which is capable of synthesizing RNA. Although it has another role in the intact cell, it can be used in vitro to assemble RNA nucleotides in a random fashion without following a DNA template. This meant that one could provide the enzyme with certain specific ribonucleotide units and obtain an RNA of known composition. Using the phosphorylase, Nirenberg and Matthei prepared a synthetic messenger that consisted only of repeating units containing uracil (polyuridylic acid). Such RNA could then be used as a messenger and supplied to the protein-making system described earlier (Fig. 12-10).

In the experiments, separate test-tube mixtures of the essential factors were followed (Fig. 12-11A). In each mixture, one of the 20 different kinds of amino acids was labeled with ^{14}C, the remaining 19 being nonradioactive. This made it easy to tell which specific amino acid was being incorporated into any insoluble polypeptide that formed in response to the messenger RNA that had been supplied; if a polypeptide formed and if it was labeled, then it must contain the particular labeled amino acid in that mixture.

When the polyuridylic acid (UUU) was used as mRNA in the cell-free system, a polypeptide was formed. Chemical analysis showed it to be polyphenylalanine, a polypeptide composed of repeating units of the amino acid phenylalanine (Fig. 12-11B). The messenger contained only repeating units of nucleotides with uracil. Therefore, if the code *is* based on triplets, the RNA triplet UUU must indicate phenylalanine during translation. Since the RNA triplet is a transcript of DNA, the DNA triplet AAA in the template strand must correspond to phenylalanine (Fig. 12-12). A sequence of three nucleotides designating an amino acid or other information was termed a codon. Thus there are DNA codons and RNA codons, such as AAA and UUU, respectively, for phenylalanine.

In the same way, polymers composed of other single kinds of ribonucleotides were prepared and supplied as messengers. Polycytidylic acid (repeating nucleotides containing cytosine) was found to induce the formation of a polypeptide composed of repeating units of the amino acid proline. Thus the RNA codon CCC and the complementary DNA triplet GGG represent proline. Polymers were then made containing two and three different bases. For example, suppose the enzyme polynucleotide phosphorylase is supplied with equal amounts of two kinds of nucleotides, half of them containing cytosine and the other half uracil. At random, the enzyme will assemble an RNA containing eight different triplets in equal amounts: UUU, UUC, CUU, UCU, UCC, CCU, CUC, and CCC. When used as messenger, this polyribonucleotide directed the incorporation of phenylalanine and proline, because the codons UUU and CCC are present. But in addition, the polypeptide that forms also contains leucine and serine. This means that these two amino acids must be coded by RNA codons that contain both uracil and cytosine. However, the specific triplets for leucine and serine (or for any amino acid coded by more than one base) cannot be determined by this procedure. This is so because the exact order (e.g., UUC, UCU, or CUU) cannot be established. Later techniques devised by Nirenberg permitted the precise assignments of codons and clarified other features of the genetic code.

In these procedures, simple trinucleotides are synthesized that are composed of just three ribonu-

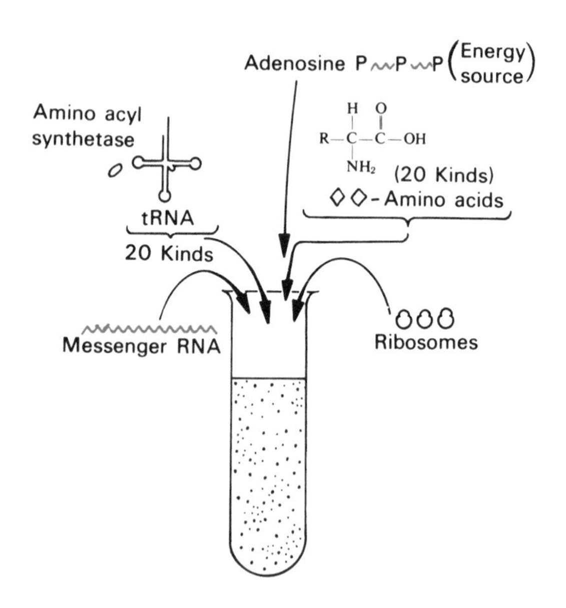

FIGURE 12-10. Cell-free system for polypeptide synthesis. Simple polypeptides can be made in vitro without the presence of intact cells. However, several basic cellular ingredients are required, as shown here.

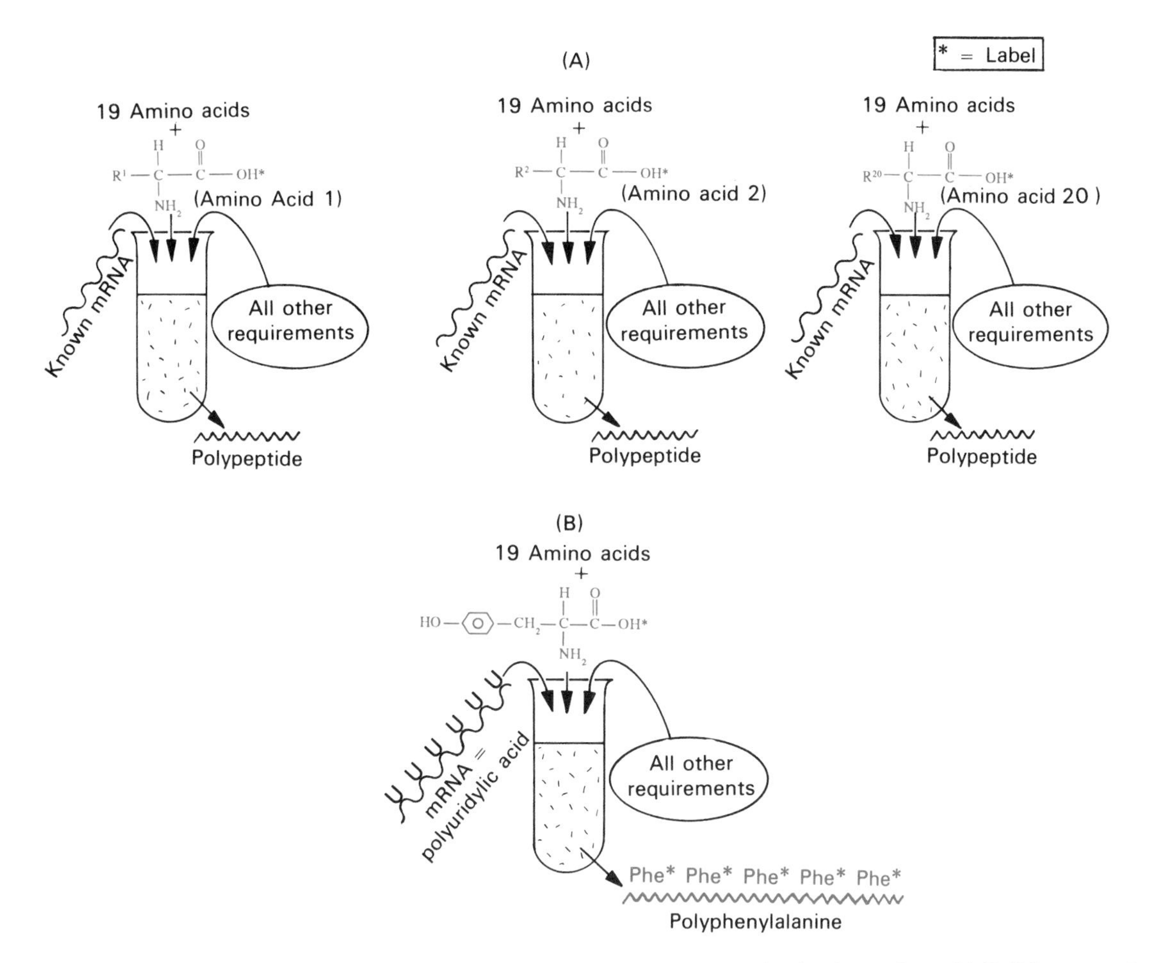

FIGURE 12-11. In vitro synthesis of polypeptides using known mRNA. **A:** To determine whether a specific codon is related to an amino acid, an mRNA of known composition is supplied to cell-free systems. Twenty separate systems must be followed. In each, one of the 20 amino acids is labeled. If the label is recovered in the resulting polypeptide manufactured by the system, it is then known that the messenger used contained a codon for that amino acid. **B:** When a synthetic mRNA composed of repeating units of uridylic acid (UUU) is supplied to each of 20 separate systems, only the system containing labeled phenylalanine gives rise to a labeled polyphenylalanine chain. Since none of the other 19 yielded a labeled polypeptide, it could be concluded that UUU is the triplet that codes for phenylalanine.

cleotides in a known sequence. It was found that each simple trinucleotide was able to act as mRNA by binding to both a ribosome and tRNA carrying its specific amino acid. For example (Fig. 12-13), the trinucleotide CCC will form a complex with a ribosome and a transfer RNA charged with the amino acid proline. Because the small trinucleotide is obviously too short to act as a regular messenger, protein synthesis does not take place. However, ribosome–transfer RNA–trinucleotide complexes can be isolated on filters along with free ribosomes. As in the other method, any one mixture contains all 20 kinds of amino acids with only one kind labeled. The ribosome fraction of the system is isolated on the filters. This also contains the ribosomes that are complexed with the charged tRNA and the trinucleotides. By determining whether the fraction is labeled, a specific trinucleotide sequence can be directly related to a specific amino acid.

Features of the genetic code

The procedure just described gives conclusive proof of the concept of a triplet code. But it has also shown that most of the triplets represent amino acids. As we will discuss later in more depth, only 3 of the 64 codons do not correspond to amino acids. This means that the code is "degenerate"; more than one triplet can desig-

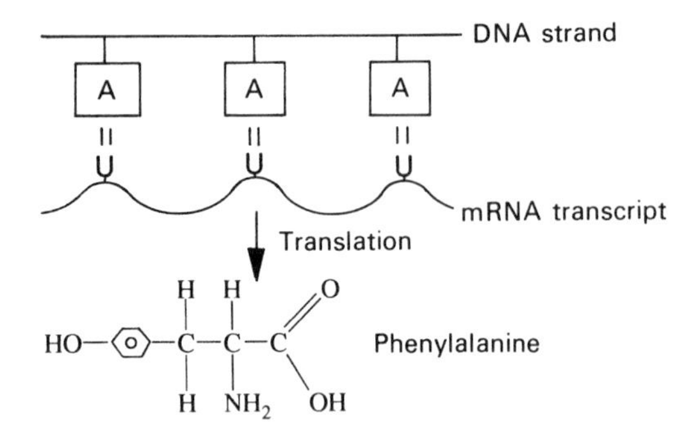

FIGURE 12-12. RNA and DNA codons. If the mRNA codon UUU specifies phenylalanine, then the DNA triplet AAA must designate UUU, because the mRNA is a transcript of DNA. Therefore each amino acid has a DNA as well as an RNA triplet or codon that is specific for it.

nate a specific amino acid. This is the case for almost all of the amino acids (Table 12-1). At first, it might seem surprising that the code is degenerate. But this in no way means that the code is haphazard; for any one amino acid is designated only by specific codons. In addition, the degeneracy of the code provides an actual advantage to living things. As can be seen by referring to Table 12-1, it is usually the third letter of the codon that differs when two or more triplets signify the same amino acid. Consider the case of arginine, which is specified by six codons. Of these, the first two letters are identical in four of them. In the remaining two codons, the first two letters are the same.

An advantage becomes apparent when we consider the codons more closely. Suppose the CGU sequence

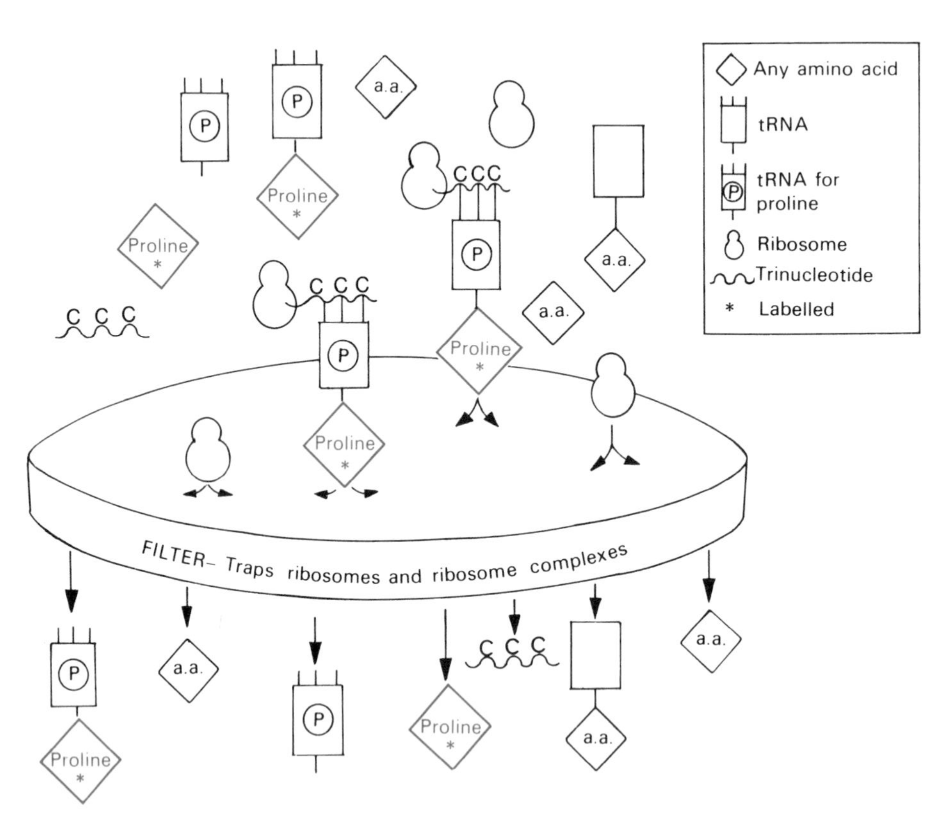

FIGURE 12-13. System for precise codon assignments. A short known sequence of three bases, a trinucleotide, can be synthesized. When added to the in vitro system, it attaches to a ribosome and also binds to a specific tRNA carrying its specific amino acid. In the procedure, only one amino acid of 20 is labeled in any one test-tube system. In this figure, it is proline, and the synthetic trinucleotide is CCC. The complex, composed of the ribosome, trinucleotide, and charged tRNA, can be isolated on filters. In this case, the ribosome fraction will be labeled because the proline is labeled. The presence of a label in the ribosome fraction of any one test tube indicates the presence of a labeled complex. This in turn indicates that the particular labeled amino acid is coded by the known trinucleotide supplied. Such procedures have shown that more than one kind of trinucleotide can code for the same amino acid, meaning that the code is degenerate.

that designates arginine is altered by mutation to CGA (Fig. 12-14). This codon also designates arginine. Actually, in four of the codons any base change at the third position will still code for arginine. A change in the CGA triplet at the first position to A produces AGA, which again corresponds to arginine. We see that the degeneracy of the code actually acts as a buffer against the effects of mutation. If only one triplet coded for one amino acid and the remaining 44 were nonsense, then a mutation would always result in an altered protein, either by causing an amino acid substitution or by bringing about premature termination of the protein chain. Degeneracy, on the other hand, results in a number of "silent mutations"—codon changes that give rise to different codons designating the same amino acid. Since no amino acid substitution would follow a silent mutation, the protein product would be normal and the very fact of the mutation would remain unnoticed. Examination of the code also shows that amino acids with similar chemical properties have similar codons that differ by only one letter. This means that if the AGA codon for arginine were changed by mutation to AAA, lysine would be substituted. Inasmuch as both are basic amino acids, the mutation might not cause too great a change in the properties of the protein and could also possibly remain undetected. Natural selection has undoubtedly operated during evolution of the genetic code to reduce the effects of gene mutation as much as possible through the establishment of a genetic code with the greatest buffering power.

Three of the codons, however, do not stand for amino acids. These are the so-called nonsense codons, UAA, UAG, and UGA (also referred to as *ochre, amber* and *opal,* respectively). As suggested by Crick's studies with the T/4 virus, these codons have been shown to provide a type of punctuation. They are better termed *chain-terminating codons* than nonsense, because the presence of any one of them on an mRNA indicates the end of the message and hence the end of translation by the ribosome. The existence of these RNA chain-terminating codons also tells us that the complementary DNA codons must be ATT, ATC, and ACT (Fig. 12-15A). The identification of the chain-terminating codons has been made possible through genetic studies with *E. coli* and phage, as well as through work with synthetic RNAs of known nucleotide sequence. For example, presence of the triplet UGA brings about termination of the reading of a synthetic messenger RNA by a ribosome. If it arises in vivo through gene mutation, we say that a nonsense mutation has arisen; translation comes to a premature halt with the production of an incomplete protein or polypeptide fragment (Fig. 12-15B).

We see from all of these studies that on the molecular level mutations may represent very different alterations of the codons. A *missense mutation* (see Fig. 12-9E), we have seen, is a change from one codon specific for a certain amino acid to another codon specific for a different amino acid. An altered protein results, and the magnitude of the effect depends on which amino acid has substituted for the original. A *nonsense mutation,* on the other hand, occurs when a codon specific for an amino acid is changed to one of the chain-terminating codons and thus brings translation to an end. An incomplete protein or fragment forms. A *silent mutation* (Fig. 12-14) represents a change from a codon specific for an amino acid to another codon specific for the same amino acid. No effect is produced on the protein product because no amino acid substitution has occurred. Figure 12-16 summarizes these classes of mutation. A deleted base or an added one can alter

TABLE 12-1. RNA codons and their designations[a]

	Second base				
First base (5′ end)	U	C	A	G	Third base (3′ end)
U	phe	ser	tyr	cys	U
	phe	ser	tyr	cys	C
	leu	ser	ter	ter	A
	leu	ser	ter	trp	G
C	leu	pro	his	arg	U
	leu	pro	his	arg	C
	leu	pro	gln	arg	A
	leu	pro	gln	arg	G
A	ile	thr	asn	ser	U
	ile	thr	asn	ser	C
	ile	thr	lys	arg	A
	met	thr	lys	arg	G
G	val	ala	asp	gly	U
	val	ala	asp	gly	C
	val	ala	glu	gly	A
	val	ala	glu	gly	G

[a]See Figure 12-1 for the amino acid symbols. "ter" designates a chain-terminating or "nonsense" codon.

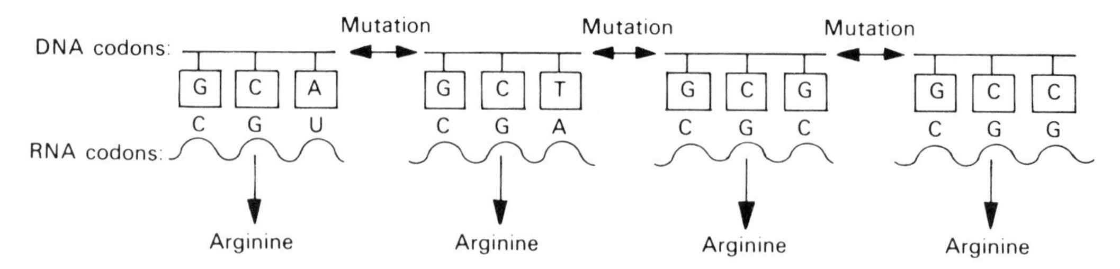

FIGURE 12-14. Silent mutations. Shown here are some of the possible silent mutations that involve codons for the amino acid arginine. If any base substitution occurs at the third position in any one of these four DNA codons shown, the result is still an RNA codon that designates arginine. Therefore, certain mutations may take place and remain undetected, because they bring about no amino acid substitutions. The buffering effect of a degenerate code is apparent; if a change in a codon always caused a change to a codon standing for another amino acid or for termination of the protein chain, every mutation would have a detectable effect, most of them harmful.

the reading frame and bring about a *frame shift mutation* in which a portion of the coded information in the gene will be out of phase (Fig. 12-9). These can result in very defective protein products with stretches of incorrect amino acids. A frame shift mutation can even create a chain-terminating codon with the consequent formation of an incomplete protein.

An outstanding feature of the genetic code is that it is almost universal. The same code is used in all groups of living organisms from prokaryotes through the highest eukaryotes. Parts of the protein-synthesizing machinery from different groups of organisms can be successfully interchanged. For example, a human mRNA can be faithfully translated by the tRNAs and ribosomes from wheat or some other distantly removed species. However, a few exceptions have now been revealed in certain prokaryotes and lower eukaryotes. Very often the exceptions are found to affect the chain-terminating codons. In a species of the ciliated protozoan Tetrahymena, the codons UAA and UAG are recognized as glutamic acid. UGA appears to be the only codon acting as a chain terminator.

Exceptions to the genetic code are found in several species including mammals when the coding dictionary of the mitochondria is considered. However, the genome of the mitochondrion, as we will see in Chapter 21, may have had a special evolutionary history, and certain apparent exceptions to the code may have another explanation. Moreover, the ciliates of today in which exceptions are found branched off from other eukaryotes very early in the evolution of the eukaryotic cell, perhaps before factors to stabilize reading of the code had come into existence. Despite occasional exceptions to it, the genetic code can nevertheless be regarded as a universal system perfected by natural selection throughout the early evolution of the cell.

One gene–one polypeptide chain

Long before any insight had been gained into the molecular basis of information transfer, many examples had accumulated showing that a gene mutation blocked a chemical step in a metabolic pathway by its effect on a specific enzyme. This led to the formation of the *one gene–one enzyme* theory, a concept that implies that each enzyme or protein is controlled by a specific gene. Today the increased knowledge of the structure of enzymes, proteins, and the gene itself has made it necessary to rephrase the theory. In many cases, an enzyme is composed of one long polypeptide chain that is under the control of just one gene. An example is the enzyme tryptophan synthetase in the red bread mold, *Neurospora*. This important catalyst is needed for the final step in the manufacture of the amino acid tryptophan. A segment of DNA, a gene, controls the formation of this specific protein or long polypeptide chain that makes up the entire enzyme tryptophan synthetase.

We can think of any gene as a functional unit, meaning that it has a specific function to perform. The function of the gene in this example is to guide the formation of a particular kind of polypeptide, tryptophan synthetase (or "t'ase"), which has a specific enzymatic action (Fig. 12-17A). A gene that contains coded information for some structural product such as the amino acid sequence of a polypeptide is called a *structural gene*. A structural gene may also code for an RNA product, as we will see later on.

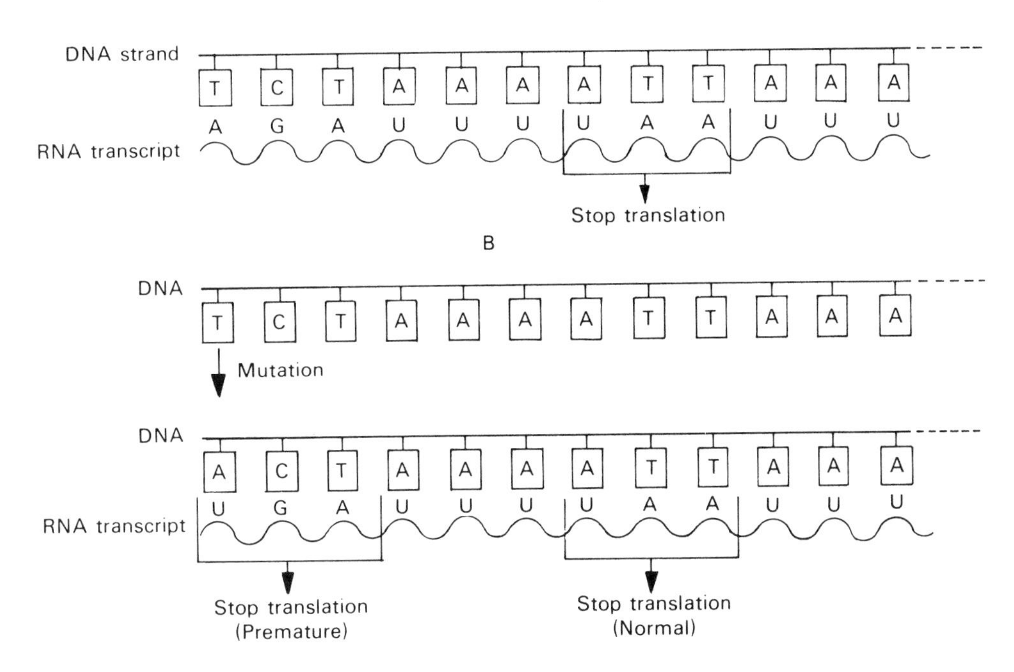

FIGURE 12-15. Chain-terminating codons. **A:** The RNA codons UAA, UAG, and UGA establish the end of translation of an mRNA. The complementary codons on the DNA would be ATT, ATC, and ACT. When the ribosome reads the messenger, the chain-terminating codons indicate the end of a polypeptide and the chain is terminated. **B:** If a nonsense mutation arises in a gene (shown here as ACT), a chain-terminating codon will be transcribed onto the mRNA in a position ahead of the normal one. A premature stop signal is thus carried on the messenger and the ribosome does not translate the message past that point. Thus that particular polypeptide chain is incomplete.

The one gene–one enzyme concept is very clear-cut in this case; however, this is not always so. In *E. coli*, the enzyme t'ase is under the control of two distinct genes (Fig. 12-17B). Two different kinds of polypeptide chains make up this bacterial enzyme. The function of each of the genes is to transfer information for the formation of a specific polypeptide. The two kinds of polypeptides then assemble to form tryptophan synthetase, essential to the final step in the formation of tryptophan. Since the completed enzyme in this case is composed of two kinds of polypeptides, each specified by a different gene, the relationship of one gene–one enzyme is not clear-cut. Because all proteins really consist of one or more polypeptide chains, however, we can modify the original statement to "*one gene–one polypeptide chain.*" This statement is basically sound, for every kind of polypeptide chain is specified by a particular gene. On the other hand, many proteins are composed of several separate chains that become associated. Indeed, the tryptophan synthetase of *E. coli* is actually composed of four, but only two different kinds are present, each one represented twice. In Chapter 17, we will reexamine the gene in relation to the way its mRNA is processed and learn that certain genes may be associated with more than one similar but nevertheless different polypeptide. Although the one gene–one polypeptide concept holds in most cases, discoveries at the molecular level present exceptions that continue to challenge established ideas.

Choice of DNA strand in transcription

One very important decision that must be made is the choice of which of the two DNA strands of a gene is to be read at the time of transcription. A moment's thought tells us that the RNA polymerase probably does not read both DNA strands when it forms mRNA. One reason why we would not expect this is that the genetic code is degenerate. If both strands were read, then two different kinds of protein would form after the translation of the two kinds of messen-

FIGURE 12-16. Three types of mutations. If the RNA codon for arginine, CGA, is changed by mutation to AGA, the altered codon still codes for arginine. No change will occur in the polypeptide governed by the gene carrying this mutation. However, if CGA is changed to UGA, the protein will be incomplete. If the codon AGA for arginine is changed to AAA, a missense mutation will have occurred, because another amino acid will be placed in the polypeptide instead of the correct one. The protein may or may not be greatly altered, depending on the particular substitution. In this case, the substituted lysine would probably not cause a pronounced effect, because its properties are similar to arginine.

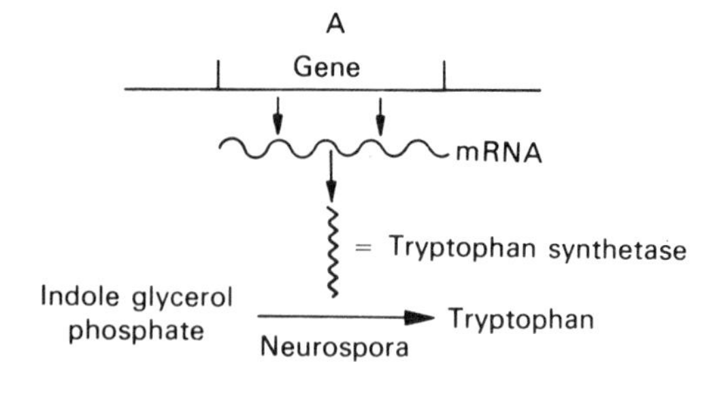

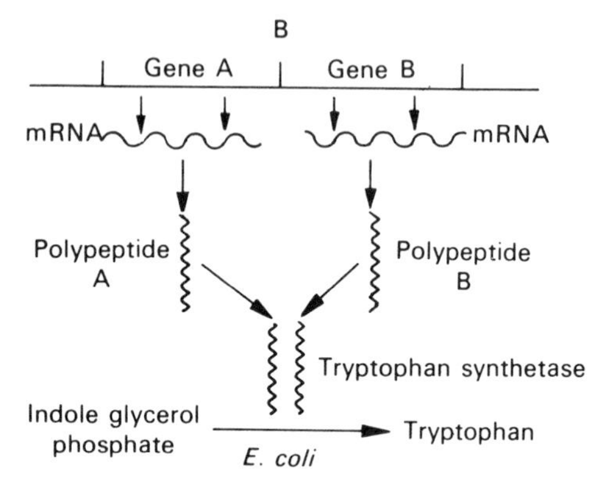

FIGURE 12-17. One gene–one polypeptide. **A:** In *Neurospora,* the enzyme tryptophan synthetase is composed of one kind of polypeptide chain that is under the control of one gene. In this case, the one gene–one enzyme relationship is direct. **B:** In *E. coli,* the enzyme t'ase is composed of two different kinds of polypeptides. Each is controlled by a separate gene. In this case, therefore, the concept of one gene–one enzyme does not strictly hold. However, in both *Neurospora* (above) and *E. coli* (below), the function of each gene is to govern the formation of a specific type of polypeptide, and the one gene–one polypeptide relationship is clear.

ger. This is not the case, for only one kind of protein or polypeptide chain is associated with a specific gene, an observation that led to the formulation of the one gene–one enzyme theory. Work with certain microorganisms has clearly shown that only one of the two strands of a gene is read.

How can it be demonstrated experimentally that a certain DNA sequence is responsible for the formation (transcription) of a certain kind of RNA? This can be accomplished by the technique of DNA–RNA hybridization, one of the most valuable tools of molecular biology and one to which we will continually refer. In any variation of this procedure, the double-stranded DNA must first be denatured, transformed into a single-stranded state. This can be achieved by an alkali treatment or by raising the temperature above the melting point of the DNA, a point at which the two strands of the double helix separate (Fig. 12-18A). If the denatured DNA is quickly cooled, the two complementary strands will not have the opportunity to reform double strands (to reanneal). RNA transcripts that are radioactively labeled may then be added to these single-stranded preparations. The mixture of DNA and RNA is incubated at a temperature that is too low to permit double-stranded DNA to form but sufficient to allow RNA to combine with any complementary DNA segments. The formation of hybrid duplexes, which can be detected by the radioactive label, indicates specific DNA segments and the RNA transcribed from them.

In one application of this procedure, denatured DNA may be trapped on a filter and then presented with labeled RNA (Fig. 12-18B). The filters are washed and exposed to an RNase that digests away any single-stranded RNA present but does not attack the RNA paired with the DNA. The presence and level of radioactivity can be detected, indicating the presence and amount of the DNA–RNA hybrids.

The technique of DNA–RNA hybridization has provided much information on the choice of DNA strands in transcription. Single DNA strands from certain sources can be separated and isolated from their complementary strands using density gradient ultracentrifugation. This separation and isolation of complementary strands is possible in those cases in which one DNA strand is lighter than its complementary strand. Complementary strands can be recognized as heavy (H) or light (L) when the two differ with respect to the percentage of purines or pyrimidines they contain. This is so because the purines adenine and guanine are heavier than the pyrimidines thymine and cytosine. Thus a single strand rich in adenine and guanine is heavier than its complement and will settle in a different region in a density gradient. This makes it possible to isolate H and L strands on separate filters. Certain very simple viruses contain a single chromosome composed of a DNA molecule in which one strand is heavy and one is light. This makes possible the separation of the two strands of the DNA molecule and the isolation of the H and L strands on separate filters. Populations of such viruses can be allowed to infect bacterial hosts that are growing in the presence of radioactive phosphorus. The viral mRNA that forms following such an infection is consequently labeled. When this labeled RNA is isolated and presented to the separate H and L filters, it is found that the single-stranded DNA on only *one* filter hybridizes with the labeled mRNA. This tells us that only one of the two strands of the viral DNA has undergone transcription. In the case of these simple viruses, it is usually the H strand.

Experiments with other types of viruses have shown that portions of the H strand and portions of the L strand hybridize with the RNA transcripts. This means that, unlike the situation in the very simple viruses, the H strand and the L strand of the DNA molecule both contain sequences that code for RNA transcripts. It appears that in all genomes except for the most simple viruses, it is not just one entire strand of the double helix that is coded for all the polypeptides. Transcription can arise from the H strand for one gene but from the L strand for another gene in the same organism. However, for a given gene it is always the same strand of the double helix that undergoes transcription to form mRNA. Therefore, strictly speaking, only one strand of a gene carries a nucleotide sequence that specifies the amino acid sequence of a polypeptide.

Initiation of transcription in prokaryotes

It is not known exactly how the decision is reached as to which of the two DNA strands will be read, but the enzyme RNA polymerase is actively involved. In a living cell, the polymerase must be able to recognize the beginning of a gene. If the enzyme could initiate transcription at any point in a gene, it would indiscriminately copy only portions of either one of the two strands into RNA, and no orderly transfer of the genetic information would occur. A definite starting region for the attachment of the enzyme has been found to exist. This is called the *promoter,* a DNA region containing start signals that are recognized by RNA polymerase. The information coded in the promoter enables the enzyme to recognize the beginning of a gene and the exact point where transcription is to begin.

Four different polypeptides make up the bacterial RNA polymerase: α, β, β', and σ. Since two α-chains are present, the complete enzyme, called the *holoenzyme,* can be represented as $\alpha_2\beta\beta'\sigma$. The σ-chain is quite easily removed from the rest of the enzyme, leaving the core ($\alpha_2\beta\beta'$). Actually, the σ portion is not involved in catalyzing the formation of RNA. This is accomplished by the core without σ. However, σ is essential for the recognition of signals in the promoter region. Without σ, as can be seen clearly in test-tube systems, transcription may begin at random anywhere along the DNA, with either of the two strands acting as the template.

When RNA polymerase is bound to any DNA region, it confers on that segment protection from digestion by the DNAases, DNA-digesting enzymes. This fact has facilitated the isolation of promoters. DNA fragments that contain them remain protected when they are complexed with the RNA polymerase and then exposed to a DNAase that digests away any unprotected region (Fig. 12-19). Their isolation has permitted analysis of promoter segments and the determination of their nucleotide sequences. Promoter mutations are studied to identify any nucleotide sequences that may be critical to the start of transcription. The isolated promoters may also be used in vitro to identify the first bases that are transcribed from the template strand of a gene.

Analyses of well over 100 *E. coli* promoters have brought to light many important features of this region and its interaction with RNA polymerase. The enzyme recognizes certain signals in the promoter; however, it does not begin transcription from the site to which it first attaches. The attachment to this recognition segment of the promoter is loose and is a distance from the point where transcription begins (Fig. 12-20A). The terms *upstream* and *downstream* are used to refer to portions of a DNA template in relation to the site of initiation of transcription. The polymerase is pictured as entering the promoter and flowing downstream into the gene. The DNA position at which transcription begins is given the designation "+1." Proceeding downstream, the + values assigned to the positions in the gene increase accordingly (Fig. 12-20B). All positions before the site of initiation are given negative values, starting with −1, next to +1.

When different promoter regions are compared, similarities in nucleotide sequences are looked for. If a particular sequence of nucleotides is critical to a certain gene function, we would expect to find it when we compare regions having similar functions such as promoters. A certain DNA segment that is vital for the

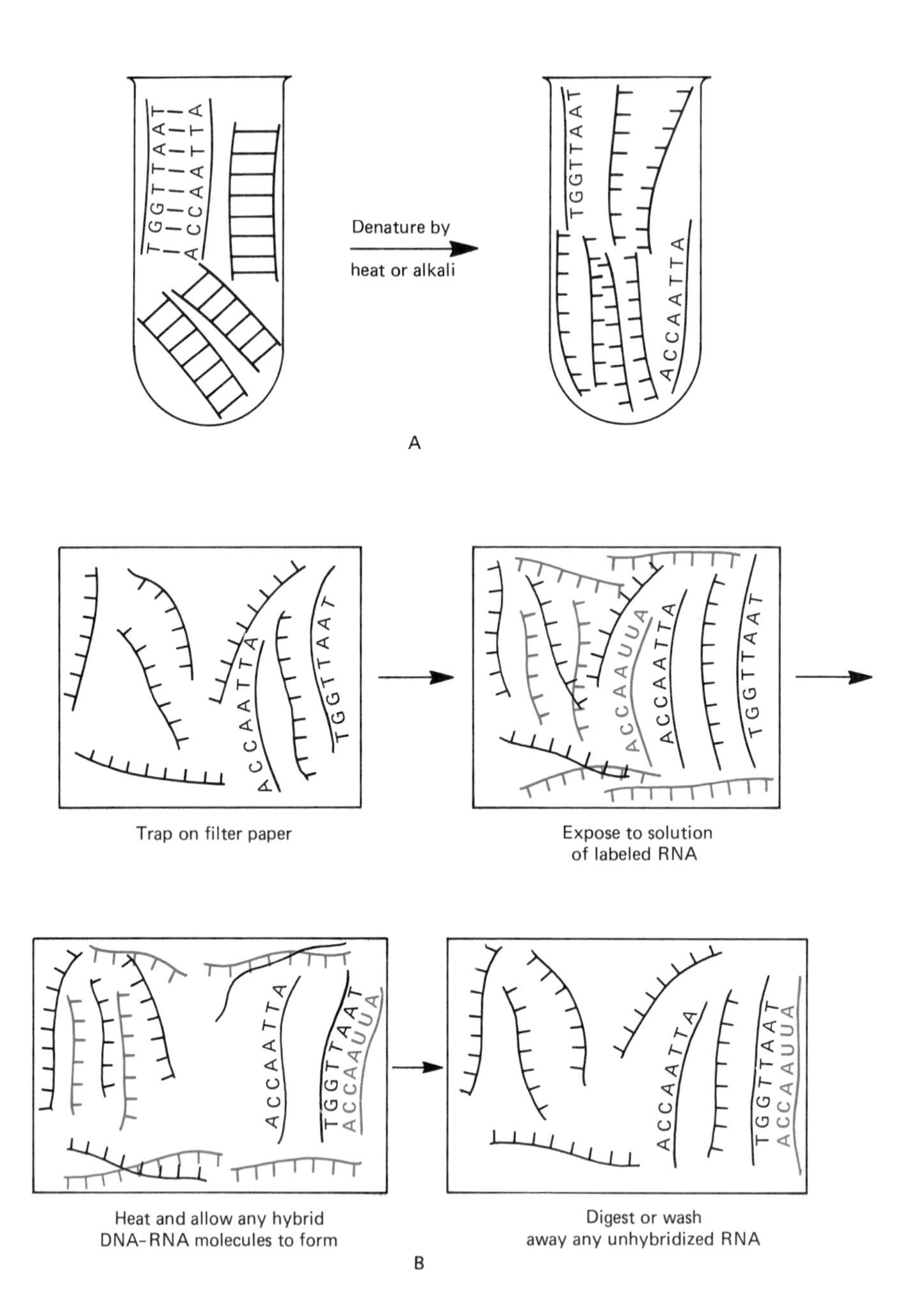

FIGURE 12-18. DNA–RNA hybridization. **A:** DNA molecules may be rendered single stranded (denatured) by heat or alkali treatment. The single-stranded preparation may then be used to identify any specific DNA sequences coded for specific RNA transcripts. **B:** The single strands may be fixed on filter paper and then exposed to labeled RNA molecules that diffuse over them. Any extended complementary sequences on the DNA and RNA will result in the formation of DNA–RNA hybrid molecules. Excess unpaired RNA is washed away. Only the labeled RNA that is part of a hybrid molecule remains. The hybrid duplexes can now be detected as a result of the label.

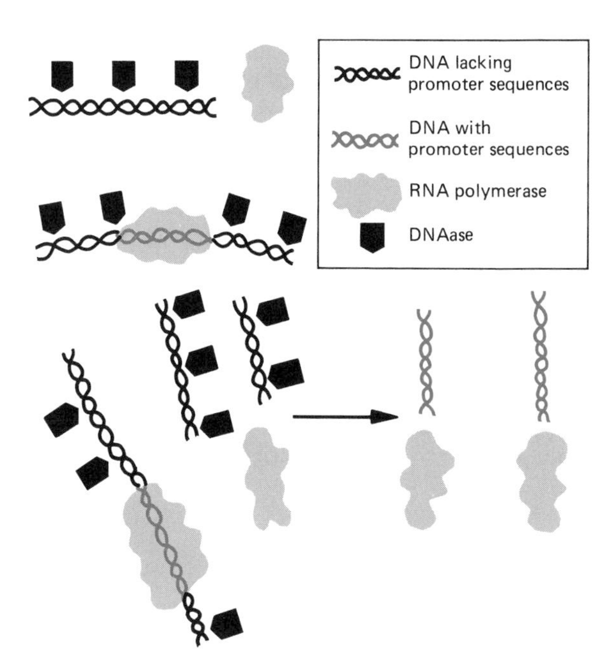

FIGURE 12-19. Isolation of promoters by DNAase protection method. One way in which promoters can be isolated takes advantage of the fact that RNA polymerase protects portions of DNA to which it is bound from digestion by DNAases (left). When the holoenzyme is added to any DNA fragments, the σ-chain enables the entire enzyme to bind tightly to any promoter that is present. Exposure to a DNA-digesting enzyme will then degrade all the DNA not covered by enzyme. The protected pieces, including the promoter sequences, may then be retrieved and analyzed (right).

recognition of RNA polymerase, for example, should be maintained or conserved from one promoter to the next. Any drastic change at a given site could upset its role. Consequently, little change would be tolerated in the region. Such a conserved sequence is known as a *consensus sequence.* It is the most prevalent nucleotide sequence found in a particular genetic region or genetic factor. There may be slight departures in any given case when a sequence of interest is compared to a consensus sequence. Nevertheless, the consensus sequence indicates the most likely nucleotides at specific positions in a given DNA segment.

Comparison of promoters shows that a characteristic sequence occurs at approximately position −35, which is the recognition region to which the polymerase first binds. The −35 sequences in the promoters are variants of the consensus sequence TTGACA. (In Figure 12-21, refer to the top strand, the strand that does not undergo transcription.) A consensus sequence is often written to reflect how often a given base occurs at a specific site. The −35 sequence is often shown as T_{82} T_{84} G_{78} A_{65} C_{54} a_{45}. The subscripts tell us the percentage of times the bases occur at those sites in the total number of promoters that have been compared. A small letter is sometimes used for those bases with percent occurrences of about 50% or less. This contrasts with those designated with capital letters, which have a higher degree of conservation. A base present at a given site almost 100% of the time is considered much more critical in its position than one that occurs with a frequency of 50%. The letter "N" is used to indicate a position in any sequence at which there is apparently no preference for a specific base so that the site appears to be occupied at random.

The polymerase leaves the −35 sequence and then proceeds to move downstream to another region that is also conserved from one promoter to the next. The consensus sequence TATAAT centers approximately at position −10 and is known as the Pribnow box (Fig. 12-21). The TA that begins the sequence and especially the final T that completes it are considered to be extremely important to polymerase binding, because they are present in the vast majority of promoters. The distance between the Pribnow box and the −35 sequence is about 16–18 bases in the great majority of cases. It is the region of the Pribnow box where the polymerase binds more firmly, and this is believed to be the location where the duplex DNA opens up so that the template strand can interact with the RNA nucleotides to form a transcript.

It should be noted that the region of the Pribnow box is rich in AT base pairs, making the region more susceptible to denaturation, because an AT pair is joined by only two hydrogen bonds as opposed to the three joining a GC pair. The RNA polymerase is believed to play a main role in causing the region to open up so that single strands are exposed. The open region extends from the Pribnow box to a few positions past position +1, the starting point of transcription. The first nucleotide in the transcript is typically one with a purine, adenine, or guanine. This means that a pyrimidine—thymine or cytosine—marks the position in the sense strand at which transcription begins. As the polymerase moves along the strand, ribonucleotides are added in the 5′ to 3′ direction, and the transcript increases in length. The σ-chain of the polymerase detaches from the rest of the enzyme after approximately 10 nucleotides have been added to the growing transcript, so that the core enzyme alone carries out the elongation of the RNA. As the enzyme moves along, the open region of DNA extends only a few base pairs in length, because the double helix reforms not far behind the enzyme. The RNA that has formed a hybrid region with the DNA in the opened-up segment is then released from its hydrogen bonding with the DNA as the two DNA strands rejoin.

A few points regarding conventions used in discussing genes must be understood at this point. As we have noted, RNA, like DNA, is assembled in the 5′ to 3′ direction. RNA is written in a way that stresses this fact, with the 5′ end to the left. As Figure 12-22 shows, the beginning of the DNA sequence of the strand that undergoes transcription starts at the 3′ end of the DNA. This is so because of the assembly of the ribonucleotides in the 5′ to 3′ direction as the RNA transcript is synthesized. Therefore, in the case of any DNA template, transcription starts toward the 3′ end of the strand that is coded for the polypeptide. Each growing RNA transcript will thus have a free 3′ end to which the 5′ end of a new nucleotide can be added to increase the length of the chain.

When referring to a gene, however, we consider the 5′ end to be the beginning of the gene, and we refer to the DNA strand that does not undergo transcription to form mRNA. The reason for this convention is readily understood when we compare this strand with the mRNA. We see that not only is the orientation (5′–3′) the same in the two, but they both have the same sense (Fig. 12-23). A triplet in the mRNA that designates an amino acid is exactly the same as the corresponding one in the nontemplate strand if we substitute Us (uracils) for Ts (thymine). Therefore it is logical to call the strand that is not the template the *sense strand*, since it reflects the picture in the messenger. The template strand on which the messenger is formed is known as the *antisense strand.* Although it is the sequence of DNA triplets in this strand that actually designates the triplets in the messenger, the messenger and its template are complementary. Consequently, the template strand and the mRNA do not have the same sense. Referring to the template strand does not indicate directly the RNA codons that designate the amino acids. However, reference to the nontemplate strand, the sense strand, gives us a direct picture of the code for the specific polypeptide or for any signals that accompany it in the particular gene. In our discussion of the Pribnow box and the −35 sequence, attention was paid to the sense strand, the one that is not transcribed (Fig. 12-21).

Promoters and effector molecules

In *E. coli,* before efficient transcription of certain genes can take place, the presence of specific regulatory molecules is required. For full transcription of these genes to occur, two factors are necessary. One is a simple molecule, cyclic adenosine monophosphate (cAMP), formed from ATP. The small cAMP molecule has been designated the *second messenger*, because of its interactions with hormones in their control of cell activity. But cAMP has also been shown to stimulate the transcription of certain genes in *E. coli*. However, it does not do this directly. It must activate a specific cell protein, catabolite activator protein (CAP). When CAP is

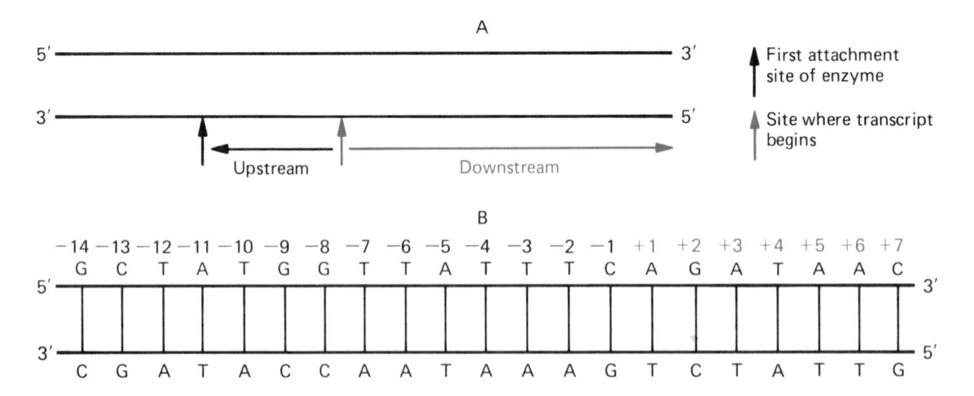

FIGURE 12-20. RNA polymerase and the start of transcription. **A:** The enzyme first attaches to the promoter at a site that is a distance before the point where transcription starts. All sequences before the starting site are designated *upstream.* Those from the site of initiation of transcription on are considered *downstream.* **B:** The position at which transcription is initiated is designated +1, and all positions downstream from it are given + values. The sequences prior to +1 are given negative values, which go in ascending order from position −1, just adjacent to the initiation site.

activated by cAMP, the CAP–cAMP complex binds directly to the DNA (Fig. 12-24). The exact location of the CAP binding site varies from one specific gene promoter to another. It may lie adjacent to it as in the case of the lactose region. However, it actually lies within the galactose promoter but is found very far upstream from the arabinose promoter. Those promoter regions in *E. coli* that require CAP and cAMP for transcription at a maximum rate are associated with DNA sequences that code for enzymes needed in the use of various sugars such as galactose, arabinose, and lactose (the lactose region in *E. coli* is discussed in some detail in Chap. 17). In the absence of the CAP–cAMP complex, the RNA polymerase becomes bound to the DNA in the region of the Pribnow box, but apparently the DNA in the area does not tend to open up. The action of the CAP–cAMP complex is not completely known, but one possibility is that the complex may destabilize the DNA in the Pribnow box region, causing it to denature. In those promoters requiring CAP–cAMP, the polymerase by itself is unable to do this and cannot proceed with transcription. With low levels of CAP–cAMP, the efficiency of such promoters may fall greatly, so that only a small percentage of the potential amount of transcription occurs.

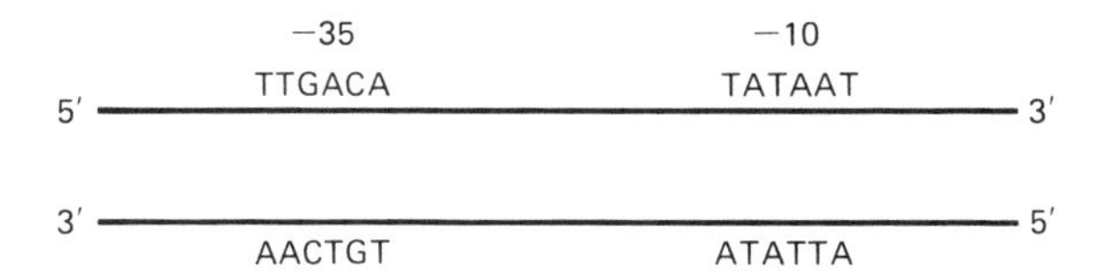

FIGURE 12-21. Conserved sequences in the promoter. Two promoter sequences in which little variation exists between different promoters can be recognized. One of these is found at about –35 and is apparently the region to which the RNA polymerase first attaches. Downstream, at approximately –10, is another conserved sequence, the Pribnow box. This is the region where the enzyme attaches firmly and the double helix opens up.

Termination of transcription

The transcript proceeds to elongate as the polymerase continues downstream, adding ribonucleotides according to the information provided by the sense strand. Obviously transcription must be brought to a halt at some point to prevent reading into the next gene. Signals to halt transcription are built into the DNA in the form of sequences called *terminators*. These have been well studied in bacteria and their viruses, but not much detail is known about them in eukaryotes. Studies of terminators reveal a few features they share. Very significant is the presence of inverted repeats. As Figure 12-25A shows, reading of an inverted repeat in the same direction (5′ to 3′, for example) on either of the two DNA strands yields the same sequence of

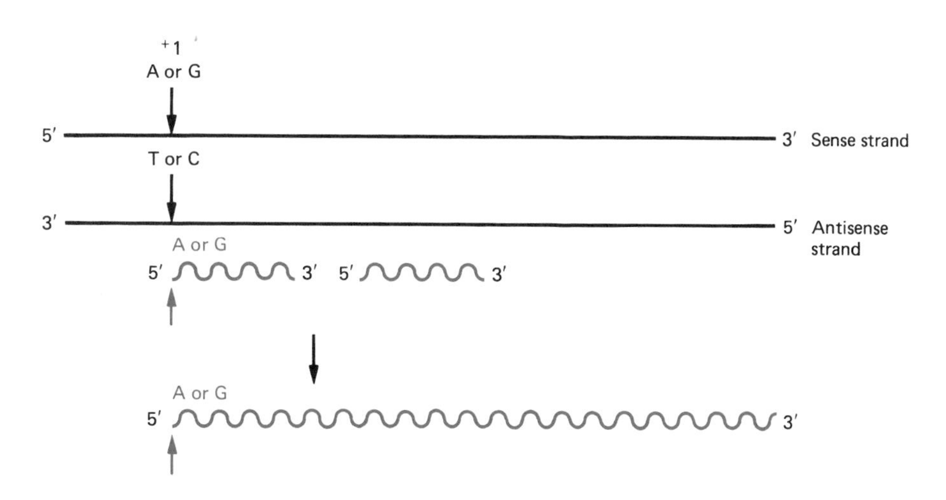

FIGURE 12-22. Relationship of transcript to gene. The first nucleotide of the transcript is typically a purine. Consequently, an A or a G will be found at the 5′ end of the transcript, which is assembled in the 5′ to 3′ direction. The beginning of the sequence that undergoes transcription on the antisense strand is toward the 3′ end of the strand. Consequently, the leading RNA (red) nucleotide will have a free 5′ end and a free 3′ end to which the next RNA nucleotide can join. The final nucleotide in the transcript will have a 3′ end.

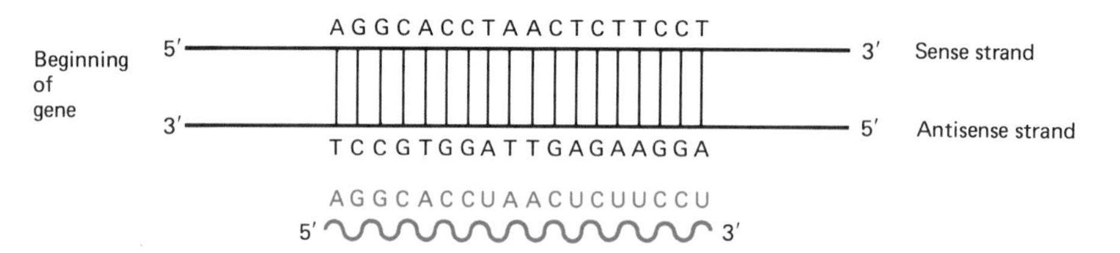

FIGURE 12-23. Orientation of gene and transcript. When reference is made to a gene, the 5′ end is typically considered its beginning, and the sequence of the sense strand is presented when the nucleotides are named going from upstream (left) to downstream (right). This convention is followed, because the 5′ to 3′ orientation of the mRNA (red) is preserved and can be related to a 5′ to 3′ orientation in the DNA. A comparison of the transcript with the sense strand shows that the two correspond exactly in nucleotide sequence, substituting Us in the transcript for Ts in the DNA strand.

nucleotides. You will note that on either of the two strands the inverted repeat provides a region where base pairing can take place between segments of the same strand if the two strands are separated (Fig. 12-25B). The importance of this in termination will be evident in a moment. An inverted repeat of a terminator is typically associated with a region rich in GC.

Another characteristic of a terminator is the presence of an AT-rich region following the inverted repeat and directly preceding the point at which transcription comes to a halt. Transcripts consequently end with a sequence of six to eight uracils as a result of transcription of the series of As on the antisense strand (Fig. 12-26A). The exact manner in which termination takes place is still uncertain, but research at the molecular level is consistent with the following idea. The evidence indicates that both the inverted repeat and the stretch of ATs provide signals to halt transcription. The RNA polymerase has been shown to pause as it approaches the region of the inverted repeat. This slowdown may start a series of reactions that bring about changes in the enzyme, leading to termination. Note from Figure 12-26B that when the polymerase proceeds past both the inverted repeat and the AT-rich region, the end of the transcript will contain a segment in which base pairing can take place. As a result, a hairpinlike configuration results. The formation of this hairpin is believed to favor the disengagement of the transcript from the template strand. Once the transcript is dislodged, reassociation of the sense and antisense strands in the region is favored. More details remain to be worked out, however. The string of ATs with the formation of the string of Us appears to be quite important in the dissociation of the transcript from the template. Evidence for this is the fact that deletions that cause a reduction in the number of Us in the transcript interfere with the termination of transcription.

The story of termination includes even additional factors at the molecular level, since certain terminators are known that require a special protein factor, called *rho,* for termination to occur. This group of terminators is known as rho-dependent terminators, as opposed to the simple or rho-independent type of terminator just discussed. Like the rho-independent terminators, those terminators that require rho protein also contain an inverted repeat that results in the formation of a hairpin in the transcript. However, there is less GC in the region of the inverted repeat and no segment that is rich in AT. Consequently, there is no string of Us at the end of the transcript.

Much remains to be learned about the action of rho, which apparently binds weakly to DNA. However, it can bind tightly to RNA, and according to one theory the rho protein attaches to the 5′ end of an elongating transcript and moves along, using its ability to degrade ATP and obtain energy. As the polymerase pauses in the terminator region, the rho factor interacts with the RNA polymerase, with the result that the transcript, enzyme, and rho dissociate from the template strand. Different terminators vary in their responses to the presence of rho protein; some are able to effect termination at low rho concentrations, and others require very high concentrations. Although two classes of terminators are recognized—simple and rho-dependent—much remains to be learned about the interplay of rho on the molecular level with RNA polymerase, template, and transcript.

Initiation of transcription in eukaryotes

In eukaryotes, three distinct types of RNA polymerase are found in the nucleus of the cell. One of these, RNA polymerase I, is confined to the nucleolus and is concerned exclusively with the synthesis of the major

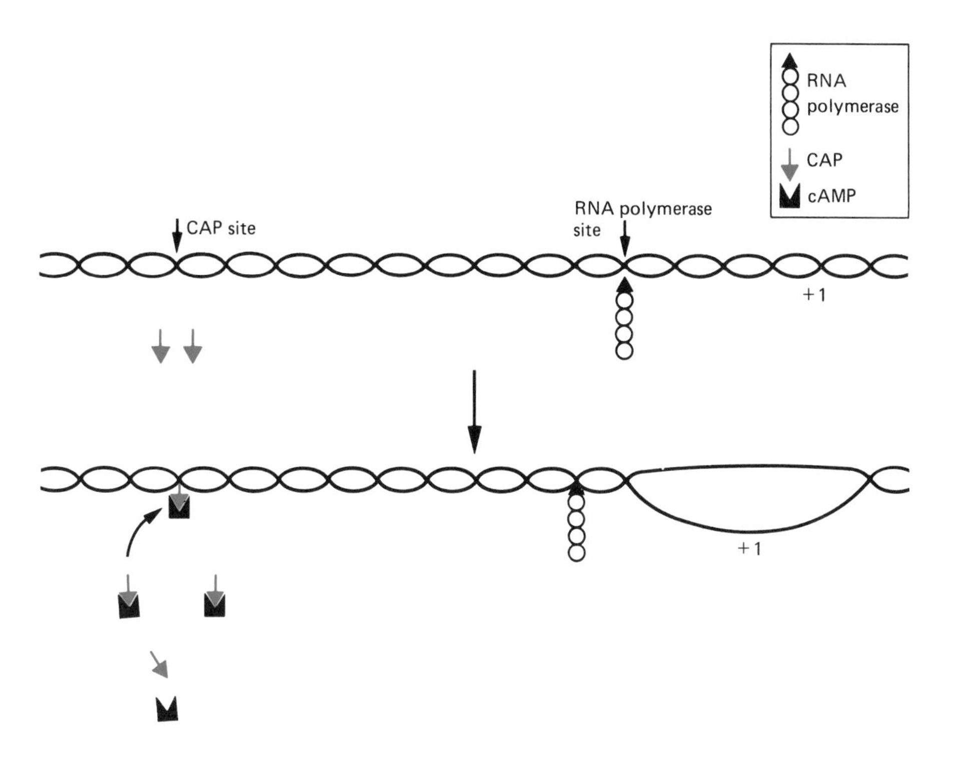

FIGURE 12-24. Promoters and effector molecules. Certain promoters require specific effectors for efficient transcription to occur. Promoters of several genes in *E. coli* interact with a CAP-binding site found upstream from the RNA polymerase binding site of the promoter. The location of the CAP site varies from one promoter to the next in its distance from the polymerase site. In the absence of binding at the CAP site (above), the double helix downstream from the polymerase site tends to remain stably paired. The presence of cAMP (below) activates the CAP. A CAP–cAMP complex can now bind to the DNA. This apparently facilitates opening up of the region downstream from the polymerase site so that the enzyme can proceed to initiate transcription.

classes of RNA of the ribosome. Most of the RNA synthesis going on in a cell is due to the activity of this enzyme (see Chap. 13). The other two species of the enzyme occur in the nuclear sap. RNA polymerase II is responsible for the formation of mRNA transcripts from all the various kinds of structural genes coded for polypeptides and accounts for most of the remaining RNA formation in the cell. RNA polymerase III is reserved for the synthesis of tRNA as well as certain other small RNA molecules, such as the smallest RNA molecule of the ribosome.

Two additional types of RNA polymerase are found in eukaryotic cells, one in the mitochondria of all cells and the other in the chloroplasts of cells capable of photosynthesis. These polymerases are smaller than their nuclear counterparts and show less resemblance to them to prokaryotic polymerases. (We will defer discussion of the organelle polymerases until Chap. 21.)

The nuclear RNA polymerases are very large proteins, each composed of 2 large subunits and about 10 smaller ones. The precise function of each subunit has still not been determined. You will recall that in prokaryotes the RNA polymerase, by virtue of its σ factor, has the ability to bind to the DNA of a promoter and to initiate formation of the transcript. In contrast, the three eukaryotic polymerases are unable by themselves to recognize and bind to specific base sequences in promoters. The initiation of transcription depends entirely on certain proteins that are not part of the enzymes themselves. These molecules, known as *transcription factors,* are the elements that recognize specific sequences in promoters and that act to initiate transcription. Transcription factors are very varied, and we can understand why after considering the following points. Certain genes contain information for functions that are common to all cells and that are essential for basic metabolic activities. These so-called *housekeeping genes* must be expressed most of the time in all cells. Promoters of housekeeping genes must contain sequences that are highly conserved from one such gene to the next. These are the sequences that must interact with general transcription factors. On the other hand, those genes that are called to activity only in certain cell types would depend for their expression on transcription factors that are tissue specific.

Almost all promoters that interact with RNA polymerase II contain a conserved sequence known as the *Hogness box,* or, more commonly, the *TATA box.* This sequence, found about 15–30 bases upstream from +1, the site of the start of transcription, has the characteristic consensus sequence TATAAAA and occurs in eukaryotic promoters from organisms as varied as mammals, insects, plants, and yeast. Investigations have revealed that four different transcription factors bind in the vicinity of the TATA box. One of these, transcription factor IID (TFIID), which is known as the *TATA protein,* actually recognizes the TATA sequence and must bind to the DNA before any other molecules can do so. Once TFIID attaches, it is then followed in a specific order by transcription factors A, then B, the polymerase, and finally TFIIE. TFIID must be a part of the general mechanism to stimulate transcription in eukaryotic cells, since the TATA sequence is found in most RNA polymerase II promoters.

While the TATA sequence apparently determines the exact site for the start of transcription, it does not work alone. It is influenced by another nucleotide sequence known as the *CAAT box*, which has the characteristic consensus GGCCAATCT and which is located farther upstream at a location of about −80 (Fig. 12-27). Its precise distance from +1 can vary greatly from one gene to the next. Mutations in the CAAT region greatly reduce the level of transcription from the promoter, even more so than do mutations in the TATA box. While mutations in the latter can reduce transcriptional activity, some of them still permit transcription, but the initiation site is altered.

In addition to the TATA and CAAT sequences, still others have been recognized that influence eukaryotic transcription. These include two GC boxes, each of which contains the sequence GGGCGG. One such box is usually found upstream and one downstream from

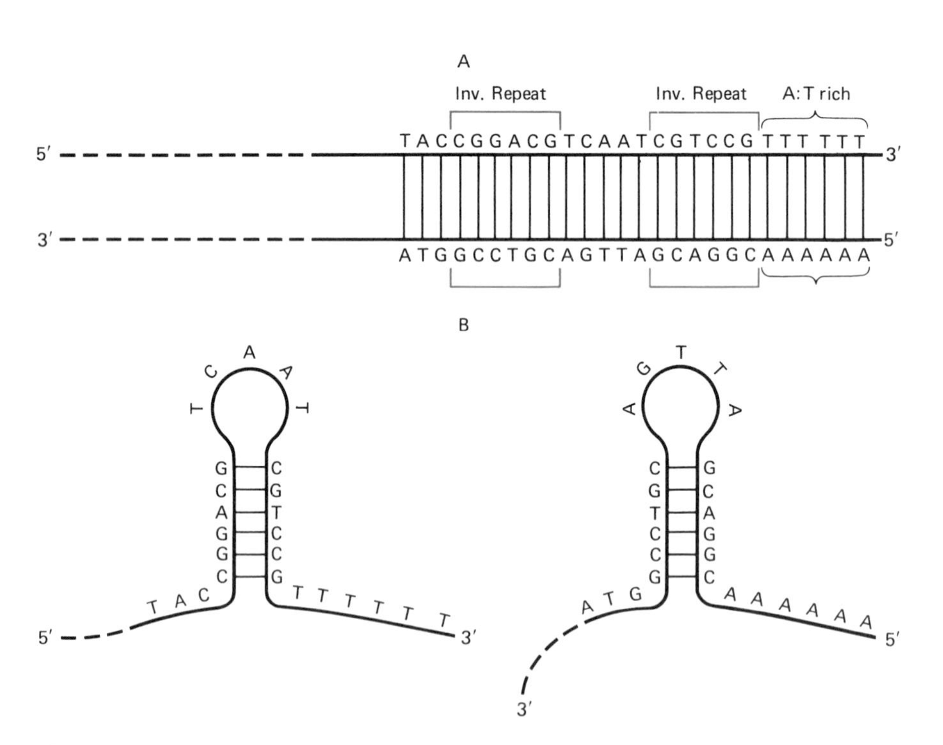

FIGURE 12-25. The terminator region. **A:** A terminator typically contains an inverted repeat (Inv. Repeat). Note that when opposite strands are read in the same direction (such as 5′ to 3′), the nucleotide sequence is identical: CGGACG. Reading will also be identical when the inverted repeat is read in the direction 3′ to 5′. The inverted repeat is rich in GC pairs. A region rich in AT pairs is often found at the end of a terminator. **B:** If the two DNA strands are separated, each single strand, as a consequence of the inverted repeat, has a region that makes possible intrastrand base pairing. As a result of this, a loop or hairpin configuration forms.

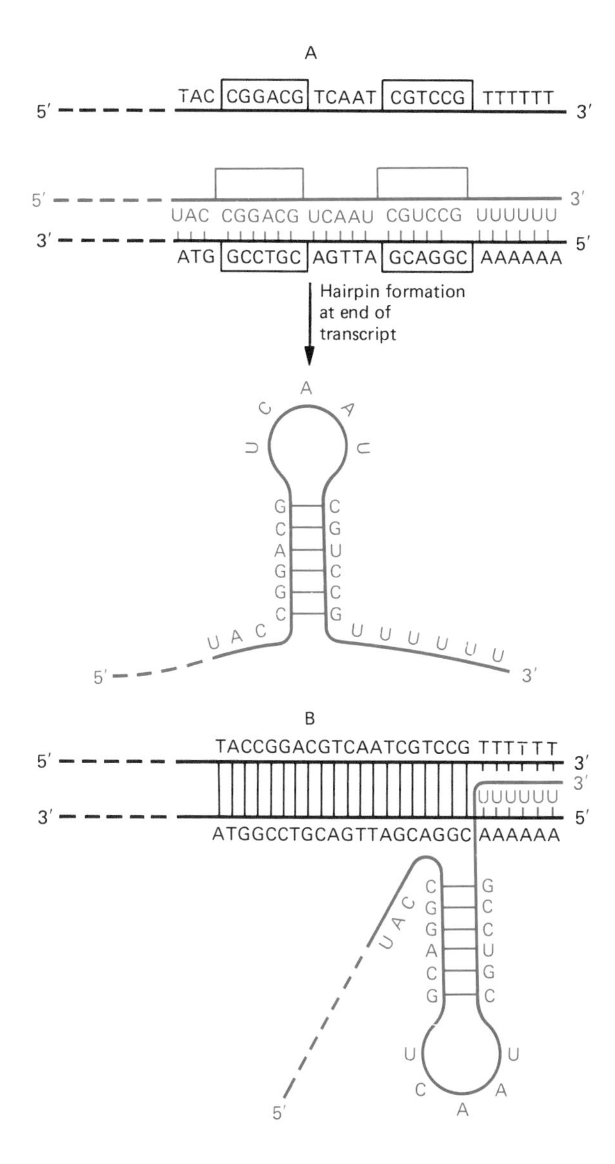

FIGURE 12-26. Events at the terminator. **A:** When transcription proceeds into the AT-rich region, the final transcript will contain at its 3′ end a copy of the inverted repeat (brackets) and a terminal string of Us because of the string of As in the template strand. As a result of the inverted repeat, a hairpin can form at the 3′ end of the transcript because of intrastrand pairing. **B:** According to one model, the formation of the hairpin at the 3′ end of the transcript favors the disengagement of the transcript from the antisense strand, allowing the two DNA strands to reassociate and reform a double helix in the region.

the CAAT box in most polymerase II promoters (Fig. 12-27). Transcription factors are also known that operate at both the CAAT and the GC sites. While the TATA and GC boxes interact with just a few defined transcription factors, certain sequences such as the CAAT box are recognized by a diversity of protein molecules, some of which apparently act as transcription factors that facilitate transcription. Other proteins play the role of repressors by combining with the sequence, thus preventing it from interacting with transcription factors. Various kinds of transcription factors have now been characterized, and we will encounter some in later discussions (see Chap. 17). These include receptors for steroid hormones and the proteins encoded by homeotic genes that play fundamental roles in the organization of body plan.

Sequences that are very far upstream can also exert an influence on the start of transcription. The evidence indicates that along with the GC and CAAT boxes these control the initial binding of RNA polymerase and that the TATA box then determines +1, the precise starting point. There are a few unusual promoters that interact with RNA polymerase II that lack one or more of the boxes or that have different combinations of them. Those lacking a TATA box do not have fixed sites for the start of transcription and depend on certain transcription factors acting along with RNA polymerase.

An unexpected finding in eukaryotes is that some promoters are greatly influenced by DNA sequences named *enhancers*. An enhancer can increase the rate of transcription from some promoters as much as 200 times. The element is unusual in that it can exert its effect on an initiation site even though it may be located as far away as 10 kb upstream or downstream but may even be found within the coding region of the gene itself! We still do not know the percentage of promoters in a cell that rely on enhancers for full transcription. It is possible that some enhancers may be active in all cells, whereas others are responsible for stimulating transcription from the promoters of just specific genes. An example of the latter case is an enhancer that has been found to stimulate the promoter of the insulin gene, but only in insulin-producing cells. Enhancers, like promoters, interact with regulatory molecules and have specific sites to which transcription factors and other proteins may bind.

Most of the discussion in this section has related to promoters that respond to RNA polymerase II. It must be noted at this point that enhancers are also known that stimulate RNA polymerase I promoters and that both polymerases I and III promoters also depend on transcription factors. Certain unusual aspects of these other two kinds of promoters are brought to light in Chapter 13. Enhancers, transcription factors, and regulatory proteins are the focus of much ongoing research, since they are so intimately tied to the prob-

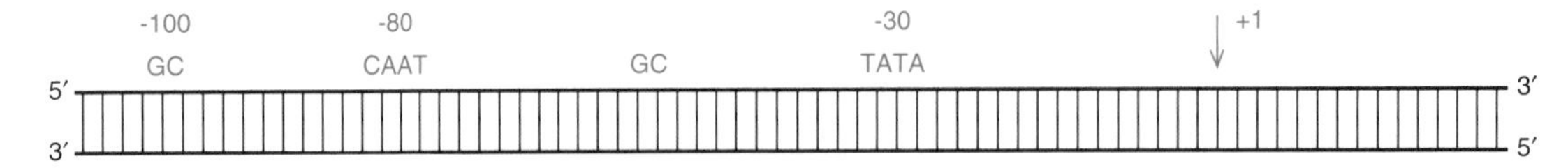

FIGURE 12-27. The eukaryotic promoter. Eukaryotic promoters contain consensus sequences required for the proper initiation of transcription. The TATA sequence appears to control the exact site at which transcription starts, +1. Proceeding upstream, a GC box occurs, followed farther upstream by a CAAT box and then another GC box. These sequences appear to control the initial binding of the RNA polymerase and to interact with other sequences still farther upstream.

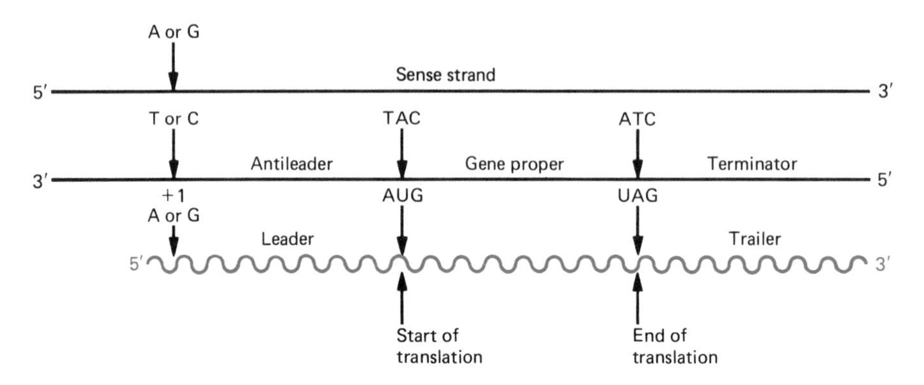

FIGURE 12-28. General relationship between a DNA template and its transcript. Only some of the nucleotide sequences of the template strand are reflected in the transcript. Moreover, only the part of the transcript that bears the information encoded in the gene proper is translated (see text for details).

lem of gene regulation and cell differentiation. More details on all of these are given in Chapter 17.

The gene and its mRNA transcript

The preceding sections tell us that the transfer of coded information from the gene to mRNA entails the interaction of many factors. Moreover, not all of the nucleotide sequences in the gene are reflected in the transcript. First of all, only one strand of a gene, the template (or antisense strand), contains portions that undergo transcription. A promoter region is present that contains signals for the binding of certain factors, such as the σ-chain of RNA polymerase. The actual mRNA transcript, coded for a specific polypeptide product, begins past the promoter region. Note (Fig. 12-28) that the mRNA contains a segment at its 5′ end (the leading end) called the *leader.* The length of this leader varies from approximately 20 to hundreds of nucleotides.

As mentioned earlier, a purine nucleotide occurs at the 5′ end of the mRNA strand. The complementary DNA segment from which the leader is transcribed is referred to as the *antileader.* It must be emphasized that the leader sequence of the mRNA does not contain coded information that designates the amino acid sequence in a polypeptide. The leader is not translated by the ribosome. Its role involves the binding of mRNA to the ribosome, so that translation of the nucleotide sequences that *do* designate amino acids can take place properly. The codon AUG usually designates the point where translation is to begin (see Chap. 13). The sequence of the transcript that codes for the amino acid sequence of a polypeptide is found in that part of the mRNA between the leader and the *trailer.* This trailer sequence, like the leader, is not translated. A chain-terminating codon (UAG, UGA, or UAA) signals the end of translation. The trailer is thus any sequence at the 3′ end of the mRNA that follows a chain-terminating codon. The terminator region of the template strand contains sequences complementary to the trailer. Trailer length can vary from one gene transcript to another, because transcription does not appear to terminate at any one specific site. The messenger transcript of eukaryotes undergoes modification or processing after the first, or primary, transcript is formed as we will learn in Chapter 13.

That part of the mRNA containing coded information for the amino acid sequence of a polypeptide lies between the two stretches that are *not* translated, the leader at the 5′ end and the trailer at the 3′ end. If we consider a gene to be a sequence of nucleotides in a DNA molecule that contains coded information for some functional product such as a polypeptide, then it is just that portion of the template (antisense) strand with this coded information that is the *gene proper.*

Still other interactions involving mRNA, tRNA, and the ribosome in translation are required to achieve the final transfer of information from the DNA that is needed to complete the formation of a specific polypeptide. These form the basis of discussion in the following chapter.

REFERENCES

Adhya, S., and M. Gottesman. Control of transcription and termination. *Annu. Rev. Biochem.* 47:967, 1978.

Avers, C.J. *Molecular Cell Biology.* Benjamin-Cummings, Menlo Park, CA, 1986.

Brenner, S., A.O.W. Stretton, and S. Kaplan. Genetic code: The "nonsense" triplets for chain termination and their suppression. *Nature* 206:994, 1965.

Burgess, R.R. Separation and characterization of the subunits of ribonucleic acid polymerase. *J. Biol. Chem.* 244:6168, 1969.

Cold Spring Harbor Laboratory. *The Genetic Code,* Vol. 31. Cold Spring Harbor Laboratory, Cold Spring Harbor, NY, 1966.

Cold Spring Harbor Laboratory: *Transcription of Genetic Material,* Vol. 35. Cold Spring Harbor Laboratory, Cold Spring Harbor, NY, 1970.

Crick, F.H.C. The genetic code. III. *Sci. Am.* (Oct.):55, 1966.

Dickerson, R.E. The DNA helix and how it is read. *Sci. Am.* (Dec.):94, 1983.

Doolittle, R.F. Proteins. *Sci. Am.* (Oct.):88, 1985.

Farnham, P.J., and T. Platt. Rho-independent terminator: Dyad symmetry in DNA causes RNA polymerase to pause during transcription in vitro. *Nucleic Acids Res.* 9:563, 1981.

Freifelder, D. *Molecular Biology,* 2d ed. Jones and Bartlett, Boston, 1987.

Goldstein, L., and D.M. Prescott (eds.). *Gene Expression: The Production of RNA's. Cell Biology, A Comprehensive Treatise,* vol. 3. Academic Press, New York, 1980.

Lake, J.A. Amino-acyl-tRNA binding at the recognition site is the first step of the elongation cycle of protein synthesis. *Proc. Natl. Acad. Sci.* USA 74:1903, 1977.

Marx, J.L. Gene control research gets a boost. *Science* 245:1329, 1989.

Nirenberg, M.W., and J.H. Matthaei. The dependence of cell-free protein synthesis in *E. coli* upon naturally occurring or synthetic polyribonucleotides. *Proc. Natl. Acad. Sci.* USA 47:1588, 1961.

Pastan, I. Cyclic AMP. *Sci. Am.* (Aug.):97, 1972.

Radman, M., and R. Wagner. The high fidelity of DNA duplication. *Sci. Am.* (Aug.):40, 1988.

Roberts, J.W. Termination factor for RNA synthesis. *Nature* 224:1168, 1969.

Rosenberg, M., and D. Court. Regulatory sequences involved in the promotion and termination of RNA transcription. *Annu. Rev. Genet.* 13:319, 1979.

Roth, J.R. Frameshift mutations. *Annu. Rev. Genet.* 8:317, 1974.

Sarabhai, A.S., A.O.W. Stretton, and S. Brenner. Colinearity of the gene with the polypeptide chain. *Nature* 201:13, 1964.

Yanofsky, C., R. Drapeau, and J.R. Guest, et al. The complete amino acid sequence of the tryptophan synthetase A protein (α subunit) and its colinear relationship with the genetic map of the A gene. *Proc. Natl. Acad. Sci.* USA 57:296, 1967.

Zubay, G., D. Schwartz, and J. Beckwith. Mechanism of activation of catabolite-sensitive genes: A positive control system. *Proc. Natl. Acad. Sci.* USA 66:104, 1970.

REVIEW QUESTIONS

1. Give three features that the various kinds of amino acids have in common.

2. In addition to the 20 kinds of amino acids, the appropriate buffers, and inorganic substances, give five major essentials that must be supplied in a cell-free in vitro system for polypeptide synthesis to occur.

3. Assume a segment of a DNA strand has the following nucleotide sequence:

3′ TACAAATCTCATTGTATAGGA 5′

A. Give the base sequence and polarity in the complementary DNA strand.

B. Give the base sequence and polarity of an mRNA strand formed by transcription from the first-mentioned 3′ strand.

C. How many amino acids would this segment represent?

4. A. Assume that exposure to a mutagen brings about a base substitution so that the first nucleotide in the third codon from the 3′ end of the DNA strand is changed to A. What would the ultimate effect be? (Consult the coding dictionary, Table 12-1.)

B. What do we call this class of mutations?

5. A. Assume exposure to a mutagen brings about a base substitution, so that the third nucleotide of the second codon from the 3′ end is changed to G. What would the ultimate effect be?

B. Name this class of mutations.

C. Assume the second nucleotide of the second codon from the 3′ end was changed to C. What would the effect be?

D. Name this class of mutations.

6. **A.** Suppose exposure to an acridine causes elimination of the third nucleotide in codon 1 and elimination of the first nucleotide in codon 3. What will this do to the reading of the gene, and what would be a likely outcome?

   B. Suppose the second nucleotide in codon 4 were removed in addition to the two already mentioned. Now what would the effect be?

   C. Suppose the first elimination, removal of the third nucleotide in codon 1, is followed by the addition of a nucleotide containing G between the first and the second nucleotides of codon 3. What would happen to the reading of the gene, and what would be a likely outcome?

   D. What do we call the class of mutations described here?

7. Of the two amino acids tryptophan and arginine, which is more likely to be substituted in a polypeptide chain by another amino acid as a result of a mutation? Explain why after consulting a coding dictionary.

8. Suppose the base ratio is determined for each of the two strands of a certain structural gene. The transcript of the gene is isolated and the base ratio is then determined for it. Using the following results, answer these questions:

   A. Which is the H strand?

   B. Which is the template (or antisense) strand, and why?

	A	T	G	C	U
DNA strand A	31.1	26.3	29.1	13.5	—
DNA strand B	26.1	30.9	13.7	29.3	—
Transcript	26.2	—	13.3	29.3	31.2

9. Synthetic trinucleotides are added to cell-free test-tube systems that contain all the ingredients for polypeptide synthesis except mRNA. In each tube, 1 of the 20 kinds of amino acids present is labeled, and the labeled amino acid type varies from tube to tube. Ribosomes and any complexes formed with them can be trapped on filters. Features of five such systems are given here. Answer the questions that follow.

Tube	Trinucleotide added	Labeled amino acid present
1	AAU	Leucine
2	AAC	Proline
3	AGU	Serine
4	AGC	Leucine
5	UGA	Serine

   A. After consulting a list of codon assignments, determine what amino acid, if any, will be bound in a complex in each tube.

   B. Determine whether the filter will be labeled in each case as a result of the trapping of labeled complexes.

   C. If we had not known the codon assignments at the outset, what would the results have indicated?

10. Slow heating of a DNA solution brings about separation of the DNA strands. The DNA is said to undergo melting or denaturation. The melting temperature is denoted by T_m, the temperature at which half of the double-stranded molecules are denatured. Of the three samples of duplex DNA represented here, which would you expect to have the highest melting temperature (T_m) and which the lowest? Explain.

Sample 1 AGAAGATTATCACAT
TCTTCTAATAGTGTA

Sample 2 AGCCATCGACAGGTG
TCGGTAGCTGTCCAC

Sample 3 AGTTGTCATTAGGAC
TCAACAGTAATCCTG

11. Of the following two double-stranded DNA molecules, which would require a higher temperature for complete strand separation, and why?

Sample 1 AGAAGTCACAAACTTTGATTC
TCTTCAGTGTTTGAAACTAAG

Sample 2 AATTAGCCGGCAAACATTATA
TTAATCGGCCGTTTGTAATAT

12. Following is a portion of a sense strand with an upstream position noted. Give the transcript that would be associated with this portion of the DNA along with its polarity.

–6
5′ TATAACGAACATAAATTT 3′

13. Below is a portion of an antisense strand with a downstream position noted. Give the transcript expected along with its polarity.

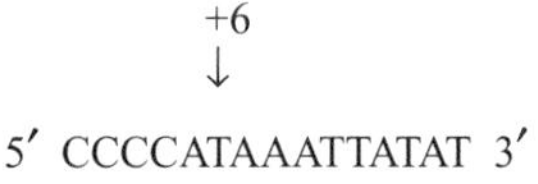

5′ CCCCATAAATTATAT 3′

14. A transcript is represented here. Give the sense and antisense strands that are related to it, along with their polarities.

5′ AGACCCUAUGCAGAG 3′

15. Following is a section of duplex DNA containing inverted repeats. Identify the repeats and show the hairpins that can form when both strands are separated. Also give the polarities of the strands.

5′ ATCAAATTATGAGCCACAACTCATAACTAGGG 3′
3′ TAGTTTAATACTCGGTGTTGAGTATTGATCCC 5′

16. In Question 15, assume that the lower strand is the antisense strand. Show the hairpin that can form from the transcript along with its polarity.

17. Assume that the following sequence represents an antisense strand for five amino acids.

3′ TACGAGCCCTCAAAG 5′

A. What must be the polarity of the strand? Explain.
B. What amino acids would the strand designate reading the transcript in the direction 5′ to 3′?

18. Following are two DNA strands composing a region of a gene that codes for nine amino acids. Determine which of the two is the antisense strand and which the sense strand. Deduce the polarity for each strand. Explain.

CCC ATTATGATCAC TGTGTAT T TAGAT
GGGTAATACTAGTGACACATAAATCTA

19. Following is a portion of a sense strand. The entire complementary sequence in the antisense strand undergoes transcription. Keeping in mind that the sense strand shown contains a segment that corresponds to a portion of the leader in the transcript, give the amino acid sequence for which this portion is coded.

5′ AATTTTATGTTCTCTATCGGG 3′

20. Following is a portion of an antisense strand of DNA. Give the amino acid sequence for which this strand is coded, numbering the amino acids 1, 2, and so on. Keep in mind that the transcript is translated in the direction 5′ to 3′.

3′ TACTATTGCGGA 5′

21. In some viruses, the genetic material is DNA, whereas in others it is RNA. On the basis of the composition shown here for four viruses, determine whether each virus is a DNA or an RNA virus and whether the nucleic acid is single or double stranded.

Nucleic acid source	Base (and percentage of total nitrogen bases)				
	A	G	C	T	U
Virus 1	25	24	18	33	
Virus 2	28	22	22		28
Virus 3	31	19	19	31	
Virus 4	22	19	26		33

22. Assume that a given sequence of eight nucleotides was studied in 200 promoters. The following gives the typical base at a given position and in parentheses the number of promoters in which it occurs there. Write a consensus sequence that reflects the picture at this region. 1) Cytosine (176), 2) adenine (140), 3) adenine (84), 4) guanine (120), 5) any base likely, 6) thymine (130), 7) adenine (124), 8) thymine (96).

Molecular Interactions in Transcription and Translation

13

The complex nature of ribosomes

Ribosomes from different organisms are quite uniform in their physical and chemical characteristics. Their size and weight can be measured by the rate at which they settle out in a centrifugal field. The rate of sedimentation is expressed in S (Svedberg) units. On this basis, two size classes can be recognized, those ribosomes from prokaryotes such as *Escherichia coli* with an S value of 70 and those from the cytosol of eukaryotes with an S value of 80. Ribosomes are also found in the chloroplasts and mitochondria of eukaryotic cells, and these resemble the 70S ribosomes of prokaryotes. These organelle ribosomes are discussed more fully in Chapter 21.

A ribosome is composed of two subunits that can be separated by lowering the magnesium concentration in the environment (Fig. 13-1). The smaller of the two subunits sits like a cap on the larger. The subunits also have characteristic S values, 30S and 50S for the smaller and larger ones of prokaryotes and 40S and 60S for the corresponding subunits in eukaryotes. The subunits can reassociate to form a complete ribosome when the magnesium concentration is again raised.

Each ribosomal subunit is composed of protein complexed with RNA, known as the rRNA and having a characteristic S value. In prokaryotes, the 30S subunit contains a 16S rRNA molecule, whereas the 50S contains two RNA molecules, a large and a small one with S values, respectively, of 23 and 5. The 40S subunit of eukaryotes contains an RNA molecule with a value of 18S. The 60S subunit has one large RNA component whose sedimentation value differs with the species (Table 13-1). For multicellular animals, the S value of the large rRNA molecule is 28S. Two smaller RNA molecules are found in the 60S subunit of all eukaryotes, a 5S molecule and one with an S value of 5.8. The latter associates very closely in the larger subunit with the larger RNA molecule. The RNA in a subunit loops around itself and assumes a secondary structure that is maintained by base-pairing between different parts of the strand.

The proteins of the subunits are very complex. In eukaryotes, 33 different kinds occur in the smaller subunit and 49 in the larger. Several of these proteins are essential for binding tRNA molecules to the ribosome. Most of the mass of the organelle, however, consists of RNA. Some of the unpaired bases of this rRNA are involved in the binding of mRNA and tRNA so that translation can begin. Since rRNA does not contain coded information for a product, its primary function has continued to intrigue molecular biologists. For many years, its role was thought to be mainly a structural one in the organization of the ribosome, and catalytic activity was believed to be associated exclusively with ribosomal proteins. Now evidence has been presented that shows that rRNA probably possesses catalytic activity and plays a primary role in polypeptide formation during translation. Later in this chapter, we discuss various examples of the catalytic activity of RNA molecules that lead us to reexamine rRNA in the dynamics of translation.

The two subunits of the ribosome actually exist in the cell as separate entities and do not associate until they engage in translation (Fig. 13-2). Once a ribosome reads an mRNA strand, the two subunits dissociate and are free to recycle. This means that the "top" of what was an intact ribosome may associate with a different "bottom" in the next round of translation. Figure 13-2 shows that single ribosomes do not read a messenger strand. Instead, several groups of them associate with a single mRNA, an association termed a *polyribosome* or *polysome.* Each single ribosome acts independently, however. One ribosome does not cooperate with another to form a polypeptide. Therefore, at any moment as a message is being read, polypeptide molecules of the same kind in different stages of completion are associated with different ribosomes along the length of the mRNA strand. Once the ribosome reaches the end of the strand, the completed polypeptide is released. The ribosome dissociates into subunits that may participate again in translating this same or a different message.

Transfer RNAs and their synthetases

Since the genetic code is degenerate, we would expect to find 61 different types of tRNA molecules, one for each codon designating an amino acid. And indeed, it has been found that most amino acids in the cell are recognized by more than one kind of tRNA. However, only 20 different varieties of aminoacyl tRNA synthetase exist. Each kind interacts with a specific amino acid and attaches it to a tRNA. This means that one type of amino acid can be recognized by more than one kind of tRNA but only by one specific synthetase. A synthetase, therefore, must be able to recognize not only its specific amino acid but all the species of tRNA molecules to which that amino acid can be attached. The interactions involved in the recognition of a tRNA molecule by a given synthetase have been the center of ongoing research. We would expect to find differences between tRNA molecules, since they attach to different amino acids, but we would also expect those tRNAs that recognize the same amino acid to have some features in common. We would expect this even though they are responding to different code words as a result of degeneracy of the genetic code. Analyses of tRNA, however, have been unable to demonstrate any significant similarities among the tRNAs for a given amino acid.

Actually all transfer RNA molecules, those from a wide assortment of prokaryotes and eukaryotes, look very much alike. Although they are all single stranded, x-ray diffraction studies show that they possess a double-helical structure in certain portions of the molecule. This results from the fact that the single RNA

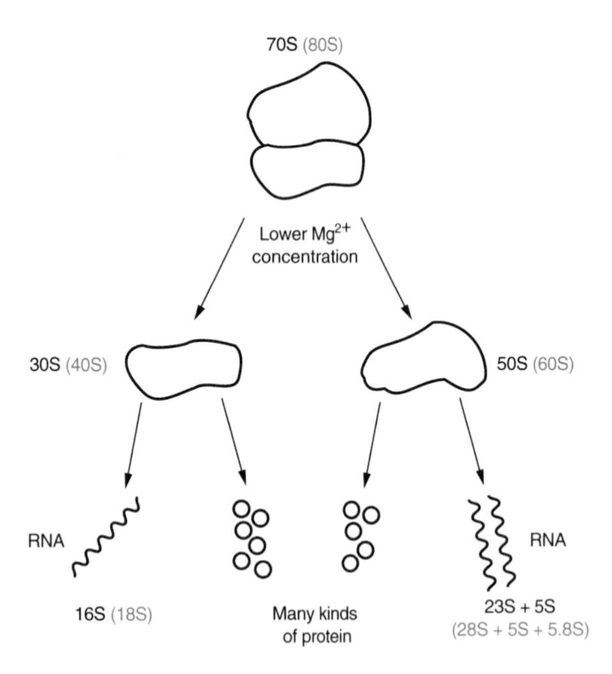

FIGURE 13-1. Ribosomal components of prokaryotes and multicellular animals. An intact ribosome consists of two subunits that may dissociate or associate depending on the Mg^{2+} concentration. Each subunit can be fractionated into its RNA and protein components. Sedimentation values are characteristic for ribosomes and their parts. Those in black are for prokaryotes, and those in red are typical of multicellular animals.

TABLE 13-1. Characteristic S values of cytoplasmic ribosomes and their components

Source	Intact ribosome	Ribosome subunits	rRNA in subunit
Prokaryotes	70S	30S	16S
		50S	23S, 5S
Eukaryotes	80S	40S	
		60S	
Protozoa		40S	18S
		60S	25–26S, 5S, 5.8S
Fungi		40S	18S
		60S	25–26S, 5S, 5.8S
Plants		40S	18S
		60S	25–26S, 5S, 5.8S
Animals		40S	18S
		60S	28S, 5S, 5.8S

strand loops to form folds, giving the molecule a secondary structure that has a stem and leaf configuration resembling a cloverleaf (Fig. 13-3A). The stems or arms are regions where pairing occurs between bases on opposite sides and maintains the shape of the molecule. The loops, on the other hand, lack pairing and are open. The number of nucleotides in a tRNA molecule has been found to vary from 75 to 95 due to variation in two of the arms.

While Figure 13-3A depicts a tRNA specific for the amino acid alanine, we can use it to bring out similarities and differences between all tRNAs studied thus far. All tRNAs possess four stems, known as the *major arms.* Note the 5′ and 3′ ends of the RNA strand. Starting at the latter, we see that it terminates in the sequence CCA, a feature of all tRNAs. It is this –OH end that is joined to an activated amino acid by a synthetase (review Fig. 12-3). Hence it is known as the *acceptor* stem or arm. When the acceptor stem is bonded to an activated amino acid, we say that the tRNA is *charged.* Proceeding away from the 3′ end, we next encounter the TΨC stem and loop, so named because of the presence in the loop of that characteristic triplet that contains the unusual base pseudouridine. The loop usually has seven unpaired bases and the sequence 5′–TΨCG–3′. Unusual bases are found in all the loops of all tRNAs. Each is derived by the modification of a typical base after it has been inserted into the RNA strand. Ribothymidine in the one loop is derived from ribouridine by methylation, and its presence in this loop presents an exception to the rule that thymine does not occur in RNA!

We next encounter the *variable* loop, aptly named because most of the variation in nucleotide number occurs at this position, which is often called the extra arm. In about 75% of all tRNAs, this loop is very small, containing only about three to five nucleotide residues. In the rest, it is very large, containing 21 nucleotides, making it the largest arm of all. The significance of this variability is unknown.

Next we find the anticodon stem and loop of the molecule. This loop plays a major part in creating differences among tRNA molecules. In its center (Fig. 13-3A), a sequence of three nucleotides occurs. This triplet, the *anticodon,* is critical in placing the amino acid that is bound to the 3′ end of the tRNA in the cor-

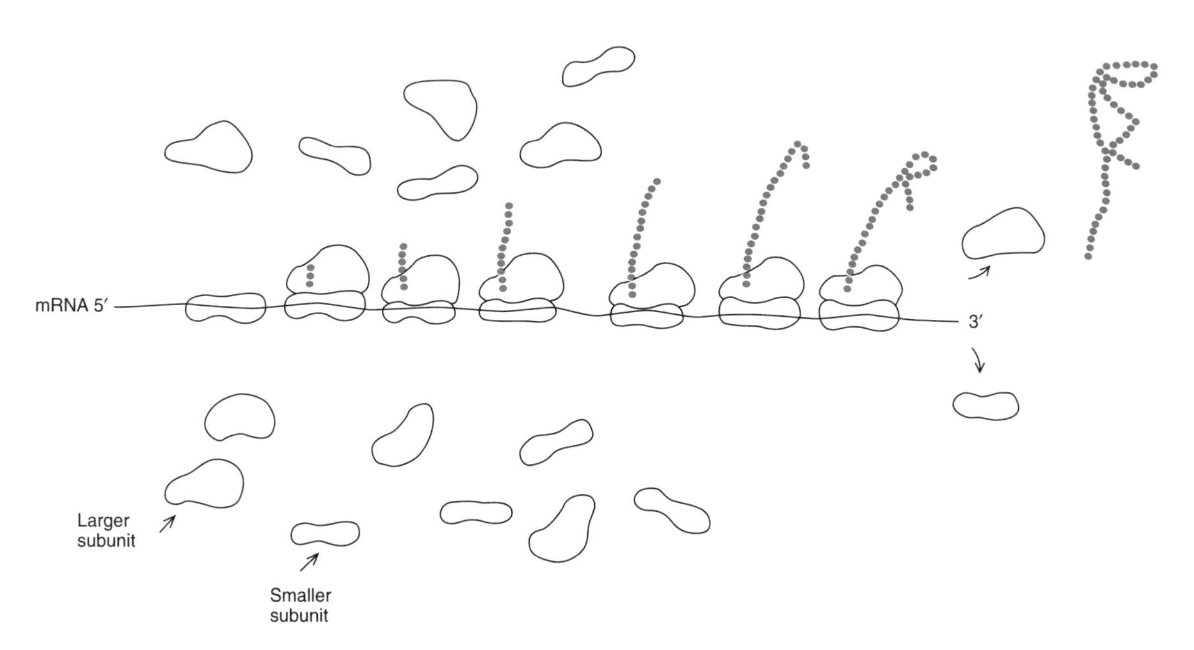

FIGURE 13-2. Polysome and ribosome. When not engaged in translation, the ribosomal subunits exist in the cell as a pool of separate entities. As translation begins, a small and a large subunit bind the mRNA toward its 5′ end. A single mRNA becomes associated with a group of ribosomes, and the entire association is termed a *polysome.* Each ribosome works independently, and, as it moves toward the 3′ end of the transcript, its polypeptide (red) grows in length. Near the end of the transcript, the completed polypeptide is released, and the two subunits dissociate to reenter the pool of subunits in the cell. Folding of the polypeptide upon itself into its characteristic three-dimensional form may begin while it is still being increased in length by a ribosome. (The diagram does not show tRNA and other details.)

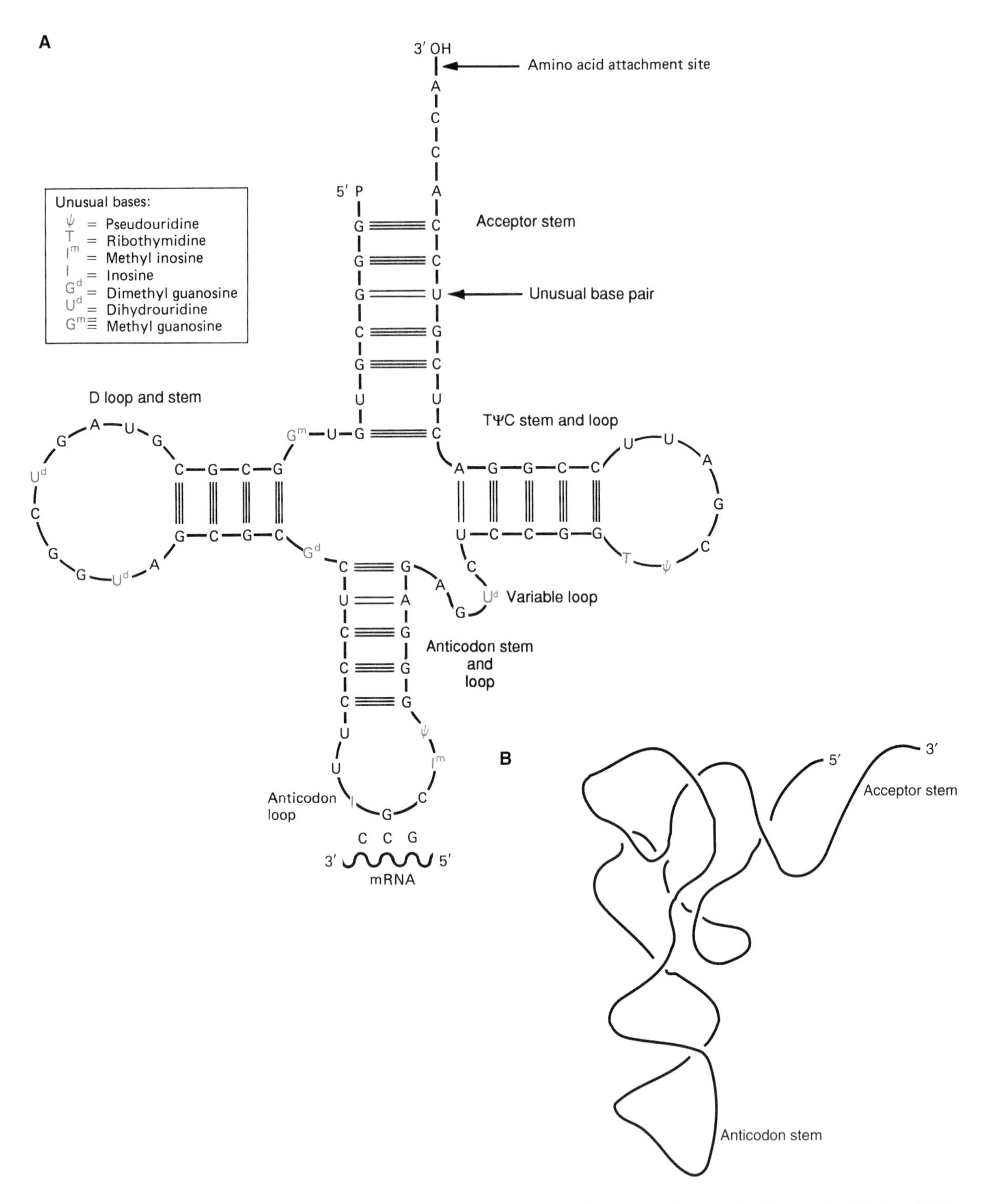

FIGURE 13-3. Structure of tRNA. **A:** The secondary structure of all tRNA molecules assumes the form of a cloverleaf as in the case of the yeast tRNA for alanine depicted here. The shape is maintained by base pairing, some of which (such as GU) is not usual. All tRNAs terminate at the 3′ end in the sequence–CCA. It is this end of the molecule to which a specific amino acid attaches. All tRNAs contain unusual bases (red) in addition to the usual ones. The anticodon loop contains a triplet that can base pair with a codon in the mRNA. This anticodon may contain the unusual base inosine (I). **B:** The tertiary structure of tRNA resembles an L with the short arm toward the top. It is this form of the molecule that interacts with the synthetases and other molecules. The anticodon and acceptor stems are at the tips of the L and are the most important regions in interactions with synthetases.

rect position in a growing polypeptide. It is so called because it is complementary to a codon in the mRNA and can form base pairs with it.

It is important here to note some points that relate to the polarity of codons and anticodons. Remember that a DNA strand and its transcript possess opposite polarities (review Fig. 12-5). The RNA codons given in Table 12-1 are written as they would occur in mRNA reading in the direction 5′ to 3′. The corresponding DNA codons are written in the reverse order, 3′ to 5′. Thus the RNA codon 5′–AUG–3′ for methionine corresponds to the DNA codon 3′–TAC–5′. The anticodon in the tRNA is complementary to the codon in the mRNA and possesses the opposite polarity and in this example would be 3′–UAC–5′, the same base sequence as in the DNA codon, substituting U for T. The ability of the anticodon to base pair with a codon in the mRNA enables the tRNA to position its amino acid properly, as we will appreciate more a bit later on. All anticodon loops contain unusual bases, and, as Figure 13-3A shows, one of these, inosine, can pair with C and may even occur in the anticodon itself.

Proceeding toward the 5′ end of the tRNA, we encounter the D loop and stem, named because of the presence of the unusual base dihydrouridine. This loop can vary in its content by as many as four nucleotides, the smallest containing only seven.

While base pairing maintains the cloverleaf structure of the molecule, we occasionally find that, in addition to the usual kinds of base pairs, an unusual one such as GU may also occur. Although the cloverleaf configuration occurs in all tRNAs, we must not think of it as the form that interacts with other molecules. Instead, it is the three-dimensional configuration that the molecule assumes that gives it an identity essential for interactions with other molecules such as the synthetases. The cloverleaf becomes folded into a shape that is roughly that of an "L" (Fig. 13-3B). This shape is also maintained by hydrogen bonds, but most of these involve unusual base pairs and also interactions of sites in the sugar–phosphate backbone.

As we noted at the outset, if different tRNAs can be recognized by one type of synthetase, we would expect these tRNAs to have certain nucleotide sequences in common. But this has not been demonstrated. This fact plus experiments that alter the structure of tRNA show that its recognition by the synthetase lies in its three-dimensional configuration. We now know that most of the nucleotide sequence of a tRNA molecule is not involved in its recognition by its enzyme. The interaction between the two molecules is complex and involves many points of contact along the inner side of the "L." Involved is the tip region, which contains the anticodon (Fig. 13-3B). This tip fits into a pocket in the synthetase. The anticodon is crucial to the recognition of the tRNA by its enzyme, which appears to react mainly with the first two anticodon bases rather than the third. Mutations in the anticodon stem can greatly affect recognition by the enzyme.

Very important as well is the other tip of the "L," the acceptor stem. Mutation in this stem can also prevent recognition by the synthetase. The stem tip, the site of formation of the aminoacyl bond between the amino acid and the tRNA, becomes inserted into the enzyme. It has also been demonstrated that there is a site here in the enzyme that binds to ATP, the supplier of energy for the bond formation (see Fig. 12-3). The acceptor and the anticodon stems seem to be the most important regions for recognition by a synthetase. Recognition appears, however, to involve just a few points of contact at each end of the "L." Many biochemical and genetic studies show that a relatively small number of nucleotides (called *identity elements*) form the basis of tRNA recognition by the enzyme. Detailed studies of the tRNA for glutamine show that its enzyme interacts with the anticodon and indicate that the middle base of the anticodon is of major importance in recognition. In the acceptor stem, it appears that the first three base pairs and the fourth nucleotide from the 3′ end are the most important.

The tRNA synthetases themselves actually compose a very diverse group of proteins. Some consist of a single polypeptide chain, others of two or four. There is not much similarity among them, and their molecular weights vary from 40,000 to 100,000. Each particular kind of tRNA may have its own specific way of interacting with its proper synthetase. There may be no one generalization regarding recognition sites that applies to all.

Initiating tRNA and its interactions

Before proceeding to the details of polypeptide formation, let us first become familiar with a few of the interactions that take place between tRNA and mRNA at the beginning or translation.

The messenger RNAs in a cell come in a wide variety of sizes. This is to be expected, since they are transcripts that reflect coded information from different genes. This information is for the construction of polypeptides whose lengths vary greatly. Consequently, longer polypeptides require longer transcripts than do shorter ones. Therefore mRNA is very diverse and cannot be characterized as precisely as is the case with rRNA and tRNA.

Since an mRNA transcript contains a leader sequence that does not designate amino acids, it must contain some kind of signal to indicate where polypeptide formation is to begin. This is usually achieved by the first AUG codon to follow the leader segment. The coding dictionary (Table 12-1) shows that this triplet designates methionine and is the only codon assigned to it. Chemical analyses of proteins tell us that methionine is the most frequent amino acid to be found at the start of a polypeptide chain (Fig. 13-4A). Moreover, it has been demonstrated in prokaryotes that the chain formed in vitro begins with *n*-formyl methionine, a form of methionine in which the –NH2 group is bonded to a formyl group and is therefore blocked, making it incapable of joining to the carboxyl group of any amino acid. However, it does have a carboxyl group of its own with which any amino acid can form a peptide linkage. In the living cell, this special form of methionine is placed in the first position to start a polypeptide, but it does not occur in any completed chain. This is so since it later becomes changed by enzyme action to an ordinary methionine with an unblocked amino end. It may also be cleaved from the chain so that the second amino acid residue in the polypeptide finally becomes the leading one (Fig. 13-4B).

In the prokaryotic cell, the *n*-formyl methionine is carried by a special kind of tRNA, designated $tRNA_f^{met}$. It is known as *initiating tRNA*. This special tRNA interacts with the initiating codon AUG to position the very first amino acid at the beginning of the polypeptide chain. Actually, it is ordinary methionine that first becomes attached to the initiating tRNA. However, this special transfer molecule makes it possible for a formyl group to be added to it by enzyme action, whereas ordinary methionine tRNA ($tRNA_m^{met}$) cannot do this, and so any methionine attached to it remains unchanged. Although the special tRNA recognizes the initiating codon AUG when it is toward the 5′ end of the mRNA following the leader segment, it cannot respond to AUG that is located elsewhere along the mRNA. Instead, the codon is recognized by $tRNA_m^{met}$ and ordinary methionine is inserted (Fig. 13-5). (The reason for this difference in response between the two tRNAs will be evident in the next section.) We see that a codon can mean different things, depending on its position in the mRNA. The first AUG triplet toward the 5′ end of the mRNA indicates that the polypeptide is to be started from that point, and it is recognized by the special initiating tRNA charged with *n*-formyl methionine. Located elsewhere, the triplet AUG is recognized by the usual $tRNA_m^{met}$ carrying methionine. Both types of tRNA possess the same anticodon, 3′–UAC–5′, which can base pair with the codon 5′–AUG–3′.

Although AUG is the usual initiating codon, the triplet GUG in prokaryotes may in some cases serve in this way. Ordinarily GUG is recognized as a valine codon, but when found at the end of the leader sequence on the mRNA it is recognized as a methionine codon by $tRNA_f^{met}$, the initiating tRNA!

In eukaryotes, the general picture is the same as just described but with some important distinctions. Two types of tRNA for methionine also occur in eukaryotes. One of these acts as initiating tRNA and has been designated $tRNA_i^{met}$. However, no formulation of methionine occurs in the cytosol of the eukaryotic cell. Ordinary methionine is always the first amino acid placed in the polypeptide chain. (In Chapter 21, we will learn that the mitochondria and chloroplasts possess an initiating tRNA as well as an enzyme that can convert methionine to *n*-formyl methionine.) As is also true in prokaryotes, methionine is the first amino acid positioned in the polypeptide chain, but it may be removed during processing of the polypeptide so that the "second amino acid" becomes the first in the chain. In eukaryotes, GUG never acts as the initiating codon, a role restricted to AUG. With these points in mind, let us now proceed to a few of the details of the process of translation.

Features of translation

Translation includes an initiation phase that must be completed before peptide linkages can be made. This phase requires the formation of an initiation complex that entails the binding of mRNA and the first tRNA to the smaller subunit of the ribosome (follow Fig. 13-6). The binding is toward the 5′ end of the mRNA, since the messenger is read in the direction 5′ to 3′, the same direction in which it was assembled on its template strand. The binding involves a sequence of six bases in the leader portion of the mRNA, a sequence that precedes the first triplet designating an amino acid. Every mRNA has such an initiation site that is essential for its proper position-

A

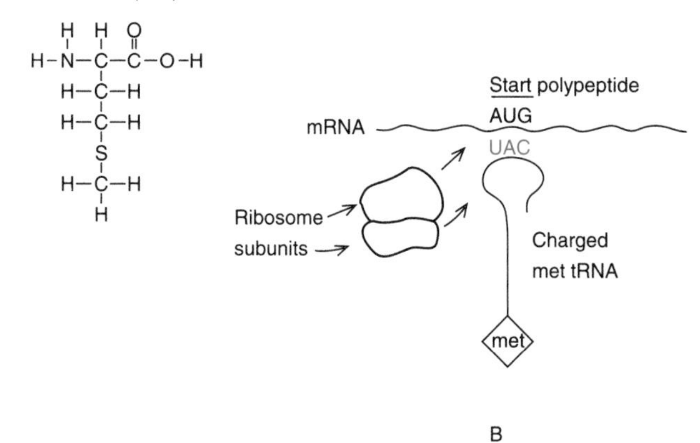

B

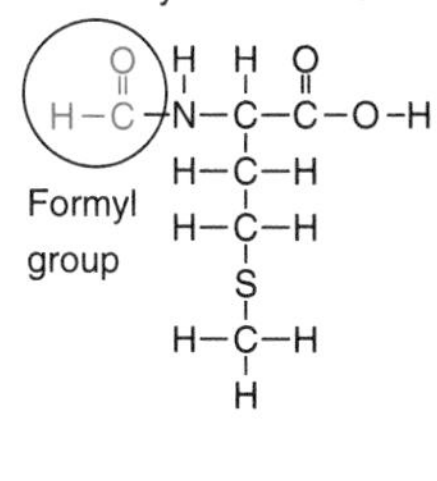

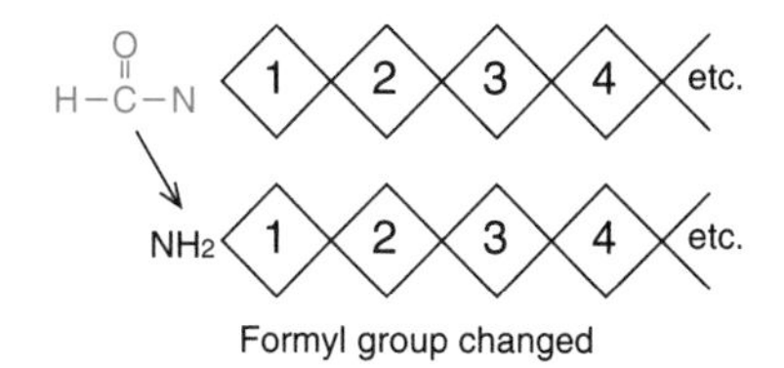

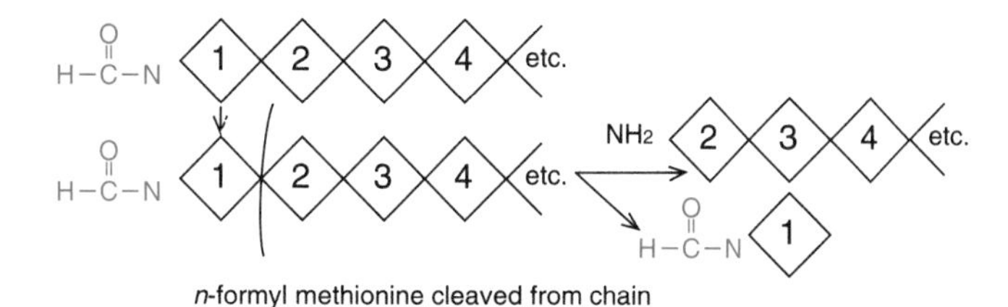

FIGURE 13-4. Start of a polypeptide. **A:** Methionine is the amino acid that occurs most frequently at the beginning of polypeptide chains. It is designated by only one codon, AUG. This codon acts as an initiating codon and signals the start of chain formation at that point in the mRNA. The specific tRNA charged with methionine is thus the first to complex with the messenger and the ribosome. The subunits of the ribosome associate before the formation of the first peptide linkage. **B:** In prokaryotes, *n*-formyl methionine, which carries a blocked amino group, is placed in the first position in the polypeptide chain. Its formyl group is later changed to an amino group. It may also be cleaved from the chain so that the second amino acid becomes the leading one.

ing on the ribosome surface. In *E. coli,* the 16S RNA of the 30S subunit has a six-base sequence complementary to the initiation site in the leader. This suggests that base pairing plays a part in the mRNA–ribosome binding.

By itself, however, the 30S subunit cannot bind to mRNA or tRNA. It requires initiation factors (IFs), three of which are found in *E. coli.* They only occur on 30S subunits from which they are released when the intact ribosome forms. Their main role is the establishment or the initiation complex. One of the IFs is specifically required for binding of the 30S subunit to the mRNA. Without it binding fails, even though the ribosome finds the initiation site on the mRNA. A different IF binds to the initiating tRNA charged with *n*-formyl methionine and takes it to the 30S subunit, enabling the tRNA to bind to it and to the mRNA. The anticodon of the initiating tRNA pairs with the initiating codon (usually AUG). A third IF binds to the initiation complex and appears to stabilize it.

In eukaryotes, the formation of the initiation complex is generally the same as that described for *E. coli.* However, more initiation factors are involved. Each one is known as an *eIF,* and at least nine participate in such tasks as unwinding the mRNA and binding it to the initiating tRNA. As noted in the previous section, formylation of methionine does not occur in the eukaryotic cytosol, and only AUG acts as the initiating codon.

After formation of the initiation complex, which includes the smaller ribosomal subunit, an mRNA, and the charged initiating tRNA, the larger subunit associates with the smaller. The 30S subunit can bind to the 50S only if all three IFs have been released and only if it is bound to the mRNA. An intact ribosome is formed that has certain special features. It possesses two cavities that can accommodate charged tRNA molecules. One is known as the *P* (*peptidyl*) site and the other as the *A* (*aminoacyl*) site. Both the 30S and 50S subunits form parts of each site. The codon that is found in position at either of the sites makes that site specific for a given charged tRNA. Note from Figure 13-7 that the A site accommodates all the charged tRNA molecules arriving after the first one. The second tRNA charged with its specific amino acid responds to the codon at the A site and is the first to move into that position. This is the case since the P site is already occupied by the initiating tRNA when the intact ribosome forms. Its placement there depended on the special IF that recognizes only the initiating tRNA. On the other hand, a charged initiating tRNA cannot enter the A site. The ability to do this requires a special protein factor, an elongation factor (EF). Any tRNA other than the initiating one can respond to this EF, which places it at the A site. The EF then leaves and recycles. Cleavage of the high-energy bond in GTP is required for the binding of the charged tRNA to the A site. This is similar in both prokaryotes and eukaryotes.

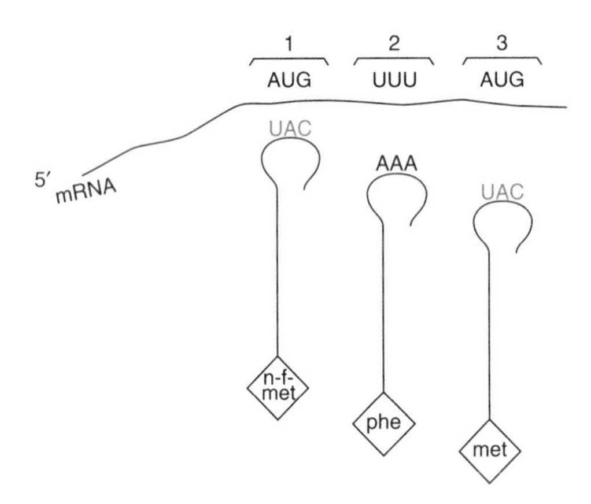

FIGURE 13-5. Initiating tRNA. In prokaryotes, a special kind of tRNA recognizes the first AUG codon and places *n*-formyl methionine in position at that point to start the polypeptide chain. When located at any other position in the coded sequence on the mRNA, the triplet AUG designates ordinary methionine and is recognized by the typical tRNA for that amino acid. This figure shows AUG in the first and third positions and the codon for phenylalanine (phe) at the second. Phenylalanine will be placed at position 2 and ordinary methionine at position 3 in the polypeptide as the ribosome reads this mRNA. Note that the anticodon (red) is the same for both forms of methionine. (The ribosome has been eliminated to permit clarity.)

After entry of the second tRNA into the A site, a peptide bond is formed between the carboxyl group of amino acid 1 at the P site and the amino group of amino acid 2 at the A site. This bond formation requires peptidyl transferase activity located in the larger subunit of the ribosome and is apparently due to the active participation of the rRNA. (More details are given later in this chapter.) Once the peptide bond is formed, there is an uncharged tRNA at the P site, and the A site is occupied by a tRNA with the growing chain of amino acids. The ribosome now moves three nucleotides along the mRNA in the 3′ direction. This causes the ejection of the uncharged tRNA and movement of the tRNA with the amino acids from the A to the P site. This movement of a ribosome one codon at a time is known as *translocation.* It requires energy from the conversion of GTP to GDP as well as another EF, one that is found in very large amounts in the cell. This EF binds to the ribosome to bring about translocation and is then released. A ribosome cannot bind both EFs at the same time. One must be released before the other can bind. An EF similar to this is also known in eukaryotes.

Translocation results in the placement of the next codon (the third in our example) in the proper position at the A site. A third charged tRNA whose anticodon can respond to the codon in position now moves to the A site, and codon–anticodon pairing again takes place. The amino group of the third amino acid, now at the A site, forms a peptide bond with the carboxyl group of amino acid 2 at the P site. The second tRNA, now uncharged, is ejected, and tRNA 3, holding a tripeptide, moves to the P site. Translocation results in placement of the next codon in proper position at the A site. A fourth charged tRRNA enters, and the process repeats until one of the chain termination codons comes into position. These codons are not recognized by any tRNAs, but they are identified by special proteins known as *release factors* (RFs). One RF responds to UAA or UAG and the other to UGA and UAA. The RFs act at the A site.

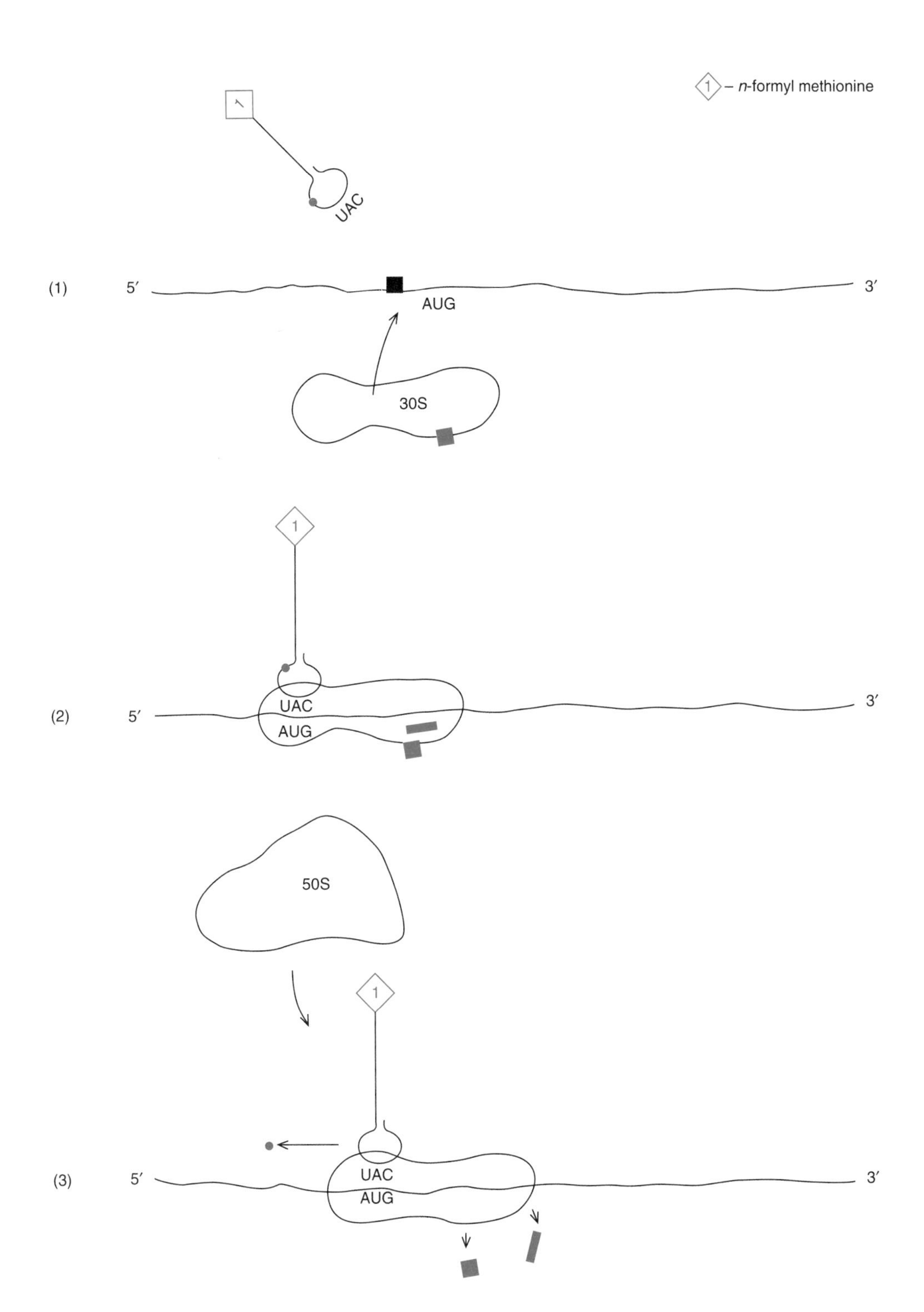

FIGURE 13-6. Formation of the initiation complex in *E. coli.* **1:** The 30S subunit can recognize the initiation site (black square) in the leader of the mRNA, but it cannot bind unless it becomes complexed with a special initiation factor, an IF (red square). A different IF (red dot) binds to the initiating tRNA to enable it to bind to the mRNA. **2:** The initiation complex includes the mRNA, the 30S subunit, and the initiating tRNA charged with *n*-formyl methionine. The tRNA's anticodon is paired with AUG, the usual initiating codon. A third IF (red bar) stabilizes the complex. **3:** Release of the IFs enables the 50S subunit to associate with the 30S one to form an intact ribosome.

Termination brings about the release of the polypeptide from the tRNA, removal of the latter from the ribosome, and the dissociation of the ribosome into its subunits. Only one release factor appears to exist in eukaryotes.

As a consequence of the manner in which peptide linkages are formed, the first amino acid in a polypeptide chain has a free amino group and hence an "amino end." Each amino acid that is added to a growing polypeptide chain is linked by its amino group to the

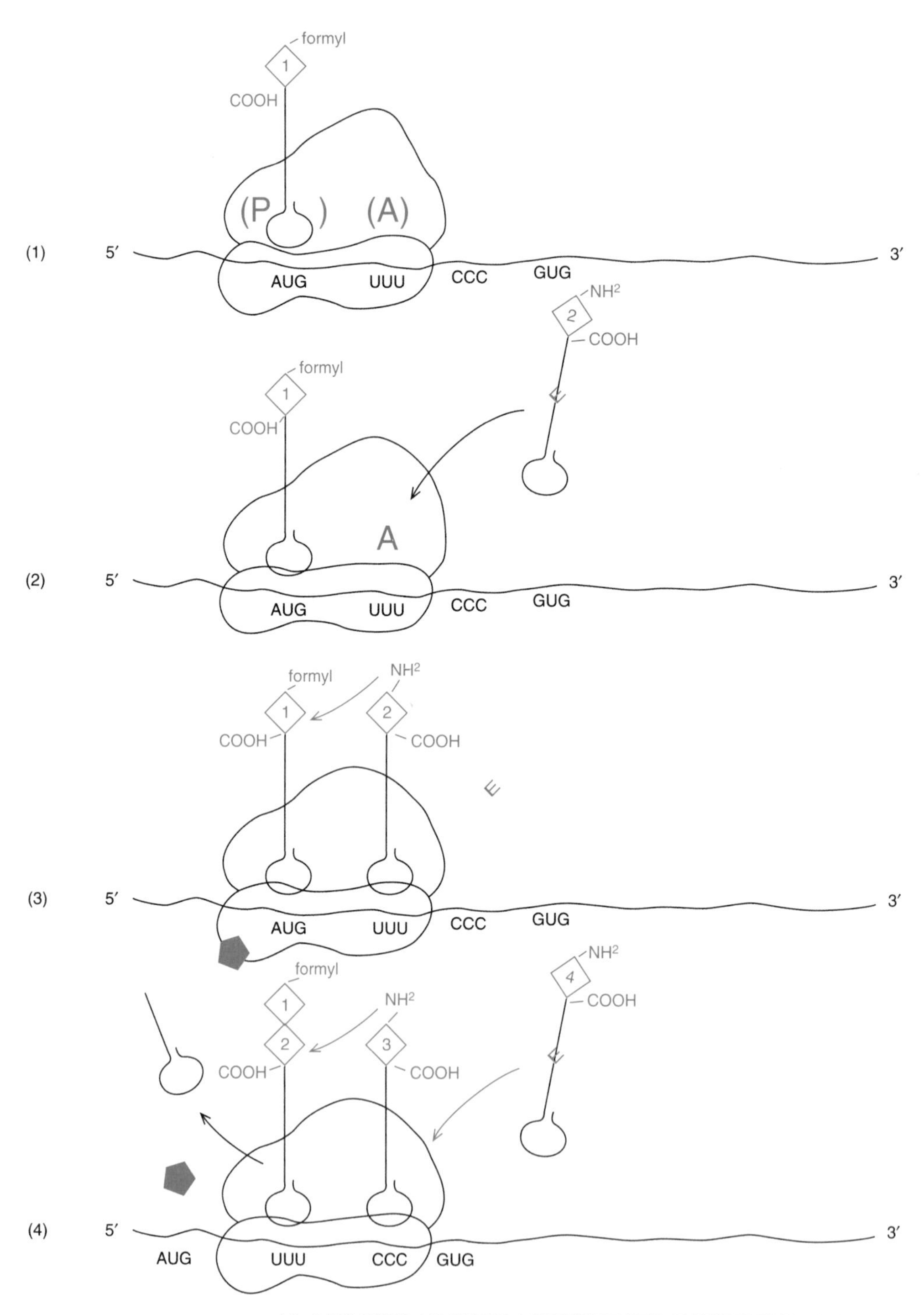

carboxyl group of the preceeding one. The last amino acid will therefore have a free carboxyl end. As a linear polypeptide forms, it may fold upon itself and begin to assume its final three-dimensional configuration (see Fig. 13-2). The complex foldings assumed by proteins result from interactions of amino acid residues along one chain as well as between different polypeptide chains in those cases in which the protein is many-stranded. The folding is independent of the ribosome that does not guide the three-dimensional architecture of a protein.

The anticodon and the wobble hypothesis

Since 61 of the 64 mRNA codons designate amino acids, it seems logical to expect to find 61 different kinds of tRNA molecules, one kind with a specific anticodon for one kind of codon. While more than one kind of tRNA is known for certain specific amino acids, a 1 to 1 correspondence does not hold. Only 40–50 different types of tRNA occur, well under the expected 61. Moreover, it has been demonstrated that one given kind of tRNA (and hence one given anticodon) can recognize more than one RNA codon. In addition, some anticodons contain the unusual base inosine (see Fig. 13-3). This base is derived by enzymatic action from adenine, which is originally in that position in the tRNA loop.

FIGURE 13-7. Steps in translation in *E. coli.* **1:** After formation of the initiation complex, the intact ribosome is ready to form peptide linkages. The two subunits of the ribosome form parts of two cavities, a P site and an A site, both of which can accommodate charged tRNAs. The P site is already occupied by the initiating tRNA charged with *n*-formyl 17 methionine. The anticodon of the tRNA is bound to AUG, the usual initiating codon. (Anticodons are not shown for sake of clarity.) **2:** To enter the A site, a tRNA must be complexed with an elongation factor, an EF (red E). This EF can recognize all tRNAs except the initiating one. All charged tRNAs other than the initiating tRNA can therefore enter the A site. **3:** The tRNA carrying the second amino acid is now at the A site. Its EF is released. A peptide bond forms between the carboxyl group of the residue at the P site and the amino group of the one at the A site. A second EF (red pointer) binds to the ribosome. It is required for the ribosome to move one codon at a time down the mRNA in the 3' direction. A ribosome can bind only one EF at a time. **4:** The next codon (CCC in this case) moves to the A site. The first tRNA, now uncharged, is ejected from the P site. The tRNA carrying two amino acids moves from the A to the P site. Another peptide bond can now be formed between amino acid 2 at the P site and amino acid 3 at the A site. The EF required for translocation is released. A fourth tRNA bound to its EF is now prepared to enter the A site. The process continues until a chain-terminating codon brings a halt to translation.

Crick proposed the wobble hypothesis, which helps explain these several points. Remember that the RNA codon in the mRNA is written in a 5′ to 3′ direction and that the anticodon that base pairs with it has the opposite polarity, 3′ to 5′. According to the wobble hypothesis, when the codon and anticodon interact, the first base pairing occurs between the bases at the 5′ end of the codon and the one at the 3′ end of the anticodon. The middle bases pair next, and finally the two at the third position, the bases at the 3′ end of the codon and the 5′ end of the anticodon. The first two pairings are specific. The base at the 5′ end of the anticodon, however, can wobble in the sense that it is freer to move about spatially than are the other two. You will recall from the discussion of tRNA and the synthetases that the latter interact primarily with the first two bases of the anticodon rather than the one in the 5′ position. This base may form hydrogen bonds with more than one kind of base at the 3′ end of the RNA codon. Consider an anticodon for arginine: 3′–UCU–5′. The base uracil is at the 5′ or wobble position. At this position, uracil can pair with either the base adenine or guanine in a codon, hence with either the codons 5′–AGA–3′ or 5′–AGG–3′. Both of these RNA codons are for arginine.

The unusual base inosine (I on Fig. 13-3A) at the wobble (5′) position may pair with A, U, or C at the 3′ position in an RNA codon. G at the wobble position may pair with U or C in the codon. U as the 5′ base in the anticodon can pair with A or G. There are restrictions, however; C in the wobble position can pair only with G (as in the case of the codon–anticodon interaction for methionine). Similarly, A at the 5′ position can base pair only with U. No single kind of tRNA can recognize four different codons.

The wobble hypothesis explains why fewer than 61 different tRNAs can exist and also why one purified kind of tRNA has been shown to interact with more than one kind of RNA codon. In addition, it can account for the presence of inosine in certain anticodons. Moreover, the hypothesis has been strengthened by the fact that certain predictions made from it have been verified. It correctly predicted, for example, the minimum number of tRNAs required for the amino acids serine and leucine. According to the wobble hypothesis, 32 tRNAs are sufficient to read all 61 codons.

In the process of translation, we see the active role of the ribosome in reading the message, forming the peptide linkages, and moving along the mRNA. We also see that tRNA is not just a passive carrier of an amino acid. Through its anticodon, it actively places its amino acid in the proper position, following the instructions coded in the messenger strand. The interactions of the mRNA, tRNA, and ribosome in translation achieve the final transfer of information from the DNA codons within the double helix—the formation of a specific polypeptide.

Ribosomal genes in prokaryotes

Cells that are synthesizing proteins at a high level are very rich in ribosomal content. In an actively growing bacterial cell, it may account for as much as 30% of the cell mass. A decrease in growth rate is followed by a drop in the number of ribosomes. Ribosome production is very closely coordinated with other cell activities and entails a variety of molecular interactions.

In *E. coli,* studies of hybridization between rRNA and the DNA of the chromosome have shown that seven genes exist for each one of the three classes of prokaryotic rRNA (16S, 23S, 5S). The genes are not found by themselves but occur in sets. There are seven sets, each one known as an rrn and designated by a letter (rrnA, to rrnG). Surprisingly, each rrn has been found to contain at least one gene for a tRNA. Figure 13–8 depicts one arrangement of genes in a set. The seven rrns are not adjacent to one another but are dispersed around the *E. coli* chromosome. Each rrn contains a promoter region and behaves as a unit of transcription so that one transcript contains information from all the genes in the set. This transcript is in the form of a precursor molecule which must be cut to yield the products of the genes in the set, the three classes of rRNA plus the tRNA. In all rrns, the arrangement of the three genes for the rRNAs is the same as shown in Figure 13-8, but there is variation regarding the number and sites of the regions coded for tRNA. Some sets contain just one tRNA gene in the spacer region between the 16S and 23S, whereas others have two. In addition, one or two tRNA genes may be found in the trailer region downstream from the 5S gene. For example, rrnC contains one tRNA gene in the spacer region and two in the trailer; rrnD contains two in the spacer and one in the trailer. The number of tRNA genes in the seven rrns varies from one to three. Each tRNA gene in an rrn is coded for a specific tRNA molecule.

Processing of the precursor RNA to yield the finished rRNA and tRNA products requires the activities of at least two enzymes: RNAase III to process the rRNA and RNAase P for the tRNA (details of the processing are not given in Fig. 13-8). As the transcript is forming, proteins begin to assemble on it. This means that, when the enzymes are doing much of their cutting, the RNA is becoming associated with the ribosomal proteins. That RNA that is discarded during the cleavage and trimming is degraded by enzymes.

In the bacterial cell, the rrns are continually undergoing transcription. As one RNA polymerase moves down from the promoter region, another attaches to the recognition site. This continual transcription of the rrns can account for the large supply of ribosomes in the cell and the fact that most of the RNA in a cell is the RNA of the ribosomes.

Locating the rRNA in eukaryotes

In eukaryotes, the cytoplasm of a metabolically active cell stains deeply with basic dyes because of the large number of ribosomes. Such a cell typically shows a nucleolus that is larger than that in a less active cell (Fig. 13-9). Conclusive evidence that the nucleolus is required for ribosome formation was presented in the 1960s in investigations with the African clawed toad *Xenopus laevis.* In this animal, the normal diploid cell contains two nucleoli. A mutation (actually a deletion of the nucleolar organizer) was discovered that interfered with the formation of nucleoli. An animal heterozygous for this change has cells that contain only one nucleolus. When heterozygotes are mated, about 25% of the offspring are homozygous for the deletion, and they die while still in the larval stage. Their cells lack nucleoli and cannot produce the major classes of rRNA. This can be demonstrated by exposing anucleolate cells and normal ones to labeled CO_2. When this is done, the anucleolate cells fail to incorporate the label into 28S and 18S rRNA in contrast to the cells having nucleoli. Those of the mutants cannot form the major RNA classes on their own. The unlabeled ribosomal RNA that does occur in cells lacking nucleoli is derived from the egg. The embryo reaches the tail bud stage of development only because of the maternal ribosomes. Once these are depleted, the larva dies.

Additional work with *Xenopus* gave more information on the genesis of the ribosomes. When DNA is isolated from cell nuclei, it can be fragmented into pieces within a certain size range and subjected to density gradient ultracentrifugation. A DNA frag-

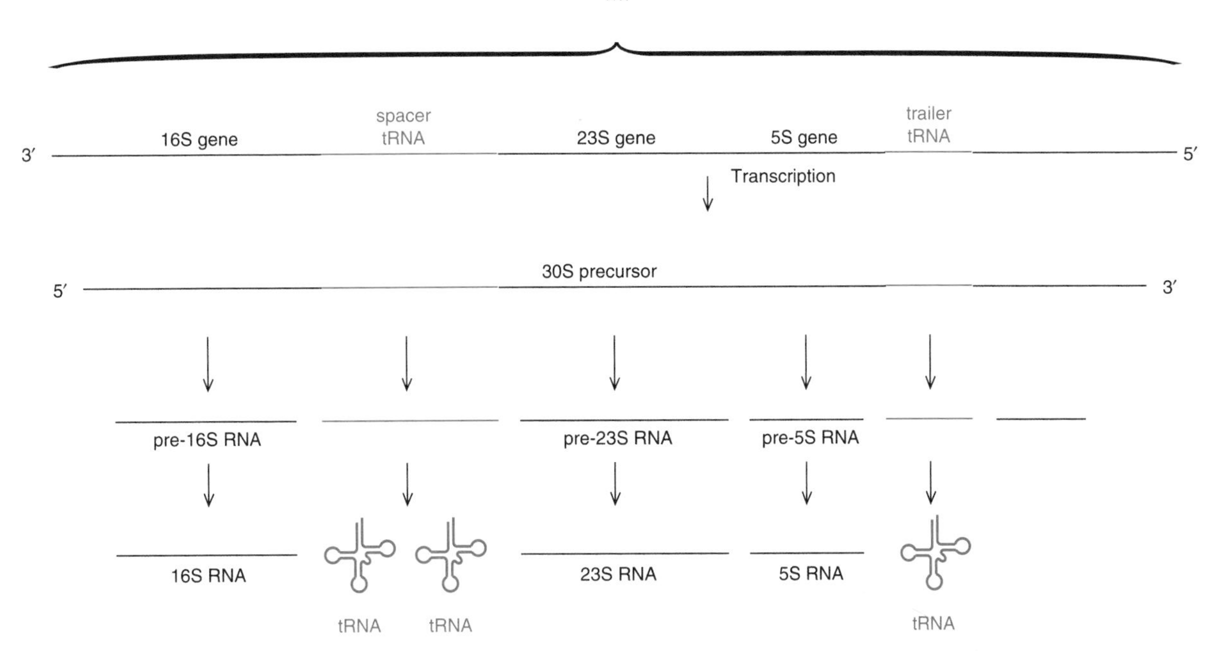

FIGURE 13-8. Processing of ribosomal RNA in *E. coli.* A set of genes known as an rrn includes sequences for the three classes of rRNA as well as for one or more tRNA molecules (red). The template strand of the rrn undergoes transcription to produce a 30S precursor, which becomes cleaved to produce smaller precursor products. These are processed further to yield mature 16S, 23S, and 5S rRNA as well as one or more tRNA molecules. Any discarded RNA is degraded. (The pattern depicted here is just one of several that exist.)

ment will move to a position in the tube at which its density matches that of the solute (review Fig. 11-15). Most of the fragments settle at a characteristic position and form a major band. This is so because most of the fragments (the bulk DNA) contain a G+C content of 40%. If fragments are denser than fragments in the bulk DNA (i.e., they contain a higher G+C content) or if they are less dense (having a lower G+C content), they will separate from the bulk DNA as smaller satellite bands (Fig. 13-10). The DNA in any given band may be isolated from that found in another band of different density. DNA–RNA hybridization (review Fig. 12-18) makes it possible to identify the DNA in a band by exposing it to RNA derived from some known source. Since ribosomal RNA is abundant in cells, it is relatively easy to determine whether a given DNA sample contains ribosomal genes (often designated *rDNA*). This is so because only the rDNA strands will hybridize with rRNA, the actual transcripts of the rDNA. Ribosomal RNA of wild-type *Xenopus* can be made radioactive by the incorporation of tritium. When such labeled RNA is used to form DNA–RNA hybrid molecules, it is found that only the DNA from a particular satellite band, not that of the major band, can form hybrid molecules with labeled 18S and 28S rRNA. The ribosomal DNA forms this satellite band as a result of its base composition, which distinguishes it from the rest of the DNA. The rDNA is especially rich in G+C content and settles in a discrete band, having a density different from that of the bulk DNA.

The genes for the 5S RNA (the 5S DNA) separate into a band that settles at a different position in the density gradient, and these can also be isolated for analysis. The rDNA for the 18S, 28S, and 5.8S RNAs band together in the density gradient, and this indicates that these ribosomal genes lie in close proximity and at a distance from the 5S genes, which are located on fragments of different density in a distinct band of their own.

The location of the 18S and 28S genes in *Xenopus* was clearly shown to be confined to the region of the nucleolar organizer, one of which is found per haploid chromosome set in *Xenopus.* This was possible with the technique known as *in situ hybridization,* a variation of the DNA–RNA hybridization procedure. In the in situ technique, the DNA is left in place in the chromosomes instead of being extracted. Cells are fixed on a slide and then subjected to a mild alkali treatment that denatures the DNA. The single strands can then be presented with single-stranded radioactive fragments of DNA or RNA from some known source. To identify the rDNA regions of the chromosomes, radioactive rRNA (18S and 28S) is used. Any DNA sequences complementary to these classes of RNA will bind to the rRNA and will hence pick up the radioactivity. Slides are washed to rid them of any unbound RNA and then coated with a photographic emulsion so that an autoradiograph can be prepared. Silver grains will be developed wherever there is a radioactive emanation from any labeled rRNA that has become bound. It has been demonstrated that the region of the nucleolar organizer exclusively binds the 18S and 28S rRNA, as shown by the fact that silver grains develop only over the nucleolar organizing region of the chromosome (Fig. 13-11). However, the 5S DNA is located elsewhere, at the tips of all the *Xenopus* chromosomes.

It was not surprising to find that the 5S DNA is located outside the nucleolar organizer, because anucleolate toads that cannot form 28S and 18S rRNA (as shown by the failure of their cells to incorporate label into these RNA types) are able to produce 5S rRNA. The 5S genes have been shown in eukaryotes to be located at a distance from the 18S and 28S genes, which are associated with the nucleolar organizer. In the human, the 5S rDNA is located near the tip of the long arm of chromosome 1. The rDNA for the other species of rRNA is localized in the short arms of chromosomes 13, 14, 15, 21, and 22, all of which possess satellites and have nucleolar organizing function.

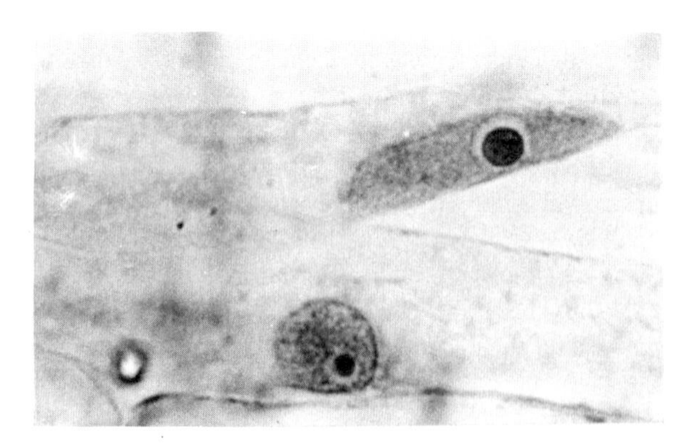

FIGURE 13-9. Epidermal cells of a grass. A cell forming a root hair is much more active metabolically than a cell that fails to produce one. Note the large nucleolus in the nucleus of the hair (above) in contrast to the smaller nucleolus in the nucleus of the hairless cell (below).

The availability of rRNA, the product of rDNA, has made it possible to estimate the number of ribosomal genes present in a set of chromosomes. Again, use is made of DNA–RNA hybridization. A sample of a known amount of single-stranded nuclear DNA is allowed to hybridize with a known quantity of pure rRNA. The larger the number of ribosomal genes in the DNA sample, the greater the amount of the rRNA that will hybridize with it. In the process (similar to that shown in Fig. 12-18), the radioactive rRNA is added to single-stranded DNA. The DNA–RNA hybrid molecules are trapped on filter paper, the remaining unbound RNA being washed away. The amount of bound RNA in the form of hybrid molecules can be measured, since the RNA is radioactive. The hybridization procedure thus acts as an assay and reveals that about 450 copies of the genes for each 18S, 28S, and 5.8S rRNA are present in a haploid set of *Xenopus* chromosomes. The corresponding figure for the human is 280. In lower eukaryotes, the number varies from about 100 to 200, but in all eukaryotes the number of gene copies for the 5S rRNA greatly exceeds that for the other classes (24,000 in *Xenopus,* 2,000 in humans).

Organization of ribosomal genes in eukaryotes

Isolation of the ribosomal genes of various species has made it possible to subject them to very detailed analyses using an assortment of chemical procedures as well as electron microscopy. We now have a very precise picture of the genes for rRNA in several species. A ribosomal DNA region that governs the synthesis of the 18S, 28S, and 5.8S RNAs was found to consist of repeating units of a given length, each unit containing certain recognizable sequences (Fig. 13-12). The unit includes DNA sequences for the three main classes of rRNA. Each unit is separated from the next by an untranscribed spacer that varies greatly in its length. The variation is due to the fact that some of the DNA sequences in the spacer can be repeated. In some species, there is more repetition of certain sequences than in others. This spacer DNA has never been found to hybridize with any cellular RNA. The units and spacers are repeated hundreds of times. The repeated rRNA genes present an excellent example of a *simple multi-*

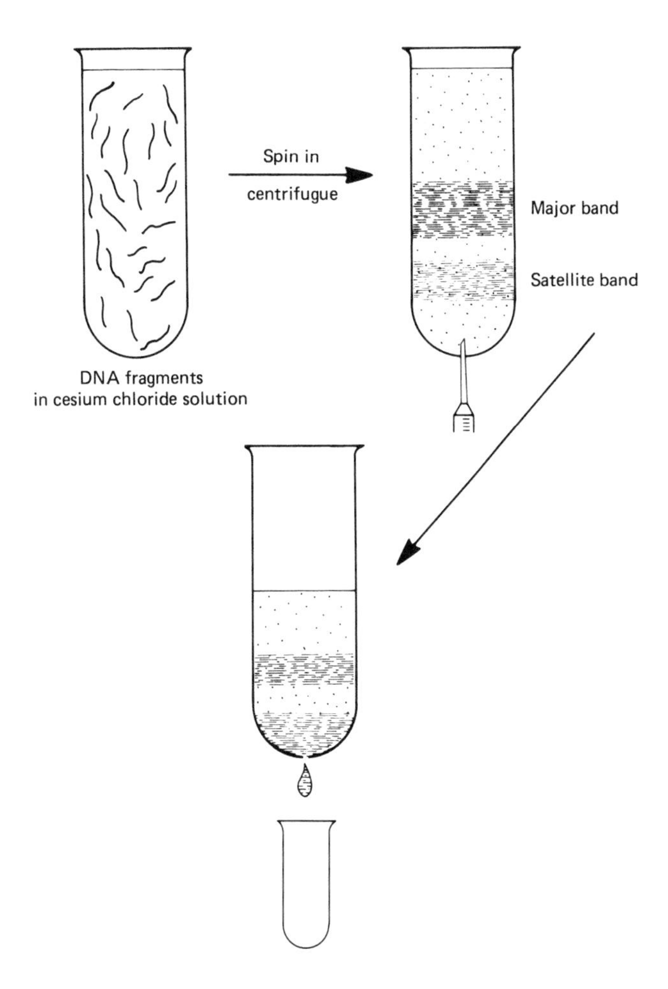

FIGURE 13-10. Bulk DNA and satellite band. If DNA from a cell has regions that contain more (or less) G + C than the major portion of the DNA, the two kinds of DNA can be identified and separated, using density gradient ultracentrifugation. When extracted from the nuclei of cells, the DNA fragments at random during the course of the procedure. In the density gradient, those fragments with a higher G + C content will separate from those having a lesser G + C content because of the density difference between them. Most of the DNA has a G + C content of 40% and settles out as a major band. Any DNA fragments with either a higher or lower density will separate from it as one or more satellites. The separation permits isolation of the bands, as seen here for the band which is denser because of its higher G + C content. The tube may be punctured and samples collected as the tube drains off drop by drop.

gene family, a set of repeated genes. In this case, the duplicated genes are clustered on a chromosome and repeated next to each other in a linear array.

When a unit of rDNA undergoes transcription in mammals, a 45S preribosomal RNA transcript is produced that contains the sequences for the 18S, 28S, and 5.8S RNAs. This transcript must be processed and becomes cleaved in several places. There are variations in the order in which cuts are made. In the scheme shown in Figure 13-12, the transcript is cut in four places. One cut removes a leader sequence from the 5′ end of the transcript. The second produces the 18S rRNA for the smaller subunit. A 36S RNA fragment remains. The next cut removes from it a small transcribed spacer region, leaving a 32S fragment. A final cut produces the 28S and 5.8S RNAs. Both of these form a close association in the larger subunit by the formation of base pairs.

RNA polymerase I, which is confined to the nucleolus, forms the preribosomal 45S transcript from each rDNA unit and is actually responsible for more than 50% of the RNA synthesis in the cell. Since this enzyme acts on a single kind of transcription unit, one might expect the promoter sequences for the units to possess many regions in common, since only one kind of transcript is involved. However, this does not prove to be the case. When different species are compared, certain homologous segments can be found when the organisms are members of closely related groups. But no conserved promoter sequences are evident in comparisons of the regions from less closely related species, such as yeast, *Drosophila, Xenopus,* and the mouse. The data suggest that nucleotide sequences located directly around the initiation site play an important role in transcription control. Species-specific transcription factors may also be involved, since successful in vitro transcription by RNA polymerase I depends on components that can be supplied from cell extracts. Ribosomal units isolated from the mouse, the human, and protozoans are transcribed accurately only if all the components supplied are from the same species. Extracts from the mouse will not support transcription of human ribosomal genes or vice versa.

Transcription from the DNA of the nucleolar organizing region can actually be visualized by studying photographs made with the electron microscope (Fig. 13-13). The untranscribed spacers are seen as regions between stretches of ribosomal DNA to which RNA transcripts are attached. The function of the untranscribed spacer remains unknown.

Analyses of the ribosomal DNA molecules for the 5S RNA have shown that the 5S DNA also contains units that repeat, making up a simple multigene family. The portion that is transcribed into 5S RNA makes up only one seventh of a unit and is separated from the next by an untranscribed spacer of variable length. RNA polymerase III, which is responsible for tRNA formation, also produces the 5S rRNA (see next section).

The ribosomal proteins are formed on the ribosomes of the cytosol. When the 45S precursor is formed, it immediately associates with proteins. Therefore, all the processing described involves ribonucleoprotein, not naked RNA. Somehow the 5S RNA, transcribed from genes located outside the nucleolar organizer, becomes packaged along with the 28S and 5.8 RNAs into the larger subunit, a process that must entail complex molecular interactions.

The very existence of mutigene families, as seen here for the genes for the rRNA classes, presents an enigma. Within a family, many copies of the same genetic sequence exist in the genome. What is the mechanism that keeps the many different copies of

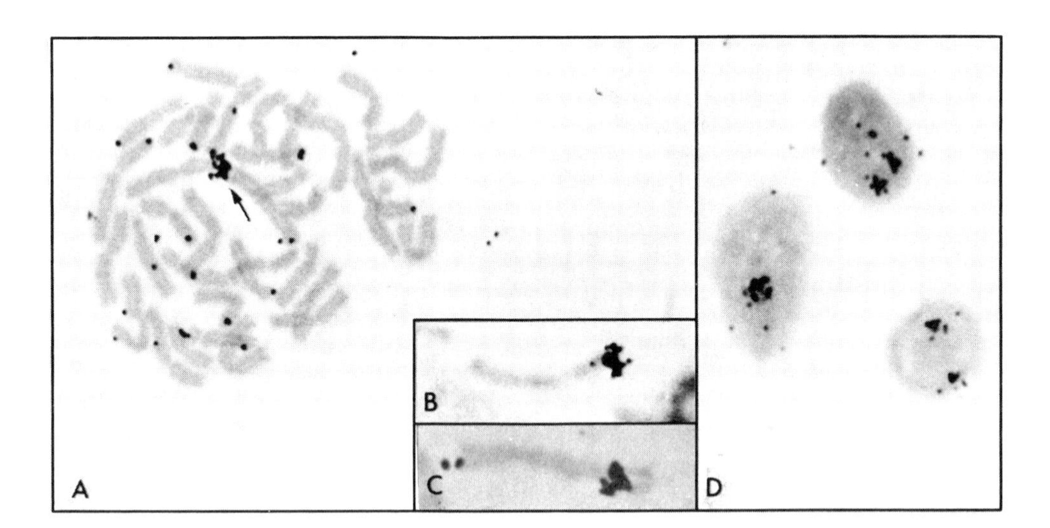

FIGURE 13-11. Autoradiographs of cytological preparations from *Xenopus mulleri* and *Xenopus laevis* hybridized in situ with labeled cRNA transcribed in vitro from DNA sequences coding for 18S and 28S ribosomal RNA. **A:** Metaphase plate from *X. mulleri* showing hybridization of the labeled RNA over the terminal secondary constrictions on a pair of the largest submetacentric chromosomes (arrow). **B:** The nucleolus organizer chromosome from *X. mulleri,* showing localization of the sequences coding for 18S and 28S ribosomal RNA at the extreme end of the shorter arm. **C:** The nucleolus organizer chromosome from a *X. laevis* cultured cell line, showing localization of the sequences coding for the 18S and 28S ribosomal RNA over the secondary constriction. **D:** Interphase nuclei from *X. mulleri,* showing hybridization over nucleoli. (From *Cold Spring Harbor Symp. Quant. Biol* 38:475, 1973. Courtesy of Dr. Mary Lou Pardue.)

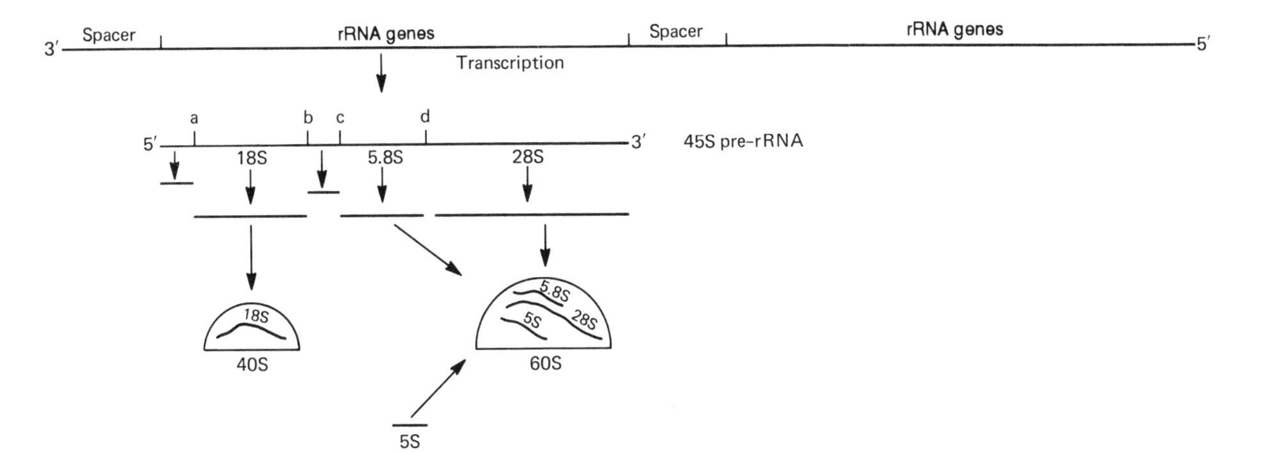

FIGURE 13-12. Processing of ribosomal RNA in eukaryotes. The rDNA region for the major classes of rRNA is found in the nucleolar organizer. The region is composed of repeating units, each one including genes for the 5.8S, 18S, and 28S rRNAs. One unit is separated from the next by a spacer that is not transcribed. When a unit undergoes transcription, a 45S RNA precursor is produced. Processing of this transcript then takes place as four cuts are made. One cut (a) removes a leader sequence from the 5′ end. The second (b) results in the formation of an 18S RNA, which becomes part of the 40S subunit of the ribosome. A third cut (c) removes a small transcribed spacer. A final cut (d) yields the 5.8S and 28S RNAs of the larger ribosomal subunit. The 5S rRNA of the larger subunit is a transcript of DNA found outside the nucleolar organizer.

each one of the rRNA genes identical and prevents them from diverging? In Chapter 10, we stressed the importance of duplications in the evolution of different genes. Once a gene becomes duplicated, each of the two copies of the original is now subject to its own history of gene mutation. Eventually, two very different genes may arise from the two copies as the forces of evolution interact with the genetic alterations that accumulate throughout time. It is even possible that one copy will accumulate deleterious alterations that eventually render it nonfunctional. Such a useless DNA sequence resembling that of an active gene is known as a *pseudogene* (see Chap. 20). Various examples of these are known in the mammalian genome. Some are found in the vicinity of functional immunoglobulin genes and histocompatibility genes. Why have not some of the copies of the ribosomal genes deteriorated into pseudogenes? Various mechanisms such as gene conversion (see Chap. 14) have been proposed as devices to preserve the identity of the many copies of a gene in a multigene family. However, there is no conclusive evidence to favor any one of the models. Whatever the mechanism proves to be, it must explain how clusters of repeated genes that may be found in different locations in the genome (such as five different nucleolar organizing regions in the human) can be kept identical. The maintenance of multigene families meanwhile remains a puzzle.

RNA polymerase III interactions and transfer RNA

As pointed out in Chapter 12, RNA polymerase III recognizes the promoters for 5S RNA genes as well as those for tRNA genes. The surprising fact came to light that these polymerase III promoter sequences are located downstream from +1, the site of transcription initiation, and are found in two separate parts (Fig. 13-14). For the 5S RNA, the promoters occur within the gene between positions +55 and +80. Each part of the promoter has a characteristic base sequence. Three different transcription factors must interact with the promoter segments before RNA polyerase III can bind and initiate transcription.

Each of the two segments of the tRNA promoters is about 20 base pairs long, one located between +8 and +30 and the other between +51 and +72. If the distance between the two segments is reduced, transcription is prevented. However, some small increase in the distance is tolerated. The nucleotide sequences of these promoter segments of tRNA promoters have been very highly conserved in eukaryotes, and they interact with two transcription factors.

Our discussion of rRNA acquainted us with structural genes that are coded for a product other than a polypeptide. The products of these genes, the classes of rRNA, become components of the ribosomes and are never translated. The genes encoded for the various tRNA molecules provide other examples of genes whose RNA product does not undergo translation. We will now see that the production of functional tRNA bears many similarities to the formation of the mature classes of rRNA.

All genes encoded for tRNA tend to lie in clusters. Each cluster is concerned with the production of several different types of tRNA. Some of the tRNA genes in a cluster form a transcript bearing a coded sequence for just a single tRNA molecule. In other cases, a long transcript arises from a single promoter region, and it carries information for several tRNAs, as many as seven. Nevertheless, in all cases, tRNA molecules arise by transcription in the form of a much larger precursor having a leader sequence and a trailer at the 5′ and 3′ ends, respectively. Those precursors that contain several tRNA sequences, as shown in Figure 13-15, contain variable amounts of RNA on either side of each tRNA portion. All tRNA precursors, whether they contain one or several tRNAs, require processing in order to generate mature tRNA molecules.

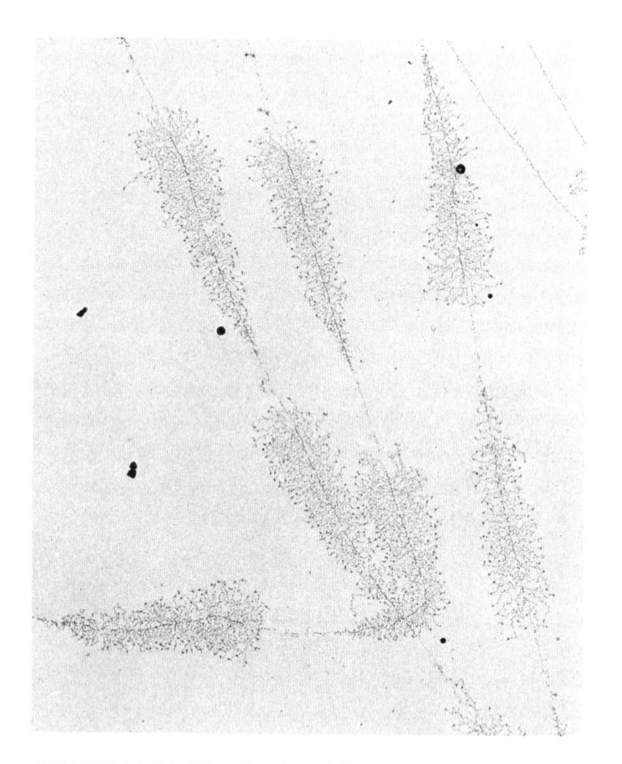

FIGURE 13-13. Visualization of ribosomal DNA from oocyte nucleoli of the spotted newt *Triturus viridescens*. The linearly repeated ribosomal units are separated by untranscribed spacer segments. Each unit actively transcribes RNA molecules that will be processed to yield the major classes of rRNA. The shorter transcripts are the more recently synthesized and are closer to the beginning of each gene. The difference in their lengths accounts for the feathery, arrowhead appearance. (From Miller, O.L., Jr., and B. Beatty. *Science* 164:955, 1969. Courtesy of O.L. Miller, Jr.)

Earlier in this chapter, we learned that in *E. coli* some tRNA sequences occur on the same transcript with sequences coded for the three classes of rRNA (Fig. 13-8). This additional way of generating tRNA does not occur in eukaryotes. Most of the information on tRNA processing that will now be described comes from prokaryotes, but similar steps are believed to be followed in eukaryotes. The activities of several enzymes are required in all tRNA processing. In *E. coli,* the enzyme ribonuclease P, an endonuclease, acts on every pre-tRNA transcript regardless of the number of tRNA sequences present. Its activity brings about the 5′ end of all tRNA molecules (Fig. 13-15). Each tRNA portion of the precursor assumes its cloverleaf shape. It is believed that this aids in the recognition of different sites by the appropriate enzymes. Ribonuclease P is quite an unusual enzyme, since it is composed of an RNA portion in addition to the protein part. Both are required for full catalytic activity. The significance of this will be discussed a bit later on in this chapter.

The enzyme RNAase D, an exonuclease, removes nucleotides on the 3′ side of each tRNA sequence in the precursor to generate 3′ ends. When it contacts the –CCA sequence common to all tRNA molecules, it stops. Other enzymes come into play during processing to modify some of the familiar bases in the pre-tRNA and to change them to such forms as inosine and dihydrouridine. After the extensive cutting, trimming, and modification are completed, mature tRNA molecules are available to perform their important role in translation.

Processing the primary transcript

Several very important distinctions between prokaryotes and eukaryotes at the molecular level center around messenger RNA. In prokaryotes, the ribosomes begin translation of the mRNA as it is still being formed by transcription of the antisense strand of the DNA. This means that the template and its transcript are in close proximity when translation begins. This is not the case in eukaryotes, because the immediate transcript (the primary transcript, or the pre-mRNA) cannot act as functional mRNA. As in the case of the preribosomal RNA, the pre-mRNA must first undergo several modifications

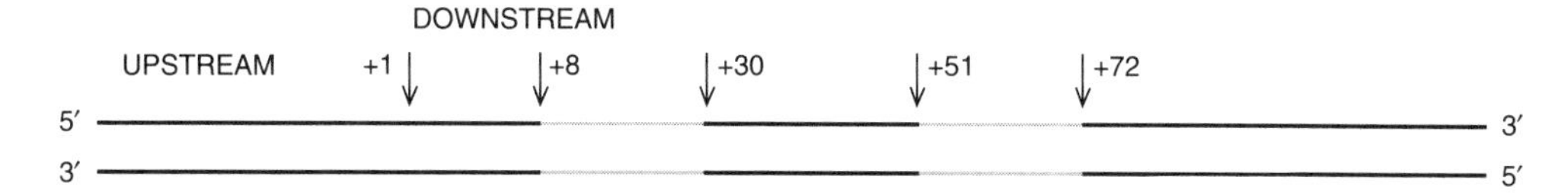

FIGURE 13-14. RNA polymerase III promoters. These promoters exist in two parts (red) and are located downstream from +1 within the region coded for the transcript.

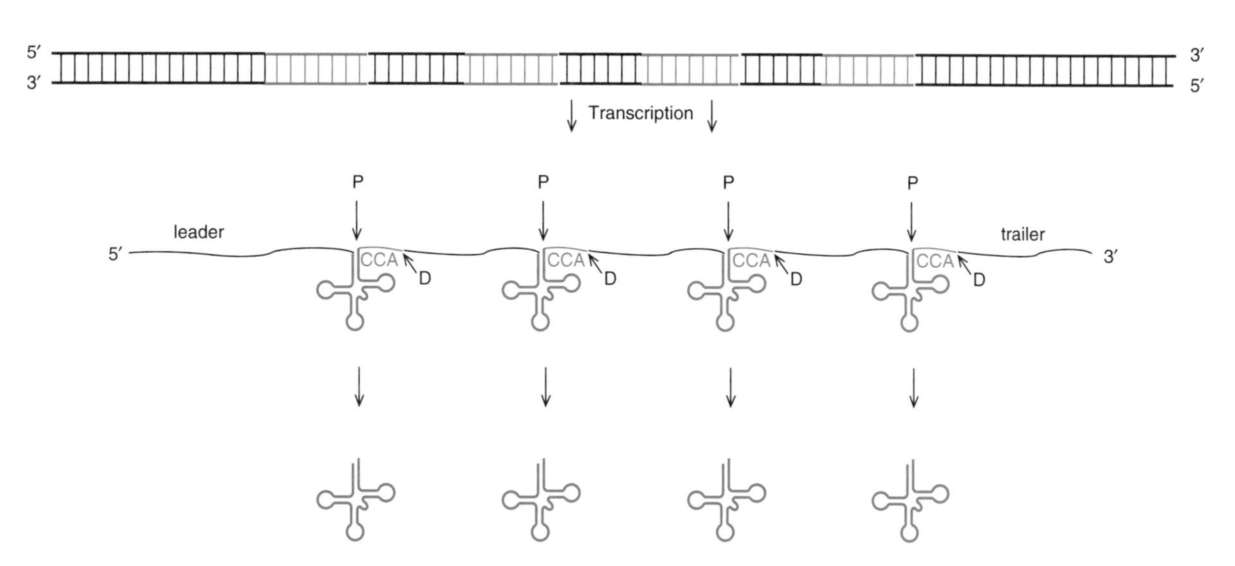

FIGURE 13-15. Processing of tRNA. In a chromosome region where tRNA genes are clustered, a single promoter may be associated with several tRNA genes (above, red). The promoter region occurs before the first tRNA sequence. A long transcript arises containing tRNA sequences and requiring extensive processing. The endonuclease RNAase P (P) generates the 5′ end of each tRNA. Exonuclease D (D) removes nucleotides between tRNA sequences but recognizes the –CCA end of each tRNA where it stops, thus generating the 3′ end of each molecule. Further enzyme activity is involved in trimming and in the modification of bases. Mature tRNA molecules are the end product of the processing.

known as *RNA processing.* The processing of the pre-mRNA alters it and converts it to the final messenger which then travels from the nucleus into the cytoplasm, where translation takes place. Actually, the first modification of the primary transcript occurs about 1 second after the initiation of transcription, while the RNA strand is fewer than 30 nucleotides long. This involves the addition of an unusual base, 7-methylguanosine, to the transcript's 5′ end. This cap becomes linked by a triphosphate connection to what was the first base of the transcript. The linkage involved is an unusual 5′ to 5′ linkage, which results from an interaction between the 5′ triphosphate end of the pre-mRNA and the triphosphate of the guanosine (Fig. 13-16). A further modification of the cap takes place in all eukaryotes, except for one-celled organisms. This is the addition of another methyl group to the 2′ position of the sugar of the first nucleotide of the primary transcript. Such a cap is called a *cap 1* type. Lacking this methyl group, the cap is called *cap O,* typical of one-celled eukaryotes. In a few eukaryotes, still another methyl group may be added to the sugar of the second nucleotide, also at its 2′ position. This is a *type 2* cap. The mRNAs of all eukaryotes and of most of the viruses that infect them have been found to have caps. The role of the cap appears to involve binding of the mRNA by the ribosome and the initiation of translation. In the test tube, mRNAs with altered caps show a slowdown in the start of translation and are not translated efficiently.

Another change in the pre-mRNA of eukaryotes occurs at its 3′ end or trailer end. This modification takes place after the primary transcript has been formed. The primary transcript contains a site known as a *poly(A) site.* The pre-mRNA is cut at this spot, and an enzyme then proceeds to add to the 3′ end a stretch of up to 200 nucleotides, each containing the base adenine (Fig. 13-17). The result is the production of a poly(A) tail at the 3′ end of a processed

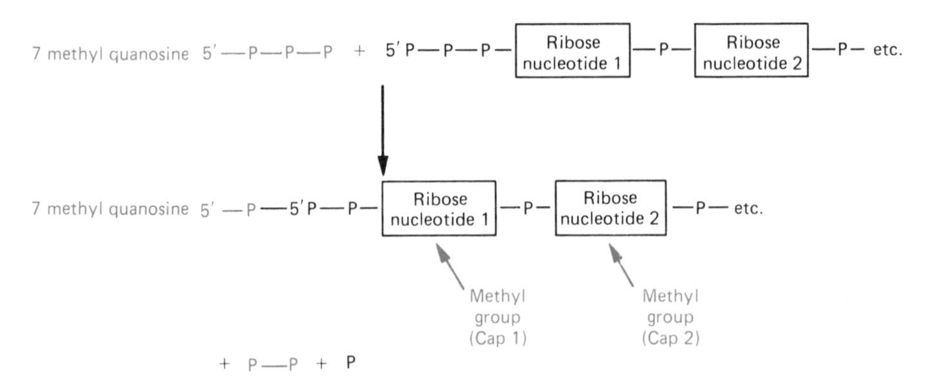

FIGURE 13-16. Capping of primary transcript. The base 7-methyl guanosine is joined by a 5′–5′ linkage to the 5′ nucleotide (nucleotide 1) of the transcript. Such a cap is called a *cap 0*. Methylation of the sugar of the first nucleotide can occur to produce a cap 1. Further methylation at the position of the second sugar produces a cap 2.

mRNA. Approximately 20 nucleotides upstream from the poly(A) site, the following nucleotide sequence occurs on the transcript: AAUAAA. This sequence appears to be involved in the recognition of the poly(A) site downstream from it. A few eukaryotic mRNAs (those for histones, for example) lack the poly(A) tail, the function of which remains unknown. It has been suggested that the tail may help to bind the mRNA to the ER of the cytoplasm, thus facilitating translation, and that it may also act to prevent degradation by the many ribonucleases present in the cytoplasm. The poly(A) tail is also typical of viruses that infect eukaryotic cells.

Figure 13-18 gives a general summary of some of the aspects of eukaryotic mRNA following posttranscriptional modification. It can be seen that the 5′ end is capped and that an untranslated leader segment, which ends with the initiating codon AUG, follows the cap. From the initiating codon to one of the chain-terminating codons, nucleotide sequences that code for the polypeptide are found. Following the chain-terminating codon is an untranslated trailer to which is attached the poly(A) tail at the 3′ end. Before messenger RNA is ready to leave the nucleus and enter the cytoplasm, several other events must take place, as are discussed in the following section.

Colinearity and fragmented genes

Both DNA and protein are essentially linear molecules. Each codon in a gene and in its mRNA is specific for an amino acid and determines the position of that amino acid in a polypeptide chain. It is tempting to conclude, therefore, that the linear sequence of the nucleotides in DNA and RNA corresponds to the linear sequence of the amino acids along a polypeptide. This was demonstrated very clearly years ago by Yanofsky and coworkers, who studied in detail a gene coded for one of the two polypeptide chains composing the bacterial enzyme tryptophan synthetase. Several separate mutations in this gene, gene A, were mapped very precisely (Fig. 13-19A). The defective polypeptide chain associated with each mutation was analyzed in detail. Each defective chain was found to have a single amino acid substitution. The substitutions brought about by the different mutations were found to be located at different positions in the defective chains. When the positions of the amino acid substitutions were related to the sites of the mutations in the gene, it became evident that a mutation mapping at one end of gene A produced an amino acid change at one end of the polypeptide, let us say the amino end. A second mutation mapping at the other end of the gene was shown to be related to an amino acid alteration at the other end of the polypeptide, the carboxyl end. A third mutation might map between the first two. The amino acid substitution associated with it would then fall at a position in the chain between the other two amino acid changes. In other words, the position of a mutation within a gene corresponds to the position of an amino acid substitution in the polypeptide chain for which it is coded. The relative position of a triplet in the DNA and in the RNA transcribed from it is related directly to the position of a specific amino acid in a polypeptide. The nucleic acid molecules and the polypeptide are colinear (Fig. 13-

19B). The sequence of nucleotides in the gene corresponds exactly with the sequence of nucleotides in the mRNA and with the amino acid sequence in the polypeptide for which it is coded. A change at a site in the DNA is reflected at corresponding positions in the mRNA and in the polypeptide formed after translation. Colinearity was found to be the case in prokaryotes and was believed to apply to all genes. One of the biggest surprises in molecular biology was the discovery that most genes of eukaryotes are not strictly colinear with their mRNA transcripts. It has been firmly established by DNA sequencing (see Chap. 19) that most eukaryotic genes as well as those of animal viruses are fragmented. The sequences that are coded for a product, such as a polypeptide, are interrupted by intervening, noncoding, sequences, designated *introns.* The coding segments of the gene are named *exons.* This means that the typical eukaryotic gene is in pieces and has a much longer nucleotide length than is required to code for the amino acid sequence of a polypeptide (Fig. 13-20).

Although genes for histone proteins and for actin lack introns entirely, most genes studied are spread out into 20 or more pieces. In lower eukaryotes, introns are shorter than those of higher forms and occur less frequently. In higher eukaryotes, the coding sequences (the exons) may represent only a small portion of the length of a gene, whereas the introns may make up the larger portion. A gene that requires a coding sequence of 1,000 base pairs could be spread out over a length 10 times as great. The gene for α-collagen in the chicken consists of over 50 exons. Some of the introns are the size of the exons, whereas others are much longer. The total length of the coding sequences (the exons) makes up only one-eighth the combined length of the exons and introns.

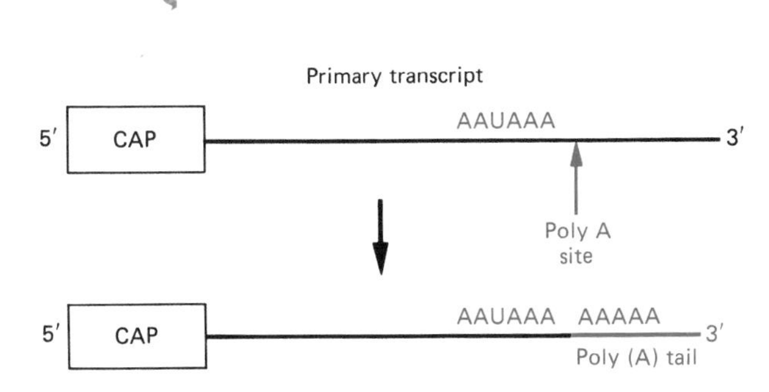

FIGURE 13-17. Processing at the 3′ end of the primary transcript. A nucleotide sequence, AAUAAA, occurs a short distance upstream from the poly(A) site, where the transcript will be cut. An enzyme then adds to the 3′ end a poly(A) tail that may include up to 300 nucleotides.

In the mouse, the gene for β-hemoglobin contains two introns whose total length is 762 base pairs, whereas the sum of the coding sequences is only 432 base pairs. The gene in cases such as these is very much longer than its mRNA, which carries mainly the coded information for the product.

When a eukaryotic gene undergoes transcription, the primary transcript contains both exon and intron sequences complementary to those in the template strand. The transcript is a precursor. Before it can serve as mRNA, it must be supplied with a cap and a poly(A) tail, but it must also undergo processing to remove the introns. This removal must then be followed by a joining of the exons (Fig. 13-20). The snipping out of the introns and the linking of the exons is called *RNA splicing.* Splicing not only achieves joining of the exons, it also maintains them in the same order as the related exons in the gene. Although distances in the gene do not correspond to distances in the mature mRNA or in the polypeptide due to the presence of introns in the gene, colinearity nevertheless exists between the exons and the amino acid sequences in the polypeptide. The position of a mutation in an exon is related to the position or an amino acid substitution in the polypeptide.

When splicing takes place, it is essential that the factors that remove the introns recognize the junctions between introns and exons. Imprecise removal will lead to defective or nonfunctional mRNA as a result of upsetting the reading frame. While a mutation in an intron does not bring about an amino acid substitution, it can interfere with splicing and the assembly of exons and consequently prevent mRNA formation. Sequencing of nucleotides has shown that at an exon–intron junction in a transcript every intron begins with GU at the 5′ end and terminates with AG at the 3′ end. The GU and AG in turn are parts of short consensus sequences that occur at the left and right exon–intron junctions (Fig. 13-21). These short sequences are definitely required for proper splicing as is another short sequence, known as the *branch sequence,* located 30 bases upstream from the 3′ end of the intron. The branch sequence shows variation and is not well defined in most eukaryotes, but its importance in splicing will become apparent shortly. Mutations in these three sequences can prevent splicing.

Clues to the events taking place during splicing have come from experiments in which removal of introns and joining of exons have been achieved in the test tube. These in vitro experiments have implicated small ribonucleoprotein particles in the splicing. Particles of

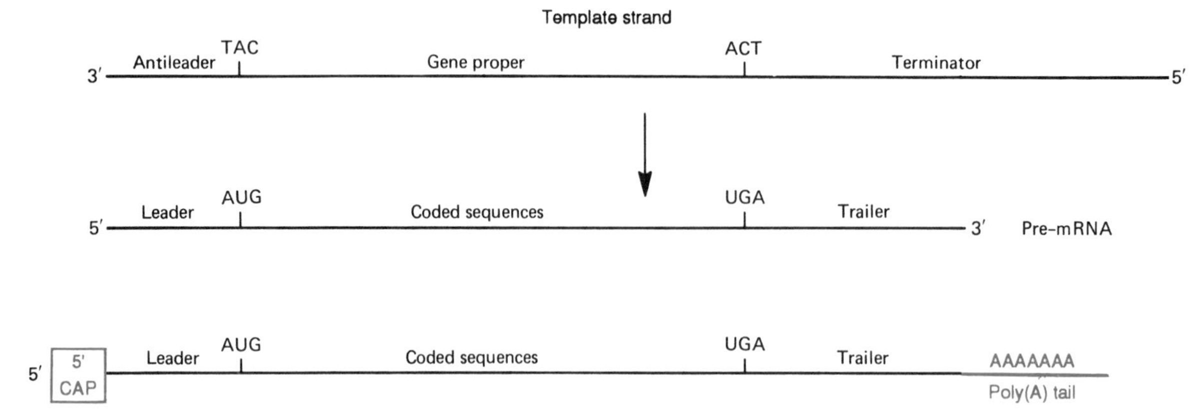

FIGURE 13-18. Posttranscriptional modification of eukaryotic messenger RNA. Pre-mRNA, the immediate transcript of the template strand, undergoes modification before the final mRNA is produced and leaves the nucleus. The 5′ end of the final messenger RNA is capped and the 3′ end bears a long stretch of nucleotides containing adenine. Prokaryotic messenger RNAs also contain a trailer and a leader but lack the 5′ cap. Very short poly(A) tails appear to be characteristic of the 3′ ends of prokaryotic mRNA. (Compare with Fig. 12-28, and see the text for details.)

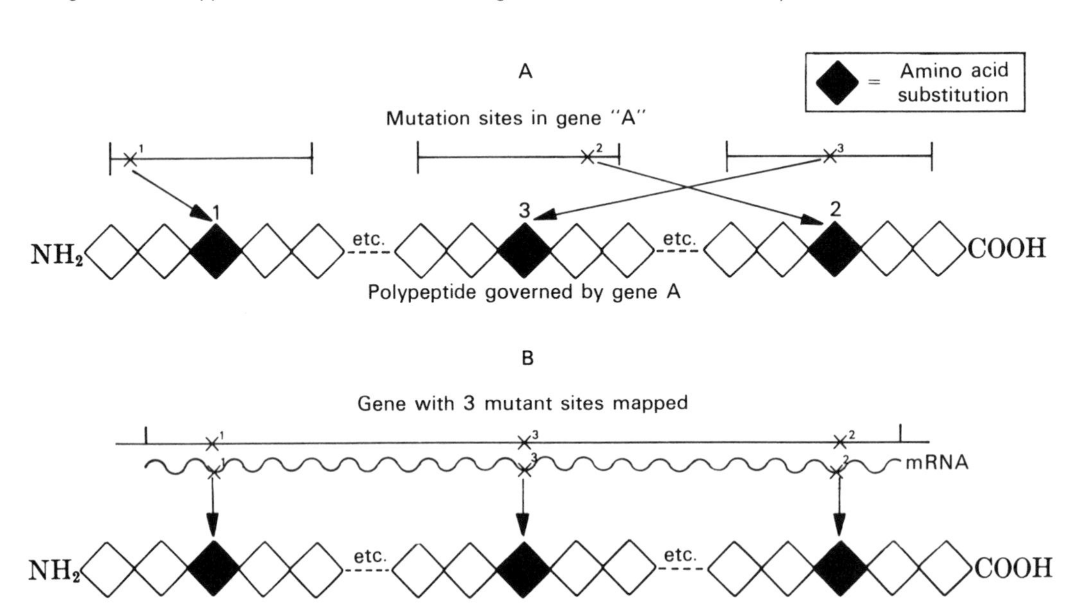

FIGURE 13-19. Colinearity. **A:** Mutations occur at different sites within gene A of *E. coli,* which determines the amino acid sequence of one of the polypeptide chains of tryptophan synthetase. The relative positions of the separate mutations can be mapped. This schematic representation shows that if a mutation maps at the beginning of a gene, an amino acid substitution is made at the beginning of the polypeptide chain. A mutation at the end of the gene (mutation 2) is reflected by an amino acid substitution near the other end of the protein, the carboxyl end. Mutations, such as mutation 3, which map between ends of the gene, are associated with amino acid substitutions between ends of the polypeptide. The gene and the polypeptide chain it governs are thus colinear. **B:** Messenger RNA, carrying the information for amino acid sequence, is a transcript of the DNA. Therefore alterations in the nucleotides of the DNA are reflected at corresponding positions in the mRNA. These in turn are responsible for the amino acid substitutions at positions in the protein. These amino acid changes therefore correspond to altered nucleotide sites in the mRNA and in the DNA. The gene, its RNA transcript, and the polypeptide are thus colinear.

this kind composed of RNA and proteins are found in the cytoplasm and the nucleus. The nuclear ones are referred to as *snurps* and the cytoplasmic ones as *scyrps.* RNA polymerase II synthesizes the RNA portion of some of these particles, whereas polymerase III is the catalyst for others. Six different types of snurps, U1–U6, occur in vertebrates. These have been named with respect to variations in their RNA, known as *U-*

RNA, which is relatively rich in uracil. Such particles are abundant in eukaryotic cells, especially those of mammals. Some types may be present in a million copies. U1 snurps are the most abundant. A typical snurp consists of just a single RNA molecule, 100–300 nucleotides long, complexed with about 10 proteins. Five different kinds of snurps are involved in splicing. The RNAs of these particles have nucleotide sequences complementary to those at the exon–intron junctions or to the branch sequence. In the human, a U1 snurp, which contains eight proteins, has an RNA sequence complementary to the consensus sequence at the left intron junction. Base pairing between the U1 snurp RNA and the left intron junction is definitely involved in splicing as shown by studying effects of mutations in the U1 RNA sequence or in that of the consensus sequence. The U1 RNA apparently binds to the complementary sequence at the beginning of the intron. The binding seems to depend on the entire snurp, since U1 RNA by itself cannot accomplish this. Once the transcript is bound to the particle, a cut is made at the left exon–intron junction (Fig. 13-22). The intron is then folded in such a way that it assumes the form of a lariat. Another particle, a U2 snurp, is believed to be involved in lariat formation in which the nucleotide with G at the free 5′ end of the intron becomes linked by a 5′–2′ bond to a nucleotide with the base A in the branch sequence. Following lariat formation, a cut, believed to be accomplished by U5 snurp, takes place at the right junction. The intron is now removed, and it assumes a linear form in which it is soon degraded. Exons are finally joined by 5′ to 3′ linkages.

The splicing mechanism includes a variety of proteins as well as the snurps. The various parts assemble on the precursor RNA as it is ready to be spliced. This aggregate forms a 50–60S particle that has a

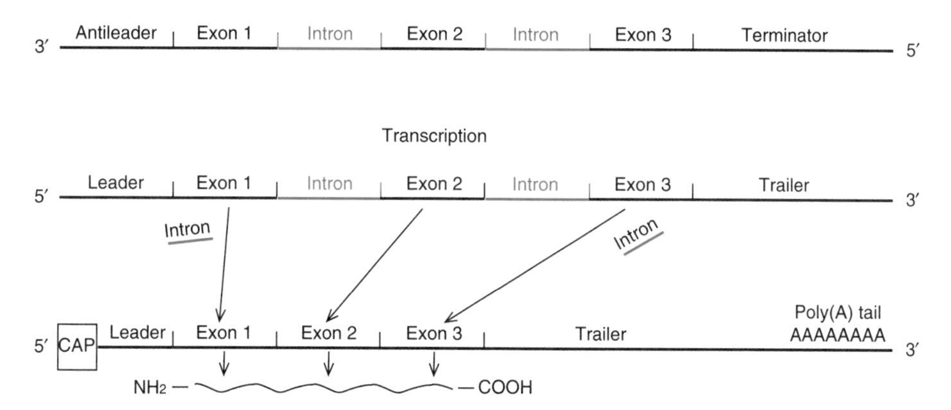

FIGURE 13-20. The fragmented gene of the eukaryote. **Top:** The region of the gene containing the information coded for a polypeptide is fragmented. The segments coding for amino acid sequences (exons) are interrupted by noncoding segments (introns). **Middle:** Both exons and introns are transcribed and are thus represented in the immediate transcript, the pre-mRNA. **Below:** The introns are snipped out enzymatically and the exons joined together during processing of the pre-mRNA to form the final mRNA containing an uninterrupted region coding for the amino acid sequence of a polypeptide. The exon order in the mRNA is the same as that in the DNA template. The exons in the gene and the mRNA are colinear with amino acid sequences in the polypeptide.

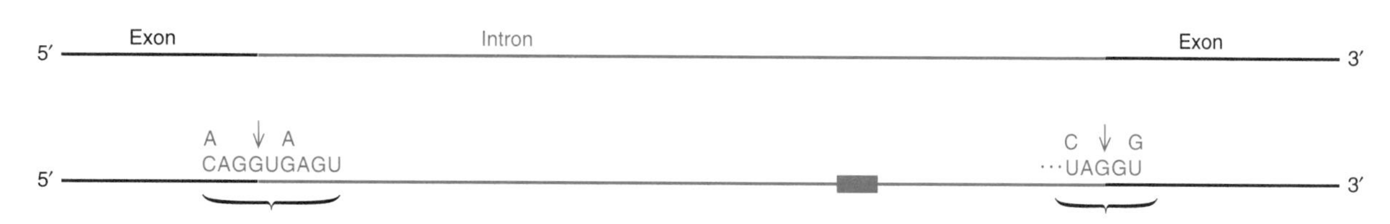

FIGURE 13-21. Exon–intron junctions of transcript. **Top:** A portion of a primary transcript is depicted bearing two exons separated by an intron. **Bottom:** An intron begins with the sequence GU at the 5′ end (left arrow) and ends with AG (right arrow) at the 3′ end. These sequences are part of two consensus sequences (brackets). Toward the 3′ end, a short sequence occurs that varies in its constitution and that is known as the *branch sequence* (red bar).

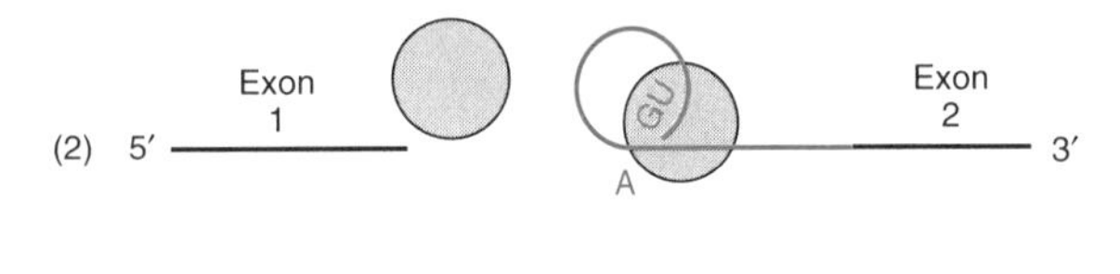

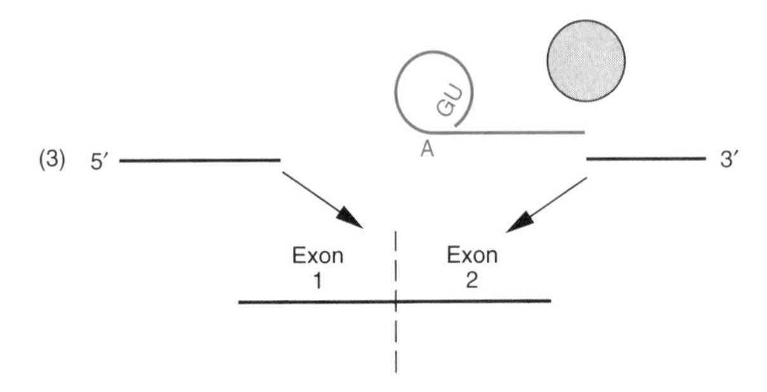

FIGURE 13-22. Splicing a transcript. **1:** A snurp binds to the 5′ exon–intron junction. **2:** A cut is made at the 5′ end of the intron that is then folded by a different snurp to produce a lariat. A bond is made between a G nucleotide at the free end of the intron and one with A in the branch sequence. **3:** Another snurp makes a cut at the 3′ end of the intron, which is then removed, and the exons are joined together.

characteristic shape and is known as a *spliceosome.* Many details remain to be filled in regarding the interactions among the parts of the spliceosome and the transcript that lead to a mature RNA ready for transport through the nuclear membrane to the cytoplasm where it can be translated. Splicing and processing of primary transcripts are related to the problem of gene regulation, and we will discuss them in this context in Chapter 17.

A very significant finding has been made regarding intron removal from eukaryotic precursor tRNAs. The anticodon loops of these molecules contain small introns that are removed by a process in which an endonuclease cuts at the two sites that mark the junctions of each intron with the exons found on either side of it. Following intron removal, ligase activity joins the cut ends of the anticodon loop. It has been clearly demonstrated in yeast and *Xenopus* that the intron in the anticodon loop of the precursor tRNA plays an active role in the splicing reaction. Recognition of the cutting sites by the endonuclease requires specific base-pairing between a base in the anticodon loop and one in the intron. The importance of the intron in the splicing reaction of eukaryotic pretRNA raises questions regarding the significance of introns, a discussion of which follows.

Introns and their significance

It is important to note that there is a difference between introns and spacers, such as those found along with genes for rRNA and tRNA. We have just seen that introns are found *within* a gene. Spacers, on the other hand, do not interrupt a sequence coded for a product. The transcribed spacer encountered in the rRNA unit (see Fig. 13-12) does not interrupt the sequences coded for the major classes of rRNA. We do know, however, that introns *do* occur within certain genes coded for tRNA and rRNA. In two lower eukaryotes (the protozoan, *Tetrahymena* and a slime mold), a single intron interrupts the sequence coded for the large rRNA. In yeast, some genes for tRNA contain a single intron that is always located near the start of the sequence for the anticodon loop.

Introns are now known to occur in both the Eubacteria, the prokaryotic line containing most of today's bacterial species, and the Archaebacteria, the ancient bacterial line represented today by only a few groups that live under extreme environmental conditions. The two prokaryotic lines as well as that of the eukaryotes are believed to have shared a common, distant ancestor. The discovery of introns in the two prokaryotic lines is therefore extremely pertinent to theories relating to the origin of introns. Introns are also found in the genes of organelles (see Chap. 21). We see that introns do occur in various kinds of genes, but they are most characteristic of eukaryotic genes coded for mRNA transcripts. The question of the significance of these sequences that interrupt the coding sequences of genes becomes a challenge to answer.

The discovery of introns has helped to clarify several puzzling observations about eukaryotic genes and their transcripts. In contrast to prokaryotes, eukaryotes contain DNA far in excess of the amount needed to code for all their proteins, even considering noncoding DNA essential for chromosome structure and gene regulation. It has been calculated that about 90% of the human DNA does not contain coding sequences for polypeptides! Equally puzzling was the observation that not only is RNA also present in excess but that large RNA strands occur that never leave the nucleus. These strands were designated *heterogeneous nuclear RNA* (hnRNA) to reflect their varying sizes. Their significance was unknown, and only much shorter RNA strands were found to enter the cytoplasm. We now know that some of the excess DNA of eukaryotes is found in introns and that hnRNA represents primary transcripts containing introns and exons, as well as transcripts in various stages of processing and splicing.

While the discovery of introns and RNA splicing answers certain questions, it fails to explain the spreading out of a gene over a distance many times longer than that needed to code for its product. One suggestion is that such an arrangement has value for evolutionary progress. The presence of noncoding, intervening sequences stretch out the coding sequences of a gene. Each exon, it is argued, carries information for a part of the polypeptide that performs a distinct functional role in the completed protein. This has been demonstrated for many genes. For example, each exon of the genes coding for antibodies, hemoglobin, and certain enzymes is coded for a portion of the protein that has a distinct function to perform. Being spread out in pieces, it is easier for the coded sequences to undergo recombination by crossing over than if they were together in one coded segment. Spreading out of the exons, each coded for an important part of the polypeptide, makes it easier for them to assemble into new combinations. This can result in the formation of polypeptides possessing advantages over those with older arrangements of the functional parts. The genes with their coding regions spread out would thus be selected by the forces of evolution.

The proponents of this idea consider introns to be additions to the gene introduced into the chromosomes during evolution of the eukaryotic cell. There is now reason to question this concept, since introns have been discovered in both the Archaebacteria and the Eubacteria, even though their occurrence is not extensive. Among the Eubacteria, an intron has been found in seven species of cyanobacteria, commonly known as blue-green algae. This intron occurs at the same position in the anticodon loop of a gene coded for a transfer RNA that is specific for leucine. Moreover, studies of RNA chemistry support the concept that the very first genes were RNA in nature. Chains of RNA can form spontaneously under conditions of a high concentration of salt and RNA molecules. One can argue that, before the origin of cells, the first genes were short chains of RNA coded for very simple proteins. The useful information coded into these primitive genes probably would have been interrupted by stretches that contained no useful information. These stretches would be the equivalent of the introns. Thus, according to this idea, introns were present from the beginning and were not introduced at random as the eukaryotic gene evolved.

There is good reason to believe that at the earliest time RNA may have achieved the removal of introns by self-splicing as a result of its intrinsic catalytic ability, examples of which are now known (see next section). According to the RNA-first concept, the gene eventually became DNA instead of RNA. How this transfer may have taken place is highly speculative, but there is support for the idea that the enzyme reverse transcriptase (see Chap. 16) may have been present very early in cellular evolution. Through the action of this enzyme, the genetic information found in RNA would have been converted to DNA, a molecule more stable than RNA and that would then have been selected as the genetic material. This DNA would have included both the exons and introns present in the RNA from which it was derived. As bacteria evolved, introns would have been almost completely eliminated in contrast to the eukaryotes which retained them.

While various theories regarding the significance of introns have been proposed, the picture becomes more complicated as new observations are made. We now know that certain introns that are located in some mitochondrial genes carry coded information. We pursue the topic of the mitochondrial genome in some detail in Chapter 21. Quite unexpected have been the findings of coded information in introns of two human chromosomal genes. The first case to come to light concerned the gene whose product, human clotting factor VIII, is required to prevent hemophilia A. Embedded in one of its introns is another gene, the function of which is still unknown. It has been shown, however, that transcription from the gene in the intron is not correlated with the clotting factor gene in which it is embedded.

The second case involves the gene associated with neurofibromatosis, a disfiguring neurological disease associated with many benign tumors that frequently return after removal and some of which at times may become malignant. This is one of the most common dominant autosomal disorders, affecting about 1 person in 3,000 in all ethnic groups equally. After being assigned to chromosome location 17q 11.2, the gene has finally been identified and described. It was found to be unusually large (from 500,000 to 2 million base pairs in length) and to have embedded within an intron at least three small genes! The exact functions of these are not as yet clear, but they are not believed to play a significant role in the affliction. Genes embedded in introns have also been reported in the fruit fly. The various examples of exceptional introns bearing functional genes or coding sequences prevent us from reaching conclusions on the origin of introns and their role in the economy of the cell.

RNA as a catalyst

With their highly diversified structures, proteins have been considered to be the only molecules with sufficient variation and complexity to act as enzymes in biological systems. We must reexamine this concept, however, for we now know that RNA can behave as a true catalyst as the following examples show.

The unusual enzyme ribonuclease P of *E. coli* discussed earlier in this chapter contains both an RNA and a protein component, each encoded by its own separate gene. This enzyme is responsible for generating the 5′ end of all tRNA molecules in *E. coli* by making appropriate cuts in tRNA precursors (Fig. 13-15). It is possible to dissociate the enzyme into its protein and RNA portions. Under in vitro conditions, it has been demonstrated that the RNA part by itself (which makes up the greater part of the enzyme) can make accurate cuts in a tRNA precursor. In vivo, however, both portions, RNA and protein, are required for enzyme activity as demonstrated by the fact that a mutation in either the gene for the protein or that for the RNA part of the enzyme can abolish ribonuclease P activity. Nevertheless, it is the RNA part of the intact enzyme, not the protein, that provides the catalytic activity.

Introns that occur in the rRNA genes of a species of slime mold are also capable of self-splicing (autocatalysis), as are those in mitochondrial genes of fungi. By itself, the isolated RNA can carry out self-rearrangement by autocatalysis and needs no protein, although the latter probably aids the process in vivo. In Chapter 16, we encounter viroids. These tiny, naked RNA molecules, which can infect plants, have all been shown in vitro to be capable of autocatalysis.

In the ciliate *Tetrahymena,* a 26S ribosomal RNA is formed from a 35S precursor following removal of a single intron. Excision of the intron and splicing of the RNA to form the 26S rRNA has been shown to result from the intrinsic splicing ability of the RNA itself. Furthermore, it has been demonstrated that the intron, when present in vitro with certain polynucleotides, is able to cleave and to rejoin them. The short RNA acts as a true enzyme with RNA polymerase activity. At a higher pH, it has been shown to possess ribonuclease activity. The catalytic capabilities of RNA have been shown to be even more extensive as the result of further experiments performed with this self-splicing intron from *Tetrahymena.* The intron RNA was altered in the laboratory to permit it to bind to an amino acid–tRNA fragment. The engineered RNA was found to possess a slight but significant amount of tRNA synthetase activity that clearly demonstrated its ability to make and break those bonds that join amino acids to tRNA.

Of primary importance to the field of molecular biology is the presentation of very strong evidence that the peptidyl transferase activity of the ribosome resides in the RNA component of the organelle and not in the protein. Following extraction of all its known proteins, the bacterial ribosome still retains most of its ability to form peptide linkages. Work is in progress to eliminate the slight possibility that a bit of protein remains following the rigorous extraction procedure.

The versatility of the various catalytic RNAs, *ribozymes*, has many implications regarding RNA activities in the cell, and causes us to reflect on the possibility that an ancient world existed in which RNA performed roles now associated with DNA and protein. The discoveries that RNA can make and break amino acid–tRNA bonds and that it most likely possesses peptidyl transferase activity lend strong support to this concept. We know that proteins cannot exist without DNA and that DNA cannot exist without certain specific proteins required for its replication and several of its other functions. The RNA world hypothesis offers a way out of this dilemma. Such a world—one that operated solely with RNA molecules—would have preceded our familiar one in which DNA is the primary molecule of genetic information. Living things would have stored their information in RNA and performed all the reactions required for a living system with RNA molecules exclusively. Both DNA and protein would be derivatives of this RNA world. Recognition of the range of activities of ribozymes places the hypothesis of an RNA world on a firm foundation. We now appreciate RNA as an active participant in living systems rather than a passive one. The RNAs of the ribosome may yet prove to be active molecules in the initiation, translocation, and termination activities that are essential to the process of translation.

It is very significant that the intron found in the tRNA gene of certain eubacteria (the cyanobacteria discussed above) belongs to a class of introns that possesses self-splicing ability. This ability has actually been demonstrated for the intron in several species of the cyanobacteria. Moreover, it has also been shown that chloroplasts contain an intron in a tRNA gene for leucine that appears to be the same as that found in the cyanobacteria. As we discuss more fully in Chapter 21, there is strong evidence that chloroplasts are descendants of ancient cyanobacteria that established a symbiotic relationship with a primitive ancestor of the eukaryotic cell. The finding of introns in cyanobacteria gives strong support to the argument that introns are very old and predate the origin of the eukaryotic cell. One can argue that the introns we see today in prokaryotes are remnants of an RNA world and that they have been lost from most prokaryotic genes but are retained in the eukaryotic line. The presence of the intron in the gene of the chloroplast would reflect the ancient origin of the intron. The intron would have already been present in the tRNA gene of the ancestor of the chloroplast and not added to the chloroplast gene later, after it took up residence in the primitive eukaryotic cell.

However, one can still argue in favor of the idea that introns arose later in cell evolution and point out that finding introns in a few bacterial species does not show that they were ever widespread in prokaryotes. Proponents of the "introns later" view propose that introns may have the ability to spread from the genome of one species to the next in much the same way that certain DNA elements are known to move about (see Chap. 17). Such an intron would insert at the same position in the DNA as the one it occupied in the previous genome. Therefore, according to this concept, the introns are invaders of bacterial genomes and do not represent ancient DNA sequences inherited from an ancestor common to both prokaryotes and eukaryotes.

Perhaps the question of the origin of introns may become clarified in future studies of other species of eubacteria as well as of very primitive eukaryotes. The DNA of such organisms has not been examined in detail until now and may provide clues to the route of evolution of the eukaryotic cell.

Heterogeneous nature of eukaryotic DNA

We have noted that much more DNA exists in a eukaryotic than in a prokaryotic cell. Moreover, the percent GC content of the DNA in lower forms, especially in prokaryotes, shows great variability from one species to the next, whereas in eukaryotes the value is much more stabilized—about 40% in mammals.

Another distinction between prokaryotic and eukaryotic DNA is seen when DNA is isolated from cells, fragmented into segments of comparable size, and then subjected to density gradient ultracentrifugation. Handled in this way, the prokaryotic DNA yields one band following centrifugation, indicating that the separate fragments of comparable size are approximately the same in density and hence in their base composition. We have seen in this chapter that, when eukaryotic DNA is processed in this way, centrifugation reveals that the DNA is heterogeneous and consists of a main band (including the bulk of the DNA) and one or more satellite bands. This means that the GC and AT pairs are not uniformly distributed on the fragments but may be clustered in certain regions, yielding DNA fragments of quite different densities. Some of the satellite DNA represents the DNA of the nucleolar organizer. Moreover, certain satellite bands represent the DNA found outside the nucleus in organelles and chloroplasts. However, the term *satellite band* is generally taken to refer only to those minor bands that are nuclear in origin.

A most unusual feature of eukaryotic DNA is its inclusion of base sequences of varying lengths, which are repeated throughout the bulk of the DNA. This difference between prokaryotes and eukaryotes can be demonstrated as follows. DNA extracted from a prokaryotic cell may be fragmented into pieces, all approximately the same size, perhaps 500 nucleotide base pairs long. The DNA is then subjected to a high temperature so that the DNA strands separate (denature), and single-stranded DNA fragments arise. The material is then cooled, permitting the single strands to rejoin and form double-helical DNA segments (renaturation). If only one copy of a stretch of a nucleotide sequence is present, then renaturation can occur only when a single-stranded DNA fragment encounters its original partner or the same complementary stretch from another cell. This means that renaturation tends to occur slowly. The course of the renaturation of the DNA is expressed in the form of a Cot curve in which the fraction of DNA molecules that have reassociated is plotted against the log of Cot. A Cot value is equal to Co $\times$ t, where Co is the starting concentration of single-stranded DNA in moles per liter and t is the time, in seconds, that the incubation has proceeded. In *E. coli* (Fig. 13-23A), renaturation occurs slowly, progressing at a constant rate for the first 10 minutes, then slowing down even more. This is so because as strands manage to find their complements, fewer and fewer single strands are present, and the chance that a complementary single-strand encounter will occur is reduced.

When this same procedure is followed with DNA from eukaryotic cells, a different picture emerges. At first a very fast renaturation takes place, as indicated by the very low Cot values (Fig. 13-23B). This tends to level off and the appearance of the curve then comes to resemble that of *E. coli,* except that the DNA reassociations occur at higher Cot values. Renaturation studies have revealed that the DNA of most eukaryotes is heterogeneous in the sense that it includes very many copies of certain nucleotide sequences and just single copies of others. Human DNA, for example, contains approximately 2.5×10^9 base pairs, and about 20%–30% of this consists of multiple copies of certain nucleotide sequences. The DNA of eukaryotes can be classified into three general categories on the basis of the kinetics of DNA renaturation. The DNA that renatures slowly (at a high Cot value) is nonrepetitive, composed of single-copy sequences. When the eukaryotic DNA is fragmented, a given stretch of nonrepetitive DNA, represented as just a single copy on a given chromosome, can unite only with its original complementary partner or with the same complementary stretch from another cell. This DNA renatures slowly, since most chance encounters do not bring together complementary strands.

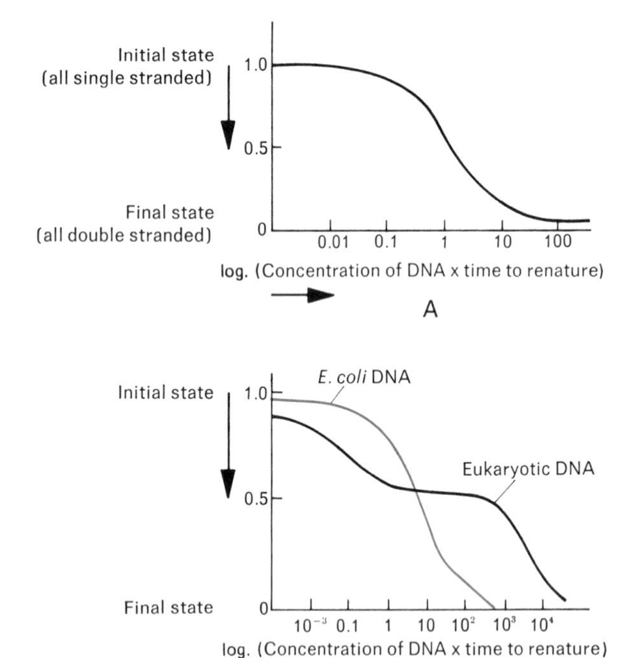

FIGURE 13-23. DNA renaturation curves. **A:** Generalized renaturation curve expected for DNA in which stretches of nucleotide sequences occur only in single copies. Curves for prokaryotes, such as *E. coli,* approach this idealized curve. **B:** DNA renaturation curves of *E. coli* and a typical eukaryote compared.

A renaturation experiment can be monitored to permit the isolation of fragments that have reassociated to produce duplexes at a particular Cot value. This is accomplished by passing the DNA preparation over hydroxyapatite columns at a given incubation time. The fragments that are double stranded bind to the hydroxyapatite and can then be isolated from those that have not renatured and used in further analyses.

The slow renaturing, single-copy DNA has been shown to include the bulk of the structural genes, those sequences coding for polypeptides. Not all the single-copy sequences in this class consist of such genes, however. Many other single-copy sequences occur that are not associated with any specific product.

A second class that can be recognized by the kinetics of reassociation is the moderately repetitive DNA. In this group of fragments, the nucleotide sequences are not all identical by any means. The fragments include various stretches that are sufficiently alike to permit a single strand to pair not just with its exact complement but also with other strands that bear similar, though not completely identical, sequences. Within this kinetic class, one can recognize families. The repeated sequences recognized as a family, though clearly similar, are not necessarily identical due to effects of mutation that cause them to diverge. The most frequently occurring base sequence is the one that is taken to represent a given family. Within a family, the related single-stranded fragments are able to pair during a renaturation experiment and to form stable duplex DNA that can bind to the hydroxyapatite. The conditions of the experiment can be so altered (by changing the temperature, for example) that fragments that are less related in nucleotide sequence can hybridize. The duplex DNA formed under these conditions would thus contain several regions that are matched but some regions that remain unpaired. Depending on the experimental conditions, these may be sufficiently stable to permit their binding to the hydroxyapatite and thus to fall into the middle repetitive class. This DNA class, therefore, includes an assortment of fragments from different families, and duplexes may form that are not completely complementary through their length.

A third class is the highly repetitive DNA. Short sequences occur in this group, which may be repeated in tandem a million or more times! The fragments bearing these highly repeated sequences therefore reassociate at very low Cot values. The term *satellite DNA* is often used to refer to this group of sequences, since the highly repetitive DNA frequently forms a satellite band in a cesium gradient. This is because of the differences in its A + T:G + C ratio compared with the bulk of the cellular DNA.

The distinctions between highly repetitive, moderately repetitive, and single-copy DNA are not clear-cut but are somewhat arbitrary. Other classes or subclasses can even be recognized for certain eukaryotic DNAs. One additional type of DNA has now been identified and can be considered a fourth kinetic DNA class. This is referred to as *fold back, snap back,* or *palindromic DNA.* The naming and the behavior of this DNA are based on the fact that a fragment in this class contains an inverted repeat (refer back to Fig. 12-25). When a DNA fragment with an inverted repeat is rendered single stranded, each strand can fold back on itself very rapidly during a renaturation experiment, since a collision with another strand is unnecessary. The fold-back DNA can thus bind almost immediately to the hydroxyapatite just a few seconds after the reassociation experiment has begun.

What is the significance of these different kinetic classes that are so typical of eukaryotic DNA? What are the roles of these different classes, and how are they arranged along the intact, unfragmented DNA of the chromosome? Many investigations have been performed to answer these questions. DNA representing a given kinetic class that has reassociated at a certain Cot value can be prepared for in situ hybridization studies to determine its chromosome distribution. The unique or single-copy DNA has been shown to be rather evenly distributed along the chromosome, not concentrated in any particular locations. In the moderately repetitive fraction, which represents approximately 20% of the total cellular DNA, some families have been shown to be dispersed, as is true of the unique DNA, whereas other families are clustered in tandem repeats at certain chromosomal regions such as the centromeres and telomeres. In this moderately repetitive fraction, some genes are found whose functions are well known. These include genes for the histone proteins, tRNA, ribosomal RNA, β-globins, and others. However, these sequences with known functions represent only a small fraction of the moderately repetitive DNA and constitute less than 1% of the total chromosomal complement. Most of this kinetic class of DNA, as is true of the highly repetitive DNA, is made up of sequences that have no known function. These moderately repetitive sequences are dispersed all over the genome and may be repeated up to 10^5 times.

Some of the highly repetitive DNA sequences are dispersed, but most of them occur in tandem arrays concentrated at the centromeres and telomeres. The amount of this highly repetitive or satellite DNA can vary greatly in amount when different eukaryotes are compared. In mammals the satellite DNA constitutes approximately only one-third of the total repetitive DNA. In contrast to the moderately repetitive DNA, this highly repetitive class does not contain as many different families, and most of it is not transcribed. When two closely related higher eukaryotic species are compared, similarities are found in the sequences; however, the sequences of the one species are always distinct from those of the other. These decided differences evident in the nucleotide sequences of satellite DNAs taken from related species indicate that, once a change occurs in a repeat family, it spreads very rapidly throughout the genome. Within any one species, however, the sequence in a family of satellite DNA is exceptionally uniform. This indicates that some device must keep these repeated sequences similar. This is reminiscent of the situation that exists with respect to the repeated units containing genes for rRNA (discussed earlier in this chapter).

Patterns of repetitive DNA in eukaryotes

Unlike the highly repetitive DNA sequences, which tend to occur as clusters following each other in tandem, the moderately repetitive DNA contains sequences that are scattered throughout the genome, some of them occurring in many thousands of copies. A significant portion of this moderately repetitive DNA is composed of short interspersed sequences (SINES). When genomes from very different eukaryotic species are compared, a characteristic arrangement of SINES with unique sequences is seen. SINES measuring about 100–300 base pairs in length are typically interspersed with single-copy sequences approximately 1,000–2,000 base pairs long (Fig. 13-24). Note that a repeat may be direct or inverted, a distinction that could be detected as snap back DNA in a renaturation experiment.

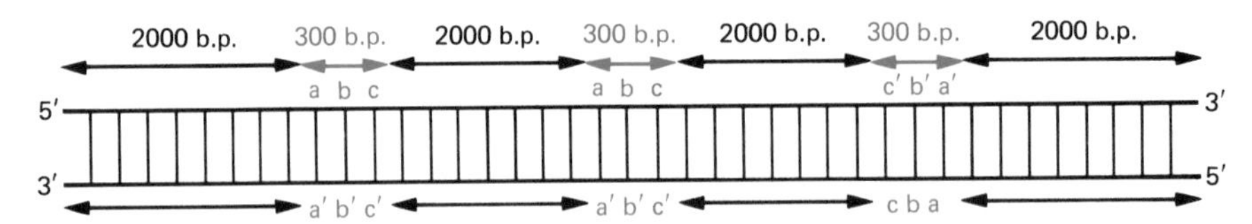

FIGURE 13-24. Pattern of dispersed repeats in eukaryotic DNA. A portion of a double-stranded DNA is depicted. Along the length of the DNA, in eukaryotes, dispersed repeats (red) approximately 100–300 base pairs in length are typically interspersed with unique sequences approximately 2,000 base pairs long. A repeated sequence may occur as a direct or as an inverted repeat (right).

This pattern of SINES is common in eukaryotic species as diverse as sea urchins, amphibians, and mammals. In the sea urchin, however, the short repeats belong to many unrelated families; no one kind of repeat is most numerous. On the other hand, the DNAs of rodents and primates contain a family of SINES that is much more abundant by far than any other in the genome. This is the Alu family, which in human DNA consists of 300 base pair long sequences repeated between 3×10^5 and 5×10^5 times and accounts for about 3% of the total human DNA. Approximately 300,000 copies or more are found in the haploid chromosome complement. The name *Alu* derives from the fact that nucleotide sequences occur in these repeats that can be cut by a certain restriction enzyme, AluI (see Chap. 16).

The Alu sequences make up a large population of the total interspersed 300 base pair repeats in human DNA. The distance between Alu repeats throughout the human genome varies when different chromosome regions are compared. Members of this abundant, widely dispersed Alu family are not necessarily identical, although they are related in sequences, and the characteristic stretch of nucleotides identifying a human Alu sequence has been recognized. Although evolutionary divergence has occurred within members of the Alu family, there has been at the same time a great deal of conservation in nucleotide sequence over the length of the 300 base pair stretch. The DNAs of other primates and rodents contain a prominent Alu-equivalent family of interspersed repeats, but, unlike human Alu, the length is only about 130 base pairs. Although the Alu family is indeed a prominent one, several other very abundant repeats are now being described in mammalian DNA. It is estimated that Alu sequences, along with two other family members, make up about half of the repeat families in human DNA.

In addition to the SINES, in which the repeat is less than 500 base pairs long, there are long interspersed sequences (LINES) in the eukaryotic genome that measure more than 5,000 base pairs in length and may be repeated as much as 100,000 times in the human. These LINES, along with the SINES, account for the largest amount of the moderately repetitive eukaryotic DNA. Recognition of the very repetitive nature of eukaryotic DNA raises the question of the significance of repeated sequences. Highly repetitive DNA does not appear to undergo transcription to any significant degree. This observation has led to the suggestion that it is not coded for any polypeptide product and that any role it may perform is involved with the structural organization of the chromosome. In contrast, SINES may undergo transcription. Some Alu family members appear to be transcribed by RNA polymerase II, but the transcripts appear to be part of the hnRNA and to be removed before transport of mature mRNA to the cytoplasm. Members of the Alu family may also be transcribed as short RNA molecules by RNA polymerase III. However, only a small proportion of any interspersed repeat family ever undergoes transcription, indicating that most of it is inactive. As is the case for the highly repetitive DNA, any function of the moderately repetitive DNA, SINES as well as LINES, remains unknown.

According to one school of thought, repetitive DNA may actually have no function at all. Indeed, it is highly probable that most of the DNA of higher forms does not code for polypeptides. For many years, several perplexing facts about eukaryotic DNA have been apparent. When the amount of DNA in a haploid chromosome set (the *C value*) of one eukaryote is compared with that of others, we can see that the range is extreme. Quite amazingly, the degree of complexity of an organism does not seem to be related directly to its C value. Although the amount of DNA in the human is 800 times that in *E. coli,* certain plant species and amphibians have 30 times as much DNA as humans. Very perplexing is the fact that the DNA content varies greatly among certain closely related species. In amphibians, DNA content among species can vary by a factor of 100. In buttercup species of similar karyotype, the content between some types may differ by a factor as great as 80. How can we explain the fact that the karyotypes of the plants are similar, as are the morphological resemblances, and yet their C values are quite different? In some cases, species differing in C value can be crossed to produce offspring. Bivalent formation in the hybrid looks quite normal, implying that the extra DNA is fairly evenly distributed over the chromosomes.

Various estimates suggest that in mammals not more than 1% of the genome may be involved in regulation or in coding for proteins. The mammalian genome has enough DNA to code for about 3 million proteins, but it does not seem likely that it is coded for more than 30,000 of them. While some of the noncoding DNA may indeed be involved in the structure and organization of the chromosome, there still appears to be too much DNA. According to the "selfish DNA" concept, many of the repeated DNA sequences have no function whatsoever and are not essential for cell survival. Their persistence in the genome results from the fact that the sequences have been selected for their ability to spread and duplicate efficiently in the genome without harming the host organism. We will return to this subject in Chapter 17, where the ability of the dispersed sequences to spread throughout the genome will be discussed in terms of the possibility that they may represent mobile DNA elements.

Allelism and the functional unit at the molecular level

Our knowledge of the gene and its interactions at the molecular level has provided us with a sound physical basis for some established genetic concepts. This is well demonstrated by a return to the subject of allelism.

In Chapter 5 (Fig. 5-1), we discussed the multiple allelic series that included vg^+ (wild), vg (vestigial wing), and vg^a (antlered wing). Since vg and vg^a represent separate recessive mutations in the same gene at the vg locus, a cross of vestigial (vgvg) and antlered (vg^avg^a) flies results in mutant offspring ($vgvg^a$). The antlered and vestigial mutations cannot complement each other, because a fly of either type lacks a wild-type allele at the vg locus.

Now let us consider another recessive mutation, dumpy (dp), which also alters the shape of the wing, making it inadequate for flying. The dp locus is located on chromosome II, which is also the case for the vg locus. When dumpy flies are crossed to vestigial (or antlered flies), the F_1 is found to consist of all wild-type flies. We can readily understand this, because the dp and vg loci, though they both affect the wing and are on chromosome II, are nevertheless separate and distinct. Genes residing at the two loci are different genes. In other words, dp and vg are *not* allelic to each other. When two wing mutants are crossed (Fig. 13-25A), each parent contributes a normal form of the "wing" gene.

Here again we see that two separate genes at different loci can affect the same characteristic. Since the loci dp and vg are far apart, we can expect crossing over to occur between them. As this does indeed take place, the dihybrid with the trans arrangement can produce chromosomes with both mutant alleles on the same chromosome and both wild alleles on the other (Fig. 13-25B). As a result, dihybrid flies that have the cis arrangement can arise.

Although these points are simply a review of basic principles, it is important to realize here that, when we are dealing with two recessive mutations at different loci, the F_1 heterozygote in cis or trans will be wild. This is so because both heterozygotes contain a normal dominant allele for each of the defective ones. We say that mutations such as dp and vg complement each other. This means that, when two mutant forms are crossed, the recessive defect carried by either mutant parent can be "covered up" by the corresponding normal allele contributed by the other mutant parent. The F_1 resulting from a cross between dp and vg is normal, since the dp parent contributes a functional vg^+ allele and the vg parent a normal dp^+ allele. Two completely normal wing alleles are present, and thus complementation takes place. This is in contrast with the cross between vestigial and antlered, in which the two mutations do not complement each other (Fig. 5-1).

With these points in mind, we see that it is possible to tell if two recessives are allelic. If a cross between two organisms that show mutant phenotypes produces normal offspring, the mutations being followed complement each other, and the defects must be in two separate genes. However, if a cross between two mutant forms yields offspring that are also mutant instead of wild, we know that the two mutant genes are not complementary. They must be allelic to each other and therefore represent defects in the same gene. If we are

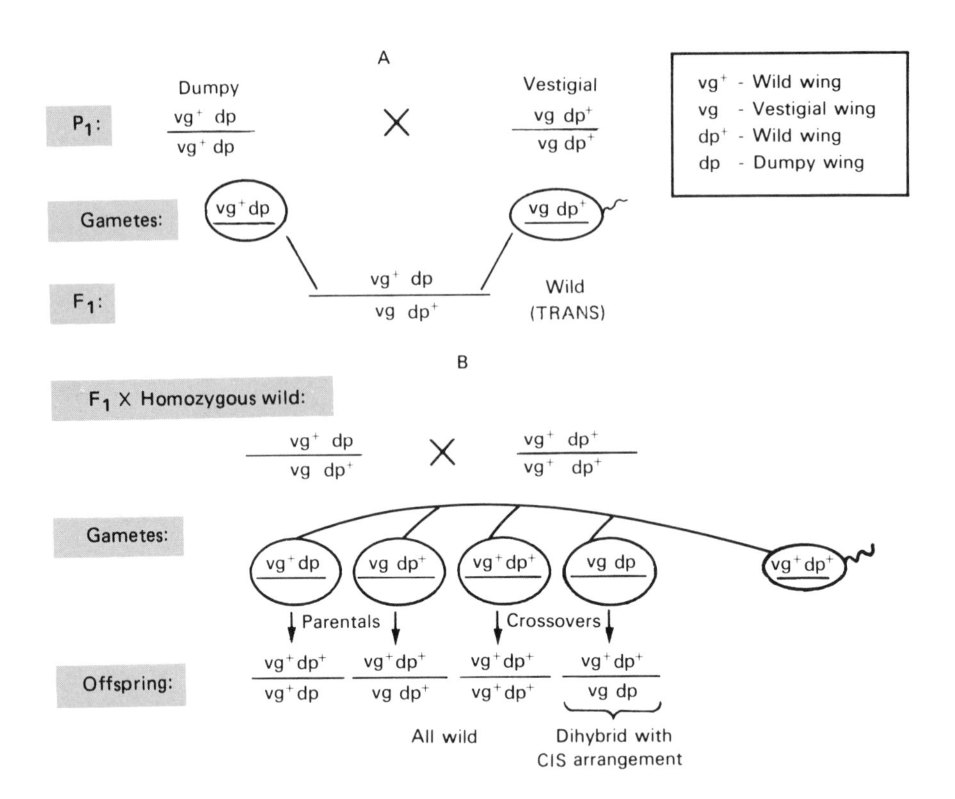

FIGURE 13-25. Cis and trans arrangements in *Drosophila.* **A:** A cross of two mutants, homozygous for recessives at two different loci, produces an F_1 that is wild and carries alleles in the trans arrangement. The F_1 is wild in phenotype, even if the parents showed a mutant trait for the same characteristic. This is so in this example because the dumpy parent contributes the wild allele of the vestigial, and the vestigial parent contributes the wild of dumpy. We say that such alleles complement each other. **B:** Since vestigial and dumpy are separate loci, crossing over may occur between them. This can give rise to offspring that are dihybrid and have the cis arrangement. These dihybrids are wild, as are those with the trans arrangement.

following mutations that represent alterations at two separate and distinct sites within the same gene, it should be possible to detect crossing over between them. Although intragenic crossing over occurs in both higher and lower forms, it can be detected much more readily in microorganisms. The reason for this is simply a reflection of what we have already learned about linkage and chromosome mapping. Genes that are closely linked on the DNA of the chromosome tend to stay together in the parental combination. There is less crossing over between them than between two genes that are widely separated. And, obviously, different sites within the same gene are very closely linked, usually more so than two sites in two different genes. Therefore, to detect crossing over within a gene, we need to raise a very large number of offspring, because the new combinations (the recombinants) will be very few. To detect intragenic crossing over, even in a species as prolific as the fruit fly, requires the breeding and examination of thousands of offspring. With microorganisms (viruses, bacteria, and certain molds), it is possible to raise millions of progeny in a very short time. And techniques are available for the detection and scoring of recombinants. This extremely rapid rate of multiplication is one of the main reasons that crossing over within genes was fully analyzed in microorganisms. Intragenic mapping is discussed in more detail in Chapter 16.

Cases of crossing over within the gene came to light slowly in *Drosophila,* and their true nature had to await clarification from studies with microorganisms. Since the mapping of sites within a gene entails vast numbers of offspring, intragenic maps in higher organisms are few compared with those in microorganisms. In the fruit fly, several genes have been mapped for a few mutant sites, and progress has been made in mapping certain genes in mice that influence coat color (the yellow gene) and the development of the embryo (the tailless gene).

Let us now see how allelism can be explained in light of our knowledge of molecular interactions. We have considered the gene in relationship to its function at the molecular level, the transfer of information to direct the formation of a specific polypeptide chain. Any gene, because it has a certain function to perform, therefore may be considered as a *unit of function*. Each gene, or unit of function, contains smaller units within it, the nucleotides. Mutation may occur at any one of these sites within a gene, and, as we will see in more detail in Chapter 16, crossing over can take place between nucleotides within the gene. Therefore the gene, the unit of function, is composed of smaller units of mutation and crossing over, the nucleotides. Each gene is a stretch of DNA with a specific function to perform, and each includes separate sites of mutation and crossing over. The term *cistron* has been coined to refer to the gene as a unit of function. The words *gene* and *cistron* mean the same thing insofar as both designate a genetic unit with a specific function. The term *cistron,* however, implies that the functional unit, the ordinary gene, is divisible. It represents a genetic region within which separate mutations can occur and crossing over can take place. It is equally correct to say "one gene–one polypeptide" or "one cistron–one polypeptide."

We can now examine these ideas in terms of gene function. Suppose two separate recessive mutations arise that cause the blocking of a metabolic step controlled by a certain enzyme (Fig. 13-26A). The blocking is shown to result from the production of a defective enzyme by the mutant types. The genetic defects in each mutant are mapped and are shown to lie at closely linked, but nevertheless separate, sites on the chromosome. The question now arises, "Are these defects separate mutations within the same gene or are they in different genes?" A cross is made between the two mutant types, each carrying a mutation that maps at a different site. When the offspring are produced, they are also found to be mutant, defective for this same enzyme.

What does this mean in terms of the unit of function (Fig. 13-26B)? Since the mutations are not complementary, no normal enzyme can be produced by the F_1 offspring. This is so for the following reasons. One parent contributed a genetic factor with a defect at a certain site, and this results in the production of a defective polypeptide, one incapable of catalyzing the needed metabolic step. The other parent also contributed a defective genetic factor—a form of the same gene that is defective in the first parent—and so the same polypeptide is affected. The only difference between the two mutants is that they carry defects at different sites within the same gene. The two factors are allelic to each other. Therefore, the trans arrangement produces mutant types, because no functional polypeptide governed by that one gene is being produced. No intact enzyme can be formed.

Crossing over in the F_1 offspring with the trans arrangement can yield a doubly defective allele and one that is normal (Fig. 13-26C). Any individual carrying both a normal allele and a gene form with two defects within it would produce normal enzyme. This is so because the individual with the cis arrangement possesses one completely normal allele (B in Fig. 13-26C). Contrast this with the F_1 heterozygote with the trans arrangement. The latter carries no completely normal unit of function for the formation of the necessary polypeptide and is consequently mutant. Therefore, when crossed, if two recessive mutant types produce an F_1 that is defective for the same characteristic (here a specific enzyme), the two separate mutations are allelic; they affect the same unit of function. The trans arrangement gives a mutant phenotype, whereas the cis yields a wild type. This is a demonstration of the cis–trans effect.

Now let us apply exactly the same kind of reasoning to another case of two separate recessive mutations (Fig. 13-27A). In this example, however, a cross of two mutants defective for the same enzyme gives an F_1 that is wild, one that can produce normal enzyme and carry out the essential step. This means that the two mutations can complement each other in the trans arrangement. The defects must be in separate units of function, separate cistrons. They complement each other because gene A controls a polypeptide chain that is distinct from the polypeptide chain controlled by gene B. For the enzyme to be functional, there must be at least one good chain A and one good chain B. Mutant A contributes a defective A but a good B. Mutant B contributes a good A but a defective B. As a result, the F_1 has some good A and some good B and can therefore form functional enzyme. The hybrid with the cis arrangement will also form functional enzyme (Fig. 13-27B). No cis–trans effect is apparent.

The cis–trans test is a fundamental one used to determine allelism. The different results in the cis and trans arrangements are easily understood on the basis of gene function and molecular interactions.

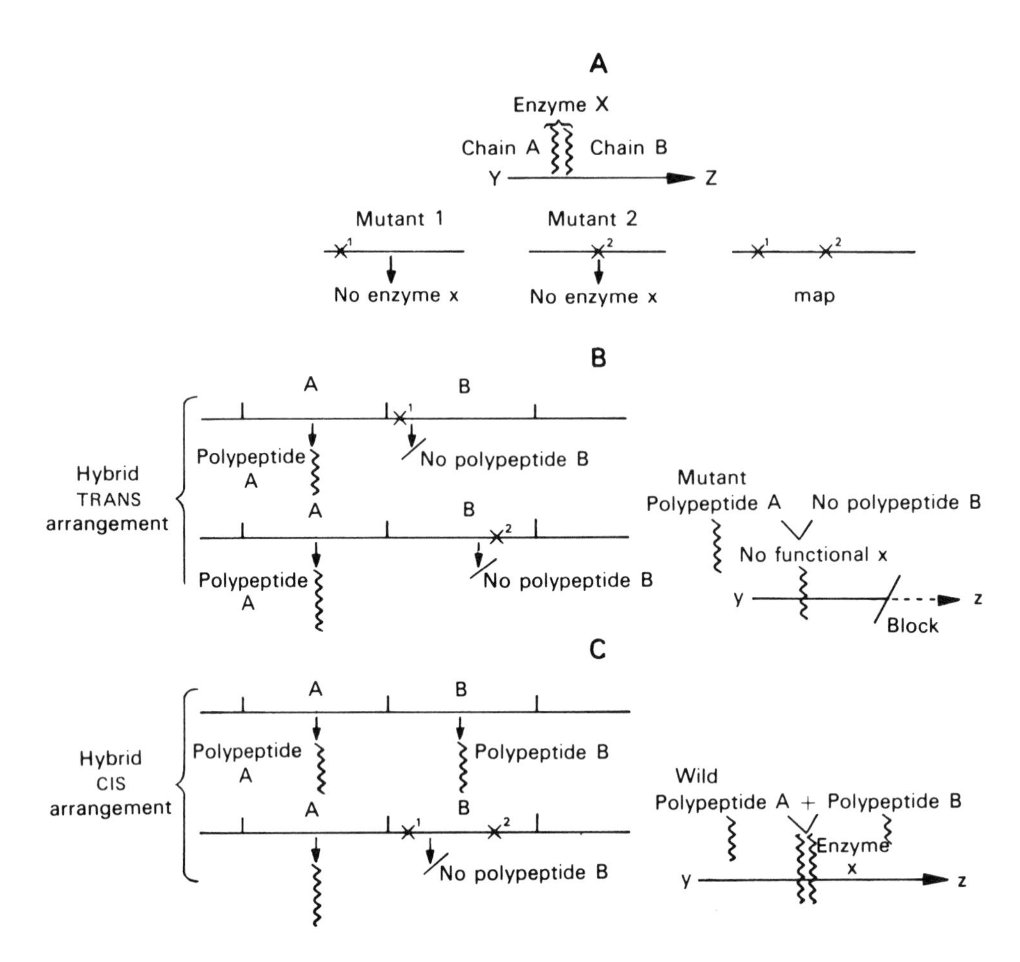

FIGURE 13-26. Lack of complementation. **A:** Assume enzyme X, which is composed of two kinds of polypeptide chains, is needed to catalyze the step in which Y is changed to Z (above). Two separate mutations occur that block formation of enzyme X, so that the step does not proceed (below). When the sites of the mutations are mapped, they are found to occur at separate, closely linked positions. **B:** When the two mutants are crossed in this example, the combination of the two separate mutations gives no enzyme. This is to be expected if the two defects are in the same gene. In this diagram, both are represented in gene B. If the normal enzyme structure depends on both polypeptide A and polypeptide B, the F_1 hybrid will not be able to form normal enzyme, because no normal polypeptide B can arise. This follows from the fact that the F_1 received a defective B allele from each parent. The two separate defects or mutations do not complement each other. **C:** An individual with the mutations in the cis arrangement will be able to produce normal enzyme. This is so because at least one completely normal B allele is present to form the required polypeptide. Both defects are within the same stretch of DNA, giving a doubly defective allele, but the completely normal one can carry out the essential function.

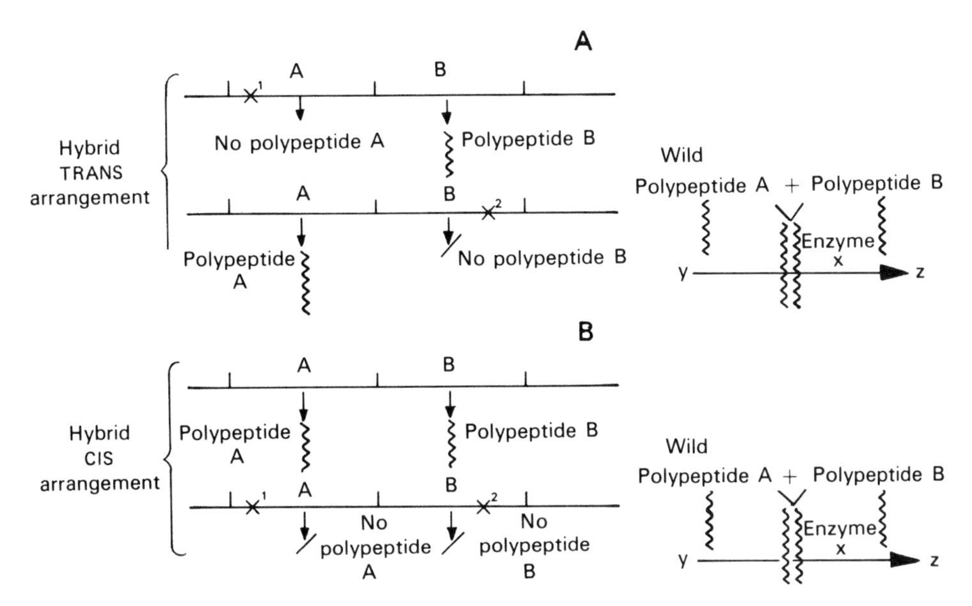

FIGURE 13-27. Cis–trans arrangements. Complementation. **A:** In this example, an F_1 hybrid is formed between two mutants (A and B), each with a defect in a separate gene. Consequently, one functional A allele and one functional B allele are present to allow formation of the enzyme. **B:** This cis arrangement is wild, because one chromosome contains both the functional A and the functional B alleles of two different genes.

REFERENCES

Alberts, B. et al. *Molecular Biology of the Cell,* 2d ed. Garland, New York, 1989.

Baldi, M.I. et al. Participation of the intron in the reaction catalyzed by the *Xenopus* tRNA splicing endonuclease. *Science* 255:1404, 1992.

Been, M.D., and T.R. Cech. RNA as an RNA polymerase: Net elongation of an RNA primer catalyzed by the *Tetrahymena* ribozyme. *Science* 239:1412, 1988.

Benner, S.A., A.D. Ellington, and A. Tauer. Modern metabolism as a palimpsest of the RNA world. *Proc. Natl. Acad. Sci. USA* 86:7054, 1989.

Britten, R.J., and D.E. Kohne. Repeated DNA *Sci. Am.* (April) :36, 1970,

Cech, T.R. The chemistry of self-splicing RNA and RNA enzymes. *Science* 236:1532, 1987.

Cech, T.R. RNA as an enzyme. *Sci. Am.* (Nov) :64, 1986.

Chabot, B., D.L. Black, D.M. Le Master, and J.A. Steitz. The 3′ splice site of pre-messenger RNA is recognized by a small ribonucleoprotein. *Science* 230:1344, 1985.

Chambon, P. Split genes. *Sci. Am.* (May) :60, 1981.

Clark, B.F.C., and K.A. Marcker. How proteins start. *Sci. Am* (Jan) :36, 1968.

Crick, F.H.C. Split genes and RNA splicing. *Science* 204:264, 1979.

Crick, F.H.C. Codon-anticodon pairing: The wobble hypothesis. *J. Mol. Biol.* 19:548, 1966.

Darnell, J. E., Jr. RNA. *Sci. Am.* (Oct) :68, 1985.

Doolittle, W.F. What introns have to tell us: Hierarchy in genome evolution. *Cold Spring Harbor Symp. Quant. Biol.* 52:907, 1987.

Fabrizio, P., and J. Abelson. Two domains of yeast U6 small nuclear RNA required for both steps of nuclear precursor messenger RNA splicing. *Science* 250:404, 1990.

Federoff, N.V. On spacers. *Cell* 16:697, 1979.

Gall, J.G. Chromosome structure and the C-value paradox. *J. Cell Biol.* 91:3s, 1981.

Gall, J.G., and M.L. Pardue. Formation and detection of RNA–DNA hybrid molecules in cytological preparations. *Proc. Natl. Acad. Sci. USA* 63:378, 1969.

Gilbert, W. The exon theory of genes. *Cold Spring Harbor Symp. Quant. Biol.* 52:901, 1987.

Gilbert, W. Genes in pieces revisited. *Science* 228:823, 1985.

Jelinek, W.R., and C.W. Schmid. Repetitive sequences in eukaryotic DNA and their expression. *Annu. Rev. Biochem.* 51:813, 1982.

Jukes, T.H., et al. Evolution of anticodons: Variations in the genetic code. *Cold Spring Harbor Symp. Quant. Biol.* 52:769, 1987.

Korenberg, J.R., and M.C. Rykowski. Human genome organization: Alu, Lines, and the molecular structure of metaphase chromosome bands. *Cell* 53:391, 1988.

Kozak, M. How do eukaryotic ribosomes select initiation regions in messenger RNA? *Cell* 15:1109, 1978.

Kuhsel, M., R. Strickland, and J.D. Palmer. An ancient group I intron shared by eubacteria and chloroplasts. *Science* 250:1570, 1990.

Lake, J.A. The ribosome. *Sci. Am.* (Aug) :84, 1981.

Lerner, M.R., and J.A. Steitz. Snurps and scyrps. *Cell* 25:298, 1981.
Lewin, R. Repeated DNA still in search of a function. *Science* 271:621, 1982.
McSwiggen, J.A., and T.R. Cech. Stereochemistry of RNA cleavage by the *Tetrahymena* ribozyme and evidence that the chemical step is not rate-limiting. *Science* 244:679, 1989.
Miller, O.L. The nucleolus, chromosomes, and visualization of genetic activity. *J. Cell Biol.* 91:15s, 1981.
Noller, H.F., V. Hofforth, and L. Zimniak. Unusual resistance of peptidyl transferase to protein extraction procedures. *Science* 256:1416, 1992.
Orgel, L.E., and F.H.C. Crick. Selfish DNA: The ultimate parasite. *Nature* 284:604, 1980.
Padgett, R.A., S.M. Mount, J.A. Steitz, and P.A. Sharp. Splicing of messenger RNA precursors is inhibited by antisera to small nuclear ribonucleoprotein. *Cell* 351:101, 1983.
Piccirilli, J.A., J.S. McConnell, A.J. Zang, H.F. Noller, and J.R. Cech. Aminoacyl esterase activity of the Tetrahymena ribozyme. *Science* 256:1420, 1992.
Rich, A.R., and S.H. Kim. The three dimensional structure of transfer RNA. *Sci. Am.* (Jan) :52, 1978.
Roberts, L. Down to the wire for the NF gene. *Science* 249:236, 1990.
Rould, M.A., J.J. Perona, D. Soll, and T.A. Steitz. Structure of *E. coli* glutaminyl-tRNA synthetase complexed with tRNA Gln and ATP at 2.8 A resolution. *Science* 246:1135, 1989.
Ruskin, B., and M.R. Green. An RNA processing activity that debranches RNA lariats. *Science* 229:135, 1985.
Schmid, C.W., and W.R. Jelinek. The Alu family of dispersed repetitive sequences. *Science* 216:1065, 1982.
Sharp, P.A. RNA splicing and genes. *J. Am. Med. Assoc.* 260:3035, 1988.
Sharp, P.A. Splicing of messenger RNA precursors. *Science* 235:766, 1987.
Smith, M.M. Concerted evolution in multigene families. *Science* 251:308, 1991.
Spiegelman, S. Hybrid nucleic acids. *Sci. Am.* (May) :48, 1964.
Steitz, J.S. "Snurps." *Sci. Am.* (June) :56, 1988.
Wallace, R.W., et al. Type I neurofibromatosis gene: Identification of a large transcript disrupted in three NF1 patients. *Science* 249:181, 1990.
Waldrop, M.M. Did life really start out in an RNA world? *Science* 246:11248, 1989.
Watson, J.D., N.H. Hopkins, J.W. Roberts, J.A. Steitz, and A.M. Weiner. *Molecular Biology of the Gene.* Benjamin-Cummings, Menlo Park, CA, 1987.
Xu, M-Q., et al. Bacterial origin of a chloroplast intron: Conserved self-splicing group I introns in cyanobacteria. *Science* 250:1566, 1990.
Yanofsky, C., B.C. Carlton, J.R. Guest, D.R. Helinski, and U. Henning. On the colinearity of gene structure and protein structure. *Proc. Natl. Acad. Sci. USA* 51:226, 1964.
Zaug, A.J., and T.R. Cech. The intervening sequence RNA of *Tetrahymena* is an enzyme. *Science* 231:470, 1986.

REVIEW QUESTIONS

These questions are based on information in Chapters 12 and 13.

1. Match the number of the term on the right with the appropriate description on the left. A number may be used more than once or not at all.

A. Region of gene that contains recognition signals.	1. Promoter
B. Needed for normal termination of transcription of some genes.	2. σ factor
C. Signal for the start of translation.	3. cAMP
D. Is formed only on initiating tRNA.	4. AUG
E. Actively recognizes the signals in the gene for the start of transcription.	5. *n*-formyl methionine
F. Needed to link an amino acid to tRNA.	6. Rho factor
G. Signal to end a polypeptide.	7. Methionine
H. Found at position 6 in the β-chain of hemoglobin A.	8. Valine
I. Protein needed to stimulate transcription of certain genes before the binding of RNA polymerase.	9. Glutamic acid
J. Needed to form linkages between carboxyl and amino groups of amino acids.	10. UUU
	11. UGA
	12. CTT
	13. CAP
	14. Peptidyl transferase
	15. Aminoacyl-tRNA synthetase

2. When pure DNA of the nucleolar organizer is hybridized with 18S and 28S RNA from the same cell type, only about 25% of the DNA present hybridizes with the RNA. Explain.

3. Answer the following questions in relation to this DNA strand:

3′–TACAAATCTCATTGTATAGGA–5′

A. How many individual tRNA molecules are required to translate the transcript of this segment?
B. What are possible anticodons that could be used for translation of the transcript of this segment?

4. For each of the following anticodons, give the possible corresponding RNA codons and the DNA triplets in the template strand. Include the polarities.

A. 3′–UUA–5′.
B. 3′–UCU–5′.

C. 3′–AAU–5′.
D. 3′–CCI–5′.

5. Consult the genetic dictionary (Table 12-1), and give the amino acid designation in each of the four cases in Question 4.

6. From the genetic dictionary, note the RNA codons for the amino acid serine. What would be the *minimum* number of tRNAs needed to insert serine into a growing polypeptide chain during translation?

7. Answer Question 6 for the amino acids leucine and arginine.

8. Consider an mRNA carrying ribonucleotides numbered 1, 2, 3, and so on, from the beginning of the message for a specific polypeptide. The normal polypeptide has 300 amino acids.

   A. Ribonucleotide 14 undergoes a change resulting in a missense mutation. At which position from the NH_2 end in the polypeptide will an amino acid substitution occur?
   B. Assume ribonucleotide 23 changes and a "nonsense" mutation arises. How many amino acids would you expect in the peptide fragment?

9. A certain polypeptide chain is composed of 100 amino acids. During its synthesis, a methionine molecule was picked up by a specific tRNA and placed in position 1 in the growing polypeptide. An asparagine molecule was placed in the second position, a leucine in the third, a histidine in the fourth, and so on, until a tryptophan is placed in the last position. Identify these five particular molecules for each of the following situations:

   A. The tRNA carrying this amino acid was the first to move to the A site of the ribosome.
   B. This amino acid will have a free carboxyl group.
   C. The tRNA carrying this amino acid did not associate with the A site of the ribosome.
   D. The carboxyl group of this amino acid formed a peptide linkage with the amino group of leucine.
   E. This amino acid has a free amino group.

10. Concerning ribosomes:

   A. Give the characteristic S value of the intact eukaryotic ribosome.
   B. Give the S value of the subunit in prokaryotes that attaches to the mRNA and to the first tRNA.
   C. Give the S value of the subunit of *E. coli* that contains A and P sites.
   D. Give the S value of the eukaryotic subunit that contains the enzyme required for the formation of peptide linkages.
   E. Name the ion required for the cohesion of the two subunits.
   F. Give the S value of the eukaryotic rRNA whose genes are not in the nucleolar organizer.

11. The following represents a tripeptide: NH_2 met-glu-trp COOH. Give the double-stranded sequences that could code for it, giving the polarity as well as the sense and the antisense strands.

12. Assume that a particular DNA template normally undergoes transcription and that the transcript is in turn translated to produce a polypeptide. A series of alterations arises that affects the template. For each of the following, give that part of the template that has most likely been altered.

   A. Transcript produced but is unable to bind to ribosome. No polypeptide is produced.
   B. No transcript formed because RNA polymerase cannot start transcription. Therefore no polypeptide is produced.
   C. Polypeptide of normal length is produced but is nonfunctional because of an amino acid substitution.
   D. Defective polypeptide of abnormal length is produced in which additional amino acids occur at the carboxyl end.
   E. Only a nonfunctional peptide fragment is produced, consisting of the first four amino acids of the normal polypeptide.

13. Indicate the polarity of each of the following as either 5′ or 3′.

   A. Translation of mRNA begins at this end.
   B. The beginning of the gene proper is toward this end of the template.
   C. The base at this end of the anticodon is in the wobble position.
   D. The first nucleotide in the mRNA has this end free.
   E. The amino acid with the carboxyl end was designated by a codon toward this end of the transcript.

14. The DNA associated with five different structural genes can be represented as shown here, where H and L designate the heavy and light strands:

H5′_3′	H5′_3′	H5′_3′	H5′_3′	H5′_3′
L3′_5′	L3′_5′	L3′_5′	L3′_5′	L3′_5′
(A)	(B)	(C)	(D)	(E)

Indicate for each structural gene whether transcription will occur in the direction left to right (→) or right to left (←) after considering the following for each DNA region.

   A. The core enzyme transcribes nucleotides of the gene proper in the L strand.
   B. The antileader is toward the 3′ end of the H strand.
   C. The terminator ending in a series of A-containing nucleotides is in the H strand.
   D. The first 5′–ATG–3′ sequence is in the H strand.
   E. The first 5′–TAG–3′ sequence is in the H strand.

15. In a certain human blood disorder, one of the β-chains of hemoglobin is absent. However, transcripts have been isolated from the cells of afflicted persons. These transcripts form hybrid molecules with the DNA known to code for the β-chain. Offer explanations for the nature of the genetic disorder.

16. DNA analyses of three different mammalian species shows the following base percentages:

Species	A	T	G	C
A	29.6	29.6	20.4	20.4
B	29.5	29.5	20.5	20.5
C	29.8	29.8	20.2	20.2

Denaturation–renaturation experiments are performed with two species at a time. In such an experiment, the separate DNA chains of one species are left intact; in the other, they are fragmented. It is found that almost 100% of the DNA from species A and C hybridize. Only about 5% of the DNA of species B hybridizes with A or C. Offer an explanation.

17. Next to each of the following statements, place a "P" if it is more applicable to prokaryotes, an "E" if it is more applicable to eukaryotes, and "P + E" if about equally applicable to both.

A. ____Little variation in GC value.
B. ____5′–3′ growth of DNA chains.
C. ____Three types of RNA polymerase.
D. ____*n*-formyl methionine in cytosol.
E. ____mRNAs terminating in long stretches of A-containing nucleotides.
F. ____Initiating type of tRNA.
G. ____Satellite DNA.
H. ____Very fast renaturing DNA fraction.
I. ____hnRNA.
J. ____Trailer and leader.
K. ____AUG-initiating codon.
L. ____tRNA genes within rDNA.
M. ____7-methyl guanosine.
N. ____Degenerate code.
O. ____Translation accompanies formation of transcript.

18. Answer the following:

A. The anticodon ACG will pair with what codon or codons for what amino acid?
B. The anticodon ACA will pair with what codon or codons for what amino acid?
C. The codon UGG for tryptophan will pair with what anticodon, and what is its polarity?

19. Answer the following:

A. Considering the fact that two specific anticodons, as noted in Question 18, are used for cysteine, why is there no anticodon at all with the sequence ACI, since this, too, could pair with the two codons for cysteine?
B. There is no ACU anticodon, yet it could pair with the codon for tryptophan. Explain.

20. In sickle-cell anemia, the mRNA coded for the β-chain of hemoglobin contains the codon GUG at the sixth amino acid coding position. In the case of normal hemoglobin, this position in the transcript is occupied by the codon GAG. Show this position in duplex DNA, indicating the sense and antisense strands along with their orientation. Indicate the specific change at the site in the DNA that is responsible for sickle-cell hemoglobin.

21. In a renaturation experiment, rank the following four fractions from lowest to highest expected Cot values: 1) satellite DNA; 2) sequences encoded for polypeptides; 3) snap back DNA; 4) moderately repetitive sequences encoded for histones, tRNA, and rRNA.

22. The term $Cot_{1/2}$ refers to that Cot value required for the reassociation of half of all the single DNA strands in a renaturation study. $Cot_{1/2}$ is related directly to the amount of DNA present. In a certain renaturation study, the same concentration of bacterial DNA and DNA from a higher eukaryote was used. When the renaturation curves are compared, would you expect the eukaryotic $Cot_{1/2}$ to be higher, lower, or the same as that for the bacterial $Cot_{1/2}$?

23. Does the Cot curve shown here represent a Cot curve from a prokaryotic or eukaryotic source? Explain.

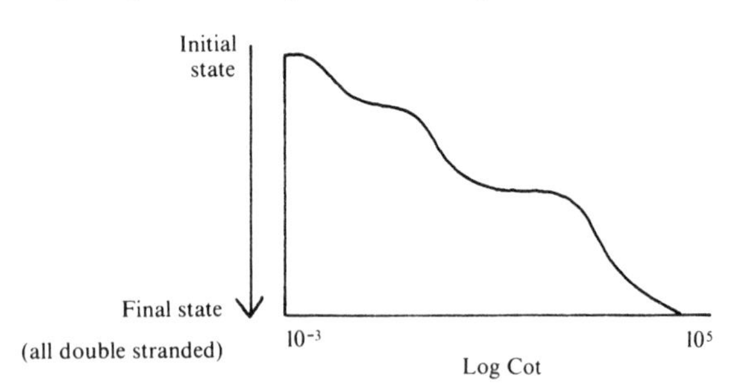

24. A gene in the chicken that codes for a specific egg protein was isolated. In one set of experiments, the duplex DNA was denatured and allowed to hybridize with mature, processed messenger RNA, which was isolated from cells that secrete the protein. The electron microscope revealed heteroduplex molecules formed between the mRNA and single DNA strands. Such a heteroduplex is shown here. Answer the following questions regarding this heteroduplex:

A. Is the DNA represented by the dark or the light strand? Explain.
B. How many introns appear to be present?
C. The longer strand shows an unpaired segment at each end. What do these represent?
D. The shorter strand shows an unpaired segment at one end. What does this represent, and is it at the 3′ or 5′ end of the strand?

25. In a certain fungus, several recessive mutations occurred independently, each of which prevents the manufacture of adenine. Normal wild-type fungi of this species have the ability to make adenine. The several adenine-less stocks have been named ad_1, ad_2, ad_3, and so on, to distinguish them according to their order of occurrence. Various crosses among some of the stocks were made with the following results:

1. $ad_1 \times ad_3$ (wild)
2. $ad_1 \times ad_8$ (wild)
3. $ad_1 \times ad_{16}$ (wild)
4. $ad_8 \times ad_{16}$ (adenine-less)
5. $ad_3 \times ad_8$ (wild)
6. $ad_3 \times ad_{16}$ (wild)

How many loci are being followed in these crosses, and which of the adenine mutations seem to be allelic to each other?

26. Assume that a, b, and c are recessive mutations in a certain mold and that the sites of the mutations are very closely linked. The following combinations give the results indicated for each:

$$\frac{a^+\ b^+}{a\ b} = \text{wild} \qquad \frac{a^+\ b}{a\ b^+} = \text{mutant}$$

$$\frac{b^+\ c}{b\ c^+} = \text{wild} \qquad \frac{b^+\ c^+}{b\ c} = \text{wild}$$

Do any of the mutations complement each other? Are any of them allelic?

27. Assume that in a certain biological step substance A is converted to substance B by enzymatic action. The step can be blocked if an individual is homozygous for the mutations a^1, a^2, a^3, or a^4. These separate mutations map at closely linked but different sites that are separable by crossing over. A series of crosses gives the following results. Explain with the concept of the cistron, and diagram the crosses.

A. $a^1 \times a^1$, step blocked.
B. $a^1 \times a^4$, step blocked.
C. $a^1 \times a^2$, step proceeds.
D. $a^1 \times a^3$, step proceeds.
E. $a^2 \times a^3$, step blocked.
F. a^1a^4 double mutant × normal, step proceeds.
G. a^1a^2 double mutant × normal, step proceeds.
H. a^2a^3 double mutant × normal, step proceeds.

28. Two married persons suffer from an anemic condition that follows an autosomal recessive pattern of inheritance in each family line. Fingerprinting of their hemoglobin A shows that the female has one amino acid substitution in the α-chain and the male has one amino acid substitution in the β-chain. Their children, however, do not suffer from the anemia. Explain.

Gene Mutation

14

Mutations and their consequences

Progress in evolution depends on the ability of a gene to undergo occasional mutation. Indeed, all the hereditary variation in living things, good as well as bad, stems from mutations in the genetic material. Changes in the DNA arise in somatic cells as well as in the germ line. Certain somatic mutations are responsible for various clinical disorders in humans (see Chap. 20). They also result in mosaic individuals: those with phenotypically different patches of tissue. Some cases of mosaicism of the human iris are the result of somatic mutation (Fig. 14-1). Examples in plants are variegated flowers with their segments of different colors.

However, in diploid organisms a somatic mutation tends to remain undetected, because the wild form of a given gene is present on the homologous chromosome and probably exerts a dominant effect that offsets that of the newly arisen mutant form. Why mutant alleles are usually recessive is understandable when we consider that most wild alleles have been selected for their dominant effects. Natural selection has screened alleles over millions of years and has selected those that are most beneficial to a given population in a specific environment. Part of the benefit of an allele is its ability to express itself in the presence of any mutant form that may prove harmful.

Although somatic mutations may be significant to the individual, germinal mutations are of greater consequence to the population, as these can be transmitted from one generation to the next. Harmful alleles are generally kept at a low level in the population. However, if a rare recessive allele should arise that happens to carry an advantage superior to that of the wild, natural selection will then favor the mutant gene form. It will also favor forms of the mutant gene that arise later to make it even more dominant. Eventually the mutant allele may supplant the original and become established as the new wild type.

A spontaneous mutation, one that arises for no apparent reason, can occur at any site on a chromosome. However, the frequency of mutation at different genetic loci can vary greatly. In the course of his pioneer investigations of eight genes in corn, Stadler noted great variation in spontaneous mutation rates (Table 14-1). Mutation analyses in microorganisms have also shown that certain sites in the genome are more likely to undergo mutation than others. While controlled studies of mutation rate are difficult to conduct in higher species, variation in genetic change from one locus to the next has been demonstrated in several, including the fruit fly. It is even apparent in the human that certain genes have higher mutation rates than others. The neurofibromatosis (NF) gene (mentioned in Chap. 13) has an unusually high mutation rate. Half of the cases of this dominant disorder actually stem from a mutation that occurs in the NF gene in a cell of that afflicted person rather than from an inherited NF gene. The high mutation rate may be associated with the very large size of the NF gene, which is approximately 500,000 to 2 million base pairs long.

The overwhelming majority of gene mutations produce harmful effects on the individual because spontaneous mutation is a random event. Unplanned change

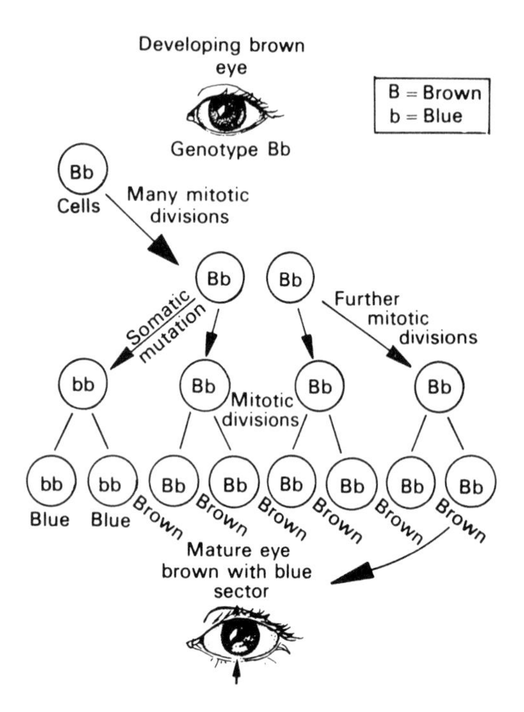

FIGURE 14-1. Effects of somatic mutations. If an individual is heterozygous, a somatic mutation can affect the dominant allele and generate a cell that is homozygous for the recessive. Further cell divisions will give rise to two populations of cells that differ both genotypically and phenotypically.

TABLE 14-1. Spontaneous mutation rates of eight genes in corn

Gene	No. of mutations per million gametes
R	492
I	106
Pr	11
Su	2.4
Y	2.2
Sh	1.2
Wx	0
C	2.3

in a system as complex as a living organism is likely to be harmful in contrast to a plan that shows thought and foresight. By far the most common mutations are detrimentals, changes that entail just a slight degree of harm. This is understandable since many genes influence the expression of a character and many biochemical pathways are interrelated in cellular metabolism. Some genes control chemical steps that are more critical than others, but by far the majority of them influence a character only slightly. Therefore any mutation, being random, will more likely occur in a gene with a slight effect, such as a modifying gene or a gene with a small quantitative effect, than in one that can alter a character drastically.

Well-controlled studies with the fruit fly have shown that gene mutations giving rise to alleles with a visible effect on a character are in the minority and that the more pronounced the visible effect, the greater the detriment that accompanies it. However, more common than these mutations are the complete lethals that have no visible effect but can remove the individual before the age of sexual maturity. Most frequent of all, however, are the detrimentals, those mutant alleles with no detectably visible effect but that contribute a slight degree of harm.

Forward and back mutations

A mutation that changes a gene from its wild to a mutant form is called a *forward mutation.* Once a gene has mutated, the new allele may continue to duplicate itself unless it is quickly eliminated. The new gene form, as well as the original, may mutate further, resulting in still additional allelic forms to produce a multiple allelic series. A mutant allele may even revert to the original wild form of the gene. This is called a *reverse mutation* or *back mutation.* Although the frequency of back mutation is characteristic for a specific gene, back mutation is rarer than the forward change from wild to mutant. We can see that this is so since a true reversion or true back mutation entails an exact reversal of the change that produced the forward mutation. For example, suppose that a forward mutation results in the substitution of the base pair GC for the pair AT and produces a defective allele. In a true reversion, the same position in the gene would be involved and the GC pair would be replaced by the original AT, restoring the wild gene form. Since a true reversion requires a second specific change at one specific site, it is a much rarer event than a forward mutation, which can involve any base pair change at a site within the gene. Other kinds of reversions can occur, however. These are more common than the true reversions and are due to suppressor mutations, mutations at sites other than the original, which can restore the wild phenotype. We have already been introduced to certain mutations of this type. One example is the frameshift mutation (see Chap. 11). A deletion or addition of a base in a gene can shift the reading frame, producing a mutant allele that can no longer code for the normal polypeptide. A second frameshift mutation nearby in the same gene, however, can restore function, such as a + mutation following a − one. This is an example of intragenic suppression or *second-site reversion,* a suppressor mutation that occurs at a different site in the same gene as the first mutation and restores function.

Second-site reversions can occur in other ways besides frameshift mutations. For example, as a result of a missense mutation, an amino acid that now substitutes for the original one may alter the folding of a polypeptide. This can render the polypeptide defective by disrupting the position of some active site such as one involved in the catalytic action of an enzyme. A second missense mutation at a different position in the gene may cause the substitution of still another amino acid at a second site in the polypeptide. By itself, this second mutation could very well be harmful, because it, too, can alter the folding of the polypeptide chain. However, the two missense mutations in the same cell may have the effect of substituting two amino acids that can now interact to restore normal or near-normal folding of the polypeptide chain. Consequently, normal activity may be restored to the enzyme by bringing the active catalytic site back into proper position.

Another way in which a reversion can take place and restore normal function to a gene stems from a mutation that, instead of being intragenic, occurs at a site in another gene. Such *intergenic mutations,* which restore normal function and thus suppress the deleterious effects of a first mutation, are usually called *intergenic suppressors.* Let us see how these suppressors may be able to cause reversion from mutant to wild. In the Simian virus SV40 (see Chap. 16), a certain mutation causes a base pair substitution near the site of initiation of DNA replication of the viral genome. This alteration prevents the effective binding of a specific protein to the DNA area around the initiation site. The binding of the protein to the DNA is apparently needed to trigger DNA replication. Hence the rate of replication of the mutant virus is greatly reduced, because the protein can no longer bind efficiently to the required site. However, a mutation in another gene, the one coded for the protein, can cause suppression of the mutant effect. The altered mutant protein, unlike the normal one, *is able* to recognize the altered initiation site very effectively and can bind well to the DNA, restoring the normal replication rate.

Suppressor mutations that occur in genes coded for tRNA have been very well studied. You will recall that a point mutation can change a codon that designates a specific amino acid to a chain-terminating codon: UAG, UAA, and UGA (see Chap. 12). Such a nonsense mutation will bring translation to a premature halt, resulting in an incomplete protein. Let us suppose that the RNA codon 5′–UGG–3′ for the amino acid tryptophan is converted to 5′–UAG–3′ as a result of a mutation in the template strand of a structural gene. The mutation converts the DNA triplet ACC to ATC. This means that the normal tryptophan tRNA in the cell no longer recognizes this codon (Fig. 14-2A). Instead, it is recognized only by release factors that bring translation to a halt. Later, a second mutation may arise, this time in a gene coded for a tRNA—let us say one for leucine, $tRNA^{leu}$. This tRNA normally picks up leucine and recognizes one of the mRNA codons for leucine, 5′–UUG–3′. This means that its anticodon would be 3′–AAC–5′. The mutation may bring about a change in the anticodon to 3′–AUC–5′. The mutation does not alter the part of the tRNA that enables it to pick up leucine. However, the mutant tRNA can now only recognize the nonsense codon UAG and can add leucine at that site (Fig. 14-2B). Although leucine is substituting for tryptophan at a specific site in the amino acid chain, and the completed polypeptide may not be as efficient in its activity as the wild type, some activity has been restored. The very damaging effects of the nonsense mutation are averted, and the second mutation has acted as a suppressor mutation. Certain intergenic mutations, however, cannot prevent the damaging effects of a first mutation if the substituted amino acid is one that disrupts a crucial site in the polypeptide.

It is fortunate that multiple copies of most tRNA genes are found in cells. Otherwise a suppressor mutation in a transfer RNA would probably have lethal effects. This is so because a change in the anticodon, from ACC to AUC in our example, means that the one altered $tRNA^{leu}$ can no longer deposit leucine at the correct locations in the mRNA. It can now deposit leucine only at sites of UAG, a chain-terminating codon. Every protein containing the amino acid leucine would be incomplete, and the cell would not survive. But does this mean that the normal termination of translation will now be altered, since the mutant tRNA can deposit leucine at a normally positioned UAG codon? This does not happen, because the binding between release factors and UAG is stronger than the binding of the mutant $tRNA^{leu}$ to UAG. This means that UAG is not always read by the altered tRNA. Therefore polypeptides with UAG as a stop signal to translation will be terminated properly most of the time.

The site of mutation in any tRNA gene can change any one of the three bases in an anticodon. And so, there are several mutations that cause a change in the tRNAs of amino acids other than the one for leucine and that also enable these altered tRNAs to recognize the UAG codon and to deposit their specific amino acids at the UAG site. While our discussion here has centered around the UAG codon, the same applies to the other two chain-terminating codons, UAA and UGA.

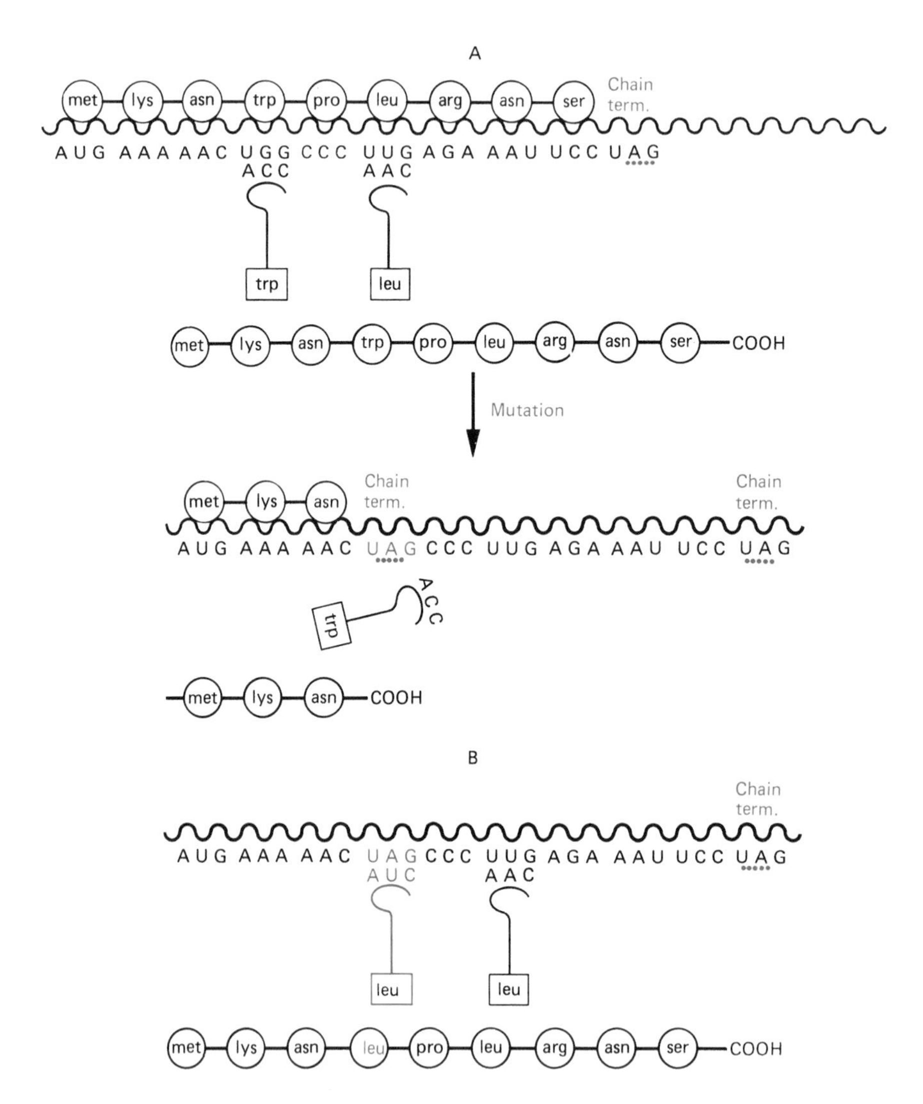

FIGURE 14-2. Nonsense mutations and suppressor mutation. **A:** A portion of an mRNA strand is depicted with the corresponding amino acids above the codons. One chain-terminating codon, UAG, is shown bound to release factors (red dots). Two charged tRNAs, $tRNA^{trp}$ and $tRNA^{leu}$, are shown with their anticodons. (The other tRNAs, as well as ribosomes, are omitted for clarity.) The amino acids will be joined together until the chain-terminating codon brings translation to a halt. A nonsense mutation changes the codon UGG for tryptophan to UAG, a chain-terminating codon. The $tRNA^{trp}$ can no longer recognize that site, which is now recognized instead by release factors, bringing translation to a premature halt. An incomplete, nonfunctional polypeptide results. **B:** A mutation in a gene for $tRNA^{leu}$ changes the anticodon in the tRNA from ACC to AUC. This is complementary to the chain-terminating codon UAG. The mutant tRNA can now position leucine at the site normally occupied by tryptophan. There is more than one gene for $tRNA^{leu}$, and so the cell will contain $tRNA^{leu}$ molecules with the normal anticodon. These will continue to position leucine at its normal sites. The resulting polypeptide, though normal in length, will carry a substituted amino acid. Nevertheless, some function may now be restored because of the mutation in the $tRNA^{leu}$ gene, which has acted as an intergenic suppressor.

Mutation rate and some factors that influence it

Although spontaneous mutation rates for individual loci are very low (a probability, on average, of 1–10 per million gametes), the frequency of spontaneous mutation should not be underestimated, as any higher organism contains thousands of genetic loci. Because so many genes are present, the chance that *one* of them will undergo a change is quite significant. Studies on the total mutation rate in *Drosophila* indicate that in one generation the probability is that 5% of the gametes will contain a mutation that arose in that generation. Calculations of total mutation rate in other organisms, including the human, give a comparable figure.

It is important to understand what this 5% value means in the life cycles of different species. Note that the value for the overall mutation frequency is expressed *not* in relation to definite time units such as months or years, but in relation to generation time. Different species obviously have different generation spans, approximately 2 weeks for the fruit fly in contrast to approximately 30 years for the human. But when different species with their different generation times are compared (excluding microorganisms), the overall mutation rates expressed in *generation time* are comparable. Obviously, in terms of weeks or months, this would mean that a species with a short life cycle of just a few weeks would have a mutation rate hundreds of times higher than humans or any other species with a much longer life cycle. The very fact that the figure is similar when expressed in generation time implies that natural selection has somehow geared the spontaneous mutation rate to the life cycle. Evolutionary progress depends on the ability of the gene to change. However, a very high or uncontrolled amount of gene mutation would prove fatal to a species, because most mutations are harmful.

Thus it seems logical to expect that mutation rate itself is a genetic feature, a characteristic adjusted to the life cycle of the species. If this is so, there must be genes that can influence the mutation frequency of other genes. This brings us to a discussion of factors that can affect the typical mutation frequency. The genetic composition itself has indeed been shown to play an important role. In the fruit fly, spontaneous mutation rates may vary as much as 10-fold among different stocks. Such marked increases in mutation frequency have been attributed to definite genes, *mutator genes,* that can increase the mutation rate of other genes. Normally, we would not be aware of the presence in the genotype of genes that keep the mutation frequency adjusted to the life cycle. However, if a mutation occurs in such a gene, the adjustment is interrupted. An increase in mutation frequency follows, and a mutator gene can be recognized.

The first factor found to alter mutation frequency was temperature. Morgan and his students noted that fruit flies raised at 27°C had two to three times the mutation rate per generation than those raised at 17°C. Later studies showed that not only high temperatures but also abnormally low ones elevated the mutation frequency. Increased mutation rate also followed sudden temperature shifts back and forth from high to low. This increase was noted even if the shift was from a high to a low temperature within the normal range for the species. Thus it would appear that departures from temperature conditions to which the organism is normally adapted may increase the mutation rate.

Observations made on some plant and animal groups indicate that aging can increase the mutation rate. An increase is seen after the aging of seed and pollen in plants and after the aging of sperm in *Drosophila.* There is even evidence that in humans a similar effect pertains. As stated before, a mutation has occurred in humans if a condition, known to be inherited as a dominant defect, suddenly appears in an offspring of two unaffected parents. Certain dominant disorders, such as achondroplastic dwarfism, have been found to appear with a greater frequency when the male parent is considerably older than the average.

The importance of malnutrition as a causative agent in mutation is indicated by the doubling of the mutation frequency in plants, such as the snapdragon, when they are deprived of certain mineral elements. Such effects as those of malnutrition, aging, and other environmental influences on mutation rate may be related to the failure of repair mechanisms in the cell, a topic discussed more fully later in this chapter.

Radiation and mutation rate

An increase in the mutation frequency in the fruit fly following exposure to X-rays was reported in 1927 by H.J. Muller, who had been one of Morgan's students. Muller was impressed by the fact that the gene, a very stable entity that goes through countless accurate duplications, can suddenly change to another form that in turn can duplicate itself accurately. These rare, sudden changes in a gene were found to be random: One environment did not favor a particular kind of mutation over another, and, when a rare gene change occurred in a cell, it was always just one change in one gene. If a gene mutates in a diploid cell, the same gene on the homologous chromosome is not affected. Since such observations indicated to Muller that mutation was some kind of chance disturbance on the molecular level, he reasoned that if a mutation actually were the result of a chance pointwise excitation, then high-energy radiations should be able to increase the frequency of mutation. This would be expected, because radiations were known to cause chance disturbances as they passed through matter. The radiations pertinent to the topic of mutation are those that can travel through space that is empty of matter and do not therefore include sound waves.

Of the two major types of radiation, the electromagnetic ones are probably the most familiar (Table 14-2). In one sense, they may be considered waves that form a continuum extending in length from the long radio waves to the extremely short cosmic rays. It is only waves with lengths shorter than those of visible light that are effective mutagenic agents. Visible light and longer wavelengths do not possess enough energy to disorganize subatomic particles, whereas X-rays possess sufficient energy to be absorbed by the electrons of atoms. Atoms containing electrons with excess kinetic energy are said to be in an *excited state.* If sufficient energy is absorbed, an energized electron may break away from the nucleus to which it belongs, creating an ionization, and may knock out additional electrons in its path. When the energy is finally dissipated, free electrons attach to other atoms, resulting in positively and negatively charged particles. The overall effect of excitations and ionizations is to make material more reactive. Ultraviolet light with its relatively long wavelength does not cause ionizations (its effects are discussed more fully later in this chapter).

The other major class of radiations includes the corpuscular radiations (Table 14-3). Unlike the electromagnetic ones, these radiations are bodies possessing a definite mass. Streams of protons, neutrons, or electrons affect matter in a way similar to that described for the other class of radiations. By actually hitting subatomic particles, they can impart sufficient energy to cause excitations and ionizations. Since both electromagnetic and corpuscular radiation can bring about chance, pointwise disturbances in matter at the level of the atom, we can appreciate why Muller suspected that high-energy radiations could induce mutations.

Muller designed an experiment with *Drosophila* that clearly demonstrated that the number of sex-linked lethals produced was directly proportional to the X-ray dosage he applied as measured in roentgen (R) units (Fig. 14-3). The direct relationship existing between radiation dosage and genetic effect has been found to apply in general to the production of mutations other than sex-linked lethals and to hold for effective wavelengths other than X-rays.

The corpuscular radiations (neutrons and protons) can also induce genetic change, but their ionization distribution is much more condensed than that produced by electromagnetic radiations. Consequently, certain differences are apparent between the two classes in the relationship between dosage applied to the number of mutations produced.

In addition to their ability to induce gene mutations, ionizing radiations are very effective in the production of all types of structural chromosome changes. However, the general rule of direct propor-

TABLE 14-2. Spectrum of electromagnetic radiations

Wave type	Wavelength	Å = 1/10,000 μ
Radio	10^7–0.04 cm	Insufficient energy to excite or ionize atomic electrons
Infrared	20,000–7,800 Å	
Visible light	7,800–3,800 Å	
Ultraviolet	3,800–150 Å	Energy to excite
X-rays	150–0.15 Å	Sufficient energy to excite and ionize
γ-Rays	0.15–0.005 Å	
Cosmic rays	0.005–0.00008 Å	

TABLE 14-3. Some of the corpuscular radiations

Particle	Weight (atomic mass units)	Charge
Electron (β-particle)	0.00055	Negative
Positron (positive β-particle)	0.00055	Positive
Proton (nucleus of common isotope of hydrogen)	1.007	Positive
Deuteron (nucleus of heavy isotope of hydrogen)	2.013	Positive
α (nucleus of helium)	4.002	Positive
Neutron	1.009	Neutral

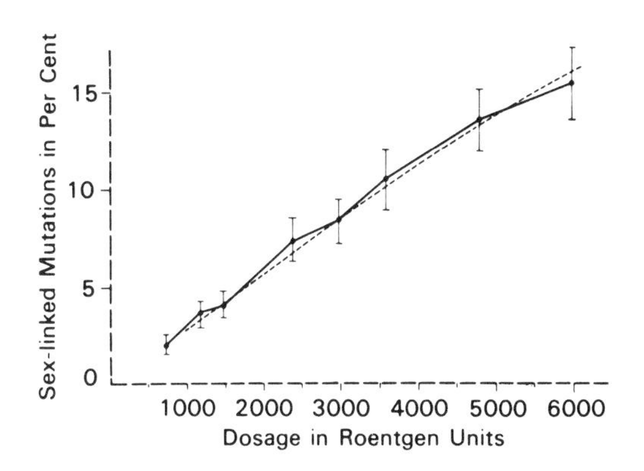

FIGURE 14-3. Sex-linked mutations and X-ray dosage in *Drosophila.* The number of point mutations is directly proportional to the dosage of X-rays as measured in roentgens. At higher doses, the relationship falls off as the radiation decreases the viability of the flies. (Reprinted with permission from E.W. Sinnott, L.C. Dunn, and T. Dobzhansky, *Principles of Genetics,* 5th ed. McGraw-Hill, New York, 1958.)

tion between mutation and radiation dose does not apply for gross structural changes, since two breaks are required for a structural change to take place. A given dosage corresponds to a certain probability for a break to occur in a chromosome. The chance that two breaks will occur together is equal to the product of the chances of the singles. Therefore the production of gross structural chromosome aberrations does not increase in a linear relationship to dose but in proportion to the square of the dose (Fig. 14-4). The higher the dose, the greater the chance of producing chromosome aberrations. Very small deletions and inversions might be caused by just one hit; these would be proportional to the dose, as seems to be the case. And the intensity of the dose seems to play a role in the production of structural changes. More of them are obtained if a large dose is given at one time rather than distributed in small packages. This can be explained by assuming that, when a dose is weak, the probability that two breaks will be produced in the same nucleus is less. The break tends to be repaired before another one was formed. Thus, if an intense dose is applied at one time, there will be more single breaks overall at one time that can engage in new arrangements.

Later work with mice has shown that these animals react differently from *Drosophila* in response to the intensity of the radiation. The intensity exerts an effect in the production of point mutations similar to that in the production of structural aberrations. Such differences among species in response to radiation suggest that the mechanism of inducing gene mutation is not as direct as it appeared at first.

Several factors can affect the number of mutations that are associated with a given radiation dose, even within the same organism. Generally, cells with chromosomes in a condensed state, such as sperm and cells at nuclear divisions, are much more susceptible to both point mutations and chromosome damage than are those with chromosomes in an extended condition. This accounts in part for the greater damage by radiation to actively dividing cells such as those in the bone marrow and to cancer cells.

Moreover, the number of mutations that arise after a given radiation dose can vary depending on the stock or strain. In fruit flies, just as some stocks have a higher spontaneous mutation frequency, others vary in their sensitivity to the radiation. The influence of environmental conditions on the yield of induced mutations has also been definitely demonstrated by the effect of oxygen. Fewer mutations for a given dose are found at lower oxygen tensions. Among microorganisms, those forms that can exist either aerobically or anaerobically undergo a higher rate of both spontaneous and induced mutation under aerobic conditions.

The temperature of the surroundings at the time of applying the radiation has an influence that is probably related to the oxygen effect. Colder conditions strengthen the induction of chromosome alterations, as well as gene mutations; this is most likely associated with the presence of more oxygen at lower temperatures. Also connected to the amount of oxygen is the action of certain reducing agents present in the environment at the time of irradiation. The number of chromosome breaks is reduced in the presence of –SH compounds, alcohols, and various strong reducers. These compounds may afford a certain amount of protection by removing oxygen dissolved in tissues.

Ultraviolet light and the induction of mutations

The amount of energy associated with ultraviolet wavelengths is too small to produce ionizations. The relatively long waves are not highly penetrating and do not usually reach cells that give rise to gametes. Nonetheless, it has been demonstrated in several higher organisms that ultraviolet has a definite mutagenic effect once it does reach the DNA. Exposure of microorganisms to ultraviolet showed that wavelengths in the range of 2,600 Å are the most mutagenic and that these are most strongly absorbed by the genetic material. Little effect is caused by those wavelengths only weakly taken up by the DNA.

Later work with isolated DNA demonstrated ultraviolet-induced changes in nucleic acid. It is the bases in the DNA that are primarily responsible for the absorption of the ultraviolet. The pyrimidine bases thymine and cytosine are particularly sensitive. A major consequence of the absorption is the formation of stable bonds between two thymine bases or between two units of cytosine. Such associations between two identical molecules or units of a molecule are called *dimers.* Production of thymine dimers seems to be a very important means by which ultraviolet alters the DNA (Fig. 14-5).

Ultraviolet absorption may also cause stable bond formation between two different primidine bases, giving a "mixed" dimer (e.g., cytosine–thymine). When dimer formation occurs, it is two pyrimidine bases adjacent to each other on a DNA strand that become stably bonded. At very high doses, ultraviolet has been shown to be capable of breaking the sugar–phosphate backbone of isolated DNA. Whether it operates this way in the intact cell, ultraviolet can cause chromosome breakage in the fruit fly and some other species. Its ability to induce breakage, however, is much less than that of the high-energy ionizing radiations.

The Watson-Crick model enables us to understand how ultraviolet causes some of its mutagenic effects. Replication of DNA depends on clean separation of the two polynucleotide chains, each of which acts as a template for the assembly of new, complementary chains. Any process that interferes with this, such as the formation of stable bonds by dimerization, can result in DNA unable to transmit properly the information stored within it.

One unusual aspect of ultraviolet is that its mutagenic effect may be reduced by white light. A simple way to demonstrate this (Fig. 14-6) is to spread each of two agar plates with the same number of bacteria and subject each to a dose of ultraviolet known to have killing activity. The plates are then incubated, one completely in the dark but the other after a 30 minute exposure to a strong source of white light. Observation of the plates the next day will show a much better growth on the plate treated with the white light. The killing effect of the ultraviolet is reduced so that only about 40% of the original killing activity remains. Therefore, if a certain dosage can kill half the bacteria on a plate, exposure to white light can reduce this lethal action from 50% to 20% (0.50 × 0.40), so that 80% of the bacteria will survive. This reversal of the lethal effect of ultraviolet light by white light, called *photoreactivation,* has also been demonstrated in *Drosophila* and higher plants. The mechanism of photoreactivation remained a mystery until the discovery of repair enzymes, which can undo DNA damage and which are the subject of the following section.

Radiation dose (arbitrary units)	Chance of one break	Chance of two breaks together
1	1/10	1/100
2	2/10	4/100
3	3/10	9/100
4	4/10	16/100

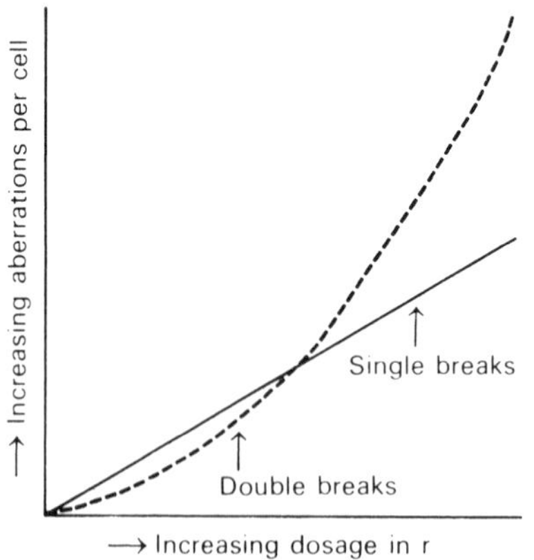

FIGURE 14-4. Radiation dosage and production of chromosome breaks. Single breaks in a chromosome show a direct proportion to dosage applied, as is true for the production of point mutations. The chance of two breaks occurring in the same nucleus is equal to the product of two single breaks. Consequently, large structural chromosome alterations increase in proportion to the square of the dose. This means, as can be seen on the right, that at lower doses more single breaks are produced, but with increased dosage, the chance is greater for two breaks, and hence structural aberrations, to take place.

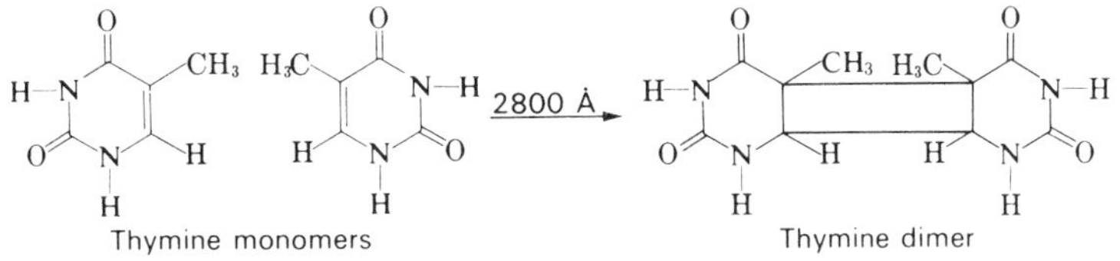

2800 Å

Thymine monomers

Thymine dimer

FIGURE 14-5. Dimerization by thymine. Ultraviolet light may cause stable bonds to form between two identical molecules of thymine or cytosine. The result is a dimer. If dimers form between two bases located along the same strand of a stretch of DNA, replication of the DNA molecule could become impaired.

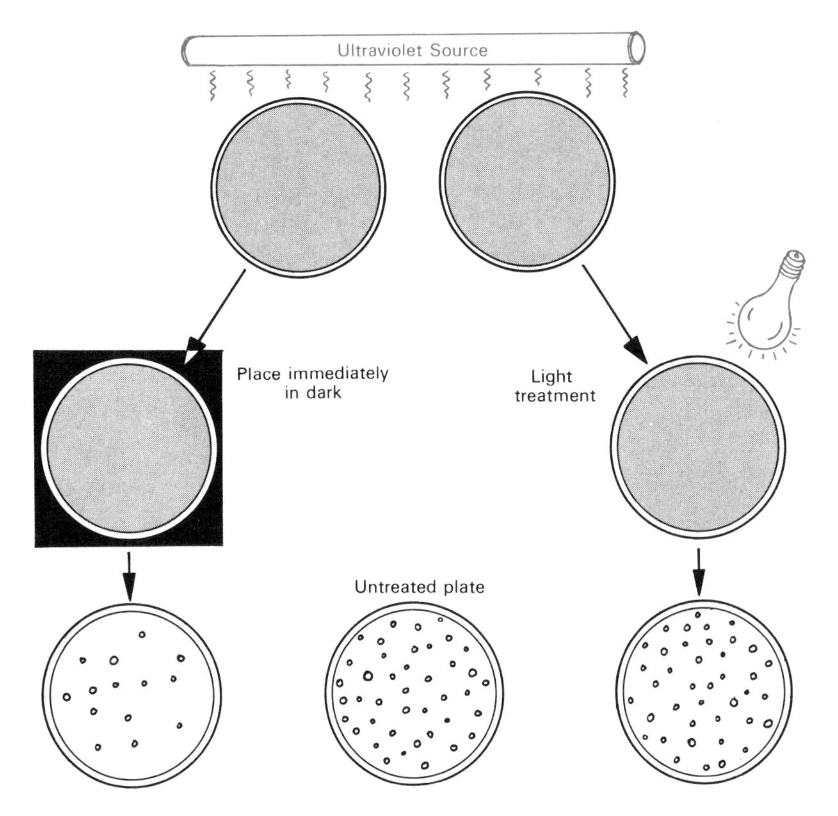

FIGURE 14-6. Photoreactivation. Two plates spread with the same number of bacteria are subjected to a dose of ultraviolet radiation. One is placed immediately in the dark. The other is kept in the light for at least 30 minutes. After incubation of the plates, the light-treated one is found to have a much larger number of bacterial colonies than the plate placed immediately in the dark. The number may approach that found on a plate not exposed to the ultraviolet at all. This untreated plate can be placed in the dark or given a light treatment, but the number of colonies on it will not be altered. The photoreactivation operates on the damage induced by the ultraviolet.

DNA repair

The prokaryotic cell has several kinds of systems to deal with damage to the DNA. Damage includes actual structural distortions such as those caused by dimer formation that can interfere with replication and transcription. It also includes mismatches between bases on opposite DNA strands that prevent proper pairing. While these do not interfere with replication or transcription, they lead to point mutations unless corrections are made.

The enzyme involved in photoreactivation is an example of a repair enzyme that can recognize a distortion in the DNA caused by dimer formation after exposure to ultraviolet light (Fig. 14-7). Although the enzyme can bind to the altered site in the dark, it requires the energy in visible light (actually in those wavelengths in the blue portion of the spectrum) to become activated. It is then able to cleave the bond between the two bases, thus restoring the DNA to its original state.

The prokaryotic cell contains other repair systems that do not depend on the energy of light. One general category is *excision–repair* in which a DNA segment that includes damage is removed and replaced by a new segment. This repair entails a battery of enzymes including those with endonuclease and exonuclease activities. Several distinct types of excision–repair systems exist in the prokaryotic cell, each dependent on a set of genes coded for proteins that interact to achieve correction of the damage. Figure 14-8 is an overall picture of excision–repair and brings out features that the different types have in common. The initial step requires an endonuclease that can recognize the damaged region as a result of the distortion resulting from dimer formation. The endonuclease produces nicks in the region on either side of the damage. An exonuclease then digests away the damaged part of the strand, including the dimer. Once the damage is removed, DNA polymerase synthesizes the missing segment in a 5′ to 3′ direction using as a template the complementary region that lies exposed on the other strand. Finally DNA ligase seals the remaining nick by catalyzing a phosphodiester linkage between the 3′ end of the newly formed segment and the 5′ end of the strand from which the damage was removed. Keep in mind that the pathway just described is a general scheme. In *Escherichia coli,* many genes and enzymes are involved, not just four. A single mutation in any one of the genes can result in a deficiency in excision–repair.

You will recall that DNA polymerase can act as a proofreader at the time of replication and can correct an error resulting from the incorporation of an incorrect nucleotide into the DNA strand being assembled on the template (see Fig. 11-26). At times, this editing fails, and a "wrong nucleotide" is not removed. As a result, a site is created within the double helix where mismatching exists, such as that between a thymine and a guanine. A mismatch can also arise following recombination when heteroduplex DNA regions form (see Fig. 11-29). In these regions where the two DNA

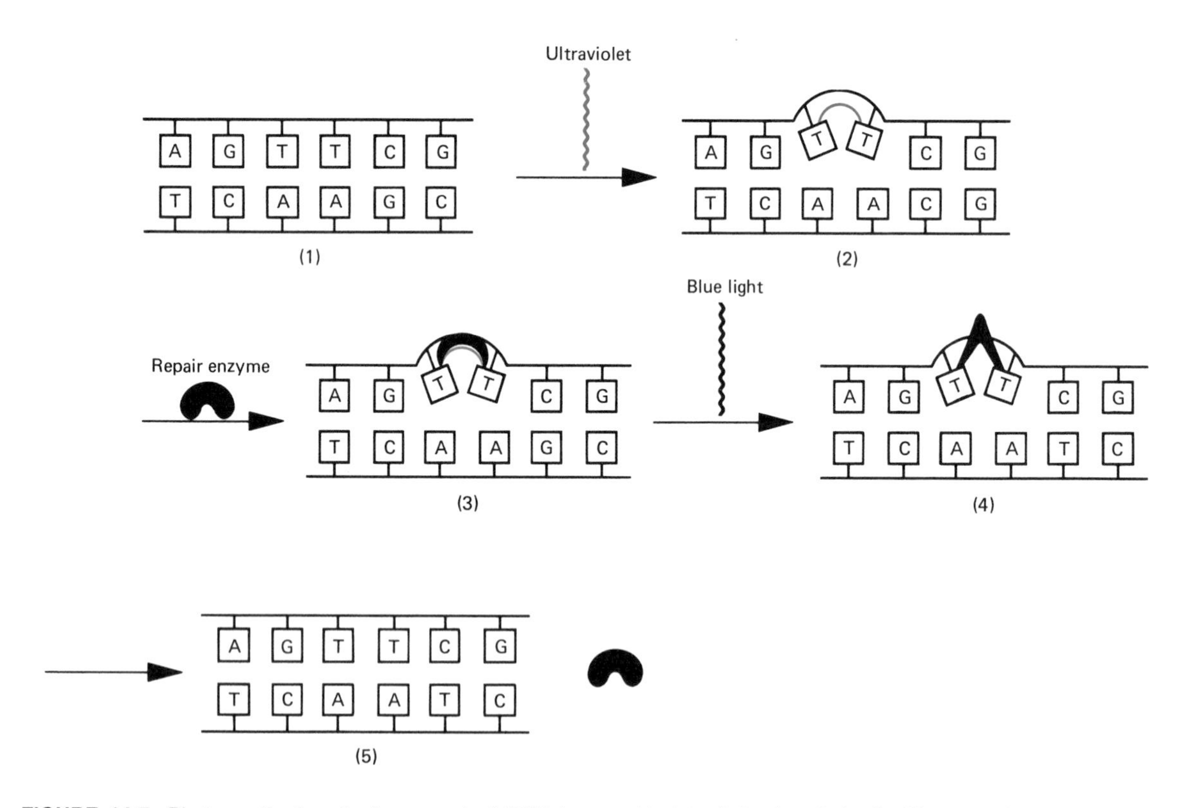

FIGURE 14-7. Photoreactivation. **1:** A segment of DNA is subjected to ultraviolet light. **2:** A dimer forms between two thymines. **3:** This is recognized by a repair enzyme that can bind to it in the dark. **4:** Blue wavelengths activate the enzyme, which then cleaves the bond between the bases. **5:** The segment of DNA is restored, and the enzyme is released.

strands had their origins in different parental duplexes, the strands may not be perfectly complementary, and therefore certain base pairs may be mismatched.

Still another way in which duplexes can have mismatched base pairs results from tautomeric shifts. Suppose thymine exists in its rare state at the time of replication so that it can pair with guanine but not with adenine as depicted in Figures 11-13 and 11-14. The unusual base pair TG results and is not recognized by the proofreading mechanism, because hydrogen bonding will be satisfied at the time, and pairing relations will appear to be correct. Since the tautomeric shift is a rare event, however, the base (thymine in this example) returns to its normal state. A mismatched site arises in which hydrogen bonding is not correct.

At least two possible outcomes are in store. At the next replication, two duplexes can form (see Fig. 11-14), but the two will differ from each other at the site that had been mismatched. The base pair change actually represents a mutation. Another possible outcome is actual removal of the mismatch and a small area surrounding it by a mechanisms called *mismatch repair* (Fig. 14-9), a type of excision–repair. In this repair process, the mismatched region is recognized and removed as a result of the activities of several proteins encoded by several different genes. A mismatch correction enzyme scans the DNA when replication is taking place. The segment with the wrong nucleotide is recognized and removed. DNA polymerase then fills in the gap, replacing the incorrect nucleotide (guanine, in this example) with the correct one, thus restoring the duplex with the correct base pair.

A moment's thought tells us that the mechanism just described has the same chance of removing the "correct nucleotide" on the template strand as it does of removing the "wrong one" on the new strand. This would produce a mutation. Fortunately, the cell has another mechanism that guards against this outcome and makes it more than likely that the wrong nucleotide will be removed in mismatch repair. The template strand contains certain adenine and cytosine nucleotides to which methyl groups have become attached (methylation is discussed further in Chap. 17). Along a DNA strand we find positions occupied by the sequence 5′–GATC–3′. Wherever this sequence occurs, it can be recognized by a specific enzyme, a methylase, which modifies the adenine (Fig. 14-10). The methylation of the adenine occurs very shortly after synthesis of a GATC sequence at a replication fork, but it does not take place immediately. There is a slight delay, which is usually sufficient to allow a correction enzyme to detect the difference in degree of methylation and to distinguish the newly synthesized region from the parental one. Consequently, the chance is greater that the wrong base rather than the correct one in the template will be removed following mismatch repair. The mismatch repair system backs up the proofreading capacity of the DNA polymerase at the time of replication and reduces the frequency of spontaneous mutations. If mismatch repair does not occur shortly after the wrong base has been added, however, there may be sufficient time for the adenine in the GATC sequence in the new strand to become methylated. Discrimination between parental and daughter strands is then not possible, and the correct segment on the parental strand may be removed, generating a mutation.

In some bacterial strains, the ability to methylate bases has been lost or the enzyme mechanism in mismatch repair is defective. Such strains have a mutation frequency far above that of other strains. Not only does the mismatch repair system correct mismatched bases, it also detects small deletions and duplications and as a result can reduce the effects of certain frame shift mutations.

Another category of repair mechanism studied mainly in *E. coli* is known as *postreplication repair.* This system, which also operates in more than one way, depends on the interactions of many kinds of proteins. The RecA protein involved in recombination plays a vital role in postreplication repair, so named because correction takes place after replication of the DNA containing the damage due to the presence of dimers. It is also known as *recombination repair,* since it entails features of crossing over.

In one kind of postreplication repair, a recombination event takes place that generates an undamaged duplex DNA. According to one model (Fig. 14-11), when the DNA altered by the presence of dimers undergoes replication, it encounters difficulty because of the distorted regions that cannot act as templates to form new DNA strands. DNA polymerase is blocked at these positions, but it does manage, after a lag period, to resume replication at a site beyond the distorted area. New DNA strands are formed, but these contain *postreplication gaps.* Excision repair systems cannot take care of these; however, a recombination event may take place between the two sister duplexes. An intact, undamaged duplex arises if segments of sister duplex strands are switched during the recombination event. As shown in Figure 14-11, damaged DNA is still present, and

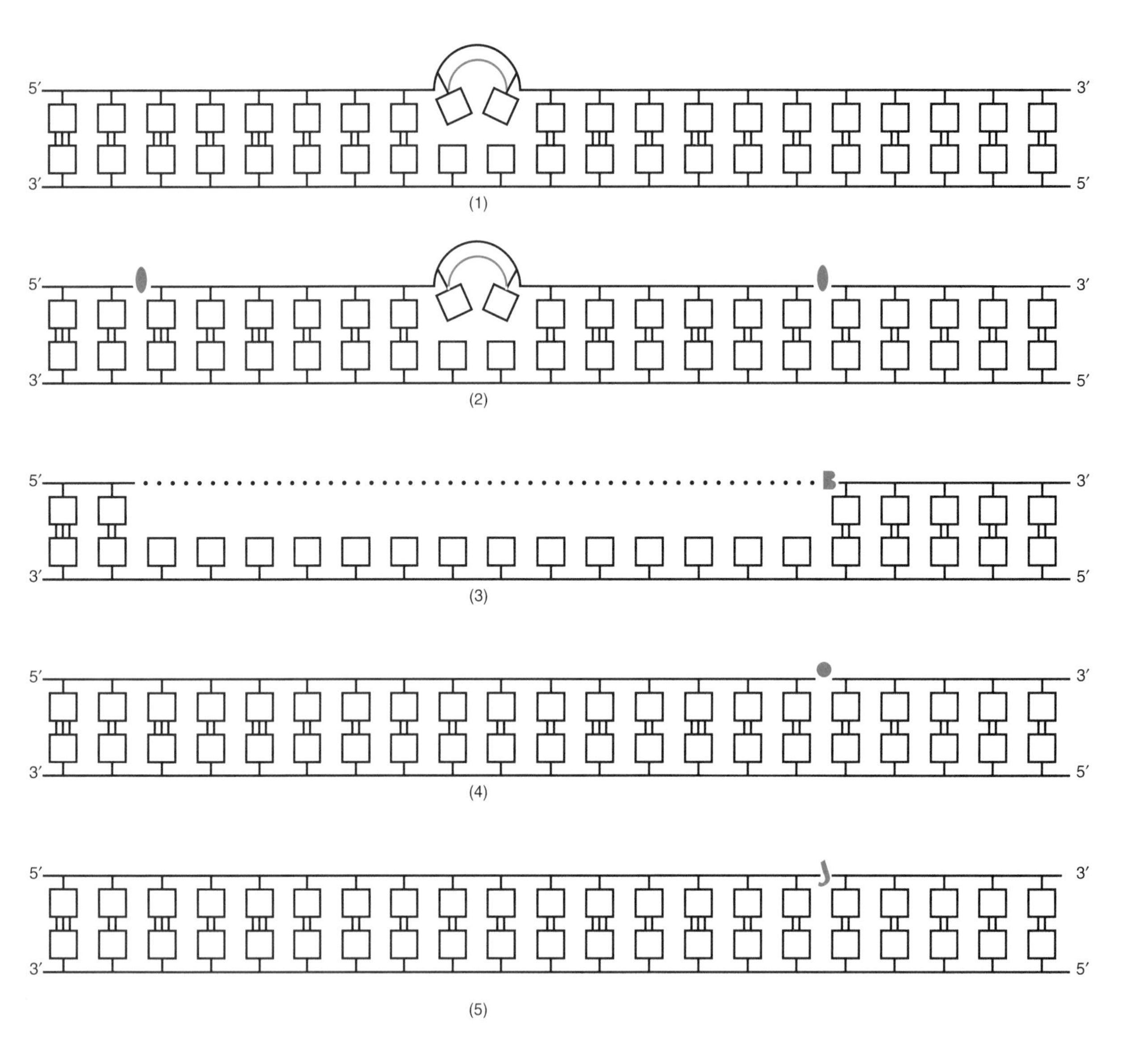
5′
3′
3′
5′
(1)
5′
3′
3′
5′
(2)
5′
3′
3′
5′
(3)
5′
3′
3′
5′
(4)
5′
3′
3′
5′
(5)

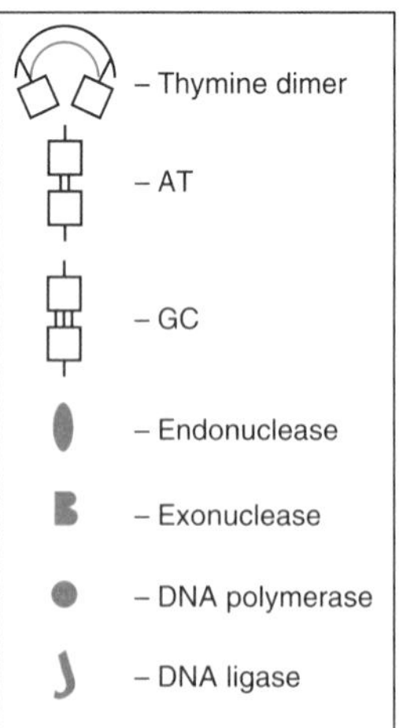
– Thymine dimer
– AT
– GC
– Endonuclease
– Exonuclease
– DNA polymerase
– DNA ligase

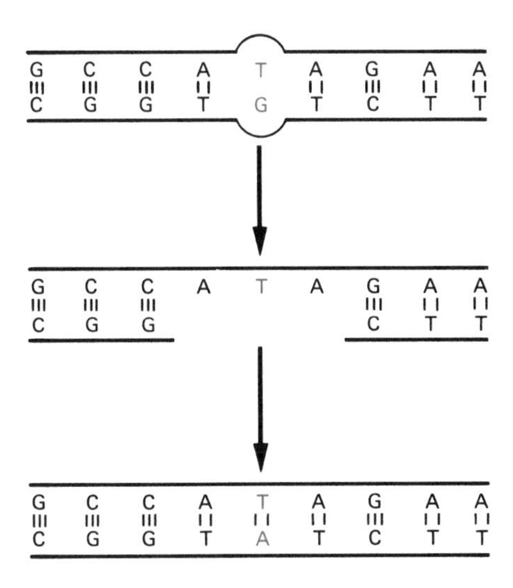

FIGURE 14-9. Mismatch repair. A mismatched base pair, TG in this example (above), can arise in more than one way. Such a mismatch may be corrected by a mismatch repair mechanism. A small segment of one of the strands containing a mismatched base is removed by nuclease activities (middle). The gap is then filled in by DNA polymerase using the complementary strand as a template, and correct pairing is thus restored (bottom).

this may not survive. The repair has nevertheless produced one intact, undamaged duplex. Postreplication repair mechanisms appear to play a very important part in the repair of DNA in *E. coli,* since cells deficient in this type of repair are exceedingly sensitive to ultraviolet light, more so than mutant cells that are defective in the other repair mechanisms.

There are other types of postreplication repair systems. One requires DNA polymerase III and is known as *SOS repair,* since the response is rapidly initiated and results in an increased ability to repair damaged DNA. SOS repair is also referred to as *error-prone repair,* because the repair of damaged regions is accompanied by the generation of mutations. In *E. coli,* an error-prone response may be called into play following various treatments that damage DNA or interfere with its replication. The response to ultraviolet light exposure has been studied in greater detail than has the response to other treatments such as exposure to alkylating agents.

In error-prone or SOS repair pathways, the proofreading capacity of DNA polymerase becomes relaxed. As the DNA polymerase assembles nucleotides on the template strand during replication and encounters a region distorted by dimers, it nevertheless continues to insert nucleotides, even if these are not the correct ones! This insertion of incorrect nucleotides constitutes the origin of gene mutations: The mutation frequency increases as a result of the relaxation of proofreading function of DNA polymerase during SOS repair.

SOS repair systems involve the interactions of several gene products. The genes coded for these products that participate in repair are held in check by a very stable protein coded by a gene known as *LexA.* The LexA protein acts as a repressor and prevents the expression of many genes involved in repair. When an SOS response is triggered, the repression is lifted. Again, a pivotal role is played by the RecA protein, which acts as a switch to turn on the SOS repair mechanism in response to damage by ultraviolet light. If the radiation has created more dimers than the excision repair mechanism can accommodate, the RecA protein starts the chain of events leading to SOS repair. It does this by cleaving the LexA protein. As a result, all the genes that were inactive due to the repressor are rapidly called to expression. Since the SOS mechanism is error prone, mutations can accumulate. It would therefore be detrimental to the cell to permit SOS repair to go on indiscriminately in response to every DNA damage that arises. Hence various genes in the system are coded for products that repress the expression of certain genes in the SOS response and thus prevent the system from activating when excision repair is sufficient. Only when DNA damage caused by ultraviolet light signals an emergency is SOS repair triggered.

The significance of repair enzymes cannot be overestimated. There is no doubt that both prokaryotic and eukaryotic cells have various devices to restore damaged DNA to its original state in the face of mutagens both within and outside the cell. The development of such mechanisms was critical to the evolution of life on land, where cells are exposed to doses of radiation higher than those reaching cells in an aquatic environment. This is demonstrated by the serious inherited skin disorder xeroderma pigmentosum. Persons with the condition are extremely sensitive to sunlight and develop skin cancers at an early age. Their cells have

FIGURE 14-8. General pathway of excision–repair. A stretch of DNA damaged by ultraviolet light contains a thymine dimer that produces a distortion (1). An endonuclease recognizes the region and produces nicks in the vicinity on either side of the damage (2). An exonuclease digests away the segment that includes the damage (3). DNA polymerase in a 5′ to 3′ direction then synthesizes the portion that has been eliminated (4). DNA ligase links the newly formed segment to the original strand, thus restoring the DNA to normal (5).

been shown to be deficient in the excision–repair process that excises ultraviolet-damaged DNA. The defect results in the accumulation of damaged DNA and the development of malignancies.

In mammals, excision–repair appears to involve a large number of different genes and enzymes, and there is also evidence for the existence of recombination repair mechanisms. Repair systems undoubtedly are of prime importance in the maintenance of normal cell activities. It has become evident that crossing over entails enzymes that also participate in DNA repair, as evidenced in certain mutant *E. coli* strains in which the capacity for recombination as well as the ability to repair ultraviolet damage to the DNA have been lost. Further support for the involvement of repair enzymes in recombination is that DNA synthesis and the production of short DNA segments take place at pachynema, that stage of meiosis when crossing over is believed to occur. Moreover, at the outset of meiotic prophase I, enzymes such as DNA polymerase, DNA ligase, and endonuclease start to appear, increasing in levels until pachynema. We have also seen that the RecA protein appears to be vital to crossing over (Chap. 11) as well as to postreplication and SOS repair systems. The versatility of the repair enzymes may thus extend from the preservation of DNA structure to the generation of new genetic combinations essential for evolutionary progress.

Mutation and variation in populations

Without knowing its basis, Darwin recognized inheritable variation among individuals in a population as the basis of evolution. We realize today that spontaneous mutation is the ultimate source of new alleles. Independent assortment along with recombination through crossing over produce new combinations of the hereditary material, providing the variation that is the raw material for evolutionary progress. According to Darwinian concepts, advantageous modifications become established through the force of natural selection, which raises the population to a greater adaptive level in its environment. The variation present in a population, however, was not considered to be very widespread. Except for some uncommon variants carrying mutant alleles, population members were regarded as rather standard types, homozygous at most loci for wild-type alleles. This picture of somewhat limited variability is now rapidly changing as molecular biology approaches are applied to the study of natural populations.

Using protein variation in individuals as the guide, one can estimate the amount of variation in the genetic material, since variation in a given protein indicates variation in the genetic material that codes for it. One can select a number of proteins to represent a random sample of genes in an individual. The proteins are then

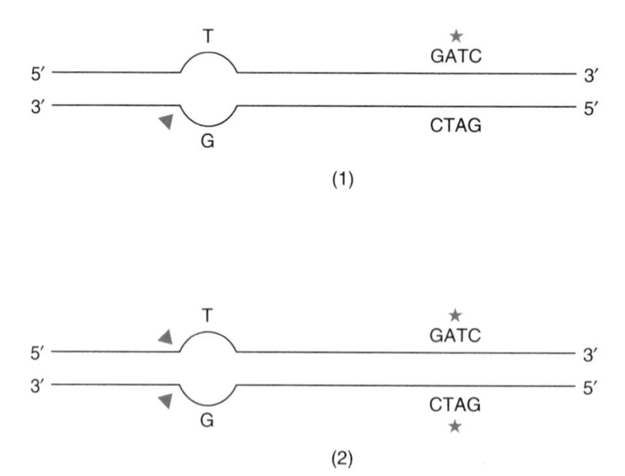

FIGURE 14-10. Distinguishing the parental DNA strand. The adenine in the sequence 5′–GATC–3′ becomes methylated (asterisk) at the time of replication after a short delay. The lack of methylation in the newly synthesized strand permits a scanning enzyme (red wedge) to distinguish the new from the template strand and to set in motion steps in the removal of the wrong nucleotide (1). If the new strand succeeds in becoming methylated before detection by the repair enzyme, the latter cannot distinguish the new from the parental strand. Either the correct or the wrong nucleotide may be removed (2).

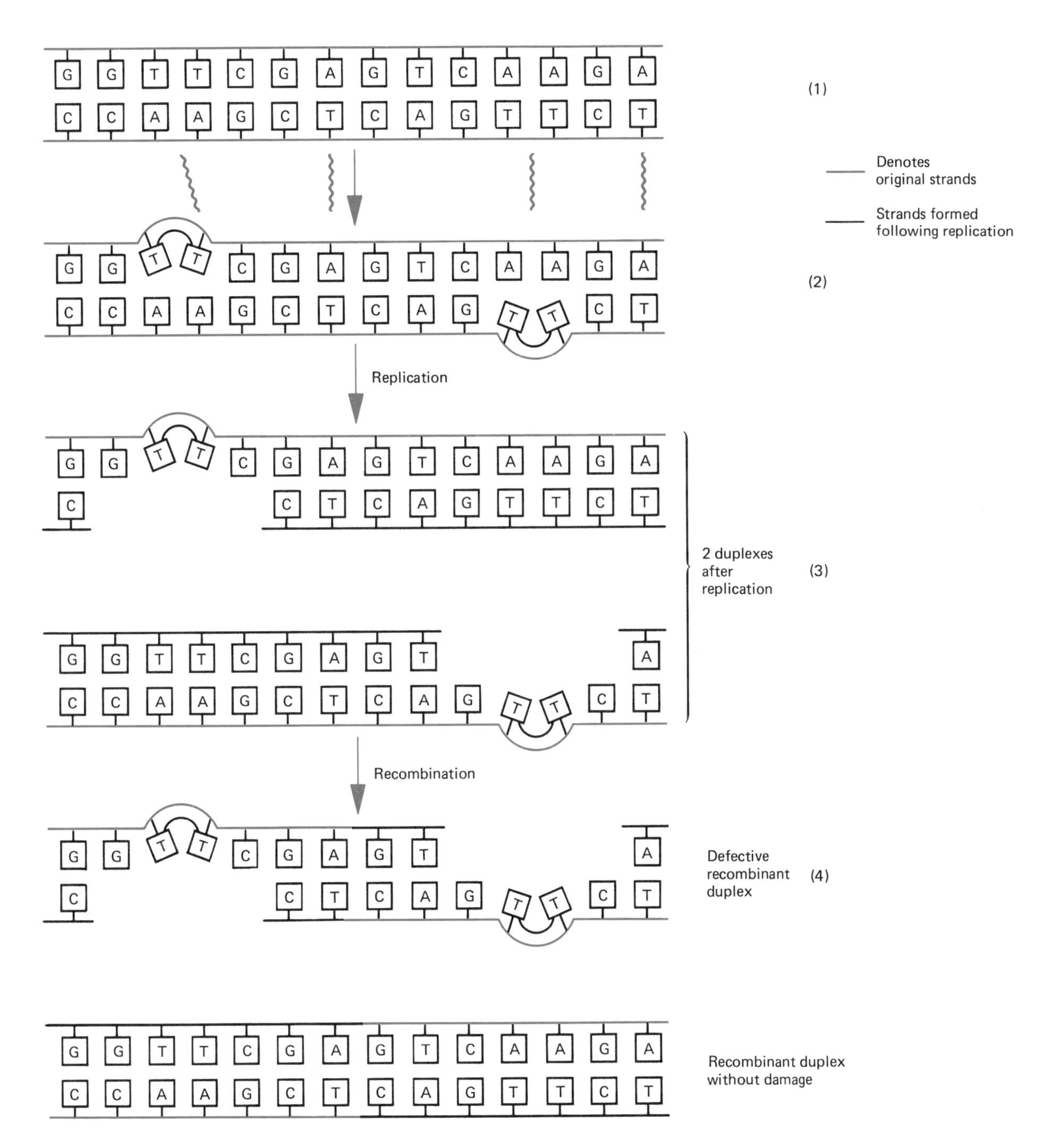

FIGURE 14-11. Postreplication repair. **1:** A segment of DNA is subjected to ultraviolet light. **2:** Dimers are produced in the irradiated strands. **3:** When the damaged duplex replicates, complementary strands are produced, but these contain postreplication gaps. Therefore the two new duplexes are doubly defective. **4:** If recombination takes place, the four strands of the two duplexes may rearrange to form a defective duplex as well as one with no damage.

subjected to gel electrophoresis, a technique so sensitive that it can detect one difference out of 100 amino acids when two proteins are compared. In the procedure, protein samples from several populations can be run side by side on a gel, allowing comparisons of various specific proteins among individuals. Suppose blood samples are taken from different population members and that soluble proteins are to be studied to estimate the degree of genetic variability. The blood is treated to remove cellular components from the serum containing the proteins. Serum samples are then placed on a gel and subjected to an electric current. Exposure to the electric field causes a protein to migrate through the gel in a direction and at a rate that depends on its net electric charge and molecular size. Each type of protein will come to assume a characteristic position. After completion of the run, the gel is treated with a solution that will stain a specific protein and reveal its location (Fig. 14-12).

Suppose a given polypeptide chain is encoded by a certain gene and that two of the chains unite to form an enzyme. As a result of spontaneous mutation, different allelic forms of the gene specifying slightly different versions of the polypeptide can exist within a population. Polypeptide chains can therefore combine to produce different versions of the enzyme. Forms of an enzyme resulting from different allelic forms of a gene are referred to as *allozymes.* Allozymes may differ sufficiently in molecular structure and charge to have different mobilities in an electric field. An individual homozygous for one allelic form of the gene (A1A1) will produce two identical polypeptides and only one form of the enzyme, visualized as one band in a gel. An individual homozygous for another allelic gene form (A2A2) will also produce one enzyme form, but the resulting band may assume a different position in the gel because the A2 polypeptide is slightly different from the A1. An individual heterozygous for these alleles can produce three different kinds of polypeptides and thus three different molecular forms of the enzyme. The three allozymes are identified as three bands (A1A1, A1A2, A2A2). Some genetic loci may be associated with genes that exist in more than two allelic forms. In such cases, more than two kinds of homozygotes can be identified as well as a greater number of heterozygotes and allozymes. Gel electrophoresis and its modifications can be used in this way to study various kinds of tissue samples taken from a number of individuals in a population and thus to estimate the variation at a given genetic locus. The procedure has revealed unsuspected amounts of genetic variation in populations. Studies of more than 250 species have shown that between 10% and 60% of the genes in each species exist in more than one allelic form and hence are *polymorphic.* On the average, between 1% and 36% of the genetic loci of an individual are heterozygous, leading to the conclusion that natural populations contain a much greater amount of genetic polymorphism than that anticipated by classic Darwinian theory.

It appears therefore that population members are heterozygous at a significant proportion of their genetic loci. The implications of this for the generation of new combinations of alleles are staggering. Heterozygosity in the human has been estimated to occur at about 6.7% of the genetic loci. Assuming there are 100,000 human loci, this indicates that a person is heterozygous at approximately 6,700 loci and could potentially produce 2^{6700} kinds of gametes! Most of the genetic variation found in a population stems not from the occurrence of

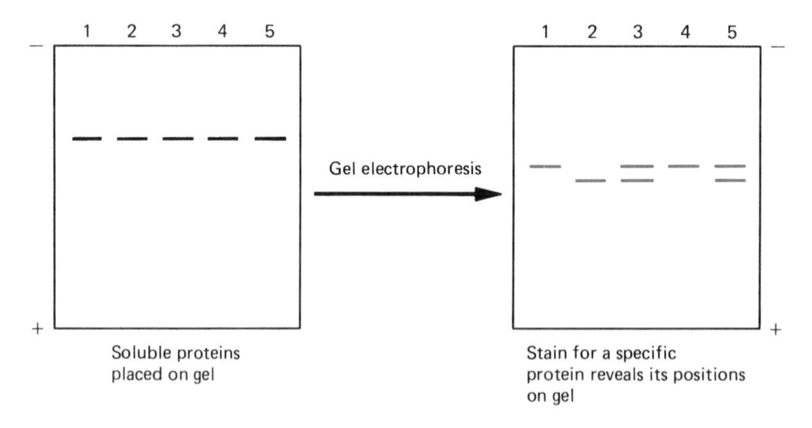

FIGURE 14-12. Detection of genetic variability in a specific protein. Soluble protein samples (black) from five different individuals are subjected to an electric field that will separate the mixture of proteins in each sample. After the separation, a specific strain is applied to reveal the position in the gel of a specific protein (red), such as an enzyme. If the gene for the enzyme exists in two allelic forms (A1 and A2) in a population, and if the enzyme is composed of two polypeptide chains encoded by the gene, some individuals may be genotype A1A1 (individuals 1 and 4), some A2A2 (individual 2), and others A1A2 (individuals 3 and 5).

new mutations in each generation, as believed by classic Darwinists; rather, it is generated through recombination. The reshuffling of stored alleles reveals the amount of variation that lies hidden in a population over generations. While gene mutation is the ultimate source of genetic change, it is a rare event that adds only a small amount of variation at each generation. A given generation does not depend on recurrent mutation to provide it with needed variation.

Neutral theory of evolution

According to the *neutral theory of evolution* (popularly known as *non-Darwinian evolution*), the mutant alleles detected in populations by chemical procedures are neutral. A mutant allele giving rise to a protein variant would be neither more nor less advantageous than the gene form from which it arose. Most evolutionary changes that take place at the molecular level in populations are brought about by recurrent spontaneous mutation (mutation pressure) and genetic drift (random or chance fluctuations in allelic frequencies independent of natural selection and mutation). Proponents of the neutral theory consider the Darwinian force of selection to be operative mainly on phenotypes shaped by the activities of many genes. The phenotypic level is considered to be the level of form and function produced through the operation of many genes.

Adherents of the neutral theory recognize the role of the environment in the selection of particular phenotypes over others and point out that selective forces are not concerned with *how* genotypes determine these phenotypes. They envision the forces of molecular evolution to be quite different from those operating at the level of the phenotype of the individual and to be driven to a large degree by genetic drift. To the proponents of the neutral theory, findings at the molecular level are incompatible with what is expected according to a strict interpretation of selectionism or neo-Darwinism. For example, when different vertebrate species are compared with respect to a specific protein (e.g., the α-chain of hemoglobin), it is found that amino acid substitutions occur at about the same rate in many of the now divergent genetic lines. Rather than following a pattern, the substitutions seem to be random. The rate does not appear to depend on such factors as generation time, living conditions, or population size. In the α-hemoglobin chain of 141 amino acids, the substitution rate is about 1 amino acid per 7 million years or 1 substitution per 1 billion years for any specific amino acid site.

As discussed earlier, the procedure for detecting small differences among proteins revealed a great deal of genetic variability, in turn revealing that a significant proportion of the genes in different groups of creatures is polymorphic, that is, such genes occur in various allelic forms. Many of the protein variants seem to have no visible phenotypic effect and appear to have no correlation with environmental conditions. To leading proponents of the neutral theory, such as M. Kimura, this suggests that most of the nucleotide substitutions in the population gene pool during the course of evolution do not stem from the force of Darwinian selection. Rather, they result from the random fixation of essentially neutral mutants. These neutral mutant alleles are as effective as the original gene forms in promoting survival and reproduction of the individual. Moreover, the observed protein polymorphisms would be neutral with respect to selection. They would be maintained in the population as a result of mutational input and random elimination. Thus a neutralist maintains that some mutants can spread through a population by themselves, even though they may impart no selective advantage to the population. On the other hand, the rate of evolution of those alleles that actually do carry a selective advantage depends on that specific advantage, the rate of mutation, and population size.

Neutralists also point out that certain DNA regions between genes and even within genes do not participate in protein formation. Therefore, these should be less subject to natural selection. There should be a lack of selective constraint in these cases, unlike the selective pressures for other regions of the DNA. DNA sequencing supports this concept. For one thing, it has been shown that silent mutations are quite common. It also reveals much more variability in that DNA that is not coded for a product as opposed to the coding DNA as seen in the higher incidence of nucleotide substitutions in introns and spacer regions between genes. Genetic changes such as these would appear to be neutral, and it would seem that some of them are present in populations at their current frequencies due to mutation and random genetic drift. Moreover, when a gene that is coded for a specific protein is studied for nucleotide substitutions, it is found that much more variation is present in some of its parts than in others. Protein analyses show that those parts of a protein that are involved in critical functions such as catalysis or binding show much less diversity in amino acid sequences than do those portions with less or very little critical functions. All the observations indicate that those portions of a gene coded for protein parts having important roles may be greatly influenced by natural selection, whereas those portions coded for parts of lesser relevance are subject to neutral evolutionary changes.

That selection and neutral evolution are both involved at the molecular level is also supported by comparisons of the rates of evolution of whole protein molecules. Histone proteins have changed very little throughout the entire evolution of eukaryotes (see Chap. 17). They are apparently so perfected for their role in chromosome organization that a change anywhere in a histone molecule is likely to upset organization of the chromosome. Natural selection has undoubtedly been responsible for the conservative history of the histone genes, which have the slowest rates of evolution of all. However, DNA sequencing of histone genes has revealed that silent mutations affecting the third position of codons is not uncommon. These mutations, which bring about no amino acid substitution, may very well represent the operation of neutral evolution. A dramatic contrast to the histone is revealed by studies of fibrinopeptides, proteins that have little function after they are cut off from fibrinogen in the blood-clotting reaction. These are the most diversified of the proteins and show the most rapid rate of evolution, in contrast to the histones, in which amino acid substitutions are exceedingly rare. Studies of both protein and DNA polymorphisms show that a great deal of genetic variation exists in natural populations. Some appear to be the result of natural selection and some of neutral evolution. In a given environment, a particular allele associated with a particular form of a protein may be selected. As the environment changes, another allelic form may be favored. An allele that is neutral under one set of conditions may prove to be valuable under another. The force of natural selection along with the process of neutral evolution may provide populations with sufficient diversity to evolve as they meet new environmental challenges.

According to a strict selectionist interpretation, a mutant allele must have some selective value to spread through a population and to become fixed. Occasionally a neutral allele may persist for a while if its locus is closely linked to that of an allele that is being for. According to selectionists, the molecules that appear to be evolving rapidly would indeed have some important function that is not yet apparent. The rapid evolution would indicate the accumulation of beneficial changes leading to a higher degree of adaptation.

Various tests have been carried out to help distinguish between the opposing neutral and selection theories. Unfortunately, all the data are sufficiently ambiguous to preclude clear-cut support for either concept. Many of these tests have involved studies of protein polymorphisms. For example, the polymorphic enzyme alcohol dehydrogenase (ADH) was the focus of one investigation. This enzyme can detoxify the ethanol found in rotting fruit eaten by *Drosophila* larvae. The gene coded for ADH is found in several allelic forms that in turn are associated with different allozymic forms of the enzyme. Two of the alleles together are by far the most common in different fly populations. One of them codes for a form of the enzyme, the slow (S) allozyme, which is much more stable at higher temperatures than is the fast (F) form coded by the other common allele or the ADH gene. The S allozyme has a greater heat stability than the F form, and thus from the selectionist's point of view, it would occur at a higher frequency than the F allozyme in populations closer to the equator. We would also expect to find a higher frequency of the S form correlated with the high maximum monthly temperatures of the areas in which the fly populations live. Examination of populations in North America, Asia, and Australia do show that the S form increases in frequency in populations found closer and closer to the equator. This provides support for the concept of selection of the S allele of the ADH gene. However, when one looks for a correlation in frequency between the S allozyme and the maximum monthly temperature of the immediate population environment, the results conflict. A correlation exists when North American populations are examined, but not in the cases of the Asian and Australian populations. The data allow no choice to be made between the selection and the neutral theories.

Strict Darwinians continue to maintain that selection operates at all levels and is primarily responsible for the frequency of an allele. The environment is a decisive factor, determining what is to be selected. However, the neutralists contend that much evolutionary change is driven by random genetic drift, not by environmental change. Although the matter remains unsettled, the controversy between neutralists and selectionists has propelled geneticists to reexamine ideas and has resulted in the design of better experiments and in more reasoned models of the forces that shaped the biological past.

Mutation and molecular clocks

If the neutral theory of evolution is correct, most of the changes that we observe in DNA and in proteins have accumulated at a constant rate in populations over eons. Consequently, the rate of evolutionary change in genes or their protein products provides the basis for a molecular clock. If the rate of change over generations is indeed constant, then a comparison can be made of different populations to reveal differences among them with respect to a given protein or a gene. An example of how the clock can be expressed in actual time units is the examination of the fossil record to see when a certain species first appears. This information is added to that on the number of amino acid substitutions by which a given gene or a protein now differ between populations of that species. An estimate is then made of the rate at which genetic change has taken place in a gene or protein from the appearance of the species up to the present. One can calculate backwards to estimate the actual time in years at which the two populations began to diverge. The longer two groups have been separated, the longer the time for random mutations to occur in a given gene and the greater the number of amino acid substitutions in the gene's protein product. If two species show very little difference in a given gene, one could conclude that the two groups have been recently separated. In contrast, two groups showing many differences in the gene would be considered separated for a longer period of time. Comparing human proteins with those of the chimpanzee or gorilla reveals little difference in contrast to the great difference seen between the proteins of the human and those of the lemur. We conclude from these observations that the human species has been separated from the lemur for a much longer period than that separating it from the higher apes. By running the molecular clock backwards in chronological units, some molecular biologists have placed the time of separation of human and apes at 5 million years.

The molecular clock would appear to provide a valuable approach to the tracing of origins of populations and their relationships. Selectionists, however, emphasize natural selection and claim that it weeds out many genetic changes while favoring others so that the differences observed among populations are not simply a reflection of a constant rate of genetic change. This argument has been countered by assertions that over millions of years the different rates of substitution in various genes and their proteins due to selection average out so that an apparent constant rate of genetic change becomes evident. In one approach to calibrate the molecular clock, amino acid substitutions in four proteins taken from over a dozen vertebrate species were compared to the fossil records of the animals involved. The results show that the molecular clock is not constant in the short term but that over long periods of geologic time the erratic substitutions of nucleotides in DNA is eliminated and the average rate of change becomes constant.

Various calculations indicate that molecular clocks are fairly accurate. Each gene (and hence its product) produces its own molecular clock, since genes differ in the rate at which neutral mutations occur in them. A major goal is to accumulate data from many genes to yield a reliable clock that can be used in evolutionary studies to trace times of divergence of related groups. Many questions regarding the use and reliability of molecular clocks remain to be answered to the satisfaction of biologists using different approaches to problems of phylogeny. We will learn in Chapter 21 that the molecular clock concept is key to the current controversy regarding human origins.

Gene conversion

In all of our discussions of linkage and recombination, we have assumed that when a crossover event takes place a reciprocal exchange occurs between two homologous chromatids. A review of Figure 8-16 depicts the expected results of a single crossover between markers when a^+ and b^+ are linked on one chromosome and a and b on the homolog. We see that four kinds of gametes are generated in equal numbers. A wealth of data from pioneer genetic studies have left little doubt that crossing over entails breakage followed by the mutual exchange of chromosome material between two homologous chromatids. However, studies of linkage and crossing over in certain organisms brought to light many exceptions to the rule. Fungi such as *Neurospora* permit detailed analyses of spores that are the products of a single meiosis in a cell, thus enabling the precise events of crossing over between two markers such as a and b (Fig. 8-16) to be examined. When very closely linked markers were being followed, a single crossover event between two of them such as a and b did not necessarily produce reciprocal products. Instead of generating the crossover products a^+b and ab^+, the recombinant ab^+ may not occur. Consequently, following the crossover between a and b, instead of the expected four types of meiotic products the results would be a^+b^+, a^+b^+, a^+b, and ab. The allele a^+ would be represented three times instead of two, and the alternative form a only once, as if the one gene form (a) were being converted to the other (a^+). The name *gene conversion* was given to the process that generates nonreciprocal recombinants.

Gene conversion occurs when markers are very closely linked; they may even reside within the same gene. At times the number of meiotic products showing conversion between such closely linked markers exceeded that showing the expected pattern. Gene mutation could not explain conversion. Not only did it occur with too high a frequency, but the phenomenon was always associated with crossing over and always resulted in a change from one form to the form present on the homolog (e.g., a might be changed to a^+ but not to a completely new allele such as a′).

These unexpected conversions can be explained on the basis of mismatch repair accompanying genetic recombination. A review of Figure 11-29 shows that heteroduplex DNA is generated if crossing over occurs according to the widely accepted Holliday model. As explained in the section on DNA repair, in a heteroduplex region the two strands may not be perfectly complementary, and a mismatch may be present. In a dihybrid cross in which the alternative genetic forms are found at very closely linked sites, even within the same gene, a crossover event can generate heteroduplex segments containing mismatched pairs in the region where the markers reside (follow Fig. 14-13). This heteroduplex is found in those two of the four chromatids that participated in the crossover event. A mismatch repair may follow that excises a portion of the DNA in one of the strands of the heteroduplex. In Figure 14-13, excision takes place in both strands where heteroduplexes are found. (However, this need not be the case. Neither strand, or only one of them, may be repaired. In such cases the out-

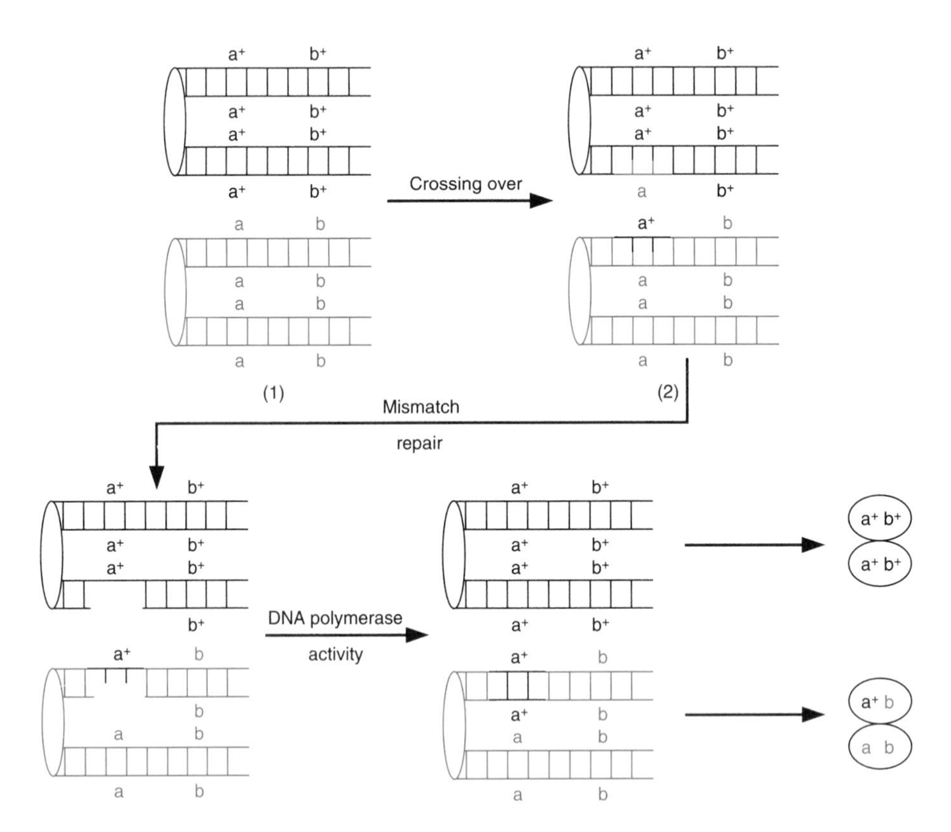

FIGURE 14-13. Gene conversion by mismatch repair. **1:** The duplex DNA in four chromatids of a bivalent are depicted. Two chromatids are a^+b^+ and the other two are ab with respect to two very closely linked markers. **2:** Following crossing over, heteroduplex DNA is generated. The two chromatids that participated in the crossover event now have heteroduplex segments in the a region, where a mismatch is now found. Mismatch repair mechanisms may excise a portion of one of the strands of a heteroduplex. **3:** Excision in this case is shown to take place in both chromatids. **4:** Filling in of the gaps by DNA polymerase brings about a gene conversion. **5:** Meiosis is completed, but the products are not reciprocal. Although the ratio for the b marker is $2b^+$:2b (1:1), the proportion of the a marker is $3a^+$:1a.

come would be different from those depicted in the figure.) DNA synthesis fills in gaps resulting from excision. The final result is a departure from the expected pattern. In this case, instead of the expected $1a^+$:1a, we find $3a^+$:1a. As noted previously, this is not the only possible outcome. For example, if the segments excised had been the two carrying a^+ instead of a, the outcome would have been 3a:$1a^+$.

Chemical mutagens and their effects at the molecular level

In 1941, Auerbach and Robson discovered the first chemical mutagens, mustard gas and related compounds. Like high-energy radiations, the mustard compounds produce a high incidence of gene mutations and structural chromosome aberrations in plants and animals. They can also greatly damage rapidly growing tissues by producing chromosome changes and have been used as well as radiations to combat malignancies.

The mutagenic action of mustard gas, however, differs from that of radiation in a very important way. The effect of treatment with mustard gas is frequently delayed; a mutation is not necessarily induced in the cell that was exposed but may arise in a descendant of that cell, as if the treated DNA had somehow become unstable. The mutation may be delayed for two generations, and when it does occur it may arise in a body cell as well as in a germ cell. A mutation occurring during the development of a fruit fly, for example, could take place in just one body cell. This mutant cell could then give rise, through cell division, to a patch of cells surrounded by nonmutant cells. The result is a mosaic individual. This delayed effect makes mustard gas a more serious mutagenic agent in many ways than one whose action is expressed directly.

After the discovery of the first mutagenic chemicals, many other substances were also found to possess mutagenic activity, and the list is still increasing. Few have the strong effect of the mustard compounds, and the mechanism by which many of them operate to increase the frequency of mutations is not known. The chemical structures of the many mutagens have nothing in common. Furthermore, a substance found to be mutagenic in one organism may be ineffective in another. Caffeine, for example, is highly mutagenic in some organisms, but is apparently broken down in others before it can exert any action. Formaldehyde markedly increases the mutation rate in the fruit fly, but only if applied to the food of the larval male. It is ineffective at other stages of the life cycle and is entirely nonmutagenic in the female! Most chemical mutagens have also been found to be capable of inducing structural chromosome changes as well as point mutations.

The interesting discovery was made that some substances have an antimutagenic action—that is, they tend to suppress the mutagenic activity of other agents. Antimutagens were first detected by Novick, working with bacteria. The strong mutagenic action of purines, such as caffeine and adenine, was suppressed by the presence in the medium of purine ribonucleosides, such as adenosine and guanosine. The antimutagens were ineffective toward the mutagenic action of ultraviolet and γ radiations. Even the spontaneous mutation rate was shown to be reduced when the antimutagens were present. It is significant that some purine nucleosides were found to have a high antimutagenic activity and that many of the nontoxic chemical mutagens were purines or their derivatives. Indeed, the mutation rate seemed to depend in part on the physiological state of the organism.

Thus observations indicate that the study of mutation can be approached on a biochemical basis and lead us to suspect that the spontaneous mutation rate, which is geared toward the generation time of a species, results from an interaction between mutagenic and antimutagenic substances arising during the course of metabolic activities in the cell. Let us examine the kinds of events at the molecular level that produce gene mutation.

As a result of tautomeric shifts, we know that the usual pairing relationships among the bases may be altered. Instead of adenine pairing with its complementary base, thymine, the rare form of adenine may pair with cytosine to give an AC pair instead of the normal AT. Conversely, the occurrence of a rare tautomeric form of cytosine may result in a CA pair instead of CG. Likewise, instead of TA or GC, the unusual pairs CA and AC may form. The ultimate effect of these pairing errors is shown in Figure 14-14. It can be seen that the AC error results in the formation of a GC nucleotide pair. The original position of the purine adenine on one of the two strands of the double helix is now occupied by the purine guanine. In other words, one purine–pyrimidine pair is now represented by a different purine–pyrimidine pair. It can be seen from Figure 14-14 that, in each case, the purine–pyrimidine relationship is maintained—one purine–pyrimidine pair substitutes for another *or* one pyrimidine–purine pair substitutes for another. Base pair substitutions of this type have been called *transitions.*

In contrast to this type of change is the *transversion,* the substitution of a purine–pyrimidine pair by a pyrimidine–purine pair (or vice versa; Fig. 14-15). In the transversion, there is a new orientation of purine and pyrimidines. A purine on one strand is replaced by a pyrimidine; a pyrimidine on the complementary strand is replaced by a purine. The significant fact is that, in the case of either a transition or a transversion, a different sequence arises at a site on the DNA. This alteration thus modifies the genetic information at that point and can cause a mutant effect as a consequence.

It was reasoned that if spontaneous mutation did actually result from such transitions and tranversions, then it should be possible to induce mutations with chemicals whose activities are known. An example of such a chemical is nitrous acid, which has the ability to remove free $-NH_2$ groups from the bases found in nucleic acids (Fig. 14-16A). After deamination by nitrous acid, adenine is converted to another base, hypoxanthine. In hypoxanthine, the arrangement of atoms involved in base pairing is different from the adenine (A), but similar to guanine (G). This means that if adenine is changed to hypoxanthine (H) at a point on the double helix, a transition from AT to GC can occur. By deamination, nitrous acid converts cytosine (C) to uracil (U, Fig. 14-16B), whose pairing characteristics are like those of thymine (T). Again a transition can result, this time from CG to TA. Since nitrous acid can cause such changes in nucleic acid, it is therefore not surprising that it is a very powerful mutagenic chemical. Several other chemical agents are known to cause alterations in the bases of nucleic acids, and these too show mutagenic activity. Most of these investigations have been performed on microorganisms. Data from phages strongly support the concept of induced mutation arising from changes in base pairing at the time of gene replication.

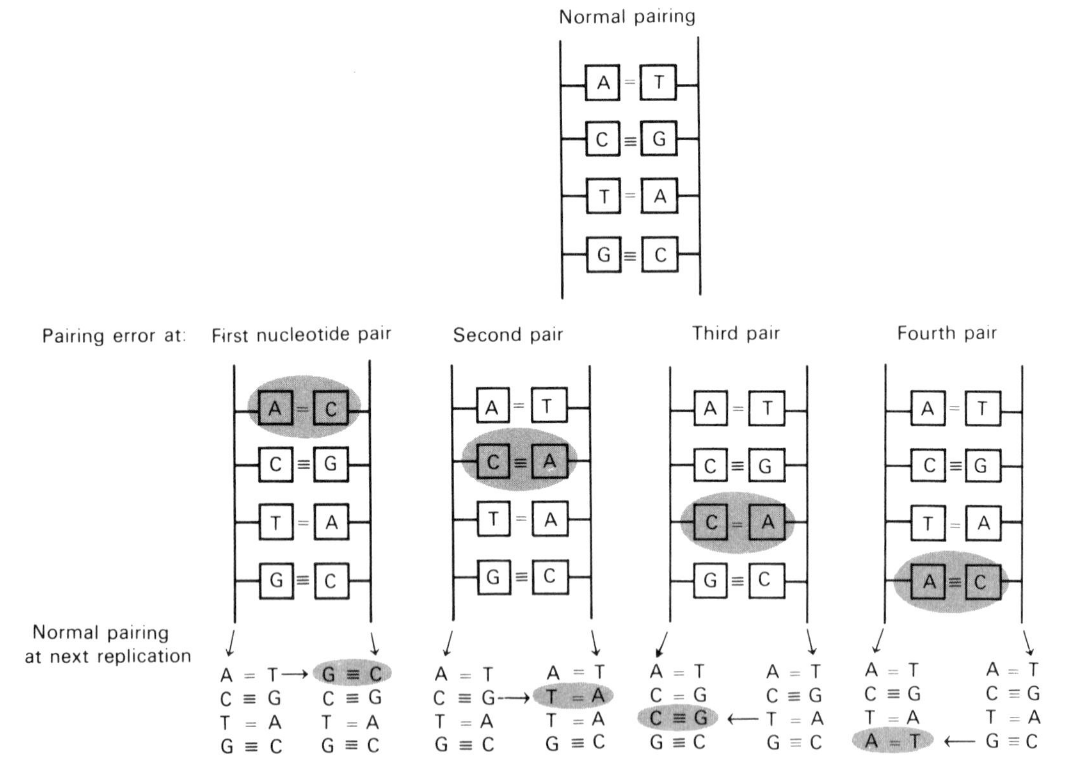

FIGURE 14-14. Transitions. A stretch of four nucleotides with normal pairing is shown above. Abnormal pairs may form at the time of replication because of the occurrence of a rare tautomeric state of one of the bases. If adenine and cytosine form a pair, instead of the normal AT the result is a transition, a substitution of one purine–pyrimidine pair by another or the substitution of one pyrimidine–purine pair by another. In this example, we see **1:** GC substitutes for AT, **2:** TA for CG, **3:** C:G for TA, and **4:** AT for CG.

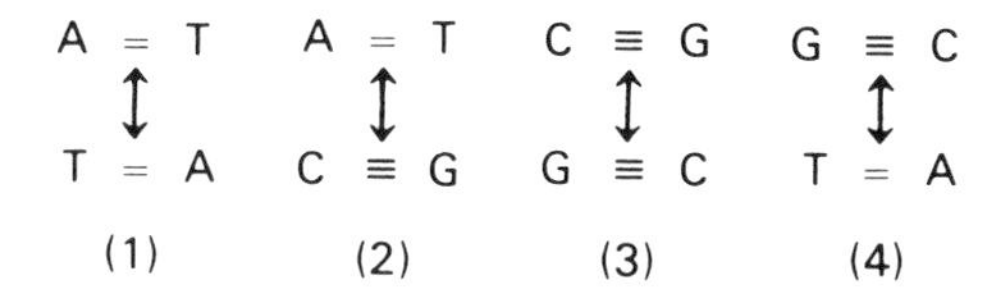

FIGURE 14-15. Summary of types of transversions. In an alteration of this kind, the end result is a substitution of a purine–pyrimidine pair by a pyrimidine–purine pair (or vice versa). For example, in 1, the purine–pyrimidine pair AT is replaced by the pyrimidine–purine pair TA (or vice versa), and so on for the other substitutions.

Another class of mutagenic substances is known as the *base analogs,* so called because they resemble the normally occurring nucleic acid bases both in structure and activity. For example, thymine has a base analog, 5-bromouracil (5-BU), so similar to thymine that it can substitute for it. This means that it can be attracted to adenine and form hydrogen bonds with adenine in much the same manner as thymine.

One important difference between thymine and 5-BU, however, is that the latter undergoes tautomeric shifts more often than the thymine. In its rarer form, 5-BU behaves like cytosine and tends to pair with guanine. Therefore the 5-BU could be in its "cytosine-like" state when it is being taken into the DNA, as shown in Figure 14-17. Or it may be in this form later at the time of replication, *after* it has been incorporated into a strand of the nucleic acid. The latter is called a *mistake in replication;* the former is a *mistake in incorporation.* Again the end result is a transition, GC to AT (from the error of incorporation) or AT to GC (from the error of replication). Highly mutagenic base analogs such as 5-BU undoubtedly exert their mutagenic activity in this manner.

The alkylating agents, which include the nitrogen mustards and ethyl sulfonate, constitute another class of mutagenic agents. In contrast to substances such as nitrous acid, which are effective mainly in prokaryotes, these are very efficient mutagens in eukaryotes. The alkylating agents can impair the normal hydrogen bonding of guanine and thymine by adding an alkyl group to the hydrogen-bonding oxygen of these two bases. As a consequence, mispairing of G with T takes place, leading to transitions (AT to GC and GC to AT). Alkylation can also bring about the loss of guanine from the DNA by breaking the bond that joins the base to the sugar. Mutation can result from this depurination (Fig. 14-18), but more likely excision repair mechanisms will restore the correct nucleotide. An error-prone repair system may come into play, however, to repair positions in the DNA that have been alkylated. An alkylated nucleotide is removed, but errors of incorporation are often made, frequently bringing about transversions.

Acridine dyes are mutagenic substances frequently used in mutation analysis. Acridines can insert themselves into the double helix between two adjacent nucleotides. Following exposure to acridines, there is an increase in frame shift mutations in which there is either the addition of an extra base, a + mutation, or the deletion of a base, a – mutation. Frame shift mutations, which cause very serious consequences by upsetting the proper amino acid sequence of polypeptides, are also known to occur spontaneously as well as following exposure to ultraviolet light. We have seen (in Chap. 12) that mutations of this type were very important in deciphering the genetic code. No matter what the exact mode of action, all known chemical mutagens bring about the same result—a change in the coded information through some type of alteration in nucleotide sequence.

Although we know that certain chemicals can cause characteristic molecular changes, this information must not be misconstrued to mean that the investigator can bring about a specific kind of gene mutation by using a specific chemical, for these agents raise the entire mutation rate, and the four bases—adenine, thymine, guanine, and cytosine—are found throughout the DNA. It is known from work with microorganisms that one mutagen may be more likely than another to induce a certain *class* of mutation. For example, Novick found that ultraviolet treatment of *E. coli* caused a higher mutational change to resistance to a kind of virus, phage T/6. In contrast to this, certain purine mutagens caused a greater probability of mutation to resistance against a different virus, T/5. Comparison of some chemical mutagens has shown that some genes may be more susceptible to mutation by one chemical than to another. The chemical 2-amino purine favors a certain kind of back mutation (from mutant to wild) at a site in a gene controlling the enzyme tryptophan synthetase in *E. coli.* However, no mutagen specific for just one kind of gene is known. Certain techniques now being used with viruses have an important bearing on this point and are discussed in Chapter 18.

Bacteria and carcinogens

Bacterial genetic systems have valuable applications in procedures designed to yield information on the potential dangers that certain environmental factors present because they are carcinogens, or cancer-inducing agents. Although the steps triggering the transformation of a normal cell to the malignant state are far from clear, there is little doubt that alteration of some kind in the cellular DNA is a key factor (see Chap. 17). When a cell is transformed to the malignant state, it has somehow been released from the normal genetic controls that regulate its DNA replication and its interactions with neighboring cells. Once the cell is changed, its malignant state is transmitted to all those cells that trace back to it.

Today we are surrounded by an ever-increasing number of chemical compounds in the food we eat, the substances we handle, the air we breathe, and in many other ways. As chemicals continue to be poured into the environment, it is essential to have a quick, inexpensive way to detect any that may be carcinogenic. The use of mammalian test systems, however, to screen only a fraction of the chemicals we contact is not only very expensive but also very time-consuming. Fortunately, tests with microorganisms have been devised that reveal a chemical's mutagenic potential. The most widely used of these was devised by Bruce N. Ames of the University of California at Berkeley. One might wonder whether the response of bacteria to any test substance can be equated to that of a cell from a mammal. However, the results have shown that the bacterial tests distinguish between *known* carcinogens

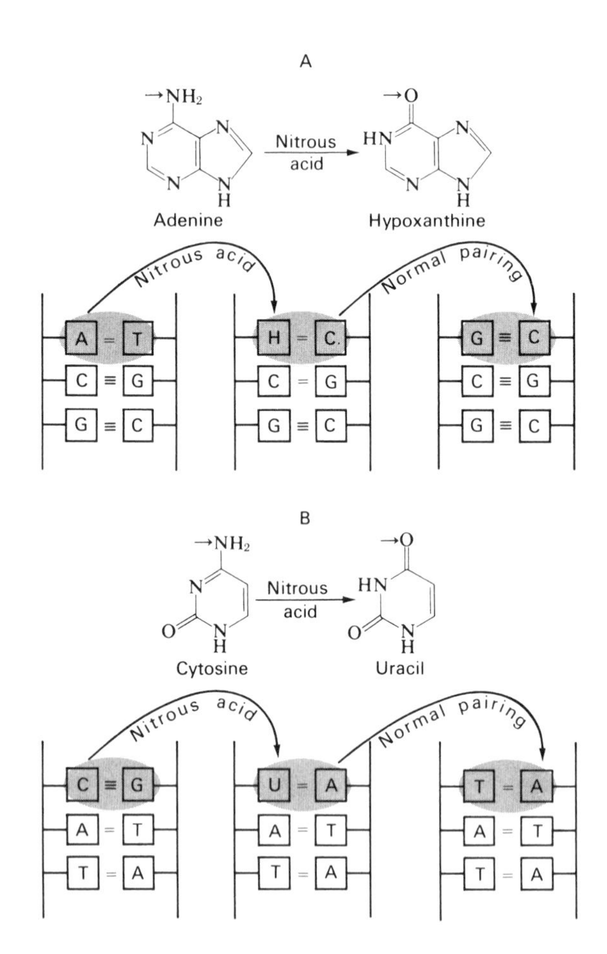

FIGURE 14-16. Mutation induction by nitrous acid. **A:** Nitrous acid is able to remove free amino groups from bases in nucleic acid in the conversion of adenine to hypoxanthine. The latter has pairing properties similar to those of guanine and tends to pair with cytosine. The end result is a transition, the substitution here of GC for the original AT. **B:** Cytosine is converted to uracil, which behaves like thymine and thus pairs with adenine. Again the result is a transition, in this case a change from CG to TA. Nitrous acid also causes deamination of guanine and its conversion to xanthine, which behaves like guanine in its pairing preference for cytosine.

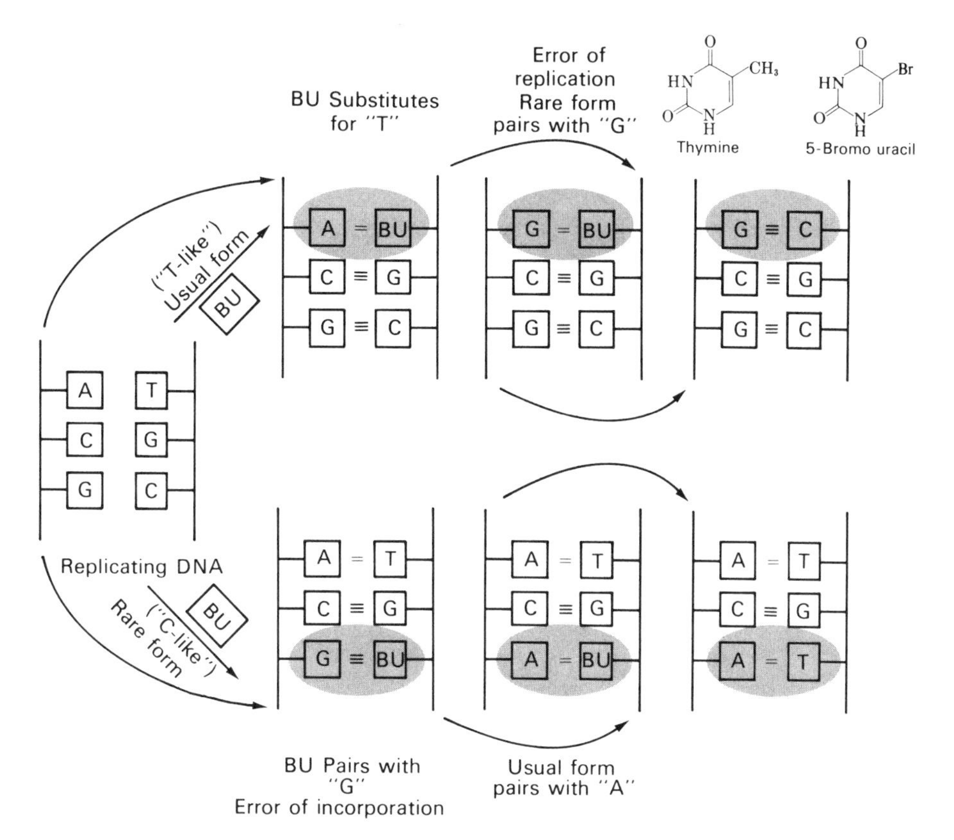

FIGURE 14-17. 5-bromouracil as a mutagen. This chemical closely resembles thymine in structure (above), and it can substitute for it at the time of DNA replication. The upper part of the figure shows this. At a following replication, it may be in its rarer form and behave like cytosine. At the next replication, the incorporated guanine pairs with cytosine. The result is a substitution of a GC pair for the original AT. The lower part of the figure indicates that 5-bromouracil may be in its rarer C-like state before it enters a DNA chain. It can thus pair with guanine. At the next replication, it would behave typically (like thymine) and attract adenine. At the next replication the adenine attracts thymine. The result is a substitution of an AT pair for the original GC.

and *known* noncarcinogens with an accuracy over 90%. Moreover, they have singled out certain previously unsuspected substances as potential carcinogens. When tested later in animals, these chemicals were indeed shown to have cancer-inducing ability.

The values of the Ames test are manyfold. It is much less expensive than those conducted with animals. The results are available much more quickly—in days versus years! It can also detect substances that may be only weak carcinogens. Once a chemical is shown to be suspicious, on the basis of the Ames procedure, it can then be followed further in mammalian systems. The bacterial test thus makes it possible to check a large number and variety of compounds that otherwise could not be screened by the animal tests. Once we are alerted to the potential danger of a substance, it can be studied further.

The basis of the Ames test is the demonstration that a gene mutation has occurred in response to a specific substance. This is then taken as evidence of DNA damage and hence the ability of that substance to effect a malignant transformation. Although it is true that not all mutagens have been found to be carcinogens, the correlation is extremely high. If a substance is shown to be mutagenic, the probability is about 90% that it is also a carcinogen.

The Ames test employs strains of a particular colon bacillus, *Salmonella typhimurium*. The bacteria used in the test carry a mutation (his$^-$) that makes them nutritionally deficient because they cannot synthesize one of the enzymes needed to manufacture histidine. Some of the his$^-$ strains used in the Ames test carry base pair substitutions (transitions or transversions); others are defective because of frame shift mutations.

Cells that are his$^-$ can, of course, undergo back mutation. However, the spontaneous reversion to his$^+$ occurs at a very low rate. If some agent can increase the rate of mutation in the cells, it will consequently increase the chance that his$^-$ will revert to his$^+$. Since different his$^-$ tester strains are used (those with base pair substitutions and those with frame shift mutations), a chemical can be scrutinized for its ability to achieve one or more kinds of genetic alterations. The ability to cause reversion from his$^-$ to his$^+$ is thus taken as a measure of an agent's mutagenic, and hence carcinogenic, potential.

To use the his$^-$ strains, however, it was essential to circumvent the fact that many substances are unable to pass through bacterial cell walls. The strains used therefore carry a mutation that makes the wall more penetrable to chemicals. The strains also carry another mutation that causes a defect in the DNA repair mechanism that excises defective DNA (see Fig. 14-8). Such a mutation permits easier detection of various kinds of alterations in the DNA, since a defect would not be removed. Very important consideration was also given to the need to create a system that has certain mammalian characteristics. This was necessary, because it is known that chemical carcinogens are not necessarily active in their original forms. Metabolic processes may take place in the mammal that would change the chemical somehow and render it carcinogenic. Conversely, cells may convert a potentially harmful compound to an innocuous one. Ames therefore added an extract of rat liver to the bacterial system. In this mixture, a test chemical is thus exposed to mammalian metabolic processes.

In the Ames test, the test substance is added to the his$^-$ bacteria–rat liver mixture. The mixture is then plated on solid, minimal medium, which contains only enough histidine to support a few cycles of replication, but not enough to permit colonies to form.

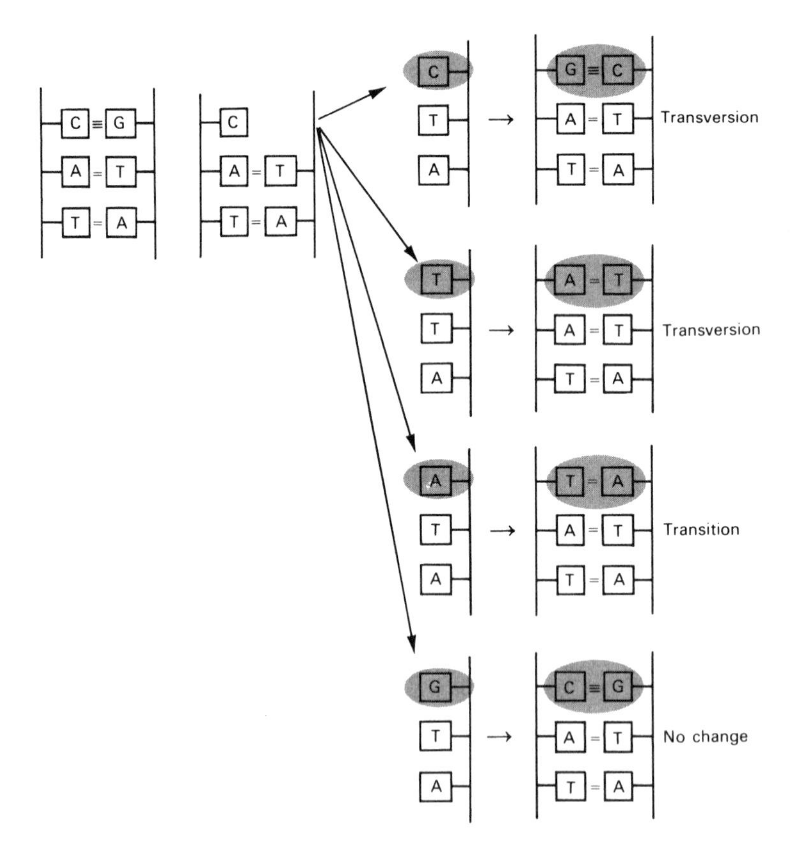

FIGURE 14-18. Effect of mustard compounds. Sulfur and nitrogen mustards can cause the base guanine to be eliminated from a strand of DNA. The missing base may later be replaced by any one of the four bases. When this strand replicates, a transition or a transversion may occur, depending on which base substitutes for the original guanine.

After about 2 days of incubation, the plates are examined. Those cells that have experienced a his$^+$ reversion will give rise to colonies on the minimal medium. The number of these can be related to controls and to the amount of the test substance used, to give a quantitative result. The value of the Ames test has been recognized by government and private industry and is currently in use in thousands of laboratories throughout the world. The speed, accuracy, and inexpensive features have already resulted in the screening of thousands of chemicals, a figure that would not have been approached by the more cumbersome animal tests.

REFERENCES

Ames, B.N., R. Magaw, and L.S. Gold. Ranking possible carcinogenic hazards. *Science* 236:271, 1987.

Avers, C.J. *Process and Patterns in Evolution.* Oxford University Press, New York, 1989.

Ayala, F.J. On the virtues and pitfalls of the molecular evolutionary clock. *J. Hered.* 77:226, 1986.

Chu, G., and E. Chang. Xeroderma pigmentosum group E cells lack a nuclear factor that binds to damaged DNA. *Science* 242:564, 1988.

Denniston, C. Low level radiation and genetic risk estimation in man. *Annu. Rev. Genet.* 16:329, 1982.

Devoret, R. Bacterial tests for potential carcinogens. *Sci. Am.* (Aug.) :40, 1979.

Drake, J.W. Comparative rates of spontaneous mutation. *Nature* 221:1132, 1969.

Futuyma, D.J. *Evolutionary Biology,* 2d ed. Sinauer, Sunderland, MA, 1986.

Haseltine, W.A. Ultraviolet light repair and mutagenesis revisited. *Cell* 33:13, 1983.

Hollaender, A., and F.J. de Serres (eds.). *Chemical Mutagens Vol. 5: Principles and Methods for Their Detection.* Plenum, New York, 1978.

Howard-Flanders, P. Inducible repair of DNA. *Sci. Am.* (Nov) :72, 1981.

Kenyon, C.J., and G.C. Walker. DNA-damaging agents stimulate gene expression at specific loci in *Escherichia coli. Proc. Natl. Acad. Sci., USA* 77:2819, 1980.

Lindahl, T. DNA repair enzymes. *Annu. Rev. Biochem.* 51:61, 1982.

Little, J.W., and D.W. Mount. The SOS regulatory system of *E. coli. Cell* 29:11, 1982.

Marx, J.L. DNA repair: New clues to carcinogenesis. *Science* 200:518, 1978.

McDowell, R.E., and S. Prakash. Allelic heterogeneity within allozymes separated by electrophoresis in *Drosophila pseudoobscura. Proc. Natl. Acad. Sci. USA* 73:4150, 1976.

Muller, H.J. Artificial transmutation of the gene. *Science* 66:84, 1927.

Nelson, K., and L.B. Holmes. Malformations due to presumed spontaneous mutations in newborn infants. *N. Engl. J. Med.* 320:19, 1989.

Novick, A., and L. Szilard. Experiments on spontaneous and chemically induced mutations of bacteria growing in a chemostat. *Cold Spring Harbor Symp. Quant. Biol.* 16:337, 1951.

O'Brien, S.J., D.E. Wildt, and M. Bush. The cheetah in genetic peril. *Sci. Am.* (May) :84, 1986.

Setlow, R.B. Repair deficient human disorders and cancer. *Nature* 271:713, 1978.

Singer, B., and J.T. Kusmierek. Chemical mutagenesis. *Annu. Rev. Biochem.* 51:655, 1982.

Stadler, L.J. Mutations in barley induced by X-rays and radium. *Science* 68:186, 1928.

Thoday, J.M. Non-Darwinian evolution and biological processes. *Nature* 255:675, 1975.

Vogel, F., and R. Rathenberg. Spontaneous mutations in man. *Adv. Hum. Genet.* 5:223, 1975.

West, S.C., E. Cassuto, and P. Howard-Flanders. Postreplication repair in *E. coli:* Strand-exchange reactions of gapped DNA by Rec A protein. *Mol. Genet.* 187:209, 1982.

Wilson, A.C. The molecular basis of evolution. *Sci. Am.* (Oct) :164, 1985.

REVIEW QUESTIONS

1. The allele W for chlorophyll development is dominant over the recessive w for albino in a certain plant species. As a result of spontaneous mutation, which of the following plants would be the most likely to exhibit mosaicism if mutation arises in cells of a developing leaf? Which would be the least likely?

A. WW.
B. Ww.
C. ww.
D. WWWw, a tetraploid.

2. Suppose a certain human disorder that is inherited as a dominant trait is associated with the allele B. The normal allele is the recessive b. It is found that in a given period of time five abnormal children are born among a total of 100,000 births to normal parents with no family histories of the disorder. From these figures, what appears to be the mutation rate per germ cell per generation from b to B?

3. In corn, the genetic factor R has a spontaneous mutation rate of 492 per million gametes, whereas the factor Pr has a rate of 2.4 per million gametes. What is the chance that a plant of genotype RRPrPr will produce a gamete of the constitution rpr?

4. A mutation converts the DNA triplet ATA in the template strand of a structural gene to the triplet ATT. As a result of the mutation, translation comes to a premature halt with severe consequences for the cell. A second mutation arises in a gene encoded for a glutamine transfer RNA and brings about a change in the anticodon. Some activity is now restored as a result of the second mutation. Explain. What was the nature of the first change? What anticodon change must have been involved in the suppressor mutation? (Consult the codon dictionary in Table 12-1.)

5. A portion of a double helix has a segment with the following nucleotide sequence:

GGATCC

CCTAGG

Suppose that at the time of replication a tautomeric shift causes A in the upper strand to pair with C. For the duplex DNA with the mismatch, answer the following:

A. What is the outcome at the next replication if no mismatch repair occurs?

B. What is the more likely outcome if mismatch repair does occur shortly after the error is made?

C. What would be the outcome in a mutant cell that has lost the ability to methylate bases?

6. In *E. coli,* assume that the amino acid tyrosine occurs at amino acid position 70 in a certain polypeptide that is 110 amino acids long. A mutant strain arises spontaneously in which the normal polypeptide is no longer formed. Instead, a short polypeptide 69 amino acids long occurs. The mutant strain is subjected to mutagenic chemicals. Among the cells derived from the treated cells are those that are wild and produce the normal polypeptide of 110 amino acids.

A. Explain the nature of the spontaneous mutation, indicating the codon or codons involved.
B. Would you expect all of the wild-type cells that arose following chemical treatment to be true revertants? Explain.

7. Suppose a wild-type male fruit fly is treated with X-rays. It is then crossed to an ordinary wild female carrying no known mutant alleles. If a recessive, X-linked lethal is induced in any of the radiated X chromosomes, what would be the effect on the sex ratio:

A. In the first generation?
B. In the second generation?

8. A certain dose of X-rays applied to a culture of actively growing cells is found to induce single chromosome breaks at random with a frequency of 1/50.

A. How many single breaks would be expected in the cell culture by increasing the dosage three times and then four times?
B. How many structural chromosome changes (inversions, reciprocal translocations, etc.) would be expected by increasing the dosage three times and then four times?

9. In *Drosophila,* the percentage of sex-linked lethals is directly proportional to the X-ray dosage as measured in roentgens. Suppose that, among the offspring of males treated with 600r, the frequency of sex-linked lethals was 3%. Among males irradiated with 1,600r the frequency of sex-linked lethals found among the offspring was 8%. What is the percentage of sex-linked lethals expected among offspring of males exposed to a dosage of 2,500r?

10. A. Suppose male fruit flies were irradiated with 1,000r and that 72 sex-linked mutations were found among 720 offspring. In another group of males irradiated with 1,700r, 136 sex-linked mutations were produced among 800 offspring. How many mutants would you predict among 780 offspring of males irradiated with 2,250r?

B. Suppose a group of male fruit flies were given a dose of 1,250r in one experiment and a dose of 1,000r in another. How many mutations would you expect among 780 offspring of these males?

11. In *Neurospora,* a single gene mutation can block a step in a biochemical pathway by causing a defect in an enzyme that governs that specific step. A mutant mold with such a defect can grow on a simple, basic growth medium only if the medium is supplemented with one or another growth requirement: substance A, B, C, D, etc. Four nutritionally deficient strains were studied in relation to the substances that when added to the basic medium can satisfy their growth requirements. In the results given in the following table, + = growth and 0 = no growth. From this information, determine the sequence of reaction steps in the pathway and the specific step at which each mutant is blocked because of an enzyme defect.

	Substance added			
Strain	A	B	C	D
1	+	+	0	0
2	0	+	0	0
3	+	+	+	0
4	+	+	+	+

12. Suppose a fifth mutant *Neurospora* strain is found that can grow on none of the substances mentioned in Question 11 except substance B. However, it can also grow if supplied with another substance, substance E, which the other strains do not require for growth on the basic medium. Where is the block in the case of this mutation?

13. H.J. Muller estimated that the total background radiation from natural sources is equivalent to about 0.045r for a 4 week period. Would you expect this background radiation to be more significant in the production of spontaneous mutations for a species like the human or one like the fruit fly with a life span of about 4 weeks? Why?

14. Assume that a bacterial suspension is diluted so that a given volume contains approximately 200 cells. The same amount of suspension is placed on several plates divided into three groups: A, B, and C. Nothing is done to group A. Groups B and C are exposed to a dosage of ultraviolet light sufficient to kill 30% of the cells. Group C plates are given a half-hour exposure to strong white light. The three groups are then placed in an incubator. Since about 60% of the killing effect of ultraviolet can be neutralized by photoreactivation, on the average about how many colonies would you expect to see the next day on plates of groups A, B, and C?

15. Suggest why exposure to excess radiation might cause lower male fertility as well as severe anemia resulting from destruction of blood cells.

16. Certain human cells are found to accumulate DNA containing thymine dimers, particularly after exposure to doses of ultraviolet light. When samples of the cells are exposed to ionizing radiation, which can induce chromosome breakage, the chromosomes show an ability to undergo repair. Explain.

17. Cells of a certain bacterial strain are exposed to ultraviolet light. A sample is exposed immediately to white light. A second sample is placed in the dark. The cells in the first sample show a very high rate of survival; only a small percentage dies. In the case of the second sample, a very high mortality rate results, which is far in excess of that shown by other strains that received the same ultraviolet exposure and dark treatment. Explain.

18. Consider the following sequence of nucleotide pairs along a molecule of DNA:

AGT
TCA

Exposure to nitrogen mustard causes G to become less firmly linked to the sugar–phosphate backbone, and it is eventually eliminated. Give the possible base sequences after the missing base is replaced and the strand replicates. Classify each change as a transition or a transversion.

19. Assume that the polynucleotide strand with the sequence AGT is replicating but that it has been stretched between G and T because of an acridine dye. On the growing complementary strand a nucleotide with thymine is inserted in the stretched region. What nucleotide pairs will result when this complementary strand has undergone a replication producing its own complementary strand?

20. A polynucleotide strand with the segment

AGT
TCA

is exposed to nitrous acid, and the base cytosine is changed to uracil.

A. What strand will be formed that is complementary to this one in which the change has occurred?
B. What will the base pairs at the next replication be, and will this outcome be a transition or a transversion?

21. The amino acid sequence of a portion of a certain polypeptide in *E. coli* is as follows: NH_2-met-pro-ser-ser-asp-lys-tyr-arg-leu-. . . . Cells were exposed to an acridine, and mutants arose in which the same segment of the polypeptide had the following amino acid sequence: NH_2-met-pro-ser-gln-ile-asn-thr-asp-leu-. . . . The mutant cells were again exposed to an acridine, and offspring cells arose in which the same polypeptide segment was found to be as follows: NH_2-met-pro-ser-gln-ile-lys-tyr-arg-leu-. . . . Determine the following:

A. The most likely kind of mutation that has arisen. Using the dictionary of codons, determine the nucleotide sequence in the original mRNA, and show the change produced as a result of the first treatment with the mutagen.
B. The second change produced by the acridine in the mRNA from the first mutant cells.

22. Artificial mRNA can be synthesized and then used in an in vitro translation system. Four different kinds of artificial messenger RNA can be made that contain only a single kind of ribonucleotide. These are poly(U), poly(A), poly(C), and poly(G). As indicated in the following table, each homopolymer can be translated into a polypeptide containing only one kind of amino acid. As the table also shows, when streptomycin is present in the system, each kind of homopolymer is translated into a polypeptide containing three or four kinds of amino acids. From the information in the table, and with the aid of the dictionary of codons, answer the following:

A. How does streptomycin appear to influence translation and the reading of codons?
B. Why is it that some polypeptides that are synthesized in the presence of the drug contain only three kinds of amino acids whereas other polypeptides contain four?

	Streptomycin	
Homopolymer	Absent	Present
Poly(U)	phe	phe, leu, ile, val
Poly(A)	lys	lys, gln, glu
Poly(C)	pro	pro, ser, thr, ala
Poly(G)	gly	gly, trp, arg

23. A certain food additive is supplied to bacteria in the same test situation as that in the Ames test except for the elimination of the rat liver extract. It is found that a statistically significant number of his^+ cells arise in relation to controls. When the chemical is tested again in the presence of the liver extract, there is no significant increase in the number of his^+ cells over the controls. Explain.

24. Suppose soluble protein samples extracted from the blood of animals from population A are placed on a gel and exposed to electrophoresis. The gel is then exposed to a chemical solution, which contains a stain specific for one particular enzyme. The different patterns detected are shown here in A.

When the same procedure is performed on protein samples from animals of population B, the results shown in B are always obtained. Explain.

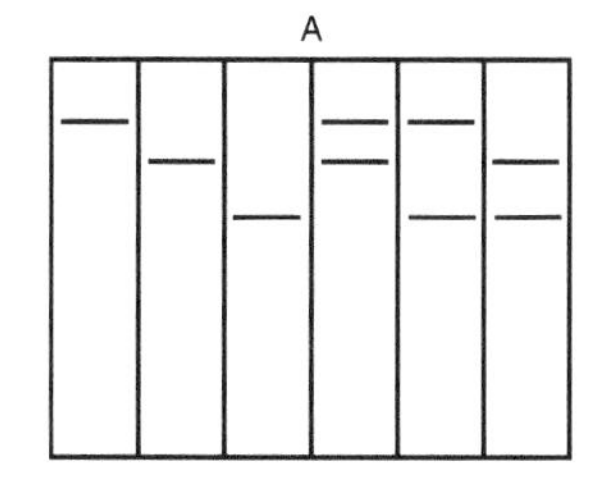

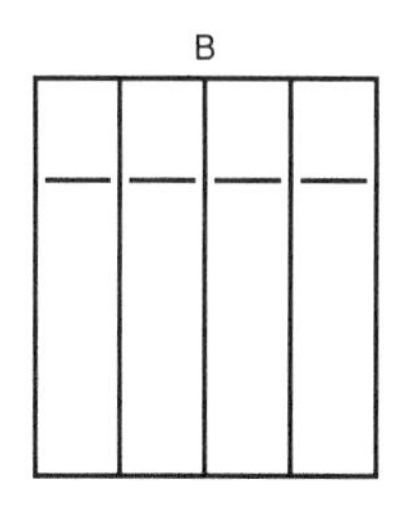

Bacterial Genetic Systems

15

The fluctuation test and replica plating

Until 1943, it was not certain that true genetic mutations occur in bacteria as they do in higher organisms. Although variations in microorganisms were well known, it was unclear whether they were the result of spontaneous genetic change or whether they were a direct adaptive response by the cells to environmental factors. The matter was settled by Luria and Delbruck with the design of the fluctuation test, which laid the foundation for all later research into the genetics of microorganisms.

The procedure was designed to distinguish between changes in population structure through 1) direct modification by an environmental agent or 2) spontaneous mutation and subsequent selection by an environmental agent. The specific bacterium used was *Escherichia coli* strain B, which is sensitive to the bacteriophage T/1. If the bacteria are exposed to the T/1 virus on solid media, most of the cells are destroyed, but a few colonies manage to develop. In broth, the turbid culture clears and later becomes cloudy. These observations indicate that lysis is rampant at first, killing a majority of the cells. But a few survivors, resistant to the virus, remain and grow. When these in turn are exposed, they show resistance to the virus. This is the same situation as that resulting from exposure of bacterial cells to a drug, and it raises the same questions: Did the agent (the virus here) cause some of the cells to change and become resistant by direct contact? Or were some cells already resistant to the virus before it was applied, so that only these cells were able to survive and form the basis of a resistant culture?

It was known that in a culture of 10^8 *E. coli* cells, an average of approximately 100 could be expected to show phage resistance. To answer the question of the origin of these resistants, the experiment was set up in two different ways. In the first (Fig. 15-1A), a culture containing a large number of cells was divided into 10 parts. For the sake of simplicity, let us say that a culture of 10^8 was divided into 10 parts of 10^7 cells each. Therefore, in this part of the investigation a culture of cells was simply divided into equal parts and then tested for any virus-resistant cells that might be present. No further growth of the samples from the original large culture was allowed up to this point. The divisions were plated on phage-impregnated media and then incubated to permit growth of any resistant cells. The results showed that a small portion, approximately 100 cells per 10^8 (or 10 cells on the average in each subdivision), were phage resistant. How can this be explained?

When the results were subjected to statistical analysis, it was found that the *mean* number of resistants in the subdivisions was of the same order of magnitude as the variance. The variance is equal to the square of the standard deviation. Statistical analysis has demonstrated that, in any random distribution, the value of the variance is close to that of the mean. Therefore, the results here are to be expected and can be explained by either of the two contrasting ideas. One could argue that, in a culture of 10^8 cells, spontaneous mutations occur from time to time, and some of these genetic changes result in phage resistance. Some resistant cells have recently arisen in the culture, whereas others have descended from mutant

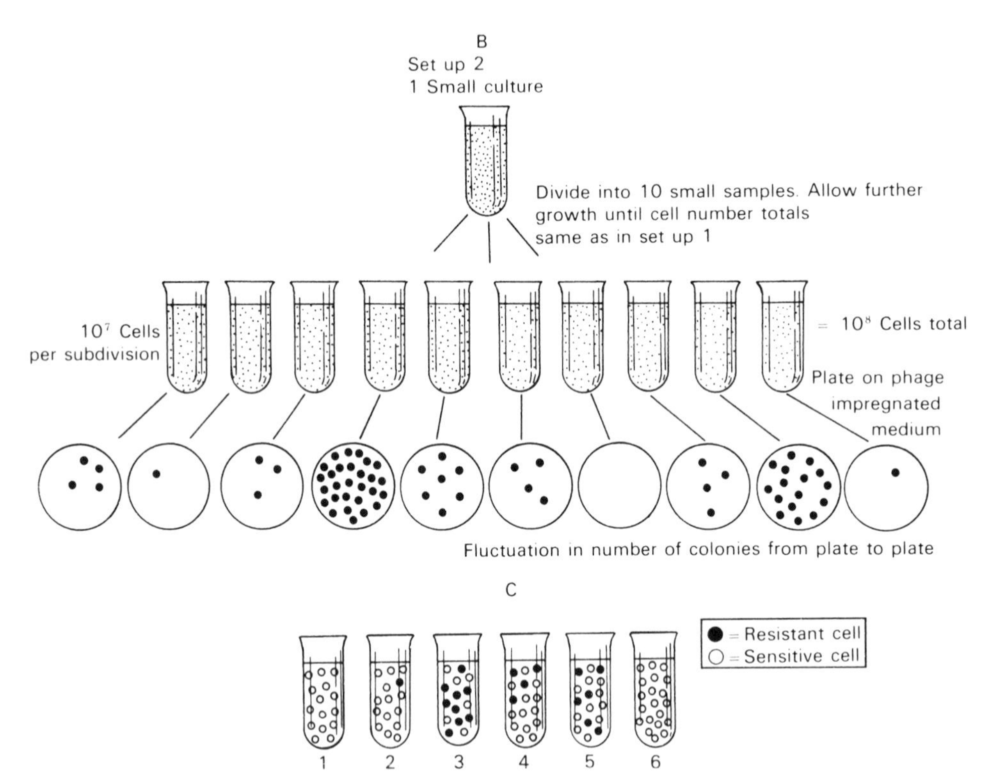

FIGURE 15-1. The fluctuation test (see text for details).

cells that arose earlier. When the culture is divided and tested for mutants, one would obtain a random sampling of the culture and therefore a distribution of mutants with a variance equal to the mean. Alternatively, one could argue that no mutations existed in the culture as it grew to the original total of 10^8 cells. When the culture of 10^8 was divided and plated out in contact with the phage, most of the cells lysed. A few, however, adapted or were induced to mutate by the phage. This occurred at random in the samples, and so the variance should be close to the mean.

Therefore, this setup does not permit a distinction between the two hypotheses. However, the second part of the experimental design allowed a clear choice between the two arguments. A small culture of bacteria was taken and divided (Fig. 15-1B). Each of these contained 50–500 cells and was allowed to grow until the total number of cells was equal to the total number in the first setup, in our example 10^8 cells. Note the difference between the two setups. In the first, a culture of 10^8 cells was subdivided and allowed no further growth before being brought into contact with the phage. In the second approach, each subdivision grew up independently from a small sample until a total of 10^8 cells was reached. According to the adaptation hypothesis, there is little difference between the two setups. In both cases, we have 10^8 cells. It should not matter that in the one case a large culture of 10^8 was subdivided into 10 smaller ones and in the other case 10 smaller cultures were grown separately until the same total of 10^8 was reached. There would be no relevant difference between them, because the adaptation (or induced mutation) hypothesis assumes that the resistants appear *only* when exposed to the phage. Thus, when the samples are tested, the variance should reflect the randomness and should be about the same as the mean.

According to the spontaneous mutation concept, however, we would expect a wide distribution in the number of mutants in the separately grown cultures. This is so because these will differ in the number of mutations that arose within each of them during growth, and they will also differ in the stages of culture growth at which the mutations occurred. In Figure 15-1C, we see that certain tubes (1 and 6) may have no mutants at all. If a single mutant arises at the end of the growth of the culture (tube 2), there will be fewer mutant cells present than if the mutation occurs early. In the latter case (tube 3), many descendants of the mutant cell will be present in the culture, because the bacteria have the opportunity to divide further. Therefore, even in cultures in which the same number of mutations take place, the number of resistants recovered would be very different depending on whether the mutants arose earlier or later in the growth of the culture (compare tubes 2–5). This is not the same as taking random samples from one large cell culture, and one would not expect the variance to be close to the mean. Since the number of resistants in the second setup fluctuates from one culture to the next, the procedure was named the *fluctuation test.* Its application to studies of variations in microorganisms leaves no doubt that spontaneous mutations do occur in lower forms and that inherited variations can arise independently of an external agent.

A procedure now routinely used in the laboratory also shows that spontaneous mutations can arise in microorganisms without the influence of a specific environmental factor. This is the technique of replica plating, devised by Lederberg and Lederberg. Figure 15-2 demonstrates the procedure in reference to streptomycin resistance in bacteria. Many strains of *E. coli* cells are sensitive to streptomycin, and colonies developing from them will not grow on a solid medium that contains the drug. However, cells at times arise that produce resistant colonies.

It can be shown by the replica plating that this change to drug resistance is completely independent of the drug itself. A piece of velvet is secured over a block. The velvet is gently applied to the surface of the colonies growing on the plate. The nap of the velvet picks up sufficient cells to transfer to other plates containing various media. Therefore, imprints or replicas of the original colonies can be easily made onto a plate containing streptomycin. This second plate must be oriented in the same way as the original one, so that any colonies that develop on it can be related back to colonies on the first plate. A colony may arise on the plate with the drug. Since the two plates are oriented, the colony on the first plate that produced this drug-resistant colony can be identified. Cells taken from the colony on the plate without the drug can then be exposed to a tube of broth containing streptomycin. These cells will grow in the presence of the drug, whereas other cells from neighboring colonies will not. Because no streptomycin was present on the first plate, the cells had never come into contact with the drug until they were placed in the broth. We would not have known that streptomycin-resistant cells had arisen and were present unless we had identified the colonies by replicating onto the second plate that has the drug. The procedure leaves no doubt that the drug-resistant cells were there on the first plate before any contact had been made with the streptomycin.

Detection of sexual recombination in bacteria

Until the mid 1940s there was no conclusive proof of a sexual mechanism in bacteria. The first clear-cut studies were those of Lederberg and Tatum working with different *E. coli* strains, normal ones, and those biochemically deficient in one or more ways. Any wild-type bacterial cell can grow on a minimal or basic medium containing just a few simple substances from which it can manufacture all the complex organic molecules needed for its metabolism. Such cells, which thrive on a minimal medium, are called *prototrophs.* In contrast, other cells may be nutritional deficients lacking the ability to make one or more essential growth factors from the minimal source. These are known as *auxotrophs,* cells that can grow only on a

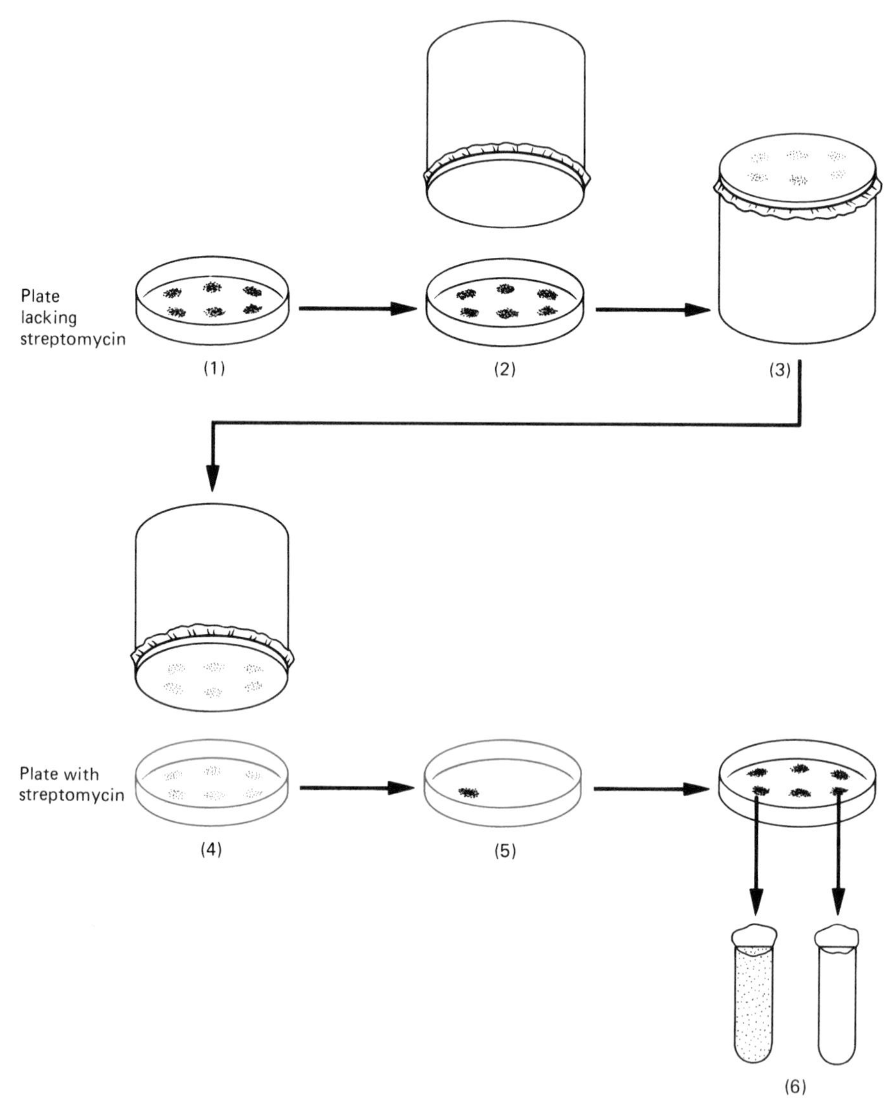

FIGURE 15-2. The replica plating technique. **1:** Separate bacterial colonies, each originating from a single cell, are allowed to grow on a nutrient plate lacking streptomycin. **2:** A block covered with a piece of velvet is brought into contact with the colonies on the plate. **3:** The velvet picks up cells from the colonies. The cells will be in the same orientation as they were on the plate. **4:** A plate containing streptomycin is oriented with the first plate, and the velvet on the block is allowed to contact the surface of the second plate. Cells are transferred from colonies on the first plate. **5:** In this example, one colony grows on the plate with the antibiotic. The two plates are compared, and this one colony is related to a colony on the plate without the streptomycin. **6:** Cells are taken from this colony on the plate lacking streptomycin and placed in broth containing the antibiotic. The cells grow in the broth. A sample taken from another colony on the same plate proves to be sensitive to the streptomycin and cannot grow in the broth.

supplemental medium, a minimal medium to which one or more defined factors have been added.

Lederberg and Tatum mixed together different auxotrophic strains. Strain 1 might require biotin and methionine and can be represented as B^-M^-. Strain 2, needing threonine and proline to supplement the minimal medium, can be designated T^-P^-. After the two were mixed together, samples were incubated and finally plated on minimal medium. A small number of colonies was found to arise on the plates (Fig. 15-3A). Since only cells that are $B^+M^+T^+P^+$ can grow on minimal medium, the origin of the prototrophs that gave rise to the colonies required an explanation. The frequency of prototrophic cells that arose following the incubation was only about 1 in 10, and thus we might assume the colonies are composed of prototrophs that resulted from back mutations, for example, $B^-M^-T^+P^+$ (strain 1) to $B^+M^+T^+P^+$. Identical studies were then performed with strains, each one of which was nutritionally deficient for three factors. Again prototrophs arose with the very same frequency as when doubly deficient auxotrophs were used. Back mutation cannot explain such results, because the frequency of prototrophs should be much lower when triply deficient strains are used than when doubly deficient strains are handled in the same way.

However, if we assume that the loci are linked, we can picture the mixing of the two doubly deficient auxotrophic strains as a cross and the appearance of the prototrophs as the result of cell fusion followed by crossing over and the production of recombinant cells (Fig. 15-3B). The parental cells cannot grow on minimal medium and neither can any recombinants other than $B^+M^+T^+P^+$ prototrophs, with the ability to carry out all the biochemical steps required for nutrition.

Various other possible explanations were offered to explain the origin of the prototrophs, but only the concept of sexual recombination was upheld. Further studies clarified the sexual system of *E. coli* and showed it to entail various unusual features, which we will encounter in the following sections.

The fertility factor

During the course of linkage analysis in *E. coli,* it became apparent that the production of recombinants depends on the survival of just one of the parents. Suppose strain 1 is $B^-M^-T^+P^+$ and is also streptomycin sensitive. Strain 2 is $B^+M^+T^-P^-$ but is resistant to the drug. Following mixing of the strains, prototrophs can be isolated on minimal medium containing streptomycin. Exactly the same procedure can be followed in a second cross in which the only difference is that strain 1 is streptomycin resistant whereas strain 2 is drug sensitive (Fig. 15-4). In such a cross no recombinants at all arise. In this example, the recombinants must be coming from strain 2, for if it is killed, no recombinant prototrophs are formed. Strain 1 is transferring genetic material to strain 2, but, once the transfer is achieved, strain 1 cells are no longer needed for recombinant formation. Studies such as this clearly demonstrated that the appearance of recombinants requires two cell types, a donor to transfer genetic material and a recipient to receive it. The latter then acts as a "zygote" that must survive to give rise to recombinant cells.

It was also discovered that cells lacking the ability to act as donors can arise in a culture of donor cells. Some factor appeared to be required for a cell to transfer genetic material. Following its loss, a previous donor then behaves as a recipient. This fertility factor was designated F, while donor and recipient cells were named F^+ and F^-, respectively.

Observations then showed that the F factor could be transferred with relative ease to F^- cells, thus converting them to the donor state. Simply bringing donor and recipient cells into contact for 1 hour or so would change the F^- cells to the F^+ state. The following summarizes what was now evident about the sexual system in *E. coli:*

1. Sexual recombination depends on the presence of two cell types, a donor or male (F^+) and a recipient or female (F^-). $F^- \times F^-$ crosses are ineffective. Mixing together F^+ cultures with other F^+ cultures yields some recombinants because of the presence of F^- cells that arise spontaneously, but the amount of recombination is much less than in crosses of $F^+ \times F^-$.

2. The F^+ or donor state results from the presence of F, the fertility factor, which can be transferred by cellular contact to F^- cells, which are then converted to F^+ donors.

3. The appearance of recombinants depends on the continued viability of the F^- parent.

It must be realized at this point that, although $F^+ \times F^-$ crosses yield recombinants, the actual number of the recombinant cells is low, in the vicinity of 10^{-6}. In other words, although the F factor is transferred with a high frequency or efficiency, the chromosome itself is transferred only in a small minority of the cells. Hence the chromosomal genes undergo recombination infrequently.

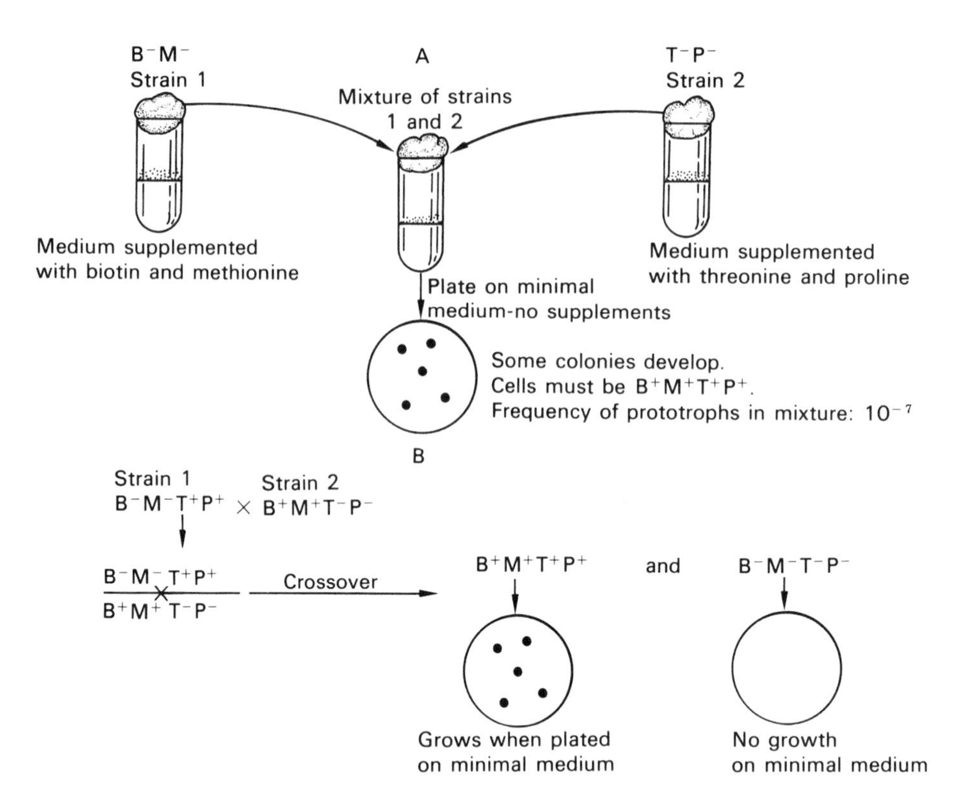

FIGURE 15-3. Prototrophs from auxotrophs. **A:** An auxtrophic strain, such as strain 1 or 2, cannot grow on minimal medium. Essential substances must be added. When the strains are mixed and samples of the mixture plated on minimal medium, some plates develop colonies. The cells in these must be prototrophs, because no essential metabolites have been added to the medium. **B:** The mixture can be depicted as a cross of bacterial strains in which the genes are all linked. The prototrophs could arise through a crossover event. Cells with the reciprocal crossover product cannot develop, because they need to be supplemented with four essential metabolites. Other crossovers are possible, but these cannot develop because the cells will require one or more essential substances. Only the prototrophs ($B^+M^+T^+P^+$) can grow on the minimal medium.

Partial genetic transfer

Another unorthodox characteristic of the *E. coli* sexual system was revealed. It was noted that when genetic material was transferred from F^+ to F^- the transfer was usually partial; only a portion of the genetic material of the F^+ seemed to enter the F^-. Certain genes were not transferred at all from donor to recipient. Evidence for this was obtained from reciprocal crosses between strains. For example, in *both* strain 1 and strain 2, F^+ and F^- types were isolated. This made possible reciprocal crosses, and the results of these proved to be quite different (Fig. 15-5). Only some of the genetic material of the F^+ parent entered the F^- cell to contribute to the offspring. The F^- parent, on the other hand, contributes all of its genetic material. Recombinants therefore will contain a greater percentage of their genes from the F^- cells than from the F^+. This is quite a different situation from that in higher forms in which both parents make equal donations of the genetic material to the zygote.

High frequency and the transfer of markers

Some clarification of the unusual features of recombination in *E. coli* was provided through the discovery of still another donor or male type. These donors behave as "supermales." When F^- cells are crossed with them instead of with ordinary F^+ males, the percentage of recombination is increased approximately 1,000 times. Therefore the supermales were termed high frequency (Hfr), denoting their ability to bring about genetic transfer.

Various Hfr males arose independently in cultures of ordinary F^+ donors. When different strains of Hfr were examined, it was noted that they could become F^+ again. This suggested that the sex factor (F) was present in the Hfr cell. But some puzzling facts emerged when different Hfr × F^- crosses were compared with one another and also with ordinary $F^+ \times F^-$ crosses. In the latter type, it will be recalled, the F factor itself is transferred quite readily to the F^- cells, converting them into donor types. And although a low frequency of recombinants is obtained in $F^+ \times F^-$ crosses, *any* portion of the genetic material can be transferred from F^+ to F^-. In contrast to this, when any specific Hfr strain is used as the donor parent, only a restricted portion of its linkage group is readily transferred. Some genes rarely go from the Hfr to the F^- parent. Moreover, the F factor that must be present in the Hfr cell usually is *not* transferred. Consequently, in a cross of Hfr × F^-, the offspring are almost always F^-. For example, consider the cross in Figure 15-6. After the strains are mixed, samples are plated out on a medium lacking threonine and leucine but containing streptomycin. By doing this, we are selecting prototrophs that are resistant to streptomycin. Therefore, only T^+L^+ colonies will grow. These cells must be recombinant, because the drug will have killed the Hfr parent; the F^- parent, being an auxotroph (T^-L^-), cannot grow on minimal medium. The recombinant colonies on the plates will thus all be T^+L^+. They can subsequently be tested for any other markers transferred by the Hfr parent.

In one strain of Hfr, the streptomycin marker is rarely transmitted, whereas the segment of genes T through λ is transmitted with a high frequency. Figure 15-6 shows that very curious ratios are found among the recombinants. Of the T^+L^+ offspring, 90% are azide sensitive (meaning that 90% received the Az marker from the Hfr parent), 70% received the T_1 marker, and so on for the others. It would seem that not all the marker alleles have the same chance of entering the F^- cells from the Hfr. Moreover, the recombinants are usually F^-, not F^+ or Hfr, as would be expected if the F factor had a good chance of entering the recipient cell.

An insight into the basis of this odd mating system was provided in an ingenious procedure employed by

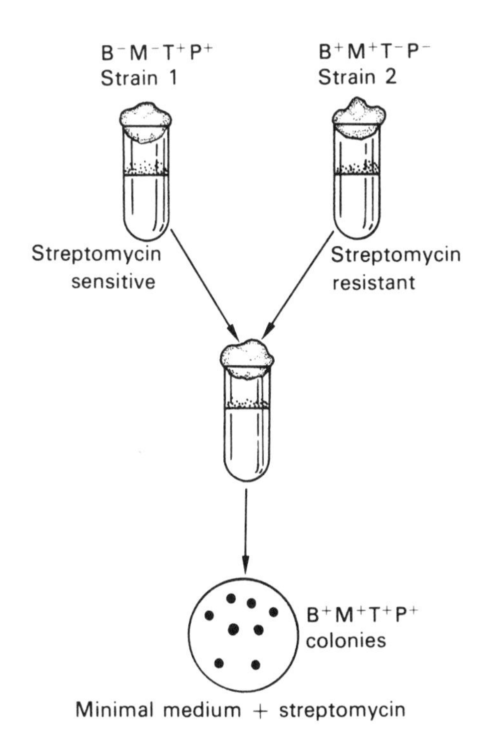

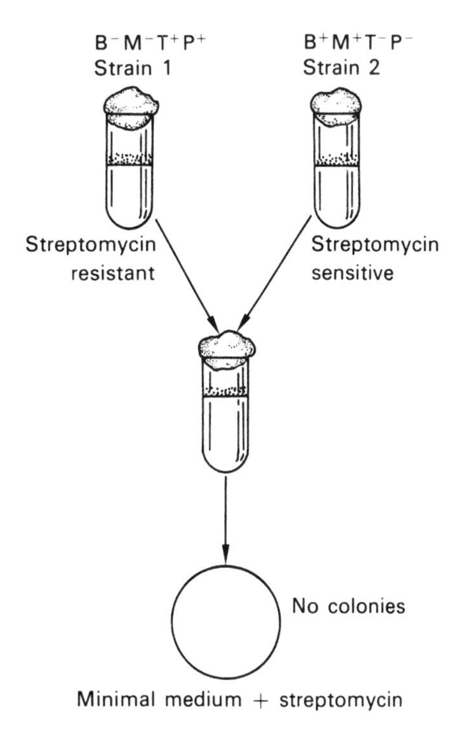

FIGURE 15-4. Survival of one parent. To obtain any recombinants, the survival of one of the parental strains must be ensured. One way to demonstrate this is by performing two types of crosses between two auxotrophic strains. The only difference between the crosses is that a parental strain is sensitive to streptomycin in one cross and resistant to it in the other. As the diagram shows, one of the two strains must not be killed if recombinants are to be obtained (strain 2 must survive here). This would appear to be the recipient strain. Strain 1 would be a donor that transfers genetic material to a recipient.

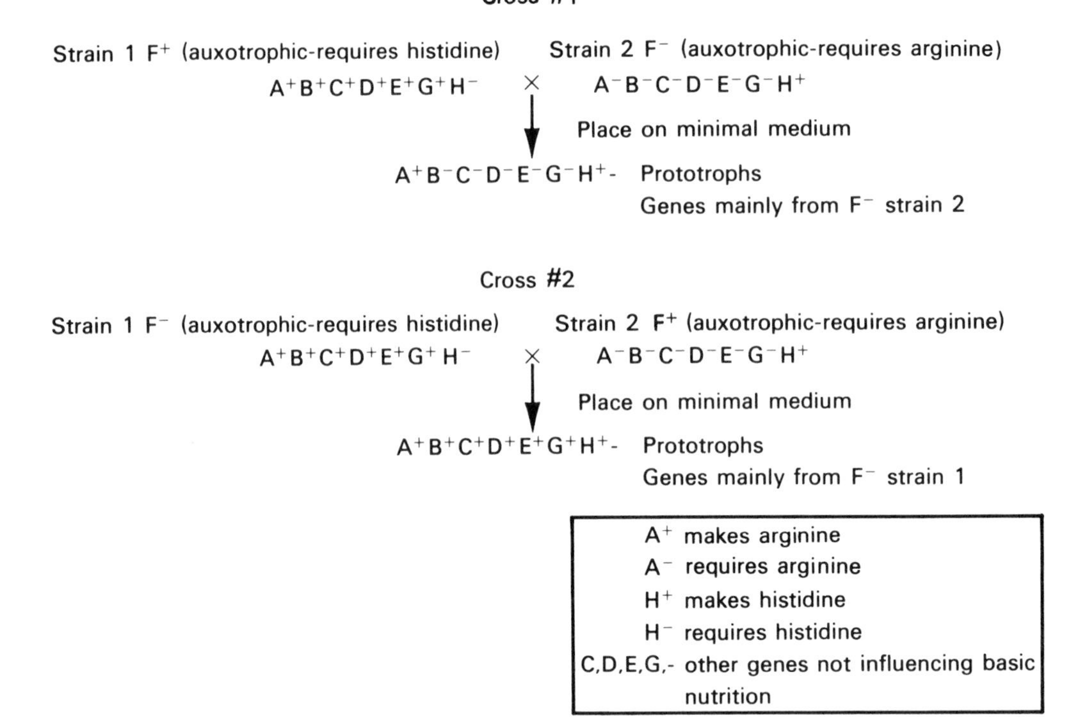

FIGURE 15-5. Partial transfer. F^+ and F^- types were found within strains. This makes possible reciprocal crosses. In this example, strains 1 and 2 are both auxotrophs, because each lacks the ability to manufacture a specific amino acid (histidine in 1 and arginine in 2). The other loci indicated are for traits that do not influence the ability to grow on minimal medium. When the strains are crossed and the cells placed on minimal medium, only the prototrophs (A^+H^+) will grow. These prototrophs can be tested for the other markers. It is found that the reciprocal crosses differ. Although various combinations are possible, most of the alleles in the A^+H^+ prototrophs will have been derived from the F^- parent. The transfer is only partial from the F^+ to the F^-.

Wollman and Jacob. Using strains similar to those just discussed, they mixed together Hfr and F^- and then interrupted the mating at various time intervals. They did this by subjecting the mixture to a blender, which can pull apart pairs of cells without damaging them. They then plated out samples and tested for the presence of Hfr markers in the recombinant cells. They found (Fig. 15-7) that no prototrophs were formed before an 8 minute period. Then the threonine marker was transferred and shortly thereafter the leucine marker.

Although only 9 minutes of contact between cells was needed for entry of the Az marker, 25 and 26 minutes were necessary for Gal and λ. If mating was interrupted at 18 minutes, for example, some cells would have received Lac, but none would be Gal^+. This suggested to Wollman and Jacob that the genetic material or chromosome of the Hfr parent entered the F^- cell at a definite rate and in an established order (Fig. 15-8A), in this case, the order T through λ. If the mating was interrupted and the chromosome broken at a certain time, some genes would not have yet been able to penetrate into the F^- cell.

The breaking of the chromosome does not prevent the integration of the donor genes into the recipient by recombination. The peculiar recombination ratios that occur under normal conditions of mating can be explained by the spontaneous breakage of the donor chromosome as it is being transferred. The closer a gene is to the origin—the chromosome end that penetrates the recipient first—the greater its chance of being transferred. The spontaneous breakage point would vary from one mating pair to the next. The result would be different percentages of recombination for the different genetic markers carried by the male parent. The bacterial mating system therefore has unusual features that permit us to map genes in two ways: by recombination percentages and by time of entry into the female parent (Fig. 15-8B). We will see later that the donor cell does not lose genetic material, since DNA replication takes place as genetic material is transferred.

The *E. coli* mating types

Wollman and Jacob were able to demonstrate that Hfr cells originate as occasional mutants in F^+ cultures. It is only these Hfr cells in an F^+ culture that are able to transmit the chromosome. The F^+ cells are ineffective in transferring genes from male to female cells. Pure Hfr strains were obtained from single Hfr cells that arose independently in Hfr cultures. Crosses were then made between each Hfr type and F^- cells. Analyses using the interrupted mating method clearly showed that different Hfr types have a different gene at the anterior end of the chromosome (Fig. 15-9A). Moreover, some blocks of genes that are not transferred with a high frequency in one strain are transferred readily in another. (The segment M through S is transferred in strain 1 but not in strain H; the segment Az through λ is transferred with high frequency in strain H but not in strain 1).

The data show that only one linkage group is present in a bacterial cell and that it exists in the form of a circle (Fig. 15-9B). The gene arrangement is the same in the various Hfr strains, but the order in which the genes are transferred to the F^- parent differs from one Hfr type to the next. Moreover, each Hfr strain has a characteristic segment of the circular chromosome which it usually does not transfer at all. When the circular chromosome map was first proposed, the physical shape of the *E. coli* chromosome was unknown. As noted in Chapter 11, Cairns demonstrated the actual circular nature of the *E. coli* chromosome, a superb correlation between genetic analysis and cytology.

The peculiarities of the *E. coli* mating system make perfect sense in light of all the findings. In the F^- cell, the circular chromosome remains in this form, and no genetic material is transferred. This is generally the case in the F^+ cell, which in addition to the circular chromosome contains an independently existing F element, an entity that is also circular and composed of double-stranded DNA. When an F^+ cell contacts an F^- cell, the free existing DNA of the F

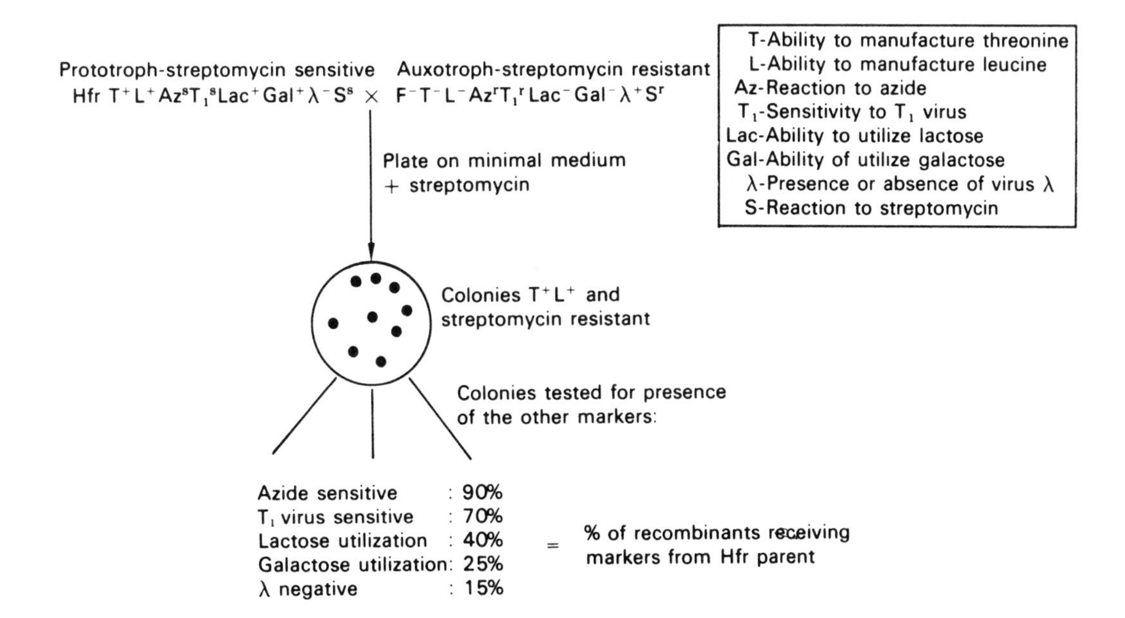

FIGURE 15-6. Cross of Hfr and F^-. Prototrophs are selected by plating on minimal medium. These cells will all possess the T^+L^+ markers, which come from the prototrophic Hfr parent. The Hfr parent, however, is destroyed by the presence of the streptomycin in the minimal medium, so all the prototrophic colonies are truly recombinant. The F^- parent, being an auxotroph, cannot grow on the minimal medium. These recombinant offspring are then tested for the presence of the other markers from the F^+ parent. It can be seen that the markers have very different chances of being transferred to the F^- cells to enter into the formation of recombinants.

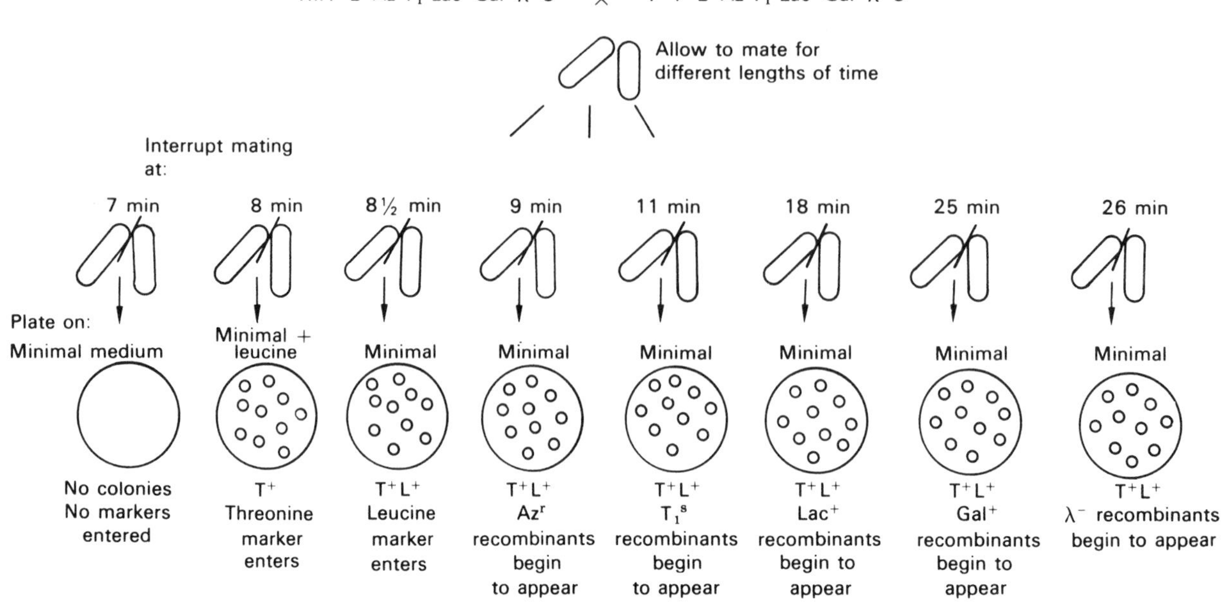

FIGURE 15-7. Interrupted mating. The same two strains as shown in Figure 15-6 are allowed to mate for different periods of time. Samples of a mating mixture are removed at specific times and subjected to a blender, which pulls the cells apart. The sample is then tested for recombinants. It is found in this cross that no markers enter if the cells are separated before 8 minutes. The markers being followed in the cross are then found to enter at different characteristic times.

element is replicated and transferred to the F^- cell, which then becomes F^+.

In an occasional F^+ cell, the F factor may insert itself into the chromosome of the bacterial cell, thus converting it to an Hfr type. The inserted element is generally replicated along with the bacterial chromosome as if it were a natural part of it. However, contact with an F^- cell sets off a cycle of DNA replication in the Hfr cell that results in transfer of chromosome material to the F^- cell. To appreciate the events taking place in the transfer of genetic material from both F^+ and Hfr cells to F^-, we must become acquainted with the rolling circle method of DNA replication described in the following section.

Rolling circle method of replication

In Chapter 11, the replication of the circular *E. coli* chromosome was discussed. Another method by which a circular DNA molecule can replicate is the rolling circle mechanism. Rolling circle replication (summarized in Fig. 15-10) is initiated by a nick in only one strand of the circular DNA molecule by an endonuclease. This cut produces a free 5′ end and a free 3′ end. The intact circle then proceeds to act as a template strand as nucleotides are added to the free 3′ end of the nicked strand. Growth of the new DNA (always in the 5′ to 3′ direction) causes the 5′ end to become displaced and to move away from the circle. We can imagine that the growing strand rolls or revolves around the circular template as new nucleotides continue to be added to the 3′ end. As new growth continues at the 3′ end and proceeds around the circle, the 5′ end of the strand rolls out and appears as a tail, which in turn continues to grow in length. After the growth has completed one turn around the circular template, the tail length represents the whole length of the original nicked strand that was displaced.

The tail, even before it reaches full length, may serve as a template for the assembly of DNA and be converted to the double-stranded state. It is the equivalent of a lagging strand, so that the 5′ to 3′ replication would be discontinuous (recall Fig. 11-23). If the replication continues and two revolutions of the circle take place, then two whole, connected tail lengths would be generated, as actually occurs in the replication of some viruses (Chap. 16). On the other hand, if replication ceases before one revolution is completed, then the tail length that is displaced will be less than the full length of the original strand.

Rolling circle and the *E. coli* sexual system

The transfer of sex factor (F) from F^+ to F^- entails rolling circle replication, as outlined in Figure 15-11. Conjugation triggers a nick in one of the strands of F, producing free 5′ and free 3′ ends. A cycle of replication is initiated. Growth at the 3′ end displaces the 5′ end, which moves through the conjugation tube from the F^+ to the F^- cell. Replication at the 5′ end takes place in the F^- cell.

After DNA growth has completed one revolution, both the transfer and the synthesis of DNA ceases. Transfer of the strand takes about 2 minutes. The strand transferred to the F^- cell is completely replicated and becomes circularized. The original F^- cell has now been converted to F^+.

When an Hfr cell contacts an F^- cell, the F factor that is inserted into the bacterial chromosome becomes activated, and a round of DNA replication is

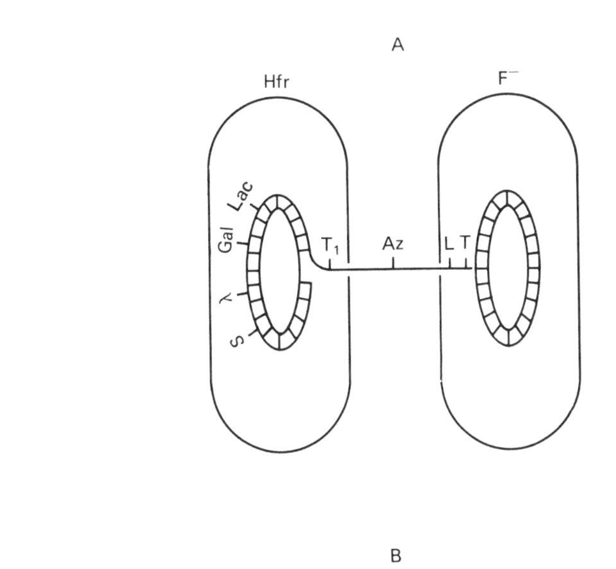

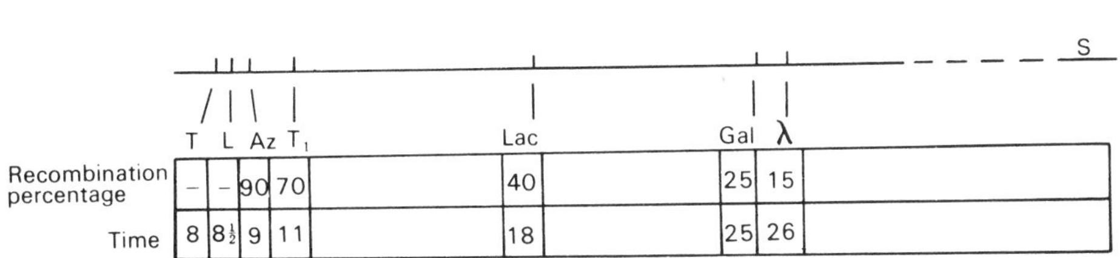

FIGURE 15-8. Entry of Hfr DNA. **A:** The results depicted in Figures 15-6 and 15-7 can be explained in part as shown here. The chromosome of *E. coli* is known to be circular. At the time of mating, the circle in the Hfr cell generates a linear chromosome strand that penetrates into the F^- cell at a specific rate and in a definite order. The donor cell loses no genetic material, as the material transferred is replaced by replication of DNA. Only portions of the donor chromosome material enter the F^- cell to engage in recombination, since mating cells usually break apart before the transfer of an entire chromosome strand. The closer a locus is to the beginning of the Hfr chromosome, the greater the chance that it will be transferred. The T and L loci, as shown here, have a greater chance of entry than those following them. Some loci, such as the one for streptomycin tolerance, are so far toward the end that breakage usually occurs in front of them, and they rarely enter. **B:** According to this concept, the chromosome may be mapped in either of two ways, by recombination percentages or by times of entry. In this example, all of the cells would be T^+L^+, because only prototrophs were selected in the cross. Different percentages of these prototrophs carry the other markers. We see that 90% of them will receive the azide marker, only 25% the Gal marker. The difference is explained by the different times of entry, 9 versus 25 minutes. The sooner a marker enters, the closer it is to the beginning and the greater the number of cells that will receive it. The whole chromosome may be mapped in time units. At 37°C, approximately 90 minutes are required for any cells to receive the entire Hfr chromosome.

Hfr Type	A Leading End ←
H	T L Az T_1 Pro Lac T_6 Gal λ
1	L T B_1 M Mtol Xyl Mal S
2	T_1 Az L T B_1 M Mtol Xyl Mal S
3	T_6 Lac Pro T_1 Az L T B_1 M Mtol Xyl Mal S
4	B_1 M Mtol Xyl Mal S λ Gal
5	M B_1 T L Az T_1 Pro Lac T_6 Gal λ

B

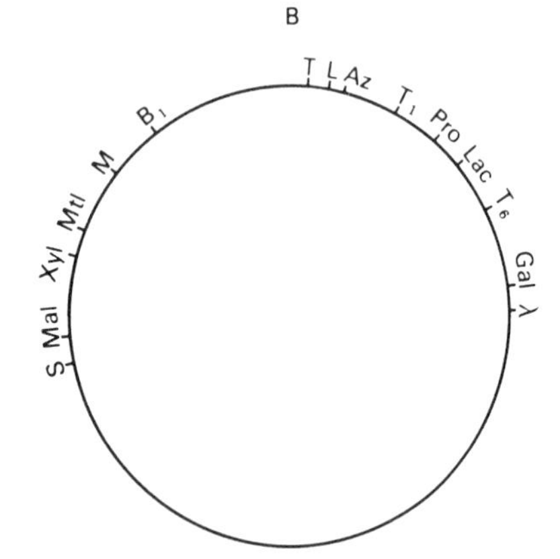

FIGURE 15-9. Hfr types and partial linkage map. **A:** Several Hfr strains were isolated (only six are shown here). When mated to F⁻, the strains were found to differ in relation to the genes that were transferred with a high frequency and to the order in which the markers enter the F⁻ cell. **B:** The observations on the six Hfr types can be explained on the basis of a circular linkage group. When genetic material is transferred to the F⁻ cell, a linear chromosome is derived from the circle. The anterior end that is established for the linear chromosome would differ among the Hfr strains, so that the first loci to enter the F⁻ would vary from one Hfr to the other. The arrangement of the loci, however, remains the same; only the order of entry is different. Those genes near the leading end are transferred with a high frequency. Those farther back go over to the F⁻ with a low frequency or not at all.

initiated that takes place by means of the rolling circle mechanism. When Hfr transfers chromosome material to F⁻, the Hfr parent remains viable; it does not lose genetic material, because rolling circle replication takes place. An endonuclease-induced nick occurs within the sex factor itself (Fig. 15-12) and affects just one of the two strands of the inserted F element. Note that the nick splits the F factor. Consequently, the 5′ end that enters the F⁻ cell contains only some of the genetic material of F. The remaining portion of F is at the 3′ end and can be transferred to the recipient cell only if replication by the rolling circle mechanism completes one revolution. However,

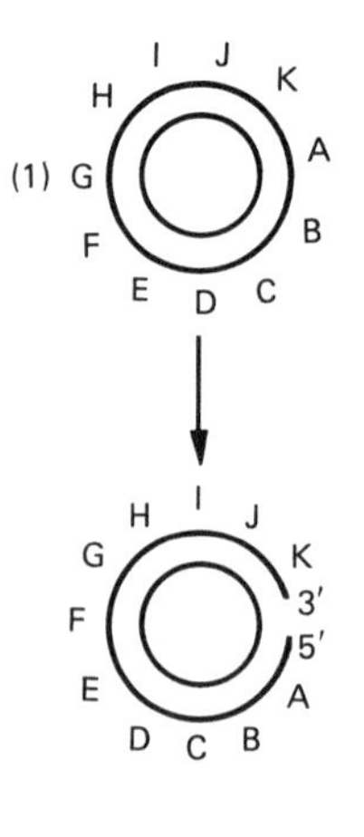

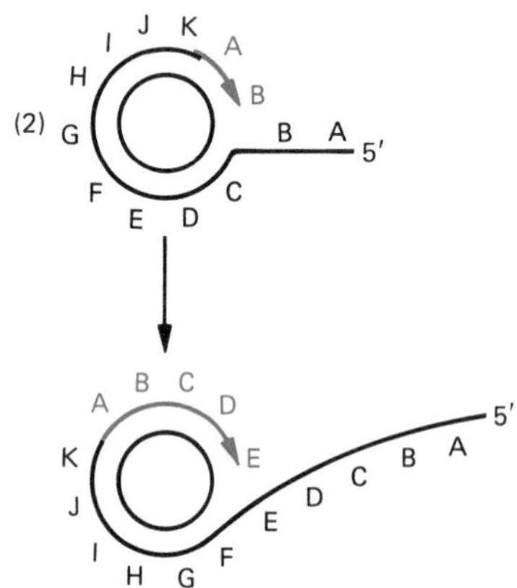

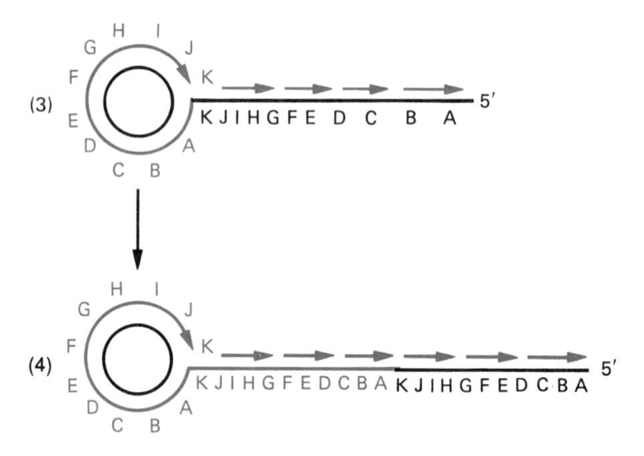

FIGURE 15-10. Rolling circle replication. **1:** A circular DNA molecule is nicked in one of the strands by an endonuclease. This produces a free 3′ end and a free 5′ end. **2:** Nucleotides are added (red) to the 3′ end, bringing about growth of the strand. This addition of nucleotides causes displacement of the strand at its 5′ end. As growth continues at the 3′ end, the tail at the 5′ end increases in length. **3:** After one revolution, the tail length represents the length of the original strand that was displaced. Discontinuous replication may take place on the tail to form double-stranded DNA (arrows). **4:** Two revolutions produce two tail lengths, whereas less than one revolution produces a tail shorter than the original strand (2).

breakage usually occurs before the entire amount of DNA is transferred. Once inside the F⁻ cell, the transferred strand acts as a template for DNA synthesis, and the portion becomes double stranded. It is this double-helical DNA, representing a portion of the Hfr genetic material, that can then engage in crossing over with the F⁻ chromosome.

It should now be apparent why the F⁻ cell usually remains F⁻, since the breakage of the transferred strand before one revolution of replication is completed results in transfer of only a part of the F factor to the F⁻ cell. Those Hfr genes in the region away from the leading 5′ end (i.e., near the 3′ end) will have little chance of entering the recipient. Note (Fig. 15-13) that the different Hfr strains simply represent cells with the sex factor integrated at different positions in the circular chromosome. When replication is triggered by conjugation and the F factor is nicked in one strand, a leading end is established, and this will obviously differ from one Hfr strain to the next, depending on the site of insertion of F and its orientation. We can now understand why the various Hfr strains differ with respect to which chromosomal segments they transfer efficiently and which they rarely transfer at all to F⁻ cells.

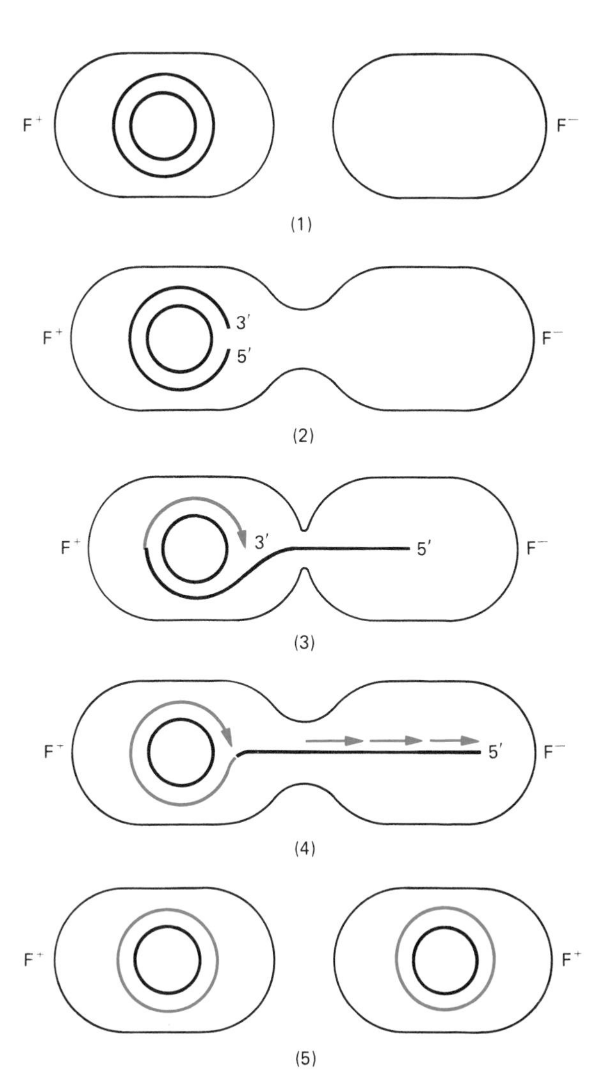

FIGURE 15-11. Transfer of sex factor by rolling circle replication. **1:** An F⁺ cell contains an F element that the F⁻ cell lacks. The F factor exists free of the chromosome (not shown here for clarity). **2:** Contact with an F⁻ cell brings about a nick in one of the strands of F, and free 3′ and 5′ ends are produced. **3:** A round of replication by the rolling circle mechanism is initiated, and the displaced 5′ end enters the F⁻ cell. **4:** As the 5′ end moves, DNA replication takes place in the F⁻ cell, and the transferred strand becomes double stranded. Transfer ceases after one round of replication. **5:** Synthesis of the complementary strand is completed. The two cells separate, and circularization of the linear strand occurs. DNA ligase seals all nicks. Both cells are now F⁺.

Chromosome transfer and recombination

Once a portion of the Hfr chromosome has penetrated the F⁻ parent, it may undergo recombination with the genetic material in the F⁻ cell. If any recombinant bacterial cells are to arise, the introduced Hfr segment must pair with the homologous chromosome region of the F⁻ cell. Genes are then reciprocally exchanged between the recipient chromosome and the introduced segment (Fig. 15-14). Note that all the chromosomal genes contributed by the Hfr parent do not necessarily enter into the formation of the recombinants (e.g., A in Fig. 15-14). The free fragment that results is apparently discarded from the recombinant cell along with the portion of the sex factor associated with the introduced segment.

Rarely, a recombinant cell is found that has received a gene that is typically *not* transmitted by a particular Hfr strain. When this does occur, the recombinant may then prove to be Hfr instead of F⁻. We now understand how this can come about, because the entry of a chromosomal gene at the extreme end means the entire chromosome strand must have been transferred. Therefore there is a good chance that the other end of the split F factor may have entered the F⁻ cell as well; thus all of the genetic material of the F factor is now present. In such an unusual case, in which premature strand breakage does not take place, an intact sex factor can integrate into the chromosome, converting the F⁻ to Hfr (details on insertion are discussed in the following section). The Hfr recombinants derived in this way may also revert to the F^+ state, indicating that the F factor can exist in a free state or attached to the chromosome.

For the entire chromosome strand to be transferred from Hfr to recipient cell, about 100 minutes are required. As noted earlier in this chapter, genes may be mapped according to time of entry from Hfr to F^- cell (see Fig. 15-8). The *E. coli* map is therefore scaled on the basis of 100 minutes (Fig. 15-15). Note that the time, in minutes, starts arbitrarily at 0 with threonine. All the other loci are related to the threonine locus on the basis of their entry times, as revealed by interrupted mating experiments.

An ordinary population of F^+ cells contains occasional Hfr cells that have arisen independently by insertion of the F factor at different positions. So when F^+ and F^- populations are mixed, only the occasional Hfr types in the population are able to transfer the chromosome. But because the point of insertion of the F factor varies from one Hfr type to another, the linear chromosomes that arise in the several Hfr types at the time of conjugation have different anterior and different terminal ends. Consequently, different segments are transferred from one mating pair to another. The F^+ population is thus a mixed one containing an assortment of Hfr types, in contrast to an isolated population pure for one Hfr strain. Thus, when F^+ and F^- are mixed, *all* genes seem to be transferred at a low frequency. The F particle itself goes over at a high frequency because it exists free in most of the cells in the F^+ population.

F factor as an episome

Jacob and Wollman coined the term *episome* to designate any factor with a behavior such as that of the F factor. An episome is a genetic entity that may exist free in a cell or be integrated into the chromosome. The episome is also a self-replicating unit, a replicon (see Chap. 11). Different techniques have demonstrated that

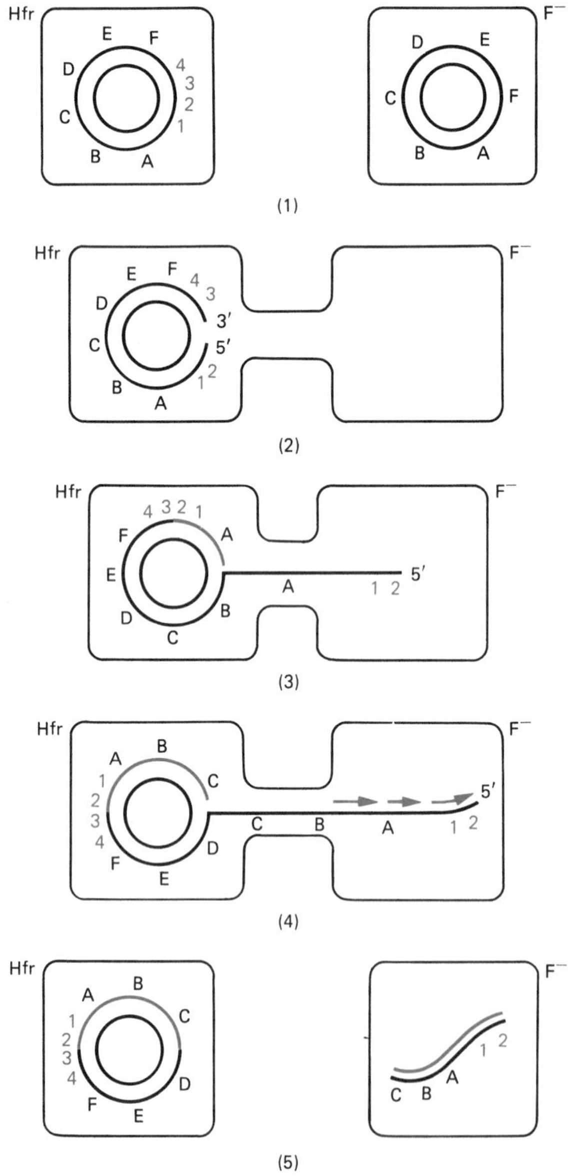

FIGURE 15-12. Transfer of chromosome from Hfr to F^-. **1:** An Hfr cell contains an F element (1, 2, 3, 4) integrated into its circular chromosome. An F^- cell lacks the F element. (Chromosomal genes are designated A through F.) **2:** Contact between an Hfr and an F^- cell triggers a round of DNA replication. (The chromosome is not shown in the F^- cell, for clarity.) An endonuclease nicks one strand of the F element, establishing free 3′ and 5′ ends. The nick splits the F factor. **3:** The 5′ end enters the F^- cell through the conjugation tube. The leading 5′ end carries a part of the genetic material of the F factor (2, 1). Chromosomal genes enter behind the portion of the sex factor. **4:** As more of the chromosome strand is transferred, it is replaced in the Hfr by rolling circle replication. DNA replication also occurs in the F^- cell (red arrows), with the transferred segment acting as a template. **5:** Breakage of the strand usually occurs before the entire chromosome strand is transferred. The transferred segment (now double stranded) can participate in recombination with a homologous chromosome region of the F^- cell. The F^- cell remains F^- because it carries only a portion of the F factor (2, 1). The Hfr cell has lost no genetic material, because rolling circle replication replaces any DNA that is transferred (segment carrying 2, 1, A, B, C).

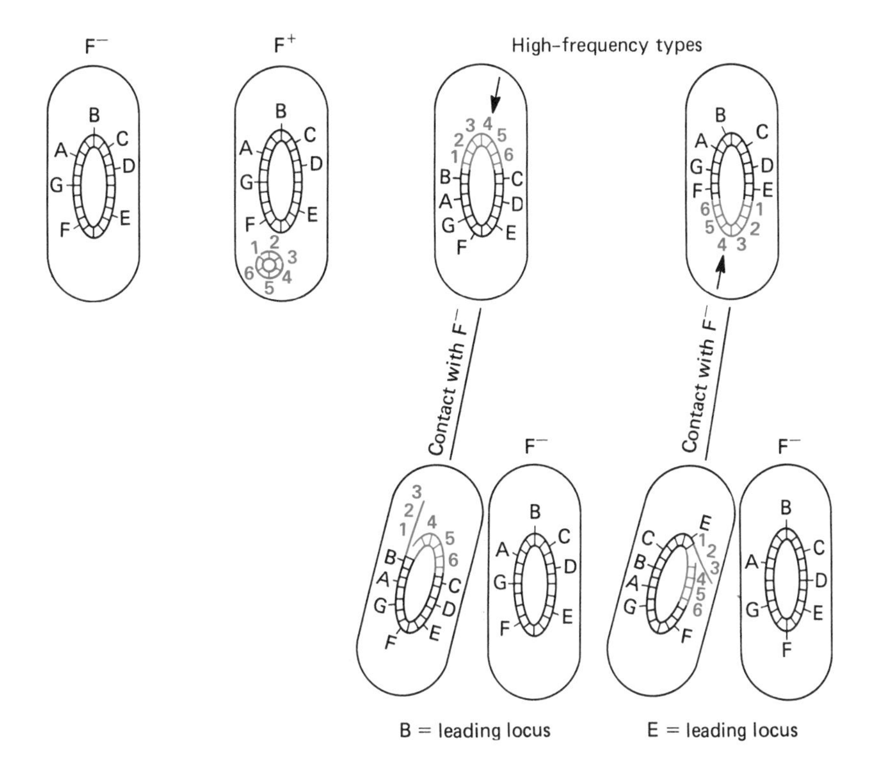

FIGURE 15-13. Differences in mating type. The F⁻ cell contains a circular chromosome, as do the F⁺ and Hfr types. In the F⁺, however, the fertility factor, F, is present as a free entity. The Hfr type contains F in an integrated state on the circular chromosome. The various Hfr types differ in the position at which F is located (arrows) and its orientation. Contact with an F⁻ cell triggers DNA replication in the Hfr, and a break is produced within F. This establishes a leading point and an endpoint. Because F can be located at different positions and in different orientations on the chromosome, the order of entry of the genes varies from one Hfr strain to the next.

the F factor of *E. coli* is a circular DNA element 94,500 base pairs long, approximately 2% the size of the bacterial chromosome. As we have seen in the case of the F factor, an episome may be present or absent in a given kind of cell. The F factor also illustrates that an episome may have several remarkable attributes. The F factor of *E. coli* confers an assortment of properties on the bacterial cell. Not only does the episome convert the cell from a nondonor to a potential donor cell (sometimes called a *male cell*), its presence has also been shown to bring about the production of particular types of appendages. Termed *sex pili* or *sex fimbriae,* these appendages are necessary for conjugation. Electron micrographs of mating cells reveal cytoplasmic connections or bridges between them. Each such bridge may represent one of the sex pili required to establish a connection for the transfer of the chromosome between mating cells (Fig. 15-16).

Presence of the F factor renders a cell sensitive to certain single-stranded DNA and RNA phages, the so-called male-specific phages. It also causes the cell to become resistant to certain other phage types that can infect F⁻ cells. Presence of a single F factor also renders the *E. coli* cell immune to the entry of or the maintenance of any additional F factors, making the relationship 1:1 between F factor and circular chromosome. Since the F factor confers several traits on the cell that harbors it and since it is a replicon of appreciable size, it is not surprising that several genes are now recognized that occur on the episome. Both mutant and nonmutant F elements have been isolated and analyzed with respect to the genetic regions carried on the F factor itself. A cluster of 13 genes (*tra* genes) is known to be needed to establish sex pili and transfer DNA during conjugation. One genetic sequence is known to be required for the origin and initiation of transfer. Other known genetic regions of the F factor are responsible for the reaction of the host cell to specific phages and for its resistance to the establishment of additional F elements in the cell. Other genes are also present that are required to replicate the F element itself.

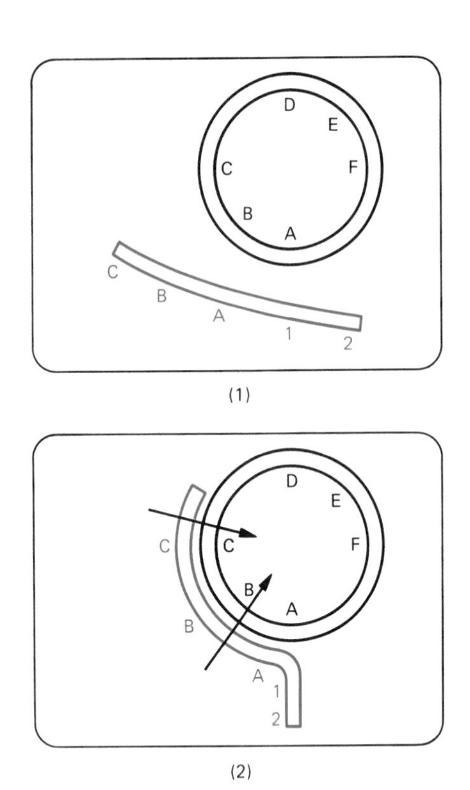

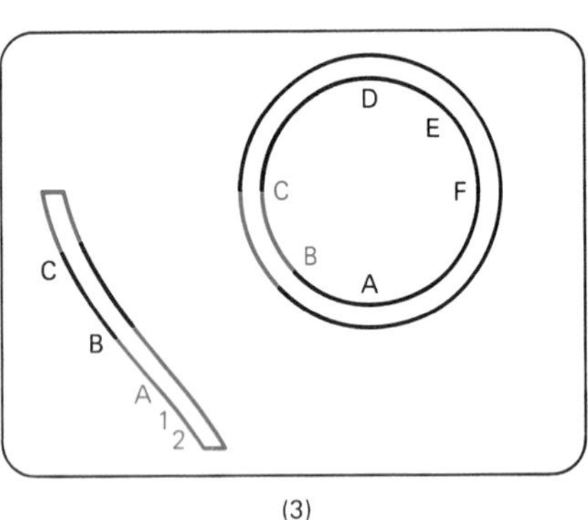

FIGURE 15-14. Formation of a recombinant F⁻ cell. **1:** Chromosomal segment transferred from Hfr to F⁻ includes chromosomal genes (A, B, C) and a portion of the F factor (2, 1). **2:** Pairing takes place between segment and homologous region of the chromosome. The F⁻ cell carries no region homologous to the portion of the F factor. Pairing is followed by crossing over (arrows). **3:** The reciprocal crossing over results here in exchange of portion CB and the formation of a recombinant chromosome in the F⁻ cell and a recombinant fragment. The latter is discarded along with the portion of the F factor.

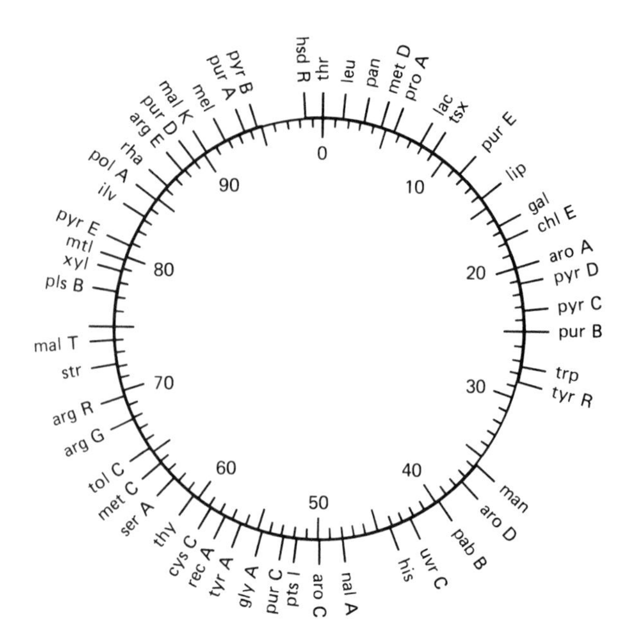

FIGURE 15-15. The 100 minute linkage map of *E. coli*, showing some important genes. (Reprinted with permission from B.J. Bachmann, K.B. Low, and A.L. Taylor. *Bacteriol. Rev.* 40:116, 1976.)

Very important is the presence on the F factor of certain short sequences called *insertion sequences* or *IS elements*. These range from 800 to 1,400 base pairs in length, and four of them are known to exist on the F factor. These IS elements are responsible for the integration of the episome into the *E. coli* chromosome. The *E. coli* chromosome also contains IS elements, some identical to those in the episome. It is the positions of the IS elements in the different F factors and in the circular chromosome of the different strains of *E. coli* that are evidently responsible for determining the different sites of integration of the episome as the cell is converted from F⁺ to Hfr. The IS elements are able to bring about the recombination between the episome and the chromosome (Fig. 15-17A), an event that results in insertion of the episome into the host chromosome. By a reversal of the process, it can again become free, converting the cell back to an F⁺. The IS elements have great bearing on other phenomena that have come to light; these are discussed later in this chapter.

When the F factor removes itself from the chromosome, it may, at rare times, become associated with a portion of the host chromosome and carry it away (Fig. 15-17B). As a result, an F factor can originate that contains a marker allele, such as Gal⁺. The piece of the chromosome now replicates along with the genetic material of the episome and is part of a replicon containing genetic material of both F factor and the chromosome.

Harboring pieces of genetic information in this way, F factors can enter F⁻ cells. This means that genes may be transferred from one cell to another by an episome, which can thus confer a new genetic property on a recipient cell. Suppose that an F^- cell is Gal^- and thus cannot ferment galactose (Fig. 15-18). This cell may contact an F^+ cell that contains an F factor carrying a Gal^+ allele. The F^-Gal^- cell in turn can be converted to an F^+Gal^+ type by receiving the fertility factor, which not only makes it a potential donor cell but also gives it the capacity to ferment galactose. This transfer of genetic material from one cell to another by way of a sex factor has been called *sexduction*. The term *F genote* was coined to designate a sex factor that has incorporated a portion of the chromosomal DNA. Those bacterial strains that carry F genotes are known as F′ strains. A bacterial cell therefore can acquire the ability to perform a function as a result of an allele contained in an episome it harbors and not as a result of an allele on its chromosome. Such a cell is haploid except for a small chromosome portion. In this short region, it can possess a pair of alleles and would therefore be partially diploid as well as partially heterozygous. A *merozygote* is any partially diploid bacterial cell resulting from conjugation or a process such as sexduction. The term *heterogenote* describes a cell heterozygous for just a segment of the genetic material, as opposed to *heterozygote,* which is reserved for diploid cells.

Plasmids and drug resistance

The discovery of the many remarkable properties of the F factor focused attention on the possible occurrence of similar particles in various kinds of cells. Of great importance to humans are certain replicons that have been recognized in bacteria and can confer drug resistance on the cells harboring them. Like episomes, these nonchromosomal genetic entities may be absent from a cell but when present they can replicate autonomously. However, they do not integrate with the chromosome of the cell as do F factors. The more general term *plasmid* is used to define any replicon that may exist in a cell independently of the chromosome. Episomes are plasmids that *do* have the ability to integrate with the chromosome. Most plasmids are not

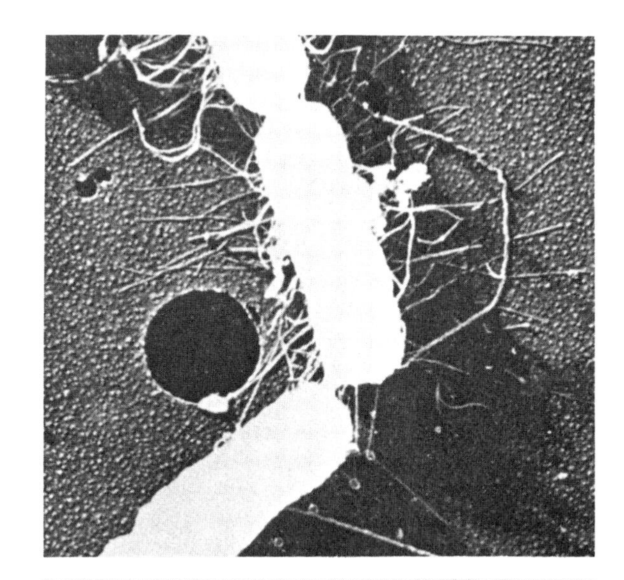

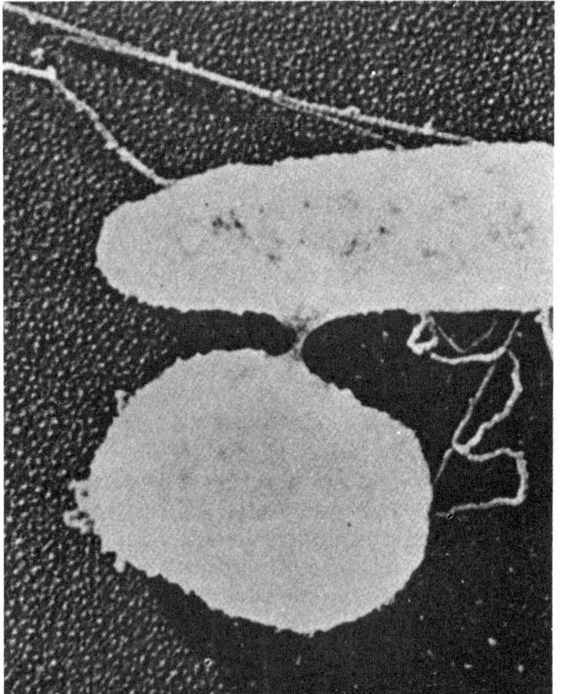

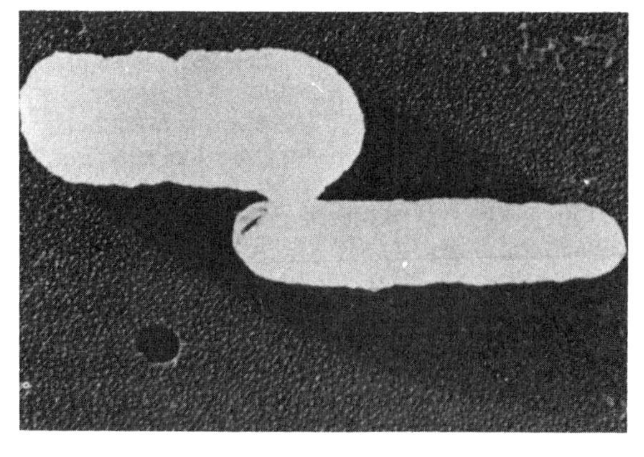

FIGURE 15-16. Electron micrograph of conjugating *E. coli* cells. (Reprinted with permission from T.F. Anderson, E.L. Wollman, and F. Jacob, *Sur les processus de conjugaison et de recombinaison genetique chez* E. coli. *III. Aspects morphologiques en microscope electronique.* Annales Institut Pasteur, 1957.)

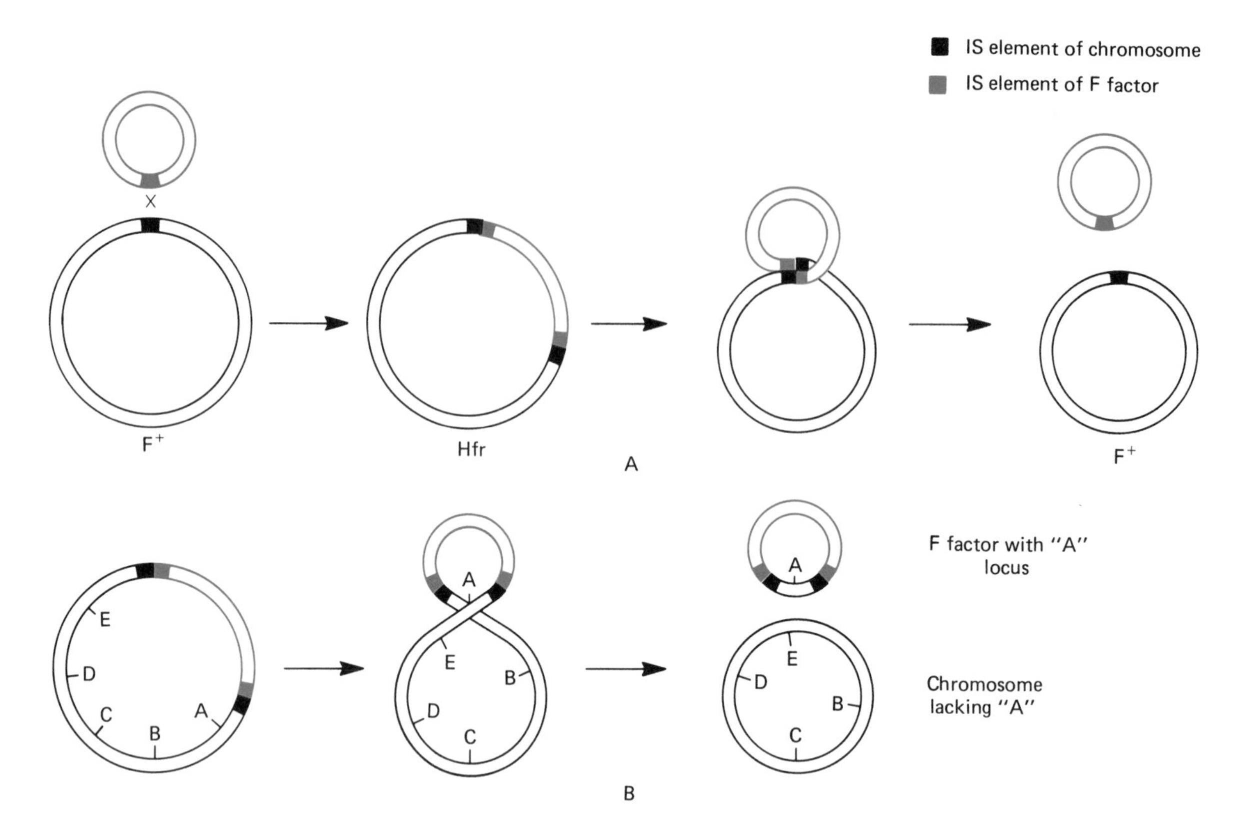

FIGURE 15-17. Insertion and removal of F factor. **A:** The factor possesses certain regions, IS elements, that are homologous to segments of the chromosome. This enables the fertility factor to pair with the chromosome at different points. A recombination event inserts F into the chromosome of an F^+ cell, thereby converting it to Hfr. At times, the F factor may remove itself from its integrated state and become free, converting the cell to an F^+. Normally this would occur by a reversal of the integration process. Homologous regions would again pair, and a recombination event would take place. **B:** At rare times, the F factor may carry away a portion of the chromosome. This is believed to come about when small homologous portions of the chromosome on either side of the inserted F factor pair and a crossover event takes place. (Only one IS element is shown in the F^+ chromosome, for simplicity.)

usually essential to the survival of the cell in which they occur.

Most strains of bacteria are known to harbor some kind of plasmid. Plasmids have been isolated from cells by density gradient procedures and then subjected to detailed analysis. They have also been visualized with the electron microscope. All plasmids are naked, circular, double-stranded DNA molecules. They vary greatly in the number of genes they carry, from just a few to hundreds. Plasmids also vary in the number of copies in which they exist in bacterial cells. Certain plasmids occur in only a single copy, others are found in a few copies, and some exist in large numbers, even up to 100. The replication of the plasmid depends on many of the enzymes of replication coded by the host cell. However, the plasmid itself is coded for sequences that regulate the initiation of replication and the number of copies of the type of plasmid that may occur in a given cell. Regulation of copy number may involve repressor protein substances that can bind to DNA sequences and prevent the initiation of DNA replication (see Chap. 17 for details of repressor action). In the case of a plasmid like F, with low copy number, the level of repressor in the cell is sufficient to inhibit replication of the plasmid, but, as the cell grows, repressor becomes less concentrated, and the plasmid can replicate. A plasmid type that has high copy number requires more of its specific repressor protein to inhibit its replication. Therefore, if the plasmid number is low in this case, so is repressor concentration. Plasmid replication therefore continues until sufficient plasmids of that type, with their repressor genes, are present to raise the repressor concentration to a level that effectively inhibits further plasmid replication.

The concept of repressor involvement in control of copy number explains why certain plasmids cannot coexist in the same cell. Assume that two plasmids, A and B, are identical or very much alike. They would then be coded for the same kind of repressor. The level of repressor would consequently double when they are present in the same cell. The repressor from one can bind to the DNA of the other and inhibit replication. Two plasmids that cannot coexist in the same cell are said to be *incompatible.* Presence of one F factor in a cell precludes the maintenance of another F factor in the same cell. If, however, F and some unrelated plasmid are present together, each plasmid would code for its own kind of repressor, and replication of one plasmid would not be regulated by the other's repressor. Two such plasmids that can coexist in the same cell are called *compatible.* In addition to repressor proteins, various other mechanisms are believed to be involved in the control of plasmid copy number and compatibility among plasmids.

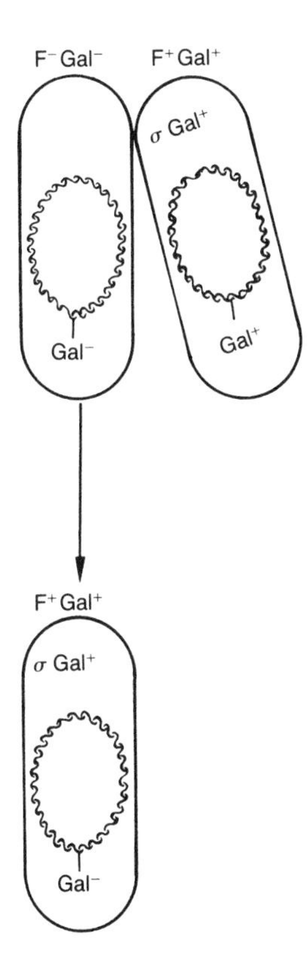

FIGURE 15-18. Sexduction. This diagram shows an F^- cell that cannot ferment galactose and an F^+ cell that carries the allele for galactose use on its F factor and also on its chromosome. After the two cells make contact, the fertility factor may be transferred to the F^-, which is therefore converted to an F^+ with the ability to use galactose. It carries the allele for this ability on its F factor, whereas its chromosome carries the allele that does not enable the cell to use the sugar.

One way in which plasmids are classified relates to the kind of genetic information coded in their DNA. The fertility factor, F, and the F′ factors derived from it constitute one class of plasmid. Bacteria may also carry plasmids that enable them to produce proteins or toxins that kill sensitive bacterial cells. Of extreme importance is a class of plasmid-carrying genes that confer resistance to one or more antibiotics. Let us explore this topic more closely.

Along with the use of antibiotics, the number of bacterial strains that have become resistant to one or more drugs has increased with amazing rapidity. Several loci on the bacterial chromosome are known to control drug resistance. Antibiotics such as chloramphenicol commonly exert their lethal effect on the bacterial cell by associating with the ribosomes, thus interfering with translation. Chromosomal gene mutations that confer drug resistance generally bring about changes in the ribosome itself, so that the antibiotic can no longer combine with it; this permits normal translation and protein synthesis in the presence of the drug. Another well-known class of drug-resistant determinants does not represent genetic information native to the chromosome. Instead, these genes have originated outside the chromosome, as part of a plasmid known as an *R plasmid* (previously called a *resistance-transfer factor,* or *RTF*). Some of the R plasmids are conjugal, meaning that, like the F factor, they can bring about conjugation between two cells and in this way promote their own transfer to cells that lack them. This transfer is independent of F, which can coexist in a cell with a conjugal R plasmid (Fig. 15-19). Since the plasmid may carry several genes for drug resistance, conjugation can confer multiple drug resistance on a previously sensitive cell.

Conjugal plasmids have been shown to cause a cell to produce sex fimbriae resembling those under the control of the F factor. The cells with the R plasmid–induced fimbriae also become sensitive to phages that can lyse male cells. Both the R plasmid and the F factor may be removed from cells after treatment with acridines. These are intercalating agents that inhibit the replication of F and other plasmids but do not affect the replication of the chromosome. Loss of F factor and R plasmid is accompanied by the

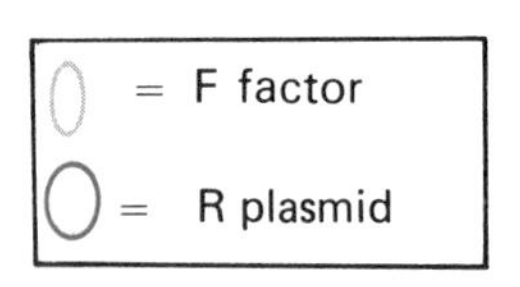

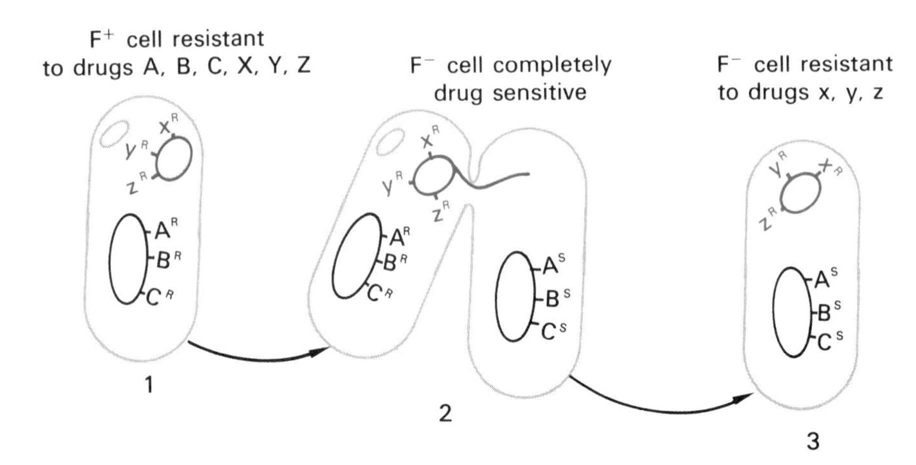

FIGURE 15-19. R plasmids. A bacterial cell may carry genetic determinants that make it resistant to certain drugs (1). Some of these determinants may be on the chromosome, but others may be on plasmids. R plasmids can exist in a cell independently of the F factor. A conjugal R plasmid can promote conjugation with a cell (2). In this way, a previously drug-sensitive cell can be converted to one that is multiply drug resistant if it receives a plasmid that carries resistance to two or more drugs (3). It is possible for a cell to carry a resistance determinant to a specific drug, say streptomycin, on either its chromosome or on an R plasmid (see the text for more details).

inability of the cells to produce sex fimbriae and to adsorb the male-specific phages. We see from this that these two types of replicons have many features in common. The fertility factor, however, serves mainly to bring about transfer of the chromosome once it becomes integrated, whereas the R plasmid is responsible for its own transfer and for that of the resistance genes it may be carrying.

While nonchromosomal genes for drug resistance may exist as part of a conjugal R plasmid, they may also occur on R plasmids that are not conjugal. For example, in *E. coli,* a gene that confers tetracycline resistance may be integrated with a conjugal R plasmid that regularly transmits itself and the genetic factor for tetracycline resistance. However, other genes for resistance, such as those for resistance to streptomycin and ampicillin, may exist on other replicons that are nonconjugal plasmids. By themselves, nonconjugal plasmids cannot be transferred from one cell to another; their transfer requires the presence of a conjugal plasmid. If a conjugal R plasmid is also present in a cell with nonconjugal plasmids, one or more of the latter may be transferred (Fig. 15-20). A variety of R plasmids is known, and these carry different combinations of resistance-determining genes.

The existence of such an assortment of R plasmids, both conjugal and nonconjugal, raises questions concerning their origin. Unlike the chromosomal genes, which can confer drug resistance by altering the structure of the ribosome, the nonchromosomal determinants exert their effect by directing the synthesis of products that act directly on the drug and inactivate it. The ribosome remains unaltered. This different method of conferring drug resistance suggests that the genes of the plasmid are distinct from their chromosomal counterparts. As we will learn in the following section, R plasmids have been found to possess several very remarkable properties.

Plasmids and transposons

A great deal of information on plasmids has come to light throughout the course of investigations with microorganisms. Unexpected discoveries have been made that are pertinent to eukaryotic cells as well and have significant bearing on such important matters as gene regulation and cellular differentiation. Here we discuss only those findings that relate to certain facets of plasmid behavior in microorganisms. Further discussion of plasmids and transposons appears in later chapters.

A class of mutations discovered in *E. coli* produced unusual properties. When such a mutation arose, its effect extended beyond the site of the gene that had mutated. It was found that the mutated DNA had become longer than the normal DNA, indicating that the mutation resulted from a piece of DNA that had become inserted into the gene that had mutated (see Fig. 17-14)! Such DNA segments were found to be able to insert into many different sites on the bacterial chromosome. When they did so, the activity of a gene at that position was abolished. These segments were called *insertion sequences* or *IS elements*—the same IS elements discussed in reference to insertion of the F factor into the *E. coli* chromosome.

A curious feature of the IS elements became apparent. Evidently they could insert at a large number of chromosomal sites. This indicated that nonhomologous recombination was involved, because an IS element could not be expected to contain DNA sequences homologous to so many different sites on the chromosome. Other unusual observations were being made at this time by those studying antibiotic resistance in bacteria. It was found that certain antibiotic resistance genes (genes that code for proteins that can confer resistance to certain drugs) can move about in the cell from one site to another. For example, the gene for chloramphenicol resistance can move from one replicon to another in the cell, from an R plasmid to the chromosome to a virus. A relationship to the IS elements became evident when it was found that transfer from one replicon to another, such as from one plasmid to another, was always accompanied by an increase in the length of the DNA of the recipient

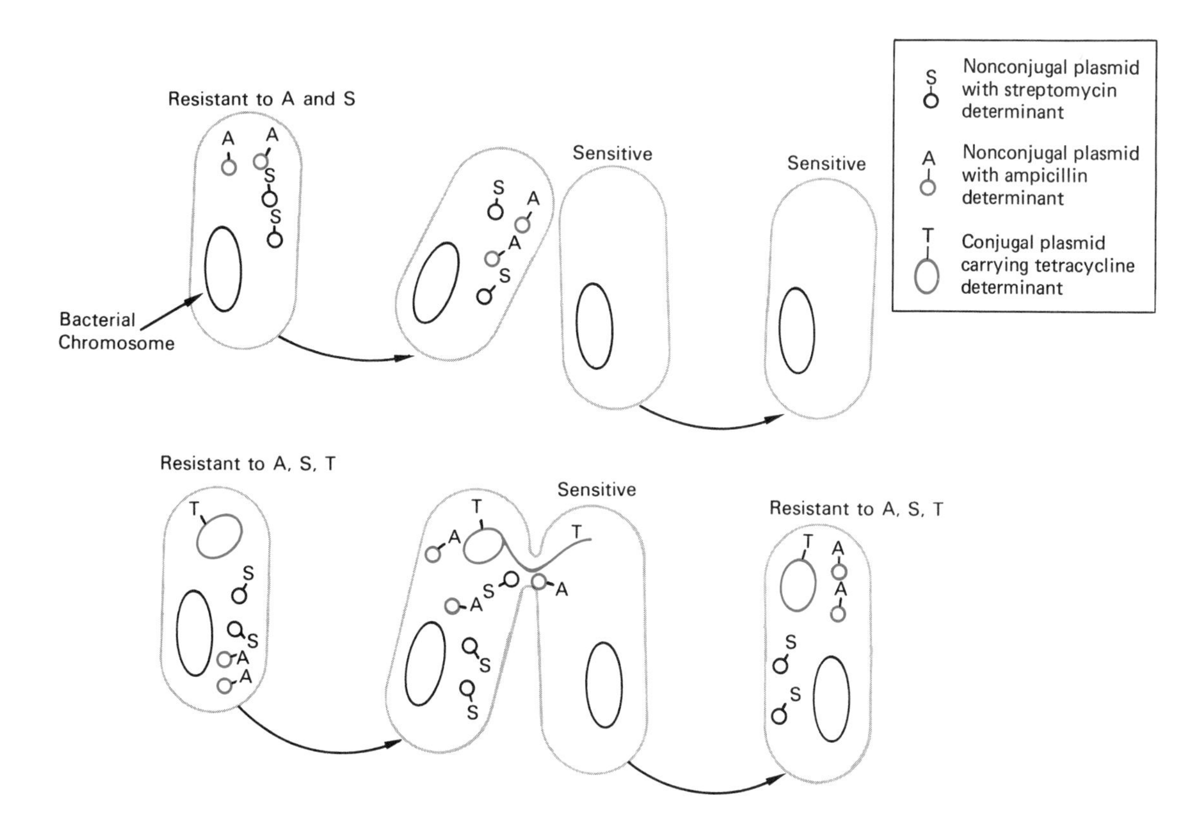

FIGURE 15-20. Conjugal and nonconjugal plasmids. Certain determinants to drug resistance do not reside on conjugal plasmids but are found instead on other R plasmids that are nonconjugal and that can replicate independently. Nonconjugal plasmids cannot bring about their own transfer to another cell (above). A conjugal plasmid carrying one or more resistance determinants can bring about its own transfer. When this takes place (below), nonconjugal plasmids, if present in the same cell, may also be transferred.

replicon. Apparently the gene for antibiotic resistance was being carried by a DNA sequence that could move from one molecule to another. The name *transposon* was given to a unit that could move about in this way.

It was found that transfer by a transposon could occur in bacterial cells that were mutant for a particular gene, rec A, whose protein product was known to be required for recombination in the cell. This, too, implied that a kind of nonhomologous recombination event was taking place, enabling the transposon to insert at nonhomologous positions in different molecules. Although transposons are apparently able to insert by nonhomologous recombination, the process is not completely random, because some transposons are known to have certain preferential sites of insertion on the DNA molecule. Some transposons seem to prefer one or a small number of specific sites; others seem to be less discriminate and insert at a large number of sites, whereas most are intermediate. Certain parts of the *E. coli* genome, known as "hot spots," seem to be selected for insertion over and over again, whereas other regions are excluded by transposons for residence.

IS elements are the simplest of the transposons and contain no known genes related to any properties other than transposition. Transposons that carry genes for resistance are larger and more complex. They may even include IS elements. The IS elements were the first class of transposons to be discovered. Now, by convention, all transposons, regardless of the number of genes they contain, are designated by the symbol *Tn* plus an identifying number, such as Tn10 or Tn55 (Fig. 15-21). A distinguishing feature of all transposons is that the DNA of the transposable unit has inverted repeats at its ends. This means that a copy of the same sequence is present at each end of the double-stranded DNA, but in reverse order (Fig. 15-22). (We encountered inverted repeats in our discussion of the terminator region in Chapter 12.) As a result of the inverted repeat, each *single* strand carries nucleotide sequences that are complementary but in reverse order. When the two strands of the transposon are separated by denaturation, each separate strand can reanneal with itself. A loop forms, which can be visualized with the electron microscope.

The investigations with transposons have given us a very clear insight into how R plasmids are constructed. A conjugal R plasmid is a compound entity, built up from discrete units. A conjugal plasmid contains a segment that carries those genes necessary for the conjugal or transfer properties of the plasmid. These are genes similar to the *tra* genes of the *E. coli* F element. This segment of the conjugal plasmid is called the *RTF (resistance transfer factor)* segment. The RTF segment also carries genes that regulate copy number and DNA replication. The RTF is flanked at its two ends by IS elements. The R plasmid contains, in addition to the RTF, a segment bearing the genes for drug resistance—a segment known as the *R determinant,* or *resistance determinant,* segment. Note from Figure 15-21 that an R segment may contain several genes for drug resistance and several transposons. Two or more R determinant segments may be present in a complex R plasmid. Conjugal plasmids reveal their compound nature by their ability to dissociate reversibly at various sites, yielding a conjugal and a nonconjugal plasmid. At these sites, IS elements are found and evidently account for the dissociation. The process, as seen in Figure 15-23, is reversible, as a larger plasmid may be built up from smaller ones by the reverse process.

The RTF segments of different plasmids have been compared by DNA–DNA hybridization procedures. These show that RTFs from different sources are quite similar. The R determinant segments appear to be collections of various transposons carrying genes for resistance to antibiotics. All the evidence indicates that large R plasmids are constructed from separately derived R determinant segments that become bound to a basic RTF segment. The separately evolving R determinants are joined to the RTF in different combinations, generating a great variety of R plasmids, which has important medical implications. Also of medical concern is the fact that identical transposons are not only found in related species of bacteria but also move from one unrelated species to another.

Transposition

The transposition process requires proteins coded by genes of the transposon. One of these, the enzyme transposase, is able to recognize the ends of transposons and can cleave DNA at target sites for the insertion of the moveable element. However, the frequency of transposition is quite low and is due to the very small amount of transposase produced in the cell. Less than one molecule may be made in a cell generation, giving an overall rate of 10^{-3}–10^{-4} per generation for a given transposon to move.

In some transposons, the nucleotide sequence 5′–GATC–3′ is involved. (Recall that the same sequence played a role in strand recognition, as shown

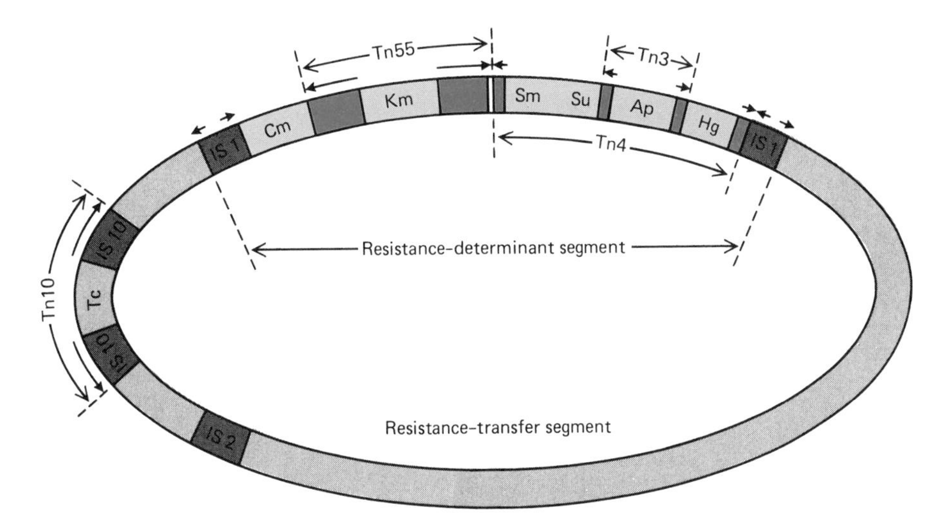

FIGURE 15-21. Transposons and conjugal R plasmids. IS elements contain only properties related to insertion. Other transposons may carry genes for drug resistance. Some of these transposons may be compound and include other transposons. A conjugal R plasmid is constructed from smaller units. A segment is present, the resistance transfer segment, which carries genes required for the conjugal properties of the plasmid. A conjugal R plasmid also contains a segment bearing genes for drug resistance. This resistance determinant segment may contain several transposons. Note that transposon 3 (Tn3) is within Tn4. Each transposon can be transferred independently to another molecule. The small arrows pointing outward indicate inverted repeats at the ends of each transposon. Note the presence of IS elements at the junctions of the resistance transfer segment and the resistance determinant segment. Note that Tn10, a transposon encoding resistance to tetracycline (Tc) occurs here on the resistance transfer segment and that it includes IS elements (see the text for further details). (Reprinted with permission from S. Cohen and J. Shapiro, *Transposable Genetic Elements.* Copyright © 1980 by Scientific American, Inc. All rights reserved.)

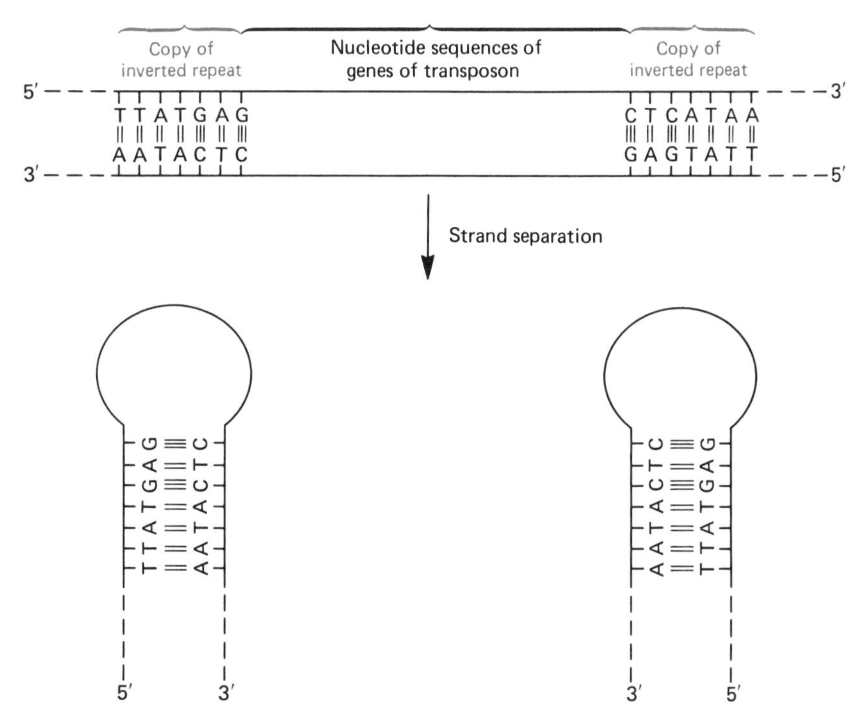

FIGURE 15-22. Inverted repeats. A transposon carries at each end nucleotide sequences that are identical but in the reverse order. When DNA bearing a transposon is denatured and each single strand is allowed to anneal with itself, the complementary bases in the inverted repeat form pairs. This results in single strands that bear a stem and a loop. The loop contains the genes carried by the transposon. The stem–loop configuration can be detected with the electron microscope and indicates the presence of a transposon.

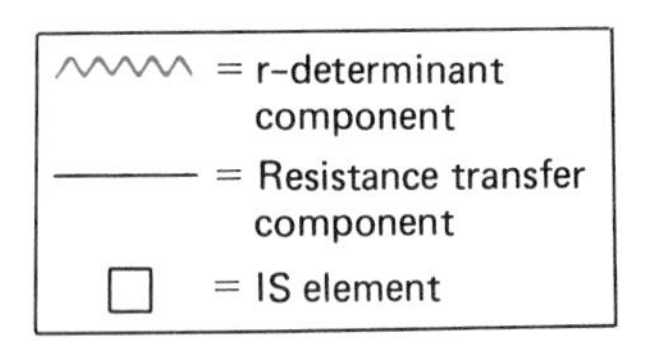

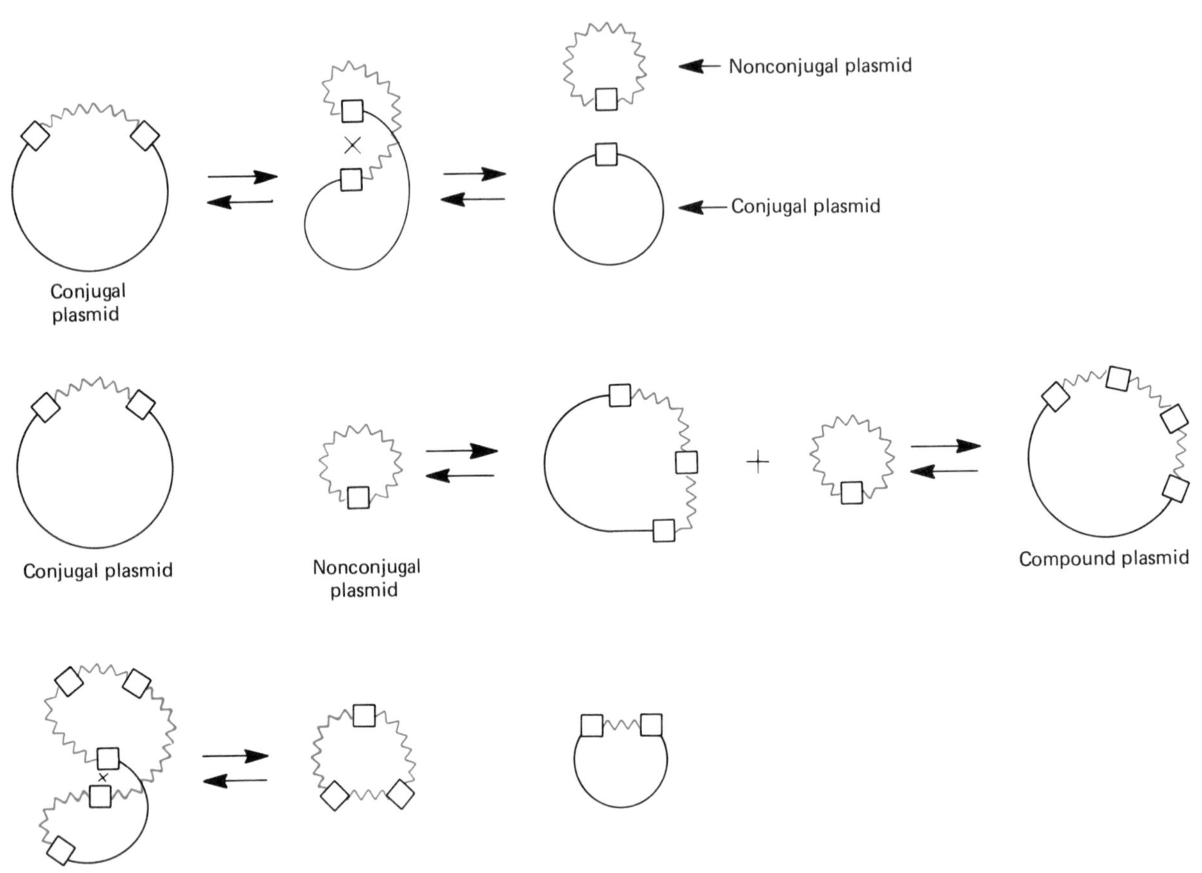

FIGURE 15-23. Reversible dissociation of a conjugal R plasmid. A conjugal R plasmid is a compound entity that can dissociate reversibly at sites where IS elements reside. Dissociation may result in a conjugal and a nonconjugal plasmid (above). A large, compound plasmid in turn can be built up from smaller ones by the reverse process (middle), and this can dissociate (bottom).

in Figure 14-10.) The promoters of some transposase genes contain this sequence, which becomes methylated in *E. coli* at the adenine position. When both strands of the promoter have the methylation at this site, the promoter is inactive. However, when replication takes place, GATC in the new strand is not methylated immediately, and the promoter is more active in this hemimethylated state. Therefore a small amount of transposase may be made just after replication. Consequently, transposons with this type of promoter tend to transpose immediately following replication.

A characteristic feature of transposon insertion relates to the site at which the transposon takes up residence. As noted previously, some sites may be preferred over others by some transposons. Nevertheless, insertion of a transposon always involves the duplication of a short nucleotide sequence of the recipient DNA. These duplications range in size from about 3–12 base pairs and flank the inverted repeats on either side of the transposon (Fig. 15-24). The sequence that becomes duplicated in the recipient is known as the *target sequence,* and it is believed to become repeated in the manner shown in Figure 15-25. The transposase brings about two single-stranded breaks that are staggered, one in each strand, at the ends of the selected target sequence. The transposon DNA becomes linked to the ends of the cut strands in such a way that two gaps are produced, one on each strand. A DNA polymerase fills in these gaps, and DNA ligase seals the nicks. The result is duplication of the target sequence, the length of which is characteristic for a given transposon. This length in turn is determined by the degree to which the transposase staggers the single-stranded nicks. Duplication of a target sequence is so characteristic of transposons that duplication of the direct repeat type flanking certain DNA sequences in a genome is taken as a clue that a transposon has been inserted at that region (see Chap. 17).

When transposition takes place, it may promote various kinds of rearrangements of the DNA such as deletions and inversions. Two major classes of transposition can be recognized. In simple (or conservative) transposition, cuts are made at each end of the transposon, which then leaves its position and moves to a new site. Staggered cuts are made, and insertion then occurs in the manner depicted in Figure 15-25. The original site now contains a gap, and its fate is unknown. It is possible that the gap becomes filled in by the repair mechanisms of the cell. Certain IS elements and some composite transposons move only by simple transposition.

The second type of movement is replicative transposition in which the transposon becomes duplicated. The transposon at the original site is not cut out. Instead, a copy of it is made that then inserts at the new site. Therefore, a transposon moving in this manner increases in number in a cell. Some noncomposite transposons are known to move exclusively in this way, and certain IS elements can move by either type of transposition.

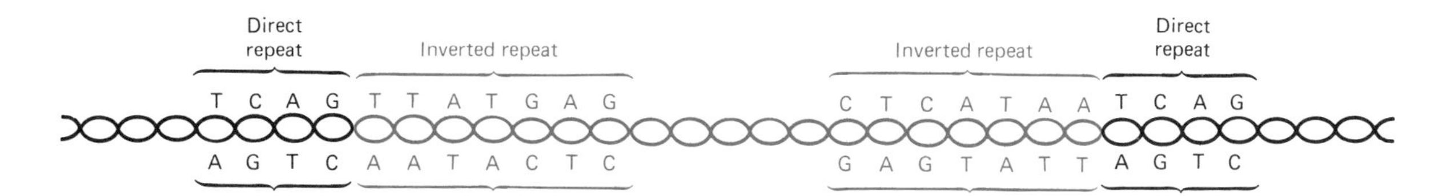

FIGURE 15-24. Features of transposon insertion. Insertion of a transposon (red) is accompanied by the direct repeat of a short sequence of nucleotides in the recipient DNA. At the ends of the transposon are nucleotide sequences that are inverted repeats of each other.

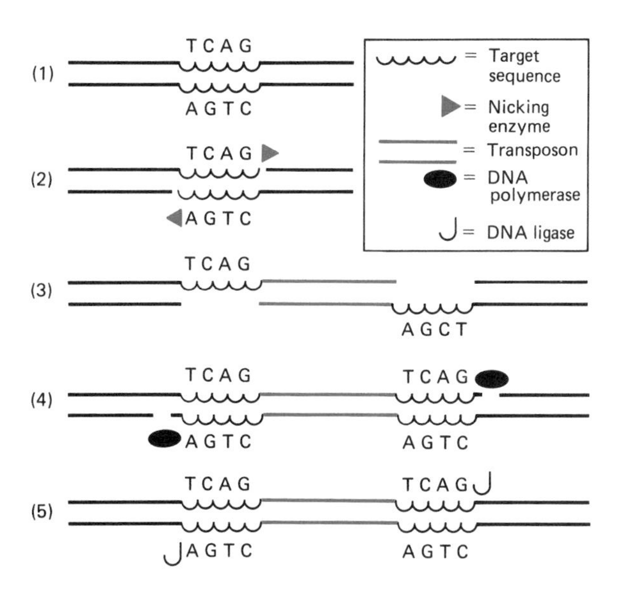

FIGURE 15-25. Generation of direct repeats. **1:** A target sequence is present in one copy in the recipient. **2:** The enzyme transposase produces two single-stranded staggered breaks. **3:** The transposon links to staggered ends of cut strands, producing two gaps. **4:** The gaps are filled in by DNA polymerase, using target sequences on the separate strands as templates. **5:** DNA ligase seals the nicks. The target sequence is now present in two direct copies.

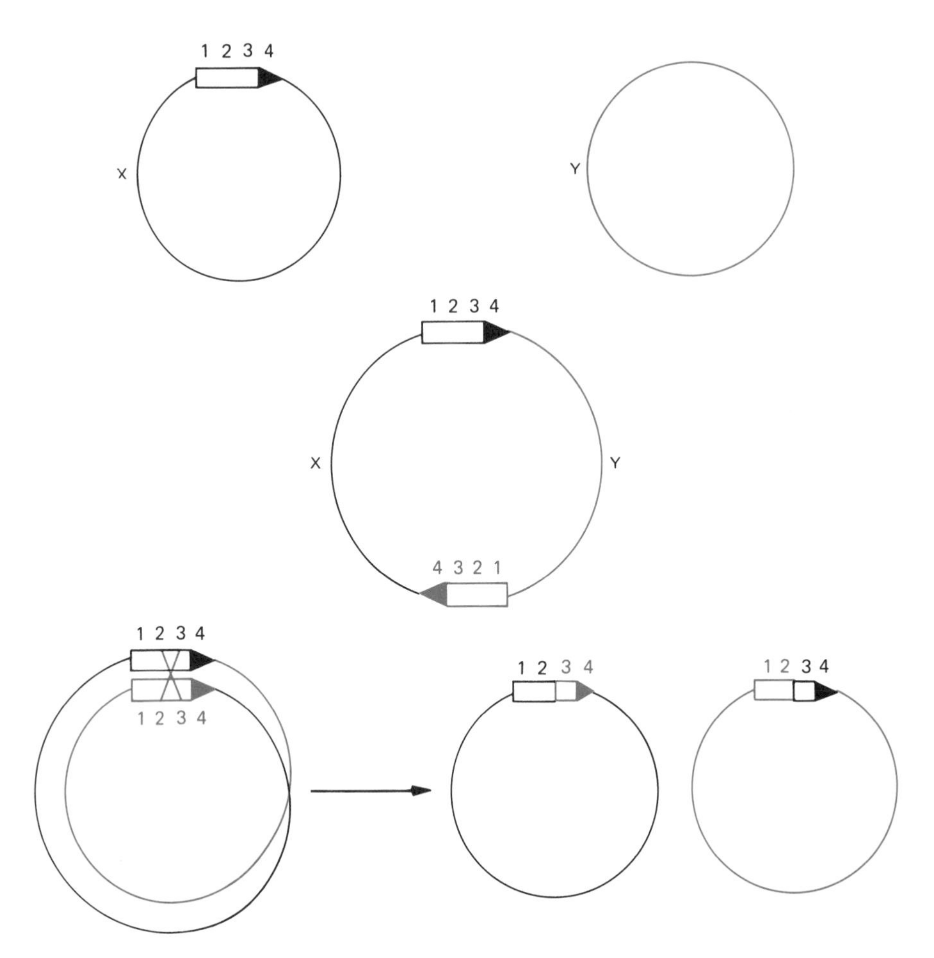

FIGURE 15-26. Cointegrate formation and transposition. **Top:** Two replicons in a cell are represented, plasmid x with the transposon and plasmid y without it. **Middle:** Replicon fusion produces a cointegrate, a composite of the two plasmids. In the process, a copy of the transposon has been generated. Each copy is at the junction uniting the two replicons, and each of the copies is a direct repeat with the same orientation. **Bottom:** A recombination event that depends on the enzyme resolvase resolves the cointegrate into two replicons, each with a copy of the transposon.

The replicative transposition of some transposons involves the formation of a cointegrate, which depends on the enzyme transposase. A cointegrate is a composite replicon formed by the fusion of the replicon containing the transposon with a replicon lacking one. Note from Figure 15-26 that the cointegrate contains two copies of the transposon, indicating that replication of the transposon DNA has accompanied the fusion process. The transposon copies are located at the sites where the two replicons are joined, and each is in the same orientation. After many generations, cells containing cointegrates may give rise to other cells that no longer contain the cointegrate but now contain two plasmids, each with a copy of the transposon. The two plasmids are believed to arise as a result of a crossover event between the two transposons in the cointegrate, a step that depends on an enzyme called *resolvase* which is coded by a gene of some transposons. This enzyme brings about the reduction of the cointegrate with the formation of two plasmids or replicons.

A well-known transposon, Tn3, found in bacteria is 5,000 nucleotides long and contains three genes. One of these is coded for transposase, the second for resolvase, and the third for the enzyme β-lactamase (Fig. 15-27A). The latter inactivates ampicillin and thus confers resistance to that antibiotic on cells harboring Tn3. The expression of both the transposase and the resolvase genes is regulated by the resolvase enzyme itself. Besides its role in transposition, the resolvase possesses repressor activity and can bind to a region between the transposase and resolvase genes, thereby preventing their expression and a high rate of transposition. The transposase and resolvase genes are transcribed in opposite directions, as shown in Figure 15-27B, because their antisense strands are opposite. Therefore, when the repressor binds to the binding site between the two genes, transcription from initiation sites in both genes is blocked. We will learn about still other kinds of transposons in Chapter 17, where their relationship to gene regulation will be emphasized.

Transformation and recombination

Before leaving this discussion of bacteria, let us reexamine one method of genetic change in bacteria that may have implications for higher forms of life. In Chapter 11, the phenomenon of transformation in *Pneumococcus* was discussed in relation to the identification of the chemical nature of the gene. It will be recalled that, in transformation, DNA preparations from a donor strain may be applied to a recipient. After a period of incubation, the latter may incorporate some of the donor DNA and acquire a new trait. When completed, this transformation is stable and represents a true genetic alteration.

At first, transformation was considered more or less a curiosity in the pneumonia organism, but it now seems to be rather widespread. Transformation has been recognized in at least 17 bacterial species, among them such well-known groups as *Streptococcus, Bacillus,* and *Hemophilus.* Many characteristics of these microorganisms are known to be susceptible to transformation—drug resistance and ability to synthesize specific enzymes, among others. It seems that any genetic locus may be transferred from donor to recipient in transformation. We must emphasize that the DNA donated to the recipient does not require cell

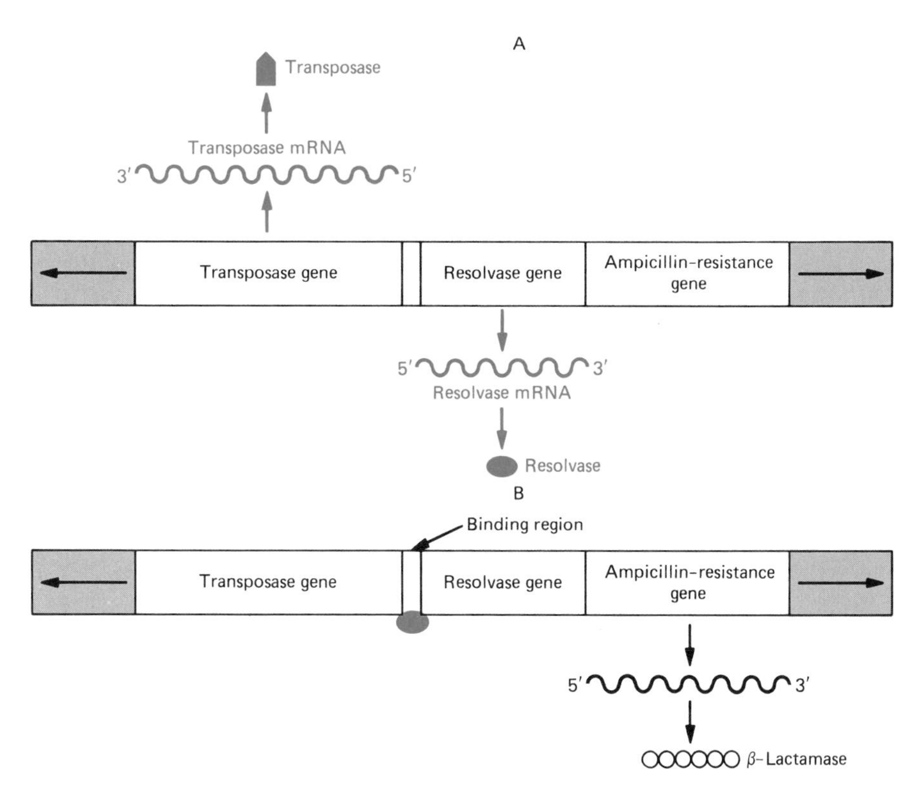

FIGURE 15-27. Transposon Tn3 and regulation of transposition. **A:** Tn3 contains three genes between its inverted repeats. The two involved in transposition are transcribed in different directions. The transposase and resolvase result from translation of the corresponding mRNAs. **B:** A high frequency of transposition is apparently prevented by the ability of the resolvase to act as a repressor when its concentration reaches a certain level. It can bind to a region between the two genes that contains the sites for the start of transcription of both transposition genes. The repressor does not affect transcription of the gene for ampicillin resistance.

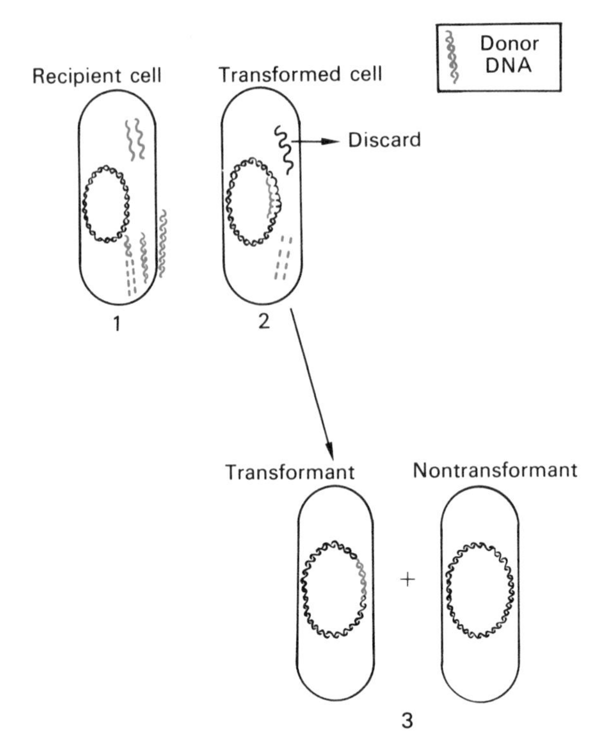

FIGURE 15-28. Steps in transformation. To bring about transformation, the donor DNA must be double stranded; otherwise it will not penetrate the recipient cell. Once inside the recipient, some of the donor DNA is degraded and some transformed to the single-stranded condition (1). This single-stranded form may associate with the region on the chromosome with which it is homologous. It may then replace a single-stranded portion of the recipient DNA (2). This recipient segment is eventually lost. The cell, now carrying the hybrid DNA, is transformed. After it divides (3), it gives rise to nontransformed and to transformed cells. In the latter, the region involved in the transformation has both strands identical to the original donor DNA. This results because the inserted donor strand forms a complementary copy of itself at the next replication.

contact; pure DNA preparations extracted from donor cells are effective. Nor does the transfer depend on any vector, such as an episome or virus. Bacterial cells appear to have an affinity for uptake of DNA from the environment during a certain phase of the growth cycle of the bacterial population, near the end of the log phase. This receptive stage is of brief duration in the recipient cells and entails physiological changes that are not completely understood. A cell in the receptive stage that can take up extraneous DNA and be transformed by it is said to be *competent. Competence* may also refer to the portion of the growth cycle of the bacterial population when most of the cells are physiologically capable of transformation.

Several ingenious experiments revealed many facts about transformation that are of general biological significance. In the early 1960s, studies were conducted in which labeled donor DNA was followed from the moment of contact with recipient cells until the latter were transformed. It was found that, immediately after entering the recipient cells, some of the donor DNA is converted to a single-stranded state, while the rest is degraded. This labeled, single-stranded donor DNA is then inserted into the recipient DNA (Fig. 15-28). Once inserted, it replaces the homologous segment of the recipient DNA, which is then discarded. It is this newly inserted DNA that is responsible for the transformation. It is able to effect a genetic change and become a part of the genetic material of the recipient cell. This cell in turn may be used as a donor in a subsequent transformation. Thus the labeled DNA introduced in the first transformation can be followed further to see how it behaves in later transformations or in later cell divisions.

Extremely valuable information has come from following DNA throughout all phases of transformation in a single cell and throughout two or more divisions of transformed cells. One of the most important discoveries, with great implications for genetic theory, was the demonstration that the introduced donor DNA is physically integrated into the recipient DNA. This observation supports the concept of crossing over as a result of synapsis followed by breakage and reunion, a matter that has been controversial for decades. The studies with the labeled donor DNA clearly show that the introduced transforming DNA penetrates the cells and becomes associated with the homologous DNA segment of the recipient. It is then incorporated into the recipient chromosome by an actual physical insertion, which is accompanied by the genetic transformation of the cell.

As noted, only one strand of the donor DNA participates, and the evidence indicates that *either* strand of the donor double helix can bring about the transformation. It is interesting that the DNA that is picked up by the recipient cell in the first place *must* be double stranded. It must also be above a certain minimal molecular weight (5×10^5). Single-stranded DNA is completely ineffective in transformation. But, once inside the recipient cell, the double-stranded donor DNA is altered, and it is only a single strand that engages in the transformation process. This means that when this single donor strand replaces one of the two strands of the recipient, heteroduplex DNA results. One strand of the double helix in the involved region is a strand of the recipient, the other the strand of the donor. Therefore, just one strand, either of the two of the original double helix, can cause a transformation. At the cell divisions that follow the transformation, the DNA of the transformed cells will come to have both strands of the *original* donor DNA in the region of the transformation (Fig. 15-28).

Although most studies of transformation employ DNA preparations that are extracted from donor cells by an investigator, it is now known that transformation may take place spontaneously. Aging cultures of some bacterial strains contain cells that are undergoing self-digestion and release DNA into the medium. This DNA can cause transformations of other cells in the population. *Pneumococcus* transformations may occur spontaneously when two different bacterial strains are injected into mice. Such observations raise the possibility of transformation in higher species. Actually, this has been reported in the fruit fly. Eggs of *Drosophila* containing very young embryos have been treated with DNA from other stocks carrying certain marker alleles. The treated embryos have given rise to stocks of flies that are true genetic transformants, whose cells have incorporated some of the genetic material applied to them. However, in the flies, the donor DNA does not replace the resident DNA of the recipient chromosome, as it does in bacteria, but manages somehow to persist along with it in the cell. Moreover, the donor DNA does not always undergo transcription. Its expression is sporadic. Whether the information contained in the DNA of the donor or in that of the host will be transcribed seems to vary from cell to cell, so that the flies are always mosaics. The mosaic flies pass down to their offspring the information introduced from the donor stock, but these offspring are in turn mosaics. Transformation, a genetic phenomenon discovered in bacteria, may prove to have many implications for various higher life forms.

REFERENCES

Adelberg, E.A., and S.N. Burns. Genetic variation in the sex factor of *Escherichia coli*. *J. Bacteriol.* 79:321, 1960.

Cold Spring Harbor. *Movable Genetic Elements,* vol. 45. Cold Spring Harbor Laboratory, Cold Spring Harbor, NY, 1981.

Cohen, S.N., and J.A. Shapiro. Transposable genetic elements. *Sci. Am.* (Feb):40, 1980.

Fox, A.S., S.B. Yoon, and W.M. Gelbart. DNA-induced transformation in *Drosophila:* Genetic analysis of transformed stocks. *Proc. Natl. Acad. Sci. USA* 68:342, 1971.

Gurney, T., Jr., and M.S. Fox. Physical and genetic hybrids in bacterial transformation. *J. Mol. Biol.* 32:83, 1968.

Hayes, W. *The Genetics of Bacteria and Their Viruses,* 2nd ed. Wiley, New York, 1968.

Kingsman, A., and N. Willets. The requirement for conjugal DNA synthesis in the donor strand during F'Lac transfer. *J. Mol. Biol.* 122:287, 1978.

Lacks, S. Molecular fate of DNA in genetic transformation of *Pneumococcus*. *J. Mol. Biol.* 5:119, 1962.

Lederberg, J. Genetic recombination in bacteria: A discovery account. *Annu. Rev. Genet.* 21:23, 1987.

Lederberg, J., and E.L. Tatum. Gene recombination in *Escherichia coli*. *Nature* 158:558, 1946.

Luria, S.E., and M. Delbruck. Mutations of bacteria from virus sensitivity to virus resistance. *Genetics* 28:491, 1943.

Novick, R.P. Plasmids. *Sci. Am.* (Dec):102, 1980.

Shapiro, J.A. Molecular model for the transposition and replication of bacteriophage mu and other transposable elements. *Proc. Natl. Acad. Sci. USA* 76:1933, 1979.

Taylor, A.L., and C.D. Trotter. Linkage map of *E. coli* K12. *Bacteriol. Rev.* 36:504, 1972.

Vapnek, D., and W.D. Rupp. Identification of individual sex factor DNA strands and their replication during conjugation in thermosensitive DNA mutants of *Escherichia coli*. *J. Mol. Biol.* 60:413, 1971.

Willets, N., and R. Skurray. The conjugation system of F-like plasmids. *Annu. Rev. Genet.* 14:41, 1980.

REVIEW QUESTIONS

1. In one part of an experiment with a strain of penicillin-sensitive cells, a sample of bacteria is taken and allowed no further growth. The sample is divided into several equal parts and then each is plated immediately on a medium containing penicillin. In another part of the experiment, a small sample of the sensitive culture is taken and divided further into several subdivisions. Each subdivision is then allowed to grow until it contains the same number of cells as each division in the other part of the experiment. From the following data, tell which part of the experiment is represented by the data in column A and which by that in column B. Explain.

	A	B
Total number of cells	10^9	10^9
Mean number of resistant cells	22	15
Variance	4,480	17

2. Colonies derived from eight different cells are grown on a plate supplemented with all the necessary growth substances. Colonies 2, 5, 6, and 8 are prototrophs. The others are auxotrophs. Colonies 3 and 7 require leucine; 1 and 4 require arginine. Answer the following:

   A. What colonies will grow following incubation after a replica is made onto a plate with minimal medium?
   B. What colonies will grow if a replica is made onto a plate with minimal medium supplemented with leucine?
   C. What colonies will grow on a replica plate containing minimal medium plus arginine?

3. Let A^+, B^+, C^+, D^+, and E^+ represent markers enabling an *E. coli* cell to manufacture amino acids required for growth on a minimal medium. *E. coli* strain 1 is $A^+B^+C^-D^+E^-$. Strain 2 is $A^-B^-C^+D^-E^+$. The two strains are mixed and incubated together. After plating on minimal medium, a small percentage of prototrophs is selected.

   A. What would the genotype of the prototrophs be with respect to these markers?
   B. Two other *E. coli* strains, 3 and 4, can manufacture all the amino acids required for growth on minimal medium. Strain 3 is Lac^+ and can consequently use lactose, a sugar that can be used as an accessory energy source. Strain 4 is Lac^- and cannot use lactose when this sugar is provided. Which of these strains is auxotrophic? Explain.

4. In *E. coli,* five different single-point mutations resulted in the origin of five different auxotrophic strains, each with a genetic block in the biochemical pathway leading to the synthesis of an essential substance. An auxotrophic mutant can grow if the minimal medium is supplemented with one or more substances: A, B, C, D, E, and F. From the following data, determine the sequence of steps in the biochemical pathway and indicate the step at which each auxotrophic strain is blocked (+ = growth; 0 = no growth).

	Medium supplement					
Strain	A	B	C	D	E	F
1	+	+	0	0	+	0
2	0	+	0	0	0	0
3	0	+	+	0	+	0
4	0	+	0	0	+	0
5	0	+	+	0	+	+

5. Two reciprocal crosses are made:

	Strain 1		Strain 2
1.	$A^+B^+C^-D^+E^-$ streptomycin resistant	×	$A^-B^-C^+D^-E^+$ streptomycin sensitive
2.	$A^+B^+C^-D^+E^-$ streptomycin sensitive	×	$A^-B^-C^+D^-E^+$ streptomycin resistant

After incubation, plating is made on minimal medium containing streptomycin. From cross 1 colonies develop, but none arises from cross 2. Explain.

6. Explain the differences between the results of the following two crosses in *E. coli:*

   A. F^+ cell having chromosome markers $T^+L^+Pro^+Lac^+ \times F^- \rightarrow \ldots F^+$ cells with no chromosome markers transferred.
   B. Hfr cell having chromosome markers $T^+L^+Pro^+Lac^+ \times F^- \rightarrow \ldots F^-$ cells with chromosome markers transferred.

7. Let A^+ and B^+ represent markers for essential growth substances needed for survival on minimal medium. The other markers, L^+, M^+, and so on, are for traits that do not affect the minimal nutritional requirements. From the cross and the information given here, arrange the marker alleles in the proper order of entry.

Hfr $A^+B^+L^+M^+N^+O^+Q^+$ × $F^-A^-B^-L^-M^-N^-O^-Q^-$
strep. sensitive strep. resistant

Colonies on minimal medium with streptomycin: A^+, 100%; B^+, 100%; L^+, 25%; M^+, 88%; N^+, none; O^+, 73%; Q^+, 35%

8. Markers are transferred from five Hfr strains to F^- cells in the following order:

Hfr strain	Order of entry
1	BKARM ←
2	DLQEOC ←
3	OEQLDN ←
4	MCOEQLDN ←
5	RAKBN ←

A. Draw a map showing the sequence of these markers on the chromosome.
B. For each strain, indicate on the map the site of insertion of the fertility factor by placing an arrowhead so that the first gene to be transferred by a strain is behind the arrowhead.

9. Assume that Hfr recombinants are desired and that crosses are made with the Hfr strains in Question 8. For each strain, indicate which marker should be selected to derive the highest number of recombinants that will also be Hfr. Explain.

10. Assume a DNA of an Hfr *E. coli* has the following arrangements of nucleotides:

5′–AGCTAT–3′
3′–TCGATA–5′

Assume that Hfr cells have been growing on medium supplied with ^{14}C and ^{15}N so that all its DNA is heavy. Mating of the Hfr with F^- cells is performed on ordinary medium. In the mating, the lower DNA strand is transferred from its 5′ end.

A. Show this segment of the double helix in the donor cell after DNA replication, indicating light (L) and heavy (H) strands.
B. Do the same for the recipient, assuming this whole strand is incorporated.

11. A certain Hfr strain normally transmits the Lac^+ marker as the last one during conjugation. In a cross of this strain with an F^-Lac^-, some recombinant Lac^+ cells received the Lac marker too early in the mating process. When these Lac^+ cells are mixed with F^-Lac^- cells, the majority of the latter are converted to F^+Lac^+, but other markers are not transferred. Explain. What can you say about the genotypes of the cells from the F^- strain that became Lac^+?

12. Suppose strain A of *Pneumococcus* is streptomycin resistant and strain B is streptomycin sensitive. In a transformation experiment, competent B cells are incubated with DNA from strain A. DNA of strain A carrying the resistant markers enters some B cells and associates with the B DNA. The latter cells become transformed. Answer the following:

A. Will the DNA of the transformed recipient have A or B DNA for the marker region?
B. When any transformed cell divides, what will be the nature of its two cell products with regard to streptomycin resistance, and what will the DNA of these cells be like regarding the A and the B DNA?

13. In a transformation experiment, donor DNA is randomly broken up into fragments in the process of extraction. Therefore genes closely linked will be included on the same fragment more often than they will become separated onto different fragments. Genes that are far apart from each other will tend to become separated onto different fragments during extraction of the donor DNA.

DNA from a donor strain of a species of Streptococcus was used to transform recipient cells. The donor cells are a^+b^+ in relation to two growth requirement substances, whereas the recipients are doubly defective nutritional deficients, a^-b^-. Three types of transformants arose in the following frequences:

Type of transformed cell	No. of transformed cells
a^+b^+	650
a^-b^+	155
a^+b^-	95
	900

A. Based on this information, would you say that the two genes are closely linked or far apart? Explain.
B. What appears to be the distance in map units between the two genes?
C. Why do the recombinant classes occur in unequal numbers?

14. Suppose the following sequence is a target sequence in a recipient cell:

5′ – A T C G T – 3′
3′ – T A G C A – 5′

Let the following represent sequences in a transposon that recognizes the particular target sequence (the transposon inserts in the sequence X Y Z as shown):

5′ – X Y Z – 3′
3′ – X′ Y′ Z′ – 5′

Determine the arrangement of the DNA after the transposon is inserted.

15. Allow A and B to represent determinants for drug resistance that are found on a conjugal plasmid in certain *E. coli* strains. C and D represent drug-resistance determinants that occur on two different nonconjugal plasmids. E and F are genetic drug-resistance determinants found on the chromosome. For each of the situations given here, tell the likely maximum number of resistance determinants transferred to any cell that completely lacks them:

A. F^+ ABCDEF mixed with F^-.
B. F^- ABCDEF mixed with F^-.
C. F^- CDEF mixed with F^-.
D. Hfr ABCDEF mixed with F^-.

16. A strain of bacterial cells carries genetic factors for streptomycin and tetracycline resistance. Both resistance factors are regularly transferred together, on contact, to sensitive cells. Cells are isolated from the original strain that are still resistant to both antibiotics but do not always transmit both determinants. Some recipient cells become resistant only to streptomycin and in turn transmit that drug resistance determinant by itself to other sensitive cells. Other cells isolated

from the strain are resistant only to tetracycline, and these no longer transmit resistance of any kind to other cells. Explain these observations.

17. In a certain F^+ *E. coli* strain, a factor for streptomycin resistance is transferred along with the factor for chloramphenicol resistance on a compound R plasmid. Hfr cells arise in this F^+ population, and these are isolated and found to be resistant to both antibiotics. However, it is found that some of the Hfr cells contain an R plasmid that transfers only the determinant for chloramphenicol resistance. The factor for streptomycin resistance is transferred only when chromosomal markers are transferred to an F^- cell. Explain these findings.

Viruses and Their Genetic Systems

16

Viral diversity

Much of what we know about the cell and its chemistry does not apply to viruses. Since they lack cell structure, viruses can be considered neither prokaryotes nor eukaryotes. Indeed, the mature virus particle, the virion, lacks organelles and all the precursors required to build its structure. Each kind of virus is thus completely dependent on its specific host cell to supply the machinery and substances needed for its reproduction. Nevertheless, the virus is able to enter its host cell and direct it to produce new viral particles. Some infected cells may continue to release viral particles for generations without apparent harm. Others, however, may become destroyed or altered in some way, even transformed to the malignant state.

Although viruses occur in a variety of shapes, the virion is essentially nucleic acid encased in the center of a protein shell. The protective protein covering surrounding the genetic material is called the *capsid.* The capsid is simple in some viruses and consists of just a single kind of protein. In others, the capsid is more complex, composed of different layers of proteins to which lipid and carbohydrates may be attached. Regardless of the type, all capsids contain protein that occurs in many identical copies. The capsids of certain bacteriophages (see Fig. 11-2) can be very complex and include a polyhedral head and a tail associated with fibers that enables it to attach to a host cell. Such tails are never found in animal viruses, which, unlike phages, do not inject their genetic material into the host cell. Instead, animal viruses gain entry into the cell by fusion or endocytosis, interactions that take place between the surface of the virion and receptors found on the host cell membrane. Once inside the cell, the virion sheds its outer covering and thus releases the genetic material, enabling it to interact freely with molecules in the cellular environment.

In addition to the capsid, some viruses possess an envelope. This additional covering, which surrounds the capsule, is derived from the cell membrane of the host but also contains proteins and glycoproteins that are encoded by the genome of the virus itself. Those viruses that lack an envelope are classified as *naked.*

Most viruses are smaller than cells, but there is great range in their size, from about 20 nm in diameter (smaller than a ribosome) to about 200 nm (about the size of the smallest bacteria). It is therefore not just size that sets them apart from cellular organisms. In addition to their lack of cell structure and inability to replicate on their own, the viral particle never contains both RNA and DNA. This indicates something about the diversity among viruses in the nature of their genetic material. Double-stranded DNA composes the genetic material of many animal and human viruses, such as the herpes, chickenpox and Epstein-Barr viruses. It also makes up the genomes of a large number of bacteriophages, such as the T-even viruses discussed later in this chapter. However, among viruses one finds many exceptions to the rules followed by cellular organisms. In certain viruses, the DNA of the virion is single stranded, not the familiar double helix of a cell. The single- or double-stranded DNA exists as a single molecule and may be linear or closed in the form of a circle. Single-stranded DNA is found in the virions of several groups of phages (discussed later this chapter) and even in those of some animal and human viruses. However, DNA may not be the genetic material at all. Instead, it may be single-stranded or even double-stranded RNA! In those viruses with double-stranded RNA, the genome consists not of one but of several separate RNA molecules and is said to be *segmented.* In these viruses, each molecule carries a separate gene, unlike the situation in DNA viruses where all the genes are carried as part of a single polynucleotide.

An interesting difference between DNA and RNA viruses relates to genome size. While genomes of DNA viruses vary widely from about 3 to 200 kb pairs, the size range of RNA viral genomes is not nearly so great, extending only from 2 to 20 kilobase pairs. One reason for this difference probably relates to the fact that cells contain mechanisms for DNA proofreading (see Fig. 11-26) but lack ways to correct errors arising during RNA replication. Actually, the mutation rate that accompanies RNA virus replication is quite high when compared with that associated with the replication of DNA viruses. Any RNA genome approaching the size of the largest DNA virus genome would accumulate so many errors that infective virions would probably not be produced.

Most DNA viruses replicate in the nucleus, where they have access to the cell's RNA polymerase and the enzymes involved in mRNA processing. However, a ready supply of the precursor substances and enzymes needed for DNA replication is restricted to the S phase of the cell cycle when the cell divides. In a nondividing cell, such molecules are present only in low concentrations. As a result, certain DNA viruses replicate only in the S phase of the cell cycle when they can take advantage of the molecules of replication manufactured by the cell. Some of the more complex DNA viruses, however, are able to stimulate nondividing cells to start DNA synthesis, thus giving the virus the supplies it needs to replicate. In contrast, almost all RNA viruses are independent of the host cell nucleus. The human polio virus, for example, undergoes its entire reproductive cycle in the cytoplasm. (The exception is the influenza virus, which requires the nucleus of the host cell if it is to replicate.)

Are such curious entities as viruses to be considered living? We do not think of a chromosome or piece of a chromosome to be a living thing. The same reasoning would apply to any virus that is thought to be a small piece of genetic material surrounded by a protein shell that enables it to travel from one host cell to the next. Although viruses appear simpler than any cell, their replication and relationships to their hosts have proved to be quite complex. The rest of this chapter will provide some examples illustrating how studies of viruses and the cells they infect provide insights into the mechanism of gene regulation in both the normal and abnormal cell. We will see below how the rapid growth rate of phages and their bacterial hosts has made possible a refinement of genetic analysis that would be impossible with higher forms.

Events following infection by phages

Phages have the capacity to destroy the bacterial hosts they infect. Virulent phages are those viruses that undergo a lytic cycle resulting in death and lysis of the host cell. Phages T2 and T4 of *E. coli* are examples of virulent double-stranded DNA phages, all of which must encounter a supply of host cells to continue their reproduction. Figure 11-2 shows the virulent virus T4. The DNA is stuffed into the protein of the headpiece during maturation of the virus in the host cell. The tail piece, needed to attach the infecting phage to the cell, is associated with a plate equipped with prongs. The contraction of the tail piece causes the viral DNA to pass through the hollow core into the bacterial cell. An enzyme (lysozyme) that can break down the cell wall enables the phage to inject its DNA into the cell.

Once inside the cell, the phage particle takes control of the cellular machinery. One of the first "new" substances to appear after infection is an mRNA that is complementary to the DNA of the virus, not to that of the host. This is followed by the appearance of proteins that represent certain enzymes needed for the activities of the virus. Meanwhile, the normal activities of the host cell cease. The bacterial DNA becomes degraded and enters a pool in the cell. Substances from the cell medium are added to this pool, from which virus-specific substances will be made (refer to Fig. 12-6). By the ability of its nucleic acid to undergo transcription in the host cell, virus-specific mRNA is formed. When this is translated by the ribosomes of the host, specific viral enzymes are constructed that are needed to build new viruses. Following the appearance of the enzymes, products foreign to the cell then form in the host cell. All of this is accomplished by means of the ability of the virus DNA to be transcribed and its mRNA to be translated in the host cell.

In the first few minutes after entry, no mature virus particles can be detected in the bacterial cell, because the preliminary activities needed for virus construction are going on. After the degradation of the nucleic acid of the host, construction of the phage DNA begins. This occurs approximately 6 minutes after infection. After the appearance of the protein, which has an enzymatic function, another class of protein begins to appear. This is the protein that will form the envelope of the completed viruses. Finished phage particles may appear in the cell approximately 12 minutes after the infection, and their number will increase until lysis of the cell occurs. The exact time of this cell burst depends on the particular bacterial host and the strain of the virus. Approximately 30 minutes after infection by a phage, the cell bursts and releases 100 or so mature phage particles.

The lysis results not just from pressure of virus particles inside the cell but also from the activity in the viral tail of the enzyme lysozyme, which can break down the cell wall. This is the same enzyme that enables the phage to inject its DNA into the cell. After lysis, the released phage particles are free to enter other susceptible cells and repeat the process. It is evident that the virus is much more than just a simple association of nucleic acid and protein. Indeed, its ability to use cell machinery for the complex reactions necessary for its own reproduction implies a high degree of specialization. This must have involved a long history in which the forces of evolution acted on both the virus and the host.

Not all phages, however, are virulent like phage T4. Many can infect cells and take up residence in them without causing lysis or cell destruction. Such a nonvirulent virus residing in a bacterium is called a *symbiotic* or *temperate phage.* The bacterial cell remains unharmed by the presence of the phage and may actually receive some benefits from the particle it contains. If we artificially burst open a bacterial cell that harbors a temperate phage, we cannot detect any mature phage particles. The virus seems to have disappeared. Yet its presence in the cell can still be demonstrated. Assume we have a strain of bacterial cells that contains a temperate virus. Although we find no mature virus in the cells, the fluid in which the cells are growing will always contain some mature phage particles. The reason for this is that, in a few rare cells of the bacterial strain, the virus becomes virulent, enters a lytic cycle, and causes lysis of its host. We can detect these released particles if we bring the culture fluid into contact with an indicator strain of bacteria. The latter would be a strain that is susceptible to the phage. If a drop of the culture fluid is added to a cloudy tube containing the susceptible cells in broth, the broth clears because of the destruction of the sensitive cells by the virus. Many bacteria normally contain unsuspected virus of some kind whose presence is revealed by contact with a sensitive strain.

Phage lambda and temperate phages

Phage lambda (λ) of *E. coli* has been extensively studied, and a great deal of information has been assembled on events following its infection of a bacterial cell. After entry, the viral DNA may undergo a lytic cycle in the cell or it may behave as a temperate phage. Which pathway it follows depends on a balance of several regulatory proteins, a topic discussed in some detail in Chapter 17.

The viral DNA is replicated at first as a circle during a lytic pathway. Intermediate θ-shaped DNA molecules arise, as when the *E. coli* chromosome replicates (see Fig. 11-19). When rapid replication of the viral DNA ensues later in the lytic cycle, replication takes place through the rolling circle method (see Figs. 15-10 and 15-11). The λ DNA that is to be packaged into the empty protein envelopes occurs in the form of a concatamer (Fig. 16-1A). This is a chain or linear series of genomes linked together. Figure 15-10 shows how more than one genome's worth of DNA can arise using the rolling circle method. An enzyme recognizes certain nucleotide sequences of the λ DNA known as *cos sites* that define the ends of each genome in the concatamer. Consequently one genome's worth of λ DNA is snipped off from the concatamer for packaging into the head of the phage. Snipping of the enzyme may result in slightly more than one genome's worth of DNA being inserted into a protein envelope in phages such as T4 (Fig. 16-1B). Concatamers occur in the replication and maturation of certain other phages (T1, T7), and they may arise in ways other than the rolling circle method of replication.

The DNA that is packaged into a λ envelope is linear, and it enters the *E. coli* cell in this form. However, it soon assumes the form of a circle. This circularization can occur because the linear DNA bears 12-nucleotide-long, single-stranded projections at its 5′ ends (Fig. 16-2). These are the cos sites mentioned above. We say the ends are sticky or cohesive, because they are complementary and base pairing can take place between them. DNA ligase activity seals the nicks, and the circular form of the virus is stabilized. The sticky ends are generated as a result of staggered, single-stranded cuts made by the enzyme involved in snipping.

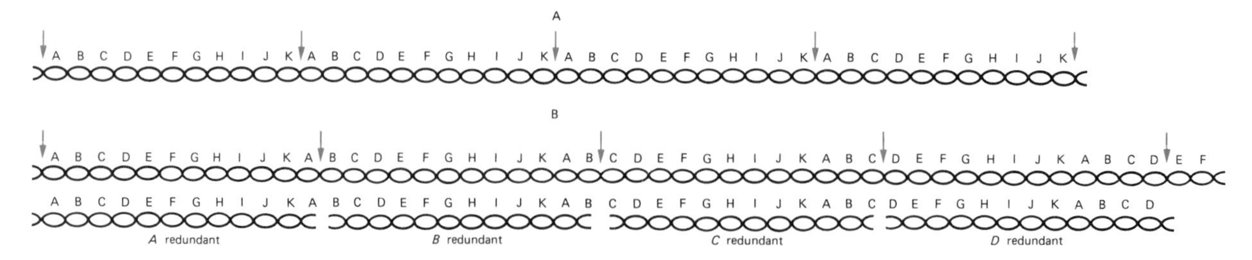

FIGURE 16-1. Concatamers and their processing. **A:** The concatamer consists of a series of genomes (each A–K) linked together. Snipping (arrows) by an enzyme results in the formation of separate genomes. Only one genome's worth of DNA (A–K) will be packaged in phage λ into a protein envelope. **B:** The snipping of the concatamer in phage T4 can start at a random point and produce slightly more than one genome's worth of DNA. There is thus terminal redundancy, a repetition of genes at each end. As a result, certain genes are represented twice in the T4 viruses, and the duplicated genes differ from one mature virus to the next.

In its circular form inside the *E. coli* cell, the λ DNA is replicated if a lytic pathway is followed (Fig. 16-3). If phage λ behaves as a temperate virus after entering the cell, however, the circular form of the DNA does not undergo replication. Instead, it inserts into the chromosome by a recombination process, in much the same way that the F factor of *E. coli* becomes inserted into the Hfr strain, although IS sequences are not involved (see Fig. 15-17). Instead, specific attachment sites play a role. One of these is located on the bacterial chromosome and is designated att B. This site is located next to the bacterial gal locus, a region concerned with the ability of the cell to produce an enzyme needed for using galactose. The corresponding attachment site, at P, occurs on the phage DNA (Fig. 16-4). Each attachment site—att B on the bacterial DNA and att P on the phage DNA—contains a homologous segment, and it is within this homologous portion of the DNA that recombination takes place, resulting in the insertion of the phage into the chromosome as a linear sequence. An enzyme coded by a phage gene can recognize the attachment site. This phage protein, along with a protein factor coded by the host DNA, is able to bring about the recombination event that results in the integration of phage λ. We see that a bacterial cell harboring an integrated virus gives no visible evidence of the virus. In its reduced state, the temperate phage is called *prophage.* The cell contains only one intact prophage of a given kind, such as λ, because the presence of an integrated virus renders the cell immune to further infection by the same kind of virus.

In its integrated state, the temperate virus behaves in every way as if it were part of the bacterial replicon. In an occasional cell, the virus may come out of the chromosome by a reversal of the process of insertion (Fig. 16-4). This excision process depends on the two proteins that are also required for insertion, as well as a third protein that is coded by a phage gene. Excision from the chromosome may come about spontaneously, or it can be stimulated to do so by certain agents, such as ultraviolet light. Most of the time, the temperate, integrated virus is prevented from leaving the chromosome and thus from becoming detrimental to the cell by a repressor substance, coded by the virus itself (see Chap. 17). If this repressor is somehow prevented from acting, as from the effects of ultraviolet radiation, the genes of the virus that are concerned with viral replication will be expressed. The result is replication of the virus, cell lysis, and the entry of mature phage particles into the medium. These may then take up residence in other appropriate host cells, or they may destroy cells that are sensitive to them.

Since the bacterial cell with the temperate phage contains information to make virus that can cause lysis of sensitive cells, it is called *lysogenic.* The establishment of an integrated state between viral and bacterial DNAs is known as *lysogeny.* A lysogenic strain therefore has the potential to make virus and to bring about lysis of cells from an indicator strain. On the level of microorganisms, we find many different strains of cells that vary in their viral susceptibility as well as different strains of viruses that vary in their virulence. A spontaneous mutation may arise that changes a sensitive bacterial cell to one resistant to a specific virus. In turn, mutation in a phage gene can make the virus still more virulent to the bacterial strain. A mutation in a phage gene may even change the virus to a less virulent or even temperate one. The establishment of a symbiotic relationship between a bacterium and its temperate phage is the product of mutation operating along with the force of natural selection.

The closeness of the phage–bacterium relationship becomes apparent when we realize that a given virus such as λ, in its prophage state, resides not at just any chromosome location, but at a very specific site, unlike the F factor, which can insert at any of several sites. As noted in Figures 15-8B and 16-4, the attachment site for phage λ is near the bacterial marker Gal. Phage λ may or may not be present in a cell at this site, depending on the bacterial strain. An Hfr lysogenic strain harboring λ at the attachment site near Gal is designated *E. coli* (λ^+). An Hfr cell that is λ^- can transmit the absence of λ to a λ^+ cell, just as if the λ state (presence or absence of λ phage) were inherited as a specific bacterial gene (Fig. 16-5A). If an Hfr λ^+ cell is mated to an $F^-\lambda^-$, the chromosome of the former may penetrate into the F^- cell and transfer phage. However, the prophage is immediately activated when transmitted in this way to an $F^-\lambda^-$ cell (Fig. 16-5B). The virus removes itself from the chromosome and causes destruction of the F^- cell. Thus some phages can be induced to become virulent as a result of their transfer during conjugation to cells lacking them. The activation of the virus to a virulent state by transfer during conjugation is called *zygotic induction.* Not all phages, however, may be induced. A host cell may contain one or more different kinds of viruses in their prophage state, each inserted at a characteristic location on the chromosome and each behaving as if it were a bacterial locus.

Phages and transduction

When the virus in its prophage state removes itself from the chromosome, it may carry with it an adjacent piece of the host chromosome. Phage λ on rare occasions (1 in 10^6 cells) picks up a Gal^+ marker from a cell (follow Fig. 16-6 in this discussion). As it does this, it leaves behind a portion of itself that remains inserted in the host chromosome. The virus, now containing a bacterial gene, is released into the

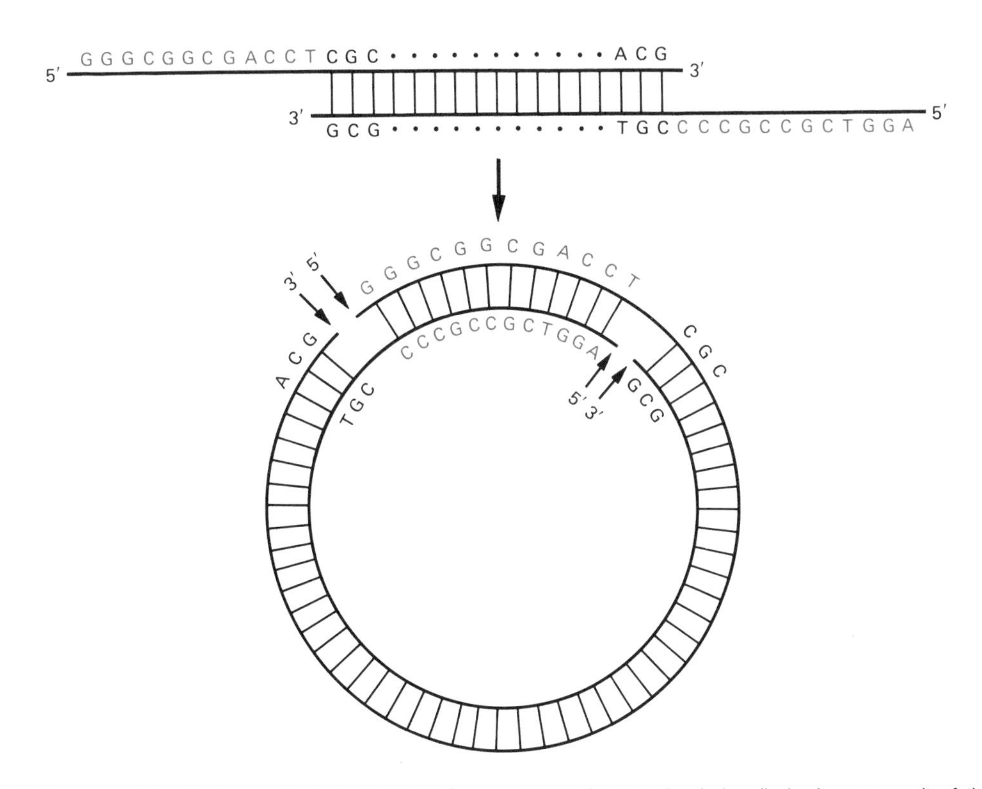

FIGURE 16-2. Linear and circular forms of phage λ. Phage λ enters the *E. coli* cell as a linear molecule with single-stranded extensions (red) at the 5′ ends (above). The linear molecule can circularize (below) as a result of the complementarity of the ends. DNA ligase seals at the positions of nicks (arrows) to form an intact circle.

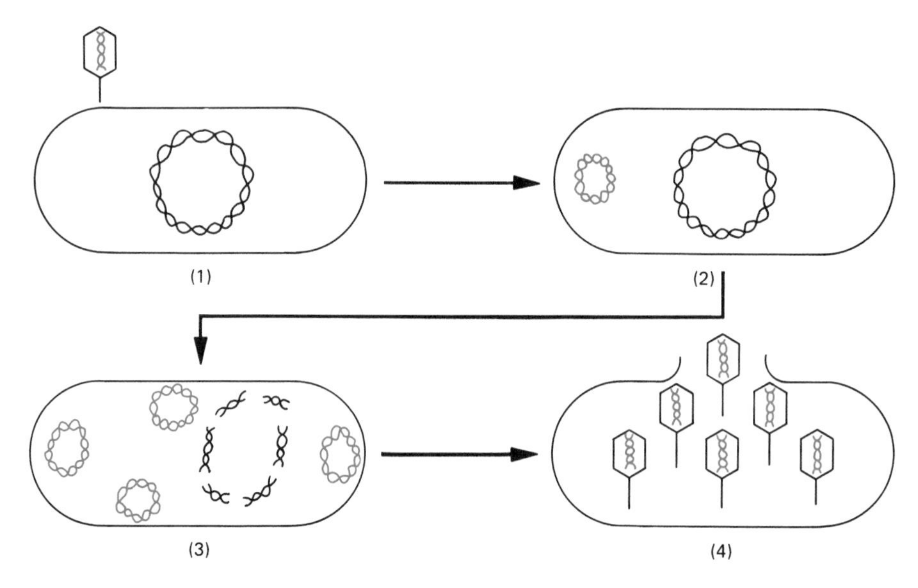

FIGURE 16-3. Lytic cycle of phage λ. **1:** Phage λ occurs in a linear form in the infective virus. **2:** The injected DNA circularizes after entry into the cell. **3:** The virus replicates as circles, while the host DNA is degraded. **4:** The infective phages released from a lysed cell contain the DNA in its linear form.

medium, where it is free to enter an appropriate cell, containing no λ prophage. The virus can insert itself as prophage in the second cell. As a result, it confers a new property on the host. If the latter, for example, is Gal^- on the chromosome, it acquires the ability to use galactose because of the presence of the Gal^+ allele associated with the prophage that it is now harboring. This transfer of genetic material from one cell to another by way of a virus is called *transduction.* The phenomenon of transduction was discovered in 1952 by Zinder and Lederberg while working with the bacterium, *Salmonella typhimurium,* and the phage, phage P_1.

It is now well known that variations occur in transduction, depending on the host and the phage under consideration. Phage λ, which has been extensively studied in *E. coli,* illustrates specialized transduction. The behavior of this virus brings to light several interactions that can take place between the host cell and the virus. When transduction takes place through a virus such as λ, which has existed as a prophage in a previous host cell, the second host becomes diploid for a very small segment of the genetic material, the Gal region in our example. The cell is a partial diploid, heterozygous for just a small portion of the genome. We again use the term *heterogenote* for such a condition.

Since the virus left a part of itself inserted in the chromosome of the first host as it picked up the Gal marker, we might expect the virus to show some alteration in its behavior. And indeed, the virus is now changed. This is seen in its inability to become infective again; it cannot cause cell lysis or be released into the cellular environment. The defective prophage can be liberated again, however, if the cell also contains an intact λ. A transduced bacterial cell (one that has received a marker by means of a virus) may contain both the defective transducing phage with the marker allele and an intact phage containing all of its own genetic information (Fig. 16-7). The latter is able to supply the information missing in the former and can bring about the lysis of an occasional cell. Such transducing cells are immune to further infection by λ particles. Bacterial cells from a strain that has been transduced may thus undergo lysis and contribute to the culture medium both kinds of phages, defective ones carrying a bacterial gene and intact ones with all the needed virus information to effect lysis. Both types of particles, in turn, would be able to enter other cells of the same host strain and again take up residence.

It should be apparent that transduction by a virus clearly resembles sexduction—transfer of a bacterial gene by a sex factor. Indeed, the two processes are so similar that they may be considered aspects of the same phenomenon. The F factor, as well as the phage, may also leave a piece of itself behind when it picks up a bacterial gene. We may consider the phage to be an episome, because it satisfies the definition in the same way as the F factor.

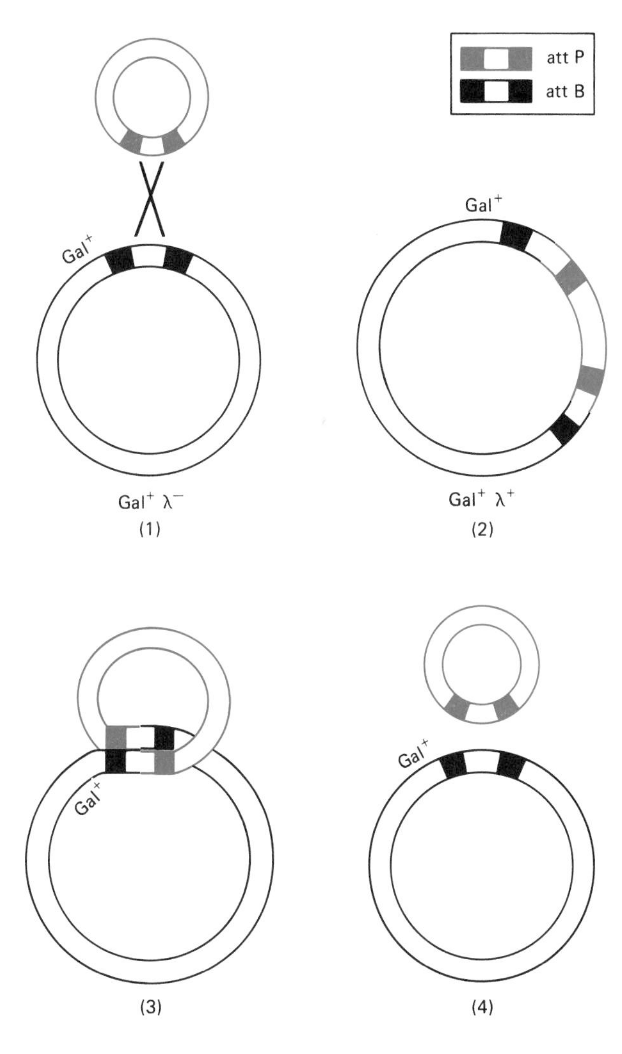

FIGURE 16-4. Integration and excision of phage λ. Phage λ DNA (red) and the host cell DNA (black) contain corresponding attachment sites, att P and att B, respectively. Both sites contain a core region that is a DNA segment homologous to both sites. It is within this region that recombination takes place (1). The attachment sites are close to the Gal locus of the cell. Integration converts a λ^- cell to one that is λ^+ (2). The integrated virus (a prophage) on rare occasions undergoes a reversal of the process of integration (3). Pairing between the two attachment sites, followed by recombination, produces a chromosome free of the virus plus an intact phage (4). The latter can now replicate and enter the lytic pathway, ending with destruction of the host cell.

Another kind of transduction is generalized transduction, which can be effected by phage P_1 in *E. coli* and by phage P_{22} in *Salmonella.* As the name implies, any region of the host chromosome may be involved and thus be carried over to another bacterial cell by the phage particle. This is quite different from specialized transduction in which a specific phage particle such as phage λ is able to carry away to another cell only a specific bacterial marker gene, such as Gal.

Another difference relates to the manner in which the bacterial DNA segment is incorporated into a phage particle. Recall that in specialized transduction, the phage carries away a portion of the chromosome during the process by which it emerges from the chromosome. In the case of generalized transduction, the fragment of host chromosome becomes incorporated into the headpiece of the virus during the lytic cycle of the phage. Remember that when a virulent virus is completing its lytic cycle, the host DNA is broken down and enters a DNA pool. A piece of bacterial DNA that has not been completely degraded to its constituent nucleotides may be packaged accidentally into the headpiece. It appears that P_1 particles capable of generalized transduction carry very little, if any, phage DNA. As a matter of fact, experimental evidence indicates that at times only bacterial DNA may become packaged into the headpiece of certain kinds of phage particles.

When such a phage particle encounters another bacterial cell of the proper strain, the DNA fragment of the first bacterial host is injected. This fragment would contain a region homologous to a region on the second host's chromosome. If a crossover event takes place between the two regions, the second host may become a recombinant cell if the donor DNA carries an allelic form different from that present on the homologous region of the recipient.

Unlike specialized transduction with phage λ, the P_1 transducing particles do not contain defective viral DNA integrated with bacterial genes. These particles do not undergo multiplication in the presence of any helper phages, and the transduced cells, the transducants, are not immune to infection by other P_1 particles, as in specialized transduction.

It has been demonstrated that the very presence of a phage in a bacterium may impart a new characteristic to the cell, one that is not due to a gene introduced from another bacterium. For example, certain nontoxic strains of *Corynebacterium diphtheriae* gain the

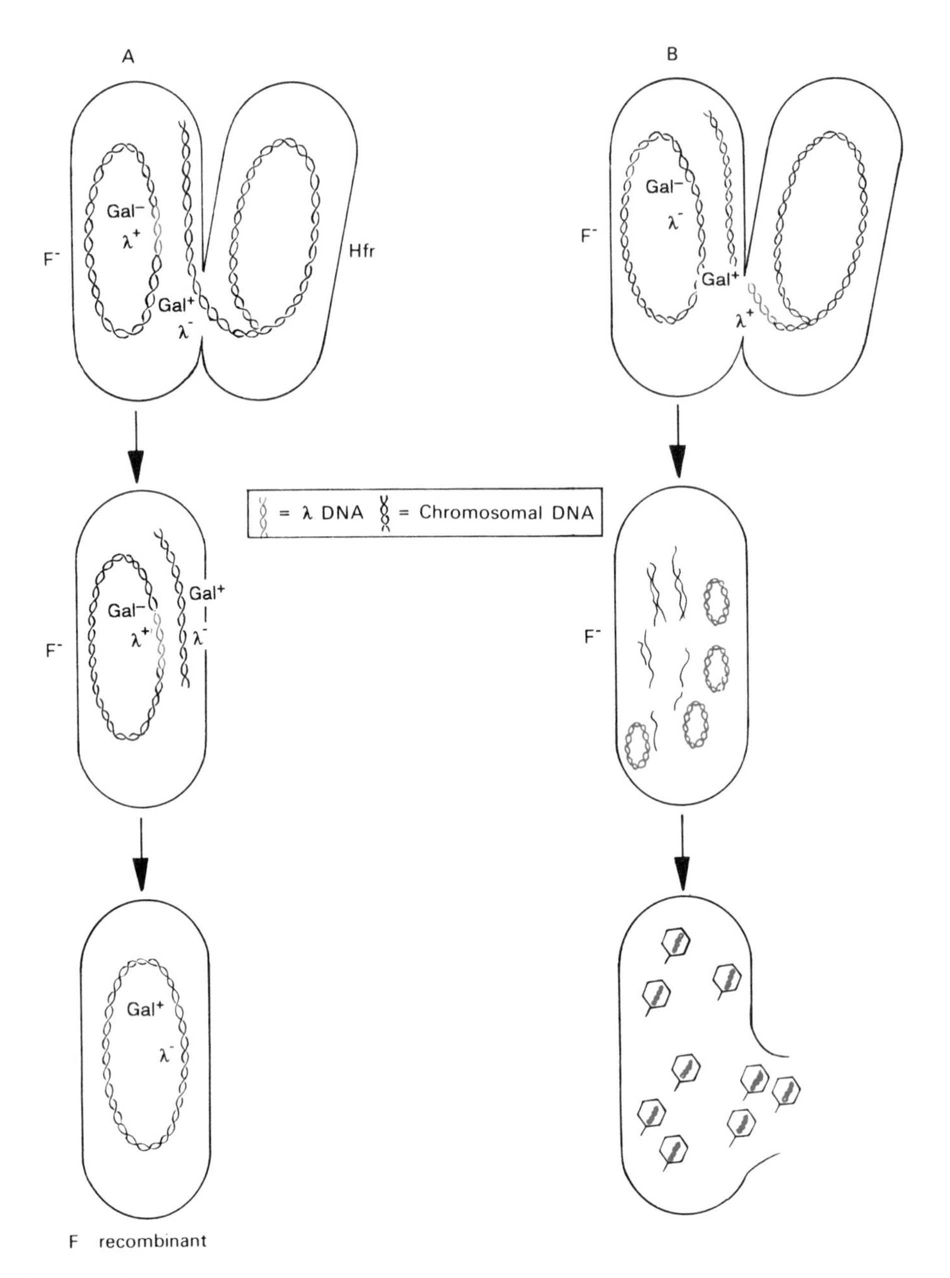

FIGURE 16-5. Transfer of λ prophage. **A:** A prophage occupies a definite location on the bacterial chromosome and may behave like an ordinary bacterial gene. The phage λ occupies a position adjacent to the marker, Gal. In conjugation, the absence of λ (λ^-) state may be transferred to the F^- cell, just as any other marker. Recombinants can arise that are λ^-. **B:** The prophage cannot be transferred from the Hfr in the λ^+ condition without causing lysis of the F^- cell. This is because of zygotic induction, in which the phage, on entering the F^- cell, becomes activated to manufacture more λ virus. Lysis of the cell results with release of mature λ phage.

ability to produce the virulent diphtheria toxin when they are infected with particular phages. The production of the toxin by the cell does not result from the presence of any bacterial genes that have been carried by the virus; rather, it depends on the genetic machinery of both the cell and the phage. If the phage is lost from the cell, so is the ability to form the toxin. This is an example of *conversion*, the acquisition of an inherited cell property as the result of the interaction of host and virus genes.

The genetic system of phages

Since mutations occur in viral genes just as they do in higher forms, the study of the inheritance of a mutant allele and its standard form permits genetic analysis of the virus and the mapping of its genes. We have noted that viruses grow only in certain host strains. Phages thus have characteristic host ranges. A mutation, however, may widen or narrow that range. For example, the *T-even* phages can infect and lyse *E. coli.* A variant form of these viruses can infect and lyse strain B of *E. coli* but cannot cause lysis of strain K. Host range is thus a genetic characteristic that can be recognized in viruses. Another common one is the type of plaque produced by a virus when it infects a lawn of bacteria spread on an agar dish (Fig. 16-8). Each plaque, a clear zone on the bacterial layer, represents an area where a single infecting virulent phage originally entered a bacterial cell and was reproduced, followed by cell lysis and the release of new phage particles. These in turn infected other bacterial cells in the area and continued the process, resulting in the formation of a visible plaque or hole in the bacterial lawn. The wild-type plaque is quite distinct for a given type of virus. It may be small with smooth edges, whereas a mutant form may be large. Some plaques are cloudy; others are clear. Another trait that is easily followed is the rapidity with which the virus bursts the host cells.

Studies conducted in the late 1940s brought out the features typical of phage crosses. Suppose host range and speed of lysis are being followed. The wild allele (h^+) enables the virus to grow on *E. coli* strain B but not on *E. coli* B/2. The mutant allele (h) increases the host range to include the B/2 strain. The wild allele (r^+) determines small plaques, whereas its allele (r) results in large plaques as a consequence of very rapid lysis. To determine whether recombination can take place between the loci h and r, we may proceed as follows (Fig. 16-9A). The two bacterial strains B and B/2 are mixed and then spread on agar plates. We know that if a virus is h^+r, it can grow only on strain B and will produce large plaques. We can easily identify plaques produced on a plate containing both B and B/2 bacteria. The virus will lyse only the B cells, but the large plaques will be turbid or cloudy because they

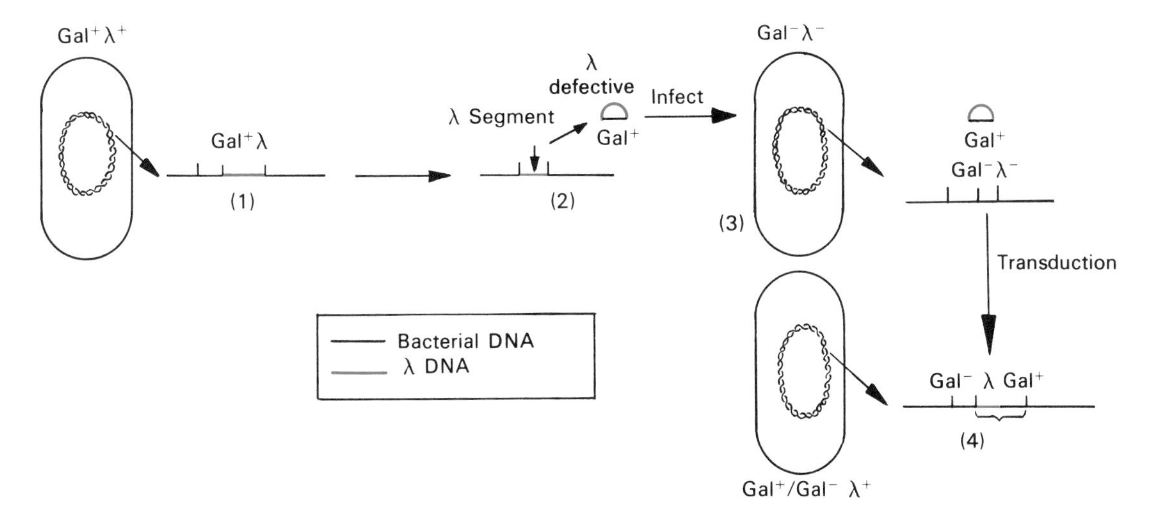

FIGURE 16-6. Prophage and transduction. Phage λ resides adjacent to the Gal locus (1) in a λ^+ cell. When it comes out of the prophage state and deintegrates from the chromosome, it may carry the Gal marker with it (2), but, in so doing, it leaves a portion of itself behind. The phage is now defective, having lost some of its abilities. If the defective virus, carrying a Gal^+ marker, infects a Gal^- cell that does not contain λ (3), it may insert itself in the Gal region. The cell is transduced (4) and contains the defective λ plus the Gal locus in the diploid condition. The cell is a heterogenote. The defective nature of λ is seen in the fact that, by itself, it cannot remove itself from the chromosome and reproduce. This function was lost when the phage left a segment of itself behind (2) and incorporated the Gal locus.

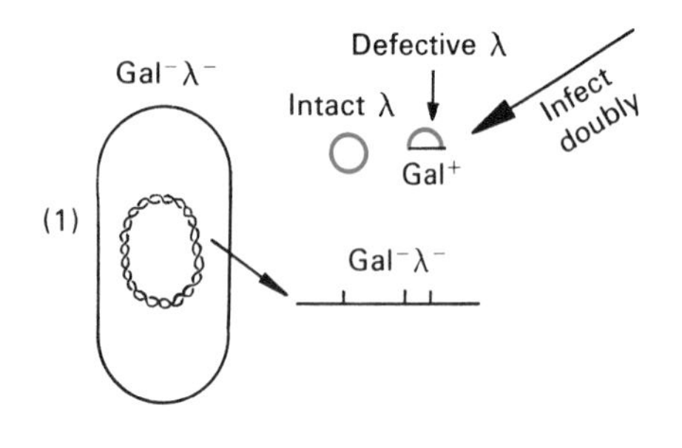

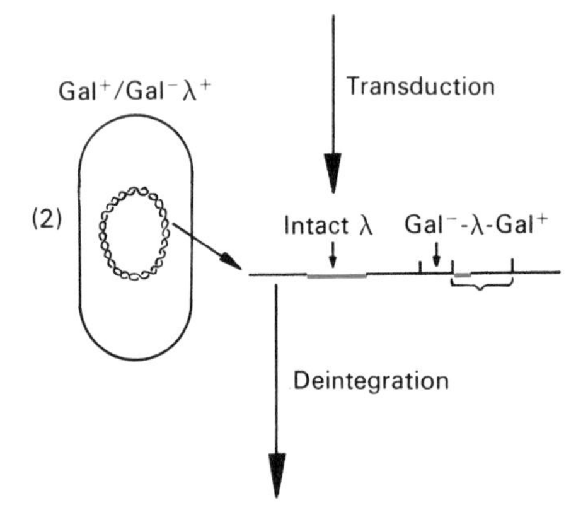

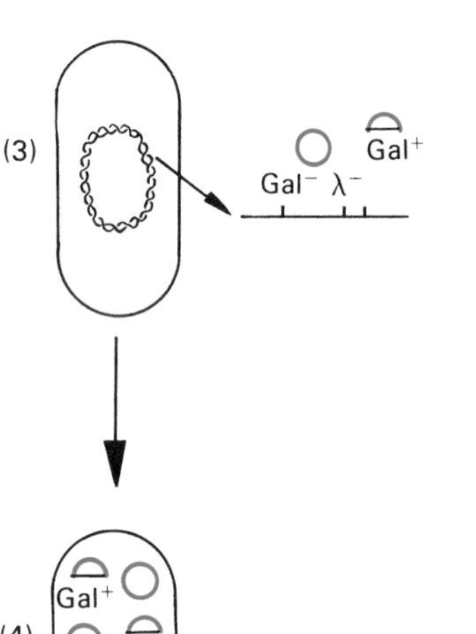

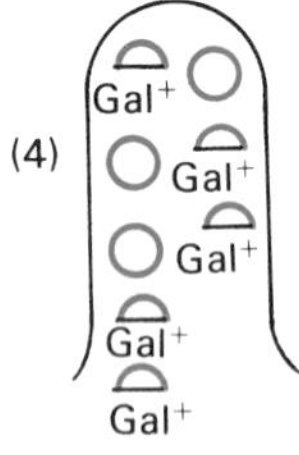

FIGURE 16-7. Double infection. A bacterial cell may become infected simultaneously with an intact virus and a defective one carrying a marker (1). Both viruses take up residence on the chromosome, and the cell is transduced and becomes a heterogenote (2). In a rare cell, the two viruses may deintegrate (3). This is possible only because the normal intact virus is present to supply the function missing in the defective phage. Both types of phages are able to reproduce, and the cell lyses (4), releasing the two types of particles.

will contain unlysed cells of strain B/2. On the other hand, a virus that is hr^+ will produce small clear plaques because r^+ determines slower (wild) lysis and the h allele enables the phage to lyse both B and B/2 cells. The resulting plaques will contain no intact cells and therefore will be clear.

A mixture of the two parental virus types (h^+r and hr^+) is allowed to infect a liquid culture of strain B host cells (Fig. 16-9B). Both parental viruses can infect and lyse the B cells. This B culture is then diluted by liquid broth and allowed to undergo lysis by the phages. The viruses that are released after lysis of the B cells are then plated on a mixture of fresh host cells, this time both B and B/2. By plating on this mixture, we can detect virus offspring with normal host range (turbid plaques) and mutant (clear plaques). The plaques that form are then examined for size and clearness. The two types caused by the two parental phages are seen (large, turbid; and small, clear types). But we also find some that are large and clear and others that are small and cloudy. These latter two types must result from the presence of viruses with the respective genotypes hr and h^+r^+. This tells us that recombination has occurred. We can express the percentage of recombination by using the same reasoning as was used to determine crossover frequency in higher organisms. The proportion of the number of recombinant offspring, hr and h^+r^+ (as measured by features of the plaques), to the number of total progeny (the total of all the plaques) is obtained. The percentage value tells the recombination frequency, just as it would if we were studying eye color and body color in fruit flies.

The classic genetic principles used in making three-point testcrosses (see Chap. 8) may also be applied to phages. Suppose that three loci are being mapped: m, r, and tu (Table 16-1). Again, plaque characteristics are examined. From the data, it can be seen that the most numerous type of plaque and the least numerous represent the parental types and the double crossovers, respectively. The locus r is established as the one in the middle, because it appears to have switched position. Establishment of regions 1 and 2 allows us to determine the corresponding crossovers in the two regions. Again we must add the frequency of doubles to each class (3.3 here). And thus the map becomes

m 12.9 r 20.8 tu

By applying detailed linkage analysis to phage T/2 of *E. coli,* it was found that all of the genes fall

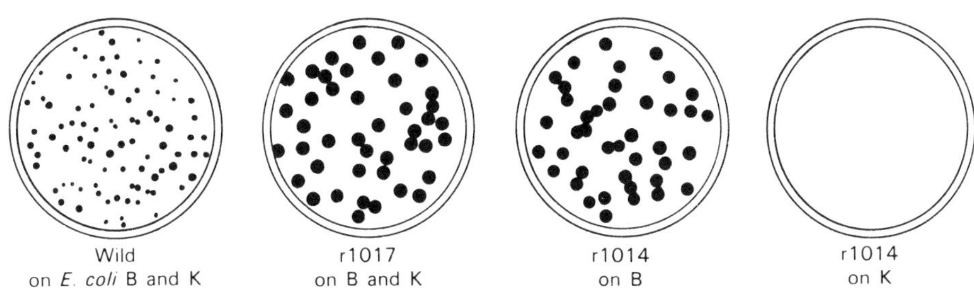

FIGURE 16-8. Plaque formation. Viruses may differ in the type of plaque they form and also in the range of hosts in which they can reproduce. Phage T4 produces the small, wild-type plaques on *E. coli* strains B and K. Certain mutants (r mutants) produce large plaques on B and K. Still others, the rII mutants, produce the large mutant plaques on B but no plaques at all on K.

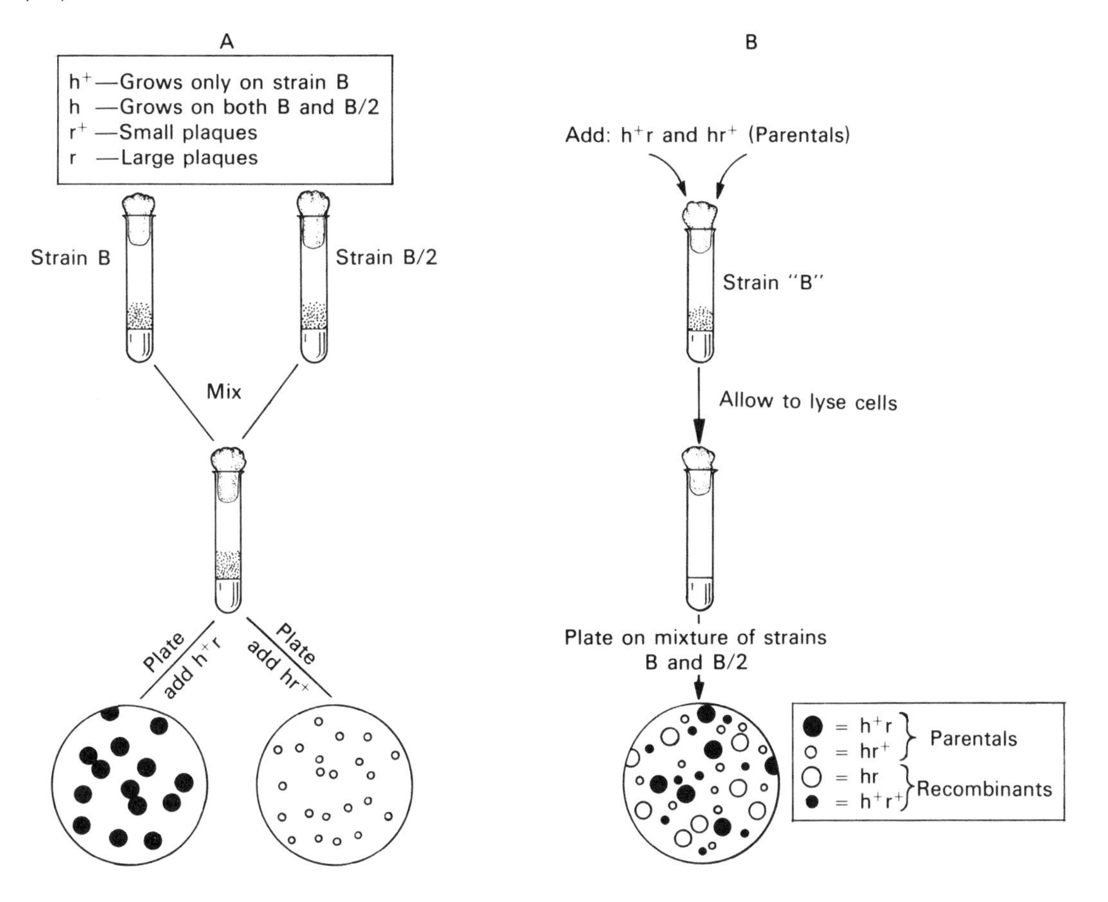

FIGURE 16-9. Recombination in phages. **A:** Mixture of strains B and B/2 can be used to identify the kinds of phages present. The h^+r type produces large, turbid plaques, because it produced rapid lysis on strain B but cannot grow in B/2. Therefore, the plaques are cloudy because of the unlysed B/2 cells. The hr^+ produce wild-type (small) plaques, but they are clear, because both B and B/2 cells are lysed. **B:** One can determine the amount of recombination in this procedure by looking at the type of plaques. Again a mixture of B and B/2 is used as an indicator, and the four types of plaques are easily identified. The amount of recombination is determined by counting the total number of plaques and expressing the number of recombinants as a percentage.

into one linkage group and that the map is circular. The circularity of the map, however, does not mean that the DNA of the virus "chromosome" is necessarily in the form of a circle as in *E. coli.*

Actually, the T/2 and T/4 chromosomes are rod shaped in the mature virus particle, even though the map of the chromosome is circular. The reason for this is that the DNA that is to be packaged, as in the case of phage λ, occurs in the form of a concatamer. As a result of random snipping of the concatamer, however, terminal redundancy occurs at each end of the chromosome (see Fig. 16-1A,B). The part of the chromosome that is repeated varies from one phage particle to the next. Since mapping the chromosome entails a large population of phage particles, it will always seem as if any gene in the sequence is next to the following one. No ends will be detected as a consequence of the terminal redundancy.

Repeated mating in phages

Although we may build up genetic maps employing the familiar logic applied to higher organisms, close inspection reveals several unusual features peculiar to phage systems. When the crosses are performed in the manner described earlier, a tremendous number of bacterial cells and phage particles are obviously involved. Techniques are available, however, that enable us to make critical dilutions. From these, we may study the offspring phages coming from the burst of a single bacterial cell that had been infected by two types of phage particles, let us say h^+r and hr^+. When such dilutions are made, it is found that the recombinant classes hr and h^+r^+ are not present in equal amounts. On the basis of our understanding of crossing over, we would expect the recombinant classes to be reciprocal and hence equal in number. In the case of phages, these are usually found to be very unequal. Moreover, if a cell is infected with three different kinds of virus particles, such as mr^+t^+, m^+rt^+, and m^+r^+t, we might find offspring that are mrt. For such a combination to arise, three parents must be involved! How can such an observation be explained?

Some clarification is provided by bursting the infected cells artificially several minutes before they would lyse naturally. When this is done, it can be demonstrated that recombinant viruses exist in the cell many minutes before the natural cell burst. As time increases toward natural burst, the number of recombinants also increases. In other words, the acts involved in recombination start well in advance of bursting of the host cell. This means that, before lysis, more than one round of mating and recombination can take place. According to the theory of repeated mating, the phages enter the cell and then multiply. The phage units in the cell now mate at random repeatedly in pairs, so that several rounds of mating and multiplication take place. Exchange of genetic material by crossing over takes place and produces recombinants. If reciprocal recombinants are formed early in the mating process, they may be changed by participating in later mating. Therefore the phage offspring coming from a lysed cell represent several generations of multiplication. We can now understand how three parents may be involved in the production of a phage genotype (Fig. 16-10). A mating of m^+rt^+ in one round of multiplication can give $m^+r^+t^+$ and mrt^+. If the latter mates with m^+r^+t in the next round of mating, the genotype mrt can arise.

When we study the products of many cell bursts after cell infection by two different phage types (h^+r and hr^+), we are unaware that multiple rounds of mating have taken place. Equal numbers of reciprocal recombinants are found among the collection of offspring from the bursting of the entire population of cells as a result of randomness. Since an extremely large number of phage units is involved, the recombinant types will be produced in comparable amounts.

The recognition of the cistron

One of the greatest contributions of phage analysis is the revised concept it has given us on the nature of the gene. Previous to the 1950s, classic ideas held the gene to be an indivisible unit. It performed a certain

TABLE 16-1. Progeny from three-factor linked cross in phage T/4[a]

Category	Class	Genotype	Total plaques	Percentage
Parental	A	mrtu	3,467	33.5
Parental	A	$m^+r^+tu^+$	3,729	36.1
Single crossover	B	mr^+tu^+	520	5.0
Single crossover	B	m^+rtu	474	4.6
Single crossover	C	$mrtu^+$	853	8.2
Single crossover	C	m^+r^+tu	965	9.3
Double crossover	D	mr^+tu	162	1.6
Double crossover	D	m^+rtu^+	172	1.7

[a]A, noncrossovers; B, crossover between m and r; C, crossover between r and tu; D, double crossover; m, minute; r, rapid lysis; tu, turbid.
Source: Reprinted with permission from A.H. Doermann, *Cold Spring Harbor Symp. Quant. Biol.* 18:4–11, 1953.

function. As a unit, the gene could mutate and engage in crossing over. When the gene mutated, all of it was somehow changed. Mutation at different sites within the gene was not considered possible. Nor was crossing over within a gene recognized. Because it was an intact unit, breakage and crossing over could occur on either side of it but not within it. Contradictory cases arose in the fruit fly as well as in other well-studied organisms, but these apparent exceptions were explained by coining new terms and adapting them to fit the classic, accepted idea of the gene.

The work of Benzer, undertaken in the 1950s, revolutionized the model of the gene as the indivisible unit of function, mutation, and recombination. Benzer worked with the T-even viruses that infect *E. coli.* He was one of the first investigators to try to relate genetic units, such as units of crossing over and units of mutation, to the actual physical basis of DNA. This procedure requires the construction of very detailed chromosome maps. In turn, constructing such maps relies on the examination of extremely large numbers of offspring. We have seen that microorganisms are ideal for such purposes, especially phages, in which millions of progeny can be examined in a short time.

Benzer studied in detail the genetics of plaque formation in phage T4 of *E. coli.* The wild-type plaque, as seen on an agar plate, is a small, clear area with rough edges. These wild-type plaques are produced when phage T4 is plated on either strain B or strain K of *E. coli.* Certain mutants arising in phage T4 are called *rII* mutants. When plated on strain B, they form a so-called r plaque, one that is larger than the wild type (r^+) and has smooth edges. These rII mutants, however, produce no plaques on strain K. (Fig. 16-8 summarizes these points.) The rII mutants infect K cells and start the synthesis of phage DNA and other viral products, but they cannot produce infective offspring. As a result of this defect, they cannot cause cells of strain K to lyse.

This growth defect of the rII viruses on strain K provided Benzer with a very critical tool. One of his goals was to take different rII mutants that had arisen independently and cross them in various combinations. The point was to determine whether the numer-

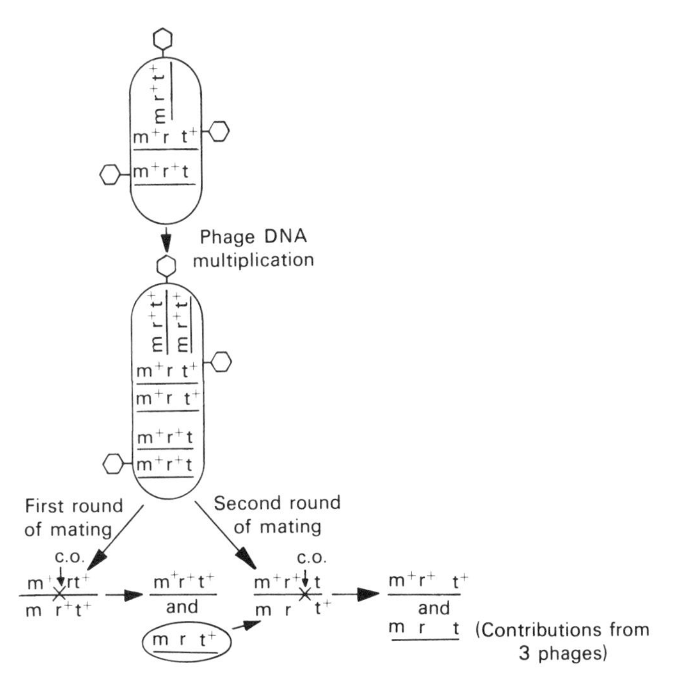

FIGURE 16-10. Repeated mating. When phage DNA enters a cell, it multiplies. The resulting phage units may undergo several rounds of random mating during which recombination can take place. If a cell is infected with phages of three different genotypes, as shown here, offspring may arise that carry contributions from three separate phage units (in this case m, r, and t were originally in three separate particles). This is easily understood when it is realized that more than one mating takes place in the infected cell before mature viruses are released. In this example, one of the products of the first round of mating (mrt^+) participates in a second mating. The genotype mrt can thus be found in a mature phage released upon lysis of the cell.

ous independent mutants represented separately occurring mutations at the same or at different sites on the genetic material. For example (Fig. 16-11A), assume that two r mutations, r_1 and r_2, are actually alterations at two different positions within the DNA. Being rII mutants, the phages produce a mutant plaque on strain B and none on strain K. However, a strain B cell, infected with the two types, could give rise by recombination to wild, $r_1^+r_2^+$. This is possible, because the r_1 mutant is carrying a normal r_2 region and would have the genotype $r_1r_2^+$. Similar reasoning tells us that the r_2 mutant would be $r_1^+r_2$. Recombinant wild-type phages, $r_1^+r_2^+$, could arise by crossing over. The resulting wild phage would be able to produce wild plaques on both strains B and K. But if the two mutant sites, r_1 and r_2, are sites within the same gene and are therefore extremely closely linked, the wild recombinants would be exceedingly rare. This means that the investigator would be forced to search through perhaps millions of mutant plaques on strain B to find the one wild plaque. This is a feat that would be truly impossible. Fortunately, the growth defect of the rII virus on strain K can avoid the need to do this. It can act as a screen to uncover the rare recombinant among the millions of others.

In the procedure, rII mutants are therefore recognized by the r plaques they produce on strain B and their inability to produce any plaques on K. Strain B is then infected with two rII mutants that arose independently (Fig. 16-11B). The infected *E. coli* B are allowed to lyse. Samples of the resulting phages are then plated on agar dishes containing strain K. Only the wild recombinants will form plaques. Their numbers will be few, because the recombinants are rare. But this number of wild plaques on K can now be compared with the number of r plaques produced by samples plated on strain B. From these counts, the frequency of recombination is easily determined; the logic is again the same as for the estimation of crossover frequency in higher organisms.

Benzer's method was so sensitive that he could have detected one recombinant among 10^8. The wild plaque that this recombinant can form would be lost among the r plaques if only strain B were available. The rare recombinant would never be found if only strain B were used; therefore, no maps could be constructed. With his method, Benzer found that the rII mutations cluster in a small portion of the T4 map. Within that small region, the positions of different rII mutations can be established. Actually, 300 separate sites have been recognized within the small map segment (Fig. 16-12). Well over 2,000 separate mutations were studied, and some of these were found to be mutations at the very same site. As a matter of fact, Benzer recognized "hot spots," sites where large numbers of the mutations map in contrast with other locations where only one or a few mutations have arisen. The smallest recombination frequency was approximately 0.02%, meaning that two different rII mutant sites may be only 0.02 map units apart. Benzer translated this into terms of nucleotides along a stretch of DNA. Making certain assumptions, the 0.02% value indicated that a distance of not more than six nucleotide pairs would separate two mutant sites along a strand of DNA. From the wealth of genetic analyses of microorganisms that followed Benzer's studies, we now know that two mutant sites may be adjacent to each other, representing two nucleotide pairs side by side.

Does the fact that two r mutant sites may be so close together mean there are different mutant sites within the gene and they can undergo recombination by crossing over within the gene? Consequently, is the unit of function different from the unit of mutation and recombination? We have already discussed these points in relation to the gene and molecular interactions (Chap. 13), but it was Benzer's work that reformed thought on the nature of the units of heredity. Using strain K of *E. coli*, Benzer demonstrated that two genes of function are present in the rII region of phage T4. He discovered

FIGURE 16-11. Crossing of rII mutants. **A:** Recognition of phage types. rII mutants are recognized by the fact that they produce a mutant plaque (large) on *E. coli* strain B and no plaques on strain K. If strain B is infected with two rII mutants that have alterations at different sites within the gene, a very rare recombinant may be formed, $r_1^+r_2^+$. This can produce wild (small) plaques on either B or K. On strain B, both the rII mutants and the wild type can grow. The rare wild plaque would be lost among the mutant ones. Strain K, however, acts as a screen, because only the wild type can grow on it. **B:** Recombination frequency in the rII region. The different rII mutants are recognized as shown in A. Two types of rII mutants are allowed to infect sensitive strain B. When B cells lyse, they will release phages that are primarily mutant, because recombinants within the short rII region under study are rare. However, by plating on strain K and observing for wild plaques, the number of these can be established. These can then be compared with the number of mutant plaques produced on strain B. A comparison of the number of mutant plaques on B with the number of wild ones on K allows one to calculate the frequency of crossing over between the two mutant sites. (The figure does not attempt to reflect actual numbers of plaques.)

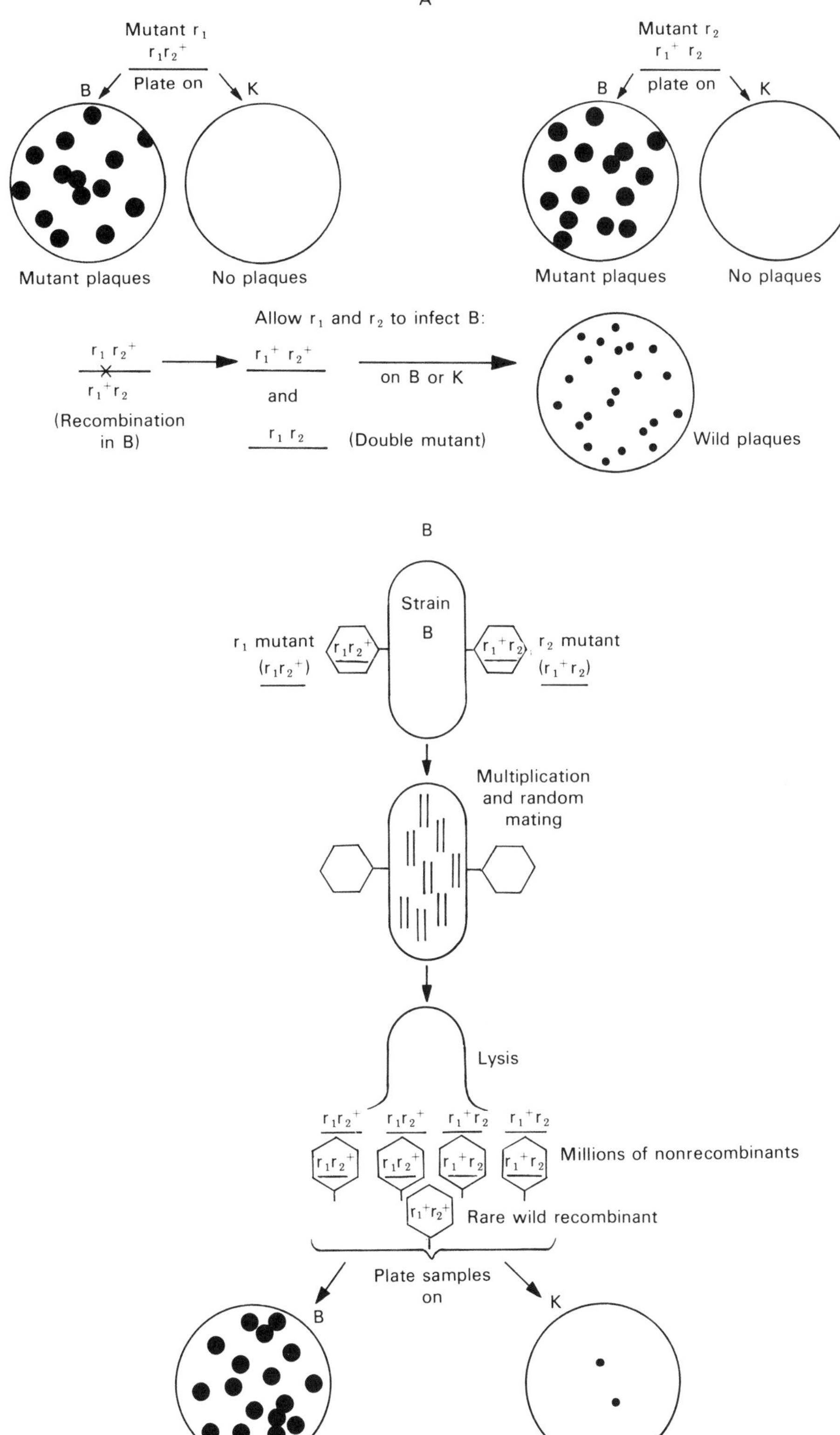
A
Mutant r_1
$r_1 r_2^+$
Plate on
B
K
Mutant plaques
No plaques
Mutant r_2
$r_1^+ r_2$
plate on
B
K
Mutant plaques
No plaques
Allow r_1 and r_2 to infect B:
$r_1 r_2^+$
$r_1^+ r_2$
(Recombination in B)
$r_1^+ r_2^+$
and
$r_1 r_2$
(Double mutant)
on B or K
Wild plaques
B
Strain B
r_1 mutant ($r_1 r_2^+$)
r_2 mutant ($r_1^+ r_2$)
Multiplication and random mating
Lysis
Millions of nonrecombinants
Rare wild recombinant
Plate samples on
B
K
Mutant plaques
Rare wild plaques

that a mixed infection of strain K with a wild-type phage plus any rII mutant results in the lysis of the cell and the reappearance of both wild-type and mutant viruses (Fig. 16-13). This means that the presence of the wild phage makes possible the reproduction of the mutant type. In a sense, it is as if the wild (+) form is dominant over a recessive r allele.

Benzer then infected strain K with different pairs of rII mutants (follow Fig. 16-13). From the mapping he had done, he knew how far apart any two rII mutants would be. He now wanted to see if two rII mutants could interact in some way to cause lysis of strain K with the production of offspring phages; this is an effect that no rII mutant can bring about on its own. Benzer found that certain pairs of rII mutants could interact to effect lysis on strain K, whereas others could not. The small rII region was shown to be divisible into two subregions, A and B. A combination of any two mutants in the same region—both in A or both in B—cannot cause lysis. However, a combination of two mutant types—one with a mutation anywhere in A and the other with one anywhere in B—can cause lysis.

This observation told Benzer that all of those sites within subregion A had more in common with one another than with any site in B. The same relationship would hold for the B sites. The common feature that distinguishes all of the sites within one of the subregions is evidently a specific function to which they are all related. In this case, the function is one that is necessary for cell lysis and the production of phage. Thus in the rII region we have two subregions, A and B, which are units of function. For lysis to occur, each unit must be carrying out its specific function. When a wild phage is crossed with any rII mutant, the wild particle contributes two normal functional units, A and B, and so lysis can occur. Similarly, a cross of any A mutant with any B produces lysis, because a mutant defective in A contributes an intact B, and vice versa. If a cell of strain K is infected with two B mutants, however, there is no normal, functional B unit in the cell (and likewise for any two A mutants). Because one of the two functions is lacking, lysis cannot occur. The work with rII phage clearly demonstrates that the unit of function is larger than the unit of mutation and crossing over. Benzer coined the term

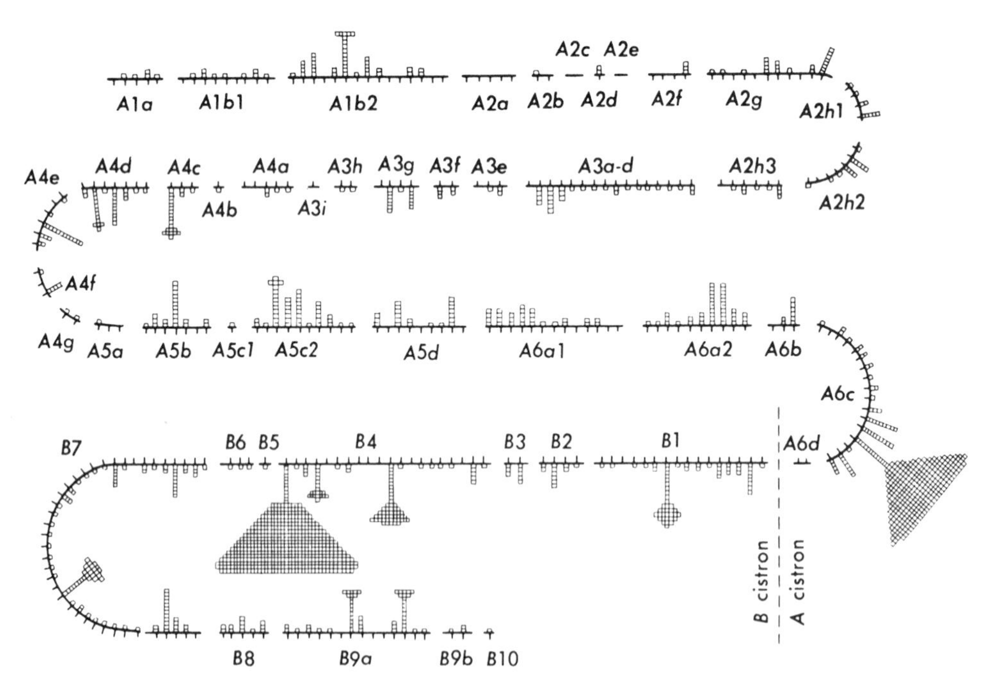

FIGURE 16-12. Map of rII region. A large number of different mutations that originated independently have been assigned to the rII region. The exact order of the mutations in a small segment is not known in most cases. Each square represents an independent mutation whose site has been mapped. It can be seen that more mutations occur at certain sites than at others. (Reprinted with permission from S. Benzer, *Proc. Natl. Acad. Sci. USA* 47:410, 1961.)

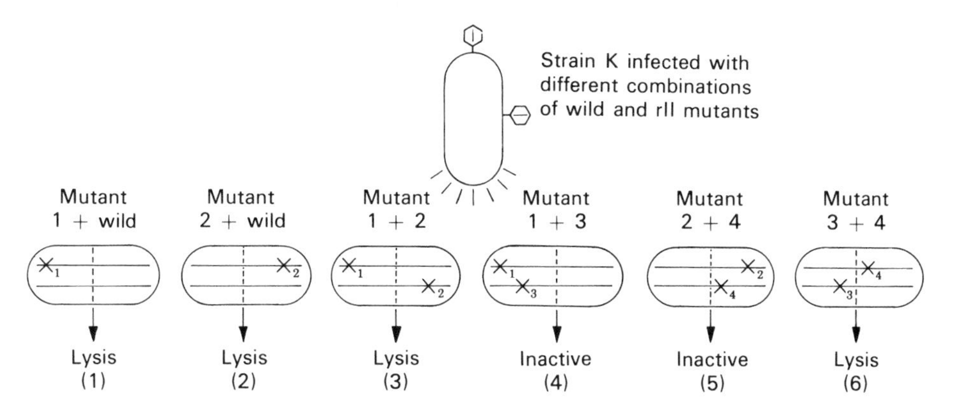

FIGURE 16-13. Demonstration of the functional unit. Strain K can be infected with wild-type phages plus one of the rII mutant types. Any combination of a wild and a mutant results in lysis of strain K (1 and 2), indicating that the wild-type phage is able to carry out the function that is deficient in the mutant (inability to grow on strain K by itself). Certain combinations of two different rII mutants on strain K result in lysis (3 and 6), whereas other combinations do not (4 and 5). It is found that the rII region can be divided into two segments (indicated by the dotted line). Any combination of mutants whose mutant sites are within the same region fails to cause lysis. This indicates that two such mutants carry defects in the same functional region, even though the defects are not at exactly the same site. When the defects are in separate regions, the one mutant contains a normal, functional unit that the other lacks (3 and 6). Lysis occurs because the two functional units needed for growth in K are present in the cell.

cistron to designate the unit of function. And the concept of the cistron, as we have seen, is applicable to higher species as well as to microorganisms.

In Chapter 13, the cistron, or gene of function, was discussed in relation to its control of a polypeptide chain. In light of the knowledge assembled by molecular biologists since Benzer's original work, the concept of the gene as a unit of function (a stretch of DNA) composed of smaller units (nucleotide pairs that can mutate independently and engage in crossing over) agrees perfectly with the known physical basis of the gene itself. No confusion need exist about the terms *cistron* and *gene.* They are equivalent if we picture the "familiar gene" as a stretch of DNA controlling a specific function (formation of a specific polypeptide chain) and composed of smaller sites that may mutate independently and engage in crossing over (the nucleotide pairs). The expressions *muton* and *recon* were also coined by Benzer to refer to these sites of mutation and crossing over within a cistron. However, they are now less frequently used, because they are known to represent the same thing in physical terms, one nucleotide pair.

A cistron, or gene of function, can be recognized in any group of organisms by the performance of a cis–trans test (see Chap. 13). When any two mutants are crossed, the mutant sites in the F_1 offspring are found in the trans arrangement. In our discussion of allelism in higher forms, we saw that a cross of two recessive mutants produces mutant offspring if the two parents are carrying mutations within the same gene. Benzer was actually performing cis–trans tests in his work with strain K and the rII mutants. For a cross of any rII mutants gives a trans arrangement (see Fig. 16-13). If the resulting phenotype is wild (here lysis and plaque formation), the mutations complement each other and hence belong to separate genes or functional units. If a mutant phenotype (no lysis) follows from a mixed infection by two rII mutants, then the mutations do not complement each other but are in the same gene or cistron.

Phages with single-stranded DNA

One virus that has been a valuable tool in molecular genetics is ϕX 174. After its discovery in 1958, chemical analysis of ϕX 174 revealed an unexpected feature. Although its DNA contains the familiar building blocks of phosphate, sugar, and bases, the proportions of adenine to thymine and of guanine to cytosine are not 1:1. At first, this seemed to contradict the rule of preferential base pairing. The ϕX 174 DNA was subjected to a variety of tests, including ultracentrifugation and exposure to enzymes. The data showed clearly that the DNA of this virus is single stranded rather than in the form of a double helix and that it exists in the form of a closed circle.

The ϕX 174 virus is now known to be just one of several DNA viruses that can infect bacterial, human, and animal cells. The manner in which ϕX 174 replicates has been worked out in detail. It was found that the single-stranded form of the phage is infective, capable of invading its *E. coli* host. Once inside the cell, however, the single-stranded DNA (the + strand) can direct the formation of a complementary strand (the – strand), producing a double-stranded form of the virus, the so-called replicative form. The latter is thus composed of the infective strand (+) and its complementary strand (–) made inside the cell (Fig. 16-14A). The formation of this first double-stranded replicative form (designated *RF1*) depends entirely on enzymes present in the host cell. A short primer RNA is synthesized. DNA polymerase III adds nucleotides to its 3′ end and continues to lengthen the strand until the template + strand is followed completely. The RNA primer is removed, the gap is filled in, and ligase seals the nick to produce a circular – strand and consequently a replicative form (RF1).

The – strand that is now present acts as the template strand for the few genes of ϕX 174. Actually, this virus was among the first forms in which it was demonstrated that only one strand of a gene undergoes transcription (see Chap. 12). Transcription of a specific gene from the – strand of RF1 results in the formation of the protein Rep A. This protein brings about a nick in the + strand, always at the replication origin (Fig. 16-14B). The protein remains attached to the 5′ end of the nicked strand. The viral form with the intact – and nicked + strands is referred to as *RF2*. Rolling circle replication then ensues. Nucleotides are added to the free 3′ end of the + strand as it is displaced. After one replication cycle is completed, the Rep A protein cuts the + strand off from the tail and then circularizes it. Depending completely on host cell enzymes, it can now also act as a template for the synthesis of a new – strand to form another RF1, which in turn can be converted to RF2 and undergo the cycle just described.

After it cuts and circularizes the + strand, the Rep A protein then links to the 5′ end of the new + strand made in the first round of replication. Another cycle of replication can then ensue, with the Rep A protein again cleaving and sealing the second new + strand. The cycle can keep continuing in this fashion as the intact – strand continues to act as a template for + strands that, as noted, may serve to generate more RF1 forms. However, some + strands start to be packaged as protein phage envelopes begin to appear in the cell. More and more of the + strands are packaged as phage heads arise, until no more RF1 forms are produced. New single-stranded ϕX 174 phages, each with a + strand, are ready for release from the cell and eventual infection of other *E. coli.*

Animal viruses with single-stranded DNA replicate in ways that are different from those described for φX 174. Some depend on rapidly dividing cells. Others will grow only in cells also infected with more complex viruses.

Overlapping genes

Later investigations with the phage ϕX 174 have yielded some unexpected results that demand a new look at the nature of the gene and its reading. The single DNA strand of the phage has been subjected to intense investigations, many involving the cleavage of the DNA with enzymes and then separation of the resulting DNA fragments. The length of the ϕX 174 DNA has been found to be 5,375 nucleotides, and their exact sequence has been completely determined. A dilemma, however, presented itself regarding the length of the DNA of the phage and the number of proteins for which the DNA is apparently coded. The virus can bring about the production of nine different proteins whose molecular weights are known. The amount of DNA required to code for these is far in excess of the amount of DNA present in the phage.

The answer to the problem was quite unanticipated. Research teams in the laboratory of F. Sanger in Cambridge, England, studied in detail several genetic regions of ϕX 174, one of which contains two genes, D and E, which code for proteins D and E, respectively. The D protein, whose exact amino acid sequence was known, is produced in great amounts in the host cell and is required for the production of single-stranded viral DNA, although its exact role in the process is unclear. The E protein is required for the lysis of the host cell when hundreds of new viruses have been produced and are to be released. The two genes, D and E, on all counts seem like any other separate and distinct genes. Nonsense mutations are known to occur in each one, without apparently affecting the function of the other.

The exciting find was made that a certain stretch of the phage DNA contains the coded information for both the D and the E proteins and that the D and E genes overlap! The start of the E gene is in the middle of the D gene. Both genes end close to the same spot. However, the E gene is not just a segment of the D gene or just the end part of the D gene. The two proteins, D and E, are very different in amino acid sequence. The DNA, however, in the D and E genetic region is read in different frames. Recall frame shift mutations (see Chap. 12), which were recognized in Crick's pioneer studies on the triplet code. Reading of a gene begins at a fixed starting point, three nucleotides at a time. A – or a + mutation may cause the entire reading of a gene to shift. In the case of the D and E genes, the starting sequence for the E gene was found to be within the D gene. The D gene and the E gene within it, however, are read in different frames (Fig. 16-15).

Another interesting finding is that the termination codon of the D gene overlaps by one nucleotide the start codon of the next gene, gene J. Moreover, overlapping of reading frames is not confined to just this one region of ϕX 174 DNA. Gene A, which carries coded information for another protein needed for viral construction, contains gene B, coded for a protein that produces a nick in the DNA when new copies of the DNA are needed. Even more unexpected was the discovery that the coding sequence for gene K, which is involved in cell lysis, begins in gene A, following gene B, and ends in gene C. It thus overlaps both genes A and C. Moreover, the boundaries between A and C actually overlap slightly.

The discovery of overlapping genes is quite an exception to the general model of the gene. The genetic code was found to be nonoverlapping, each

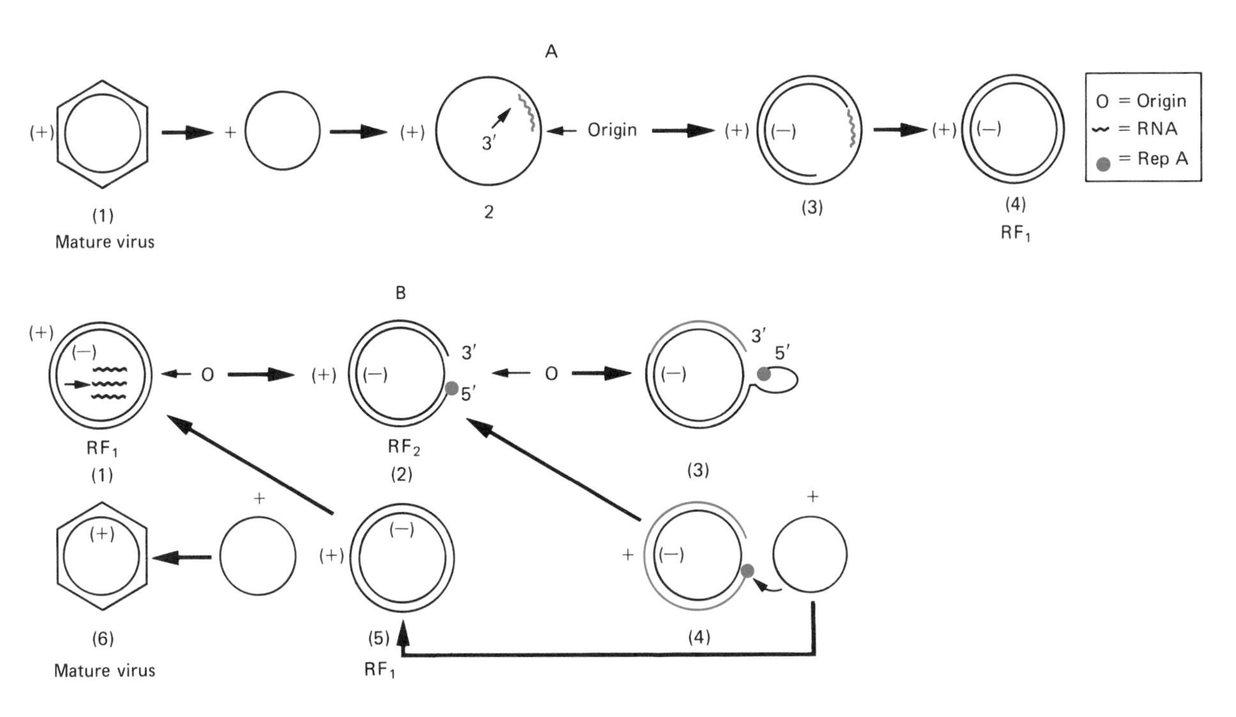

FIGURE 16-14. Stages in replication of ϕX 174 DNA. **A:** An infective + strand of a ϕX 174 virus enters a cell (1). Using enzymes of the host, an RNA primer (red) is assembled to a 3′ end of the origin (2), and a – DNA strand complementary to the + strand is started (3) and then finally completed (4). The primer is removed, the gap left is filled, and the new strand is circularized. A replicative form, RF1, is now completed. **B:** The – strand of the replicative form acts as the template strand for all the genes of ϕX 174. Transcripts of the genes (1) are coded for assembly of a protein, Rep A, which can nick the (+) strand at its origin and bind to it at the 5′ end (2). This form is referred to as *RF2,* and from it, rolling circle replication may ensue (3). Nucleotides in a new round of replication (red lines) are added to the 3′ end of the nicked + strand as the intact – strand acts as a template. The + strand is displaced from the circle. The + strand then is cleaved from the new DNA to which it is linked by the Rep A protein, which also circularizes it (4). The Rep A protein then attaches to the 5′ end of the newly assembled + strand. The + strand may act as a template for assembly of a new – strand (5), as depicted in A. This new RF1 can then be converted to RF2, and the cycle is repeated. Eventually + strands will be captured by phage envelopes and packaged (6) to form mature virus. The – strand with the newly assembled + strand, shown in step 4, can serve as an RF2, which then starts another cycle of replication, forming another + strand. The cycle can keep repeating.

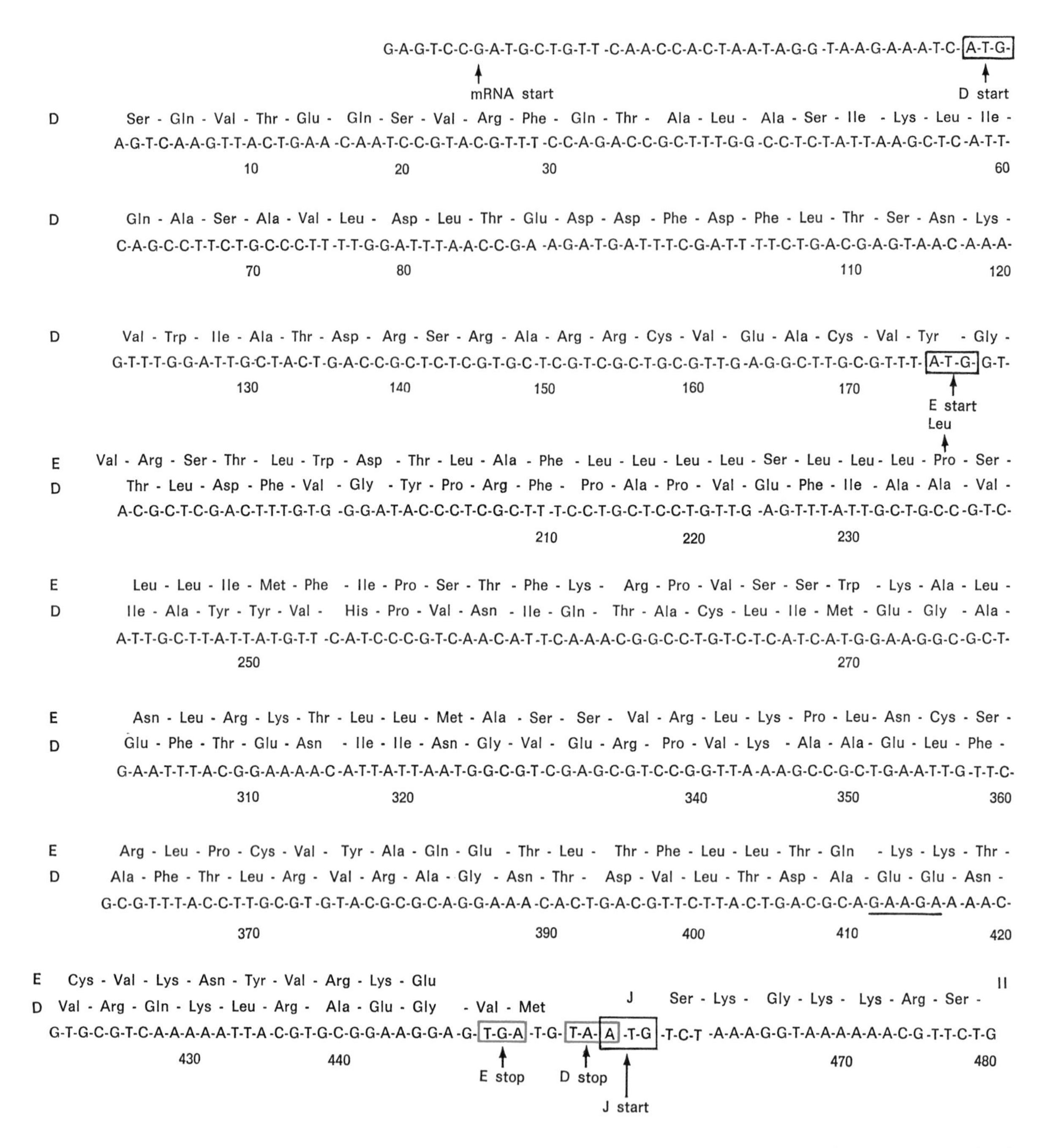

FIGURE 16-15. Overlapping genes in ϕX 174. The starting sequence and the end sequence of the E gene are within the D gene. In the overlapping genes, codons are read in different frames. The amino acid sequences of the D and E proteins, as well as that of the J protein (product of the gene following gene D) are aligned with the DNA sequences. Note the overlapping between the terminating codon of gene D and the start codon of gene J. The DNA triplets shown here have the same sense as the RNA triplets of the messenger RNA. Therefore, the mRNA sequence can be immediately deduced by changing all the Ts to Us.

single nucleotide being part of only one code word or codon. However, in certain viruses a single nucleotide may be part of two code words, depending on the reading frame. Only in viruses has complete overlapping of genes been found. The finding of the overlapping genes helps solve the problem of fitting the nine genes of ϕX 174 into a supposedly less than adequate length of DNA. The overlapping may have evolved in viruses as a device to accommodate a larger number of genes than would otherwise be possible in so small an amount of DNA. The discovery of the phenomenon focuses attention on such matters as genetic controls and gene expression and also raises the possibility that overlapping genes may yet be found in other organisms. While complete gene overlapping has not been found in bacteria, slight overlapping is known to occur between the beginning of one coding region and the end of another (see tryptophan operon, Chap. 17).

Cellular defenses against foreign DNA

When the foreign DNA of a phage enters a bacterial cell, a cellular defense mechanism comes into operation to degrade the DNA of the virus, thus protecting the cell against destruction. This *restriction activity* is accomplished by a certain class of enzymes, the *restriction endonucleases.* Restriction endonucleases fall into two main classes. Let us consider first those designated type II enzymes, which are the more common type. These are the enzymes at the very core of genetic manipulation that have made possible the incredible advances in molecular genetics (see Chap. 18).

Recall that an endonuclease is capable of breaking an internal phosphodiester bond. Unlike other endonucleases that are nonspecific as to where they cleave a DNA strand, these restriction endonucleases produce internal double-stranded nicks only within certain nucleotide sequences—target sites they are able to recognize. Protective enzymes such as these have been isolated from various species and strains of bacteria. The enzymes in this group are characteristically dimers, each composed of two identical subunits. When these endonucleases recognize a target site on the foreign DNA, they produce a double-stranded, highly specific cut at the site of recognition. The DNA sequences recognized by most of these enzymes contain a point or axis around which a symmetrical arrangement of base pairs occurs (Fig. 16-16A). Note that in such a stretch of DNA the reading in the 5′ to 3′ direction on either of the two paired strands produces the same base sequence. Such a symmetrical arrangement is known as a *palindrome.* A palindrome is equivalent to inverted repeats that are side by side. Some restriction endonucleases produce breaks in the two DNA strands that are both at the axis of symmetry. This results in the formation of DNA segments with blunt ends (Fig. 16-16B). Other restriction enzymes produce breaks in the complementary DNA strands at points actually several nucleotides apart in each strand. As a result of this staggered cleavage, DNA regions are generated that have single-stranded projections.

When the foreign DNA of a virus enters a bacterial cell, it may be broken up into fragments as a result of cleavage by restriction enzymes at specific sites they recognize on the viral DNA. The fragments may then be degraded by exonucleases of the cell. Because any one of these restriction enzymes cleaves only at a specific target site, the number of fragments resulting from attack by that enzyme depends on the number of specific target sites found on the viral DNA that the enzyme can recognize. This means that such an enzyme can cleave a given kind of DNA into a specific number of fragments of characteristic size. As we will see in Chapter 18, this specific property of these restriction enzymes has provided the molecular biologist with a tool to probe into the very nucleotide sequence of a given gene.

Since cells of a bacterial strain possess enzymes that can cleave foreign DNA, thus restricting the growth of certain phages within them, it is necessary for the cells to protect their own DNA from the cleaving potential of their own restriction enzymes. If a given site occurs on the bacterial DNA as well as on the viral DNA, a particular restriction endonuclease of the cell would be able to attack targets on both kinds of DNA.

The cell, however, has a device to prevent this, because it possesses methylase activity. This gives the cell the ability to modify a target site that can be recognized by an endonuclease. In the case of the type II restriction enzymes, a corresponding *modification enzyme* is found that can recognize the same nucleotide sequence as a given restriction endonuclease. It can modify the particular DNA sequence recognized by both and thus protect it from cleavage. A modification enzyme protects the cell's DNA by methylating certain bases at the site of recognition (Fig. 16-16C). The modification methylase transfers methyl groups to particular bases after these bases have been incorporated into a DNA strand. Following modification by the methylase, the site in the cell's restriction endonuclease is now immune to cleavage.

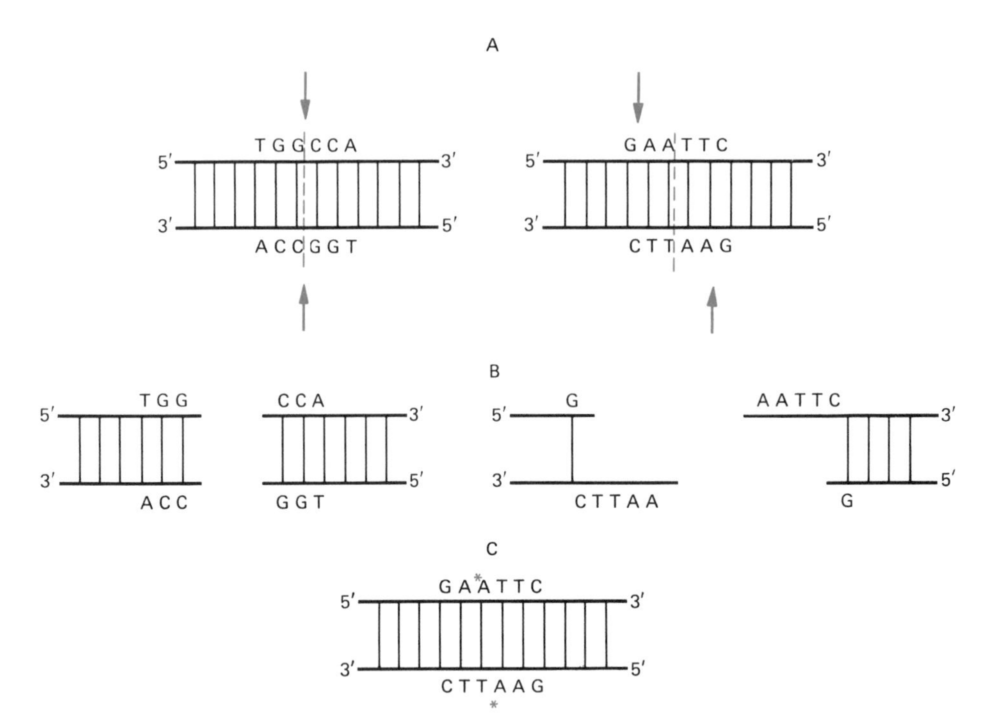

FIGURE 16-16. Restriction enzyme sites. **A:** A restriction endonuclease can recognize a specific nucleotide sequence that occurs in a symmetrical arrangement or palindrome. Two such sequences are shown. (The axis of symmetry is indicated by the broken red line.) A restriction enzyme produces very specific cuts at sites in each DNA strand (arrows). **B:** Cleavage in both strands at the axis of symmetry (left) yields DNA segments with blunt ends. Staggered cleavage (right) results in DNA segments that bear complementary single-stranded projections. **C:** Modification methylases protect the DNA of the cell from restriction enzymes by methylating certain bases (asterisks) in a recognition sequence.

The entering phage DNA, however, is not necessarily modified and therefore is vulnerable to the restriction enzyme; it thus has a reduced chance of establishing itself in the cell. Usually, entering phages become restricted in the presence of restriction endonucleases. At times, however, sequences at recognition sites in the viral DNA do manage to become methylated by the cell's modification enzymes before the cell's restriction enzymes can cleave the viral DNA. In such a case, the viral DNA becomes resistant to the cell's restriction endonucleases and can therefore multiply in the cell and effect lysis. We see here an excellent example of the interplay between a host cell and its virus.

The second class of restriction endonucleases is composed of enzymes, each of which possesses *both* endonuclease and methylase activities. Two subclasses are recognized. In one of these (type I enzymes), each enzyme is multimeric, composed of three types of subunits—one that is needed to recognize the target sequence, another type capable of restriction, and a third type that is responsible for methylation. Since such an enzyme is capable of both restriction and modification, what determines whether a target site will be cleaved or methylated? The decision is apparently made in response to the condition of the target sequence. If the nucleotide sequence is completely methylated, the enzyme may bind to the target, but is released without acting further. If only one of the strands of the duplex DNA at the site is methylated, the enzyme will then proceed to bring about complete methylation of the target site. On the other hand, if the site is not methylated at all, cleavage of the DNA takes place following binding.

Unlike the type II enzymes, these type I enzymes, when they cleave the DNA, do so at sites that are more than 1,000 base pairs away from the sites of recognition! Moreover, the site that is cut is not a specific sequence. Certain sites, however, appear to be preferred over others by the enzymes in this group. The second subclass of this group (type III enzymes) consists of two kinds of subunits, one for restriction and one that accomplishes both recognition of a target and methylation. Only a few enzymes in this subclass have been studied in detail.

Plant viruses

The genomes of plant viruses, like those of animals, may be composed of either DNA or RNA. The genetic material of the cauliflower mosaic virus consists of double-stranded DNA. One unusual group of plant viruses is made up of particles that are actually two capsids held together, each one containing a single-stranded circular DNA molecule! In some of these, the two DNA molecules are identical, whereas in others they are different. Plant viruses such as the wound tumor virus possess genomes that are segmented double-stranded RNA molecules. However, most plant viruses that have been studied fall into the + single-stranded RNA category (see next section). These viruses vary in shape, but the best known plant virus of all, the tobacco mosaic virus (TMV), assumes the shape of a rod.

TMV is a single-stranded RNA + virus that can also attack members of the sunflower and nightshade families. It holds historical significance, since it clearly demonstrated that RNA rather than DNA serves as the genetic material. The genome of TMV is an RNA strand 6,395 nucleotides long and contains four genes that are coded for polypeptides. Two of these appear to be involved in the replication of the virus, but the function of the third one is still poorly understood. The fourth, however, codes for the coat protein, which is composed of 158 amino acids. The capsid of the virus is made up of over 2,100 copies of the same protein molecule. These subunits become assembled around the linear RNA genome. We see here an excellent example of the repetition of one kind of protein in the make up of a viral capsid. Being limited in size, a viral genome is restricted in the number of structural genes that may be involved in protein production for the formation of a capsid. This is particularly the case in RNA viruses in which genome size is even more restricted due to the lack of RNA proofreading.

In the classic work with TMV, the protein coat and RNA core were separated. It was shown that the coat alone cannot infect a cell and bring about the production of either new virus protein or RNA. On the other hand, the RNA by itself can do this. Experiments were performed using different TMV strains. For example, in addition to the standard strain (S), a more virulent strain (H) also occurs. After separating the protein coats from the RNA of viruses of each strain, hybrid virus particles can be reconstituted, those with H-RNA and S-coat and also those with S-RNA and H-coat protein. When these hybrid particles are allowed to infect plant cells, the progeny viruses that arise are always typical of the strain that provided the RNA. As seen in Figure 16-17, H-RNA with S-coat governs the formation of typical H viruses, which consist of both H-RNA and H-protein. Long before the foundations of molecular genetics were firmly established, pioneer experiments of this kind clearly demonstrated that RNA is not only the genetic material of the virus but that it also governs the formation of virus-specific protein as well as RNA, much as DNA governs the formation of more DNA and specific protein.

TMV was also the first virus to illustrate that the nucleic acid and protein components of a mature virus particle can assemble spontaneously to produce accurate offspring particles. Enzymes are often unnecessary for the aggregation and assembly of the different viral parts that compose a virion. Work with TMV demonstrated that only weak secondary bonds are required in viral assembly. If TMV is completely broken down to its RNA core and individual coat protein subunits, complete virions will still reassemble when the separate ingredients are mixed!

RNA viruses and the flow of genetic information

Bacterial and animal cells as well as those of plants can serve as hosts for RNA viruses. The events that take place in a cell following infection by an RNA virus differ depending on the virus and the cell type. Those viruses with double-stranded RNA carry within the virion on enzyme, a transcriptase. This is needed following infection of animal cells, since normal cells produce no kind of enzyme that permits the replication of RNA. Once inside the cell, the packaged transcriptase is released along with the RNA genome. The enzyme is then able to bring about the formation of mRNA transcripts using the entering strands as templates. More of the transcriptase must be formed by the end of the viral reproductive cycle so that it can be packaged along with the newly made RNA. In this way, progeny viruses are formed that are equipped to repeat the replication cycle in the next host cell. Double-stranded RNA viruses are known that can cause a form of diarrhea in the human.

Animal viruses with genomes composed of just a single RNA strand fall into two major categories.

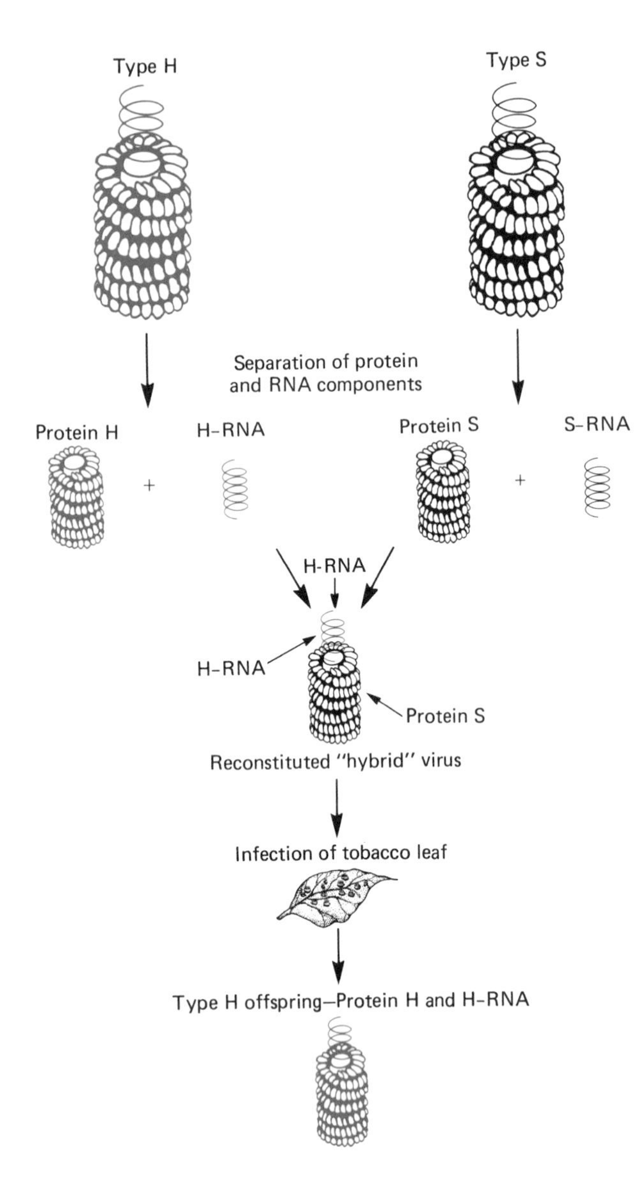

FIGURE 16-17. Reconstitution of tobacco mosaic virus. The RNA of one strain of virus can be combined with the protein coat of another strain. When such hybrids infect the tobacco plant, the viruses that are later isolated are typical of the one from the strain that supplied the nucleic acid. The RNA of the virus must therefore direct the production of more specific RNA and also of the specific protein surrounding it.

Members of one of these groups carry prepackaged transcriptase into the cell as described above for the double-stranded RNA viruses. This is necessary in this viral group since the single RNA strand cannot act as mRNA and thus undergo translation. It is therefore referred to as a – strand. The transcriptase uses this strand as a template to form a complementary strand, which then acts as the mRNA. Among such viruses we find those that are the agents of measles, mumps, and influenza.

In contrast to the – RNA viruses are the + RNA viruses in which the single-stranded RNA of the virion can function as mRNA and thus interact directly with the cellular enzymes of transcription without the need to carry in a transcriptase. However, the + strand is coded for a replicase, an enzyme needed for replication of the virus once it is finally inside the animal cell. Such an enzyme, as in the case of the transcriptase, is not produced in uninfected cells, since normal cells never contain RNA molecules that are capable of serving as templates for forming more RNA. Examples of + single-stranded RNA viruses are those that cause the common cold, polio, German measles, yellow fever, and encephalitis.

The genomes of the different types of RNA viruses are all coded for the protein coverings of the mature virus particle. These coverings eventually assemble around the RNA to form mature virions, which are finally released from the cell in ways that vary among the different viruses. Nevertheless, the process of information transfer in these various types of RNA viruses is essentially similar to the manner in which DNA transfers information (Fig. 16-18A). The flow of genetic information shown in Figure 16-18B was first proposed by Crick. It seemed to be the universal process for transferring coded information from DNA, and so the concept became established as the central dogma of molecular biology. The central dogma was taken by many biologists to mean that the flow of genetic information from DNA to protein could occur only in the one direction, as indicated in the figure. Protein could not impart information to other protein or to RNA. RNA could not impart information to DNA. Any such reverse flow of information, as from RNA to DNA, would be a violation of the central dogma.

Therefore, it was with a great deal of excitement that certain findings were reported from studies with a group of RNA viruses; these appeared to violate the central dogma. The particular assortment of viruses includes some that are capable of inducing tumors in several animal species: rats, mice, and chickens, for example. The Rous sarcoma virus has been one of the most widely studied members. After infecting a chicken cell, the virus can transform it into a cancer cell, which can in turn divide and produce new virus particles. The first clue to the unexpected way in which this virus replicates was found by Temin, who worked with the antibiotic actinomycin D as part of his analytical procedure. This substance had been shown to prevent transcription of DNA into RNA by blocking the DNA as a template. Therefore, it has become a commonly used tool for determining DNA-dependent RNA synthesis. If applied to a cell infected with an RNA virus, RNA would still be produced (Fig. 16-19A). This is so because the RNA template of the virus is not blocked. Therefore RNA-dependent RNA synthesis would still go on, because only DNA-dependent RNA synthesis is shut off by the antibiotic.

When Temin applied the drug to cell cultures infected with the Rous sarcoma virus, he found that all RNA production ceased. This was perplexing and suggested that DNA might somehow be involved in the production of the RNA virus. A variety of experiments was then performed by Temin's group and others, and the data strongly supported the idea that new DNA is produced in a cell infected with the virus. This DNA is actually viral DNA and is formed on an RNA template, the RNA of the infective virus (Fig. 16-19B). If actinomycin D is applied to a cell containing the Rous virus, the viral DNA formed from the viral RNA template cannot undergo transcription into RNA, and so no new viral RNA is made by the cell.

Temin's group succeeded in demonstrating the presence of a DNA polymerase in the Rous sarcoma virion as it occurs outside the cell. This polymerase can form DNA from an RNA template as well as from a DNA template. It has been designated *RNA-directed DNA polymerase.* In popular usage, this polymerase is called *reverse transcriptase.* In the Rous sarcoma virus, this DNA polymerase preferentially uses RNA as a template for the assembly of DNA, whereas most DNA polymerases use a DNA template. It is apparent that a strict interpretation of the central dogma must be modified to include the reverse flow of information from DNA. The findings, as we will now see, also demand a new outlook on the interactions between the genetic material of a virus and the host cell.

Retrovirus replication

Since the discovery of the Rous sarcoma virus, many other RNA viruses were found to use reverse transcriptase to generate a duplex DNA viral form that integrates into the host cell DNA. The term *retrovirus* designates this unusual class, which includes several viruses of birds and mammals, among them the human AIDS virus. The retrovirus virion is not naked but possesses an envelope in addition to the capsid. The genome of a retrovirus is actually diploid, since inside the capsid we find two identical single-stranded RNA molecules, each 7–10 kb in length. Packaged along with the genome is the enzyme reverse transcriptase as well as a tRNA taken from the host cell in which the virion was formed. Once inside the cell, all of the contents are released from the protein covering to permit further steps in the cycle of replication.

The RNA genome of a typical retrovirus is capped at the 5′ end and carries a poly(A) tail at the 3′ end (see Chap. 13). The genome includes three genes: gag, which is coded for the capsid proteins; pol, which con-

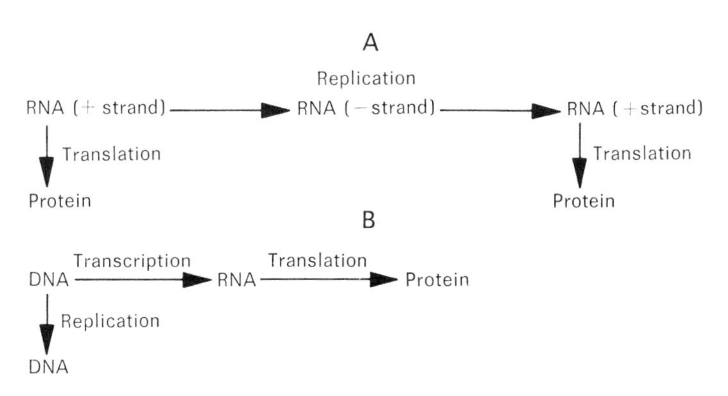

FIGURE 16-18. Flow of genetic information. **A:** In many RNA viruses, the nucleic acid can bring about the formation of more RNA. Some of the RNA can act as mRNA that is translated into protein. **B:** DNA contains information that can guide the formation of more DNA or of RNA. The latter contains information imparted to it by the DNA, and this is translated into protein. The concept of the flow of genetic information in only one direction, from DNA to RNA to protein, is known as the *central dogma.*

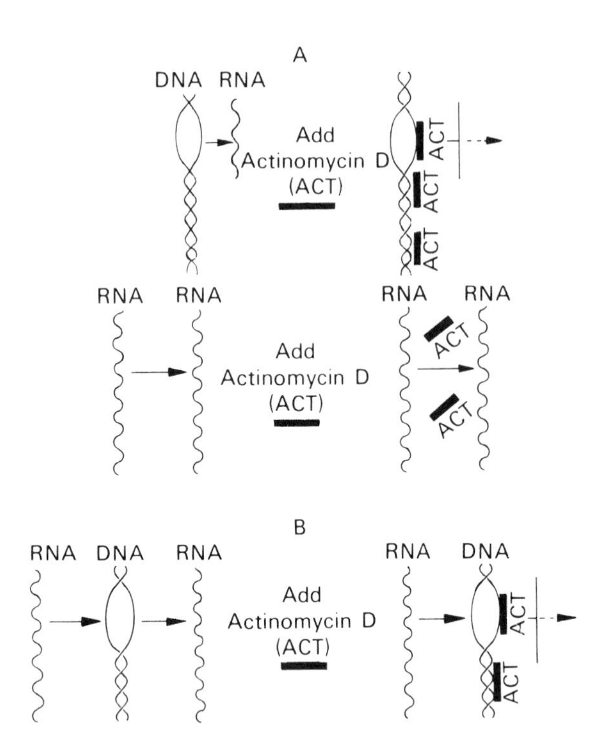

FIGURE 16-19. Effect of actinomycin D. **A:** The drug blocks DNA as a template, so that no RNA transcript can be formed when it is applied to a cell (above). When it is applied to a cell harboring an RNA virus (below), more viral RNA is made, because the drug cannot block an RNA template. **B:** Reproduction of the RNA of the Rous sarcoma virus entails DNA that evidently depends on the RNA template. This DNA in turn gives rise to more viral RNA. If actinomycin D is applied, production of the viral RNA ceases, because the DNA template, which is needed to guide the formation of the RNA of the virus, is blocked.

tains coded information for reverse transcriptase; and env, which is coded for glycoprotein of the envelope (Fig. 16-20). Close to the 5′ and 3′ ends of the genome, a short sequence occurs that is a direct repeat (R). Associated with each one of these is a unique stretch of nucleotides, U5 at the 5′ end and U3 at the 3′ end. In addition, the tRNA is attached to the 5′ end. This interaction results from a complementarity that exists between a small portion of the viral RNA at that end and a sequence at the 3′ end of the tRNA molecule. The latter was lifted from the last host cell to act as a primer for the DNA synthesis that will follow infection. Reverse transcriptase is no exception to the rule that all DNA polyerases require a primer to provide a 3′ end for the extension of the DNA polynucleotide chain.

Although the retroviral genome is diploid, only one double-stranded DNA form of the virus arises for each virion that infects the cell. This DNA stage is produced in the host cell cytoplasm and is known as the *provirus*. The provirus, which arises as a result of the activity of the remarkably versatile enzyme reverse transcriptase, is not an exact copy of the virus genome in DNA form. It is actually a bit longer at each end as a result of the process in which the provirus is formed. Since this is a very complex situation, only a few of its features will be presented here (follow Fig. 16-21).

Reverse transcriptase adds deoxynucleotides to the 3′ end of the tRNA primer to begin production of the – DNA strand. When the latter is about 200 nucleotides long, the reverse transcriptase nicks the RNA template strand, and the portion of the viral RNA that has now been copied into DNA is degraded. Once the nick is made and the degrading begins, the nicked end of the viral RNA acts as a primer for + strand DNA synthesis, during which – strand DNA is used as a template. The reverse transcriptase continues to degrade the viral RNA as it becomes copied to form the completed – DNA strand. The enzyme is able to accomplish the degrading, since part of the transcriptase molecule possesses activity of RNAase H, an enzyme capable of degrading RNA in an RNA–DNA hybrid.

The final result of the degrading and synthesizing ability of reverse transcriptase is the provirus (Fig. 16-22A). In addition to the genes gag, pol, and env, the provirus carries at each end a long terminal repeat (LTR). Each one of these consists of the sequences U3–R–U5 and arises from the R, U3, and U5 sequences found in the RNA genome of the retrovirus. The provirus is thus longer than the RNA genome, since the complex process of provirus formation adds a U3 sequence to the 5′ end and a U5 sequence to the 3′ end. Once formed, the linear proviral DNA enters the nucleus, where it assumes a circular form and then integrates into the host DNA. In the process of integration, the provirus itself loses two base pairs from the U3 at the 5′ end as well as two base pairs from the U5 at the 3′ end. The provirus also brings about a short, direct repeat of four to six base pairs of the host DNA (Fig. 16-22B). This feature is characteristic of the behavior of transposable genetic elements, a major topic of Chapter 17. Although the RNA form of the virus has now disappeared from the cell, all the information required to make more viral RNA is present and is carried along in the provirus with the host's own genetic material. The provirus of the typical retrovirus does not kill the infected cell. It remains integrated and does not remove itself from the chromosome as does prophage in the infected bacterial cell.

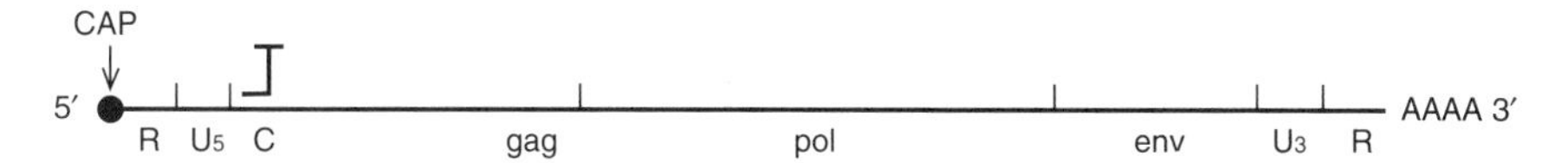

FIGURE 16-20. A typical retrovirus genome. The RNA genome is capped at the 5′ end and carries a poly(A) tail at the 3′ end. A genome sequence (C) toward the 5′ end is complementary to a tRNA sequence and enables pairing. (See text for details.)

Proviral DNA is capable of acting as a template for RNA synthesis. Each LTR carries a promoter in its U3 region. The left one initiates proviral transcription. The right LTR is capable of starting transcription from any host genes that lie adjacent to it. Enhancers (Chaps. 12 and 17) that act to stimulate transcription are also found as well in each U3 region. Signals in both the left and the right LTRs establish the length of the transcript, which arises through the activity of the host RNA polymerase II.

Once RNA transcripts arise, they become capped at the 3′ end and gain a poly(A) tail at the 3′ end. These then leave the nucleus and enter the cytoplasm, where they are utilized in different ways. Some transcripts will become packaged as genome RNA into new virions that bud off from the cell. Other transcripts will serve as mRNA for the gag and pol genes (Fig. 16-22B). Translation of gag mRNA yields a protein that becomes cleaved into four smaller ones that form the capsid. Translation of the pol mRNA produces a protein that is cleaved to give reverse transcriptase as well as an enzyme that plays a role in the integration of the provirus into the host chromosome. A small number of the full-length transcripts become cleaved and spliced to produce mRNA for the gene env, which upon translation results in production of glycoproteins for the virion envelope.

The first retrovirus to be isolated from the human and the very first virus to be linked to a human cancer is HTLV-I, a human T-cell leukemia virus. Although retroviruses were known to cause a variety of cancers and other diseases in animals, none had ever been found in the human until the isolation of HTLV-I. This virus is capable of infecting T lymphocytes (T4 cells) and transforming them to the malignant state. The AIDS virus, originally designated HTLV-III but now named HIV (human immunodeficiency virus), shares a number of properties with HTLV-I such as the ability to infect T4 cells. Despite certain similarities, HIV differs from HTLV-I in many respects. For example, infection by HIV does not transform lymphocytes to a cancerous state, but it does bring about cell death.

HIV differs significantly from the typical retrovirus that has been depicted here. Its genome has proved to be much more complex than first expected. Its gag gene codes for units of a dense protein core that encases the two RNA molecules composing the HIV genome. As virus particles bud off from the surface of an infected cell, each virion acquires a lipid envelope that surrounds the core. The particle carries within it several enzymes encoded by its genome that are required for viral replication, such as reverse transcriptase.

The genome of the provirus is about 10 kb in length and includes the flanking LTRs, each containing regulatory sequences required for replication of the virus. Also present are the typical retroviral pol and env genes, coding respectively for the reverse transcriptase, and the envelope glycoproteins. In addition to these, HIV also contains overlapping reading frames carrying sufficient information to code for seven additional proteins. Research continues in order to clarify the exact manner in which they operate in viral infection and replication. Three of the proteins, encoded by genes art, tat, and rev, are critical to viral replication. The consensus is that the rev protein is needed for the transport of viral mRNA from the infected cell nucleus to the cytoplasm. The product of the art gene appears to regulate the expression of env and gag. The protein encoded by gene tat can trigger a tremendous amount of viral replication, and it is believed to do so by activating the promoter of the viral genome, which in turn activates other viral genes. A gene known as sor plays a role critical to the formation of infectious particles, whereas the gene designated 3′ orf appears to downregulate expression of the virus.

The proviral DNA arising in the host cell from the viral genomic RNA by reverse transcriptase activity can exist in either a linear or a circular form. Unlike most other retroviruses, not all of the retroviral DNA becomes integrated into the host cell DNA. Instead, large amounts of unintegrated viral DNA accumulate in the nucleus, a factor that may be important in the lethal effect of HIV on the infected cell.

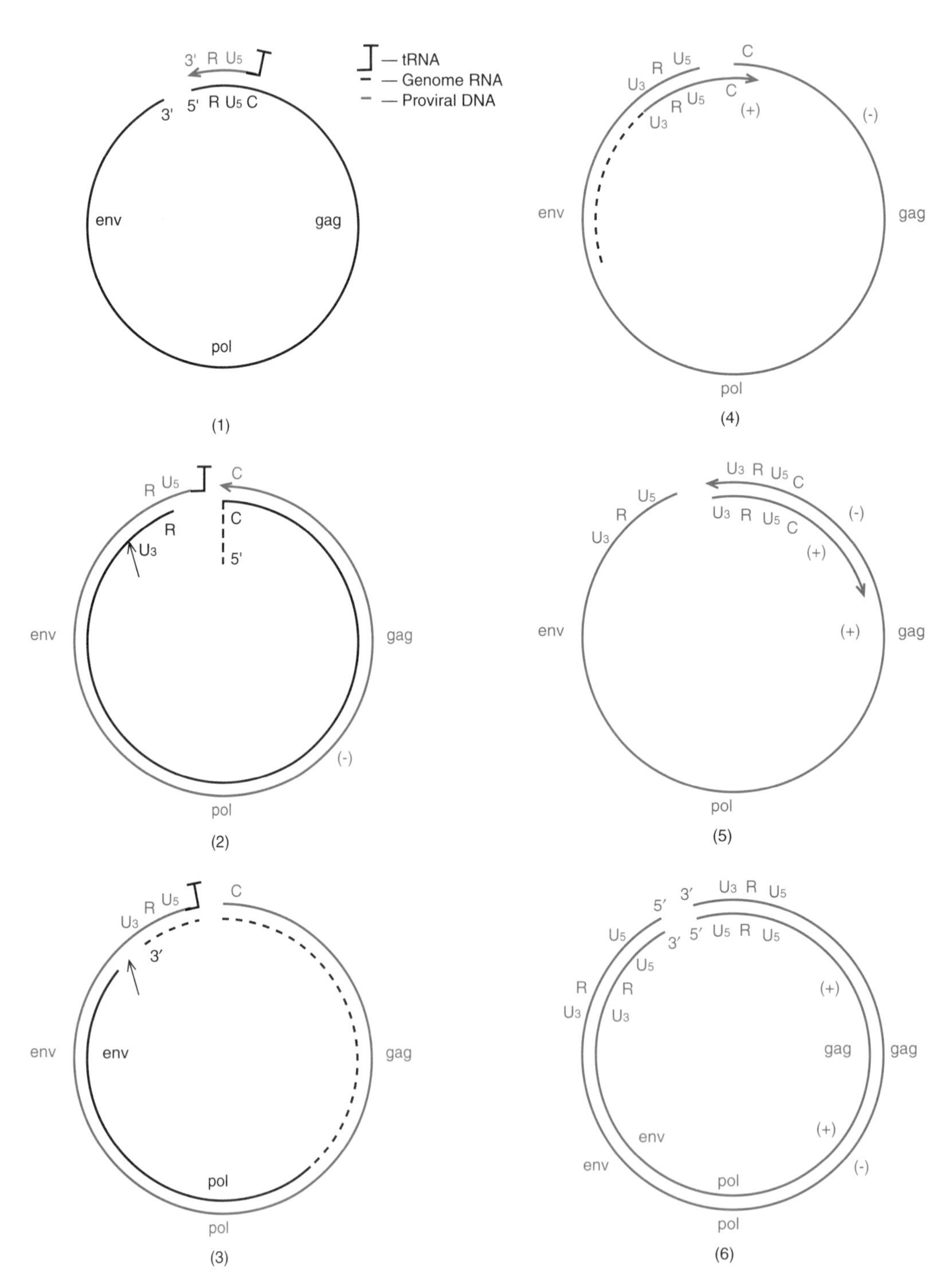

Ongoing investigations are actively pursuing the many complexities and unexpected features of the HIV virus with the goal of finding its vulnerable aspects. Once these are well characterized, pathways may be opened toward finding both curative and preventive approaches to defeat the replication of the AIDS virus, one of the most serious pathogens of the human.

Endogenous proviruses

Surprisingly, most proviruses do not harm the host cell. The stability of the retrovirus in its integrated state becomes vividly apparent when we consider endogenous proviruses, retroviruses that infect sperm and eggs and therefore can be passed down through the germ line and inherited as if they were a native part of the host genome. Such viruses occur in a variety of vertebrates, including the human. In the mouse, where they may constitute as much as one-half of 1% of the cellular DNA, thousands of endogenous proviruses are integrated into the genome at many different chromosomal sites. These proviruses usually do not undergo transcription but appear to be turned off as a result of DNA methylation (see Chap. 17). However, if infected cells are treated with certain chemicals, they may produce infective virions, and at least one example is known from the mouse, where an integrated provirus can cause gene inactivation. The existence of proviruses raises the question of their significance in the cells of vertebrates. It is a matter of particular concern to the investigator who uses animal cells as research tools. Data obtained from experiments utilizing cells that prove to be harboring endogenous viruses are rendered suspect, since they can lead to erroneous conclusions. This is seen dramatically in the case of the AIDS virus (HIV). It has been demonstrated that when the HIV virus is passed through mouse cells and then allowed to infect human cells in culture, the infective activity becomes altered. The virus acquires the ability to reproduce much more rapidly than is typical, and it becomes able to infect a number of cell types that it does not normally infect. The acquisition of these new properties appears to be due to interaction between HIV and endogenous mouse viruses. Besides its importance to interpretation of data, the observations raise many implications, not the least of which is the potential for producing viruses that are more virulent due to interactions between endogenous viruses and a virus that has entered a cell that is not normally its host.

Moreover, reverse transcriptases were believed to be a unique feature of the world of retroviruses. This no longer appears to be the case, as evidence continues to accumulate that reverse transcriptases are widespread among higher and lower eukaryotes and possibly even prokaryotes. Activities of these nonviral reverse transcriptases in cells are pursued in Chapter 17.

FIGURE 16-21. Steps in provirus formation. **1:** The tRNA pairs with the genome RNA near its 5′ end due to the complementary region (C). Using the tRNA as primer and the genomic RNA as a template, reverse transcriptase starts to build the – DNA strand and moves toward the 5′ end of the template RNA in the direction of the arrowhead. It will next switch over to the 3′ end of the same strand and continue to increase the – strand in length. **2:** U5 and the 5′ R of the template become degraded (broken line). The 3′ R of the template pairs with the R of the – DNA strand, which continues to increase in length. The template RNA becomes nicked at the site of the arrow. **3:** Degradation of template RNA continues. The site of the nicked RNA (arrow) provides a primer end for synthesis of the + DNA strand. **4:** The RNA genome and tRNA become completely degraded. + strand synthesis continues. The + strand includes a C site copied from the tRNA. Pairing takes place between the – and + DNA strands at the C site. The arrowhead indicates the direction of growth of the + strand, which is opposite that of the – strand. **5:** DNA synthesis continues as the – and + strands use each other as templates. **6:** DNA synthesis ends with completion of the – and + strands.

Oncogenes and RNA tumor viruses

Many retroviruses are now known to be capable of causing tumors and are thus said to be oncogenic (having the ability to transform the cells they infect to a state in which they grow uncontrollably). Oncogenesis, tumor induction, occurs among several groups of DNA viruses, but within RNA viruses it is associated only with some of the retroviruses. Most oncogenic retroviruses fall into one of two major categories on the basis of the genome. Each virion in the first group possesses an intact retroviral genome and is capable of undergoing the typical cycle of infection and replication depicted in Figure 16-21. The viral particles are highly infectious, but they produce tumors only after a long period in the cell. Members of this group cause leukemia in cats, mice, and chickens. Their oncogenic ability is a result of various changes that take place in their genomes, such as point mutations. The specific change imparts to such a virus the ability to activate a cellular gene.

Rous sarcoma virus (RSV), the first retrovirus to be isolated, falls into the second category of oncogenic retroviruses. In addition to the three genes typical of most retroviruses (Fig. 16-20), RSV carries one other. This is src, a gene that codes for a protein required for oncogenesis and that is classified as an oncogene, a genetic sequence capable of tumor induction. In almost all oncogenic retroviruses in this second category, known as the *acute transforming viruses,* the tumor-inducing ability of the virus is due to the single oncogene it carries. Approximately 30 genes have now been isolated from retroviruses oncogenic in various birds, rats, the mouse, the cat, and the monkey. Acute transforming viruses can transform cells to the malignant state in culture and can cause tumors in vivo in a matter of days.

In the mid-1970s, the surprising discovery was made that the src viral oncogene is not truly viral but is an almost exact copy of a gene found in all chicken cells! Each of the various oncogenes identified has been shown to be very closely related to a normal gene in the genetic complement of the animal. Oncogenes are designated by three letters (*src* for sarcoma; *fes* for feline sarcoma virus, and so forth). The prefix *v,* as in v-src or v-fes, indicates a viral oncogene (v-onc), whereas *c,* as in c-src and c-fes, refers to the cellular counterpart (c-onc). There is little doubt that an oncogenic virus has somehow captured its oncogene from a cellular sequence. A clue to the way in which the retrovirus picks up the c-onc, also known as a *protooncogene,* comes from the observation that the latter usually contains introns that interrupt the coding information found in the exons. On the other hand, the corresponding v-onc lacks introns. This suggests that v-onc has arisen through proviral transcription following an infection. According to one idea, a provirus formed in an infected cell may become integrated with the cellular DNA very close to the site of a protooncogene. The provirus and the cellular gene could become fused as a result of a deletion arising between them. Consequently, when the provirus undergoes transcription a transcript could be produced that consists of viral sequences followed by those of the cellular gene, the c-onc. Following processing of the transcript, the introns of the protooncogene in the transcript would be removed. Such a transcript could then become incorporated into a virion. The cellular information is now intimately associated with the virus and has become a v-onc, which will be copied into double-

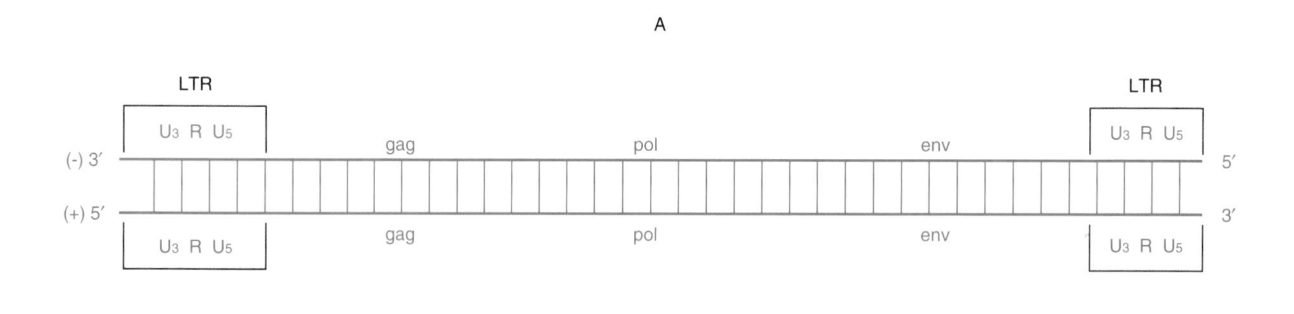

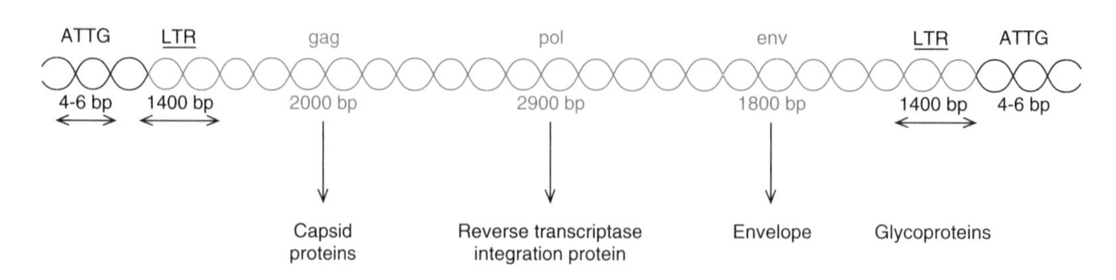

FIGURE 16-22. The provirus. **A:** Due to the manner in which it arises, the provirus is longer than the original RNA genome since it possesses an LTR at each end. **B:** When it integrates, the provirus brings about a direct repeat (indicated here as ATTG) of a few base pairs of the adjacent host DNA. The viral genes become expressed as new virions form. (See text for details. Lengths are not in proportion for sake of diagrammatic clarity.)

stranded DNA along with the adjacent viral genome during the next cycle of provirus formation.

The same protooncogene can be found in different acute transforming viruses. In the cell, any one of the 30 or so known oncogenes appears to be coded for protein products that must play important roles in cell development. The protooncogene obviously is not designed to harm the cell by evoking tumor formation, but it can become oncogenic in various ways. It can become activated if a retrovirus integrates nearby. Once it is picked up by a retrovirus and becomes a v-onc the uninterrupted message of the cellular gene becomes part of the viral transcript and is expressed at a very high level in contrast to the c-onc counterpart, which is typically expressed at a low level in its normal cell location. Overproduction of a normal cellular product or its production in an inappropriate cell type are ways that a v-onc might evoke tumor formation. This, along with the subject of protooncogenes, is pursued more fully in Chapter 17 in relation to cell regulation.

When the two are compared, the v-onc and the c-onc are usually found to differ. In some cases, it appears that an entire protooncogene has been picked up by the virus and has remained essentially the same. Frequently the v-onc and its cellular counterpart exhibit differences attributable to the accumulation of point mutations. The incorporation of the cellular gene into the viral genome often entails some loss of genetic material from the protooncogene. The latter may lose sequences from either or from both of its ends.

In the process by which a retrovirus acquires a protooncogene which then becomes a v-onc, the virus itself typically becomes defective due to the fact that cellular sequences replace part of the viral genome. For example, all or most of the env gene of the virus could be lost (Fig. 16-23A). Exactly what region is lost and its length vary with the virus. The defective virus can nevertheless be packaged into a virion as long as a portion of its 5′ end is present for recognition by the virion proteins. Following infection of a cell by a single particle with a defective genome, the latter can be copied by reverse transcriptase into a proviral form as long as it has an intact primer-binding site at its 5′ end to enable the start of – DNA strand synthesis and also the 3′ end for the origin of the + strand (refer back to Fig. 16-21). If its LTRs are intact, the provirus can integrate and undergo transcription. It can even transform a cell into the malignant state by expression of the oncogene. Being defective, however, such a retrovirus cannot complete a cycle of replication. Nevertheless, the lost function may be supplied by an intact virus that infects the cell along with the defective one. The helper virus can bring about replication of both viral types, the intact helper and the defective one with the v-onc (Fig. 16-23B). The Rous sarcoma virus is the only member of the acute transforming viruses that is not defective. Its oncogene, src, has been added to the three genes, gag, pol, and env, without replacing any part of the genome.

Many aspects of the interactions between an oncogenic retrovirus and a cell are reminiscent of transduction by bacteriophages, discussed earlier in this chapter. Indeed, the term *transduction* may be appropriately extended to include the transfer of eukaryotic genetic material from one cell to another by a virus.

DNA tumor viruses

Unlike the RNA tumor viruses, the oncogenic ability of DNA tumor viruses depends on the viral genome itself, which in some way brings about an upset in cellular regulatory devices. The DNA tumor virus does not carry an oncogene such as the retroviral v-onc, which has a cellular counterpart, the c-onc. Oncogenic DNA viruses are found among the adenoviruses and a family that includes the papilloma viruses and the very well studied simian vacuolating virus (SV40). The DNA of SV40 is double stranded, and its exact sequence of 5,226 base pairs has been determined as well as the locations of various genes (see Chap. 19). This virus is another example of one in which overlapping genes are found.

SV40, which lacks an envelope, can follow either of two kinds of cycles, depending on the host cell. After it infects cells growing in culture that are derived from certain monkey species, a lytic cycle is initiated that leads to cell death. The monkey cells become infected by a linear viral DNA, which circularizes inside the cell. Both an early and a late phase of the viral lytic cycle can be recognized. Shortly after infection, the virus migrates to the cell nucleus, where viral replication and transcription take place. Two of the proteins encoded by the virus are found during the early stage; these are known as the T and t antigens or proteins. Both are required for replication of viral DNA, which starts about 12 hours or more after a cell is infected. SV40 is able to stimulate a nondividing cell to start DNA synthesis, thus providing itself with a supply of precursors and other molecules needed in DNA replication. The replication of the viral DNA in turn appears to be necessary for stimulation of its late genes to undergo transcription. The late proteins finally assemble with the viral DNA to form thousands of mature virions that are released 2 days or so after infection, resulting in death of the monkey cells.

If SV40 is allowed to infect mature mouse cells in culture, a different scenario is followed. Approximately 24 hours after they are infected, the mouse cells begin to show some of the characteristics of malignant cells as the T antigen is produced in the cell nuclei. Late genes of the virus are not expressed, nor do mature virions appear in these cells. Most of the infected cells revert back to the normal state, but a number of them becomes permanently transformed and grow into fatal tumors when injected into mice.

The difference in the responses of the two types of host cells illustrates a feature of oncogenic virus interaction with a cell. A cell that responds as does a monkey cell to SV40 is said to be *permissive,* meaning that it enables a specific virus to multiply and complete its cycle of replication. Release of offspring viruses takes place from permissive cells and is usually followed by cell death. A cell that behaves as does one from the mouse to SV40 is said to be *nonpermissive,* to indicate that, after entry, the virus cannot grow. Instead, the virus may bring about the transformation of some infected cells to the malignant state. A given cell type, a mouse cell, for example, may be permissive with respect to one viral species and nonpermissive to another.

During the conversion of normal to cancer cells by SV40 and various other DNA tumor viruses, the viral DNA is integrated into a host chromosome. This can be demonstrated by turning to the technique of DNA–DNA hybridization. If we demonstrate that viral DNA hybridizes with the cellular DNA, we can conclude that viral DNA sequences are present along with certain cellular DNAs. To determine whether this is the case, we extract the DNA from the nuclei of transformed cells and fragment it with the aid of a specific restriction endonuclease (Fig. 16-24). The resulting fragments are then placed on an agarose gel and subjected to electrophoresis, a procedure that separates the mixture of fragments on the basis of size, smaller fragments traveling faster down the gel. Fragments of a given specific size can be visualized in the gel as a band. A band at a given location reflects a fragment of a specific size. (Chapter 19 contains details on procedures used in genetic manipulation.) Since our goal is to detect any hybridization between viral and cellular DNA, we must render both DNAs single stranded. The two DNAs must then be incubated together to permit hybridization to take place. Because hybridization cannot take place when one of the DNAs is in a gel, it is necessary to transfer this DNA from the gel to a surface, where it can come into contact with DNA from the other source. Moreover, transfer from the gel must be done in such a way that the order of the fragments in the gel is preserved.

A technique developed by E. Southern, known widely as *Southern blotting,* enables us to do just this. The gel containing the separated bands of fragments is placed in a solution that denatures the DNA so that all of the fragments consist of single strands. The gel is then laid on a large piece of nitrocellulose paper so that the gel and the filter paper are in direct contact. A buffer is allowed to flow through the gel toward the filter, at right angles to the direction in which the fragments were separated by the electrophoresis. The denatured DNA is carried by the buffer out of the gel onto the filter to which the single-stranded DNA binds tightly. The original positions of the fragments in the gel will correspond almost exactly with their positions on the filter. This routinely used Southern blotting procedure, which enables us to transfer denatured DNA fragments on a gel to a filter, is invaluable in molecular biology. (The procedure can be altered to permit blotting of RNA from a gel to another medium. It is then referred to as *Northern blotting.*)

Following Southern blotting, the denatured fragments are then exposed to denatured labeled viral

FIGURE 16-23. Defective and helper viruses. **A:** A provirus inserts into the genome of a cell near a protooncogene, c-onc (1). In acquiring the c-onc, rearrangement of DNA occurs; the cellular genetic material becomes a v-onc and replaces part of the viral genome (2). The virus in this case loses part of gene pol and all of env. If essential 5′ and 3′ ends of the virus remain intact, the RNA transcripts of the provirus can be packaged into virions (3). These can infect other cells, but the resulting provirus is defective and by itself cannot complete a new cycle of replication, and no new virions arise (4). **B:** If an intact virus (left) is present in the genome of a host cell, it acts as a helper and can generate transcripts that are intact and carry the information missing in the virus with the v-onc. Two types of virus particles, defective and nondefective, can arise from the cell.

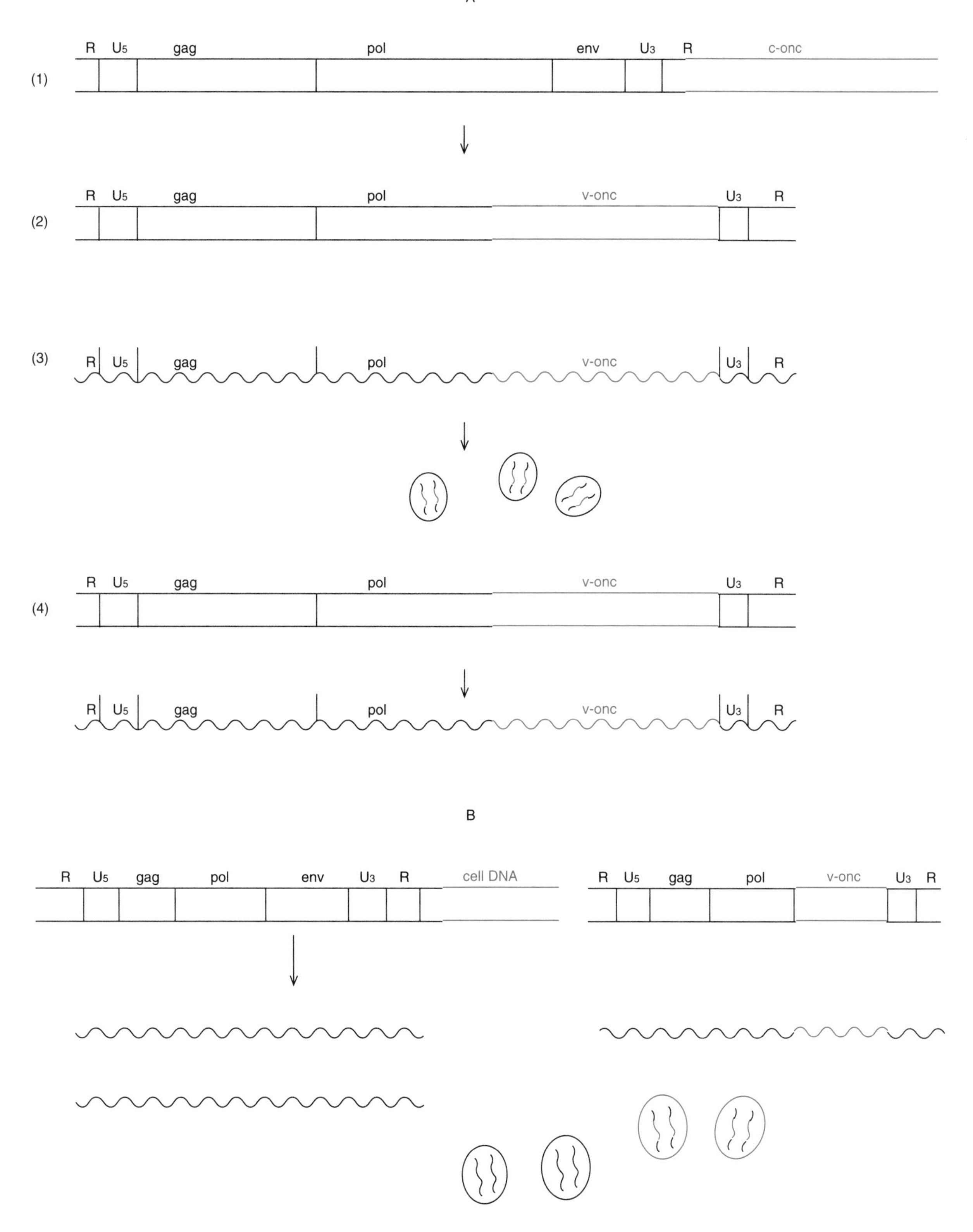
A
R
U5
gag
pol
env
U3
R
c-onc
(1)
R
U5
gag
pol
v-onc
U3
R
(2)
(3)
R
U5
gag
pol
v-onc
U3
R
(4)
R
U5
gag
pol
v-onc
U3
R
R
U5
gag
pol
v-onc
U3
R
B
R
U5
gag
pol
env
U3
R
cell DNA
R
U5
gag
pol
v-onc
U3
R

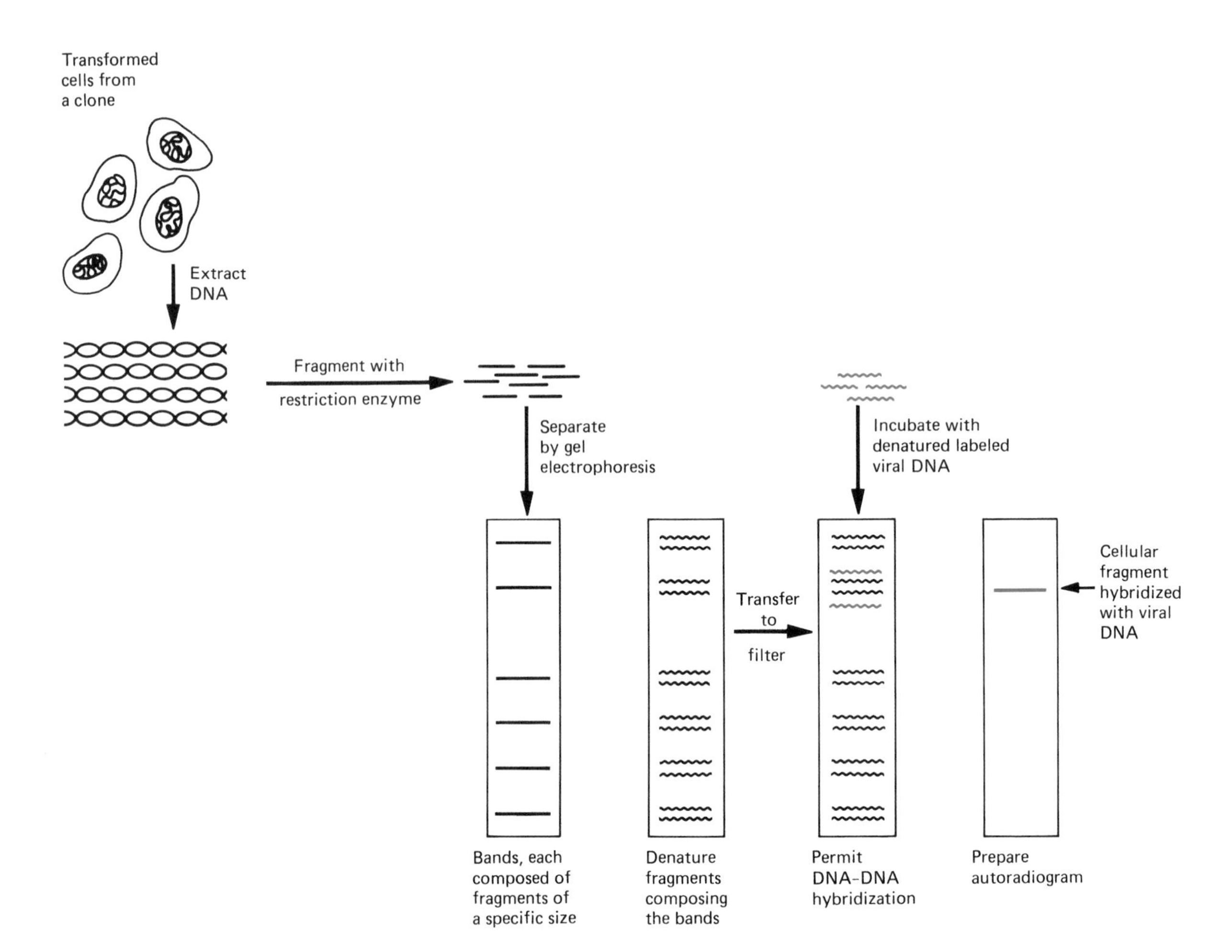

FIGURE 16-24. Demonstration of SV40 integration. DNA is extracted from cells composing a population that traces back to one cell that was transformed by SV40. The DNA is exposed to a restriction enzyme that fragments the DNA into pieces of specific sizes. These fragments can be separated into bands by gel electrophoresis, the larger fragments toward the top of the gel. Each band represents a fragment of a specific size. The DNA is rendered single stranded and then transferred to a filter, where it can be used in a DNA–DNA hybridization procedure. After exposure to labeled, single-stranded SV40 DNA, a certain fragment hybridizes with the viral DNA, indicating that a viral DNA sequence is present. The band, representing the specific fragment, can be detected when an autoradiograph is prepared, because of the labeled DNA of the virus. If this same procedure is repeated with DNA from cells of another clone of transformed cells, the fragment that pairs with the viral DNA may be different from the one shown here.

DNA. Pairing between the viral DNA and any DNA on a chromosome fragment can be detected by autoradiography and indicates that strands complementary to the viral DNA are integrated with the host DNA. The integrated viral DNA thus behaves as inherited elements in transformed cells. When this procedure (Fig. 16-24) is repeated using DNA from different clones of infected cells, the results continue to show that the viral DNA is integrated, but the DNA is not necessarily associated with the same fragments from one clone to the next. This indicates that SV40 virus, unlike phage λ, does not have a single preferential site of integration but that it integrates at random. There must be no sequence homology between the viral and host DNA to narrow the sites of integration. It is possible to find clones in which two viruses are integrated close to each other so that occasionally more than one fragment on the filter may hybridize with the viral DNA. The genomes of 1–20 SV40 viruses can integrate, depending on the cell line.

The T antigens of SV40 and other DNA tumor viruses are now known to be able to transform normal cells to the malignant state. As in the case of SV40, some DNA viruses may be able to transform the cells of some species in cell culture but not the cells of others. Moreover, although the viruses may transform the cells, these cells, when injected into healthy animals, do not cause tumors to form. This is true even though the cells express T antigen and have malignant characteristics when growing in cell cultures. If the same cells are injected into animals with suppressed immune systems, however, tumors result. This sort of observation indicates that the ability of cells that were transformed in culture to produce tumors depends on interaction between the virally transformed cells and the immune system of the host animal. A low resistance of the transformed cells to the immune system of the host results in their rejection; a high resistance results in the growth of fatal tumors. There is good evidence that the early-appearing T antigens of the tumor viruses, by interacting with the host DNA, regulate the level of resistance that transformed cells show to the host immune system and are responsible for the pathway followed. In Chapter 19, we present more details on the T antigen and genes of SV40.

REFERENCES

Alwine, J.C., D.J. Kemp, and G.R. Stark. Method for detection of specific RNAs in agarose gels by transfer to diazobenzylomethyl-paper and hybridization with DNA probes. *Proc. Natl. Acad. Sci. USA* 74:5350, 1977.

Arya, S.K., C. Guo, S.F. Josephs, and F. Wong-Stall. Transactivator gene of human T-lymphotropic virus type III (HTLV-III). *Science* 229:69, 1985.

Benzer, S. On the topology of the genetic fine structure. *Proc. Natl. Acad. Sci. USA* 47:403, 1961.

Bishop, J.M. Oncogenes. *Sci. Am.* (March):80, 1982.

Chang, D.D., and P.A. Sharp. Messenger RNA transport and HIV regulation. *Science* 249:614, 1990.

Cold Spring Harbor. *Replication of DNA in Microorganisms,* vol. 33. Cold Spring Harbor Laboratory, Cold Spring Harbor, NY, 1968.

Contreras, R., R. Rogiers, A. Van de Voorde, and W. Fiers. Overlapping of the VP_2-VP_3 gene and the VP_1 gene in the SV40 genome. *Cell* 12:529, 1977.

Fauci, A.S. The human immunodeficiency virus: Infectivity and mechanisms of pathogenesis. *Science* 239:617, 1988.

Fiddes, J.C. The nucleotide sequence of a viral DNA. *Sci. Am.* (Dec.) :54, 1977.

Fraenkel-Conrat, H., and B. Singer. Virus reconstitution II. Combination of protein and nucleic acid from different strains. *Biochim. Biophys. Acta* 24:540, 1957.

Gajdusek, D.C. Unconventional viruses and the origin and disappearance of kuru. *Science* 197:943, 1977.

Gallo, R.C. The first human retrovirus. *Sci. Am.* (Dec.) :88, 1986.

Gallo, R.C., and L. Montagnier. AIDS in 1988. *Sci. Am.* (Oct) :40, 1988.

Gallo, R.C. The AIDS virus. *Sci. Am.* (Jan) :40, 1987.

Haseltine, W.A. Molecular biology of the human immunodeficiency virus type 1. *FASEB J.* 5:2349, 1991.

Landy, A., and W. Ross. Viral integration and excision: Structure of the lambda att sites. *Science* 197:1147, 1977.

Lusso, P., F. Di Marzo Veronese, B. Ensoli, G. Franchini, C. Jemma, S.E. DeRocco, V.S. Kalyanaram, and R.C. Gallo. Expanded HIV-1 cell tropism by phenotypic mixing with murine endogenous retroviruses. *Science* 247:848, 1990.

Moyzis, R.K. The human telomere. *Sci. Am.* (Aug) :48, 1991.

Nash, H.A. Integration and excision of bacteriophage lambda: The mechanism of conservative site specific recombination. *Annu. Rev. Genet.* 15:143, 1981.

Nathans, D. Restriction endonucleases, simian virus 40, and the new genetics. *Science* 206:903, 1979.

Olsen, H.S., P. Nelbock, A.W. Cochrane, and C.A. Rosen. Secondary structure is the major determinant for interaction of HIV rev protein with RNA. *Science* 247:845, 1990.

Pratt, D. Single stranded DNA bacteriophages. *Annu. Rev. Genet.* 3:343, 1969.

Ptashne, M., A.D. Johnson, and C.O. Pabo. A genetic switch in a bacterial virus. *Sci. Am.* (Nov) :128, 1982.

Razin, A., and A.D. Riggs. DNA methylation and gene function. *Science* 210:604, 1980.

Sanger, F., G.M. Air, B.G. Barrell, and N.L. Brown, et al. Nucleotide sequence of bacteriophage ϕX 174 DNA. *Nature* 265:687, 1977.

Simons, K., H. Garoff, and A. Helenius. How an animal virus gets into and out of its host cell. *Sci. Am.* (Feb) :58, 1982.

Smith, H.O. Nucleotide sequence specificity of restriction endonucleases. *Science* 205:455, 1979.

Smith, M., N.L. Brown, G.M. Air, and B.G. Barrell, et al. DNA sequences at the C termini of the overlapping genes A and B in bacteriophage ϕX 174. *Nature* 265:702, 1977.

Southern, E.M. Detection of specific sequences among DNA fragments separated by gel electrophoresis. *J. Mol. Biol.* 98:503, 1975.

Stahl, F.W. Genetic recombination. *Sci. Am.* (Feb):90, 1987.

Stent, G.S., and R. Calendar. *Molecular Genetics. An Introductory Narrative.* Freeman, San Francisco, 1978.

Temin, H.M. RNA-directed DNA synthesis. *Sci. Am.* (Jan) :24, 1972.

Temin, H.M. Function of the retrovirus long terminal repeat. *Cell* 28:3, 1982.

Tjian, R. T. Antigen binding and the control of SV40 gene expression. *Cell* 26:1, 1981.

Varmus, H.E. Form and function of retroviral proviruses. *Science* 216:812, 1982.

Varmus, H.E. Retroviruses. *Science* 240:1427, 1988.

Wong-Staal, F., and R.C. Gallo. Human T-lymphotropic retroviruses. *Nature* 317:395, 1985.

zur Hausen, H. Viruses in human cancer. *Science* 254:1167, 1991.

REVIEW QUESTIONS

1. Show how the following can be converted into a duplex DNA with extended 5′ sticky ends that would allow the molecule to form a circle.

5′–TTACCTTCATAAAGCGCACATTACCTTCAT–3′

3′–AATGGAAGTATTTCGCGTGTAATGGAAGTA–5′

2. Let A, B, and C represent three strains of *E. coli*. Drops of culture fluid are interchanged, as shown in the following. The recipient tubes are examined in each case for clearing + or no change –. Explain the results.

Drop derived from	Recipient	Clearing (+)
A	B	+
A	C	+
B	A	+
B	C	+
C	A	–
C	B	–

3. The order of entry of markers in a certain Hfr strain is

T L Lac λ S

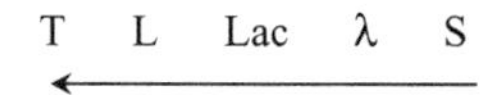

The following three crosses are made. Answer the questions that follow, and offer an explanation.

(1). $HfrT^+L^+Lac^+Gal^+\lambda^+S^s \times F^-T^-L^-Lac^-Gal^-\lambda^-S^r$
(2). $HfrT^+L^+Lac^+Gal^+\lambda^-S^s \times F^-T^-L^-Lac^-Gal^-\lambda^+S^r$
(3). $HfrT^+L^+Lac^+Gal^+\lambda^+S^s \times F^-T^-L^-Lac^-Gal^-\lambda^+S^r$

A. From which of the crosses would it be possible to obtain recombinants that are $Lac^+Gal^+\lambda^-$?
B. From which of the crosses would it be possible to obtain recombinants that are $Gal^+\lambda^+$?
C. From which of the crosses would it be possible to obtain rare recombinants that are S^s?

4. A. In what way does transformation differ from specialized transduction?
B. How does conversion resemble and differ from transduction?

5. A culture of *E. coli* cells that are λ^-Gal^- is infected with λ phages, which have been residing in cells that are λ^+Gal^+. The concentration of virus used to infect the *E. coli* is very low. A small percentage of Gal^+ cells is isolated. However, none of these transduced cells is found to be lysogenic, and none can be made to liberate virus. Explain.

6. *E. coli* cells that are λ^-Gal^- may become transduced by phage λ to Gal^+ cells. These cells can give rise to colonies. It is then found that 1%–10% of the cells in most of the colonies cannot use galactose. Explain.

7. Circle any answer that is correct. In generalized transduction:
A. A helper phage enables the transducing particle to replicate in the host cell.
B. The transduced cell becomes lysogenic.
C. The transduced cell is vulnerable to further infection by the same virus.
D. The transducing virus can be activated in the bacterial cell by zygotic induction.
E. The viral particle can integrate at any one of various sites on the host chromosome.

8. Formation of a phage particle capable of generalized transduction is a very rare event. Consequently, if particles capable of transduction frequently bring about cotransduction of two genes, this is taken to mean that the two genes are closely linked. Suppose a donor strain of bacteria is c^+d^+ and the recipient is c^-d^-. Phages capable of generalized transduction are allowed to infect donor cells. Phage particles from this infection are then allowed to infect recipient cells, and recombinants are formed among the recipient cells. The following kinds and numbers of transductants arise. Give the recombination frequency and map unit distance between the genes.

c^+d^- – 150
c^-d^+ – 300
c^+d^+ – 650
1,100

9. Suppose you are studying three genes, r, s, and t, in relation to cotransduction in generalized transduction experiments. It is found that r and s are cotransduced fairly often. The same is true for r and t. However, s and t are cotransduced with a much lower frequency. What appears to be the gene order?

10. Bacterial strain 1 cannot grow on a minimal medium, because it is unable to form the essential growth substances m and n. However, it can form the essential metabolite o. The strain can be represented as $m^-n^-o^+$. This strain is infected with a phage capable of generalized transduction. The offspring phages are used to transduce strain 2. Strain 2 is also auxotrophic, since it is unable to form substance o, and can be represented as $m^+n^+o^-$. After the second infection, strain 2 cells are plated on minimal medium. A reciprocal experiment is performed in which strain 1 is the recipient and is transduced with phage particles derived from an infection of strain 2. From the results given here, tell the order of the three genes and explain.

Experiment 1: strain 1, donor; strain 2, recipient	Experiment 2: strain 2, donor; strain 1, recipient
200 wild-type cells per 10^8	20 wild-type cells per 10^8

11. One type of T/2 virus is h^+r^+ and another is hr. The two types are allowed to infect strain B of *E. coli*. Plating of viruses from the lysate is made on a mixture of strains B and B/2 of *E. coli.* Plaques are then scored for size and cloudiness. Based on the following information, give the percentage of recombination and the viral genotypes represented by the plaques.

Small, clear	42
Small, cloudy	2,195
Large, clear	2,230
Large, cloudy	33

12. When an experiment is performed such as that described in Question 11, the virus particles emerging from a single cell may be followed further. It is often found that one recombinant class, h^+r or hr^+, may not be found at all. Explain.

13. Assume that a, b, and c are recessive mutations and are very closely linked. Suppose the following double heterozygotes have been constructed and the resulting phenotypes are as indicated here:

$\frac{ab^+}{a^+b}$ = mutant $\frac{ab}{a^+b^+}$ = wild $\frac{bc^+}{b^+c}$ = wild

$\frac{bc}{b^+c^+}$ = wild $\frac{ac^+}{a^+c}$ = wild $\frac{ac}{a^+c^+}$ = wild

Which of the following would appear to be true? Circle any answer that is correct.

A. b and c are engaged in different functions.
B. All three are engaged in the same function.
C. a and b are complementary to each other.
D. a and b are alleles.
E. The three sites are all part of the same cistron.
F. b and c are alleles.

14. Assume that a DNA segment of the infective form of the virus ϕX 174 has the nucleotide sequence AAGTTACCA. The single-stranded DNA infects an *E. coli,* and mRNA is formed by the replicative form RF1. Give the sequence of bases in the mRNA that would relate to the preceding segment.

15. The sequence in the following gene has the same sense as the mRNA. Consulting a genetic dictionary, if needed, determine where it is possible in the sequence for two genes to overlap, assuming a difference in reading frames.

AAATTAGCTCCCGGAGCGTGATGTCTAAAGGT

16. Suppose the genome of a DNA virus contains overlapping genes. The same base sequence can be read in three different frames, depending on the starting point of the messages. The DNA segment depicted here is from the sense strand of the virus. What would you predict as the amino acid sequence in the translation products if the codons are read three at a time beginning with 1) the first base; 2) the second base; 3) the third base?

5′–GCTGCTGCTGCTGCTGCTGCT–3′

17. Assume that the + strand of an RNA virus enters a bacterial cell. A stretch of the RNA in the + strand is as follows:

AACAGGACGCAG

A. What will be the nucleotide sequence in the nucleic acid that attaches to the ribosomes of the host?
B. What RNA sequence is required for replication of the preceding sequence?
C. Name the enzyme required for replication of the viral RNA.

18. The following is part of the base sequence of a particular viral mRNA:

5′–ACGCGUUAAUCAAA–3′

A. An in vitro system is provided with reverse transcriptase, and a DNA strand is synthesized. Give the base sequence in this DNA reading in the direction 5′ to 3′.
B. What will the results be if actinomycin D is also added to the system?

19. A. In each of the following nucleotide sequences pick out the palindrome and indicate the axis of symmetry.

1. 5′–CTTGGAAGCTTAATAC–3′
   3′–GAACCTTCGAATTATG–5′
2. 5′–AACGTTAACTTCCTA–3′
   3′–TTGCAATTGAAGGAT–5′

B. The palindrome in sequence 1 is cleaved by a restriction enzyme at staggered sites. In the upper strand, the cut is two nucleotides to the left of the axis of symmetry. In the lower strand, it is two nucleotides to the right. Show the fragments that will be generated after endonuclease action.
C. The palindrome in sequence 2 is cleaved exactly at the axis of symmetry in both strands. Show the fragments that will be generated by endonuclease action.

20. A particular type II restriction enzyme can attack DNA with the palindrome found in the following DNA sequence. However, a particular methylating enzyme can methylate all the cytosines in the palindrome. Indicate by asterisks in this sequence all the cytosines that would become methylated.

5′–CCACCGGTCCTCATACCGGTC–3′
3′–GGTGGCCAGGAGTATGGCCAG–5′

21. A particular oncogenic virus, virus A, is allowed to infect mouse cells growing in culture, and a number of infected cells is transformed to the malignant state. From single, transformed cells, clones are established: 1, 2, 3, and so on. DNA is extracted from cells constituting a clone and is then subjected to a specific restriction enzyme. The resulting fragments are placed on a gel, subjected to electrophoresis, and then denatured. The denatured fragments are transferred to a filter by Southern blotting and then exposed to denatured labeled DNA from the oncogenic virus A. DNA is also extracted from nontransformed cells and handled in exactly the same way as described. Autoradiographs are then prepared for each DNA source. When the autoradiographs are compared, the following picture is seen. Explain the results.

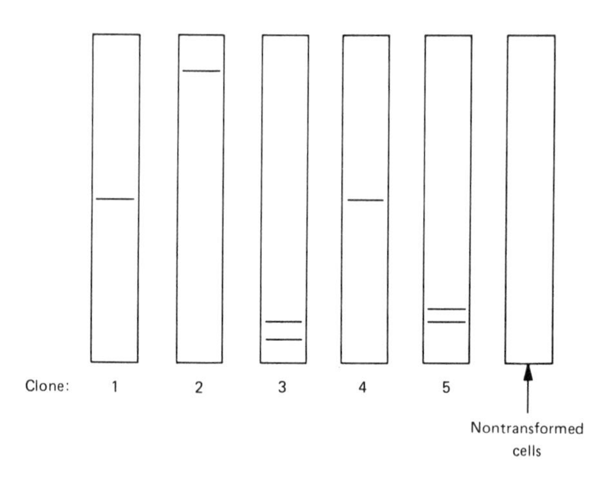

22. A. In the procedure discussed in Question 21, why was it necessary to transfer the denatured fragments from the gel?

B. Suppose oncogenic DNA virus B is allowed to infect other mouse cells in culture. Exactly the same procedure is followed as was discussed in Question 21. Explain the following results of a comparison of autoradiographs.

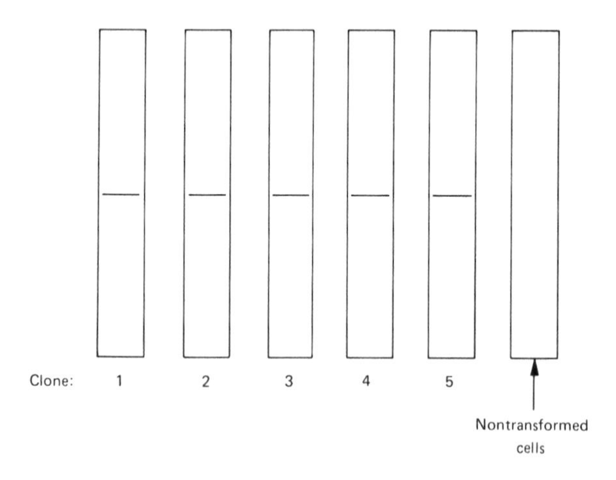

23. The MS_2 virus of *E. coli* is a single-stranded RNA phage in which the RNA serves as both mRNA and the genetic material of the virus itself. The RNA codes for a few genes, one of which governs the formation of the coat protein. This protein consists of 129 amino acids, and the exact amino acid sequence has been established. Work was undertaken to determine the precise nucleotide sequence of the gene encoded for the amino acid sequence of the coat protein. In the procedure followed, the RNA of the virus was fragmented and the fragments isolated on gels. The nucleotide sequence was then determined for each fragment. Here we show a sequence of 15 amino acids in the coat protein along with the nucleotide sequences of four fragments.

NH_2–pro-phe-tyr-ala-thr-ala-asn-ser-
1 2 3 4 5 6 7 8

gly-ile-tyr-arg-gly-gly-val–COOH
9 10 11 12 13 14 15

Fragment 1: U U U C U C U G U A A A A A G
A U G

Fragment 2: G C A A A C U C C G G U A U C
U A C C G U

Fragment 3: U G U U G G C A U U U U G A U
C A C C U C C A U

Fragment 4: U U U U A U G C A A C U G C A
A A C U C C G G U A U C U A C

With the aid of the genetic dictionary, determine which fragments relate to this portion of the gene for coat protein, and establish as much of the nucleotide sequence as possible for the gene. (Assume each fragment starts at the 5′ end of a complete codon.)

Control Mechanisms and Differentiation

17

Genetic control of enzyme induction in bacteria

The first few mitotic divisions of an embryo produce cells that are essentially identical. How do diverse cell types arise from a population of cells that all contain the same genetic information? What devices permit only certain genes to come to expression in specific cells so that specialization or differentiation occurs? Pioneer genetic studies with microorganisms have addressed these questions and thus contributed basic information that has served as a point of departure for investigations into the regulation of gene expression and differentiation of cells in higher species.

Several critical investigations depended on the isolation from bacteria of certain mutants in which control mechanisms have gone awry. For example, a normal *Escherichia coli* cell can use lactose, a β-galactoside sugar, when it is supplied in place of glucose. The ability to split lactose, a disaccharide, into the monosaccharides glucose and galactose depends on a specific enzyme, β-galactosidase, which the normal cell can manufacture. However, in the absence of lactose, this enzyme is not needed, and fewer than five molecules are present in a cell. It would be a waste of energy for the cell to manufacture large amounts of any protein it does not require. Normally, therefore, the enzyme is present in significant quantities only when lactose, the substrate of the enzyme, is present. About 5,000 molecules are formed within minutes when this sugar is supplied. Such a cellular reaction is called *induction,* the formation of a specific enzyme in response to the presence of a substrate. However, certain mutant *E. coli* cells arise in which the enzyme is formed indiscriminately. These are called *constitutive mutants,* and they produce large amounts of the enzyme in both the presence and absence of substrate. Such mutants possess some defect in a control mechanism, and they consequently waste energy by producing enzyme when it is not needed.

Associated with β-galactosidase are two other enzymes, β-galactoside permease and galactoside transacetylase. The former is involved in the transport of lactose across the cell membrane into the cell, where it is cleaved by the galactosidase into glucose and galactose. The function of the transacetylase is still unclear. All three enzymes are formed by the constitutive mutants in the absence of the substrate lactose. Chemical analysis of these three protein products in the constitutive mutants shows them to be normal. Therefore, nothing has gone wrong to alter the composition of the protein. Instead, a mutation has interfered with the cell's ability to produce the three enzymes only when they are needed. The constitutive mutant cell cannot control the level of enzyme production; it produces large amounts of normal enzyme regardless of the cell's need for it.

The lactose system is one of several in which mutations are known that can upset controls, and it has been studied in great detail. The information gained from it applies to the other systems as well and permitted Jacob and Monod to formulate a model of genetic control. This model is based on the results of crossing various kinds of mutants in different combinations. As in the case of any other proteins, there are mutations that can alter the structure of the three enzymes of the lactose system. From crossing cells containing mutations of this sort, it has been possible to map the sites of the genes that code for the formation of the three enzymes. Mapping has shown (Fig. 17-1A) that the three loci are clustered together in a sequence on the chromosome (Z, Y, and A for the galactosidase, the permease, and the transacetylase, respectively). A mutation in any one of these can alter the structure of its polypeptide product. Since these three genes direct the amino acid sequences in polypeptides, they are structural genes.

Mutations that affect the regulation of enzymes and produce constitutive mutants do not occur in their structural genes. Instead, they map in other locations. One kind of constitutive mutation maps in a portion of the chromosome that is a distance away from the three clustered structural genes. However, these mutations result in the unregulated production of the three enzymes of the lactose system. Such mutations enabled Jacob and Monod to identify a *regulatory gene*, a gene that is not concerned with

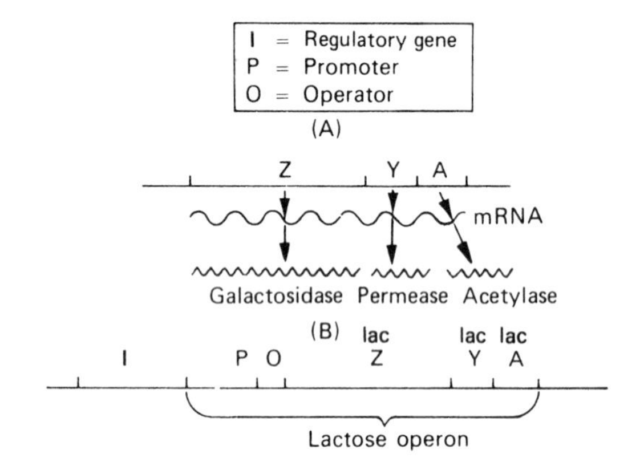

FIGURE 17-1. The lactose operon. **A:** Three enzymes are involved in the metabolism of lactose. Each is controlled by a structural gene, and the three genes map together in a portion of the *E. coli* chromosome. These genes undergo transcription to form mRNA when the cell is induced by the presence of lactose. A mutation within any one of the genes can alter the structure of the protein it governs. **B:** Other genes, the operator and the regulatory gene, can affect the lactose system. However, a mutation within either of these does not alter the structure of any one of the three enzymes. Instead, the mutation can result in a failure of the enzymes to be produced at the appropriate time. The operator is located adjacent to the structural genes. The promoter region to which RNA polymerase attaches to begin transcription is next to the operator. The structural genes for the enzymes plus the operator and promoter compose an operon. All the genes and sites within an operon are transcribed on the same mRNA. The regulatory gene (I) maps on the other side of the promoter.

the formation of an enzyme but rather with a regulator substance (Fig. 17-1B). But the regulatory gene is not the only genetic factor involved in control of the three genes of the lactose system. A genetic region called the *operator* has also been recognized as a result of mutations that have arisen within it. It has been mapped and shown to be located adjacent to structural gene Z. Like the regulatory gene, it is concerned with control and not with the amino acid sequence of an enzyme. Gene Z, along with the promoter and operator, was isolated in 1964 using transducing phages (Fig. 17-2). The isolated gene has been shown to be 1.4 mμ in length.

Jacob and Monod were able to show that the regulatory gene is responsible for the production of a repressor substance that is produced by the gene in the absence of the substrate lactose (Fig. 17-3A). When lactose is not present in the cell, the repressor can combine with the operator. As a consequence, RNA polymerase that attaches to the promoter site cannot transcribe the information of the three genes into messenger RNA, even in the presence of cylic AMP and CAP (see Chap. 12). Transcription is thus effectively blocked by the interaction of the repressor with the operator site.

Now let us suppose that lactose is added to the medium and enters the cell (Fig. 17-3B). The substrate lactose has the ability to act as an inducer or effector which can bind to a specific repressor molecule. The latter is normally present only in very small amounts. As a result of the interaction between repressor and inducer, the operator is freed. The RNA polymerase bound at the promoter encounters no barrier and can proceed with transcription. However, if a constitutive mutation arises in the regulatory gene, the latter may be unable to produce a normal repressor. An altered repressor may be unable to combine with the operator. Consequently, RNA polymerase is free to transcribe the structural genes at any time. Transcription can thus take place even in the absence of the substrate, when the three enzymes are not needed (Fig. 17-3C).

While lactose, the substrate of β-galactosidase, is commonly thought of as the inducer, this is not the case in the living cell. A metabolite of lactose, allolactose, is formed after the entry of the sugar. The very small amount of β-galactosidase found even in an uninduced cell brings about the formation of the allolactose, which is the natural inducer of the enzyme.

If a constitutive mutation occurs in the operator gene, again the enzymes are produced without regard to the presence of substrate. This is so because the

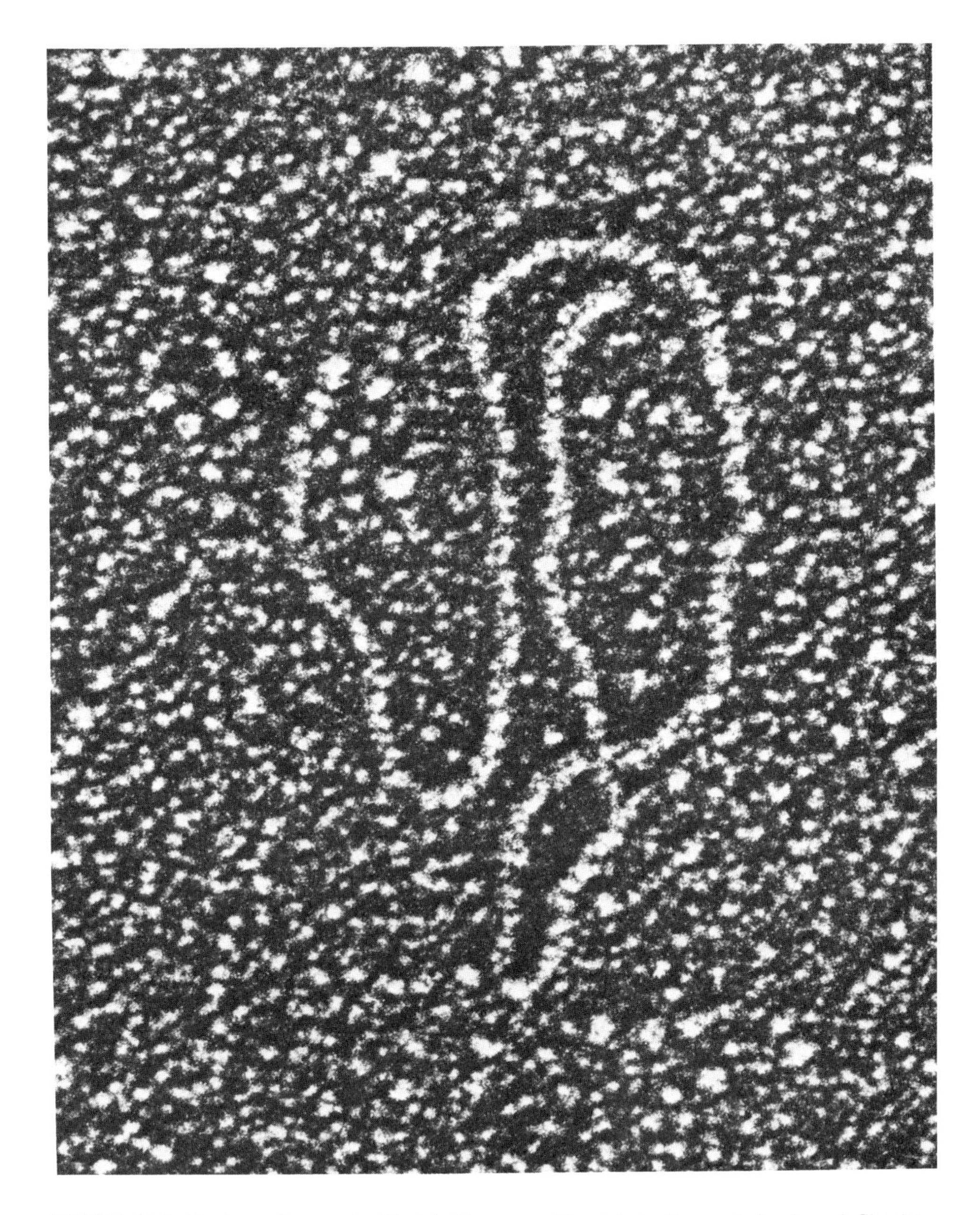

FIGURE 17-2. Electron micrograph of isolated lac gene. (Reprinted with permission from J. Shapiro, L. MacHattie, L. Eron, et al. *Nature* 224:768–774, 1969.)

mutation in the operator renders it incapable of binding with the normal repressor (Fig. 17-3D). Consequently, the RNA polymerase is again free to transcribe the structural genes. One should note that the three structural genes are undergoing transcription together on the same stretch of messenger RNA. Thus the promoter, the operator, and the adjacent genes define a unit—a unit of transcription called the *operon.* An operon is composed of an operator, a promoter, and associated structural genes (or cistrons), which are all transcribed on the same piece of mRNA. A piece of mRNA that contains the coded information for two or more genes is polycistronic; it carries the information that can direct the formation of more than one protein. In transcription of the operon, the promoter region is not included in the transcript, but the operator or a portion of it may be. We can now appreciate more fully the need for punctuation codons in the mRNA to distinguish between two or more polypeptide chains controlled by two or more different genes in an operon. The RNA codons—UAA, UAG, and UGA—along the messenger are needed to prevent a run-on protein. Without them, the ribosome would be unable to complete one protein and start the next. The three enzymes in the lactose operon would all be hooked together and would not exist as separate functional entities.

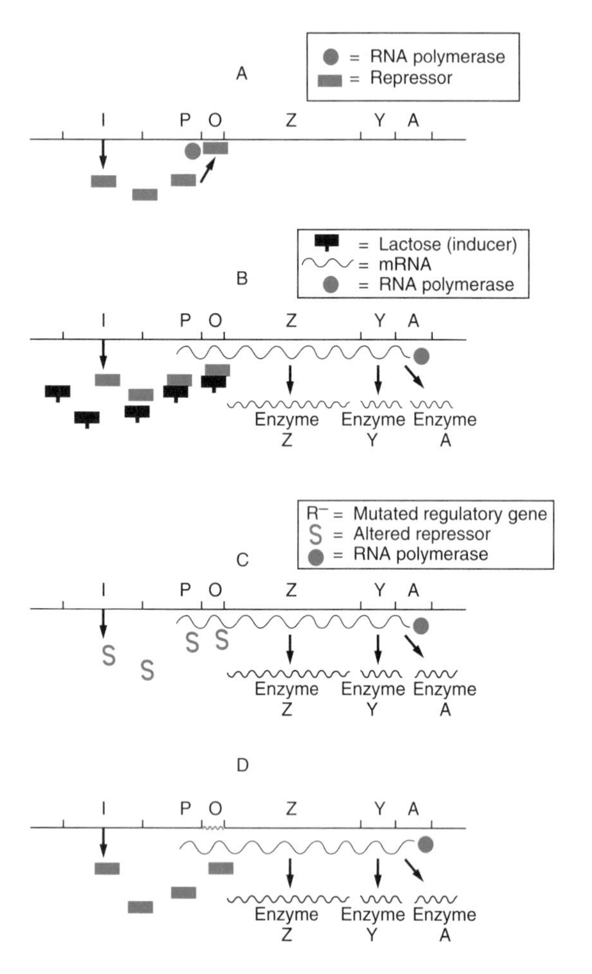

FIGURE 17-3. The repressor. **A:** The regulatory gene produces a substance that can combine with the operator. In the absence of the substrate, lactose in this example, combination of the repressor and the operator prevents the RNA polymerase from moving along the DNA to bring about transcription. **B:** When an inducer is added (the lactose), it binds to the repressor. The operator site is freed. The polymerase can now move along the DNA to transcribe it into mRNA. **C:** If a mutation arises in the regulatory gene, the repressor may be altered in structure so that it cannot combine with the operator. Therefore, even in the absence of the lactose, the RNA polymerase is free to transcribe genes in the operon. The mRNA is then translated, and the three enzymes are formed in the cell, even when they are not needed. **D:** A mutation can affect the operator in such a way that it no longer can combine with the normal repressor. The result, again, is freedom of the RNA polymerase to transcribe genes in the operon. The three enzymes are formed constitutively, without regard to presence of the substrate.

The lactose repressor and its interactions

When Jacob and Monod announced the operon model in 1962, they did not identify the repressor, but they had sufficient evidence to propose its existence. For example, it was possible through sexduction (see Chap. 15) to obtain *E. coli* cells that were diploid for the lactose region of the chromosome (Fig. 17-4). Partial heterozygotes were derived that possessed a normal regulatory gene and a normal operator on the sexduced fragment; the chromosome contained a defective regulatory gene. Such a cell behaves normally; it is not constitutive but produces enzyme only when needed. Figure 17-4 shows clearly why this is so. The defective repressor cannot bind with the operator. If the cell were haploid, lacking the F factor with the chromosome segment, enzyme would be produced indiscriminately (Fig. 17-3C). However, the presence of a normal regulatory gene anywhere in the cell, even on the episome, gives rise to the production of normal repressor. This must be a substance that can diffuse throughout the cell and bind with both normal operators, the one on the chromosome and the one carried on the sex factor. The structural genes on both DNAs are therefore not transcribed unless an inducer can tie up the normal repressor. In effect, the constitutive regulatory gene mutations are behaving as recessives to the normal gene form.

In contrast to this, the constitutive operator mutations were shown to be dominant. A study of Figure 17-5 shows why this is expected. In the diploid state, one of the DNA segments has a normal operator (the F factor in Fig. 17-5). The repressor substance produced by either regulatory gene can bind to it. But this is not so with the mutant operator (the one on the chromosome). Since it cannot bind with the repressor, the structural genes adjacent to it (in the cis arrangement) are free to be transcribed in both the absence and presence of substrate.

In the mid 1960s, genetic repressors were isolated by the team of Gilbert and Muller-Hill, working with the lactose operon, and by Ptashne, who was studying a repressor of phage in *E. coli.* The lactose repressor turned out to be a rather ordinary acidic protein, an aggregate consisting of four identical polypeptide sub-

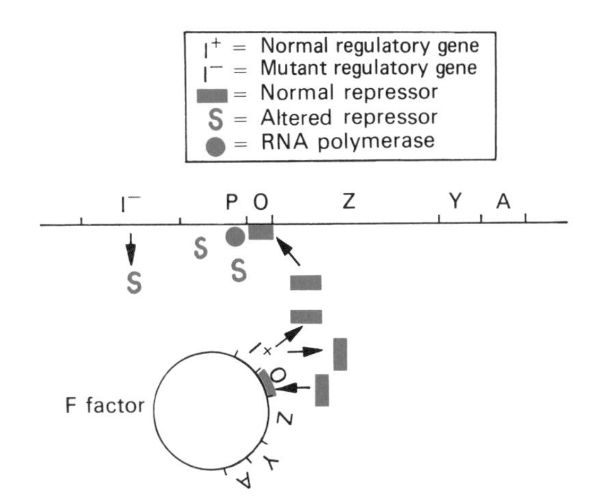

FIGURE 17-4. A heterogenote for the regulatory gene. A bacterial cell may possess a fertility factor that carries a lac operon and a regulatory gene. The regulatory gene on the chromosome may be defective and produce ineffective repressor. However, if a normal regulatory gene is present in the sex factor, it will cause normal repressor to be formed. This combines with the operator on the chromosome and also with the operator associated with the lac genes on the F factor itself. The cell consequently behaves normally and does not produce enzyme in the absence of the substrate. The regulatory gene mutation is thus behaving as a recessive to the normal.

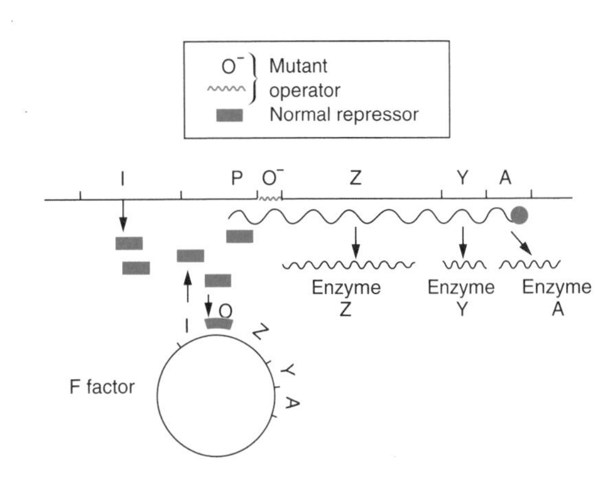

FIGURE 17-5. A heterogenote for the operator. If a mutation occurs in an operator gene, it may not bind with the normal repressor. This means that those structural genes adjacent to it (in the cis arrangement) will be freely transcribed. The figure shows a normal operator in the sex factor combining with the repressor, which is being produced by both regulatory genes, one on the chromosome and one on the episome. The operator on the chromosome does not combine with the repressor, so that the three enzymes are produced in the presence or absence of substrate. The operator mutation therefore behaves as a dominant.

units. Each of these is composed of 147 amino acids whose exact sequence is known. Since the regulatory gene is coded for the amino acid sequence of a polypeptide, it too is a structural gene, even though its polypeptide product is regulatory rather than enzymatic.

In one of the experiments, isolated repressor was made radioactive and mixed with the appropriate DNA. For example, lac repressor was placed in a centrifuge tube along with DNA known to contain the lactose genes. When spun in a density gradient, the repressor bound tightly to the DNA and sedimented with it. In contrast, when the repressor was spun in the centrifuge tube with DNA from cells containing lactose-constitutive mutations, the repressor did not bind. These results provided strong support for the model of the operon and the concept of a repressor that physically blocks genetic transcription.

Extensive investigations have been made on the lactose operon over the years. When the Z gene was isolated using transducing phages (see Fig. 17-2), the adjacent operator and promoter regions were obtained along with it. The exact nucleotide sequence of the operator region is now known. This has been accomplished by subjecting *E. coli* DNA to nuclease treatment. Lactose repressor protein, when added to the DNA before exposure to the nuclease, protects from digestion the operator region to which it binds. Analysis of the protected portion has shown it to consist of about 26 base pairs and to contain two regions where there is a degree of symmetry:

$$\overrightarrow{\text{5}'\text{–AATTGT}}\text{--------ACAATT–3}'$$
$$\text{3}'\text{–TTAACA-------}\overleftarrow{\text{TGTTAA–5}'}$$

(Recall that such inverted repeats were discussed in relation to terminator sequences in Chapter 12 and to restriction enzymes in Chapter 16. We will continue to encounter such DNA regions throughout the following discussions. Inverted repeats and palindromes apparently act as signals in various kinds of molecular interactions.) The particular portion in the case of the lac operon can apparently be recognized by the repressor, two subunits of which bind to the regions of symmetry. Alteration within the 26 base pair portion as a result of mutation can result in defective binding of the repressor. Not more than 20 copies of the lac repressor are normally found within a cell.

The 26 base pairs of the operator region are now known to contain the starting point for synthesis of lac mRNA. A bit of overlapping of nucleotides exists between the end of the lac promoter region and the beginning of the operator. When repressor is bound, it inhibits transcription of the operon by preventing RNA polymerase from beginning transcription at +1, the site of initiation of the mRNA (Fig. 17-6). The operator and promoter have been referred to as *controlling elements,* because they respond to environmental signals and thus control transcription of the gene or genes with which they are associated.

When the lac repressor is removed, the rate of transcription of genes in the operon may nevertheless still be low. The operon requires an activator, a regulator molecule that acts in a positive fashion to control transcription as opposed to a repressor that acts in a negative way. CAP is the activator, and it must bind to the region of the promoter of the lac operon if RNA polymerase is to bind at a high rate. As a positive regulator molecule, CAP is required for maximum transcription of those genes that code for enzymes needed to metabolize sugars, such as lactose, arabinose, and galactose. If glucose is present in the environment, certain microorganisms such as *E. coli* will use it preferentially, even if other sugars are present. The presence of glucose brings about inhibition of the synthesis of enzymes required for metabolism of the other sugars, a phenomenon known as the *glucose effect.* A metabolite of glucose is somehow able to reduce the level of cAMP that CAP requires if it is to stimulate the promoter. By itself, CAP is ineffective. Once it becomes complexed with cAMP, however, it is able to bind to the promoter region (review Fig. 12-24) and increase the rate at which RNA polymerase molecules bind to the promoter and initiate transcription. When glucose is present, cAMP levels fall; fewer of them are available to activate CAP molecules, and consequently transcription of the operon is reduced. Conversely, when glucose amounts decrease, cAMP levels increase and so do the positive effects of the CAP–cAMP complex as a stimulus of transcription.

CAP occurs as a dimer composed of two identical subunits. A single cAMP is sufficient to activate the dimer and bring about binding to the CAP site, which is adjacent to the promoter and which is very short, only about 22 base pairs long. A consensus sequence of five base pairs has been identified at the binding site. It is apparently this sequence that CAP recognizes, since mutations that occur in it prevent CAP binding.

In various microorganisms, the maximum amount of energy is provided by glucose as the carbon source. The lactose operon and others that show the glucose effect illustrate very clearly the interactions between positive and negative controlling factors for the highest degree of cell efficiency.

The structural gene coded for the amino acid sequence of CAP and the one coded for the amino acid sequence of the enzyme required for cAMP formation have also been identified. A mutation in either of these structural genes greatly reduces the amount of transcription from those operons such as lac that show the glucose effect. Moreover, operons are known in bacteria (the L-arabinose and L-maltose operons, e.g.) in which a specific regulatory gene associated with a specific operon is coded for a substance required for the activation of the structural genes in the particular operon. Both kinds of control, positive and negative, are undoubtedly involved in regulating most structural genes.

Repression of enzyme synthesis

The amino acid tryptophan is synthesized in five steps. The completion of these steps requires five different polypeptides, encoded by five structural genes that lie adjacent to each other on the *E. coli* chromosome (Fig. 17-7A). Three enzymes, essential to the five steps leading to tryptophan synthesis, are derived from these

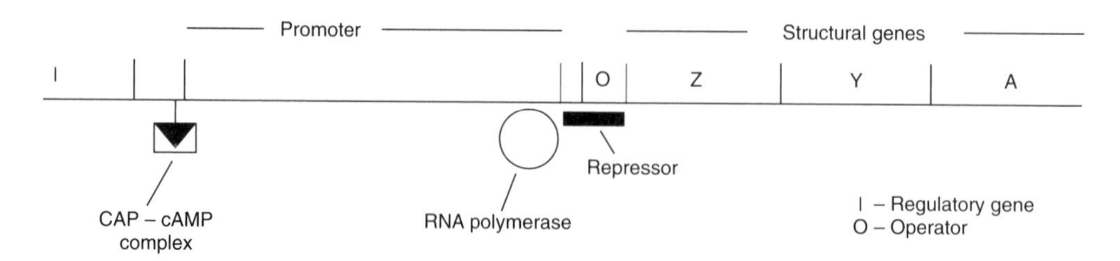

FIGURE 17-6. Blockage of transcription by repressor in the lac operon. Adjacent to the lac promoter is a CAP site to which CAP, activated by cAMP, binds and stimulates transcription. The site in the promoter at which RNA polymerase initiates transcription overlaps with the beginning of the operator. If repressor is attached to the operator, the polymerase that is bound cannot initiate transcription.

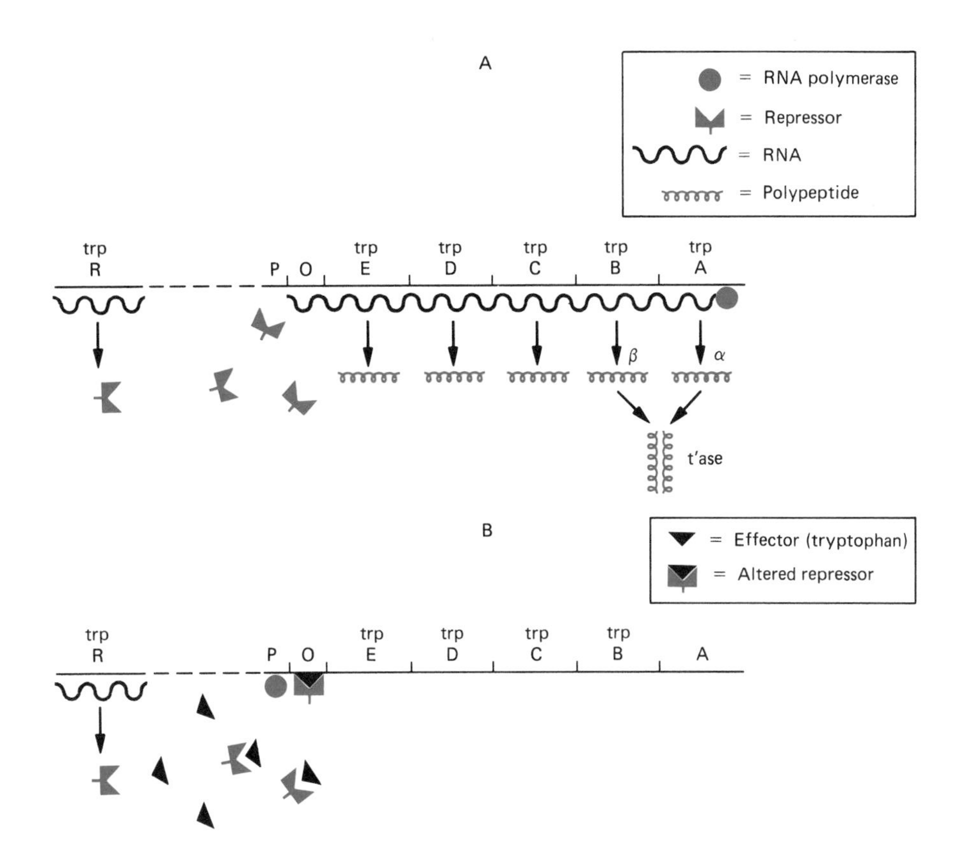

FIGURE 17-7. The tryptophan operon and repression of enzyme synthesis. **A:** The trp operon includes five structural genes lying next to each other, each coded for a polypeptide needed as a catalyst to form tryptophan. Two of these, β and α, encoded by trp B and trp A, associate to form tryptophan synthetase (t'ase). The regulatory gene, trp R, is responsible for producing a repressor that cannot by itself combine with the operator (O). Consequently, transcription is allowed from the promoter (P). The five polypeptides are formed, β and α combining to form t'ase. **B:** If tryptophan is added to the cell medium, t'ase and the other proteins needed to catalyze the synthesis of tryptophan are no longer necessary. The tryptophan acts as an effector or corepressor, which alters the repressor. This altered repressor can now combine with the operator and block transcription.

polypeptides. The last step in forming tryptophan (the union of indole and serine) requires the enzyme tryptophan synthetase. This enzyme is composed of two of the polypeptides (β and α) encoded by the structural genes B and A, respectively. The five contiguous structural genes, associated with an operator and promoter, constitute the tryptophan operon.

While complete overlapping between adjacent genes does not occur in bacteria as it does in viruses (Chap. 16), slight overlapping may be found at gene boundaries in cases of bacterial operons containing several structural genes. For example, the sequence TGATG occurs at boundaries between trp E and trp D, as well as between trp B and trp A. TGA is read as a stop codon (UGA in the mRNA) in one frame and the overlapping ATG (AUG in the mRNA) as an initiation codon in the other frame.

The regulatory gene trp R, associated with the trp operon, in contrast with the lac operon, lies an appreciable distance away from the operator with which it interacts. trp R is responsible for the production of the trp repressor, which, like the lac repressor, functions as a tetramer composed of four identical subunits.

Unlike the lac operon, the trp operon in *E. coli* is repressible and illustrates the repression of enzyme synthesis. The bacteria need tryptophan for growth and usually must synthesize it. Therefore they manufacture tryptophan synthetase and the other catalysts involved in the synthesis of tryptophan. However, if tryptophan is supplied to the cells, production of the five polypeptides ceases as the five structural genes are repressed. This is economical, because it would be wasteful for the cell to make all five polypeptides and the end product, tryptophan, when a ready supply of the required substance exists. The operon model explains this repression in a manner similar to that used for the lactose operon (Fig. 17-7A,B). The main difference between the tryptophan and lactose control systems is that the operator in the former does *not* react with the repressor in the absence of the effector, in this case the tryptophan. But, when tryptophan is added to the cell, it acts as a *corepressor* and alters the repressor. It is this changed repressor that can bind to the operator and block transcription. In a repressible system, as seen in the example of tryptophan, therefore, the effector (the end product in a reaction) alters the repressor so that it *can* combine with the operator and block transcription. In inducible systems, on the other hand, as in the lactose operon, the effector (substrate of an enzyme) alters the repressor so that it *cannot* combine with the operator, thus allowing transcription to take place (see Fig. 17-3B).

Choice between lysis and lysogeny in phage λ

An excellent example of gene control that entails more than one operon is provided by phage λ. Whether λ establishes itself as prophage after entry into the cell or enters a lytic cycle depends on the interaction and balance among six regulatory proteins that form in the host cell and are coded by the DNA of viral genes. Establishment of the virus as prophage depends ultimately on the presence of the main λ repressor protein, which is formed following transcription of one of the viral genes, cI. As noted in Chapter 16, phage λ is linear in the phage particle, but it becomes circularized shortly after the *E. coli* cell is infected. It is in this form before it becomes integrated to take up residence as prophage.

When a genetic map of λ is presented, the letters R and L (for right and left) are used to designate the direction of synthesis of two mRNAs, transcripts of genes N and cro, which are produced very early after the entry of λ (Fig. 17-8). The designation also indicates that mRNA synthesis is taking place on opposite strands of the λ DNA. A map of λ (Fig. 17-9), shows that genes with similar or related functions tend to cluster together. Gene clustering often occurs with related genes in organisms from bacteria to diverse eukaryotic species, including the human.

Let us first follow the chain of events that take place when a lytic cycle results following λ infection. In the lytic cycle, three sets of genes and three stages are recognized to indicate the timing of events in the cycle. These are *immediate early, delayed early,* and *late.* Right after the entry of λ, the RNA polymerase of the *E. coli* cell recognizes two promoter–operator regions, P_L/O_L and P_R/O_R. Transcription from these two promoters occurs in opposite directions, as noted earlier and as the names imply. Transcription from P_R leads to the formation of mRNA from gene cro. (Follow Figure 17-10 throughout this discussion.) Transcription of this gene gives rise to cro protein, which can interact with the two operators. The importance of this will become clear a bit later. Transcription from P_L leads to formation of an mRNA from gene N and the formation of N protein. This protein is necessary for the next set of genes, the delayed early genes, to be read. The N protein combines with the host RNA polymerase and modifies it so that it ignores termination signals at the end of both the N and the cro genes (refer to Fig. 17-9 for positions of genes). Lacking N protein, transcription would halt at these points. N product is thus an activator that enables transcription to continue and to yield transcripts of two delayed early genes, cII and cIII. These transcripts are not involved in the lytic cycle, and so we will return to them a bit later.

N protein also enables the delayed early genes O, P, and Q to undergo transcription. Genes O and P are required for lysis, because their products are needed for replication of λ DNA. The fourth delayed early gene, Q, is coded for Q protein, which, like N protein, behaves as an activator or antiterminator. It is required for the late genes to undergo transcription. These tran-

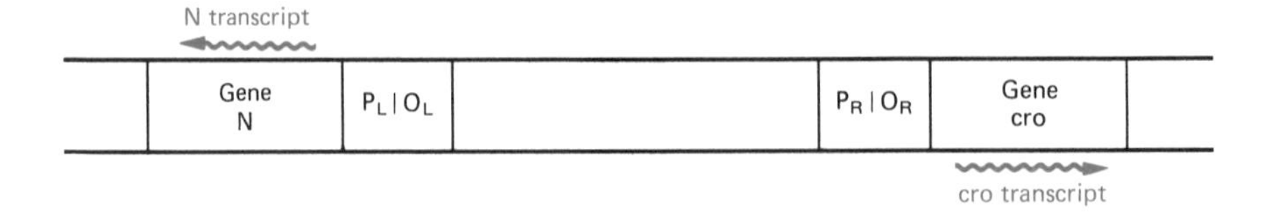

FIGURE 17-8. Direction of transcription from operators in phage λ. Not all of the genes of phage λ have the same template strand. When a map is presented, the symbols R and L are used to distinguish two different promoter/operator regions. Hence a left promoter/operator (P_L/O_L) and right one (P_R/O_R) can be recognized for genes that are expressed very early after phage infection (genes N and cro). The transcripts of these genes are seen to be derived from opposite strands of the DNA.

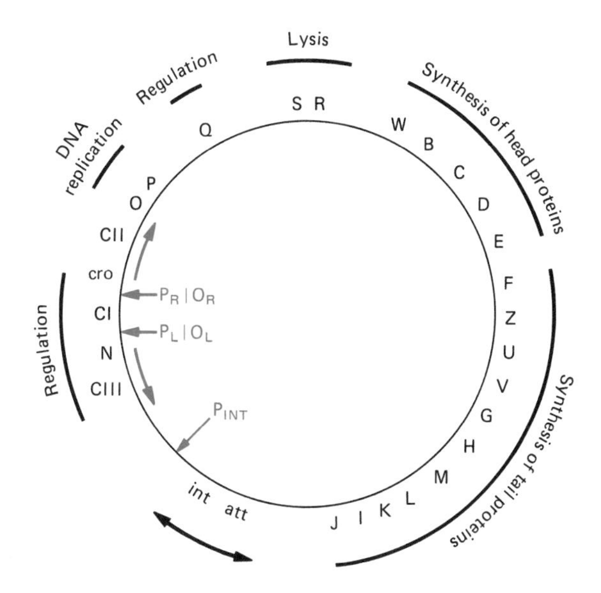

FIGURE 17-9. Partial map of phage λ showing the relative positions of some genes involved in the lytic and lysogenic cycles. Note that genes that have similar functions tend to occur in clusters. Two promoter/operator regions, P_L/O_L and P_R/O_R (red), are found on either side of gene cI. The red arrows indicate the different directions in which transcription occurs from these two regions. A third promoter, P_{INT}, is associated with genetic sequences for integration of phage into the chromosome of the host.

scripts of the late genes are coded for the protein of phage heads and tails and also for the enzyme, a lysozyme, which is needed for the bursting of the host cell. We see here in phage λ a neat regulatory system that enables critical events in the lytic cycle to follow an orderly time sequence.

Let us now examine the relationship of these events to the lysogenic cycle of λ. We noted that two delayed early genes, cII and cIII, were activated by N protein (Fig. 17-10). Both of these genes encode products that also have activator properties. The two products can bring about transcription from a promoter, P_{RE}, the promoter for the establishment of the main repressor protein. When formed, this strong repressor has an affinity for the two operators, O_L and O_R. These operators overlap the corresponding promoters, P_L and P_R. When the λ repressor is bound to the operators, transcription can no longer occur from P_L and P_R. Thus, N protein is no longer produced. Consequently, the genes involved in the delayed early and late parts of the lytic cycle are not activated. Note from Figure 17-10 that the cII and cIII products also activate another promoter, P_{INT}, the promoter for the integration of the phage (see also Fig. 17-9). Remember that phage integration requires a crossover event between phage and host-cell DNAs. The crossing over requires two proteins, one coded by a viral gene and one by a gene of the cell (see Chap. 16). In this case, the viral gene coded for the crossing over is under the control of P_{INT}, which in turn requires activation by cII and cIII products.

But what determines whether lysis or lysogeny occurs, since we see that N protein activates not only genes O, P, and Q of the lytic cycle but also cII and cIII, which are required for lysogeny? This takes us back to the very beginning of the story (Fig. 17-10). Recall that right after infection the immediate early gene cro is read. This gene encodes cro protein, which also has repressor properties. Like the λ repressor protein, it too can combine with operators O_R and O_L. It is not as strong a repressor, however, as the main λ repressor, does not bind as well, and is less stable. Thus, although it slows down transcription from O_R and O_L, it does not completely prevent it. Therefore, some N protein is produced, even though the cro repressor is bound. Just how much cro binds is central to the decision whether lysis or lysogeny will occur. Above a certain level, cro causes a decrease in the amount of N protein, which causes lysis to be favored. This is so because at the low level less activation of cII and cIII occurs. Their products are required for activation of the promoter for the repressor protein, P_{RE}. A low level of the cII and cIII products is insufficient to activate P_{RE}, and hence strong repressor is not formed to any extent. However, this lower level of N *can* sufficiently stimulate genes O, P, and Q, which are required for the lytic cycle. Even though a lesser amount of N and even Q products is produced, this is still sufficient for completion of the lytic cycle. Therefore gene cro is the critical gene in the choice between a lytic or a lysogenic pathway. Above a certain level of cro protein, a lytic cycle will be followed; below this level, a lysogenic one will occur. The amount of cro protein can be influenced by such factors as temperature, the genetic constitution of both λ and the host, and the metabolic state of the host cell.

A moment's thought tells us that once the lysogenic state is established, no more N protein will be produced. Therefore lysis does not occur, but neither does stimulation of cII and cIII, whose products are needed to stimulate the promoter of the repressor gene. A small amount of repressor must be synthesized to replace any that breaks down and thus to maintain lysogeny. This small amount results from transcription from a promoter other than P_{RE}—one called P_{RM}, for *promoter for repressor maintenance.* The control of this promoter depends on the concentration of λ

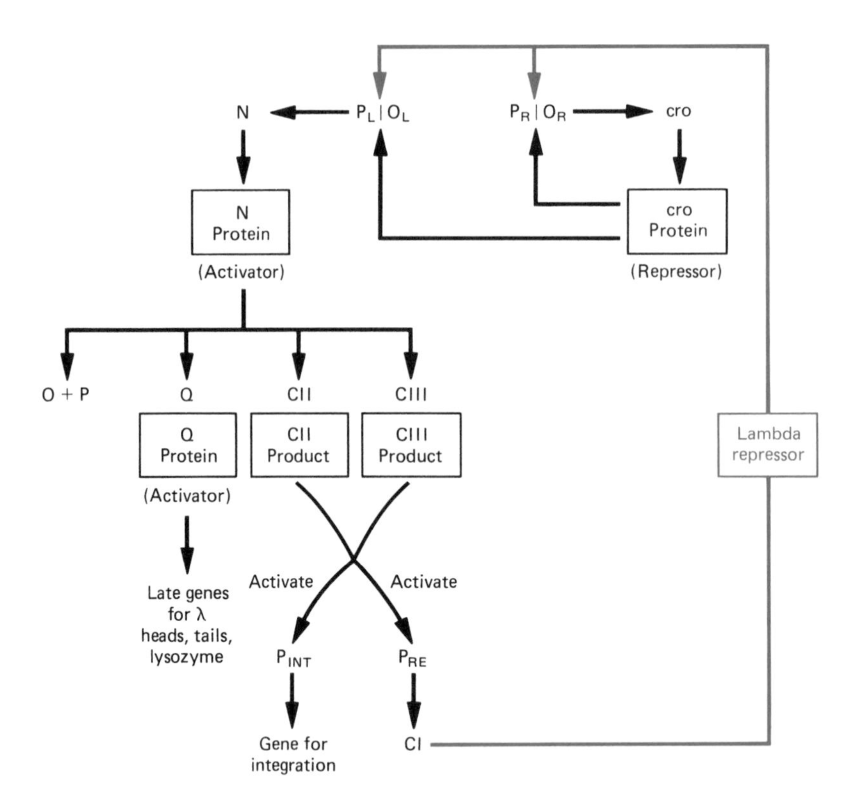

FIGURE 17-10. A scheme showing some factors involved in the lytic and lysogenic cycles of phage λ. Whether the virus establishes itself as prophage or follows a lytic cycle depends on a balance of several factors. Critical to the choice is the balance between cro protein, a repressor, and the main λ repressor (red), encoded by gene cI.

repressor itself. When λ repressor falls to a low level, the repressor somehow stimulates its own synthesis, whereas it inhibits transcription when it is present in higher amounts.

The monomer of the λ repressor, isolated by Ptashne, has a molecular weight of 26,000, and it may bind to O_R and O_L as a dimer or a tetramer. Ptashne has succeeded in isolating both operators and has found that each is really composed of three regions separated by spacers (Fig. 17-11). Mutation analysis indicates that these spacer regions are promoter sites. Recall that the operator and promoter regions overlap in the lac operon. The arrangement is more striking in the case of phage λ and accounts for the fact that the two regions are referred to as P_R/O_R and P_L/O_L.

There are mutations that affect λ that are comparable to the constitutive operator mutations in the lac operon. Certain mutant phages cannot establish themselves as prophage because of the inability of the mutant operator to bind to the repressor efficiently. As a result, N and cro genes remain active, and a lytic cycle is favored. Mutations are also known in the promoter regions that overlap with the operators.

In Chapter 16, the phenomenon of zygotic induction was discussed briefly. The activation of the prophage after entry into an $F^-\lambda^-$ cell results from the lack of repressor in the λ^- cell. When the chromosome segment of the Hfr cell bearing λ enters the recipient, transcription can start from both operators, to begin a lytic cycle. It should also be clear why a lysogenic cell is immune to another λ particle. If another λ enters such a cell, the repressor being made by the integrated prophage will immediately bind to O_L and O_R of the entering phage and prevent it from starting the lytic cycle.

Treating lysogenic *E. coli* with ultraviolet light, nitrogen mustard, or most carcinogens will activate the prophage with destruction of the host cells. These agents apparently can damage the host DNA and thus reduce the efficiency of binding of the repressor to the host chromosome.

Transposons and gene expression

Transposable genetic elements (Chap. 15) are very relevant to the subject of control of gene expression. It is now well established that certain transposons may act as biological switches. For example, the insertion of a particular IS element (IS3) into a particular plasmid can turn off the expression of a nearby gene, which can confer tetracycline resistance. Excision of the IS element from the region of the gene restores its expression for drug resistance.

IS elements are also known that can insert at multiple sites on bacterial and phage chromosomes and produce polar mutations. In these cases, insertion of the IS element not only switches off the function of the gene into or near which the element inserts but also affects the expression of other nearby genes. For example, the element IS1 can insert into the gal operon of *E. coli* (Fig. 17-12). The gal operon includes three structural genes coded for enzymes involved in galactose metabolism. When the IS1 element inserts next to the operator–promoter region, expression of genes distal to the insertion site is turned off. When the IS element leaves the site, transcription is restored and the three structural genes can be expressed upon induction. IS elements have also been shown to produce a similar effect in the lac operon.

Moreover, it is known that in the case of some IS elements the orientation or direction in which the element inserts is significant. Inserted in one direction, the IS element turns off genes nearby; inserted in the opposite direction, a nearby gene that is unexpressed may be turned on! A particular IS sequence (IS2) can insert in either of two orientations in the operator–promoter region of the gal operon (Fig. 17-13). When it is inserted in a direction opposite to the direction in which the gal structural genes are transcribed, the expression of the genes is decreased. This means that the three gal enzymes are produced in lower than normal amounts upon induction of the enzymes.

If, however, the element is inserted in the same direction in which transcription of the structural genes takes place, it then acts like a positive control element. The structural genes are turned on, and the enzyme production becomes constitutive! The element IS1 mentioned previously does not show any difference in its effect related to its orientation when it is inserted into the gal operator–promoter region. Transcription of the genes distal to its site of insertion is abolished regardless of its orientation.

It can be shown that the mutated DNA associated with these polar effects has increased in length compared with the corresponding nonmutant DNA. One way to illustrate this entails the use of transducing phages, one group carrying a specific unmutated DNA sequence, the other carrying the same sequence but showing a mutant polar effect. DNA is extracted from the phages and rendered single stranded (Fig. 17-14). When denatured single strands lacking the mutation are incubated with strands carrying the polar mutation, a loop can be seen with the electron microscope. This loop represents an additional DNA sequence that has been inserted into the normal sequence. The loop reflects the base pairing between the inverted repeats flanking the transposon. The length of the loop is a direct measure of the length of the inserted DNA.

There is no question that transposable elements such as IS sequences may alter gene expression. We have already seen (in Chap. 15) that they are actually needed in *E. coli* for the conversion of an F^+ cell into an Hfr type, and hence for the transfer of genes from one cell to another. Perhaps they actually play some basic role in gene regulation. We can speculate that reversal of orientation of certain IS elements such as IS2 can determine whether certain genes are expressed. Transposons, or "jumping genes" as they are often called, are the focus of much current research at the molecular level and may prove to have very important implications for normal gene control and expression.

The unusual virus mu

The unusual entity known as mu was discovered in 1963 and possesses properties of both a virus and a transposable element. It must be classed as a virus, however, since its DNA becomes packaged into a viral

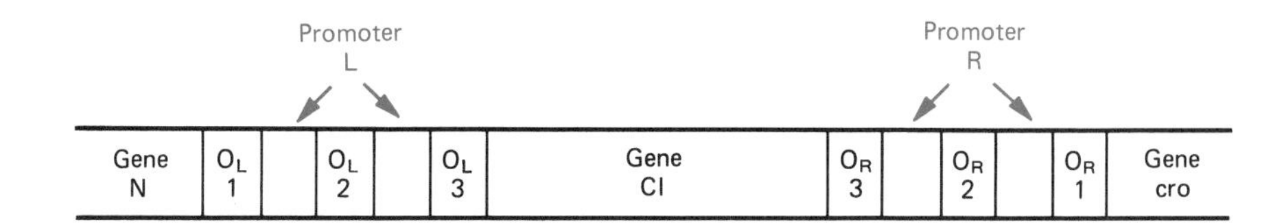

FIGURE 17-11. Operator regions of phage λ. Each operator is composed of three binding sites. Mutations in these alter the binding of λ repressor. Promoter mutations map between these operator sites, indicating that the promoters and operators in each of the two operons overlap.

particle. Moreover, it behaves as a temperate phage, resembling phage λ in certain respects. However, it can integrate just about anywhere on the bacterial chromosome, and integration is necessary for its replication. These are two features of transposons. Mu contains genes for its movement as well as genes coded for proteins that package its DNA. Its possession of transposase enables it to undergo transposition, which it does much more frequently than do other transposons.

The packaged virus contains segments of bacterial DNA from its previous host at each of its ends, 50–100 base pairs at its left and 1,000–2,000 at its right. When it integrates into a new host to take up residence as prophage, it sheds the bacterial DNA. The mu DNA continually remains in a linear form and does not circularize when it integrates. Mu DNA carries inverted repeats at its ends and brings about a five base pair duplication of a target site. A host gene that becomes interrupted by the integration of mu becomes inactive.

Mu is able to move by both simple and replicative transposition (see Chap. 15). After entering the cell, it inserts by simple transposition, and, if enough repressor is then made, it remains as a prophage. Otherwise,

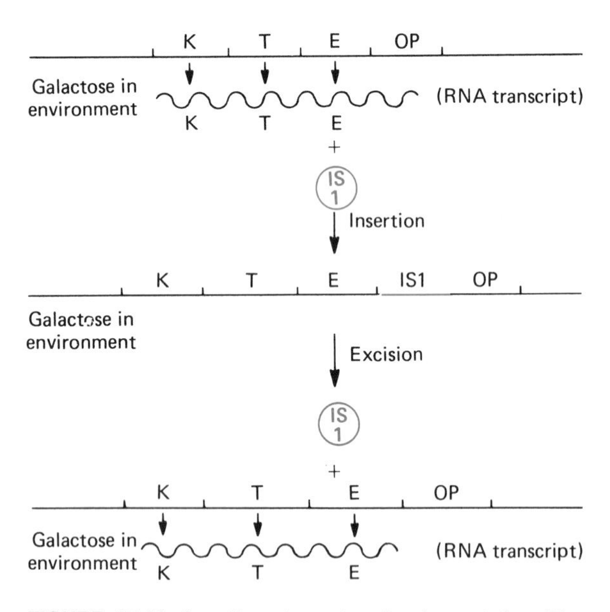

FIGURE 17-12. Insertion element and polar mutation. The three structural genes (K, T, and E) in the normal inducible galactose operon are coded for three enzymes required in galactose metabolism. The operon undergoes transcription when galactose is in the environment. If the insertion element IS1 inserts in the operator/promoter region (OP) of the operon, the three genes in the operon fail to undergo transcription when galactose is present and the three enzymes are required. When the IS element leaves the OP region, normal and inducible activity is restored to the operon.

a lytic cycle follows during which mu becomes duplicated by replicative transposition into different sites of the bacterial DNA. One copy always remains at the "old site," and another copy appears at another site. Consequently, mu keeps increasing in number of copies, up to about 100. However, resolvase is not made in the replicative part of the cycle, and cointegrates are not formed (see Fig. 15-26). When mu becomes packaged, some of the adjacent host DNA is carried along with it. Since the mu particles can insert at random in the host chromosome during the replicative phase, the various packaged mu particles will usually have different bacterial DNA flanking them.

This curious entity was named *mu* for *mutator,* since it causes various kinds of genetic alterations while it is established as prophage in the host cell. It is able to bring about the fusion of two different replicons. For example, it can cause a λ virus that has entered the cell to integrate with the cell's chromosome. It can cause portions of the host chromosome to be transposed to a plasmid. In addition, it can bring about deletions and inversions of the host chromosome (Fig. 17-15). Following such changes, the particular rearranged DNA segment is found to have a copy of mu at each of its ends. The recognition of other transposable elements that have properties similar to mu and that can alter gene expression and bring about chromosome rearrangements is highly significant to studies of cellular control mechanisms.

Transposable elements in corn

The very existence of transposable genetic elements was first deduced in the 1940s in a brilliant series of genetic studies by Barbara McClintock, who was studying the inheritance of color and pigment distribution in kernels of the corn plant. Certain kernels were mottled or variegated in appearance, because they contained different-sized patches of pigment. In these plants, certain genes were being turned off at abnormal times. These plants had also experienced chromosome breakage. McClintock followed many generations of such plants throughout a complex genetic analysis. She concluded that transposable genetic elements (which she called *controlling elements*) were responsible for the mottled pattern and that these could move about the genome from one corn chromosome to another. As they did so, they acted as switches. Genes could be turned on or off by particular transposable elements. More than one gene could be affected at the same time. Examination of

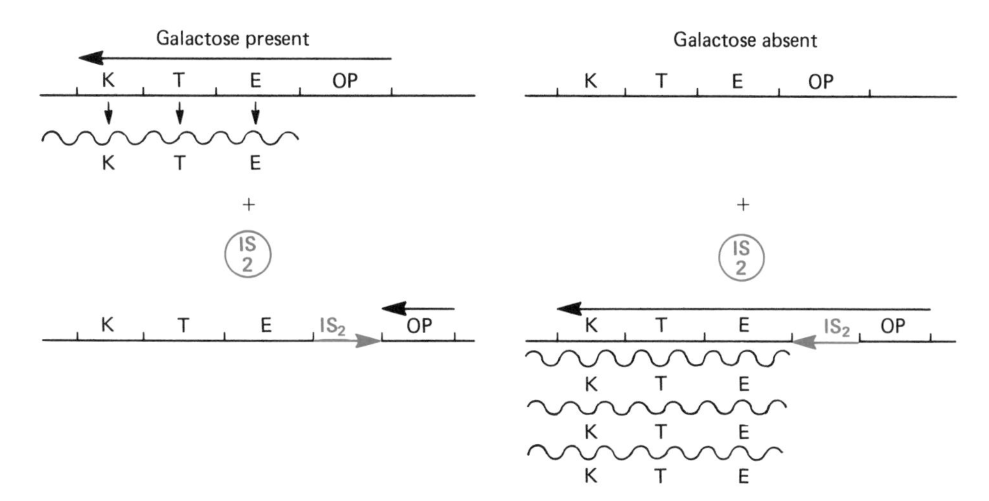

FIGURE 17-13. Effect of orientation of IS element. The three structural genes of the galactose operon normally undergo transcription only when induced by galactose. If the insertion element IS2 inserts into the operon in a direction opposite to that of transcription of the structural genes (left), transcription is greatly reduced when galactose is present and the three enzymes are required. If IS2 inserts so that it is oriented in the same direction as that of transcription of the operon (right), transcription then occurs indiscriminately, and enzyme synthesis is thus constitutive.

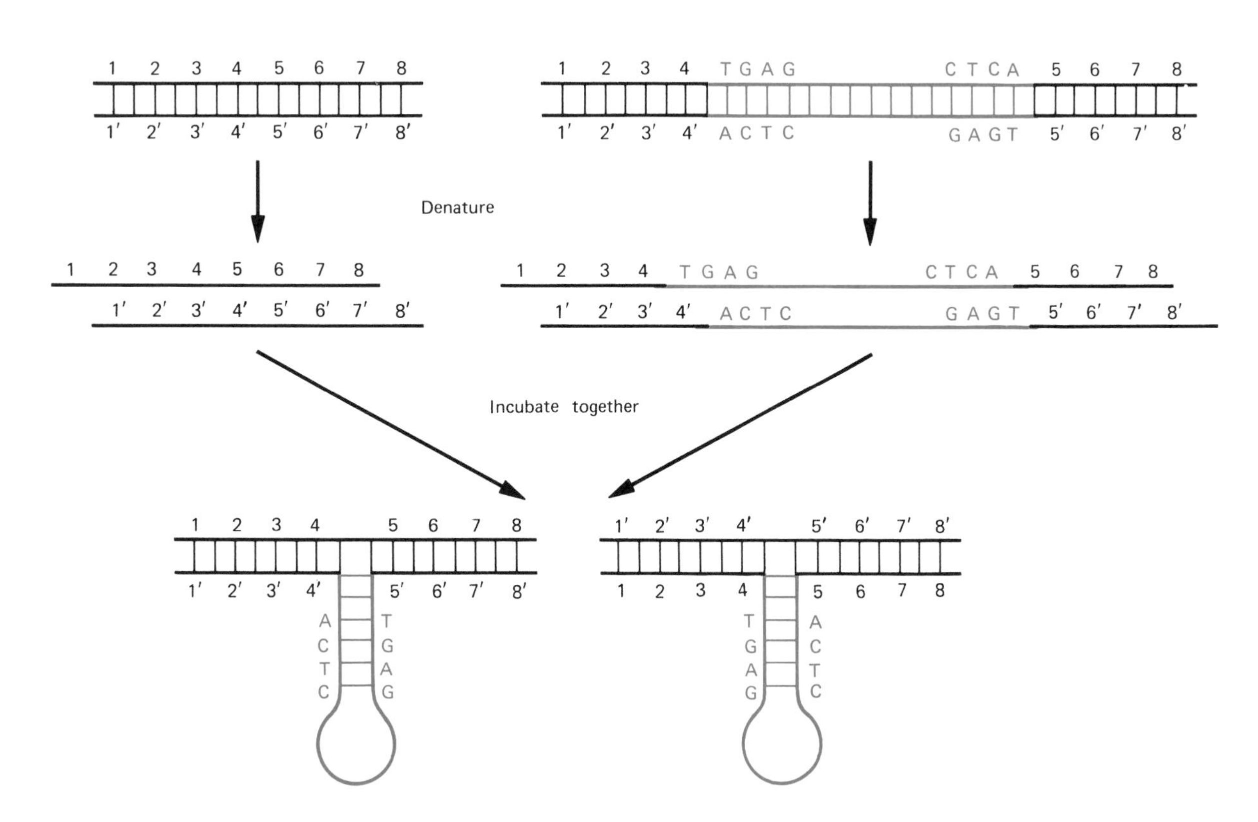

FIGURE 17-14. Evidence for an inserted transposon. DNA from phages carrying a specific genetic sequence is denatured, as is the DNA from other phages that carry the same sequence and that show a polar mutant effect. When single-stranded DNA from the two sources is brought together and allowed to anneal, DNA that contains loops can be distinguished. A loop represents an inserted sequence with the base of the loop paired as a consequence of the inverted repeats at the ends of the transposon.

corn chromosomes with the microscope showed that chromosomes with transposable elements could indeed undergo breakage and rejoining with the production of alterations in chromosome structure.

McClintock's report in 1947 that a genetic element could move about the genome was considered a peculiarity probably confined to a few very special cases. However, well over 20 years later, with the description of transposable elements in viruses and bacteria, McClintock's discovery was recognized as one of general importance, not just an isolated phenomenon. For her discovery of transposition, now considered a major discovery of the twentieth century, McClintock was awarded the Nobel Prize in 1983.

The two most thoroughly studied transposable elements in corn are Ds and Ac. The Ds element was the one McClintock first found to be capable of transposition. Associated with the movement of Ds was a breakage of the chromosome at the site or locus occupied by Ds. Hence McClintock named the locus *Dissociator* (Ds). However, it became clear to her that although the Ds site is the point where the chromosome breaks, Ds by itself cannot bring about the breakage. Only if another element at a different locus is present will the breakage occur at the Ds position. This second locus was named *Activator,* or Ac, to denote its ability to trigger chromosome breakage at the Ds site. Ac may occur on the same chromosome as Ds or on a different one. Nevertheless, in either case, Ac can activate breakage at the Ds site (Fig. 17-16).

McClintock soon realized that, in a small percentage of the offspring after certain corn plants were crossed, Ac could move to a new location (Fig. 17-17). This movement could be to the same or to a different chromosome. Moreover, Ac could even disappear. She showed that Ds could also move about, but, just as Ds depends on Ac for chromosome breakage, it also depends on Ac for its movement from one site to another (Fig. 17-18). The movement of Ac or Ds entails loss of the element from the original site and its transfer to a new site. It is therefore an example of simple transposition as described for transposons in Chapter 15. Because Ac can move about by itself, it is classified as an autonomous element in contrast to the nonautonomous element Ds, which depends on the presence of Ac for its transposition. As we will discuss later, Ac contains genes coded for proteins that apparently act as enzymes of transposition. Lacking these genes, the nonautonomous Ds depends on Ac for the proteins required for movement from one chromosome site to the next.

Early in this century, before McClintock's studies, it had been noted that a certain mutation in corn could affect the synthesis of reddish pigment in the layer of cells surrounding and protecting each individual kernel. The unusual feature of this mutation was that, although it prevented formation of pigment, it was very unstable and appeared to revert back to the normal pigment allele very often as each corn kernel was developing. In those cells in which the mutation had

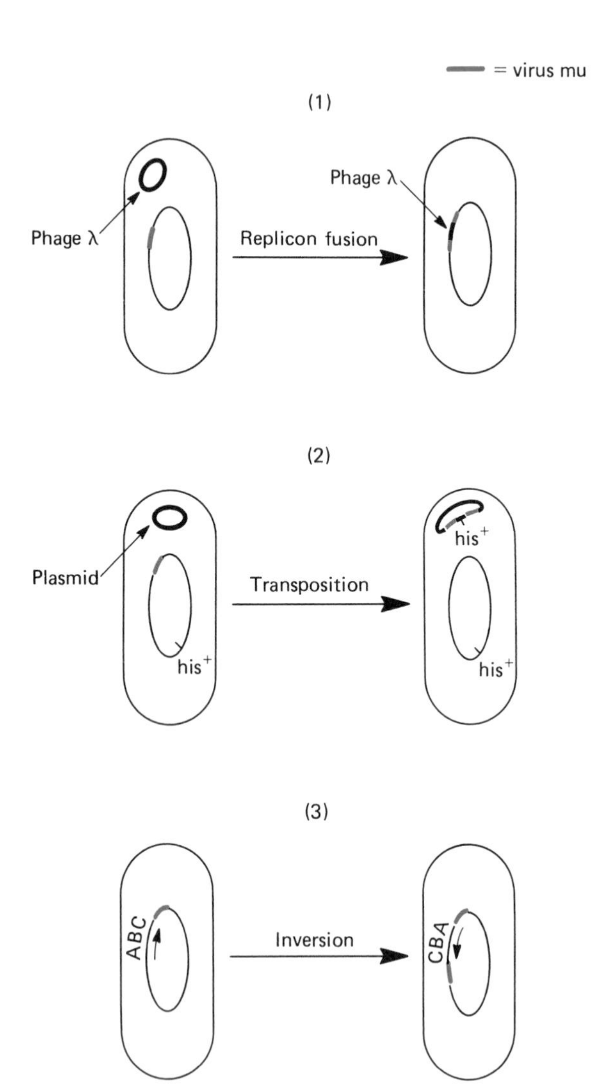

FIGURE 17-15. The virus mu. **1:** If phage mu is inserted into a bacterial chromosome, it can bring about the fusion of the chromosome with a phage λ, which enters the cell. Phage λ is then found between two copies of phage mu. **2:** Inserted into the chromosome, phage mu can cause transposition of a bacterial gene (his$^+$) to a plasmid in the cell. The transposed gene is then found between two copies of mu. **3:** Phage mu can bring about inversion of a segment of the chromosome. A copy of mu is then found at each end of the inserted segment.

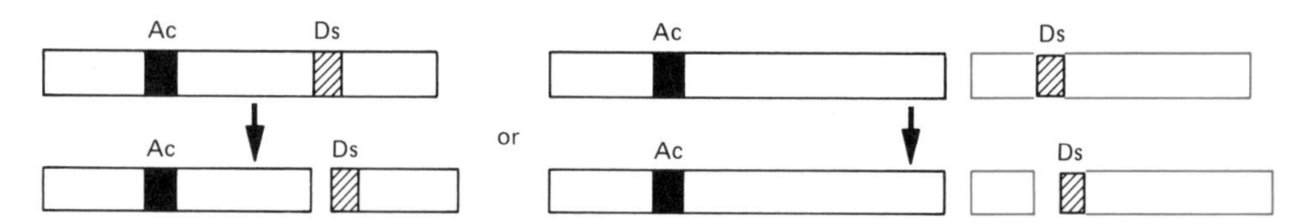

FIGURE 17-16. The Ac and Ds elements in corn. The Ds element may bring about chromosome breakage at the site it occupies, but it can do so only if the Ac element is present in the same cell. Ac and Ds can occur on the same chromosome (left) or on different chromosomes (right). In both cases, breakage is at the Ds site.

reverted, the pigment could be produced. The overall effect was mottling or variegation, a kernel with regions of pigmented and unpigmented cells.

Studies by M.M. Rhoades in the 1930s had shown that certain apparently stable gene mutations in corn can become unstable when a particular gene is present. This results in many reversions from the mutant allele back to the normal. The mutation in this case was one that affected the synthesis of purple pigment in the outermost layer of cells in the endosperm of the kernel, the layer just under the protective covering of the kernel. As a result of the mutation, no synthesis of purple pigment occurs, and a kernel may be white. However, if a certain gene is present in cells carrying the mutation, the mutation reverts frequently, so that the kernel has patches of purple amid the white background of the kernel. Experiments of McClintock with the Ac–Ds system clarified the story of the unstable mutations. McClintock found a case in the course of her studies in which Ds moved from its original site to the site of a gene designated *C*. The allele C is required for the synthesis of purple pigment in the outermost layer of the endosperm of the kernel. A mutation at the locus occupied by C can result in lack of pigment. McClintock was able to show that a certain new mutation at C giving colorless endosperm was the result of the insertion of the Ds element into C. The kernels in such a case are white, but only if Ac is absent from the rest of the chromosome complement (Fig. 17-19). If Ac remains present in the genome along with Ds, then chromosome breakage occurs at the C locus, just as it did at the site of Ds before it moved into C. Consequently, the mutation to colorless then reverts back to pigment production in some cells when Ac is present. The result is a kernel with splotches of pigment against a colorless background.

If Ds becomes lost from the C locus in germ cells of a plant, then offspring plants would have completely pigmented kernels, having reverted completely. The C gene functions perfectly well again after Ds removes itself. If Ac is present in the genome after Ds is gone, no further breakage occurs at the site of C. These unstable mutations studied by McClintock were similar to those studied earlier by Rhoades and other investigators and cast them in a new light.

The techniques of molecular biology have now made it possible to isolate and characterize Ac and Ds elements and to gain more details on their behavior. Ac has been shown to be a DNA element about 4,500 nucleotides in length flanked by inverted terminal repeats 11 base pairs long. When Ac inserts at a site, a target sequence of eight base pairs becomes duplicated. Most of the length of Ac appears to represent a single gene that produces an mRNA transcript that contains five exons. Ds elements have been found to vary in size as well as in DNA sequence. Most resemble Ac, but all are shorter due to the presence of deletions of different lengths. Most Ds elements appear to be defective Ac elements missing a nucleotide sequence in the gene that may code for a transposase. The inverted repeats that they all retain enable them to respond to transposase, however. There seems to be little doubt that these inverted repeats are essential for the recognition of transposase supplied by the autonomous element. Some Ds elements vary in structure from Ac, and these may not be derived from it, but they carry the same inverted terminal repeats found in Ac.

It appears that when an Ac or Ds element moves from one site to another it usually has been replicated. This means that when a chromosome becomes composed of two chromatids the element will at first be present in both of them at a particular site. If transposition should take place, the element will become lost from that site on one of the sister chromatids (Fig. 17-20). As noted, transposition is simple or conservative. Consequently, the chromosome will contain one chromatid with the element at that site and one lacking it. The element may move into a different site on the same chromatid or into a different chromosome. The chromosome site into which it moves may or may not have replicated.

The corn genome has been scanned to determine the frequency of elements such as Ac and Ds. It appears that

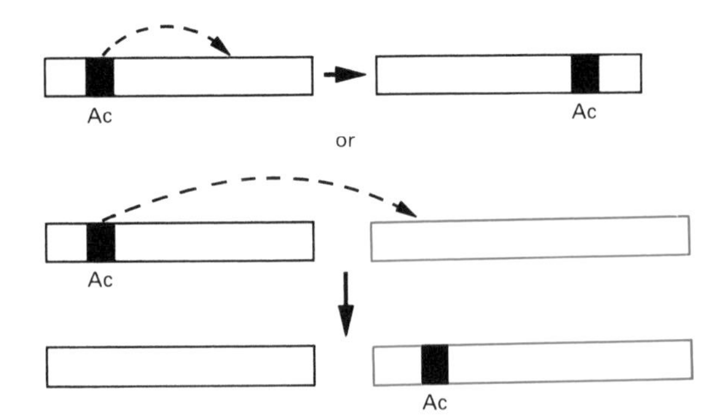

FIGURE 17-17. Transposition of Ac. Ac can move to a new location, either to one in the same chromosome (above) or to one in another chromosome (below). In either case, Ac is lost from its original site. Ac may even be lost from the cell entirely.

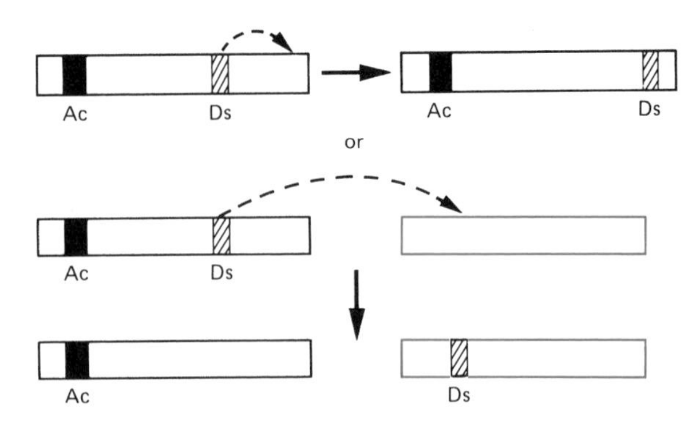

FIGURE 17-18. Transposition of Ds. Ds can move to another site on the same chromosome (above) or to a site on another chromosome (below). However, its movement depends on the presence of Ac in the same cell. In the absence of Ac, Ds will not undergo transposition.

corn may contain as many as 100 Ac or Ac-like sequences scattered throughout the chromosomes. However, the Ac–Ds system represents only one category of transposable elements in corn. Additional families have been identified by McClintock and others. Each family includes an autonomous element and several kinds of nonautonomous ones. The latter can be activated only by the specific autonomous element of that given family. (Ds, for example, can be activated by Ac but not by Spm, an autonomous element of another family.) Within each family, the elements appear to depend on family-specific enzymes and recognition signals for their movements. Nonautonomous elements all lack certain internal sequences, and the extent of the deletions may be great, but all still retain the inverted terminal repeats. It is believed that all nonautonomous elements are derived from autonomous ones following structural changes that inactivate the gene coded for the transposase but that leave the inverted terminal repeats intact.

As more transposable elements in eukaryotes are identified, their possible role in normal gene regulation will continue to be a focus of study. Very intriguing is the significance of such elements in organic evolution. The ability of some transposable elements to break and rearrange chromosomes makes them good candidates as important agents in rearranging the genome, thereby providing another source of genetic variability.

Mobile elements in *Drosophila* and yeast

In 1976, a certain class of DNA was recognized in *Drosophila melanogaster* that undoubtedly represents an assortment of transposable elements. This DNA consists of repeated sequences dispersed throughout the chromosomes. Some 20–30 families of dispersed repeated sequences are known in the fruit fly, and they may constitute as much as 5% of its DNA. The best studied of the families has been named *copia* to stress the presence of very large amounts of RNA transcripts of the repeated sequences of this family in the mRNA fraction of the cell. The behavior and structure of copia can be used to illustrate the general characteristics of several other families, even though the nucleotide sequences differ from one family to the next. Copia is present in about 50 dispersed copies per cell. Chromosome mapping has shown that the repeats vary in their locations in different strains of the fly, even among individuals of the same strain.

Each copia sequence is 5,000 base pairs in length, although smaller and larger sequences occur in other families. A copia element is flanked by direct repeats, each 276 base pairs long, reminiscent of the LTRs of retroviruses (review Fig. 16-22 for retroviral structure and replication). Each repeat is itself flanked by inverted repeats, each 17 base pairs in length (Fig. 17-21). On either side of the copia element, a short, five base pair direct repeat is found, suggesting the insertion of the copia element at a target site. The characteristics of copia in its structure and locations all imply that these are transposable elements. Supporting this is the observation that several different gene mutations in the fruit fly are the result of the insertion of copia sequences. Moreover, when cells of the fly are grown in tissue culture, the number of copia repeats per cell increases up to three times, and the new copies of copia appear at new chromosome sites.

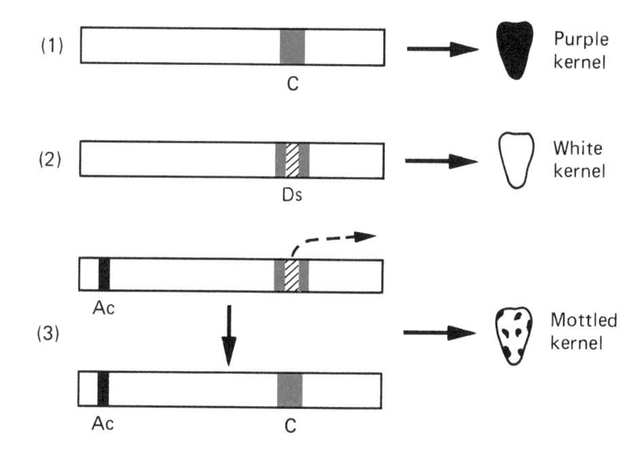

FIGURE 17-19. Mutation at the C locus in corn. **1:** The allele C is necessary for the production of purple pigment in cells of the kernel. **2:** If the Ds element inserts into C, the allele is no longer functional in pigment production, and cells of the kernel are white. To move, Ds requires the presence of Ac in the genome. If Ac is lost, Ds will no longer move, and all cells of the kernel will be colorless. **3:** If Ac remains present in the genome, Ds will undergo transpostion in some cells of the kernel as it develops.The mutation for colorless will thus revert to the gene form for pigment production in such cells. The kernel will appear mottled, showing patches of purple (pigmented cells in which Ds has moved out of C) against a white background of cells (those in which Ds is still inserted in C).

Culture conditions somehow appear to stimulate the rate of transposition of copia.

The DNA of the copia element has now been sequenced and shown to include one very long reading frame from which the abundant transcripts arise. The way in which these transcripts are handled in the cell has not been determined. However some of the gene sequences have been shown to be related to sequences that are found in retroviral genes coding for reverse transcriptase and the proteins involved in viral integration into the host DNA. These and the overall similarities in structure between copia and retroviruses suggest that copia's origin is related to these viruses and that its movement may involve reverse transcriptase activity and an RNA intermediate.

In yeast, we find a family similar to copia in many ways. It is also composed of repeated DNA sequences dispersed throughout the genome and found at different chromosome locations in different strains. The family consists of elements known as TY, the commonest being TY-1. This 6 kb pair element is found in about 35 copies in the yeast genome. In many respects it shows a very strong resemblance to retroviruses. At each of the ends of the TY element is a long terminal repeat (LTR), each containing a promoter. The element carries two genes that possess sequences that are homologous to those in retroviral genes coding for reverse transcriptase and other viral proteins. Transcription from the TY element is very frequent. Its transcripts can make up as much as 10% of the mRNA in a yeast cell. Experiments involving DNA manipulation have demonstrated that the transposition of TY elements involves an RNA intermediate in much the same way as does integration of a retrovirus. The DNA of the element was manipulated in such a way that it was placed on a plasmid under the control of the Gal promoter, and an intron was inserted into it. Transposition was induced, and it was found that the new copies of the element which became inserted at different sites in the yeast genome all lacked introns. There is no other way known for an intron to be removed other than by splicing of RNA. When transposition occurs, a TY element must form an RNA intermediate that is used as a template for reverse transcriptase, which forms a DNA copy that then inserts into a chromosome site. TY elements have been called *retroposons* to distinguish them from transposons that move from one DNA site to another without the formation of an RNA intermediate.

Mobile elements in the mammalian genome

Not only are members of the copia family thought to be related to retroposons, but there is also strong evidence to suggest that a good portion of the moderately repetitive DNA of mammals is also made up of them. In Chapter 13, we discussed moderately repetitive DNA and noted that within this kinetic class two general fractions may be recognized, the short and the long interspersed sequences (referred to as *SINES* and *LINES*, respectively).

The most abundant of the SINES in mammals is the Alu family. There are several reasons for suspecting that these repeats are capable of moving about in the genome and are related to transposons. For example, analysis of their DNA shows that each Alu sequence is flanked by direct repeats that range in size from 7 to 20 base pairs. Moreover, the repeat on each side of an Alu sequence is unique to that specific Alu. In other words, when one Alu repeat is compared with others, the short flanking repeats are found to differ. This is again reminiscent of the direct repeats that flank transposons and inserted retroviruses. A good argument can be made that the direct repeats flanking an Alu sequence have resulted from duplication of the chromosomal DNA at the target site where the mobile

Alu inserted (refer to Figs. 15-24 and 15-25). RNAs transcribed from these inserted sequences would then function to transpose them to other sites throughout the genome in much the same way that a retrovirus enters the genome of the host cell. Duplex DNA would be formed from the transcripts using reverse transcriptase, and the DNA sequences would then become inserted into the chromosomes at target sites.

Excellent evidence that Alu sequences do move about the genome is provided by the discovery that Alu became inserted into two genes, a cholinesterase gene and gene NF1 associated with neurofibromatosis F1, an autosomal dominant with a high mutation rate (refer to Chaps. 13 and 14). Along with its transcripts, the cloned NF1 gene has been studied in some detail with techniques of molecular biology such as DNA fingerprinting and the polymerase chain reaction described in Chapter 20. The procedures were applied to study the DNA of a young man afflicted with NF1, neither of whose parents showed any sign of the disorder. It was demonstrated that an Alu element was present in an NF1 gene of the patient at a location in an intron found 44 base pairs upstream from one of the gene's exons, exon 6. Analysis of the patient's RNA showed that his NF1 gene transcripts were of two types: a larger one containing all of the exons represented in the NF1 gene and a short type lacking exon 6. RNA from normal controls and from other NF1 patients showed only the larger transcripts containing exon 6. It appears that in this case normal mRNA splicing was disrupted in the transcript of the NF1 gene containing the Alu insertion (see Ch. 13 for a review of RNA splicing). As a result of the disruption, exon 6 was lost from the transcript of that NF1 gene, and a transcript was produced containing exons 5 and 7 joined together. The loss of exon 6 also generated an upset in the reading frame of the transcript that, when translated, results in the formation of an incomplete NF1 protein, one that is missing almost 800 amino acids at its C-terminal end. Analyses of DNA from both parents of the patient strongly indicate that the Alu element inserted into the patient's NF1 gene by retrotransposition in the germ line of his father.

Since Alu sequences lack the means for transposition, the mechanism permitting their movement is unknown, although an explanation may now be provided by some very important findings involving a certain long interspersed sequence known as *LINES-1*. This repeat is found in 20,000–50,000 or more copies interspersed throughout the mammalian genome. In the human, LINES-1 (the L1 element) makes up about 5% of the genome, occurring in 50,000–100,000 copies. The 6,500 bp L1 sequence has been shown to contain a reading frame possessing regions homologous to those found in the reverse transcriptase genes of retroviruses. This strongly suggests that these repeated regions are related to mobile elements coded for an enzyme of transposition. However, most of the L1 elements contain large deletions. Only about 3,500 of them are full-length, and many of these harbor point mutations that may render them defective.

Although conclusive proof has been lacking, it has been considered likely that some L1 elements are capable of retrotransposition. Very strong evidence to support this concept has come to light with the discovery that an L1 element disrupted the gene for blood-clotting factor VIII. Recall (from Chap. 6) that hemophilia A stems from defects in the X-linked gene

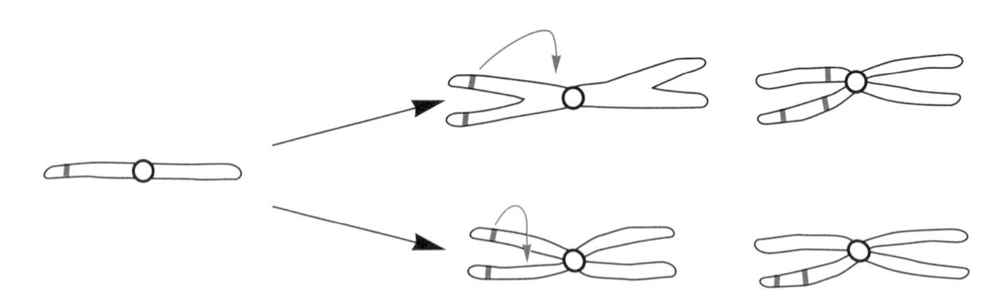

FIGURE 17-20. Transposition and replication. A Ds or Ac element (red) may be found at a specific site on a chromosome (left). The element usually becomes replicated before it moves to a new location. If it moves into a site on the same chromosome that has not yet replicated (above), the chromosome will become composed of two chromatids, one containing a single element at a new site and the other containing two elements, one at the original and one at the new site. If an element moves into a site that has also replicated (below), the result can produce a chromosome with one chromatid lacking the element entirely and the other with two elements. In both cases, daughter cells will be produced that differ in genotype from each other and from the original cell. (An element can move to locations other than those shown, but it will always be lost from the original site on one chromatid.)

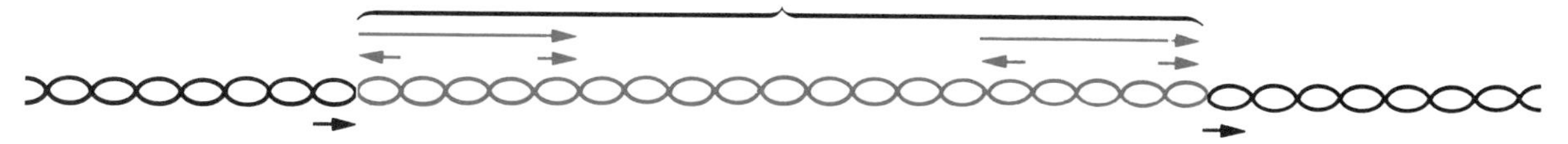

FIGURE 17-21. The copia element in *Drosophila.* A copia element (red) consists of 5,000 base pairs. The element carries at each of its ends direct repeats, DNA sequences in the same orientation (large red arrows). Each of these repeats includes, at each of its ends, inverted repeats, sequences that have opposite orientations (small red arrows). On either side of the copia element is a small direct repeat of a chromosomal sequence (dark arrows), apparently representing a target sequence.

for clotting factor VIII. About one-third of the cases of hemophilia A are known to result from the origin of new mutations. Analyses of the factor VIII gene of many children with hemophilia A revealed two cases in which foreign DNA was present in the middle of the gene. This DNA was identified as an L1 element that had become inserted into exon 14 of the factor VIII gene and rendered it inactive. This and the cases of the inserted Alu elements are the first examples of diseases caused by transposable elements.

Since there are about 100,000 LINES in the human genome, an attempt was made to discover the actual source of the particular L1 element that had inserted into the clotting factor gene. Sequences of several different L1 elements were compared, and a consensus sequence was established. The L1 element that had moved was found to depart from this consensus sequence by 3 bp over a 20 bp sequence. This made it possible to probe for this specific element in human DNA and thus to distinguish it from all the others scattered about the genome. In both parents of an afflicted child an L1 element was found on chromosome 22. The element inserted into the clotting factor gene proved to be a shortened version of the chromosome 22 element, but both agreed in the sequence of all those bases that had not been deleted. It was not possible to determine if the L1 element that moved came from the male or the female parent. It is clear, however, that this particular element is present at that position in chromosome 22 in all persons. It has even been found in the same position in the gorilla, implying that it has been in that location throughout millions of years of primate evolution.

Extremely significant was the demonstration that the specific L1 element could make reverse transcriptase. This was accomplished by splicing the region of the element apparently coded for the enzyme into a yeast TY retroposon. The latter, containing the L1 region in place of the comparable yeast region, was shown to possess reverse transcriptase activity. Moreover, the production of a point mutation in the region rendered the protein nonfunctional.

This demonstration of a moveable L1 element with reverse transcriptase activity leaves little doubt as to the retroposon nature of much of the moderately repetitive mammalian DNA. SINES such as Alu that lack the machinery for transposition may be provided with it by LINES capable of producing reverse transcriptase. However, extremely provocative questions still remain to be answered. How did these elements enter the mammalian genome in the first place? Do they move about frequently, and if so how often are they responsible for causing disease? What has been their role in the evolution of the mammalian genome? Do retroviruses, transposons, and much of the repetitive DNA of eukaryotes have a common origin?

Still another significant finding concerning reverse transcriptase in eukaryotes has been made in relation to the replication of the DNA in telomeres. Telomeres are the specialized ends of eukaryotic chromosomes and are required for the integrity of a chromosome and its normal behavior at nuclear divisions (Chap. 10). The DNA of a telomere includes simple repetitive sequences such as 5′–TTAGGG–3′ in the human. A special enzyme known as *telomerase,* which has been demonstrated in the human and several other eukaryotes, is required for the synthesis of the telomeric repeats. Telomerase is another example of an unusual enzyme composed of an RNA as well as a protein portion (Chap. 13). It has now been established that a nucleotide sequence in the RNA portion of the enzyme acts as a template for the synthesis of the telomeric repeats. For example, in the ciliated protozoan *Euplotes,* the telomeric sequence 5′–TTTTGGGG–3′ is formed from the sequence 3′–AAAACCCC–5′, which is part of the RNA of the enzyme. It is thus clear that the telomerase is acting as a reverse transcriptase, since a segment of its RNA component is the template for the DNA repeats of the telomere. Telomerase, however, differs from the typical reverse transcriptase, since the RNA template is an actual part of the enzyme itself.

Hybrid dysgenesis

An interesting phenomenon associated with transposable elements in *Drosophila* is known as *hybrid dysgenesis*. This term refers to a collection of correlated genetic abnormalities that arise spontaneously in hybrid offspring after certain strains of flies are crossed. Transposable entities known as P elements and P factors are responsible for hybrid dysgenesis. Strains of flies found in the wild called *P strains* harbor both of these. The P elements can be found at multiple sites throughout the genome. They range from 0.5 to 1.4 kb pairs in length and are flanked by inverted repeats of 31 base pairs. The P elements are believed to be derived from the P factor as a result of internal deletions of varying size. The P factor is composed of about 3 kb pairs and contains four open reading frames all of which are required for transposition.

Those fly strains lacking P elements and P factors are known as *M strains*. Strains of flies that have been established for years in the laboratory are M strains, as if the transposable entities have been lost from them. Over 30 P elements may be scattered throughout the genome in a P strain, but the P factors in them are present in much smaller numbers.

When males from a P strain are mated with females from an M strain, the P entities become activated and undergo transposition. This is believed to result from the transposase activity provided by P factors in the male. The transposase is apparently coded by the reading frames in the P factor, since all four frames have been shown to be necessary for transposition. The P elements depend on the transposase of the P factor for their mobilization since the deletions they carry have rendered them defective in transposase production. Transposition, followed by insertion of the many P elements into new locations, brings about the production of gene mutations and structural chromosome aberrations. However, the transpositions occur primarily in germ cells rather than somatic ones. Consequently, the offspring resulting from the cross usually appear normal but are almost always completely sterile.

The reciprocal cross, P female with M male, does not bring about the dysgenesis and gives rise to completely normal offspring, as is also the case with the other crosses, P with P and M with M (Fig. 17-22). It appears that the cytoplasm of a P strain fly contains a repressor substance that inhibits the expression of the P factor. The cytoplasm of cells of strain M flies lack the repressor. Crosses of P × P are normal, because P elements cannot move about and insert at new locations due to the presence of repressor in the cytoplasm. In contrast to this, when a P male is crossed to an M female, the cytoplasm of the egg lacks repressor, and the genetic information of the P factor can be expressed. Transposase is produced, followed by movement of the many P elements and the resulting dysgenesis. In the reciprocal cross, M male with P female, the egg contains sufficient repressor to prevent expression of the information coded in the P factor. No transposase is produced to mobilize the P elements, and thus dysgenesis does not occur.

While activation of the P elements occurs primarily in the germ line, the P factors actually undergo transcription in both somatic and germ line cells. The four reading frames of the P factor are in effect exons separated by introns. A transcript is produced, but this transcript is handled differently in the two lines, somatic and germinal. In the former, the first three exons or reading frames are joined together, but an intron remains between them and the fourth exon in the transcript. In the germ line cells, the intron is removed so that the reading frames are joined in the transcript. This splicing together of the exons brings about the formation of a polypeptide that provides the transposase activity required for activation of the P elements. The reason for the difference in splicing in the two lines is not known, but differential processing of mRNA is well known in eukaryotes, as we will see a bit later on in this chapter.

The story of the P elements and factors is highly significant from an evolutionary point of view. Let us suppose that two populations of a given species come to possess different families of transposable elements such as the P elements. The possession of the different kinds of elements can act to isolate the populations since breeding between the two would lead to an appreciable number of sterile hybrid offspring. New gene mutations arising independently in the two populations would have a chance to accumulate, leading to further separation down the pathway to speciation. The recognition of transposable elements such as those in the P group raise many provocative ideas concerning alteration of the genome and the evolution of species.

The proteins of chromatin

Any approach to the problem of the control of gene expression in eukaryotes must take into consideration the assortment of proteins associated with the DNA of the chromosomes. These proteins fall into two major classes, the histone and the nonhistone proteins. Histones are relatively small (MW 10,000–20,000) and have a very basic nature because of the inclusion of large amounts of the basic amino acids lysine and arginine. This high proportion of positively charged amino acids enables them to bind tightly to the DNA, which carries negatively charged phosphate groups. The total mass of the histones is about equal to that of the DNA to which it is bound.

Five different classes of histone proteins occur in chromatin. Four of these—H2A, H2B, H3, and H4—are exceptionally constant in their composition from one cell type to another and from one species to the next. Of the 102 amino acids composing H4, 100 are identical in chromatin from the cow and in that from the pea plant! Very little change in the amino acid sequence of these four histone classes has taken place over the course of evolution of eukaryotes. Such a high degree of conservation implies that the histones were perfected for an important role they must play in the structure of the chromosome, a function so crucial that almost no amino acid substitution can be tolerated.

Except for one central region, the amino acid sequence found in the fifth histone class, H1, shows much less conservation. Variation in H1 is found when chromatin from different organisms is compared. Moreover, several closely related varieties of H1 are found in each cell of an organism.

Enormous quantities of histone proteins exist in a cell, about 60 million copies of each type! A tremendous amount of histone must be synthesized during the S period of the cell cycle, so that the histones can bind to the DNA as it is replicated. The demand for such large amounts of histone during S may be met as a consequence of the repetition of the genes for the five histone types. As is true of the ribosomal RNA genes (the rDNA), the histone genes also occur in clusters and are found in the moderately repetitive DNA fraction, as revealed by renaturation kinetics (see Chap. 13). Each gene in a cluster (or gene family) is transcribed separately on its own mRNA. Unlike most eukaryotic genes, sequences coded for the five histone proteins lack introns.

The size of the spacers that separate the histone genes, as well as their order in a cluster, varies from one species to the next, as is evident in Figure 17-23. For example, in the sea urchin, the order is H1, H4, H2B, H3, H2A, whereas in *Drosophila* it is H1, H3, H4, H2A, H2B. In *Drosophila,* two of the genes (H4 and H2B) are transcribed left to right on the strand opposite the template strand for the other three genes. In contrast, all transcription comes from the same strand in the sea urchin.

A great deal of variation among species is also seen in the number of clusters found per genome. Several hundred copies of each cluster occur in sea urchins. *Drosophila* possesses about 100 repeats of a histone cluster, whereas in the mammal only about 20 repeats are found. In the chicken, the number of repeats is only 10. The gene clusters in a given species are located at specific chromosomal regions. In the human, in situ hybridization shows them at a certain region in the long arm of chromosome 7.

Although the histone proteins make up a very well-defined conservative class of proteins, the nonhistone proteins of chromatin compose an assortment of thousands of different proteins with many roles. Included

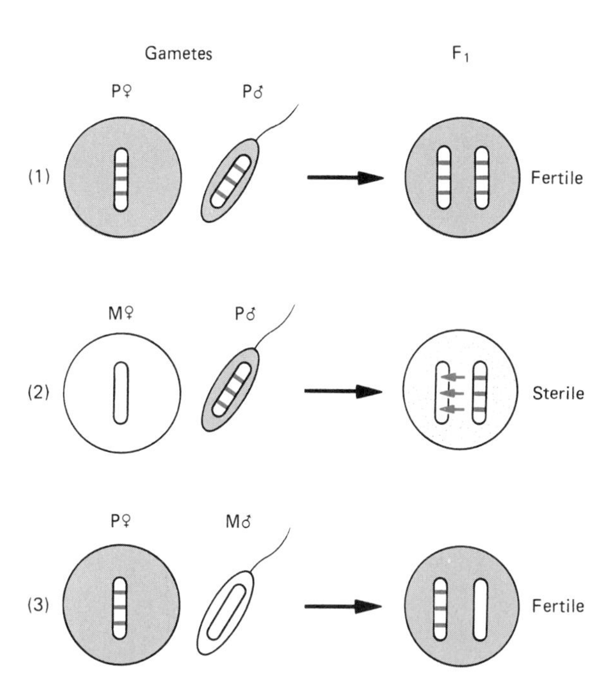

FIGURE 17-22. Hybrid dysgenesis in *Drosophila.* **1:** Crossing flies taken from a strain that harbors P factors and P elements (red) yields fertile offspring because repressor in the cytoplasm (pink) prevents their transposition and hence disruption of the genome in the F_1. **2:** The cross of M females with P males results in sterile offspring, because the cytoplasms of the egg (colorless) lacks repressor. In the hybrids, insufficient repressor is present. Transposase can be produced by the P factors, enabling the P elements in the chromosomes from the male to transpose to new locations, disrupting the genome. **3:** The cytoplasm of the egg contains large amounts of repressor in P females. The P elements contributed by the female cannot undergo transposition in the F_1 because sufficient repressor is present in the F_1 cytoplasm to prevent expression of the genetic information for transposase production in the P factor. Consequently, the hybrids are fertile.

are various acidic proteins and the functional proteins such as the RNA and DNA polymerases. Most of the nonhistone proteins are present in very small amounts, making it difficult to isolate each kind for study. In contrast with the histones that are very highly conserved, the nonhistones are variable in kind. Evidence shows variation from one species to the next, as well as from tissue to tissue within a species. The nonhistones also vary greatly in relation to the mass of the DNA in chromatin, from less than 50% of the DNA in chromatin to more than 100% of the DNA amount. We will return to a discussion of certain kinds of nonhistone proteins following a look at the organization of the chromatin fiber.

The chromatin fiber and the nucleosome

The unreplicated chromosome in an interphase nucleus consists of a single continuous chromatin fiber composed of DNA associated with histones and nonhistone proteins. Various procedures have shown that only one double-stranded DNA molecule runs the length of the continuous fiber. Following replication at S of interphase, one duplex is present per chromatid. The chromatin fiber of each chromatid folds back on itself repeatedly to form the compact chromosome of metaphase (Fig. 17-24).

In the 1970s, the chromatin fiber was analyzed in detail to gain an insight into its organization. Roger Kornberg pioneered the biochemical studies, and Olins and Olins produced superb electron micrographs of chromatin fibers. The investigations have resulted in the formulation of the *nucleosome model* of chromatin fiber organization. After its extraction from cells, chromatin can be treated to remove its nonhistone proteins. When such chromatin is then exposed to different salt concentrations, its organization is seen to vary. At very dilute salt concentrations, which do not exist in the living cell, the chromatin fiber is seen at its thinnest and appears to have a diameter of about 10 nm. This continuous fiber has a repeated, beaded appearance (Fig 17-25). Many details of fiber organization have come from biochemical studies that make use of an endonuclease obtained from micrococci. The fiber, composed of the DNA and histone, is subjected to different concentrations of the digesting enzyme for various periods of time. A brief exposure to the enzyme yields particles that are rather uniform in length and shows that the continuous fiber consists of repeating units. These were called *nucleosomes.* The nucleosome includes an average of 200 nucleotide base pairs, 146 of which are wrapped in two turns around an octamer of histone proteins that form a bead (Fig. 17-26A). The octamer consists of two molecules each of histones H2A, H2B, H3, and H4. Each nucleosome also includes one molecule of histone H1 located at a position where the DNA leads into and leaves the bead.

Further exposure to the nuclease yields the nucleosome core particles. Each core particle consists of 146 nucleotide base pairs wrapped around the octamer of histones. H1 protein is no longer present (Fig. 17-26B). These results come from the fact that a brief enzyme treatment attacks very sensitive sites on what is called the *linker* DNA, the DNA that joins together the nucleosome core particles. Units consisting of about 200 nucleotide base pairs, one H1 molecule, and the octamer of histones are recovered. Once the linker is cut by the enzyme, it is digested away rather rapidly. The H1 protein is lost along with it in the process. What is

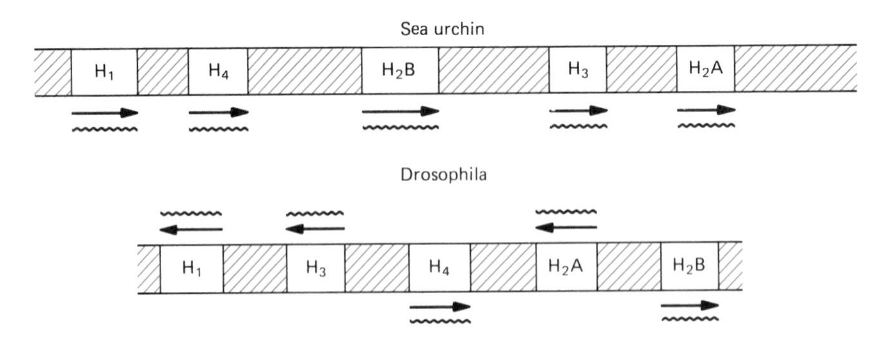

FIGURE 17-23. Histone gene clusters. The genes for the five classes of histone proteins occur in clusters repeated over and over. Variation is seen in the order in which the specific genes occur in a cluster and in the size of the spacers separating them. Each gene is transcribed separately on its own mRNA. Transcription may be all from one strand, as in the sea urchin, or from both strands, as in *Drosophila,* in which three genes are transcribed from right to left and two from left to right.

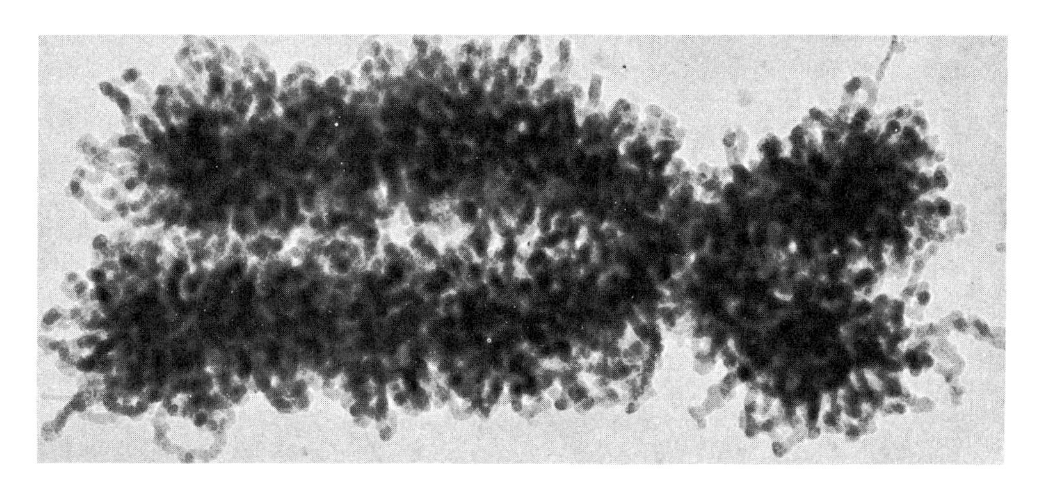

FIGURE 17-24. An electron micrograph of a whole mount of human chromosome 12 at metaphase. Each of the two chromatids is composed of a compactly folded 30 nm fiber, which in turn arises from a narrower 10 nm fiber. (Courtesy of E.J. DuPraw.)

left is the core particle, 146 base pairs wrapped around the histone octamer. The DNA of the core receives some protection from digestion as a result of its association with the eight histone molecules. While the core particle, also called the *nucleosome bead,* always includes 146 base pairs, the length of the linker DNA can show considerable variation, from approximately 10 base pairs per nucleosome to more than 100. This variation becomes apparent when different species are compared, and even when nucleosomes from different cell types within an organism are compared.

The very high degree of conservation of the amino acid sequence of the nucleosome core proteins fits in well with the fact that nucleosome organization of chromatin is found in almost all eukaryotes, the only exception being found among the fire algae or dinoflagellates. This indicates a very crucial role for nucleosome organization in the biological world. The consensus is that nucleosome structure represents the first level of packaging of the DNA. The fundamental need to pack DNA to form the compact chromosome at nuclear divisions becomes obvious when we realize that the DNA in each human chromosome would be about 5 cm long if it were stretched out! Packing such a long, flexible duplex into a compact body that permits distribution at nuclear divisions is a critical process.

The function of the H1 protein appears to be to generate further condensation of the chromatin. Kornberg and his associates have shown that the H1 protein brings about folding of the 10 nm chromatin fiber to form a fiber 30 nm wide twisted further into a supercoil (Fig. 17-27). They have named this 30 nm structure a *solenoid,* a helix with about six nucleosomes per turn. Under experimental conditions, the 10 and 30 nm fibers can be interconverted, but, if H1 is removed, only irregular chromatin clumps result. It has been proposed that H1 brings about coiling of the 10 nm fiber into the 30 nm one through contact of adjacent H1 proteins with one another. As mentioned previously, the 10 nm fiber is seen when chromatin is exposed to very dilute salt concentrations. At the concentration found in the living cell, the 10 nm fiber is converted to the 30 nm fiber. It is this 30 nm fiber that is considered to be the chromatin fiber of interphase and the fiber of the metaphase chromosome (see Fig. 17-24). The chromatin is thought to adopt the extended "beads-on-a-string" form only rarely. When cells are lysed following a very gentle treatment, the chromatin is seen in electron micrographs as a fiber 30 nm in diameter, not as a 10 nm fiber.

Histone and nonhistone proteins in gene regulation

Much attention has been focused on histones as potential regulators of gene expression. Histones can undoubtedly influence the regulation of genes, since DNA associated with the core of the nucleosome would not be as exposed as linker DNA to regulatory molecules in the cell. This brings us to the question of *nucleosome phasing.* If nucleosomes are phased in a given region, this means that the very same DNA sequences would be in the same position in all cells of the same type. Therefore each specific DNA sequence would always be found in a given position in relation to the nucleosome. The same sequences would be available for interaction with regulatory molecules. The linker DNA sequences would all be the same. If, on the other hand, the nucleosome beads are randomly placed along the DNA, then the linker sequences would differ from one cell to the next. Different DNA sequences would be associated with the bead from one cell to another, thus making different sequences available for interaction with regulatory molecules (Fig. 17-28). There is evidence for nucleosome phasing in at least some parts of the chromatin. One idea is that, in different cell types, the nucleosomes are arranged in different phases. This means that a given gene could be bound to a bead in one cell type, whereas in a cell type with different phasing the same gene would be unbound to the histone octamer of a bead and thus more accessible to regulatory substances.

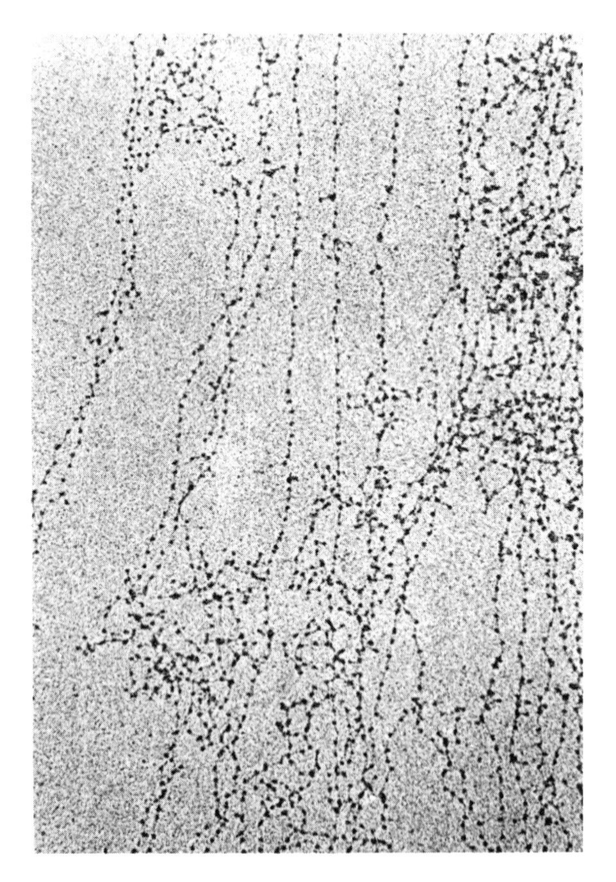

FIGURE 17-25. Electron micrograph of 10 nm chromatin fiber. Note the repeated, beaded appearance. (Courtesy of Dr. Barbara Hamkalo.)

Whatever the case, it seems quite clear that DNA remains associated with its histones, even during transcription and replication. It appears that the histone octamer rarely, if ever, leaves the DNA region to which it binds. It remains somewhat of a puzzle to explain how the DNA bound up in a nucleosome can lend itself to transcription by RNA polymerase. Some temporary change in the nucleosome would seem necessary. It is even harder to envision how transcription can occur without some packing alteration in the 30 nm fiber, where the chromatin is even more condensed than in the 10 nm fiber.

Evidence exists that the H1 protein, whose role appears to involve packing the nucleosomes together, is less tightly bound to chromatin that is actively undergoing transcription, although the histones of the bead appear unchanged. However, certain modifications of the histones of the core particle are known to occur. A certain enzyme can add acetyl groups to some of the lysines. Highly active chromatin appears to have highly acetylated histone octamers. Acetylation of the histones has been shown in vitro to decrease the tendency for the nucleosomes to pack together. Both the increased acetylation and reduced binding of H1 may be factors in altering the conformation of the 30 nm fiber, making certain DNA regions more accessible to substances in the nuclear sap. The pattern of acetylation constantly changes in the chromatin. As acetyl groups are added, others may be removed because of the activity of another enzyme.

Besides acetylation, the histones of the bead can be modified in at least two other ways. Serines may become phosphorylated and dephosphorylated as a result of the activities of still other enzymes. The other kind of modification involves certain lysines in the H2A histone. These become complexed with a protein called *ubiquitin,* which consists of 74 amino acids. The H2A–ubiquitin complexes seem to occur only in the interphase nucleus, and there is some evidence that the ubiquitin-modified histones are associated with actively transcribing genes. Ubiquitin may exert its effect by causing the chromatin to unpack, thus favoring gene transcription in that region.

A large assortment of nonhistone proteins is also associated with chromatin, and, unlike the histones, they are quite variable. Since most of them are usually present in only small amounts, isolating a given kind for study is difficult. Only those nonhistone types present in very large quantities are available for detailed investigations. These nonhistone proteins are the HMG proteins. They are relatively small, charged molecules that

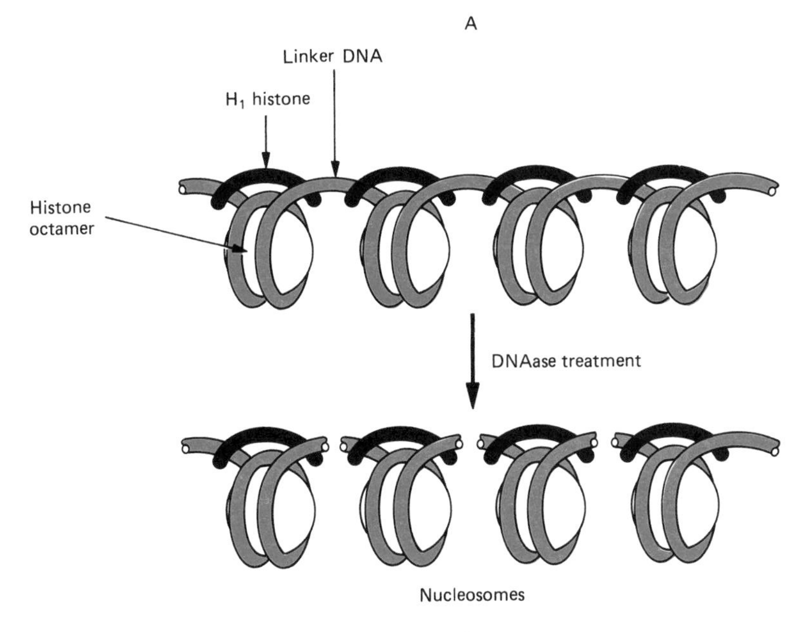

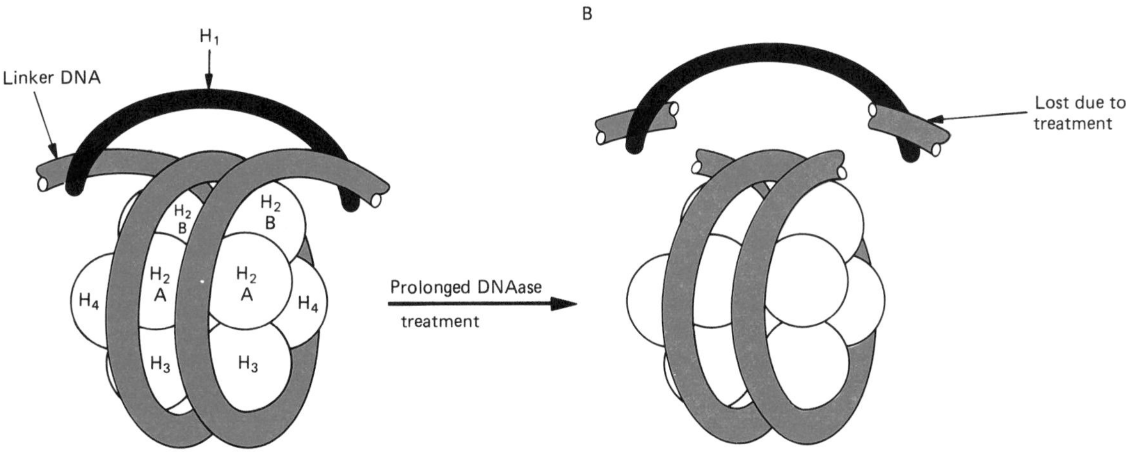

FIGURE 17-26. The nucleosome model of chromatin fiber organization. **A:** According to the model, the chromatin fiber is continuous and consists of repeating units, the nucleosomes. Each nucleosome includes a bead containing an octamer of histone proteins around which DNA is wrapped in two turns. Also included is an HI protein found outside the bead. **B:** Detail of a portion of a chromatin fiber. A nucleosome (left) arising after a brief enzyme treatment can be exposed to further enzyme digestion. The linker DNA and H1 protein are lost, yielding the nucleosome core particle or bead.

move rapidly when electrophoresis is performed. This characteristic gives them their name–HMG for *high-motility-group* proteins. Two HMG proteins have been implicated with nucleosome beads that are associated with active genes (see the next section). Most of the nonhistone proteins, because of their small amounts, are yet to be characterized and studied for their possible interactions with chromatin. It is considered very possible that a number of these nonhistones may act as regulatory proteins by binding to specific DNA sequences, thus determining which genes undergo transcription.

The bulk of the nonhistone proteins are yet to be studied for their effects on chromatin. They may produce changes in gene activity in a number of ways. Possibly they may sometimes interact with the histones; at other times they may interact directly with enzymes, such as the polymerases, which are also components of the chromatin.

Very good evidence indicates that, when eukaryotic genes undergo transcription, the activation is accompanied by changes in the chromatin structure. This is clearly seen by differences in sensitivity of genes in response to various digestive enzymes. For example, compared with inactive genes, active ones are much more sensitive to DNAase I, S1 nuclease, and restriction endonucleases. DNAase hypersensitive sites are located toward the 5′ end of the gene, anywhere from a short distance to hundreds of base pairs upstream from the starting site of transcription. Hypersensitive sites have also been found in or near DNA regions that serve as binding sites for proteins. These hypersensitive sites have been shown, in some cases, to be present at the time the gene becomes active and, in other cases, before transcription begins. One suggestion is that certain critical controlling regions of the DNA that interact with regulatory molecules are exposed to the nuclear environment and are not bound up in nucleosome beads. Hence, in cell differentiation, parts of the DNA may become unassociated with a nucleosome bead, thus becoming exposed to various enzymes and factors that can initiate transcription. In an inactive gene, on the other hand, such regions would be bound to the nucleosome core and not exposed to controlling molecules.

It has been shown that hypersensitive sites may be passed down from one cell generation to the next, and the suggestion has been made that such sites could possibly act as signals in development. It is known that certain hypersensitive sites become associated with two types of nonhistone proteins (two HMG proteins) that are necessary for maintaining the sensitivity of the genetic region. The information indicates that, in gene activation, a region undergoes a change that enables it to bind to a nonhistone protein necessary to keep the site sensitive and hence available for interactions with other molecules. Sites bound to the nonhistone protein could thus serve during embryonic development as signals indicating which genetic regions are available for interaction with the enzymes of transcription and possibly various controlling molecules as well.

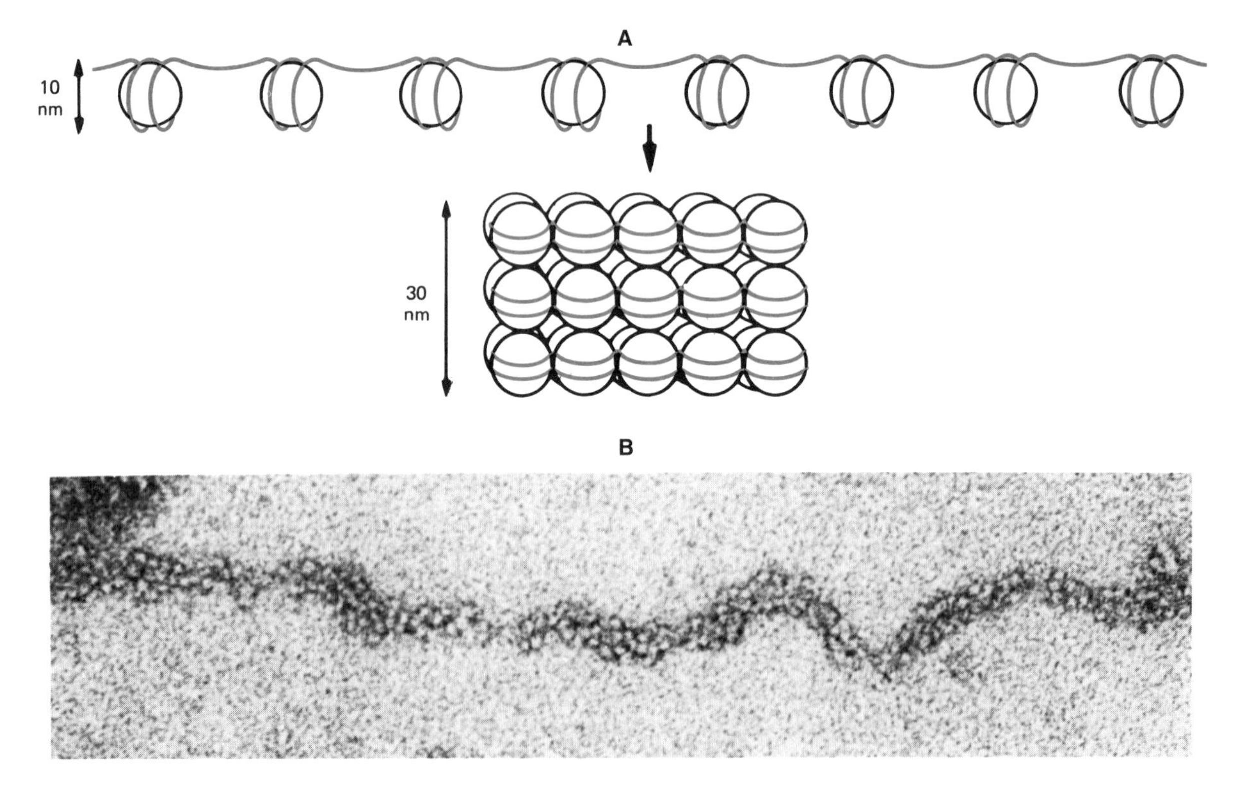

FIGURE 17-27. The 30 nm thread. **A:** The 10 nm thread folds to produce a thicker, 30 nm thread, the chromatin thread of interphase. Histone HI is responsible for the condensation. The 30 nm thread is twisted further into a supercoil and is also called a solenoid. **B:** Electron micrograph of 30 nm thread. (Courtesy of Dr. Barbara Hamkalo.)

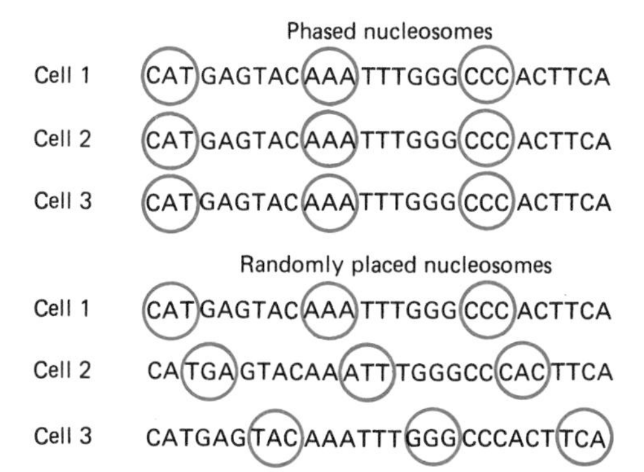

FIGURE 17-28. Phased and randomly placed nucleosomes. If nucleosomes are phased (above) in a given region of the chromatin, specific DNA sequences would always be found in the same position when different cells are compared. The linker regions would be the same from one cell to the next, and the same sequences would be associated with the nucleosome bead. If nucleosome beads are placed at random, then different sequences are found in the linker DNA from cell to cell when this chromatin region is compared (below).

Enhancers

In Chapter 12, we noted that several promoters in eukaryotes are known to be greatly influenced by other DNA sequences named *enhancers.* Actually, these genetic elements were first discovered in several animal viruses, such as the monkey tumor virus SV40. In SV40, transcription starts from two promoters that lie next to each other on the circular DNA. Transcription of the early genes from the early promoter proceeds in the direction opposite that from the late promoter (see Fig. 19-5). It was demonstrated that removal of the enhancer sequence from the early promoter region decreases early gene transcription to less than one-one hundredth its normal amount.

Following their discovery in viruses, enhancer sequences were recognized as activators of several different eukaryotic genes. Investigations of enhancers have revealed that they possess several distinguishing features that set them apart from other controlling elements known to influence the rate of transcription. For one thing, experimental manipulation shows that the enhancer's stimulating effect can be felt at different distances and locations in relation to the gene. It may be located as far as 5–10 kb upstream or downstream from the gene it influences. In some cases, as in immunoglobulin genes, the enhancer is located *within* the segment that undergoes transcription and therefore lies downstream from the promoter it stimulates. Experiments in which enhancers are moved about also reveal that an enhancer will increase the rate of transcription of any promoter that is placed nearby, showing that a given enhancer is not limited in its effect to just one promoter of a given gene. In addition, an enhancer can work efficiently even if it becomes inverted and thus oriented in a direction opposite to its usual one.

Enhancer sequences have been discovered that greatly increase transcription from promoters of specific genes in specific cell types. For example, they are known to activate the insulin gene only in insulin-producing cells of the pancreas, immunoglobulin genes only in B lymphocytes, and genes coded for chemotrypsin enzyme only in certain secretory cells. It is clear therefore that some enhancers are active only in those cells in which specific genes are being expressed.

It is not yet clear how enhancers exert their stimulating effect since they vary so greatly in their locations from the promoters they influence. Moreover, while we know that enhancers interact with regulatory molecules such as transcription factors, we do not know what these factors actually do when they bind to specific consensus sequences in enhancers. Mutations are known that affect these sites, and some of them reduce the stimulating activity of the enhancer. Others, however, appear to have no effect.

Hormones are known to be involved in the activation of some enhancers. Hormones are chemical messengers that stimulate the synthesis of specific proteins in target cells, those cells receptive to the particular hormones. Most of the details of hormone action on the molecular level are known about certain steroids, but other hormones, such as the thyroid hormone, appear to act in a similar fashion. When given to animals in vivo, steroids increase the rate of RNA synthesis. A specific hormone enters the target cell and binds to a specific protein receptor. Once bound to the hormone, the receptor becomes activated and acts as a transcription factor. The complex, which is found only in the nucleus, binds to a consensus sequence of an enhancer and activates it. This in turn stimulates the promoter associated with the enhancer. It seems that steroid hormones may be able to regulate various genes by activating transcription factor proteins that in turn stimulate specific enhancers. These curious enhancer elements and their interactions continue to be the focus of many studies relating to the problem of regulation of gene activity.

Methylation and gene activity

The chromatin structure associated with active genes has also been shown to be related to the degree of methylation. It was noted in the late 1940s that certain cytosines in the DNA were converted to 5-methylcytosine as a result of enzyme action. The modification involves the covalent linkage of a methyl group ($-CH_3$) to carbon 5 in cytosine. This alteration does not interfere with replication or transcription, because it does not influence in any way the hydrogen bonding involved in these processes. The 5-methylcytosine acts as a fifth base, since it is chemically distinct from the cytosine.

When methylation of cytosine takes place, it is always at positions where the cytosine is followed by guanine in the nucleotide sequence. Thus it can be seen from Figure 17-29A that both DNA strands at any one site are symmetrically methylated. Once methylation takes place, this state is inherited from one cell generation to the next.

The observations that about 80% of mammalian DNA is methylated and that cellular methylation patterns are inherited prompted investigators to observe cells for any connection between gene activation and methylation. And indeed it was found that in many cases hypermethylation of cytosine within or near a coding sequence in a structural gene is associated with gene suppression. Those genes known as *housekeeping genes,* on the other hand, are permanently active in all cells and lack methyl groups. Moreover, experiments with promoter regions of certain viruses showed that the promoter does not function when it is methylated but that absence of methylation is associated with promoter activity. The pattern that emerged indicated that genes are active only if hypomethylated.

Transfection experiments in which eukaryotic cells take up foreign DNA (explained in Chap. 19) have shown that only those genes that are nonmethylated express themselves after being introduced. Moreover, the methylation pattern of the introduced DNA was kept for many cell generations. Further experiments of this type support the concept that demethylation is a factor in switching genes on selectively in the course of cell differentiation. For example, when a methylated gene, let us say the insulin gene, is inserted into a specialized cell type in which that gene is normally expressed (in this case, a pancreatic cell), the introduced gene becomes demethylated and is expressed. On the other hand, the same gene in the methylated state remains methylated and unexpressed when it is introduced into cells that are undifferentiated. In all those examples in which demethylation takes place in differentiated cells, DNA synthesis is not going on. This suggests that the mechanisms involved are not dependent on replication but may depend on molecules that are active only in specific cell types.

There is another kind of experimental evidence that shows that altering methylation patterns can switch genes on or off. This involves treatment of DNA with the drug 5-azacytidine. DNA that has incorporated the drug cannot be methylated, because a nitrogen blocks the carbon atom that normally accepts the methyl group. Many cases could be presented to show that demethylation of a gene with the drug is associated with its activation. For example, manipulation of the methylation pattern has been carried on with the inactive X chromosome of the female mammal. The X that appears to have most of its genetic information turned off in a cell has been treated with the drug. When DNA fragments of such a treated X are introduced into other cells, the genes located on them are expressed. However, genes on a fragment from an inactive X that was not exposed to the drug remain inactive after introduction into a cell.

Very impressive results were obtained when the drug was applied to fibroblast cells that were immortalized (discussed in Chap. 18) and growing in culture. Normally these connective tissue cells reproduce true to type and do not switch to some other kind of cell. However, following drug treatment some of them changed into muscle cells and others into fat-storing cells. The differentiation into muscle was shown to be accompanied by demethylation in a regulatory gene. The implications from such an outcome strongly imply that altering the methylation pattern of a cell's DNA can steer it in a particular direction during development.

The maintenance of a particular methylation pattern from one cell generation to the next is kept as a result of the activity of an enzyme called a *maintenance methylase.* The enzyme, which ensures that methylated sequences are inherited as a cell divides, is believed to act only on that DNA that is hemimethylated, DNA in which methylated sequences occur on one strand but not on its complement (Fig. 17-29B). At the time of replication, a fully methylated sequence will at first give rise to two hemimethylated double helixes. The very fast-acting methylase recognizes these and swiftly methylates the unmethylated strand. The enzyme does not recognize those DNA sequences that are completely unmethylated. Consequently, both types of sequences, methylated and unmethylated, are inherited in the original state as cell divisions take place.

Delineation of the entire mechanism of methylation in gene regulation, however, awaits additional findings. While the maintenance of a methylation pattern seems

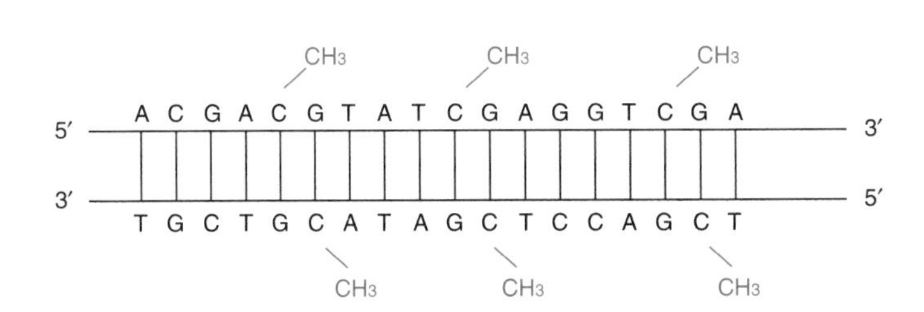

B

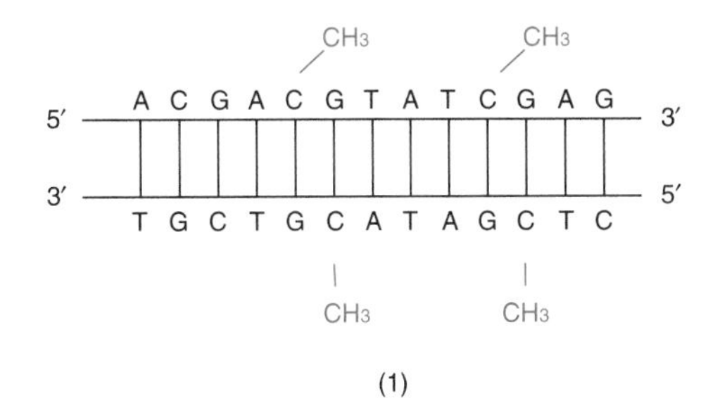

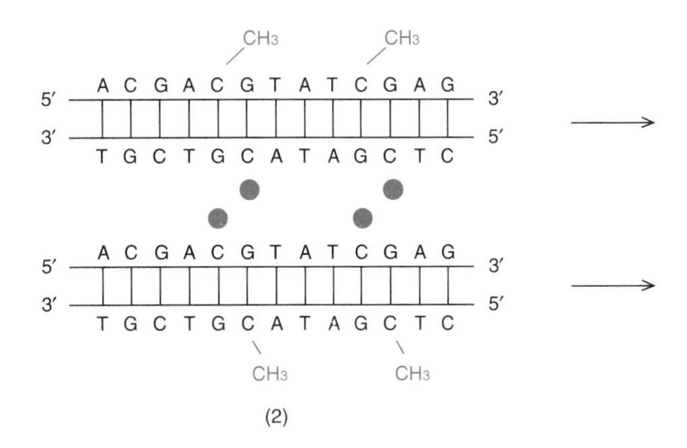

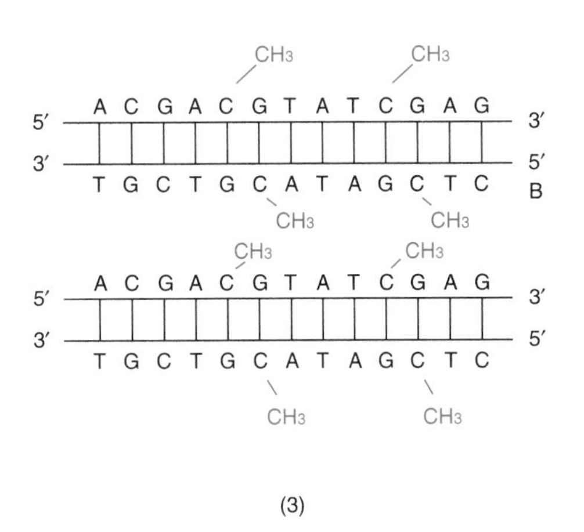

FIGURE 17-29. DNA methylation. **A:** Any cytosine on a DNA strand that is methylated is always one followed by a guanine. Since both strands are methylated, the methylation is symmetrical at that site. Not every C followed by a G is methylated, as shown here on the left. **B:** When a symmetrically methylated region of a duplex is replicated (1, 2), hemimethylated regions arise. These are quickly recognized by the maintenance methylase, which adds methyl groups to the appropriate cytosines. The enzyme does not recognize the CG sequence on the left, since it responds only to hemimethylated sites, and this site was unmethylated from the start. The original pattern of methylation is thus preserved through a cycle of replication (3).

fairly well understood, it is still not known exactly how genes can be switched on and off during cell differentiation. DNA apparently contains far too many methylated CG regions for them all to be involved in switching. Only a minority of them may participate, and these are likely to be found within large DNA sequences that interact with regulatory molecules. The completion of the switching story depends heavily on the recognition of an enzyme or enzymes in addition to the methylase. Some enzyme is necessary to add methyl groups in the first place to unmethylated sequences and also to remove them from methylated regions. Switching of a pattern could occur, for example, if an enzyme recognizes some unmethylated CG sequence in a regulatory gene and attaches a methyl group to a cytosine on one of the strands. The maintenance enzyme would then methylate the complementary strand. In this way a pattern could be switched. Conversely, a specific enzyme might recognize a sequence that is already methylated and remove a methyl group before DNA replication. The maintenance enzyme, since it only recognizes hemimethylated regions, would add no methyl groups. Such a switch in a region from a methylated to an unmethylated state could steer the cell down a certain pathway of differentiation.

One proposal on the role of methylation is that it keeps certain nucleosomes in a fixed position in the DNA. Those gene regions that are involved in control would not be wound in nucleosome beads. Since DNA in the bead is much more highly methylated than regions between the beads, the idea is that the methyl groups stabilize the nucleosome beads, preventing them from associating with other DNA regions. If methyl groups are moved, then a bead is free to shift and associate with another DNA region. This in turn influences gene expression.

It has been suggested that certain proteins, determinator proteins, may interact with the methylation. These would first fix the nucleosome beads in a certain pattern. Then DNA methylation would keep them in that pattern permanently. Methylation, according to this idea, is a secondary rather than a primary device in gene regulation. It could help to explain the seeming contradiction that in those vertebrates lower than mammals, only about 20% of the DNA is methylated. In the invertebrate *Drosophila,* no methylation is found at all. Perhaps an organism such as the fruit fly, having few cell divisions in its life cycle compared with mammals, would not require methylation. It could make the proteins to fix the nucleosome pattern every time a cell division occurs. Since so few cell cycles are involved, methylation would not be required in such a species.

Mammalian cells must have many different ways to accomplish differential gene activity, methylation being one of them. It may prove to be a device superimposed on others at times to ensure that basic patterns of gene expression are preserved, as would be the case in establishing a nucleosome bead pattern. In other cases, demethylation may set the stage for a gene to come to expression. Whether the gene actually is ever expressed may depend on other regulatory factors operating in the specific cell type. Additional information continues to implicate DNA methylation in unexpected ways in the control of gene expression, as seen in the following section.

Differential imprinting between the sexes

A basic concept of Mendelian genetics is that it does not matter whether a given gene is inherited from the male or female parent. An increasing body of information from experiments and from clinical observations indicates that this fundamental principle is not quite correct. It now appears that it does make a difference whether a gene has been transmitted by the mother or the father. This knowledge may help to explain the fact that while an unfertilized frog egg can be artificially stimulated to complete normal development, such a feat has never been accomplished with a mammalian egg. It appears that normal mammalian development requires contributions from both a paternal and a maternal parent. Studies on the molecular level are now showing that, in the human, the mouse, and various other animal species some genes are inactivated when received from one parent but not when received from the parent of the opposite sex. This differential activation and deactivation of parental genes has been termed *parental imprinting.* The pattern set by imprinting may last throughout the life of the individual receiving the genes. However, preceding the formation of gametes from that individual, all imprinting is somehow erased. Reimprinting then takes place, the pattern depending on the sex of the individual.

Results from several laboratories have given evidence that imprinting is the result of different patterns of methylation placed on the maternal and paternal genes. Usually those genes inherited from the mother contain more methyl groups than those received from the father. The more methylated the genes, the lower is their activity. We see once more support for the concept that methylation is important in gene expression.

Differential methylation and its effects on gene activity may explain various observations made over several years. For example, mouse eggs can be treated in such a way that they come to possess either two maternal or two paternal pronuclei (the haploid nuclei of gametes). This can be accomplished by removing one pronucleus from a mouse zygote just before nuclear fusion and replacing it with a pronucleus from another zygote. The two pronuclei then fuse, in effect producing a fertilization nucleus. In this way, it is possible to obtain diploid zygotes with two female-derived or two male-derived genomes. Although embryo development begins, it comes to a halt midway in the gestation period. It has been found that when only the male genome is present in the diploid do the placenta and extraembryonic membranes form, but the embryo proper fails to develop to any degree. In contrast to this picture, in absence of the male genome, a small embryo arises, but no extraembryonic membranes form, and again development comes to a halt. Development of the embryo proper clearly depends on the presence of the genome inherited from the mother. The picture that emerges is that the genomes received from the male and female parents perform different functions in development. In the human, abnormal pregnancies (hydatidiform mole) are known in which an egg comes to possess two paternal nuclei and none from the maternal parent. An excess growth of extraembryonic membranes takes place with little or no development of the embryo proper.

One group of investigators injected newly fertilized mouse eggs with foreign genes (transgenes). Some of the animals that developed from these eggs were found to have the foreign genes incorporated into their chromosomes and were able to transmit those genes to their offspring. A given foreign gene could be followed in the tissues of the different animals. Studies of several different mouse lines showed that the transgenes integrated into the chromosomes at different sites. The methylation patterns were determined, and it was found that the degree of methylation of genes in a given mouse line depended on the sex of the parent contributing them. If a given gene was received from a male parent, it always proved to be less methylated than when it came from the mother. In addition, the methylation pattern of the genes was found to change from one generation to the next, again depending on the sex of the parent. When a male received the gene from its mother, the gene was usually in a highly methylated state. Nevertheless, in both the male and female offspring of this male, the gene would be in a low methylated form (Fig. 17-30). Conversely, a sister of such a mouse would transmit the highly methylated gene to both her male and female offspring in the highly methylated form. She would also transmit a low methylated gene received from her father in a highly methylated form to all her offspring. Such observations leave no doubt that parental imprinting is imposed on a gene as it is transmitted from one generation to the next.

After investigations were made with transgenes, imprinting of naturally occurring endogenous genes was found. While the newer studies with the native genes reinforce the work with transgenes, they also caution us about reaching premature conclusions. In one study with mice, three endogenous genes were followed: a gene for insulin-like growth factor 2; the gene for the receptor of this factor; and gene H19, which encodes an RNA of unknown function. In the case of each gene, either the maternal or paternal allele was expressed, never both. However, the preliminary studies indicate that it is not always maternal genes that become more highly methylated and less likely to be expressed when imprinting takes place. Moreover, there is some reason to believe that methylation is not the primary change taking place during imprinting. Evidence indicates that the chromatin is more tightly wound on a chromosome known to carry imprinted genes. Perhaps the chromatin structure dictates which genetic regions are to be methylated. A search is underway for winding proteins that may be involved in setting the pattern of imprinting.

Regardless of the mechanism of imprinting, the work with endogenous genes supports earlier studies that suggested that those genes that become imprinted are ones that function early in development. The three endogenous genes noted above are all expressed during this time. If either the maternal gene for the receptor or the maternal H19 gene is deleted, a mouse embryo dies by day 15. In one study, cells that were genetically either all male or all female were implanted into mouse embryos. The "all female" cells were found to retard embryo growth by 50%, whereas the "all male" cells were found to enhance it by the same amount! It was further found that the female and male genomes contribute differently to specific tissues. The male-derived genomes have a decided effect on skeletal parts. Female-derived genomes made no contribution to skeletal muscle but contributed to the development of the brain.

Knowledge of imprinting may help to explain some curious observations on genetic disorders as well as on development. Wilm's tumor, a kidney tumor discussed in Chapter 18, is associated with loss of genetic material from chromosome 11 in the malignant cells. It appears that the affected chromosome is usually the one received from the female parent. Huntington's disease (see Chap. 20) follows an autosomal dominant inheritance pattern. The observation has been made that when the dominant allele is received from the father, the disorder is more severe and its onset is at a

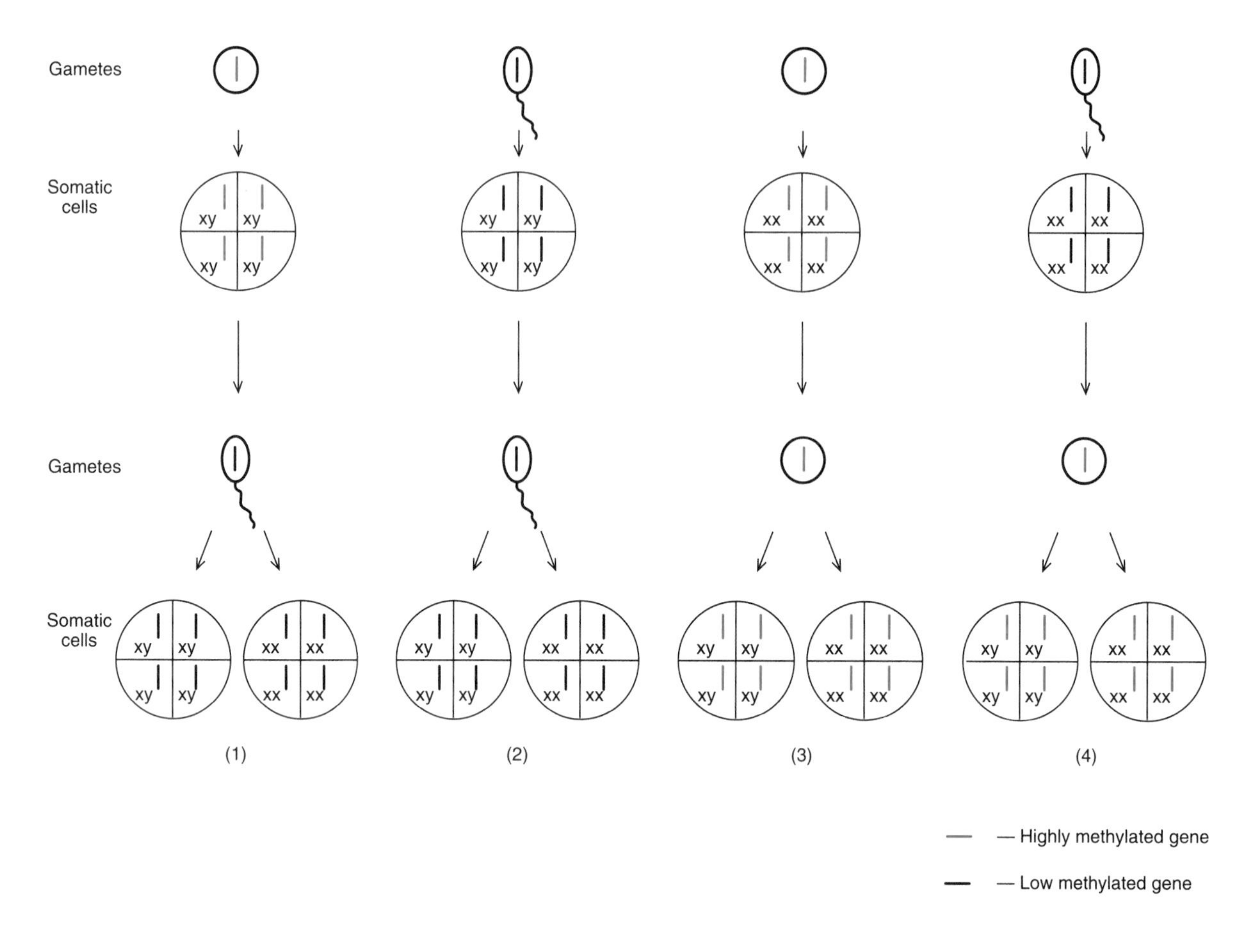

FIGURE 17-30. Differential imprinting. A foreign gene that has been introduced into a line of mice can be followed in the tissues of the animals from one generation to the next. If a male receives a highly methylated gene from his mother, the gene persists in this state in his somatic cells (1). However, the same gene is transmitted by him to both his male and female offspring in a low methylated state in which it persists in their somatic cells. A low methylated gene coming through a sperm (2) persists in this condition in a male offspring, which in turn transmits it in a low methylated state. A female animal that receives a gene in a highly methylated state from her mother (3) will transmit it in this state to all her offspring, male and female. If she receives a low methylated gene from her father (4), it persists as such in her somatic cells, but she transmits it to all her offspring in a highly methylated state.

much earlier age than if it comes from the mother. In the case of myotonic dystrophy, the reverse is true. Perhaps such differences in these and other disorders are somehow related to differential imprinting.

The problem of inactivation of one X chromosome in the somatic cells of female mammals is being actively pursued on the molecular level. There is evidence that methylation of cytosines may play an important role in the phenomenon, and various models have been proposed to explain it. We expect more clues on this intriguing aspect of DNA inactivation to be forthcoming.

Gene regulation and DNA plasticity

As noted in Chapter 11, more than one form of the DNA molecule can exist. The B form discovered by Watson and Crick is the one we have depicted in our discussions. However, it was apparent even at the time of Watson and Crick's announcement of DNA structure that another form, known as the A form, also existed.

Today, improved techniques permit the synthesis of short DNA molecules composed of any nucleotide sequences. Pure preparations of a single kind of DNA can now be analyzed with precision using X-ray diffraction methods. A most significant outcome of these studies is the realization that the DNA molecule is far from a rigid, inflexible molecule, as might be assumed from the depiction of its familiar B form. The evidence indicates that the DNA can undergo changes in conformation and thus assume a form other than the familiar one in certain localized areas. This flexibility undoubtedly enables the DNA to interact in different ways with the assortment of cellular molecules that regulate gene expression. An acquaintance with some of the DNA configurations is therefore in order.

Three basic types of duplex DNA have been defined by the improved methodology. The familiar B form is stable under the conditions of high relative humidity (92%) that occur in cells and is undoubtedly the most typically occurring form in vivo. The A form mentioned earlier can be assumed by the DNA when the humidity falls to about 75%. Changes in ion concentration also play a role in changes from the B to A form. The A form of the molecule is shorter and fatter than the B form, and its bases are tilted differently in relation to the axis of the double helix (Fig. 17-31A). It is very likely that the A form is assumed when DNA–RNA hybridization takes place in regions undergoing transcription. Both the A and B forms of DNA are molecules with a right-handed twist.

Studies of short synthetic DNA molecules composed of dinucleotides of alternating purines and pyrimidines (GC, CG) led to the discovery of an unexpected DNA form, the Z form (Fig. 17-31B). Unlike the A and B forms, the Z form does not have a right-handed twist. The sugar–phosphate backbone follows a zig-zag course (hence Z DNA) and is left handed. In the Z form, a single, deep minor groove replaces the major and minor grooves found in B and A DNAs. The Z DNA is thin and elongated and has 12 base pairs in each turn of the helix—more than the other forms.

The question of whether the Z form exists in the living cell is still unsettled. For the Z form to arise, the alternating purine–pyrimidine sequence appears to be essential. It is considered possible that certain stresses imposed on the molecule may cause the B form to unwind and favor the formation of the Z form. This Z configuration could provide recognition signals for certain important regulatory molecules.

The information from in vitro studies has enabled us to look at the DNA molecule in a new light. Far from being inflexible, the commonly occurring B form, in response to certain environmental changes may undergo transitions to the A or Z forms in certain limited regions, whereas the B configurations of the molecule on either side of the altered site are preserved.

You will recall that inverted repeats occur throughout the DNA. When the two strands of the double helix separate, either strand may form a hairpin as the result of intrastrand pairing (see Fig. 12-25). It is considered possible that factors favoring the unwinding of specific regions of the DNA may result in the formation of these hairpins and, as in conversion to the Z form, may serve as recognition sites for regulatory molecules.

The ability of the DNA to assume localized changes in its form is now being considered as an important factor in the interactions between the DNA and regulatory molecules. It is known that the lac repressor recognizes nucleotide sequences in the operator in preference to other sequences in the bacterial DNA. Hydrogen bonds link certain amino acid side chains of the repressor to bases in the nucleotides of the operator. It is the formation of these specific hydrogen bonds that is largely responsible for the specificity of the repressor for the operator. Electrostatic forces involving positively charged amino acids of the repressor and negative charges in the sugar–phosphate backbone contribute to the strength of the binding of the repressor.

The cro repressor protein that was discussed in this chapter and that plays a central role in the lysogenic cycle of *E. coli* fits into the major groove, where it forms hydrogen bonds with certain bases in the DNA. Again, it is the hydrogen bonding between specific amino acids of the repressor and bases of the DNA that provides the specificity for the protein–DNA interaction. The ability of the DNA molecule to alter its shape is probably a factor in its interactions with regulatory proteins such as cro and lac. There is good evidence that deformation of the DNA *does* take place in some interactions. This is seen in the case of the interaction of CAP protein with the lac promoter. When free of the protein, the lac DNA is straight, but when the CAP–cAMP complex binds to it, the DNA bends. This can be detected by comparing differences in the way the free and the compressed DNAs travel in a gel electrophoresis experiment.

An important fact that has emerged about DNA is that all naturally occurring DNA (as mentioned in Chap. 11) exists in a supercoiled conformation rather than as an extended, linear duplex. Remember that the duplex DNA is right handed. The twisting that produces the supercoil, however, runs in the direction opposite to that of the turns of the duplex DNA. The DNA is said to be negatively supercoiled (review Fig. 11-28). To gain an insight into the generation of such a conformation, consider the following (see Fig. 17-32). Let us suppose that we take a circular duplex DNA molecule and cut both strands at a particular site. We then take the ends of the two strands and rotate them about each other in the direction *opposite* to the direction of the twist of the helix. This results in an unwinding of the helix. If the cut ends are then joined to reform a closed circle, the DNA will assume the form of a twisted circle. It assumes the negatively supercoiled configuration to relieve the strain that was imposed on it as a consequence of the unwinding.

In both bacteria and eukaryotes, the DNA is organized into loops, each one held together at its base by an RNA–protein complex. A loop behaves as if it were a closed circle and is found in the supercoiled state. In bacterial cells, the enzyme DNA gyrase, a topoisomerase, is responsible for bringing about the formation of supercoils and maintaining them. When DNA is replicating, the unwinding of the DNA strands can generate positive supercoils in the part of the molecule that has not yet undergone replication. These twists would become very tight and interfere with replication. The DNA gyrase prevents this by introducing negative supercoils.

The supercoiling is believed to exert an important influence on transcription. The negative supercoiling favors strand separation, which is a necessary preamble to transcription. An increase in negative supercoiling in a region would favor the transcription of certain genes. Negative supercoiling also stabilizes Z DNA, as well as regions with hairpins formed by intrastrand pairing. By favoring changes in the conformation of DNA, supercoiling may play a significant part in the interactions between the DNA and regulatory molecules. We see that the DNA molecule is one of unexpected flexibility, which enables it to interact in various ways with diverse cellular components as it transfers its information to the rest of the cell.

Gene expression and differential processing

In Chapter 13, we learned that cellular genes are characteristically fragmented and that their transcripts undergo processing and splicing (see Figs. 13-17 and 13-22). These events are pertinent to the subject of the control of gene expression. We can envision the primary transcript of a gene being processed and spliced in two or more different ways so that, in effect, more than one kind of mRNA would result. This means that one gene could carry information for more than one kind of mRNA and hence for more than one kind of product. How the transcript becomes processed in a given cell type could be the basis for how that gene is expressed. We now know that several such genes do exist among eukaryotes and viruses. A gene of this kind is known as a *complex transcription* unit, and it can give rise to different species of mRNA, depending on the cell type and several other factors.

As Figure 17-33 shows, a primary transcript could possibly contain more than one poly(A) site as well as several exons. Depending on where the cut is made at the 3′ end and how the splicing takes place, more than one kind of final mRNA could result. Transcripts from complex transcription units are known in mammals, and these appear to be processed in different ways, depending on the cell type. For example, the primary transcript of a particular gene in one type of muscle cell may be processed and spliced in a certain way. In another type of muscle cell, the same gene's transcript may be processed in another way. Since the transcripts are different, their translation will result in similar but nevertheless different proteins. Even though splicing can occur in more than one way, as in this example, it is still essential that a transcript, to be an active mRNA, retain its 5′ cap and be fitted with a poly(A) tail.

A particular complex transcription unit in the rat has been shown to be associated with at least two different mRNAs—one typical of the pituitary and one found in the thyroid. However, the same primary transcript is found in both thyroid and pituitary. The difference between the mRNAs is explained as a result of differential processing of the primary transcript, which contains two poly(A) sites (Fig. 17-34). In thyroid cells, this primary transcript is cleaved at the first site, and the first four exons are then spliced together to give a prehormone protein. In the pituitary, cleavage occurs at the second poly(A) site. Splicing then results in elimination of the introns along with one of the exons that is found

FIGURE 17-31. Forms of the DNA molecule. **A:** In the A form of DNA (left), the base pairs (represented here as planks) are titled and pulled away from the axis of the double helix. In B DNA (right), the base pairs are perpendicular to the axis with one base of a pair on each side of it. The sugar phosphate backbones are represented as ribbons. Both forms are molecules with a right-handed twist. (Reprinted with permission from R.E. Dickerson. The DNA helix and how it is read. *Sci. Am.* (Dec):94, 1983. Copyright © 1983 by Scientific American, Inc. All rights reserved.) **B:** The B form of DNA (left) is clearly seen to have a right-handed twist as well as a major and a minor groove, neither of which extends into the helix axis. The smooth heavy line traces the backbone of the molecule, going from phosphate to phosphate. The Z form of the DNA molecule (right) lacks a major groove but possesses a deep minor groove that extends to the axis of the left-handed helix. The backbone, as shown by the heavy line, is irregular and follows a zig-zag course.

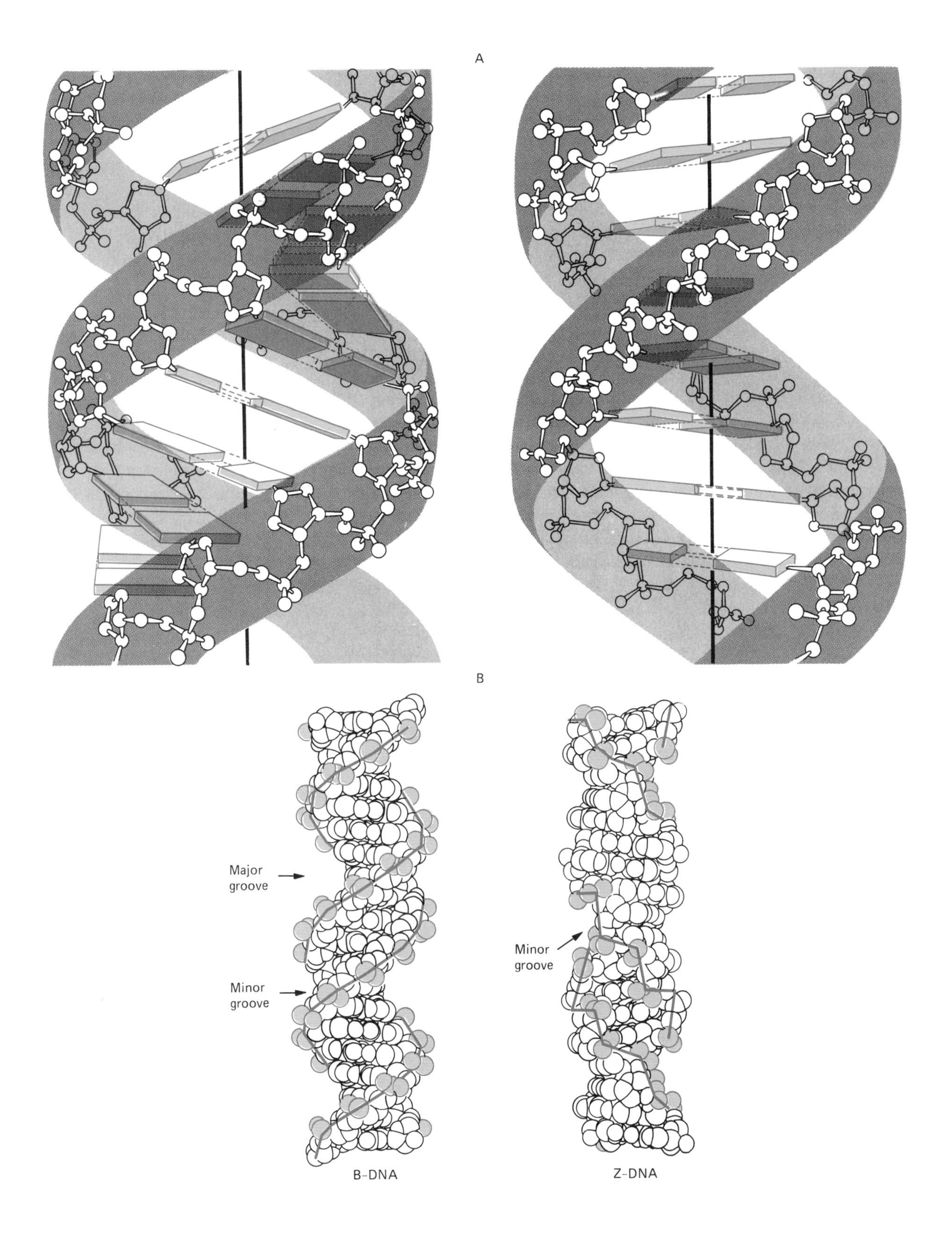
A
B
Major
groove
Minor
groove
B-DNA
Minor
groove
Z-DNA

in thyroid mRNA. However, the coded region at the end of the primary transcript is then joined to the first three exons. The outcome is production of an mRNA for a different prehormone in the pituitary.

These complex transcription units cause us to pause in our definition of a gene, because in these cases a given gene is associated with more than one kind of final product. Although differential processing of the primary transcript can determine which particular products may be produced in different cell types from transcripts of a specific gene, the primary way in which genes are regulated is still at the level of transcription. This is seen in that a given gene, though its transcript may be handled differently depending on the cell types, is not actually expressed at all in most cells of the organism. Housekeeping genes are indeed expressed in all cells; however, only a small percentage of the genome is being expressed in any given cell type. We now know conclusively that genes that are *not* expressed in a cell are *not* transcribed. Any expression of a gene depends on its undergoing transcription. Although the transcript may be handled differently in different cells, as in the case of complex transcription units, it is still the factors that switch transcription on and off that are dominant in gene control.

Giant chromosomes and gene activity

Some excellent studies of differentiation have made use of giant polytene chromosomes, described in Chapter 2. Such giant chromosomes can be seen in the salivary gland nuclei of the larval *Drosophila* (see Fig. 2-17). They are also found in other fly genera and can be seen in cells of other organs as well—the midgut, rectum, and excretory organs. These unusual structures provide a unique opportunity to study gene action. Each giant chromosome is a linear body composed of about 2^{10} single-unit chromatin fibers. As a result of the juxtaposition of so many individual chromosome threads, the polytene structure reflects features of the single chromosome that would otherwise go unnoticed. For example, the bands or chromomeres along the salivary chromosomes are regions where DNA appears to be concentrated because of tight packing of the individual chromatin fibers; the interbands, on the other hand, are more extended. We can see the band and interband arrangement, because more than 1,000 separate fibers are closely associated next to each other.

The nuclei in which these giant chromosomes occur are in a permanent interphase state. They are not preparing for cell division, but are concerned only with the synthesis of products needed in the nondividing nucleus. They are thus in a more active, stretched out condition than they would be if they were preparing to divide. Structural chromosome changes (see Chap. 10) have enabled us to associate some of the bands with specific genes. An inversion, duplication, or deletion is often related to a comparable change in a portion of one of the giant chromosomes. Each band may represent a site occupied by one gene, but this matter is still not resolved. Besides DNA, the chromosomes contain all of the other molecules typically associated with chromatin, such as RNA and protein.

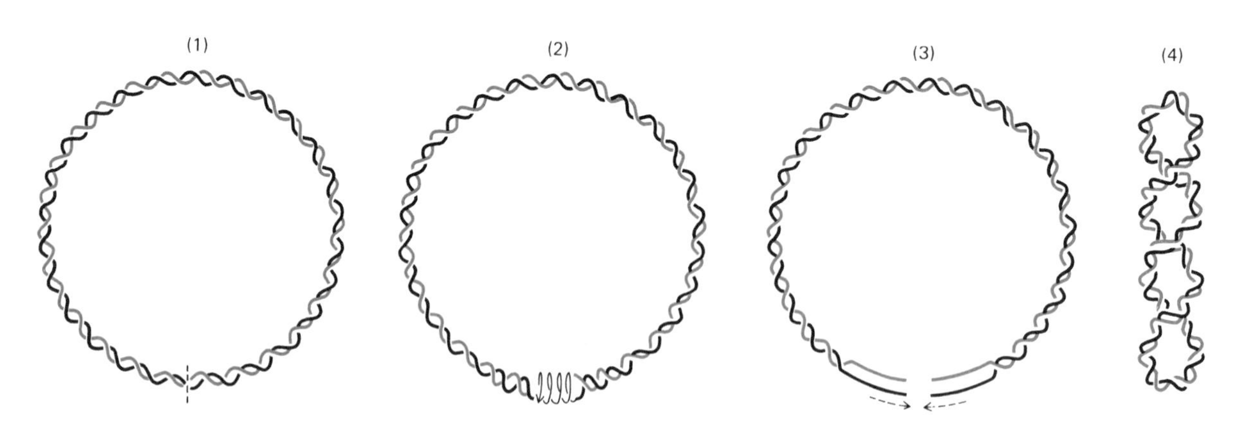

FIGURE 17-32. Generation of supercoiled DNA. **1:** The two strands of a circular duplex DNA are cut at a particular location (dotted line). **2:** The strands are then rotated about each other in the direction opposite to that in which the helix is twisted, and the helix begins to unwind. **3:** The number of times one strand goes around the other is reduced as a result of the unwinding. The cut ends are then rejoined (arrows). **4:** The DNA now assumes the form of a twisted circle, relieving the strain imposed on it by the unwinding. (Reprinted with permission from J.C. Wang, DNA topoisomerases. *Sci. Am.* (July):94, 1982. Copyright © 1982 by Scientific American. All rights reserved.)

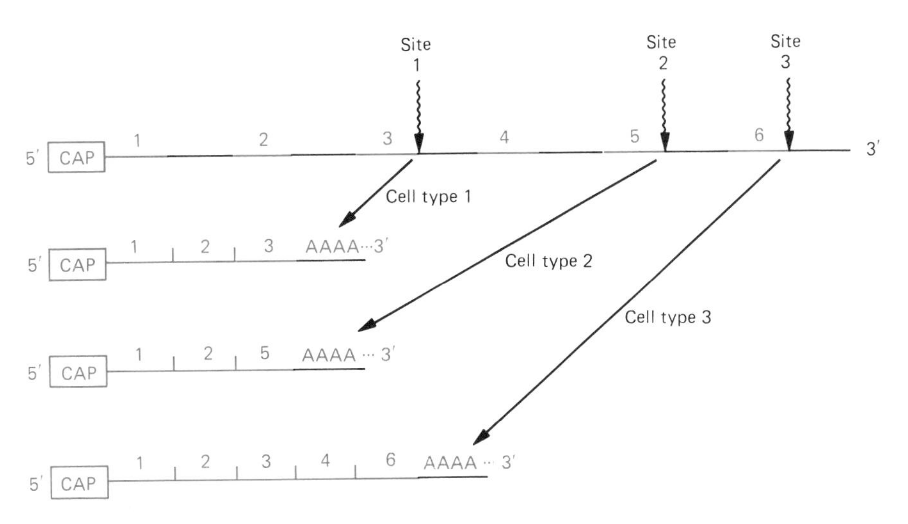

FIGURE 17-33. A complex transcription unit. Assuming a transcript of a specific gene contains more than one poly(A) site (arrows 1, 2, and 3), the transcript could be cut at site 1 in one type of cell, site 2 in a second type, and site 3 in a third. The 5′ cap must be retained, and a poly(A) tail must be added. In the processing, certain exons (red) may be discarded along with the introns in a given cell type. In cell type 1, exons 4, 5, and 6 are discarded along with the introns. In cell type 2, exons 3, 4, and 6 are discarded, whereas in type 3, only exon 5 is discarded. The one gene has thus given rise to three different mRNAs as a result of differential processing from one cell type to another.

In the early 1950s, Beerman, working with *Chironomus,* noted that some bands of the polytene chromosomes appear in a swollen or puffed condition. (These puffs are also found associated with giant chromosomes of several fly species, including *Drosophila.*) Comparison of the giant chromosomes in different tissues of the same individual also showed puffing, but significantly, the pattern of puffing differed from one kind of cell to another. Moreover, the pattern seen in a tissue is characteristic of a certain stage of development; the pattern changes at a later stage (Fig. 17-35).

Cells with polytene chromosomes were supplied with labeled amino acids and labeled RNA precursors. The results clearly showed that swollen bands are actively engaged in RNA synthesis but are not incorporating amino acids at an increased rate. All of the evidence taken together strongly suggests that the bands are sites of genes and their puffed state indicates that a gene or genes have become active. The gene activity is reflected by both the puffing and the increased rate of RNA formation. Observations with the electron microscope indicate that the puffing itself results from an unwinding of chromosome fibers in the more condensed regions of the bands. This unpacking exposes a much greater length of the individual chromosome fibers, and this makes them available as templates for RNA transcription (Fig. 17-36). Beerman and his colleagues have actually been able to isolate RNA associated with different puffs. If the RNA formed at separate puffs represents different kinds of mRNA coded by different genes, we would expect the base composition of the isolated RNA samples to vary. And this is what has been found. Not only have different types of RNA been demonstrated, but the various puffs show different rates of RNA synthesis. There is little doubt that the polytene chromosomes, with their puffing patterns and associated species of RNA, afford an exceptional opportunity to study differential gene activity at various stages of development.

If the puffs represent genes that have been switched on at specific times by an unpacking of the DNA, an important question arises concerning the nature of the mechanism that accomplishes this. A clue was provided by the discovery that a specific hormone, the insect molting hormone ecdysone, can induce the formation of very specific puffs. When ecdysone is released by the larva at a certain stage in development, the puffs that were already present decrease in size, and new ones quickly appear, eventually more than 100 of them, until the pupal stage is reached. Each puff has a characteristic time of appearance and duration.

The hormone ecdysone is a steroid, and it is believed that the first puffs appearing after its release are caused directly by the hormone's ability to promote transcription. Various steroid hormones, as noted earlier in this chapter, are believed to exert their effects by

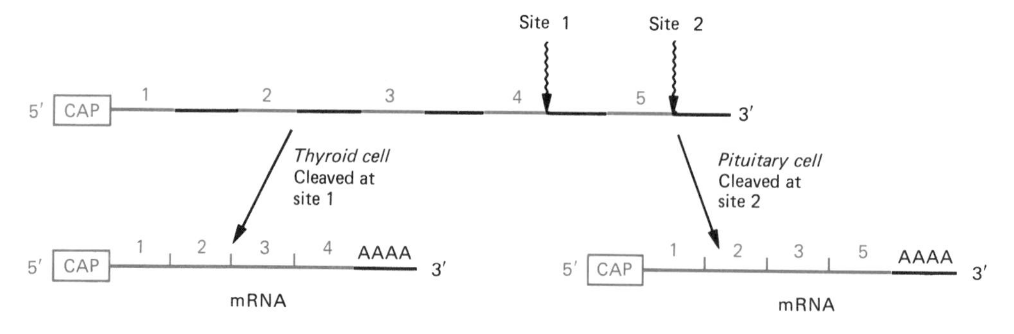

FIGURE 17-34. Differential processing in two cell types. The same primary transcript (above) of a given gene occurs in both the thyroid and the pituitary of the rat. However, the same transcript is processed differently (below). In the thyroid, cutting of the transcript occurs at site 1. Introns (black) are discarded, the first four exons (red) are joined, and a poly(A) tail is added. In the pituitary, cutting takes place at site 2. One of the exons kept in the thyroid (exon 4) is discarded, but the one at the 3′ end is included. Two different mRNAs have resulted from the same transcript and hence from the same gene.

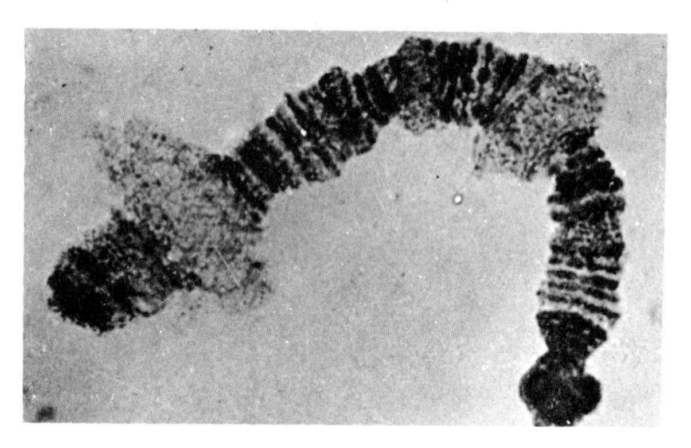

FIGURE 17-35. Chromosome puffs in *Rhynchosciara angelae.* The chromosome shows puffs that were not present at earlier stages of larval development but appear at a characteristic time. (Reprinted with permission from W.V. Brown. *Textbook of Cytogenetics,* p. 27. C.V. Mosby Co., St. Louis, 1972.)

directly promoting the transcription of specific genes. Ecdysone will cause the earlier puffs to form even if protein inhibitors are applied. This indicates that synthesis of new proteins is not essential for the formation of these early puffs but that the hormone is acting directly on the chromatin. Later puffs, however, do seem to depend on proteins made as a result of the earlier puffing, because they will not arise if protein synthesis is blocked. The hormone level also seems to be important, because several puffs may appear at ecdysone concentrations that bring about no changes in other bands along the length of the chromosome. Certain puffs may actually decrease in size, whereas others increase in response to the hormone when it is given experimentally.

Observations indicate that the various genetic regions along the length of the chromosome may react quite differently to a given hormone level. Although the changes in puffing patterns are taking place, the histone content remains constant, again indicating that these proteins are not removed from the DNA when it becomes active in transcription. Any modifications that may take place in histone or nonhistone components of the chromatin of the polytenes to permit transcription at a specific time await clarification.

Cytoplasm and external environment in cellular differentiation

The importance of the cytoplasm must not be overlooked when considering cellular differentiation. Very striking experiments that show the involvement of the cytoplasm are those that entail the transplantation of nuclei in amphibians. In such an experiment, the nucleus is removed from an egg cell. Another nucleus is then excised from a different kind of cell and placed into the egg cytoplasm. Shortly after this is done, synthesis of RNA and DNA takes place in the cell with the transplanted nuclei. This occurs even when the nucleus introduced into the egg is from a highly differentiated cell type. A brain cell nucleus, for example, does not normally divide again; yet when inserted into the egg, it can reacquire this ability. Several cell divisions may ensue. The nuclei taken from some cells trigger more divisions than others. Nuclei taken from the gut not only can divide but also can give rise to normal tadpoles! Such results demonstrate that the genetic material in differentiated as well as nondifferentiated cells possesses all of the information needed to direct the development of a complete individual. Cytoplasmic substances undoubtedly play a regulatory role by signaling the expression of specific genes at the proper time.

Superb experiments with plants have shown both the ability of a single differentiated cell to develop into a new individual and the importance of the external environment in cellular control. Most cells in the root of a carrot or of the underground stem (tuber) of the potato no longer divide when the plant reaches a mature state. It is common knowledge, however, that such quiescent cells express their capacity for cell division when the organ is wounded by cutting and the cells become exposed. Once the wound is healed, division again comes to a halt. Evidently, the information for cell division and growth is present in these cells. There must be some kind of signal that indicates when these processes are to start and stop. Based on observations with healing, it seemed likely that some signal could come from outside the cell itself. Steward and his associates, after exposing fragments of carrot root to a variety of substances, found that coconut milk acts as a potent growth stimulator. Fragments exposed to the milk increased in weight about 80-fold in approximately 3 weeks. The coconut milk must contain something that triggers the mechanism for cell division in these typically quiescent cells.

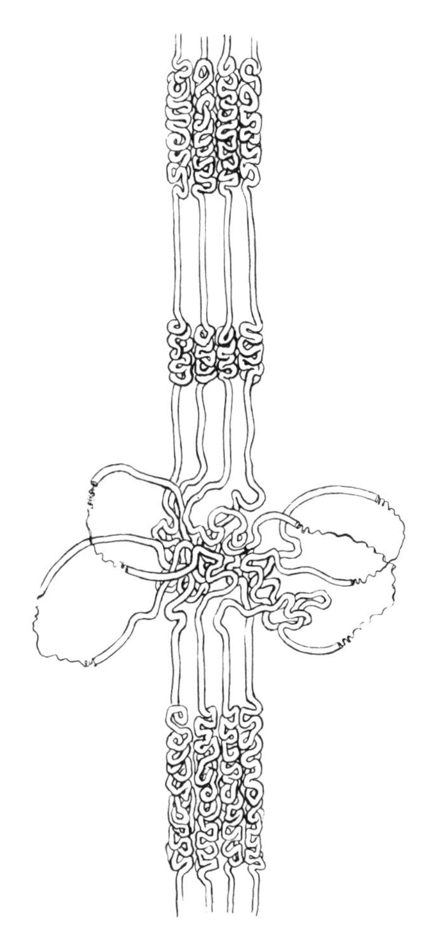

FIGURE 17-36. Nature of a polytene chromosome. The diagram shows that the giant chromosome consists of many unit chromosome threads or chromatin fibers side by side. Tight folding in all the threads at a specific site produces a band. A puff arises when the threads at a site become unpacked. The threads in their extended state can then engage in RNA synthesis. RNA isolated from puffs at different sites has been shown to be different in its composition, indicating that the RNA is mRNA. Different bands along a chromosome would thus include different genes that form different mRNAs during the puffed state. (Reprinted with permission from E.J. DuPraw and P.M.M. Rae. *Nature* 212:598, 1966.)

A very graphic illustration of the potential of a mature cell to express all of its coded genetic information was shown when individual root cells were freed completely from their neighbors. Apparently, surrounding cells in the intact root can restrict the growth of other cells in the vicinity. Separate root cells cultured in the coconut milk can develop into complete, normal plants. This is clear-cut evidence that all of the genetic information for directing the formation of another individual resides in the nucleus of the mature root cell. In the intact root tissue, however, the ability to express all of this information is suppressed by surrounding cells. When freed from this inhibitory effect and subjected to substances that can stimulate growth, the full potential of the mature nucleus is realized.

Another striking demonstration of this was seen when separate cells were obtained from an intact carrot embryo and then were cultured in coconut milk. Thousands of such individual embryo cells developed into thousands of other embryos, just as if each separate cell had regained the power of the fertilized egg (Fig 17-37). Moreover, the stages of development through which these secondary embryos passed were similar to those of embryos that develop in the usual way in the intact plant.

We see that, although an adult, differentiated cell contains all the coded information of the fertilized egg, it is normally restricted from expressing the entire set of instructions. Apparently, this suppression takes place during normal development by the inhibitory effect of certain environmental substances, as well as by the influence of surrounding cells. Other substances may trigger the information for division and growth at the proper time or "turn on" the tran-

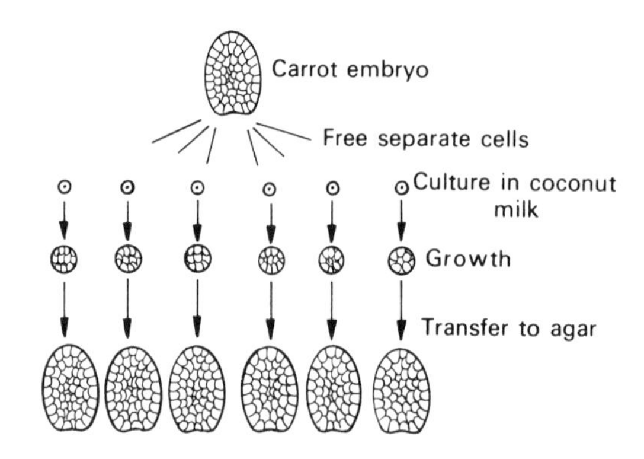

FIGURE 17-37. Embryos from embryos. Separate cells can be freed from a growing carrot embryo. When these separate cells are cultured in coconut milk, they can multiply to form a small group of cells. Each of these in turn can form a carrot embryo when raised on agar medium.

scription of the appropriate genes when their products are needed. By a series of coordinated inhibitory and stimulatory signals, the genetic information is so controlled that essential gene products are present at the needed time. When not required, superfluous gene products could be detrimental and would waste valuable cell energy, as in the case of the constitutive bacterial mutants.

It is of course imperative that cells do *not* express all of their genetic information; the result would be masses of identical cells. Normal differentiation from the fertilized egg on requires that only a fraction of the total coded information be translated. Expression of extra information in a growing cell or a mature cell may cause chaotic or abnormal growth.

REFERENCES

Amara, S.G., V. Jonas, M.G. Rosenfeld, E.S. Ong, and R.M. Evans. Alternative RNA processing in calcitonin gene expression generates mRNAs encoding different polypeptide products. *Nature* 298:240, 1982.

Ashburner, M., C. Chihara, P. Meltzer, and G. Richards. Temporal control of puffing activity in polytene chromosomes. *Cold Spring Harbor Symp. Quant. Biol.* 38:655, 1974.

Bauer, W.R., R.H.C. Crick, and J.H. White. Supercoiled DNA. *Sci. Am.* (July) :94, 1982.

Beerman, W., and U. Clever. Chromosome puffs. *Sci. Am.* (April) :210, 1964.

Bingham, P.M., M.G. Kidwell, and G.M. Rubin. The molecular basis of P–M hybrid dysgenesis: The role of the P element, a P-strain-specific transposon family. *Cell* 29:995, 1982.

Blackburn, E.H. Telomeres and their synthesis. *Science* 249:489, 1990.

Brosius, J. Retroposons—Seeds of evolution. *Science* 251:753, 1991.

Brown, D.D. Gene expression in eucaryotes. *Science* 211:667, 1981.

Bukhari, A.I., J. Shapiro, and S.L. Adhya. *DNA: Insertion Elements, Plasmids, and Episomes.* Cold Spring Harbor Laboratory, Cold Spring Harbor, NY, 1977.

Cattanach, B.M., and M. Kirk. Differential activity of maternally and paternally derived chromosome regions in mice. *Nature* 313:496, 1985.

Cedar, H. DNA methylation and gene activity. *Cell* 53:3, 1988.

Cohen, S.N., and J.A. Shapiro. Transposable genetic elements. *Sci. Am.* (Feb.) :40, 1980.

Darnell, J.E., Jr. Variety in the level of gene control in eukaryotic cells. *Nature* 297:365, 1982.

De Robertis, E.M., and J.B. Gurdon. Gene transplantation and the analysis of development. *Sci. Am.* (Dec) :74, 1979.

Dickerson, R.E. The DNA helix and how it is read. *Sci. Am.* (Dec) :94, 1983.

Dickerson, R.E., H.R. Drew, B.M. Conner, R.M. Wing, A.V. Fratini, and M.L. Hopka. The anatomy of A-, B-, and Z-DNA. *Science* 216:475, 1982.

Dickson, R.C., J. Abelson, W.M. Barnes, and W.S. Reznikoff. The lac control region. *Science* 187:27, 1975.

Doerfler, W. DNA methylation and gene activity. *Annu. Rev. Biochem.* 52:93, 1983.

Dombroski, B.A., S.L. Mathias, E. Nanthakumer, A. Scott, and H.H. Kazazian Jr. Isolation of an active human transposable element. *Science* 254:1805, 1991.

Dunsmuir, P., W.J. Brorein, Jr., M.A. Simon, and G.M. Rubin. Insertion of the *Drosophila* transposable element *copia* generated a 5 base pair duplication. *Cell* 21:575, 1980.

Elgin, S.C.R. DNase I–hypersensitive sites of chromatin. *Cell* 27:413, 1981.

Emerson, B.M., C.D. Lewis, and G. Felsenfeld. Interaction of specific nuclear factors with the nuclease-hypersensitive region of the chicken adult β-globin gene: Nature of the linking domain. *Cell* 41:21, 1985.

Federoff, N.V. Transposable genetic elements in maize. *Sci. Am.* (June) :84, 1984.

Federoff, N., S. Wessler, and S. Shure. Isolation of the transposable maize controlling elements Ac and Ds. *Cell* 35:235, 1983.

Felsenfeld, G. Chromatin. *Nature* 271:115, 1978.

Felsenfeld, G. DNA. *Sci. Am.* (Oct) :58, 1985.

Felsenfeld, G., and J. McGhee. Methylation and gene control. *Nature* 296:602, 1982.

Gurdon, J.B. Transplanted nuclei and cell differentiation. *Sci. Am.* (Dec) :24, 1968.

Herendeen, D.R., K.P. Williams, G.A. Kassavetis, and E.P. Geiduschek. An RNA polymerase-binding protein that is required for communication between an enhancer and a promoter. *Science* 248:573, 1990.

Herskowitz, I., and D. Hagen. The lysis–lysogeny decision of phage lambda: Explicit programming and responsiveness. *Annu. Rev. Genet.* 14:399, 1980.

Hoffman, M. How parents make their mark on genes. *Science* 252:1250, 1991.

Holliday, R. Ageing: X-chromosome reactivation. *Nature* 327:661, 1987.
Holliday, R. A different kind of inheritance. *Sci. Am.* (June) :60, 1989.
Hunter, T. The proteins of oncogenes. *Sci. Am.* (Aug) :70, 1984.
Jacob, F., and J. Monod. Genetic regulator mechanisms in the synthesis of a protein. *J. Mol. Biol.* 3:318, 1961.
Jagadeeswaran, P., B.G. Forget, and S.M. Weissman. Short interspersed DNA elements in eucaryotes: Transposable DNA elements generated by reverse transcription of RNA P1 III transcripts? *Cell* 26:141, 1981.
Khoury, G., and P. Gruss. Enhancer elements. *Cell* 33:313, 1983.
Korge, G. Direct correlation between a chromosome puff and the synthesis of a larval saliva protein in *Drosophila melanogaster. Chromosoma* 62:155, 1977.
Kornberg, R.D., and A. Klug. The nucleosome. *Sci. Am.* (Feb) :52, 1981.
Laybourn, P.J., and J.T. Kadmaga. Role of nucleosome cores and histone HI in regulation of transcription by RNA polymerase II. *Science* 254:238, 1991.
Lewis, A.M., Jr., and J.L. Cook. A new role for DNA virus early proteins in viral carcinogenesis. *Science* 227:15, 1985.
Maniatis, T., and M. Ptashne. A DNA operator-repressor system. *Sci. Am.* (Jan) :64, 1976.
Marx, J.L. Z-DNA: Still searching for a function. *Science* 230:794, 1985.
Mathias, S.L., A. Scott, H.H. Kazazian Jr., J.D. Boeke, and A. Gabriel. Reverse transcriptase encoded by a human transposable element. *Science* 254:1808, 1991.
McClintock, B. The control of gene action in maize. *Brookhaven Symp. Biol.* 18:162, 1965.
McGhee, J.D., J.M. Nichol, G. Felsenfeld, and D.C. Rau. Higher order structure of chromatin: Orientation of nucleosomes within the 30 nm chromatin solenoid is independent of species and spacer length. *Cell* 33:831, 1983.
Olins, D.E., and A.L. Olins. Nucleosomes: The structural quantum in chromosomes. *Am. Sci.* 66:704, 1978.
O'Malley, B.W., and W.T. Schrader. The receptors of steroid hormones. *Sci. Am.* (Feb) :32, 1976.
Palo, C.O., et al. The lambda repressor contains two domains. *Proc. Natl. Acad. Sci. USA* 76:1608, 1979.
Ptashne, M., and W. Gilbert. Genetic repressors. *Sci. Am.* (June) :36, 1970.
Richmond, T.J., J.T. Finch, B. Rushton, D. Rhodes, and A. Klug. Structure of the nucleosome core particle at 7 Å resolution. *Nature* 311:532, 1984.
Sapienza, C. Parental imprinting of genes. *Sci. Am.* (Oct) :52, 1990.
Schippen-Lentz D., and E.H. Blackburn. Functional evidence for an RNA template in telomerase. *Science* 247:546, 1990.
Smith, G.R. DNA supercoiling: Another level for regulating gene expression. *Cell* 24:599, 1981.
Solter, D. Differential imprinting and expression of maternal and paternal genomes. *Annu. Rev. Genet.* 22:127, 1988.
Steward, F.C. The control of growth in plant cells. *Sci. Am.* (April) :104, 1963.
Wallace, M.R., et al. A de novo Alu insertion results in neurofibromatosis type 1. *Nature* 353:864, 1991.
Wang, J.C. DNA topoisomerases. *Sci. Am.* (July) :94, 1982.

REVIEW QUESTIONS

For Questions 1 through 4, allow R^+ and O^+ to stand for regulatory gene and operator and R^- and O^- to represent their deficient alleles. Z^+, Y^+, and A^+ represent normal alleles of structural genes for enzyme production and Z^-, Y^-, and A^- their alleles for absence of enzyme. In normal cell types, the enzymes are inducible unless otherwise stated.

1. In cells of the following genotypes, indicate whether enzyme production will be inducible or constitutive:

A. $\underline{R^+O^+Z^+Y^+A^+}$ **B.** $\underline{R^+O^-Z^+Y^+A^+}$

C. $\frac{R^+O^+Z^+Y^+A^+}{R^+O^-Z^+Y^+A^+}$ **D.** $\frac{R^+O^-Z^+Y^+A^+}{R^+O^-Z^+Y^+A^+}$

E. $\frac{R^+O^+Z^+Y^+A^+}{R^-O^+Z^+Y^+A^+}$ **F.** $\frac{R^-O^+Z^+Y^+A^+}{R^-O^+Z^+Y^+A^+}$

2. For each of the following, tell which enzymes will be produced constitutively and which by induction:

A. $\frac{R^+O^-Z^+Y^-A^-}{R^-O^+Z^-Y^+A^+}$ **B.** $\frac{R^+O^+Z^-Y^+A^+}{R^-O^+Z^+Y^-A^-}$

3. Let O^0 represent an operator mutation causing the operator to bind irreversibly with the normal repressor. Let R^S represent a regulatory gene mutation that causes the formation of an altered repressor, which cannot react with the inducer even though it can bind with the normal operator. For each of the following, tell which enzymes will be produced constitutively and which by induction:

A. $\underline{R^SO^+Z^+Y^+A^-}$ **B.** $\underline{R^+O^0Z^+Y^+A^+}$

C. $\frac{R^+O^0Z^+Y^+A^-}{R^-O^+Z^-Y^-A^+}$ **D.** $\frac{R^-O^+Z^+Y^+A^-}{R^SO^-Z^-Y^-A^+}$

4. Assume that A^+ and B^+ govern enzyme formation in an operon that is a repressible system, such as in the case of tryptophan synthetase. In such a system, the effector (end product) must combine with the regulatory gene product for the operator to block transcription. For each of the following, tell what enzymes would be expected from the cells in the absence of effector and in the presence of effector:

A. $\underline{R^+O^+A^+B^+}$ **B.** $\underline{R^+O^-A^+B^+}$

C. $\underline{R^-O^+A^+B^+}$ **D.** $\frac{R^+O^+A^+B^-}{R^+O^-A^-B^+}$

5. Merozygotic *E. coli* may be partially diploid for genes in the lac region. From the results given in the table (1) determine the dominance relationships among the I alleles, and (2) indicate evidence showing that the lac repressor is a diffusible substance (+ = enzyme synthesized; 0 = no enzyme synthesis).

Strain	Genotype (Chromosome/F factor)	Lactose	
		Present	Absent
1	$I^+O^+Z^+/I^sO^+Z^-$	0	0
2	$I^sO^+Z^+/I^+O^+Z^-$	0	0
3	$I^sO^+Z^+/I^-O^+Z^-$	0	0
4	$I^-O^+Z^+/I^sO^+Z^-$	0	0
5	$I^+O^+Z^+/I^-O^+Z^-$	+	0
6	$I^-O^+Z^+/I^+O^+Z^-$	+	0

6. Suppose the lac repressor were not diffusible. Show the results that would then be expected for the six strains shown in question 5 when in the presence and absence of lactose.

7. The operator–repressor system is considered to be a negative control, whereas the cAMP–CAP mechanism is considered to be positive. Why is this so?

8. Some bacterial cells harbor a plasmid known as R6-5. The plasmid does not confer drug resistance. From a population of such cells, tetracycline-resistant cells have been isolated, and the resistance determinant has been found to be associated with the plasmid. Analysis of the plasmid from the resistant cells shows that it is about 1.4 kb shorter than R6-5. Explain these observations.

9. A certain inducible operon includes three structural genes coded for enzymes A, B, and C, respectively. Cells have been isolated that produce the three enzymes constitutively. Other cells have been isolated that cannot produce the enzymes at all and are thus completely uninducible. DNA analysis of both kinds of cells shows that the region of the operon is about 2.5 kb longer than the same region in cells of the wild type. How can this be explained?

10. In some variant *E. coli* cells, all the structural genes of the gal operon are permanently turned off, in contrast with those in the wild-type operon, which is inducible. Viruses capable of transducing the operon were isolated, and the DNA was extracted. DNA was also isolated from transducing viruses carrying the mutant operon. The two kinds of DNA were denatured, and the single strands were separated. Single strands from the normal and the mutant cells were then allowed to hybridize. The electron microscope reveals DNA that is duplex but contains an unpaired segment. This segment appears as a stalk or stem with a loop at the top of the stem. Account for all these observations.

11. A particular IS element bears the following sequence at the end of one of its DNA strands: 3′–GGTGATGTA. This is followed by nonduplicated sequences that can be represented as XYZ. The sequence at the 3′ end exists as an inverted repeat at the 5′ end.

A. Give the double helical structure of this IS element, using XYZ and X′Y′Z′ to represent the regions not involved in the inverted repeat.

B. If the strands are separated, show what should be expected as a result of base pairing in each of the isolated strands.

12. Following are nine strains of phage λ, each with the specific mutant defect noted. In which of these strains will lysogeny be favored over lysis?

A. Gene O deleted.
B. Gene cII deleted.
C. P_{RE} deleted.
D. Gene cro deleted.
E. P_{RM} deleted.
F. Overexpression of cro.
G. Underexpression of N.
H. O_R with strong affinity for lambda repressor.
I. P_{INT} deleted.

13. Red kernel color in corn depends on the dominant allele R, whereas its recessive counterpart r is associated with white kernels because of lack of pigment production. Assume that certain plants of genotype RR are mated to plants with white kernels, genotype rr. Plants of genotype RR produce gametes of the following five types:

1. A gamete with the Ds element located on a chromosome other than the one with the pigment locus. No Ac element is present.
2. A gamete with Ds inserted into the pigment locus. Ac is on the same chromosome.
3. A gamete with Ds inserted into the pigment locus. Ac is on another chromosome.
4. A gamete with Ds inserted into pigment locus. Ac is lost.
5. A gamete with Ac near pigment locus. Ds is lost.

How would the kernels be expected to appear in the next generation in each case, when the gamete described unites with a gamete carrying the recessive r?

14. The transposable element "P" is found in P strains of the fruit fly and is associated with hybrid dysgenesis in certain crosses. M strains lack the element.

A. Suppose highly fertile flies with normal chromosome behavior arise from a cross involving a P and an M strain. From which parental strains were the females and males in this cross?

B. With respect to M and P strains, what other crosses will yield fertile offspring with normal chromosome behavior?

15. Assume that a certain nucleotide sequence is known to occur in the DNA of the rat. Chromatin is extracted from cells of different tissues and then exposed to an endonuclease treatment that digests away linker DNA, leaving the nucleosome beads. When the beads derived from chromatin of pancreatic cells are subjected to further analysis, it is found that the particular nucleotide sequence can always be recovered. However, when chromatin taken from cells of the skin, liver, and other sites is treated in the same way, the same DNA sequence is found to be lost, recovered only in part, or recovered at times in its entirety. Explain these findings.

16. A particular gene is known to be highly active in most tissues of the rat. Would you expect the promoter region at its 5′ end to be bound to the histone octamer? Explain. How could you determine whether the sequence is tightly bound?

17. In a particular DNA virus, a certain early gene becomes very active shortly after infection of the cell. Its transcripts are needed to turn on genes that become active later in the viral cycle. A strain of the virus is isolated in which transcription from the early gene has decreased by well over three-fourths, greatly interfering with replication of the virus. Analysis of the viral DNA shows the entire early gene to be perfectly intact along with its promoter. However, a deletion of a DNA segment is found more than 1 kb away from the gene. Explain.

18. Would you expect a gene associated with each of the following to be active or inactive: 1) associated tightly with histone octamer; 2) associated with acetylated histones; 3) associated with a high number of methyl groups; 4) associated with ubiquitin; 5) associated with receptor–steroid hormone complex; 6) associated with increase in negative supercoiling?

19. Indicate by an asterisk any site at which methylation may take place in the following DNA sequence:

5′– CACGTCACGTACC – 3′
3′– GTGCAGTGCATGG – 5′

20. Assume in a certain animal species that the pancreas, pituitary, and thyroid each secretes a particular prehormone. Certain missense mutations that cause amino acid substitutions toward the amino end of the three different prehormones are known. When such a missense mutation alters the thyroid prehormone, the other two prehormones remain unaffected. However, a missense mutation that brings about an amino acid substitution toward the amino end of the pancreatic protein always has the same effect on the pituitary protein though never altering the thyroid prehormone in any way. Molecular analysis reveals that the pancreatic and pituitary prehormones are identical for about the first 50% of their amino acids, but the latter portion toward the carboxyl end is very different. The amino acid sequence of the thyroid prehormone does not resemble those of the other two. How many genes appear to be involved in the case of these three prehormones? Explain the observations.

21. A *Drosophila* larva is fitted with a bond in such a way that portions of the salivary glands become tied off from the rest of the glands. The effect is such that one portion continues to be exposed to ecdysone, whereas the rest does not receive the hormone. The polytene chromosomes in cells receiving hormone continue to undergo puffing, and the cells continue to develop. The chromosomes in cells cut off from the hormone show little puffing, and tissues do not differentiate.

   A. Using radioactively labeled precursor molecules, how might you show that chromosome puffs undergo transcription whereas unpuffed regions do not?
   B. What can you suggest as to the role of ecdysone, based on information regarding steroid hormone–chromatin interactions?

22. In each of the following, an Ac element moves from site 1 to site 2 in the same chromosome. For each case, tell how many Ac elements will be present at sites 1 and 2 when replication is complete and the chromosome consists of two sister chromatids. Also give the total number of Ac elements present in the chromosome. 1) Ac replicates at site 1, and one copy moves to site 2, which has not replicated when Ac inserts. 2) Ac moves from site 1 to site 2 before replication of either site. 3) Ac replicates at site 1. Site 2 has replicated, and one Ac inserts into site 2 of the sister chromatid.

Controls Throughout the Life Cycle

18

Genetic regulation of body plan

The larval form of the fruit fly is composed of a series of segments separated by grooves (Fig. 18-1A). Besides the head segments, three of these (T_1–T_3) are thoracic and eight (A_1–A_8) are abdominal. These segments, each of which undergoes its own pattern of differentiation, correspond to those in the adult insect. The organization of the body plan of *Drosophila* can actually be traced all the way back to genetic factors which come to expression in the unfertilized egg. These are the maternal genes, and approximately a dozen of them have been identified. They are essential to development of an egg that will permit normal embryo development. If the female carries a mutant form of any one of them, the proper dorsoventral polarity of the egg fails to become established, and an abnormal embryo arises. Posterior body regions may develop as anterior ones, and parts that should be ventral can develop in a dorsal location. It is immaterial if the male parent contributes a wild-type counterpart of the defective allele from the female. The effect of these genes is thus strictly maternal.

Following fertilization, another group of genes comes to expression that are critical to normal organization of the body. These are the segmentation genes required for the correct number and location of the body segments as well as for their proper polarity. It now seems likely that the visible or morphological segments that come to be defined in the larva and the adult are actually not the basic units of organization of the body of the fly. Instead, the basic unit is the parasegment (Fig. 18-1B). Each parasegment extends from the middle of one morphological segment to the middle of the next one. As development progresses, the constriction that forms in the center of a parasegment deepens so that each is composed of two compartments, P followed by A. The boundary between each morphological segment is really the center of a parasegment. Each visible segment is composed of an anterior (A) and a posterior (P) compartment.

About 20 segmentation genes are known that regulate segment organization, and they fall into three main classes. Very early in embryo development, those in the gap gene class are called into action. Their proper expression requires the establishment of normal egg polarity, which in turn depends on the action of the maternal genes before fertilization. A gap gene mutation results in the deletion of several adjacent segments from the body of the larva. Following gap gene activity, genes come to expression that are known as the pair-rule genes. A larva homozygous for a mutant allele of a gene in this class has only half the number of segments present in a normal larva. Every other segment found in a normal larva is missing. The third class of segmentation genes is activated after expression of the previous group. These are the segment polarity genes. If a normal gene in this group is not present, half of each body segment is missing, but the missing portion is replaced by a mirror image of the segment that remains (Fig. 18-2A–C).

In situ hybridization has made it possible to detect transcripts of the three classes of segmentation genes in the embryo and thus to determine where the specific genes are being expressed as the body becomes organized. The results indicate that the segmentation

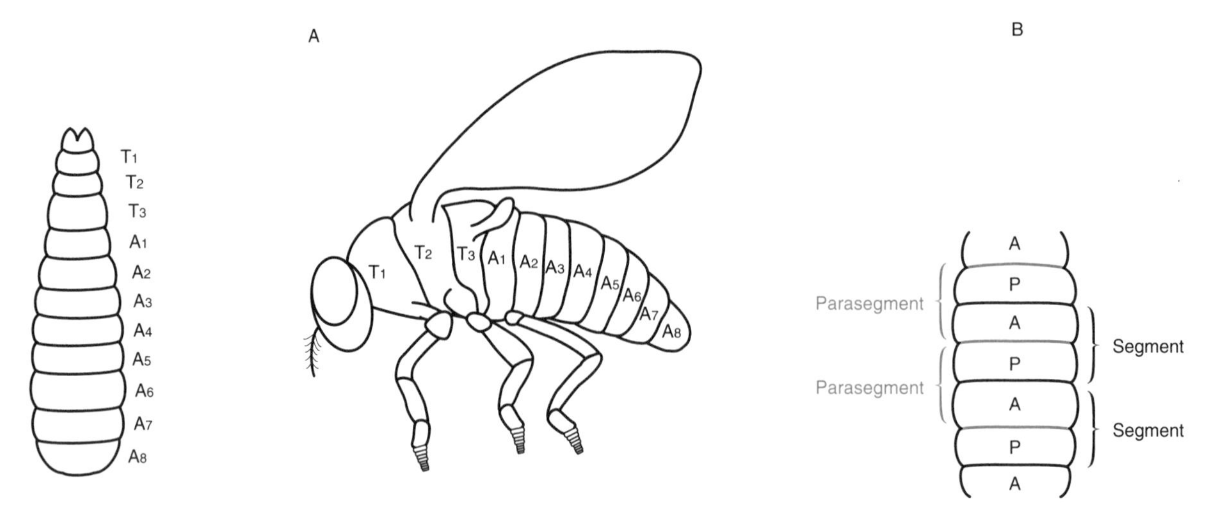

FIGURE 18-1. Segmentation in the fruit fly. **A:** The larval form of the fly (left) possesses eleven segments other than those involved with the head. There are three thoracic segments (T_1–T_3) and 8 abdominal ones (A_1–A_8). All these segments can be traced to their final differentiation in the adult (right). **B:** The basic units of organization of the fly body appear to be the parasegments. Each of these is composed of a P compartment followed by an A compartment. The morphological segments of the larva and adult are composed of an A compartment followed by a P. The middle of each parasegment (black) marks the boundary of a visual segment.

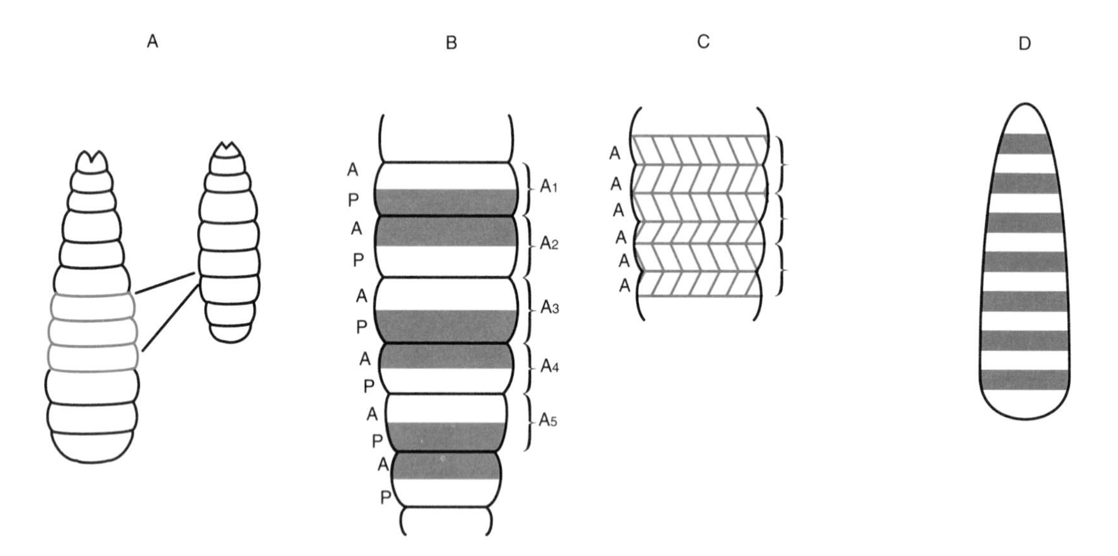

FIGURE 18-2. Segmentation gene mutations. A gap mutation (A) results in the elimination of adjacent segments normally found in the larva (red segments). A pair-rule mutation (B) causes the elimination of certain P and A parasegment compartments (red) in such a way that a mutant larva contains one-half the normal number with every other visual segment apparently missing. The remaining compartments fuse. (Only a few abdominal compartments are shown here.) A segment polarity mutation (C) causes loss of a compartment, usually the P. This is replaced by a mirror image of the compartment that remains. Transcripts of a particular pair-rule gene can be detected in the embryo by in situ hybridization and are visualized as bands (D, red). When a mutant allele of this gene replaces the wild type, the transcripts are absent, and hence the bands cannot be demonstrated. The embryo develops into a larva with missing parasegments and only one-half the normal number of visual segments as described for B.

genes undergo transcription in defined patterns and that the expression of these genes determines the pattern of development of the compartments that form the parasegments and segments of the larva. Studies of embryos with mutant alleles of these genes reveal altered patterns of gene expression followed by alteration in body organization. For example (Fig. 18-2D), a mutant allele of a well-studied pair-rule gene alters the pattern of gene expression and the ensuing pattern of normal development in such a way that the regions that correspond to even-numbered parasegments are missing, producing a larva with only one-half the normal number of segments.

Once the segmentation genes have established the normal body segments of the larva, other genes become activated. These will now determine the specific course of development of the segments. The battery of genetic factors involved in this important aspect of differentiation is the topic of the following section.

Homeotic genes

For years, geneticists have been aware of mutations in the fruit fly that result in bizarre deformities. Such a mutation might give rise to a fly with four wings instead of two or to one bearing a pair of perfectly formed legs attached to its forehead in place of antennae, a mutation known as antennapedia (Fig. 18-3). These kinds of mutations, which produce developmental upsets in which one body part may be replaced by another, have been called *homeotic mutations.*

Genetic analyses showed that a homeotic transformation could result from a mutation in just a single gene. This was a most important finding, since it is well known that many genes interact in the formation of any body part or organ. The very fact that just a single gene mutation can produce a dramatic effect on the organization of the body tells us that a homeotic mutation must occur in a master gene, that is, a gene that controls the activity of other genes.

Two clusters of homeotic genes have been recognized on the right arm of chromosome III of *Drosophila,* and these are the focus of many developmental studies in the fly. One of the clusters makes up the antennapedia complex (ANT-C), which guides development of the head and three segments of the thorax. The other is known as the bithorax complex (BX-C), which regulates development of the eight segments of the abdomen and two of the thorax. Between them, these two complexes regulate the formation of a great deal of the anatomy of the fruit fly. A mutation in any one of the genes in these complexes usually produces such a developmental upset that the embryo dies. A few malformed ones survive, however, typically with misplaced or transformed body parts.

The methodology of molecular biology that we will encounter in the following chapters made it possible to isolate several homeotic genes in the fruit fly in the early 1980s. DNA sequences of two developmental genes were compared, and this led to an amazing discovery: A certain portion of the sequences was identified that was very highly conserved, corresponding almost exactly base pair for base pair. This meant that the conserved sequence could be used as a probe (see Chap. 19) to locate any other genes in a DNA sample that contain the same or a very similar base pair sequence. Armed with the probe, molecular biologists isolated other homeotic genes. The conserved DNA region in each of these genes was named the *homeobox.* As noted several times in our previous discussions, a conserved sequence implies a segment that must serve a function so vital that it has been kept almost intact throughout evolution. The homeobox region is a 180 base pair sequence that encodes a stretch of 60 amino acids in the protein product of each homeotic gene. This very similar sequence from one homeotic gene protein to the next has been designated the *homeodomain* (Fig. 18-4). We now know that its function is to recognize a specific DNA sequence of any gene that is under the regulation of the homeotic gene. Since homeodomains are very much the same from one homeotic gene protein to the next, this means that the DNA segment to which they bind are all also very similar.

Highly unexpected was the finding that the homeobox DNA from the fly could be used as a probe to locate homeotic genes in vertebrates. This is again a reflection of the amazing degree of conservation of the homeobox throughout animal evolution. Genes with homeoboxes are now known in fungi, amphibians, the mouse, the human, and even in the plant kingdom. When the protein products of the various homeotic or homeobox genes are compared, it is found that they differ greatly from one another except for the homeodomain region. Some homeodomains are more similar than others, and it is interesting to note that the homeodomains of some fly proteins are very similar to those of mammals. We see that the variation in the amino acid sequence of the homeodomain is indeed small even when the homeotic gene proteins are compared from species as different as the insect and the human.

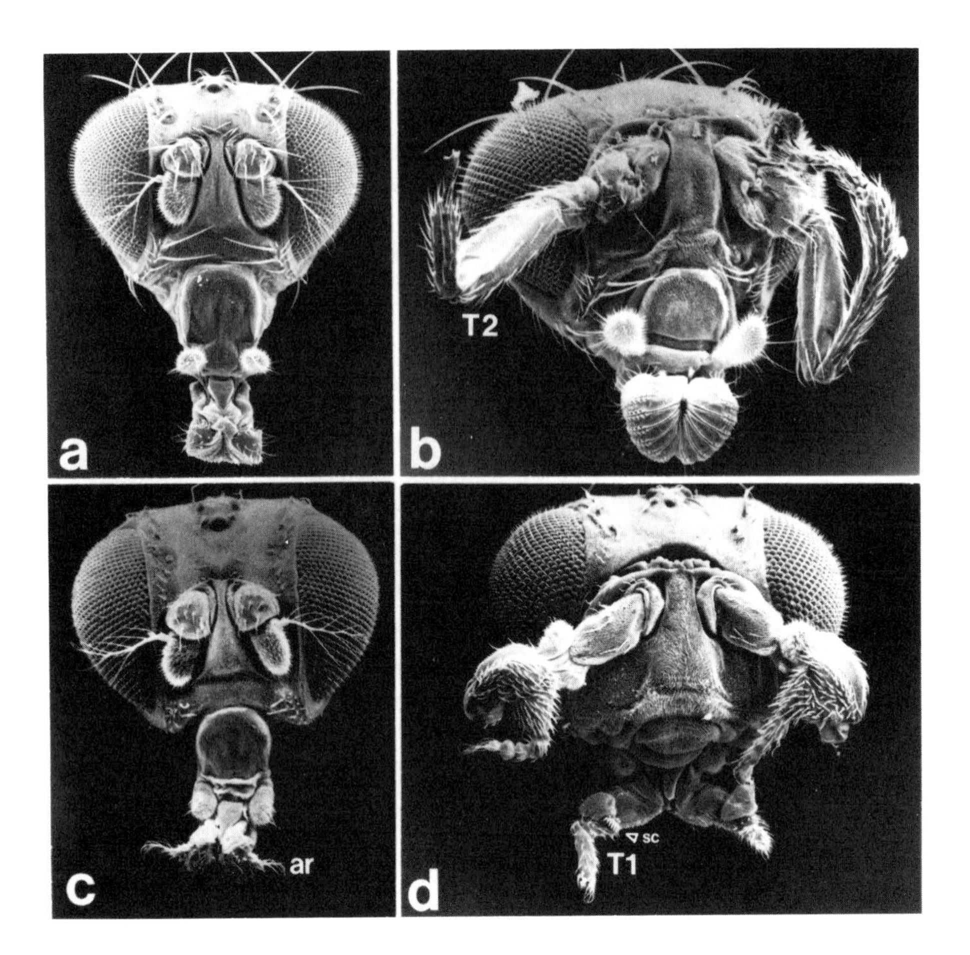

FIGURE 18-3. Normal fly head and homeotic mutants (ANT-C). **A:** Normal head front view. **B:** *Antennapedia* mutant showing legs (T2) growing in place of the antennae. **C:** A *proboscipedia* mutant showing antennae (ar) replacing the mouthparts. **D:** A double *(Antp pb)* mutant showing legs replacing the antennae and the mouthparts. (Courtesy of Dr. Thomas C. Kaufman, Indiana University, Bloomington, Indiana.)

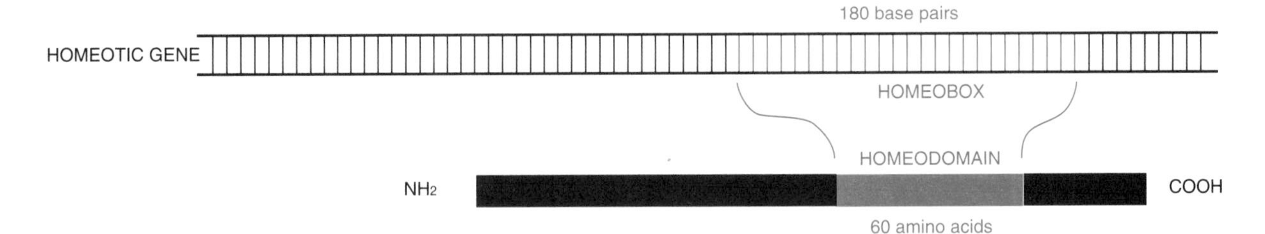

FIGURE 18-4. Conserved sequences. The homeobox is a conserved sequence of 180 base pairs in a homeotic gene (above). It codes for a 60 amino acid sequence known as the homeodomain. While homeodomains are very similar from one homeotic gene to the next, even those of different species, the rest of the homeotic gene protein is very variable.

Recall (from Chap. 12) that the activity of a promoter or of an enhancer may be increased by the binding of a transcription factor. Analyses of these proteins, which usually act to increase rather than decrease the expression of genes, revealed that some of them contain homeodomains. This means that the genes that code for these transcription factors are homeotic and leaves little doubt that homeotic genes regulate the activities of other genes and that they do so by way of proteins that they encode.

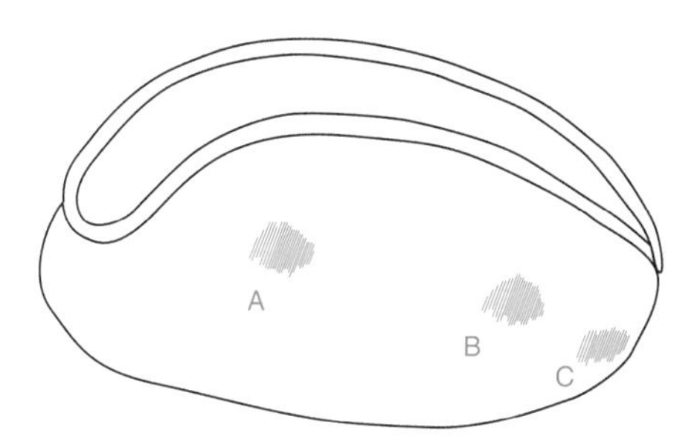

FIGURE 18-5. Fields of organization. In the vertebrate embryo, groups of cells can be identified in which homeotic genes are expressed and that contain specific homeobox proteins. From these areas, specific body parts will arise, such as forelimb (A), hindlimb (B), or tail (C). The embryo is thus divided into developmental fields well in advance of the formation of the specific structures that they designate.

As is the case in the fruit fly, genes with homeoboxes in the frog, mouse, and human have been found to come to expression during development of the embryo. This indicates that such genes play a role in controlling the body organization of many very different species. Extremely pertinent is the detection of the proteins of homeotic genes in different locations of the amphibian (*Xenopus*) embryo at different points in development. One specific protein can be detected in distinct bands around the head region at a certain stage. This protein is found in the nuclei of cells composing ectodermal and mesodermal tissue. All cells that give rise to the forelimb express the gene controlling this protein, which is present well before any structure has developed. The same is true for the distribution of still other proteins encoded by homeobox genes and that appear in bands in other parts of the developing embryo. Since homeotic genes for these proteins are being expressed in the embryo well before the appearance of identifiable structures or organs, we can view the embryo as a body divided into various fields, each area identified or marked out by the protein of a homeotic gene and having the potential to form a specific structure such as a forelimb or hindlimb (Fig. 18-5). Further observations lend support to the concept that the expression of homeotic genes gives identity to cells. When structures finally do arise in the regions where bands of homeodomain proteins accumulate, the same homeotic genes expressed earlier become very actively expressed again. It is as if the identity of the cells is being kept intact while a body part develops. In a precursor to a structure, a particular homeodomain protein is not uniformly distributed but occurs in a concentration gradient, more abundant on one side, less so in cells on the opposite side. For example, as a limb develops, a given protein continues to display a concentration gradient, as if guiding portions of the limb to develop in a specific fashion. The protein products of other homeotic genes also appear, and these too show characteristic concentration gradients. It seems that different concentrations of a given protein can cause cells to have different fates. For example, a certain amount of a protein may steer cells toward the development of a thumb, whereas a lower concentration may lead to production of a smaller digit.

We noted at the outset that in the fruit fly homeotic genes occur in two separate clusters, each one called a *complex.* The homeotic genes have a similar arrangement on the chromosomes of vertebrates, where they occur in four clusters. The genes in a vertebrate cluster have been named *Hox* genes, and these have been very closely studied in the mouse. In the human, they occur on chromosomes 2, 7, 12, and 17. As many as eight genes may be found in a vertebrate cluster, which can extend in length anywhere from 20 to 100 kb. Genes of two of the Hox clusters correspond to those of the antennapedia group of the fly and show a 70% similarity in their homeoboxes, an amazing amount of conservation. Complexes of homeobox genes appear to be another example of the role of duplication in evolution. A single group of complexes such as that found in the fly may have undergone a history of duplication. Therefore the human and other vertebrates now have four copies of each gene related to one found in a fly complex, such as four in the ANT-C group and four in the BX-C group. Another remarkable similarity has been found between the mouse and fly genes. The order of the genes in a cluster is not at random. Instead, the genes in a cluster occur in the order in which they are expressed going along the body axis from anterior to posterior (Fig. 18-6). As we go from one gene to the next starting at one end of a cluster, we find that the genes are expressed in embryo parts that go progressively from more anterior to more posterior. There is evidence that this specific sequential arrangement of the genes in a cluster has been selected during evolution, since the genes may have to be activated in a specific order. The addition of different growth factors to cells growing in culture has demonstrated that a specific substance may selectively activate a group of genes in a cluster that are expressed in a posterior body region. Another compound may activate only those genes expressed in anterior portions. Moreover, it has been found that the order in which homeotic genes are activated as limb development proceeds corresponds to their order in a cluster. Many lines of evidence support the concept that homeotic genes give identities to embryonic cells and by so doing shape the body and guide the form of specific body parts. For example, in one kind of experiment, antibodies against a specific homeotic gene protein was prepared. In *Xenopus,* expression of the specific gene takes place in tissues that normally become part of the spinal cord. Injection of the antibody at the time of gene expression binds the protein and renders it inactive. In such a case, the animal develops abnormally. The spinal cord portion becomes replaced by hindbrain portions, structures that are more anterior ones.

The exciting work with homeotic genes appears to have uncovered genetic factors that act as regulators in development. They appear to do so by coding for proteins that bind to the DNA of sets of genes in cells in defined regions of an embryo and bring about expression of those genes. In this way, they give identity to the cells in a specific region. They are called to expression at crucial times to maintain cell identity as a structure proceeds to develop fully. The order of activation of the homeotic genes corresponds to the order in which they are found in their cluster on a chromosome. The unbelievable amount of conservation displayed by homeoboxes and hence homeodomains of invertebrates and vertebrates suggests that a most effective way of establishing body plan arose eons ago in some invertebrate form and that the basic plan has been maintained with some variation throughout animal evolution. If this is indeed the case, molecular geneticists may have discovered the strategy that is common to organisms from worms to mammals and that has operated throughout geologic time to guide the development of body plan in a specific fashion. As more is uncovered about the molecular basis of development, additional information may become available allowing a deeper understanding of clinical defects in the human that may prove to be the result of defects in homeotic genes.

Very exciting is the announcement of the discovery of homeobox-containing genes in the plant kingdom. A gene known as *Knotted-1* (Kn1) was isolated from corn and was found to contain a region coded for the amino acid sequence of a polypeptide. The sequence of amino acids in the Kn1 polypeptide was deduced from the nucleotide sequence of gene Kn1. The amino acid sequence was then related to polypeptides encoded by different genes from various other eukaryotes. A 61 amino acid region in the Kn1 polypeptide was recognized that corresponded to homeodomain sequences in homeotic genes from a variety of eukaryotes. The similarity between the Kn1 corn protein encoded by the Kn1 gene and the proteins of the other eukaryotes was restricted to the region recognized as a homeodomain. The corn region was found to be most similar to the homeodomains of a certain human gene and to one from yeast. Comparison of the sequences of the corn gene with those eukaryotic genes known to contain homeodomains suggests that the Kn1 protein acts as a transcription factor. The Kn1 gene of corn appears to be involved in the regulation of leaf development, since mutant forms of the gene alter the developmental pattern.

The portion of the Kn1 gene containing the homeobox sequence was used as a probe to isolate other genes in corn that also contain homeobox sequences. The gene appears to be just one member of a class of plant genes containing homeoboxes. The discovery of homeobox-containing genes in the plant kingdom promises further insight into the action of genes involved in regulation of body plan. The implications of the discovery are especially profound for theories of cellular evolution. The similarities of the homeoboxes in yeast, animals, and plants suggest that they are derived from a sequence found in genes of ances-

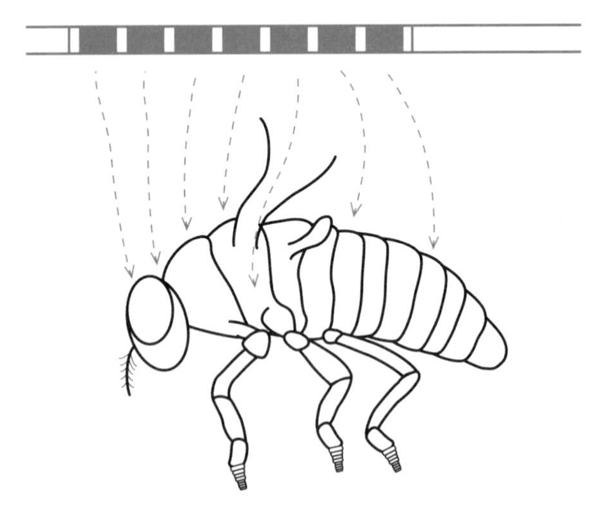

FIGURE 18-6. Order of expression of homeotic genes. The homeotic genes in a cluster on a chromosome are arranged in the order in which they are expressed along the body axis. Genes affecting more anterior body parts are followed by those influencing more posterior ones.

tral cells about 1 billion years ago, before the divergence of plants, animals, and fungi. The homeobox-containing genes we find today may regulate patterns of development in multicellular eukaryotes by modifications of regulatory devices that existed in unicellular ancestors.

The best understood organism

Drosophila melanogaster no longer holds title as the best understood multicellular organism. This distinction is now reserved for the round worm or nematode, *Caenorhabditis elegans.* After years of exacting and tedious observations, the announcement was made in 1983 that the complete cell lineage from egg to adult had been determined for this organism. What this means is that from the very first cleavage of the fertilized egg, the origin and destiny of every single cell has been determined. Such an accomplishment requires the tracking of every cell as it arises and pursuing the many cell divisions that can follow. Therefore, any cell in the adult has a lineage that can be traced back farther and farther to the fertilized egg. This amazing feat makes it possible to follow each cell from its origin through its differentiation into a specific cell type. It also permits study of the effects of various mutations on the maturation of specific kinds of cells.

C. elegans possesses several features that lend it to studies of this kind. While it contains all the major types of animal cells (nerve, muscle, reproductive, and so forth), the worm is only about 1 mm long (Fig. 18-7). Its body plan is quite simple, basically that of a tube within a tube typical of round worms. The transparency of the animal makes it possible for a microscopist, with the aid of an instrument equipped with special lenses, to follow cellular events even through the body wall. Moreover, the generation time is short. In about 1 week, 10,000 offspring can arise from one animal! While there are two sexes, one is male whereas the other is a self-fertilizing hermaphrodite. Any eggs left unfertilized can be fertilized by sperm that the male deposits in the hermaphrodite. The adult animal contains 959 nuclei. In development, some cells fuse, whereas in other cases mitosis takes place without cytokinesis, giving rise to cells with more than one nucleus. Therefore the cell number is 945, less than the number of nuclei.

The fantastic feat of tracing the origin and fate of each cell has shown that the cell lineages are invariant in this animal. The history of any adult cell is always the same from one worm to the next. Preceding the differentiation of any cell of interest, for example, are the very same events, such as a defined number of cell divisions or certain characteristic movements and rearrangements of cells. Any one adult worm is exactly like any other with respect to cell history, having precisely the same number of cells preceded by the same events. The anatomy of the nervous system, which is composed of 302 neurons, was reconstructed from thousands of serial section electron micrographs. Each nerve cell and its connections to any other is known in detail and has resulted in a "wiring diagram". Such information opens the door for studies of aspects of behavior with a precision never before achieved.

Experiments have been performed utilizing a laser beam to destroy specific cells of the worm and to ascertain the consequences. (The laser beam only affects the target cell.) The results indicate that each cell acts as if it possesses a very precise set of programmed instructions to which it adheres. The various cells behave quite independently. The differentiation of a given cell does not seem to depend on any neighboring cells for its maturation. It is as if various lines of cells develop independently and then assemble to form the multicellular adult. For example, all cells that compose the intestine are in a lineage that can be traced back to the specific cell that established it (Fig. 18-8). It is only cells in this line that give rise to the intestine, and no other lineages appear to influence it. Such a pattern is referred to as *mosaic development* and is in contrast to *regulatory development* in which differentiation of cells and tissues depends on interactions with neighboring cells. In higher organisms, it is the regulatory pattern that is the rule, not the mosaic one. Studies of vertebrate development present many examples of tissue interactions, such as the development of nervous tissue from ectoderm in response to underlying mesoderm. The only regulatory development that appears to play a role in *C. elegans* is restricted to the differentiation of certain germ cells and parts of the reproductive tract.

Among the many significant findings in this nematode is the recognition that the normal fate of many cells in the course of development is death; specific cells appear to be programmed to die! For example, 407 cells are involved in the differentiation of nerve cells, but 105 of these always die, leaving 302 mature nerve cells in the adult. Besides being important to studies of life span (discussed later in this chapter), gene action that brings about cell death may also have

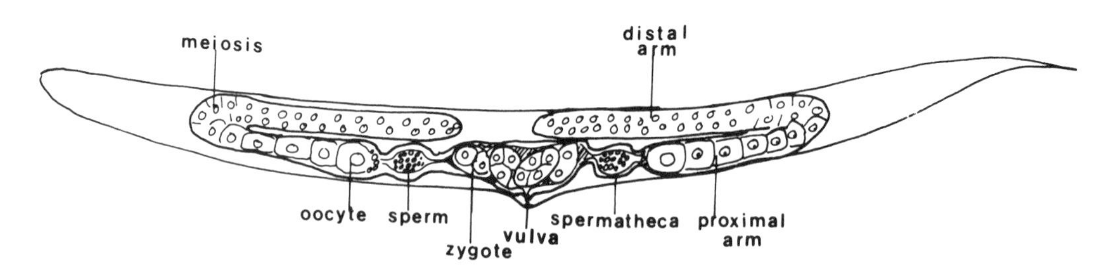

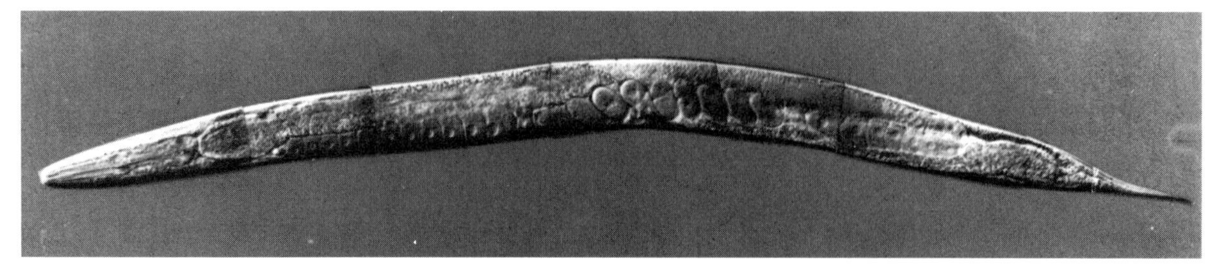

FIGURE 18-7. The round worm, *Caenorhabditis elegans.* Hermaphrodite. (Courtesy of Dr. Lois Edgar, University of Colorado, Boulder, Colorado.)

evolutionary significance. This is seen when *C. elegans* is compared with a very closely related species, *C. redivivus.* The ovary shape is very different in the two species. In the former, the ovary of the hermaphrodite possesses two arms, whereas in the latter there is but a single arm. It appears that the ancestor of these two species had an ovary with two arms and that a mutation arose causing loss of one of them. During the course of evolution, the uterus grew into the space left by the arm and then enlarged, so that *C. redivivus* now retains its eggs, which hatch in utero, while *C. elegans* evolved from the line lacking the mutation and continues to lay its eggs. Strong support for this concept comes from the finding that in *C. redivivus* a specific cell in the gonad dies at a specific point in development. The death of this cell prevents the origin of one of the ovary arms. In *C. elegans,* each developing ovary arm has a cell known as a *distal tip cell.* If a laser beam destroys one of these, an arm fails to arise from that part of the ovary. We can speculate that a mutation caused the death of a distal cell in the ancestor of the two species and that this led to a basic anatomical change in a species.

Further observations on *C. elegans* revealed genes that appear to be part of a genetic program that determines the death of specific cells. Two genes come to expression at a specific point in the development of 131 nondividing cells. When these genes act, chromosomes and other cell parts begin to break down. Other genes then become activated, resulting in the disintegration of a dying cell and its digestion by a neighboring cell. We see here an actual program that triggers cell death as a part of normal development. Perhaps after serving their function up to a certain stage, these cells become superfluous to later development of the animal and therefore have been eliminated by natural selection. In a mutant strain of *C. elegans,* a mutation has been found that brings about destruction of a cell that normally is not scheduled to die. This illustrates how a spontaneous mutation could have arisen that brought about death of a distal cell in the ovary and set the stage for an anatomical difference between two species of nematodes, *C. elegans* and *C. redivivus.* Genes that control development have undoubtedly played leading roles in the evolution of animal species.

Research with *C. elegans* continues at a rapid pace, and it remains at the forefront of developmental studies. A map of the entire genome of the worm is just about complete, and efforts are underway to learn the sequence of the 100 million base pairs making up the entire set of genetic instructions of the animal (see Chap. 19). In the not too distant future, the geneticist may be able to turn to *C. elegans* and select for study any one of its 2,000 or so genes. Knowing the precise function of all the genes of a multicellular creature will provide an insight into gene action and development in higher organisms. Those genes that trigger death are

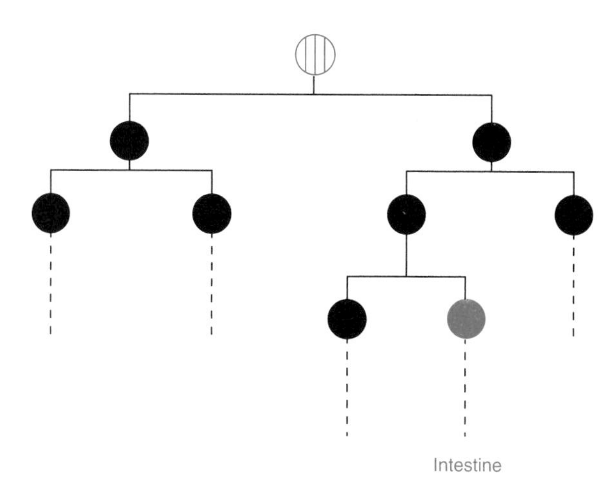

FIGURE 18-8. Establishment of cell lineages. Any cell in the adult worm is part of a lineage that can be traced back to the fertilized egg (red hatching). Conversely, the establishment of cell lines can be followed starting with the fertilized egg and leading to the formation of adult structures. After division of the egg, the two daughter cells, their division products, and their positions can be tracked in the developing animal. The precise cell (red) that gives rise to that lineage of cells that forms the intestine can be traced back here directly to the egg. The other cells shown are precursors to other lineages that give rise to other body parts.

especially pertinent to the human, where it appears that essential nerve cells are dying off in such disorders as Alzheimer's and Parkinson's diseases. Clues from the lowly nematode may fill in essential gaps in our understanding of the causes of these afflictions.

Control mechanisms and the cancer cell

The problem of cancer is related directly to the subject of cell regulation and differentiation. A normal differentiated cell that was once performing certain functions in cooperation with its neighbors can somehow undergo a fundamental change that converts it to a malignant state. It is released from various restraints that hold cell division in check. When it is transformed to a cancerous state, a cell that was noncycling (see Chap. 2) reenters the cell cycle and proliferates wildly to produce billions of altered cells, which constitute the tumor.

Uncontrolled growth is just one feature a cell acquires when it is transformed to the malignant state. The cancer cell also gains the property of immortality in the sense that it can multiply indefinitely in cell culture, unlike normal cells that lose the ability to divide after about 50 cell generations and then die (discussed later in this chapter). Cancer cells also become less responsive to feedback mechanisms that control the growth of normal cells both in culture and in vivo. For example, normal cells divide and form a neat, single-cell layer when grown on a culture dish. Cell division shuts down when the entire surface of the dish is covered. Malignant cells, on the other hand, are not confined to an orderly single layer but grow on top of one another in disordered piles.

Cell division has not shut down in response to the presence of other cells. The malignant cell has lost the property of *contact inhibition,* a term that may be somewhat misleading. Experiments suggest that it is not the actual contact among normal cells that is the basis for shutting down cell division but rather the degree to which cells spread out. A cell that is able to spread out on a surface has a shorter growth cycle. The rate of protein synthesis decreases in the more rounded-up cell, even in the absence of contact with other cells. Apparently, when a normal cell contacts other cells, it cannot spread out to its fullest extent. This reduces the rate at which it makes proteins, one or more of which may be needed for cell division. Cancer cells, even when they cannot flatten out on a culture dish, still appear unaffected and continue to proliferate. This may be related to the fact that cancer cells require fewer protein factors than normal cells for their growth and survival in cell culture. Moreover, the cancer cell usually differs in shape from its normal counterpart. Unlike normal cells, cancer cells engage in anaerobic respiration to a large extent. In addition, the cell membrane of a cancer cell displays differences from that of the normal cell, among these the possession of special antigens that can be detected by immunological procedures. Cancer cells also frequently have chromosomes in excess of the normal diploid number.

These are just a few of the features that distinguish a cancer cell from the normal type from which it arose (Fig. 18-9). Such a list of changes accompanying the malignant transformation indicates that the cellular control mechanisms have been profoundly altered in the transition from the normal to the cancerous state. There is reason to suspect that a genetic alteration of some kind is the basis of the malignant state, because carcinogenic agents such as high-energy radiation and various chemicals are known to have the potential to damage DNA and cause gene mutations. However, for many years no evidence existed for any specific genes in the cancer cell that had undergone change leading to transformation and maintenance of the malignant state.

The first clue that cells of higher eukaryotes might contain specific genes that can cause cancer came from studies with viruses. In Chapter 16, we discussed that some retroviruses carry, along with other genes in their genomes, a single gene, an oncogene, that can evoke tumor formation. Those retroviruses carrying oncogenes, the acute transforming viruses, are the fastest acting carcinogens known and can efficiently convert susceptible cells to the malignant state very shortly after infection. They have been isolated only from animals with tumors. The retroviral oncogenes are the only known genes that can initiate and maintain the cancerous state. The viral oncogene (v-onc) has been acquired by the virus from the cell. This cellular counterpart is the protooncogene or c-onc. Once in the genome of the retrovirus, the protooncogene is activated to become an oncogene, which can then transform a susceptible animal cell to the cancerous state.

Does this mean that the normal cell carries genes that can potentially trigger a malignant transformation? Evidence for this comes from gene transfer experiments. In one type of experiment, DNA is extracted from cancer cells, such as mouse fibroblasts. The DNA is next precipitated with calcium phosphate and then applied to normal mouse fibroblasts in culture. (Procedures used in gene transfer are described in Chapter 19.) Human DNA from tumors of the lung, colon, and bladder, as well as from leukemic cells, was also applied to normal mouse fibroblasts. It was demonstrated in all cases that the diverse donor DNA from the various cells could transform normal mouse cells in culture. When DNA from normal cells is used as the donor and is applied to normal fibroblasts, no transformation takes place. Clearly, the donor malignant cells of the human or mouse must have some common elements not present in normal cells capable of transforming the fibroblasts (Fig. 18-10). It does not matter that the donor DNA may come from a different species or tissue. The donor DNA, evidently, must contain some genetic information capable of evoking the transformation in normal cells, even those of a different species or tissue type.

Other evidence indicated that the DNA sequences responsible for the transformation must be present in just a small DNA region. This becomes apparent, because the DNA from the first transformed cells (Fig. 18-10) can be extracted and used to transform additional normal cells. The procedure can be repeated through several cycles of DNA extraction and infection of normal cells. Transformation still occurs throughout the course of several transfers. Only short pieces of the transforming DNA from the original malignant cells used at the beginning can be

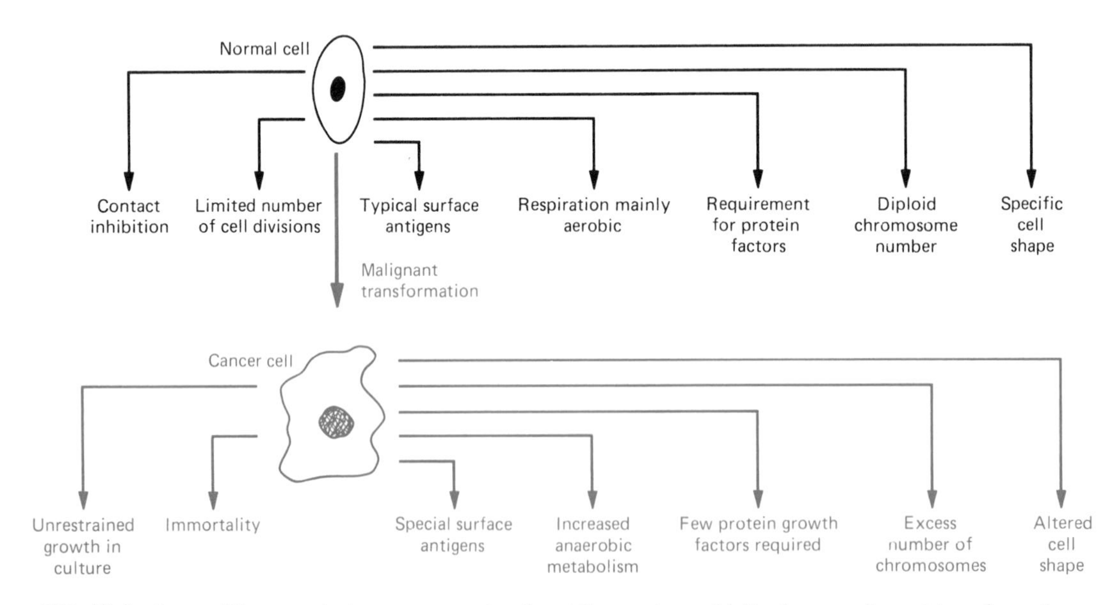

FIG. 18-9. Some differences between a normal cell and its counterpart following a malignant transformation.

expected to persist throughout several repetitions of the procedure. It becomes apparent that the genetic sequences responsible for the malignant change must be present on a very small DNA segment. This eliminates the possibility that many genetic elements must work together to produce a cancerous change, because very few small segments, at most, could survive the several transfers and enter the normal cells.

Attention was next focused on identification of the sequences responsible for the oncogenesis. Techniques of recombinant DNA technology involving cloned DNA made this possible (see Chap. 19 for details). In each case, several teams of investigators were able to recognize a single gene from the malignant donor DNA that was responsible for transforming the normal cells. Oncogenes or cancer genes were thus demonstrated in tumors from the mouse and the human. The next question to be answered concerned the origin of the oncogenes in mammalian cells. Again using the techniques of recombinant DNA analysis,

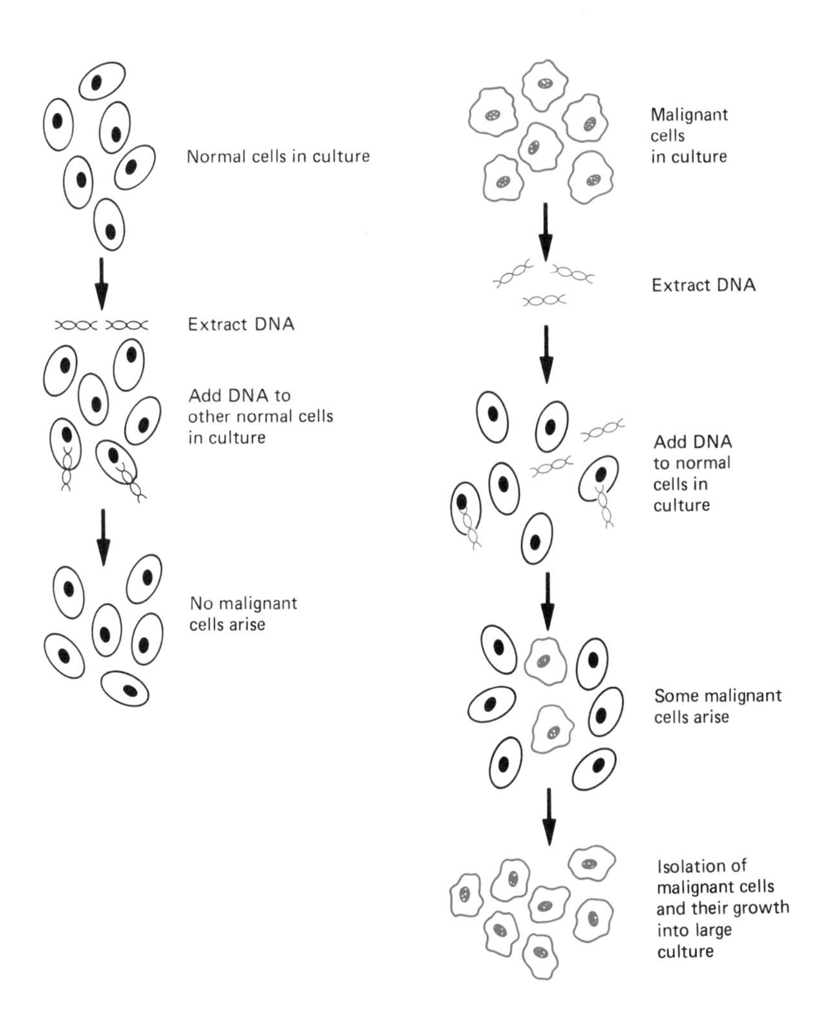

FIGURE 18-10. Demonstration of genetic factors in malignant transformation. (Left) DNA taken from normal cells can be added to cultures of other normal cells. The extracted DNA has been treated to enable it to enter cells more readily than untreated DNA. No malignant cells arise. The same procedure can be followed using DNA extracted from malignant cells growing in culture (right). When this DNA is applied to normal cells, some of the latter are transformed to cancer cells. These transformed cells can then be isolated from the culture and used to produce a large culture of cancer cells. DNA can, in turn, be isolated from these cells, and the entire process repeated again and again. After several repetitions, the results are the same as shown. Some normal cells are still transformed. This indicates that the factor or factors responsible for the transformation occur on a small segment of DNA, since repeating the process over and over causes loss of more and more of the original transforming DNA.

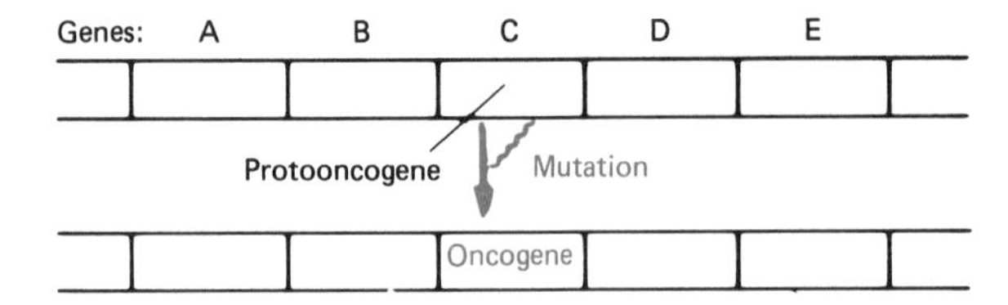

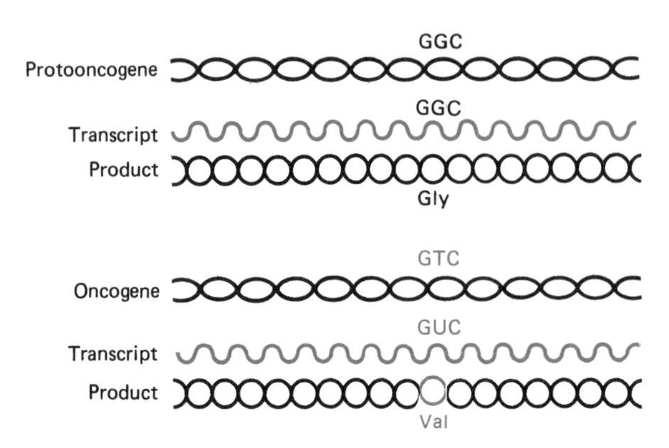

FIGURE 18-11. Oncogene and protooncogene. Recombinant DNA methodology has permitted oncogenes to be characterized. An oncogene has been found to have a counterpart, a protooncogene, in a normal cell. The protooncogene (gene C, above) can experience a point mutation and give rise to an oncogene. In one known mutation, the only difference between oncogene and protooncogene is a change in one codon that substitutes valine for glycine in the polypeptide product.

several research teams were able to characterize specific oncogenes isolated from malignant cells. For example, the oncogene associated with a human bladder tumor was isolated from malignant cells and was shown to be almost identical to a normal gene present in human cells. The oncogene can bring about malignant transformation, but its normal counterpart, the protooncogene, cannot do so when transferred to normal cells. DNA sequencing (see Chap. 19) revealed that the protooncogene is converted into the oncogene as a result of a point mutation that changes a codon for glycine (GGC) to a codon for valine (GTC). This is the only change in the 5,000 base pair long gene associated with the bladder tumor (Fig. 18-11).

What brings about the change converting a protooncogene to an oncogene? A spontaneous mutation or a carcinogenic agent might be responsible for the point mutation. (Recall that most carcinogens are also mutagenic.) As a result, any product encoded by the protooncogene would be altered. As we have seen, a protooncogene may also be activated as a result of its integration into a retrovirus. Most viral oncogenes are hybrids of coding regions of protooncogenes linked to coding regions of genes of the retrovirus. A few viral oncogenes are composed of coding regions from a protooncogene linked to regulatory regions (promoter, enhancer) of the retrovirus. It appears that the same protooncogene can be activated in two ways: by mutation in a human cell and by integration with the retroviral DNA.

There appear to be still other ways in which a protooncogene can be activated to bring about a cellular transformation. A protooncogene can become amplified in some way so that a large number of copies are present in the cell instead of the normal diploid amount. As a result, an unusually high level of the product is found to be associated with the gene. This amplification of a protooncogene seems to be quite common in tumor cells, where the gene may have about 30 copies. A normal protein that is necessary for proper cell regulation and is associated with the gene might very well cause a malignant transformation when it is present in excessive amounts.

Structural chromosome changes may also be responsible for the activation of protooncogenes. A reciprocal translocation between human chromosomes 8 and 14 has definitely been shown to be associated with a malignancy of B cells in the human immune system, Burkitt's lymphoma (Fig. 18-12). The translocation moves the protooncogene, *myc,* whose oncogenic counterpart has been found in human and mouse tumors, from its normal location on chromosome 8 to chromosome 14. In its new location, it is situated next to genes responsible for the synthesis of antibodies. As a result, the protooncogene is released from controls that normally keep its expression at a low level in antibody cells. The translocation apparently has an activating effect only in cells that produce antibodies and not in other kinds of cells. Enhancer sequences have been shown to occur within the regions that are coded for antibodies, and it has been suggested that the translocation of myc from chromosome 8 to 14 places the protooncogene next to an enhancer. Released from normal controls, it would then be expressed along with the antibody genes as if it were part of the activity of these specialized cells. The abnormally high level of the myc product would somehow upset regulatory mechanisms, producing a malignancy.

A somewhat similar situation holds for the reciprocal translocation associated with chronic myelogenous leukemia (CML). The structural change in this case involves a mutual exchange of the end segments of the long arms of human chromosomes 9 and 22 (see Chap. 10). In a band located near the end of its q arm, chromosome 9 carries a prototooncogene (c-abl). The application of procedures from molecular biology established that the translocation results in the fusion of c-abl from chromosome 9 with a chromosome 22 gene of unknown function called the *breakage cluster region* (*bcr*). This fusion creates a hybrid gene, bcr/abl, in which coding sequences of both original genes are brought together on the Philadelphia chromosome. This fused gene undergoes transcription to produce an exceptionally long pre-mRNA. After processing, the transcript is translated into a hybrid protein, $p210^{bcr/abl}$. The hybrid protein at its amino end is composed of amino acids coded by the bcr gene, and these are followed by those coded by the abl gene. This fusion protein has proved to be a phosphoprotein possessing tyrosine kinase activity. Such proteins have been associated with several cellular oncogenes (see the following section).

In vitro, protein $p210^{bcr/abl}$ is able to transform various kinds of circulatory cells to the malignant state. Moreover, it has been introduced into mouse bone marrow cells by way of a retrovirus (see Chap. 19), and these infected cells were then transplanted into mice. Some of the recipient mice proceeded to develop symptoms associated with human CML. Now under suspicion as a factor in the induction of CML, the hybrid protein is a focus of attention in studies of this human malignancy.

Products of oncogenes

What is the function of protooncogene products in the normal cell? We can reason that it must be extremely fundamental to normal cellular activities, since protooncogenes have remained practically unchanged during the course of evolution. Protooncogenes related to those in the human occur not only in other mammals but in all vertebrates, and even in *Drosophila*. It appears that they must be indispensable to the cells of major eukaryotic groups. Still, the number of protooncogenes in cells is not expected to prove large. Only about 30 are now known, and it has been estimated that not more than 100 of them occur out of a total of about 30,000 genes in the human genome.

How can we account for the fact that products of protooncogenes have the potential to harm the cell? Studies of viral oncogenes (v-oncs) have shown that their related protooncogenes (c-oncs) are genes associated in some way with normal growth. Some alteration in an essential gene product of such a gene could upset cell growth by interfering with the orderly pattern of gene expression. We have seen that a protooncogene product may be changed by a single point mutation. Moreover, as a result of the incorporation of the gene into a retrovirus, the product may become part of a larger protein that now carries a beginning segment coded for viral functions. That portion coded for the cell product may have been shortened or altered further in some way by mutation. Even if the cell product remains unaltered, it could be produced in

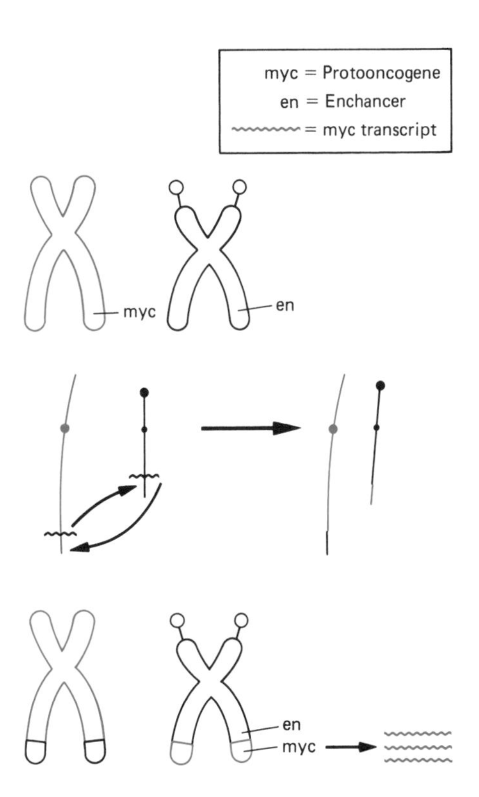

FIGURE 18-12. Reciprocal translocation and Burkitt's lymphoma. Normal chromosome 8 (above) carries a protooncogene, myc, near the tip of its longer arm. Chromosome 14, which bears satellites, contains an enhancer region associated with antibody production in its longer arm. A reciprocal translocation (middle) between the two chromosomes places segments of the longer arms in new locations. The result of the translocation (below) is the production of mature chromosomes with the rearrangement in both chromatids. The myc gene from chromosome 8 is now placed near the enhancer for antibody production. This can release the protooncogene from controls imposed on it in its normal location, enabling it to undergo transcription at a highly increased rate.

abnormally high amounts as is the case when its gene is placed under the control regions (LTRs) of a virus or when it is moved to a new location following a structural chromosome change. Let us examine these points in light of what is known about some of the products associated with protooncogenes and their viral counterparts.

The protein products of several oncogenes have been found to have protein kinase activity. A protein kinase transfers energy-rich phosphate from ATP to a particular protein. Most such kinases add the phosphate to serine or threonine residues in the protein. Unlike these, the protein kinases of most oncogenes phosphorylate tyrosine. In normal cells, low levels of the cellular counterparts that are tyrosine specific have been detected, and phosphorylation of tyrosine does occur, but infrequently. Almost all of the transferred phosphate is linked to serine or threonine. However, the amount of phosphotyrosine in cells transformed by viruses carrying oncogenes with kinase activity increases 10-fold. How could phosphorylation of tyrosine residues in cell proteins bring about a malignant transformation? Although the answer has not been established, we can suggest that such phosphorylation may alter proteins playing key roles in various regulatory pathways that control cell division, cell shape, and an assortment of vital metabolic processes. The kinases of oncogenes, being similar to those of the corresponding normal cellular kinases, could well mimic the effect of the latter. Since tyrosine protein kinase activity in a normal cell is low, a critical balance between phosphorylated and nonphosphorylated proteins could be disturbed, upsetting control mechanisms.

A normal gene product may be produced in excess in cells as a result of protoncogene amplification or indiscriminate production by a retrovirus. Present at excessively high levels, the normal product might effect a malignant transformation by too much protein kinase activity.

One protooncogene, whose v-onc counterpart is the oncogene of simian sarcoma virus, is known to code for one of the polypeptide chains of a growth factor. This is PDGF, platelet-derived growth factor, which is normally secreted by platelets and certain white blood cells during the formation of clots. It is also believed to stimulate growth of cells in the area of a wound and in this way promote healing. Two oncogene products that possess tyrosine kinase activity have been shown to act as receptors for cell growth factors. In the case of one of these v-oncs, the protein product lacks most of the amino terminal end of the receptor. This is a result of loss of the corresponding part of the protooncogene when it becomes inserted into the viral genome. The shortened product, which is also altered at its carboxyl end as a result of point mutations, can no longer bind to its normal growth factor. Highly significant is the fact that it no longer carries out its normal control function but keeps on transmitting signals for the cell to divide, even though the normal growth factor does not bind to it.

At least two oncogenes are coded for proteins that can bind closely to guanosine nucleotides and to GTP. The corresponding protooncogene products possess GTPase activity which is very much reduced in the oncogene proteins. As a result, the oncogene products may possibly remain in an active form and keep on interacting with cell proteins, resulting in deranged growth. Four oncogenes and their cellular counterparts produce products that are found inside the nucleus, unlike those discussed so far found in the cytoplasm or associated with the plasma membrane. The protooncogene myc, which is associated with Burkitt's lymphoma (Fig. 18-12) is one of these. While their role in the cell is not clear, observations indicate they are involved in cell division.

Research with oncogenes has provided many new approaches to the cancer problem, which can now be studied more closely on the molecular level using many of the procedures of molecular biology. However, it is important to note that oncogenes are only a part of the entire picture, which entails an assortment of genetic and environmental factors. For example, activated protooncogenes are not always associated with tumors. Moreover, in some cases activation appears to occur after the tumor has started, suggesting that it is a consequence rather than a cause of the malignant change. In the case of Burkitt's lymphoma, the protooncogene myc is not always translocated as shown in Figure 18-12 but remains at its normal site.

Finally, it must be realized that malignant transformation is not a one-step process. Introduction of a single oncogene into cultured cells taken directly from a normal animal, as opposed to cells growing in culture for a period of time, does not effect a malignant change. It is clear from such evidence that cancer induction in mammals entails not just one but many steps, a matter we will now pursue further.

Retinoblastoma and tumor suppression

Although it occurs only in 1 of 20,000 births, retinoblastoma is the most common eye tumor in children, striking victims anytime from birth to age 4 years. Two forms of the malignancy have been recognized, the inherited form and the sporadic or spontaneous one. Children who are victims of the inherited form usually develop tumors in both eyes at about age 9 months. They are also hundreds of times more prone than others to develop a bone cancer, osteosarcoma, when they are in their teens. These children typically have a parent who also had the affliction. The pattern of inheritance of retinoblastoma is that of an autosomal dominant with a penetrance of about 90%.

In over half of the cases, however, a child with the disorder comes from a family with no history of the disease. Such children do not develop the malignancy until about age 18 months, and in these sporadic cases only a single tumor develops in one eye. Moreover, they are not prone to bone cancer. The explanation for the differences between the two forms of the disease is now at hand. The first clue came with the observation that in certain retinoblastoma (RB) patients a deletion was present in chromosome 13. This was eventually narrowed down to the q14 band. The application of techniques from molecular biology eventually resulted in the isolation and mapping of a 30 kb length of DNA in this band region. A transcript 4.7 kb long was found in normal cells that hybridizes to this region. Such a transcript is absent from both RB and osteosarcoma cells, which carry the q14 deletion. The RB gene has now been isolated and cloned (see Chap. 20). Its product, a protein 920 amino acids long, is found in the nucleus of most normal cells but is absent or altered in some way in RB and osteosarcoma cells.

We now know that the normal individual carries two doses of the normal RB allele (RB^+), one on each chromosome 13, band q14. In inherited retinoblastoma, the afflicted child receives from one parent a normal allele, but a defective allele (RB^-) from the other. In less than half the inherited cases, the defects are due to deletions, which vary and may affect any part gene. The rest of the time mainly point mutations either prevent gene expression and formation of the product or else alter the transcript so that an ineffective product arises. Those who have inherited one defective allele will have this defect in every body cell. By itself, the one defect is insufficient to cause tumors. However, during the divisions of the millions of cells forming the retina, a somatic event may take place in one cell causing a defect in the one remaining RB^+. This can be a deletion, a gene mutation, or even at times mitotic nondisjunction that completely removes the chromosome from a cell. Lacking any functional RB allele, this cell can now initiate a tumor (Fig. 18-13A). A loss of a functional RB^+ gene due to a second somatic event in still another cell is able to initiate a second tumor. It is important to note that while inherited retinoblastoma behaves as an autosomal dominant, it is really an autosomal recessive at the cellular level, since both copies of the RB^+ allele must be rendered functionless for the condition to develop.

In contrast to a child inheriting one defective allele, the one who develops the sporadic form of the disease begins life with two functional alleles. For a tumor to develop, two somatic events must occur in one retinal cell to produce two nonfunctional RB alleles (Fig. 18-13B). The chances of this happening even once are extremely low, and so the probability for the events to take place in yet another retinal cell are even more remote. We see why multiple tumors are not apt to develop and why tumor formation starts at a later age in the sporadic as opposed to the inherited form of RB. Moreover, a child with the inherited form, having a defective allele in every body cell, runs a greater risk of osteosarcoma, since a defective RB is already present in bone cells, and a later single somatic event may eliminate the functional allele in an occasional cell. In an individual with the sporadic form, all the bone cells possess two normal RB alleles, making it highly improbable that two more somatic events will occur again to eliminate them in yet another cell.

The story of RB has many implications. RB was the first cancer found to be recessive at the cellular level. In contrast, oncogenes behave as dominants. The product of an oncogene can initiate a malignant change, whereas the RB gene acts as a tumor suppressor that produces a product needed to prevent tumor development. Several lines of evidence suggest that the RB product acts to inhibit a cell from progressing through the cell cycle and that its inhibitory effect can be weakened by phosphorylation and interactions with oncogenic viral products such as the SV40 T antigen. The occurrence of tumor suppressor genes and the actions of their products provide a focus of much active research and have given us a new concept of cancer as we will see in the sections that follow.

Cancer, a genomic disease

Many lines of evidence indicate that cancer is typically a multistep process resulting from an accumulation of as many as 10 genetic changes in a single cell. The activation of oncogenes is only one step in the process, which includes point mutations, structural

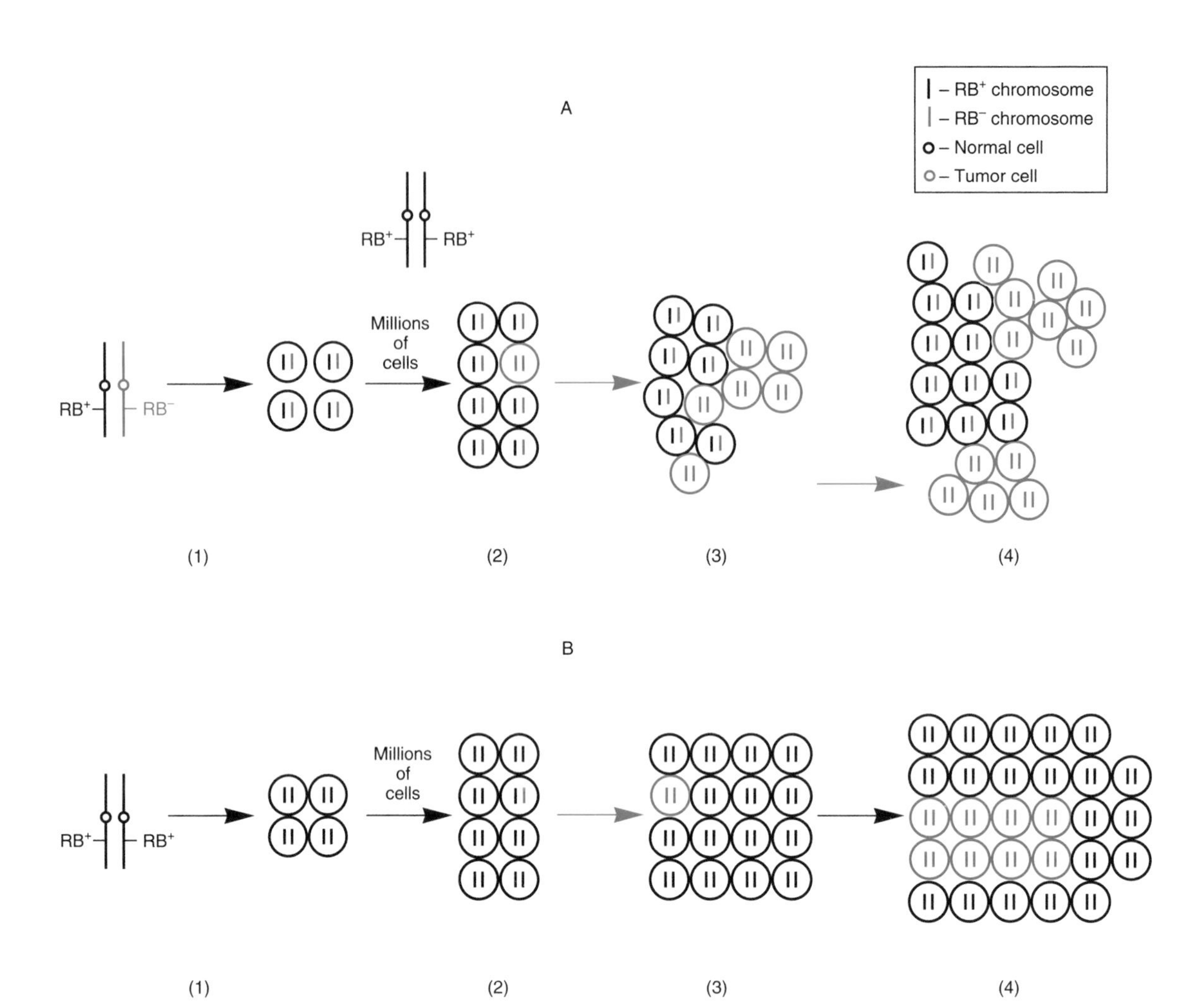

FIGURE 18-13. Development of retinoblastoma (RB). **A:** The normal cell possesses two doses of the RB^+ gene (above). In the inherited form of RB, an individual acquires a normal RB^+ allele from one parent and from the other a chromosome with the allele, RB^-, which is defective due to a deletion or to gene mutation. All the cells of the individual carry this defect (1). Heterozygous cells are nonmalignant, since one RB^+ dose is sufficient to maintain regulated growth. However, such an individual is vulnerable to any somatic event that may take place in any one of the millions of cells dividing to form the retina (2). A loss of the other RB^+ gene in a cell releases it from regulatory controls, and a tumor arises (3). Since all the cells already carry one defective RB^- allele, a second somatic derangement in still another cell can generate an additional tumor of the eye (3, 4). The person is also susceptible to losses of RB^+ in other cells including those of the bone. **B:** In the sporadic form of RB, the individual starts off with two functional RB^+ alleles in all somatic cells (1). A somatic event may generate a defective RB-allele (2), but the affected cell is nonmalignant. A second mitotic event in the same cell is required to eliminate the other RB^+ allele (3). This cell is now capable of producing a tumor (4). Since all the other cells possess two functional RB^+ alleles, it is highly improbable that two more somatic events will occur to produce a second tumor in the eye or one in the bone.

chromosome aberrations, and the amplification of genetic sequences. Another important alteration is believed to be the inactivation of tumor suppressor genes, genes such as the retinoblastoma (RB) gene, which normally act to curb cell growth. Cells normally control their growth very closely by a host of mechanisms under genetic control. A cell does not change from normal to malignant in just one step. Cancer is a genomic derangement entailing many separate genetic upsets.

At the moment, we have considerable information on the series of genetic alterations leading to cancers of the colon and rectum due to the fact that these malignancies develop slowly and therefore can be studied over a period of years. The most lethal tumors are those found to have acquired more genetic changes. It is the accumulation of changes over a period of time that is most important. While they usually occur in the order shown in Figure 18-14, this is not always the case. An early change is a deletion in chromosome 5q, also seen in the inherited disease familial adenomatous polyposis. In this condition, benign polyps arise, but some of them are highly likely to become cancerous. Among the next changes leading to colorectal tumor is the loss of methyl groups from DNA. Such a loss may alter gene expression and raises interesting questions regarding the kinds of gene activity in the cancer cell. About midway down the pathway there is an activation of the oncogene ras (first identified as a v-onc that causes rat sarcoma).

The bulk of the other alterations that have been detected are deletions, especially those in chromosomes 17 and 18. These occur in 70% of the colorectal cancers, and it is believed that the affected regions are sites of tumor suppressor genes. A sequence located in 18q21 is an excellent candidate for such a gene. It is expressed in most normal tissues, but its activity has been found to be either absent or greatly reduced in most colorectal cancers. Cells of these tumors may carry several kinds of alterations that include a homozygous deletion of the 5′ end of the gene, a point mutation in an intron, and an insertion of a DNA sequence. There is evidence that the normal gene plays some role in cell to cell-surface interactions involving contact inhibition. A deleted segment of chromosome 17p carries a gene known to have tumor-suppressing activity. Its loss may be involved in the development of certain lung and breast cancers as well as colorectal tumors.

One particular kind of genetic alteration, such as the 17p deletion, may be common to cancers of different organs or tissues. However, a given type of genetic change is not necessarily common to all cancers. The RB gene is defective in some way in almost all small cell lung cancers as well as certain other kinds of lung tumors. However, there is no evidence of its involvement in colorectal cancer. Deletions from chromosomes 3 and 11, also common in lung cancers, are not found in colorectal malignancies. Chromosome 5 and 18 deletions typical of the latter are not found in the former.

Since the bulk of evidence indicates that many genetic changes must take place in a cell before it becomes malignant, we can appreciate why many cancers (colorectal, lung, breast) are more frequent among older persons. Those tumors that afflict younger persons (retinoblastoma, leukemia, lymphoma) may require fewer gene changes. There is the hope that understanding the specific alterations involved in a given kind of cancer may aid in prognosis and guide the treatment of the patient.

Tumor suppressor genes

Cancer undoubtedly results from cell alterations that upset the checks and balances normally in force to regulate orderly cell growth and development. Two kinds of genes apparently interact to preserve the norm. One group acts in a positive way to facilitate growth; the other operates negatively to curb it. Overexpression of a growth-promoting gene as in the case of an oncogene, could provide one step down an abnormal growth pathway. We now know that chromosome losses are major factors in the development of cancers. The critical regions lost are considered to be the locations of tumor suppressor genes. A tumor suppressor gene (TSG) is believed to be coded for a protein that regulates normal growth and suppresses tumor development. Such a gene typically acts recessively. Therefore two separate mutations must arise in a somatic cell to inactivate both copies of the gene, thus eliminating completely its normal product and function. We saw this in the case of the RB gene, regarded as an excellent example of a TSG. The sequences in chromosomes 17 and 18, that are involved in colorectal cancer are believed to contain TSGs. Such genes are undoubtedly involved in a variety of cancers, among them those of the lung, breast, and colon, which by themselves constitute 40% of all human cancers.

Only a few TSGs have been cloned and sequenced, but all the evidence indicates that many more exist. Since the effects of TSGs are negative, unlike the positive role played by oncogenes, their detection is more difficult. Refined cell fusion studies are aiding in this direction. In certain cases, when mouse cells derived from various kinds of tumors were fused with normal mouse cells, most resulting hybrid cells were often like the normal parent. As the hybrid cells become established as clones in cell culture, chromosomes were lost in some of the clones, and the malignant phenotype reappeared. It has now been shown that in these cases the reversion to tumor cell type was due to loss of chromosomes carrying TSGs.

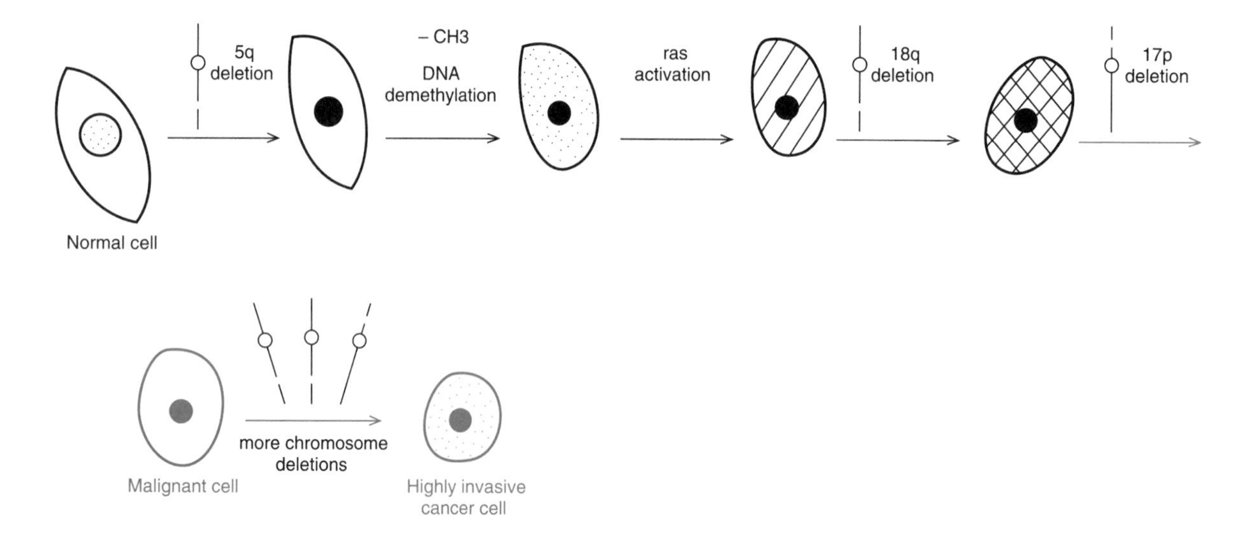

FIGURE 18-14. Some steps in the development of a malignant colorectal cancer cell from a normal cell. A series of genetic changes accumulates, altering the cell which remains benign (black) until it is finally released from normal controls as a result of the upsets and becomes malignant (red). The greater the number of genetic alterations, the more malignant the cell.

Microcell transfer now makes it possible to transfer by membrane fusion one or a few chromosomes from normal to tumor cells (see Chap. 9). If suppression of the tumor phenotype occurs, identification of the chromosome carrying the suspected TSG can be made. This type of approach has made it possible to identify several human chromosomes that are apparently carrying TSGs. For example, transfer of chromosome 11 from normal human cells was followed by suppression of malignancy in kidney cells from a Wilms tumor, but this did not happen when chromosome 13 or the X were transferred. This and other observations have strongly indicated that a TSG resides on chromosome 11 band p13 and that loss of its activity is a factor in the development of Wilms' tumor, a malignancy responsible for 85% of kidney cancers in children. It has been demonstrated that the gene normally functions in the formation of the kidneys where it expresses itself at 7 and 18 weeks of fetal development, times when its transcripts can be detected. Both copies of this gene, which has now been isolated and cloned, were missing in some tumor cell lines that showed little or no mRNA from this gene. Like the RB gene, the Wilms gene behaves as a recessive at the cellular level. However, while only one gene has been identified so far in retinoblastoma, evidence from molecular biology and cytogenetic analysis strongly suggest that in Wilms tumor, as in other cancers, more than just one gene is involved. Still other suppressor genes on chromosome 11 and elsewhere in the genome are implicated in the kidney tumor.

Analysis of restriction fragment length polymorphisms (RFLPs) has been used to detect loss of alleles and its association with various kinds of cancers. (RFLP analysis and its applications are major topics of Chapter 20.) This procedure has shown that deletions have eliminated both copies of a single chromosome region in a variety of human tumors, suggesting that elimination of TSGs at these sites has been a factor in the generation of the tumors along with other derangements of the genome. Among several tumors in which TSGs are believed to be operating as shown by RFLP analysis are malignant melanoma, a bladder cancer, cervical cancer, several kinds of breast and lung cancers as well as retinoblastoma, Wilms tumor, and colorectal cancer. It is significant that tumor suppressor genes have been found in other mammals and even in invertebrates. The fruit fly genome includes several regulatory genes that are involved in orderly development. Mutations in these bring about uncontrollable, tumorous growth leading to death. These mutations are recessives that apparently cause a lack of specific products needed for normal development. One such mutation affects the lethal giant larvae gene, which has now been cloned. Absence of its normal product has been shown to produce the lethal affect. When the gene is introduced back into developing larvae, the lethal affect is prevented.

There seems to be little doubt that tumor suppressor genes play key roles in regulating growth and preventing tumor formation in the human and other organisms. As more TSGs continue to be characterized and their products identified, the chances increase for understanding the complexities of normal growth and differentiation and for devising preventions and treatments to alleviate the toll taken by the genomic disorder, cancer.

Differentiation in relation to time and aging

Differentiation in a many-celled organism begins with the development of the fertilized egg, and it does not cease upon completion of the individual but continues throughout adulthood until death terminates the process. The very earliest signs of differentiation occur even before zygote formation. (Recall that some maternal genes are active in the unfertilized egg.) No active form of DNA polymerase can be found in oocytes until they are ready for fertilization. Once an oocyte becomes receptive, the polymerase appears in the cytoplasm and moves to the nucleus where it can activate DNA synthesis.

In the zygote, most of the structural genes are repressed. During ontogeny, only certain genes will be called to expression at specific times in specific cells. Genes such as those for development of limbs and digits are turned on at the appropriate embryonic stages. After birth, the DNA segments coding for the mature eye pigments undergo transcription. At puberty, the information for beard or breast development is switched on. Throughout the normal life span, the whole array of genes may have come to expression in the appropriate cells at the correct time. Indeed, aging and life span itself may be a genetic characteristic of all living things. We realize that, as one gets older, the chances of dying increase. A person of 50 years has a greater probability of death than a 30-year-old. The 20-year-old has a lower probability of dying than the other two. Certainly more people reach an older age today than was true years ago. But has the true limit to the span of life really been increased? Or have our modern medical practices and technology allowed more people to reach the upper age limit, which cannot be exceeded? An intriguing series of experiments by Hayflick suggests that the latter case may be correct and that life span and aging are genetic characteristics, just as much as eye color or blood groups.

Hayflick has employed cell cultures of human fibroblasts, cells that produce collagen and fibrin and continue to divide in the body of the adult. If there is no restriction on the capacity of these cells to divide, then one might expect them to continue forever in tissue culture from one cell generation to the next. However, immortality does not seem to be a property of normal human cells. Fibroblasts taken from 4-month-old embryos divided over and over again for approximately 50 cell generations. Then the population died. That the number of doublings is somehow genetically controlled was suggested by the amazing fact that the *total number* of doublings was not changed, even if cell division was interrupted for different periods of time. For example, cells were allowed to undergo 30 population doublings and were then suspended in cold storage for years. Upon thawing, they went on to divide an average of 20 times more!

Fibroblasts were taken from persons in different age groups. These divided fewer times than the average of 50, which is typical of embryos. Moreover, cells from older donors underwent fewer doublings than those from younger persons. A limit to length of life span through genetic restriction on cell division was strongly suggested by observations made on cells from species with short life spans. Fibroblasts taken from rat embryos undergo no more than 15 doublings in cell culture, compared with the 50 divisions for the human. And those from the adult rat undergo even fewer.

Single human embryonic cells were isolated and allowed to divide. Any cells arising from a common cell would all be genetically identical. From these, groups of genetically identical cells, *clones,* can be established (Fig. 18-15). The ability of these genetically identical populations to divide was then followed. It was shown that a gradual decline in the capacity for cell division took place. More and more of the clones lost the ability to divide as the fiftieth doubling generation was approached. Finally, division ceased in all the clones. The clonal studies strongly supported the idea that aging was determined largely by factors internal to the cell rather than by substances in the outside environment.

Studies of cell lineages in the nematode *Caenorhabditis* (discussed earlier in this chapter) show that certain cells are programmed to die as differentiation of body parts proceeds. We also know that certain kinds of body cells in vertebrates cease active

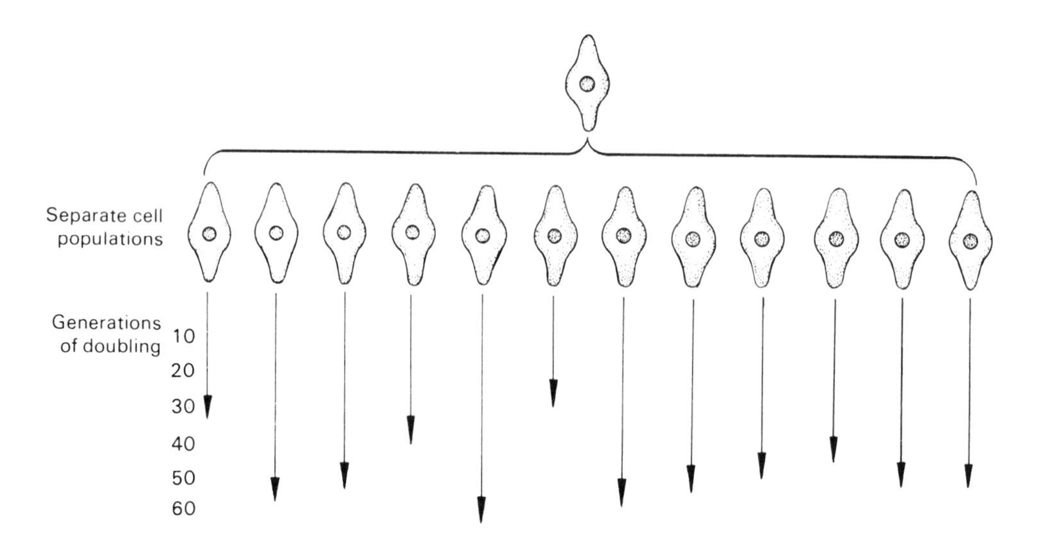

FIGURE 18-15. Clonal aging. Groups of cells that trace back to the same ancestral cell are genetically identical and compose a clone. When populations of identical cells are followed through cell divisions, it is found that the separate populations cease division at different times but that more and more of them lose the capacity for division as they approach the fiftieth doubling generation. Eventually all the cell populations cease to divide further and die.

division as a normal part of their development. Such a restriction on division is seen in the pronephros or metanephros during embryology. Moreover, at old age, many organs weigh less and contain fewer cells. This may be a reflection of a limit imposed on the number of cell doublings that is somehow coded in the genetic material. If so, the DNA would then determine the limit to the life span of normal cells.

Molecular aspects of senescence

The problem of cell aging is now being approached on the molecular level, mainly with human fibroblasts. As we have just noted, a characteristic of normal cells is their limited ability to keep on dividing. Normal human fibroblasts in culture undergo a mitotic period followed by a gradual, irreversible decline in cell divisions, ending in cell enlargement and death. This progression of events has been termed *cellular senescence* and takes place anywhere after 20 to 60 doublings. The age at which senescence occurs varies inversely with the age of the donor supplying the cells. Senescence, however, is not equivalent to cell death, since senescent cells may remain viable for months as they continue to synthesize RNA and proteins. However, senescent cells do not incorporate thymidine and cannot be stimulated to enter S, no matter what the mitotic stimulator (mitogen) or what growth factors are supplied. On the other hand, younger, nondividing cells in the G_0 state (see Fig. 2-9) do respond to such factors.

In contrast to normal cells, many tumor cells can be grown indefinitely in culture. They would appear to have escaped senescence and are termed *immortal.* It is extremely rare for human cells in culture to give rise spontaneously to cells that are immortal. They are also very resistant to malignant transformation by chemical carcinogens, radiation, and oncogenic viruses. Many carcinogenic agents can immortalize normal cells, but this by itself is not enough for malignant transformation. However, most immortal cells unlike normal human cells have an increased tendency to undergo transformation spontaneously, by oncogenes or by treatment with carcinogens. Escape from senescence may very well be a change preceding and disposing cells to malignant transformation. Cellular senescence has been viewed as one way in which tumor suppression occurs. There are two major ideas regarding cellular senescence. According to one of them, an accumulation of random damage or mutations over time is responsible. The other concept is that senescence is genetically programmed. There have been several lines of support for the latter.

Evidence has accumulated that indicates that senescent human fibroblasts express one or more dominant genetic inhibitors of cell division and that the finite life span is a dominant phenotype. For example, when somatic hybrids are formed between normal human cells and immortal hamster cells, the majority are found to undergo senescence, indicating dominance of the senescent phenotype over the immortal. When human–rodent hybrid cells are formed, there is a tendency for human chromosomes to be lost (see Chap. 9). Karyotype analyses of those hybrid clones that escaped senescence and became immortal revealed that all of them lost both copies of human chromosome 1. All other human chromosomes were present in one or two copies in at least one of the immortal hybrids. Transfer of a single copy of chromosome 1 to hamster cells by microcell fusion brought about symptoms of cellular senescence. Transfer of another chromosome, such as 11 instead of 1, had no effect on the growth of the immortal cells. Transfer of chromosome 1 into endometrial cancer cells brought about senescent changes in them. The gene or genes responsible for the senescence are located on the q arm of 1. Alteration of this arm occurs in a variety of cancers (intestine, breast, ovary, colon). Alteration in 1q has also been associated with the acquisition of immortality in cultured cells from benign colorectal growths.

The somatic cell hybrid studies also show that escape from senescence involves not just one gene but several. Immortality actually seems to require defects in a relatively small number of genes. These defects can be corrected in cultured immortal cells so that they again proceed to undergo senescence. This can be accomplished by fusing an immortal line of cells with normal diploid ones, by introducing a single chromosome 1, and by fusion between certain immortal cell lines (Fig. 18-16). In the latter case, we see an example of complementation. The defects in one line (the inactivation of genes causing senescence) are covered by the active senescence genes operating in the other. Both the activation of oncogenes and the inactivation of senescence genes may be ways by which a cell can escape senescence. Indeed, observations also support the concept that senescence may be a process of terminal differentiation and an escape from malignant transformation.

In another approach to the problem of cellular senescence, a comparison of gene expression was made among senescent cells, actively growing cells, and younger, quiescent (nondividing) cells arrested in the G_0/G_1 interval before DNA synthesis in the cell cycle. Of seven genes studied, the expression of three was found to be greatly reduced in the senescent cells. The results indicate that senescent fibroblasts do not behave like younger, nondividing cells arrested in the G_0/G_1 interval. While they express and regulate some genes in the manner typical of younger fibroblasts, they show striking, irreversible changes in the expression of others. Some of the latter are genes known to have a role in the blocking of cell proliferation. Very striking is the repression of transcription of the protooncogene, c-fos on chromosome 14q. This inhibition of c-fos transcription in the senescent cells has features of dominant inhibition of cell division and is believed to be a cause in stopping proliferation. Normally c-fos is expressed in the G_0/G_1 interval preceding S in the cell cycle and can be induced in the quiescent younger cells.

Other studies have called attention to changes in methylation of DNA as factors in cell aging. In Chapter 17, we noted the importance of methylation patterns in normal differentiation. Changes in these patterns could result from some failure at mitosis to maintain methyl groups or from failure to replace them by enzymes during DNA repair. Undermethylation could lead to the activation of certain genes which are normally inactive in specific tissues and eventually to derangements leading to senescence and death. Evidence to support the involvement of demethylation with aging comes from studies with mice. In these animals, it has been noted that a certain gene, when located on the inactive X chromosome in the female, becomes reactivated with a higher and higher frequency as the animals become older. In a similar fashion, a gene that is coded for pigment formation becomes turned off when it is inserted into an inactive X resulting in an albino animal. However, as the animal ages, coat color becomes more and more pigmented, indicating that the gene is becoming progressively activated in the cells.

It has also been demonstrated that as human cells in culture continue to divide, the number of methylated DNA sites in them declines. This progressive loss of methylation is associated with the eventual loss of the cell to proliferate. Another indication that loss of methylation is a factor to consider in aging comes from experiments in which cells in culture are treated with azacytidine. Young cells at first show no effects when exposed to this drug, which removes methyl-

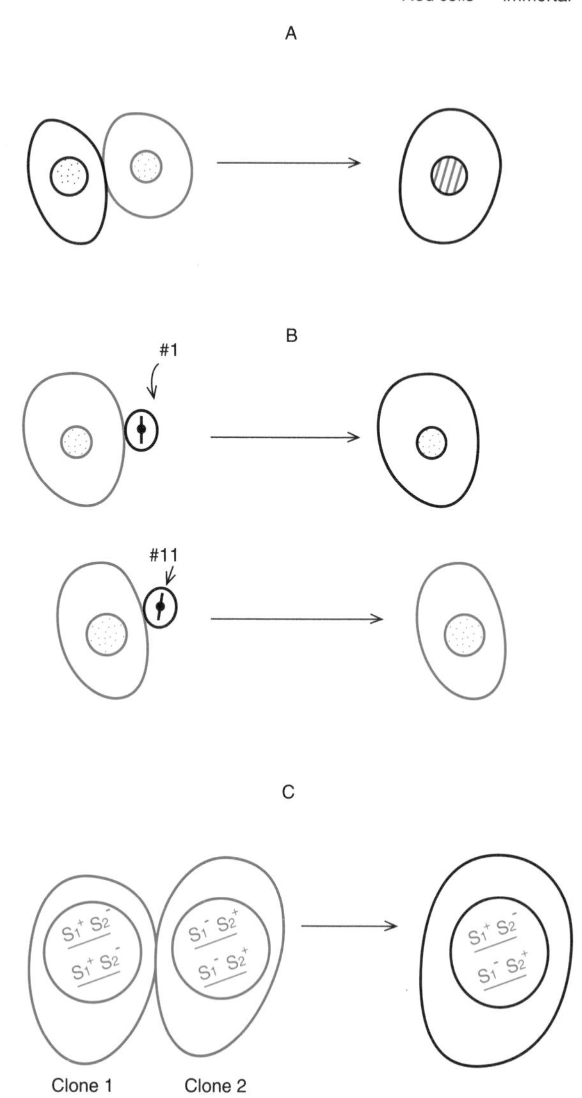

FIGURE 18-16. Correcting for defective senescence genes. **A:** Fusion between normal human cells and immortal hamster cells usually yields hybrid cells that are no longer immortal but that will undergo senescence. The normal cell supplies functional senescence genes defective in the immortal cells, and thus the hybrid loses the ability to proliferate indefinitely. The senescent phenotype behaves as a dominant. **B:** Introduction of chromosome 1 by itself from a normal cell into an immortal cell brings about senescence (above), whereas introduction of any other chromosome alone other than 1 has no such effect, and the immortal phenotype persists (below). Chromosome 1 supplies normal senescence genes which are defective in the immortal cells. **C:** Fusion between cells from different clones of immortal cells may yield hybrids that have lost the immortal phenotype and will undergo senescence. This is to be expected if each line is defective for two or more different senescence genes. Complementation can occur, and hybrid cells can arise in which the defects have been overcome, and the cells gain the dominant senescence phenotype.

ated DNA sites in them declines. This progressive loss of methylation is associated with the eventual loss of the cell to proliferate. Another indication that loss of methylation is a factor to consider in aging comes from experiments in which cells in culture are treated with azacytidine. Young cells at first show no effects when exposed to this drug, which removes methyl groups. However, they eventually fail to undergo as many cell divisions as untreated cells and die well before them.

Further studies of senescence on the molecular level are certain to reveal other differences between younger and aging cells. It is quite possible that aging involves chromosome aberrations in addition to mistakes made during transcription, translation, and replication. Failure of repair could well add to the accumulation of genetic errors leading to derangements in biochemical activities necessary to maintain normal cell functions. Research on cell aging is certain to add to our understanding of malignant transformation, a process interrelated with the phenomenon of aging.

REFERENCES

Aaronson, S.A. Growth factors and cancer. *Science* 254:1146, 1991.

Bishop, J.M. The molecular genetics of cancer. *Science* 235:305, 1987.

Bookstein, R., J.-Y. Shew, P.-L. Chen, P. Scully, and W.-H. Lee. Suppression of tumorigenicity of human prostate carcinoma cells by replacing a mutated RB gene. *Science* 247:712, 1990.

Croce, C.M., and G. Klein. Chromosome translocations and human cancer. *Sci. Am.* (March):54, 1985.

Daley, G.Q., R.A. Van Etten, and D. Baltimore. Induction of chronic myelogenous leukemia in mice by the P210 bcr/abl gene of the Philadelphia chromosome. *Science* 247:824, 1990.

De Robertis, E.M., G. Oliver, and C.V.E. Wright. Homeobox genes and the vertebrate body plan. *Sci. Am.* (July):46, 1990.

Fearon, E.R., *et al.* Identification of a chromosome 18q gene that is altered in colorectal cancers. *Science* 247:49, 1990.

Gehring, W.J. Homeo boxes in the study of development. *Science* 236:1245, 1987.

Gehring, W.J., and Y. Hiromi. Homeotic genes and the homeobox. *Annu. Rev. Genet.* 20:147, 1986.

Goldstein, S. Replicative senescence: The human fibroblast comes of age. *Science* 249:1129, 1990.

Graham, A., et al. The murine and *Drosophila* homeobox gene complexes have common features of organization and expression. *Cell* 57:367, 1989.

Haluska, F.G., Y. Tsujimoto, and C.M. Croce. Oncogene activation by chromosome translocation in human malignancy. *Annu. Rev. Genet.* 21:321, 1987.

Hanahan, D. Dissecting multistep tumorigenesis in transgenic mice. *Annu. Rev. Genet.* 22:479, 1988.

Hansen, M.F., and W.K. Cavenee. Retinoblastoma and the progression of tumor genetics. *Trends Genet.* 4:125, 1988.

Harbour, J.W., et al. Abnormalities in structure and expression of the human retinoblastoma gene in SCLC. *Science* 241:353, 1988.

Hayflick, L. The cell biology of human aging. *Sci. Am.* (Jan) :32, 1980.

Ingham, P.W. The molecular genetics of embryonic pattern formation in *Drosophila. Nature* 335:25, 1988.

Jaynes, J.B., and P.H. O'Farrell. Activation and repression of transcription by homeodomain-containing proteins that bind a common site. *Nature* 336:744, 1988.

Kinsler, K.W., *et al.* Identification of a chromosome 5q21 gene that is mutated in colorectal cancers. *Science* 251:1366, 1991.

Klein, G. Specific chromosomal translocations and the genesis of B-cell derived tumours in mice and man. *Cell* 32:311, 1983.

Knudson, A.G., Jr. Genetics of human cancer. *Annu. Rev. Genet.* 20:231, 1986.

Levine, M., and T. Hoey. Homeobox proteins as sequence specific transcription factors. *Cell* 55:537, 1988.

Marx, J. Many gene changes found in cancer. *Science* 246:1386, 1989.

Nusslein-Volhard, C., G. Frohnhofer, and R. Lehman. Determination of anteroposterior polarity in *Drosophila. Science* 238:1675, 1987.

Oliver, G., C.V.E. Wright, J. Hardwicke, and E.M. De Robertis. A gradient of homeodomain protein in developing forelimbs of *Xenopus* and mouse embryos. *Cell* 55:1017, 1988.

Roberts, L. The worm project. *Science* 248:1310, 1990.

Rose, M.R. *Evolutionary Biology of Aging.* Oxford University Press, New York, 1991.

Ross, J. The turnover of messenger RNA. *Sci. Am.* (Apr.) :48, 1989.

Sachs, L. Growth, differentiation and the reversal of malignancy. *Sci. Am.* (Jan) :40, 1986.

Sager, R. Tumor suppressor genes: The puzzle and the promise. *Science* 246:1406, 1989.

Seshadri, T., and J. Campisi. Repression of c-fos transcription and an altered genetic program in senescent human fibroblasts. *Science* 247:205, 1990.

Stanbridge, E.J. Identifying tumor suppressor genes in human colorectal cancer. *Science* 247:12, 1990.

Sugawara, O., M. Oshimura, M. Koi, L.A. Annaab, and J.C. Barrett. Induction of cellular senescence in immortalized cells by human chromosome I. *Science* 247:707, 1990.

Sternberg, P.W., and H.R. Horvitz. Genetic control of cell lineage during nematode development. *Annu. Rev. Genet.* 18:489, 1984.

Sulston, J.E., E. Schierenberg, J.G. White, and J.N. Thomson. The embryonic cell lineage of the nematode *Caenorhabditis elegans. Dev. Biol.* 100:64, 1983.

Volbrecht, E., B. Veit, N. Sinha, and S. Hake. The developmental gene knotted-1 is a member of a maize homeobox gene family. *Nature* 350:241, 1991.

Weinberg, R.A. Tumor suppressor genes. *Science* 254:1138, 1991.

Weinberg, R.A. The action of oncogenes in the cytoplasm and nucleus. *Science* 230:770, 1985.

Weinberg, R.A. A molecular basis of cancer. *Sci. Am.* (Nov) :126, 1983.

Weisbrod, S. Active chromatin. *Nature* 297:289, 1982.

Yunis, J.J. The chromosomal basis of human neoplasia. *Science* 221:227, 1983.

Yunis, J.J., and A.L. Soreng. Constitutive fragile sites and cancer. *Science* 226:1199, 1984.

REVIEW QUESTIONS

1. A. A female fruit fly produces an egg mutant for a maternal gene (m^-) but wild for all gap genes (gap^+), pair-rule (pr^+), and polarity genes (po^+). It is fertilized by a sperm from a male homozygous wild for all these genes. What kind of embryo is expected?

B. A second female produces an egg that is wild for all the genes. The egg is fertilized by a sperm that is m^-, gap^+, pr^+, po^+. What kind of embryo is expected?

C. A female contributes a gamete that is m^+, gp^+, pr^-, po^+.This is fertilized by sperm which is m^-, gp^+, pr^-, po^+. What kind of offspring is expected?

D. A female gamete is m^+, gp^+, pr^-, pol^- and is fertilized by a male gamete that has the same constitution. What kind of offspring is expected?

2. It is unusual to find a very young person afflicted with colorectal or lung cancer, although young people develop other types of cancers such as retinoblastoma, leukemias, and lymphomas. Offer an explanation.

3. In cancer, cells gain the ability to undergo division without restraints. However, many cancers are associated with the loss of genetic material. Offer an explanation.

4. Fibroblasts from rat embryos undergo an average of 15 doublings in cell cultures. Suppose some cells are allowed to undergo 6 doublings and others 10 doublings before they are suspended in cold storage for 3 years and 2 years, respectively. About how many more doublings on the average are to be expected when both groups of cells are thawed and permitted to divide?

5. In normal mouse cells the protooncogene (p-onc) resides at a certain chromosome site. The DNA in the region of the protooncogene can be broken into specific fragments by a restriction enzyme, as indicated here by arrows:

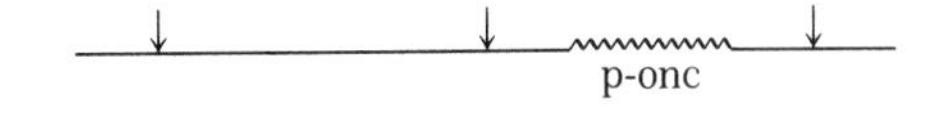

In the course of an experiment, normal mouse cells growing in culture are transformed into tumor cells following infection with a retrovirus. You wish to determine if the malignant transformation is the result of viral integration near p-onc. DNA is extracted from both normal and transformed cells. The DNA from both sources is exposed to a restriction enzyme, which fragments the DNA and attacks the sites indicated previously by arrows. Following gel electrophoresis, DNA denaturation, and Southern blotting, the fragments are exposed to 1) denatured labeled p-onc DNA and 2) DNA of the specific virus. (The DNA from these two sources is being used as probes to detect the presence of that specific DNA.) Autoradiographs are then finally prepared. Interpret the comparison of the four autoradiographs shown here:

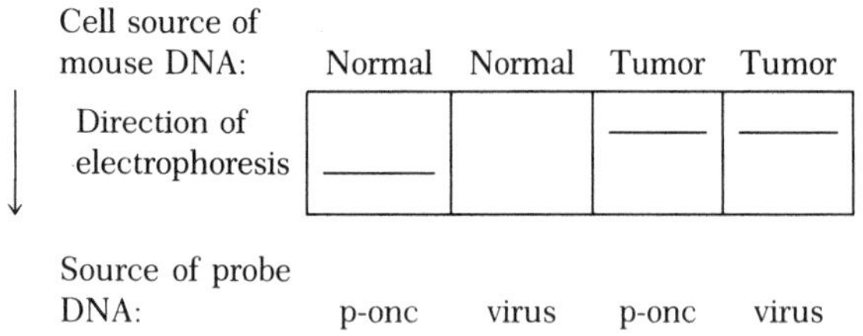

Cell source of mouse DNA:	Normal	Normal	Tumor	Tumor
↓ Direction of electrophoresis	____ (lower)		____ (upper)	____ (upper)
Source of probe DNA:	p-onc	virus	p-onc	virus

6. Suppose in the case of another virus, the following results were obtained in a situation exactly like that described in Question 5. What would you conclude?

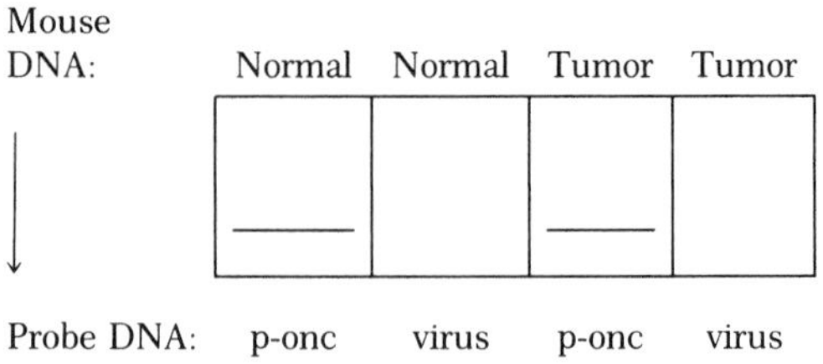

Mouse DNA:	Normal	Normal	Tumor	Tumor
↓	____		____	
Probe DNA:	p-onc	virus	p-onc	virus

The Impact of Genetic Manipulation

19

A completely artificial gene

In 1977, the announcement was made that Khorana and his associates at MIT had succeeded in synthesizing the first truly artificial gene, one that can function in a living cell. Khorana's group synthesized a gene by chemical procedures alone, without relying on a natural DNA template. The gene that was synthesized is one found in phage λ, which can infect *Escherichia coli* cells. The naturally occurring viral gene codes for a new kind of tRNA that appears in the bacterial cells after viral entry, a tRNA specific for the amino acid tyrosine. (Certain viruses carry genes that code for new types of tRNA. These new tRNA species may be necessary for viral replication in the cell, because some viruses, after entry, are known to alter the ribosomes of the host. Perhaps the new types of tRNA are needed to work efficiently with the altered ribosomes.)

The sequence of nucleotides in the gene proper (the part that undergoes transcription into RNA) includes 126 nucleotide pairs. This sequence had been deduced by English investigators from analyses of the RNA transcript. For reasons still unknown, the transcript is cleaved by enzyme action after its formation. Forty nucleotides are removed in the host cell, leaving a tRNA composed of 86 nucleotides. In addition to the gene proper, which is transcribed into RNA, the gene has a promoter region of 56 nucleotide pairs and, at its other end, a stop signal that is 25 nucleotide pairs long. Khorana and his group determined the nucleotide sequences of these other gene parts. They also artificially synthesized short gene fragments, each containing a dozen or so nucleotide pairs. In the procedure, each segment that was used bore a short piece of single-stranded DNA extending from each end (Fig. 19-1). These extensions act as joints or splints, which enable them to complex more readily as a result of the complementarity of bases in the extensions. The segments are then joined together by DNA ligase.

By joining the short segments, the entire gene (gene proper, promoter, and terminator regions) composed of 207 nucleotide pairs was constructed without the use of any natural DNA product. Khorana then tested the ability of the synthetic gene to function. A certain mutant strain of phage λ cannot produce the new tyrosine tRNA in the host cell because of a defect in the gene. The defect appears to be in the terminator region. As a result of this defect, and hence the defective tRNA, the mutant strain produces short, nonfunctional proteins in the cell and cannot multiply in the host. The artificial gene was incorporated into mutant viruses, and these were then allowed to infect *E. coli* cells. The originally defective viruses could now multiply in the host cells as well as normal viruses. This conclusively showed that the artificial gene was functioning. The new and necessary tyrosine tRNA was proven to be present in the host, because normal translation took place with the formation of normal viral proteins.

The construction of totally artificial genes opens up many new pathways in genetic research. One extremely important one can lead to a better understanding of a gene, how it is regulated, and how it is expressed. The investigator can produce alterations in the gene at will and observe any ensuing effects. Information of this type is of profound significance in studies of cell differentiation and pathological cell changes. Although the

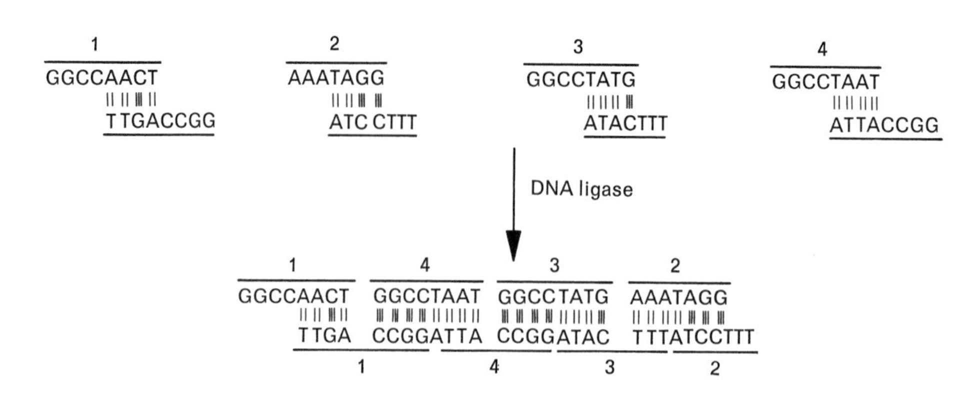

FIGURE 19-1. Joining of DNA fragments. Each fragment is synthesized bearing single-stranded projections to act as splints so that stretches of complementary bases on the extensions are exposed and can complex, bringing the segments together. After the segments are properly positioned, DNA ligase activity (arrow) joins them together by forming phosphodiester linkages. (The sequence shown here is purely illustrative.)

accomplishments of Khorana and others were with microorganisms, equally incredible strides are now being made in the manipulation of genes of higher forms, as will be presented later in this chapter.

Cleavage maps

Restriction endonucleases (see Chap. 16) have made it possible to construct cleavage or restriction maps of DNA using the fragments resulting from enzyme digestion. Such maps have been extremely important in deciphering the nucleotide sequences of DNA fragments and even of entire genomes. They also perform an important role in the localization of genes on chromosomes, and we will encounter their many applications in this and the following chapters.

Recall that a given restriction enzyme of type II recognizes certain specific sites on DNA and can break up any specific long fragment into a characteristic set of shorter pieces. These can be separated from one another by gel electrophoresis, a technique used routinely in molecular biology (see Chap. 16). In this process, a sample of the material is placed on top of a narrow column of a polyacrylamide or an agarose gel, which is then subjected to an electric field. The concentration of the ingredients can be varied to prepare gels having different pore sizes. This governs the size range of the molecules that can then move into the gel. The more concentrated the gel, the greater the restriction on entry of larger molecules. Under the influence of the electric field, fragments of proper size may enter the gel. The smaller the molecule, the faster its migration. Consequently, smaller molecules come to rest at points farther away from the point of origin, the position in the gel where the DNA sample was applied, whereas the larger fragments come to rest at sites nearer to it.

The procedure can be modified for the production of autoradiographs. This requires the labeling of the DNA with ^{32}P and the preparation of the gel as extremely thin sheets. After electrophoresis, the paper-thin gel is placed in contact with X-ray film. This is finally developed to reveal bands that represent the different positions of the labeled DNA fragments. Each band consists of a group of molecules found in the original sample that are of the very same length and that came to rest in the gel at a specific site. In addition to radioactivity, special stains and the use of ultraviolet optics can reveal the bands in a gel. The actual size of the molecules in a band can be determined by running them in a gel in a lane adjacent to one that contains DNA fragments of known size. In this way, the gel becomes calibrated, and one can determine the sizes of fragments by comparing the bands they produce in the gel with the bands of known size (Fig. 19-2A).

Let us see how a physical map of a DNA segment or even of a genome can be prepared with restriction enzymes. We will assume in this example that we are starting with the circular genome of some virus. The DNA will be subjected not to just one, but to several restriction endonucleases. This treatment will be followed by gel electrophoresis and the preparation of autoradiographs. Suppose we find that one of the enzymes (R1) produces from the viral DNA a single band that migrates to a position

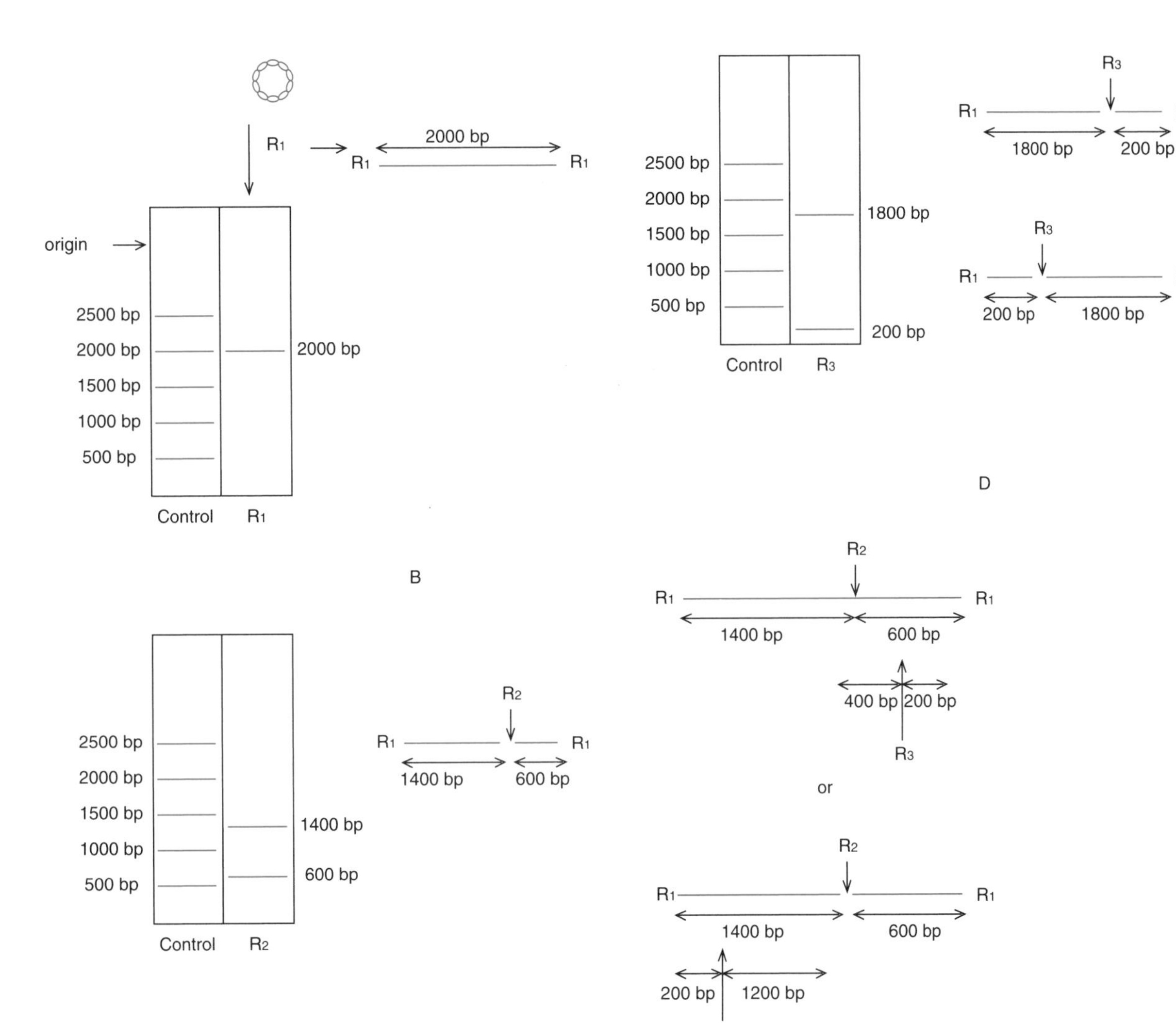

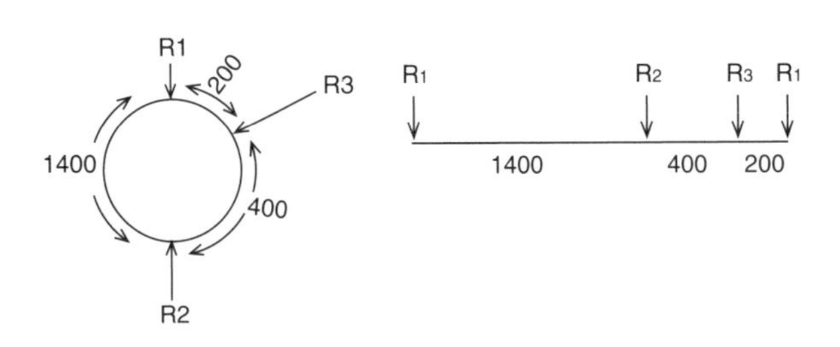

FIGURE 19-2. Preparing a cleavage map. **A:** A gel can be calibrated by separating a mixture of fragments of known size. These come to rest at different distances from the point of origin, the small ones travelling the farthest. The unknown length of DNA in an adjacent lane can be directly determined by comparing the positions of any DNA bands in that lane with those positions assumed by fragments of known length. In this example, circular DNA is treated with enzyme R1, which renders it linear. In the gel, it comes to rest at a position corresponding to that of a 2,000 bp long fragment. **B:** The linear 2,000 bp fragment is cleaved by enzyme R2 at one site, producing fragments 1,400 and 600 bp long. **C:** The 2,000 bp fragment is cut by enzyme R3 into fragments 1,800 and 200 bp long. However, we do not know at which side of the fragment R3 cuts in reference to R2. **D:** If the R3 cut is to the right of the R2 cut, the 1,400 bp fragment will be left intact while two smaller fragments arise from the 600 bp fragment. If the R3 cut is to the left, the 600 bp fragment is left intact, and two smaller fragments are generated from the fragment 1,400 bp long. **E:** A cleavage map of the circular and linear forms of the DNA shows sites of cleavage of the three restriction enzymes and the physical distances between these sites in base pairs.

that corresponds to a 2,000 base pair (bp) fragment. This tells us that R1 recognized a single restriction site on the circular molecule and cut it there, thus generating a linear molecule (Fig. 19-2A). The 2,000 bp DNA can be extracted from the gel and then subjected to another restriction endonuclease (R2). When this is done, it is found that fragments of two sizes are produced, 1,400 and 600 bp. This means that R2 cut at a site 1,400 bp from one end of the linear molecule and 600 bp from the other (Fig. 19-2B).

A third restriction enzyme (R3) also cuts the 2,000 bp DNA at just one site, yielding fragments of 1,800 and 200 bp. However, we do not yet know where this site is found in reference to the site cut by R2. It could be either to the right or to the left of the R2 site (Fig. 19-2C). To resolve this problem, a double digest is prepared in which both enzymes R2 and R3 are allowed to cleave the 2,000 bp molecule simultaneously. If the R3 site is to the right of the R2 site, then the double digest should produce a pattern on the gel in which the 1,400 bp fragment is left intact but in which the 600 bp fragment is replaced by fragments 400 and 200 bp long (Fig. 19-2D). If the R3 site is to the left, then the 600 bp fragment will be found intact but the 1,400 bp piece will be cut into fragments 1,200 and 200 bp in length. If the autoradiograph reveals a pattern with fragments of 1,400, 400, and 200, we then know that the first alternative is the correct one.

Now that the R3 site has been established, the three restriction enzyme sites can be placed correctly on a physical map of the circular genome or on a map of the linear fragment (Fig. 19-2E). The map indicates the sites at which a specific restriction enzyme cuts the DNA. The distances between sites are measured in base pairs. The map shows the DNA divided into various regions, each of a definite length and each between restriction enzyme recognition sites. Continuing with the reasoning in this example and using still other restriction enzymes, the map can become more and more detailed.

The cleavage map of SV40

The DNA of simian virus 40 (SV40) has been analyzed in detail with the aid of restriction enzymes. When the viral DNA is subjected to a particular restriction enzyme isolated from the bacterium *Hemophilus,* 11 fragments (A–K) are produced that are visualized following gel electrophoresis and the preparation of an autoradiograph (Fig. 19-3A). (The enzyme was called *Hind* but is now known actually to be a combination of two enzymes HindII and HindIII.) The sizes of the 11 fragments, from the smallest one, A, through the largest, K, were determined from their positions in the gel and were checked by electron micrograph measurements. Steps were then taken to identify these fragments precisely and to produce a cleavage map showing their exact order in the intact circular chromosome. This was achieved in a fashion similar to that described in the previous section.

Viral DNA can be treated with Hind and the resulting fragments exposed to a restriction enzyme from *E. coli,* EcoR1. The fragments produced by the activities of both enzymes are then separated by gel electrophoresis. It is now found that fragment F is not present in its typical position. Instead, two smaller fragments are generated, and these together equal the size of fragment F. From this we can deduce that the F fragment carries a site that is recognized and cleaved by EcoR1. Only one such site must be present in the entire genome, because all the other fragments are characteristically present, as if they had been treated by Hind alone. One can then turn to still other restriction enzymes and study their effects on the viral genome. Treatment of the SV40 DNA with the enzyme HpaI, for example, produces three large fragments, Hpa-A, -B, and -C. Each of these can be isolated by itself from the gel, exposed to other restriction enzymes, and finally subjected to electrophoresis to establish the order of digest products in a single fragment. When isolated Hpa-C is exposed to Hind, two bands are produced (Fig. 19-3B). These are identical in position to fragments Hind-B and Hind-I. This tells us that Hind fragments B and I must be found right next to each other. Together they form the fragment Hpa-C, produced following treatment of the chromosome with HpaI.

Using an assortment of restriction endonucleases, D. Nathans and associates proceeded in this way, isolating separately from gels the partial digest products of one enzyme, such as the product Hpa-C. Each such partial digest was then subjected to further digestion with another enzyme. The electrophoretic patterns of the products of the second digestion were studied and compared with known standards. The result has been the production of cleavage maps of the SV40 chromosome. The initial map was based on cleavage sites generated by three enzyme treatments, EcoR1, HindII plus HindIII, and HpaI plus HpaII.

Let us make sure we understand this simple map (Fig. 19-4). The outer circle shows the sites of cleavage of the Hpa endonucleases. The inner circle shows

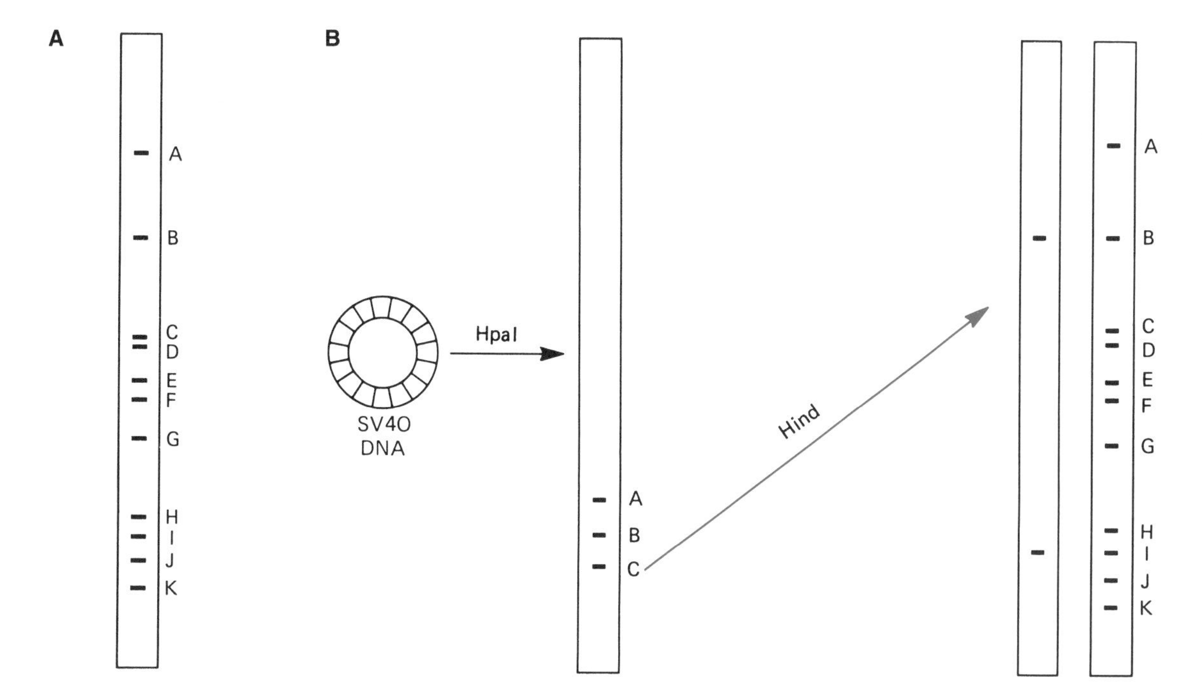

FIGURE 19-3. Some steps in preparing a cleavage map of SV40. **A:** When labeled SV40 DNA is subjected to digestion by restriction enzyme Hind, 11 fragments are produced. These are visualized following gel electrophoresis and preparation of an autoradiograph, which is depicted here showing the longest fragment, A, at the top and the shortest, K, at the bottom. **B:** Treatment with other restriction enzymes can help to establish the order of the fragments in the chromosome. When SV40 DNA is exposed to the enzyme HpaI, only three large fragments are produced. Each of these can be isolated from the gel and exposed to other enzymes. When fragment C is isolated and exposed to Hind, two fragments are generated that can be separated by electrophoresis. The pattern that results can then be compared with that obtained when SV40 DNA is treated with Hind alone (right). Comparison shows that the two fragments are identical in position to fragments Hind-B and Hind-I. We now know that fragments B and I are adjacent in the intact chromosome.

the 11 Hind fragments, A through K, which were mentioned earlier in this section. The true positions of these fragments were established by the reasoning we have just presented. Note that the inner circle also shows the site of cleavage of restriction enzyme EcoR1. This single cleavage site, indicated by the arrow, was designated coordinate 0. The other map units (0.1, 0.2, etc.) represent the fraction of the DNA length of SV40 from that EcoR1 site in an arbitrary direction around the map.

$$\text{Map units} = \frac{\text{distance from 0 coordinate}}{\text{length of SV40 DNA}}$$

More detailed maps have been produced using the initial one as a reference. To obtain more detailed cleavage maps, still other restriction endonucleases were used. Since the SV40 genome can be cleaved by these enzymes at any of several different sites, one can generate large or small fragments from any part of the genome. In the more detailed maps, the sites of cleavage of all the various additional restriction enzymes are localized and indicated.

Localization of genes in SV40

The cleavage map and the sequencing of the entire SV40 genome have permitted identification of the positions of structural genes, as well as the site of origin of viral DNA replication. DNA replication in SV40 is known to begin at a definite site, map coordinate 0.67 in segment C (Fig. 19-4), and to proceed bidirectionally until termination of replication occurs in segment G. RNA transcripts that are present early in an infected cell before the start of viral DNA replication were isolated. These were then exposed to denatured restriction fragments of the viral DNA that had been transferred to a filter following Southern blotting (see Chap. 16). DNA–RNA hybrids were formed between the early RNA and approximately half the viral genome between 0.17 and 0.67 (Fig. 19-5). The transcription was found to follow a counterclockwise direction from an early promoter. The late mRNA, originating from a late promoter adjacent to the early one, was shown to be coded by the other half of the genome and to be synthesized in a clockwise direction. (Recall that the early promoter is known to be activated by an enhancer.)

Nathans and his colleagues were able to locate the positions of certain structural genes. For example, they studied temperature-sensitive viral mutants (ts mutants), viruses that cannot replicate at certain temperatures that permit replication of wild type. One way in which the mutant genes were located on the viral DNA included heteroduplex DNA and mismatch correction. The circular DNA from a mutant strain was denatured to single strands. Wild-type DNA was then exposed to a restriction enzyme to yield specific fragments, which were then also denatured. The denatured DNA fragments were allowed to incubate with the mutant, single-stranded circles. If a fragment is used that corresponds to the region containing the site of the mutation causing temperature sensitivity, a partial heteroduplex will form (Fig. 19-6A). Two possibilities now follow. Replication of the DNA can result in the formation of two daughter homoduplexes after separation of the two imperfectly matched strands. The formation of wild virus (non-ts) indicates that the fragment *did* cover the mutant site. If it did not, then only mutant viruses would arise after two rounds of replication (Fig. 19-6B).

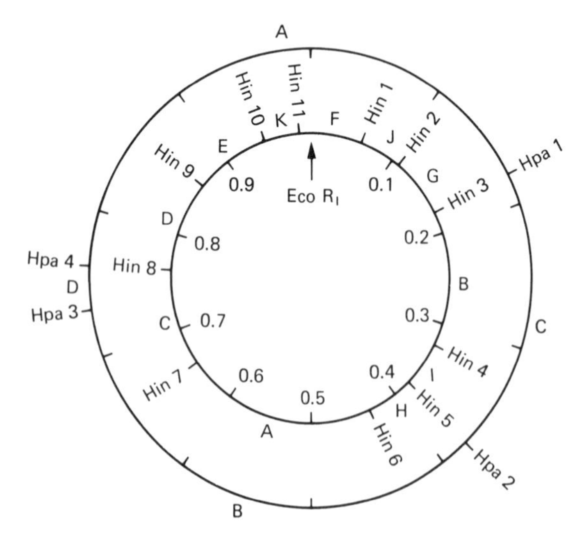

FIGURE 19-4. Initial cleavage map of SV40 chromosome (see text for details). (Reprinted with permission from K.J. Danna, G.H. Sack Jr., and D. Nathans, *J. Mol. Biol.* 78:363, 1973. Copyright by Academic Press, Inc. [London], Ltd.)

Another possibility entails mismatch repair (Fig. 19-6C). If a heteroduplex forms, digestion of part of one of the strands takes place. The intact strand is then copied. Again, wild-type viruses can arise. If the strand did *not* cover the mutant site, no mismatch repair will take place and hence no wild-type viruses will be generated (Fig. 19-6D). Therefore, use of the various restriction fragments to determine which specific ones cover up specific mutations enables one to identify the location of mutant sites. This is possible because wild-type viruses arise as a result of replication or mismatch repair if the specific restriction fragment corresponds to (and hence can cover up) a region carrying the mutation.

In this way, it was shown that all the ts mutants that were defective in the initiation of DNA replication mapped in the early region of the viral genome. From various mutation analyses, it became clear that the early region of the SV40 genome codes for T antigen and that this region is sufficient for viral DNA replication and the transformation of the host cell.

The T antigen is now known to have several functions. Besides initiating the replication of the viral DNA and transforming the host cell, the T antigen also plays a role in the control of expression of the SV40 early and late genes. It is responsible for the ability of the virus to bring about the formation of replication enzymes and to initiate a cycle of host DNA replication. The role of the t antigen, the other early protein, is still not clear. Both the T and t proteins are encoded by the same stretch of DNA, and their mRNAs are both derived from the same primary transcript following differential processing. While the two proteins are identical in 80 amino acids at the amino terminal end, they differ at the carboxyl end as a result of the way in which processing of the transcript takes place. Actually the mRNA for the T antigen, the larger of the two proteins, is smaller than that for the t antigen! This stems from the size of an intron sequence removed during processing of the primary transcript. The length removed is smaller in the case of the RNA, which becomes mRNA for t protein. This resulting mRNA, however, contains a terminator sequence that brings translation to a halt. The mRNA for T protein has a longer piece excised, but this includes the terminator sequence. Consequently translation of this smaller mRNA results in a T antigen that is larger than the t protein.

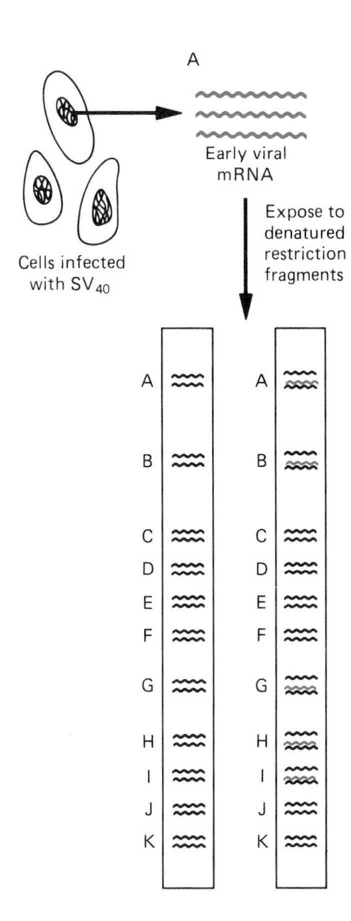

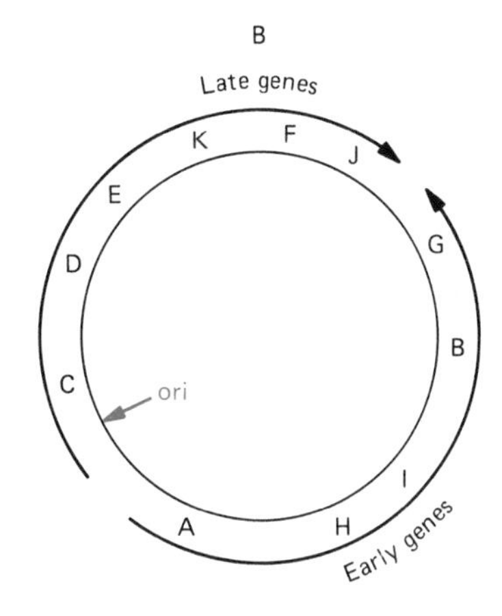

FIGURE 19-5. Localization of early and late genes in SV40. **A:** Early-appearing viral mRNA is extracted from cells and allowed to hybridize with specific restriction enzyme fragments (A–K). Results with the early mRNA show that only certain fragments, composing about half the genome, contain sequences that were transcribed into the early RNA. The same procedure followed with late mRNA shows hybridization with the other fragments. **B:** Hybridization of the early and late transcripts to the specific fragments enables the genes from which they were transcribed, and which were coded for them, to be related to the SV40 cleavage map. The early genes, composing about half the genome, are transcribed in a counterclockwise direction. The late genes are transcribed from the opposite strand in a clockwise direction. The site of viral DNA replication is found in fragment Hind-C.

Sequencing DNA

Restriction endonucleases have been a major factor in the progress achieved in DNA sequencing. Exposure to one of these enzymes yields fragments of characteristic sizes that can be separated and then studied for their exact nucleotide sequence. Sequencing fragments obtained by cutting a source of DNA with several such enzymes has made it possible to determine the nucleotide sequences for larger and larger stretches of DNA encompassing entire genes and even entire genomes. Two main methods, one designed by Maxam and Gilbert and the other by Frederick Sanger, have permitted rapid study of DNA fragments and have made possible the accumulation of many thousands of sequences. Stored in computer data banks, this information on specific nucleotide sequences continues to lead to a greater insight into the roles of specific genes and the manner in which genetic controls operate. The available data have also facilitated improvements in biotechnology in such areas as the synthesis of artificial sequences for use as probes in locating desired genes from various DNA sources.

Besides utilizing specific restriction fragments, both sequencing methods rely on gel electrophoresis, a procedure that is now so refined that it can separate on a single gel DNA fragments that differ in length by just a single nucleotide. Let us familiarize ourselves with some of the major steps in the Sanger method, which has become the more widely used in recent years.

Critical to the Sanger procedure are nucleotides that differ slightly from their four counterparts that are found in natural DNA. These are the dideoxynucleotides, so called because they lack not just the one –OH group at the carbon 2 position in the sugar component but also the one at the carbon 3 site. This means, for example, that in its triphosphate form, a dideoxynucleotide (dd) can be added to a growing DNA chain. However, since it lacks an –OH at the carbon 3 position, once it is incorporated into a chain it

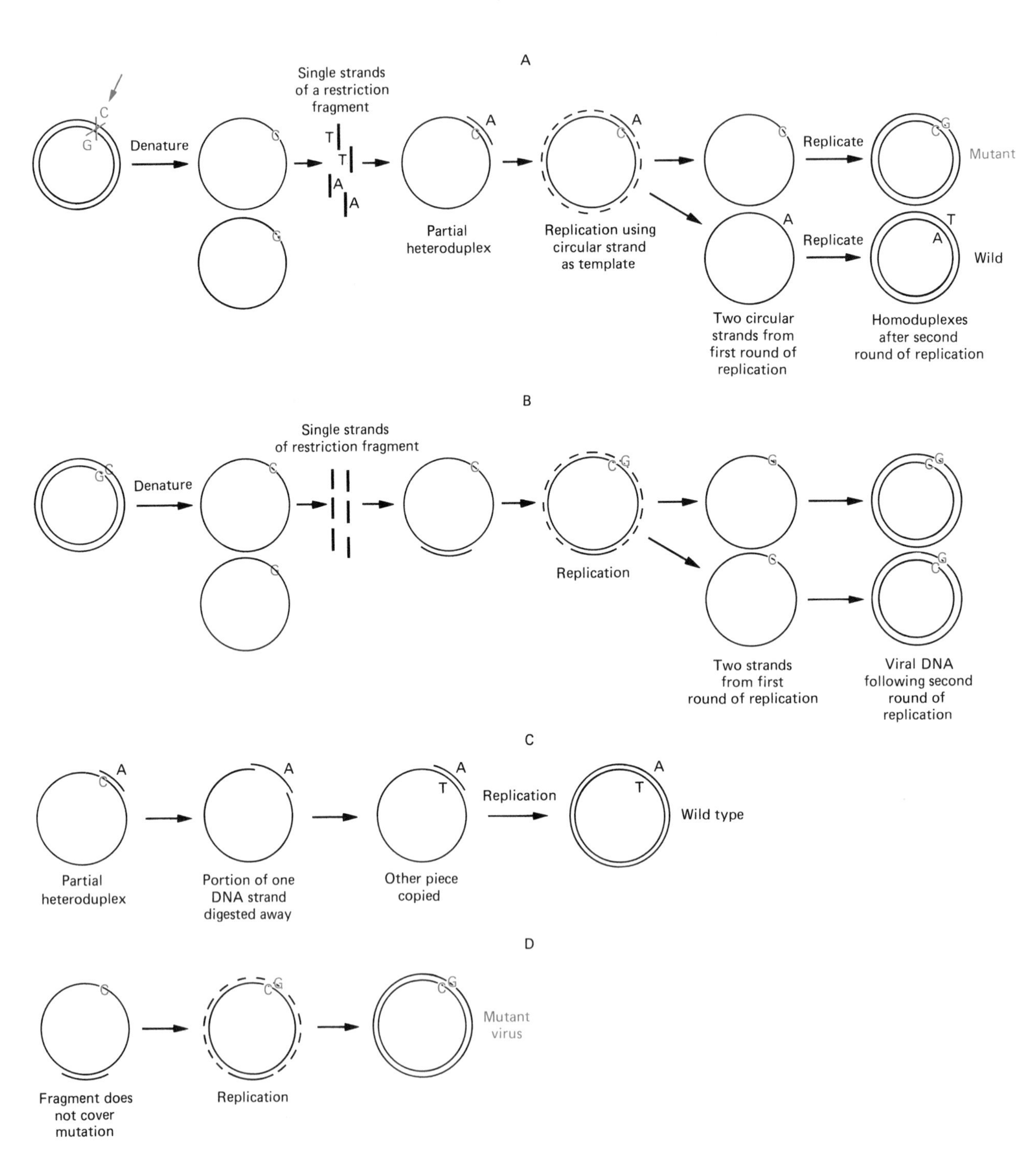

cannot form a phosphodiester band with any other activated precursor. Therefore a dd acts as a terminator at the site that it occupies, bringing the synthesis of a growing DNA segment to a halt.

The incorporation of a dd into a growing DNA chain is pivotal to the Sanger method, as an outline of the steps in Figure 19-7 shows. At first DNA from some source such as that of a cloned gene is subjected to a restriction enzyme, and a preparation containing fragments of a specific size, perhaps 100 base pairs long, is prepared for sequencing. The DNA is then denatured to yield pure preparations of the single strands. This can be accomplished by taking advantage of the fact that one of the two strands may be heavier because it contains more purine nucleotides than the other, which is lighter and contains more pyrimidines. Gel electrophoresis enables us to separate the two kinds of strands. We can then proceed to analyze either one of these for determination of its nucleotide sequence, the heavy (H) or the light (L).

Assuming that we select a pure preparation of the L strand, we then divide it among four test tubes, designated ddG, ddA, ddC, and ddT. Each of the four portions contains all the ingredients needed for DNA synthesis. In addition to the L strand, which we have selected for analysis and which will act as the template, the following are present: DNA polymerase, a short known DNA sequence to act as a primer, the four usual deoxynucleoside triphosphates (dG, dA, dC, dT), plus a small amount of one kind of dd. The primer is radioactively labeled to permit the preparation of an autoradiograph later on. Each of the four set-ups is identical to the other except for the one specific dideoxynucleotide it contains, ddG, ddA, ddC, or ddT.

Under the proper incubation conditions, the synthesis of new strands will begin in each test tube as nucleotides in their activated form are added to the primer. In the ddT tube, for example, when a thymine nucleotide is required to pair with adenine on the template strand, the ordinary deoxy (d) form will be competing for incorporation with the dd form, but usually it is the ordinary d form that will be added to the chain at a given position. This is so because more of the d form is present in the tube. Therefore strand elongation will tend to continue past a given site that requires T to pair with A. However, at times the dd form will be inserted at a given site, and this will terminate DNA synthesis on the template strand. Since synthesis will be halted at all the possible sites where dT should be incorporated and where ddt has substituted for it, new strands differing from one another in length will be formed, each ending with ddT. Every A on the template strand will be represented by an incomplete chain that terminates in ddT. The new strands of varying lengths are radioactive due to the label in the primer.

Following denaturation, the products in each of the four tubes are placed in four lanes side by side on the same gel and separated by electrophoresis. An autoradiograph is finally prepared. The nucleotide sequence of the newly made strand can be read directly from the band positions that are now resolved in the gel and that represent the newly made DNA segments of varying lengths. The smallest segment is represented by the band at the bottom, since it travels fastest in the gel. The nucleotide sequence of the template strand (the L strand) can be deduced directly from that of the new H strand. To confirm the sequence deduced in any one run, another run can be made using the strand complementary to the one used as template in the first run, in this case the H strand. In this way the sequence of the duplex DNA can be verified.

Improvements in technology have increased greatly the resolution of bands that are seen following gel

FIGURE 19-6. Locating positions of mutant genes in SV40. **A:** Viral DNA containing a mutant site (arrow) is extracted from a temperature-sensitive mutant and rendered single stranded. (Only one of the two strands is followed here, for simplicity.) Single circular strands are exposed to a specific, denatured restriction fragment from wild-type virus. Assuming the fragment from the wild-type virus is homologous to the region in which the mutant site is found, it will pair with that region following incubation to form a partial heteroduplex. In the heteroduplex, there is no correspondence at the position of the base substitution (C vs. A). A second round of replication can result in formation of homoduplexes. Wild, double-stranded viral DNA can arise: **B:** If a fragment is used that does not cover the mutant site, no wild-type virus can arise, because completion of replication of the fragment will copy the mutant site in the circular DNA. (Only one of the two strands is followed here.) After a second round of replication, two mutant viral DNA types will arise from the original circular DNA. **C:** If the fragment covers the mutant site, as it also did in A, mismatch repair can take place. A stretch of DNA in the vicinity of the mutation is digested away. Digestion can remove either a piece of the fragment or a piece of the circular DNA. (Only the latter is shown here.) Completion of replication of the fragment, using the rest of the circular DNA as a template, produces wild-type viruses. Even if the nonmutant sequence is removed at times by mismatch repair, wild-type viruses will also be produced, in addition to mutant ones, because the picture depicted here will also be operating. **D:** If the fragment does not cover the mutant site, no wild-type virus can possibly be formed by mismatch correction, since, as in B, completion of replication of the fragment always results in copying the mutant site.

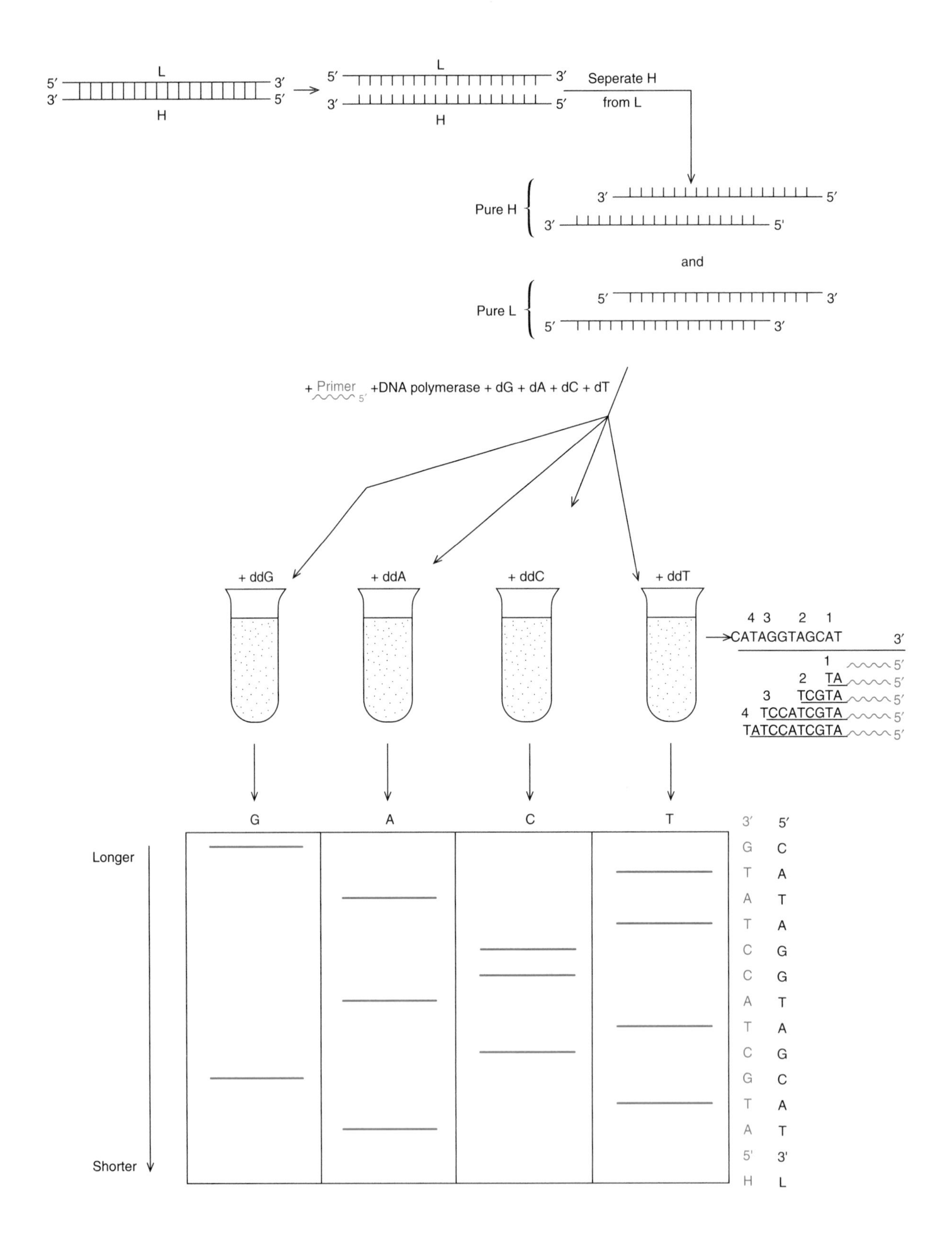
L
5′
3′
H
Seperate H
from L
Pure H
and
Pure L
+ Primer
+DNA polymerase + dG + dA + dC + dT
+ ddG
+ ddA
+ ddC
+ ddT
4 3 2 1
CATAGGTAGCAT
TA
TCGTA
TCCATCGTA
TATCCATCGTA
G
A
C
T
Longer
Shorter
H
L

electrophoresis, making it possible to sequence on just one gel a fragment 300–500 nucleotides long (Fig. 19-8). The sequencing of large stretches of DNA, entire genes, and even entire genomes has been accomplished by piecing together sequences determined for fragments that were produced by subjecting the DNA source to different restriction enzymes. For example, suppose enzyme 1 produces a series of small fragments and that these are sequenced. The same is then done for fragments generated by treatment with enzyme 2. We then compare the sequenced fragments and look for sequences that overlap. Suppose the following sequence has been worked out for a particular fragment generated by enzyme 1: 5′–TGGAGGACC-CGGAT–3′. A fragment generated by enzyme 2 is found to have the following sequence: 5′–AGAATT-TAGTGGAGGACC–3′. When the two are compared, an overlap is evident, as indicated by the underlining. From this, it can be deduced that the two fragments have been generated from the same region of the chromosome. A larger sequence can now be established: 5′–AGAATTTAGTGGAGGACCCGGAT–3′.

Proceeding to look for overlapping sequences among fragments in this way, investigators completed the entire genome of 5,375 nucleotides for ϕX 174 and the 5,226 nucleotide pairs of SV40 in the late 1970s. Progress in DNA sequencing continues to advance at a rapid rate. The complete genomes of viruses 10 times the size of the simpler ones became sequenced in the early 1980s: viruses such as phages T7 and λ with its 48,513 base pairs. Now the 100,000 or so base pairs of the Epstein-Barr virus as well as the DNA sequences of several chloroplast genomes have been worked out. At the moment, the largest DNA to be sequenced is that of a herpes virus, cytomegalovirus, which consists of about 250,000 base pairs. The sequences of many specific genes have also been established, making the total number of base pairs that are now presently known to be more than 37 million! Within 10 years, it is believed that the entire *E. coli* genome of 4.8×10^6 base pairs will be worked out through the cooperation of various laboratories in the United States and Japan. Already 800,000 of its base pairs are known.

The rewards of DNA sequencing have indeed been many. It revealed the unexpected fact that overlapping genes occur in some forms as seen in ϕX 174 and SV40. It established the fact that the genes of eukaryotes and their viruses are typically in pieces, consisting of introns and exons. Extensive portions of the human DNA have been sequenced showing the arrangement of genes in a chromosome region (see Chap. 20 for the arrangement of the globin gene clusters). Sequencing techniques have been an impetus to what is known as *reverse genetics,* the approach to a genetic investigation that starts with a cloned gene that is then handled in various ways to gain information on its function. This is in contrast to the classic or *forward genetics* approach, which begins with studying effects of gene mutations on the phenotype, works toward identification of the gene product, and then zeroes in on the gene. Indeed, with today's new approach, rather than striving to obtain the polypeptide product of a gene so that its amino acid sequence can be determined, it is now often simpler to start with the gene's nucleotide sequence. From this, one then deduces the amino acid sequence of the product. This approach has been used in gaining information on the products associated with certain import human genes, such as the one associated with Duchenne muscular dystrophy (Chap. 20).

Especially exciting are the advances being made in the development of automated DNA sequencing machines. A DNA sequenator has been designed that is able to sequence nucleotides at a rate at least 10 times greater than that which can be achieved by more costly and time-consuming manual methods. Now that automatic DNA sequencing technology permits about 8,000 nucleotides a day to be read off, molecular biologists are able to turn their attention to certain projects that would not be feasible by manual methods. Among these is what is perhaps the most ambitious goal ever

FIGURE 19-7. Sequencing DNA by the Sanger method. A pure preparation of a restriction fragment of a given length is denatured. One kind of strand (H or L) is selected to act as a template. A preparation of the selected strand type (L in this case) is then divided among four test tube set-ups. Added to each tube are all the other ingredients for DNA replication such as a known primer (which is radioactively labeled), DNA polymerase, and (in their activated forms) the four usual nucleotides (dG, dA, dC, and dT). Each tube also contains a small amount of one specific dideoxynucleotide (dd) that will act to terminate replication if it becomes incorporated at a given location. In the tube shown at the right, ddT can become paired with A in the template at every possible site, four of which are shown here. A similar situation pertains in all the tubes, each of which will yield labeled DNA strands that vary in length. The products in each tube are denatured and the strands separated side by side in four lanes by gel electrophoresis. The labeling that was in the primer permits the preparation of an autoradiograph that reveals positions of the newly made strands as discrete bands. These are easily read from the 5′ bottom (shortest) end to the top (3′ end), giving the sequence of the newly made H strand from which the sequence of the original template (L) is directly deduced.

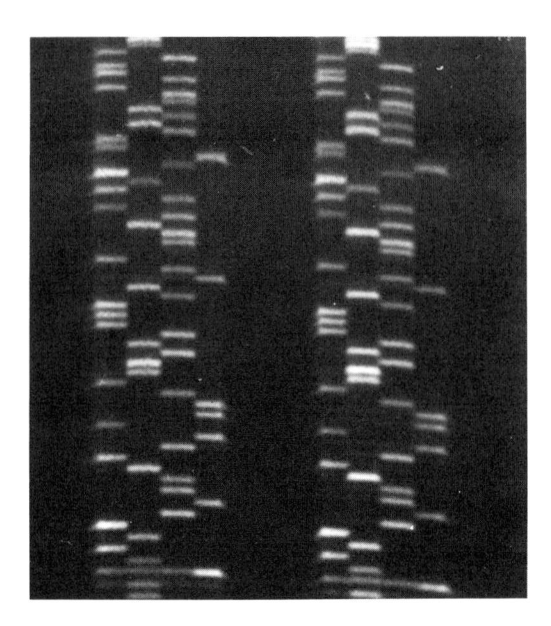

FIGURE 19-8. Autoradiogram of sequencing gel showing resolution of DNA fragments. (Courtesy of Dr. June Polak.)

conceived in the biological world, that of sequencing the entire human genome, which is composed of 3 billion base pairs! In the United States, the human genome project is a national objective, organized and supported by the National Institutes of Health (NIH) and the Department of Energy. Among its many activities, the NIH Office of Genome Research coordinates ongoing research in laboratories throughout the country and assembles all the data for analysis. Human genome projects have been announced by Japan and several European countries. A number of prominent scientists have formed the Human Genome Organization (HUGO) with the hope of establishing international cooperation with the sharing of data and coordination of efforts, all with the aim of establishing the human genome sequence by the year 2005.

The enormity of the goal staggers the mind when one realizes that in addition to its very size, the genome varies from one person to the next and contains vast regions of repetitive and noncoding DNA. To formulate the data into a meaningful whole will demand efforts previously unknown. However, the rewards reaped may be greater than from any other scientific adventure. At the molecular level, we will become able to understand how a human functions and exactly how defects in specific genes bring about afflictions. Armed with such information, strides can be taken to alleviate suffering from disorders that take a severe human toll, such as diseases of the circulatory system, cancer, diabetes, and mental disturbances.

Mutation at preselected sites in SV40

As was stressed in Chapter 14, the application of a mutagen raises the overall mutation rate. It does not enable the investigator to predetermine which specific gene or genetic region will be altered. The possibility of bringing about a mutation at a site selected in advance has now been realized in SV40 through a series of brilliant studies by Nathans and his associates. Some of the mutations produced were deletions; others were base pair substitutions. Production of the latter is summarized here.

First, the circular viral DNA is nicked in one of its two strands at a specific site, with the aid of a specific restriction endonuclease let us say at the short sequence CCGG. With the aid of an exonuclease, a small gap is then produced that causes a nucleotide sequence to be exposed on the intact, complementary strand (Fig. 19-9). A mutagen that can react with single-stranded DNA is then applied. Sodium bisulfite is a mutagenic chemical that can cause deamination of cytosine to uracil. If a mutation is produced in the exposed region, the sequence GGCC on the complementary strand could be altered to GGUC. Next, DNA polymerase and ligase are added to repair the gap. As a result of these steps, a double-stranded circular DNA with a sequence that was originally

CCGG
GGCC

is now changed to

CCAG
GGUC

Realize that if the original sequence occurs at only one site in the viral DNA, then only this one spot will be attacked by the restriction enzyme, and any alteration will occur only at this spot. Also, the location of this region on the cleavage map is known at the outset. Therefore, if a change has been produced as a result of the procedure, its location is known.

Any changes that are induced would be expected to occur only in a certain percentage of the treated viral genomes. Therefore circular molecules that still have the original sequence would also be present in the preparation. These molecules are unaffected by the mutagen and are easily eliminated by exposing the preparation to the restriction enzyme employed at the outset. The enzyme attacks only the sequence

CCGG and destroys all those molecules with the original sequence. Only the double-stranded circular molecules carrying the mutation will remain. These can now be allowed to infect cells for study of changes in viral characteristics. Nucleotide sequence analysis can also be done on restriction fragments of the mutant viral DNA, to pinpoint precisely the site of the base substitution in the area.

Such local mutagenesis procedures have been used to construct mutant viruses that carry base pair substitutions associated with the point of origin of DNA replication—a region that is palindromic and that, you will recall, maps to a coordinate of 0.67. Specific restriction enzymes were used to nick the SV40 DNA in the region of this palindrome (Fig. 19-10). Local mutagenesis was carried out, and base pair substitutions were produced. The mutant viruses with these substitutions were then studied for any alterations in their ability to replicate. A base pair change from GC to AT at the point of axial symmetry was found to have no effect on the rate of DNA replication. However, other changes shown in Figure 19-10 led to the production of either mutants with definite decreases in DNA replication or ts mutants. The evidence indicates that the T antigen plays a critical role in the initiation of viral DNA replication and that it must first bind to a portion of the SV40 genome that includes this palindromic segment associated with the point of origin. A single base pair change in the region may alter the ability of the T protein either to bind in the proper amount or to interact properly with the origin site.

The ability to select within the genome a region for mutations is a monumental step in refining genetic procedures that permit the manipulation of DNA. Through studies of DNA regions involved in the con-

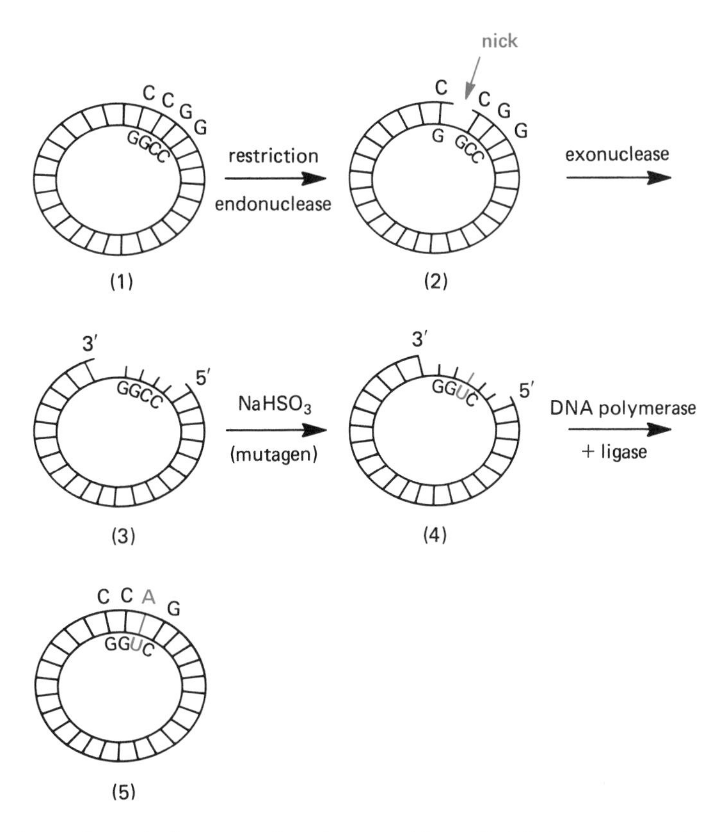

FIGURE 19-9. Summary of steps in producing a mutation at a preselected site. **1:** A restriction enzyme is selected that will attack a sequence of nucleotides occurring at one site in the viral DNA. The position of this site is known as a result of the construction of a cleavage map and DNA sequencing. **2:** The particular restriction enzyme nicks one strand of the duplex, circular DNA. **3:** An exonuclease attacks the free ends and digests away DNA, producing a gap. **4:** A mutagenic chemical is applied that can cause a base change in the exposed area of the intact circular strand. **5:** DNA polymerase is added and makes a complementary copy of the missing region using the intact strand as a template. It thus inserts a nucleotide that is the complement of the one changed. Ligase seals the DNA, resulting in a double-stranded, circular molecule carrying a base pair substitution at the preselected site.

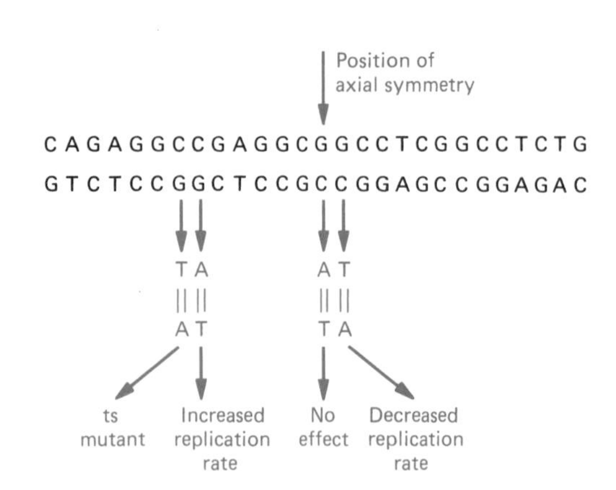

FIGURE 19-10. Mutagenesis of a palindromic sequence at the site of origin of SV40 DNA. Following local mutagenesis, specific base pair substitutions were made in the palindrome. A GC pair occurs at the position of axial symmetry in the palindrome. Transition to an AT pair does not affect replication of the viral DNA. However, other base pair substitutions were shown to have effects. A similar change to an AT pair at the next position to the right results in a decrease in replication rate. The change from CG to TA on the left leads to a temperature-sensitive mutant. The change to AT at the position next to it leads to an increased rate of viral DNA replication.

trol of transcription or replication, it may eventually be possible to establish the precise location of sites critical to normal gene expression, growth, and development in higher forms.

Antisense molecules

An unexpected way to regulate transcription has been uncovered in viruses and bacteria. This entails the formation of antisense RNA, RNA that is formed on the sense DNA strand, that strand of the double helix that does not undergo transcription to form mRNA. As Figure 19-11 shows, antisense RNA is complementary to mRNA, which can be called *sense RNA* since it carries the sequence of triplets coded for a polypeptide. The sense and antisense RNAs therefore are able to hybridize. The formation of the double-stranded RNA prevents translation of the mRNA. There is evidence that this interference can come about in one or more ways. The duplex may be unable to leave the nucleus and may be degraded by enzymes to which it is very susceptible. If it does get to the cytoplasm, the ribosomes will be unable to translate any of it. There is even evidence that the antisense RNA in some cases can inhibit translation without even forming a hybrid RNA molecule. It may somehow act in the nucleus to prevent the normal processing of the messenger, such as its splicing or capping (see Chap. 13). Regulation of gene activity by antisense RNA production has not as yet been reported in eukaryotic cells, but a variety of genes in viruses and bacteria appears to be under such control. The list includes genes involved with DNA replication as well as those associated with different aspects of metabolism.

The recognition of the control of gene activity by antisense RNA has led to the development of procedures that may have wide application in medicine as well as in studies of gene regulation. Different techniques are being explored for producing antisense molecules. In one approach, short strands of specific antisense RNA are synthesized and injected directly into cells, such as frog oocytes. In one such experiment, antisense RNA was made that was complementary to the sense RNA (mRNA) for actin, the vital protein that is a component of the cytoskeleton of the cell. When this RNA directed against actin was introduced into oocytes, the cells became flat and possessed defective cytoskeletons.

Keep in mind that the DNA strand that acts as the template for mRNA is the antisense strand. Short DNA strands 15–25 bases long that carry antisense sequences can be constructed in the laboratory. When these short strands (oligonucleotides) are introduced into cells, they act in the same way as the antisense RNA (Fig. 19-12). They inhibit translation of that mRNA that contains a stretch of nucleotides complementary to the antisense sequence they carry. It has been found that some cells can absorb sufficient numbers of DNA antisense molecules directly from the environment to bind to some mRNAs. It is important, of course, that a given sequence complementary to an antisense molecule occur in only one kind of mRNA. Otherwise a short antisense DNA strand would not be specific and could inactivate the mRNA transcribed from more than one gene. However, it has been calculated that a given sequence of 12 nucleotides should occur only once in an mRNA of a vertebrate cell, making such an antisense oligonucleotide specific against the RNA product of a target gene in higher organisms.

Antisense DNA is being manipulated and modified in various ways to make it more effective against specific target genes. For example, an acridine molecule was attached to an antisense DNA strand. The antisense was complementary to a nucleotide sequence which is found on most mRNAs of a certain parasitic African trypanosome. Applied to cell cultures of the organism, the antisense DNA carrying the acridine was able to kill the parasite. It is clear that antisense molecules have great potential in the treatment of disease-causing agents. This is especially so with respect to viruses against which antibiotics are ineffective. Success has already been reported in tissue culture studies in which antisense molecules have impeded infection by several viral types (influenza, herpes, HIV).

Antisense technology is making it possible to gain a better insight into the control of expression of oncogenes. In one investigation, Rous sarcoma viruses were developed that were incapable of transforming cells but that carried antisense RNA against src, the oncogene of the retrovirus. When chicken cells were infected with these viruses and then exposed to viruses capable of causing a malignant change, the cells were able to resist transformation to the cancerous state. One goal of studies of this type is to be able to target oncogenes and prevent cancer development without interfering with expression of protooncogenes, the normal counterparts of the cancer-inducing genes.

Antisense molecules are also aiding in our understanding of gene action in cell growth and development. In one experiment, antisense DNA was prepared against the mRNA of the gene for cyclin, a protein that accumulates during certain portions of the cell cycle. Introduction of the specific antisense molecules brought a halt to cell growth and division, an important observation indicating that cyclin is vital to these normal cell processes.

Antisense molecules can be utilized to mimic specific gene mutations. If antisense RNA or DNA directed against the mRNA of a specific gene is introduced into cells, the antisense molecule binds to the target mRNA and renders it inactive. The antisense molecule therefore suppresses gene activity and produces the effects of a specific gene mutation. Antisense technology clearly provides a new approach to the study of the consequences of specific mutations and has already been used in fruit fly and mouse experiments and may become adapted to studies of

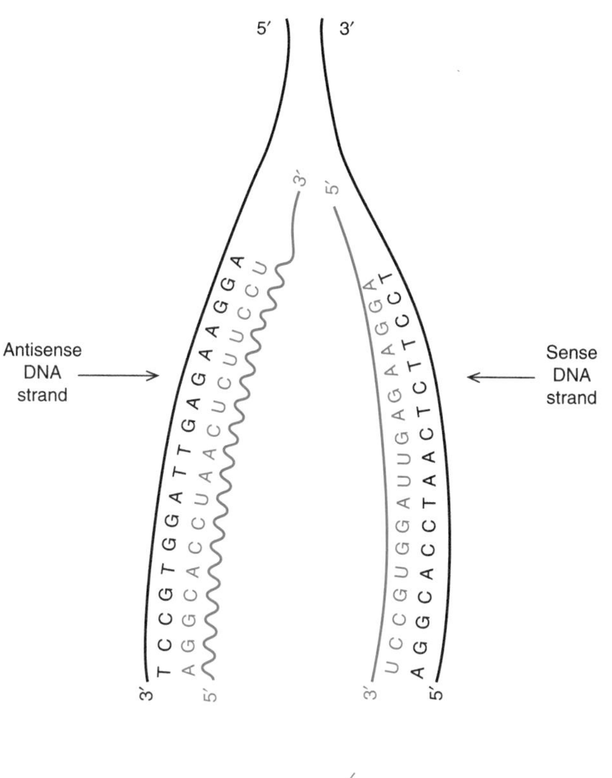

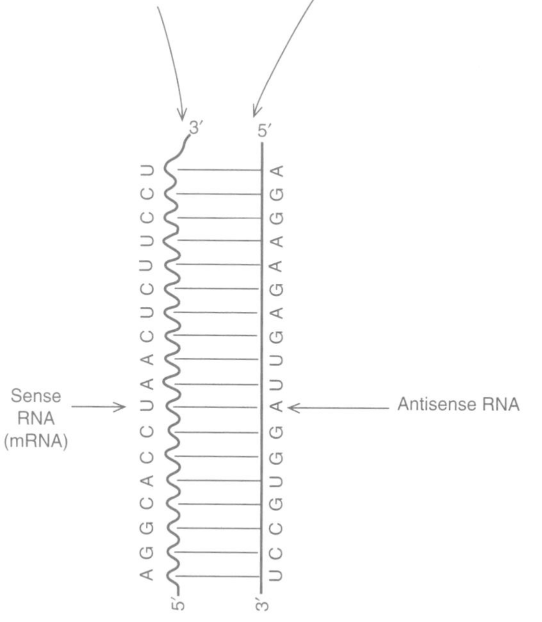

FIGURE 19-11. Sense and antisense. The antisense DNA strand of a gene is the template for the formation of the sense RNA (mRNA), which carries the code for the amino acid sequence of a polypeptide. If the sense DNA strand also acts as a template for the assembly of RNA, that RNA formed on it is antisense RNA since the nucleotide sequences are complementary to the corresponding sequences in the sense RNA. The two kinds of RNA, sense and antisense, form duplex RNA molecules that are unavailable for translation.

various human diseases. Moreover, experiments with antisense molecules in plants hold promise for the recognition of specific genes that may produce more productive and disease-resistant crops.

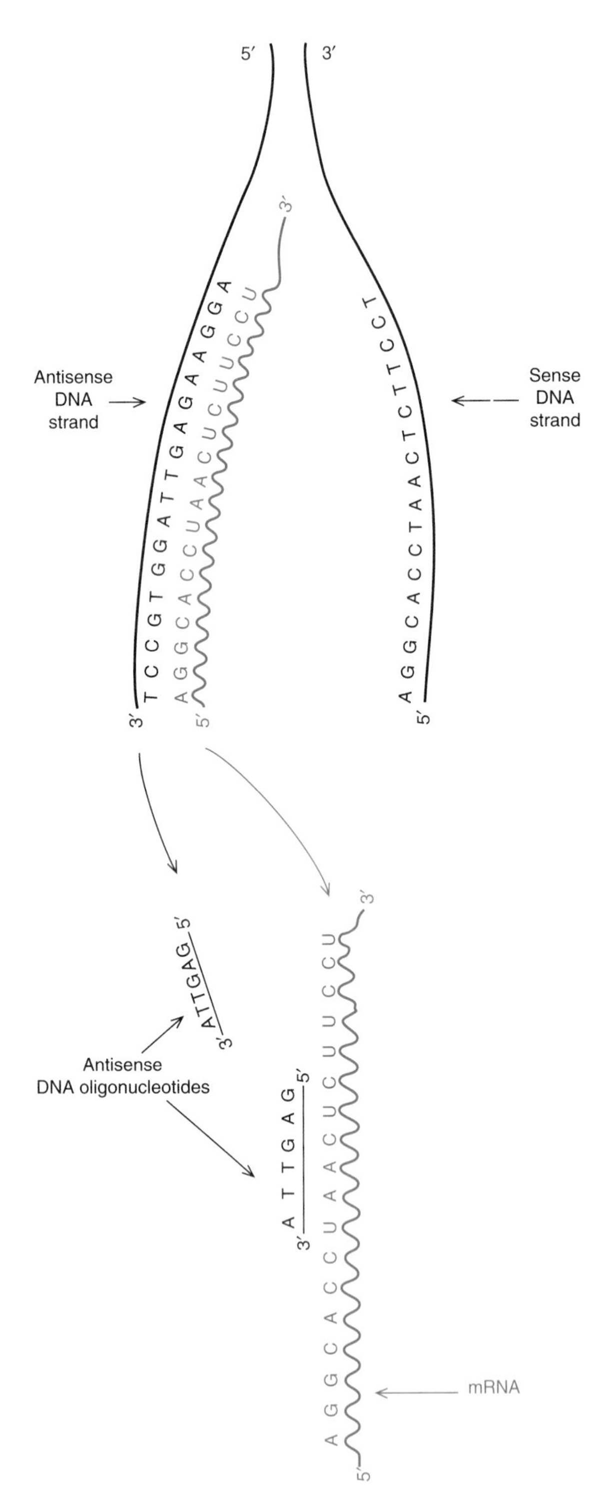

Genetic manipulation

The strides made in genetic manipulation, in which genetic materials from different sources are combined to produce a recombinant DNA, are at the forefront of the most dramatic scientific research in modern times. So rapid has been the breakthrough in recombinant DNA methodology that the widest bridges between species have been spanned. Let us familiarize ourselves with some of the basic aspects of recombinant DNA technology and then examine some of its refinements and applications.

Recall that in their synthesis of the first truly artificial gene, Khorana and associates used single-stranded projections as splints to join together separate DNA fragments (see Fig. 19-1). Recombinant DNA procedures commonly use such single-stranded "sticky" ends. These single-stranded projections are generated by many restriction enzymes that produce staggered nicks within DNA sequences that are palindromes and contain an axis around which a symmetrical arrangement of base pairs occurs. Figure 19-13A shows the site of cleavage of the widely used restriction enzyme EcoR1. This enzyme produces breaks in complementary DNA strands at points that are actually several nucleotides apart. As a result of this property, DNA regions are generated with single-stranded projections that will facilitate their joining to strands bearing the complementary nucleotide sequences.

Remember that fragments or segments of DNA do not usually possess the ability to replicate unless they are part of an intact replicon. Therefore, if any DNA is to be introduced into a cell, it must be part of a suitable replicon. The first vehicle selected to carry foreign

FIGURE 19-12. Antisense DNA strands. Short DNA oligonucleotides (usually twice as long as those shown here) can be constructed that have a nucleotide sequence found in the antisense DNA strand of a gene. The sequence is therefore complementary to a corresponding sequence in the mRNA transcribed from such a DNA strand. The synthetic antisense DNA can hybridize with the complementary portion of the mRNA and inhibit it from undergoing translation. If a short antisense DNA sequence is unique to a specific gene, then the antisense DNA can be directed specifically against a selected target gene.

DNA into a cell, and a vector still commonly used, was an R plasmid, which may contain factors providing antibiotic resistance (see Chap. 15). R plasmids can be isolated from cells by density gradient ultracentrifugation. It was discovered that when bacteria are treated with calcium chloride their cell membranes become permeable to extraneous DNA. When salt-treated cells are exposed to R plasmids, about 1 in 1 million becomes transformed by taking up an R plasmid.

This information and the available techniques set the stage for combining DNA from different sources to produce recombinant DNA. One kind of genetic engineering combines plasmid DNA from two distinct species of bacteria. For example, *E. coli* and *Staphylococcus aureus* are distinct species and cannot exchange genetic material. Either one may carry its own kind of R plasmid. Assume that the *S. aureus* plasmid carries a factor for penicillin resistance and that the *E. coli* plasmid carries one for resistance to tetracycline. From cultures of cells carrying known R plasmids, the DNA may be extracted and subjected to density gradient ultracentrifugation. Certain chemical treatments are used to impart a greater density to the plasmids, which are then easily separated from the chromosomal DNA in the gradient.

Once isolated, the two kinds of plasmids are mixed and subjected to a restriction endonuclease such as EcoR1, which can recognize a specific palindrome and cleave the DNA in the region of axial symmetry at staggered sites. The *E. coli* plasmid used as a carrier in the experiments had only one such site, so that the circular plasmid becomes linear but is not broken up into short fragments (Fig. 19-13B). Fragmentation occurs in a plasmid, or in a stretch of DNA if the palindrome is repeated in various regions, since the enzyme will attack each of them. The same sequence can occur in DNAs from different sources. The particular *E. coli* plasmid used as a carrier has the same palindrome as the *S. aureus* R plasmid. In the latter, however, the palindrome may occur more than once, so that fragments are produced from the *S. aureus* R plasmid as a result of the enzyme activity (Fig. 19-13C). Keep in mind that the action of enzymes such as EcoR1 also causes broken DNA to have projecting single-stranded ends and that these ends are complementary. Complementary projections on any DNA pieces, no matter what the source, permit base pairing. If this pairing occurs between DNA pieces from different sources, exposure to DNA ligase can join them by forming phosphodiester links and thus give rise to a recombinant DNA (Fig. 19-13D).

Plasmids treated in this way are then mixed with *E. coli* cells, which have been treated with calcium chloride to render them permeable to DNA. Since the majority of salt-treated cells will not be transformed and since some of those that have been transformed will contain two plasmids of the same kind joined together, it is necessary to screen out cells that carry a composite plasmid, composed of DNA from the two bacterial species. It is possible to do this by growing cells in the presence of the antibiotics for which the two types of plasmids carry resistance factors—in our example penicillin and tetracycline. Only doubly resistant cells can grow and therefore must carry antibiotic-resistance factors from both species. This permits isolation of those cells containing integrated foreign DNA combined with DNA native to the *E. coli.* The foreign DNA inserted can be from any source, from another bacterial species as described here to pieces of human DNA, and each case can be adapted for the screening out of just those cells with the recombinant DNA. The foreign DNA continues to replicate as if it were native to the cell. The procedure bridges the natural barrier to genetic exchange between species. Most of the genetic information expressed is that of *E. coli,* but the bacteria also carry DNA derived from a genetically separate species, and this DNA may become expressed in the cells as well.

In later sections of this chapter, we will encounter other recombinant DNA procedures now in use. Advantages and disadvantages accompany each one. The method described above is widely used and carries the advantage of sticky ends to facilitate joining of pieces. Moreover, as Figure 19-13 shows, the recombined plasmid possesses an EcoR1 site at either end of the foreign DNA. This makes it easy to retrieve the inserted DNA from the plasmid just by using EcoR1 to cut at the plasmid sites on either side of it. However, a disadvantage to the procedure is that a sticky end formed following EcoR1 treatment can pair with any other sticky end in the preparation. Consequently, some of the plasmids that become opened up simply reform by coming together again before any foreign DNA becomes inserted. On the other hand, when other plasmids become opened, many foreign fragments joined end to end can insert into them, creating a complicating factor. It is therefore necessary to weed out just those plasmids carrying only the single foreign DNA piece that is desired.

Further applications of genetic manipulation

The term *cloning* is applied to those recombinant DNA procedures in which a foreign stretch of DNA is replicated in cells, using a carrier such as a plasmid or virus. The cells continue to grow, proliferating the specific recombinant DNA. A pure culture or population of identical cells, a clone, is thus estab-

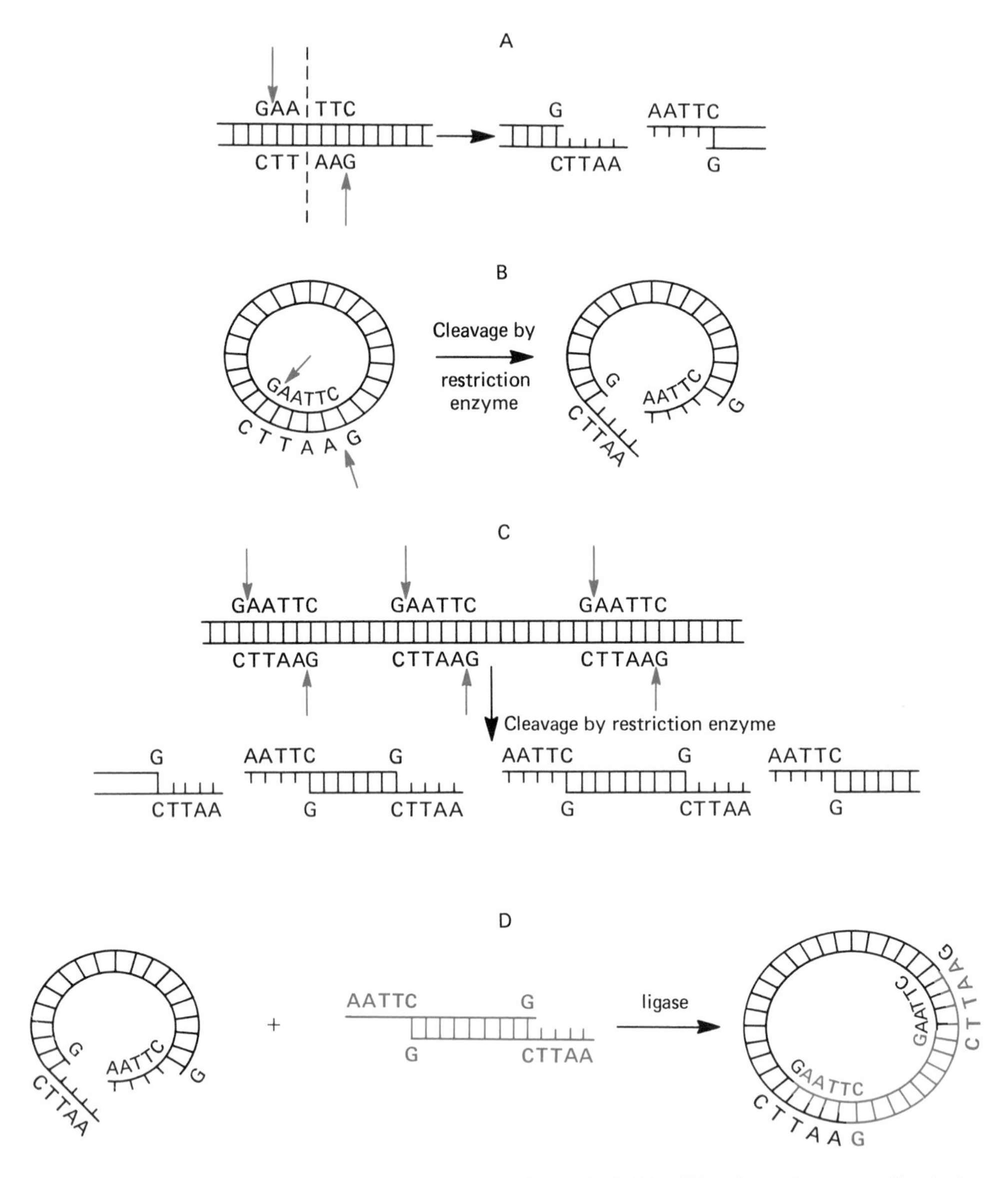

FIGURE 19-13. Steps in genetic manipulation. **A:** A palindrome may occur in a stretch of DNA and be recognized by a specific restriction endonuclease. The enzyme EcoR1 cleaves at staggered sites four nucleotides apart on each complementary strand. Consequently, the resulting fragments possess single-stranded projections. **B:** If the circular DNA of a plasmid contains only one such palindrome that can be recognized by EcoR1, it is broken at a specific site in each strand (arrows) and becomes linear with single-stranded ends projecting. **C:** If the same palindrome exists more than once in DNA, the DNA will be cleaved at a specific site in each region (arrows) with the production of many fragments, each with projecting, single-stranded ends. Note that these extensions in the fragments are complementary to those in the plasmid (B). **D:** A fragment of DNA from some source, and with the proper complementary base sequences on the ends (red), can pair with exposed bases in the plasmid. DNA ligase can then join the ends together, resulting in the production of a recombinant DNA molecule.

lished. The introduced DNA is multiplied thousands of times in the clone. Plasmids carrying the foreign DNA can easily be isolated from bacterial cells and then subjected to further study. The DNA carried on a plasmid is thus always readily available for various lines of investigation.

One goal of genetic manipulation is to introduce into bacteria those genes known to control the formation of some rare but vital substance, such as insulin. If an introduced gene can function and produce the valuable product in the cells, vast amounts of costly and rare substances might become readily available to help alleviate human suffering. However, recombinant DNA procedures are often rather hit or miss, in the sense that the DNA from some given source, such as the human, is broken enzymatically into a large number of fragments, from a few hundred to a few thousand nucleotide pairs. Each piece is then inserted into bacteria to establish clones. The term *shotgun experiment* is often used to denote the collection of a large sample of cloned DNA fragments representing the DNA of a given species. The collection can serve as a reserve of cloned DNA, and from it one can later select clones of interest for further studies. An important aim of genetic manipulation is to be able to place into a plasmid for cloning *only* the DNA that carries some desired genetic sequence, such as the DNA coded for the amino acid sequence of insulin.

How can one insert only the gene or sequences one specifically wants for cloning and thus avoid the need to examine many thousands of clones to find the desired one? One approach is applicable in those cases in which pure RNA transcripts of the gene in question are available at the outset. The mRNAs for various human globin chains are examples of specific transcripts that can be isolated in pure form. The availability of the globin transcripts has been responsible for the wealth of information we now possess on the arrangement of sequences that code for these polypeptides.

We can make good use of the fact that the information for the amino acid sequence of a polypeptide is carried in the messenger RNA and this coded information is complementary to the DNA strand that underwent transcription to produce it. One can convert the RNA information into DNA information by the use of the enzyme reverse transcriptase (Fig. 19-14A). DNA that is formed on an RNA template by reverse transcriptase is cDNA (copy DNA). Once a cDNA is obtained, the mRNA is destroyed. The single-stranded cDNA can then be rendered double stranded with DNA polymerase. The two DNA strands, however, are joined at the end by strong covalent linking. Another enzyme (S_1) solves this problem by breaking the linkage to give a normal double helix.

In a commonly used procedure, this double helix is treated further to provide it with projecting ends. Returning to Figure 19-13, recall that both the plasmid DNA and the DNA segments to be inserted bore single-stranded projecting ends and that those of the plasmid were complementary to those of the fragments. Projecting complementary ends are not always present, as is the case here where the double-helical cDNA bears no projections at all. The properties of still another enzyme—terminal transferase—can be called on to provide sticky ends. This enzyme has the ability to add to the 3′ ends of DNA strands a series of identical nucleotides (Fig. 19-14B). The cDNA is exposed to the enzyme in the presence of, let us say, molecules of guanosine triphosphate. This results in the production of an extension at each 3′ end of the cDNA composed of a sequence of repeating bases, all with G. The plasmid that is to be the carrier is processed separately. It is first rendered linear by treatment with a restriction enzyme. In our example, it would then be subjected to terminal transferase in the presence of molecules of cytidine triphosphate. Complementary ends of repeating sequences of C are thus added to the plasmid to produce projecting ends.

The plasmid and the cDNA are then mixed together (Fig. 19-14B) to permit complementary base pairing. Any gaps are filled in by adding DNA polymerase and extra nucleotides. DNA ligase finally seals breaks between the inserted DNA and the plasmid to produce a recombinant DNA. The plasmid selected as a carrier contains at least one genetic factor that confers on a bacterial cell resistance to a specific antibiotic. As described earlier, this enables the investigator to screen out any transformed cells that have arisen following exposure of the bacterial cells to calcium chloride and then to the R plasmid.

An advantage of the terminal transferase procedure described in this section is that it makes it possible to insert any DNA into a vector and eliminates the worry of finding a restriction enzyme that will provide sticky ends without cutting up the foreign DNA into tiny pieces. This can often present a problem. In this method, a restriction enzyme can be used that produces only blunt ends, as shown in Figure 16-16. If the selected enzyme produces fragments of the proper size for insertion, projecting ends can then be added on to both the vector and the foreign fragments in a manner similar to that shown in Figure 19-14. However, a disadvantage to this procedure is that the restriction enzyme site used to open up the plasmid in the first place becomes eliminated when the foreign DNA is inserted. Note from Figure 19-14 that the foreign DNA has poly(G) and poly(C) at either end, and thus it becomes difficult to retrieve the inserted DNA.

In still another recombinant DNA protocol, DNA ends may be joined together without the presence of any sticky ends. This method, known as *blunt end ligation,* utilizes the ability of the DNA ligase from phage T4 to join together ends lacking extensions. A great advantage to this method is that it enables one to join together directly any pairs of ends, and no additional material need be added between them. However, a drawback is that there is no control over which blunt ends join together. Therefore it becomes necessary to retrieve the recombinant DNA that is desired from a large mixture of unwanted products.

Each process used today in recombinant DNA studies has its advantages and disadvantages, as well as its variations. Which method is used depends on the particular experiment, in addition to the circumstances under which it is conducted.

Establishing gene libraries

Restriction enzymes have been responsible for many highly significant advances in molecular genetics and cell biology. Among these is the progress made in the molecular analysis of the DNA of entire genomes of species such as the human and the mouse. For example, total DNA is extracted from human fibroblasts and then exposed to a given restriction enzyme such as HindIII, which cleaves it into a certain number of specific fragments. These fragments are now ready for insertion into a suitable plasmid containing only one site vulnerable to the restriction enzyme. The general procedure is identical to that described in Figure 19-13.

A very popular plasmid used as a vector in cloning is the one designated *pBR322.* The reason for its wide use is easily understood. The plasmid carries two genes that confer antibiotic resistance, one to ampicillin (Amp^r), the other to tetracycline (Tet^r). Fortunately, each gene contains sites that can be recognized by several restriction enzymes. Certain of these endonucleases recognize sites only in the Tet^r gene, others only in the Amp^r gene. The value of this fact is that the plasmid may be cut selectively in either of these two genes. The cuts are staggered so that projecting, sticky ends are generated. If a fragment of foreign DNA, such as from the human, is inserted into either resistance gene, the activity of that gene is lost. For example, cleavage by endonuclease HindIII produces staggered cuts in gene Tet^r at one site, rendering the plasmid linear. (The restriction enzyme EcoR1 can also cut the plasmid, but its specific site is found outside both of the resistance genes.) Human DNA cleaved by HindIII will produce many fragments, each with sticky ends complementary to those in the plasmid. Endonuclease-treated plasmids and the endonuclease-generated human fragments are incubated together. Following DNA ligase treatment, the material is mixed with calcium chloride–treated bacterial cells and then screened to detect those cells that are resistant to ampicillin only. Loss of resistance to tetracycline because of insertion of the human DNA enables us to screen out those cells that have been transformed and that contain plasmids into which a fragment of human DNA has been inserted (Fig. 19-15). In this way, pieces of human DNA may be inserted into bacterial cells, which in turn produce a number of clones. Each clone that is established harbors a specific piece of human DNA.

Recall that the steps in the cloning of human DNA began with the extraction from cells of the total DNA content. This total DNA was then fragmented by the endonuclease and exposed to the plasmid. It is obvious that the procedure is a shotgun approach, hit or miss as to which fragment enters a particular plasmid. It is possible to follow the procedure using a more purified type of DNA. For example, a cesium gradient may supply us with a small amount of a particular band of DNA. If this DNA contains a region vulnera-

FIGURE 19-14. Summary of major steps in recombinant DNA production, starting with transcripts of a specific gene. **A:** The mRNA is used as a template in the presence of reverse transcriptase to form a complementary DNA strand. The RNA is then destroyed, leaving the single-stranded cDNA. This is then used as a template to form a complementary DNA strand. An enzyme is required to separate the two strands that are linked. A cDNA duplex results that is now available for further processing. **B:** The cDNA and the plasmid are handled separately to add single-stranded projecting ends that are complementary. This is accomplished by terminal transferase, an enzyme that can add strings of identical nucleotides to 3′ ends. The plasmid must first be opened up before the application of the transferase. This is accomplished by a restriction enzyme specific for a palindrome in the plasmid. Finally, the cDNA and plasmid are brought together. Extensions on the cDNA may pair with their complements on extensions of the plasmid. Filling in of gaps and sealing yield a circular duplex recombinant DNA carrying a copy of the specific gene.

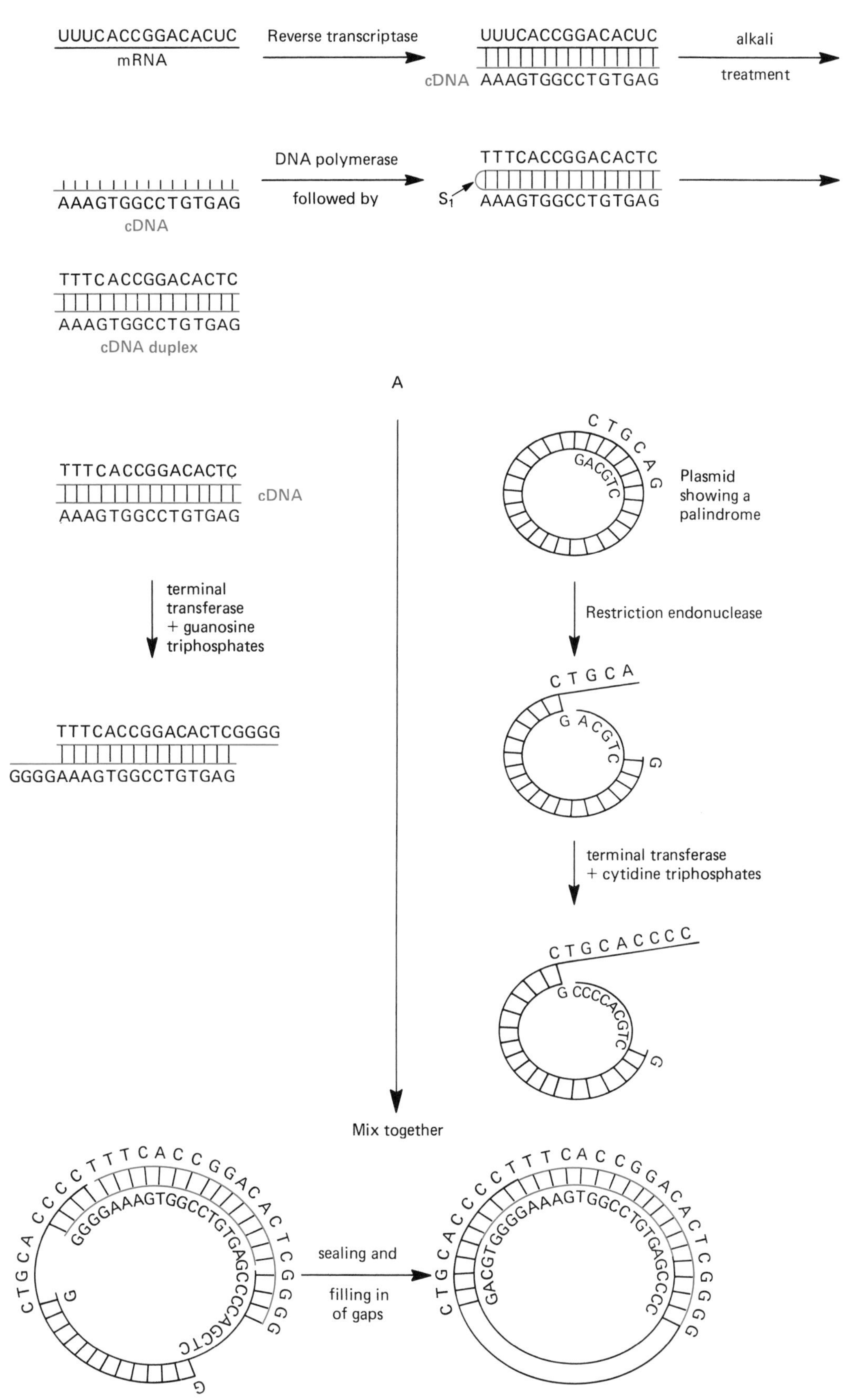

UUUCACCGGACACUC
mRNA
Reverse transcriptase
UUUCACCGGACACUC
cDNA AAAGTGGCCTGTGAG
alkali
treatment
AAAGTGGCCTGTGAG
cDNA
DNA polymerase
followed by
TTTCACCGGACACTC
S1
AAAGTGGCCTGTGAG
TTTCACCGGACACTC
AAAGTGGCCTGTGAG
cDNA duplex
A
TTTCACCGGACACTC
cDNA
AAAGTGGCCTGTGAG
terminal
transferase
+ guanosine
triphosphates
TTTCACCGGACACTCGGGG
GGGGAAAGTGGCCTGTGAG
CTGCAG
GACGTC
Plasmid
showing a
palindrome
Restriction endonuclease
CTGCA
GACGTC
G
terminal transferase
+ cytidine triphosphates
CTGCACCCC
G CCCCACGTC
G
Mix together
sealing and
filling in
of gaps
B

ble to the proper restriction enzyme, then it may be handled in the manner just described. The result is the establishment of clones covering a smaller but much more specific portion of the genome.

By joining fragmented DNA from the human or some other species to plasmids and then establishing clones, we can construct a gene library. The library consists of all the different clones, as if each one were a volume. At this point, however, we do not know what is in any of the clones or volumes, because our procedure has not enabled us to take a specific gene and place it in a given plasmid. Somehow, we must now try to determine what gene or sequences are located in a given clone. If we can achieve this, then we can turn to the clone at any time to study further a particular gene that is of interest to us. To solve our problem, we must find a probe to locate the clone with the desired gene.

The importance of probes in cloning and genetic manipulation cannot be overestimated, and indeed advances in these and related areas depend on the availability of the proper probes. Probes are specific, labeled stretches of RNA or DNA that can hybridize with complementary nucleic acid stretches. A probe for a specific nucleic acid sequence may be a pure mRNA species, cDNA that has been prepared from the mRNA, or a stretch of DNA. A probe need not match perfectly with the target sequence in order for hybridization to take place between them. Moreover, those parts of a probe that do match a desired gene need not be very long as we will learn in the following section. Having a probe for a specific gene or DNA segment makes it possible to screen out the gene or segment that otherwise might remain undetected among the numerous other pieces of DNA in the many clones constituting a gene library.

Let us assume that we have available the purified transcripts of some specific gene, such as the one for β-globin. Since these are at hand, our task becomes greatly simplified, and excellent use can be made of DNA–RNA hybridization. First of all, separate bacterial colonies, each representing a clone carrying unidentified human DNA, are maintained on nutrient agar (Fig. 19-16). Copies of these clones are transferred to nitrocellulose filters using the replica plating technique. The bacterial cells are lysed, and their DNA is released onto the filter paper and then denatured. Pure radioactive transcripts of the gene being searched for are presented to the denatured DNA. Any single-stranded DNA present on the filters that carries nucleotide sequences complementary to the RNA will hybridize with it. Those DNA–RNA molecules that form are radioactive and are easily detected by autoradiography after the filters are washed. Since replicas of the clones have been maintained, the investigator can return to the original plate and identify any clone on the agar that corresponds to one on the filter. In this way, using the mRNA as a probe, one can identify from a library those clones that have incorporated vec-

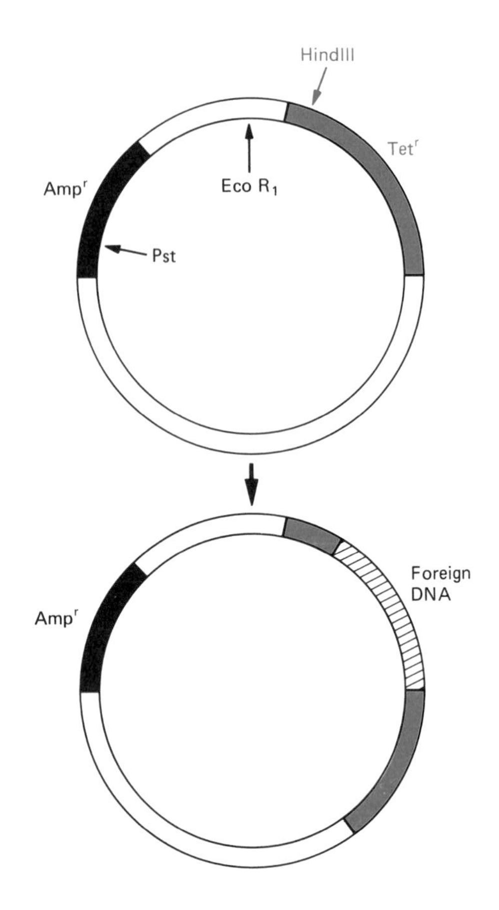

FIGURE 19-15. Plasmid pBR322 as a vector. **Top:** This plasmid contains a gene for resistance to ampicillin (Amp^r) and one for resistance to tetracycline (Tet^r). Various restriction enzymes recognize sequences in both of these genes. Others, such as Pst, recognize sites found only in the Amp^r gene. Still others, such as HindIII, recognize sites found only in the Tet^r gene. The EcoR1 site is found outside the resistance genes. Using recombinant DNA procedures, we can insert a fragment of foreign DNA into one of the two resistance genes, making use of a restriction enzyme that will cleave within one of the genes but not the other. In this example (**bottom**), the foreign DNA has been inserted into the Tet^r gene after the plasmid was cut by HindIII in the Tet^r gene. Carrying the foreign DNA, the Tet^r gene no longer confers resistance to tetracycline. Loss of resistance to this antibiotic, but not to ampicillin, makes it possible to screen out those cells that have been transformed and also carry an inserted foreign DNA.

tors carrying the gene sequences that are desired. Such clones are now available for further study.

Labeled cDNA derived from the transcripts of known genes are also very valuable as probes in genetic manipulation. After cDNA denaturation, DNA–DNA hybridization may lead to the identification of a specific gene in the same way as described here using the transcript itself.

Once a clone has been identified as a population of cells that contains a specific structural gene, further genetic manipulation is generally required if that gene is to express itself in the bacteria. We know that the normal expression of a genetic message depends on regulatory signals. If a foreign gene, lacking appropriate signals, is inserted into a plasmid, there is no reason to expect expression of the gene and formation of its associated product. The gene could simply be replicated along with the plasmid DNA, without transcripts being formed from it. If transcripts are formed that lack the proper "start" and "stop" signals, they could fail to undergo proper translation.

The task of providing a gene with the proper regulatory signals to permit its transcription, followed by translation of the transcript, is being approached in various ways. Using restriction enzymes and ligase, the investigator may insert a foreign gene *into* a bacterial gene residing on the plasmid. The introduced gene is thus provided with the regulatory signals of a resident bacterial gene. The result of such a manipulation is the production of a hybrid genetic region that starts off as a bacterial gene, changes to the introduced gene, and ends as the bacterial gene. The hybrid region has, however, the signals required for the beginning and ending of transcription.

If all the necessary steps are followed, the cells of the clone may secrete into the medium a polypeptide that begins as a bacterial protein, then changes to the desired one, and ends as a bacterial protein again. The desired portion may then be retrieved using protein-digesting enzymes. Relying on the various procedures for clone production, identification of the proper clone, and production of a desired protein, several teams of investigators have demonstrated that recombinant DNA technology can indeed be used to obtain from bacteria biologically active proteins typically secreted by specialized eukaryotic cells. Such valuable products as human insulin and growth hormone have been produced using such methods and may be made available in ever larger amounts as cloning techniques become even more refined. Yields of exceptionally scarce substances in increased quantities, such as the potential anticancer protein interferon, may enable the design of critical experiments to evaluate their biological properties and their value in medicine.

Searching and probing for the desired clone

Unfortunately, RNA transcripts are not available for most of the genes one wishes to clone. Therefore the task of producing and locating a clone that contains a desired gene can become much more laborious than the steps depicted in Figure 19-16, although the same general procedure is followed.

In one approach to the problem, one uses the tissue that is specialized to produce the product of the gene one is seeking. If insulin is the product of the desired gene, one would turn to those pancreatic cells specialized for production of the hormone. The reason for this is that these cells produce mRNA transcripts for the protein, whereas other cells, such as those of the kidney, do not, since genetic information for producing insulin is repressed in the kidney. Of course, the insulin-producing cells contain other mRNAs in addition to that for insulin, because all cells need to produce products for basic cell survival. This means that the insulin mRNA is mixed in with significant amounts of other kinds of RNA.

The next steps in the procedure are the same as those described previously and in Figure 19-14. However, one now has cDNAs of various kinds, not just those desired. Different clones are thus produced that carry everything from fragments of various kinds to the structural information for various genes, including those with complete information for the gene desired, in this case the one for insulin. Without a specific pure RNA probe to locate the desired clone or clones, additional steps must be taken.

The clones, whose complete genetic contents are unknown, are numbered, and a DNA sample is taken from each. This DNA is denatured in a test tube to which is added that RNA taken from the specialized cells at the start of the procedure. This is the very RNA that was used to make the cDNA that was inserted into the bacteria. This RNA is of course unpurified, a mixture of various species of RNA. Any of this added RNA with nucleotide sequences complementary to a DNA sequence present in a clone will hybridize with that DNA. This fact of hybridization can now be used to identify the desired clone(s). This stems from the fact that RNA that has formed a hybrid molecule with DNA is unavailable for translation. Therefore no polypeptide can be formed from hybridized RNA when hybrid DNA–RNA is placed under conditions favoring translation. To find out whether a specific hybrid DNA–RNA is present, all the requirements for polypeptide formation are added to the test tube (Fig. 19-17; review also Fig. 12-10).

The amino acids that are added are labeled with radioactivity. Polypeptide is formed as a result of translation of any *unhybridized* mRNA. To tell whether the system is making the desired polypeptide, one adds antibodies to that polypeptide, the polypeptide specified by the gene one wishes to clone. The antibodies will react specifically to the polypeptide and will bind to it, causing it to precipitate out of the solution if it is present. If the desired gene is present in the system, the mRNA for that gene will be bound up in hybrid DNA–RNA and will not be available for translation. The antibody, being specific, does not react with any other proteins formed by translation of free, unhybridized mRNAs, and thus no precipitate is formed.

We are now in a position to identify those clones that contain the sought-after DNA. In a clone in which the precipitate forms, the DNA of the desired gene *cannot* be present. If it were, it would have bound the mRNA complementary to it. Since the gene is not present, the transcript is available for translation, and the protein (insulin in our example) precipitates out, because it is recognized by the anti-

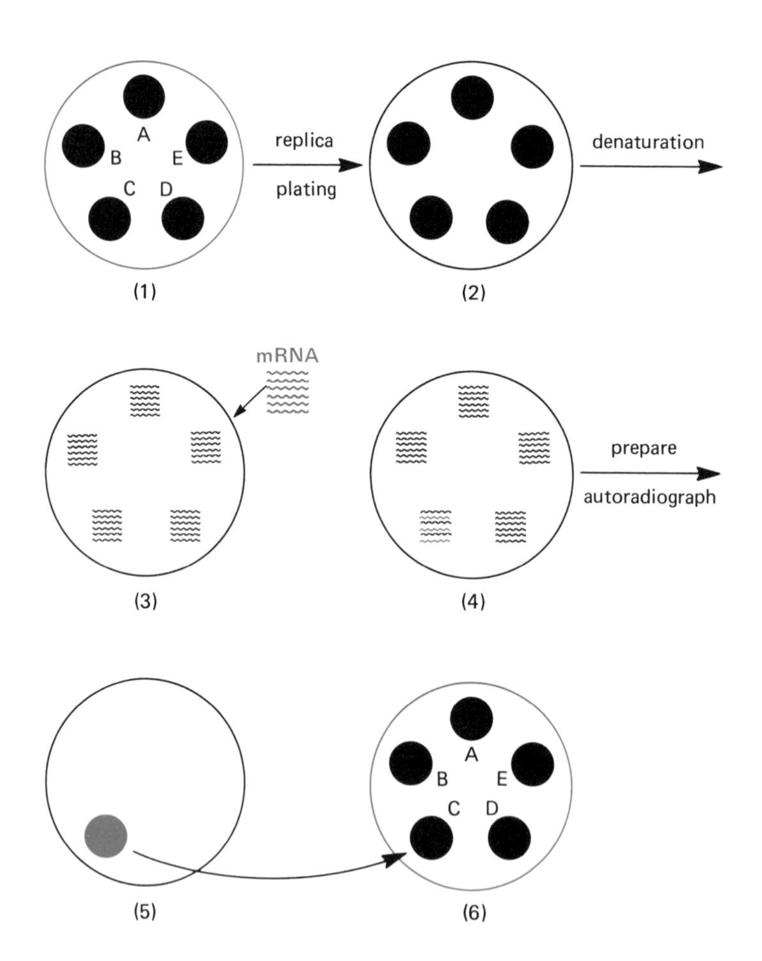

FIGURE 19-16. Identifying the proper clone. Clones are maintained on master plates (1). To determine which, if any, carry the desired gene or genetic sequence, a copy of the master is made by replica plating (2). The DNA on this plate is then treated to denature it. Labeled transcripts of the gene in question are added to serve as a probe. (3), DNA from one or more clones may form hybrid molecules with the added RNA if the DNA carries nucleotide sequences complementary to ones in the probe. Hybrid formation indicates that desired gene sequences are present in that clone (4). To detect any DNA–RNA hybrid molecules, the plate is covered with a photographic emulsion to produce an autoradiograph (5). The clone with the desired gene (C) is identified by relating it back to the clone held in reserve on the master plate (6).

body. Other proteins are also formed, since much of the added mRNA would not be hybridized. However, this protein is *not* precipitated; only the insulin is. We then know that this clone does *not* contain the structural gene for insulin.

In a clone that *does* contain the structural gene for insulin, the DNA–RNA hybrids are formed between the DNA of that gene and its messenger. Consequently, translation is blocked. In that system, no precipitate forms, because no insulin could be synthesized to permit a reaction with the specific antibody against it. Again, other proteins are formed here, but they are not precipitated out. In this way, a desired clone can be recognized.

A further check can be made on the identity of a clone. The DNA–RNA hybrid molecules can be separated from the unbound RNAs present in the system. The RNA bound in the hybrid molecules can be released from the DNA and utilized for various purposes. It can be used to direct the synthesis of protein in another translation system and in this way identify the protein. Since in this case we now know that this retrieved RNA is a transcript of the insulin gene, we can use it as a probe in further investigations or even use it to derive cDNA probes for insulin gene sequences.

Another way to obtain specific probes is to turn to a specific differentiated cell type, as we did in the case of the pancreatic cells in our search for the insulin mRNA. Mixed mRNAs are isolated from the specialized cells and used to form a mixture of cDNAs after a treatment with reverse transcriptase. Again, the procedure is similar to that shown in Figure 19-14, except that we would be starting with many different transcripts of unknown identity and would generate a mixture of cDNAs, also of unknown identity. These different cDNAs are inserted into a suitable plasmid type, such as pBR322. We now have a mixture of plasmids carrying different cDNAs. Bacterial cells are then exposed to the plasmids. Any transformed cells are identified, and clones are finally established from single bacterial cells. In essence, we are constructing a cDNA library (Fig. 19-18). Each bacterial cell carries a certain unknown cDNA, which is being multiplied in each clone.

We must now try to identify the cDNA in a given clone. To do this, the cDNA is extracted from cells in a clone, rendered single stranded, and then attached to a nitrocellulose filter. The bound cDNA is next exposed to mixtures of mRNA. If any mRNA in a mixture hybridizes with this cDNA, it can be located on the filter and then isolated by itself. Although this RNA represents a single kind of transcript, we still do not know what information it carries. To ascertain this, we may proceed to place the mRNA in a cell-free system, as explained in Chapter 12 (see Fig. 12-10). Any protein that is made would reflect information carried in the mRNA used, and hence would indicate the specific gene that gave rise to the transcript.

The particular protein must now be characterized using one or more procedures. Once its identity is established, we have also identified a probe for the gene that specifies this protein. The probe is the cDNA that was able to bind to the mRNA coded for the protein and that has been multiplied in the bacterial cells. Armed with the cDNA probe, we can now return to the volumes in our DNA library to locate the clone that carries genetic sequences corresponding to the probe and hence identifying the presence of a given gene.

Improvements in protein sequencing have made it possible to determine the amino acid sequences of many polypeptides, and this has proved to be most valuable in the construction of probes for specific genes. Let us suppose that a particular polypeptide is known to be associated with the activity of a certain gene that we wish to isolate from a genomic library and that we have determined the amino acid sequence of a section of this molecule. With this information, we can attempt to deduce the mRNA sequence related to this stretch of amino acids. Technology makes it possible to link together nucleotides in a specific sequence to form a short molecule, called an *oligonucleotide.* Assuming we know a sequence of 12 amino acids in the polypeptide, we can relate these to RNA triplets as they might be found in the mRNA coding for the polypeptide, for example, methionine (AUG) or leucine (CUU). Each RNA triplet has a corresponding DNA triplet, and thus we can proceed to construct an oligonucleotide of 36 DNA nucleotides (12 triplets) to be used as a probe. Of course, the degeneracy of the code must be taken into consideration, since most amino acids are designated by more than one triplet. Consequently any probe we construct may have an "incorrect" triplet at one or more positions in the oligonucleotide, for example, CTT for leucine instead of the correct CTC for that specific position. To overcome this problem, several oligonucleotide probes may be constructed in relation to a given sequence, each one with a different triplet at a given location, say CTT for leucine at triplet position 3 in one molecule and CTC at that same position in a second potential probe. Fortunately, an oligonucleotide

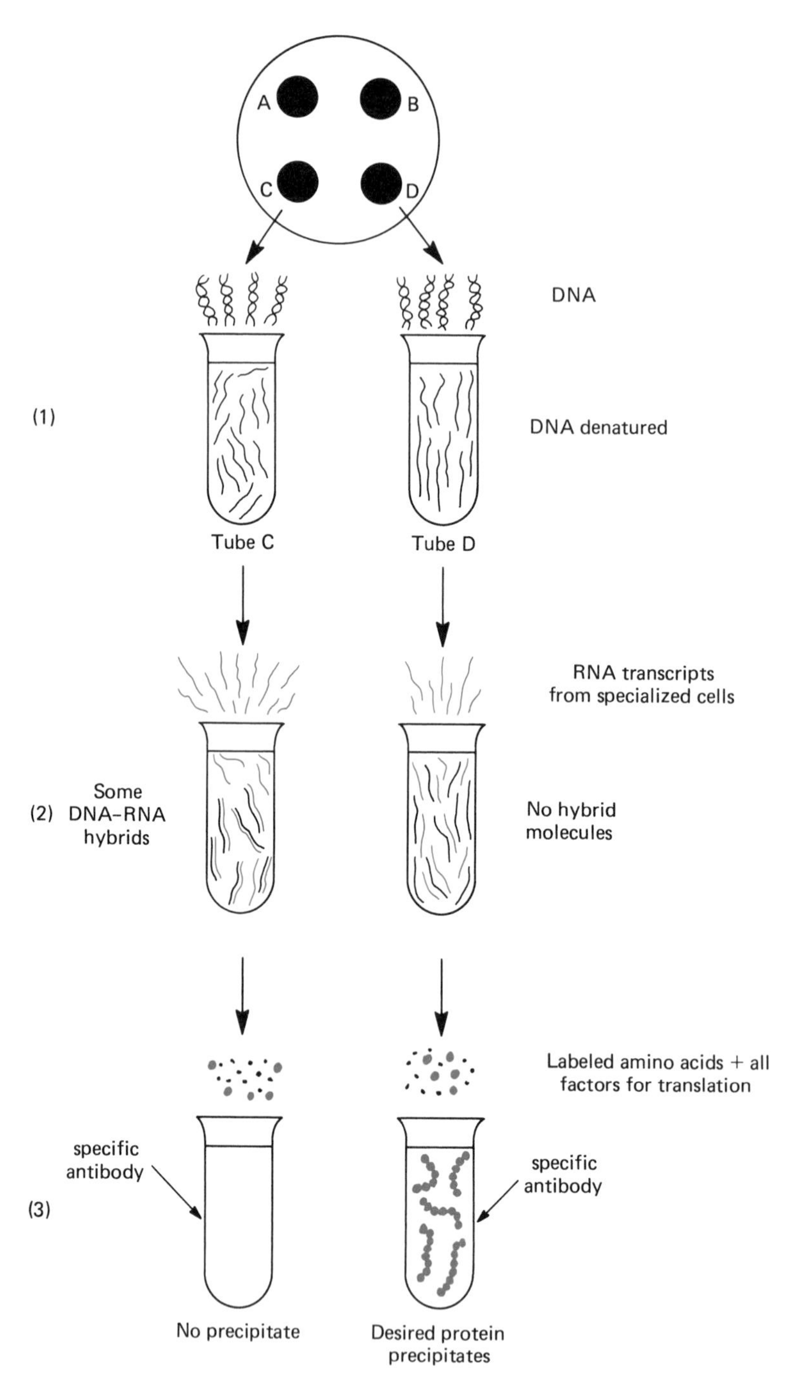

FIGURE 19-17. Identifying the proper clone in the absence of a pure RNA sample. **1:** From clones held in reserve (C and D here), DNA samples are taken and denatured. **2:** Messenger RNA that has been isolated from cells known to produce transcripts of the desired gene is then added. These transcripts will be mixed with other transcripts from the cells. However, if the desired gene sequences are present in any of the denatured DNA, hybrid molecules will form (tube C). This hybridization will tie up the transcripts of the gene, making them unavailable for translation. **3:** In a translation system, the protein product of the desired gene can form only if the mRNA coded for its amino acid sequence is present. This is possible in tube D but not in tube C. Antibody precipitates the protein out if it is present. The absence of the protein in tube C tells us the transcript is tied up and the coded information for that protein is present. Hence clone C, held in reserve (above), carries the desired gene.

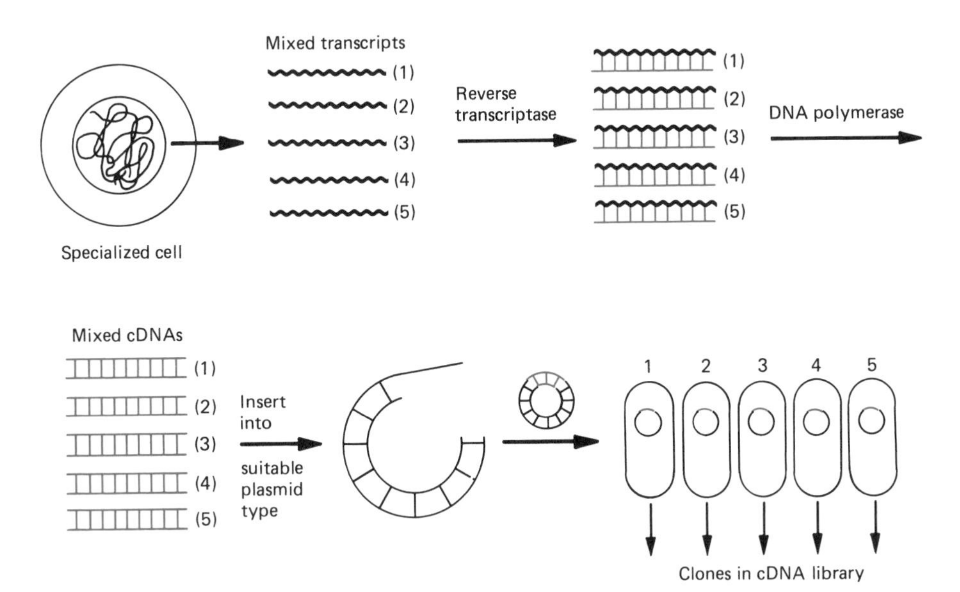

FIGURE 19-18. Establishing a cDNA library. Transcripts are isolated from a differentiated cell type. These mRNAs are mixed and represent transcripts of many different genes. They are used as templates in the formation of cDNA in a fashion similar to that shown in more detail in Figure 19-14. The mixture of cDNAs that are derived represent many different genes. These are then inserted into a suitable plasmid type such as pBR322, which in turn is used to transform bacterial cells. Those cells that are transformed and that contain inserted cDNA can be selected and allowed to form clones. A library is thus constructed. Each clone will harbor a cDNA representing a genetic sequence whose identity remains unknown until a suitable probe is found.

need not correspond exactly in order to be useful as a probe. Just a few "incorrect" nucleotides can still enable it to bind sufficiently to a nearly complementary sequence and thus permit recognition of a desired sequence from a genomic library. As our discussions continue, we will encounter still additional ways to obtain probes for specific genes.

Cloning in bacteriophages

While DNA fragments from 5–10 kb pairs in length can be readily cloned in plasmids such as pBR322 (Fig. 19-15), difficulties arise with the cloning of larger pieces. Plasmids carrying large amounts of inserted DNA become unstable and tend to lose portions of the foreign DNA. The larger the plasmid, the slower its rate of replication so that the smaller plasmids that have lost parts of the inserted DNA come to outnumber greatly those that remain intact. Luckily, larger pieces of foreign DNA may be inserted into a suitable bacteriophage and then cloned successfully. Phage λ has been used extensively as a cloning vector for pieces up to 15 kbp long and has been the vehicle of choice for establishing human DNA libraries.

The procedure is similar to that in which foreign DNA is inserted into a plasmid. Phage λ is subjected to the restriction enzyme EcoR1, which cuts its DNA into pieces (Fig. 19-19). Among the pieces generated are two large ones, and these are the only portions of the virus needed for its replication in the *E. coli* host. The end pieces can be separated from the smaller fragments due to their size and then mixed with DNA from some other source, such as human DNA, which has also been cut by EcoR1. Some of the foreign fragments will become inserted between the end segments of the phage. A fortunate aspect of λ phage maturation now comes into play. Unless DNA is about 45 kb in length, it will not be packaged by the λ system to form mature phage. Pieces both above and below the critical length remain unpackaged. And thus the only DNA pieces formed in vitro that can be successfully inserted into phage heads are those that are formed from a combination of the ends of the phage and a foreign DNA in the 15 kb range.

A recombinant phage carrying a foreign DNA of appropriate length is able to replicate in *E. coli* and give rise to a clone. A plate covered with bacteria can be infected with large numbers of different phage particles carrying different human DNA fragments. Each plaque produced has been derived from a single infecting phage and represents a clone of phages (Fig. 19-20). The problem now is to locate clones carrying specific DNA fragments. The DNA from a clone of phages can be denatured and exposed to a specific probe to determine if the cloned human DNA carries sequences homologous to those in the probe. In this way, an entire phage library can be screened by a probe to determine the location of a given gene.

However, the procedure just described places a size limit of 15 kb on the fragment to be cloned. While this often presents no problem, various genes are known that greatly exceed this size, 40 kb, for example. The problem in cloning such large pieces has been overcome by the creation of special vectors to permit the packaging of large fragments. A review of Figure 16-2 will show that phage λ in its linear form carries at both ends single-stranded projections that are complementary and that enable the virus to circularize. These ends, known as *cos* sites, also define one genome's worth of λ DNA for packaging into the head of the phage. It has been possible to clone cos sites into plasmids of various kinds. For example, they may be cloned within the ampicillin-resistance gene of plasmid pBR322. Except for the cos sites, the rest of the phage λ DNA has been deleted. Plasmids carrying the λ cos sites are known as *cosmids,* and they can be handled in the same way as any other plasmids used as cloning vectors.

DNA from some eukaryotic source can be subjected to a restriction enzyme to produce fragments in the 35–45 kb range. Cosmids are also opened up by the enzyme and are treated to remain in a linear form (Fig. 19-21). Eukaryotic fragments generated by the enzyme may become inserted between cos sites of linearized cosmids in the preparation. Viral packaging extract is then added, and the enzymes that it contains recognize cos sites. Hybrid DNA composed of the eukaryotic DNA flanked by the cos sites may then become packaged into a phage head. However, any eukaryotic fragments less than 35 kb that become flanked by the sites will not be packaged, since the total DNA length is too small.

Those phage heads carrying the successfully packaged hybrid DNA in the proper size range are able to infect *E. coli.* Infected cells can be selected out from

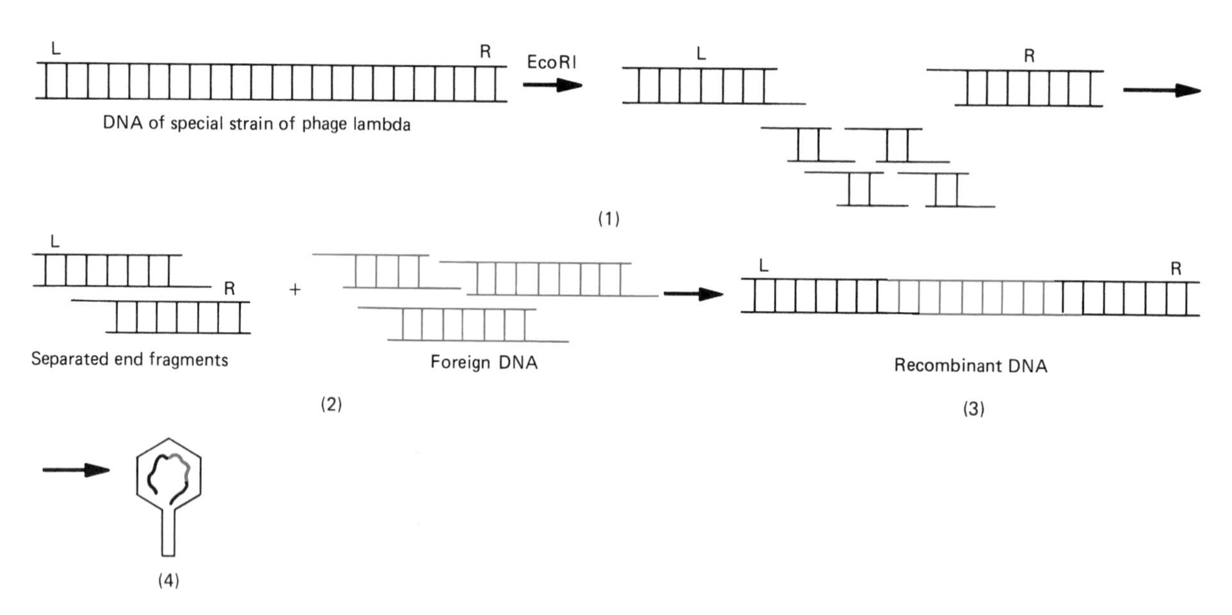

FIGURE 19-19. Inserting foreign DNA into phage λ. **1:** The DNA of certain λ strains, when exposed to EcoR1, is cut into two large fragments representing the left (L) and right (R) ends of the λ DNA plus several small fragments. The DNA of the latter is not required for the replication of λ in *E. coli.* **2:** The two large end fragments, which bear sticky ends, are separated from the small ones. Foreign DNA (red) that has also been cut by EcoR1 will also bear sticky ends that are complementary to those of the L and R fragments of λ. **3:** Foreign DNA can thus be inserted between the two end fragments, yielding recombinant DNA. **4:** A particle bearing foreign DNA can now be packaged in the protein coat of the virus. Such a virus can replicate in *E. coli.*

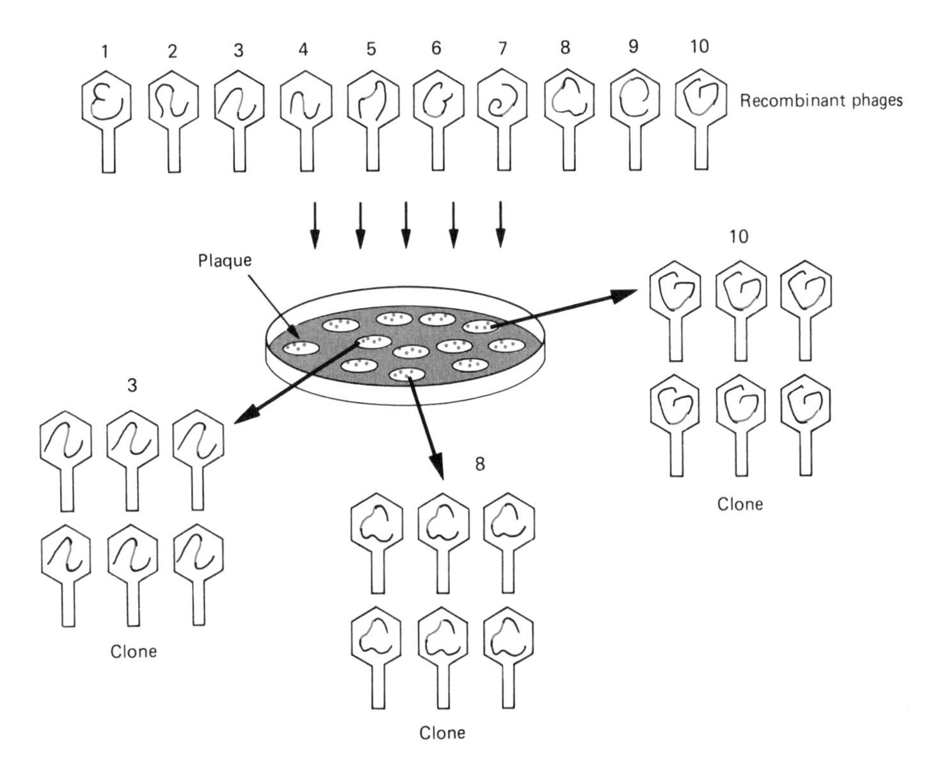

FIGURE 19-20. Establishing a phage library. Recombinant virus particles bearing different portions of a foreign genome (e.g., 1, 2, 3) are allowed to infect a lawn of *E. coli* growing on a plate. The phages infect the bacterial cells, multiply in them, and lyse them, resulting in the formation of plaques on the lawn. Each plaque represents a clone of phages, each one derived from a single infecting recombinant phage. Consequently, each plaque is equivalent to a clone of recombinant DNA. An assortment of clones such as these covering the entire genome of a species constitutes a genomic library.

the others, since they will be tetracycline resistant in the case of a cosmid formed from plasmid pBR322. Since only the cos sites of the phage are present in the *E. coli* cell along with the foreign DNA, no active virus is present. However, due to the cos sites and the packaging treatment, circularization can now take place; the hybrid DNA that includes the large eukaryotic fragment can continue to be propagated as a plasmid in the *E. coli* cell.

Cloning in YACs

Cosmids and vectors utilizing phage λ have been extremely valuable in the isolation and cloning of many portions of the genome of various eukaryotes. However, there is a limit to the size of the DNA that these vectors can accommodate. Phage λ vectors can maintain DNA inserts to a limit of 24 kb in size and cosmids up to 45 kb. However, many human genes are much larger and therefore cannot be cloned as a single fragment in these vectors. For example, the gene associated with Duchenne muscular dystrophy exceeds 1,800 kb! The size restriction of cloning vectors poses quite a problem, since a very large gene cannot be cloned as a single piece but requires several clones, each with a portion of the gene. The problem is especially serious for projects whose goal is the physical mapping of an entire genome. The smaller the cloned fragments, the larger the number of clones needed for analysis. For example, using cosmids with an average DNA insert 33 kb in length requires at least 10,000 clones for analysis of the entire *Drosophila* genome. The problem facing the project to clone and sequence the human genome with its 3 billion base pairs would appear to be overwhelming in its requirement for different clones. Methodology is required that will permit a larger size of the cloned fragment and thus a reduced number of separate clones required for analysis.

A significant step in this direction has been taken with the development of yeast artificial chromosome (YAC) vectors. These are plasmids constructed from portions of plasmid pBR322 (Fig. 19-15) combined with chromosome segments derived largely from yeast. The vector is assembled to contain all the elements needed to replicate as a plasmid in *E. coli.* It also contains portions required for orderly chromosome maintenance and replication such as centromere and telomere regions as well as autonomous replication sequences that possess the properties of replication origins (follow Fig. 19-22A). The vector also carries restriction enzyme sites, one of which provides a location for the insertion of exogenous DNA for cloning. This site is found within a gene whose function becomes interrupted by the DNA insertion. The effects of the loss of function can be detected visually, and, as we will see shortly, this provides a way to detect those vectors that have incorporated DNA for cloning.

The basic cloning procedure requires exposure of the circular YAC vector to restriction enzymes that linearize it (follow Fig. 19-22B). This treatment yields

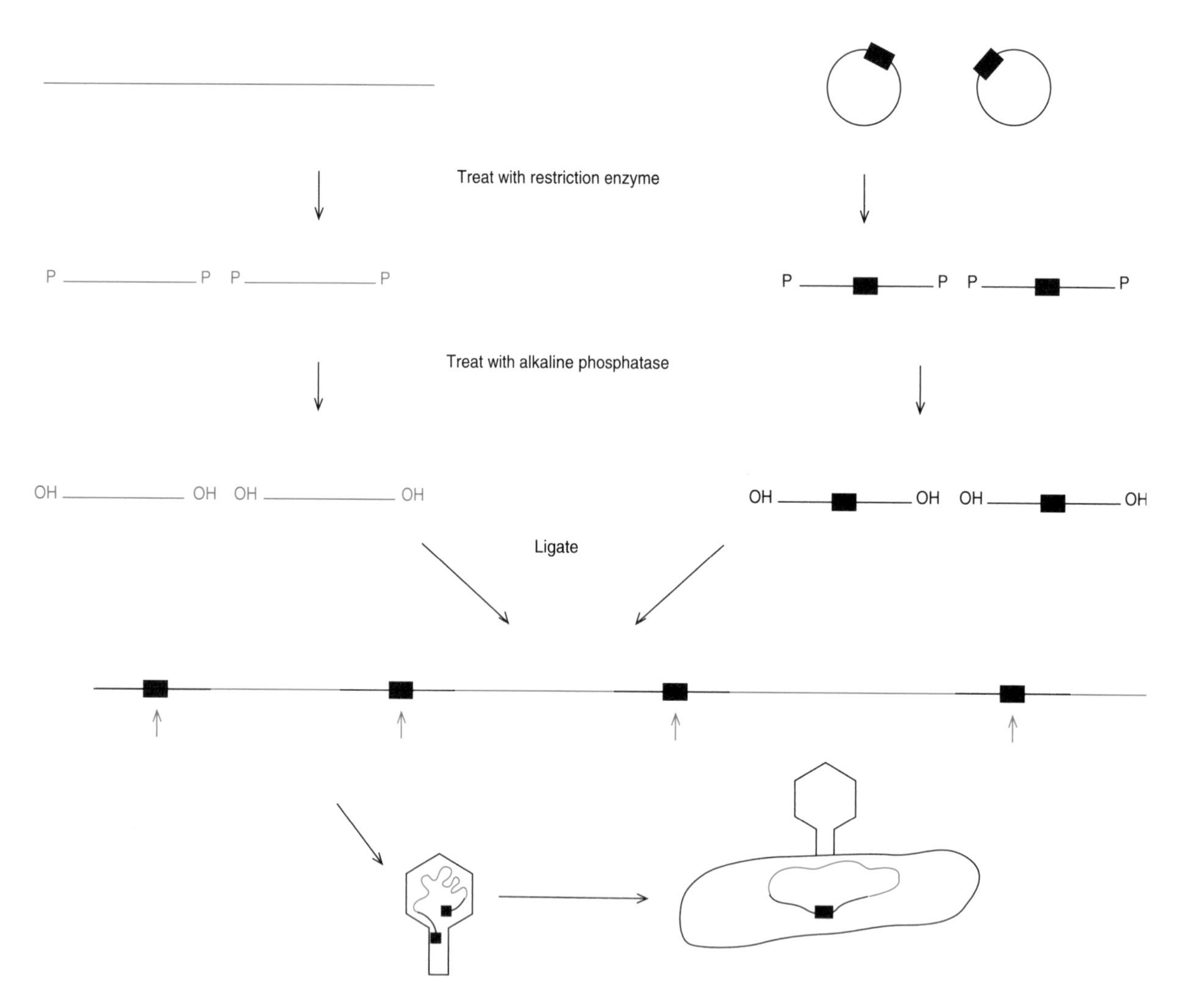

FIGURE 19-21. Outline of cosmid cloning. Eukaryotic DNA (red) is cleaved to yield large fragments in the 35–45 kb range. Cosmids are also opened up by enzyme treatment (cos sites shown as black bars). All the DNA is treated with alkaline phosphatase to remove terminal phosphate groups (P). This prevents recircularization of the cosmid as well as any indiscriminate end-to-end joining of DNA segments, such as many linearized cosmids linking together. The preparation can be treated to produce overlapping ends on the DNA pieces, which are then ligated. Eukaryotic fragments may become joined between different cosmids. A high concentration of the eukaryotic DNA and an excess of cosmids favors the formation of long chains, with the former alternating with the latter. Packaging enzymes (red arrows) recognize the cos sites. Only those pieces of DNA within the proper size range will become inserted into phage heads. *E. coli* cells are then infected by the phage. Inside the cells, circularization can now occur, since the packaging treatment generated cohesive ends.

three parts from each YAC vector: a left chromosome arm that includes a telomere, the centromere, and sequences to act as replication origins; a right arm with a telomere; and a portion between the telomeres of the vector that is discarded. The arms are treated to prevent their joining together. An outside source of DNA that is to be cloned is then subjected to restriction enzymes to yield relatively large pieces in the 200–500 kb range. Such a piece may become inserted between two arms utilizing the technique of blunt end ligation. As noted earlier, the DNA insertion destroys the activity of a gene in the YAC vector. This will now be put to use to detect those YACs that contain an inserted piece of DNA.

Yeast spheroplasts (cells with much of the wall material removed and thus that assume a spherical shape) are exposed to the linearized products of the YAC vectors. Those cells that have become transformed by picking up the inserted exogenous DNA can be detected visually due to the loss in gene activity caused by the insertion. The transformants are then screened further to select out only those cells carrying exogenous DNA that has been successfully joined to two different arms. It is these transformants that carry the DNA as part of a chromosome capable of replication and maintenance in a cell. Their identification is possible, since each arm of a YAC carries a different marker gene that permits selection. These cells with the intact YACs are now available to establish clones that can provide many copies of the inserted exogenous DNA for further analysis.

Success in cloning large DNA fragments in YACs is being currently reported in analyses of genetic regions in a variety of organisms. Work with the *Drosophila* genome suggests that most sequences of eukaryotes can be successfully cloned utilizing the advantages of YACs. Construction of complete mammalian DNA libraries in YACs is now feasible. Libraries containing overlapping sequences can be used in the same way as phage λ genomic libraries or cosmid vectors. The methodology of YACs holds great promise in the project to characterize the entire human genome.

Gene cloning and chromosome mapping

In Chapter 9 we discussed advances in the mapping of human chromosomes made possible through somatic cell genetics. This procedure, although responsible for the assignment of well over 100 genes to specific chromosomes, entails a definite limitation. Recall that the cell hybridization procedure depends on the ability of the specific gene being mapped to express itself in the hybrid cell. Failure of a gene to produce some detectable product or effect in the hybrid cell makes it impossible to assign it to a chromosome using the procedure of somatic cell genetics by itself. Let us see how gene cloning, coupled with somatic cell procedures, has been able to overcome this severe limitation and to make it possible to detect the presence in a hybrid cell of a gene that is not being expressed.

Once any specific gene has been successfully cloned, it becomes available for a host of studies and uses. Among these is the use of the cloned gene as a probe, such as a globin gene or the insulin gene. Let us suppose that we have as a probe a labeled cDNA of the gene for β-globin. We are now in a position to locate the β-gene sequence on a specific chromosome. To accomplish this, we first perform somatic cell hybridization. Human and mouse cells are fused, as described in Chapter 9, and stable clones containing just a few human chromosomes are established. Because the β-gene is not expressed in the cell hybrid, we were previously prevented from screening the clones, since no detectable evidence of the globin gene in the form of a product would be present in any hybrid cell containing the gene. Now, however, we can proceed further. DNA is extracted from human cells and then subjected to a particular restriction enzyme. Mouse DNA is similarly treated, and so is the DNA taken from the hybrid cell containing, let us say, human chromosomes 5, 8, and 11. The three DNA samples that have now been cut by the specific restriction enzyme are placed on an agarose gel and subjected to electrophoresis, which separates the fragments on the basis of their sizes. Our ultimate goal is to hybridize fragments with a probe. To do this, it is necessary to transfer and bind the fragments to a nitrocellulose filter. This can be accomplished using the Southern blotting procedure (see Chap. 16). The filter with the denatured fragments is now available for hybridization with our probe.

The filter contains many separated DNA fragments—mouse in one lane, human fragments in another, and mouse DNA plus DNA for human chromosomes 5, 8, and 11 in a third lane. Our labeled probe of denatured cDNA of the globin gene is then brought into contact with the filter. Any single-stranded DNA fragments on the filter that contain sequences complementary to sequences on the probe will form hybrid duplex DNA with it. Such complementary sequences can exist in rodent as well as human DNA fragments because the rodent also carries genes coded for globin proteins. Since our probe is labeled, any fragments on the filter that hybridize can be detected by autoradiography. As Figure 19-23A shows, signals from the radioactive probe are now evident. We see that the lane with human DNA alone contains two fragments that hybridized with the probe, indicating that those fragments contain DNA sequences complementary to sequences in the probe and hence contain sequences of the β-globin gene. The mouse lane gives a different pat-

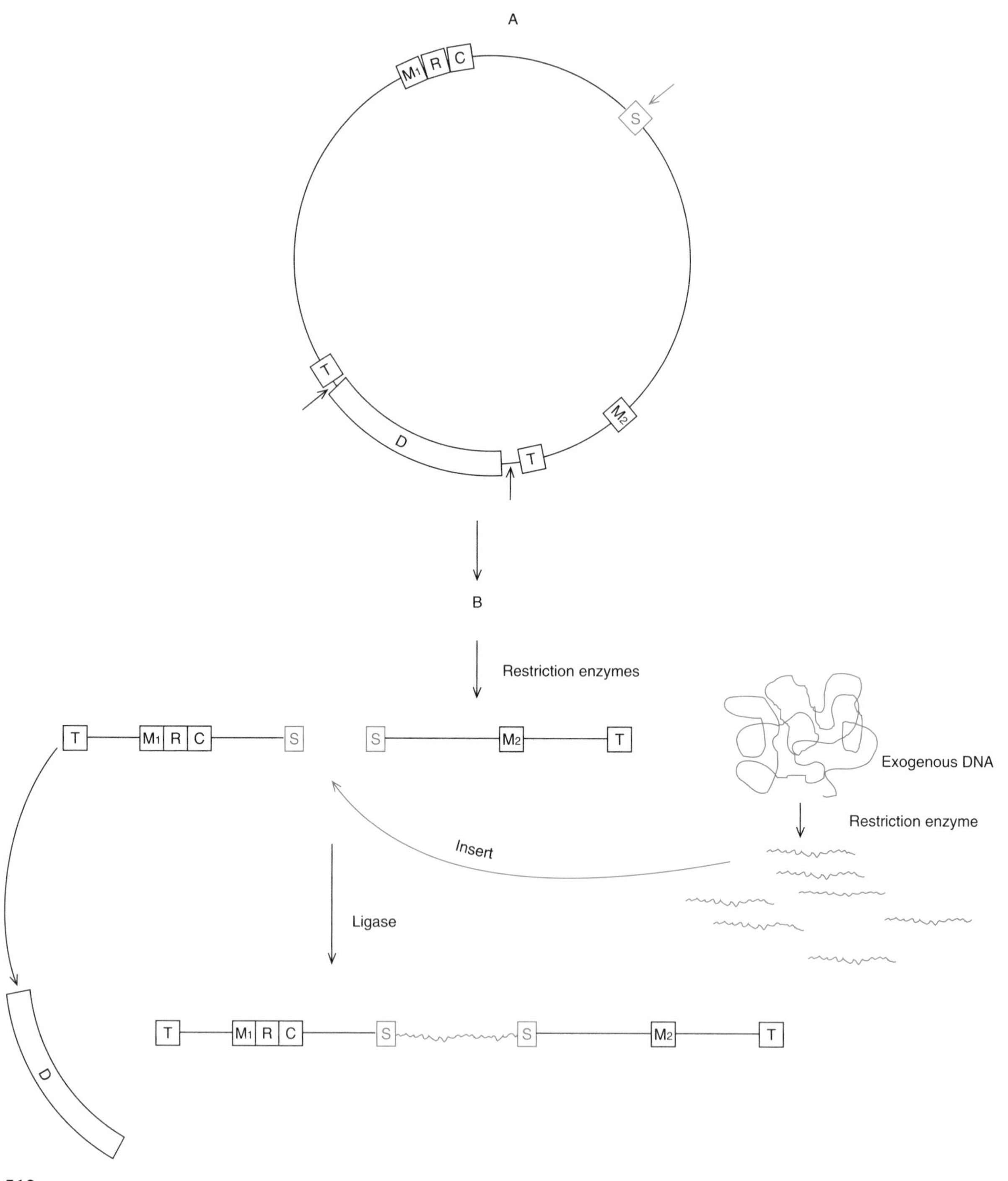

tern. One fragment contains the sequences for the gene. The DNA from the hybrid cell shows a pattern that combines both the human alone and the mouse alone. We now know that human β-globin DNA sequences are present on chromosome 5, 8, or 11.

To ascertain which human chromosome actually carries the β-globin gene, we must screen clones containing other combinations of human chromosomes. It is found that only when chromosome 11 is present do we find the pattern shown in Figure 19-23A. DNA from any hybrid cell lacking chromosome 11 shows only the mouse pattern (Fig. 19-23B). We can thus deduce that the gene for β-globin is on chromosome 11.

This procedure enables us to assign a gene, even though that gene is not expressed in the hybrid cell. What we are detecting are specific fragments, generated by restriction enzymes, that contain stretches of the gene. Our probe enables us to detect these fragments. Coupling this procedure with the procedure of somatic cell genetics, we can now relate the fragments to a specific chromosome. We simply look at our filter to see which chromosome, when absent from a hybrid cell, is associated only with the mouse pattern and when present is always associated with the human plus mouse pattern. The reasoning is the same as that used in Chapter 9 when screening panels for a gene associated with a detectable product. Using the procedure made possible by strides in molecular biology, human chromosome mapping is advancing at a rapid rate. Not only does it enable us to assign to a chromosome a gene that is not expressed in a hybrid cell, it also allows us to assign such a gene to a specific chromosome band. The basic reasoning is again the same as was described in Chapter 9 and is illustrated in the following case of the insulin gene.

Earlier in this chapter, we discussed procedures permitting the identification of clones containing insulin cDNA. Once clones with cDNA of any gene have been identified, they provide a source for further investigation of a specific gene. Using insulin cDNA as a probe, investigators were able to assign the insulin gene to chromosome 11. It was then assigned to a specific region of chromosome 11. This was made possible through the use of a human cell culture containing a specific reciprocal translocation. Cells from individuals carrying specific translocations have been used to establish hundreds of different lines, each representing a particular translocation. Each such reciprocal translocation can be identified, because it alters the banding pattern of the chromosome. Held in reserve, such cell lines are available for further use. In the case of the insulin gene that had been assigned to chromosome 11, it was possible to pinpoint its location on the chromosome. Figure 19-24 shows that only when the very end portion of the short arm of chromosome 11 is present does the hybridization give the human plus mouse pattern. In the absence of that specific chromosome segment from chromosome 11, only the mouse pattern is revealed following hybridization with the human cDNA probe, indicating that the probe is hybridizing only with complementary sequences on the mouse DNA.

Use of structural aberrations such as translocations and deletions along with the advances in molecular procedures is rapidly increasing our knowledge of the precise chromosome positions of genes that previously would not have been assigned to a chromosome at all because of their failure to express themselves in the human–rodent hybrid cell.

FIGURE 19-22. Essentials of a YAC cloning vector and a yeast artificial chromosome. **A:** The YAC cloning vector is constructed from DNA sequences derived from different sources. Those whose source is plasmid pBR322 are represented by thin lines. Most of the other genetic regions shown in boxes are from yeast. The vector contains all the elements necessary for chromosome replication such as a centromere (C), replication origins (R), and telomeres (T). The latter are derived from the protozoan *Tetrahymena.* The gene represented here by S contains a restriction enzyme site (red arrow) where exogenous DNA may be inserted for cloning. Insertion interrupts the gene's function, producing an observable effect. Also present are two marker genes (M1, M2), which are used to identify intact YACs containing inserted DNA fragments. The region D between the two telomeres (T) is flanked by two restriction enzyme sites (small arrows) and become discarded during insertion of the exogenous DNA. **B:** In the presence of the appropriate restriction enzymes, the vector is opened up at the cloning site and the portion (D) between the telomeres (T) is discarded. The treatment yields from each vector a left arm region with a marker gene (M1) as well as a right arm region also carrying a marker gene (M2). The two arms together contain all the sequences needed for chromosome replication (C, T, R). DNA from some source (right) may be cut into fragments with an appropriate restriction enzyme, and a fragment may become inserted between two arms, which are then ligated to the introduced DNA. The YAC that is produced is then transferred to yeast cells by standard transformation procedures. Cells transformed by picking up recombinant DNA can be identified, since disruption of gene S function produces a visible effect on colonies of cells containing the inserted DNA. Transformed cells can be screened further to make certain that they contain YACs with two different arms and therefore can undergo stable chromosome replication. Isolation of such intact YACs is possible, since each arm carries a marker gene, thus permitting selection for both arms. An intact artificial chromosome can continue to replicate inside yeast cells, yielding clones containing copies of the exogenous DNA that is now available for further analysis.

Good use can also be made of somatic cell hybrids to obtain probes. For example, mouse–human hybrid cells were formed in which the only human chromosome present was chromosome 11. A DNA library was constructed using the DNA from the hybrid cells. Most of the DNA fragments in this library would represent mouse DNA, but a small number of the clones would contain fragments of human chromosome 11. These can be identified by taking *total* human DNA that is fragmented and labeled to screen the mouse–human library for human fragments (Fig. 19-25). In this way we can identify DNA fragments that represent sequences of a specific chromosome. These fragments are then available for further investigations, such as

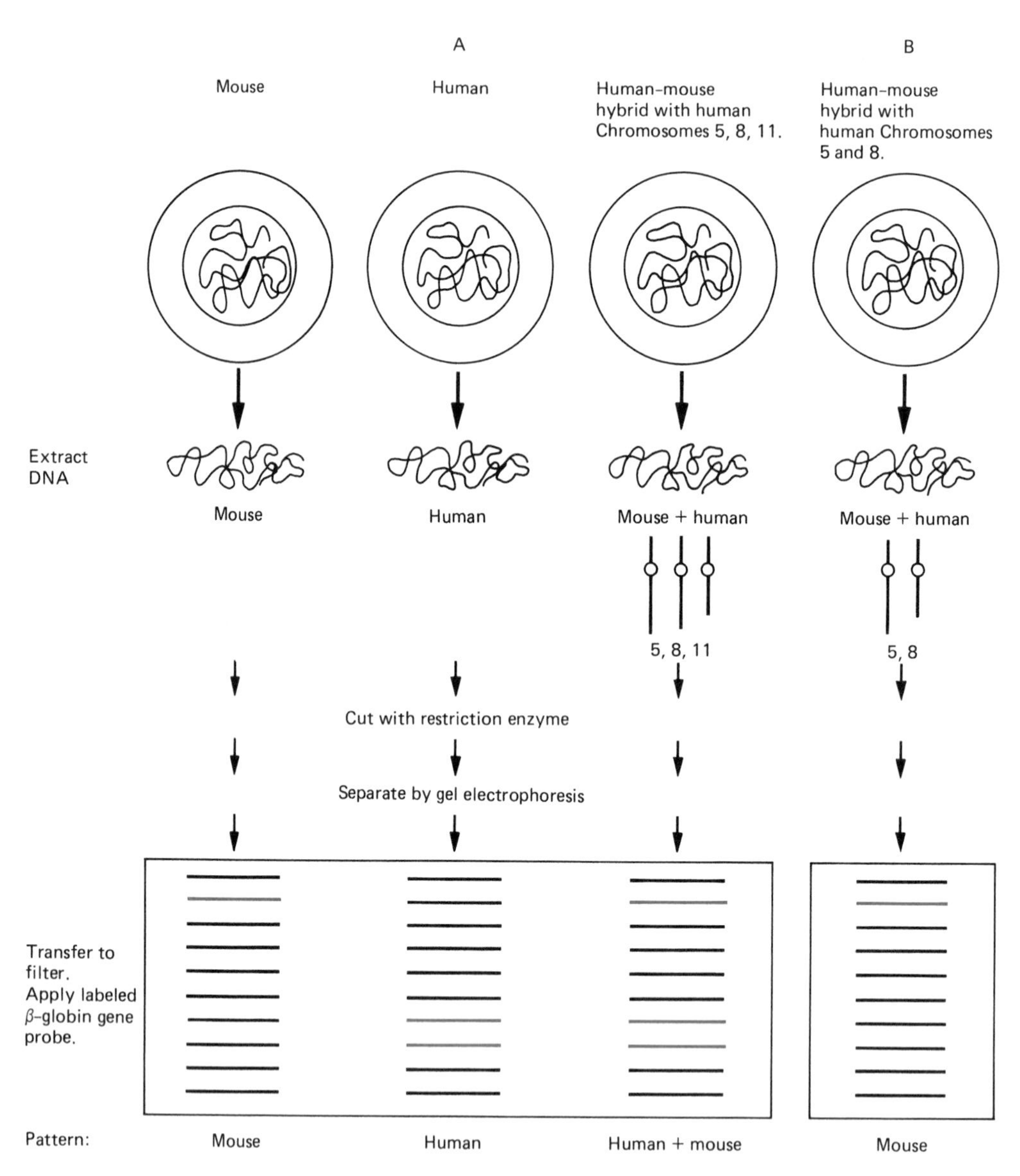

FIGURE 19-23. Assigning the gene for β-globin. **A:** Total mouse DNA, when subjected to a particular restriction enzyme, yields one band (red) that hybridizes with the labeled probe, indicating that sequences for the globin gene are present on that fragment. Similarly treated total human DNA yields two bands that hybridize with the probe, hence carry β-globin gene sequences. When DNA from human–mouse cell hybrids is similarly treated, the mouse plus human pattern is found when chromosomes 5, 8, and 11 are present. **B:** In the absence of chromosome 11, only the mouse pattern is found. The gene for human β-globin must reside on chromosome 11.

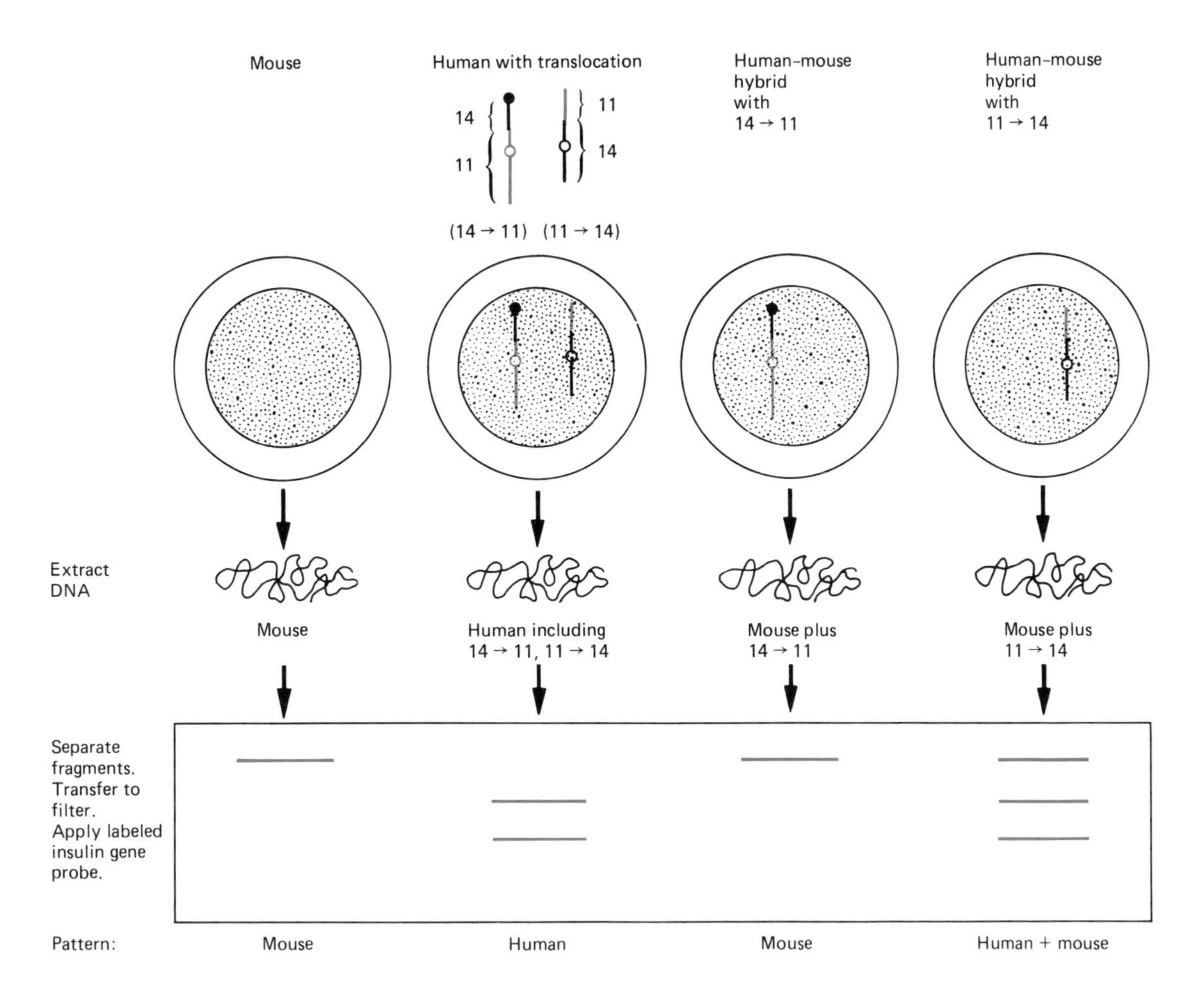

FIGURE 19-24. Localization of the gene for insulin. The gene for insulin had been assigned to chromosome 11 in a manner similar to that shown in Figure 19-23. Using a reciprocal translocation involving chromosomes 11 and 14, we can show, following somatic cell hybridization, that the end portion of chromosome 11 carries the insulin gene. Only when the end of human chromosome 11 is present in a cell do we find hybridization with the insulin gene probe, giving the mouse plus human pattern.

DNA sequencing, and may even be used as probes themselves for screening other DNA samples for corresponding homologous sequences now known to be on chromosome 11.

Searching for an oncogene

Phage λ libraries were used to search for a genetic sequence that can bring about a malignant transformation (see Chap. 17). In one approach, DNA was extracted from cells of a human bladder tumor. This DNA was then used to transform mouse cells to a malignant state. DNA from these malignant cells was extracted in turn and used to transform a second set of mouse cells. From these cells in the second cycle of transformation, a phage library was constructed.

The following reasoning was applied. The amount of human DNA in the second set of transformed mouse cells must be very small, because during the course of the gene transfer, starting with the original amount of human DNA through the second mouse transfer, much human DNA would have been lost. *Only* those cells that had been transformed were being selected on the plate (Fig. 19-26). In effect, only that DNA that had transforming ability was being selected for. Much of the DNA *not* involved in the malignant transformation would be discarded along with the cells that were not transformed. It was also reasoned that any human DNA fragment had a high probability of being associated with Alu sequences. Recall from Chapter 13 that these are widely scattered throughout the human chromosomes. A probe for the Alu sequence was available. The DNA in the phage library from the second round of infection could now be screened using the Alu probe.

The phage library would contain only a small number of phage particles carrying the transforming gene. This would be so because the DNA incorporated into the phages was extracted from mouse cells that had been transformed by a small amount of human DNA. The library was plated on *E. coli* cells (Fig. 19-27). The resulting plaques represented DNA clones, and these were replicated onto filter paper. A labeled Alu probe was then allowed to contact the plaques. Autoradiography revealed the location of any plaques containing DNA with Alu sequences. DNA from such a plaque was then used to infect mouse cells. It was found that this DNA could indeed transform cells to the malignant state. The DNA containing the genetic sequence for the human bladder cancer oncogene had thus been isolated.

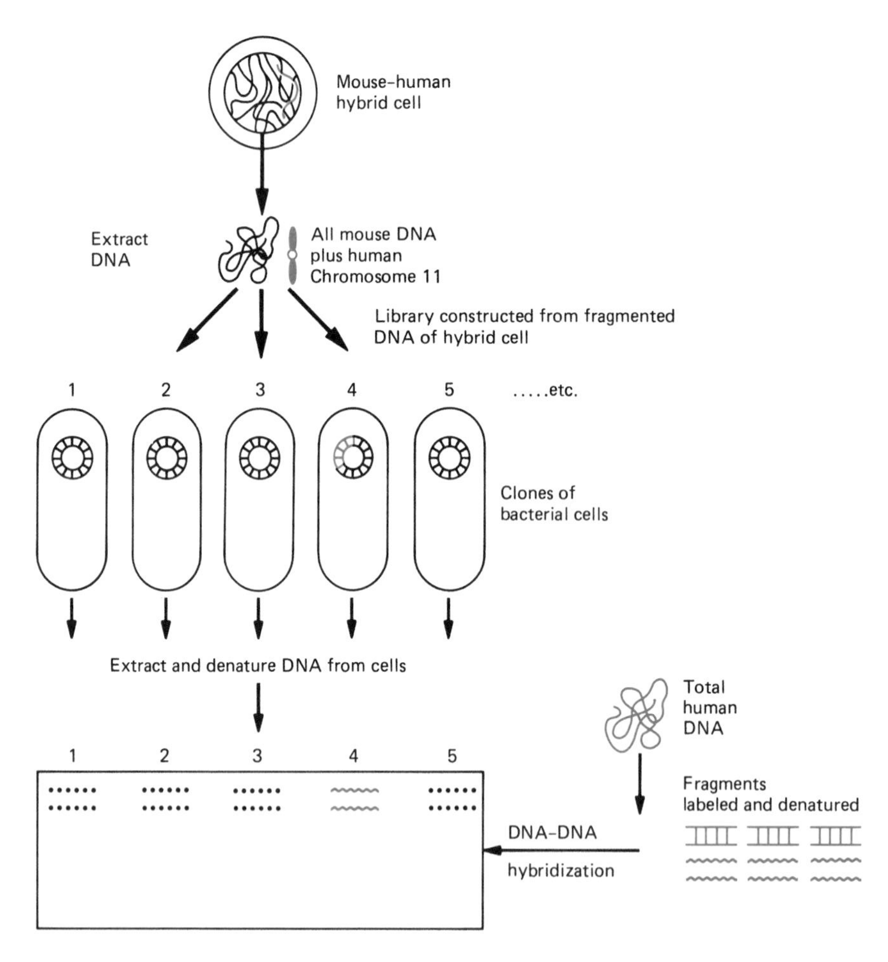

FIGURE 19-25. Obtaining a probe for human chromosome 11. DNA is extracted from mouse–human hybrid cells that carry, in addition to the mouse DNA, only human chromosome 11. A library is constructed from the DNA of the hybrid cells. Most of the clones of the transformed bacterial cells contain plasmids with mouse DNA only. The occasional clone with a plasmid containing an inserted piece of chromosome 11 can be recognized by performing DNA–DNA hybridization procedures between DNA from the clones and labeled denatured fragments from total human DNA. Only when a clone is harboring a piece of chromosome 11 does hybridization occur between the DNAs.

Gene transfer and DNA uptake

Especially exciting is the progress being made in gene transfer experiments. Several methods are available to introduce foreign DNA (donor DNA) into nuclei of recipient cells, frequently those of another species. One of the goals of such work is to obtain more information on the mechanism of gene control and expression. Another goal is to develop efficient methods for gene therapy to correct the genetic defect in individuals handicapped by certain inherited disorders. The amazing advances in recombinant DNA technology are greatly accelerating research in gene transfer and are making possible more efficient and precise ways of placing specific genes or DNA segments into recipient mammalian cells.

Three main methods are used. One of these, commonly referred to as transfection, entails incubation of a monolayer of recipient cells with donor DNA that has been precipitated with calcium phosphate. Although the reasons are not clear, it was noted in 1973 that DNA treated with calcium phosphate had a much greater chance than untreated DNA fragments of being taken up by the recipient cells. Although the calcium phosphate treatment enhances DNA uptake, the number of cells that acquire the donor DNA is nevertheless low. One problem is to locate those few cells that have incorporated the donor DNA. One way around this is to use donor DNA, which carries a gene enabling only those cells that have picked up the DNA to survive under certain conditions. For example, recipient cells that cannot produce thymidine kinase (tk^-) cannot grow on HAT medium (refer to Fig. 9-5). In a pioneer experiment, calcium phosphate–precipitated DNA carrying the tk^+ allele from the herpes simplex virus was incubated with tk^- mouse cells on HAT medium. Only transfected cells (those that took up the donor DNA carrying the tk^+ allele) could grow, permitting them to be isolated. If proper selection methods are available, a variety of marker genes in addition to the tk gene can be used to transfect cells.

It was soon discovered that *cotransformation* occurs in mammalian cells. Most cells, for example, that take up DNA with the tk^+ allele can also be shown to take up other genes at the same time—genes that were *not* being selected for by the medium. This was shown in the following way. If gene transfer is to be effective at all, a certain minimum amount of DNA must be used to treat donor cells. The efficiency of the DNA uptake is increased by using more DNA precipitate to layer over the recipient cells. Unexpectedly, it was found that tk^+ cells that had been selected on HAT medium also contained the carrier DNA. Obviously the selector marker (tk^+ in this case) and the unselected marker do not have to be linked on the same DNA fragment to be taken up together. In one experiment, two unlinked genes, tk and the rabbit gene for β-globin, were tied together after entering the cell. Once inside, the joined genes could integrate into a cell's chromosome.

It was found that certain cell lines are unstable and tend to lose the foreign genes they have taken up. These unstable lines lose the joined DNAs before integration takes place. Other lines are stable and retain the new DNA. In these lines, it was shown that the donor DNA has become integrated into a chromosome, a different chromosome in each of the cell lines. It was demonstrated that two originally separate genes are joined together *before* they are integrated. This is demonstrated by the fact that two added DNAs are always found next to each other on the chromosome. If each had integrated with the chromosome separately, each one would usually be found by itself, bounded on each side by the cellular genes of the recipient instead of next to each other. Moreover, it is now clear that the introduced DNA becomes part of a large molecule, a concatamer. Apparently any DNA in the precipitate (DNA with a marker, part of the carrier DNA, as well as specific genes from still other sources) can be taken up by cells, then joined together as a large molecule that may then be integrated into the chromosome (Fig. 19-28). We thus see that cells can be readily cotransformed. Evidently, any gene can be introduced into a mammalian cell growing in culture, along with a suitable selectable marker gene, although a given nonselectable gene is not necessarily expressed.

Gene transfer and micromanipulation

A second method used to introduce foreign genes into a cell entails direct injection of donor DNA into the nucleus of a recipient cell by means of a micropipette drawn out to a fine tip with a diameter of approximately 0.5 μm. Although this procedure requires special equipment and expertise in manipulation, its efficiency is much greater than that of direct uptake of calcium phosphate–precipitated DNA. Moreover, it permits insertion into the recipient of any DNA of choice. The recipient cell may be a body cell, or even a zygote. After being injected with foreign DNA, a mouse zygote, for example, may be inserted

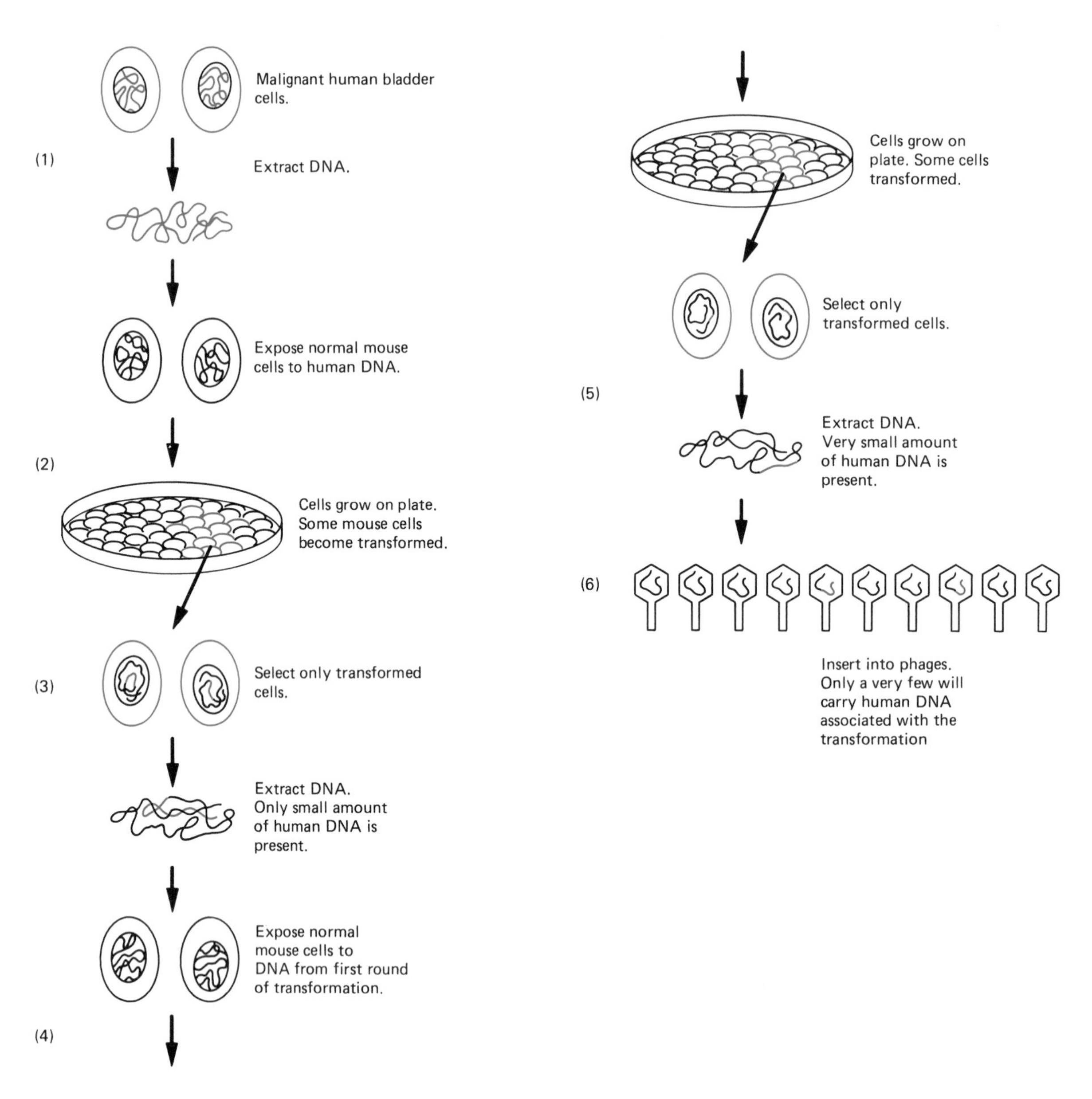

FIGURE 19-26. Transforming cells to the malignant state. Human DNA (red) is extracted from cells of a malignant tumor (1). This DNA is applied to normal mouse cells that take up the DNA. Some of them undergo a malignant transformation (2). These cells are selected, and their DNA is extracted (3). Less human than mouse DNA is present, because only malignant cells are selected. These must contain the human sequences for the malignant transformation. Any other human DNA in the untransformed cells is ignored. This extracted DNA is used to transform more normal mouse cells (4). Malignant cells are again selected, and their DNA in turn is extracted (5). Very little human DNA is now present, and again it must contain the sequences for the transformation. Any human DNA in the untransformed cells is again ignored. The DNA from the second round of transformation is used to establish a phage library (6). Most of the inserted pieces will be mouse DNA. Only a very few phages will carry the small amount of human DNA. This DNA is associated with transforming ability, because most of the rest of the human DNA was not selected during the transfers.

into the reproductive tract of a foster mother to complete its development.

A major goal of such experiments is to study gene expression during development. Recipient cells are typically injected with a solution containing plasmids into which cloned donor DNA has been inserted. Injecting a recipient cell with plasmids rather than with pure copies of the linear gene favors multiplication or amplification of the transferred donor sequence in the recipient cell. Some of these amplified sequences may be retained as the embryo develops, and some may become integrated into the host genome and even transmitted through the germ line. Although foreign DNA can be introduced into cells of experimental animals, getting the proper expression of the transferred gene is difficult. Therefore, an attempt is made to introduce foreign DNA that has a chance of being expressed not only in the recipient but also in the proper tissues in a regulated way.

The dramatic story of the "giant mice" illustrates many aspects of microinjection techniques. The investigators selected the rat growth hormone gene for introduction into mouse zygotes. Since regulated gene expression was desired in the recipient mouse, the rat growth hormone (GH) gene was modified. It was realized that achieving regulated expression would depend on the proper regulatory sequences associated with the introduced gene. Since the structural gene for the GH was associated with rat regulatory sequences, it might not respond properly to mouse controls in the recipient mouse cells. Therefore, using recombinant DNA procedures, the rat gene was altered by genetic engineering. The regulatory sequences at the 5′ end (the end of the gene where transcription begins) was replaced with a corresponding region from a mouse gene, the metallothionein (MT) gene.

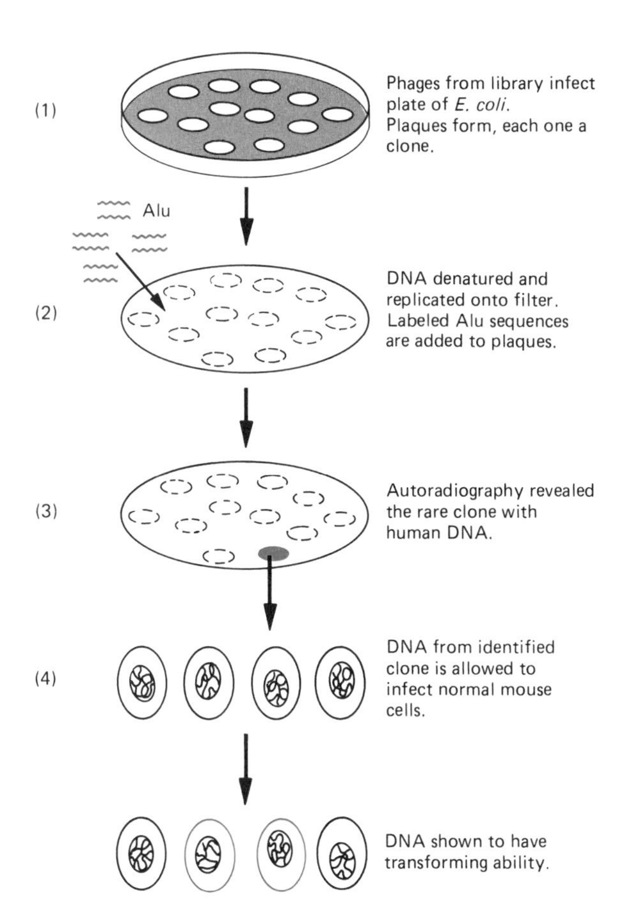

FIGURE 19-27. Detecting DNA with an oncogene. Phages from the library (Fig. 19-26) are allowed to infect plates of *E. coli* cells (1). Plaques form, each one a DNA clone. DNA is denatured and transferred to a filter (2). Each DNA clone is exposed to a labeled probe for the repetitive human Alu sequence. The rare clone with the human DNA can then be identified by autoradiography (3). DNA from this clone, when added to normal mouse cells, has transforming ability (4). The oncogene for the human bladder cancer must be in the DNA of this clone. Other clones lack transforming ability.

The MT story is of great interest by itself. Actually two linked structural genes, MTI and MTII, are found at the MT locus. These genes are coded for two distinct proteins known as *metalloproteins.* These low-molecular-weight proteins are found in species varying from blue-green algae to humans, a fact that underscores the importance of their role in the cell. These proteins have a strong affinity for various trace metal ions (copper, zinc, cadmium) and play key parts in regulating the level of essential trace elements in the cell. By binding to cadmium, metalloproteins prevent it from accumulating in a toxic form. Each MT gene is associated with a strong promoter that acts as an active switch or control region to be turned on or off, depending on the concentration of certain metallic ions in the environment. For example, an appropriate concentration of Zn^{2+} induces metalloprotein synthesis.

After the switch that regulates the MI gene of the mouse was located, it was joined to several other unrelated genes by recombinant DNA methods and then used in various gene transfer experiments. It was demonstrated that the expression of the recombinant genes could be induced by exposing cells to proper doses of Zn^{2+} and Cd^{2+}. Because it can respond to metals and certain other inducers, transcriptional regulation of the recombinant structural genes is possible in these systems. The MT promoter or switch proved to be an excellent means of manipulating the expression of other genes joined to it.

Armed with the modified rat GH gene, which was now combined with the mouse MT switch (Fig. 19-29), the investigators injected about 600 copies of the fusion gene into each of almost 200 recently fertilized eggs. These were then implanted into foster mothers, and 21 mice developed, 7 of which had one

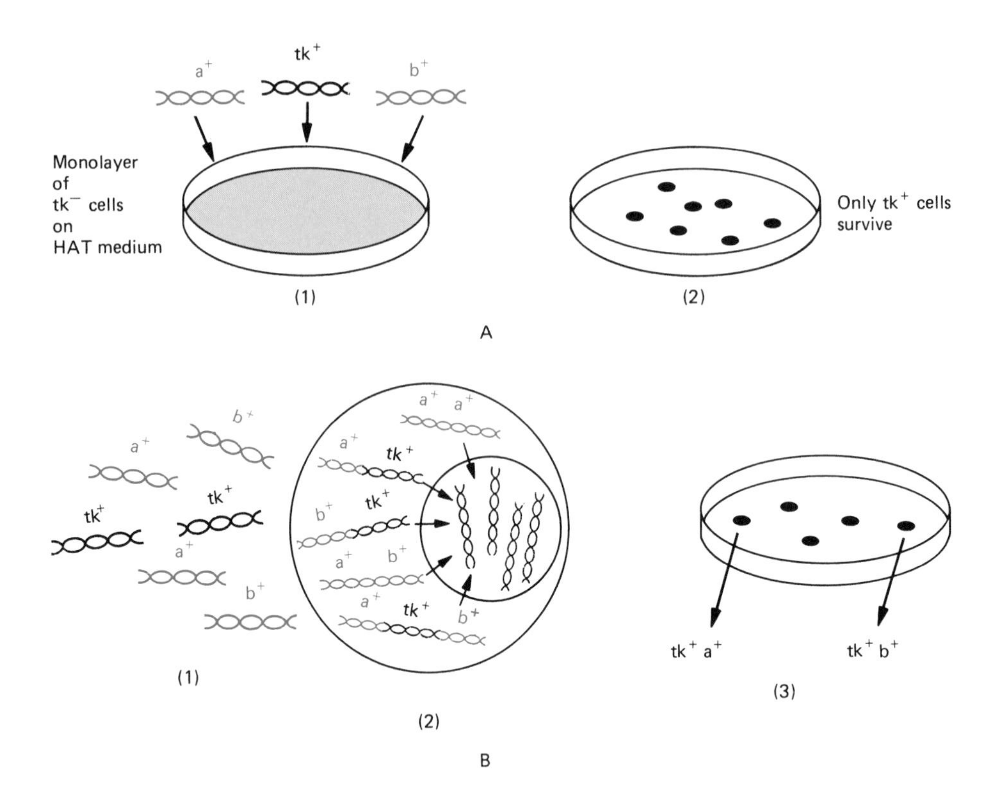

FIGURE 19-28. DNA uptake and cotransformation. **A:** (1) A monolayer of tk$^-$ cells may be incubated with calcium phosphate–precipitated DNA, which carries the tk$^+$ allele. However, a certain minimum amount of DNA must be present in the precipitate if DNA uptake is to be efficient. The DNA of the carrier DNA (red) used to increase efficiency may also contain certain alleles, e.g., a$^+$, b$^+$. (2) The only cells that survive on HAT medium are those that have taken up the DNA carrying the tk$^+$ allele. Any carrier DNA that entered along with the tk$^+$ allele is not selected for by the medium, and so it is not now apparent if any genes associated with the carrier DNA (a$^+$, b$^+$) are present in the tk$^+$ cells. **B:** (1) Actually, DNA from several sources (the DNA with the selected marker as well as any other DNA in the precipitate) may enter a cell. (2) DNAs from different sources are joined together after entering the cell to form a larger molecule. This means that carrier DNA can be joined to DNA with the marker. (Only some arrangements of the alleles in the large molecule are shown here.) Once formed, the larger molecule may become integrated into one of the chromosomes of the host cell. (3) Those tk$^+$ cells that survive on the HAT medium can then be treated further. Most tk$^+$ cells can be shown to contain the a$^+$, b$^+$ or some allele that was not selected for on the HAT medium.

or more copies of the modified GH gene. At a young age, the mice were placed on a diet containing zinc to stimulate the MT switch combined with the GH gene. Six of the seven mice grew significantly faster than did the others in the litter. Greater than normal levels of GH were produced in these mice, resulting in gigantism. The high level of GH in the animals corresponded to a high level of GH mRNA in the liver and increased levels of GH in the serum. All the data indicate that the altered mouse phenotype results from the integration and expression of the modified GH gene. The gene was transmitted and inherited by 10 of the 19 offspring of one mouse and has been followed to at least the seventh generation.

Although the transfer and expression of the recombinant gene have been successful, we must note that the normal location of expression of the GH gene is the pituitary, whereas here it is the liver. Moreover, certain controls are apparently not operating properly, because higher than normal hormone levels were produced, resulting in gigantism, which indicates the failure of some feedback mechanism in liver cells. Nevertheless, such an exciting accomplishment has many implications besides those for studies of development and gene expression. The approach may lead to a means of correcting genetic defects, a way of accelerating animal growth, and a way of injecting desirable genes into farm animals to produce new breeds. Controlled, tissue-specific expression of genes in transgenic animals still remains a central problem in this approach. However, some degree of success has been reported in several experiments of this type. In one case, a rat gene for a pancreatic enzyme was transferred by microinjection into mice. In four out of five of the transgenic animals, the enzyme was expressed selectively in the pancreas at levels characteristic of the differentiated state of the gland.

Viral vectors and gene transfer

A third method employed in gene transfer involves retroviral vectors to deposit foreign DNA into animal cells. Such success has been achieved with this procedure that it is now being applied to humans. We know a great deal about how a given virus enters and leaves a cell, and it is a simple task to infect cells with many virus particles. If the virus has been altered and is carrying a chosen foreign DNA, this makes it likely that many copies of the desired DNA will be introduced into recipient cells along with the virus. Retroviruses are especially attractive as vectors. In their double-stranded DNA form, they integrate into the host's chromosome and behave as a part of it. Moreover, integrated retroviruses have very strong promoters located in the long terminal repeats at either end of the virus. (We have just seen, in the preceding section, the value of a strong promoter in the case of the MT gene.) If the promoter of the virus is activated to bring

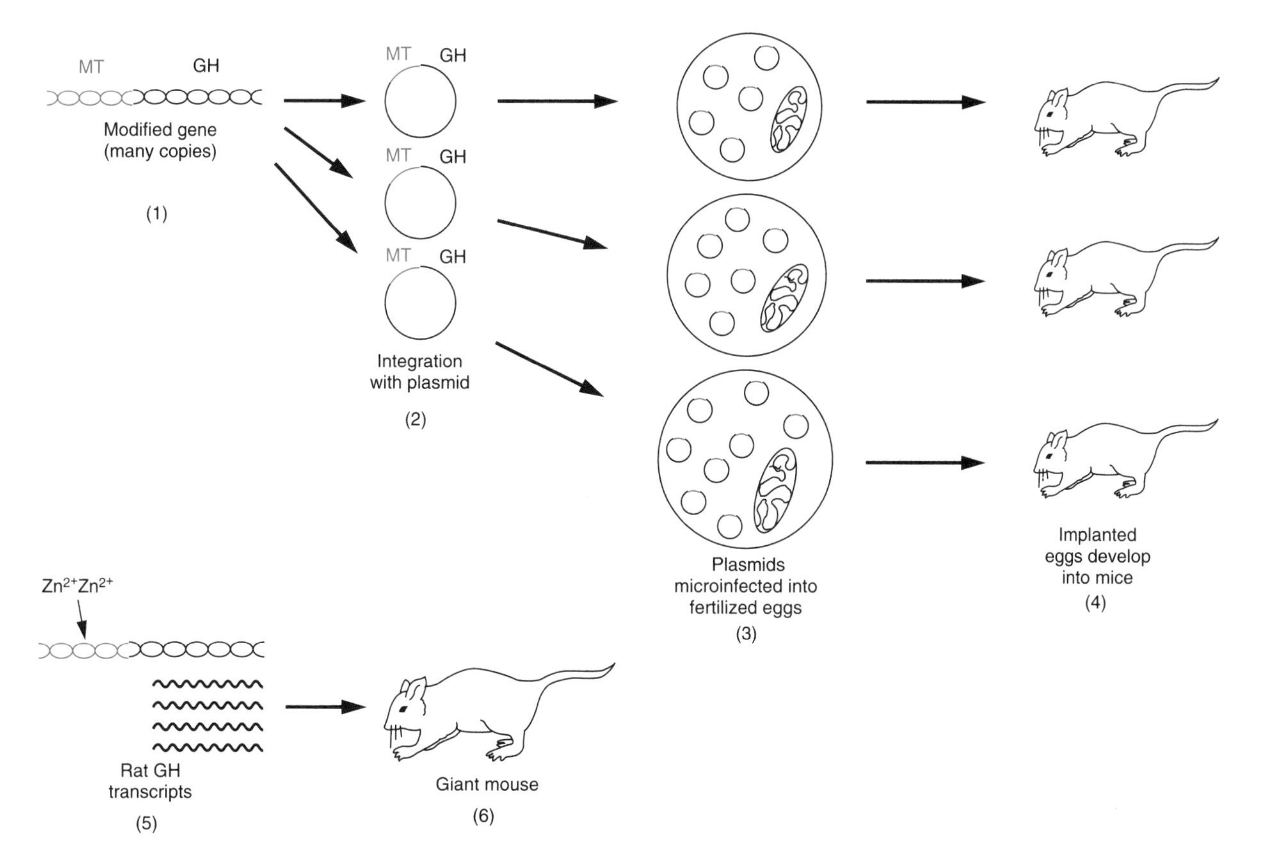

FIGURE 19-29. Transfer of a gene and its expression in host. **1:** With the use of recombinant DNA procedures, the rat growth hormone (GH) gene can be combined with the mouse MT switch, a strong promoter that responds to Zn^{2+}. **2:** The modified gene is then integrated with a plasmid. **3:** Plasmids containing the modified gene are then microinjected into newly fertilized eggs. Integration of the donor DNA into a plasmid favors its amplification in the recipient cells. **4:** After implantation into foster mothers, some of the treated eggs develop into mice that contain copies of the modified gene. **5:** Zn^{2+} in the diet of the mice with the altered gene causes stimulation of the MT promoter. **6:** A high level of transcripts of the rat GH is produced, which results in a high level of rat growth hormone in some mice, which are conspicuously larger than those lacking high levels of the hormone.

about transcription, then any foreign gene spliced into the virus will also be transcribed.

Retroviruses that have been integrated into cellular DNA can be isolated. A genetic region may be cut out using the proper enzymes and a cDNA of a specific foreign gene inserted in its place. This recombinant virus can then be cloned in bacteria using standard cloning procedures (Fig. 19-30). The goal is to establish a cell line that harbors a recombinant virus carrying a desirable gene and that can be used as a ready source of the virus to infect other cells, such as bone marrow cells.

But a problem with the recombinant virus becomes evident. Although it can perform all the necessary retroviral functions, it remains noninfective. It cannot produce protein envelopes that will enable it to leave the cell and enter another cell. Somehow we must enable noninfective engineered viruses to leave cells from an established stock and to enter other cells (e.g., bone marrow cells), insert the desired gene, and *do no more.* A helper virus can be called on to enable us to establish a cell line carrying such an engineered retrovirus. A helper virus in the same cell with the defective recombinant virus will supply the coat needed by the defective virus to replicate and to enter other cells. However, another problem becomes obvious. The helper virus supplies coats for itself and the defective virus. Both of these are then free to leave and enter other cells indiscriminately. This is not at all desirable; it means that the vector can enter human cells, leave, and transfer itself and the inserted gene into any tissue, perhaps with fatal consequences. Therefore, the vector must be able to enter a cell but not reproduce.

Amazingly, this problem has been solved with the discovery that a specific segment of the retroviral genetic material is required for the virus to package or wrap itself in its protein envelope. Missing the genetic segment with the packaging gene, such a retrovirus, although it has the gene to produce protein for the envelope, cannot be wrapped in its coat and cannot leave the cell. Employing molecular manipulation, a team at MIT was able to construct a retrovirus-packaging mutant that can be used to establish a cell line yielding a source of helper-free defective retroviruses (follow Fig. 19-31). The helper supplies the protein coat for the recombinant, defective virus, but cannot itself be wrapped in the coat, because it lacks the packaging gene. Therefore it cannot leave the cell or enter others. The recombinant virus can get into another cell and insert into the chromosome, but it is unable to leave, because it lacks the genes required to make the protein coat, even though it does have the packaging gene. But now there is no helper virus to supply the protein. Such systems utilizing retroviruses to transfer genetic material into cells have been designed in several laboratories and are now in current use in human gene transfer studies.

In 1989, the first authorized gene transfer experiment was performed on a human, a patient with malignant melanoma, the deadliest form of skin cancer. The patient received an infusion of tumor-infiltrating leukocytes (TILs) that had been genetically engineered. In an approach to the treatment of melanoma, TIL cells are surgically removed from a patient's tumor and are cultured in the laboratory for weeks with a potent growth factor that stimulates the immune system, interleukin-2 (IL-2). The TILs are then introduced back into the patient. A significant feature of TILs is that they are selective and tend to attack tumors from which they were obtained. When returned to a patient's body, IL-2–treated TILs have the ability to infiltrate specific tumors and in some cases successfully shrink them.

It was important for clinical researchers to find out exactly where TILs go when they are returned to the body. To do this, they used a retrovirus as a vector to deliver to the TILs a marker gene, one for resistance to the antibiotic neomycin. In a manner similar to that explained above, TILS growing in culture were infected with the engineered virus that carried the marker gene and deposited it into the DNA of the TILs. Being defective, the virus was unable to replicate itself. In this pioneer experiment, the transferred marker gene provided no therapy for the patient but gave the scientists a way to track the journey of the TILs in the body. This first test also showed that foreign genes can safely be introduced into a person's lymphocytes where it becomes expressed. It also paved the way for actual gene therapy to cure disease.

Now in progress are attempts to treat some very serious human afflictions utilizing engineered retroviruses. One of the trials involves mouse leukemia retroviruses carrying the human gene for tumor necrosis factor (TNF). This factor, when injected into mice, brings about an almost immediate shrinkage of tumors by its ability to shut off the blood vessels that supply them. Unfortunately, the success in animals has not been duplicated in humans. The goal is to deliver larger doses of TNF to patients with malignant melanoma by infecting their IL-2–cultured TILs with retroviruses carrying the TNF gene. The TILs are then transferred back into the patient. The hope is that sufficient numbers of the TILs, armed with the TNF gene, will zero in on the tumors and that sufficient amounts of TNF will be produced to destroy them. The actual trials have thus far shown that the TILs are indeed going to the patients' tumors and that normal tissue is left intact. Some success with tumor shrinkage has also been reported. One fear of this type of gene therapy has been that a retrovirus used as a vector could conceivably integrate at a site in a cell's DNA near an oncogene and then turn it on, triggering a malignant transformation. In all the trials conducted thus far, the virus has proved harmless to the patient.

The focus of another gene therapy trial is the very rare but fatal immune disorder adenosine deaminase deficiency (ADA). ADA is a form of SCID (see Chap. 5) and makes up about one-half of these severe immune system disorders, which are inherited on an autosomal recessive basis. The ADA gene is critical to the development of B and T lymphocytes. Those with ADA are immune deficient because the ADA enzyme is needed to degrade nucleic acids. In its absence, adenosine and deoxyadenosine accumulate, a buildup that for unknown reasons proves especially toxic to B and T cells.

In the gene therapy protocol, T cells are extracted from the young victim of ADA and then exposed to engineered mouse leukemia retroviruses carrying the ADA gene. About 1 billion of the T cells are then infused into the child. The amount of ADA needed for a healthy immune system is very variable. Once the ADA gene is incorporated into the T cell's DNA, it need only be expressed in moderate amounts. Since no complications are involved with regulatory mechanisms, ADA is an excellent choice for a gene therapy study.

It would be ideal to add the ADA gene to stem cells of the bone marrow, those cells that give rise to the red and all the white cell types. If a retrovirus were to insert the gene into stem cells' DNA, repair would be permanent, since those cells and all of their progeny would possess the normal gene. Unfortunately, stem cells are evasive, very rare, and thus unavailable in large quantities; they are also very difficult to infect with the viruses. Until such problems can be overcome, gene therapy is being pursued with other cell

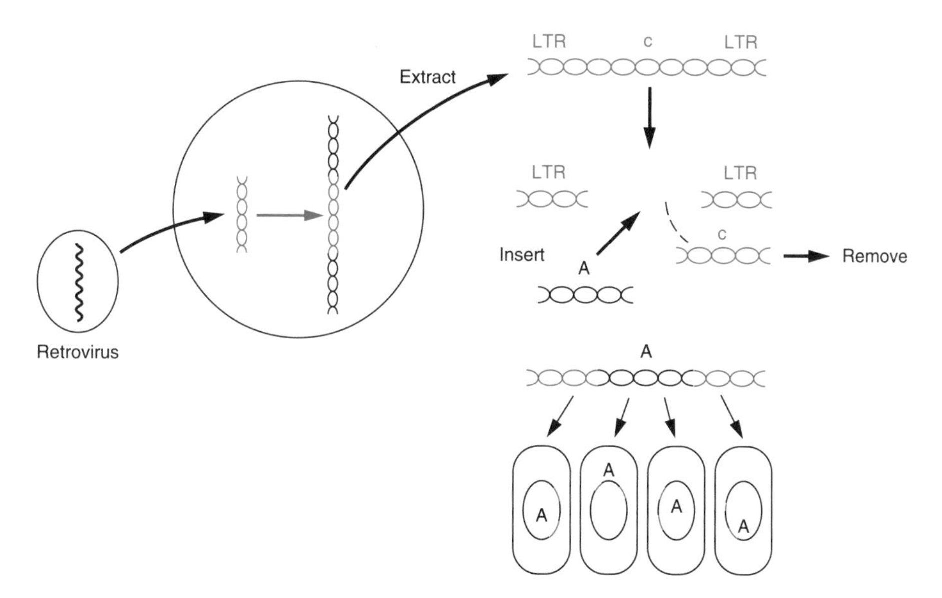

FIGURE 19-30. Cloning a desired gene in a retrovirus. A retrovirus infects a mouse cell, and its RNA acts as a template for synthesis of a DNA duplex copy of the viral genome (red), which inserts into the chromosome of the cell. Viral DNA can be extracted from infected cells and then processed using recombinant DNA procedures. A number of viral genes can be snipped out, including gene c, required for the formation of the protein coat of the virus. The virus has long terminal repeats (LTR) at each end, which contain promoters. A desired gene, A, is inserted between the segments with the promoter regions. The engineered recombinant virus with the desired gene is cloned in bacteria for use as needed. The virus, however, is now defective. Because it lacks gene c, it cannot make coat protein, and therefore cannot produce infectious viral particles.

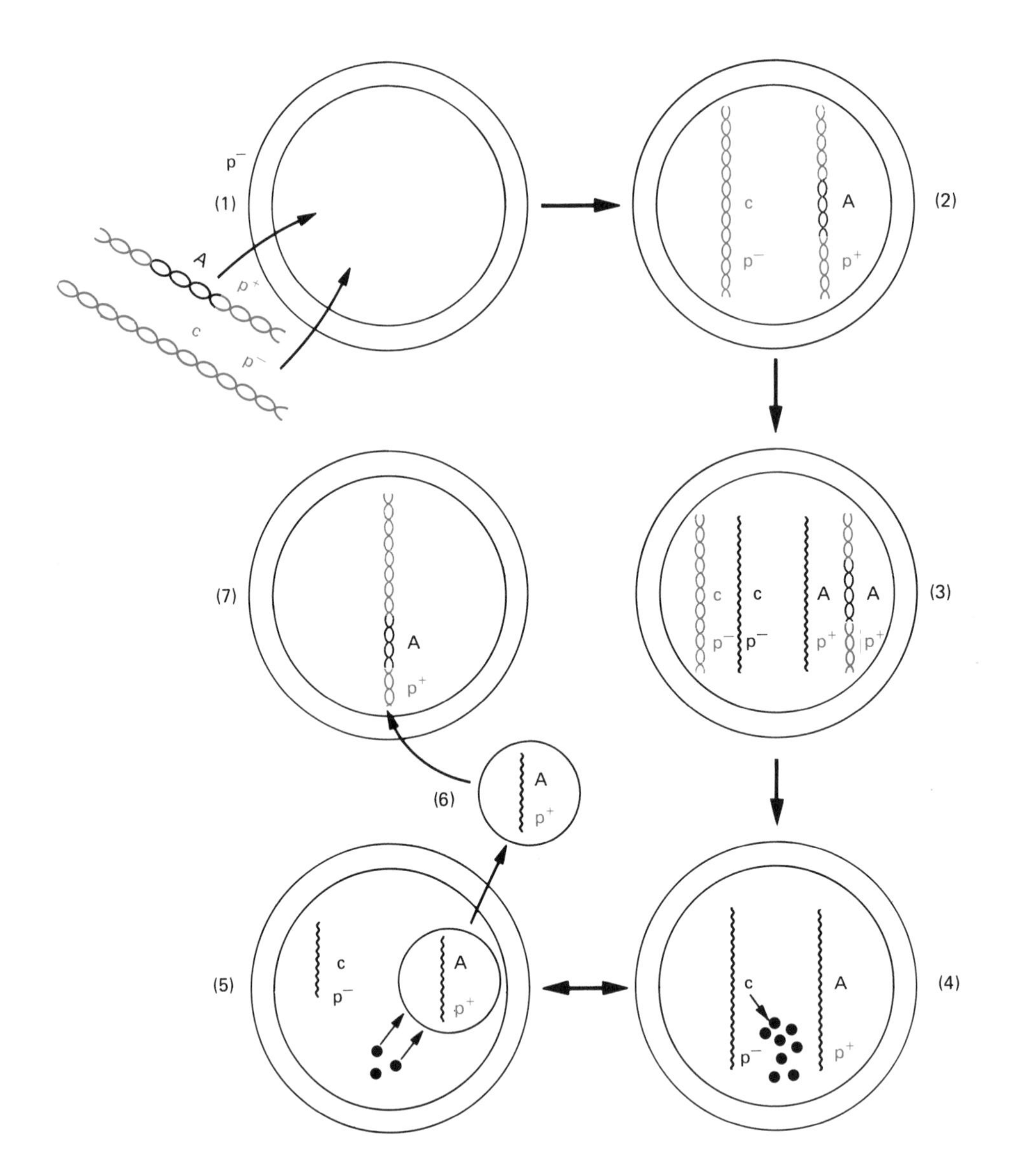

FIGURE 19-31. Establishment of cell lines to supply noninfective retrovirus free of helper. **1:** Cells are incubated with many copies of the engineered gene and also with DNA copies of a helper virus that contain gene c to make coat protein but lack the gene (p^-) necessary for wrapping the protein around the virus. The DNA of the engineered virus carries this gene (p^+), even though it lacks the coat protein gene c. **2:** Some DNA of both viruses is taken up by cells and inserted into the chromosomes (not shown). **3:** Both the helper virus and the one with the desired gene (A) undergo transcription, making RNA copies (single dark lines). **4:** RNA copies of the helper are translated and coat protein is produced (dark dots). **5:** The helper cannot package the protein and so remains noninfective. The viral RNA with the desired gene can wrap itself in the coat protein, produce infective particles, and leave the cell. **6:** Such a virus from this cell line can then readily enter other cells, carrying the desired gene along with it. **7:** Once inside another cell, it again makes a duplex DNA copy that inserts into a chromosome. However, the virus can no longer become infective and leave the cell, because it cannot bring about the formation of coat protein. Free of helper virus, the cell contains the desired gene (A), which will be replicated along with the rest of the chromosomal DNA.

types such as TILs and T cells to deliver the desired genes. This means that a patient receiving the engineered T cells with the ADA gene is not cured but must receive periodic infusions. Various kinds of cells in addition to lymphocytes are being considered as candidates to deliver engineered genes to proper sites in the body. Endothelial cells, for example, have shown their potential since they interact with many components of the circulatory system.

Thus far the trials have demonstrated the feasibility of gene therapy, and already a host of other protocols designed to treat a variety of diseases is awaiting approval. Disorders, such as emphysema and Parkinson's disease, that have no clear-cut genetic basis are also targets for gene therapy, and high on the list for this approach is AIDS. Still in the experimental stage with mice, the procedure to outwit the AIDS virus calls for the transfer of a cloned gene to lymphocytes by way of harmless retroviruses. The gene to be incorporated is coded for the soluble protein CD4. In its attack on cells, the AIDS virus binds to specific sites, CD4 receptors, found on the cell surface. The soluble CD4 protein is able to block the virus from entering cells. Perhaps delivery of sufficient amounts of the protein to susceptible cells can thwart an AIDS infection.

In the not too distant future, complex disorders that affect large numbers of the human population and that probably have a complex genetic basis (atherosclerosis, diabetes) may yield to treatment provided by advances in gene therapy.

REFERENCES

Botstein, D., and D. Shortle. Strategies and applications of in vitro mutagenesis. *Science* 229:1193, 1985.

Burke, D.T., G.F. Carle, and M.V. Olson. Cloning of large segments of exogenous DNA into yeast by means of artificial chromosome vectors. *Science* 236:806, 1987.

Culliton, B.J. Gene therapy: Into the home stretch. *Science* 249:974, 1990.

Culliton, B.J. Mapping terra incognita (humani corporis). *Science* 250:210, 1990.

Danna, K.J., G.H. Sack, and D. Nathans. Studies of simian virus 40 DNA. VII. A cleavage map of the SV40 genome. *J. Mol. Biol.* 78:363, 1973.

D'Eustachio, P. Gene mapping and oncogenes. *Am. Sci.* 72:32, 1984.

D'Eustachio, P., and F.H. Ruddle. Somatic cell genetics and gene families. *Science* 220:919, 1983.

Friedman, T. Progress toward human gene therapy. *Science* 244:1275, 1989.

Garza D., J.W. Ajioka, D.T. Burke, and D.L. Hartl. Mapping the *Drosophila* genome with yeast artificial chromosomes. *Science* 246:641, 1989.

Gilboa, E., M.A. Eglitis, P.W. Kantoff, and W.F. Anderson. Transfer and expression of cloned genes using retroviral vectors. *BioTechniques* 6:504, 1986.

Gordon, J.W., and F.H. Ruddle. Gene transfer into mouse embryos: Production of transgenic mice by pronuclear injection. *Methods Enzymol.* 101:411, 1983.

Green, P.J., O. Pines, and M. Inouye. The role of antisense RNA in gene regulation. *Annu. Rev. Biochem.* 55:569, 1986.

Khorana, H.G. Total synthesis of a gene. *Science* 203:614, 1979.

Kornberg, A. The synthesis of DNA. *Sci. Am.* (Oct) :64, 1968.

Kucherlapati, R.S. Introduction of purified genes into mammalian cells. *ASM News* 50:49, 1984.

Maniatis, T., E.F. Fritsch, and J. Sambrook. *Molecular Cloning: A Laboratory Manual.* Cold Spring Harbor Laboratory, Cold Spring Harbor, NY, 1982.

Maniatis, T., R.C. Hardison, E. Lacy, J. Lauer, C. O'Connell, D. Quon, G.K. Sim, and A. Efstratiadis. The isolation of structural genes from libraries of eucaryotic DNA. *Cell* 15:687, 1978.

Mann, R., R.C. Mulligan, and D. Baltimore. Construction of a retrovirus packaging mutant and its use to produce helper-free defective retrovirus. *Cell* 33:153, 1983.

Maxam, A.M., and W. Gilbert. A new method for sequencing DNA. *Proc. Natl. Acad. Sci. USA* 74:560, 1977.

Merriam, J., M. Ashburner, D.L. Hartl, and F.C. Kafatos. Toward cloning and mapping the genome of *Drosophila. Science* 254:221, 1991.

McKusick, V.A. Mapping and sequencing the human genome. *N. Engl. J. Med.* 320:910, 1989.

Myers, R. M., T. Tilly, and T. Maniatis. Fine structure analysis of a β-globin promoter. *Science* 232:613, 1986.

Nathans, D. Restriction endonucleases. Simian virus 40 and the new genetics. *Science* 206:903, 1979.

Prober, J.M. A system for rapid DNA sequencing with fluorescent chain-terminating dideoxyribonucleotides. *Science* 238:336, 1987.

Robins, D.M., S. Ripley, A.S. Henderson, and R. Axel. Transforming DNA integrates into the host chromosome. *Cell* 23:29, 1981.

Ruddle, F.H. A new era in mammalian gene mapping: Somatic cell genetics and recombinant DNA methodologies. *Nature* 294:115, 1981.

St. Louis, D., and I.M. Verma. An alternative approach to somatic cell gene therapy. *Proc. Natl. Acad. Sci. USA* 85:3150, 1988.

Sanbrook, J., E.F. Fritsch, and T. Maniatis. *Molecular Cloning—A Laboratory Manual,* 2nd ed. Cold Spring Harbor Laboratory, Cold Spring Harbor, NY, 1989.

Sanger, F. Determination of nucleotide sequence in DNA. *Science* 214:1205, 1981.

Sanger, F., and A.R. Coulson. A rapid method for determining sequences in DNA by primed synthesis with DNA polymerase. *J. Mol. Biol.* 94:441, 1975.

Smith, M. The first complete nucleotide sequencing of an organism's DNA. *Am. Sci.* 67:57, 1979.

Stephens, J.C., M.L. Cavanaugh, M.I. Gradie, M.L. Mador, and K.K. Kidd. Mapping the human genome: Current status. *Science* 250:237, 1990.

Swift, G.H., R.E. Hammer, R.J. MacDonald, and R.L. Brinster. Tissue-specific expression of the rat pancreatic elastase gene in transgenic mice. *Cell* 38:639, 1984.

van der Krol, A.R., J.N.M. Mol, and A.R. Stuitje. Modulation of eukaryotic gene expression by complementary RNA or DNA sequences. *BioTechniques* 6:958, 1988.

Van Embden, J. The use of cosmids as cloning vehicles. In *Techniques in Molecular Biology.* MacMillan, New York, 1983.
Verma, I.M. Gene therapy. *Sci. Am.* (Nov) :263:68, 1990.
Weintraub, H.M. Antisense RNA and DNA. *Sci. Am.* (Jan):401, 1990.

REVIEW QUESTIONS

1. Khorana's artificial gene is a DNA unit 207 nucleotide pairs long. Suppose this DNA were hybridized with the tRNA for which it is coded. About what percentage of the DNA would you expect to form hybrids with the RNA? Why?

2. A viral DNA is subjected to the restriction enzyme Alu. Five fragments are produced and separated by gel electrophoresis. These are designated Alu A, B, C, D, and E, respectively, in order of decreasing size. When the same viral DNA is subjected to the restriction enzyme Hpa, only two fragments are produced, Hpa A and Hpa B. Hpa A is isolated and subjected to Alu. Two fragments are found on the gels that correspond in position to Alu B and Alu D. When the viral DNA is subjected to Alu and then to the restriction enzyme Hae, the gels show fragments that correspond in position to Alu A, B, C, and D. Two additional very small fragments are found. What information do we now have about fragments Alu A, B, C, D, and E?

3. A DNA sequence 10 kb long contains one site at which it may be cut by the restriction enzyme PstI. The same 10 kb long sequence also contains one site that can be cut by the restriction enzyme BglII. Three separate fractions of this DNA were treated separately, as follows: 1) digestion by PstI; 2) digestion by BglII; 3) digestion by both enzymes. The three treated fractions were then subjected to gel electrophoresis. The following table shows the lengths of the restriction fragments associated with each treated fraction. From the results, draw a restriction map that shows the PstI and the BglII cutting sites in the 10 kb DNA sequence.

Enzyme in mixture	Fragment lengths (kb)
PstI	6.4, 3.6
BglII	8.8, 1.2
PstI + BglII	5.2, 3.6, 1.2

4. A particular virus has a duplex DNA genome in the form of a circle. This circular molecule can be attacked by the restriction enzymes HpaI, HindIII, and EcoR1. Each enzyme attacks the circular DNA at only one site, which is specific for that enzyme. The circular genome is cut first with HpaI, yielding a linear duplex. This linear molecule was then subjected to further enzyme treatment with the production of restriction fragments. The results are shown in the following table:

Enzyme treatment	Fragment lengths (kb)
HpaI	5.6
HpaI + HindIII	3.0, 2.6
HpaI + EcoR1	4.8, 0.8
HpaI + EcoR1 + HindIII	2.6, 2.2, 0.8

From these results, do the following:

A. Draw a restriction map of the genome after it has been linearized, showing the HindIII and the EcoR1 restriction sites relative to the HpaI site.
B. Draw the restriction map as a circle, indicating all three restriction enzyme cutting sites.

5. Let us suppose that a ts (temperature sensitive) mutant strain arises from the virus described in Question 4. How could you establish where the mutant site is located on the circular restriction map derived? (See answer 4B.)

6. The restriction enzyme BamI recognizes the following sequence in which the arrows indicate the cut sites:

↓
5′–GGATCC–3′
3′–CCTAGG–5′
↑

The enzyme was allowed to attack a cloned DNA sequence from a mouse genome. Two fractions were obtained following BamI digestion of this genomic sequence. Each of these fractions contained a different duplex restriction fragment. The duplex fragment digest in each fraction was melted to single strands, and only one strand from each fraction was sequenced by the Sanger method. The gel patterns of the two fragment digests were as follows:

Digest 1

ddG	ddA	ddT	ddC
		—	
		—	
—			
			—
	—		
—			
			—
		—	
		—	
			—
			—
		—	
	—		
—			

Digest 2

ddG	ddA	ddT	ddC
	—		
		—	
	—		
—			
		—	
			—
			—
		—	
	—		
—			

Determine the following:

A. The DNA sequence from digest 1.
B. The DNA sequence from digest 2.
C. The nucleotide sequence of the cloned duplex DNA before it was cut by the enzyme. (Show the 5′–3′ orientation.)
D. The nucleotide sequences of the two duplex fragments after the cut. (Show the orientation.)

7. Suppose the strand complementary to the one sequenced in digest 2 (Question 6) had been the strand that was analyzed. What pattern would result when an autoradiograph is prepared?

8. The following is an RNA transcript of a segment of DNA you wish to clone:

5′–CCAUUUCGAAUGGAGCAA–3′

Double-stranded DNA can be derived from this transcript with reverse transcriptase and DNA polymerase.

A. Show the derived DNA and how it can be provided with projecting ends employing the appropriate enzymes and thymidine triphosphate.
B. Show how you would then process the plasmid selected to carry this DNA.

9. Following is a segment of DNA containing a palindrome.

C T A C G A A T T C G T C T
G A T G C T T A A G C A G A

A restriction enzyme that recognizes this produces breaks four nucleotides apart from the axis of symmetry in each strand. The break is to the left in the top strand and to the right in the lower one. Show the fragments that can be generated with complementary single-stranded projections.

10. Following is a segment of a circular plasmid.

G A C T A T T A C G T A A T A G C A
C T G A T A A T G C A T T A T C G T

A restriction enzyme specific for the palindrome present produces breaks four nucleotides apart from the axis of symmetry, to the left in the inner or upper strand, and to the right in the outer, or lower one.

A. Show the ends that can be generated.
B. Could the fragments generated in Question 9 be inserted? Explain.

11. Assume that you have a number of bacterial clones, each of which contains human DNA inserted into a plasmid. The goal is to locate any clones that contain the entire gene for the production of human enzyme Z. DNA is taken from clones 1 through 5 and is rendered single stranded. It is then added to systems containing all the requirements for polypeptide formation, including labeled amino acids. The RNA transcripts present were taken from cells of the tissue in which the enzyme is produced. Following are the observations made after addition of anti-Z antibody to each of the five systems (+ = precipitate; – = no precipitate):

Clone or system				
1	2	3	4	5
–	–	+	+	+

A. What clones contain the gene coded for enzyme Z? Explain your reasoning.
B. Are any polypeptides formed in systems 1 and 2? Explain.

12. A. 3′–CATTGGCA–5′ represents a nucleotide sequence found on an antisense DNA strand. Give the corresponding sequence and its polarity on the antisense RNA.
B. 5′–GTTAGACA–3′ is a sequence on a sense DNA strand. Give the corresponding sequence on the mRNA and its polarity.
C. 5′–GAACUCUG–3′ is a sequence on a mRNA transcript. Give the corresponding sequences and their polarities on 1) the sense DNA strand, 2) the antisense DNA strand, and 3) the antisense RNA.

13. A type of plasmid used frequently as a vector in recombinant DNA procedures carries gene markers for resistance to tetracycline (Tet^r) and to ampicillin (Amp^r). In one experiment, such plasmids were exposed to a restriction enzyme that cuts within the Amp^r gene but leaves the Tet^r gene intact. Plasmids exposed to the enzyme were next mixed with genomic restriction fragments from a cloned library of mouse DNA. The aim was to produce and detect recombinant DNA among transformed drug-sensitive *E. coli* cells. Following their exposure to the treated plasmids, *E. coli* cells were plated on nutrient media and allowed to form colonies. Replica plates of these colonies were made on selective drug-containing media to find bacterial colonies that carry recombinant DNA. The results are as shown:

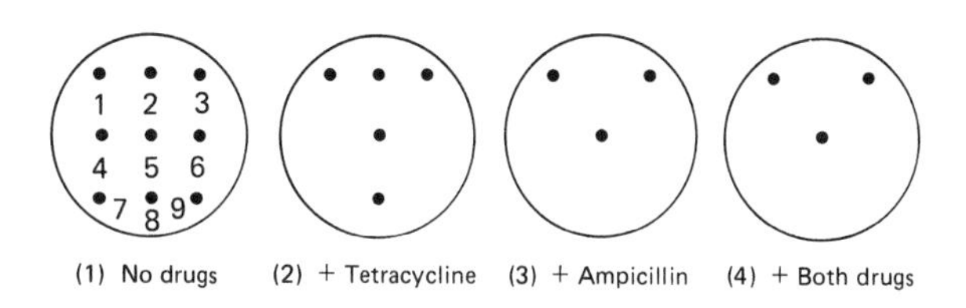

Answer the following questions, giving a brief explanation:

A. Which colonies have neither plasmid nor recombinant DNA?
B. Which colonies carry intact plasmid DNA but no recombinant DNA?
C. Which colonies carry recombinant DNA?

14. Suppose you have a plate on which several transformed *E. coli* colonies are growing. You wish to identify any colony that carries the mouse gene that codes for thymidine kinase. Briefly outline steps that would enable you to locate such a colony.

15. Let us suppose that one wishes to assign to a particular human chromosome the gene coded for a certain critical enzyme. DNA is extracted from human cells, mouse cells, and hybrid human–mouse cells that carry different combinations of human chromosomes. The DNA from the various sources is cut by a specific restriction enzyme, and the fragments are separated by gel electrophoresis on the basis of size. Following Southern blotting and denaturation of the fragments on a filter, the fragments are exposed to a labeled cDNA probe of the gene. Autoradiography is performed with the following results. (The numbers represent the human chromosomes present in the hybrid cells.)

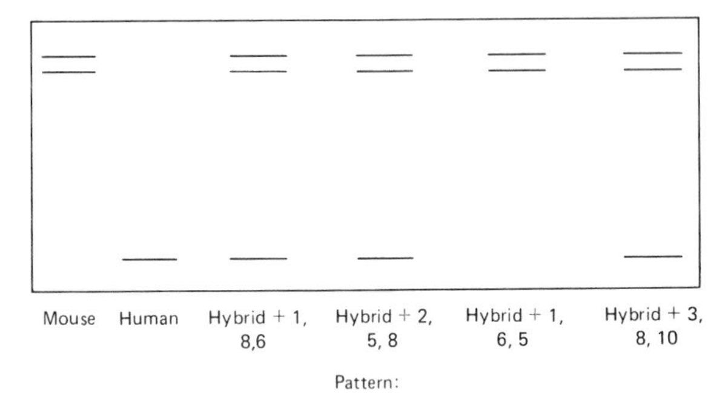

On which human chromosome does the gene for the human enzyme reside?

16. A. It is possible for two right arms to combine and produce a YAC (consult Fig. 19-22). What would be the outcome when such a YAC replicates?

B. What would be the outcome if two left arms combine to form a YAC?

Further Applications of Molecular Biology

20

Molecular biology and prenatal diagnosis

Although most of our early knowledge of molecular interactions came from elegant experiments with microorganisms, it was actually a human disorder, sickle-cell anemia, that provided the first experimental demonstration that the amino acid sequence in a polypeptide is under the control of a gene. The unfortunate effects of the disease had been shown to stem from some sort of change in the hemoglobin molecule, but it was not known whether the change was an alteration in the amino acid content or in the folding of the molecule.

An answer to the question demanded the detailed analysis of the amino acid sequence in hemoglobin. In the adult, most of the hemoglobin in a red blood cell is hemoglobin A, composed of two α-chains combined with two β-chains (more details are given later in this chapter). In sickle-cell anemia, normal hemoglobin A is not present in the red blood cell, since a change has occurred in the β-chain.

The analysis of hemoglobin was greatly facilitated by a shortcut procedure known as *fingerprinting,* which was used by Ingram in his pioneer determination of amino acid sequence in the protein. In this method, a protein that is to be analyzed is broken up into short fragments by cleaving it with a protein-digesting enzyme such as trypsin (Fig. 20-1). The resulting fragments, the peptides, then must be separated from one another. This is accomplished by using the technique of paper chromatography coupled with that of electrophoresis. The products of the enzyme action are placed near one edge of a piece of filter paper. The paper is then exposed to a specific solvent, which flows across it in one direction. The peptide fragments travel along in the direction of the solvent, but because they do not all move at the same rate in a particular solvent they can be partially separated. To obtain a more complete separation of the fragments, the paper is then turned 90 degrees and exposed to an electric field. Because of differences in their net electric charges, the fragments move at different rates to the + or − pole. The overall result is a good separation of peptide fragments, which then may be stained by methods that color proteins.

The technique is called *fingerprinting* because a particular protein gives a characteristic picture. When the fingerprints of normal and sickle-cell hemoglobin were compared, they were found to be identical except for one fragment that was out of place. This indicated the presence of some difference that alters the rate of migration of the specific fragment in the solvent and electric field. Because the peptide fragments are small (approximately eight amino acids each), their exact amino acid composition can be determined with relative ease.

Ingram was able to show that the peptide fragment difference between normal and sickle-cell hemoglobin was a result of a difference in just a single amino acid (Fig. 20-2). Position 6 in the β-chain of normal adult hemoglobin (numbering from the amino acid with the free amino group) was shown to be occupied by glutamic acid, whereas this position was occupied by valine in sickle-cell hemoglobin. After Ingram's attack on the problem, the different peptide fragments were eventually pieced together to show the exact sequence

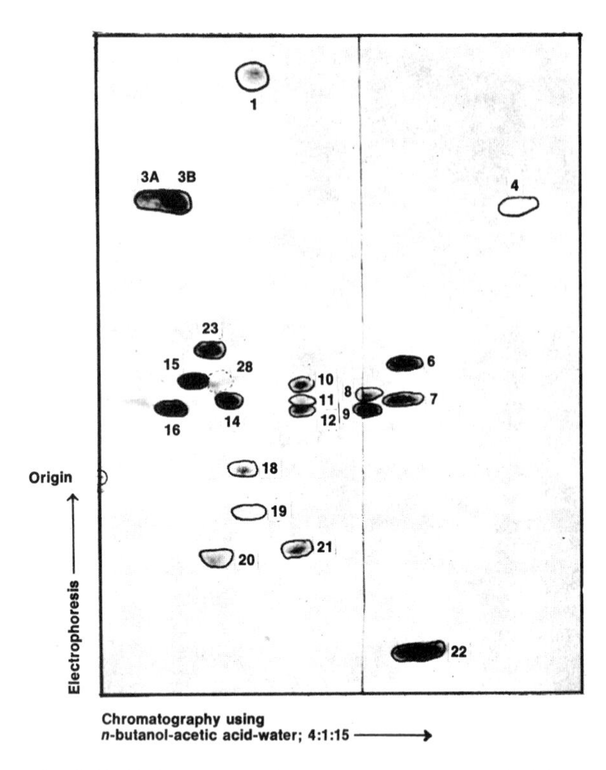

FIGURE 20-1. Fingerprinting technique. The figure shows a fingerprint of the enzyme ribonuclease from the sheep pancreas. The enzyme was exposed to trypsin. A portion of the digested material was then applied to the small spot at the left. It was next subjected to paper chromatography, as indicated by the arrow below. After allowing the solvent to evaporate, the paper was moistened with buffer solution and subjected to electrophoresis. The sheet was sprayed with ninhydrin solution, which stains the areas containing peptides. These areas can be cut out and the peptides washed from the paper. The amino acid composition of each peptide can then be determined by further analysis. (Reprinted with permission from C.B. Anfinsen, *The Molecular Basis of Evolution*, p. 145. Wiley, New York, 1959.)

of amino acids in hemoglobin. It was finally shown that this single amino acid substitution demonstrated by Ingram is the only difference between normal adult hemoglobin (hemoglobin A) and sickle-cell hemoglobin (hemoglobin S).

Shortly after this was established, still another abnormal hemoglobin—hemoglobin C—was analyzed that causes an anemia. It results from a recessive mutation, but the cells do not sickle, and the effects are not as severe as those of sickle-cell anemia. However, hemoglobin C was shown to differ from the normal hemoglobin A at exactly the same position as hemoglobin S (see Fig. 20-2). In hemoglobin C, position 6 is occupied by lysine. The two amino acids (valine and lysine) substituting at this position for the glutamic acid in normal hemoglobin have electric properties that differ from each other and from those of glutamic acid as well. These differences may cause changes in the shape of the β-polypeptide chain. An alteration in the shape of this chain may in turn affect the architecture of the entire hemoglobin molecule and so cause a series of metabolic disturbances.

A normal β-chain for normal hemoglobin A depends on an allele that can be represented as Hg_{β}^{A}. The alleles for β-chains, which result in sickle-cell hemoglobin and hemoglobin C, can be represented respectively as Hb_{β}^{S} and Hb_{β}^{C}.

Obtaining fetal blood for prenatal detection of a deleterious allele entails a risk almost 10 times that associated with amniocentesis. Consequently, use of fetal blood to check for the presence of the sickle-cell allele has been limited. Molecular biology now offers a way to overcome this problem by precluding the need for blood to detect the defective allele. Human β-globin mRNA has been isolated and used to obtain double-stranded cDNA following treatment of the transcript with reverse transcriptase. Sequencing the cDNA has indicated that the mRNA codon corresponding to the glutamic acid residue normally found at position 6 in the β-globin polypeptide is GAG. This RNA codon, if changed to GUG as the result of point mutation, would code for valine. As Figure 20-3 shows, the sickle-cell mutation at this position is accounted for by a transversion in which a base pair alteration in the β-globin gene has changed an AT pair to a TA pair.

This single base pair change happens to affect a region in the gene that can be recognized by a certain restriction enzyme, Dde. The enzyme recognizes the sequence CTNAG (in which N represents any nucleotide). As a result of the mutation, the enzyme can no longer identify the site and will therefore not cut the DNA at that location. Armed with this knowledge, we can now screen DNA from fetal cells obtained by amniocentesis to determine whether the DNA has been altered at that restriction enzyme site. If it has, we know the fetus is carrying the sickle-cell allele.

Suppose that DNA is extracted from human cells known to contain the normal β-globin allele on each of the two homologous chromosomes (chromosome 11). The DNA is treated with the restriction enzyme, which recognizes the site in the normal gene and consequently cuts the DNA at that location. DNA extracted from cells of a fetus at risk for sickle-cell anemia is also exposed to the enzyme. Let us assume that the DNA from the fetal cells carries the AT→TA transversion and that the enzyme will therefore not cut the DNA at that site. Both DNA samples, the one with the normal allele and that with the sickle-cell gene form, are subjected to gel electrophoresis and Southern blotting. Many fragments are generated, because the enzyme will cut the DNA wherever the recognition sequence occurs in the two total DNA samples. To find and visualize the β-globin DNA fragments among the many others that have been separated on the gel, we must turn to a suitable probe. One such probe is a labeled cDNA specific for the 5′ end of the β-globin gene. This probe enables us to locate the fragments on the filter containing those DNA sequences that code for the beginning of the β-globin polypeptide.

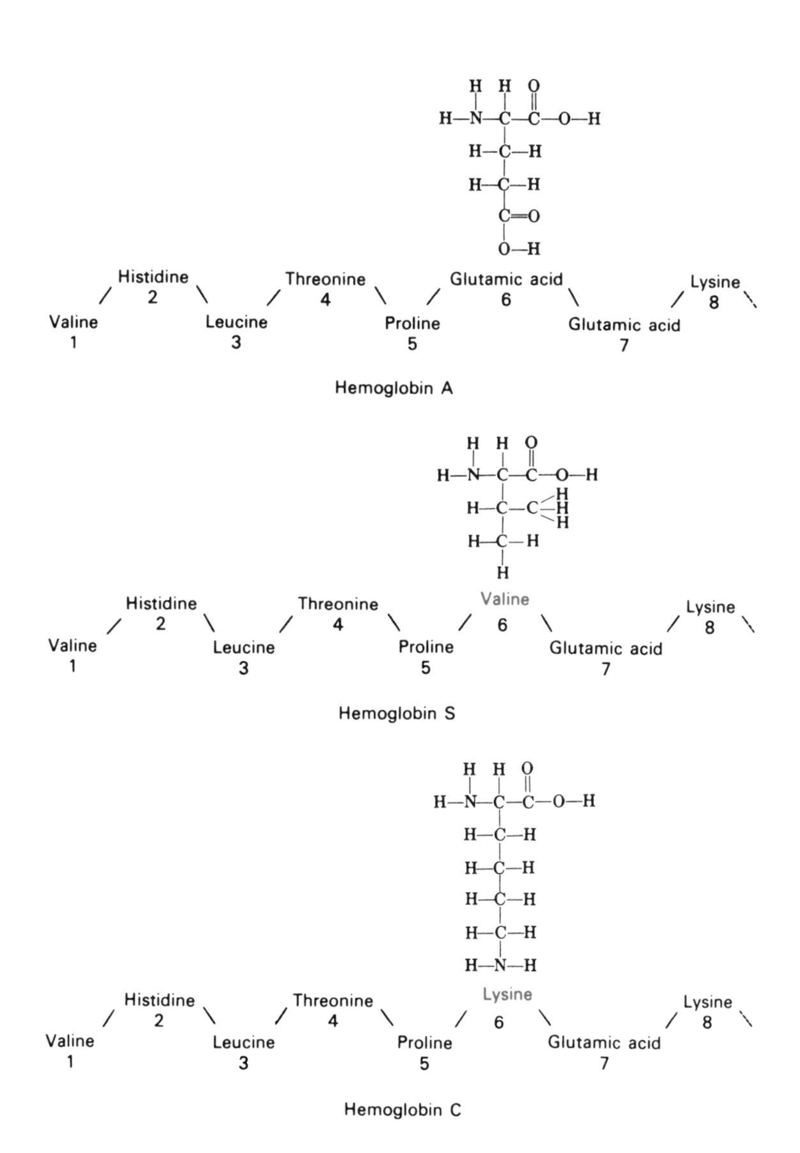

FIGURE 20-2. The difference between normal hemoglobin and two mutant types. The precise amino acid sequence was determined for normal hemoglobin (hemoglobin A) and for sickle-cell hemoglobin (hemoglobin S). A difference between them was found at one position. When the exact amino acid sequence was established for the entire β-chain of the hemoglobin, this was shown to be the only difference in the entire chain. At position 6 (near the NH_2 end), the glutamic acid in hemoglobin A is replaced by valine in hemoglobin S. Another mutant hemoglobin, hemoglobin C, is also altered at position 6. In this case, lysine is substituted.

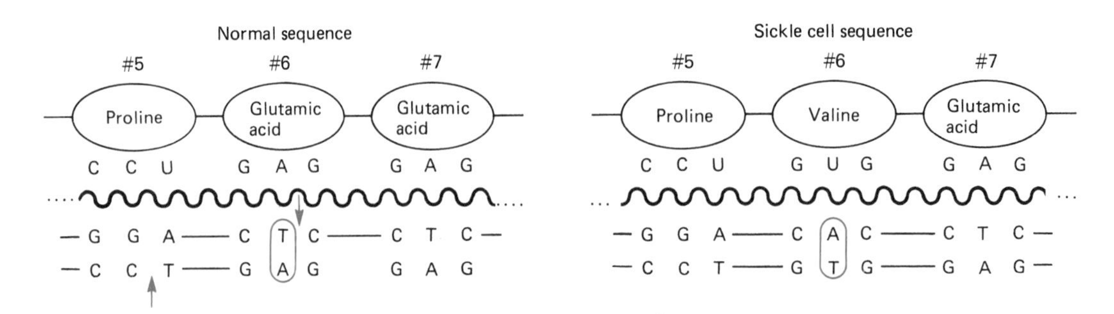

FIGURE 20-3. The sickle-cell mutation. On the left, the normal DNA sequence is shown, as well as the specific sites of cutting by enzyme Dde (red arrows). This restriction enzyme recognizes the sequence CTNAG. A transversion (red circling) that results in the sickle-cell allele eliminates the palindrome that the enzyme recognizes. The cut, therefore, cannot take place.

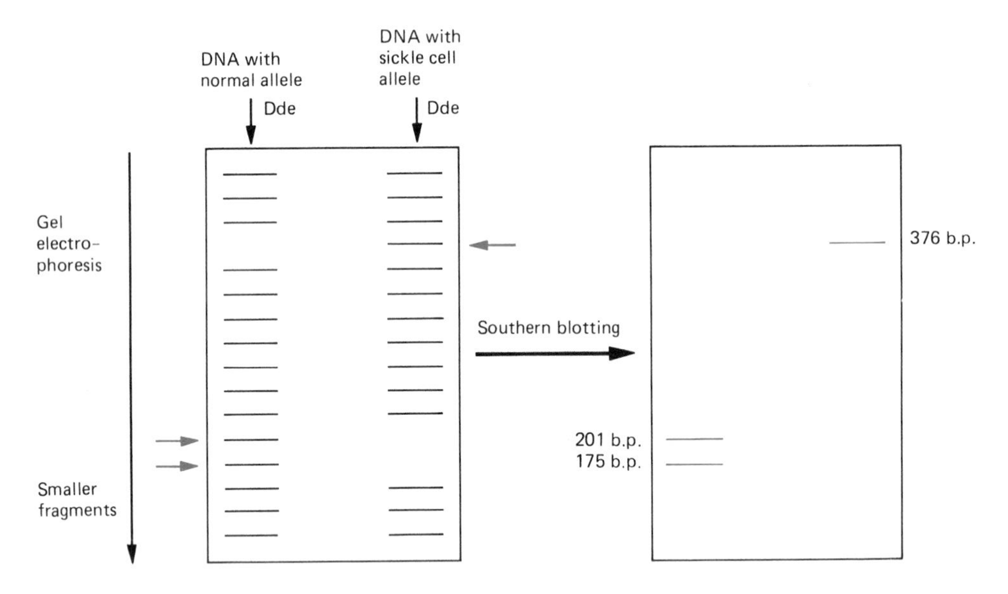

FIGURE 20-4. Restriction fragments formed from DNA of normal β-globin gene and the sickle-cell allele. DNA from cells with the normal allele and cells with the sickle-cell allele is subjected to the restriction enzyme Dde. Many fragments are produced, because the sequence CTNAG occurs at many sites along the DNA in addition to the site within the start of the β-globin gene. To detect fragments containing the sequence at the beginning of the gene (red arrows), Southern blotting and a suitable probe are required. The labeled probe used can visualize the 201, 175, and 376 base pair fragments when an autoradiograph is prepared. The red arrows indicate the specific fragments on the gel that are undetected until they are exposed to the probe (right).

Comparison of the fragments formed from the DNA of the normal β-globin gene with that carrying the sickle-cell allele shows a difference. It is seen that a larger fragment of the sickle-cell DNA hybridized with the probe than is the case for the normal DNA (Fig. 20-4). The reason, of course, is that elimination of the restriction site in the sickle-cell DNA eliminated a cut, allowing a larger than normal fragment to be generated. In this case, the normal hemoglobin will produce two bands that hybridize with the probe, a 175 base pair band and a 201 base pair band. On the other hand, sickle-cell DNA produces only a 376 base pair band. The 175 base pair fragment cannot form but becomes part of the 376 base pair fragment (Fig. 20-5).

Other enzymes besides Dde can be used to resolve fragments and show a difference between normal and sickle-cell DNAs. Different probes are available that vary in the length of the DNA sequences they cover. Therefore the fragment size that can be detected varies, but the procedures are nevertheless the same as that outlined here and can resolve differences between normal hemoglobin A and hemoglobin S. The need for taking fetal blood and the dangers it entails are avoided.

Hemoglobin chains and their genes

Recombinant DNA procedures have provided us with an excellent picture of the arrangement of the various genes coded for globin chains of hemoglobin and have also led to some unexpected findings. To appreciate the accomplishments in this area, we must examine a few details on human hemoglobin.

As noted in the previous section, most of the hemoglobin of the adult, hemoglobin A, is composed of four polypeptide chains, two α combined with two β. However, more than one type of hemoglobin appears during the human life cycle. Nevertheless, all hemoglobin molecules consist of a tetramer composed of four globin chains. Two of these polypeptides are α-like and two are β-like, a distinction made on the basis of their structure. Each kind of polypeptide chain is encoded by its own specific gene, and at least seven genes occur for the globin polypeptides of the human. Throughout the life span, different globin genes are called to expression, and some of these become repressed later on.

As Table 20-1 shows, the first hemoglobin to arise in the human is composed of two epsilon (ϵ)-chains, which are β-like, and two zeta (ζ)-chains, which are α-like. This hemoglobin, $\zeta_2\epsilon_2$, is known as Gower I. As the embryo reaches the eighth week of development, the genes encoded for these chains become repressed as other globin genes are called to expression. These are genes for the α-chains and for two different types of gamma (γ)-chains, which are β-like. These chains, ${}^G\gamma$ and ${}^A\gamma$, are encoded by two separate genes and are identical except for amino acid position 136. A glycine is present at that site in ${}^G\gamma$, whereas it is an alanine in ${}^A\gamma$.

A combination of two α-chains and two of the γ-chains ($\alpha_2{}^G\gamma_2$ or $\alpha_2{}^A\gamma_2$) forms hemoglobin F (general formula $\alpha_2\gamma_2$), which gradually replaces the embryonic hemoglobin and becomes the main hemoglobin throughout the life of the fetus. (During the transition from embryonic to fetal hemoglobin, one can detect other hemoglobin types such as $\alpha_2\epsilon_2$ and $\zeta_2\gamma_2$.) The two γ-genes, in turn, start to become repressed shortly before birth as the globin chains characteristic of adult hemoglobin appear. Genes coded for β-globin and delta (δ)-globin polypeptide are now called to expression. At 6 months of age, the child has a hemoglobin profile typical of the adult. About 97% of the hemoglobin is hemoglobin A ($\alpha_2\beta_2$), the remainder consisting mainly of hemoglobin A_2 ($\alpha_2\delta_2$). Approximately 1% fetal hemoglobin ($\alpha_2\gamma_2$) continues to be produced. The α-globin chain is composed of 141 amino acids. The β-, δ-, and γ-chains all contain 146 amino acids.

The nucleotide sequences of the different globin genes have been worked out using recombinant DNA techniques. All the α-like and β-like globin genes have been shown to be split, each containing two introns. Moreover, all the globin genes have been mapped. The α-like genes compose a family that occurs in a cluster on chromosome 16. Two α-globin genes, α_2 and α_1, typically occur in a cluster (Fig. 20-6A). The β-like genes constitute another family clustered on chromosome 11 (Fig. 20-6B).

One reason that such rapid advances have been made in our understanding of globin genes is the

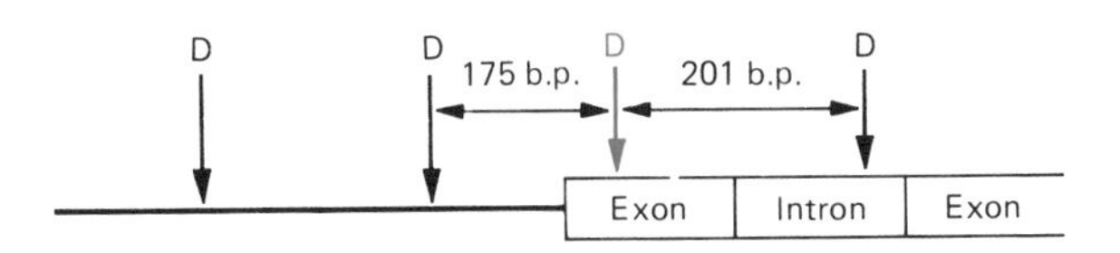

FIGURE 20-5. Variation in restriction fragment length because of gene mutation. The reason for the results seen in Figure 20-4 is the elimination of the Dde restriction site occurring in the first exon of the gene (red D) by the sickle-cell mutation. Therefore, the 201 base pair fragment cannot form. Along with the 175 base pair fragment, it becomes part of a 376 base pair fragment. Other fragments are generated, because other Dde sites occur along the DNA and are not associated with the 5′ end of the β-globin gene.

TABLE 20-1. Human globin chains and hemoglobins

Stage	Globin chains		Hemoglobin class
	α-like	β-like	
Embryonic (up to 8 weeks)	ζ_2	ϵ_2	$\zeta_2\epsilon_2$ Gower I
Fetal	α_2	${}^G\gamma_2$ or ${}^A\gamma_2$	$\alpha_2\gamma_2$(general formula) HgF
Adult (after birth)	α_2	β_2	$\alpha_2\beta_2$ HgA
Adult (after birth)	α_2	δ_2	$\alpha_2\delta_2$ HgA_2

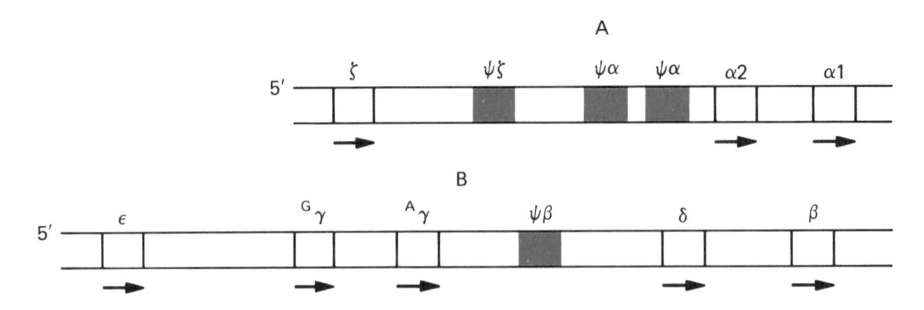

FIGURE 20-6. The globin gene clusters. Sequence of genes and pseudogenes (red) in the α-like cluster on chromosome 16 (A) and the β-like cluster (B) on chromosome 11. The 5′ end of each gene is toward the left. The direction of transcription for each gene (arrows) is from left to right.

ready availability of their mRNAs, as noted previously. In relation to other kinds of mRNA, the immature red blood cells manufacture an abundance of transcripts for the globin chains. Transcripts isolated from the immature red blood cells can then be used in various kinds of investigations. For example, isolated human transcripts have been placed in cell-free translation systems composed of ingredients obtained from nonhuman sources—wheat cells in some cases. Even though the tRNA, ribosomes, and other factors are from nonhuman cells, the human globin transcripts are nevertheless translated into human globin chains. This is an important bit of information; it tells us that whatever factors regulate the transcription of the globin genes must not operate, to any large extent, at the level of translation. Instead, they must exert their major influences at the level of transcription or of processing the RNA.

The globin gene transcripts have been used to make cDNA, which is then inserted into bacterial plasmids. All the human globin genes have been cloned and are now available for use in various kinds of experimental approaches. One obvious use of the cDNA is as a probe to locate globin DNA sequences in any kind of cell or from among any group of DNA fragments. Use of probes led to the discovery of the α- and β-gene clusters and the sequence of the genes in each cluster. Several very interesting observations have been made on these α- and β-clusters. In each of them, the template strand of the separate genes proved to be the same DNA strand. With the 5′ end of each gene written on the left, the reading of each gene is from left to right (Fig. 20-6). For example, the strand used as the template to produce γ-mRNA is the same as the one used for the δ- and β-mRNAs.

As Figure 20-6 also shows, the arrangement of the genes in each cluster corresponds to the order in which they are expressed (∈ before γ before β and δ in the β-group). In addition, Figure 20-6 shows that both the α- and the β-clusters contain nucleotide sequences known as *pseudogenes*. These sequences have been detected with labeled DNA from globin genes as probes. Neither the β-like nor the α-like pseudogene sequences have ever been shown to be associated with any known globin polypeptide. Apparently, they are not expressed, even though they contain large stretches of DNA homologous to the functional genes in each cluster (about 75% homology). All the globin gene clusters studied thus far have pseudogenes between the embryonic or fetal genes and the adult genes (Fig. 20-6). Pseudogenes are considered to have arisen from functional genes as the result of gene duplications.

Duplications and the globin genes

Duplications are considered to be one class of structural chromosome change that has played a highly significant role in the evolution of the genome (see Chap. 10). Once a gene duplication arises, each gene copy becomes subject to its own history of point mutation and structural chromosome change. For a while, both genes may retain the same function. One of the duplicates, however, may accumulate mutations that render it incapable of forming a functional product. Pseudogenes are believed to represent cases of this. The detrimental mutations can accumulate because the gene involved originated as a duplication, and the other copy, lacking detrimental mutations, is able to perform the function associated with the gene. The genetic alterations in the pseudogenes of mammals include deletions that shift the reading frame, point mutations that alter intron–exon boundaries, and missense mutations that would substitute amino acids at critical positions in any associated polypeptide chain. Taken together, the alterations in the pseudogenes would be detrimental to both normal transcription and translation, even if a transcript should arise. Pseudogenes are believed to be common in vertebrates and possibly in other organisms as well. Many gene families are now known to include them other than the globin cluster—for example, the sequences coded for the immunoglobulins and those for the histocompatibility antigens.

The two adjacent genes for the α-globulin chains on chromosome 16 are clear-cut examples of very recent duplications, as are the two almost identical γ-globulin genes. In both cases, the members of the pair have clearly maintained the same function. As evolution proceeds, however, one of the two genes in a pair of duplicates may maintain the original function while the other evolves to perform a somewhat different but related function. For many years, the globin genes have been considered a primary example of a group of related genes, originating from the duplication of ancestral genes. Before recombinant DNA procedures were developed, excellent evidence for this concept was provided by the striking similarity of the amino acid sequences of the several globin chains that are encoded by several separate genes. The β- and δ-chains differ by only 10 amino acids; the two γ-chains differ from each other by only a single amino acid. The β-, δ-, and γ-chains are all the same length. The β-chain differs from the γ-chain by 30 amino acids, but it differs from the α-chain by 84.

The β- and δ-genes are now known to have over 90% identical nucleotide sequences. Close homology between the β- and δ-genes has been apparent to investigators for many years, since crossing over can occur between them. In Chapter 10, we saw that, as a result of tandem duplications in the Bar region of *Drosophila,* crossing over can take place in such a way as to yield unequal crossover products. A similar situation pertains to the β- and δ-genes. Since the two genes may have stretches of homology, a certain amount of mispairing can occur, giving rise to unequal crossing over. As Figure 20-7 shows, pairing between β- and δ-genes, followed by a crossover event, leads to the formation of two dissimilar or unequal products. A variant gene is generated that starts off as a δ-gene and ends as a β-gene. Such a gene, known as a *Lepore gene,* is responsible for the production of a variant globin chain that starts at the amino end as a δ-chain and ends at the carboxyl end as a β-chain. Since crossing over between the mispaired genes may take place at more than one site, more than one type of Lepore gene and hence Lepore globin is expected, and this is the case. Since neither a complete δ- or β-gene is present, persons with Lepore hemoglobin show symptoms of anemia. Note from Figure 20-7 that the reciprocal product yields a chromosome region that contains a δ-gene, an anti-Lepore gene (which begins as a β-gene and ends as a δ-gene), and finally a complete β-gene. The person with the anti-Lepore gene has a full set of normal β- and δ-genes plus the β–δ fusion gene and does not experience severe anemia.

These and other observations lend support to the theory that the genes for the four globin chains have been derived from a single gene, one that was also ancestral to the gene encoded for myoglobin, an oxygen-binding muscle protein. The myoglobin gene, like the genes for the globin units, is also interrupted by two introns. Myoglobin, though having many differences from the amino acid sequences of the globin polypeptides, is comparable to them in length, consisting of 153 amino acids. The α-chain is composed of 141 amino acids, and the β-, γ-, and δ-chains are made up of 146 of them. According to the proposed relationships among the globin genes and that for myoglobin, an ancestral gene duplicated. One copy proceeded down the pathway leading to the present-day myoglobin gene. The other copy underwent another duplication. One of these evolved into a gene that was ancestral to the α-cluster; the other was ancestral to the β-cluster (Fig. 20-8). Further duplications in the more recent past set the stage for the origins of the various α-like and β-like genes, as well as the pseudogenes associated with them.

All the gene members of the β-group are more similar to one another than any one of them is to the α-sequence. Very interesting is the fact that the two introns that split the five β-like genes occur at the same location in each gene in relation to the reading frame. For example, in all five genes, the first intron occurs between codons 30 and 31, the second one between codons 164 and 165. The two introns that interrupt the genes in the α-like cluster also split the reading at exactly the same location, the first between codons 31 and 32, the second between codons 99 and 100. Such observations lend strong support to the concept that gene members of the β-cluster resemble one another more than they do members of the α-gene cluster, since they stem from duplications that occurred in the more recent past. Not enough time has elapsed for major differences in nucleotide sequence to have arisen among them. Nor have they yet been separated onto different chromosomes by structural rearrangements. In contrast, the α-sequence stems from a more ancient duplication of an ancestral gene and has diverged appreciably from those in the β-group, to which it is no longer linked but is now on a separate chromosome.

Molecular basis of several globin chain disorders

Certain mutations are known in the human that alter chain-terminating codons so that normal termination signals become erased. Polypeptide chain variants of hemoglobin have been identified that are called *chain-elongation variants* because of the presence of additional amino acids at the carboxyl end of the polypeptide chain. The codon UAG is now known to be the signal for termination of the α-hemoglobin chain. If the codon is altered to CAG as a result of a mutation, it is then read as glutamine by the ribosome, which continues to translate the transcript into the trailer region until it eventually encounters another signal to cease transcription. Several chain-elongation variants are known, and these affect both the α- and β-hemoglobin chains. Some of the variants are believed to have resulted from frame shift mutations that have shifted the reading of the transcript near the 3′ end of the coding sequence and have eliminated the normal chain-terminating signal. The normal length of the α-polypeptide chain is 141 amino acid residues. As many as 172 amino acids compose some of the variants as a result of translation of sequences in the trailer, which are normally not read and are not coded for the normal hemoglobin chain. We see here the importance of the three chain-terminating codons for orderly information transfer.

Recombinant DNA procedures have supplied a wealth of information on the molecular level concerning several human disorders related to variant hemoglobin molecules. Two main classes of human blood disorders can be recognized. In the first type, a globin chain is altered, usually because of an amino acid substitution resulting from a point mutation. Sickle-cell anemia is a typical example of this class. In the second type of disorder, the globin chain is not altered in any qualitative way. Instead, there is a quantitative change. The globin chains that are produced usually have the normal amino acid sequence, but they are present in reduced amounts. The thalassemias are disorders that fall into this class. Both α- and β-thalassemias occur in which production of either α- or β-chains is reduced.

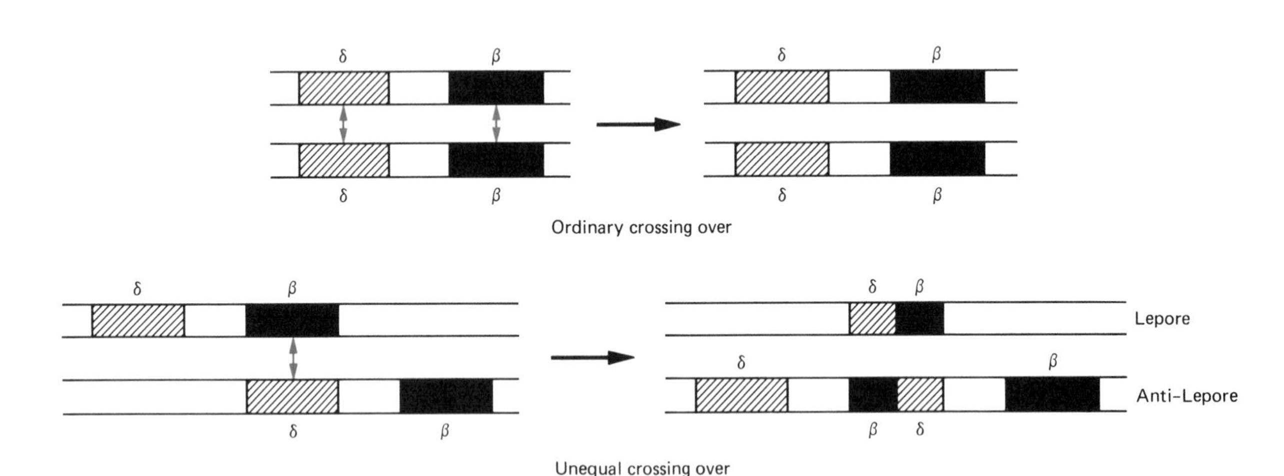

FIGURE 20-7. Equal and unequal crossing over in the δ–β region. If normal alignment of the homologous chromosomes takes place (above), the two δ-genes pair with each other, as do the two β-genes. A crossover event (red arrows) anywhere along the length of the paired region will produce equal crossover products with intact δ- and β-genes. However, if misalignment occurs (below) and crossing over takes place between a δ- and a β-gene, unequal crossover products are generated. A Lepore gene is formed, composed of the left portion of the δ-gene and the right portion of the β-gene. Depending on the point of crossing over, Lepore genes will differ in the proportion of δ- and β-sequences they contain. The other product of the unequal crossing over is a chromosome that contains an anti-Lepore sequence, the left part of the β-gene, and the right part of the δ-gene. A normal δ- and a normal β-gene are also present.

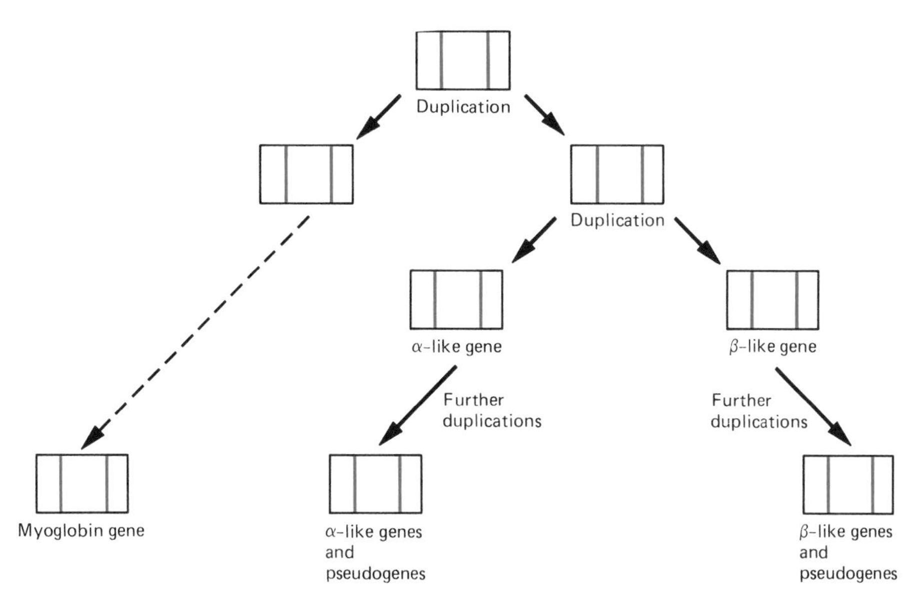

FIGURE 20-8. Duplications and the globin genes. An ancestral gene is believed to have duplicated. One copy evolved into the present-day gene for myoglobin. The other copy underwent another duplication. One of these evolved down a pathway leading to an α-like gene. Several later duplications eventually gave rise to the α-like gene cluster. The other gene copy evolved in the β-like direction, and further duplications also took place in this pathway leading to the origin of the β-like cluster. Two introns (red) are present in all the genes involved.

Let us first concentrate on the β-thalassemias, which are responsible for the deaths of hundreds of thousands of children yearly throughout the world. This very serious health problem, like sickle-cell anemia, occurs with a high frequency in malaria-infested regions. It is known that the high incidence of the sickle-cell allele in certain populations is a consequence of the advantage imparted to the heterozygote against the malarial parasite. The situation with the β-thalassemias may again be one in which heterozygosity for a blood disorder confers resistance to disease. Heterozygotes exhibit thalassemia minor, a mild anemic condition, whereas individuals with the very severe anemia, β-thalassemia major (formerly known as Cooley's anemia), usually die before puberty. There are various forms of β-thalassemia stemming from different types of defects at the molecular level. The effects of the thalassemias therefore vary in their severity in homozygous individuals. Nevertheless, in all of them, β-globin chains are significantly reduced in number and may be entirely absent.

Various kinds of point mutations are known to occur at different locations in the β-globin gene. Mutation in the promoter region interferes with the initiation of transcription. Therefore, in persons homozygous for this type of defect there is a decrease in the number of transcripts and consequently of β-globin chains, resulting in anemia. Mutations are known that alter nucleotides in the introns and thus interfere with normal splicing of the mRNA. In homozygotes, no normal β-globin chains are found.

A mutation is known that changes a codon for an amino acid near the 5′ end of the transcript to a chain-terminating codon. Since the mRNA cannot be translated to produce β-globin, severe anemia results. Frame shift mutations are also known, and these result in absence of normal β-globin chains. A mutation is even known that alters the poly (A) site with the result that the poly (A) tail of the mRNA is added farther toward the 3′ end, yielding an mRNA that is less stable than the normal and that produces a mild anemia. Some persons may actually be heterozygous for two different kinds of genetic defects and show the severe anemic symptoms associated with one of them.

There are two other rarer conditions related to the β-thalassemias. These are the hereditary persistence of fetal hemoglobin (HPFH) and δ–β-thalassemia. In persons with these disorders, no δ- or β-chain synthesis occurs at all. However, fetal hemoglobin (HgF) is produced in increased amounts. As a result, δ–β-thalassemia is a mild condition, and in HPFH anemia is absent. In the δ–β-condition, the fetal hemoglobin

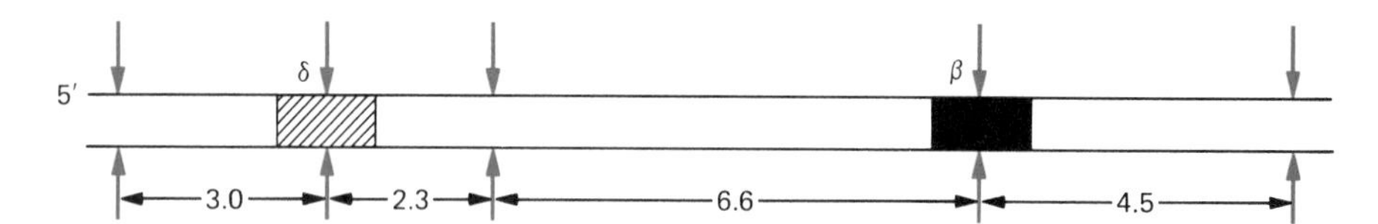

FIGURE 20-9. Organization of the δ–β region. Molecular details on the arrangement of the δ- and β-genes and the regions surrounding them were obtained using restriction enzymes and labeled cDNA probes. When human DNA from persons with normal hemoglobin chains is exposed to EcoR1, the restriction enzyme attacks the DNA at several sites (arrows) in the δ–β region. As a result, four fragments are generated. The four fragments differ in size, expressed in kilobases, and were distinguished from all other DNA fragments in the enzyme-exposed DNA by the labeled cDNA probes. The sequence of the four fragments was worked out, and the 3.0 kb fragment was found to contain the 5′ end of the δ-gene, whereas the 5′ end of the β-gene is in the 6.6 kb fragment. The 2.3 and 4.5 kb fragments contain, respectively, the 3′ ends of the δ- and the β-genes. The distance between the δ- and β- genes is approximately 7 kb.

compensation is not quite sufficient to prevent a mild anemia. However, in HPFH, the production of HgF is greatly increased and eliminates anemia. How can this be explained? The level of HgF is also increased in the β-thalassemias. However, HgF production is not raised sufficiently as in the δ–β and HPFH conditions. Consequently, serious anemia results, because compensation for the decreased or absent hemoglobin A is inadequate.

In a series of investigations, cDNA probes were used to detect globin genes in DNA obtained from persons with no blood disorders and from persons with HPFH, δ–β-, and β-thalassemias. The organization of the δ- and β-globin gene sequences and the region surrounding them was determined. With the use of restriction enzyme EcoRl, four fragments were identified that hybridize with the appropriate cDNA probes, and the sequence of the fragments was worked out for the normal condition (Fig. 20-9).

DNA from persons with normal hemoglobin was subjected to a series of different restriction enzymes. Appropriate cDNA probes were then used to identify fragments containing DNA sequences found in the δ–β-region. The same was done with the DNA from persons with variant hemoglobins. The results indicated that in persons homozygous for δ–β-thalassemia, the 5′ end of the δ-gene plus sequences immediately surrounding it are present, but the rest of the δ–β-sequences are absent. In persons homozygous for HPFH, all the δ–β-fragments normally found in this chromosome region are entirely absent (Fig. 20-10)! Hemoglobin Lepore, discussed in the previous section, is an example of an unusual β-thalassemia that entails deletion, since neither a complete δ- or β-chain is present. There is no evidence for any extensive deletions in the other β-thalassemias.

It is very significant that in the δ–β-thalassemias, where symptoms are not too severe, and in HPFH, where there is no anemia at all, extensive deletions exist in the δ–β-region. The production of HgF compensates for lack of intact genes for the δ- and β-chains; the compensation is greater in HPFH, in which the entire δ–β-region has been deleted! The increase in HgF production above the normal in β-thalassemias (in which there is no evidence for extensive deletions) is not at all sufficient to prevent severe anemia. One interpretation of these observations is that the δ–β region of the chromosome contains sequences that regulate the expression of the γ-genes and decrease the level of their expression before birth. In the case of HPFH, these sequences are lost because of deletion, and γ-chains can consequently be produced at high levels. In δ–β-thalassemia, the deletion does not remove all of this controlling region, so that some reduction in expression of the γ-genes occurs. In the β-thalassemias, in which little if any deletion exists, the controlling region is present and tends to suppress the γ-genes. As a result, HgF levels are kept too low to compensate for a decrease in the normal amount of HgA.

Although many details are yet to be uncovered, the information now available on the β-gene cluster and the changes that can alter it at the molecular level has provided us with a deeper understanding of the nature of the β-thalassemias and offers hope that ways to lessen the severity of these conditions may be found.

As noted in Figure 20-6A, two α-globin genes may occur on chromosome 16. In most human populations this is the case. Since α-genes exist in duplicate copies, this again makes possible the occurrence of unequal crossing over, as discussed for the δ- and β-genes. As the result of an unequal crossover event, a

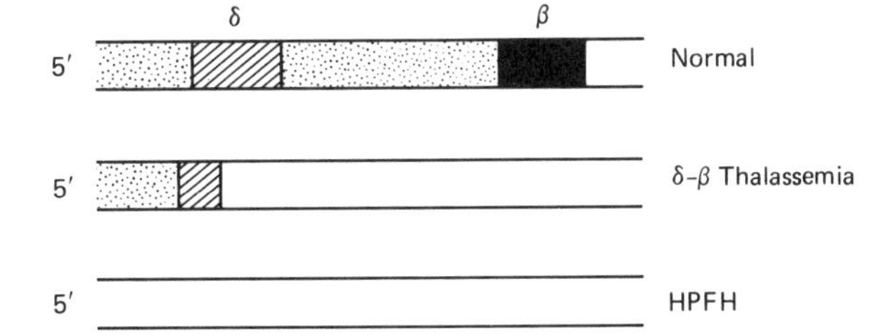

FIGURE 20-10. Deletions in the δ–β region. Labeled cDNA probes show that persons homozygous for δ–β-thalassemia have a deletion in the δ–β region. Only the sequences toward the 5′ end of the δ-gene and immediately to the left of it are present. The rest of the δ–β region is missing. In HPFH, the entire δ–β region has been deleted.

deletion of an α-gene can occur (Fig. 20-11). α-thalassemias are known that may have resulted from different deletions of α-globin genes and may have been generated by unequal crossing over. Such deletions result in a decrease in the number of α-globin chains. These α-gene deletions have been detected using α-globin cDNA probes. Four different possibilities exist regarding the number of α-genes. Deletion of just one gene gives an individual three α-globin genes, two on one chromosome and one on the other. This results in a decrease in α-globin synthesis, but no anemic symptoms are present. When two α-globin genes are deleted, there is a slight anemic effect. If three of them are deleted, the β-chains accumulate, because an insufficient number of α-chains are being produced to combine with them. The result is the formation of unstable tetramers of four β-chains, which break down, resulting in a mild anemia associated with hemoglobin H disease. If all four α-chains are deleted, the fetus dies as a result of the very severe anemia *hydrops fetalis* (Fig. 20-12 summarizes some of the possibilities in the number of α-globin genes). Human populations differ in the incidence of hydrops fetalis. Knowing what we now know about the α- and β-gene clusters, and armed with rapidly advancing techniques, it may be possible in the near future to transform human marrow cells with isolated, nonvariant globulin genes, thus preventing the severe effects associated with aberrant or reduced globin chains.

Restriction fragment length polymorphisms

We have encountered, throughout our discussions, many examples of genes whose protein products are known. Knowing the product that is encoded by a particular gene enables us to associate a defect in the protein with a specific defective allele of that gene. It may also permit detection of carriers of a defective allele (e.g., Tay-Sachs, sickle-cell anemia). Knowing the protein product also makes it possible to locate the gene and to clone it (globin genes, insulin gene). Unfortunately, in the case of most genetic disorders, the protein product associated with the gene in question is unknown. One way around this dilemma is to evaluate the inheritance pattern of the defective allele and to look for markers to which it may be linked. We discussed this approach in Chapter 9 (review Fig. 9-4).

For many genetic disorders, however, the defective allele has not been associated with any other gene that can serve as a marker. Consequently, the location as well as the linkage relationships of such a defective gene remain unknown. Now methodology from molecular biology and recombinant DNA procedures is enabling us to overcome this barrier. In our discussion of sickle-cell anemia earlier in this chapter, we saw that a gene mutation can alter a specific restriction site within a gene. Such a mutation, which alters a restriction site, can occur anywhere along the DNA. If two individuals vary in a particular DNA sequence at a given site, a specific restriction enzyme will produce fragments of different sizes when it cuts the DNA from the two sources (Figs. 20-4, 20-5). In other words, a mutation, by affecting the cutting site of a restriction enzyme, will produce variations in the DNA pattern. Pairs of variations in the DNA are inherited as traits, just as if they were variant forms of a given gene (Fig. 20-13).

In the same way that we can detect differences in a specific protein because of genetic variation, we can now detect differences in DNA sequences by observing fragments of different sizes generated by cutting with a given restriction enzyme. Whereas traditionally we used protein variations (protein polymorphisms) as markers to indicate the presence of a particular allele, we can now use DNA polymorphisms (restriction fragment length polymorphisms [RFLPs]) as genetic markers. Since the DNA polymorphisms can affect any type of DNA sequence, this means that an alteration producing a RFLP can occur within a coding sequence of a gene (as was the case with the β-globin gene), noncoding sequences (introns), sequences between genes, and even DNA with no known function, such as repetitive DNA. Because the DNA polymorphisms can affect any part of the genome and they need not be expressed as a protein product, they are extremely valuable markers.

One great advantage of the use of RFLPs stems from the fact that in the human DNA complement a very large amount of polymorphism exists. When two homologous chromosomes are compared, a difference in base pair sequence is found about every 200–500 base pairs. This means that along a pair of homologs, various restriction enzymes can recognize these many sites. At a given DNA location, the appropriate restriction enzyme can detect a variation in the DNA sequence between two chromosomes, as shown in Figure 20-5 for the sickle-cell location. The result is an RFLP.

A particular enzyme will generate a number of fragments that is dependent upon the number of times a given recognition site (specific sequence of base pairs) is found in the genetic complement. Electrophoresis must then be employed to separate the fragments. To identify the fragments resulting from cutting at a given site, the appropriate probe must be found. The probe may happen to be a piece of DNA taken at random from a library of cloned human DNA fragments (see Chap. 19, Fig. 19-18). The denatured, labeled probe DNA from the library is then applied to denatured Southern blots. These represent fragments generated by exposure of the DNA to the restriction enzyme. The probe DNA will hybridize with DNA complementary to it, and a pattern of bands may be evident when an autoradiogram is prepared following the hybridization. If DNA from two different individuals is compared, it may be found that the band pattern between them is different with respect to that particular given site, as shown in Figure 20-4. An RFLP would have been detected, and it may now be used as a marker and a reference point for later studies.

Every person will have this DNA marker in one form or another (one allele or another). It may be found that in a certain family one form of the marker (say form A) is being transmitted along with a certain genetic disease. This tells us that the marker (the RFLP) and the defective gene occur in the same region of the chromosome. Note that the marker region occurs in *all* persons, those with and those without the disorder. In a second family in which the same disease occurs, the affliction may be associated with another form of the marker, perhaps allelic form B rather than A.

An outstanding value of RFLPs as markers is that they provide many variant forms or versions. Some RFLPs are a consequence of a change in just one base pair at a given region of the DNA. If this is the case, then just two versions or allelic forms of the RFLP would exist. This would be the same as the situation discussed in Chapter 9 (see Fig. 9-4B). At times, two variable sites may occur nearby along a stretch of DNA. This means a single probe may detect four versions of the marker (see next section). Fortunately, many different alleles or versions of a pattern can be detected with respect to many RFLPs. Recall that human DNA contains many tandem repeats (Chap. 13). For any given location along a chromosome, there may be variation in the number of times that a stretch of base pairs is repeated in tandem, from a small number to a few hundred times. If a restriction enzyme cuts close to a region where tandem repeats occur, there may be a great degree of variation from one chromosome to the next in the number of times the repeat is represented at that specific chromosome location. The variation in the number of repeated stretches of DNA leads directly to variation in the size of fragments generated and thus gives rise to RFLPs (Fig. 20-14). Any one individual in the population has a high probability of possessing two homologous chromosomes, each with a different repeat number at a specific location. This means that the chance is very great in the population that any person will be heterozygous for an RFLP at that site. Such a variable site, therefore, has great potential as a marker. If such a site occurs near a gene associated with a particular disorder, it then

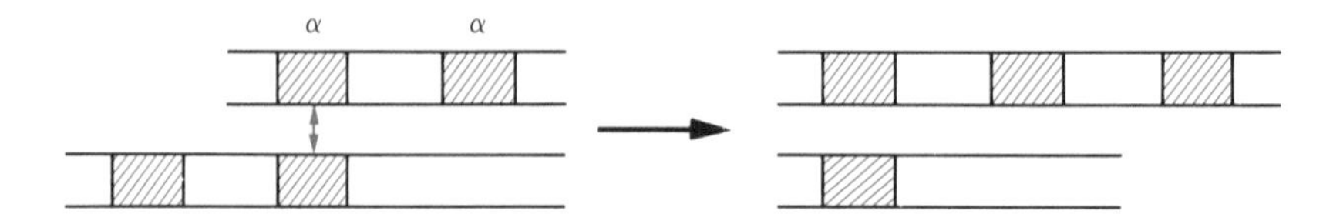

FIGURE 20-11. Deletion of an α-gene from one chromosome. Unequal crossing over in the α-gene cluster can result in the deletion of one of the α-genes from a chromosome and the origin of a chromosome with three α-genes.

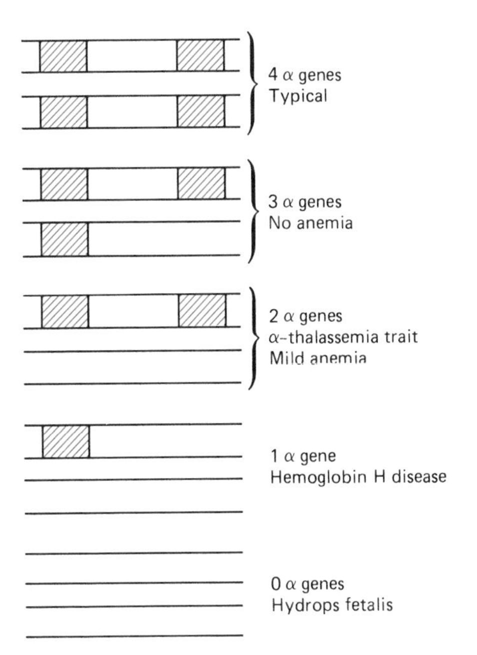

FIGURE 20-12. Different numbers of α-genes and their effects. The number of α-genes in the human can vary from four (typical) to zero (hydrops fetalis). Some of the possible arrangements are depicted here.

becomes extremely valuable. The situation is the same as that discussed earlier in terms of linkage (see Chap. 9, Fig. 9-4C).

RFLPs and Huntington's disease

A prime example of the use of RFLPs to detect the presence of a gene whose primary protein defect remains unknown is Huntington's disease (HD). This devastating disorder is inherited as a dominant, but its chromosome location remained unknown until the advent of recombinant DNA techniques. HD is a neurological disorder in which a person with an apparently normal central nervous system experiences premature death of nerve cells. Its frequency in the white population is about 1 in 10,000. The afflicted individual, whose disease may progress over a 15–20 year period, becomes completely disabled. Since the primary biochemical defect in the disorder remains a mystery, no successful approach to treatment has been found to prevent the deterioration and ultimate death of the victim of HD.

Moreover, ignorance of the gene product did not permit assignment of the Huntington's gene to a specific chromosome location or the development of a test to identify persons at risk. This was particularly significant in the case of this disorder, because the Huntington's allele is 100% penetrant. Although the disorder can occur at almost any time in the life span, it typically manifests itself between the ages of 35 and 45 years. This means that when the symptoms start, the individual can be a parent. Since the disease behaves as an autosomal dominant, the occurrence of HD in one parent gives each child a 50% chance of inheriting the defective allele and then developing the disorder, usually well after the age of sexual maturity. It is evident that the disorder is an especially insidious one that can be passed down by an unsuspecting person to one or more offspring.

For years, investigators attempted to find a linkage relationship between the HD locus and some other genetic locus that is associated with a detectable product, as was the case with myotonic dystrophy. This would make it possible to calculate the risk that a person will inherit the defective allele and would also lead to its eventual assignment to a chromosome.

After years of failing to achieve these goals, researchers in HD turned their attention to the techniques provided by molecular biology. They were also able to study a large number of families in the United States in which HD occurs. In addition, they were alerted to a large number of Venezuelan families that were interrelated. Many members of these families were at appreciable risk for HD, because they included 100 HD victims as well as 1,100 children. Tissue samples were collected from the U.S. and South American families. DNA was then extracted in a search for RFLPs that might be associated with the HD locus. DNA samples from the families were subjected to various restriction enzymes, and fragments of different lengths generated by the enzymes were then inserted and cloned in phage vectors. These cloned fragments representing different portions of the genome were now available as potential probes with which to search for any DNA fragment polymorphisms that might be associated with the HD locus. Not long after the search was started, a particular RFLP was identified that was linked to the HD locus. A fragment of this marker region, cloned in a phage vector, was available to act as a probe.

As Figure 20-15 shows, four different DNA patterns can be generated from a DNA sample representing genetic material from this chromosome region (now known to be on chromosome 4). Each pattern (A, B, C, and D) had been designated a *haplotype.* The haplotype depends on whether the variable sites

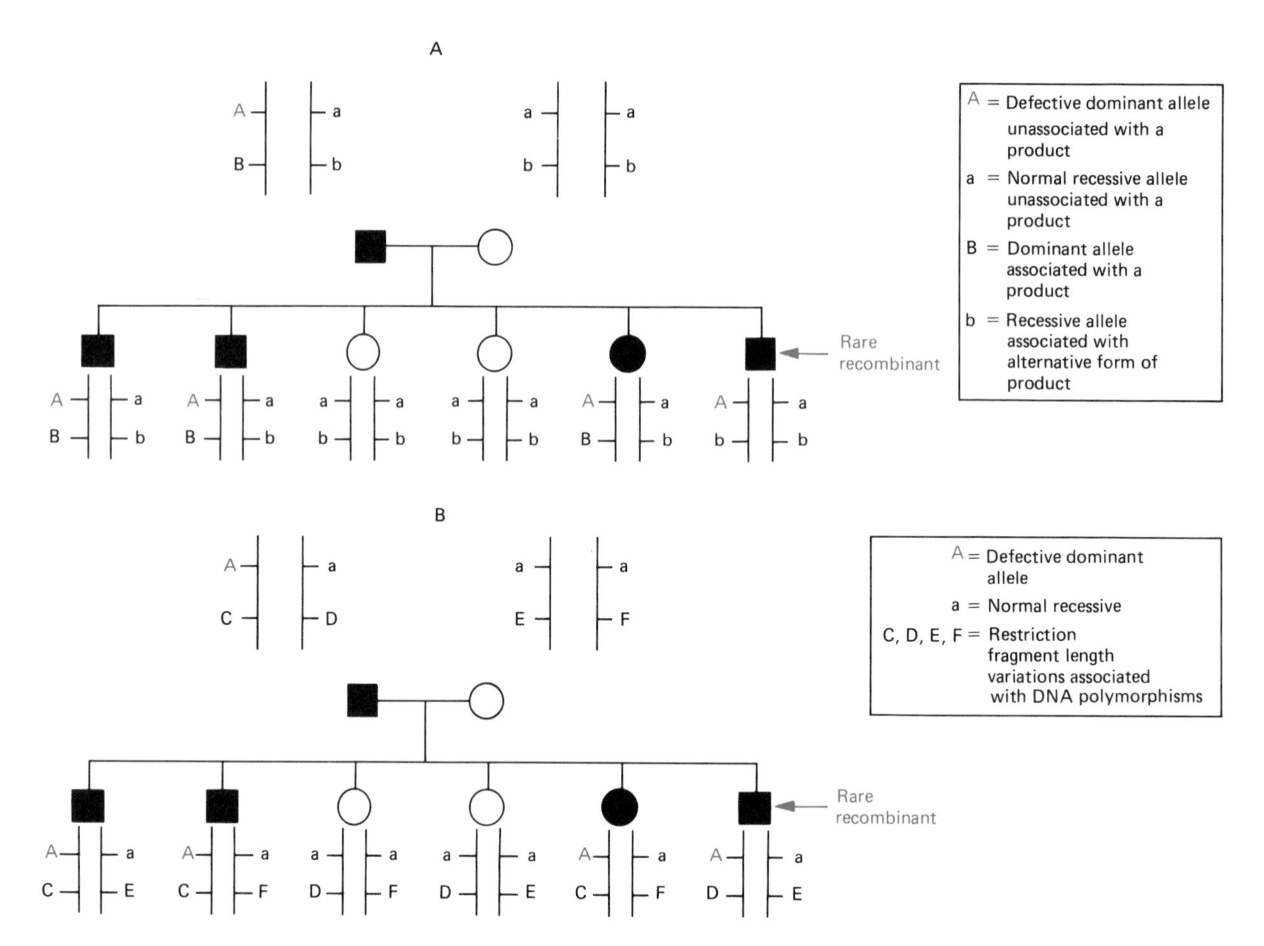

FIGURE 20-13. Inheritance of a defective allele in relation to markers. **A:** The A and a alleles are unassociated with any protein product, but their locus, a, is linked closely to the b locus occupied by alleles B and b, which *are* associated with products. Even though the dominant allele A may not be expressed until later in life, a person carrying it can be detected because of the close linkage between the two loci. The dominant defect in this pedigree is traveling with allele B. The presence of B product indicates that allele A is probably present. Therefore the b locus acts as a marker for alleles at the a locus. A recombinant resulting from crossing over may arise with a frequency depending on the distance between the a and b loci. If the two loci are 2 map units apart, then there is a 2% chance that a recombinant will arise at any birth. **B:** This pedigree is exactly the same as the pedigree shown in A, only here the marker is a DNA region on the chromosome that is variable in its response to a certain restriction enzyme because of DNA sequence variations. Consequently, restriction fragments of different lengths are generated, and these can be identified as different patterns following hybridization with an appropriate probe. Of the four patterns, pattern C is associated with the defect in this pedigree. Again, a rare recombinant can arise.

(shown by the red arrows in Fig. 20-15) allow recognition and cutting by the enzyme HindIII at that region of the DNA. If a change in base sequence is found at a variable site, the enzyme will fail to cut at that location. The probe can detect fragments of 17.5, 3.7, 4.9, and 15 kb pairs. Thus 10 different genotypes are possible because one could have a combination of haplotypes, having inherited one haplotype from one parent and the same or another haplotype from the other parent. Figure 20-16 represents autoradiographs of these different genotypes. The autoradiographs are produced following DNA digestion, gel electrophoresis, Southern blotting, and hybridization with the radioactive probe.

When DNA samples of 51 persons from unrelated families were analyzed, it was found that more than half were heterozygotes in relation to the DNA markers (the RFLPs) in this chromosomal region. When the U.S. and Venezuelan families known to be at risk for HD were studied, it was found that the HD locus *was* indeed associated with the RFLP in this region. For example, in a U.S. family, the HD gene was found to be associated with a chromosome carrying the A haplotype. In a Venezuelan family, it was found to be associated with the C haplotype. In other words, we now have a marker (the DNA region that generates the RFLPs) known to be linked to the HD locus. We can use the RFLPs to assess the risk that a person may have received the dominant HD allele from a parent. If it is shown, for example, that the allele is traveling through family members and that any member with the disease always has haplotype C, we then know that the disorder is traveling with haplotype C in that family line (Fig. 20-17).

Suppose a female parent with HD carries haplotypes A and C, and the C haplotype of that parent is known from pedigree studies to be linked to the HD allele. The male parent is homozygous for haplotype B (BB). There are two offspring. Analysis of DNA from one child shows the AB pattern, whereas the DNA of the other child shows the BC pattern (see Fig. 20-16). We now know that the former is probably free of the allele (barring recombination, estimated to be about 2%), whereas the latter probably carries the genetic defect and will later become afflicted with HD.

Human–mouse hybrid cells carrying different combinations of human chromosomes have now been screened. It is finally possible to assign the polymorphic DNA marker, and hence the haplotypes, to chromosome 4. Hybrid cells containing only chromosome 4 or only a portion of the short arm of chromosome 4

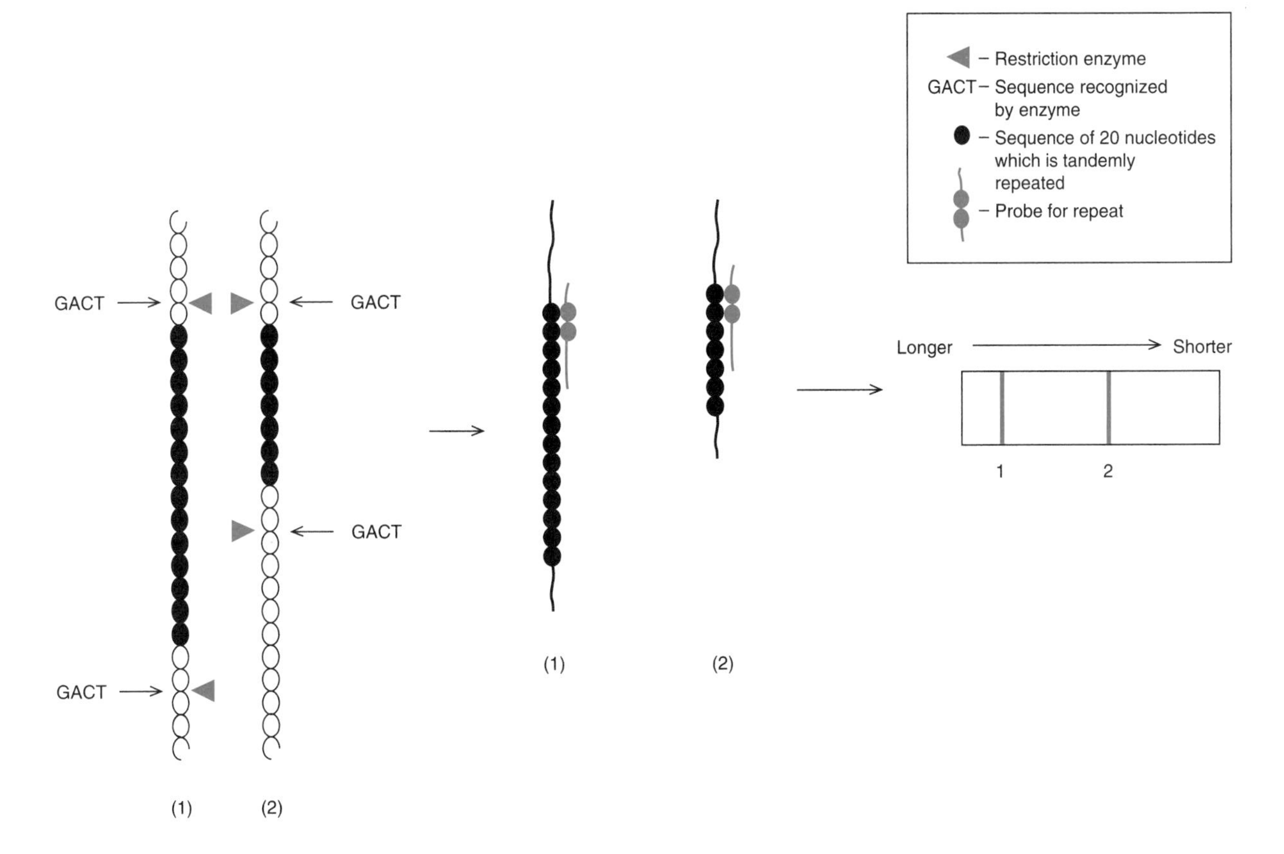

FIGURE 20-14. Tandem repeats as marker sites. In this example, a certain restriction enzyme can recognize the sequence GACT and will cut wherever it is found along the DNA. Tandemly repeated sequences may occur between two enzyme cutting sites. In a given chromosome (1), a large number of repeats can occur between two such sites, whereas only a small number of them may be present at the corresponding position on its homolog (2). Fragments of two different sizes are therefore generated by the enzyme cutting. Following electrophoresis and Southern blotting, these fragments of different size (1 and 2) can be recognized by an appropriate probe for DNA in the repeat region.

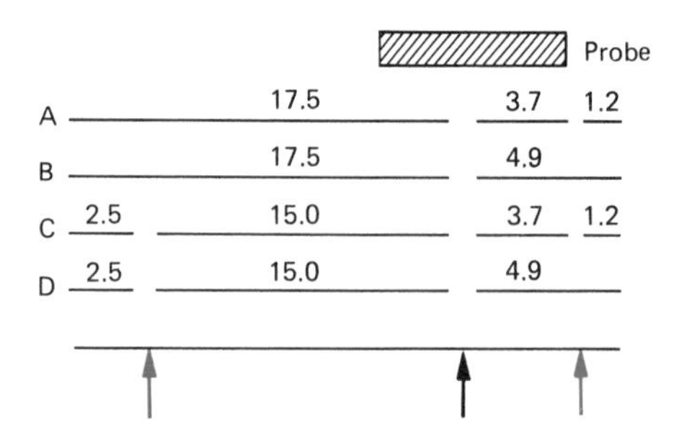

FIGURE 20-15. HindIII restriction sites in a region of human chromosome 4 linked to the HD locus. This region of chromosome 4 contains three sites (arrows) that can be recognized by restriction enzyme HindIII. Two of these sites are variable (red arrows). Consequently, when DNAs from different persons are analyzed, fragments of different lengths can be generated, depending on whether the enzyme cuts at both of these variable sites, one of them, or neither. The probe used to detect the variability identifies fragments of 17.5, 3.7, 4.9, and 15.0 kb in length. The probe does not overlap the two smallest fragments (2.5 and 1.2 kb) and therefore cannot detect them. Nevertheless, four different haplotypes or patterns can be distinguished: A (17.5 + 3.7 kb fragments), B (17.5 + 4.9 kb fragments), C (15 + 3.7 kb fragments), and D (15 + 4.9 kb fragments). Each haplotype is inherited as if it were a combination of tightly linked alleles on a chromosome. Each person carries two chromosomes 4 and thus can have any combination of the haplotypes.

yield DNA that reveals the presence of the marker. Since the marker is associated with the HD locus, the HD locus must also reside on chromosome 4.

The result of this new information on HD makes it possible to inform certain persons in whose families HD occurs if they are likely to have inherited the allele. When a routine test does become available, it will not, of course, be able to answer the question for everyone. A person would need to have cell samples tested (blood, skin) not only from himself or herself but also from the parents and even grandparents. Unfortunately, such a test will raise many additional problems. How will an otherwise healthy young person cope with the knowledge that he or she is probably carrying the HD allele and will most likely suffer from the devastating disease? Should a person at risk be required to take such a test in order to prevent transmitting the disorder to an offspring? Should a spouse have the right to insist on having such knowledge?

The Huntington's gene has now been associated with several chromosome markers found in the short arm of chromosome 4. Several teams of geneticists and molecular biologists are actively collaborating as they strive to identify the HD gene itself in this region of the human genome. Once the gene is isolated and cloned, its nucleotide sequence can be determined. This may lead to identification of the basic genetic defect in HD and the recognition of the product associated with the HD gene. The ultimate goal of the molecular procedures is to rid families of the suffering and fear this dreadful disorder inflicts. Perhaps it will someday be possible to prevent the onset of the disease in those known to carry the HD allele and to prevent its transmission through counselling of those known to be at high risk.

Chromosome walking and the search for specific genes

It is important to reemphasize the need to find a suitable probe to identify RFLPs. Lacking a probe to identify the DNA marker (to reveal the specific DNA fragments on an autoradiogram following Southern blotting), a DNA region will remain unknown no matter how variable it might be in its production of restriction fragments. Once a probe is found that reveals the variation, it still does not readily indicate where the variable site is located. However, if the RFLP, as revealed by the probe, is shown to be linked to a gene associated with a certain genetic disease, valuable information can nevertheless be obtained for diagnostic purposes to determine the degree of risk in a given case (Fig. 20-13B). Moreover, the probe may eventually be related to a specific chromosome. Once the probe is shown to hybridize only with DNA from a particular chromosome, it would then be apparent that the disease gene is also associated with that specific chromosome. Conversely, if one knows from the outset that a certain disease gene is found on a specific chromosome, one may then proceed to look for DNA markers known to be associated with that very chromosome.

To find the precise location of a disease gene, it is necessary to identify a marker that is very closely linked to it (one that shows a very low recombination frequency with the gene in question). Suppose that a marker recombines at a frequency of 5% with a gene. On the molecular level, this means that the marker and the gene are as many as 5 million base pairs apart. A recombination frequency of 1% between a gene and its DNA marker would imply that the two are about 1 million base pairs apart. This greatly increases the chance of finding the precise DNA segment on which the gene resides and hence of cloning it.

One procedure that has been used to characterize a chromosome region and zero in on a gene is known as *chromosome walking*. This method enables us to locate a marker that is much closer to the gene than any we may already have and may lead us to the very gene itself. To walk along the chromosome, we start with a probe that is known to be closely linked to the gene we wish to locate, let us say, a human disease gene. We then turn to a library of cloned DNA fragments that represent the entire genome. Using the probe, we scan the library, searching for a DNA sequence that overlaps a portion of the sequence in our probe (probe 1). If the probe happens to be a long stretch of DNA, we may simply subclone one of its ends (Fig. 20-18A). Once the probe

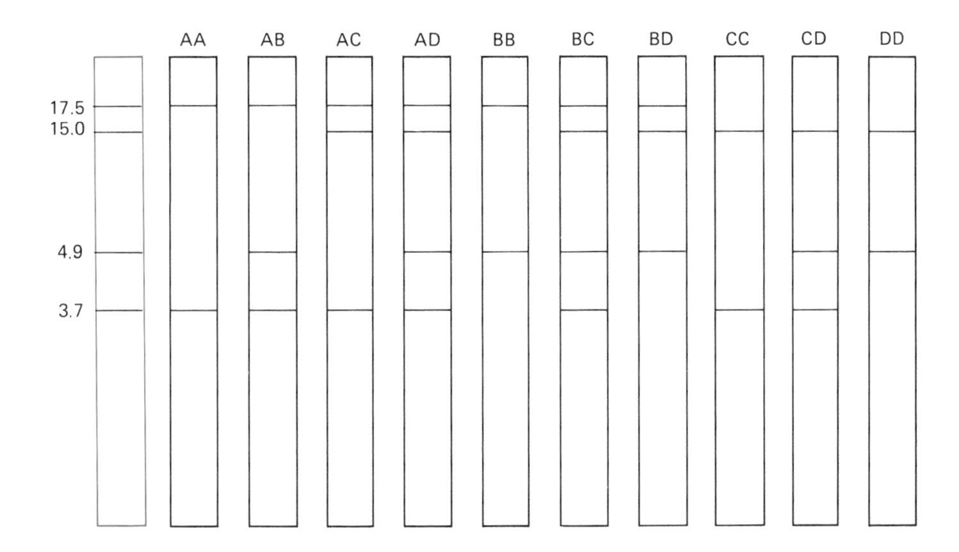

FIGURE 20-16. Depiction of autoradiograph patterns of the 10 genotypes possible from the different combinations of the A, B, C, and D haplotypes. The pattern on the left is the one seen when the genotype permits detection of the four fragments of different kilobase pair lengths. Reference to Figure 20-15 shows that these patterns are to be expected because of the different combinations of the haplotypes and because the radioactive probe will identify on the gel fragments of four different kilobase pair lengths (17.5, 3.7, 4.9, and 15 kb).

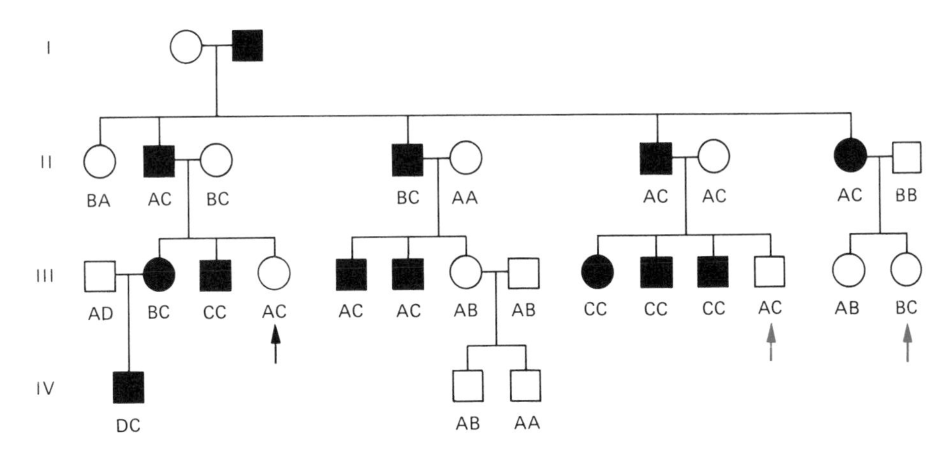

FIGURE 20-17. Pedigree showing inheritance of HD. This pedigree shows that, wherever it was possible to determine the haplotypes, each HD victim (shaded) also has haplotype C. Those persons with haplotype C who marry into the family do not have this rare allele (persons II-3 and II-7). An offspring in the pedigree who has received haplotype C from such a parent will remain free of the disorder, barring a rare crossover event (person III-4, indicated by arrow). Person III-12 (first red arrow) is at risk because the C haplotype may be derived from the afflicted or the nonafflicted parent. Person III-14 (second red arrow) is at extremely high risk, because the C haplotype has been transmitted by the afflicted female parent, and the recombination frequency is only 2%.

identifies a fragment that it overlaps, we now have available a second probe (probe 2). We can now subclone its end and in turn use this to scan the library to locate a fragment that overlaps with our probe 2. We have thus recognized probe 3 and can continue on in this way identifying more and more overlapping sequences. However, it is obvious that as we continue we do not know the direction in which we are walking along the chromosome, piecing together a long stretch of DNA from overlapping fragments. Are we going in a direction closer to or farther away from the gene of interest?

To keep sight of the direction in which we are travelling, we construct a restriction map from the fragments. If we are moving in one direction (toward the right in Fig. 20-18), then the restriction fragments that are present at the end of the map of the DNA of the first clone (probe 1) should be present in the map of the second clone (probe 2); those present at the end of the second should be in the third and so on. In this way a larger restriction map builds up that characterizes that region of the chromosome starting with probe 1 and extending in one direction. This can lead to the identification of markers (RFLPs) more and more closely linked to the gene of interest or even to the identification of nucleotide sequences within the gene itself. Chromosome walking is a very laborious process, but facilitating it are refinements such as pulse field gel electrophoresis (discussed later in this chapter) that permit the separation of larger fragments than is possible by ordinary methods.

Finding the Duchenne muscular dystrophy gene

As noted several times, sex-linked genes are the easiest ones to assign to their chromosome. Consequently, genes on the X chromosome that are associated with genetic diseases were the first to be pinpointed with the aid of technology from molecular biology. Let us now see how chromosome walking and some other procedures were used to find the gene associated with Duchenne muscular dystrophy (DMD). This fatal disorder, which entails a progressive weakening and wasting of the muscles, is inherited as an X-linked recessive and occurs in about 1 in 3,600 male births. Afflicted females are rare, but they proved to be very valuable in the identification of the region where the DMD gene resides. All afflicted females were found to have translocations involving chromosome band Xp21 in which a piece was translocated to an autosome. The particular autosome varied from case to case. As noted in Chapter 10, when a portion of an X chromosome in a female is translocated to an autosome, the remaining intact X becomes preferentially inactivated in all the somatic cells. Thus while these females carried an intact X in each cell, the normal DMD gene located on it would not be expressed. In effect, the other X would have a deletion. Since such females suffered from DMD, attention was focused on band Xp21 as the site of the DMD gene, and a search was undertaken to determine the extent of any deletions that might be present in other DMD victims.

Probes that were known to represent DNA of the X chromosome were used to determine whether they would hybridize with DNA from male patients with DMD and from persons free of the affliction. Cloned fragments from region Xp21 were identified, and these were shown to be deleted from DNA of patients but not from other persons. The deletions, however, were not the same from one patient to the next (Fig. 20-19A). Chromosome walking was then used to generate a restriction map, starting with a probe that identified region Xp21. Studies were made of DNA from a number of males with DMD, and these showed that large DNA deletions were present. Some of them extended to the left and others to the right of the region (see Fig. 20-19B). One deletion was found to be right in the middle of Xp21, extending neither to the right nor the left. This was an important observation, since it indicated that the region is most likely within the DMD gene and that it must have a major role in normal function of the gene. The assemblage of information left little doubt that the region of the DMD gene had been located, but knowing we are in the region of a gene is not enough. We must be able to piece together the DNA sequences in the region and establish which are introns and which are exons. A very interesting approach was taken to ascertain the coding and non-coding portions of the gene. Recall that Ohno's Law (see Chap. 6) recognizes that genes on the X chromosome are highly conserved. A gene that is sex linked in one mammal is almost certain to be sex linked in others. If genes on the X are indeed highly conserved, the coding sequences (exons) for the protein product of a given gene should be quite similar from one mammalian species to the next. In other words, the reading frames should have related nucleotide sequences. Therefore, if we take a probe representing a short sequence of human DNA from the region of an X-linked gene and find that this probe hybridizes with DNA fragments from the genomes of several other

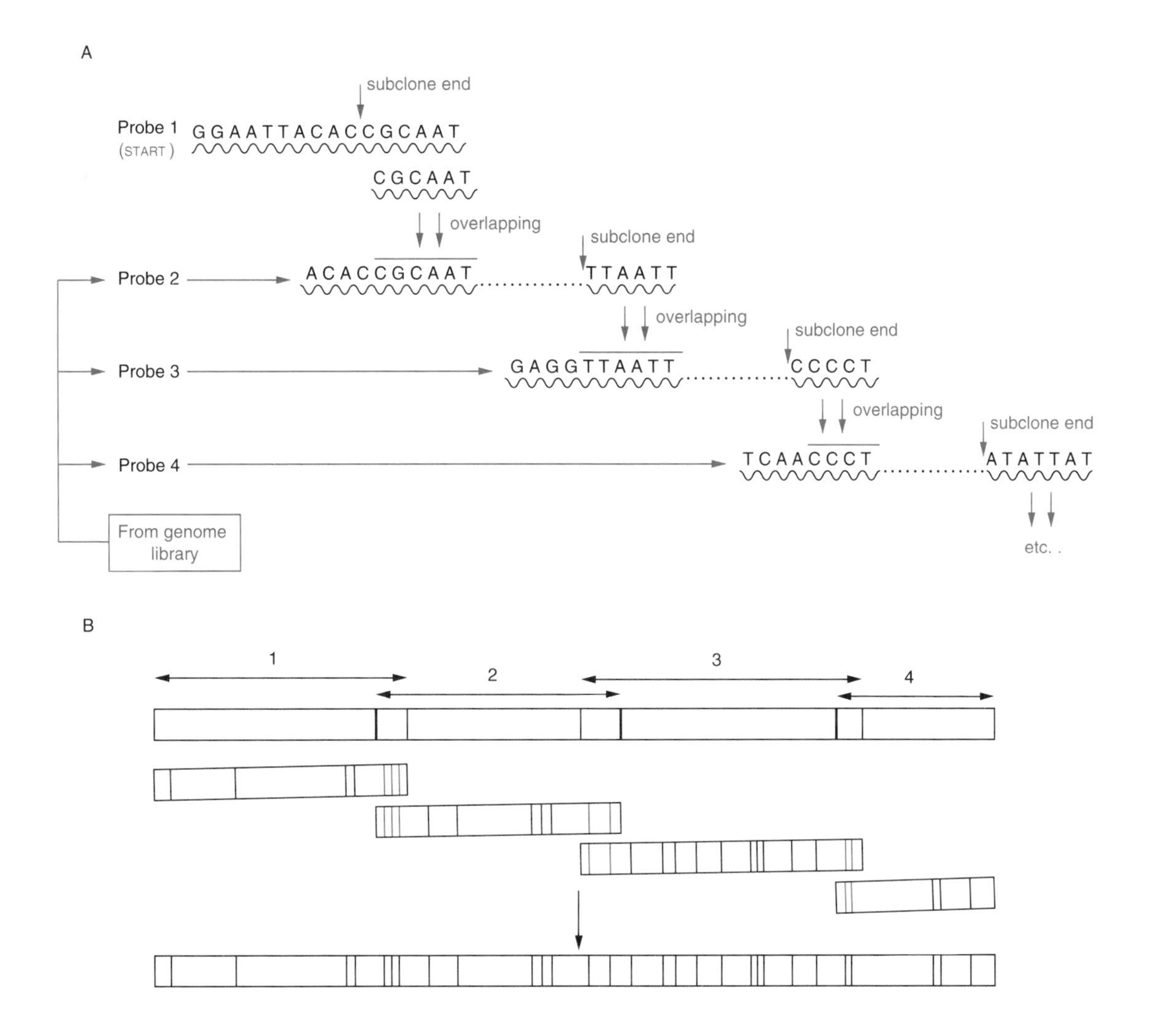

FIGURE 20-18. Chromosome walking. **A:** This procedure makes it possible to characterize a portion of the genome, to provide markers closely linked to a specific gene, and to lead to sequences within the gene itself. A probe (probe 1) that identifies sequences close to the gene is used to locate a fragment (probe 2) from a genome library that overlaps with it. (The ends of large fragments are subcloned and used as probes.) This fragment in turn is used to identify a third fragment (probe 3) from the library that overlaps with it. Probe 3 may be used to identify still another fragment and so on. The procedure identifies a series of overlapping segments that make it possible to characterize a long DNA segment. **B:** A restriction map is made of each fragment composing this portion of the genome. Restriction fragments present at the end of one fragment (shown in red) must be present in the following fragment if we are travelling in one direction. A larger restriction map is generated that may lead to the identity of a marker very close to the gene if we are walking toward it. (Sequences and fragments shown are only for diagrammatic clarity.)

species, we know that the sequence has been conserved in evolution and that it very possibly could be part of an exon. The fragment is then sequenced and the sequence studied to determine if it could be part of an open reading frame (a sequence of triplet nucleotides coding a sequence of amino acids in a polypeptide). Knowledge of nucleotide sequences at intron–exon junctions (see Fig. 13-18) also helps to distinguish the coding from the noncoding parts of the interrupted gene.

Various fragments in the region of the X chromosome suspected to be part of the DMD gene were studied in this way. Fragments believed to be part of the gene were then used as probes of an available cDNA library. This cDNA library had been prepared from mRNA derived from muscle tissue. Hybridization with the probes led to the identification of an exceptionally large cDNA and hence the mRNA from which it had been derived (Fig. 20-20). Hybridization was then carried out between cloned genome DNA representing the region of the walk and led to the identification of the exons and the entire DMD gene itself. The gene proved to be incredibly large, 10 times longer than any other gene known, containing about 2 million base pairs and 60 exons! This unusual size may account in part for the high mutation rate of the DMD gene. The majority of the alterations in the DMD gene are actual deletions of different portions of the genetic material rather than point mutations.

Once open reading frames in the gene were recognized, the amino acid sequences of parts of the product of the gene could be predicted, and short polypeptide fragments were synthesized. These were injected into animals to elicit antibody formation. The antibodies were then added to muscle preparations taken from normal persons. This procedure led to the identification of dystrophin, a protein that is present only in low levels in unafflicted persons. In almost all DMD patients, however, it is absent. The function of dystrophin, which is also present in other tissues such as the brain and kidney, is not understood, although the suspicion is that it protects muscle fibers from damage incurred during contraction.

While there is still much to learn about the nature of DMD, the identification of the gene is sure to accelerate our understanding of it and lead to steps in its prevention. Improvement in detection of female carriers has already occurred. We now have available RFLP markers around the dystrophy locus. These can enable us to recognize carrier mothers with an accuracy of about 95%. This is far from simply knowing that a boy born to a woman at risk has a 50% chance of having the fatal affliction.

Finding the gene for cystic fibrosis

Inherited as an autosomal recessive, cystic fibrosis (CF) is the most common genetic disorder in the Caucasian population, occurring about once in 2,000 births. The frequency of carriers in the population is high, about 1 in 20. The disorder, which greatly decreases life span, affects various organs including the pancreas, sweat glands, and lungs. The underlying problem appears to be some defect in the regulation of ion transport in secretory epithelial cells. This leads to the production of abnormal, thick mucus that blocks airways of the lung and ducts of the pancreas. The result is chronic lung infection and impaired nutrition due to pancreatic insufficiency. In addition to these and various other symptoms, the sweat glands secrete abnormally high concentrations of sodium and chloride.

Locating the gene associated with CF was an extremely difficult undertaking. Not only was there no information on the protein product encoded by the CF gene, but the gene was not associated with any apparent structural chromosome change as was the case with DMD. In the absence of such indicators as deletions or translocations, the gene location was more difficult to spot in the human DNA, which includes about 100,000 genes.

A big step forward in finding the CF gene was made with its definite assignment to the long arm of chromosome 7, band q3. This was done on the basis of a large number of polymorphic DNA markers. Among these were tightly linked markers that flank the sides of the CF gene and thus serve as points of departure for zeroing in on it. However, the flanking markers are about 100 kb apart, a very large distance to be travelled by chromosome walking. Fortunately, new techniques have been devised that speeded up the laborious search for the CF gene. One of these is an improvement in conventional gel electrophoresis and is known as pulsed-field gel electrophoresis (PFGE). In the ordinary procedure, the voltage applied to the gel is constant and in one direction. It does not, however, usually permit movement and separation in the gel of DNA fragments larger than 50 kb. Many genes exceed this in size and may reach 1 million base pairs or more in length. In PFGE, on the other hand, short pulses of electricity are applied to the gel in different directions. Since the electric fields are turned on and off for a matter of seconds or minutes, they are referred to as *pulsed fields.* Treating a gel containing DNA fragments in this way has made it possible to separate very large pieces, those from 50,000 to more than 5 million base pairs long! Originally designed to study DNA from entire yeast chromosomes, several variations of PFGE are now in use. The procedure makes it possible to define very large restriction fragments and pieces of a human chromosome. It is being applied more and more in analyses of the human genome on the molecular level and was utilized to study fragments throughout the search for the CF gene.

A technique known as *chromosome jumping* was devised to be used in conjunction with chromosome

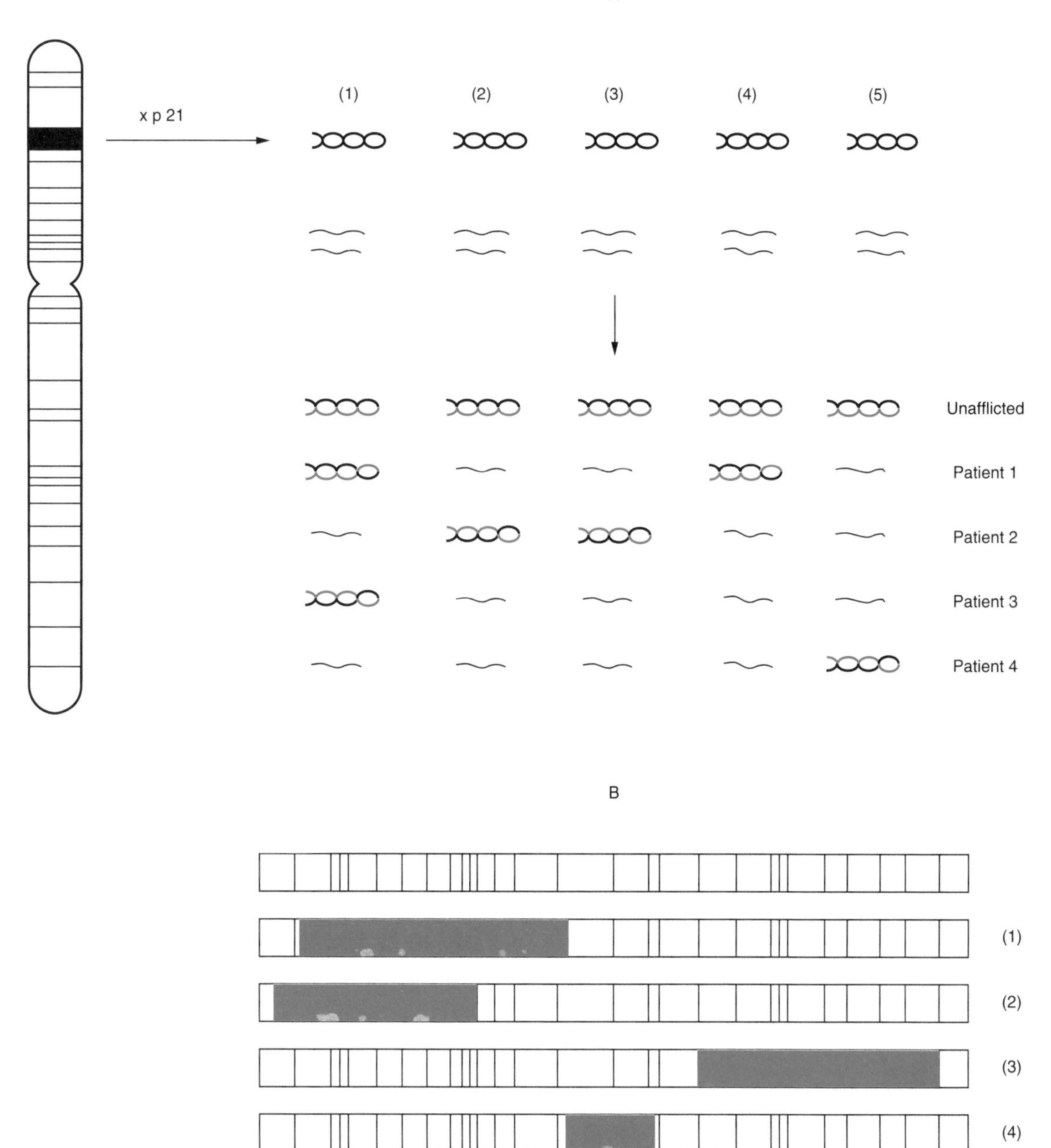

FIGURE 20-19. Deletions in the DMD gene. **A:** DNA fragments from region Xp21 were cloned and then used as probes (1–5). Denatured probe DNA (black) was allowed to hybridize with denatured DNA from various persons (red). DNA from the X chromosomes of healthy individuals hybridized with all of the fragments. In persons with DMD, the presence of deletions was indicated, since not all the fragments were able to form hybrids. The fragments that appear to be missing differ from one DMD patient to the next. **B:** A chromosome walk established a restriction map in the Xp21 region (shown diagrammatically here at the top). Studies of healthy persons showed no deletions in this region. Different DMD patients (1–4) showed large deletions (red) affecting different portions of the region.

walking in order to speed up the process of travelling along the chromosome. To form clones for use as probes in chromosome jumping, DNA is cut with certain restriction enzymes (rare cutter enzymes) that recognize sequences that occur only infrequently in the human genome (Fig. 20-21). This partial digestion yields large fragments that can be separated by PFGE. Fragments of a certain length can be selected, let us say in the 70 kb range. These linear molecules are then placed in a dilute solution along with a suitable vector that carries a marker gene. Under conditions that allow ligation, the linear molecules will eventually circularize. The circular molecules are then exposed to a restriction enzyme such as EcoR1, which will make many cuts producing many small fragments. Any small fragment with a vector and its marker can be selected out. Note that the DNA on either side of the vector will be 70 kb apart in this example. A probe of this construction can therefore pick out from a genome library fragments that are 70 kb apart. Many clones are established from these small fragments to form a jumping library, each clone representing a 70 kb jump. Jumping permits skipping over long regions of the genome and is consequently much faster than walking. It has a decided advantage in enabling the investigator to pass over many regions of the human DNA that contain stretches of repetitive DNA that are unclonable and that can end a chromosome walk.

Starting from the flanking markers, a combination of chromosome walking and jumping yielded DNA segments from genome libraries. These segments were then used to produce a restriction map (Fig. 20-22). The direction of walking with respect to the flanking markers had to be established. Hybridization

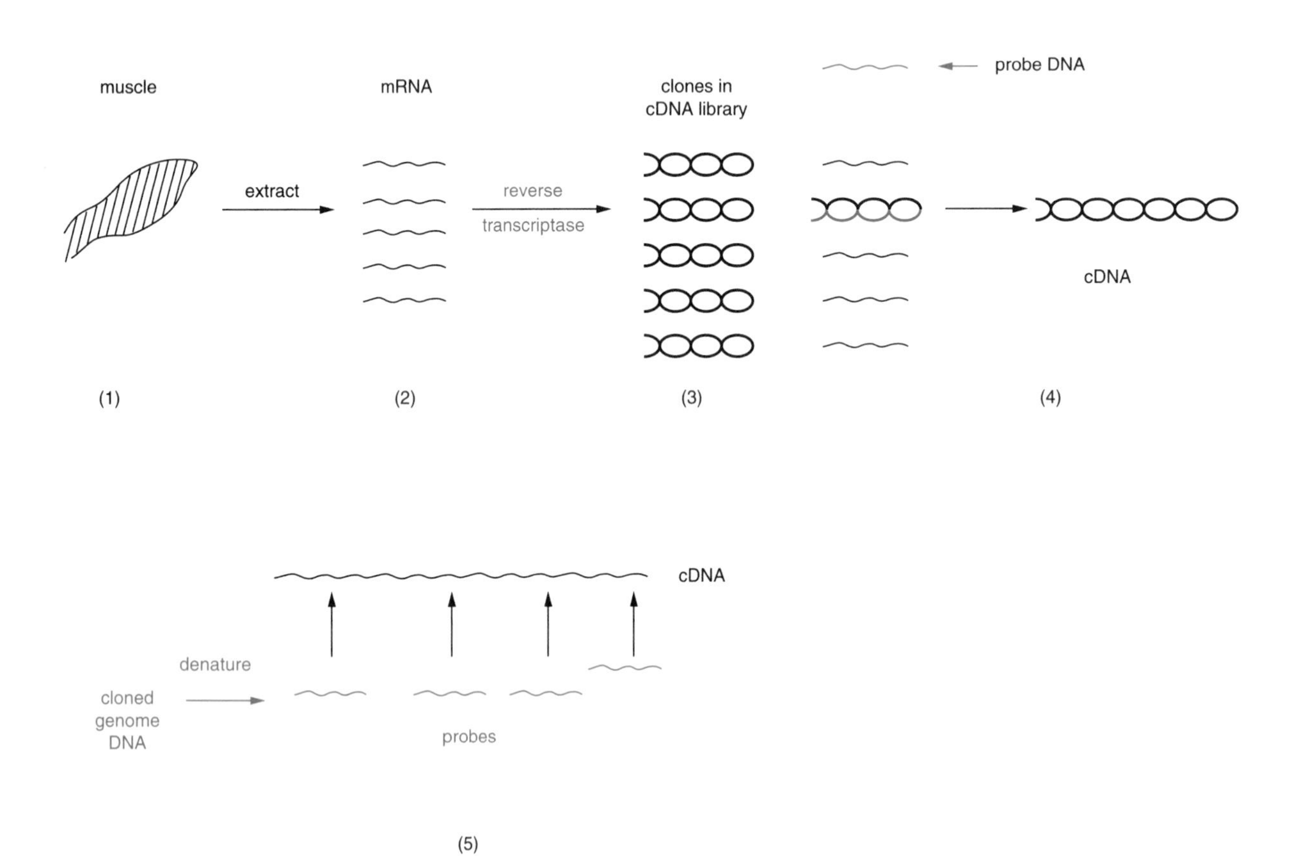

FIGURE 20-20. Some steps in locating the DMD gene. An extract from normal muscle tissue (1) yields different mRNAs (2) that are converted to cDNA to form a cDNA library (3). Denatured DNA of suspected exons of the desired gene were used as probes (red) with denatured cDNA (black). Hybridization led to the identification of a large cDNA (4). The denatured cDNA was then subjected to probes from parts of the genome thought to represent portions of the gene (5). From the hybridization between genome DNA and cDNA, the structure of the DMD gene was pieced together.

analysis of each segment was performed with human–rodent somatic hybrid lines to make sure the segments were localized on chromosome 7. The molecular cloning led to the identification of a large DNA segment thought to contain the CF gene. As the jumping and walking proceeded, a search was made for possible genes in the accumulated DNA segments. As in the case of the DMD gene, DNA sequences were compared with those of other animals to locate any fragments containing conserved sequences that might be part of a reading frame of a gene. Four such conserved sequences were found, but only one of them was to lead to the CF gene. Identification of the gene was greatly speeded up due to the fact that cDNA libraries were available, among them those derived from mRNA of sweat gland epithelial cells taken from both healthy persons and those with CF. Such cells are believed to be one type in which the CF gene is expressed. The conserved sequence was then used to probe the cDNA libraries. It identified one single clone in a cDNA sweat gland library from a healthy individual (Fig. 20-23A). This clone was sequenced and found to have a possible long open reading frame. The sequence overlap between the clone and the probe that located it was very short and was later found to be the first exon of the CF gene. With the identification of the first clone, the stage was set for the screening of the cDNA libraries and the isolation of overlapping cDNA clones. The DNA segments corresponding to the cDNA clones were isolated from a genome library, and the two DNAs, the cDNA and the genome DNA, were aligned and compared (Fig. 20-23B). Many of the cDNA sequences were from unprocessed mRNA and thus contained introns. Analysis of the cDNA and genomic DNA fragments enabled the investigators to work out intron–exon boundaries and to recognize nucleotide sequences in the gene apparently coded for a polypeptide. The gene was found to span 250 kb and to contain 24 exons. Sequencing of the coding regions led to the prediction of a protein composed of 1,480 amino acids. Once the triplet sequences of the coding regions were worked out, a comparison was made of these regions in cDNA derived from persons with and without CF. The most striking difference between the two was a three base pair deletion in the coding sequence of the CF gene. The triplet deletion corresponds to a loss of the amino acid phenylalanine at position 508 in the protein. This mutation is found in about 70% of the CF patients. More than 165 other mutations have been found to compose the rest of the genetic changes in the CF gene. A gene smaller than the large CF gene was anticipated from the absence of large chromosome rearrangements and the involvement of only a rather limited number of different types of mutations.

The structure of the predicted protein suggests a membrane protein, since its sequence resembles those of several proteins known to take part in the transport of substances across cell membranes. The thick mucus characteristic of CF may stem from a protein membrane defect that results in the inability of the cells to secrete chloride and water into the mucus. The phenylalanine deletion is in a region of the protein that appears to contain a binding site for ATP. Perhaps loss of the amino acid interferes somehow with transport of chloride ions as a result of the loss of ATP binding to the protein thus depriving it of energy needed for transport.

A bothersome question that remains regarding CF is "What accounts for the high frequency of the defective allele in the Caucasian population?" Since very few afflicted persons reproduce, is the high incidence of the defect due to an unusually high mutation rate of the CF gene? Perhaps the defect gives the heterozygote some reproductive advantage, as is true in sickle-cell anemia. Perhaps the high incidence of CF results from the founder effect, a phenomenon in which the few individuals who happen to establish a new population also happen to include a high number of individuals who carry an atypical allele that is then transmitted to their descendants among whom it persists. The identification of the CF gene and its major defect eliminates the suggestion that many loci in the genome are involved in the disorder. While various aspects of CF are still to be worked out, identification of the gene and its product are giant steps forward. Sometime in the future it may become possible to incorporate the normal gene into defective cells such as those of the lungs and help alleviate some of the dire effects of the CF mutation. One immediate benefit of the discovery of the CF gene and its mutations is in the detection of carriers. Up to now, CF carriers could be detected only in those families with affected children. It was necessary to look for markers known to be inherited with the CF gene. This requires a great deal of information from family pedigrees, as demonstrated with the Huntington's gene. Now that the CF gene and its major defect have been identified, testing is now available to any individual in the population.

Locating the genes for CF and DMD are outstanding examples of the power of an approach to a genetic problem that has been designated *reverse genetics.* It should be apparent that in locating the CF and DMD

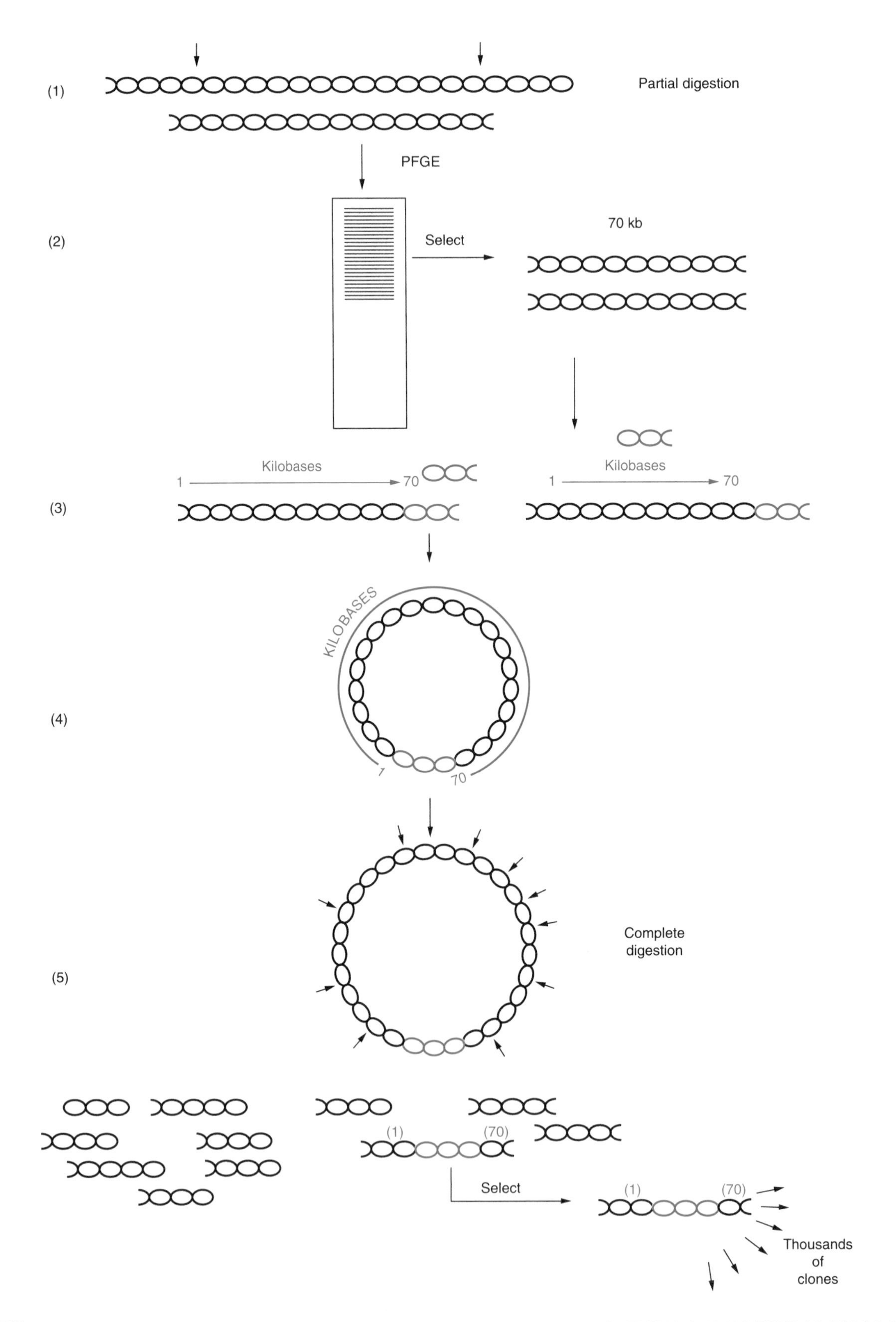

Partial digestion
(1)
PFGE
(2)
Select
70 kb
Kilobases
1
70
(3)
KILOBASES
(4)
1
70
Complete digestion
(5)
(1)
(70)
Select
(1)
(70)
Thousands of clones

genes no information on the gene product was available at the outset. The methodology of molecular biology has made it possible to approach a problem by starting with the genetic material itself. In the examples of CF and DMD the gene was pieced together, and the associated product was then deduced from knowledge of the gene. This is quite different from *forward genetics.* In this traditional approach, a phenotype (let us say sickle-cell anemia) is related to a defect in a product (the β-chain of hemoglobin) that is then related to the specific gene that is involved. The direction between the two approaches is thus reversed. Several prominent molecular biologists prefer the terms *positional cloning* and *functional cloning* to designate *reverse genetics* and *forward genetics,* respectively, and these designations are becoming widely adopted.

Localizing the genetic basis of Down syndrome

Trisomy for chromosome 21 is the most common of the human trisomies and is directly associated with the assortment of features that describe Down syndrome (see Chap. 10). Chromosome 21, the smallest of the human chromosomes, contains only 45 million of the 3 billion base pair total composing the human genome. Only about 1,500 genes are believed to reside on chromosome 21, and less than 20 are actually known at the present time. Of those persons with Down syndrome, 5% have a normal chromosome number, but nevertheless they are actually trisomic for a portion of 21 as the result of a translocation of a portion of chromosome 21 into another chromosome (typically a member of the D or G group). Studies on individuals with the translocation have shown that only the bottommost third of the long arm (q arm) of chromosome 21 is necessary to produce the characteristics of Down syndrome. The methodologies of molecular biology, coupled with cytogenetic analyses, are making possible fine analysis of that portion of chromosome 21 and have permitted the identification of specific genes in that part of the q arm.

FIGURE 20-21. Preparing a chromosome jumping library. **1:** A partial digest of high-molecular-weight DNA by a rare cutter enzyme yields very large fragments. **2:** PFGE separates large fragments in a gel. Fragments that position at a certain size zone (say 70 kb) can be isolated. **3:** These 70 kb fragments are then placed into a solution containing a vector (red) that carries a selective marker under conditions that favor joining of the marked vector to fragments. **4:** In time, the conditions enable fragments to circularize. The ends of the vector are now attached to beginning and end segments of the fragment that are 70 kb apart (as indicated by numbers). **5:** The circular molecules are subjected to an enzyme that cuts them into many small pieces. The vector, carrying pieces on each end, is selected out on the basis of the marker it carries. The piece of DNA attached to one end of the vector is 70 kb away from the piece attached to the other end as indicated by the red numbers. The selected vectors are then cloned to form jumping libraries of DNA probes to be used in conjunction with chromosome walking (see Fig. 20-18).

Particularly useful in the search for the genetic basis of Down syndrome have been cell hybrids formed between human cells and those from a strain of Chinese hamster ovary cells (CHO). These particular CHO cells carry a mutation that inactivates a particular gene known as *Gart,* which resides on human chromosome 21 and codes for a specific enzyme (phosphoribosylglycinamide synthetase). Such mutant CHO cells are deficient in purine metabolism. In the absence of the enzyme, the cells die on a medium lacking purines. When hybrid cells are formed between these hamster cells and human ones, only those hybrid cells with chromosome 21 will grow on a medium lacking purines. This is the case, since the human cells contain a functional Gart gene. Growing hybrid cells on a medium devoid of purines thus permits selection of hybrid cells containing human chromosome 21. The Gart gene has now been pinpointed to a specific portion of chromosome 21 known as 21q22.1 (long arm of chromosome 21, region 2, band 2, subband 1) (refer to Fig. 20-24 and Chap. 2).

It has been possible to narrow the gene down to this chromosome region by forming somatic cell hybrids between the hamster cells and human cells carrying a translocation of just a portion of chromosome 21. The procedure has utilized naturally occurring translocations involving various segments of chromosome 21. It has also made good use of chromosome 21 fragments resulting from exposure of cells to heavy radiation doses. Human cells with different fragments resulting from the treatment were then hybridized with hamster cells. Cells with hamster chromosomes plus a human chromosome fragment can then be screened to detect the presence of a specific DNA sequence or of a specific gene such as Gart. The localization of genes in the lower region of the q arm of chromosome 21 has also been greatly aided by PFGE, which has permitted analysis of large pieces of DNA (containing one or more genes) from chromosome 21.

Maps of chromosome 21 have now been prepared utilizing somatic cell hybridization as well as RFLPs,

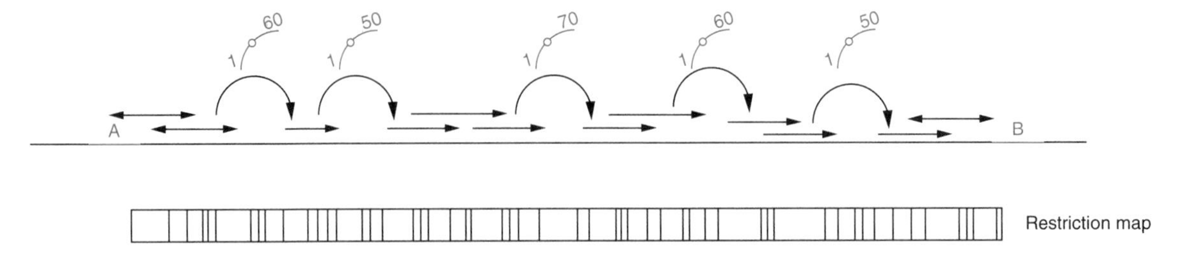

FIGURE 20-22. Searching for the CF gene. Flanking markers (red segments A and B) tightly linked to the CF gene were identified. Since the CF gene must be between them, they served as points of departure to characterize the region. Starting from the markers, chromosome walks were taken in each direction (straight horizontal lines with arrowheads). The walking was carried on in conjunction with chromosome jumping (curved lines). Probes from a jumping library (red) speeded the process up by identifying fragments from the genome that could be apart by 50 kb or more. (Note that pieces on either side of the vector are many kilobases apart, as indicated by the numbers.) A restriction map of the region was constructed on the basis of fragments from the genome identified by the combination of walking and jumping. The region represented by the map must contain the CF gene somewhere.

in situ hybridization, and PFGE. In addition, clinical and cytogenetic studies have provided valuable information that has tied together the data from the various sources. All the information has led to the identification on the q arm of chromosome 21 (specifically, band 21q22) of five genes, including Gart. This region is the one most likely responsible for the symptoms of Down syndrome. Another gene found in band 21q22 is the ets-2 gene, which is particularly interesting, since it may shed some light on the fact that Down individuals have a very high risk of developing leukemia. Studies of persons without Down syndrome but with acute myelogenous leukemia have revealed that about 20% of them possess leukemic cells carrying a translocation that involves movement of genetic material from chromosome 21 to chromosome 8. It is only the leukemic cells that show the translocation. The chromosome fragment involved in the translocation was localized to 21q22. Moreover, trisomy 21 is the most frequently occurring chromosome anomaly that occurs in the leukemic cells of children with the acute leukemia. It now appears that the ets-2 gene may be a protooncogene. Its presence in three doses, as in trisomy, or its movement to a new location following a translocation may somehow alter expression of the gene and lead to the leukemic state.

Persons with Down syndrome exhibit elevated purine levels in the blood. The Gart region of chromosome 21 codes for three different enzymes that are involved in purine synthesis. Elevation in purines is known to be associated with serious metabolic upsets, including mental retardation. A favored idea is that trisomy of the Gart region may upset purine levels and thus be responsible for many of the problems that accompany Down syndrome.

Other genes in the 21q22 region are also candidates for involvement in Down syndrome. One of these encodes for a protein found in the lens of the eye, an important finding, since persons with Down syndrome suffer from lens defects and have an increased risk of cataract development. Particularly interesting is a gene in this region that codes for an enzyme (superoxide dismutase [SOD-1]) that protects cells from highly reactive free radicals. There has been a suspicion that these radicals are involved in the process of aging. Down persons age rapidly. It is therefore valid to speculate that an extra dose of the SOD-1 gene may lead to upsets in enzyme levels that alter the normal aging rate. The long arm of chromosome 21, therefore, holds possibilities for studies on aging. This is especially so since one form of Alzheimer's disease has also been associated with the q arm of chromosome 21 (see next section).

Utilizing the methodologies of molecular biology, important insights into the direct causes of the disorders associated with Down syndrome may be made in the near future. Further experiments employing genetic manipulation may clarify the roles of the products of the genes in the 21q22 region during normal metabolism and show the exact way in which they are related to Down syndrome. Armed with such knowledge, the effects of trisomy 21 may ultimately be prevented in utero or successfully treated in a Down individual.

Insight into Alzheimer's disease

Two general forms of Alzheimer's disease are recognized: late onset, which strikes after age 70 years, and early onset, which occurs before age 50 years. Family studies indicate that the early onset form is transmitted on the basis of an autosomal dominant factor. The use of RFLPs as markers showed that the genetic defect in certain cases of early onset Alzheimer's occurs on the q arm of chromosome 21, close to the centromere.

A characteristic feature of Alzheimer's disease is the accumulation of a protein, β-amyloid protein, that

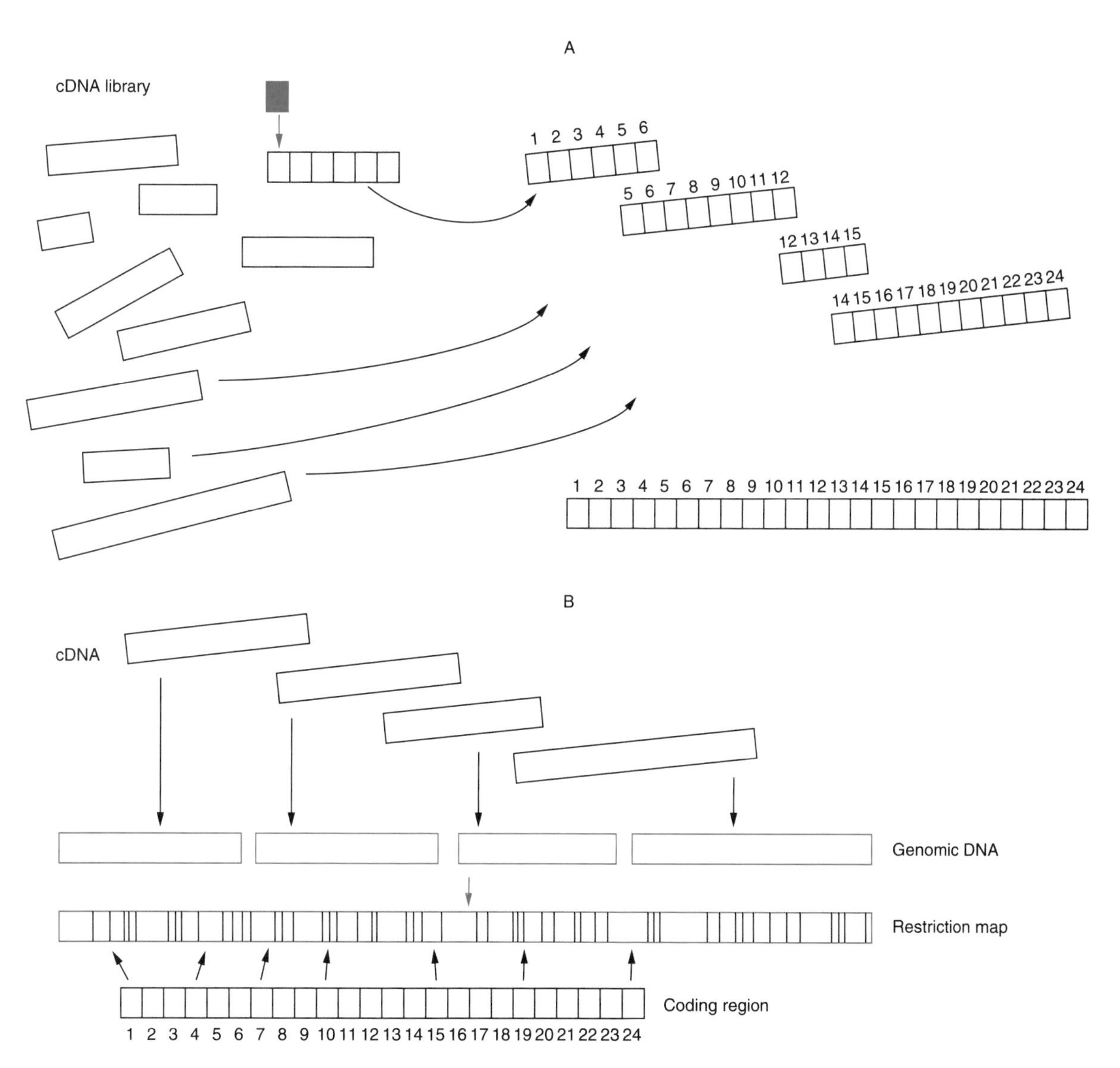

FIGURE 20-23. Zeroing in on the CF gene. **A:** A fragment from the CF region found to contain a conserved sequence (red) was used to screen a cDNA library. The latter had been prepared from mRNAs in cells of the sweat gland. The fragment identified a larger fragment from the library. Since this is cDNA, the sequence it carries must be part of a transcript from this region, which is known to contain the CF gene. The identification of the first fragment led to the identification of other overlapping fragments from the cDNA library. From these overlapping fragments, it was possible to deduce a coding region of the CF gene consisting of 24 exons. **B:** The cDNA fragments were also hybridized to DNA fragments in a genome library. The DNA (red) recognized by the cDNA sequences was characterized, and a restriction map was built. The coding regions (1–24) deduced from the cDNA were then related to the genome. (The figure is diagrammatic and does not attempt to represent the actual number of fragments and restriction sites involved.)

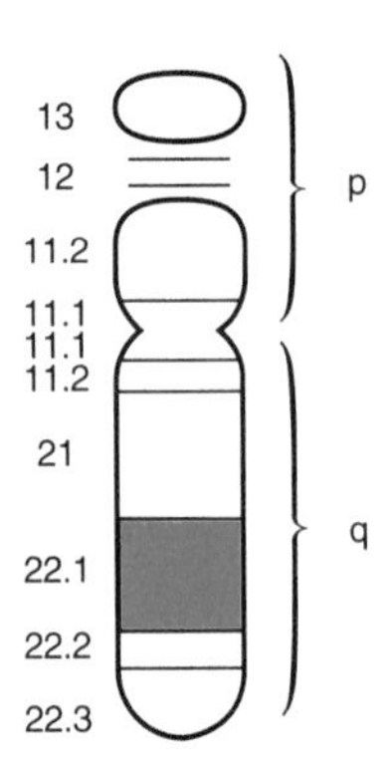

FIGURE 20-24. Localization of genes implicated in Down syndrome. Five genes in the 21q22 region (which includes bands 22.1, 22.2, and 22.3) may be associated with the metabolic upsets associated with Down syndrome. The Gart gene is involved in purine metabolism and has been localized to 21q22.1 (red) along with the SOD gene. The ets-2 gene resides in band 22.2. Two others are found in band 22.3.

piles up in brain tissue where it forms abnormal plaques. β-amyloid protein is derived from a larger protein, amyloid protein precursor (APP), a normal cell constituent. Copy DNA for the APP gene was cloned, and this in turn was used as a probe in conjunction with RFLPs. The results showed that the APP gene resides on chromosome 21 in the same region linked to early onset Alzheimer's disease, suggesting that the APP gene might very well be the gene for the familial form of the disease.

In one investigation, the nucleotide sequence of a portion of the cloned APP gene was determined. This segment was studied in a family having members who had early onset Alzheimer's disease. When the nucleotide sequence of this gene segment was compared between affected and nonaffected family members, it was found that the DNA of the former contained a specific point mutation. This genetic alteration brings about a substitution of the amino acid isoleucine for valine at a specific site in APP. Unaffected family members did not show this change, nor did persons with the late onset form of the disease. The same mutation also appears to have been found in afflicted members of still another family with a history of early onset Alzheimer's disease.

Such findings support the concept that the APP gene and the Alzheimer's gene are the same, but the picture has proved to be more complicated. The APP gene may well be the Alzheimer's gene in certain early onset families, but other studies show that it cannot be the only gene involved. For example, genetic linkage analyses of other families reveal that hereditary Alzheimer's is not always associated with chromosome 21, the site of the APP gene. All evidence strongly suggests that hereditary Alzheimer's is a heterogeneous disease. In some families, it may result from a mutation in the APP gene, whereas in others a mutated gene located elsewhere in the genome may be responsible. Nevertheless, the association of mutation in the APP gene with some cases of Alzheimer's plus the fact that a characteristic feature of the disease is the abnormal accumulation of β-amyloid protein, a derivative of the APP gene product, is very suggestive. Perhaps the various genetic and environmental factors operating to produce the disorder may all somehow bring about their effects through the APP gene, resulting in the abnormal deposition of β-amyloid protein after causing its release from the larger APP molecule.

Investigations in this direction are certain to be pertinent to the study of the aging process. In addition to aged brains of humans, those of several species of mammals (monkey, orangutan, dog, polar bear) display an abnormal accumulation of the β-amyloid protein. This indicates that the APP gene has been highly conserved throughout mammalian evolution and implies that its product, APP, performs a pivotal role in the normal cell. In normal nerve cells, APP is embedded in the cell membrane, where it may play a role in communication between neurons. Tied in with this idea is the interesting observation that mRNA for the protein occurs in all types of tissues examined. However, abnormal accumulations of the β-amyloid protein that is derived from APP are found only in the brain! This excess deposition appears to be a typical feature of the aging process.

Very provocative is the fact that β-amyloid protein also accumulates in brain tissue of all individuals with Down syndrome over age 35. Moreover, it has been demonstrated that mRNA for APP occurs at higher than normal levels in the brains of fetuses with Down syndrome, suggesting that the effects of an extra dose of the APP gene may somehow lead to an accumulation of β-amyloid protein in Down persons. Still remaining is the question concerning the factor(s) that cause the release of the β-amyloid protein from the APP and its abnormal deposition in persons with Alzheimer's and those with Down syndrome. Conceivably, the accumulation of the β-amyloid protein is an effect rather than an underlying cause of brain disorders. Also to be considered is the possibility that some abnormality in the processing of the APP protein may be involved in the disease process. Perhaps a mutation in a gene disrupts the normal function of APP and causes it to release β-amyloid protein, which then accumulates in certain cells. The association of certain hereditary cases of Alzheimer's disease and Down syndrome with the long arm of chromosome 21 is an intriguing fact that will continue to spur on investigations to clarify the underlying causes of these two major human afflictions.

One investigation has identified a site on chromosome 14 that appears to be associated with the majority of those inherited Alzheimer's cases having the very early onset age of 45. The eventual identification of the gene at this location is certain to lead to a fuller understanding of the molecular interactions operating in Alzheimer's disease.

The fragile X syndrome and mental retardation

From the early 1900s, an excess of males has been noted among mentally retarded persons. Studies of many family pedigrees suggested that a sex-linked mode of transmission is involved in certain cases of mental retardation. Estimations tell us that there are about 69 different X-linked syndromes involving mental retardation. Since it has been calculated that as many as 2%–3% of the general population of industrialized countries are mentally retarded, interest has increased in research related to the causes of retardation.

In 1969, a description was made of the fragile X site, located in the long arm of the X chromosome in subband Xq27.3. This site appears in cells growing in culture as a nonstaining gap of varying size in the terminal part of the X (Fig. 20-25). The gap is revealed by Q-, G-, and R-banding procedures and often involves both chromatids. Occasionally chromosome fragments lacking a centromere (acentric fragments) are associated with the fragile X site.

Several years after its recognition, the fragile X was determined to be associated with X-linked mental retardation in various families and is believed to be responsible for the condition that had been called the *Martin-Bell syndrome.* The latter name is now a synonym for *fragile X syndrome.* A great deal of interest has centered on this condition since it apparently occurs in all ethnic groups with equal frequency and has now been recognized as a major cause of mental retardation in all populations, second only to Down syndrome. The incidence of the fragile X syndrome is approximately 1/1,500 among males and 1/2,500 for females.

A mentally retarded male with the fragile X may possess certain characteristic physical features, including a long and coarse-appearing face, protruding ears, and unusually large testes. The syndrome shows reduced penetrance, and its expressivity is quite variable. The degree of retardation varies from person to person and is the only symptom in at least 10% of the males. Retarded males may show the typical facial features as well as the large testes. The latter trait, however, may be absent in as many as 30% of the fragile X males.

Only a fraction of the cells in a person with fragile X syndrome actually show the fragile X site on the chromosome, and its demonstration requires special culture medium. The inclusion of folic acid, for example, will inhibit expression of the fragile site, whereas various procedures that can induce it all involve inhibition of DNA synthesis. The number of cells in affected males that show the fragile X varies from less than 4% to 50%. Fragile X expression in each individual is rather constant, and the degree of expression is correlated with the degree of mental retardation in affected males.

The female carriers of the fragile X have been very difficult to explain. One-third of them are affected to some extent. For any X-linked disorder, this is a very high proportion. Some of the heterozygous females are mentally handicapped, whereas others have normal intelligence. An interesting observation is that females heterozygous for the fragile X have a high fertility rate and a two- to fourfold increase in twinning! Different degrees of the physical features typical of the syndrome are found among the normal and retarded females heterozygous for the fragile X. In at least 50% of the normal heterozygotes, it has not been possible to demonstrate the fragile X. The reasons for these differences among them is not clear, but as we will see a bit further on in our discussion, they may be related somehow to events taking place on the molecular level.

While the fragile X syndrome has many unusual aspects, particularly perplexing has been its mode of transmission. The syndrome can be transmitted through so-called transmitter males, men who show no features of the syndrome and in whom the cytological fragile X occurs not at all or only in an occasional cell. One-fifth of males who carry the fragile X fall into this category. These phenotypically normal men transmit the fragile X to their daughters, who are also typically free of phenotypic symptoms and who usually do not show the fragile X at all in their cells or at the most only in a very small percentage of them. However,

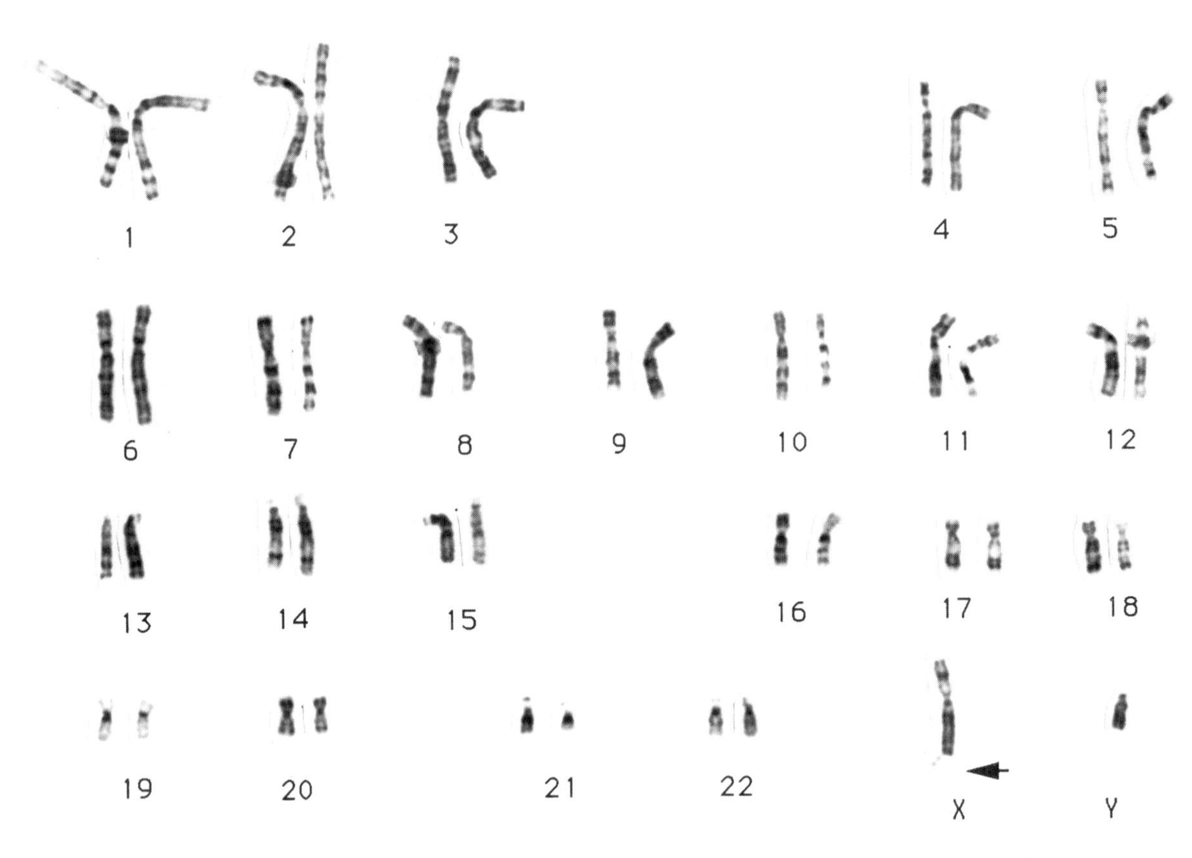

FIGURE 20-25. Karyotype with X chromosome showing the fragile X site (arrow). (Courtesy of Steven A. Schonberg, Ph.D., University of California, San Francisco.)

these daughters of transmitter males are at a very high risk of transmitting mental retardation and the physical features of the fragile X syndrome to their children, who in addition will show the cytological marker in an appreciable number of their cells (Fig. 20-26). Note that the penetrance increases in succeeding generations, beginning with the mother of a transmitter male. The mothers and the daughters of such men produce offspring who express the fragile X differently. Another unusual aspect to the story concerns those heterozygous females who receive a fragile X from their mothers rather than from transmitter fathers. Such females are at an appreciable risk of being mentally retarded and showing physical features of the syndrome. However, those heterozygous females who are sisters of transmitter males are usually free of symptoms, but their chances of transmitting the fragile X syndrome to their offspring is about the same as that of the daughters of transmitter males (in Fig. 20-26, refer to females in generations II and III).

The fragile X syndrome has presented a major challenge to both the clinician and the research biologist. Prenatal detection and recognition of carriers have been beset with many problems due to the complexities of the fragile X story. However, approaches from molecular biology have led to the development of procedures that have characterized the fragile X site and that can detect the presence of the fragile X in persons who are asymptomatic.

A region very rich in cytosine and guanine (a C–G island) was found to be closely linked to the fragile X site. Clones were established that cover 9 kb of this region, and these were used as probes to analyze DNA from asymptomatic persons as well as from those with symptoms of the fragile X syndrome. The probes were able to detect structural rearrangements that occur at the fragile X site of the chromosome. Asymptomatic transmitter males were shown to have a 150–500 bp insertion in the region. This DNA insertion was inherited by their daughters either unchanged or with only a very small size difference. However, in the following generation, those persons who were fragile X positive possessed much larger fragments in the area. In other words, the size of the fragile X region increased greatly in size in affected persons! These offspring of the carrier daughters of transmitter males showed an amazing increase in the size of the fragment, which could contain as much as 20 times the genetic material present in their asymptomatic mothers. Whenever a change in the site takes place, it is always when the fragile X mutation is passed by a female to her children. However, the size change is not obligate and is not always transmitted by the carrier female. An important observation is that the size change, when it does occur, differs among those sibs who received increased fragments. This heterogeneous pattern indicates that the fragile X mutation is highly unstable. We can recognize the insertion seen in the transmitter male as a "premutation," a DNA alteration that passes down with little or no change to his daughters. However, the insertion becomes much larger as it passes from the daughters of the transmitter males to their afflicted offspring and represents the full mutation. The size change among the afflicted is variable. The fragile X mutation is highly mutable, unlike the inherited changes of classic genetics.

Further analysis at the molecular level has associated these unusual features of the fragile X story with a specific nucleotide sequence in the fragile X gene. In unafflicted persons, the 5′ end of the gene has been found to contain 6 to about 50 tandem repeats of the triplet CCG. In persons with the fragile X syndrome, the triplet is repeated more than 200 times. Mild retardation may be present in persons having 50–200 repeats. Such individuals, however, have a high risk of transmitting the severe form of the disorder to their children or grandchildren. Severely affected persons may possess anywhere from about 230 to 1,000 CCG repeats. We have here an example in the case of the fragile X syndrome of a disorder whose severity increases as it becomes transmitted down the generations within a family line. Such an increase in severity of disease symptoms in successive generations is known as "genetic anticipation." The studies at the molecular level now tell us that the anticipation typical of the fragile X syndrome is associated with increase in gene size due to amplification of a CCG repeat. When the CCG triplet becomes repeated more than approximately 52 times, it becomes unstable, especially during the meiotic process in a female.

In its mode of transmission (anticipation, high mutability, increase in size), the fragile X gene departs widely from the gene of classic genetics. However, such unexpected features, may prove to be typical of the inheritance of a number of human genes, as we will see later in this chapter.

In addition to the size changes occurring as the fragile X mutation is transmitted, changes in methylation also take place. The fragile X site in affected persons contain more methyl groups than those in asymptomatic carriers or persons who are free of the mutation. The small 150–500 bp insertion (the premutation) in the transmitter male shows no methylation in the C–G island near the fragile X site. The same is true of this site on the active X of his asymptomatic daughters. However, in the full mutation, the C–G island is methylated in the male and on the active X of the female. According to one interpretation, the determin-

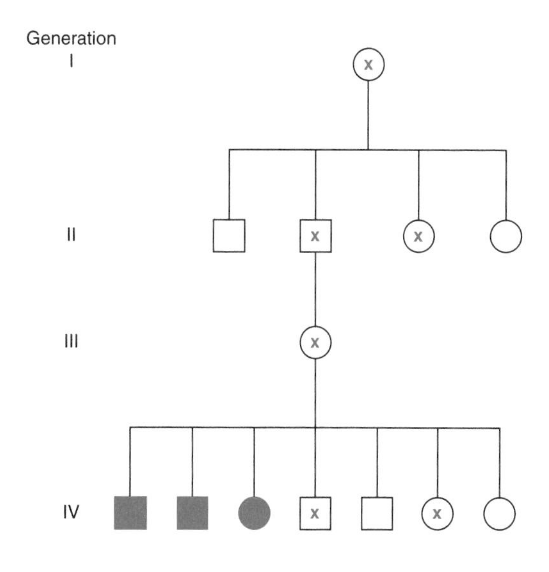

FIGURE 20-26. General mode of transmission of the fragile X. (Circles and squares represent females and males, respectively.) A heterozygous woman (generation I) who is the mother of a transmitter male may transmit the fragile X to 50% of her sons and daughters (generation II) who are usually normal. A transmitter male (generation II) will transmit the fragile X to all his daughters (generation III), who will also usually be normal. However, these carrier females are at an increased risk of producing affected offspring (more affected males than females) in the next generation (generation IV; affected shown in red). Any carrier female in generation II who received the fragile X from her mother along with a transmitter brother is at the same risk of producing affected offspring as are the generation III daughters of the transmitter males. This risk is higher than that in her mother (generation I), who produced transmitter sons. (This general scheme does not represent a pedigree and does not depict actual proportions of carriers or affecteds in families.)

ing factor in expression of fragile X symptoms is increased methylation, which occurs during imprinting. Imprinting in a female would inactivate genes in the fragile X area, turning them off and thus resulting in fragile X symptoms. The size changes that take place would be secondary. However, another idea emphasizes the amplification of the fragile X site as the primary trigger leading to fragile X symptoms and places little importance on imprinting. Support for this belief comes from the observation that the fragile X site can vary in size from cell to cell in a single individual. This indicates that size changes can occur at the site whenever DNA replicates and that there is no need for it to pass through a female for imprinting. Now that the fragile X site is available for detailed study, the issue may become resolved. Moreover, DNA analysis enables detection of the previously hidden fragile X defect in asymptomatic individuals. This is a giant step forward that makes possible more informative counseling regarding the prevention and transmission of this mysterious disorder.

Molecular aspects of myotonic dystrophy

The unusual picture presented by the fragile X syndrome at the molecular level is not unique. A somewhat similar story applies to myotonic dystrophy (DM). This serious genetic disorder is the most common form of adult muscular dystrophy, afflicting 2–14 individuals per 100,000. DM displays great variation both within and between families regarding the age at which it strikes and the severity it manifests. Genetic anticipation is a characteristic feature of DM, increasing in severity as it passes from one generation to the next (refer to Chap. 9 for a review of linkage analysis and prenatal detection of DM).

Although the biochemical basis of DM remains unknown, the gene has been mapped to a 200 kb region at chromosome 19q13.3. This region contains the DM gene as well as the DNA markers on either side of it and has been cloned in phage λ, cosmids, and YACs. Analyses utilizing restriction enzyme EcoR1 followed by separation of the resulting DNA fragments by pulsed-field gel electrophoresis revealed an unusual situation at the DM region that is similar to that described for fragile X. Normal persons were shown to possess a two allele DNA polymorphism in respect to digestion of the region by EcoR1 which generated fragments of 9 and 10 kb. Normal persons could be homozygous or heterozygous in respect to these two DNA alleles. When persons afflicted with DM were studied in this way, the DM region of most of them yielded a fragment larger than 10 kb along with the 10 or 9 kb fragment typical of the normal picture. Lack of detection of a fragment larger than 10 kb in some afflicted persons may be due to the inability of the gels to resolve certain small size increases of a fragment. It is significant that those afflicted persons who show no large fragment have mild cases of the disease. Moreover, afflicted offspring of such afflicted persons showed a large fragment!

The largest fragment detected has been 15 kb, an increase of 5 kb. No offspring with DM have been seen to have a fragment on the gel smaller than the large one seen in an affected parent. However, a size change was not always detectable. Very significant is the finding that the variable DNA region shows an increase in size as it passes down the generations. Studies of family pedigrees reveal increases in fragment size as the dystrophy is transmitted from grandparents to grandchildren. The size increase is directly correlated with the severity of the disease and its earlier onset, an excellent correlation of genetic anticipation with changes on the molecular level. Expansion of the fragment occurs whether transmission is from either the mother or the father. This is unlike the fragile X story in which fragment enlargement is usually seen in female transmission, whereas in male transmission there is no or only slight size change.

Further analysis of the cloned region containing the DM gene has revealed that, as in the case of fragile X, the genetic anticipation is accompanied by an increase in gene size resulting from amplification of a trinucleotide repeat, CTG in the case of DM. Among unafflicted persons in the population, CTG repeats ranging from 5 to 30 are found, 5 and 13 being the most common. These repeats are found in a noncoding region of what appears to be the last exon of the DM gene. Sequence analysis of the DNA containing the DM gene indicates that it is coded for a serine–threonine protein kinase.

An increase in the number of the CTG repeats is responsible for the expansion that takes place in the region. Persons with mild symptoms of DM have been found to have at least 50 CTGs tandemly repeated in the last exon of the gene. As many as 2,000 copies are found in persons with severe symptoms. Those few families (2%) in which expansion of the DM genetic region cannot be demonstrated probably represent pedigrees in which unique kinds of gene mutations are operating, genetic changes that do not involve an increase in the number of CTG repeats.

As depicted in classic Mendelian genetics, a hereditary factor remains unchanged, retaining its constant particulate nature as it passes down the generations. The discoveries of the changing genetic regions in the transmission of fragile X and myotonic dystrophy tell us we must reexamine certain established concepts. A third inherited affliction, sex-linked spinal and bulbar atrophy (a neurological disorder also known as *Kennedy disease*) exhibits genetic anticipation and has also been shown to be associated with a fragment size increase as the disorder is transmitted. Again, amplification of a trinucleotide is involved. In this disorder, the triplet CAG becomes repeated within the coding sequence of the gene, an androgen receptor gene. Such DNA expansion may prove to be a typical feature associated with various genetic diseases. Attention is being directed to other inherited afflictions such as Huntington's disease to determine whether amplification of DNA sequences is involved. It is not possible at the moment to say how expansion of segments of a gene brings about a particular clinical picture, and the molecular processes responsible for the DNA changes remain unknown. Nevertheless, knowledge of the dynamics occurring at the level of DNA in certain genetic disorders offers new approaches to a full understanding of the molecular nature of their clinical pictures and may aid greatly in both prenatal diagnosis and detection of persons at risk well in advance of the appearance of symptoms.

The power of the polymerase chain reaction

One of the major scientific achievements of the past few years has been the development of the procedure known as the polymerase chain reaction (PCR). PCR provides new approaches to problems in many diverse disciplines, since it permits the amplification of specific stretches or segments of DNA. The DNA from just a single cell or hair or even that present in trace amounts in semen and blood can be multiplied over 1 million times to provide amounts sufficient for further analyses! Since its invention by K. Mullis and the first reports of its development in 1985, PCR has undergone many refinements that have increased its efficiency and adapted it to an assortment of scientific problems. PCR can greatly amplify specific DNA segments through a series of repeated cycles entailing DNA denaturation by heat, DNA primers, and the extension of primers by a DNA polymerase.

Critical to the procedure are the two primers, each an oligonucleotide synthesized from approximately 18–30 nucleotides. The primers are complementary to known sequences that mark the ends of the segment of DNA (the target sequence) that is to be amplified (Fig. 20-27). By their 5′ ends, the primers define the ends of the two strands of the target DNA. Moreover, the primers provide 3′ ends to which the DNA polymerase can add nucleotides in the 5′ to 3′ direction, thus building new chains on each template. After one cycle, each new chain will possess a defined 5′ end but will have at its 3′ end an extension composed of nucleotide sequences that are outside the specific target DNA or interest. However, note (in Fig. 20-27) that subsequent cycles of DNA denaturation followed by primer extension on a template that *itself* resulted from extension in a previous cycle, yield DNA with both ends defined, the 3′ as well as the 5′. These defined segments actually increase exponentially as the cycles of denaturation and primer extension continue, resulting in an enormous amplification of the target region. The other products in the reaction, those chains that are not precisely defined at both ends, increase in a simple linear fashion.

The DNA polymerase used in earlier PCR reactions was heat labile. This necessitated the addition of more enzyme each time newly synthesized DNA strands were heat denatured in the course of the procedure. The efficiency of the reaction has now been greatly increased through the substitution of a heat-stable polymerase for the heat-sensitive one. The enzyme now in use is stable at the high temperatures required in the cycles of amplification. Known as Taq polymerase, this DNA polymerase is derived from the bacterial species *Thermus aquaticus,* which lives in hot springs. Being heat stable, Taq polymerase need not be replenished after each cycle. Its enzymatic activity persists cycle after cycle despite the requirement for many heating steps. With Taq polymerase it is now possible to amplify just a single copy of a target DNA sequence into microgram amounts in just a few hours!

The applications of PCR are far reaching since the basic procedure can provide a ready source of large quantities of a specific DNA sequence. It can substitute for the more laborious cloning procedures that utilize bacteria or phages. Indeed, it may actually replace cloning as the procedure to yield the amounts of DNA needed for sequencing. PCR can supply the investigator with large amounts of DNA representing a gene or portions of a gene. This facilitates the detection of defects at the molecular level that are associated with clinical conditions. Since trace amounts of DNA in body fluids and hair can be amplified, PCR is playing an increasingly important role in criminal investigations, since DNA samples taken from a crime scene can be compared with those from suspects. Such an approach has already led to acquittals and convictions (discussed in the next section).

The difficulty in culturing certain disease-causing agents such as the Lyme disease organism becomes circumvented by PCR, which can amplify just trace amounts of the genetic material of the pathogen for further study. Indeed PCR may even reveal the presence of a pathogen in just an occasional cell. For example, only a few copies of HIV, the AIDS virus, may be present in the circulation of a recently infected person. If only one cell in 100,000 harbors the virus, PCR, as a result of its power to amplify DNA segments, can reveal the presence of the pathogen well in advance of other procedures.

Since PCR can be used to amplify cDNA, this means that mRNA can be extracted from a cell, converted to cDNA, and finally amplified. This cDNA is then available for analysis to determine which specific genes may or may not be expressed in a given kind of cell. Such a determination is of utmost importance in cancer research. PCR is especially valuable in prenatal diagnosis, in which different pairs of primers can be combined in a single reaction to detect DNA deletions. In the case of the sex-linked disorder DMD there is information on the DNA sequences flanking exons. Using different pairs of primers, amplified DNA may reveal whether deletions associated with the dystrophy are present in fetal cells. The simplicity of the PCR procedure may come to favor its use over others in prenatal diagnosis.

Very exciting is the amplification by PCR of tiny bits of DNA extracted from mummies and extinct animals. Studies of the genetic material from such sources yield valuable insights into evolutionary relationships.

One important variation on the basic PCR is *inverse* or *inverted* PCR. This modification can be used to gain information on that DNA on either side of a target sequence. A return to Figure 20-27 shows that the orientation of the primers is such that a target sequence becomes defined and then greatly amplified. However, any regions flanking that area remain in amounts insufficient for study. The DNA on either side of the original target can be amplified by reverse PCR. In this approach, primers are constructed that prime DNA synthesis in both directions away from the original target sequence. In other words, the DNA that lies outside two primer sites can become amplified rather than that lying between two of them. Reverse PCR is very valuable in giving information on unknown regions of the genome found on either side of a known gene and that may possibly influence its expression in some way. The application of PCR is certain to widen as variations on the basic procedure continue to be made.

The power of DNA fingerprinting

In our discussion of RFLPs earlier in this chapter, we pointed out that the most valuable markers are those that occur in several versions. This is so since variability increases the chance that any two persons will be heterozygous for different forms of that marker. We saw that in particular tandem repeats produce a great degree of variability at different DNA sites (Fig. 20-14). They thus provide us with locations along the chromosomes that can serve as markers, each marker site existing in many forms. Although tandemly repeated sequences have not been associated with any products and although their full significance remains unknown, they are nevertheless extremely valuable as markers and are playing an increasingly important role in legal matters. The value of tandem repeats as molecular aids in identification becomes apparent when we reconsider the fact that at any one chromosome site the number of times that a sequence repeats itself can vary greatly, from just a few times to many hundreds. Since there is great variability from one human chromosome to the next in the number of times a sequence is repeated in tandem at a given site, many versions of that region exist in the human population. Not only are the chances great that any one person is heterozygous at a specific chromosome region (locus) for a tandem repeat, but the chances are exceedingly remote that any two unrelated persons will possess the same variation (same allele) at that region. Remember also that the same situation pertains to the same repeat as it occurs elsewhere in the genome. Therefore any one person is apt to be heterozygous for a given tandem repeat at various sites throughout the genome, sites that are not allelic to one another.

A review of Figure 20-14 shows a chromosome location in which a sequence is repeated a few times on one chromosome and many times on the homolog. If a restriction enzyme cuts sites in the vicinity of the repeat, the variation in the number of repeats produces fragments of different size. (Note that the restriction enzyme used must not cut the repeat unit itself.) The distance between two enzyme cutting sites is different when two homologous chromosomes are compared.

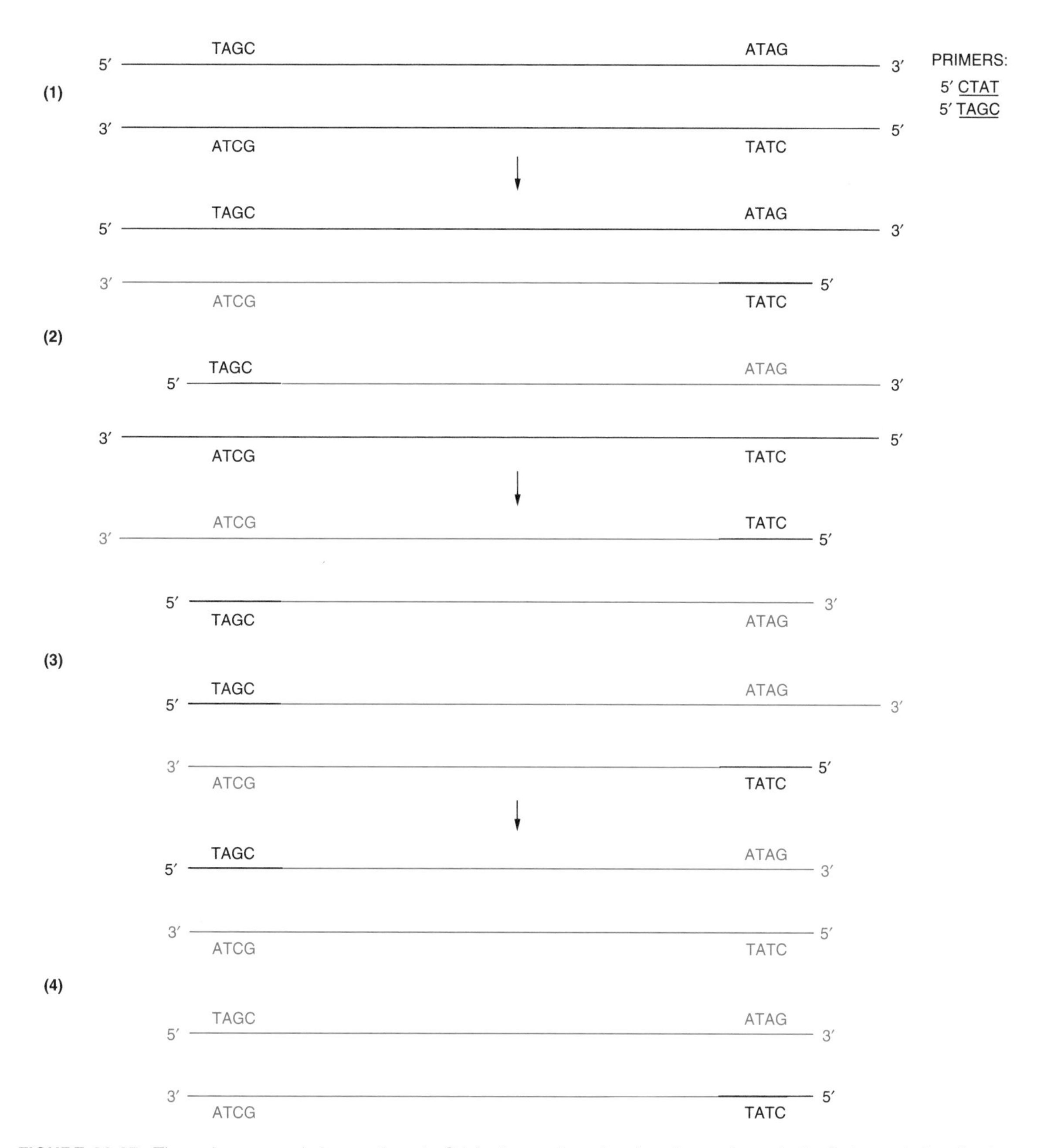

FIGURE 20-27. The polymerase chain reaction. **1:** Original duplex DNA showing nucleotide sequences that will define the target region to be amplified. **2:** DNA is denatured. Primers are added along with DNA polymerase and nucleotides. Newly synthesized strands are formed from each primer that defines a 5′ end. The newly made strands are extended at the 3′ end beyond the target. **3:** Strands formed by extensions from primers in the first round of replication act as templates in another round. Primers define 5′ ends of strands being newly synthesized on these templates. As a result, strands are produced with distinct 5′ and 3′ ends which define the target region. **4:** Defined target strands continue in successive cycles of replication to be extended by primers and act as templates amplifying the target regions.

Following Southern blotting, the difference in fragment size can finally be detected with the aid of a probe that recognizes the repetitive DNA.

We noted above that a given repeated sequence occurs at more than one location in the genome (Fig. 20-28). Therefore just one restriction enzyme will generate fragments that vary in length. Since the fragments are formed as a result of cutting at different chromosome spots, most of them are *not* allelic. In other words, the different fragments visualized by the probe do not result from variation at just one single chromosome location. This means that an enormous degree of fragment size variation exists between two unrelated persons and that this can be detected by a single enzyme and a single probe. This fact was recognized by Alec Jeffreys of the University of Leicester in England who developed the now highly publicized technique of DNA fingerprinting. Jeffreys realized that DNA from any person will produce a pattern of fragments due to variation in tandem repeats which is distinct from that of any one else. Since the chance of two unrelated persons having the identical fragment pattern has been calculated to be about one in several billion, each person's pattern can be considered unique. The pattern can be used to identify the person and is indeed the equivalent on the molecular level of a fingerprint. Even two ordinary sibs are highly unlikely to have the same pattern with respect to a given tandem repeat. Remember that the number of times a given sequence is repeated in tandem varies greatly at a given locus between two homologous chromosomes. Consequently any person in the population is almost always heterozygous with respect to that locus or chromosome site. This means that two parents, each heterozygous for a DNA marker at a repeat site, will have four different alleles between them. Excluding identical twins, any two children of the same parents have a 50% chance of sharing one of the alleles (a fragment). However, remember that cutting with a single enzyme and using a single probe generates fragments from many different nonallelic sites throughout the genome. Suppose a probe can recognize 20 of these sites. This means that the chance that two sibs will possess exactly the same DNA pattern with respect to 20 different chromosome regions is $(1/2)^{20}$.

The importance of DNA in legal matters has increased with the development of the PCR. We noted in the previous section that PCR has the power to amplify the DNA from just a single hair. Amounts measured in nanograms can be multiplied to microgram amounts, which in turn can be used to obtain a DNA fingerprint. Jeffrey's DNA fingerprinting method was first used in a rape–murder case in England. The fingerprint of DNA from semen left at the scene of the crime was compared with those fingerprints of DNA taken from white blood cells of various men living in the region. The procedure was able to exclude certain suspects and to identify the guilty person beyond any reasonable doubt.

In the United States, DNA fingerprinting has been instrumental in several court decisions. However, criticisms regarding the accuracy call attention to certain technological problems that must yet be addressed. For example, the power of PCR is so great that trace amounts of DNA of a contaminant may be amplified. Therefore steps must be taken to eliminate any contaminant DNA from blurring the results. Moreover, precise alignment of gels bearing patterns from two different DNA sources must be guaranteed so that it can be unequivocally stated that two bands on two different gels are in the same position and represent the same fragment rather than two different fragments of slightly different lengths. Despite such problems, the power of DNA fingerprinting as a tool in various legal matters has been recognized. As refinements continue to be made in the overall procedure, DNA fingerprinting is certain to assume a role of major importance and to be routinely used in matters requiring precise identity of individuals.

Interactions of cells of the immune system

The cells that manufacture antibodies (immunoglobulins) are *B lymphocytes,* small white blood cells that are formed and mature in the bone marrow. *T cells* constitute another class of lymphocytes, also found in bone marrow, which complete their maturation in the thymus gland. Although both types of cells appear very much alike, they perform differently in the immune response. A T cell participates in a direct attack, killing any cell bearing foreign antigens. T cells also act against viruses. The T cell recognizes the discarded protein of the virus that remains attached to an infected cell's membrane. The cytotoxic killer T cell then destroys the virally infected cell. It is the killer T cell that is involved in directly attacking cells following a tissue transplant. There are also *T helper cells* that, after recognizing an antigen, stimulate other T cells on guard against the same antigen. Moreover, the helper T cells appear to be necessary for the alerted B lymphocytes to undergo divisions to complete their differentiation. Another class of T cells, the *T suppressor cells,* are required to cut down on the activity of active T killer cells and hence keep the response within the proper limits. Acting together, the T helper and T suppressor cells act in a regulatory capacity.

When a B lymphocyte that has recognized a specific antigen is activated by a T helper cell, it goes on

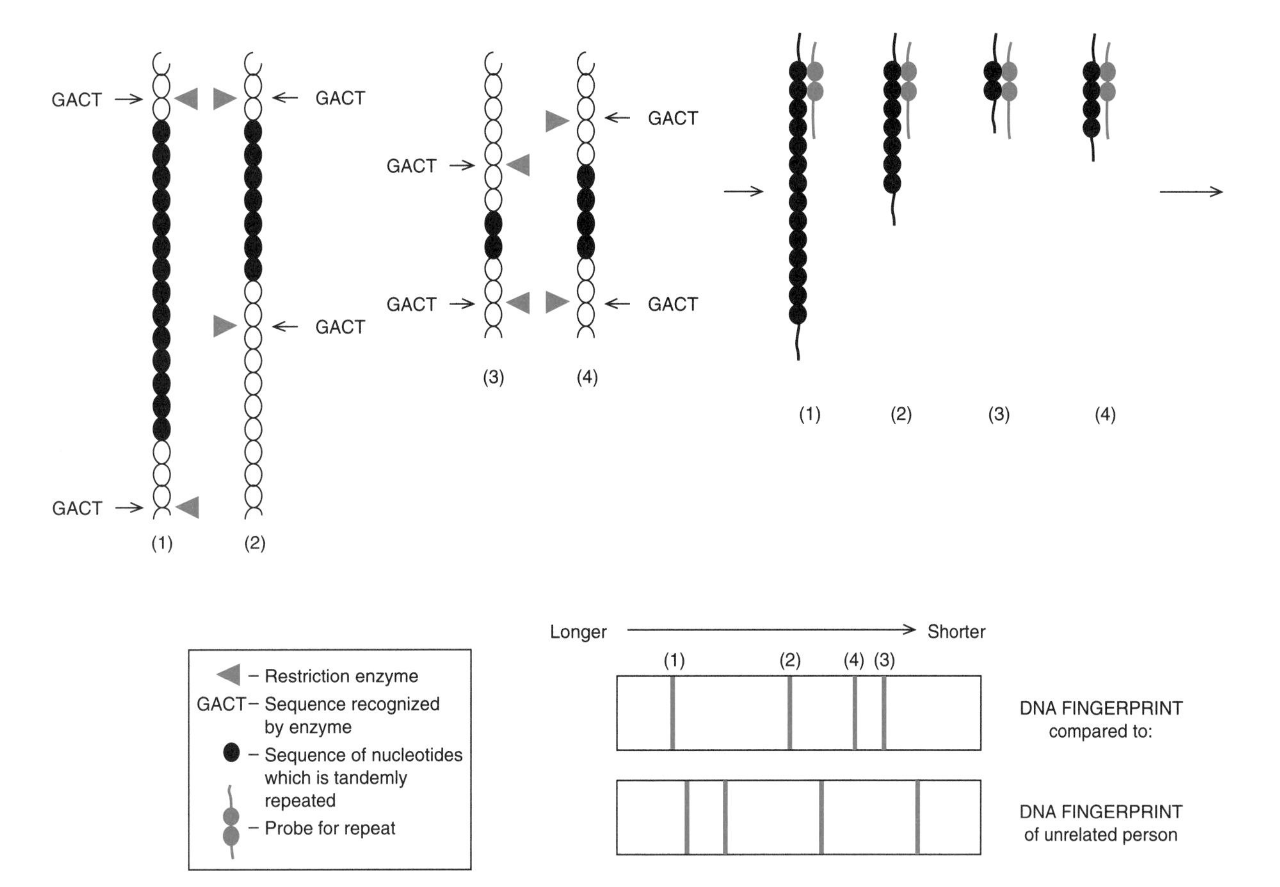

FIGURE 20-28. Tandem repeats and DNA fingerprinting. A tandemly repeated sequence may occur at various sites within the genome. Depicted here are two pairs of homologous chromosomes showing two different nonallelic locations at each of which a DNA sequence is repeated in tandem (contrast this with Fig. 20-14). A restriction enzyme recognizes the sequence GACT, which occurs on either side of the repeats. Cutting by the enzyme yields fragments of different sizes due to the difference in the repeat number. Fragments 1 and 2 result from cutting at corresponding sites and are allelic to each other. In the same way, fragments 3 and 4 are also allelic to each other. While 1 and 2 are not allelic to 3 and 4, the four fragments have been generated by the one enzyme and visualized by the same probe to give a DNA fingerprint. Since two unrelated persons are apt to be heterozygous for different alleles at each site (possess different numbers of repeats), comparison of their DNA fingerprints would show that the fragments do not match and that the fingerprints are different. (The diagram is highly simplified since the probe would typically reveal more than four fragments.)

to proliferate. Some of the resulting B cells then proceed to act as *memory cells.* These B lymphocytes continue to circulate around the body, alert to the specific antigen. They are able to bring about a swift response following an infection by the same antigen and account for the speed with which the immune system reacts to a second infection by a given antigen. Those B lymphocytes that do not become memory cells after stimulation by the T helper no longer divide but undergo a terminal differentiation to become *plasma cells*—B cells that live only a few days but secrete very large quantities of antibodies. A given B cell is committed to produce a single kind of specific antibody when stimulated by the antigen for which it is on the alert. Although the antibodies secreted by plasma cells interact with their specific antigens as they circulate in the blood system, they do not destroy them directly. Rather, they mark them so that the antigens are recognized by other components of the immune system, such as *macrophages,* which do the actual destruction.

Antibody diversity

The different kinds of foreign antigens that can infect an individual number in the millions. The body must therefore have the ability to produce a corresponding number of different antibodies to interact with them. It would seem that the genome of a mammal must include many millions of genes set aside for the coding of millions of different antibodies. However, estimates of gene number in mammals indicate that the total number of genes actually found in the genome can certainly not be more than 1 million. Only a small percentage of these are coded for specific antibodies. Yet this fraction of the genetic complement nevertheless specifies countless kinds of antibodies to counteract the effects of countless types of antigens. Laborious research by geneticists and immunologists over the years has sought an explanation for this puzzling situation. Although certain details are yet to be clarified, the basis for antibody diversity has now been determined. The fruits of the research have also demonstrated the unexpected versatility of the mammalian genome and have forced us to reexamine some established concepts, such as the one gene–one polypeptide theory.

An antibody is a Y-shaped molecule composed of four polypeptide chains held together by disulfide bridges (Fig. 20-29). Two of the chains are called *light chains.* These are identical, each composed of about 200 amino acids. The other two chains—the *heavy chains*—are longer, composed of more than 300–400 amino acids; these are also identical in any one antibody molecule. Antibodies fall into five different types or classes. Each type of antibody or immunoglobulin (Ig) is distinguished by the kind of heavy chain it possesses: M, D, G, E, or A. All five types may be produced by lymphocytes responding to the very same antigen. Consequently, the five different types of antibodies—IgM, IgD, IgG, IgE, and IgA—are specific for that one antigen. Just how the particular antibody operates against the antigen, however, differs from one type to another. For example, the IgG antibody circulates in the blood; IgD antibody remains attached to its cell's surface.

When we compare the heavy and light chains from antibodies specific for different antigens, we find that certain differences exist in the amino acid sequence of the chains. These differences are found only in one portion of each chain. This is the so-called *variable region,* which makes up about half of the light chain and one fourth of the heavy chain (Fig. 20-29). The remainder of the chains are called *constant regions,* because they have the same amino acid sequences in a given type of antibody, *regardless* of their differences in specificity for different antigens. For example, all IgG_3 antibodies have the same constant heavy chain region, even though two IgG_3 molecules may be specific for different antigens. Note from Table 20-2 that there are four kinds of constant γ-chains, which makes four subclasses of IgG possible. It is the constant region of the heavy chain—μ, δ, γ_3, and so on—that determines the type of the antibody and the way it will carry out its job.

Although there are five kinds of heavy chains (remembering that the γ type actually divides into four subclasses of its own), there are two kinds of light chains, κ and λ (Table 20-2). Either two κ or two λ will be present along with two identical heavy chains of a given type in every antibody. As in the case of the heavy chain, the constant region of any kappa is identical to that of all other κ light chains in all other antibodies of a given type. The same is true for the constant region of the λ-chain. The variable regions are those parts of an antibody molecule that can recognize a specific antigen and interact with it. The constant regions do not participate by reacting directly with the antigen, although, as we have noted, the constant heavy region determines how that type of antibody carries out its job against the antigen.

The variable regions of the heavy and light chains are found at that part of the polypeptide near the beginning of the NH_2 end of the chain. Each variable region of a heavy and a light chain contains within it three *hypervariable regions,* segments where the amino acid sequences are especially different when antibodies of different specificities are compared.

Theories of antibody diversity

The presence of a variable region in each kind of antibody chain makes possible a large number of different kinds of light and of heavy chains—at least 1,000 of each kind. According to one theory, each haploid genome contains 1,000 genes coded for the 1,000 different kinds of light chains and 1,000 genes coded for the different heavy chains. Besides the large number of genes required by this theory to code for the vast assortment of specific antibodies, another difficulty accompanies this concept. Let us say that there are 1,000 genes for light chains. What then is the device that keeps all 1,000 constant regions identical in each of the 1,000 genes? Some mechanism would be required to prevent the effects of gene mutation, and thus to conserve perfectly the constant region in each of the chains. And yet the variable region of each of these genes would be permitted to mutate and change rapidly. To explain how such a system could be maintained requires the surveillance of some unknown kind of biological device and argues against the one gene–one specific antibody chain idea.

A different theory proposed in the mid-1960s suggested that the genome contains many genes that code for all the variable regions of a chain of a given type but only one gene for a constant region of a given type. For example, in the case of heavy chains, there would be one gene for the constant region of type μ-chain, one for the constant region of a δ-chain, and so on. There would, however, be many genes for the different variable portions. In the case of the light chain—κ let us say—one gene would exist for the constant portion and perhaps hundreds for the variable. According to this theory, as an antibody-producing cell matures, the DNA is shuffled, and a constant gene is combined at random with one of the variable genes to produce an active antibody gene. In another cell, the same constant region gene might combine with a different variable gene to give another active antibody gene. The proposal is a radical one, because it implies that two separated genes or DNA segments can be brought together to form one segment or gene, which is then coded for a polypeptide chain (Fig. 20-30). Nevertheless, the basic concept of this unusual proposal has now been shown to be essentially correct. According to this scheme, as a lymphocyte matures, one constant region gene can be brought into combination with any one of many variable genes that lie a distance from it. This recombination then yields an active gene with a variable region and a constant region, which is the same for every polypeptide chain of a given specific type.

Evidence supporting this DNA rearrangement theory has been provided by experiments using recombinant DNA procedures. Mouse DNA coding for the variable (V) and constant (C) regions of the light chain, λ, were cloned in phage λ. DNA was obtained from two cell types—embryonic and mature, antibody-secreting cells. Restriction enzymes were then used to cut the DNA obtained from the two sources. If some sort of shuffling or recombination of the DNA does take place as an antibody-secreting cell matures, a restriction enzyme site could be altered. This would mean that when DNA containing V and C genes from an embryo is compared with the corresponding DNA in a mature cell, different restriction fragments would be obtained. And this proved to be the case, as seen clearly in the mouse. In the embryonic cells (or in any cells that are not antibody secreting), the DNA sequences for the V and C regions of the light chain are much farther apart than they are in a cell that is secreting antibodies. This was clearly demonstrated by the size of the restriction fragment bearing the C segment (Fig. 20-31). Such experiments leave no

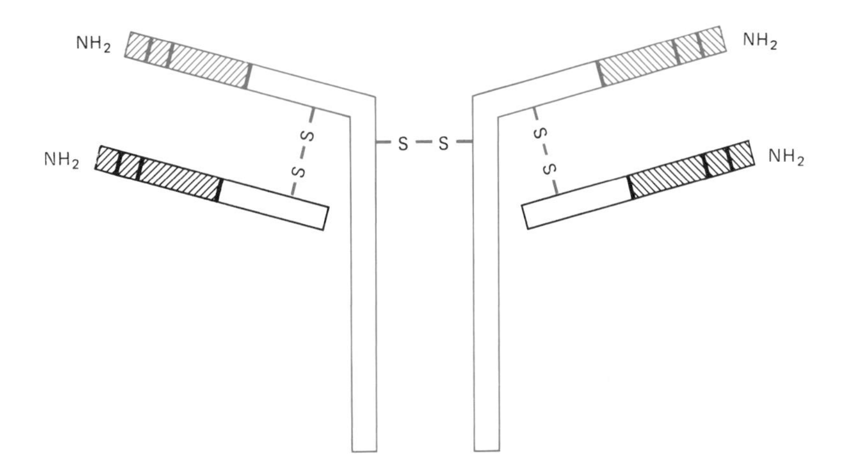

FIGURE 20-29. Representation of antibody molecule. Each antibody molecule is composed of two identical light chains (black) and two identical heavy chains (red) held together by disulfide links (S–S). A portion near the NH_2 ends of each chain, the variable region (shaded), is composed of amino acid sequences that vary when antibodies of the same type are compared. The constant region (unshaded) is the same for all antibodies of the same type. Within each variable region are three hypervariable regions.

TABLE 20-2. Classes of antibodies and their chains[a]

Class	Heavy chain	Light chain
IgM	mu (μ)	kappa (κ) or lambda (λ)
IgD	delta (δ)	kappa (κ) or lambda (λ)
IgG	gamma (γ)	
Subclasses:	gamma-3 (γ_3)	kappa (κ) or lambda (λ)
	gamma-1 (γ_1)	kappa (κ) or lambda (λ)
	gamma-2b (γ_{2b})	kappa (κ) or lambda (λ)
	gamma-2a (γ_{2a})	kappa (κ) or lambda (λ)
IgE	epsilon ($\in$)	kappa (κ) or lambda (λ)
IgA	alpha (α)	kappa (κ) or lambda (λ)

[a]The heavy chain defines the antibody class. The gamma class is divided into four subclasses of heavy chain. Any one kind of heavy chain can be combined with either a kappa or a lambda light chain to form the antibody molecule.

doubt that maturation of a B lymphocyte into an antibody-producing cell is accompanied by rearrangement of the DNA. A rearrangement of this type, which takes place during the course of differentiation, is called *somatic recombination.*

Before proceeding with details of the arrangement of antibody genes, it must be pointed out that in any antibody-producing cell, the antibody genes on *only one* of the two chromosomes undergo somatic recombination. It is these recombined segments that constitute the DNA that will undergo transcription and then be transcribed (Fig. 20-32). The DNA corresponding to these antibody genes on the homologous chromosome does not undergo somatic recombination but remains in the original arrangement found in embryonic or in nonantibody-producing cells. This DNA will not undergo transcription, and therefore does not produce an mRNA for translation. This phenomenon is known as *allelic exclusion.* Allelic exclusion is the rule, and only rarely do both corresponding segments in a lymphocyte undergo shuffling.

The generation of antibody genes

DNA sequencing, combined with other methods from molecular biology, has filled in many details on the arrangement of the DNA sequences coded for antibodies. It was demonstrated that even when the variable and constant regions are brought close together during differentiation, they still remain separated by a distance of about 1,500 nucleotides. This is because of a joining segment known as *J.* As shown for the human κ light chain in Figure 20-33A, there is a single C gene for the constant region plus five J segments and a large number of V genes. Both the J and the V sequences are encoded for the variable sequences of the chain. Each V sequence is separated from a leader by a short intervening sequence. A long noncoding sequence separates the J sequences from the cluster of V genes.

During the differentiation of a lymphocyte to a plasma cell, somatic recombination can shuffle a V gene with its leader into a combination with one of the J segments. This recombination brings about an arrangement that can act as a κ-gene and undergo transcription. When a V segment joins to a J, the intervening DNA to the left is lost in the somatic recombination. A signal to the left of the J segment indicates the site of recombination between the specific J and V sequences. Note from Figure 20-33B that pre-mRNA forms in the cells, and this contains the recombinant V–J segment and also any J regions that remained to the right of the J that combined with the V. A signal to the right of that J indicates where splicing is to occur. For example, in cell 1 in Figure 20-33B, the mRNA processing brings J4 together with C. The J5 is eliminated from the pre-mRNA along with introns. A poly(A) tail is added, and the transcript is ready for translation into a κ light chain. The leader sequence of the transcript results in the assembly of a leader portion of the polypeptide chain composed of about 20 amino acids. This leader portion is part of the original polypeptide formed and is thought to be important for the transport of the newly formed antibody chain through the cell membrane. As the chain passes through, the leader is snipped off.

The story of the λ-chain is similar to that just described for the κ-chain, with some significant differences. There are six constant genes, not just one, for this light chain. Each of the six is closely associated with its own J segment, forming a pair. Any one of the V genes can be brought into combination with any J–C pair (Fig. 20-34). The exact number of V segments for both the λ- and κ-chains in the human is still uncertain and is roughly estimated to be less than 300 for each. The mouse appears to have experienced a deletion in its V λ region where only two V segments remain! In the mouse, only 5% of the light chains are λ type, whereas in the human 40% are λ. The genes for the λ- and κ-chains have been assigned to human chromosomes 2 and 22, respectively. In each case, the cluster of V segments constitutes a gene family.

A picture similar to that for the light chains holds for the genes coded for the variable and constant regions of a heavy chain. An active heavy-chain gene results from the shuffling of four sets of sequences, V (with the leader), C, J, and D. Thus much greater diversity is possible by shuffling the heavy-chain sequences than is the case for the light chain. These sequences are located on chromosome 14, where they constitute another immunoglobulin gene family. The D (for diversity) genes or segments, each composed of about 13 nucleotides, are responsible for the greater diversity of the different kinds of heavy chains. A D segment in the finished heavy chain gene is located in a position corresponding to that of one of the hypervariable regions of the heavy chain.

In the embryonic or "nonshuffled" state, the human DNA containing antibody genes includes about 100 V segments, approximately 20 D segments, and 4 J segments. In the assembly of a functional heavy chain gene, somatic recombination brings one of the V genes (with its leader) together with a D segment and a J segment, which also code for the variable region of the heavy chain. The segments for the constant sequences are grouped together at a distance from the V genes (Fig. 20-35). Unlike the situation with light chains, several different separate C genes exist, each designating the type of heavy chain (μ, δ, γ, etc.). Since there are four different classes of γ chains, eight different C sequences exist. These occur in the order μ, δ, γ, $\in$, and α. Actually, the situation is even more complex, because each of the constant genes is separated by introns into three to six separate coded segments.

In the development of a mature B lymphocyte, a stage is recognized in which the cell is called a *pre-B lymphocyte.* Such a cell produces a heavy chain of the μ type. The heavy chain synthesis occurs before light chain synthesis. Following the formation of light chains (κ or λ in a given cell), the pre-B lymphocyte produces a δ-chain, and the cell produces completed IgM and IgD molecules. In other words, all maturing B lymphocytes first produce heavy chains of the μ class. Then δ-chains are produced. This comes about as a result of both DNA shuffling and RNA processing (Fig. 20-36). In the pre-B lymphocyte, the somatic recombination joins a V, a D, and a J segment close to the constant region. This produces an active gene, which forms an RNA transcript carrying the information for a μ constant sequence. Processing of this transcript removes introns within the μ-gene as

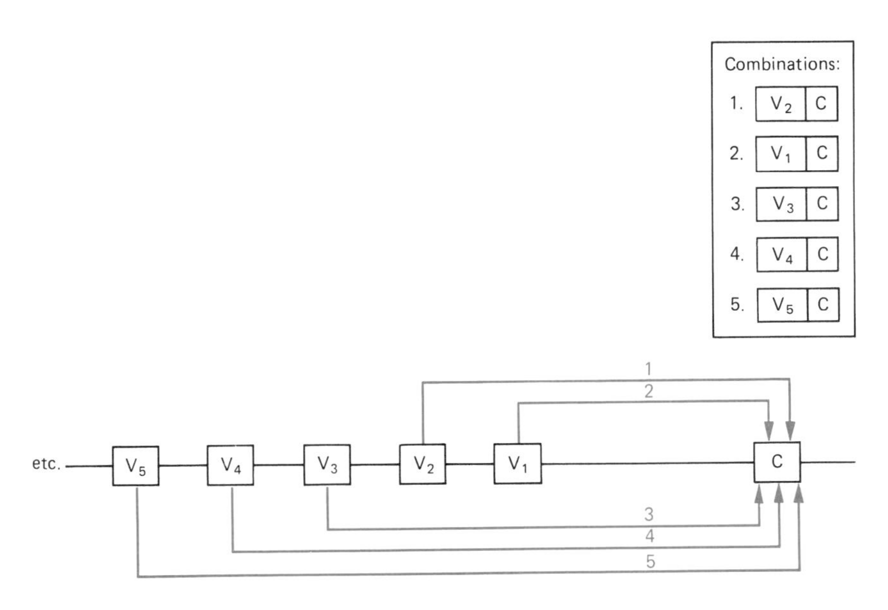

FIGURE 20-30. Theory of gene shuffling. According to one proposal, diversity of antibody genes can be accounted for by assuming that many genes or segments code for variable regions (V1, V2, etc.). Only one or a few segments would code for the constant region (one C segment is depicted here). In one cell, an active gene can be formed by combining the C segment or gene with the segment coded for V1, which lies a distance away from it. In other cells, different combinations of C and V segments occur to produce many different kinds of active antibody genes, which are then able to form transcripts.

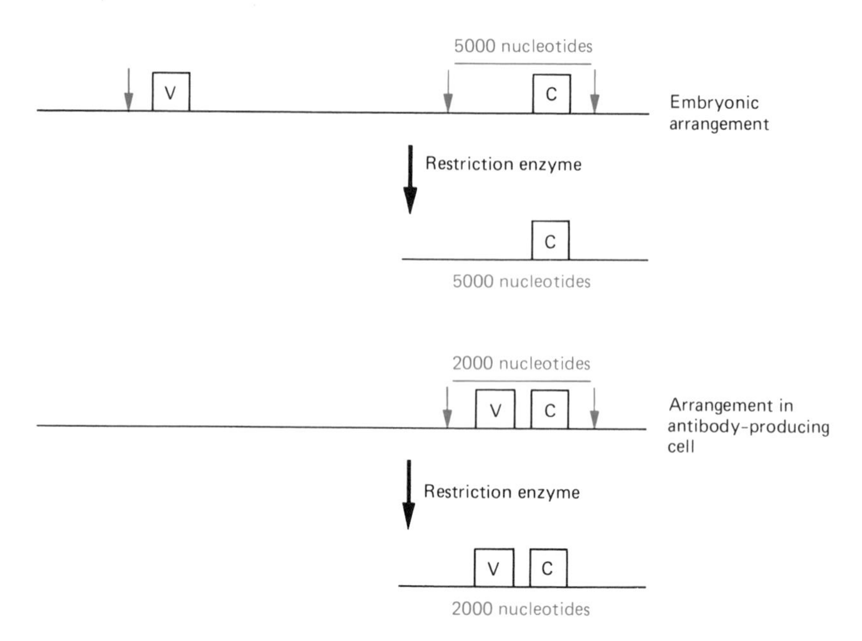

FIGURE 20-31. Detection of DNA shuffling. Suppose (above) that a C gene in an embryonic cell or in a nonantibody-producing one is located between two DNA sites (arrows) that are sensitive to a particular restriction enzyme and are 5,000 nucleotides apart. Located a distance away is a V gene with a sensitive site close to it. When the cell is exposed to the enzyme, C and V can be located with probes. C is found on a fragment that is 5,000 nucleotides long and does not contain V. If shuffling of DNA occurs, the restriction site to the left of the gene may be eliminated, and V may be brought close to C. Cleavage with the restriction enzyme along with probing now shows that C is associated with a much smaller fragment that also carries V. A piece of DNA originally between V and C, which contains the middle sensitive site, must have been eliminated.

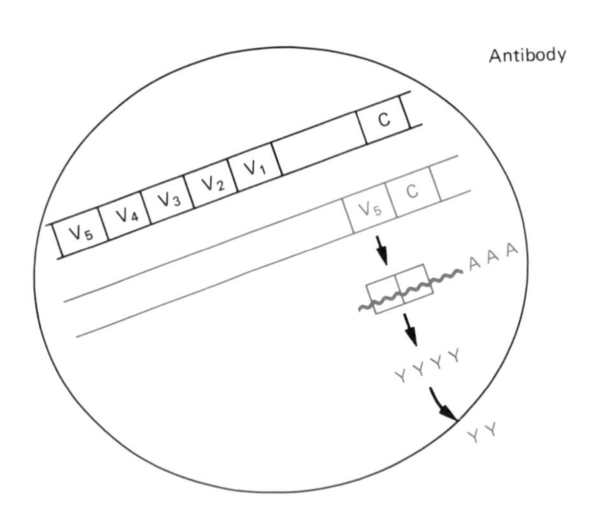

FIGURE 20-32. Allelic exclusion. In an antibody-producing cell, the DNA of only one of the two homologous chromosomes undergoes somatic recombination to produce an active antibody gene (red). Only the rearranged DNA on the one chromosome will form an active gene that will produce transcripts coded for a specific antibody. The other chromosome retains the gene sequence of the immunoglobulin family in the original (embryonic) condition.

well as other intervening sequences, to give a finished IgM heavy chain transcript. A primary transcript can also include information from a δ-gene. Splicing may then remove the μ-gene sequences and connect the segment carrying the δ information to the V–D–J segment. Therefore, differential splicing, as well as DNA shuffling, is involved in the production of IgM and IgD heavy chains in the pre-B lymphocyte. Both immunoglobulins IgM and IgD at first appear on the surface of the prelymphocyte cell. This simultaneous production of IgM and IgD is the only exception to the statement that a mature lymphocyte synthesizes *only one* type of immunoglobulin.

Further development of the B lymphocyte depends on its interacting with some specific antigen. When a specific antigen becomes bound to the cell, the lymphocyte completes its development. It goes on to divide to produce a clone of mature antibody-producing B lymphocytes. During this transition, both IgM and IgD disappear from the cell surface, and the cell then secretes one of the immunoglobulins: IgM, IgG, IgE, or IgA. This means that the *same V region* can now become associated with a heavy chain constant region *other than* the constant μ-region, a process called *heavy-chain class switching.* This switching involves additional DNA shuffling and differential processing of the mRNA (Fig. 20-37). Between the constant region genes are nucleotide sequences that seem to act as signals in switching. These signal sequences have some similarity to a sequence located between the

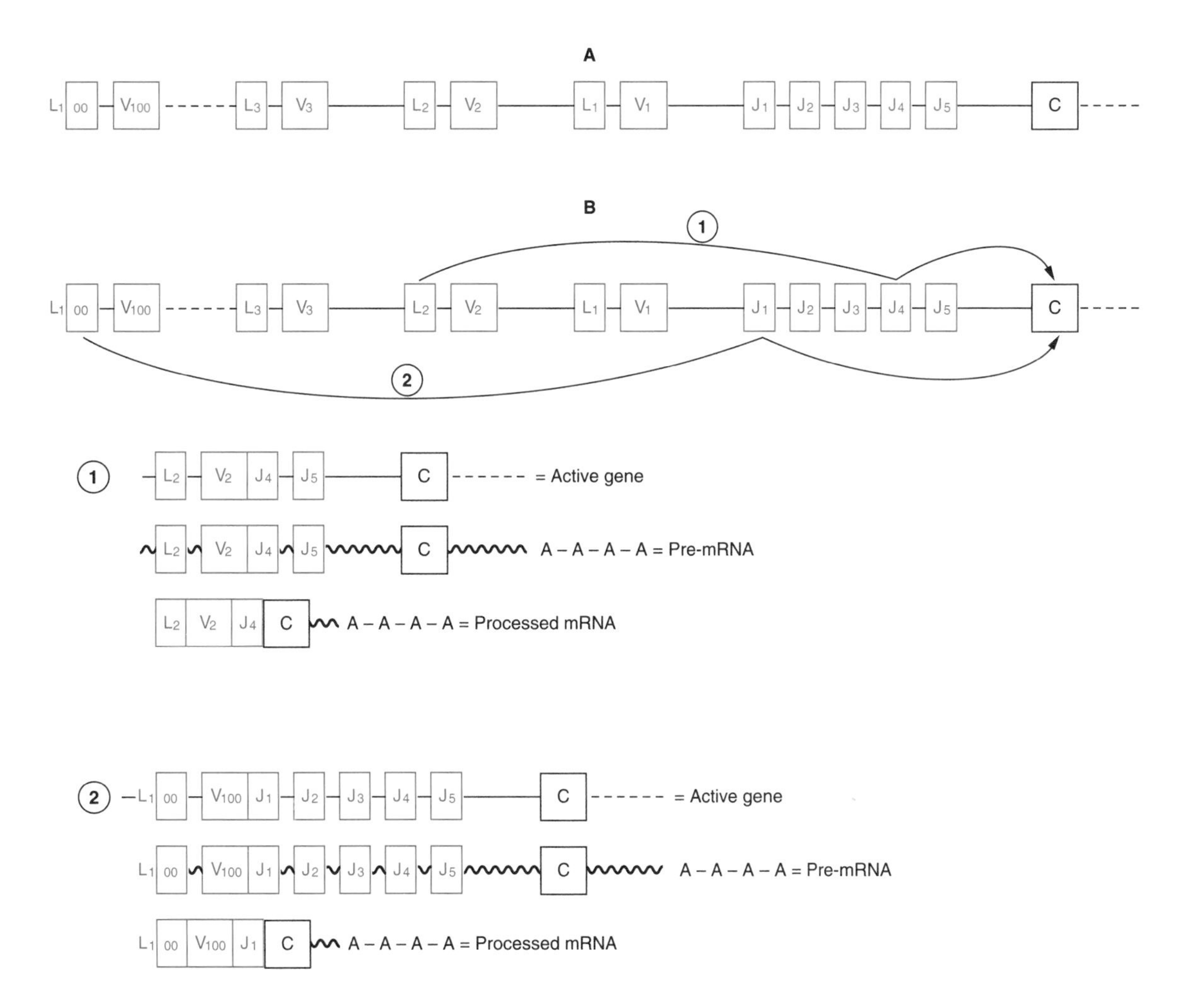

FIGURE 20-33. The formation of active κ-genes. **A:** The κ light chain is encoded by a single C gene, five J sequences, and many V genes. Each V gene has a leader segment separated from it by an intron. **B:** In the formation of an active κ-gene, somatic recombination can combine any V gene with a J segment. In cell 1, the combination joins V2 and J4. In cell 2, the combination is V100–J1. Many combinations are possible from one lymphocyte to the next. The active κ-gene undergoes transcription to form a pre-mRNA containing sequences between the V–J combination and the C. Processing of the transcript results in an mRNA with the specific V–J combination now combined with the C. The sequences to the right of the J in the pre-mRNA are eliminated along with introns in the splicing that forms the functional mRNA.

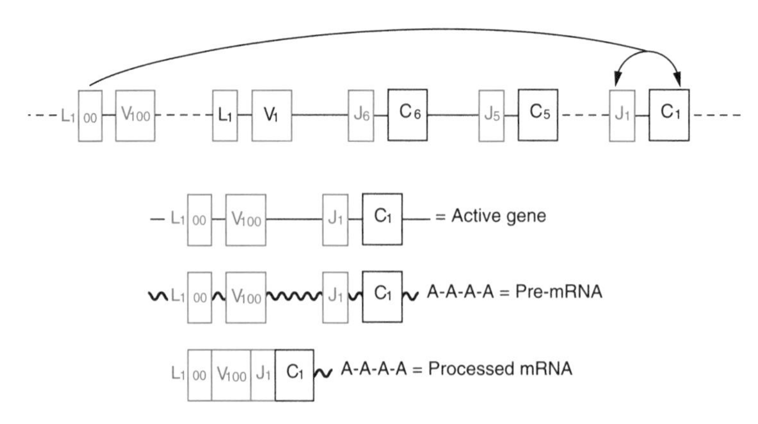

FIGURE 20-34. The formation of active λ-genes. Six genes are coded for the constant region of the λ-chain (only three are shown here). Each is paired with a J segment. Any one of the V genes can combine with any J–C pair to form an active gene that can undergo transcription. The primary transcript is processed. Introns and other intervening sequences are removed to produce a functional mRNA.

μ-gene and the variable genes. These signal sequences may be involved in bringing about a recombination event that joins the V–D–J variable sequences to one of the other constant regions. After this next DNA recombination occurs, the same variable region is joined to a different C region so that the variable region is directly adjacent to one of the C genes. After the DNA has undergone this second reshuffling, it is transcribed. The mRNA is processed to form a messenger that codes for a γ, an ∈, or an α constant region, whichever one is adjacent to the variable sequence. Signal sequences similar to those associated with the DNA coded for the heavy chains are also known to occur in the DNA coded for the light chains. In the case of the κ-chain, one of these is found near each V gene and one near the J segment.

It should be noted that the IgM that can be secreted by a mature lymphocyte is a different form of the IgM that appears bound to the surface of the prelymphocyte along with IgD. The bound form of μ contains an amino acid sequence at its end that is responsible for binding it to the membrane. The μ that is secreted lacks this amino acid chain. The two different forms of μ arise from translation of two slightly different kinds of mRNA transcripts. The primary transcript for the bound form of μ arises as a result of the complete reading of six separate DNA segments of the μ-gene. The last two segments are coded for the tail portion, which anchors μ to the membrane. The primary transcript for the secreted form results from the termination of transcription after the first four segments are read. Hence the final mRNA, after processing of the primary transcript, lacks the two segments coded for the tail. Consequently a secreted form of μ arises after translation of the shorter messenger.

The amount of diversity possible from the somatic recombination that joins different constant and variable genes for both the light and heavy chains is astronomical. In the mouse, more than 10,000 possible combinations for the heavy chain have been calculated. Each κ or λ light chain in turn can occur in a large number of different variations, since each combines different V with different C regions. Therefore, combining light chains with heavy chains yields more than 10 million possibilities! Added to the possible diversity resulting from these different combinations is the role played by somatic mutation. For reasons that are not clear, the genes in antibody-producing cells appear to be unstable and to undergo mutation at a rate far in excess of that in other kinds of cells. Some of these mutated genes in lymphocytes have been shown to differ from the corresponding variable gene before it is shuffled by just a single nucleotide.

It has been estimated that one change in the V region occurs in every 3–30 cell divisions. The B cell may contain some factors for inducing such a high mutation rate. Supplementing mutation and adding to the potential diversity is the fact that the joining of V and J and V–D–J segments can occur in more than one way. The two segments do not necessarily always recombine at the same site. The point of recombination can vary over several nucleotides. As Figure 20-38 shows, if V and J segments always combine at site 1, then sequences coded for proline and tryptophan would always result at the site of recombination. If they sometimes recombine at site 2, however, the sequence would be for proline and arginine. Since recombination can occur at different sites, new codons are produced designating different amino acids. Moreover, base pairs can be lost when recombination takes place at these junctions, but they may also be inserted, probably due to a special enzyme activity. This flexibility at the V–J and the V–D–J combining sites makes even more variation possible in sequences coded for variable regions of the antibody.

Taking all the possibilities for antibody genes into consideration, the somatic recombination along with mutation and different ways of joining segments, molecular biologists have estimated the possible number of different antibodies as high as 18 billion. Such a fantastic number of diverse molecules is encoded by only a few hundred separate genes or segments, a far cry from the 18 billion that would be called for if each antibody type were encoded by a separate gene!

Many answers must still be found for events taking place during the maturation of B cells and the formation of antibodies. The unexpected finding of gene shuffling or somatic recombination requires machinery to bring together the different V, D, C, and J segments. Perhaps the signal sequences mentioned earlier, which are associated with genes for the heavy and light chains, are involved with enzymes that must

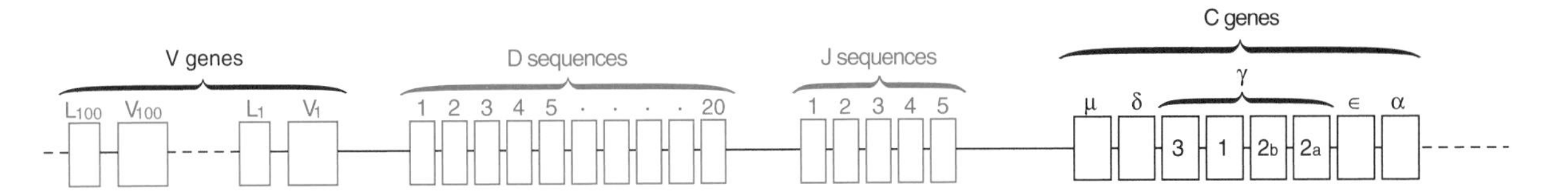

FIGURE 20-35. Unshuffled DNA sequences coded for the heavy antibody chain. In the human, the sequences include about 100 variable genes, 20 D sequences, 5 J sequences, and 8 different constant sequences, each of which designates a specific class of heavy chain. There are five main classes, but the γ class includes four subclasses. The α-gene sequence is present in two copies adjacent to each other. Only one of these is shown.

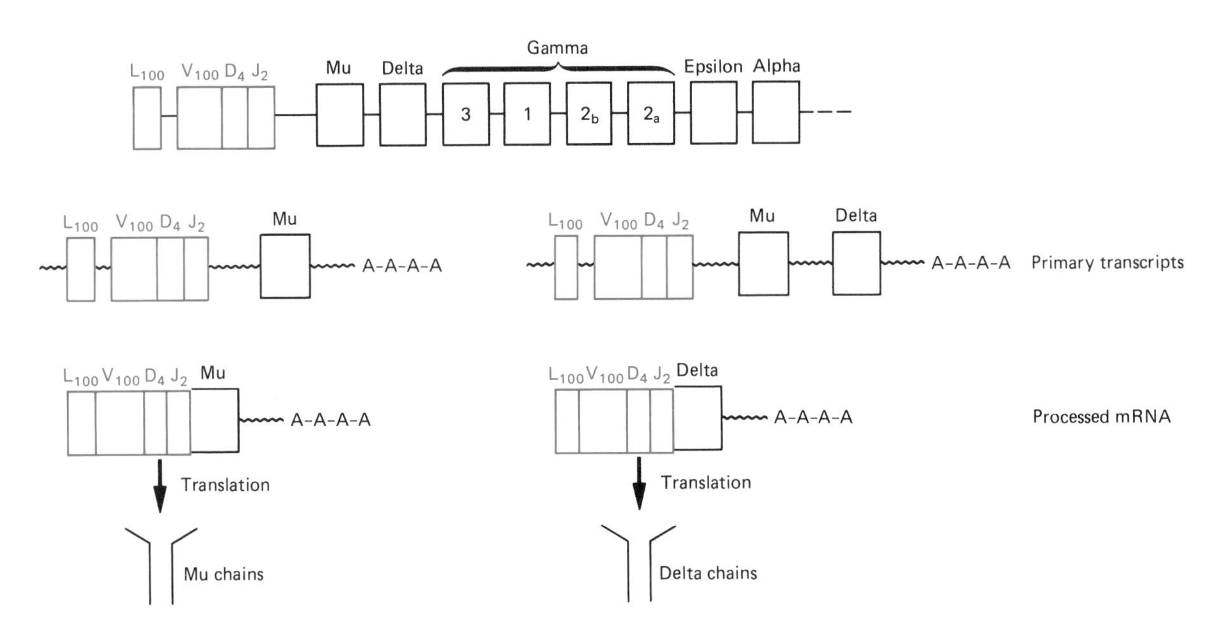

FIGURE 20-36. Somatic recombination in the prelymphocyte. DNA shuffling brings together a combination of a V gene and a D segment and a J segment (in this example, V100, D4, and J2). This combination produces an active gene. Primary transcripts arise (left) coded for the mu constant gene. Processing results in the formation of mRNA, which can then be translated. Some primary transcripts (right) include information for both mu- and delta-chains. The processing eliminates the information in the transcript for the mu-chain. An mRNA forms, which can be translated for a delta-chain. In this cell, therefore, the same V region will be associated with two different constant regions, and the cell will produce both M and D chains. (Two alpha-genes are present in the human and one in the mouse.)

be able to cut the DNA during the shuffling process. The complex story of antibody diversity makes us look at the gene in another light and reconsider how many genes may be involved in forming a polypeptide gene. We still need an explanation regarding the mechanism that controls the recombination of the coded segments or genes to produce an active gene, one that can form a transcript to carry information for the assembly of an antibody chain. The knowledge that gene shuffling occurs in the formation of active antibody genes raises the possibility that similar reshuffling may occur in cases of other DNA regions to generate diversity of genes and hence proteins involved in certain critical developmental pathways.

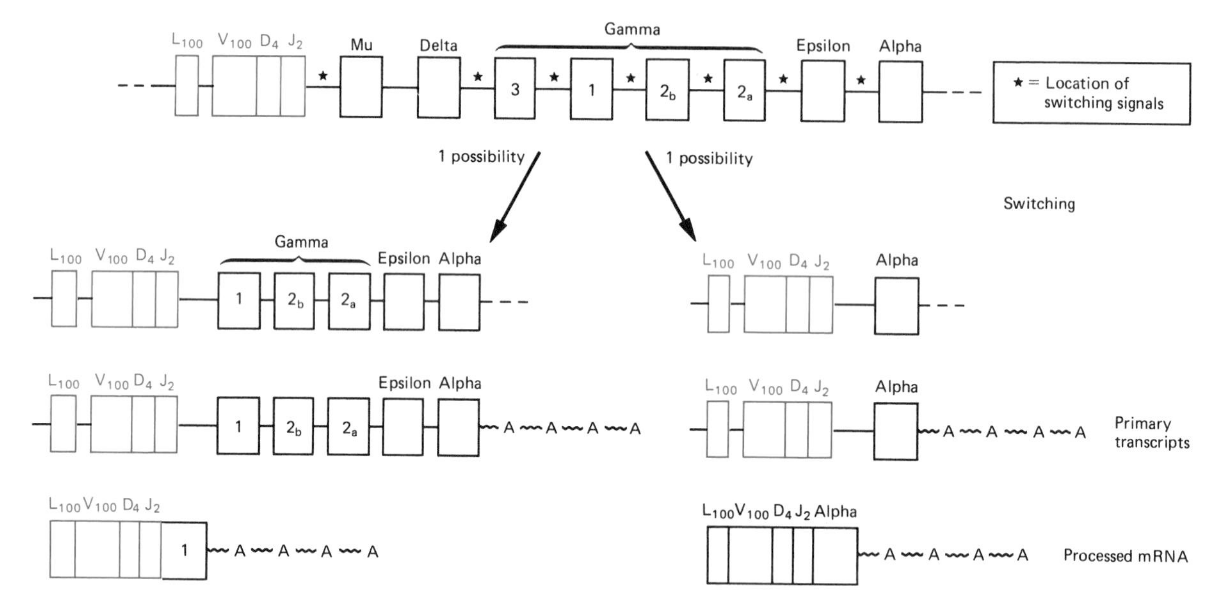

FIGURE 20-37. Heavy-chain class switching. Further DNA shuffling as the lymphocyte matures can place the same V–D–J sequence (refer to Fig. 20-36) directly next to still another C gene. A gamma-1 gene, for example, could be placed next to V–D–J (left) with the deletion of the genes to the left of it. Any one of the other C genes could be placed adjacent to V–D–J, for example, alpha (right). All constant genes still present along with the one directly next to V–D–J will undergo transcription with it, and a primary transcript is produced. We see here on the left that the primary transcript includes information from gamma-1 through alpha . Differential processing, however, produces an mRNA that carries information only for gamma-1, the C gene placed directly next to the V–D–J segment following heavy-chain class switching.

FIGURE 20-38. Joining of V and J segments. When recombination takes place to join the V and J segments, it does not always occur at exactly the same site. If joining or recombination between a given V and a given J always occurred at point 1 (arrow 1), the two triplets, as depicted in the figure, would always designate proline–tryptophan in the recombinant DNA at the position where the recombination took place. If, however, there is flexibility allowing recombination to take place at another point (arrow 2), a new sequence, proline–arginine, can be generated.

REFERENCES

Allore, R., et al. Gene encoding the B subunit of S100 protein is on chromosome 21: Implications for Down syndrome. *Science* 239:1311, 1988.

Anderton, B.H. Alzheimer's disease: Progress in molecular pathology. *Nature* 325:658, 1987.

Antonarakis, S.E. Diagnosis of genetic disorders at the DNA level. *N. Engl. J. Med.* 320:153, 1989.

Aslandis, C., et al. Cloning of the essential myotonic dystrophy region and mapping of the putative effect. *Nature* 355:59, 1992.

Bank, A., J.G. Mears, and F. Ramirez. Disorders of human hemoglobin. *Science* 207:486, 1980.

Beam, K.G. Duchenne muscular dystrophy: Localizing the gene product. *Nature* 333:798, 1988.

Brook, J.D., et al. Molecular basis of myotonic dystrophy: expansion of a trinucleotide (CTG) repeat at the 3′ end of a transcript encoding a protein kinase family member. *Cell* 68:799, 1992.

Buxton, J. et al. Detection of an unstable fragment of DNA specific to individuals with myotonic dystrophy. *Nature* 355:547, 1992.

Cavalli-Sforza, L.L., et al. DNA markers and genetic variation in the human species. *Cold Spring Harbor Symp. Quant. Biol.* 51:411, 1986.

Chang, J.C., and Y.W. Kan. A sensitive new prenatal test for sickle-cell anemia. *N. Engl. J. Med.* 307:30, 1982.

Clark, S.M., E. Lai, B.W. Birren, and L. Hood. A novel instrument for separating large DNA molecules with pulsed homogeneous electric fields. *Science* 241:1203, 1988.

Collins, F. Positional cloning: Let's not call it reverse anymore. *Nature Genetics* 1:3, 1992.

Davies, K.E. (ed). *The Fragile X Syndrome.* Oxford University Press, New York, 1989.

Delabar, J.M., et al. Beta amyloid gene duplication in Alzheimer's disease and karyotypically normal Down syndrome. *Science* 235:1390, 1987.

Drayna, D., and R. White. The genetic linkage map of the human X chromosome. *Science* 230:753, 1985.

Efstratiadis, A., et al. The structure and evolution of the human β-globin gene family. *Cell* 21:653, 1980.

Ellison, J.W., and L.E. Hood. Human antibody genes: Evolutionary and molecular genetic perspectives. *Adv. Hum. Genet.* 13:113, 1983.

Erlich, H.A. (ed.) *PCR Technology.* Stockton Press, New York, 1989.

Erlich, H.A., D. Gelfand, and J.J. Sninsky. Recent advances in the polymerase chain reaction. *Science* 252:1643, 1991.

Fu, Y., et al. An unstable triplet repeat in a gene related to myotonic dystrophy. *Science* 255:1256, 1992.

Geever, R.F., L.B. Wilson, F.S. Nallaseth, P.F. Milner, M. Bittner, and J.T. Wilson. Direct identification of sickle cell anemia by blot hybridization. *Proc. Natl. Sci. USA* 78:5081, 1981.

Goldgaber, D., M.I. Lerman, O.W. McBride, U. Saffiotti, and D.C. Gajdusek. Characterization and chromosomal location of a cDNA encoding brain amyloid of Alzheimer's disease. *Science* 235:877, 1987.

Gusella, J.F. DNA polymorphism and human disease. *Annu. Rev. Biochem.* 55:831, 1986.

Gusella, J.F., et al. Molecular genetics of Huntington's disease. *Cold Spring Harbor Symp. Quant. Biol.* 51:359, 1986.

Gyllensten, U.B. PCR and DNA sequencing. *Biotechniques* 7:700, 1989.

Goodfellow, P. (ed.) *Cystic Fibrosis.* Oxford University Press, New York, 1989.

Harley, H.G., et al. Expansion of an unstable DNA region and phenotypic variation in myotonic dystrophy. *Nature* 355:545, 1992.

Higuchi, R., C.H. von Beroldingen, G.F. Sensabaugh, and H.A. Erlich. DNA typing from single hairs. *Nature* 332:543, 1988.

Hoffman, M. Ancient DNA. Still busy after death. *Science* 253:1354, 1991.

Hozumi, N., and S. Tonegawa. Evidence for somatic rearrangement of immunoglobulin genes coding for variable and constant regions. *Proc. Natl. Acad. Sci. USA* 73:3628, 1976.

Jeffreys, A.J., V. Wilson, and S.L. Thien. Individual-specific "fingerprints" of human DNA. *Nature* 316:76, 1985.

Jerne, N.K. The generative grammar of the immune system. *Science* 229:1057, 1985.

Kerem, B., J.M. Rommens, J.A. Buchanan, D. Markiewicz, T.K. Cox, A. Chakravarti, M. Buchwald, and L.-C. Tsui. Identification of the cystic fibrosis gene: Genetic analysis. *Science* 245:1073, 1989.

Kremer, E.J., et al. Mapping of DNA instability at the fragile X to a trinucleotide repeat sequence p(CCG)n. *Science* 242:1711, 1991.

LaSpada, A.R., Wilson, E.M., Lubahn, D.B., Harding, A.E., and K.H. Fischbeck. Androgen receptor gene mutations in X-linked spinal and bulbar muscular dystrophy. *Nature* 352:77, 1991.

Leder, P. The genetics of antibody diversity. *Sci. Am.* (May) :102, 1982.

Lewin, R. DNA fingerprints in health and disease. *Science* 233:521, 1986.

Lewontin, R.C., and D.L. Hartl. Population genetics in forensic DNA typing. *Science* 254:1745, 1991.

Little, P.F.R. Globin pseudogenes. *Cell* 28:683, 1982.

Mahadevan M., et al. Myotonic dystrophy mutation: An unstable CTG repeat in the 3′ untranslated region of the gene. *Science* 255:1253, 1992.

Marrack, P., and J. Kappler. The *T* cell and its receptor. *Sci. Am.* (Feb) :36, 1986.

Marx, J. Mutation identified as a possible cause of Alzheimer's desease [sic]. *Science* 251:876, 1991.

Marx, J. Boring in on β-amyloid's role in Alzheimer's disease. *Science* 255:688, 1992.

Milstein, C. From antibody structure to immunological diversification of immune response. *Science* 231:1261, 1986.

Monaco, A.P., and L.M. Kunkel. A giant locus for the Duchenne and Becker muscular dystrophy gene. *Trends Genet.* 3:33, 1987.

Monaco, A.P., and L.M. Kunkel. Cloning of the Duchenne/Becker muscular dystrophy locus. *Adv. Hum. Genet.* 17:61, 1988.

Murrell, J., M. Farlow, B. Ghetti, and M.D. Benson. A mutation in the amyloid precursor protein associated with hereditary Alzheimer's disease. *Science* 254:97, 1991.

Nichols, E.K. *Human Gene Therapy.* Harvard University Press, Cambridge, 1988.

Nossal, G.J.V. The basic components of the immune system. *N. Engl. J. Med.* 316:1320, 1987.

Oberle, I., et al. Instability of a 550 base pair DNA segment and abnormal methylation in fragile X syndrome. *Science* 252:1097, 1991.

Oettinger, M.A., D.G. Schatz, C. Gorka, and D. Baltimore. RAG-1 and RAG-2, adjacent genes that synergistically activate V(D)J recombination. *Science* 248:1517, 1990.

Orkin, S.H. Controlling the fetal globin switch in man. *Nature* 301:108, 1983.

Orkin, S.H., and H.H. Kazazian, Jr. The mutation and polymorphism of the human β-globin gene and its surrounding DNA. *Annu. Rev. Genet.* 8:131, 1984.

Patterson, D. The causes of Down syndrome. *Sci. Am.* (Aug) :52, 1987.

Poncz, M., E. Schwartz, M. Ballantine, and S. Surrey. Nucleotide sequence analysis of the δ/β-globin region in humans. *J. Biol. Chem.* 258:11599, 1983.

Proudfoot, N. Pseudogenes. *Nature* 286:840, 1980.

Riordan, J.R., et al. Identification of the cystic fibrosis gene: Cloning and characterization of complementary DNA. *Science* 245:1066, 1989.

Risch, N., and B. Devlin. On the probability of matching DNA fingerprints. *Science* 255:717, 1992.

Roberts, L. Huntington's gene: So near, yet so far. *Science* 247:624, 1990.

Rommens, J.M., et al. Identification of the cystic fibrosis gene: chromosome walking and jumping. *Science* 245:1059, 1989.

Saiki, R.K., et al. Diagnosis of sickle-cell anemia and B-thalassemia with enzymatically amplified DNA and nonradioactive allele specific oligonucleotide probes. *N. Engl. J. Med.* 319:537, 1988.

Sakano, H., K. Hauuppi, G. Heinrich, and S. Tonegawa. Sequences at the somatic recombination sites of immunoglobulin light-chain genes. *Nature* 280:288, 1979.

Schellenberg, G.D., et al. Genetic linkage evidence for a familial Alzheimer's disease locus on chromosome 14. *Science* 258:668, 1992.

Schwartz, D.C., and C.R. Cantor. Separation of yeast chromosome-sized DNA by pulsed field gradient gel electrophoresis. *Cell* 37:67, 1984.

Selkoe, D.J. Amyloid protein and Alzheimer's disease. *Sci. Am.* (Nov) :68, 1991.

Stamatoyannopoulos, G.A., A.W. Nienhuis, P. Leder, and P.W. Majerus. *The Molecular Basis of Blood Diseases.* W.B. Saunders, Philadelphia, 1987.

Tonegawa, S. The molecules of the immune system. *Sci. Am.* (Oct) :122, 1985.

Tonegawa, S. Somatic generation of antibody diversity. *Nature* 302:575, 1983.

White, R., and J.M. Lalouel. Chromosome mapping with DNA markers. *Sci. Am.* (Feb) :40, 1988.

Yu, S., et al. Fragile X genotype characterized by an unstable region of DNA. *Science* 252:1179, 1991.

REVIEW QUESTIONS

1. Suppose a rare and very serious blood disorder is associated with an abnormality of the α-globin chain. DNA is extracted from persons with normal hemoglobin and from those suffering from the disorder. DNA from both sources is subjected to a certain restriction enzyme. This is followed by gel electrophoresis and Southern blot hybridization. Two probes are available. One of these is a labeled cDNA specific for the 5′ portion of the α-globin gene and the other a labeled cDNA specific for the 3′ portion of the gene. Two autoradiographs are prepared, and the results are given here. Explain the differences between the autoradiographs and the molecular basis of the disorder (A = DNA from nonafflicted persons; B = DNA from those with the disorder).

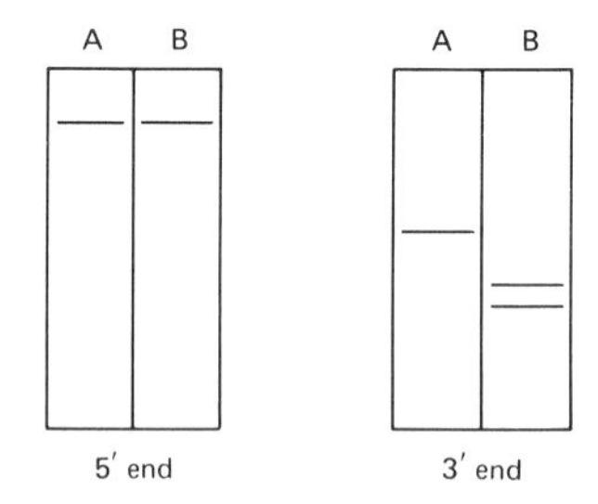

2. A 28 kb segment is known that is vulnerable to a certain restriction enzyme. The enzyme can cut the segment at three sites to yield fragments of different kilobase pair lengths as seen in the following, with arrows indicating cutting sites. However, the site at the left arrow is variable. A probe, as shown here, can detect restriction fragments resulting from the action of the enzyme.

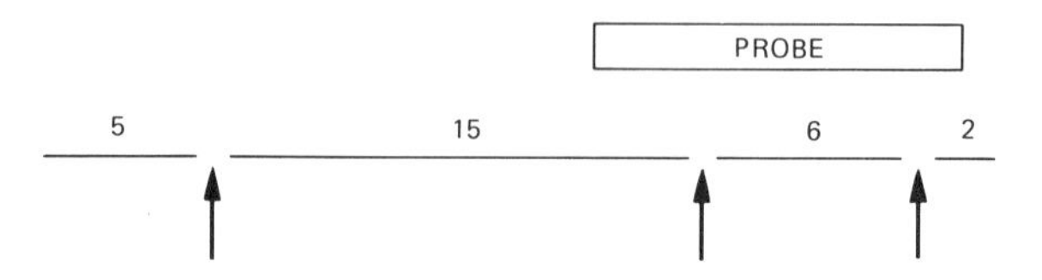

Based on the information presented, answer the following questions:

A. What fragment sizes can the probe detect?
B. What haplotypes are possible? (Use the A, B naming system to designate haplotypes.)
C. What are the possible genotypes in reference to the haplotypes?
D. What fragments would one observe on autoradiographs of the possible genotypes?

3. In reference to Question 2, suppose the sites at both the left and right arrows are variable in relation to cutting by the enzyme. Answer the following:

A. How many haplotypes are possible, and what are they in regard to fragment sizes? (Use the A, B system to name them.)

B. How many genotypes are possible, and what would they be?

4. In a particular pedigree, it is found that Huntington's disease (HD) is associated with haplotype D with respect to HindIII restriction fragments (see Fig. 20-15). Two related families are studied in which one parent is afflicted and the unafflicted parent is not related by blood to other members in the pedigree. The genotypes of the parents in each family are as follows, in reference to the haplotypes:

Family 1, afflicted parent, AD; unafflicted parent, CD
Family 2, afflicted parent, CD; unafflicted parent, CD

DNA is extracted from offspring in each family, treated with HindIII, and autoradiographs are finally prepared. The results are presented here with the pattern at left showing all possible HindIII fragments. Deduce the genotypes of the five persons involved in the two families. Give the relative degree of risk each has of developing HD, along with a brief explanation.

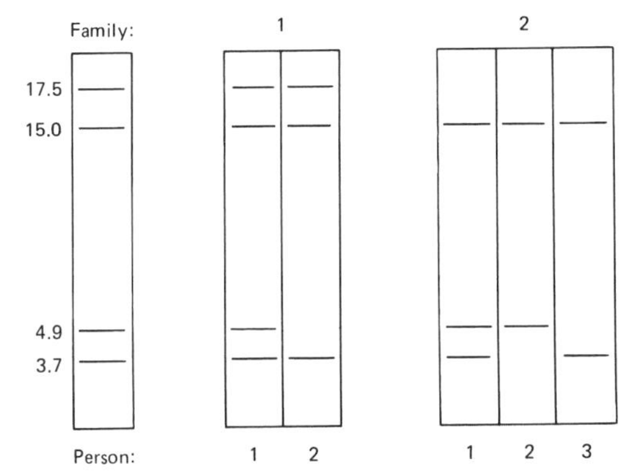

5. If T-helper cells are destroyed by a virus such as the one responsible for AIDS, the immune system breaks down. Why is this so, since the T-helper cells do not directly destroy foreign cells or produce antibodies?

6. DNA was extracted from fibroblast cells and also from antibody-producing cells. The DNA from both sources was subjected to a particular restriction enzyme, and the fragments were separated on a gel. A radioactively labeled available cDNA probe can detect a C gene for the constant region of a light chain. Following Southern blotting, the fragments were exposed to the probe, and autoradiographs were prepared. The results of a comparison of the autoradiographs are shown here, with the numbers indicating the fragment lengths in nucleotide base pairs. Explain the differences seen (F = fibroblasts; P = plasma cells).

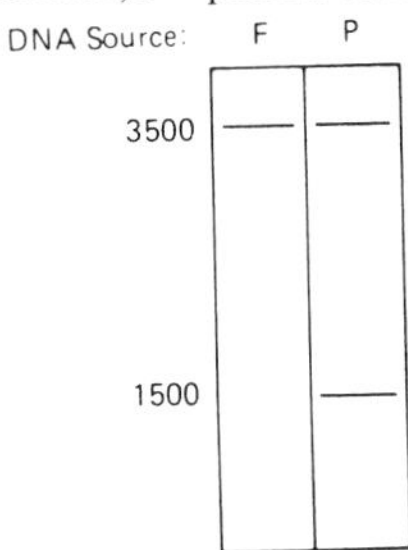

7. For each of the following, indicate whether the item is relevant to heavy or light chains or to both types: 1) κ-chain; 2) C gene; 3) μ-chain; 4) constant region; 5) γ-chain; 6) allelic exclusion; 7) λ-chain; 8) J sequence; 9) V gene; 10) D sequence; 11) α-chain; 12) somatic recombination; 13) chain class switching; 14) RNA processing.

8. As diagrammed here, with just two V (variable) genes, three J sequences, and one C (constant) gene, give the combinations possible for an active, light chain gene following DNA shuffling.

$\underline{V_{18}\ V_6\quad J_1\ J_2\ J_3\qquad C_1}$

9. As diagrammed here, with just one V gene, two D sequences, two J sequences, and two C genes, give the combinations possible for an active heavy chain following DNA shuffling.

$\underline{V_2\quad D_1\ D_2\ J_1\ J_2\qquad C_1\ C_2}$

10. During the development of immature B lymphocytes, IgM and IgD are produced and bind to the cell membrane. After antigenic stimulation, the prelymphocytes mature into antibody-secreting plasma cells, and IgM and IgD disappear from the cell surface. A mature cell may begin to secrete IgG, IgE, or IgA. Explain the basis for regulation of differential gene expression in mature plasma cells.

11. Following is the arrangement of heavy chain sequences in a pre-B lymphocyte. The symbol (S) designates a signal sequence involved in bringing about a shuffling of the DNA. (Only one γ sequence is indicated for the sake of simplicity.)

$\underline{V_{50}\ D_3\ J_2\ (S)\quad \mu\quad \delta\,(S)\,\gamma\,(S)\,\epsilon\ (S)\,\alpha}$

A. Give the arrangements that can form as mature antibody-producing B lymphocytes arise following antigenic stimulation and heavy-chain class switching.
B. For each arrangement, give the sequences that can be found in the pre-mRNA transcript.
C. Give the sequences for each that will be found in each processed mRNA and the class of chain for which the mRNA is encoded.

12. Human DNA can be obtained from the spleen or from white blood cells. When the DNA is subjected to restriction enzyme EcoR1, fragments are generated. With the use of cDNA probes, four fragments can be identified as DNA in the region containing the genes for the δ- and the β-chains of hemoglobin. These four fragments have the following lengths in kilobases: 2.3, 3.0, 4.5, and 6.6.

DNA is taken from patients with hemoglobin Lepore in which the "beginning" portion of the δ-gene is fused with the "end" portion of the β-gene. The DNA from these persons yields only the 3.0 and the 4.5 kb fragments.

DNA is then taken from persons with a very rare disorder affecting the B gene. A mutation has altered a site in the B gene that cannot now be recognized by EcoR1. The DNA from these persons gives three fragments: 11.1, 2.3, and 3.0 kb.

Give the arrangement of the four fragments generated by EcoR1 from DNA taken from persons with normal hemoglobin, starting with the beginning of the δ-gene. Explain the logic used in deducing the arrangement.

Nonchromosomal Genetic Information

21

The killer trait in *Paramecium*

Investigations with the protozoan *Paramecium aurelia* were among the first to highlight the importance of cytoplasmic factors in inheritance and to focus attention on nonchromosomal genetic determinants. In the 1930s, Sonneborn observed that at times, when two stocks of *P. aurelia* are mixed together, some of the animals become abnormal and die. It was easy to mark one of the stocks by feeding the protozoans colored food and to demonstrate that one of the stocks in each case was responsible for the killing effect. Such stocks were designated *killers,* and those affected by them were called *sensitives.* Studies conducted over the years revealed the basis of the killing action. It was shown to be due to the possession by killer cells of a cytoplasmic particle that was named *kappa* (Fig. 21-1). A cell lacking kappa is sensitive to the effects of a kappa particle when it is liberated into the immediate environment, where it exerts a toxic effect.

Fortunately, sensitive stocks are immune to kappa's killing action during conjugation, so that it is possible to cross a killer with a sensitive cell and to follow each one after it has mated. This permitted the genetics of the killer trait to be worked out. It was demonstrated that, for a cell to be a killer, it must possess the kappa particle in its cytoplasm. However, to maintain kappa, the *Paramecium,* which has diploid micronuclei, must possess a dominant allele, designated "K." A cell homozygous for the recessive (kk) is sensitive, since it cannot support kappa. Nevertheless, a cell that carries the dominant (KK or Kk) may lack kappa and is just as sensitive as a cell of genotype kk. No cell can form kappa on its own, even if it contains the dominant K needed for kappa maintenance.

When *Paramecium* cells undergo conjugation, each contains two haploid micronuclei that were formed following meiosis and that behave as gametes. Micronuclei are exchanged between mating cells by way of a conjugation tube that forms between mating pairs. The micronucleus received from a mate unites with the stationary one that remained. The union restores the diploid state to the micronucleus. Following a mating between a killer (KK) and a sensitive (kk), both F_1 cells become Kk. As each of these undergoes later divisions and self-fertilization, homozygous KK and kk F_2 cells arise. To become a pure breeding killer, a sensitive cell of the genotype KK must gain kappa particles through cytoplasmic exchange at the time of mating (Fig. 21-2A). This comes about when a mating is prolonged, and a sizeable cytoplasmic bridge forms that permits transmission of kappa from a killer to a sensitive. If a cell fails to receive kappa, it remains sensitive, even though its genotype permits the maintenance of kappa.

A mating between a killer, KK, and a sensitive of genotype kk produces F_1 cells of genotype Kk (Fig. 21-2B). If both cells following the mating are to become killers, a cytoplasmic bridge of sufficient size must form to enable passage of kappa particles to the cell that was originally sensitive. Self-fertilizations of cells descended from the mates give rise to homozygous KK and kk cells. The former will be killers only if they are derived from killer cells containing a large number of kappa particles. Cells become sensitive if they are genotype kk, even if descended from killers with a large amount of kappa.

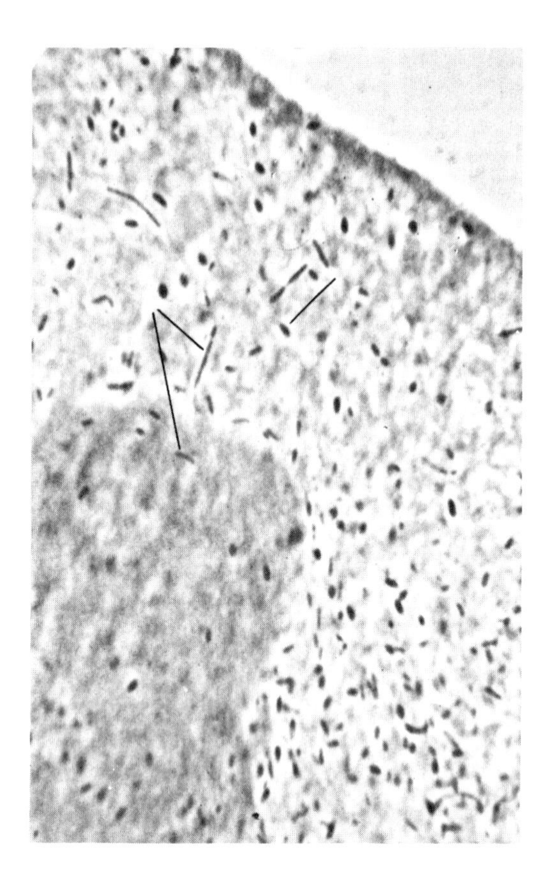

FIGURE 21-1. Kappa in a killer cell. Stained kappa particles are shown to be very numerous. The nucleus in the lower left appears slightly darker than the cytoplasm and shows kappa particles on its surface. Sensitive cells lack these particles. (Reprinted with permission from *J. Cell Sci.* 5:65–91, 1969. Courtesy of Dr. J. Preer, Jr.)

The unusual features of the killer phenomenon raised many questions regarding the origin and nature of kappa. Analysis of the particles showed that kappa is a genetic entity containing DNA and is capable of mutating on its own to give rise to mutant kappa particles, such as those that lack killing activity. Kappa is not essential to the life of the protozoan and is absent from most of them. Moreover, it cannot be made by a cell but must be introduced. These facts suggested that kappa is not native to the species but has arisen from another source. The electron microscope has shown that kappa structure resembles that of a bacterial cell in many respects. Moreover, cytochemical tests have shown that it is a Gram-negative body containing cytochromes that resemble those of bacteria, not those of any eukaryote, including *Paramecium.* Most authorities agree that kappa is an infective agent with definite relationships to bacteria.

Kappa has been found to be but one of numerous agents that can infect *Paramecium.* Indeed, any cell can be infected by some kind of extraneous entity, the maintenance of which depends on the genotype of the host cell. This fact reflects the close association that has evolved in the relationship between the cell and the infective agent and suggests some sort of advantage that may be conferred upon the host by the presence of kappa and related particles, all of which may be regarded as symbionts.

If the presence of the symbiont does confer an advantage, any genetic changes in the host that support an infective particle will be selected by the forces of evolution. However, no advantage from these particles to *Paramecium* has as yet been demonstrated other than the fact that cells with a symbiont are immune to any toxic effects it may produce. However, most cells would not seem to be vulnerable to the lethal effects, since a killing particle or released toxin would find little chance in the natural environment of contacting a cell as it swims by. Whatever the reason for such a variety of infective particles in *Paramecium* cells, it raises interesting questions related to the establishment of host–symbiont relationships and the benefits that can be derived from the coexistence of chromosomal and infective DNA in the same cell.

Extranuclear genomes and information transfer

Two organelles of eukaryotic cells, the mitochondrion and the chloroplast, both of which are essential to all aerobic life on earth, carry genetic information. All the eukaryotes that have been examined show DNA in mitochondria. Moreover, the chloroplasts from algae through flowering plants contain even larger amounts of DNA than are found in mitochondria. The electron microscope reveals that both the mitochondrial DNA (the mt DNA) and the DNA of the chloroplast (the ct DNA) are typically circular molecules. In some protozoans—*Paramecium,* for example—the mt DNA is linear. Both the mt DNA and ct DNA are double stranded and occur in their respective organelles unassociated with proteins and unbounded by any membrane.

Organelle DNA usually differs sufficiently in its G + C content from nuclear DNA that it forms a distinct satellite band when cellular DNA is extracted and subjected to density gradient ultracentrifugation. This makes it possible to isolate organelles from nuclear DNA for use in further studies. Organelle DNA under-

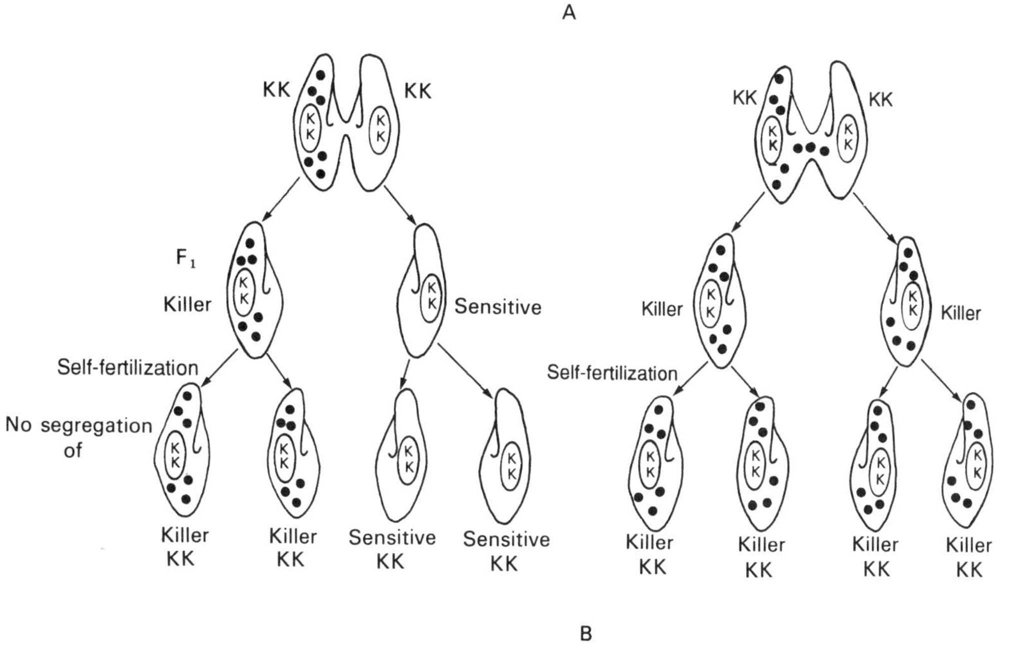

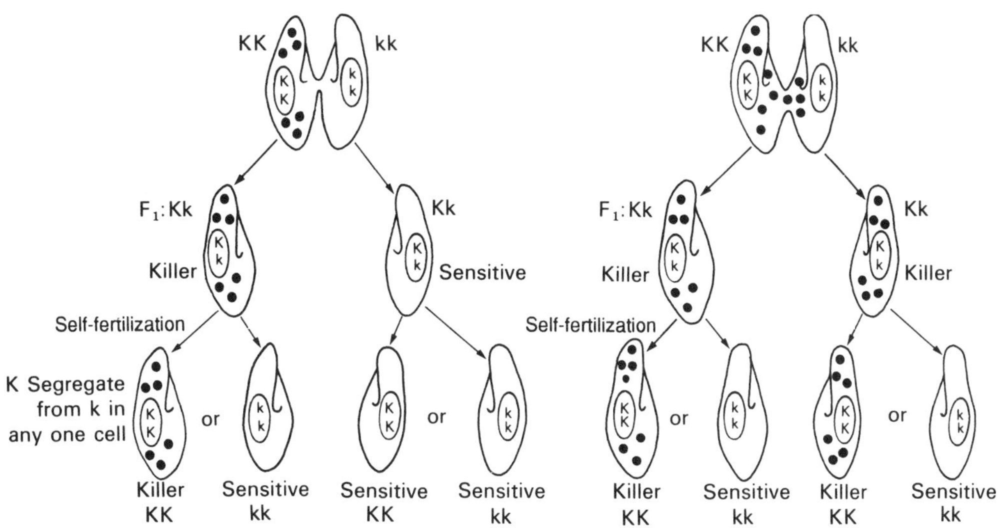

FIGURE 21-2. Genetics of the killer trait. **A:** Both strains in this cross are genetically identical, each being homozygous for the killer allele. The sensitive is sensitive only because it lacks kappa. If it receives no kappa particles from its mate at the time of conjugation (left), it remains sensitive and gives rise only to sensitive cells. If it receives kappa particles by way of a large bridge (right), it becomes a killer and gives rise to killers following later self-fertilizations. **B:** The sensitive strain (kk) in this cross is unable to support kappa in its cytoplasm in contrast to its killer mate. Following mating, both F_1 cells possess the same genotype, Kk, but only the original killer remains a killer if kappa is not contributed to the sensitive mate (left). The F_1 killer, however, will give rise to killers (KK) and sensitives (kk) following later self-fertilizations. If the sensitive cell receives kappa from its mate due to a large cytoplasmic bridge (right), both F_1 cells are killers (Kk), and both give rise to sensitives and killers at subsequent self-fertilizations.

goes semiconservative replication in the typical 5′ to 3′ manner, although the process may be modified for mt DNA. Both DNA strands do not necessarily begin replication at the same time. One strand, the heavy (H) strand, becomes displaced and forms a displacement (D) loop (Fig. 21-3). Replication begins on the light (L) strand before replication on the H strand. Consequently, the replication is staggered, as the D loop becomes more pronounced. Finally, two circular duplexes are formed, each with an old and a new strand.

In mammals, the DNA composing a genome of mt DNA consists of approximately 16,000–17,000 base pairs, a tiny fraction of that found in the nucleus. The human mitochondrial genome has been completely sequenced and consists of 16,569 base pairs, a figure that falls within the size range for animals in general (13,500–18,000 base pairs). The size of the mitochondrial genome tends to decrease as one progresses from lower to higher eukaryote. The mt DNA of yeast is about five times the size of that of the human. In higher plants, the mitochondrial genome is much larger, and great variation is found from one species to another, ranging approximately from 250,000 to more than 2 million base pairs. When algae and higher plant species are compared, the size of the chloroplast genome, the ct DNA, is seen to be much less variable than that of the mt DNA, ranging approximately from 128,000 to 180,000 base pairs.

More than one copy of mt DNA is found in a mitochondrion, where it occurs in the matrix of the organelle (2–50 copies in yeast; 4–10 in mouse, rat, and human cells). Even with the multiple copies, the amount of mt DNA in a mammalian cell is less than 1% of the total amount of the cellular DNA. In plants, however, with their larger mitochondrial genomes and in which 20–40 copies may exist, an appreciable per-

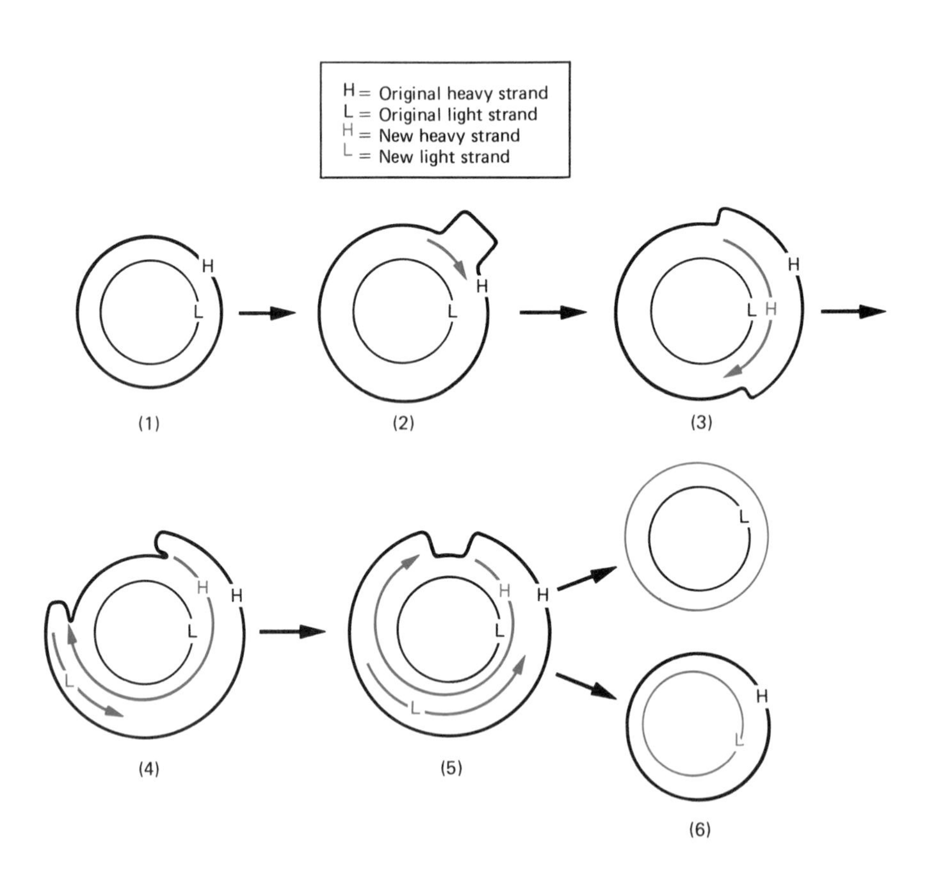

FIGURE 21-3. Replication of mt DNA. When a circular mt DNA molecule (1) composed of H (heavy) and L (light) strands undergoes replication, displacement of the H strand may occur, and a loop forms (2). Synthesis of a new H strand then begins using the original L strand as a template. Displacement of the original H strand continues (3) as the new H strand increases in length. Synthesis of a new L strand then begins, using the original H strand as the template (4). As synthesis on both original strands proceeds, H loop displacement continues (5). Completion of DNA synthesis (6) results in two mt DNA molecules, one with a new H strand, the other with a new L strand.

centage of the cellular DNA (15%) may be accounted for by the organelle.

The replication of mt DNA, unlike that of nuclear DNA, is unaccompanied by a proofreading mechanism. Since the mitochondrial DNA polymerase lacks the capacity to proofread, mutations can accumulate much more rapidly in mt DNA as opposed to the DNA of the nucleus. Consequently, mt DNA can evolve much more rapidly. The multiple copies of mt DNA in a cell may offset the effects of lack of DNA repair in the organelle. The accumulation of mutations in mt DNA has provided a basis for estimating the time of divergence of modern humans from the ancestral human population and is at the center of a current controversy on human origins (discussed later in this chapter).

Mitochondria and chloroplasts always arise by means of growth and division of the preexisting organelles. Replication of the ct and mt DNAs takes place throughout the cell cycle and is not confined to the S phase, when the nuclear DNA replicates. Moreover, in plants, the mt and ct DNAs replicate throughout the cell cycle, completely independently of each other. In all eukaryotic cells, however, there must be some mechanism that keeps replication of the organelle DNA in check, since the *total* amount of organelle DNA just doubles during a cell cycle, resulting in a constant amount of organelle DNA from one cell generation to the next.

In addition to their DNA, mitochondria and chloroplasts also possess components required for the synthesis of protein. Isolated mitochondria and chloroplasts can be supplied in the test tube with labeled amino acids. When this is done, the organelles incorporate the amino acid units, which can later be identified in certain protein fractions. Both kinds of organelles contain ribosomes, and these have been isolated and characterized. Ribosomes of mitochondria are typically composed of two subunits of unequal size. The intact ribosomes vary in sedimentation values from 55S (typical of multicellular animals) to 80S for the protozoan *Tetrahymena* (an exception whose ribosomes dissociate into two subunits of equal size). The ribosomes of chloroplasts, composed of two unequal subunits, have been found to be much less variable than those of the mitochondria. An S value of 70 has been obtained for chloroplast ribosomes from many sources. Their smaller and larger subunits have S values of 30 and 50, respectively. The chloroplast ribosomes closely resemble bacterial cell ribosomes.

The ribosomes of both kinds of organelles contain RNA, a smaller molecule in the small subunit and a larger one in the larger subunit. These ribosomal RNAs of chloroplasts are very similar to those of bacterial cells (16S and 23S). A 5S RNA is also present in the large subunit. The small and large rRNAs of the mitochondrial ribosomes, however, are much more variable in size and consequently cannot be said to be like those of bacteria. A 5S RNA has not been reported in the large subunit. The ribosomal units of both kinds of organelles contain an assortment of proteins, approximately 30 and 25 for the large and small subunits, respectively. These proteins are quite distinct from those that are constituents of the ribosomes found in the cytosol (the fluid portion of the cytoplasm, exclusive of the organelles) of the same cell.

In addition to DNA and ribosomes, the mitochondria and chloroplasts contain other parts of the machinery associated with replication and information transfer. However, these are quite distinct from their counterparts found in the cytosol. These include DNA polymerase, tRNAs for the different amino acids, aminoacyl synthetases, and RNA polymerase. Messenger RNA has also been identified in mitochondria and chloroplasts. Both organelles possess a formylase, an enzyme that can convert methionine to *n*-formyl methionine, as well as an initiating tRNA, $tRNA_f^{met}$.

The electron microscope and biochemical procedures have demonstrated the presence of polysomes in mitochondria and chloroplasts. The ribosomes of the polysomes are held together by a thin strand that has been identified as RNA as a result of digestion by RNAase. Polypeptides in various stages of growth have also been isolated from the polysome fraction of mitochondria. When the mitochondria from cells as diverse as those of *Drosophila* and human HeLa cells have been burst open and their DNA isolated, it has been found that polysomes are associated with the DNA. This indicates that translation is taking place while transcription is occurring from the DNA template. Putting the entire picture together, we can see many striking resemblances between the organelle systems and the protein synthesizing system of prokaryotes, such as the occurrence together of transcription and translation and the presence of formylase. Moreover, components of the organelle system respond similarly to those agents that inhibit nucleic acid and polypeptide synthesis in prokaryotes, whereas the counterparts in the cytosol are not affected. Very interesting is the fact that components of the organelle and prokaryotic protein-synthesizing systems (tRNAs, various enzymes) are interchangeable in in vitro investigations and can substitute for each other in translation. No such substitutions can be made between the organelle components and the equivalent parts in the cytosol. All the foregoing observations suggest that a certain degree of autonomy may reside in the mitochondria and chloroplasts of eukaryotic cells. Let us become familiar with the results of some investigations that bear on this point and have called attention over the years to the existence of extranuclear genetic determinants.

The petite mutations in yeast

Excellent genetic evidence accumulating over the years points to the mitochondrial DNA as the site of mutations responsible for certain traits that are inherited in a nonMendelian fashion. The most thoroughly studied of these are the mutations in yeast (*Saccharomyces*). Cells carrying a petite mutation grow slowly and form tiny colonies on agar, in contrast with the large ones of the wild phenotype. These petites have been shown to possess enzyme defects and to be deficient in aerobic respiration. They require a substrate containing a fermentable product such as glucose, which they use in the presence of oxygen as if they were growing anaerobically.

To understand the inheritance of the petite trait, we must be familiar with the very simple life cycle of yeast (Fig. 21-4), an ascomycete or sac fungus. It is a unicellular organism, and the haploid cells can be classified into either of two mating types, + or –. The diploid zygote formed from fusion of a + and a – cell may grow by budding to produce a diploid colony. The diploid cells can also be stimulated to undergo meiosis. The cell then enlarges and forms four haploid nuclei, each of which becomes the nucleus of a spore. The meiotic cells behave like an ascus or sac, and thus the four spores are considered to be ascospores.

We can see in Figure 21-4 that two of the ascospores will be mating type + and two will be –. This 2:2 segregation indicates that the genetic determinants of mating type are nuclear and exist as one pair of alleles that segregate at the meiosis preceding ascospore formation. Similarly, many other characteristics known in yeast behave in the same way, such as the ability to produce adenine or some other essential metabolite. In all of these cases, 2:2 segregations are found when a single pair of alleles is followed. Linkage as well as independent assortment are easily detected in yeast and have permitted the construction of chromosome maps using the same reasoning followed for other organisms.

Mutations to the petite phenotype have arisen in both mating types + and –. Some of these petites behave in the expected Mendelian fashion: A cross of a petite with a wild produces wild diploid cells; the ascospores yield a 2:2 segregation of wild type to petite. Petites such as these are called *nuclear petites.* In contrast to these are others that are decidedly non-Mendelian in their pattern of inheritance. One kind of nonchromosomal petite is known as the *neutral* or *recessive* petite. A cross of a neutral petite with a wild (Fig. 21-5) produces diploid cells that are normal in phenotype. When sporulation is induced, the ascus yields spores that produce only wild-type cells. The segregation is thus 4:0. The petite phenotype has disappeared and does not reappear when these wild cells are followed further in similar crosses. Clearly, the neutral petite is not behaving as a Mendelian trait, as indicated in the departure from the 2:2 segregation in the ascus.

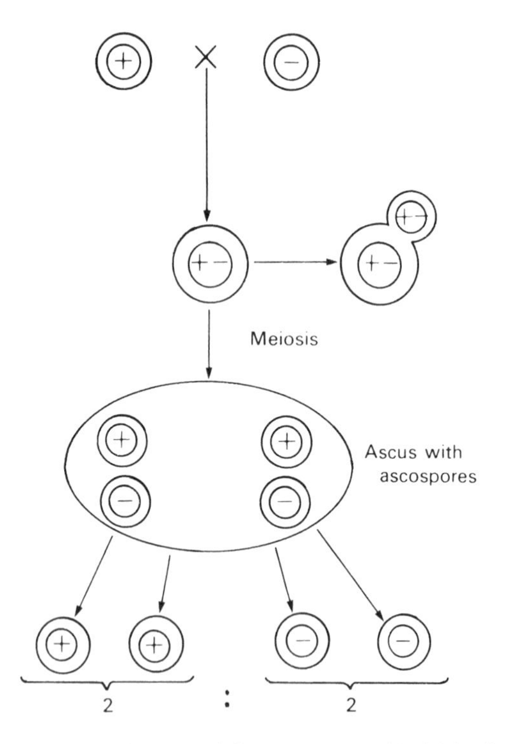

FIGURE 21-4. Life cycle of *Saccharomyces.* Haploid cells of opposite mating types (+ and –) fuse to form a diploid zygote. This can give rise to a colony of diploid cells by budding. Diploid cells may also be stimulated to undergo meiosis. The cell enlarges, and four haploid nuclei result, each the nucleus of an ascospore. The mating type as well as other nuclear alleles segregate 2:2 from the ascus.

Experiments were performed in which yeast was treated with acriflavine. (It has been found that acridine dyes can eliminate F factors from *Escherichia coli* cells.) The results showed that almost a whole population of normal cells could be transformed to petite after exposure. No known mutagen can affect nuclear genes to such an extent that every exposed cell contains an induced mutation. Other observations of this type strongly indicated that the determinants for the petite phenotype reside in the cytoplasm.

Another class of petites was discovered in which the trait also behaved in a nonMendelian fashion, but it differed in its pattern from the neutral or recessive petite. This is the *suppressive petite.* When it is crossed with a wild type, the results depend on when the zygotes are induced to sporulate (Fig. 21-6). If ascospore formation takes place very soon after the zygote forms, it is found that most of the asci will give a segregation of 0:4; that is, all the spores will give rise to petites.

The zygotes, if immediately plated out on agar after the mating, form diploid colonies that are also petite. In contrast to the neutral petite, it is as if the wild type were tending to disappear. However, different results can be obtained from the same cross. Instead of being induced to sporulate immediately, the zygotes may be subcultured in liquid medium for a period of time. If the diploid colonies are then plated out, they are almost all wild type. Similarly, if these zygotes are induced to sporulate *after* being subcultured, the ascus gives a 4:0 segregation, four wild to no petite.

Thus, depending on the treatment after mating, two very different situations can develop. The appearance of the wild type after subculture in the liquid can be explained by the fact that the liquid culture favors the survival of wild-type cells. Any cells of normal respiratory phenotype that are present immediately after mating would be selected in the liquid medium, because it is not conducive to the growth of the petites. Therefore, wild-type cells would increase and take over the culture, giving normal diploids upon plating and 4:0 segregation after sporulation.

There are many other features of petites, but it should be evident from this short discussion that the inheritance of the neutral petite and of the suppressive petite is decidedly nonMendelian. For many years it had been considered likely that the petite mutation is a result of some change in the mitochondrion, because the respiratory capacity of the petite cell is deficient. It has now been demonstrated by density gradient ultracentrifugation that the mt DNA of cytoplasmic petites

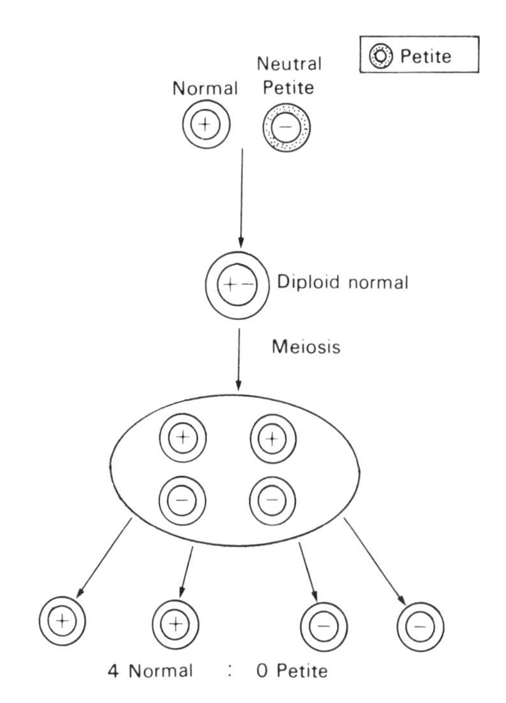

FIGURE 21-5. Wild-type yeast cells and neutral petite. In a cross between the two types diploids arise, and these are normal in phenotype. After meiosis, the mating types segregate 2:2 as expected, but all the resulting ascospores produce wild-type cells. The petite trait seems to have disappeared and is not behaving in a Mendelian fashion.

has undergone some sort of alteration. The mt DNA of suppressive petites differs in its buoyant density from the mt DNA of wild-type yeast. There is a shift to a lower G + C value. These changes generally arise as deletions of portions of the mt DNA. The deletions differ when the mt DNAs from various strains are compared. Although usually about as much DNA exists in the suppressive petites as in the wild-type yeast cell, most neutral petites have been shown to lack mt DNA altogether! We can well understand why the cytoplasmic petites do not undergo spontaneous reversion to wild type as is the case for point mutations.

Inheritance of drug resistance in yeast

Another class of nonMendelian mutations has been recognized in yeast. These bring about resistance to various antibiotics, such as erythromycin and chloramphenicol. After a cross of erythromycin-sensitive (wild type) to erythromycin-resistant (mutant) cells, the zygotes are allowed to undergo mitotic divisions.

Repeated divisions eventually yield cells that are parental types, either erythromycin sensitive or resistant. When one of these diploid cells is allowed to sporulate, the four ascospores will be identical, either all sensitive or all resistant. Such results are decidedly nonMendelian (Fig. 21-7). We can easily understand why this occurs if we consider the fact the yeast cell produces a bud when it divides mitotically. Originally, the zygote contains a mixture of both mutant (resistant) and wild-type (sensitive) mitochondria. However, when a bud is produced, only a few mitochondria enter it. Continued mitotic divisions of the diploid cells eventually lead to buds that contain either all resistant or all sensitive mitochondria. As a result of this mitotic segregation, when the diploid cell sporulates, the four spores produced will contain the same kind of mitochondria, either all resistant or all sensitive. Resistance to chloramphenicol follows the same pattern.

If a cross is made between two strains, one resistant to erythromycin and the other resistant to chloramphenicol, it can be seen that recombination can give asci containing all doubly resistant spores and asci with all doubly sensitive spores (Fig. 21-8). This tells us that recombination of nonchromosomal genetic determinants can occur as well as recombination of nuclear alleles. Such a process adds even more diversity to the combinations of inherited material possible after sexual reproduction, and its full significance in all species is yet to be assessed. There is good evidence that the determinants of drug resistance in yeast also reside in the mt DNA. For example, when mutations to petite occur and produce a physical alteration in the mitochondrial DNA, markers for drug resistance may be lost. This implies that the physical change in the organelle DNA, the elimination of a segment, has brought about the petite phenotype and the simultaneous loss of the resistance determinants. The inheritance in yeast of the petite trait and of the resistance to certain drugs offers excellent evidence for the localization of specific nonchromosomal genes in the mitochondrion.

Nonchromosomal genes in *Chlamydomonas*

Very good evidence for the storage of genetic information in the chloroplast has been provided by studies of Sager with the unicellular green alga *Chlamydomonas.* The life cycle of this species is very simple

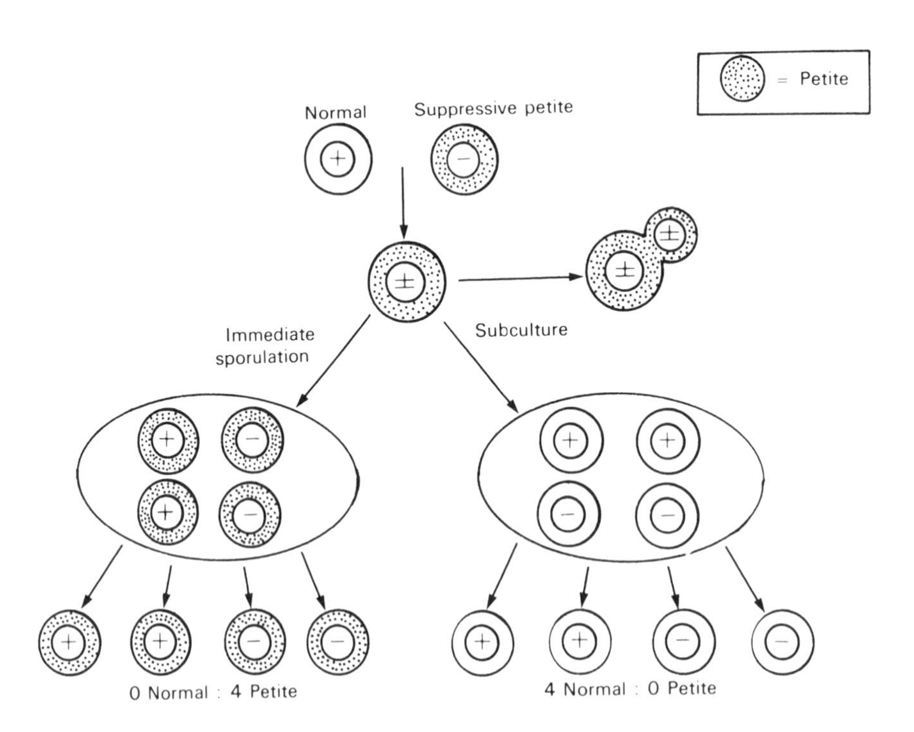

FIGURE 21-6. Wild-type yeast × suppressive petite. The suppressive petite, unlike the neutral one, tends to express itself over the wild. The diploids are petite, and, if diploid cells are induced to sporulate immediately, all the spores give rise to petite colonies. However, if diploids are subcultured and later sporulated, all the spores give rise to wild colonies.

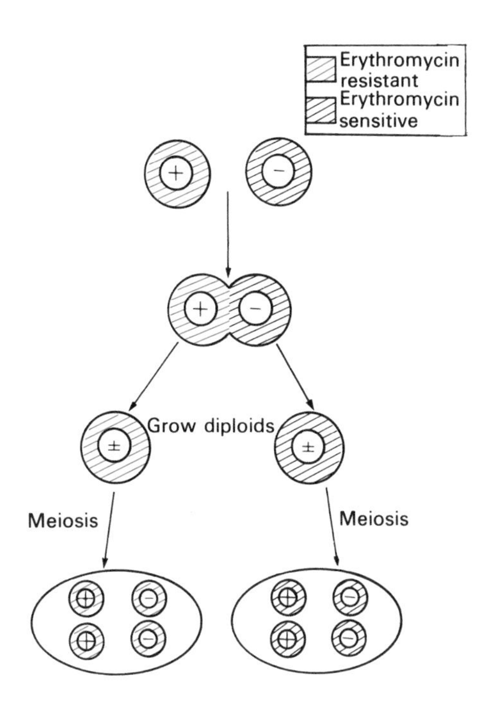

FIGURE 21-7. Erythromycin resistance in yeast. After the cross of resistant × sensitive, diploid cells can be established from growth of the zygotes. Upon sporulation, the diploids give rise either to all resistant cells or to all sensitive ones. (The two kinds of diploid cells do not necessarily occur in a 1:1 ratio.)

and is comparable to that of yeast. Again, two mating types, + and −, are recognized. Mating between cells of opposite mating type produces a zygote (Fig. 21-9). When the zygote matures, meiosis occurs, and four zoospores (the motile products of the meiotic division) arise. Half of these are mating type + and half are type −. The zoospores may undergo mitotic division to form clones of cells. Again, the mating type is seen to depend on a pair of nuclear alleles inherited in typical Mendelian fashion.

Many characteristics in *Chlamydomonas* are known to depend on nuclear genes: spore color and requirements for certain metabolites, for example. Both linkage and independent assortment have been demonstrated for these genes associated with Mendelian traits. There are, however, some genetic determinants in *Chlamydomonas* that are definitely nonchromosomal. Sager has worked extensively with certain streptomycin-resistant strains that exhibit nonMendelian inheritance. As Figure 21-10 shows, the results of a cross between streptomycin-resistant and streptomycin-sensitive strains depends on the phenotype of the mating type + parent.

Although mating type itself segregates among the zoospores in the Mendelian 2:2 fashion, the zoospores always have the streptomycin trait shown by the mating type + parent. This suggests maternal inheritance, but here the + and the − cells are identical. Neither class of gamete contributes more cytoplasm and hence more cytoplasmic genes than the other. Something is operating to prevent the transmission of the nonchromosomal genes of the mating type − parent from the zygote to the meiotic products. Since the discovery of the nonchromosomal streptomycin determinants, many others have been found. Some of these were induced by mutagens, particularly by streptomycin itself, which in *Chlamydomonas* is able to induce mutations in nonchromosomal genes but not in those of the nucleus.

It was later found that rare exceptions to this mating type + pattern of inheritance occur. In fewer than 1% of the zygotes, exceptions arise in which both sets of nonchromosomal genes, the set from the mating type − as well as the one from the mating type + parent, are transmitted. Sager also found that if the mating type + parent is treated with a dose of ultraviolet light before mating, 50% of the zygotes exhibit biparental inheritance. All of the offspring arising from such a zygote contain a complete set of nonchromosomal genes from each parent. Such cells are called *cytohets,* indicating that they are heterozygous for cytoplasmic genetic determinants (Fig. 21-11A and B).

It is possible to follow more than one pair of cytoplasmic markers in biparental inheritance. The cytoplasmic markers do *not* segregate when meiosis takes place. The haploid zoospores are therefore heterozygous for the cytoplasmic genes. When a heterozygous zoospore divides mitotically and forms a clone, the alternative forms of the cytoplasmic genes segregate and undergo recombination. This segregation and recombination continues with each additional mitotic division of the zoospores until there are no more heterozygous markers to detect. The recombination frequency of these nonMendelian genes has been measured. This has permitted the construction of a map and the assignment of several nonchromosomal genes to definite positions. It appears that the cytoplasmic genes are all part of a single circular linkage group.

Sager has presented several lines of evidence that support the hypothesis that the physical basis of the nonchromosomal genes (and hence the linkage group to which they belong) is the DNA of the chloroplast. For example, most of the cytoplasmic determinants that have been mapped were caused to mutate by streptomycin, a drug known to exert its influence almost exclusively on the development of the chloroplast, with no direct effect on other parts of the cell. Other work with *Chlamydomonas* continues to provide greater insight into the overall significance of nonchromosomal genes, as well as into the mechanism of uniparental inheritance.

Organelle genetic systems

As discussed earlier in this chapter, organelles have the ability to replicate their DNA, and they possess the machinery for protein synthesis. We may now wonder about the precise kinds of information encoded that may be in the organelle DNA and the kinds of interactions that may take place between the organelle system and the nucleocytoplasmic component of the cell.

Organelle ribosomes can be isolated and then separated into their RNA and protein components. The rRNA from mitochondria or chloroplasts may then be presented with the DNA from the same kind of organelle in the presence of RNA from some other source, such as rRNA from the cytoplasmic ribosomes. When this is done (Fig. 21-12A), it is found that all the rRNA from mitochondria forms hybrid DNA–RNA molecules only with DNA from mitochondria. Other kinds of RNA from outside the organelle do not compete with it, which demonstrates that the "foreign" RNAs do not contain a significant number of nucleotide sequences in common with the mt DNA. The same is true for the rRNA and the DNA of the chloroplast. The rRNA from an organelle (mitochondrion or chloroplast) does not form hybrid molecules with the DNA taken from the nucleus of the same cell (Fig. 21-12B). Results from such DNA–RNA hybridization studies leave no doubt that the chloroplast and the mitochondrion each contains information in its DNA coded for the specific rRNA of the organelle.

Other DNA–RNA hybridization studies that involve tRNA from yeast have indicated that all the mitochondrial tRNAs of yeast are transcripts of genes found in the mt DNA. These are tRNAs that represent all 20 amino acids. In contrast, most of the genes in yeast that code for the proteins of the ribosome and for

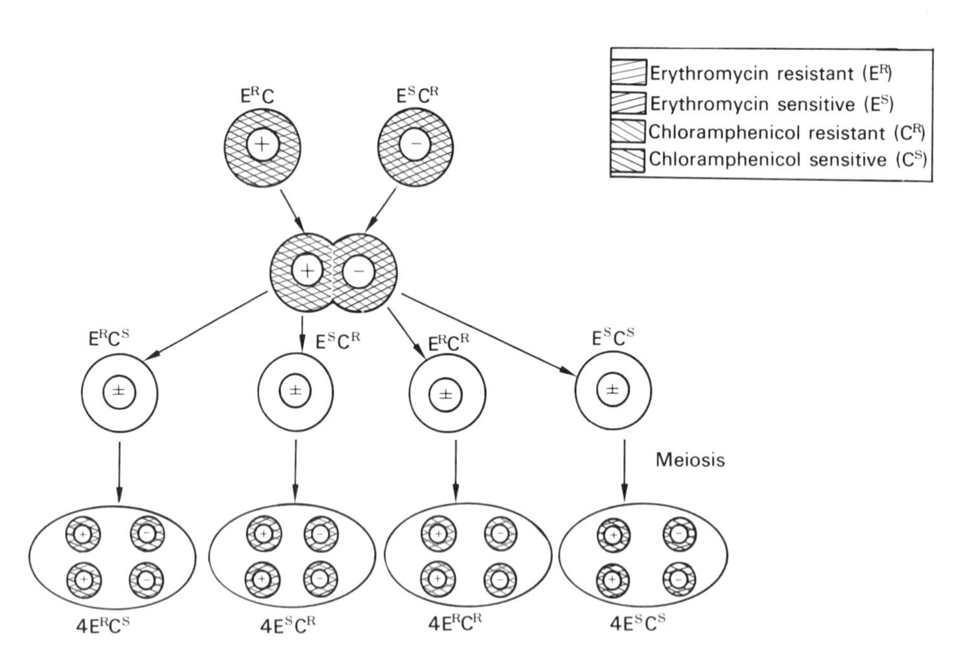

FIGURE 21-8. Two-factor cross involving drug resistance. If chloramphenicol resistance is followed alone, it shows the type of nonMendelian inheritance illustrated in Figure 21-7 for erythromycin. If chloramphenicol and erythromycin resistances are followed together in a two-factor cross, it can be demonstrated that recombination of the nonMendelian factors can take place. After the mating, diploid cells are cultured. From these, four types can be isolated (not necessarily in a 1:1:1:1 ratio). Each gives rise to ascospores, which are either all parental or all recombinant. The recombinants here are the E^RC^R and E^SC^S. (Modified slightly with permission from R. Sager, *Cytoplasmic Genes and Organelles,* p. 132. Academic Press, New York, 1972.)

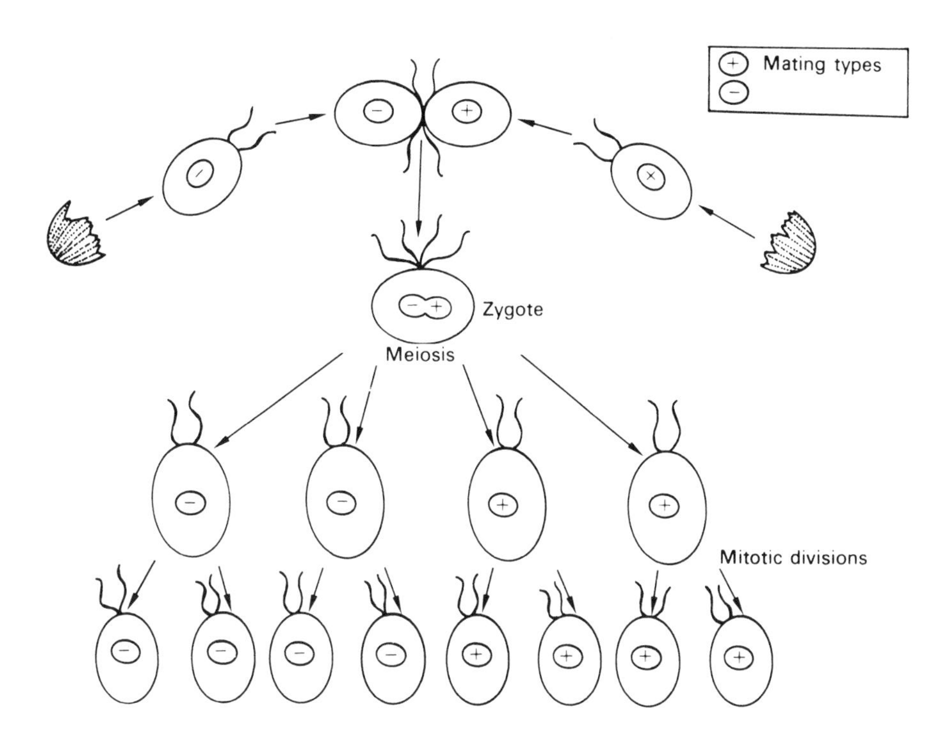

FIGURE 21-9. Life cycle of *Chlamydomonas.* Two mating types occur. After fusion of two cells of opposite types, meiosis takes place, and the mating type segregates in a Mendelian fashion. The motile cells can divide mitotically to form clones of other identical cells.

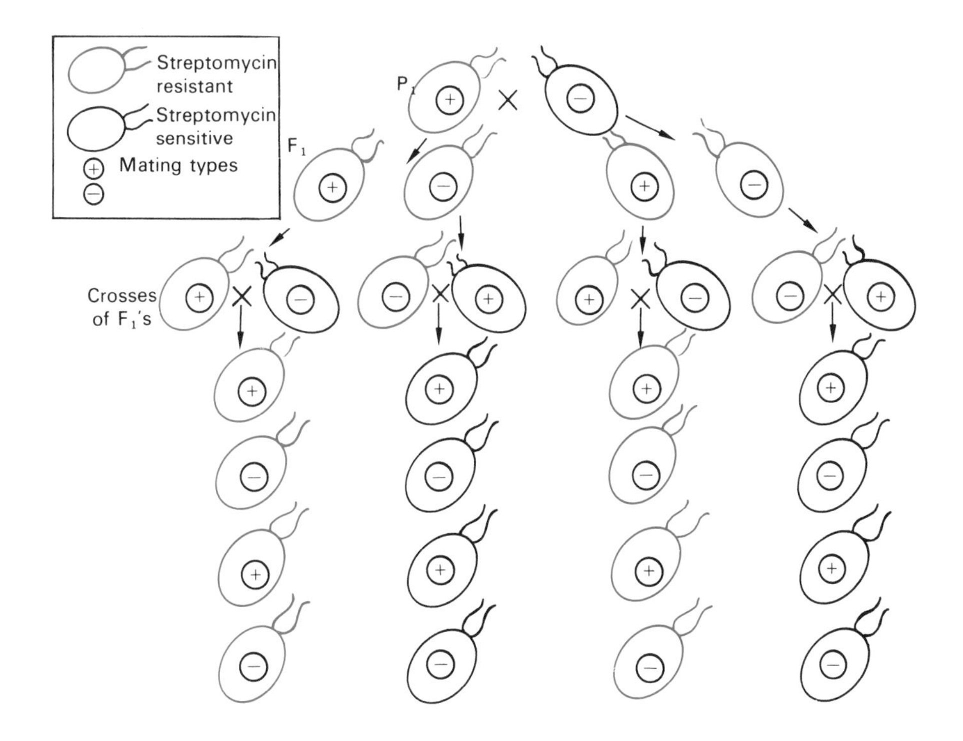

FIGURE 21-10. Inheritance of streptomycin resistance in *Chlamydomonas.* It can be seen that inheritance of sensitivity or resistance to streptomycin depends on the phenotype of the parent that was mating type +. Although mating type is segregating as expected, the cells are always like those of the + parent in relation to tolerance of streptomycin.

the aminoacyl synthetases have been found to exist in the cell nuclei. These proteins are synthesized on the ribosomes found in the cytosol and then move into the organelle. The ribosomal proteins finally associate with the rRNA made in the mitochondrion to form the complete functional ribosomes of the organelle. The same holds for the chloroplast.

It is not unexpected to find that information for many of the parts of the mitochondria or chloroplasts is coded in genes of the cell nucleus. This is so

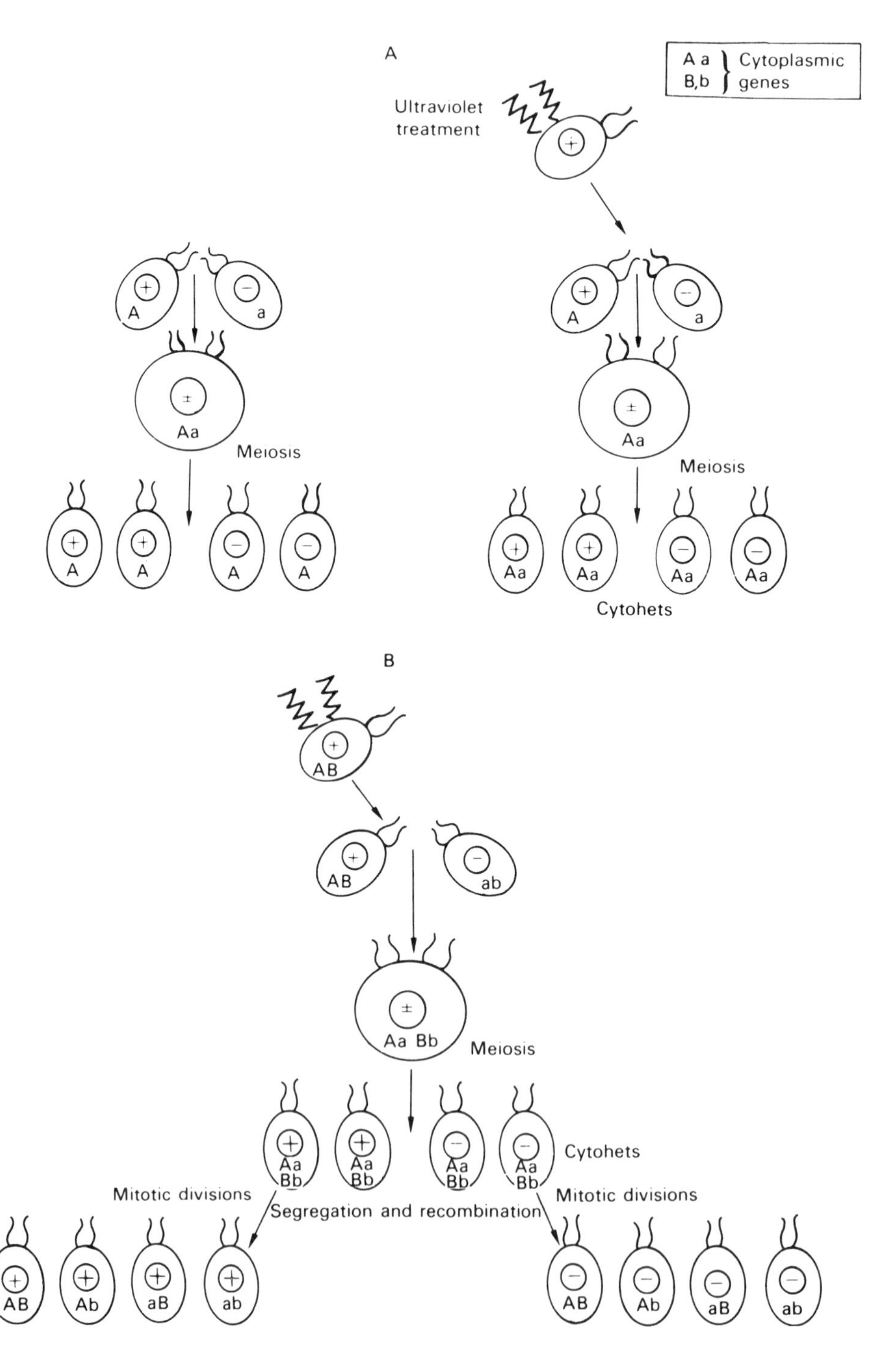

because many different nuclear gene mutations are known to affect the structure of the mitochondria or the chloroplasts. Moreover, both organelles contain hundreds of different kinds of proteins (ribosomal proteins, membrane proteins, enzymes), and the amount of organelle DNA is simply not sufficient to code for all of these.

In addition to the genes of the mt DNA in yeast that code for tRNA and rRNA, a second important class has been recognized. These genetic determinants code for polypeptides that function in electron transport and oxidative phosphorylation. Their existence was established by using mutant cells in which the mt DNA had been altered by deletions. Obviously, since neutral petites contain no mt DNA, the absence of any genes in such strains would indicate that their genetic information resides in the mt DNA. Since the amount of DNA lost in the different suppressive petites is variable, the various strains of suppressive petites can supply valuable information on the proximity of two or more genes as well. For example, if two genes are lost together when a given deletion occurs in the mt DNA, then they must have been positioned close to each other on the mt DNA (Fig. 21-13).

Recognition of extranuclear point mutations in yeast (e.g., those affecting drug resistance and the cytochrome oxidase complex) was extremely valuable in pinpointing genetic sites on the mt DNA. Unlike the petites, which cannot revert back to wild type because of deleted DNA, these mutations can revert. The point mutations exist in different allelic forms, and they have made it possible to recognize nonchromosomal loci that undergo recombination and can be analyzed genetically, as described earlier in this chapter for drug resistance in yeast. Moreover, acridine dyes can induce the origin of petites, all of which have defects of aerobic respiration. However, petites do differ in their extent of DNA loss. Yeast strains with different point mutation markers can be treated with an acridine mutagen to convert them to petites differing in the deletion of their DNA. As mentioned, the higher the frequency of simultaneous loss of two markers (or, conversely, their retention) in the different petite strains, the closer together the two genes must be. Mutant forms also known in *Paramecium* and the mold *Neurospora* are believed to be the result of alterations in mt DNA.

FIGURE 21-11. Biparental inheritance of cytoplasmic genes. **A:** Typically, the cytoplasmic genes of the mating type – parent are not transmitted from the zygote, so that the haploid products of meiosis, the zoospores, show the phenotype of the + parent (left). If, however, the mating type + parent is treated with ultraviolet light and then mated, the cytoplasmic genes from both parents may be transmitted to the zoospores (right). Although the nuclear factors (e.g., + and –) segregate at meiosis, the cytoplasmic ones do not. The F_1 cells thus have two complete sets of cytoplasmic genes and are called *cytohets.* **B:** Two or more pairs of cytoplasmic factors can be followed in cases of biparental inheritance. The haploid zoospores are cytohets. If each zoospore is then allowed to divide further to form a clone, segregation of the alleles takes place, as does recombination among the members of the different allelic pairs. Thus nuclear alleles in *Chlamydomonas* segregate and undergo recombination at meiosis, whereas these processes do not take place for the cytoplasmic genes until the mitotic divisions of the haploid zoospores, after meiosis has occurred.

Translation in organelles

Although genetic information is present in organelles, not enough of it is there to code for most organelle character-

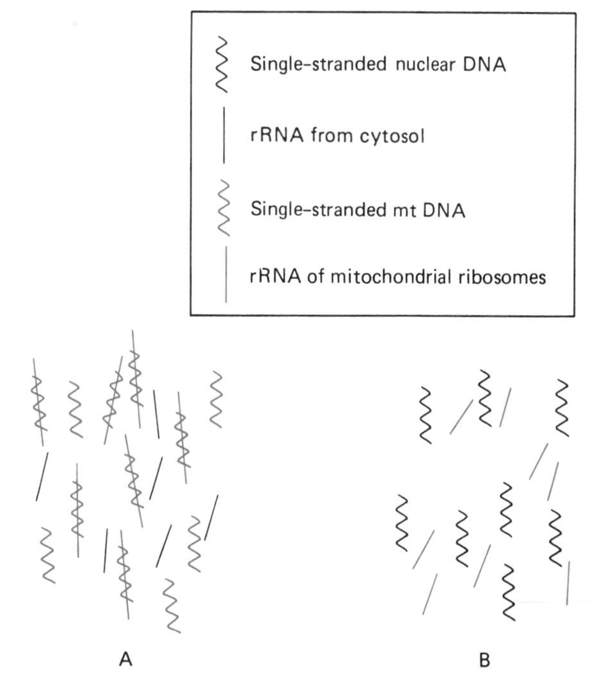

FIGURE 21-12. A: When single-stranded mt DNA is presented with RNA derived from the cytosol along with mitochondrial rRNA, the only DNA–RNA hybrid molecules that form are those between the mt DNA and the rRNA of the mitochondrial ribosomes. Here, for example, the rRNA of the ribosomes from the cytosol does not compete with the rRNA of the mitochondria for sites on the mt DNA. **B:** No DNA–RNA hybrid molecules form when the nuclear DNA is presented with rRNA from the mitochondrial ribosomes of the same cell.

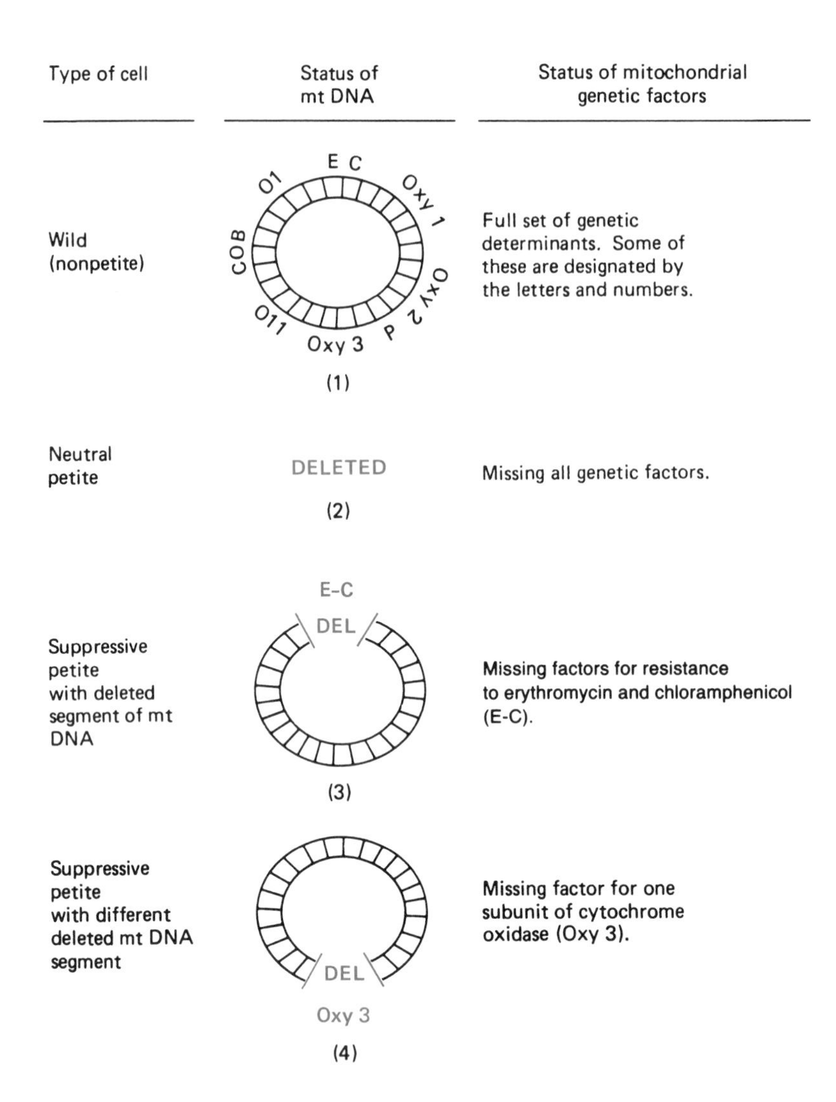

FIGURE 21-13. Nonpetite, wild-type yeast strains have an intact mt DNA and carry a complete set of genetic determinants (1). These genetic factors are absent in neutral petites, which also lack all the mt DNA, indicating that these missing genetic elements reside on the mt DNA (2). Two different suppressive petites (3, 4) carry deletions in the mt DNA that differ in size and location. The genetic factors that are missing in the two strains will be different. Those factors closer together on the mt DNA will have a greater chance of being lost together when a deletion of the mt DNA occurs. By studying different deletion mutants in this way, the positions of the mt genes can be estimated.

istics. Much information coding for features of the organelles must reside in the cell nucleus. It has been possible to sort out some of the interactions that take place between the organelle and nucleocytosol systems. When labeled amino acids are supplied to cells, the newly made proteins are radioactive and are found in all parts of the cell, including the mitochondria. How can we determine whether a protein found in the mitochondria was made in the organelle or in the cytosol and then transported into the organelle? Since the 1960s investigators have been distinguishing between these two alternatives through the use of drugs that selectively shut off the protein-synthesizing apparatus in either mitochondria or cytosol. For example, cycloheximide can inhibit protein synthesis by ribosomes in the cytosol. It does not affect the ribosomes of the organelle. Therefore, if one finds labeled protein that has arisen in cells supplied with labeled amino acids and grown in the presence of cycloheximide, it is evident that the protein was assembled on mitochondrial ribosomes. Conversely, chloramphenicol and erythromycin can affect the mitochondrial ribosomes and shut off any translation that may result from their activities. Any labeled polypeptides formed in this setup must have been made on ribosomes in the cytosol that remain unaffected by those antibiotics (Fig. 21-14).

From such selective inhibition studies, significant information has been obtained on the sites of origin of various polypeptides found in the mitochondria. The proteins of the mitochondrial ribosomes must be made by the ribosomes of the cytosol in yeast. No labeled amino acids are found in the ribosomes of yeast mitochondria following cycloheximide treatment, which shuts off the ribosomal activities in the cytosol (Fig. 21-15). Conversely, when yeast cells are treated with chloramphenicol or erythromycin, labeled amino acids appear in all the proteins of the mitochondrial ribosomes, indicating that they must have been synthesized on the ribosomes of the cytosol and then shipped into the organelle. Once there, they assemble with rRNA made in the organelle to form an essential part of the functional mitochondrial ribosomes.

Such procedures have been employed to study the complex enzyme ATPase, which is bound to the inner membrane of the mitochondria. The ATPase of the mitochondria is composed of 10 polypeptide subunits and functions in oxidative phosphorylation coupled to electron transport. Some of the ATPase subunits are made in the cytosol, others in the organelle. In yeast, four are mitochondrial in origin, coded by the mt DNA; in *Xenopus* three are mitochondrial and in *Neurospora* only two. It has also been shown that in the chloroplast some of the subunits of this enzyme are made in the organelle.

Another complex enzyme of the inner mitochondrial membrane, cytochrome oxidase, consists of seven polypeptide chains. Selective inhibition studies have shown that in yeast the three largest ones are made in the mitochondria and the four smaller ones in the cytosol (Fig. 21-16). Again we see cooperation between the cytosol and the organelle systems in the formation of a complex protein. When mt DNA is altered by a deletion, so that the genes coding for the three subunits of cytochrome oxidase are lost, as in petites, transcripts of the missing genes are consequently absent, and the three subunits cannot be formed in the mitochondria. The four smaller subunits *are* formed, however, because nuclear DNA is intact, but no functional enzyme is assembled inside the organelle (Fig. 21-17). All the results clearly show that the three larger subunits of cytochrome oxidase are coded by mt DNA and that the transcripts for them are translated in the organelle. One of the seven or eight subunits of the complex coenzyme QH_2–cytochrome c reductase is translated in the mitochondria of yeast and *Neurospora*. All the others are synthesized in the cytosol. We see another good example of cooperation between the translation systems of the mitochondria and cytosol in forming a complex mitochondrial component.

Mitochondrial genomes of the human and yeast

Several approaches have led to the construction of a genetic map of the mitochondrial genome in yeast. Various genes were localized by studies of the loss of markers associated with deletions and by the analysis of recombination of nonchromosomal alleles. Mitochondrial genes have also been positioned on a physical map with the aid of restriction enzymes, in a manner similar to that described in Chapter 19. Such analyses led to the identification and mapping of mitochondrial genes encoded for specific products. The development of methods permitting sequencing of the mt DNA has now reinforced the earlier approaches and has led to a more precise mapping of both the yeast and the human mitochondrial genomes.

Human mt DNA is composed of 16,569 base pairs, and all the reading frames have now been identified (Fig. 21-18). These include those coded for cytochrome b, three subunits of cytochrome oxidase, some ATPase subunits, and six subunits of NADH dehydrogenase. The genes are packed so closely together on the human mt DNA that there appears to be little room left for noncoding regions between them. Moreover, not only do the genes lack introns, but in some regions they actually overlap, the same nucleotides being read in different frames in two different genes. There is also a scarcity of nucleotide sequences to act as promoters.

In contrast to the mitochondrial picture, not only do most nuclear genes of eukaryotes contain introns, but the transcription of each gene is controlled by its own individual promoter that interacts with the enzymes of transcription. On the human mt DNA, only one major promoter region exists on each of the two strands of the circular mt DNA molecule. This region is localized in the D loop of both strands (Fig. 21-18). When transcription takes place, the full length of a strand is transcribed. Each long transcript is then processed to yield the tRNAs, rRNAs, and mRNAs. The transcript derived from the H strand gives rise to two rRNAs plus most of the tRNAs and mRNAs. Each mRNA carries a poly (A) tail about 55 nucleotides long. This

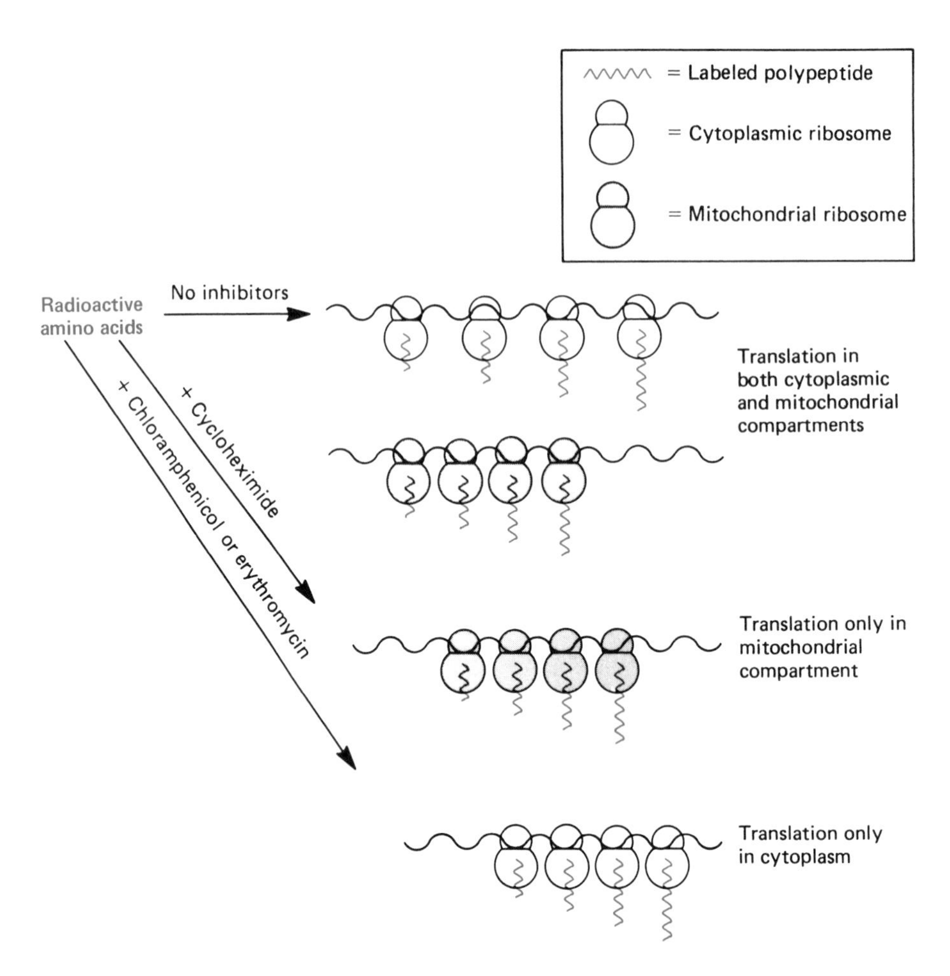

FIGURE 21-14. Top: When labeled amino acids are supplied to cells, labeled protein appears in all parts of the cell. Both the cytosol and mitochondrial translation systems are operative, and the polypeptides produced in each compartment are labeled. **Middle:** When cycloheximide is added, translation is shut off in the cytosol but remains unaffected in the organelles. Label can appear only in those polypeptides made in the organelle. **Bottom:** When the antibiotics are added, the organelle translation system is shut off, but the cytosol apparatus is intact. Label appears only in those polypeptides synthesized in the cytosol.

tail is added after the completion of transcription by the action of a mitochondrial poly (A) polymerase. The mRNAs, which all lack the 5′ cap typical of the mRNA or the cytosol, are then ready for translation.

Following processing, the long mitochondrial transcript made on the L strand yields eight tRNAs and only one mRNA, coded for a subunit of NDH dehydrogenase. About 90% of this transcript is devoid of information and becomes degraded.

The cleavage of a large mitochondrial transcript into separate smaller ones occurs while the RNA is being formed on the mt DNA template. An interesting observation is that sequences coded for tRNAs exist on each side of a longer gene and that cleavage of the long transcript takes place exactly at the beginning and end of each tRNA sequence. Since there is little noncoding DNA in the human mitochondrial genome, these tRNA sequences would seem to be acting as recognition sites for the enzymes that process each long mitochondrial transcript. The tRNA sequences may accomplish this by folding into a shape that the processing enzymes recognize.

Unlike nuclear transcripts, the mt mRNAs lack the leader sequences to which ribosomes bind at the start of translation. The ribosomes of human mitochondria are able to recognize certain codons for methionine (more on this later) as start signals and manage to bind to the transcripts without the aid of leaders. Moreover, sequencing both the human mt DNA and its transcripts has shown that most of the genes have incomplete chain-terminating codons and also lack sequences coded for the trailer. For example, an antisense strand may have the sequence–AT at its 5′ end (Fig. 21-19). Its transcript therefore terminates in–UA, an incompletion of the chain-terminating codon UAA. The problem is solved by the posttranscriptional addition of the poly(A) tail mentioned earlier. Therefore the transcript is extended to include the sequence UAA, a chain-terminating codon. In the same way, posttranscriptional extension enables a transcript ending in a single–U to include the UAA chain-terminating sequence.

Yeast mt DNA, which is about five times the length of human mt DNA, has been mapped for genes encoded for two rRNAs, about 30 tRNAs, and several membrane proteins as well as some proteins as yet unidentified. The yeast mitochondrial genome contrasts sharply with the human in many respects. Unlike human mitochondrial genes, those of yeast are scattered, separated by long noncoding regions, 95% of which are composed of AT base pairs. The mitochondrial genes of yeast *do* code for chain-terminating codons in the transcript as well as noncoding trailer sequences. The yeast transcripts do *not* have poly(A) tails added posttranscriptionally. Some mitochondrial mRNAs have been found to have leader sequences that are much longer than the sequences that code for the product. Moreover, some mitochondrial genes in yeast are split by long introns. The genes for cytochrome b and for subunit I of cytochrome oxidase contain almost 10 times the amount of DNA needed to code for the protein product. (Yeast strains vary in the number and size of their mitochondrial introns.) We can certainly conclude that much of the yeast mt DNA, unlike that of its human counterpart, is noncoding.

A very interesting discovery was made relating to one of the introns found in the yeast mitochondrial gene for cytochrome b. Mutations that arise in this intron influence the splicing of the RNA transcript of that gene. Research on the cytochrome b gene revealed the following picture (Fig. 21-20). The gene includes five introns and six exons. Transcription gives rise to a long transcript containing the introns and exons. Intron 1 is soon removed through the activity of a splicing enzyme encoded by a nuclear gene. As a result of the removal, exons 1 and 2 become linked together and to intron 2. This intron has been found to include a portion that contains a reading frame. This reading frame is now in phase with the reading sequences of the first two exons. These joined sequences (exons 1 and 2 and the reading frame of intron 2) act as an mRNA, which is translated

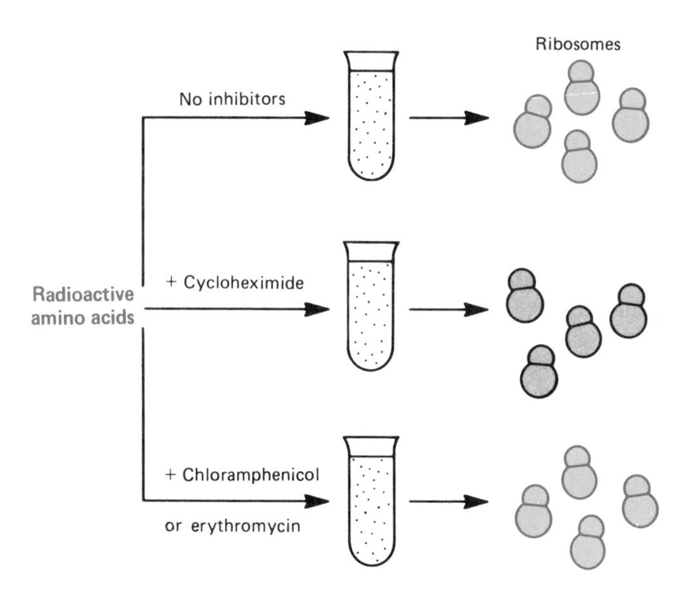

FIGURE 21-15. Top: When labeled amino acids are supplied to cells in the absence of inhibitors, the isolated mitochondrial ribosomes contain labeled protein; **middle:** whereas, no label is found in these ribosomes when cycloheximide is present. **Bottom:** however, the mitochondrial ribosomal proteins are labeled in such an experiment when one of the antibiotics is present. Because cycloheximide shuts off the translation system in the cytosol, the results indicate that the protein of the mitochondrial ribosomes is formed in the cytosol, since label appears in ribosomal protein only when the cytosol compartment is operative (compare with Fig. 21-14).

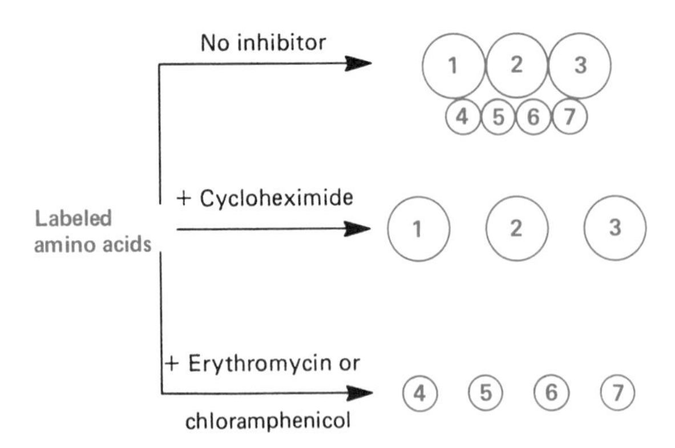

FIGURE 21-16. The enzyme cytochrome oxidase of yeast is composed of seven polypeptide subunits, three large and four small. When cells are supplied with labeled amino acids in the absence of inhibitor (top), cytochrome oxidase is formed, and all seven subunits are labeled. In the presence of cycloheximide (middle), no complete enzyme is made. Only the three large subunits are synthesized and are thus labeled. In the presence of one of the antibiotics (bottom), again no complete enzyme is made, but now the four smaller subunits are formed and are labeled. No large ones are present. Since the translation machinery of the cytosol is shut off with cycloheximide, the three subunits that appear in its presence must have been made in the mitochondria (middle). The four that appear in the absence of cycloheximide but when the mitochondrial machinery is turned off (bottom) must have been made in the cytosol.

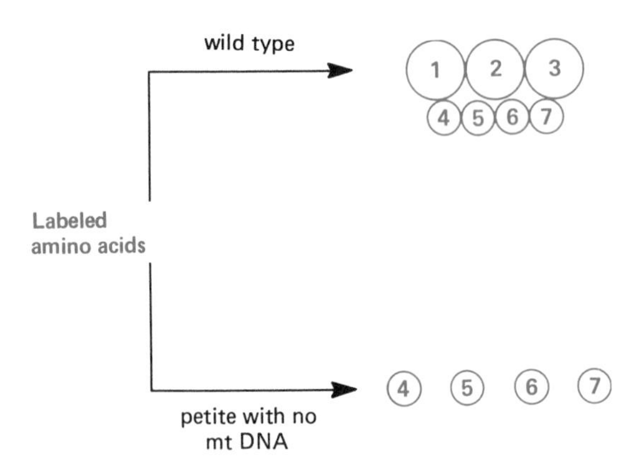

FIGURE 21-17. When wild-type yeast cells are supplied with labeled amino acids in the absence of any inhibitors, the complete cytochrome oxidase enzyme is formed, composed of its seven polypeptide subunits. In petites with deleted mt DNA, no intact enzyme is formed. The three large subunits are missing. These are also missing when wild yeast is subjected to chloramphenicol or erythromycin, as seen in Figure 21-16. The results tell us that the genetic information for the three large subunits (always absent in the petites) is coded in the mt DNA and that they are synthesized in the organelle. The four small subunits, always present in petites (and when the machinery of the cytosol is operative) are coded by the nuclear DNA and synthesized on the ribosomes of the cytosol.

into a protein. The first 143 amino acids of this protein, encoded by exons 1 and 2, are those of cytochrome b. The carboxyl end of this protein is composed of amino acids encoded by the reading frame of intron 2. This odd protein has been called an *RNA maturase.* It has the ability to aid in the catalytic removal of intron 2. By so doing, it limits its own concentration in the organelle because it destroys its own mRNA!

Introns 3 and 4 of the cytochrome b gene also contain reading frames, as do four of the introns of a gene that codes for a unit of cytochrome c oxidase. There is reason to suspect that these introns may also be partly encoded for intron-specific maturases. Introns with coding sequences such as described here are now known to occur in the mitochondrial genomes of several species of fungi and are suspected in other lower eukaryotes. Such findings again raise the question of the true purpose of introns.

Mitochondrial DNA and the human condition

Long neglected in clinical studies, more and more attention is being directed to the human mt DNA as evidence accumulates to show that the mitochondrial genome plays a role in the maintenance of a healthy phenotype. Unlike those of the nucleus, mitochondrial genes of the human are inherited exclusively through the female parent, and they are not partitioned equally when a cell divides but are distributed randomly to daughter cells. As noted earlier, there are many mitochondria per cell, and each organelle in the human contains 4–10 copies of the mitochondrial genome. Consequently, if a mutation is present in the mt DNA somewhere in one of the organelles of a cell, it is by far outnumbered by the nonmutant DNA. The appearance of any effects due to an alteration of the mt DNA by the mutation will be delayed until a sufficient number of organelles come to acquire it.

The first human disorder to become linked to a mitochondrial defect is one that shows a maternal pattern of inheritance. Known as Leber's neuropathy, it brings about loss of central vision as the optic nerve degenerates. The energy provided by the mitochondria is critical to normal stability of the optic nerve, which does not tolerate energy fluctuations. The mt DNA sequences of persons with the disorder have been compared with those from a healthy source. Observations from most patients have revealed a nucleotide substitution in which adenine replaces guanine in one of the mt DNA genes coded for a respiratory enzyme.

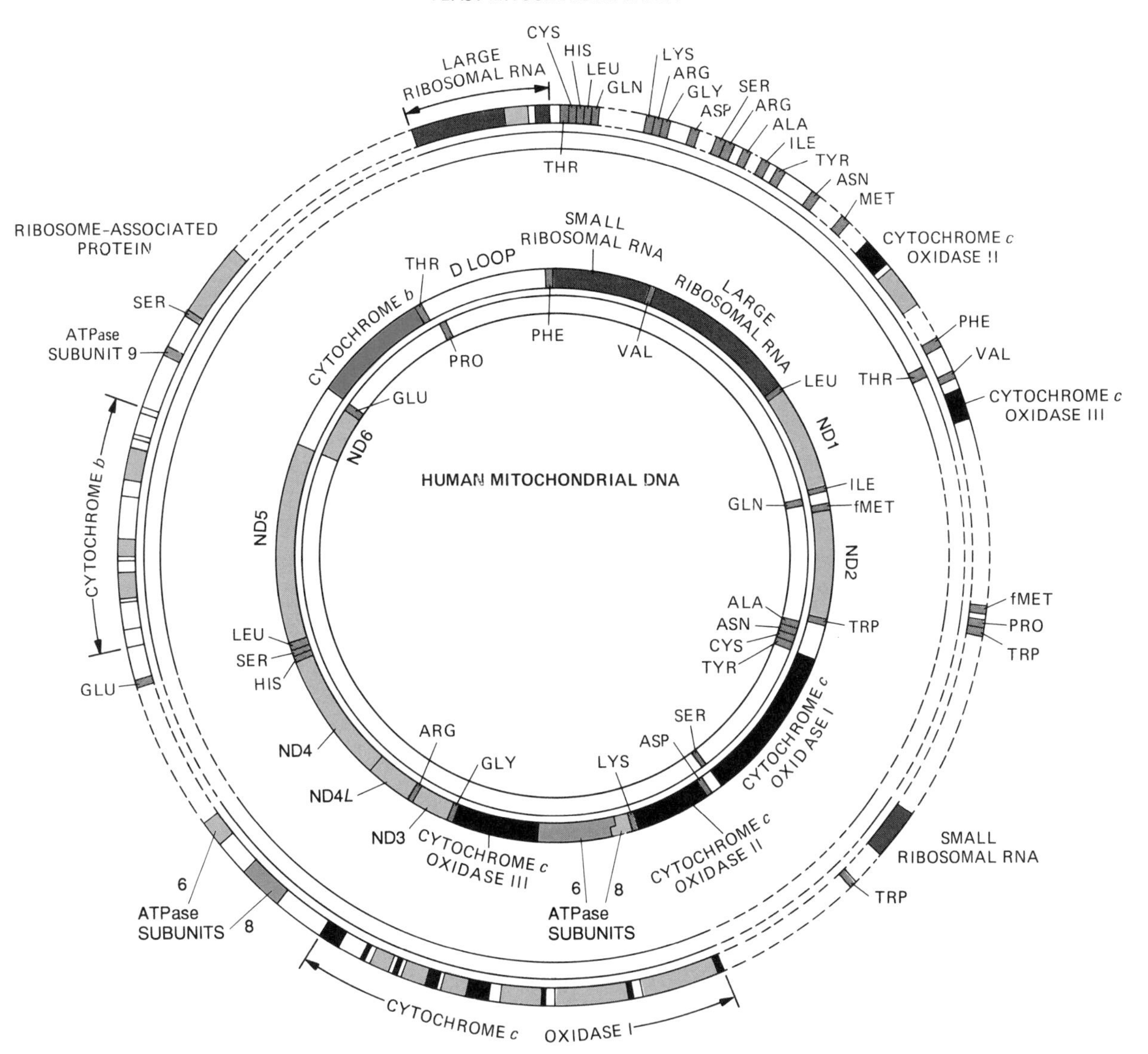

FIGURE 21-18. Organization of yeast and human mitochondrial genomes. The mt DNAs of yeast and human are shown as concentric double circles. The two single circles of each double one represent the two strands of the double helix. All darkened segments indicate coding regions, such as those for specific proteins, rDNAs, and tRNAs. Each abbreviation for an amino acid marks the location of the gene for that specific tRNA. The human mt genome contains little noncoding DNA between genes. The D loop is a region involved with the initiation of transcription. Each ND denotes a reading frame for a subunit of NADH dehydrogenase. In yeast, the mt genome contains long, noncoding (colorless) stretches, and several genes contain introns (e.g., cytochrome b). The broken lines indicate regions that have not yet been sequenced. The yeast mt DNA molecule is actually five times the length of the human one. (Modified from L.A. Grivell. Mitochondrial DNA. *Sci. Am.* (March): 78, 1983. Copyright © 1983 by Scientific American, Inc. All rights reserved.)

As a result, histidine replaces arginine in the protein encoded by the gene. All the observations lead to the conclusion that this optic disorder results from a point mutation in the mt DNA.

Another rare disease has been associated with a similar kind of mutation. This point mutation is in a gene coded for the organelle tRNA needed to carry lysine. As a consequence of the defect, there is interference with the synthesis of mitochondrial proteins and the production of abnormal-appearing mitochondria associated with a type of epilepsy.

Deletions of pieces of the DNA from the mitochondrial genome have been linked to certain disorders characterized by muscle weakness. These illnesses are not inherited, in contrast to those associated with point mutations. Instead, they appear to result from spontaneous mutations that knock out pieces of the mt DNA during the affected person's lifetime. A woman suffering debilitating defects following such losses may be too ill to produce children.

Deletions have now been associated with a disorder in which there is paralysis of certain muscles that control eye movements, muscles that contain a large number of mitochondria. Some of the cases entail severe brain and heart damage in addition to the eye muscle effect. Over 45% of the mitochondrial genome has been found to be missing in these severe cases.

It is believed that deletions of the mitochondrial genome may be taking place all the time in cells. They may prove lethal if large ones arise during prenatal development. The later the occurrence of a deletion, the later the appearance of any ill effects, since sufficient amounts of normal mt DNA would be present to prevent onset of the disorder. However, once mt DNA becomes shortened, it multiplies more quickly than the larger, intact DNA, and the shorter it is the more

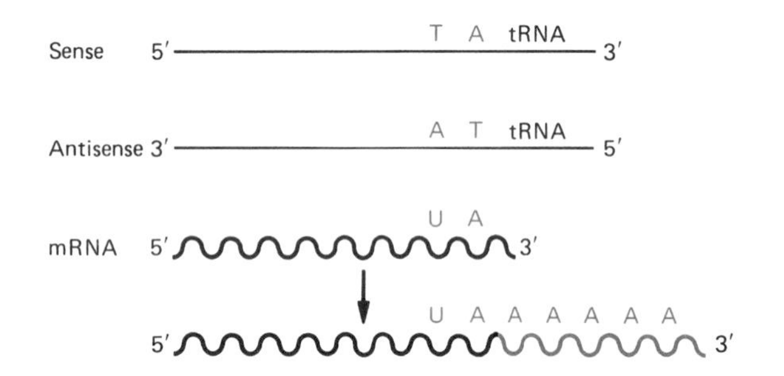

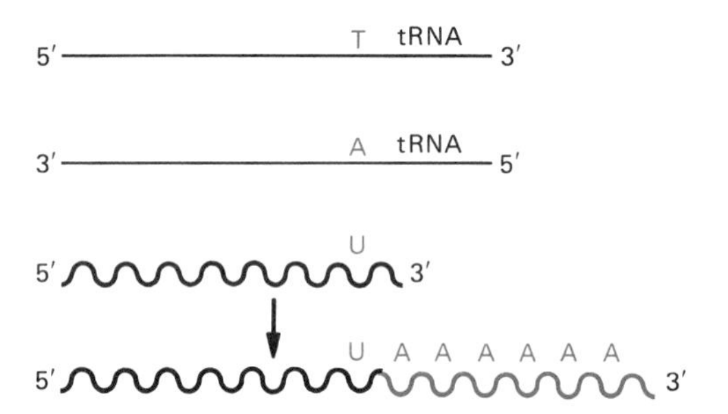

FIGURE 21-19. Incomplete chain-terminating codons of human mitochondrial transcripts. Mitochondrial genes in the human are typically very closely packed, the end of a gene next to a sequence coded for tRNA, as depicted above for the gene on the left and one on the right. Most of the genes do not contain complete chain-terminating codons. As a result, the transcript formed on the antisense strand ends in UA (left) or simply U (right). Following transcription, an enzyme adds a long poly(A) tail (red). This extension of the transcript brings about completion of the chain-terminating codon, UAA.

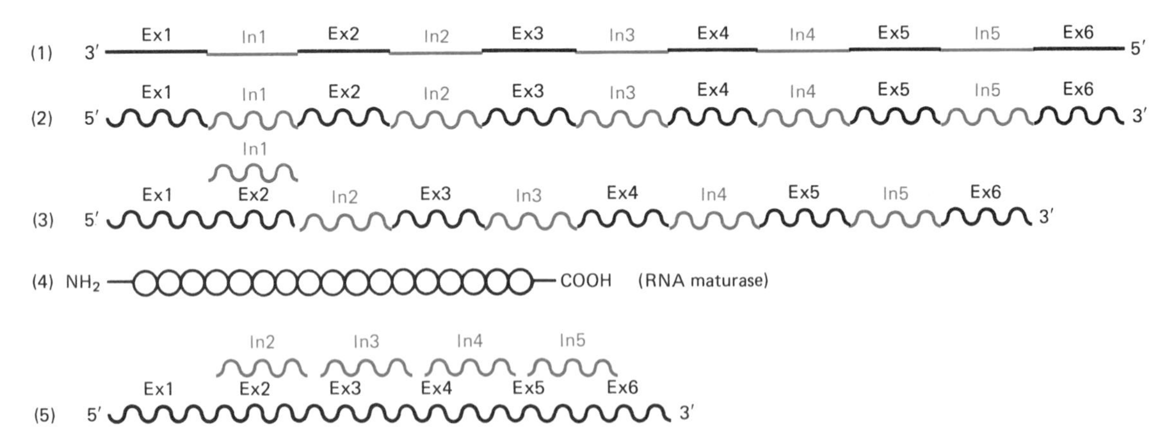

FIGURE 21-20. The cytochrome b gene of yeast mt DNA. The gene includes five introns and six exons. When the template strand (1) undergoes transcription, a transcript is formed (2) containing the introns and exons. Intron 1 is then removed (3), joining together exons 1 and 2 and putting their coding sequences in phase with those of a reading frame in intron 2. Translation of this portion of the transcript produces a short protein, RNA maturase (4), which aids in the removal of intron 2. Final excision and splicing removes the remaining introns, joins together the six exons (5), and brings about the formation of the transcript for cytochrome b.

rapidly it replicates. Eventually the deleted mt DNA will take over in the organelle population, leading to a lethal condition.

Research in the relatively new field of mitochondrial genetics is uncovering associations between certain mt DNA defects and more common human illnesses. For example, those cells of the brain that deteriorate in Parkinson's disease lack some of the respiratory enzyme activity of normal mitochondria. Very exciting is the idea that an accumulation of various kinds of mutations in the mt DNA during the life of a person may contribute to the aging process. It has been demonstrated that, with age, the mitochondrial activity in skeletal muscle declines. More attention is certain to be directed to the mitochondrial genome, which until now has been greatly overlooked while attention focused almost exclusively on the nucleus.

mt DNA and the Eve controversy

In the past several years, teams of molecular biologists have turned to mt DNA to gather information on human origins. The reasons for selecting the mitochondrial genome are several. The mutation rate of mt DNA is about 10 times that of the nuclear DNA. This is due largely to the fact that no DNA repair system is present in the organelle. This means that when one person is compared with the next, much more variation is seen in the mitochondrial genome than in the chromosomes. Nuclear DNA changes slowly and does not yield a high degree of variation in a given gene over a period of time as short as that of human history. The molecular clock runs much faster with respect to the mt DNA and provides a better basis for dating the geologically brief story of the human. Moreover, the mt DNA is more self-contained and sheltered than the DNA of the chromosomes and thus less apt to be influenced by natural selection, since little or no recombination of mt DNA is known in mammals. In addition, the organelle DNA is transmitted exclusively by the female parent. The various features of mt DNA render it in effect a set of genes that change primarily due to the accumulation of mutations and that can be traced backward in time through female lineages. Assuming a constant mutation rate, mt DNA would appear to be ideally suited to studies of human origins and the relationships among human populations. By comparing mt DNA among populations, one can estimate the number of accumulated differences and calculate the time two populations have been separated.

In the mid-1980s, mt DNA was taken from women from five different geographic regions: Asia, Africa, Europe, Australia, and New Guinea. The DNA was subjected to a dozen different restriction enzymes, and the patterns resulting from the DNA cutting were analyzed. Each pattern is known as a haplotype. The population found to contain the largest diversity of haplotypes would be considered the oldest, since it has had more time to accumulate various mutations in the mt DNA. When this was done for the five regions, the population showing the greatest amount of mt DNA differences among its members proved to be the African one. Variation in nucleotide sequences of the mt DNA was also compared to reveal differences among populations, such as African with Asian and Caucasian. All the results based on diversity of the mt DNA support the concept that African mt DNA has had the longest history and that all human populations are descended from an African one. These molecular findings are in accord with those of other students of human origins, since the oldest known human fossils are African. However, the molecular geneticists, estimating a constant rate of mt DNA mutations of 2%–4% per million years, deduced that the last common ancestor of all humans existed about 200,000 years ago. Since mt DNA is inherited maternally, the last common ancestor of all of us today would be an African woman, now popularly designated "Eve." This means that any differences now present in mt DNA between humans have accumulated since Eve lived. The figure of 200,000 years for the time of human divergence is in sharp disagreement with the ideas of scientists whose fossil studies place a time of divergence from an African ancestor about 800,000 years earlier. Moreover, the molecular data offer a good argument that all modern humans share an African heritage as opposed to the hypothesis that more archaic humans spread out from Africa to parts of Asia and Europe and later evolved to the modern human. According to the Eve hypothesis, the more archaic lineages left Africa before the emergence of modern humans, the descendants of Eve. Rather than evolving into modern *Homo sapiens,* the archaic humans became extinct after their replacement by modern humans who had already arisen in Africa and who then migrated to other parts of the world.

Some critics of the Eve hypothesis argue against the use of mt DNA as a molecular clock and claim the variation we see in mt DNA is not just the result of the accumulation of mutations that arise at a constant rate. For one thing, they maintain that the mt DNA is not immune to the effects of natural selection, which has favored certain genetic changes and has eliminated others. Thus modern humans show less variation in their mt DNA than they otherwise would, making the time of divergence from the African ancestor appear more recent than it actually is. Molecular biologists claim, however, that the mt DNA mutations that they study occur in parts of the mt genome that are neutral since they do not code for specific products. Criticism has been raised against the sampling methods used in the analysis of mt DNA, and it has been claimed that the samples used are not truly random and thus offer insufficient support for the conclusions reached by adherents of the Eve hypothesis. Other critics argue that fossils of humans found outside Africa clearly show transitional skeletal changes that take place gradually from archaic humans to modern *Homo sapiens,* a continuity very difficult to explain under the Eve hypothesis. While a rather heated controversy on the divergence of human populations continues, it has spurred on very active research that is certain to yield procedures of value in unravelling the history of human origins.

Mitochondrial translation and the genetic dictionary

The reading of RNA codons in the messenger RNA during translation has been thought to be invariable, a particular triplet always designating the same amino acid. The reading is precisely the same in viruses, bacteria, lower eukaryotes, plants, and animals. Consequently, the genetic code has been considered to be universal. Violations of this concept have now come to light from studies of mitochondrial DNA sequencing. When a nucleotide sequence for a segment of DNA has been determined, it can be compared with the amino acid sequence of the polypeptide for which it is encoded. The two sequences can be related, a triplet of nucleotides for an amino acid (Fig. 21-21A). Such comparisons between the nucleotide sequences of mitochondrial genes and the amino acid sequences of polypeptides for which they are encoded has brought to light several exceptions to the reading of the genetic code (Fig. 21-21B; Table 21-1).

In the mitochondria of yeast, invertebrates, and vertebrates, the codon UGA is not recognized as a chain-terminating signal, as it is in the cytosol. Instead, it is read as tryptophan in addition to the usual UGG codon for tryptophan. In several species, the mitochondria recognize AUA as methionine instead of isoleucine. However, not only may codon reading in the mitochondria depart from the "universal" genetic dictionary, the reading of codons can actually vary among different organisms. In corn, for example, CGG is read as an extra codon for tryptophan instead of UGG. Moreover, while the mammalian mitochondria read AGA and AGG as chain-terminating codons, yeast mitochondria translate these as arginine, following the rules of the universal code. Human mitochondria, however, follow the genetic dictionary by translating the four codons beginning with CU as leucine, but yeast mitochondria recognize these as threonine (Table 21-1).

In the mammalian mitochondria, only 22 tRNAs are found, whereas no fewer than 32 should occur according to the wobble hypothesis (Chap. 13). It appears that although certain mitochondrial tRNAs *do* follow the wobble rules, others do not. No one kind of tRNA should be able to recognize four different codons, but that is exactly what seems to be the case for certain tRNAs in human mitochondria. These particular tRNAs all have U in the wobble position. Somehow, this U may manage to pair with all four possible bases at the 3′ position in the codon. It is also possible that the anticodons of these mitochondrial tRNAs ignore the base on the codon at the 3′ position and recognize only the first two. It is not known what mechanism enables some mitochondrial tRNAs such as these with U at the wobble position to read all four 3′ positions in the codons (U, C, A, and G), whereas other mitochondrial tRNAs with U at this position fol-

TABLE 21-1. Comparison of codon recognition by mammalian and yeast mitochondrial translation systems

		Code in mitochondrial translation system	
Codon	Universal code	Mammals	Yeast
UGA	CT	trp	trp
AUA	ile	met	met
AGA	arg	CT	arg
AGG	arg	CT	arg
CUU	leu	leu	thr
CUC	leu	leu	thr
CUA	leu	leu	thr
CUG	leu	leu	thr

CT, chain terminating.

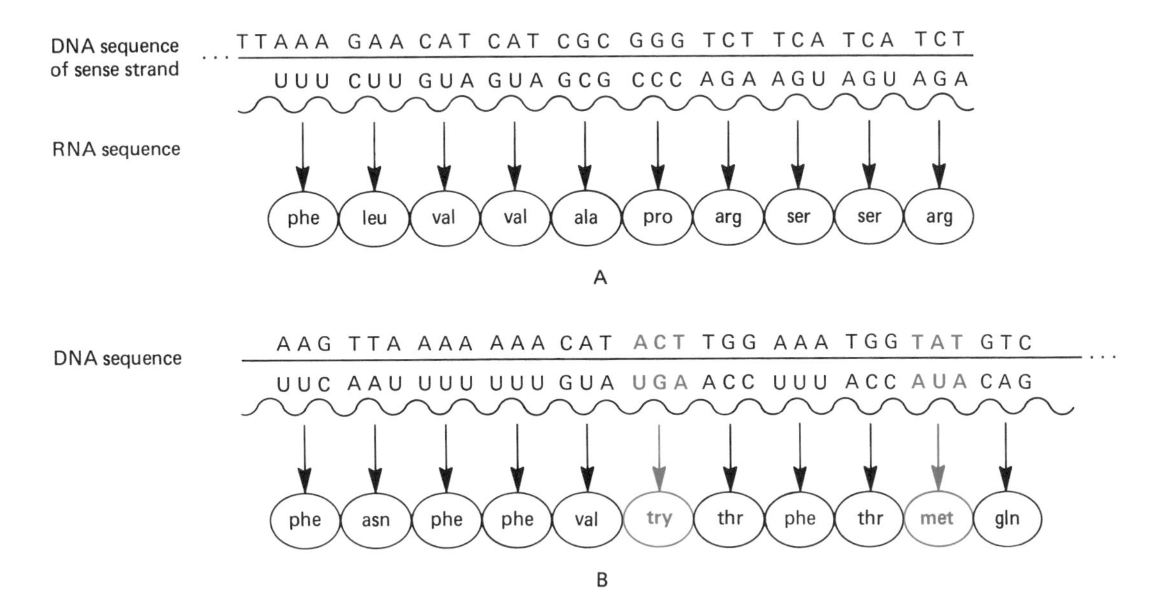

FIGURE 21-21. A: A segment of a DNA strand may be sequenced and the order of its nucleotides established. A DNA sequence can then be related to some polypeptide whose amino acid sequence is known. For example, if the DNA strand is part of the template strand for the polypeptide, it is possible to recognize a region where the sequence of DNA triplets (and hence the RNA codons complementary to them) corresponds to the sequence of amino acids in the polypeptide. If such a correspondence does exist between the two, the DNA would appear to be coded for that particular amino acid sequence. **B:** Apparent exceptions to the reading of RNA codons in translation have been uncovered. This is seen where a particular DNA triplet (and hence the RNA codon complementary to it) corresponds in position to an unexpected amino acid. The RNA codon UGA is a chain-terminating codon, but, in the case of certain mitochondrial systems, UGA appears to be read as tryptophan. Cases have also been found in which AUA (for isoleucine) appears to be read as methionine (usually associated only with the codon AUG). The sequences shown here are purely illustrative and do not depict any actual sequences in DNA or a polypeptide.

low the wobble rules and pair only with codons ending in G and A. Since many mitochondrial tRNAs have shapes that differ from the typical cloverleaf configuration, it is possible that the structure of the tRNA somehow influences the specificity of the anticodon.

Highly significant findings are coming to light that have strong implications for the subject of translation in the mitochondrion. DNA of certain genes located in the mitochondrial genome was sequenced from *Oenothera,* the evening primrose. Copy DNA (cDNA) derived from transcripts of these genes was also sequenced. Since any given cDNA is derived from a transcript of the genomic DNA, the two sequences should be identical. However, when the cDNA was compared with the genomic DNA, discrepancies were found between certain corresponding sites. For example (Fig. 21-22), at four sites in the gene for subunit III of cytochrome oxidase, the codons in the genomic DNA of the organelle differed from those at the same sites in the cDNA. Two different codons representing the amino acid serine in the genomic DNA were represented in the cDNA as codons for leucine. One for proline and one for arginine in the genomic DNA were present in the cDNA as codons for phenylalanine and tryptophan, respectively. Departures in nucleotide sequence were also noted at six different positions in the trailer region, a portion of the gene that is not coded for any product. Five discrepancies were also found in the coding region of the gene for subunit II of cytochrome oxidase. Most of the discrepancies between codons in *Oenothera* involved transitions in which C was replaced by T with respect to the sense strands.

Regarding the two cytochrome oxidase genes, one nucleotide alteration was present for every 58 nucleotides. Examples of similar alterations have been found as well in *Oenothera* for the cytochrome b gene and for a gene coding for a subunit of NADH dehydrogenase. Alterations of this type have also been reported for mitochondrial genes of other organisms and is quite extensive in the mitochondrion of a species of trypanosome. How can these discrepancies between the genomic and cDNAs be explained? It is believed that a posttranscriptional RNA editing mechanism exists in plant mitochondria and perhaps in those of other groups that alters specific codons in the mRNA. For example, codon GCC (arginine) in the template strand of the genomic DNA would be transcribed as CGG (arginine) in the mRNA. However, editing would change the RNA codon to UGG (tryptophan). Comparison of the sense strands would show at this altered site CGG (arginine) in the genomic DNA and TGG (tryptophan) in the cDNA.

These findings bear greatly on the matter of mitochondrial translation and exceptions to the genetic dictionary. Perhaps the standard code actually is used in many organelles. RNA editing, by changing one codon (let us say, arginine) to another codon for a different amino acid (tryptophan) would be noted as an exception to the genetic dictionary when the genomic DNA coding sequence is related to the amino acid sequence of the product of the gene. Actually, however, the usual codon for tryptophan would be present in the mRNA and would be translated as such.

The unusual mitochondrion, with its various exceptions to the nucleocytosol "rules," as seen in this and previous sections, causes us to reexamine many established concepts. The organelle may eventually provide us with deeper insight into the nature of the mechanisms controlling both transcription and translation.

The genome of the chloroplast

Plants contain two kinds of cytoplasmic genetic systems, that of the chloroplast and that of the mitochondrion. As noted earlier, the ct DNA is much larger than the mt DNA of yeast and animals. Consequently, information on chloroplast genome organization has been more difficult to obtain than on the mitochondrial genome. In plants, it is often unclear whether a mutant cytoplasmic genetic factor resides in the mt or the ct DNA.

In most higher plants, chloroplast inheritance is strictly maternal, because the plastids in the pollen do not enter the zygote. If chloroplast DNA from the female parent is absent or contains some sort of mutant defect, the plastids do not develop properly and may cause a lethal condition. In those plants in which plastids enter the zygote through both the pollen and the egg parents, a plastid defect in one parent can lead to the production of offspring that are variegated, containing tissue with a mixture of both normal and defective plastids, seen as streaks of green interspersed with white or yellow. Many nuclear genes are known that exert an effect on plastid development. Mutations in such genes can alter the plastid and render it defective. Actually most of the proteins of the chloroplast are coded by nuclear genes whose transcripts are translated in the cytosol. About 1,000 proteins are found in the chloroplast, and there is simply not sufficient DNA in the organelle to code for all of these. The 50 or so chloroplast proteins that are coded by the ct DNA are transcribed and translated entirely within the organelle, utilizing its own machinery for these processes. Construction of a normal chloroplast requires a coordinated interplay between the nuclear DNA and the ct DNA.

There are similarities between mt and ct DNA; however, the latter contains many more genes. The complete nucleotide sequence of ct DNA has been determined for a higher plant (tobacco) and a lower one (a liverwort). While the ct DNA of the former

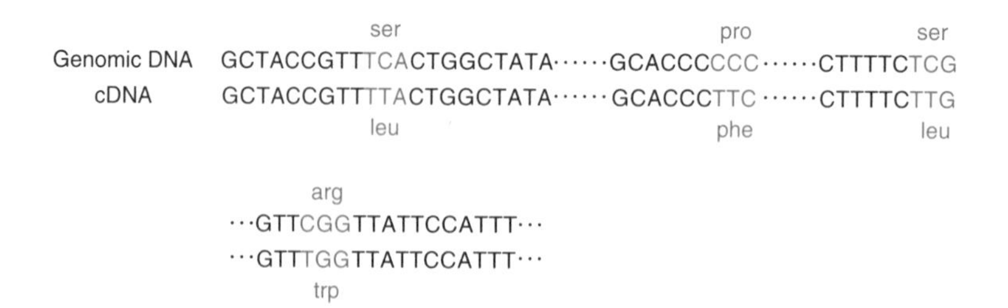

FIGURE 21-22. Genomic and cDNA discrepancies. DNA was sequenced for various regions of the *Oenothera* mitochondrial genome. The cDNA of the same regions was also sequenced and then compared to the genomic DNA. Shown here are portions of the sense stands of the genomic and cDNAs of the coding region of a cytochrome oxidase subunit gene. While the two should be identical, discrepancies at certain locations come to light, and four of these are seen (red). Two codons for serine in the genomic DNA are represented by leucine codons in the cDNA. A proline codon and an arginine codon in the genomic DNA are represented in the cDNA as codons for phenylalanine and tryptophan, respectively.

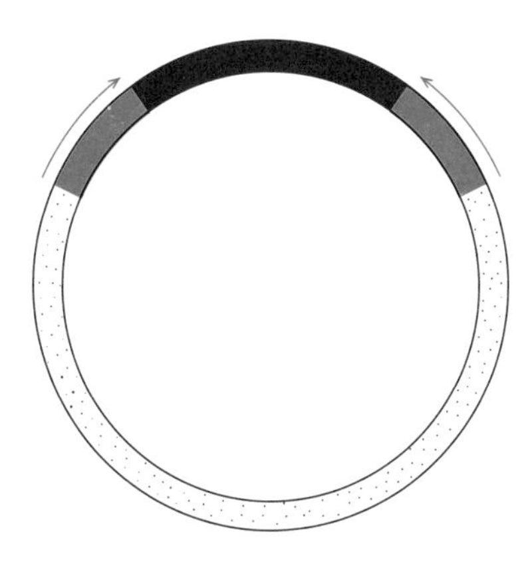

FIGURE 21-23. Regions of the ct DNA. The circular genome contains a sequence present as an inverted repeat (red) that divides the genome into two regions, SSC and LSC.

consists of more than 155,000 base pairs and that of the latter of about 121,000, the numbers of genes and their arrangements are very similar in the two despite the difference in length.

DNA length is variable among higher plant species, but it characteristically contains two copies of an approximately 25 kb sequence. The sequence is present as an inverted repeat (Fig. 21-23) and divides the circular ct DNA into two regions, known as short single copy (SSC) and long single copy (LSC). Present in each repeated sequence and therefore in two copies are genes for rRNA: 16S and 23S plus 4.5S and 5S. The ct genome contains information for all the tRNAs needed for translation and for about 50 different proteins. The latter include certain ribosomal proteins, RNA polymerase, and many essential to photosynthesis. Several nuclear genes are also coded for certain subunits composing proteins of the photosynthesis machinery. All the subunits of certain complexes critical to photosynthesis, such as photosystems I and II, are coded by the ct DNA, whereas all the subunits of others are coded by the nuclear DNA. Still other photosynthetic complexes depend on subunits coded by both genomes, nuclear and ct DNA. There are still over two dozen reading frames in the ct DNA whose products are unknown.

An interesting observation is that the ct DNA contains genes homologous to those coded for NADH dehydrogenase, which is a mitochondrial enzyme. Also interesting are the resemblances seen between certain chloroplast genes and genes of *E. coli.* These include genes for ribosomal proteins and those for the α-, β-, and β'-subunits of RNA polymerase. Unlike *E. coli,* introns are found in genes coded for proteins as well as in those for tRNAs.

The DNA polymerase that replicates the DNA of the plastid does not arise through translation of mRNA in the plastid. Actually, the polymerase forms in the cytosol as a result of translation of mRNA by the ribosomes found there. This can be shown in the following way. Cells may be treated with the drug rifampicin, which prevents the activity of the RNA polymerase of the plastid. This interference actually causes a loss of the ribosomes of the plastid, because no rRNA can be transcribed from the DNA of the plastid. The drug does not inhibit the RNA polymerase of the nucleus, and thus ribosomes of the cytosol are not affected by it. When ribosomes are lost from the plastids after treatment with rifampicin, the plastids still manage to replicate their DNA. Therefore, the DNA polymerase needed for this replication must enter the plastid, because it is a protein and cannot be made there in the plastid if no ribosomes are present (Fig. 21-24).

Although both ct DNA and mt DNA contain genes coded for parts of the structure of the organelles, it is still the nucleus that exerts the greatest degree of control over the biogenesis of the chloroplast and the mitochondrion. Most of the proteins of both organelles are imported from the cytosol, where translation of nuclear transcripts occurs. The nucleus apparently exerts control over the amount of protein the ribosomes of the organelles make. However, evidence also supports involvement of the organelles in the mechanisms that regulate the balance between the products synthesized in the organelles and those made in the cytosol for import into the organelles. Much is yet to be learned about the processes that enable the nucleus and organelle genetic systems to coordinate their activities in order to ensure efficient biosynthesis of mitochondria and chloroplasts and their orderly distribution from one cell generation to the next.

Biological implications of genetic information in organelles

The presence of genetic information in mitochondria and chloroplasts and the interactions between the organelles and the rest of the cell have led to the formulation of interesting hypotheses concerning the origin of mitochondria and chloroplasts. We have already noted several ways organelles and prokaryotic organisms resemble each other. The DNAs of the mitochondria, the chloroplasts, and bacteria are not associated with histones or bounded by membranes. The ribosomes of the organelles and of prokaryotes are inhibited by drugs such as streptomycin and chloramphenicol, antibiotics that do not interfere with the activity of the 80S ribosomes in the cytoplasm of the eukaryotic cell. The RNA polymerases of bacteria, mitochondria, and plastids are inhibited by rifampicin. Although this drug can bind to these enzymes, it does not affect the DNA-dependent RNA polymerase of the nucleus. Moreover, in bacteria the enzymes of aerobic respiration and photosynthesis form a part of the cell membrane. This is also true of the internal membrane of the plastid and the mitochondrion.

In contrast, the plasma membrane of the eukaryotic cell does not contain these enzymes, which are confined to the specific organelles. Recall that *n*-formyl methionine is the initiating unit in both bacterial and mitochondrial translation. Also the bacterial and organelle components of polypeptide synthesis can substitute for one another in translation. This cannot be done with the parallel components of the nucleocytoplasmic compartment of the cell. Translation and transcription also appear to be simultaneous in both prokaryotes and organelles, but these events do not take place in the cell nucleus simultaneously. These and other observations suggest a similarity between organelles and prokaryotes that has led some biologists to propose that the mitochondria and plastids in eukaryotic cells represent the descendants of ancient symbiotic organisms that took up residence in ancient prokaryotic host cells.

The mitochondrial ancestor would have been a primitive type of aerobic bacterium and the ancestral chloroplast a primitive type of blue-green alga. According to the thesis, their residence gave a decided advantage to the primitive prokaryotic host. Those with the mitochondrial ancestor were provided with an efficient energy source. The host cytoplasm contained the machinery for anaerobic respiration, whereas the symbiont contained the aerobic mechanism. As a result of the benefits conferred to the host, other features of the eukaryotic cell became established: the nuclear membrane, mitotic apparatus, and endoplasmic reticulum, and so on. Those cells that also carried the primitive blue-green algae evolved into eukaryotic plant cells, which had acquired the ability to trap the energy of light.

Thus, according to this concept, which actually dates back to the late nineteenth century, all eukaryotic cells have arisen from a host–symbiont relationship, and mitochondria and plastids are the descendants of the original symbionts we see today in the cells of every eukaryote. During the course of evolution, natural selection would favor any mutations in the genetic material of the host cell that helps to preserve the relationship or make it still more efficient. Thus the host cell eventually took over some of the functions of the symbiont as the relationship became more and more advantageous. This picture is reminiscent of *Paramecium* with kappa and the related particles that probably *do* represent symbionts. We saw that the host cells carried chromosomal genes that affect the maintenance and development of the particles. Symbiotic relationships between green algae and cells of *Paramecium* and *Hydra* are also well known.

It is therefore conceivable that the eukaryotic cell organelles arose in a similar fashion and that today most of the properties of the symbionts are controlled by nuclear genes of the host. Most of the proteins and the developmental stages of the mitochondria and plastids are certainly under the control of nuclear genes. And most of the essential structural components of the organelles are made outside of them in the cytosol of the cell. The amount of information in the mitochondrion is so small that not very many proteins could be coded for. Thus one can argue that, during the course of evolution, most of the original mitochondrial functions were assumed by the cell nucleus because mutations in this direction were favored by natural selection. But the annoying question still remains as to why any genetic information at all continues to reside in the mitochondrion or plastid. If the symbiont hypothesis is correct, the relationship was established countless ages ago, long before the origin of the simplest eukaryotic cell. If it were beneficial for the nucleus to assume control over the organelles, why has it not been complete? The force of evolution usually weeds out any superfluous features of an organism.

According to alternative hypotheses, mitochondria evolved gradually, not in a large step the endosymbiont theory proposes. Supposedly, an aerobic prokaryote was ancestral to the eukaryotic cell, and it contained the ingredients for aerobic respiration in its plasma membrane. Gradually, over a very long period of time, changes occurred in which the original plasma membrane was altered and formed mitochondrial membrane. This occurred as a result of selective pressure for a larger surface area needed for respiration as the organism became more and more complex. The membrane-bound organelle would have arisen as a result of infoldings of the plasma membrane of the cell. (Infoldings of the plasma membrane actually occur in bacteria,

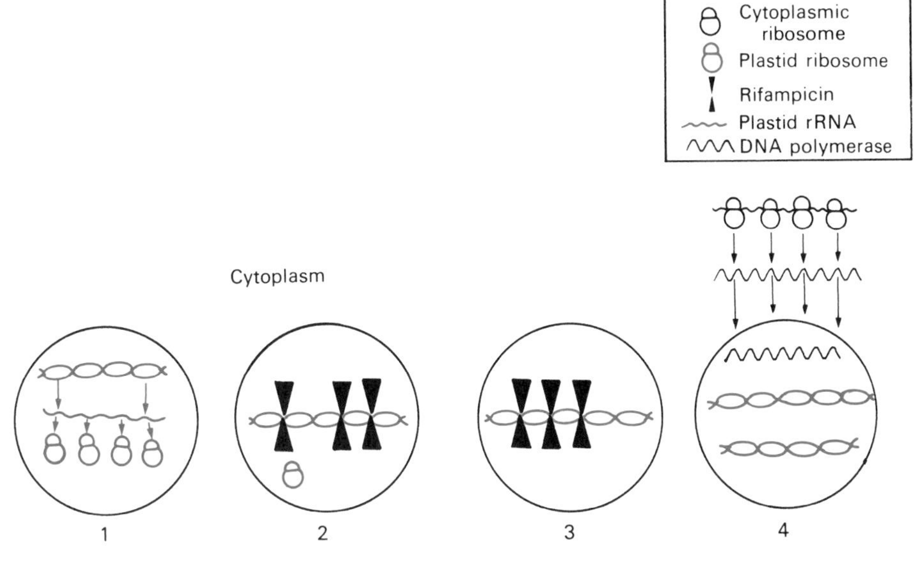

FIGURE 21-24. Interaction of chloroplast and cytosol. **1:** The normal chloroplast contains, among other things, its own DNA, RNA, and ribosomes. Some of the chloroplast DNA may undergo transcription to form the rRNA of the ribosomes of the plastid. **2:** The drug rifampicin blocks transcription of the plastid DNA. It does not block transcription of the nuclear DNA. **3:** Plastids arise that lack ribosomes, because no ribosomal RNA is formed in presence of the drug. **4:** In the plastids lacking the ribosomes, the DNA manages to replicate. This requires the enzyme DNA polymerase. The polymerase could not be formed in the plastids because they lack ribosomes, and these are needed for protein formation. The polymerase must have entered the plastid from the cytosol. The cytosol contains ribosomes because the drug does not prevent RNA formation from the nuclear DNA.

as seen in bacterial mesosomes.) The present-day machinery for protein synthesis would thus trace back to that which was present in the ancestral prokaryote. As the organism became more complex with time, the mitochondrial system and the cytosol diverged, the one in the organelle remaining more like the original bacterial ancestor.

Some alternative hypotheses propose that the DNA of the mitochondria resulted from the introduction of a plasmid into the organelle. Once this happened, the DNA, ribosomes, and other parts of the organelle evolved into the mitochondrion of today. Another suggestion is that a piece of the genetic material of the ancestral cell became separated and enclosed in the membranes that contained the ribosomes, producing the mitochondrion.

When evaluating the two main theories on the origin of organelles (endosymbiotic and gradual), we must keep in mind that mitochondria could have arisen in one way and chloroplasts in another. The chloroplasts show many similarities to blue-green algae. The amount of DNA contained in a chloroplast is appreciable compared with the very small amount in a mitochondrion. In *Chlamydomonas,* for example, the plastid DNA content is almost equal to that of a prokaryotic cell. Mitochondria differ from prokaryotes much more than plastids do. Recall the variation seen in mitochondrial ribosomes, discussed earlier in this chapter. The argument for the origin of mitochondria by endosymbiosis is not as strong as that for the origin of plastids.

Definitive explanations of the origin of organelles cannot yet be given, but adherents of one or the other of the two main hypotheses continue to offer evidence for their points of view. Clarification may someday be provided through investigations of nonchromosomal genetic determinants in a variety of species.

REFERENCES

Attardi, G. Organization and expression of the mammalian mitochondrial genome: a lesson in economy. *Trends Biochem. Sci.* 6:86, 1981.

Attardi, G., et al. Seven unidentified reading frames of human mitochondrial DNA encode subunits of the respiratory chain NADH dehydrogenase. *Cold Spring Harbor Symp. Quant. Biol.* 51:103, 1986.

Barinaga, M. "African Eve" backers beat a retreat. *Science* 255:686, 1992.

Birsky, C.W., Jr. Transmission genetics of mitochondria and chloroplasts. *Annu. Rev. Genet.* 12:471, 1978.

Borst, P., and L.A. Givell. Small is beautiful—portrait of a mitochondrial genome. *Nature* 290:443, 1981.

Butow, R.A., P.S. Perlman, and L.I. Grossman. The unusual *var* I gene of yeast mitochondrial DNA. *Science* 228:1496, 1985.

Cann, R.L., M. Stoneking, and A.C. Wilson. Mitochondrial DNA and human evolution. *Nature* 325:31, 1987.

Cavalier-Smith, T. The origin of cells: A symbiosis between genes, catalysts, and membranes. *Cold Spring Harbor Symp. Quant. Biol.* 52:805, 1987.

Crow, J.F. Sewall Wright (1889–1988). *Genetics* 119:1, 1988.

Dujardin, G., C. Jacq, and P.P. Slonismki. Single base substitution in an intron of oxidase gene compensates splicing defects of the cytochrome b gene. *Nature* 298:628, 1982.

Gillham, N.W. *Organelle Heredity.* Raven, New York, 1978.

Goddard, J.M., and D.R. Wolstenholme. Origin and direction of replication in mitochondrial DNA molecules from *Drosophila melanogaster. Proc. Natl. Acad. Sci. USA* 75:3886, 1978.

Grivell, L.A. Mitochondrial DNA. *Sci. Am.* (March):78, 1983.

Hiesel, R., B. Wissinger, W. Schuster, and S. Brennicke. RNA editing in plant mitochondria. *Science* 246:1632, 1989.

Leaver, C.J., and M.W. Gray. Mitochondrial genome organization and expression in higher plants. *Annu. Rev. Plant Physiol.* 33:373, 1982.

Ledoigt, G., B.J. Stevens, J.J. Curgy, and J. Andre. Analysis of chloroplast ribosomes by polyacrylamide gel electrophoresis and electron microscopy. *Exp. Cell. Res.* 119:221, 1979.

Lewin, R. The biochemical route to human origins. *Mosaic* 22:46, 1991.

Margulis, L. *Symbiosis and evolution.* W.H. Freeman, San Francisco, 1981.

Marshall, E. Paleoanthropology gets physical. *Science* 247:798, 1990.

Merz, B. Eye disease linked to mitochondrial gene defect. *J. Am. Med. Assoc.* 260:894, 1988.

Ozeki, H., et al. Genetic system of chloroplasts. *Cold Spring Harbor Symp. Quant. Biol.* 52:791, 1987.

Palca, J. The other genome. *Science* 249:1104, 1990.

Preer, L.B., and J.R., Preer, Jr. Inheritance of infectious elements. In *Cell Biology, A Comprehensive Treatise,* Vol. 1. L. Goldstein and D.M. Prescott (eds.), p. 319. Academic Press, New York, 1978.

Rochaix, J.D. Restriction endonuclease map of the chloroplast DNA of *Chlamydomonas reinhardii. J. Mol. Biol.* 126:597, 1978.

Sager, R. *Cytoplasmic Genes and Organelles.* Academic Press, New York, 1972.

Sapp, J. *Beyond the Gene.* Oxford University Press, New York, 1987.

Schwartz, R.D., and M.O. Dayhoff. Origins of prokaryotes, eukaryotes, mitochondria, and chloroplasts. *Science* 199:395, 1978.

Shoffner, J.M., et al. Myoclonic epilepsy and ragged red fiber disease (MERRF) is associated with a mitochondrial DNA tRNA Lys mutation. *Cell* 61:931, 1990.

Sonneborn, I.M. Kappa and related particles in *Paramecium. Adv. Virus Res.* 6:229, 1959.

Strathern, J.N., E.W. Jones, and J.R. Broach (eds.). *The Molecular Biology of the Yeast* Saccharomyces: *Metabolism and Gene Expression.* Cold Spring Harbor Laboratory, Cold Spring Harbor, NY, 1982.

Thorne, A.G., and M.H. Wolpoff. The multiregional evolution of humans. *Sci. Amer.* (Apr):76, 1992.

Vigilant, L., M. Stoneking, H. Harpending, K. Hawkes, and A.C. Wilson. African populations and the evolution of human mitochondrial DNA. *Science* 253:1503, 1991.

Wainscoat, J. Human evolution out of the garden of Eden. Nature 325:13, 1987.

Wallace, D.C. Mitochondrial genetics: a paradigm for aging and degenerative diseases? *Science* 256:628, 1992.

Wilson, A.C., and R.L. Cann. The recent African genesis of humans. *Sci. Amer.* (Apr):76, 1992.

Wurtz, E.A., J.E. Boynton, and N.W. Gillham. Perturbation of chloroplast DNA deoxyuridine. *Proc. Natl. Acad. Sci. USA* 74:4552, 1977.

Yang, D., et al. Mitochondrial origins. *Proc. Natl. Acad. Sci. USA* 82:4443, 1985.

REVIEW QUESTIONS

1. Assume that each of the following kinds of *Paramecium* cells undergoes self-fertilization. Give the genotypes of the two cells that will arise. Indicate whether they are killer or sensitive.

A. KK with kappa.
B. KK without kappa.
C. Kk with kappa.
D. Kk without kappa.

2. In each of the following, give the genotypes of the two stocks or strains of *Paramecium* that, when crossed, produce the results described:

A. A killer stock and a sensitive stock whose F_2 are all killers following cytoplasmic exchange.
B. A killer stock and a sensitive stock whose F_2 are killers and sensitives in a 1:1 ratio following cytoplasmic exchange.
C. Two killer strains whose cells, when mated, give an F_1 producing killers and sensitives in a 3:1 ratio with or without cytoplasmic exchange.

3. A. In Question 2A, assume there is no cytoplasmic exchange. What would the genotype and the phenotype of the F_1 cells arising after conjugation from the P_1 sensitive be? What would be the result from the P_1 killer?

B. In Question 2B, assume there is no cytoplasmic exchange. What would the genotype and the phenotype of the F_1 cells arising after conjugation from the P_1 sensitive be? What would be the result from the P_1 killer?

4. **A.** In yeast, would you expect the wild or the petite condition in the zygotes and from sporulation of the zygotes after crossing two neutral petites?

   B. Answer the same question for the cross of a neutral petite and a nuclear petite.

   C. How would you expect mating type to segregate in each of the preceding crosses?

5. A genetic determinant for resistance to oligomycin has been found to reside on yeast mt DNA. A cross is made between oligomycin-sensitive and oligomycin-resistant strains. Zygotes arise that are allowed to undergo mitotic divisions. Single zygotes are eventually allowed to sporulate. What is to be expected regarding sensitivity to the drug among the resulting ascospores?

6. In *Chlamydomonas,* assume the nuclear alleles C, c and D, d are being followed along with the cytoplasmic determinants P, p and Q, q. Give the genotype expected in the zygote and in the products of the zygote in each of the following crosses:

   A. Mating type +; CD; Pq × mating type −; cd; pQ.

   B. The same cross as in A, assuming that ultraviolet light was applied to the + parent.

7. In the mold *Neurospora,* there are two mating types A and a. Spores may be pigmented or nonpigmented. Some strains of the mold are slow growing (poky) in contrast to the normal, fast-growing habit. Nuclei of the mold are haploid, except for the fertilization nucleus. Meiosis occurs immediately in the ascus, giving four haploid nuclei that undergo one mitotic division. The eight haploid spores from a single ascus can be retrieved and tested.

   Two crosses are given here. From the results of testing the spores, what can you say about the transmission of mating type, spore color, and growth habit?

   Cross 1. Female, mating type A, fast growing, colored spores; male, mating type a, poky, nonpigmented spores.

   Cross 2. Female, mating type a, poky, nonpigmented spores; male, mating type A, fast growing, colored spores.

   Results of cross 1. All fast growing; 50% A, 50% a; 50% pigmented spores, 50% nonpigmented spores.

   Results of cross 2. All poky; 50% A, 50% a; 50% pigmented spores, 50% nonpigmented spores.

8. Three different petite strains are shown here. The presence (+) or absence (−) of certain marker genes is noted for each strain. What do the data suggest concerning the location of the individual markers?

	Marker					
	A	B	C	D	E	F
Suppressive petite strain 1	+	+	−	+	−	+
Suppressive petite strain 2	+	−	+	+	+	−
Neutral petite	−	−	−	+	−	−

9. After consulting the map of yeast mt DNA, give a reason why you would expect most alterations of the mt DNA to result in the origin of cells that are respiratory deficients.

10. A certain complex coenzyme consists of seven different polypeptide chains, four large (I, II, III, IV) and three small (V, VI, VII). Suppose that labeled amino acids are supplied to cell populations of a certain protozoan and of a certain mold under the conditions indicated in the following. What would the results suggest about the site of synthesis of the various chains of the enzyme in the two organisms?

	Labeled chains formed	
	Protozoan	Mold
Erythromycin added	I, IV, VI	V, VI, VII
Cycloheximide added	II, III, V, VII	I, II, III, IV
Control	All 7	All 7

11. Suppose the following sequence occurs at the end of a human mitochondrial gene that codes for a unit of a respiratory enzyme. The end of the coding sequence for the enzyme is next to a sequence that codes for a tRNA. When the transcript associated with the enzyme sequence is eventually read by the ribosomes, how can chain termination be established?

    5′–AATTTGCCGT tRNA sequence–3′

12. Suppose a random point mutation brings about a base substitution in human mt DNA. Another point mutation alters yeast mt DNA. Which mutation has the greater chance of affecting some important product—the mutation in the human or in the yeast mt DNA? Explain.

13. Following is a nucleotide sequence of a segment of a strand of mt DNA. This segment is part of a region of mt DNA that codes for the amino acid sequence of a certain polypeptide. The amino acid sequence shown is related to the given nucleotide sequence. Consult a dictionary of codons, and determine whether any exceptions to the reading of the genetic code exist.

    5′–CCTTTTCCTCATCCATGAGTTATACTA–3′
    pro-phe-pro-his-pro-try-val-met-thr

14. Which of the following is (are) true about the petite story in yeast?

    A. Following the cross of a nuclear petite with a normal, the segregation of ascospores from the zygote is 2 normal:2 petite.

    B. Following the cross of a neutral petite with a normal, the diploid cells are petite and give rise to all petites following ascospore formation.

    C. Following the cross of a suppressive petite with a normal, the diploids will be petite if they are plated on agar soon after mating.

D. Following the cross of normal with suppressive petite, wild-type cells are favored if subculturing precedes the plating on agar.
E. Petites cannot be induced by any known mutagen.

15. Which is (are) true about yeast:

A. Following the cross of an erythromycin-sensitive cell with a resistant one, the sensitive trait disappears, and all cells become resistant.
B. Following the cross of an erythromycin-resistant, chloramphenicol-sensitive cell (E^RC^S) with an erythromycin-sensitive, chloramphenicol-resistant cell (E^SC^R) asci are produced that contain spores only of the parental types.
C. The cross $E^RC^S \times E^SC^R$ produces four kinds of asci: E^RC^S, E^SC^R, E^RC^R, and E^SC^S.
D. The cross $E^RC^S \times E^SC^R$ produces asci, each of which produces four different kinds of spores, the two parentals and the two recombinant types.
E. The determinants for drug resistance may be associated with a physical alteration in mitochondrial DNA.

16. Which is (are) true about higher plants:

A. Nuclear genes do not seem to influence the development of the plastid from the proplastid.
B. There is more DNA in a chloroplast than in a mitochondrion.
C. All of the enzymes involved in photosynthesis are apparently formed from transcripts of the plastid DNA.
D. The RNA of the ribosomes of the plastid arises from transcription of the plastid DNA.
E. Replication of the plastid DNA depends on the formation of plastid DNA polymerase by the ribosomes of the plastid.

Glossary

acentric: An abnormal chromosome or chromosome fragment lacking a centromere.

acidic amino acid: An amino acid that bears a net negative charge at neutral pH.

acquired characteristics: Those features that an organism takes on during its lifetime through the effect of the environment on somatic tissue and that are not transferred to the next generation.

acrocentric: A chromosome with its centromere very close to one end, giving it one very short arm. The centromere located near one end of a chromosome.

activator: A regulatory substance that binds to a controlling element and acts in a positive fashion, stimulating transcription of a structural gene or genes.

adenine: A purine base found in DNA and RNA.

adenosine triphosphate (ATP): A nucleoside triphosphate that provides a source of high energy for metabolic processes requiring energy transfers. The main energy-storing molecule in the cell.

allele: A given form of a gene that occupies a specific position (locus) on a specific chromosome. Alternative forms of a gene occur and can thus occupy that specific locus. These variant forms are said to be alleles or to be allelic to one another.

allelic complementation: The production of wild-type or near-wild-type phenotype in an organism that carries two different mutant alleles in the trans arrangement.

allelic exclusion: The expression in plasma cells of the antibody genes found on only one member of a pair of homologous chromosomes.

allopolyploidy: A polyploid condition in which the extra chromosome sets are derived from different groups or species following hybridization.

allozymes: Different molecular forms of an enzyme coded by different allelic forms of a gene at a given locus.

Alu family: The most common set of dispersed and related DNA sequences in the human genome; each is about 300 base pairs long and repeated approximately 500,000 times.

amber codon: The chain-terminating codon UAG in mRNA.

amino acid: Any one of the 20 units that are joined by peptide linkages to form a polypeptide.

amino group: The $-NH_2$ chemical group, basic in nature, and characteristic of amino acids.

amino terminal end: The end of a polypeptide that has a free amino group.

aminoacyl–tRNA synthetase: One of a group of enzymes required to activate an amino acid and link it to its specific tRNA carrier.

amniocentesis: A procedure in which a sample of amniotic fluid is removed from a pregnant woman, so that the fluid and fetal cells present in it can be subjected to various analyses.

anaphase: The stage of nuclear division at which the chromosomes move to opposite poles.

aneuploidy: A chromosome anomaly in which the number of chromosomes in a cell or individual departs from the normal diploid number by less than a whole haploid set.

Angstrom (Å): A unit of measurement equal to 10^{-8} cm.

anneal: Form a double-stranded molecule from two single-stranded polynucleotide chains as a result of hydrogen bond formation between complementary nucleotides. The single strands that anneal may be two complementary DNA chains, two complementary RNA chains, or a DNA chain with an RNA chain complementary to it, thus forming a DNA–RNA hybrid molecule. (See also **reanneal.**)

antibody: A protein molecule, also known as an immunoglobulin, that is formed in response to a specific antigen and can recognize and react with it.

anticodon: A sequence of three nucleotides in a tRNA molecule that pairs with a specific codon in the mRNA and thus enables the tRNA to position its amino acid properly in a growing polypeptide chain.

antigen: Any substance or large molecule that can stimulate the production of antibodies following entrance into the tissues of a vertebrate.

antisense DNA: DNA composed of nucleotide sequences complementary to the mRNA.

antisense RNA: RNA that is complementary to mRNA, the sense RNA. Antisense and sense RNAs are able to hybridize.

antisense strand: The template strand on which the mRNA transcript is assembled. The template strand does not have the same sense as the mRNA but is complementary to it.

ascus: An enlarged cell in which the sexual spores are produced; typical of one major class of fungi, the *Ascomycetes.*

aster: The region that marks the poles of dividing cells and includes rays of microtubules surrounding a clear area within which two centrioles are located.

att sites: Loci occurring on both a phage chromosome and that of its bacterial host, where recombination brings about integration of the phage into, or its excision from, the host chromosome.

autopolyploidy: A polyploid condition in which the additional chromosome set(s) have been derived from within the same group or species.

autoradiography: A technique that permits the localizing of sites occupied by a radioactive substance in biological material by covering slide preparations of the material with a photographic emulsion that is later developed to reveal the darkened points at which radioactive emanations have caused decay in the emulsion.

autosomes: All the chromosomes in the complement other than the sex chromosomes.

auxotroph: A nutritional deficient that must be supplied with one or more nutrient factors that it cannot manufacture from a minimal medium.

axial symmetry: See **palindrome.**

backcross: A cross of a member of the F_1 generation to a member of one of the parental lines or to an individual that is a parental type.

back mutation: A mutation in a mutant allele that results in a change back to the wild or standard form of the gene.

bacteriophage: See **phage.**

Barr body: See **sex chromatin.**

base pair: A pair of nitrogen bases (one purine and one pyrimidine) held together by hydrogen bonds.

basic amino acid: An amino acid with a net positive charge at neutral pH.

bidirectional replication: DNA replication in which two replication forks move in opposite directions away from the same origin.

bivalent: An intimate association of two homologous chromosomes seen at first meiotic division.

C bands: Deep-staining regions that are produced following certain procedures for the visualization of the chromosome complement and that indicate sites of heterochromatin concentration.

c-onc: A protooncogene. The cellular counterpart of a v-onc, a viral oncogene.

C value: The amount of DNA found in the haploid genome of a species.

cAMP (3′, 5′-cyclic adenylic acid): A small molecule derived from ATP that interacts with specific hormones in cells and must complex with CAP before transcription of certain genes can take place.

CAP (catabolite gene activator protein): A positive control element that complexes with cyclic AMP and must bind to a region of the promoter before transcription of certain genes can begin.

carboxyl group: The –COOH chemical grouping, acidic in nature, and found in all amino acids.

carcinogen: A chemical substance or physical agent that can effect a malignant transformation in a cell. A substance or agent capable of producing cancer in an individual.

cDNA: Copied DNA; a DNA strand complementary to an RNA strand from which it was synthesized in vitro using reverse transcriptase.

centriole: A microtubular organelle marking the poles of the spindle in certain types of dividing cells.

centromere: That part of the chromosome responsible for chromosome movement. The region of the chromosome known as the primary constriction, where the sister chromatids of a prophase and metaphase chromosome are held together. (The structural part of the centromere, the kinetochore, becomes attached to the spindle fibers.)

centrosome: A clear region of the cytoplasm that contains the centrioles.

chain-terminating codon (nonsense codon): One of three nucleotide sequences (UAG, UGA, UAA) in an mRNA transcript that indicates that translation is to be halted and thus the assembly of amino acids into a growing polypeptide chain is to be terminated.

characteristic: A general attribute of an organism. In some characteristics, variation is recognizable as distinct, contrasting traits. For others, the variation is continuous, showing a gradual transition from one extreme to the other.

charged tRNA: A transfer RNA molecule attached to its specific amino acid; aminoacylated tRNA.

chiasma: A cross-like figure seen at first meiotic division and associated with the exchange between two homologous chromatids.

chromatid: A strand of a chromosome that has replicated. A chromosome is composed of two chromatids joined at the centromere region during mitotic prophase and metaphase.

chromatin: The chromosome material composed of DNA and associated proteins.

chromomere: A region of a chromosome thread that appears as a bead-like structure.

chromonema: The chromosome when it is seen as an extremely thin thread.

chromosome: A structure in the cell nucleus composed of chromatin that stores and transmits genetic information.

chromosome aberration (chromosome anomaly): Any modification that alters the morphology or number of chromosomes in the complement.

cis arrangement: That arrangement in a doubly heterozygous individual in which two linked wild alleles occur together on one homolog and the two recessive mutant alleles occur on the other.

cis–trans effect: A position effect in which two individuals, dihybrid with respect to the same genetic sites, show a difference in phenotype as a consequence of the arrangement of the mutant sites on the chromosome, the cis arrangement producing a more normal phenotype than the trans.

cistron: The gene considered as a unit of function; a segment of DNA coded for one polypeptide.

class switching: See **heavy-chain class switching.**

clone: A population of genetically identical cells.

cloned DNA: A collection of identical DNA molecules or sequences produced as the result of replication of a single molecule in a suitable cellular or viral system.

cloned library: See **gene library.**

cloning: The formation of clones or exact genetic replicas.

codominance: The expression in the heterozygote of both alleles. Codominance is generally used to refer to those cases in which two products can be detected in the heterozygote, each associated with one of the allelic forms. Also used interchangeably with incomplete dominance, as the distinction between the two is often not clear.

codon: A sequence of three adjacent nucleotides (an RNA or a DNA triplet) that designates a specific amino acid or indicates that translation is to be terminated.

coefficient of coincidence: A figure that takes into consideration interference and is used to indicate the chance that a double crossover will occur within a certain map distance.

cohesive ends: Complementary single DNA strands that extend from opposite ends of a duplex molecule or from the ends of different duplexes; can be produced as a result of staggered cutting by a restriction enzyme; also called *sticky ends.*

cointegrate: A circular molecule formed as the result of fusion of two replicons, one originally possessing a transposon and the other lacking one. The cointegrate has two copies of the transposon occurring as direct repeats, one at each of the two junctions of the replicons.

colinearity: The correlation between the sequence of codons in the DNA of a gene (cistron) and the sequence of amino acids in the polypeptide it specifies.

complementation: See **allelic complementation.**

complex locus: A cluster of very closely linked, functionally related genes.

complex transcription unit: A gene whose primary transcript is processed differently in different cell types, yielding different mRNA species coded for different products.

concatamer: A chain or linear series of genomes joined together.

consensus sequence: An idealized sequence that presents the nucleotides most often present at each position in a given DNA segment of interest.

constitutive: Unchanging or produced at a constant rate, such as a constitutive enzyme that continues to be produced in abundance, regardless of the cell's requirements.

constitutive heterochromatin: That chromatin which remains constantly condensed from one cell to the next, such as the heterochromatin associated with the centromere.

continuous variation: That phenotypic variation shown by some characteristics in which there is a gradation from one extreme to the other without any delineated categories or distinct traits.

controlling element: A genetic region, such as the promoter or operator, that can respond to an environmental signal and determine whether or not its associated gene will undergo transcription.

conversion: See **gene conversion.**

copia: A family of closely related DNA sequences in *Drosophila* that code for very large amounts of RNA and are capable of transposition.

copy DNA: See **cDNA.**

core particle: The product resulting from nucleosome digestion that consists of a histone octamer around which is wrapped 146 base pairs of DNA; also called a *nucleosome bead.*

corepressor: A small molecule needed to interact with a specific repressor before the latter can combine with an operator to block transcription.

Cot plot: A graph representing the reassociation kinetics for a given sample of denatured DNA in which the fraction of molecules remaining single-stranded is plotted against the product of DNA concentration and time (Cot value).

covalent bond: A strong interaction between atoms that share electrons.

crossing over: The event that entails a reciprocal exchange of segments between two homologous chromosomes and results in the recombination of linked alleles.

crossover: The recombinant product of a crossover event that has a new combination of linked alleles.

ct DNA: The DNA of the chloroplast.

cyclic AMP: See **cAMP.**

cytogenetics: The science that combines the methods and findings of cytology (microscopy) and genetics (breeding or molecular analysis).

cytohet: A cytoplasmically heterozygous eukaryotic cell; a eukaryotic cell containing two genetically different kinds of a specific organelle.

cytokinesis: Division of the cytoplasm producing two daughter cells from a parent cell.

cytoplasm: All the living parts of the cell outside the nucleus.

cytosine: A pyrimidine base found in DNA and RNA.

cytosol: The fluid portion of the cytoplasm exclusive of organelles.

degenerate code: The genetic code that is characterized by the fact that two or more codons may designate the same amino acid.

deletion: A chromosome alteration in which a portion of a chromosome has been lost; loss of a part of a DNA molecule from the genetic complement.

denaturation (of DNA): The separation of the two strands of the double helix as a result of hydrogen bond disruption following exposure to high temperature or chemical treatment.

density gradient centrifugation: A technique in which a mixture of substances is spun at high speed in an appropriate solution until each component of the mixture reaches an equilibrium position, where its density matches the density of the solute molecules at that position in the centrifuge tube.

deoxyribonucleic acid: See **DNA.**

detrimental allele: Any gene form that imparts a disadvantage to the individual carrying it and decreases the carrier's chances of reproduction or survival compared with that of an individual not possessing the allele.

diakinesis: Last stage of first meiotic prophase during which the bivalents are extremely condensed and well separated from one another.

dicentric: An abnormal chromosome having two centromeres.

dihybrid (cross): A cross in which two pairs of alleles are being followed; also, any individual carrying two pairs of alleles and thus heterozygous at two loci under consideration.

dimer: A compound formed between two identical molecules and thus having the same percentage composition but twice the molecular weight of one of the original molecules.

diploid: Having two complete haploid sets of chromosomes typical for the species or group.

diplonema (diplotene stage): Stage of first meiotic prophase, during which homologous chromosome threads repel and chiasmata appear.

direct repeats: Two or more identical or very closely related nucleotide sequences present in the same orientation in the same DNA molecule.

discontinuous replication: The synthesis of short DNA chains on a template strand. The short chains later become linked to form a long chain.

discontinuous variation: That phenotypic variation shown by some characteristics that can be recognized by clear-cut differences or distinct traits.

disjunction: The separation of chromosomes at anaphase of a mitotic or meiotic division.

DNA (deoxyribonucleic acid): The molecular nature of the genetic material that is composed of the four nitrogen bases—adenine, guanine, thymine, and cytosine—covalently bonded to a repeating chain of deoxyribose and phosphate residues and that occurs in cellular forms as a double helix or duplex molecule.

DNAase (DNase): Any enzyme that digests DNA to fragments composed of only a few nucleotides.

DNA–DNA hybridization: A procedure in which single-stranded DNA from one source is presented with single-stranded DNA from a second source under conditions that allow the formation of duplex DNA molecules, thus permitting an estimate of the degree of similarity between the two kinds of DNA.

DNA fingerprinting: A technique that reveals the pattern of fragments resulting from the cutting of DNA by a restriction enzyme in the vicinity of tandemly repeated DNA sequences that are scattered throughout the human genome. The great variability existing from one human chromosome to the next in the number of times a sequence is repeated produces an enormous degree of fragment size variation among unrelated persons and is the equivalent of a fingerprint on the molecular level.

DNA ligase: An enzyme that can catalyze phosphodiester linkages and can thus restore intact a polynucleotide chain containing one or more nicks.

DNA polymerase: An enzyme that catalyzes the synthesis of DNA on a DNA template from deoxyribonucleotide triphosphate precursors.

DNA–RNA hybridization: A procedure in which single-stranded DNA is presented with RNA under conditions that permit any complementary DNA–RNA sequences to combine.

dominant: Any gene form that expresses itself in some way in the presence of its allele in the heterozygote. Also, any trait that is phenotypically expressed when the responsible gene form is present singly with its allele in the heterozygote.

downstream: The direction on a DNA template from the 3′ end toward the 5′ end and toward the site of initiation of transcription.

duplex: Composed of two strands. A molecule composed of two strands, such as double-stranded DNA.

duplication: An extra copy of a gene or genetic region in the complement; refers to any situation in which one or more additional copies of a genetic region are present.

electrophoresis: The movement of charged molecules in an electrical field.

endonuclease: An enzyme that can cut phosphodiester bonds that occur internally in a DNA chain.

endoplasmic reticulum (ER): System of membranes distributed within the cytoplasm of eukaryotes which provides sites of protein synthesis and intracellular channels of transport.

enhancers: DNA sequences that can greatly increase the rate of transcription of genes and whose influence can be felt far upstream or far downstream from the promoters they stimulate.

episome: A nonessential hereditary factor that may exist either free within the cell or in a state in which it is integrated with a chromosome.

epistasis: A type of genic interaction in which a gene form at a given locus masks the expression of another genetic factor located at a different locus and therefore not its allele.

euchromatin: Those chromosome regions that are noncondensed and active in the interphase nucleus.

eukaryote: A cell or organism with a distinct membrane-bound nucleus as well as one or more membranous subcellular components (e.g., mitochondria, Golgi) that are specialized for the performance of certain specific cellular functions. (See also **prokaryote.**)

exons: Those nucleotide sequences in the gene and its transcript that are coded for a functional polypeptide.

exonuclease: An enzyme that attacks either a free 3′ or 5′ end of a polynucleotide chain and cuts phosphodiester bonds.

expressivity: The variation in the phenotypic expression of a specific allele or genotype when that genetic constitution is penetrant.

F factor (fertility factor): An episome that determines whether a bacterial cell will act as a donor (F^+ and Hfr) or a recipient (F^-) of DNA transferred by the donor cell during conjugation.

F_1: See **first filial generation.**

facultative heterochromatin: Chromatin that may behave as heterochromatin in some cells and as euchromatin in others, such as one of the X chromosomes of a mammalian female.

F-genote: A fertility factor that carries a portion of DNA that is typically found in the chromosome of a bacterium.

F′ (F prime factor): A bacterial fertility factor, F, that carries a portion of the bacterial chromosome.

fingerprinting: A technique in which a protein is subjected to enzyme digestion and the resulting peptide fragments separated by a combination of paper chromatography and electrophoresis. (See also **DNA fingerprinting.**)

first filial generation (F_1): The generation of individuals produced by the first parental generation (P_1) or the first parents being considered.

first parental generation (P_1): The first set of parents considered in any pedigree or mating.

5′ end: That end of a nucleic acid chain terminating in a free phosphate group.

5′ to 3′ growth: The synthesis of a nucleic acid chain by joining the 5′ (PO_3) end of a nucleotide to the 3′ (OH) end of the last nucleotide already in the uncompleted chain.

forward mutation: A mutation that results in the formation of a mutant gene form from the wild or standard allele.

frame shift mutation: A deletion or duplication of a base in the DNA that causes a portion of the coded information in the gene to go out of phase.

G bands: Deeply stained bands along the length of a chromosome that alternate with slightly stained regions and are revealed following a procedure using the Giemsa stain.

gel electrophoresis: A technique in which molecules are separated, because of their different shapes, sizes, and electric charges, as they migrate through a gel in an electric field.

gene: Classically a unit of inheritance occupying a specific site (locus) on a chromosome that has one or more specific effects on an organism and can both recombine with other such genetic units and mutate independently to other allelic forms. Many genes contain coded information for some functional product such as a polypeptide, rRNA or tRNA, whereas other genes serve as recognition sites for molecules involved in such processes as regulation of transcription and DNA replication.

gene conversion: A situation attributed to mismatch repair of DNA in which a pair of alleles in a diploid cell of genotype Aa segregates at meiosis in some ratio other than that expected, such as a 3A:1a segregation of ascospores instead of 2A:2a.

gene flow: The exchange of new allelic forms by migrants between two different populations of the same species.

gene library: A collection of cloned DNA fragments generated by the activity of restriction enzymes that includes all or part of the genome of a species.

gene pool: The total of all the genes carried by all the breeding individuals in a population at a given time.

gene proper: That portion of the template strand containing coded information for a functional product.

generalized transduction: The one-way transfer of genes from one bacterial cell to another by way of a generalized transducing phage, which packages bacterial genetic material instead of its own during its replication cycle in the host cell. In this process, the phage can transfer any gene from the host to the recipient cell.

genetic code: The DNA and RNA triplets (codons) that carry the genetic information specifying the 20 amino acids for polypeptide synthesis as well as the start and stop signals for translation by the ribosome.

genetic engineering: Manipulation in which the genetic material is purposefully altered in some way by combining hereditary materials from different sources or by removing a native segment and inserting one from another source.

genetic load: The sum of the mutant alleles in a population that have a detrimental effect and accumulate largely in the heterozygotes.

genetic marker: Any gene whose presence can be readily detected by its phenotypic expression and which is used to identify a cell, chromosome, or individual carrying it. Also any detectable RFLP used to identify a specific linked gene or an individual carrying a gene of interest.

genome: The genetic content of a single set of chromosomes, or a single set of bacterial or viral genes.

genotype: The genetic constitution of a cell or an individual.

guanine: A purine base found in DNA and RNA.

gynandromorph: A sex mosaic; an individual with certain body segments that can be recognized as phenotypically male and others that are female.

gyrase: A topoisomerase that can introduce negative supercoils in DNA through its nicking and closing activities.

hairpin: A double-helical region formed as a result of complementary base pairing between adjacent inverted complementary sequences in a single DNA or RNA strand.

haploid: Having one complete set of chromosomes typical for the species.

haplotype: A combination of very closely linked alleles or markers that tend to be transmitted as a unit to the next generation; also refers to a pattern of DNA restriction fragments for a chromosome region that can be recognized in an autoradiograph.

heavy-chain class switching: During differentiation of a B lymphocyte, the association of the same V region with a heavy chain constant region other than mu as a result of DNA shuffling and differential processing of mRNA.

HeLa cell: A widely studied, established cell line originally obtained from a human cervical carcinoma.

helicase: An enzyme that unwinds a double helix at the time of DNA replication.

helix: A spiral of constant diameter and pitch.

helper virus: A virus that, when present in a cell also infected with a defective virus, can provide a function missing in the latter, thus enabling it to multiply in the cell.

hemizygous: Having certain loci present in a single rather than a double dose, as the X-linked genes in a mammalian male.

heritability: The percentage of the variation seen in the expression of a character or trait that can be attributed to genetic factors.

heterochromatin: The chromatin or chromosome region that tends to remain condensed in the interphase nucleus.

heteroduplex DNA: A DNA region composed of two strands, each of which is derived from different duplex DNA molecules.

heterogametic sex: The sex in a given species that produces two kinds of gametes with respect to the type of sex chromosome. The mammalian male is the heterogametic sex because half his sperm carry an X chromosome and the other half a Y.

heterogeneous nuclear RNA (hnRNA): Those RNA strands of varying lengths that do not leave the nucleus of a eukaryotic cell.

heterogenote: A partially diploid bacterial cell heterozygous for just a segment of the genetic material.

heterokaryon: A cell that contains two or more genetically diverse nuclei.

heteropycnosis: The differential reactivity to staining displayed by a chromosome or a part of a chromosome in relation to the rest of the complement.

heterosis: Hybrid vigor; the more robust nature of a heterozygote compared with the more inbred parental lines.

heterozygous: Having alternative forms of a gene (alleles) at a given locus, one allelic form on each of the two homologs.

Hfr strain: A mating strain of *E. coli* that carries the fertility factor, F, integrated into the chromosome and that brings about a high frequency of recombination following conjugation with an F^- cell.

histocompatibility antigens: Antigens in the cell membrane that typically vary from one individual to the next and can be recognized by the immune system of a host, thus determining acceptance or rejection of a transplant.

histones: Basic proteins of low molecular weight that are complexed with the DNA of eukaryotes.

hnRNA: See **heterogeneous nuclear RNA.**

Hogness (TATA) box: A DNA segment found in those eukaryotic promoters that interact with RNA polymerase II, which contains the characteristic sequence TATAAAA and plays a role in determining the exact site at which transcription will start.

holandric: Any allele whose locus is on the Y chromosome; any trait associated only with the Y chromosome.

homeobox: The DNA region in a homeotic gene that has been very highly conserved in animal evolution and that is composed of a sequence of 180 nucleotides specifying a stretch of 60 amino acids.

homeodomain: The sequence in a homeotic gene protein product consisting of 60 amino acids and that is very similar from one homeotic gene protein to the next, even among different species.

homeotic gene: A gene that controls the activity of other genes that are involved in the organization of body plan. A homeotic gene mutation produces profound developmental upsets.

homogametic sex: The sex in a given species that produces a single kind of gamete with respect to the type of sex chromosome. The mammalian female is the homogametic sex, because every normal egg carries an X chromosome.

homologous: Having corresponding genetic loci.

homozygous: Having the same allele (gene form) present at a given locus on both homologous chromosomes.

housekeeping genes: Those genes that are expressed in all cells because they are coded for essential functions basic to all cell types.

H–Y antigen: The transplantation antigen present on the tissue of the heterogametic sex of mammals.

hybrid: A heterozygote; an individual resulting from a cross between genetically different parents.

hydrogen bond: A weak chemical interaction between an electronegative atom and a hydrogen atom that is covalently linked to another atom.

hydroxyapatite: A form of calcium phosphate that binds to double-stranded DNA.

in situ hybridization: A variation of the DNA–RNA hybridization procedure in which the denatured DNA is left in place in the chromosome and challenged with RNA or DNA chains from another source without being extracted.

in vitro: Occurring outside the living organism, for example, in a test tube or other artificially designed environment.

in vivo: Occurring within the living organism.

incomplete dominance: The expression in the heterozygote of both allelic forms; generally used to refer to those cases in which the phenotypic effect of one gene form in the heterozygote appears more pronounced than the other. Incomplete dominance and codominance are frequently used interchangeably, because the distinction between them is often not clearly defined.

incomplete sex linkage: The situation in which a specific locus is found on both the X and the Y chromosomes in that portion of the sex chromosomes that have homologous regions.

independent assortment: The segregation at meiosis of a pair of alleles independently of other allelic pairs whose loci are found on different chromosomes.

induction (of enzyme synthesis): The stimulation of specific enzyme synthesis in a cell when the specific substrate of the enzyme is supplied.

initiating tRNA: A special form of tRNA that can recognize the triplet AUG at the site in the mRNA that marks the start of translation.

interference: The effect exerted by a crossover that decreases the probability of a second crossover occurring within a certain map distance in the chromosome.

interphase: The metabolic nucleus of a cell between nuclear divisions. The condition of a nucleus that has not entered the mitotic or meiotic cycle.

introns: Those nucleotide sequences found in structural genes and their transcripts that are noncoding and interrupt those sequences containing information for the assembly of a product such as a polypeptide.

inversion: A chromosome aberration entailing two breaks in a chromosome followed by a reversal of the segment and consequently of the gene sequence in the segment. Pericentric inversions include the centromere in the inverted segment, whereas paracentric inversions do not. In an overlapping inversion a point of breakage in a particular inversion occurs in a segment that was reversed in a previous inversion.

inverted repeats: Two identical DNA sequences oriented in opposite directions on the same molecule. Adjacent inverted repeats constitute a palindrome.

ionizing radiation: Any radiation possessing sufficient energy to cause an outright separation of electric charges in the material it strikes.

IS element (insertion sequence): One of a group of small (less than 2,000 base pairs long) transposons found in bacteria that are able to move about and insert at a number of sites on the chromosome and are coded for no known properties other than those of insertion.

isozymes: Two or more enzymes capable of catalyzing the same reaction but differing slightly in their structures and consequently in their efficiencies under various environmental conditions. (A more general term than *allozymes.*)

kappa: A DNA-containing cytoplasmic particle found in certain strains of *Paramecium* that is capable of self-reproduction and can exert a killing effect on sensitive cells when it is liberated in the culture medium.

karyotype: The chromosome constitution of a cell or individual.

kinetochore: The structural part of the centromere into which spindle fibers insert.

labeled compound: A compound containing a radioactive atom that enables the compound or its breakdown products to be followed through a series of reactions or steps by detecting its radioactivity.

lagging chain: The DNA strand that is assembled discontinuously at the time of replication in the direction away from the replication fork.

leader: That portion of the mRNA toward the 5′ end that is not translated but binds to the ribosome for the proper initiation of translation.

leading chain: That DNA strand that is assembled continuously at the time of replication in the direction toward the replication fork.

leptonema (leptotene stage): The first stage of first meiotic prophase before synapsis, during which each chromosome is in an extremely extended state.

lethal allele: Any allele that can cause the death of the individual who possesses it. A complete lethal removes the individual sometime before the age of reproduction.

library: See **gene library.**

ligase: See **DNA ligase.**

linkage: The association of genes or genetic loci on the same chromosome. Linked genes tend to be transmitted together.

linker DNA: That DNA of a nucleosome in excess of the 146 base pair core DNA to which histone H1 is bound and that connects adjacent core particles.

locus: The specific position occupied by a given gene on a chromosome. At a particular locus, any one of the variant forms of a gene may be present.

long terminal repeats (LTRs): Sequences several hundred base pairs long that are directly repeated at both ends of the DNA of a retrovirus.

lysis: The bursting of a cell following dissolution of the cell membrane.

lysogenic bacteria: A strain of bacteria-harboring prophage that causes the lysis of a strain sensitive to that virus.

lysogeny: A state in which the genetic material of a virus and its bacterial host are integrated.

lytic virus: A virus that causes lysis following its multiplication in the host cell.

malignant transformation: The conversion of a noncancerous cell to the cancerous state by some trigger such as the acquisition of an oncogene.

map unit: A measure of genetic distance between two linked genes that corresponds to a recombination frequency of 1%.

marker: See **genetic marker**.

megaspore: A haploid plant cell derived from meiotic division of a megaspore mother cell that may give rise to a female gamete following mitotic divisions.

meiosis: The nuclear process that includes two divisions and results in the reduction of chromosome number from diploid to haploid.

melting: Denaturation of double-stranded DNA to single strands.

Mendelian trait: Any trait that is transmitted in accordance with Mendel's laws.

merozygote: A partially diploid bacterial cell.

messenger RNA (mRNA): The complementary RNA copy of DNA formed on the DNA template during transcription and carrying coded information for the amino acid sequence of a polypeptide.

metacentric: A chromosome with its centromere in a median position, thus having two arms of equal length; the median location of the centromere in the chromosome.

metaphase: That stage of nuclear divisions when the chromosomes are arranged at the equatorial plane of the spindle.

microspore: A haploid plant cell derived from the meiotic division of a microspore mother cell that can give rise to male gametes following mitotic divisions.

minimal medium: A medium that provides only those substances essential for the growth and reproduction of wild-type cells or organisms.

mismatch repair: Enzymatic correction of nucleotide mispairing that entails the removal of defective, single-stranded segments followed by the synthesis of new segments.

missense mutation: A point mutation in which a codon is changed to a different codon designating a different amino acid.

mitochondrion: A double-membrane cytoplasmic organelle found in eukaryotes and the site of aerobic respiration.

mitosis: The nuclear division that results in the accurate distribution of the genetic material and produces two nuclei exactly like the parent nucleus in chromosome content.

modification enzyme: An enzyme that recognizes the same nucleotide sequence as a corresponding restriction endonuclease and methylates certain bases within the sequence, thus conferring protection from the endonuclease activity.

modifier (modifying gene): A gene whose expression affects the expression of a gene or genes at other loci, which are thus not allelic to it. A modifier often has no other known effect.

molecular biology: A modern branch of biology that investigates biological phenomena at the molecular level, often using techniques of physical chemistry to pursue genetic and other biological problems.

monohybrid: An individual heterozygous at a locus under consideration, thus having a pair of alleles at that locus. Also, a cross in which a pair of alternative gene forms (alleles) is being followed.

monosomy: A chromosome aberration in which only one of a given kind of chromosome is present instead of the normal two characteristic of a typically diploid organism.

mosaicism: A situation in which the body of an individual is composed of two or more genetically distinct cell types.

mt DNA: The DNA of a mitochondrion.

multigene family: A set of genes derived by duplication of some ancestral gene followed by a history of independent mutations. Genes composing a family may be either clustered together on the same chromosome or dispersed on different chromosomes.

multiple alleles: Three or more forms of a gene, any one of which can occur at a given locus on a specific chromosome.

multiple-factor inheritance: Inheritance that entails many nonallelic genes and the complex interaction of environmental factors, many of which are unknown (see **quantitative inheritance**).

mutagen: Any agent that can cause an increase in the rate of mutation in an organism.

mutation: A sudden inheritable change that includes gene (point) mutation and chromosome aberrations in its broadest sense.

mutation pressure: Spontaneous gene mutation occurring continuously in a population.

natural selection: The differential reproduction of alleles that occurs in a population from one generation to the next and results in an increase in the frequency of certain alleles and a decrease in others.

negative control: The type of genetic regulation in which the product of the regulatory gene acts (alone or with a corepressor) to repress transcription of specific structural genes.

nondisjunction: The failure of homologous chromosomes or sister chromatids to separate at mitotic or meiotic division, resulting in aneuploid cells or individuals.

nonhistone protein: An assortment of proteins, mostly acidic in nature, that vary in kind with the species and cell type and that appear to function in controlling transcription in eukaryotes.

nonsense codon: A chain-terminating codon.

nonsense mutation: A point mutation in which a codon specific for an amino acid is changed to a chain-terminating codon.

Northern blotting: A technique similar to Southern blotting used to identify RNAs.

nuclease: Any enzyme that can break the phosphodiester bonds joining nucleotides together in a polynucleotide chain.

nucleoid: A region within a prokaryote that contains the DNA.

nucleolar organizer: That region of a specific chromosome in the complement (the nucleolar organizing chromosome) that contains genes for the RNA of the ribosome and is associated with the formation of the nucleolus.

nucleolus: A dense body within the nucleus produced by the nucleolar organizer and associated with the processing of the ribosomes.

nucleoside: A molecule composed of a five-carbon sugar linked to a nitrogen base.

nucleosome: The basic structural subunit of chromatin, consisting of an average of 200 base pairs of DNA plus an octamer of histone proteins.

nucleosome bead (core particle): An octamer of histones plus 146 base pairs of DNA wrapped around it in two turns.

nucleotide: A nucleic acid unit composed of a five-carbon sugar joined to a phosphate group and a nitrogen base.

ochre codon: The chain-terminating codon UAA.

Okazaki fragments: Short DNA chains that result from discontinuous replication on one of the template strands of a duplex DNA.

oligonucleotide: A short molecule composed of a linear sequence of nucleotides joined together by phosphodiester linkages. Synthetic oligonucleotides may be specifically constructed to serve as probes.

oncogene: A gene that can initiate and maintain a tumorous state in an organism and that arises from a protooncogene of a normal cell.

oogenesis: The entire series of events in a female in which a gamete is produced from the maturation of an immature germ cell.

oogonium: An immature germ cell in the female that can give rise to a primary oocyte.

opal codon: The chain-terminating codon UGA.

operator: That segment of DNA associated with one or more structural genes, which interacts with the product of a regulatory gene in the control of transcription.

operon: A unit of transcription composed of one or more structural genes associated with an operator and a promoter.

organelle: A subcellular component containing particular enzymes and specialized for a certain cell role.

overlapping genes: Different genes that share a common base sequence but are read in different frames so that they specify different products.

overlapping inversion: See **inversion.**

P_1: See **first parental generation.**

pachynema (pachytene stage): Stage of meiotic prophase I after the completion of synapsis and characterized by a detectable thickening of the chromosome threads. The stage of meiosis when crossing over takes place.

palindrome: In DNA, a region in which a symmetrical arrangement of bases occurs around a point in the molecule, with the result that the base sequence is the same on either side of it, reading either of the two paired strands in the 5′ to 3′ direction; adjacent inverted repeats.

paracentric inversion: See **inversion.**

parental imprinting: The differential activation and deactivation of maternal and paternal genes.

penetrance: The ability of a specific allele or genotype to express itself in any way at all when present in an organism.

peptide: A short linear stretch of two or more chemically linked amino acids.

peptide bond: A covalent linkage between two amino acids formed when the amino group of one becomes bonded to the carboxyl group of the other and water is eliminated.

peptidyl transferase: The enzyme found in the larger subunit of the ribosome that catalyzes the formation of peptide linkages between amino acids at the P and A sites.

pericentric inversion: See **inversion.**

phage: A virus with a bacterial cell as its host.

phenocopy: A nongenetic, environmentally induced imitation of the effects of a specific genotype.

phenotype: Any detectable feature of a living organism. The phenotype is a product of the interaction between the genetic material and the environment.

photoreactivation: The reversal of the deleterious effect of ultraviolet light on DNA by white light.

pilus: A filamentous, hollow appendage extending from the surface of a bacterial cell.

plaque: A clear area on an otherwise cloudy growth of bacteria or cells growing in tissue culture where cells have been lysed by a virulent virus.

plasmid: Any replicon that exists in a cell independently of the chromosomes.

pleiotropy: The multiple phenotypic effects of an allele.

point mutation: A mutation in which one nucleotide substitutes for another.

polar body: A tiny cell that receives little cytoplasm; produced by the division of a primary or a secondary oocyte.

polar mutation: A genetic alteration that affects the expression not only of the gene in which it occurs but also of other genes in the vicinity.

poly(A) tail: The stretch of adenine nucleotides at the 3′ end of eukaryotic mRNA that is added to the pre-mRNA as it is processed, before its transport from the nucleus to the cytoplasm.

polygenic inheritance: Inheritance involving alleles at many genetic loci, which interact with environmental factors. *Polygenic* refers to the many genetic factors as opposed to the environmental component. (See also **quantitative inheritance.**)

polymorphism: The existence within a population of two or more forms in a proportion that cannot be attributed to recurrent mutation alone.

polypeptide: A single long chain composed of amino acid units joined by peptide linkages.

polyploidy: A condition in a cell or individual in which one or more whole haploid sets of chromosomes are present, in addition to the normal diploid number typical for the species or group.

polysome: An assembly of ribosomes active in translation and connected by the same mRNA strand.

polytene chromosome: A chromosome composed of many intimately associated strands or unit chromatids formed as a result of endoreduplication.

population: An interbreeding group of organisms of the same species.

position effect: A phenotypic change resulting solely from the placement of a gene or genetic region in a new location in the chromosome complement, without any change within the newly located genetic material itself.

positive control: The type of genetic regulation in which a substance produced in the cell is required to activate transcription of specific structural genes.

posttranscriptional modification: Any alteration made to pre-mRNA before it leaves the nucleus as mature mRNA.

pre-messenger RNA (pre-mRNA): The immediate transcript of an antisense strand that undergoes modification in the eukaryotic nucleus and then yields the mRNA that leaves the nucleus to undergo translation.

Pribnow box: The promoter sequence TATAAT centered about 10 base pairs before the point at which transcription of bacterial genes starts, which is believed to play a very important role in the binding of RNA polymerase and initiation of transcription.

primary oocyte: A cell in the female derived from an oogonium and which undergoes first meiotic division.

primary spermatocyte: A cell in the testis derived from a spermatogonium and which undergoes first meiotic division.

primase: An enzyme required for the assembly of an RNA primer at the time of DNA replication.

primer: A short nucleic acid segment that provides a free 3′-OH end for internucleotide linkage in the 5′ to 3′ direction.

probe: Any biochemical that is usually labeled or tagged in some way so that it can be used to identify or isolate a gene, gene product, or protein.

processing: The posttranscriptional modification of pre-mRNA.

prokaryote: Organism or cell that lacks a nucleus bounded by a membrane as well as specialized membrane-bound organelles. (See also **eukaryote.**)

promoter: The segment of DNA containing start signals that can be recognized by RNA polymerase for the start of transcription of a gene.

prophage: A temperate virus devoid of its protein components, which is integrated with the DNA of the host and behaves like a genetic factor of the infected cell.

prophase: That phase of nuclear divisions following DNA replication, during which the chromosomes become evident as they gradually shorten and thicken.

protooncogene: A normal cellular gene that can be converted to an oncogene as the result of somatic mutation or recombination with a viral gene.

prototroph: A microorganism capable of supplying its own nutrient requirements from a minimal medium.

provirus: The DNA form of an RNA virus that can integrate into the chromosome of a cell.

pseudogene: A gene that closely resembles a known functional gene at another locus but has become nonfunctional because of an accumulation of defects in its structure that prevent normal transcription or translation.

purine: A nitrogen-containing compound with a double ring structure, and the parent compound of adenine and guanine.

pyrimidine: A nitrogen-containing compound with a single ring structure, and the parent compound of cytosine, thymine, and uracil.

Q bands: Those regions along the length of the chromosome that fluoresce after staining with a fluorescent dye.

quantitative inheritance (variation): That type of hereditary transmission or variation that involves many alleles at different genetic loci, each allele producing some measurable effect (often used synonymously with *polygenic* and *multiple factor inheritance*).

R plasmid: An extrachromosomal replicon carrying genetic determinants for drug resistance.

rDNA: See **ribosomal DNA.**

reading frame: A sequence of codons that can be translated, which contains a codon indicating where translation is to begin and a chain-terminating codon indicating where it is to end.

reanneal: To form duplex molecules from single-stranded complementary nucleotide chains. (*Reannealing* implies that the single-stranded chains came from the same source, whereas *annealing* implies they originated from different sources.)

reassociation: Reannealing or renaturation.

reassociation kinetics: See **renaturation kinetics.**

recessive: Any gene form that is not expressed in the presence of its allele in the heterozygote. Also, any trait that is expressed phenotypically only when the responsible gene form is present in double dose in a homozygote.

recombinant DNA: DNA molecules resulting from the union of DNA derived from different sources.

recombinant DNA technology: The techniques employed for splicing DNA molecules derived from more than one source and also for amplifying the recombinant DNA in a suitable host, to obtain large amounts of the DNA or of its products.

recombination: A new combination of alleles resulting from rearrangement following crossing over or independent assortment. (Some restrict usage of the term to those new combinations produced by crossing over.)

regulatory gene: A DNA sequence that functions primarily to control the expression of other genes by modulating the synthesis of their products. (A regulatory gene coded for a specific product is also a structural gene.)

renaturation (of DNA): The reconstitution of double-helical DNA by the reassociation of single strands derived from it by melting. (See also **reanneal.**)

renaturation kinetics: A technique that measures the rate of reassociation or reannealing of complementary single DNA strands derived from a single source that is used to indicate genome size and complexity. (See also **Cot plot.**)

repetitive DNA: Repeated nucleotide sequences that may occur in hundreds, thousands, or more copies in the chromosome complement of a eukaryote.

replication: The synthesis of a macromolecule identical to and under the guidance of a parent or template macromolecule.

replication fork: The Y-shaped region within a double-stranded DNA molecule that is undergoing replication and marks the site at which complementary strands are being synthesized at that time.

replicon: A unit of DNA replication with a specific point of origin at which replication begins.

replisome: The assemblage at the site of the replication fork of the assortment of proteins involved in replication.

repression (of enzyme synthesis): The cessation of production of a specific enzyme by a cell when the end product of the enzyme reaction is supplied.

repressor: The protein product of a regulatory gene that can combine with a specific operator and block transcription of the structural genes in an operon.

resistance transfer factor (RTF): That segment of a conjugal R plasmid carrying the genetic factors that confer on the plasmid those properties needed for transfer of the plasmid from one cell to the next.

restriction enzyme (restriction endonuclease): A type of endonuclease that recognizes and produces internal cuts only within a certain specific nucleotide sequence.

restriction fragment length polymorphisms (RFLPs): Variations found within a species in the length of the DNA fragments generated from a particular DNA region by a specific restriction enzyme.

restriction map: A physical map or depiction of a gene or genome derived from the ordering of restriction fragments produced by restriction enzymes that indicates the sites at which one or more restriction endonucleases cleave the molecule.

retroposon: A mobile genetic element that is able to transpose through the formation of an RNA intermediate.

retrovirus: An RNA virus that uses reverse transcriptase to assemble a DNA copy of its RNA genome using its RNA as the template.

reverse genetics: An approach to the study of a gene and its coded product utilizing the methodology of molecular biology, which begins with manipulation of the genetic material itself and leads to identification of the gene product. In the traditional approach (forward genetics), an aberrant phenotype is related to some change or defect in a gene product, which then leads to the gene.

reverse mutation (reversion): A change in a mutant allele that restores it to the wild type.

reverse transcriptase: A polymerase that preferentially assembles a DNA strand using an RNA strand as the template; RNA-dependent DNA polymerase.

rho factor: A protein required to halt transcription of certain genes.

ribonuclease: An enzyme that hydrolyzes RNA; RNAase.

ribosomal DNA (rDNA): DNA whose transcripts are processed into the RNA components of the ribosomes.

ribosomal RNA (rRNA): That class of RNA molecules present in both the small and large subunits of prokaryotic and eukaryotic ribosomes.

ribosome: A complex cellular component, composed of RNA and protein, that interacts with mRNA and tRNA to join together amino acids into a polypeptide chain. Found in the cytosol, mitochondria, and chloroplasts.

RNA (ribonucleic acid): A category of polynucleotides in which the component sugar is ribose and uracil takes the place of thymine. (RNA composes the genome of certain viruses and falls into three main classes in prokaryotes and eukaryotes: tRNA, mRNA, and rRNA.)

RNAase: See **ribonuclease.**

RNA polymerase: An enzyme that catalyzes the synthesis of RNA from ribonucleoside triphosphate precursors using a template DNA strand.

RNA replicase: An enzyme required for the assembly of RNA strands on an RNA template.

rolling circle: A model for the manner of replication of certain DNA molecules, in which a replication fork proceeds around a circular template for an indefinite number of revolutions and a newly synthesized DNA strand displaces the strand synthesized in the previous revolution.

rRNA: See **ribosomal RNA.**

S phase: The portion of interphase during which DNA replication takes place.

S value: A sedimentation coefficient or constant expressed in Svedberg units.

satellite: A terminal end of a chromosome arm produced by a secondary constriction, connected to the rest of the chromosome by a very narrow region, and typically associated with nucleolus formation.

satellite DNA: The DNA of a eukaryotic cell that has a different density from the bulk of the cellular DNA and equilibrates at a different position from the main band following density gradient centrifugation.

second site reversion: A suppressor mutation that restores function to a gene and occurs in the same gene as the first mutation that produced the aberrant effect.

secondary oocyte: The larger of the two cells produced by the division of a primary oocyte.

secondary spermatocyte: One of the two cells derived from the division of a primary spermatocyte that has completed first meiotic division.

selection pressure: The operation of natural selection on the allele frequency in a population resulting in an increase in the frequency of certain alleles and a decrease in the frequency of others.

semiconservative replication: The manner in which double-stranded DNA is synthesized and which produces two double-stranded molecules from the original DNA, each containing one parental strand and one strand that is newly formed.

sense RNA: mRNA. The RNA formed on the template or antisense strand of the gene.

sense strand: The one of the two DNA strand composing a gene that does not act as a template for the formation of a transcript. This strand has the same sense as the mRNA with Us in the RNA in place of Ts.

sex bivalent: An association composed of the paired X and Y chromosomes in a primary spermatocyte.

sex chromatin: A condensed or inactivated X chromosome (or part of an X) in the interphase nucleus seen as a densely straining body in cells containing more than one X chromosome.

sex chromosomes: Those chromosomes involved in sex determination (e.g., the mammalian X and Y) that show a difference in morphology or number between the sexes.

sex-influenced allele: Any allele that is expressed as a dominant in one sex and a recessive in the other.

sex-limited allele: Any allele that expresses itself in only one of the sexes.

sex-linked (X-linked) allele: Any gene form with its locus on the X chromosome. Also any trait associated with such a genetic factor.

sex pilus: An appendage on a bacterial cell produced by the presence of a fertility factor and necessary for conjugation.

sexduction: The transfer of a chromosomal gene from one cell to another by way of a sex factor, such as F, the fertility factor.

shift: A chromosome aberration in which a chromosome segment is transposed to a new location in the same or a different chromosome. The shift is a simple type of translocation.

shotgun experiment: The cloning of an entire genome in the form of randomly generated fragments to create a "gene library" for the species that is available for later studies.

sigma chain: One of the polypeptide chains composing RNA polymerase, which is essential for the recognition of signals in the DNA for the start of transcription.

silent mutation: A point mutation in which a codon specific for a given amino acid is changed to a different codon that designates the same amino acid.

small ribonucleoprotein particle (snRNP): A unit confined to the nucleus and composed of one RNA molecule and several proteins; a snurp. (Snurps differ in RNA type, the type rich in U_1 RNA playing a role in processing pre-mRNA.)

somatic cell(s): The cells of the body, in contrast to gametes and the cells from which they were derived.

somatic cell hybridization: The process of fusion between somatic cells to produce somatic cell hybrids for genetic analysis.

SOS response: An error-prone mechanism to repair damaged DNA in *E. coli* that involves the coordinate induction of many enzymes in response to irradiation or other damage to DNA.

Southern blotting: A procedure for transferring denatured DNA from an agarose gel to a nitrocellulose filter, where it can be hybridized with complementary single-stranded nucleic acid.

spacer DNA: Segments of unknown function that flank certain genes, such as the eukaryotic rRNA genes, that do not interrupt a coding sequence and that typically do not undergo transcription. The transcripts of those spacers that are transcribed, as in case of rRNA, become discarded when the primary RNA transcripts are formed.

specialized transduction: One-way gene transfer in bacteria in which a phage that had been integrated as prophage carries some of its own genome and some of the host's into a recipient cell. Only genes adjacent to the site of integration of the virus in its prophage form can be transferred.

spermatocyte: See **primary spermatocyte.**

spermatid: One of two cells derived from the division of a secondary spermatocyte that has completed second meiotic division.

spermatogenesis: The series of events starting with the origin of a primary spermatocyte and culminating with the production of sperms.

spermatogonium: A type of cell found in the wall of the testis that can give rise to a primary spermatocyte.

spermiogenesis: The portion of spermatogenesis during which a spermatid is transformed into a mature sperm.

spindle: An aggregation of microtubules essential for the positioning and distribution of the chromosomes at nuclear divisions.

spindle fiber: A microtubule composed of an assemblage of protein units.

spliceosome: The splicing mechanism that accomplishes the formation of functional mRNA and that includes snurps and a variety of proteins that assemble on the precursor RNA that is to be spliced.

splicing: Joining together separated component parts. RNA splicing in eukaryotes entails the removal of introns and the joining of exons of pre-mRNA to produce mature mRNA.

steroid: One of an assortment of complex lipids, composed of four interlocking rings of carbon atoms, that are often biologically important compounds, such as vertebrate male and female sex hormones.

sticky ends: See **cohesive ends.**

structural gene: A segment of DNA coded for the amino acid sequence of a polypeptide chain or for a tRNA or rRNA molecule. (Regulatory genes are structural genes whose products control the expression of other genes.)

structural heterozygote: An individual with a normal chromosome and some kind of change in the homolog that affects its morphology or structure.

supercoil: A coil superimposed on a coiled duplex; a conformation resulting from the coiling of a circular duplex DNA molecule upon itself, so that it crosses its own axis.

suppressor mutation: A mutation at a genetic locus separate from the one in which a first mutation occurred that reverses the effect of the first mutation and produces a wild-type phenotype in the double mutant.

symbiosis: An intimate association of two organisms of different species living together, usually for mutual benefits.

synapsis: The intimate pairing, locus for locus, of homologous chromosomes at first meiotic prophase.

synaptonemal complex: A complex structure that forms during first meiotic prophase and facilitates the intimate pairing between homologous chromosomes.

syndrome: The collection or group of symptoms or features associated with a specific disease or abnormality.

synteny: The association of genes or genetic loci on the same chromosome. Often used synonymously with *linkage.* Some restrict the term *linkage* to those genes that clearly do not undergo independent assortment, as shown by genetic analysis. *Syntenic* is used to indicate that procedures such as somatic cell hybridization have shown two genes are on the same chromosome, even if genetic analysis has not yet demonstrated this.

T cell (T lymphocyte): A white blood cell that becomes differentiated in the thymus gland and can recognize foreign antigens on cell surfaces and bring about the removal of the foreign cells.

TATA box: See **Hogness box.**

tautomerism: The reversible shifting of the location of a proton in a molecule that changes certain chemical properties of the molecule.

telocentric: A chromosome with its centromere appearing to be in a terminal location, thus having only one evident arm, the centromere located at one end of the chromosome.

telomere: A region of a chromosome marking the extreme end of a chromosome arm. Every normal chromosome possesses two telomeres.

telophase: The stage of nuclear division during which the nuclear membrane reforms and the chromosomes gradually become less and less evident.

temperate (symbiotic) phage: A bacterial virus that tends to take up residence in the host cell without destroying it.

template: A macromolecule that serves as a blueprint or mold for the synthesis of another molecule. A preformed nucleic acid chain that is required to establish the nucleotide sequence in a complementary chain.

terminator sequence: A stretch of DNA that contains a signal for RNA polymerase to bring transcription to a halt.

testcross: The cross of an individual to one that expresses a specific recessive trait or traits under consideration.

tetrad: The four nuclei or four cells that are the immediate results of meiosis in a parent cell (sometimes also used to designate a bivalent).

tetraploid: A cell or organism with four complete haploid sets of chromosomes instead of the typical two.

tetravalent (quadrivalent): An association at first meiotic division of four chromosomes that are completely or partially homologous.

3′ end: That end of a nucleic acid chain terminating in a free –OH group.

thymine: A pyrimidine base found in DNA but not in RNA.

topoisomerase: An enzyme that can convert DNA from one topological form to another.

trailer: That portion of an mRNA toward the 3′ end that follows a chain-terminating codon and does not undergo translation.

trait: A distinct alternative form of a characteristic.

trans arrangement: That linkage arrangement in a double heterozygote in which a wild allele and a recessive mutant gene form occur together on one chromosome, and the corresponding recessive and wild alleles occur together on the homolog.

transcript: The single-stranded RNA chain that is assembled on a DNA template.

transcription: The assembly of a complementary single-stranded molecule of RNA on a DNA template.

transcription factors: A varied group of regulatory molecules that are not parts of the enzymes of transcription but that recognize specific sequences in promoters and are necessary to initiate transcription in eukaryotes.

transductant: A cell that has been transduced.

transduction: The transfer of genetic material from one cell to another by a viral vector. (See also **generalized** and **specialized transduction.**)

transfection: Commonly refers to the uptake in culture by recipient cells of exogenous calcium phosphate–treated DNA.

transfer RNA (tRNA): A small RNA molecule that recognizes a specific amino acid, transports it to a specific codon in the mRNA, and positions it properly in a growing polypeptide chain.

transformation: A genetic change effected in a cell as the result of the incorporation of DNA from a virus or some genetically different cell type; also refers to the taking up of extraneous genetic material by salt-treated bacterial cells in genetic manipulation. (See also **malignant transformation.**)

transgressive variation: The quantitative variation in a characteristic of the offspring that exceeds that shown by the same characteristic in either parent.

transition: A point mutation in which a purine is replaced by a different purine or a pyrimidine by a different pyrimidine.

translation: The assembly of a polypeptide chain from the coded information in the mRNA that directs the amino acid sequence of the chain.

translocation (reciprocal): A chromosome alteration in which a chromosome segment or arm is transposed to a new location. A reciprocal translocation involves a mutual exchange of chromosome segments or arms between two nonhomologous chromosomes.

transposon: A genetic unit capable of moving from one chromosome site to another or from one replicon to another.

transversion: A point mutation in which a purine is replaced by a pyrimidine or a pyrimidine is replaced by a purine.

triploid: A cell or organism with three complete haploid sets of chromosomes instead of the normal two.

tRNA: See **transfer RNA.**

trisomy: A condition in a cell or individual of a typically diploid species in which a particular chromosome is present in three doses instead of the normal two.

trivalent: An association at first meiotic division of three chromosomes that are completely or partially homologous.

unique DNA: A class of DNA representing sequences that are present only once in a genome.

univalent: A chromosome that remains unpaired at the first division of meiosis.

upstream: The portion of the DNA template that lies away from the initiation site of transcription toward the 3′ end.

uracil: A pyrimidine base found in RNA but not in DNA.

vector: An agent that transfers material from one host to another.

virion: A complete virus particle consisting of its nucleic acid core and protein coat.

virulent phage: A bacterial virus that typically destroys the host cell it infects.

v-onc: A viral oncogene acquired by the virus from a eukaryotic cell.

wild type: The form of a gene or an individual that is considered the standard type typically found in nature.

wobble hypothesis: The concept that the base at the 5′ end of the anticodon is somewhat free to move about spatially and can thus form hydrogen bonds with more than one kind of base at the 3′ end of a codon in the mRNA.

X chromosome: In mammals and various other groups, the sex chromosome that is found in two copies in the homogametic sex and in one copy in the heterogametic sex.

X linked: See **sex linked.**

Z DNA: A form of DNA existing as a left-handed double helix whose sugar–phosphate backbone follows a zig zag course and may possibly play a role in gene regulation.

zygonema (zygotene stage): The stage of first meiotic prophase during which homologous chromosomes pair.

zygotic induction: The activation of a virus to a virulent state following its transfer as prophage from a Hfr bacterial cell to an F^- cell that lacks the virus.

Answers to Review Questions

CHAPTER 1

1. **A.** OO
   B. oo
   C. Oo
   D. oo

2. **A.** 1 OO:1 Oo. All round
   B. 1 OO:2 Oo:1 oo. 3 round:1 oblong
   C. oo. All oblong
   D. 1Oo:1 oo. 1 round:1 oblong

3. Short hair is the dominant trait. Genotypes: female 1 is Ss; female 2 is SS. The male is ss.

4. **A.** Horned and hornless in a ratio of 1:1
   B. One-fourth
   C. Three-fourths

5. **A.** Red (RR), roan (Rr), white (rr) in a ratio of 1:2:1
   B. 1 roan (Rr): 1 white (rr)

6. **A.** hhrr
   B. H–Rr
   C. HhRR

7. **A.** All with black fur and short hair
   B. One out of four; BBLL, black short; BBll, black long; bbLL, albino short; bbll, albino long

8. **A.** All blue and mildly frizzled
   B. 1 blue and mildly frizzled:1 white, mildly frizzled
   C. 1 blue, mildly frizzled:1 blue frizzled:1 white, mildly frizzled:1 white frizzled
   D. 1 black frizzled:2 blue frizzled:1 white frizzled:2 black, mildly frizzled:4 blue, mildly frizzled:2 white, mildly frizzled:1 black straight:2 blue straight:1 white straight

9. **A.** Aamm
   B. A–Mm
   C. aamm

10. **A.** Parental genotype: AaMm. Gametes: AM, Am, aM, am
    Parental genotype: Aamm. Gametes: Am, am.
    B. Offspring: nonalbino with migraine, nonalbino without migraine, albino with migraine, albino without migraine in a ratio of 3:3:1:1

11. **A.** No chance
    B. One-fourth
    C. Two-thirds

12. All the children will be genotype ss, homozygous for the defective allele, just as the parents are. The children would require treatment, because the environment has not altered the genetic material that is transmitted through the gametes. A Lamarckian would claim that the treatment has produced the normal condition and this trait can now be transmitted, so that the children would be normal eventually regardless of treatment.

13. **A.** (1) ffSs (2) FfSs
    B. Free, no sickling; free, some sickling; free, with anemia; attached, no sickling; attached, some sickling; attached with anemia. Ratio 1:2:1:1:2:1

14. **A.** An allele does not increase in frequency simply because it is dominant. The frequency is the result of the interaction of many factors. Dominance by itself does not make an allele increase. Dominance means that a gene form will express itself in the presence of its allele.
    B. The person has a 50:50 chance of developing the disorder. The afflicted parent is most likely genotype Hh, since the allele is so rare.

15. **A.** 4
    B. 16
    C. 16

16. **A.** 16
B. 8
C. 32

17. **A.** 81
B. 27
C. 243

18. (1) TG, tG
(2) TGW, TGw, tGW, tGw
(3) TGW, TgW, tGW, tgW, TGw, Tgw, tGw, tgw

19. (1) 1 tall yellow:1 dwarf yellow
(2) 1 tall yellow round:1 tall yellow wrinkled:1 dwarf yellow round:1 dwarf yellow wrinkled
(3) 1 tall yellow round:1 tall green round:1 dwarf yellow round:1 dwarf green round:1 tall yellow wrinkled:1 tall green wrinkled:1 dwarf yellow wrinkled:1 dwarf green wrinkled

20. 9 tall yellow round:3 dwarf yellow round:3 tall yellow wrinkled:1 dwarf yellow wrinkled

21. **A.** BbCc
B. bbC–
C. B–cc
D. bbcc

CHAPTER 2

1. **A.** Prophase
B. Telophase
C. Metaphase
D. Prophase
E. Telophase
F. Anaphase

2. **A.** (1) 92
(2) 92
B. (1) 46
(2) 46
(3) 46

3. **A.** 22 pairs
B. 22 pairs
C. 2
D. 2
E. 7

4. **A.** AAXX
B. AAXY

5. **1.** B
2. E
3. F
4. A
5. E
6. C
7. I
8. J
9. G
10. H

6. Prokaryotes lack mitochondria, a nuclear envelope, chloroplasts, endoplasmic reticulum, centrioles, mitotic spindles (additional distinctions are made in later chapters).

7. **A.** Locus
B. Homologous
C. Homozygote or homozygous
D. Alleles
E. DNA
F. Metacentric
G. Polytene
H. Karyotype
I. Satellite
J. Acrocentric
K. Colchicine
L. S of interphase
M. Q bands
N. C bands
O. q arm

8. **A.** +
B. +
C. –
D. –
E. –
F. +
G. –
H. +
I. –

9. **A.** LaLa
B. La^+La^+ or + +
C. La^+La or +La

CHAPTER 3

1. **A.** Diplonema
B. Zygonema
C. Diakinesis
D. S of premeiotic interphase
E. Leptonema
F. Pachynema
G. Metaphase II

2.

	No. of DNA molecules	Chromosome No.	No. of bivalents
A.	92	46	23
B.	92	46	23
C.	92	46	23
D.	46	23	0
E.	46	23	0
F.	23	23	0

3. **A.** At mitotic metaphase single chromosomes are oriented at the equator, whereas bivalents are at the equator of first meiotic metaphase.
   B. In mitosis, each chromosome moving poleward is composed of a single chromatid; at meiotic first anaphase, each is composed of two chromatids.
   C. No bivalents would be present at mitotic prophase. At the earliest meiotic prophase, the chromosome threads are much more extended than at mitosis and show obvious chromomeres.

4. **A.** Only AaBb
   B. AB, Ab, aB, ab

5. **A.** 46
   B. 46
   C. 23
   D. 23
   E. 23
   F. 23

6. **A.** AX
   B. AX and AY
   C. AAXY
   D. AAXX

7. Only in the primary oocyte

8. **A.** (1) 4000, (2) 2000, (3) 1000, (4) 8000
   B. (1) 1000, (2) 1000, (3) 1000, (4) 250

9. **A.** 10
   B. 10
   C. 11
   D. 6
   E. 5

10. **A.** 10
    B. 20
    C. 20
    D. 20
    E. 10
    F. 10

11. (1) Bo, (2) Bo, (3) Me, (4) Bo, (5) Me, (6) Me, (7) Bo, (8) Bo, (9) Me, (10) Me, (11) Me, (12) Bo

CHAPTER 4

1. The dominant E shows variable expressivity and reduced penetrance. The man with the extra toe on each foot definitely carries allele E and is most likely genotype Ee, since the allele is rare. The wife is probably genotype ee, because the allele is rare, and she does not come from her husband's family line. Both their children are genotype Ee, even the one of normal phenotype. The normal offspring's wife is almost certainly ee, again because the allele is rare, but the offspring here is again Ee, having received the dominant from the phenotypically normal parent who carried it but did not express it.

2. 60%

3. Pleiotropy

4. **A.** CcPp × Ccpp
   B. White and colored in a ratio of 5:3

5. **A.** Red
   B. White
   C. Yellow
   D. White

6. **A.** Brown × white. Offspring: 2 white:1 black:1 brown
   B. White × black. Offspring: 2 white:1 black:1 brown
   C. White × white. Offspring: 3 white:1 black

7. **A.** Deaf × deaf. No chance of a deaf child
   B. Normal × normal. No chance of a deaf child
   C. Normal × normal. One chance out of four
   D. Normal × normal. One chance out of four

8. **A.** Pea × rose. Offspring: all walnut
   B. Walnut × walnut. Offspring: 9 walnut:3 rose:3 pea:1 single
   C. Walnut × single. Offspring: 1 walnut:1 rose:1 pea:1 single

9. **A.** RrPp × rrpp
   B. RRpp × rrPp
   C. Rrpp × RrPp

10. **A.** RRppBB × rrPPbb. F_1 birds are walnut, blue: RrPpBb
    B. RrPpBb × rrppBB. F_2: black birds of the four different types of combs and blue birds of the four different types of combs

11. **A.** BBCCoo × BBccOO. F_1: BBCcOo
    B. BBCcOo × BBCcOo. F_1: 9 BBC–O–; 3 BBC–oo; 3 BBccO–; 1 BBccoo. The ratio is 9:7
    C. BbCCOO × BbCCOO. F_1: BBCCOO; BbCCOO; bbCCOO. The ratio is approximately 1:2:1.

12. The F_2 plants fall into a ratio of 9:7, a modified dihybrid ratio. Two pairs of alleles are involved, say Aa and Bb. High cyanide content requires both dominants, A and B. All those plants lacking either A or B will have a low cyanide content.

13. The ratio among the F_2s is approximately 15:1, indicating a modified dihybrid ratio. Assuming the allelic pairs Aa and Bb, the F_1 parents must have been dihybrids AaBb. The P_1s would have been AAbb and aaBB. Only double recessives, aabb, have unfeathered legs. Duplicate dominant epistasis is involved, so that the presence of either dominant, A or B, will give feathers on the legs.

14. Mexican hairless dogs are all heterozygotes, carrying an allele that can be represented as H. This acts as a dominant with respect to the hairless condition, but it also has a recessive lethal effect. Other breeds of dogs are homozygous for the recessive allele h, which permits hair growth. They thus lack the lethal. Genotypes are Mexican hairless, Hh; dogs with hair, hh; dead pups, HH.

15. **A.** Hhww
B. hhww
C. hhWW

16. **A.** 1 hairless:1 wire-haired
B. 4 hairless:3 wire-haired:1 straight-haired

17. When two plants are crossed and produce albino offspring, it is known they are heterozygotes. These plants of known genotype Ww can be crossed to those whose genotype is still unknown. If the cross of a known heterozygote and an unknown produces a large number of offspring (say, more than 20) without the appearance of any albinos, the unknown would most likely be genotype WW. Recognizing WW plants in this way and then using them exclusively in further crosses would eliminate the lethal.

18. **A.** 1 black:1 albino
B. Same results as in A because ovaries are aa
C. All black offspring

19. Pp (no PKU: no mental retardation) and pp (PKU but no mental retardation if placed on special diet).

20. Pp and pp both suffering mental retardation as a result of damage from toxic substances that crossed the placenta.

CHAPTER 5

1. **A.** S_1S_4; S_1S_5; S_2S_4; S_2S_5
B. S_3S_5; S_4S_5
C. No offspring can result, because all the pollen will abort
D. S_3S_4; S_3S_5

2. **A.** $c^{ch}c^h$ (light gray) and $c^{ch}c$ (light gray)
B. c^+c (wild) and $c^{ch}c$ (light gray)
C. $c^{ch}c^{ch}$ (chinchilla); $c^{ch}c^h$ (light gray); $c^{ch}c$ (light gray); c^hc (Himalayan)
D. c^+c^h (wild); c^+c (wild); c^hc (Himalayan); cc (albino)

3. **A.** I^Ai and ii
B. I^AI^B and I^Ai
C. I^AI^A and I^BI^B

4. **A.** $I^{A1}I^{A2}$ (type A_1); $I^{A1}i$ (type A_1); $I^{A2}i$ (type A_2); ii (type O)
B. $I^{A1}I^{A2}$ (type A_1); $I^{A2}I^B$ (type A_2B); $I^{A1}i$ (type A_1); I^Bi (type B)
C. $I^{A1}I^{A2}$ (type A_1); $I^{A1}I^B$ (type A_1B); $I^{A2}I^B$ (type A_2B); I^BI^B (type B)
D. $I^{A1}i$ (type A_1); I^Bi (type B)

5. Neither. A type O person such as man 1 cannot have AB offspring. The second man, being blood type A_2, cannot contribute the I^{A1} allele required for the child's $I^{A1}I^B$ type.

6. **A.** I^AiSese (will secrete A antigen) and I^BiSese (will secrete B antigen)
B. I^AiSeSe; I^AiSese (both types secrete A antigen); I^BiSe Se; I^BiSese (both types secrete B antigen); I^Aisese; I^Bi sese (both types are nonsecretors)
C. I^AI^BSeSe (secretes antigens A and B); I^AI^BSese (secretes antigens A and B); I^BiSeSe (secretes B); I^Bi Sese (secretes B); I^AI^Bsese (nonsecretor); I^Bisese (nonsecretor); I^AiSeSe (secretes A); I^AiSese (secretes A); I^Aisese (nonsecretor); iiSeSe; iiSese; iisese (no A or B antigen can be produced with O type, even in presence of Se).

7. **A.** 3 type A; 3 type B; 6 type AB; 4 type O
B. 3 type A; 5 type O
C. 5 type O; 1 A; 1 B; 1 AB

8. **A.** All type O; 75% for any one to secrete H antigen; 25% to secrete no H antigen.
B. $^1/_2$ chance for an offspring to be type O who will secrete no A, B, or H antigens; $^1/_8$ chance for an O who will secrete H antigen; $^1/_8$ chance for type AB (secretor of A, B, and H antigens); $^1/_8$ chance for type A (secretor of A and H antigens); $^1/_8$ chance for type B (secretor of B and H antigens).
C. All type O; 9 chances out of 16 for H antigen to be secreted; 7 chances out of 16 for no H antigen.
D. $^3/_8$ chance for type AB who will secrete antigens A, B, and H; $^3/_8$ chance for type AB who will secrete none of the antigens; $^1/_4$ chance for type O who will secrete none of the antigens.

9. Man 1 cannot be the father, because the father must contribute the L^N allele. Man 2 could be the father, but one cannot say that he is simply because he has the appropriate genotype.

10. **A.** O Rh^+ and O Rh^-
B. AB Rh^+; AB Rh^-; A Rh^+; A Rh^-; B Rh^+; B Rh^-; O Rh^+; O Rh^-
C. A Rh^+; A Rh^-; B Rh^+; B Rh^-

11. Both men could be.

12. **A.** Type O; secretes antigens H, M, and N.
B. Type A; secretes none of the antigens.
C. Type O; secretes antigens M and N.
D. Type AB; secretes none of the antigens.
E. Type O; secretes antigens D and N.

13. There would be a risk in matings B and C, because the female is Rh^- (dd) and the male is Rh^+, possessing allele D. The risk is greater in mating B, because the male is homozygous for allele D.

14. **A.** $\frac{B_2C_1A_2}{B_4C_3A_4}$, $\frac{B_2C_1A_2}{B_3C_4A_3}$, $\frac{B_1C_2A_1}{B_4C_3A_4}$, $\frac{B_1C_2A_1}{B_3C_4A_3}$
B. They can neither donate to nor accept from the parents.
C. The boy will be able to accept from either parent or his sibs, because he is deficient in the immune response, but he will not be able to donate tissue to his parents or to sibs with different haplotypes, because his tissue carries antigens that can trigger an immune response.

15. **A.** Mother: $\frac{B_9A_2}{B_8A_1}$ Father: $\frac{B_6A_3}{B_8A_3}$

Child 1: $\frac{B_9A_2}{B_8A_3}$

Child 2: $\frac{B_8A_1}{B_6A_3}$ Child 3: $\frac{B_9A_2}{B_6A_3}$

B. $\frac{B_8A_2}{B_8A_3}$ (crossover from mother)

16. **A.** All are normally pigmented; genotype AaBb
B. Chance for an albino is 7 out of 16

17. Child 2 gives the clue to the haplotypes.

Father: $\frac{5 \quad 10 \quad 8}{40 \quad 10 \quad 20}$ Mother: $\frac{17 \quad 10 \quad 20}{27 \quad 2 \quad 11}$

Child 1: $\frac{27 \quad 2 \quad 11}{40 \quad 10 \quad 20}$ Child 2: $\frac{17 \quad 10 \quad 20}{40 \quad 10 \quad 20}$

CHAPTER 6

1. **A.** AaPp
B. AapY
C. AaPY

2. **A.** aP, ap
B. Ap, AY, ap, aY
C. AP, Ap, aP, ap

3. **A.** All children with normal pigment distribution. Genotypes: AaOo, AaOY
B. Daughters with normal pigmentation; sons lacking eye pigment. Genotypes: AaOo, AaoY

4. **A.** Daughters all carriers with normal iris (Ii); sons all with normal iris (IY)
B. Daughters all carriers with normal iris (Ii); sons all with cleft iris (iY)
C. Daughters: half carriers with normal iris (Ii) and half with cleft iris (ii); sons: half with normal iris (IY) and half with cleft iris (iy)

5. Girl: Mmii; mother: mmIi; father: M–iY

6. Migraine, normal iris; migraine, cleft iris; no migraine, normal iris; no migraine, cleft iris in a ratio of 1:1:1:1 among both sons and daughters

7. **A.** ppTT × PYtt. F_1: PpTt (daughters with normal vision); pYTt (sons with red-green color blindness)
B. Normal vision, red-green color-blind, and totally color-blind in a ratio of 3:3:2 among both sons and daughters

8. **A.** All daughters with rickets (Rr); all sons normal (rY)
B. Half of the sons and half of the daughters with rickets (RY; Rr) and half of the sons and half of the daughters normal (rY; rr)

9. **A.** Females: bY (non-Bar); males: Bb (Bar)
B. Females: BY (Bar) and bY (non-Bar); males: Bb (Bar) and bb (non-Bar)

10. **A.** The females carrying the lethal would produce only half the expected number of male offspring, since no male could be born with the allele. The 1:1 sex ratio would become distorted to 2:1 in favor of females.
B. The males carrying the lethal would produce only half the expected number of female offspring. No hen could carry the allele. The sex ratio of 1:1 would become distorted in favor of males.

11. To have the affliction, a female must receive an X chromosome bearing the recessive from both parents. Since males with this affliction do not live to reproduce, offspring receive the recessive only from a carrier mother. Because a male has only one X chromosome, a male receiving the allele will express it.

12. The butterfly must be female, because the allele for whiteness can be expressed only in the female of this species. The genotype could be WW or Ww. The tortoise kitten must be female, because the genotype for tortoise must be Bb, and this requires two X chromosomes. The sex of the yellow kitten is in doubt. It could be a male (bY) or a female (bb).

13. Noncolor-blind female with three X chromosomes (PPp); noncolor-blind Turner female (PO); noncolor-blind Klinefelter male (PpY). The YO zygote would not survive.

14. Noncolor-blind Turner female (PO); noncolor-blind Klinefelter male (PPY); color-blind Turner female (pO); noncolor-blind Klinefelter male (PpY).

15. (A) 100, (B) 100, (C) 50, (D) none, (E) none, (F) 50, (G) 100, (H) 100

16. **A.** Apricot males and intermediate apricot females
B. Apricot males and females intermediate between cherry and apricot
C. Cherry males and red-eyed females
D. Males: 1 wild:1 cherry; females: 1 red:1 intermediate between cherry and apricot

17. This male must be carrying certain alleles that cause reduction in milk yield. These would be sex-limited alleles that are not expressed in the male. However, in a female these alleles can come to expression whether received from a male or a female parent. In this example, the bull is carrying them and transmitting them to female offspring, where they are expressed, causing a reduction in milk yield.

18. **A.** Half of the sons nonbald (bb) and half of them bald (Bb); daughters all nonbald (Bb and bb)
B. Sons all bald (Bb); daughters all nonbald (Bb)
C. Same as A

19. Daughters: all nonbald, but half will be color-blind. Sons: equal chances for bald and normal vision; bald and color-blind; nonbald and normal vision; nonbald and color-blind

20. (1) ll, (2) LL, (3) Ll (long fingers), (4) Ll (short fingers)

21. No early hair loss among the daughters. Early hair loss in three out of four sons.

22. (A) +, (B) –, (C) –, (D) +, (E) +, (F) –, (G) +, (H) +, (I) +

23. (1) Older sister or sister who is fraternal twin; (2) younger brother (sisters preferred in respect to HY antigen); (3) mother; (4) aunt; (5) unrelated girl

24. D, E

25. **A.** Daughters: Dd^a (normal vision); dd^a (color-blind); sons: DY (normal vision); dY (color-blind)
    B. Daughters: Dd (normal vision); Dd^a (normal vision); sons: dY (color-blind); d^aY (color-blind)

CHAPTER 7

1. **A.** AABb or AaBB
   B. AaBb
   C. aabb

2. The white parent is aabb. The intermediate is probably either AAbb or aaBB. This is so because intermediates with genotype AaBb would make it possible for children of various shades to arise.

3. Intermediate: AaBb
   Light: Aabb and aaBb
   White: aabb

4. The two parents in the first case are probably not heterozygous at both loci. Both could be AAbb; both could be aaBB; or one could be AAbb and the other aaBB. The second two parents are probably both AaBb.

5. **A.** 7 (including white)
   B. 1/4,096 red and 1/4,096 white

6. Four pairs of alleles are involved. Each effective allele contributes 0.25 inch to ear length.

7. Three pairs of alleles are involved. Each effective allele contributes 1 g.

8. Four pairs of alleles are involved. Each effective allele contributes 1 mm.

9. The two parents are obviously not extreme types. Several genotypes are possible. For example, each could be AaBbCcDd. One other possibility is that one parent is genotype AABbCcdd and the other parent is AABbccDd.

10. **A.** (Nonhygienic queen) UURR × ur (hygienic drone). UuRr, both workers and queen; workers' behavior is nonhygienic
    B. UR drones
    C. UR, Ur, uR, ur
    D. UURR, UURr, UuRR, UuRr, all nonhygienic

11. A, B, and C

12. Any cheetah would reject grafts from the cat because the two species differ genetically at the equivalent of the HLA complex. They thus possess different tissue antigens, and the cat skin is rejected as a consequence of the immune response of the cheetah. Since a cheetah tolerates a skin graft from any other cheetah, there must be a great deal of homozygosity at the genetic region governing tissue antigens. The cheetahs show the consequence of a high degree of inbreeding and have apparently lost genetic variability at the loci that govern the formation of tissue antigens.

13. (1) A greater influence of environmental factors than of genetic ones is suggested, since the concordance values for MZ and DZ are not very different.
    (2) A strong hereditary component appears to be operating, as indicated by the much higher concordance value of the MZ as contrasted with the DZ.
    (3) The large difference in concordance values between MZ and DZ suggests that a hereditary component is involved but a large environmental factor is operating, as indicated by the 73% discordance among the MZ.

CHAPTER 8

1. **A.** $\underline{r^+\ s}$ and $\underline{r\ s^+}$
   B. $\underline{r^+\ s^+}$ and $\underline{r\ s}$

2. **A.** $\underline{t^+\ u}$ (35%); $\underline{t\ u^+}$ (35%); $\underline{t^+\ u^+}$ (15%); $\underline{t\ u}$ (15%)
   B. $\underline{t^+\ u^+}$ (35%); $\underline{t\ u}$ (35%); $\underline{t^+\ u}$ (15%); $\underline{t\ u^+}$ (15%)

3. $\underline{t^+\ u^+}$ and $\underline{t\ u}$, the original combinations from the dihybrid

4. **A.** $\frac{pr\quad vg^+}{pr^+\quad vg} \times \frac{pr\quad vg}{pr\quad vg}$
   B. 15%

5. Purple flies and vestigial flies in a ratio of 1:1

6. **A.** $\frac{o^+\quad s}{o\quad s^+} \times \frac{o\ s}{o\ s}$
   B. 20%

7. **A.** $\underline{o^+\quad s^+}$ (40%); $\underline{o^+\quad s}$ (10%); $\underline{o\quad s^+}$ (10%); $\underline{o\quad s}$ (40%)
   B. Round, simple (66%); round, branched (9%); long, simple (9%); long, branched (16%)
   C. 1,000 × 9% = 90

8. P_1: $\frac{I^+\ F}{I^+\ F} \times \frac{I\ F^+}{I\ F^+}$; F_1: $\frac{I^+\ F}{I\ F^+}$

9. A. $\frac{I^+\ F}{I\ \ F^+} \times \frac{I^+\ F^+}{I^+\ F^+}$

B. $\frac{I\ \ F^+}{I^+\ F^+}$ (white, normal);

$\frac{I^+\ F}{I^+\ F^+}$ (colored, mildly brittle);

$\frac{I^+\ F^+}{I^+\ F^+}$ (colored, normal);

$\frac{I\ \ F}{I^+\ F^+}$ (white, mildly brittle)

C. 20%

10. P_1: $\frac{w\ y^+}{w\ y^+} \times \frac{w^+\ y}{Y}$

F_1: $\frac{w\ y^+}{w^+\ y}$ (red-eyed, gray-bodied females);

$\frac{w\ y^+}{Y}$ (white-eyed, gray-bodied males)

11. 15/685 = approximately 2%. Consider only the male offspring.

12. A. $\frac{d\ p^+}{d\ p^+} \times \frac{d^+\ p}{Y}$

Offspring:

$\frac{d\ \ p^+}{d^+\ p}$ (noncolor-blind daughters);

$\frac{d\ \ p^+}{Y}$ (color-blind sons)

B. $\frac{d^+\ p}{d\ p^+} \times \frac{d^+\ p^+}{Y}$

Offspring:

$\frac{d^+\ \ p}{d^+\ p^+}$ and $\frac{d\ \ p^+}{d^+\ p^+}$ (noncolor-blind daughters);

$\frac{d^+\ \ p}{Y}$ and $\frac{d\ \ p^+}{Y}$ (color-blind sons)

C. Daughters all with normal vision:

$\frac{d^+\ p}{d^+\ p^+}$; $\frac{d\ \ p^+}{d^+\ p^+}$

$\frac{d^+\ p^+}{d^+\ p^+}$; $\frac{d\ \ p}{d^+\ p^+}$

Sons: $\frac{d^+\ p}{Y}$; $\frac{d\ p^+}{Y}$; $\frac{d\ p}{Y}$ (color-blind)

$\frac{d^+\ p^+}{Y}$ (normal vision)

13. A. $\frac{o\ d^+}{o\ d^+} \times \frac{o^+\ d}{Y}$

Offspring: $\frac{o\ d^+}{o^+\ d}$ (daughters normal)

$\frac{o\ d^+}{Y}$ (sons with ocular albinism)

B. $\frac{o^+\ d^+}{o\ \ d} \times \frac{o^+\ d^+}{Y}$

Offspring:

$\frac{o^+\ d^+}{o^+\ d^+}$ and $\frac{o^+\ d^+}{o\ \ d}$ (daughters all normal)

$\frac{o^+\ d^+}{Y}$ and $\frac{o\ \ d}{Y}$ (sons normal and sons with both ocular albinism and color blindness)

14. e 4 r 7 s or s 7 r 4 e

(both maps are the same)

15. A. $0.18 \times 0.14 = 0.025$ (expected);

coincidence $= \frac{0.015}{0.025} = 0.6$

B. The locus cannot be placed with certainty on the basis of the information given. It could be located 13 units to the right of p or 13 units to the left. Information is needed on crosses involving genes at the s locus and those at m or d.

16. A. $\frac{v^+\ \ l\ \ b^+}{v\ \ \ l^+\ b}$

B. v 17.4 l 26.7 b

C. $0.174 \times 0.267 = 5\%$ (approx.)

coincidence $= \frac{40 \text{ (obtained)}}{50 \text{ (expected)}} = 0.8$

17. l 11.8 r 33.7 g

18. A. P_1 females: $\frac{x^+\ z^+\ y}{x\ \ z\ \ y^+}$; males: $\frac{x^+\ z^+\ y^+}{Y}$

B. x 6.4 z 7.3 y

C. Little, if any, interference, because five doubles were obtained, and this is about what should be expected, assuming no interference.

19. A. Would expect half of the flies to be females. Therefore, approximately 2,000 would be wild females, $r^+s^+t^+$.

B. Would expect 2,000 males. Approximately 6 should be r^+st. This is a double crossover type. The total number of doubles expected on the basis of the map is $2{,}000 \times 1.2\%$ or 24. However, the coincidence is 0.5, and so a total of 12 doubles is expected. Of these, 6 would be r^+ st and 6 would be r s^+t^+.

C. These are parental types. They are determined by subtracting the sum of the doubles (12) and the singles in region I (228) and the singles in region II (188). The singles in region I, for example, are determined in the following way: $12\% \times 2{,}000 = 240$ expected from the map information. However, the doubles (12) must be subtracted, because they were included in the construction of the map. This gives 228. The same reasoning gives 188 for region II. The parental types among the male flies, r^+s^+t and rst^+, would total about 1,572 (2,000 – 428, the sum of the doubles and the singles in I and II).

20. A. 2 wild (red eyes, straight wings):1 sepia-eyed, straight wings:1 red-eyed, curled wings

B. $\frac{se\ e^+}{se\ e^+} \times \frac{se^+\ e}{se^+\ e}$; F_1: $\frac{se\ e^+}{se^+\ e} \times \frac{se\ e^+}{se^+\ e}$

	$se\ e^+$	$se^+\ e$	$se^+\ e^+$	$se\ e$
$se\ e^+$	$\frac{se\ e^+}{se\ e^+}$ (sepia)	$\frac{se^+\ e}{se\ e^+}$ (wild)	$\frac{se^+\ e^+}{se\ e^+}$ (wild)	$\frac{se\ e}{se\ e^+}$ (sepia)
$se^+\ e$	$\frac{se\ e^+}{se^+\ e}$ (wild)	$\frac{se^+\ e}{se^+\ e}$ (ebony)	$\frac{se^+\ e^+}{se^+\ e}$ (wild)	$\frac{se\ e}{se^+\ e}$ (ebony)

The amount of recombination cannot be estimated from a cross in *Drosophila* in which the linked alleles are in the trans arrangement in the F_1s. This is so because there is no crossing over in the male fruit fly. Consequently, both the crossover and noncrossover offspring will arise in the F_2 in a ratio of 2 wild:1 mutant:1 mutant.

21. A. 25% crossovers and 75% noncrossovers. (The percentage of noncrossovers equals all the gametes formed from those meiotic cells in which no crossing over took place plus half of the gametes from meiotic cells in which crossing over did occur, in this case 50% + 25%.)

B. 50%

CHAPTER 9

1. 13%

2. 90%

3. $\frac{gd^+\ \ P}{gd\ \ p} \times \frac{gd^+\ \ P}{Y}$

Offspring:
Females, all enzyme producers with normal vision
Males, enzyme producer, normal vision, 47%; no enzyme, color-blind, 47%; enzyme producer, color-blind, 3%; no enzyme, normal vision, 3%

4. The crossover event will produce two chromosomes, in each of which the two chromatids are different genetically:

$$\frac{\frac{sn\ \ \ y^+}{}}{sn^+\ \ \ y} \text{ and } \frac{\frac{sn\ \ \ y^+}{}}{sn^+\ \ \ y}$$

An arrangement on the spindle, as shown here in some of the cells at mitosis, will give rise to cells that are homozygous for y and cells that are homozygous for sn. These cells in turn will give rise to yellow and singed patches among the cells that show the gray and the normal bristle traits.

5. The h locus must be on chromosome III, linked to sepia. The reason is that no double recessives can be found among the F_2 offspring of a cross in *Drosophila* when two mutant stocks are crossed and the recessives being followed are linked. The F_1 dihybrid offspring would have alleles in the trans arrangement:

$$\frac{se\ \ h^+}{se^+\ \ h}$$

Since there is no crossing over in the male fruit fly, only parental combinations can be passed to the gametes: se h^+ and se^+ h. When these fertilize the four kinds of gametes formed by the female, a 2:1:1 ratio results. No double recessive can be produced.

6. Enzyme X, chromosome 7; enzyme Y, chromosome 5; enzyme Z, chromosome 1.

7. A chromosome alteration such as a shift or translocation should be suspected. A karyotype analysis should be performed using such techniques as fluorescent staining to bring out the chromosome banding pattern. This person's karyotype should then be compared with that of typical persons. If a band normally present in chromosome 1 is absent from that chromosome of the atypical person, and if a similar band is now present in chromosome 5, where it is not typically seen, then good evidence can be offered for a shift. The locus governing the enzyme's production can now be more precisely assigned to a specific region of chromosome 1.

8. B, C

CHAPTER 10

1. The dicentric ——0 ABCdefba 0—— and the acentric gcDEFG

2. ——0 ABCdeFG and
——0 abfEDcg

3. HIJ 0 Klh and
nmij 0 kLMN

4. HIj 0 kLMN
hlK 0 Jimn

5. At meiosis, each chromosome will form a loop, which includes the genes T, U, and V. Crossing over anywhere between the loops, before W or before X, will lead to

——0 RSTUVWXTUVYZ
(a duplication)

and

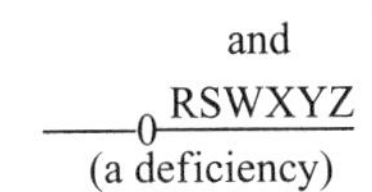

6. **A.**

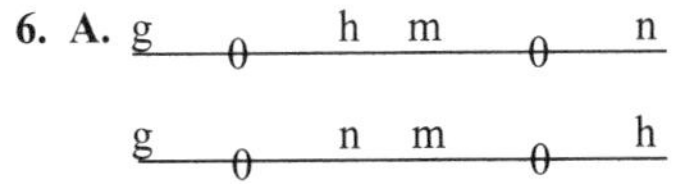

B. g–h, m–h and m–n, g–n (one possibility); m–h, m–n and g–n, g–h (the other possibility)
C. g–h, m–n and g–n, m–h

7. **A.** Half of the male embryos would die, so that the sex ratio would be 2 females:1 male. Half of the females carry the deletion.
B. One fourth of the offspring from two carrier parents will fail to survive, just as if a recessive lethal gene were present.

8. When Bar is taken out of its normal location on the X chromosome and placed next to heterochromatin of chromosome IV, a narrowing of the eye occurs. The absolute number of Bar regions in a fly is not the sole determinant of eye width. The position of the Bar locus is important. In its normal location in the X, Bar does not reduce eye size, but when Bar is placed on chromosome IV, eye size decreases, and the decrease is proportional to the number of abnormally placed Bars on chromosome 4.

9. **A.** I I II II II III III IV IV V V
B. I I I II II II III III III IV IV IV V V V
C. I I II II III III IV V V
D. I I I I II II II II III III III III IV IV IV IV V V V V

10. **A.** 24
B. 48

11. **A.** 25
B. 50

12. (1) Leaf cells fused following removal of cell walls; (2) selection of hybrid cells on medium, permitting growth of hybrid cells only; (3) grafting of plantlets, if necessary, to roots of compatible species such as a parental type; (4) development of mature plants with chromosome number of 50.

13. **A.** A reciprocal translocation involving two pairs of nonhomologous chromosomes is present.
B. A paracentric inversion is present. The bridge and fragment have resulted from crossing over within the inverted region.
C. The plant would appear to be derived as a result of autopolyploidy and would be a tetraploid, as evidenced by four chromosomes undergoing synapsis in each tetravalent.
D. The plant would appear to be an autotriploid, as suggested by the trivalents. The three univalents represent chromosomes that did not succeed in entering into trivalent formation. This also accounts for the bivalents.
E. The buckle suggests that a deletion is present in the chromosome lacking the buckle. It is also possible that the buckle reflects a duplication present in the chromosome showing the buckle.
F. Autotetraploidy is indicated by the tetravalents. The trivalent and univalent represent a potential tetravalent that failed to form.
G. Nondisjunction at first meitoic division would produce two cells—one with 13 chromosomes, the other with 11. These may complete meitoic division.
H. Complete failure of chromosome separation at first meiotic division would yield a cell unreduced in chromosome number. This may complete second meiotic division and eventually give rise to unreduced gametes.

14. **A.** 24
B. 17
C. 32
D. 15
E. 32

15. **A.** Nondisjunction of X chromosomes at first anaphase of oogenesis giving an XX egg, which is fertilized by a Y-bearing sperm. Also nondisjunction of the X and Y chromosomes at first anaphase of spermatogenesis, giving an XY sperm that fertilizes an egg.
B. Nondisjunction of sex chromosomes at spermatogenesis or oogenesis, producing gametes with no sex chromosomes. Also loss of one X chromosome at first mitotic anaphase of an XX zygote.
C. Nondisjunction of Y chromosome at second anaphase of spermatogenesis.
D. A pericentric inversion.
E. Complete failure of anaphase movement at meiosis, producing an unreduced sperm or egg that unites with a haploid gamete.
F. A translocation.
G. Nondisjunction of X chromosomes at oogenesis producing an XX egg that is fertilized by an X-bearing sperm.
H. Loss of an X chromosome during mitotic divisions of an XX embryo.
I. Loss of a Y chromosome during mitotic divisions of an XXY embryo.
J. Nondisjunction of Y chromosome during mitotic divisions of an XY embryo.

16. Turner syndrome, 2, 3, 8
Klinefelter syndrome, 2, 8
Down syndrome, 2, 6, 8
Cri du chat syndrome, 7
Philadelphia chromosome, 6 (previously considered a deletion)
XYY male, 2, 8

17. A metacentric chromosome may be formed following two breaks, one in each of two acrocentrics. In one of the acrocentrics, the break is just before the centromere, in the long arm of the chromosome. In the other, the break is just behind the centromere, in the short arm. The latter long piece with the centromere joins the other long arm piece, which lacks a centromere. The tiny pieces left will generally consist largely of heterochromatin. These may become lost or may unite to form a tiny chromosome that is later lost.

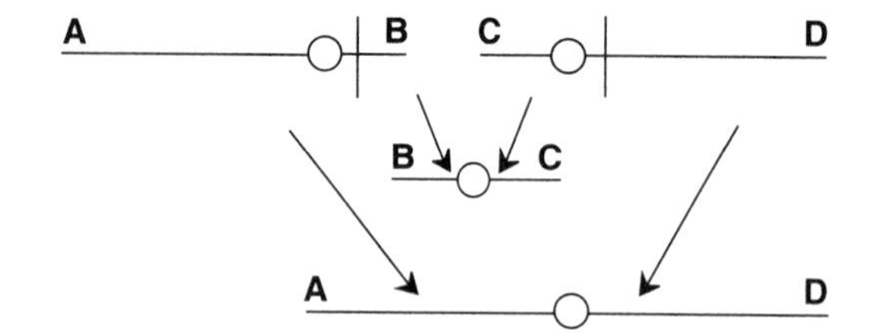

18. Testis determination seems to depend on a small Y-linked region. This could have been inserted into an X chromosome at the time of chromosome pairing at meiotic prophase. The rest of the Y chromosome could have been lost following this at meiotic divisions. The outcome could be a gamete lacking a visible Y but containing the tiny, testis-determining region.

CHAPTER 11

1. **A.** Capsulated, type III
   B. Streptomycin-sensitive
   C. Penicillin-resistant

2. Most will be in the cell-free fluid, because the radioactive sulfur was in the protein coats which did not enter the host cells.

3. Radioactivity would have been found in appreciable amounts in both the fluid and the cellular fractions. Carbon is an element common to all organic molecules, and therefore it would be incorporated into both protein and DNA. Consequently, the results would have been inconclusive.

4. **A.** Phosphoric acid, purine or pyrimidine base, sugar component
   B. Purine or pyrimidine base, sugar component

5. **A.** DNA contains thymine, never uracil. RNA contains uracil in place of thymine.
   B. The deoxyribose of DNA has one fewer oxygen at the No. 2 position in the sugar ring than does the ribose of RNA.
   C. DNA is confined mainly to the chromosomes, whereas RNA is much more widely distributed in the nucleus and the cytoplasm.

6. Sample 1, A = 15%, G = 35%, C = 35%, T = 15%
   Sample 2, A = 30%, G = 20%, C = 20%, T = 30%
   Sample 3, A = 30%, G = 20%, C = 20%, T = 30%
   Sample 4, A = 10%, G = 40%, C = 40%, T = 10%
   Samples 2 and 3 could be from the same source.

7. TAG paired with ATC (the original sequence of base pairs) and TGG paired with ACC (a GC pair now replaces the original AT).

8. **A.** 3′–TGTGGGAAATGTTTA–5′
   B. 2, or 66.6%:33.3%
   C. 1:1
   D. 1:1

9.

	Human	*Drosophila*	Yeast
A.	1.54	1.51	1.79
B.	0.99	1.02	1.00
C.	1.00	1.04	0.95
D.	0.98	0.98	1.09
E.	39.40	39.90	35.80

10. **A.** 0.2, since the amount of A + T in the complementary strand must be the same as in its partner. The same is true for the amount of G + C.
    B. 1/0.2 = 5
    C. 1 in both. The sum of the purine bases must equal the sum of the pyrimidines as a consequence of the specific base pairing.

11. **A.** 0.5, since the sum of A + T in one strand must be the same as the sum of A + T in the other strand.
    B. 0.5, since A + T = 0.5 in both strands.

12. 1/0.5 = 2

13. **A.** The one with guanine
    B. The one with thymine
    C. The 3′ or –OH end
    D. Adenine

14. No hybrid DNA band would form. At 0 generations, one heavy band. At one generation, two bands, one heavy ($^{15}N/^{15}N$) and one light ($^{14}N/^{14}N$). At two generations, two bands, one heavy and one light, but the latter about three times as wide as the former. At three generations, almost all light.

15. **A.** –A
    B. –A
    C. –B
    D. –B

16. **A.** The stem and both arms of the Y will be equal.
    B. The stem will be twice as heavily labeled as the two arms of the Y.

17. **A.** k
    B. h
    C. a, b, e, g, j
    D. a, e, j
    E. a, b, c, e, j
    F. d
    G. f
    H. i

CHAPTER 12

1. A carboxyl group, an amino group, and an R group

2. ATP, aminoacyl synthetases, tRNA, mRNA, ribosomes

3. **A.** 5′–ATGTTTAGAGTAACATATCCT–3′
   B. 5′–AUGUUUAGAGUAACAUAUCCU–3′

C. 7

4. **A.** Translation would halt prematurely
B. Nonsense

5. **A.** No effect on the protein, because both code for phenylalanine
B. Silent
C. Cysteine would be substituted for phenylalanine
D. Missense

6. **A.** The codons from the point of the first deletion would go out of phase and remain out. Likely outcome is a completely nonfunctional protein because of amino acid substitutions and also possibly one that is incomplete because of the formation of a nonsense codon.
B. Reading will come back into phase with codon 5. Polypeptide may function if the first four amino acid substitutions are not too detrimental.
C. Reading will come back into phase with codon 4. Polypeptide may function.
D. Frame shift mutations.

7. Tryptophan, because it is designated by only one codon, whereas arginine is represented by six.

8. **A.** Strand A
B. Strand A, because it is complementary to the transcript

9. **A.** (1) Aspartic acid; (2) aspartic acid; (3) serine; (4) serine; (5) none.
B. Only the contents of tube 3 will produce a labeled filter.
C. We would know that AGU codes for serine. We would also know that the other codons do not designate the particular amino acids that are labeled in the tubes to which they were added.

10. Since GC pairs form three hydrogen bonds, whereas AT pairs form only two, the melting temperature increases along with increase in G + C content. Therefore sample 2 would have the highest melting point, whereas sample 1 would have the lowest.

11. Although the G + C content is the same for both samples, sample 2 would require a higher temperature for complete strand separation because it contains a stretch of GC pairs. This G + C region would separate after the other regions surrounding it.

12. 5′–GAACAUAAAUUU–3′

13. 3′–GGGGUAUUUAA–5′

14. 5′–AGACCCTATGCAGAG–3′ (sense strand)
3′–TCTGGGATACGTCTC–5′ (antisense strand)

15. Reading in the 5′ to 3′ direction on each of the two strands, the inverted repeat is TTATGAG. Upon separation, the two strands can form the following hairpins:

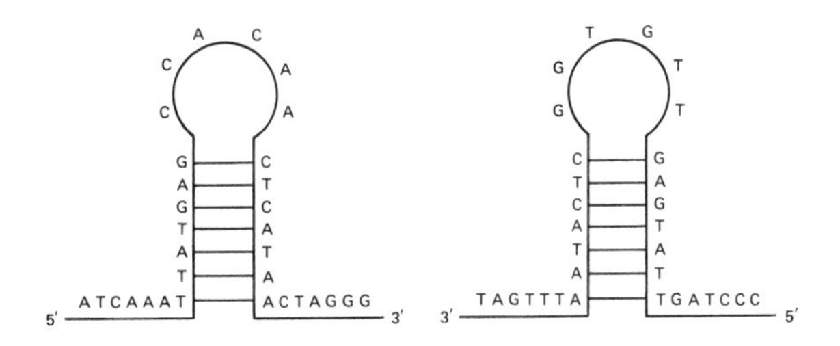

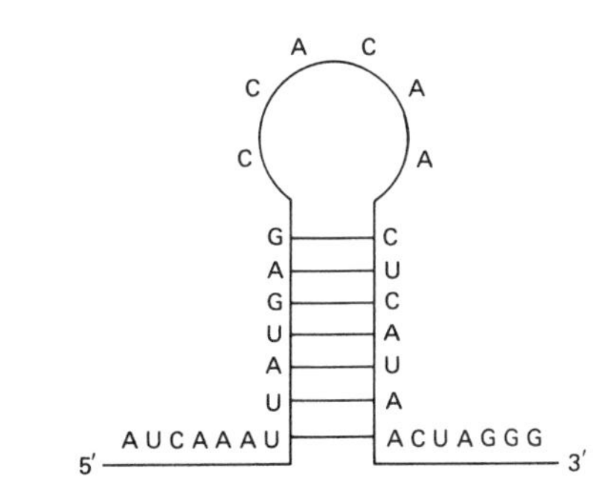

17. **A.** It must be written in the 3′ to 5′ direction, because reading in the reverse direction would produce the triplet ACT, which would yield UGA, a chain-terminating codon; thus, five amino acids could not be designated.
B. met, leu, gly, ser, phe

18. The upper strand is the sense strand and is written in the 5′ to 3′ direction. The lower strand, the antisense strand, is written 3′ to 5′. The upper strand cannot be the template strand. Reading from left to right reveals three codons—ATT, ATC, and ACT—which would be complementary to UAA, UAG, and UGA, chain-terminating codons. Reading the strand from right to left reveals ATT as the second codon, which would yield UAA upon transcription. The lower strand will not give rise to chain-terminating codons if read from left to right. If read right to left, the first codon, ATC, would be complementary to the chain-terminating codon UAG. Only in the orientation noted previously can the double-stranded DNA specify nine amino acids.

19. met, phe, ser, ile, gly

20. met, ile, thr, pro

21. Virus 1, single-stranded DNA; virus 2, double-stranded RNA; virus 3, double-stranded DNA; virus 4, single-stranded RNA

22. C_{88} A_{70} a_{42} G_{60} N T_{65} A_{62} t_{48}

CHAPTER 13

1. **A.** 1
B. 6
C. 4

D. 5
E. 2
F. 15
G. 11
H. 9
I. 13
J. 14

2. Not all the DNA of the nucleolar organizer is complementary to the 18S and 28S RNA, because a considerable amount is spacer, which is not transcribed at all. The transcribed spacer would not form part of the final rRNA, and this would also account for some of the DNA that does not hybridize with the RNA of the ribosomes.

3. **A.** 7
   B. UAC AAA UCU CAU UGU AUA GGA

4. **A.** RNA codons: 5′–AAU–3′
   DNA codons: 3′–TTA–5′
   B. RNA codons: 5′–AGA–3′ and 5′–AGG–3′
   DNA codons: 3′–TCT–5′ and 3′–TCC–5′
   C. RNA codons: 5′–UUA–3′ and 5′–UUG–3′
   DNA codons: 3′–AAT–5′ and 3′–AAC–5′
   D. RNA codons: 5′–GGA–3′; 5′–GGU–3′; 5′–GGC–3′
   DNA codons: 3′–CCT–5′; 3′–CCA–5′; 3′–CCG–5′

5. **A.** Asparagine
   B. Arginine
   C. Leucine
   D. Glycine

6. A minimum of three would be needed. Four of the degenerate codons have the same bases in the first two positions; however, no single anticodon can recognize four different codons. The base G at the wobble position can recognize U or C at the 3′ position in the codon. Similarly, U at the wobble position can recognize A or G. Thus, two tRNAs can suffice for these four. The other two codons for serine can be recognized by one transfer RNA for the same reasons.

7. Three in each case

8. **A.** The fifth position
   B. 7

9. **A.** Asparagine
   B. Tryptophan
   C. Methionine
   D. Asparagine
   E. Methionine

10. **A.** 80S
    B. 30S
    C. 50S
    D. 60S
    E. Mg^{2+}
    F. 5S

11. 5′–ATGGAGTGG–3′ (sense)
    3′–TACCTCACC–5′ (antisense)
    or
    5′–ATGGAATGG–3′ (sense)
    3′–TACCTTACC–5′ (antisense)

12. **A.** Antileader
    B. Promoter
    C. Gene proper
    D. Alteration in chain-terminating codon at end of gene proper
    E. Origin of chain-terminating codon at beginning of gene proper

13. **A.** 5′
    B. 3′
    C. 5′
    D. 5′
    E. 3′

14. **A.** ⟶ (left to right)
    B. ⟵ (right to left)
    C. ⟵ (right to left)
    D. ⟶ (left to right)
    E. ⟶ (left to right)

15. It would appear that the transcript is untranslatable for some reason. The defect could be one that permits DNA–RNA hybrid molecules to form. Some possibilities are origin of a chain-terminating codon near the 5′ end of the sequence coding for the polypeptide; defect in leader sequence; improper processing of pre-mRNA, so that introns are not properly excised.

16. Simply because the percentage values of the bases in two DNAs are the same does not necessarily mean that the distribution of the bases in the two DNAs is the same. Nucleotide sequences would be very different in species that are not closely related. Genes from more closely related species would be expected to be more similar and to be coded for related protein products. In this case, B apparently has many nucleotide sequence differences (and hence gene differences) from A and C. The latter two species would appear to be closely related, because most DNA stretches are complementary.

17. **A.** E
    B. P + E
    C. E
    D. P
    E. E
    F. P + E
    G. E
    H. E
    I. E
    J. P + E
    K. P + E
    L. P
    M. E
    N. P + E
    O. P

18. **A.** UGU and UGC for cysteine

B. UGU for cysteine
C. 3′–ACC–5′

19. A. An anticodon with the sequence ACI would also be able to pair with UGA, a chain-terminating codon. There are no anticodons for chain-terminating codons.
B. An ACU anticodon would be able to pair with the chain-terminating codon UGA. Hence no such anticodon is found.

20. Normal: 5′–GAG–3′ Sickle-cell: 5′–GTG–3′
3′–CTC–5′ 3′–CAC–5′
The sense strands are the upper ones in both cases. The change involves the substitution of a TA pair for an AT pair at the second position.

21. From lowest to highest: snap-back DNA; satellite DNA; moderately repetitive sequences; sequences coded for polypeptides.

22. Higher, because the bacterial genome, being smaller than that of a higher eukaryote, will provide a sample in which many more copies of each specific DNA sequence will be present than in the eukaryote. Eukaryotic DNA is larger and more complex than prokaryotic DNA, so that in a DNA sample equal in amount to that of a prokaryote, relatively fewer copies of each specific eukaryotic sequence are present. Therefore, because fewer copies of any given sequence are present, the time required for reassociation will be greater, giving a higher Cot½ for the eukaryote.

23. The Cot curve is typical for a eukaryotic genome and clearly shows distinct steps indicating fractions with fast annealing sequences (left) to very slow (right).

24. A. The DNA is represented by the dark strand. The mRNA represented by the light strand has undergone processing, and introns have been excised from it. These sequences are present in the DNA and form loops in the hybrid molecule because no complementary sequences are available in the mRNA to permit pairing.
B. Seven, because each loop formed by the DNA represents an intron that has been excised from the mRNA.
C. At one end is the promoter region that is not transcribed. At the other end is a segment containing sequences in the terminator that are not found in the mRNA.
D. This represents the poly(A) tail at the 3′ end of the strand. It is added after transcription is completed, and there is no complement to it in the template strand.

25. Three loci are being followed: adenine 1, adenine 3, and the locus at which the adenine 8 and adenine 16 mutant sites have occurred. The ad_8 and ad_{16} mutations do not complement each other, and so the ad_8 and ad_{16} sites are allelic.

26. Complementation is seen between b and c. However, a and b are alleles. Therefore a and c will most likely complement each other in a cross.

27. The enzyme involved is composed of two different polypeptide chains, each controlled by a separate cistron. Mutant sites a^1 and a^4 are in the same cistron that is a unit of function separate from the cistron containing both a^2 and a^3. The crosses are

A.

a^1	
a^1	

B.

a^1	
a^4	

C.

a^1	
	a^2

D.

a^1	
	a^3

E.

	a^2
	a^3

F.

a^1 a^4	

G.

a^1	a^2

H.

	a^2 a^3

28. The two different chains of hemoglobin A are coded by two different genes. The female parent may be represented as genotype aaBB (defect in the gene for the α-chain), and the male parent as AAbb (defect in the gene for the β-chain). The offspring would be genotype AaBb, having received a functional α-gene from one parent and a functional β-gene from the other. Sufficient normal hemoglobin A may form. Complementation has taken place.

CHAPTER 14

1. Plants of genotype Ww will have the greatest probability, because mutation is rare and is usually to an allele that is recessive to normal or standard. The tetraploid is the least likely to show the effect, because two dominant alleles would still be present if one of them mutates in this case.

2. 1/40,000, because 5 children out of 100,000 = 1/20,000 afflicted individuals. Each individual represents two sex cells, and the new mutation could have arisen in either the sperm or the egg.

3. $(492 \times 10^{-6})(2.4 \times 10^{-6}) = 1{,}180.8 \times 10^{-12}$

4. The first change converted the RNA codon UAU for tyrosine to UAA, a chain-terminating codon. The second mutation must have altered the anticodon GUU, which is complementary to the codon CAA for glutamine. A change at the 3′ position in the anticodon must have resulted in the anticodon AUU, which would recognize the chain-terminating codon UAA. Consequently, glutamine would be deposited at the site of the chain-terminating codon.

5. A. GGATCC / CCTAGG and GGGTCC / CCCAGG (a transition)
B. A more likely outcome is the restoration of the correct base pairing, since the mismatch repair system will recognize the region with the incorrect base pair, remove a portion of the new strand that includes the incorrect

base, and then fill in the region by DNA polymerase activity. Chances are that the system will be able to discriminate between the template and the new strand shortly after the error is made.

C. Since the cell cannot methylate bases, there will be no way for the repair system to discriminate between the template strand and the newly synthesized one with the incorrect base. Therefore it is equally possible that the correct base will be restored as described earlier, or that the correct base will be removed from the template strand. In this case a GC pair will substitute for the original AT, and a transition will result.

6. A. Two codons designate tyrosine: UAU and UAC. A single-step mutation in either of these can give rise to UAA, a chain-terminating codon.

B. Not all the phenotypically wild-type cells would be expected to be true revertants. Assuming, first, that UAU is the codon specifying tyrosine in the wild-type cell, a true revertant would be one in which UAA was changed back to UAU. However, if UAA changes to UAC, the polypeptide would still have tyrosine at position 70 and would be phenotypically wild, although the cell would not be a true revertant because a codon different from the original one has been formed. The same reasoning applies if UAC was the codon in the original wild.

7. A. Since the P_1 females are carrying no X-linked recessives, the sex ratio should be 1:1, even if any treated male X chromosomes carry a recessive, because no F_1 male would receive an X from a P_1 male.

B. If a sex-linked lethal has been induced, any F_1 female carrying the recessive and crossed to an F_1 male will produce female:male in a ratio of 2:1.

8. A. 3/50 and 4/50

B. 9/2,500 and 16/2,500

9. Since the percentage of sex-linked mutations is directly proportional to the dosage applied, there must be a constant value in each experiment for percentage of mutants/dosage in r. The constant here is seen to be 0.005 (3/600 and 8/1,600). The constant is then multiplied by 2,500 to find the expected mutations at that dosage in r, giving 12.5%.

10. A. Since 10 and 17% mutations were produced by dosages of 1,000r and 1,700r, respectively, the constant is found to be 0.01. Therefore, at 2,250r, we would expect 22.5% mutations.

B. The answer is 22.5%, because the total exposure is the same as in Part A. It does not matter in *Drosophila* whether the total dosage is given at one time or broken up into smaller doses in a series of experiments. The effect of ionizing radiations is cumulative over the life of an organism.

11. $\xrightarrow{4} D \xrightarrow{3} C \xrightarrow{1} A \xrightarrow{2} B$

The table shows that substance B permits growth of all four strains. Therefore it must be the final product in the pathway. Any block at an earlier step will be satisfied if the final substance (B in this case) is supplied. Strain 4 is the only strain that can grow with substance D. Moreover, strain 4 can grow with any of the substances. These observations indicate that substance D is the earliest of the products in the pathway, and strain 4 must be blocked in the step leading to D. Strain 2 is blocked in the last step leading to substance B, since that is the only substance it can use for growth. Strain 1 can use substance A or B, and strain 3 can use A, B, or C. This tells us that substance C comes before A and that A comes before B in the pathway.

12. Because the fifth strain can use only substance E or B, this tells us that E must be produced in an independent pathway parallel to the one described in Question 11. Since strain 5 can use the same end product—B, as all the others—it must be able to use E to make B. Substances A and E may combine to make B in the last step for which strain 2 is deficient. This is indicated as follows:

$$\begin{array}{l} C \rightarrow A \\ \qquad\qquad \searrow \xrightarrow{2} B \\ \text{Precursor} \xrightarrow{5} E \nearrow \end{array}$$

13. More significant to the human with the longer life span, because the effects of radiation are cumulative. The average fruit fly would be getting very little natural radiation. A human with a longer life span would accumulate about 15r in 30 years.

14. About 200 on plates of group A, about 140 on B, and about 176 on C. (About 40% of the killing action remains; therefore 12% killing action is left following photoreactivation.)

15. Cells in the bone marrow and in the testes are actively dividing. In such cells, the chromosomes are in a more condensed state than in nondividing cells. With their condensed chromosomes, these cells are much more susceptible to destruction by mutagens, because the chromosomes can be more readily damaged.

16. It would appear that the cells are deficient in the first step of excision repair, which requires an endonuclease to nick the DNA so that the other enzymes involved may come into action. When the DNA is broken by the radiation so that breaks or minute deletions are produced, the presence of the other enzymes can repair some of the damage.

17. The strain is normal with respect to the genetic information required to code for the enzyme of photoreactivation. However, it is deficient in one or more of the genes coding for the other enzymes of repair such as endonuclease, exonuclease, and DNA ligase.

18. AGT (base replaced with guanine, no change)
TCA
ACT (base replaced with cytosine, a transversion)
TGA
ATT (base replaced with thymine, a transversion)
TAA
AAT (base replaced with adenine, a transition)
TTA

19. AGAT
TCTA

20. **A.** AAT
B. AAT (a transition)
TTA

21. **A.** The most likely explanation is that a frame shift mutation took place following the first exposure, since too many amino acids have been changed. Missense mutations cannot reasonably account for this. We can deduce the following sequence and change for the mRNA:

			↓ Base deleted					
5′–AUG	CCA	AGU	UCA	GAU	AAA	UAC	CGA	CUU
met	pro	ser	ser	asp	lys	tyr	arg	leu
								A or
5′–AUG	CCA	AGU	CAG	AUA	AAU	ACC	GAC	UU G
met	pro	ser	gln	ile	asn	thr	asp	leu

B. The mutant cells exposed to further treatment have a polypeptide that is similar to the normal one except for the fourth and fifth amino acids. The reading frame was probably restored as a result of the second treatment, which brought about addition of a base in the sequence, as indicated here:

					↓ Base added			
5′–AUG	CCA	AGU	CAG	AUA	AAA	UAC	CGA	CUU
met	pro	ser	gln	ile	lys	tyr	arg	leu

22. **A.** Streptomycin causes a misreading of the message in the mRNA. In addition to the expected amino acid in each case, others are incorporated into the polypeptide. In each case, the first or 5′ base in the codon is the only one misread. For example, UUU (phe) can be misread as AUU, GUU, and CUU so that ile, val, and leu, respectively, can be incorporated as well as phe.
B. Misreading of poly(A) can specify UAA, a chain-terminating codon that does not specify an amino acid and would cause shorter polypeptides to be made. Misreading of poly(G) produces the codons AGG and CGG, both of which designate arginine.

23. The substance is obviously mutagenic in bacteria, but the mammalian cell can evidently convert the substance into a product or products that lack this property.

24. Genetic polymorphism exists at the enzyme locus in population A, as indicated by the variation seen in the enzyme from one individual to the next. The enzyme appears to be composed of two polypeptide chains encoded by a specific gene. The gene exists in three allelic forms, let us say A1, A2, and A3. This makes possible six different genotypes. Population B, on the other hand, is monomorphic with respect to the gene that codes for the enzyme, because only one enzyme form has been detected.

CHAPTER 15

1. The fluctuation from the mean is very great in A, as shown by the large difference between the mean and the variance. The A data must refer to the second part of the experiment, since spontaneous mutations will result in fluctuation from one sample to the next because their occurrence is random.

2. **A.** Colonies 2, 5, 6, 8
B. Colonies 2, 3, 5, 6, 7, 8
C. Colonies 1, 2, 4, 5, 6, 8

3. **A.** $A^+B^+C^+D^+E^+$
B. Neither. Both are prototrophs, because both can grow on minimal medium.

4.

$$\begin{array}{l} D \xrightarrow{1} A \\ \qquad\qquad\qquad \searrow \\ \qquad\qquad\qquad\quad \xrightarrow{4} E \xrightarrow{2} B \\ \qquad\qquad\qquad \nearrow \\ \text{Precursor} \xrightarrow{5} F \xrightarrow{3} C \end{array}$$

5. Strain 2 is the donor strain. The recipient strain is strain 1, and it must survive because the recombinants are derived from the F^- cells.

6. In cross 1, the F factor is not integrated into the chromosome but exists as a free entity that can readily transfer itself to an F^- cell and convert it to F^+, but it cannot effect chromosome transfer. In cross 2, the F factor is integrated into the chromosome and can bring about transfer of chromosomal markers. However, it is split before transfer of a strand of the chromosome. In most cases, the recipient cell will remain F^-, because only an occasional cell receives the entire chromosome strand with all portions of the F factor.

7. M–O–Q–L–N. Because the procedure selects only A^+ and B^+ cells, the order of these cannot be established here.

8. A and B:

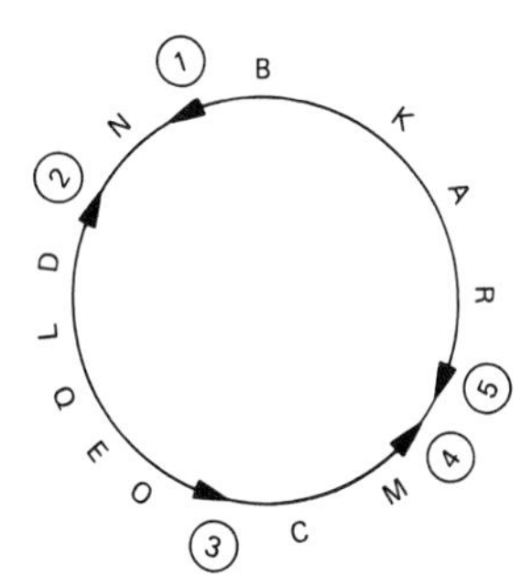

9. The last known marker to enter should be selected in each case. Select for M in the case of strain 1, C in the case of strain 2, and so on. Since the site of the fertility factor will be the last to enter, chances are greater that selection for a marker near the terminus of the transferred donor chromosome will screen out cells that have received the entire donor strand along with the fertility factor.

10. **A.** (H) 5′–AGCTAT–3′

(L) 3′–TCGATA–5′

B. (L) 5′–AGCTAT–3′
(H) 3′–TCGATA–5′

11. The cells that first received Lac$^+$ too early received the sex factor that had removed itself from its site of insertion on the chromosome along with the adjacent Lac$^+$ marker. These cells then acted as donor cells, transferring F$^+$Lac$^+$ to F$^-$Lac$^-$ with a high frequency. This is an example of sexduction. The derived F$^+$Lac$^+$ cells are heterogenotes. They carry Lac$^-$ on the chromosome and Lac$^+$ on the sex factor that is existing free in the cell.

12. A. The DNA will be hybrid: one strand A, one strand B.

B. One cell will be streptomycin resistant and the DNA will be composed of two A-type strands in the marker region. The other cell is sensitive and will have two B-type strands.

13. A. The genes a and b are closely linked, because the number of double transformants (a$^+$b$^+$) is much higher than the number of singles. If the two genes were far apart, they would tend to occur on separate fragments, and two independent events would be required to produce double transformants. The probability of doubles would be much lower than in this case and would equal the product of the separate chances for each single recombinant $(155/900 \times 95/900)$.

B. $155 + 95/900 \times 100 = 27.8\% = 27.8$ map units. When strength of linkage is estimated between two genes, the only transformant classes considered in the calculations to be recombinants are those that arose as the result of a crossover event between the two genes. In the case of the double transformants, no crossover event took place between the two genes.

C. The recombinant classes a$^-$b$^+$ and a$^+$b$^-$ are not equal because they did not result as reciprocal products from the same crossover event. Each recombinant class arose as the result of separate crossover events in which separate donor DNA fragments became incorporated in the recipient chromosome.

14. 5′–ATCGT X Y Z ATCGT–3′
3′–TAGCA X′Y′Z′ TAGCA–5′

15. A. ABCD, because the conjugal plasmid and the non-integrated F factor cannot transfer the chromosome.

B. ABCD.

C. No determinants will be transferred.

D. EF, because integrated F factor mainly brings about chromosome transfer.

16. The original strain carried a compound conjugal R plasmid with an R determinant component containing genetic factors for resistance to both drugs. This conjugal plasmid dissociated into a conjugal and a nonconjugal plasmid, the latter carrying the factor for tetracycline resistance. At times only the conjugal plasmid with the streptomycin-resistance factor is transferred. Other cells of the original strain appear to have lost the conjugal plasmid, so that the factor for tetracycline resistance is not transferred.

17. The factor for streptomycin resistance is found on a transposon that moved to another replicon, the F factor or the chromosome. It is now part of the replicon that is composed of the chromosome and the inserted F factor, and is thus transferred along with the chromosomal markers.

CHAPTER 16

1. The first and the last 10 nucleotide pairs represent a repetition or terminal redundancy. Nicking of each strand and removal of the last 10 bases toward the 3′ end yields

5′–TTACCTTCATAAAGCGCACA–3′
3′–TTTCGCGTGTAATGGAAGTA–5′

2. A and B are both lysogenic. Each harbors a different prophage, so each can be lysed by the other. C harbors no known prophage, and is thus susceptible to both A and B but cannot lyse either of them.

3. A. From either 1 or 2, since phage λ would not be transferred from the Hfr, and zygotic induction would not occur.

B. From either 2 or 3. In 2 the prophage is not transferred. In 3, both the Hfr and F$^-$ carry the prophage, so induction will not occur.

C. From either 2 or 3. Impossible in 1, because site of λ is well before the S locus and therefore λ would be transferred before S, and lysis would follow.

4. A. Transformation requires no vector. Transformation can occur by uptake of pure donor DNA alone. Transduction requires a vector; the DNA is introduced into the cell by the vector and does not enter by itself. Transformation requires that the cell be in a state of "competence." This is not so in transduction. Transduction depends on the ability of a cell to harbor appropriate prophage. This is not required in transformation.

B. Both depend on introduction of a virus into the bacterial cell, but in conversion the introduced virus does not necessarily carry any bacterial genes. The new features of the bacterial cell are not a result of any introduced genes from another bacterial strain but of the interaction of the genetic material of the virus and of the cell itself.

5. If the level of infection is very low, only one virus particle would be likely to infect a cell. Those that are transducing the cells are defective. Because no intact helper phage is present, the transducing phage cannot effect lysis.

6. The transduced cells are heterogenotes. Some of them may lose the Gal$^+$ marker and revert to the Gal$^-$ trait.

7. C

8. 40%; 40 map units

9. srt or trs

10. Gene o is in the middle, and this can be deduced in the following way. The minimal medium permits growth only of prototrophs, $m^+n^+o^+$. In experiment 1, prototrophs will arise when o^+ alone is transferred to strain 2. In experiment 2, prototrophs will arise if m^+ and n^+ but not o^- are transferred to strain 1. If o is the gene in the middle, more $m^+n^+o^+$ transductants are expected in experiment 1 than in experiment 2. This is so because the first case then requires a double crossover to get $m^+n^+o^+$. However, a much rarer quadruple crossover is required to get prototrophs in the second case. This is seen here:

1. Donor: m^- o^+ n^- 2. Donor: m^+ o^- n^+

Recipient: m^+ o^- n^+ Recipient: m^- o^+ n^-

Other arrangements of the genes require double crossovers only to derive the prototrophs in the reciprocal experiments.

11. 1.7% recombination; small clear, hr^+; small cloudy, h^+r^+; large clear, hr; large cloudy, h^+r

12. Repeated mating occurs. One recombinant class formed early in the cycle would probably engage in later matings, and thus not be represented at the time of burst.

13. A and D

14. AAGUUACCA

15. Taken three nucleotides at a time from the left, the seventh triplet is TGA, which corresponds to UGA, a chain-terminating codon in the messenger. Starting a new frame with the A in this triplet produces ATG, which corresponds to the start signal AUG in mRNA. It is thus at this point that an overlapping could occur.

16. (1) A string of alanines; (2) a string of leucines; (3) a string of cysteines

17. **A.** The same as the + strand, since the + strand acts as mRNA
B. UUGUCCUGCGUC
C. RNA replicase

18. **A.** 5′–TTTGATTAACGCGT–3′
B. The same results, because actinomycin D blocks DNA-dependent RNA synthesis, not RNA-directed DNA synthesis.

19. **A.** 5′–CTTGGAAG|CTTAATAC–3′
3′–GAACCTTC|GAATTATG–5′
(1)
5′–AACGTT|AACTTCCTA–3′
3′–TTGCAA|TTGAAGGAT–5′
(2)

B. 5′–CTTGGA AGCTTAATAC–3′
3′–GAACCTTCGA ATTATG–5′

C. 5′–AACGTT AACTTCCTA–3′
3′–TTGCAA TTGAAGGAT–5′

20. 5′–CCACCGGTCCTCATACCGGTC–3′
3′–GGTGGCCAGGAGTATGGCCAG–5′
(** marks over CC and under CC in each strand)

21. In nontransformed cells, the viral DNA has not integrated with the DNA of the cell, as indicated by lack of a signal from labeled DNA. In the transformed cells, the virus has integrated with the cellular DNA, and it has no preferential site of integration, as can be seen by the location of signals at different positions, corresponding to the fragments of different sizes (smallest at the bottom). More than one virus may integrate with the cellular DNA, as seen in clones 3 and 5.

22. **A.** Hybridization between single-stranded nucleic acids cannot be performed on a gel. The transfer is necessary to achieve this essential step.
B. Virus B, unlike virus A, has a preferential site of integration with the cellular DNA.

23. Fragments 1 and 3 do not correspond to this amino acid sequence of the coat protein. Fragment 2 corresponds to amino acids 6 through 12 in the sequence. Fragment 4 overlaps fragment 3 so that we can establish the sequence of nucleotides in the coat protein gene that corresponds to amino acid sequence 2 through 12 in this portion of the protein. The nucleotide sequence is
UUUUAUGCAACUGCAAACUCCGGUAUCUACCGU

CHAPTER 17

1. **A.** Inducible
B Constitutive
C. Constitutive
D. Constitutive
E. Inducible
F. Constitutive

2. **A.** Z constitutively; Y and A by induction
B. A, Y, and Z only by induction

3. **A.** No enzymes produced constitutively or by induction
B. Same as A
C. No enzymes produced constitutively; only A produced by induction
D. A produced constitutively; nothing else produced

4. **A.** A and B produced only in absence of effector
B. A and B produced in absence and in presence of effector
C. Same as B
D. A and B produced in absence of effector; B produced in presence of effector

5. (1) I^s is dominant to both I^+ and I^-; I^+ is dominant to I^-. (2) The altered lac repressor acts in both the cis and trans arrangement with O^+Z^+. This is seen in strains 1, 2, 3, and 4, where all are repressed. Allele I^s is expressed in both arrangements despite the presence of the + type allele in strains 1 and 2 and the I^- allele in 3 and 4. Strains 5 and 6 are both inducible. The I^+ allele is acting in both cis (5) and trans (6) in relation to O^+Z^+. In strains 1, 4, and 6, the repressor must be diffusing to the DNA in the trans arrangement in order to influence the alleles on the other molecule. If the repressor acted only at its point of formation or on the DNA on the same molecule, then it would not be diffusible, and the results presented here would be different.

6.

Strain	Lactose Present	Lactose Absent
1	+	0
2	0	0
3	0	0
4	+	+
5	+	0
6	+	+

7. In the case of a positive control system, a substance must be added to stimulate the process. A negative control mechanism requires removal of the substance to stimulate the process. When the cAMP–CAP complex is added, it binds to the promoter and stimulates transcription by influencing the binding of RNA polymerase. This is in contrast to the operator–repressor system in which the repressor must be removed from the site of the operator for transcription to proceed.

8. An IS element is present in R6-5, and its presence switches off the determinant for tetracycline resistance. Excision of the element results in a shorter plasmid but restores the activity of the determinant for tetracycline resistance.

9. The additional length suggests that an IS element has been inserted in the region of the operon in the case of both variants. The difference in expression can be explained as a result of differences in the direction of insertion of the element. Insertion in one direction causes transcription of the structural genes, even in the absence of inducer. Insertion in the opposite direction results in complete suppression of transcription from the structural genes.

10. An IS element appears to have been inserted into the region of the operon, resulting in switching off of the structural genes in the operon. The electron microscope reveals the region where the IS element is located, since the strand from the wild type does not bear the element. The "stem" results from the pairing of the inverted repeats at the ends of the single strand of the element. The "loop" contains nonduplicated sequences between the inverted repeats. No pairing of bases occurs in the loop.

11. A. 3′–GGTGATGTA . . . XYZ . . .
TACATCACC–5′
5′–CCACTACAT . . . X′ Y′ Z′
ATGTAGTGG–3′

B.

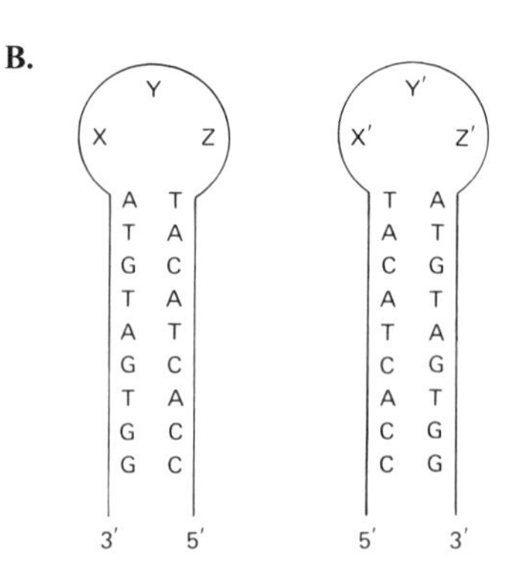

12. Lysogeny will be favored in strains A, D, and H.

13. (1) Red, (2) mottled, (3) mottled, (4) white, (5) red

14. A. The females must have been from the P strain and the males from M, since P cytoplasm has repressor activity that prevents transposition of P. Moreover, M males carry no P elements. If the females were from strain M in the cross, insufficient repressor would be present to prevent transposition of P coming from the males.

B. M × M and P × P

15. In cells of the pancreas, this sequence does not occur as part of the linker DNA. Nucleosome phasing appears to be the case here. In other types of cells, nucleosome beads appear to be randomly placed, so that the sequence may be associated entirely with the linker. It may also sometimes be associated in part or sometimes in its entirety with the octamer. Only that part of the DNA associated with the bead will be protected from digestion.

16. The promoter region of this gene would not be expected to be tightly bound to the octamer, because an active promoter undoubtedly must interact with various regulatory molecules governing transcription. If it were bound to the bead, such interactions would be curtailed. To determine whether it is tightly bound, expose the chromatin from cells in which the gene is active to a nuclease. If the region is not closely associated with the bead, it should be highly sensitive to nicking with an endonuclease. One might also attempt to detect association of the site with HMG.

17. Although the entire gene is intact, an enhancer sequence appears to have been deleted. Without the stimulating effect of the enhancer, transcription from the promoter of the gene is greatly reduced.

18. (1) Inactive, (2) active, (3) inactive, (4) active, (5) active, (6) active

19. 5′–CAC*GTCAC*GTACC–3′
3′–GTGC*AGTGC*ATGG–5′

(* marks the methylated C in each strand: asterisks printed above the C's in the top strand and below the C's in the bottom strand)

20. Two genes are involved. One is coded for the thyroid prehormone. The other gene is a complex transcription unit. Differential processing of the primary transcript occurs in pancreatic and pituitary cells. Coding sequences toward the first half of the transcript from its 5′ end are preserved in both kinds of cells, so that the amino acids toward the amino end of the two polypeptides are the same. However, the exons that are retained toward the 3′ end of the transcript must be different in the two cell types. Coding sequences discarded during processing of the pancreatic mRNA would be retained in processing of the pituitary mRNA, and vice versa. The result of translation of the two mRNAs resulting from differential processing of the primary transcript is two different proteins, identical in the first 50% of their amino acids but different in the latter portion.

21. **A.** Incorporate labeled uridine into the food. Newly synthesized RNA will be labeled, and this can be visualized by preparation of autoradiographs made of the larval salivary gland chromosomes. If RNA synthesis and puffing are associated, dots will appear only over the puffs and not over unpuffed regions.
B. Like most steroids, ecdysone binds to protein receptors. The hormone–receptor complex binds to the chromatin and turns on transcription at the site of binding. The hormone effects cell differentiation by turning on certain genes and not others, causing different gene products to be made at different times in development. Depending on these products, different kinds of cells may originate.

22. (1) Ac will be present at site 1 in one chromatid but will be present in both chromatids at site 2. A total of three Ac elements is present in the chromosome. (2) Ac will be absent from site 1 in both chromatids but present at site 2 in both of them. Total Ac is two. (3) One chromatid will lack Ac at both sites 1 and 2. The sister will have an Ac element at both sites. Total Ac is two.

CHAPTER 18

1. **A.** An abnormal embryo with upsets in polarity.
B. A normal embryo since the effect of m genes is maternal.
C. A larva with every other segment missing.
D. Same as C.

2. Cancers that occur among young persons acquire fewer genetic changes than cancers such as colorectal ones, that apparently entail many genetic changes accumulated over a period of many years.

3. A cancer may be triggered by the loss of a tumor suppressor gene as a result of chromosome alteration or damage.

4. About nine more for the first group and five for the second, since the limit of 15 doublings is not altered by the cold treatment.

5. We see that the protooncogene is present in normal cells but that the viral DNA is not. Both viral and p-onc DNA are present in the tumor cells. The viral DNA must have integrated near p-onc in the same restriction fragment, since the fragment containing p-onc is longer in the tumor cells than in the normal ones.

6. DNA of the protooncogenes is present in both normal and transformed cells, but there is no evidence that the viral DNA has integrated with the cellular DNA in the tumor cells.

CHAPTER 19

1. Somewhat under 20%. This is so because only half of the DNA would be involved in transcription. Of this amount, not all the DNA would be transcribed, such as the promoter and the terminator. Of that part that is transcribed, a portion is cleaved from the final RNA product. Only 86 nucleotides are present in the final product.

2. Alu B and Alu D are adjacent to each other. One site is present that is recognized by Hae, and it is in fragment E.

3. As indicated in the following PstI could cut toward either the right or left of the sequence to produce fragments 6.4 and 3.6 kb in length.

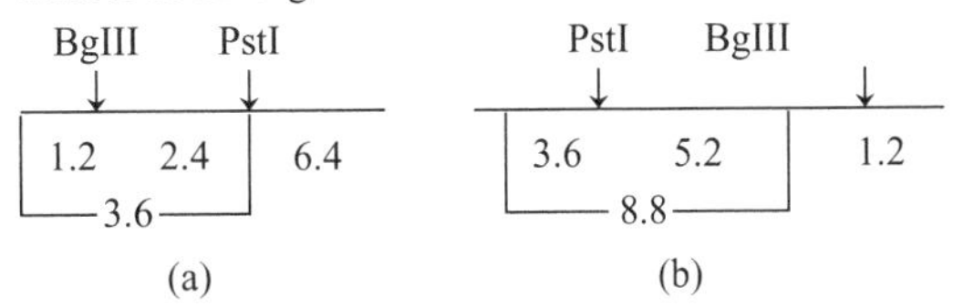

We must now locate the BglII cutting site in relation to the PstI site. Placing the BglII site nearer the left end of the sequence (a) would result in a double enzyme digest of fragments 1.2, 2.4, and 6.4 kb long. Placing the BglII site nearer the right end, the double digest will yield fragments that are 3.6, 5.2, and 1.2 kb long, as shown in b. This is what was obtained, and therefore the map shown in b is consistent with the results.

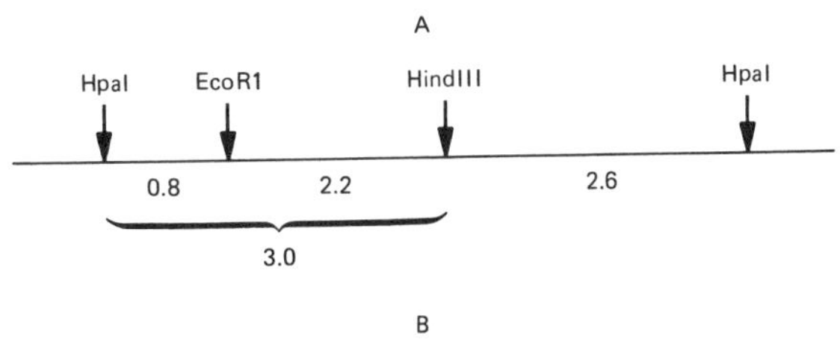

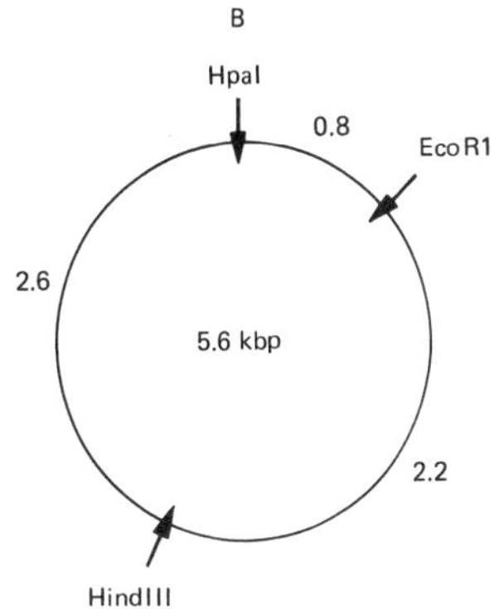

5. Denature the double-stranded circular DNA of the mutant virus to yield two single-stranded circles. Expose DNA from wild-type virus to HpaI, which will open the circle, and then to both HindIII and EcoR1. This yields the 2.6, 2.2, and 0.8 kb long fragments that can be separated from one another by gel electrophoresis. A fragment of a given length, let us say the 2.6 kb fragment, can then be denatured to the single-stranded state. These single strands are incubated with the single-stranded circles from the mutant under conditions permitting DNA replication. If the resulting viruses all prove to be mutant, we know that the fragment did not cover the mutant site, which must be located elsewhere. Conversely, origin of wild-type viruses following DNA replication would indicate that the fragment does cover the mutant site, because wild-type viruses can now arise by replication that may entail mismatch repair. The mutant site would thus be localized to the 2.6 kb region.

6. **A.** 5′–GATCCTTCGACGTT–3′
   B. 5′–GATCCTGATA–3′
   C. 5′–TATCAGGATCCTTCGACGTT–3′
   3′–ATAGTCCTAGGAAGCTGCAA–5′
   D. 5′–TATCAG–3′
   3′–ATAGTCCTAG–5′

   5′–GATCCTTCGACGTT–3′
   3′–GAAGCTGCAA–5′

7.

ddG	ddA	ddT	ddC
			—
		—	
	—		
—			
—			
	—		
			—
		—	
	—		
		—	

8. **A.** The derived DNA is
   3′–GGTAAAGCTTACCTCGTT–5′
   5′–CCATTTCGAATGGAGCAA–3′

   Terminal transferase can be used to add projecting ends of T at the 3′ ends to the DNA piece:

   3′–TTTTTTGGT . . . GTT–5′
   5′–CCA . . . CAATTTTTT–3′

   B. Render the plasmid linear by subjecting it to a restriction endonuclease that cleaves it at only one site. Then supply the plasmid with projecting ends of A at its 3′ ends using terminal transferase and adenosine triphosphate.

9. CTA
   GATGCTTAAGC
   and
   CGAATTCGTCT
   AGA

10. **A.** GACTA
    CTGATAATGCATT
    and
    TTACGTAATAGCA
    ATCGT

    B. No. Different palindromes are involved, and the ends generated in the plasmid are not complementary to those in the DNA segments of Question 9.

11. **A.** Clones 1 and 2 contain the desired gene. The absence of precipitate in these two systems indicates that the transcript for the enzyme is unavailable for translation. Therefore no polypeptide is present to react with the antibodies. The mRNA for the specific enzyme Z has complexed with its template DNA, the gene for enzyme Z. Presence of precipitate in the other three systems means the mRNA for enzyme Z is available for translation. Its template DNA, the desired gene, must not be present in systems 3, 4, and 5 to remove it by pairing with it.

    B. It is most likely, because an assortment of transcripts was added. Most of these are free to undergo translation, because their template DNA is not present. No precipitate forms in systems 1 and 2, since the antibody is specific only for enzyme Z which is not being formed. Other protein can be present that would not react with the antibodies.

12. **A.** 3′–CAUUGGCA–5′
    B. 5′–GUUAGACA–3′
    C. (1) 5′–GAACTCTG–3′
    (2) 3′–CTTGAGAC–5′
    (3) 3′–CUUGAGAC–5′

13. **A.** Colonies 4, 6, 7, and 9, since these are sensitive to both drugs. If plasmid DNA with two intact resistance genes were present, the host cells would be resistant to both drugs. If recombinant DNA is present, the host cells would be resistant to tetracycline but sensitive to ampicillin.

    B. Colonies 1, 3, and 5, since these are resistant to both drugs.

    C. Colonies 2 and 8, since they are resistant only to tetracycline, indicating that they have the plasmid with the intact Tet^r marker and that mouse DNA may have been inserted into the cut Amp^r gene, rendering it incapable of conferring resistance to ampicillin.

14. (1) Imprint colonies from the master plate onto filter paper. (2) Break open cells to extract the DNA. (3) Denature the DNA on the paper, retaining the original orientation of the colonies. (4) Add a radioactively labeled probe of thymidine kinase mRNA or of radioactively labeled single strands of the cloned thymidine kinase gene, and permit molecular hybridization to take place. (5) Wash away any unhybridized probe molecules. (6) Prepare an autoradiograph of the materials on the filter paper. (7) Any colony with radioactivity is then revealed as one with DNA sequences complementary to the labeled probe that was added (the known thymidine kinase sequences).

15. On chromosome 8. Only when 8 is present do we see the human plus mouse pattern. In the absence of 8, only the mouse pattern is seen.

16. **A.** Since there would be no replication origin and no centromere, no stable chromosome would be present at the time of replication.
B. Since two centromeres are present, orderly segregation of the product of replication could not take place.

CHAPTER 20

1. The cDNA probe specific for the 5′ portion reveals no difference between the DNA from the two sources. However, the probe for the 3′ end detects one fragment in unafflicted persons and two smaller ones in DNA from those with the disorder. The affliction appears to be associated with a gene mutation that has created a restriction enzyme site in the 3′ end of the gene. This indicates a substitution of one nucleotide base pair for another and an amino acid substitution in the α-globin chain toward the carboxyl end.

2. **A.** 20, 15, 6, and 2 kb.
B. Two haplotypes are possible. We can call one A and the other B, or vice versa. For example, A, 20 + 6 + 2 kb fragments; B, 15 + 6 + 2 kb fragments.
C. Three genotypes are possible: AA, AB, and BB.
D. An autoradiograph of genotype AA would reveal fragments of 20, 6, and 2 kb. One of genotype BB would show 15 + 6 + 2 kb, and one of AB would reveal 20, 15, 6, and 2 kb fragments.

3. **A.** Four haplotypes are possible, which can be designated as follows: A, 20 + 8 kb; B, 20 + 6 + 2 kb; C, 15 + 8 kb; D, 15 + 6 + 2 kb.
B. Ten are possible: AA, AB, AC, AD, BB, BC, BD, CC, CD, DD.

4. Four genotypes are possible among the offspring in family 1: AC, AD, CD, and DD. If haplotype D is present, fragments of both 15.0 and 4.9 kb must also be present. Therefore, person 2 cannot have haplotype D and must be genotype AC, and is consequently at low risk because the recombination frequency is low for the region. Since person 1 shows all four possible fragments, this indicates the genotype is AD. However, the person must have received haplotype D from the unafflicted parent and A from the afflicted one. Again, the person would be at little risk.

In family 2, the possible genotypes are CC, CD, and DD. Person 2 is at very high risk, because the genotype must be DD, and the person must have received one haplotype D from the afflicted parent. Person 1 is at risk, because the genotype must be CD, and there is a 50% chance that the D haplotype was contributed by the afflicted parent. Person 3 is at little risk, because the genotype must be CC.

5. T-helper cells play a pivotal role in coordinating the interactions among various cell types involved in the immune response. Helpers are needed to stimulate other T cells against a particular antigen. It is the T-helper cells that cause certain B cells to divide and complete their differentiation to circulating memory cells, which are on guard against a specific antigen. Other B lymphocytes differentiate to become the antibody-secreting plasma cells after stimulation by T helpers. T cells are also critical to the interaction between T suppressor and active T-killer cells. Lacking T helpers, for example, the suppressors may continue to cut down the activity of the killer cells. Hence without T-helper cells activities of B lymphocytes and other T cells go awry, resulting in a lack of both sufficient antibodies and killer-T cells to combat foreign antigens.

6. C gene DNA is seen to be associated with two fragments in the plasma cells, one the same size as the fragment in the nonantibody-producing (fibroblast) cells. However, C gene DNA in plasma cells is also associated with a much smaller fragment, indicating that rearrangement or shuffling of some of the DNA has taken place in the development of these cells. The shuffling may have altered a restriction site on one of the chromosomes of a plasma cell, with the result that C gene DNA becomes part of a smaller fragment. Because of allelic exclusion, the C gene DNA on the other chromosome in each plasma cell will remain in the same (original) arrangement as in nonantibody-producing cells, and will thus be associated with the larger fragment.

7. (1) Light, (2) both, (3) heavy, (4) both, (5) heavy, (6) both, (7) light, (8) both, (9) both, (10) heavy, (11) heavy, (12) both, (13) heavy, (14) both

8. $V_{18}J_1C_1$; $V_{18}J_2C_1$; $V_{18}J_3C_1$; $V_6J_1C_1$; $V_6J_2C_1$; $V_6J_3C_1$

9. $V_2D_1J_1C_1$; $V_2D_1J_2C_1$; $V_2D_2J_1C_1$; $V_2D_2J_2C_1$; $V_2D_1J_1C_2$; $V_2D_1J_2C_2$; $V_2D_2J_1C_2$; $V_2D_2J_2C_2$

10. In the maturation of plasma cells, additional gene shuffling occurs in which heavy chain class switching takes place. Before chain class switching, a specific VDJ combination has been brought adjacent to the cluster of C (constant) region genes. In the cluster, the C genes occur in the following order: μ, δ, γ-3, γ-1, γ-2b, γ-2a, ∊, α. μ and δ are immediately adjacent to VDJ, and these two classes of chains can be produced. If further rearrangement takes place, deletion may remove one or more C genes. The class of chain produced depends on the C genes that are deleted. If μ, δ, and γ-3 are deleted, then γ-1 will be directly next to the specific VDJ combination. Chains of class γ-1 will then be produced following transcription of the entire contiguous region and differential processing of the transcript. If the region μ through γ-2a is deleted in the class switching, then the ∊-gene will be placed next to the VDJ combination, and the cell will secrete ∊ class chains following formation of pre-mRNA and its differential processing to form mature mRNA. Chains of class α can be produced when all the C genes except α are deleted.

11. **A.** (1) $V_{50}D_3J_2$ (S) γ (S) ∊ (S) α
(2) $V_{50}D_3J_2$ (S) ∊ (S) α
(3) $V_{50}D_3J_2$ (S) α
B. Answers are the same as in A, because the entire C cluster adjoining the VDJ segment can be transcribed.
C. (1) $V_{50}D_3J_2$ γ codes for IgG
(2) $V_{50}D_3J_2$ ∊ codes for IgE

(3) $V_{50}D_3J_2$ α codes for IgA

12. The order must be 3.0, 2.3, 6.6, 4.5. This can be deduced because the "beginning" of the δ-gene and the "end" of the β-gene are present in persons with hemoglobin Lepore. The other persons are known to have an alteration that eliminates an EcoR1 cutting site in the β-gene only. The 3.0 fragment is present in the DNA from these individuals. This must be the start of the δ-gene. The 4.5 fragment that includes the end of the β-gene must be part of the 11.1 fragment. The 6.6 fragment must make up the rest of this large fragment, and it must include the beginning of the β-gene. The 2.3 fragment is present in these persons, and it must include the end of the δ-gene.

CHAPTER 21

1. A. KK, killers
B. KK, sensitives
C. Two KK killers or two kk sensitives
D. Two KK sensitives or two kk sensitives

2. A. KK and KK
B. KK and kk
C. Kk and Kk

3. A. KK sensitives from the sensitive and KK killers from the killer
B. Kk sensitives from the sensitive and Kk killers from the killer

4. A. Zygotes and ascospores petite
B. Zygotes wild. Ascospores 2 wild:2 petite
C. $2^+:2^-$

5. Two classes of zygotes are to be expected: those that, upon sporulation, will yield spores that are all drug resistant and those that will yield spores that are all sensitive.

6. A. Zygote: CcDdPpQq. Products of zygote: (1) CDPq, (2) CdPq, (3) cDPq, (4) cdPq. Each cell type will produce cells of the same type.
B. Zygote: CcDdPpQq. Products of zygote: (1) CDPpQq, (2) CdPpQq, (3) cDPpQq, (4) cdPpQq. Type 1 will give rise to CDPQ, CDPq; CDpQ; CDpq. Type 2 will give rise to CdPQ; CdPq; CdpQ; Cdpq. Type 3 will give rise to cDPQ; cDPq; cDpQ; cDpq. Type 4 will give rise to cdPQ; cdPq; cdpQ; cdpq.

7. The transmission of growth habit is nonMendelian. The reciprocal crosses give different results. In each case, growth habit is like that of the female parent, whereas the other traits show Mendelian inheritance.

8. D would appear to be in the nuclear DNA, because it is present in the neutral petite, which lacks mt DNA. The other markers would appear to be in the mt DNA. C and E are lost together in one strain and appear to be adjacent. The same is true for B and F in the other strain.

9. Because the genetic sequences coded in the mt DNA for the polypeptide chains of various respiratory enzymes are not localized in one part of the mt DNA. The map indicates that they are in various locations around the circular DNA.

10. In the protozoan, chains I, IV, and VI must be made in the cytosol, since erythromycin inhibits the ribosomes of the organelles. The remaining four chains are made in the mitochondria, since cycloheximide inhibits translation by the ribosomes of the cytosol. In the mold, the three small chains are made in the cytosol and the four large ones in the organelles.

11. A large transcript will be formed as a result of transcription of the entire DNA strand. Processing of the transcript will result in cleavage of the enzyme sequence from the tRNA sequence. The enzyme transcript will then terminate in a U. The transcript will next be extended by posttranscriptional addition of a poly(A) tail. The solitary U can finally be recognized as part of a chain-terminating codon, UAA.

12. The mutation affecting the human mt has the greater chance of altering some product, because human mitochondrial genes are packed closely together with little noncoding DNA present. This contrasts with yeast mt DNA, which contains an appreciable amount of DNA that is apparently not coded for specific products.

13. Three violations are seen: TGA should indicate a chain-terminating codon (UGA), ATA (AUA) isoleucine, and CTA (CUA) leucine.

14. A, C, and D

15. C and E

16. B and D

Index